Periodic Table of the Elements

Atomic masses are based on $^{12}_{6}C$. Numbers in parentheses are the mass numbers of the most stable isotopes of radioactive elements.

1 Group IA	2 Group IIA	3 Group IIIB	4 Group IVB	5 Group VB	6 Group VIB	7 Group VIIB	8 Group	9 Group VIIIB
1 **H** 1.00794								
3 **Li** 6.941	4 **Be** 9.01218							
11 **Na** 22.98977	12 **Mg** 24.305							
19 **K** 39.0983	20 **Ca** 40.078	21 **Sc** 44.9559	22 **Ti** 47.867	23 **V** 50.9415	24 **Cr** 51.9961	25 **Mn** 54.9380	26 **Fe** 55.845	27 **Co** 58.9332
37 **Rb** 85.4678	38 **Sr** 87.62	39 **Y** 88.9059	40 **Zr** 91.22	41 **Nb** 92.9064	42 **Mo** 95.94	43 **Tc** (98)	44 **Ru** 101.07	45 **Rh** 102.9055
55 **Cs** 132.9054	56 **Ba** 137.33	57 **La** 138.9055	72 **Hf** 178.49	73 **Ta** 180.9479	74 W 183.84	75 **Re** 186.207	76 **Os** 190.23	77 **Ir** 192.217
87 **Fr** (223)	88 **Ra** 226.0254	89 **Ac** 227.0278	104 **Rf** (261)	105 **Ha** (262)	106 **Sg** (266)	107 **Ns** (262)	108 **Hs** (265)	109 **Mt** (266)

58 **Ce** 140.12	59 **Pr** 140.9077	60 **Nd** 144.24	61 **Pm** (145)	62 **Sm** 150.36
90 **Th** 232.0381	91 **Pa** 231.0359	92 **U** 238.0289	93 **Np** 237.0482	94 **Pu** (244)

10 roup →	11 Group IB	12 Group IIB	13 Group IIIA	14 Group IVA	15 Group VA	16 Group VIA	17 Group VIIA	18 Group VIIIA
								2 **He** 4.002602
			5 **B** 10.811	6 **C** 12.011	7 **N** 14.0067	8 **O** 15.9994	9 **F** 18.998403	10 **Ne** 20.179
			13 **Al** 26.98154	14 **Si** 28.0855	15 **P** 30.97376	16 **S** 32.066	17 **Cl** 35.453	18 **Ar** 39.948
28 **Ni** .6934	29 **Cu** 63.546	30 **Zn** 65.38	31 **Ga** 69.723	32 **Ge** 72.59	33 **As** 74.9216	34 **Se** 78.96	35 **Br** 79.904	36 **Kr** 83.80
46 **Pd** 06.42	47 **Ag** 107.8682	48 **Cd** 112.41	49 **In** 114.818	50 **Sn** 118.710	51 **Sb** 121.760	52 **Te** 127.60	53 **I** 126.9045	54 **Xe** 131.29
78 **Pt** 95.08	79 **Au** 196.9665	80 **Hg** 200.59	81 **Tl** 204.383	82 **Pb** 207.2	83 **Bi** 208.9804	84 **Po** (209)	85 **At** (210)	86 **Rn** (222)
110 – 271)	111 – (272)							

Metals ← → Nonmetals

63 **Eu** 51.96	64 **Gd** 157.25	65 **Tb** 158.9254	66 **Dy** 162.50	67 **Ho** 164.9304	68 **Er** 167.26	69 **Tm** 168.9342	70 **Yb** 173.04	71 **Lu** 174.967
95 **Am** 243)	96 **Cm** (247)	97 **Bk** (247)	98 **Cf** (251)	99 **Es** (252)	100 **Fm** (257)	101 **Md** (260)	102 **No** (259)	103 **Lr** (262)

Introduction to Chemical Principles

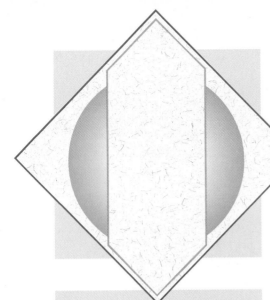

Introduction to Chemical Principles

H. Stephen Stoker
Weber State University

Fifth Edition

PRENTICE HALL
Upper Saddle River, NJ 07458

Library of Congress Cataloging-Publication Data
Stoker, H. Stephen (Howard Stephen)
Introduction to chemical principles / H. Stephen Stoker—5th ed.
p. cm.
Includes index.
ISBN 0-13-228438-3
1. Chemistry I. Title.
QD33.S86 1996
540—dc20 95-20146
CIP

Acquisitions Editor: Mary Hornby
Project Management: J. Carey Publishing Service
Manufacturing Manager: Trudy Pisciotti
Art Director: Joseph Sengotta
Cover Designer: Thomas Nery
Interior Designer: Amy Rosen
Cover Image: Thermite Reaction, Richard Megna, Fundamental Photographs

Printed in the United States of America

10 9 8 7 6 5 4 3 2 1

ISBN 0-13-228438-3

Prentice-Hall International (UK) Limited, *London*
Prentice-Hall of Australia Pty. Limited, *Sydney*
Prentice-Hall Canada, Inc., *Toronto*
Prentice-Hall Hispanoamericana, S.A., *Mexico*
Prentice-Hall of India Private Limited, *New Delhi*
Prentice-Hall of Japan, Inc., *Tokyo*
Simon & Schuster Adia Pte. Ltd., *Singapore*
Editora Prentice-Hall do Brasil, Ltda., *Rio de Janeiro*

Contents

Chapter 6 ELECTRONIC STRUCTURE AND CHEMICAL PERIODICITY 152

Chapter 7 CHEMICAL BONDS 188

Preface

Introduction to Chemical Principles is a text for students who have had little or no previous instruction in chemistry or who had such instruction long enough ago that a thorough review is needed. The text's purpose is to give students the background (and confidence) needed for a subsequent successful encounter with a main sequence college level general chemistry course.

Many texts written for preparatory chemistry courses are simply "watered down" versions of general chemistry texts; they treat almost all topics found in the general chemistry course, but at a superficial level. *Introduction to Chemical Principles* does not fit this mold. The author's philosophy is that it is better to treat fewer topics extensively and have the student understand those topics in greater depth. The very real temptation to include "lots and lots" of additional concepts in this new edition of the text was resisted. Instead the focus for the revision was on rewriting selected portions to improve the clarity of presentation.

Important Features of This Textbook

1. Because of the varied degrees of understanding of chemical principles possessed by students taking a preparatory chemistry course, development of each topic in this text starts at "ground level" and continues step by step until the level of sophistication required for a further chemistry course is attained.
2. Problem solving receives major emphasis. Nearly thirty years of teaching experience indicate to the author that student "troubles" in general chemistry courses are almost always centered in the inability to set up and solve problems. Whenever possible, dimensional analysis is used in problem solving. This method, which requires no mathematics beyond arithmetic and elementary algebra, is a powerful and widely applicable problem-solving tool. Most important, it is a method that an average student can master with an average amount of diligence. Mastering dimensional analysis also helps build the confidence that is so valuable for future chemistry courses.

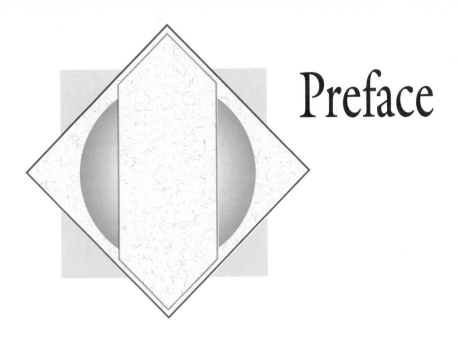

3. Significant figure concepts are emphasized in all problem-solving situations. Routinely, electronic calculators display answers that contain more digits than are needed or acceptable. In all worked-out examples, students are reminded about these "unneeded digits" by the appearance of two answers to the example: the calculator answer (which does not take into account significant figures) and, in color, the correct answer (which is the calculator answer adjusted to the correct number of significant figures).
4. Numerous worked-out example problems are found within the textual material with detailed commentary accompanying each such example. In addition an unworked practice exercise is coupled to each example. It is intended that students will work this exercise immediately after "working through" the example. For immediate feedback, the answer to each practice exercise follows the exercise.
5. All end-of-chapter exercises occur in "matched pairs." In essence, each chapter has two independent, but similar, problem sets. Counting subparts to problems, there are over 5000 questions and problems available for a student to use in his or her "struggle" to become proficient at problem solving. Answers to all of the odd-number problems are found at the end of the text. Thus, two problem sets exist, one with answers and one without answers.
6. Each end-of-chapter problem set, except for Chapters 1 and 2, is divided into three sections: (1) Practice Problems, (2) Additional Problems, and (3) Cumulative Problems. The practice problems are categorized by topic and are arranged in the same sequence as the chapter's textual material. These problems, which are always single-concept, are "drill" problems which most students will find "routine." The additional problem section contains problems that involve more than one concept from the chapter and are usually more difficult than the practice problems. The cumulative-skills section draws not only on materials from the current chapter but also on concepts discussed in previous chapters. The working of problems in this third group allows students to continue to use, rather than forget, problem-solving techniques presented earlier.

NEW FEATURES OF THE FIFTH EDITION

1. All 197 of the worked out examples in the text are new to this edition.
2. At least 50% of the end-of-chapter problems in each chapter are new. Overall, 1124 of the 1972 end-of-chapter problems and questions are new.
3. The sequence of presentation of the covalent bonding topics in Chapter 7 has been altered. The geometry of molecules (VSEPR theory) is now more closely tied to the concept of electron-dot structures.
4. Chapter 8, Chemical Nomenclature, has been extensively rewritten. An important new feature of this chapter is the introduction of nomenclature flow charts ("decision trees").
5. Chapter 9 contains an expanded discussion of the calculation of empirical and molecular formulas.

6. The material in Chapter 15 dealing with balancing redox reactions using the half-reaction method has been simplified. A new procedure for balancing half-reactions in basic solution has been added.

SUPPLEMENTS

A *Student Solutions Manual* has been prepared by Dr. Garth L. Welch of Weber State University to accompany this text. There is also an appropriate lab manual, *Prentice Hall Laboratory Experiments for Introductory Chemistry* by Charles H. Corwin of American River College. A free supplement, *How to Study Chemistry*, stressing strategies for learning and achievement in chemistry, is available to students on request from the instructor.

Also available to adopters is an *Instructor's Solutions Manual* by Dr. Welch. A test bank compiled by Dr. Stoker is available as a manual or in computerized formats, both MAC and IBM. There is an *Instructor's Manual to the Laboratory Experiments* available to the instructor, and *The New York Times Themes of the Times*, a newsprint collection of timely topics relevant to chemistry in action in our world.

ACKNOWLEDGMENTS

As always, the valuable contributions of reviewers are gratefully acknowledged: Spencer Steinberg of University of Nevada Las Vegas; Gerard Nobiling, Monroe Community College; Thomas Mincs, St. Louis-Florissant Community College; Donald Langr, North Iowa Community College.

H.S.S.

Introduction to Chemical Principles

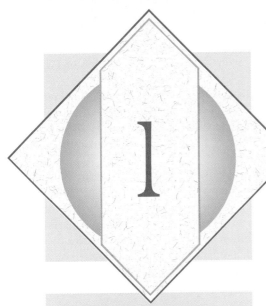

1

The Science of Chemistry

1.1 Chemistry—A Scientific Discipline

Students who are required to take two years, one year, or just one course of chemistry all often ask the same question prior to beginning their "chemistry experience." That question is: Why must I take chemistry? The answer to this simple five-word question, which involves an appreciation for the relationships between various branches of scientific knowledge, serves as our entry point into the "realm" of chemistry.

Chemistry is part of a larger body of knowledge called *science*. **Science** *is the study in which humans attempt to organize and explain, in a systematic and logical manner, knowledge about themselves and their surroundings.*

Because of the enormous scope of science, the sheer amount of accumulated knowledge, and the limitations of human mental capacity to master such a large and diverse body of knowledge, science is divided into smaller subdivisions called *scientific disciplines*. A **scientific discipline** *is a branch of science limited in size and scope to make it more manageable.* Examples of scientific disciplines are *chemistry,* astronomy, botany, geology, physics, and zoology.

Figure 1.1 shows an organization chart, with emphasis on chemistry, for the various scientific disciplines. These disciplines can be grouped into *physical sciences* (the study of matter and energy) and *biological sciences* (the study of living organisms). Chemistry is a physical science.

Rigid boundaries between scientific disciplines *do not exist*. All scientific disciplines borrow information and methods from each other. No scientific discipline is totally independent. "Environmental problems" that scientists have encountered in the last two decades particularly show the interdependence of the various scientific disciplines. For example, chemists attempting to solve the problems of chemical contamination of the environment find that they need some knowledge of geology, zoology, and botany. It is now common to talk not only of chemists, but also geochemists, biochemists, chemical physicists, and so on. The middle portion of Figure 1.1 shows the overlap of the other scientific disciples with chemistry.

Discipline "overlap" requires that scientists, in addition to having in-depth knowledge of a selected discipline, also have limited knowledge of other disciplines. Discipline "overlap" also explains why a great many college students are required to study chemistry. One or more chemistry courses are required because of their applicability to the disciplines in which the student has more specific interest.

The body of knowledge found within the scientific discipline of chemistry is itself vast. No one can hope to master completely all aspects of chemical knowledge. However, the fundamental concepts of chemistry can be learned in a relatively short period of time.

The vastness of chemistry is sufficiently large that it, like most scientific disciplines, is partitioned into *subdisciplines*. The lower portion of Figure 1.1 shows the five fundamental branches of chemistry; analytical, general, inorganic, organic, and physical. Most of the subject matter of this textbook falls within the realm of *general chemistry*, the basic laws and concepts of chemistry.

1.2 SCIENTIFIC DISCIPLINES AND TECHNOLOGY

Scientific disciplines represent abstract bodies of knowledge. The abstractness of such knowledge is modified by technology. **Technology** *is the physical application of scientific knowledge to the production of new products to improve human survival, comfort, and quality of life*. Technology manipulates nature for advantage. Technological advances began affecting our society about 200 years ago, and new advances still continue, at an accelerating pace, to have a major impact on human society. Section 1.3 considers numerous contributions of chemical technology to human well-being.

Technology, like science, involves human activities. Whether or not a given piece of scientific knowledge is technologically used for good or evil purposes depends on the motives of those men and women, whether in industry or government, who have the decision-making authority. In democratic societies, citizens (the voters) can influence many technological decisions. Therefore, it is important for everyone to be informed about scientific and technological issues.

1.3 THE SCOPE OF CHEMISTRY AND CHEMICAL TECHNOLOGY

Although chemistry is concerned with only a part of the scientific knowledge that has been accumulated, it is in itself an enormous and broad field. Chemistry touches all parts of our lives.

Many of the clothes we wear are made from synthetic fibers produced by chemical processes. Even natural fibers, such as cotton or wool, are the products of naturally occurring chemical reactions within living systems. Our transportation usually involves vehicles powered with energy obtained by burning chemical mixtures, such as gasoline, diesel, jet fuel, etc. The drugs used to cure many of our illnesses are the result of chemical research. The paper on which this textbook is printed was produced through a chemical process, and the ink used in printing the words and illustrations is a mixture of many chemicals. The movies we watch are possible because of synthetic materials called film. The images on film are produced through the interaction of selected chemicals. Almost all of our recreational pursuits involve objects made of materials produced by chemical industries. Skis, boats,

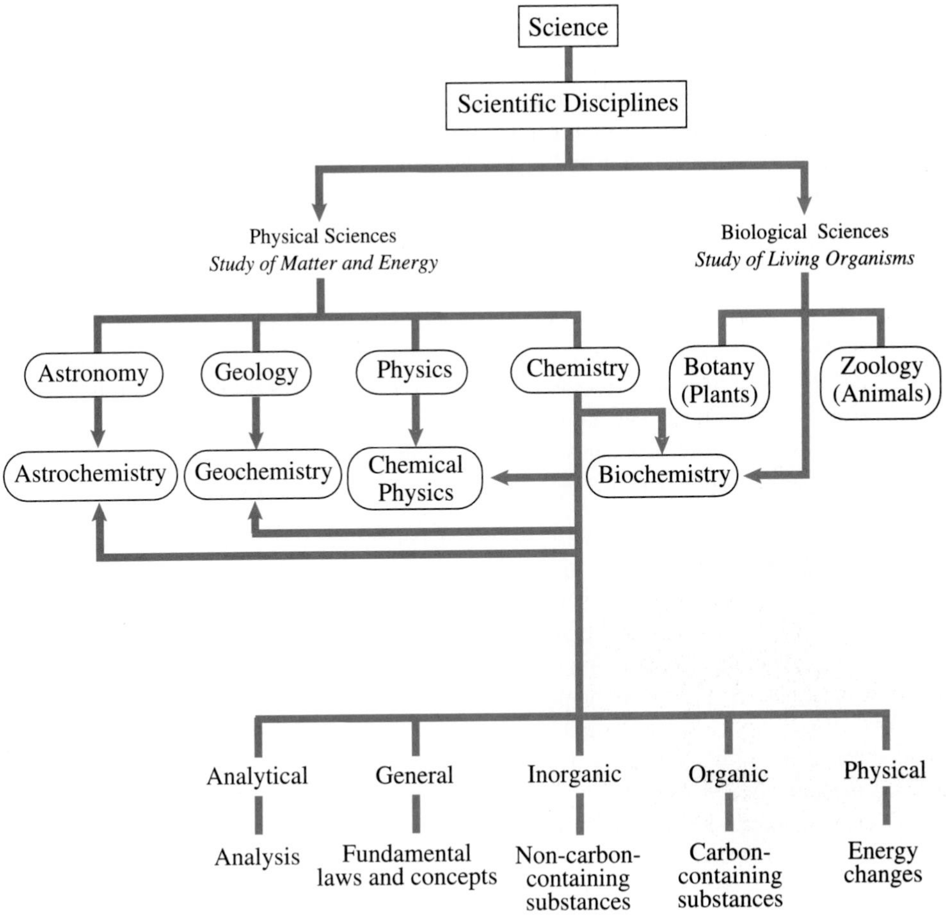

FIGURE 1.1 An organizational chart showing the relationship of the scientific discipline of chemistry to other scientific disciplines and also the substructuring that occurs within the discipline of chemistry.

basketballs, bowling balls, musical instruments, and television sets all contain materials that do not occur naturally, but are products of human technological expertise.

Our bodies are a complex mixture of chemicals. The principles of chemistry are fundamental to an understanding of all processes of the living state. Chemical secretions (hormones) produced within our bodies help determine our outward physical characteristics such as height, weight, and appearance. Digestion of food involves a complex series of chemical reactions. Food itself is an extremely complicated array of chemical substances. Chemical reactions govern our thought processes and how knowledge is stored in and retrieved from our brains. In short, chemistry runs our lives.

A formal course in chemistry can be a fascinating experience because it helps us understand ourselves and our surroundings. We cannot truly understand or even know very much about the world we live in or about our own bodies without being conversant with the fundamental ideas of chemistry.

1.4 HOW CHEMISTS DISCOVER THINGS—THE SCIENTIFIC METHOD

The word *chemistry* conjures up images of people in white lab coats peering at instruments and shaking test tubes or other similar apparatus. Why is this generally valid image associated with chemists? The reason is simple. Chemists, as well as all other scientists, discover the general principles that govern the physical world (both its seen and unseen parts) through experimentation and observation. (See Fig. 1.2.)

A majority of the scientific and technological advances of the twentieth century are the result of systematic experimentation using a method of problem solving known as the scientific method. The **scientific method** *is a set of procedures used to acquire knowledge and explain phenomena.* The procedural steps in the scientific method are

1. Identify the problem, break it into small parts, and carefully plan procedures to obtain information about all aspects of this problem.
2. Collect data concerning the problem through observation and experimentation.
3. Analyze and organize the data in terms of general statements (generalizations) that summarize the experimental observations.
4. Suggest probable explanations for the generalizations.
5. Experiment further to prove or disprove the proposed explanations.

Occasionally a great discovery is made by accident, but the majority of scientific discoveries are the result of the application of these five steps over long periods of time. There are no instantaneous steps in the scientific method; applying them requires considerable amounts of time. Even in those situations where luck is involved, it must be remembered that "chance favors the prepared mind." To take full advantage of an accidental discovery, a person must be well trained in the procedures of the scientific method.

The imagination, creativity, and mental attitude of a scientist using the scientific method are always major factors in scientific success. The procedures of the scientific method must always be enhanced with the abilities of a thinking scientist.

There are special vocabulary terms associated with the scientific method and its use.

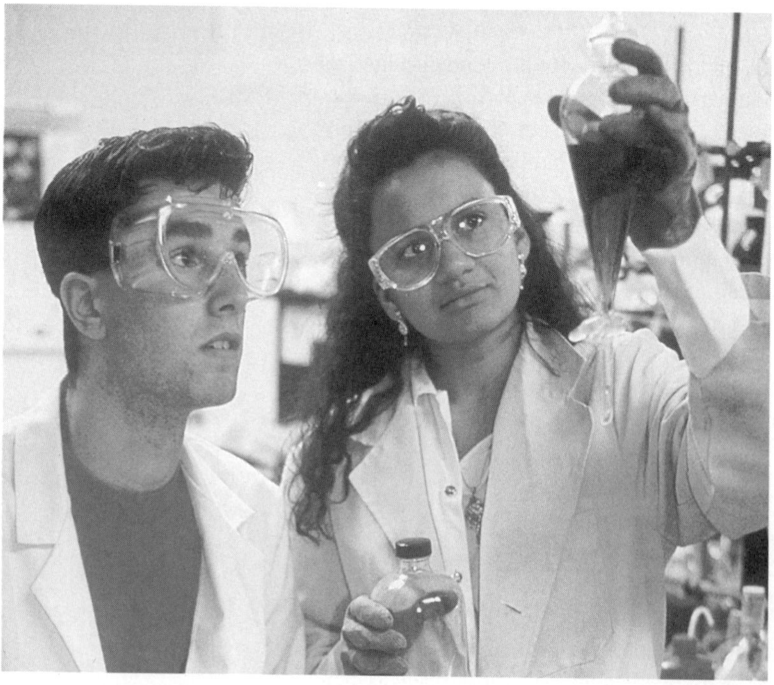

FIGURE 1.2 Chemistry is an experimental science. Most discoveries in chemistry are made through analysis of data obtained from experiments carried out in laboratories. (Jeff Greenberg/Visuals Unlimited)

This vocabulary includes the terms *experiment, fact, law, hypothesis,* and *theory*. An understanding of the relationships among these terms is the key to a real understanding of how to obtain chemical knowledge.

The beginning step in the search for chemical knowledge is the identification of a problem concerning some chemical system that needs study. After determining what other chemists have already learned about the selected problem, a chemist sets up experiments for obtaining more information. An **experiment** *is a well-defined, controlled procedure for obtaining information about a system under study.* The exact conditions under which an experiment is carried out must always be noted because conditions such as temperature and pressure affect results.

New facts about the system under study are obtained by actually carrying out the experimental procedures. A **fact** *is a valid observation about some natural phenomenon.* Facts are reproducible pieces of information. If a given experiment is repeated under exactly the same conditions, the same facts should be obtained. To be acceptable, all facts must be verifiable by anyone who has the time, means, and knowledge needed to repeat the experiments that led to their discovery. It is important that scientific data be published so that other scientists have the opportunity to critique and double-check both the data and experimental design.

It is interesting to contrast the differing ways in which scientific facts and the results of technology (Sec. 1.2) are shared. Scientists publish their observations (facts) as widely, openly, and quickly as possible. Technological breakthroughs, on the other hand, are usually kept secret by an individual or company until patent rights for the new process or product can be obtained.

As a next step, the scientist makes an effort to determine ways in which the facts about a given system relate both to each other and to facts known about similar systems. Repeating patterns often emerge among the collected facts. These patterns lead to generalizations that are called laws about how chemical systems behave under specific conditions. A **law** *is a generalization that summarizes facts about natural phenomena.*

Do not assume that laws are easy to discover. Often, many years of work and thousands of facts are needed before the true relationships among variables in the area under study emerge.

A law is a description of what happens in a given type of experiment. No new understanding of nature results from simply stating a law. A law merely summarizes already known observations (facts).

A law can be expressed either as a verbal statement or as a mathematical equation. An example of a verbally stated law is "If hot and cold pieces of metal are placed in contact with each other, the temperature of the hot piece always decreases and the temperature of the cold piece always increases."

It is important to distinguish between the use of the word *law* in science and its use in a societal context. Scientific laws are *discovered* by research (see Fig. 1.2), and researchers have *no control* over what the laws turn out to be. Societal laws, which are designed to control aspects of human behavior, are *arbitrary conventions* agreed upon (in a democracy) by the majority of those to whom the laws apply. These laws *can be* and *are changed* when necessary. For example, the speed limit for a particular highway (a societal law) can be decreased or increased for various safety or political reasons.

There is no mention in a scientific law as to why the occurrence described happens. The law simply summarizes experimental observations, without attempting to clarify the reasons for the occurrence. Chemists, and other scientists, are not content with such a situation. They want to know *why* a certain type of observation is always made. Thus, after a law is discovered, scientists work out *plausible, tentative* explanations of the behavior encompassed by the law. These explanations are called *hypotheses*. A **hypothesis** *is a tentative model or statement that offers an explanation for a law.*

Once a hypothesis has been proposed, experimentation begins again. Scientists run more experiments, under varied, but controlled, conditions to test the reliability of the proposed explanation. The hypothesis must be able to predict the outcome of as-yet-untried experiments. The validity of the hypothesis depends upon its predictions being true.

It is much easier to disprove a false hypothesis than to prove a true one. A negative result from an experiment indicates that the hypothesis is not valid as formulated and must be modified. Obtaining positive results supports the hypothesis, but it does not definitely prove it. There is always the chance that someone will carry out a new type of experiment, one that was not previously thought of, that disproves the hypothesis.

In practice, scientists usually start with a number of alternative hypotheses for a given law. Evaluation proceeds by demonstrating that certain proposals are *not* valid. A successful experiment is one in which one or more of the alternative hypotheses are demonstrated to be inconsistent with experimental observation and are thus rejected. Scientific progress is made in the same way a marble statue is: unwanted bits of marble are chipped away. Example 1.1 contains a simple illustration of this "chipping away" principle in a scientific context.

Example 1.1

Suppose you encounter a situation involving two unopened books with no identification on their covers and four alternative hypotheses about these books, which are: (1) the thinner book is a chemistry textbook, (2) the thicker book is a chemistry textbook, (3) both books are chemistry textbooks, and (4) neither book is a chemistry textbook. What evaluative information about these hypotheses can be obtained by opening the thicker book and determining that it is a chemistry textbook?

Solution

This experiment (opening the thicker book) disproves hypothesis 4; it does not prove that *only one* of the hypotheses is true, but rather demonstrates that one of them is not true. The fact that the thicker book is a chemistry textbook does not rule out the possibility that the thinner book is also a chemistry textbook.

Practice Exercise 1.1

Based on the same "two-book, four-hypotheses" situation stated in Example 1.1, what evaluative information about the hypotheses is obtained from the single observation that the thinner book is *not* a chemistry textbook?

Ans. Hypotheses 1 and 3 are disproved

As further experimentation continues to validate a particular hypothesis, its acceptance in scientific circles increases. If, after extensive testing, the reliability of a hypothesis is still very high, confidence in it increases to the extent that it is accepted by the scientific community at large. After more time has elapsed and more positive support has accumulated, the hypothesis assumes the status of a theory. A **theory** *is a hypothesis that has been tested and validated over a long period of time.* The dividing line between a hypothesis and a theory is

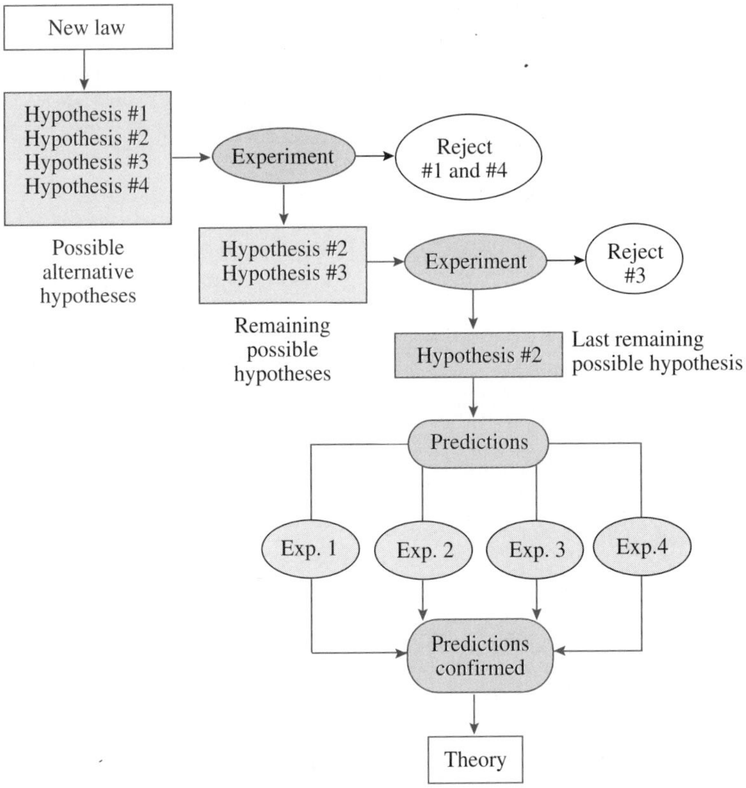

FIGURE 1.3 A number of alternative hypotheses are proposed to explain a new law; experiments are carried out to eliminate invalid hypotheses; predictions are made based on the surviving hypothesis and further experiments are carried out to test these predictions.

arbitrary and cannot be precisely defined. There is no set number of supporting experiments that must be performed in order to give theory status to a hypothesis.

Theories serve two important purposes: (1) they allow scientists to predict what will happen in experiments that have not yet been run, and (2) they simplify the very real problem of being able to remember all the scientific facts that have already been discovered. Figure 1.3 shows the interplay that must occur between hypotheses and experimentation before an acceptable theory is obtained.

Theories must often undergo modification. As scientific tools, particularly instrumentation, become more precise, there is an increasing probability that some experimental observations will not be consistent with all aspects of a given theory. A theory inconsistent with new observations must either be modified to accommodate the new results or be restated in such a way that scientists know where it is useful and where it is not. Most theories in use have known limitations. These "imperfect" theories are simply the best ideas anyone has found *so far* to describe, explain, and predict what happens in the world in which we live. Theories with limitations are generally not abandoned until a better theory is developed.

Scientists do not view scientific theories as "absolute truth." All theories in science are considered provisional—subject to change in the light of new experimental observations. A science is like a living organism; it continues to grow and change. Science develops through a constant interplay between theory and experimentation.

Facts that have been verified by repeated experiments will never be changed, but the theories that were invented to explain these facts are subject to change. In this sense, facts are more important than the theories devised to explain them. It is a mistake to believe that by knowing all the laws and theories that are derived from experimental observations, the

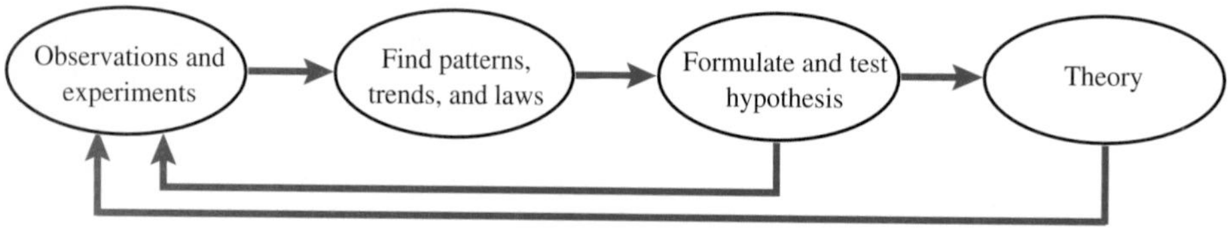

FIGURE 1.4 The scientific method, the general approach to problem solving in all sciences, has experimentation as its central focus.

experimental facts are not needed. New theories can only be developed by people who have a wide knowledge of the facts relating to a particular field, especially those facts that have not been satisfactorily accounted for by existing theories.

The term *theory* is often misused by nonscientists in everyday contexts. "I have a theory that such and such is the case" is a frequently heard comment. In this case, "theory" means a "speculative guess," which is not what a theory is. The term *hypothesis* would be more appropriate.

Figure 1.4 summarizes the sequence of steps scientists generally use when applying the scientific method to a given research problem. It also shows, as did Figure 1.3, the central role that experimentation plays in the scientific method.

KEY TERMS

The new terms or concepts defined in this chapter are

experiment (Sec. 1.4) A well-defined, controlled procedure for obtaining information about a system under study.

fact (Sec. 1.4) A valid observation about some natural phenomenon.

hypothesis (Sec. 1.4) A tentative model or statement that offers an explanation for a law.

law (Sec. 1.4) A generalization that summarizes in a concise way facts about natural phenomena.

science (Sec. 1.1) The study in which humans attempt to organize and explain in a systematic and logical manner knowledge about themselves and their surroundings.

scientific discipline (Sec. 1.1) A branch of scientific knowledge limited in size and scope to make it more manageable.

scientific method (Sec. 1.4) A set of procedures for acquiring knowledge and explaining phenomena.

technology (Sec. 1.2) The physical application of scientific knowledge to the production of new products to improve human survival, comfort, and quality of life.

theory (Sec. 1.4) A hypothesis that has been tested and validated over a long period of time.

PRACTICE PROBLEMS

Scientific Disciplines (Sec. 1.1)

1.1 Indicate whether each of the following statements is true or false.

(a) Science is the study in which humans attempt to organize and explain, in a systematic and logical manner, knowledge about themselves and their surroundings.

(b) Scientific disciplines are limited in size and scope to the extent that subdividing of subject matter within a discipline is not necessary.

(c) Boundaries between scientific disciplines are very rigid.

(d) Collectively, the knowledge in scientific disciplines constitutes the whole of scientific knowledge currently known.

1.2 Indicate whether each of the following statements is true or false.

(a) Scientific disciplines are branches of scientific knowledge limited in size and scope to make them more manageable.

(b) Scientific disciplines are defined in such a manner that each discipline is totally independent of other disciplines.

(c) Complete mastery of all concepts within a scientific discipline is difficult (but possible) because of limited size and scope for the discipline.

(d) The scientific discipline of chemistry has some overlap with other physical sciences but no overlap with biological sciences.

The Scientific Method (Sec. 1.4)

1.3 Arrange the following steps in the scientific method in the sequence in which they normally occur.

(a) Suggest probable explanations for generalizations obtained from data.

(b) Collect data concerning a problem through observation and experimentation.

(c) Identify a problem and carefully plan procedures to obtain information about all aspects of this problem.

(d) Experiment further to prove or disprove proposed explanations.

(e) Analyze and organize data in terms of general statements that summarize experimental observations.

1.4 Arrange the following terms associated with the scientific method in the order in which they are normally encountered as the scientific method is applied to a problem.

(a) law

(b) fact

(c) theory

(d) experiment

(e) hypothesis

1.5 Classify each of the following statements as a *fact,* a *law,* or a *hypothesis.*

(a) Cars rust faster during the winter months than during the summer months.

(b) The author of this chemistry textbook is bald because he chewed his food too fast as a child.

(c) A sample of oxygen gas expanded when it was heated.

(d) The boiling point of water is always 100°C at sea level.

1.6 Classify each of the following statements as a *fact,* a *law,* or a *hypothesis.*

(a) A man's hair turned gray because of the driving habits of his teenage children.

(b) All samples of gaseous substances expand when heated.

(c) The force of gravity upon an object depends on the color of the object.

(d) The diameter of the moon is 3476 kilometers.

1.7 Indicate whether each of the following statements is true or false.

(a) A theory is a summary of experimental observations.

(b) A hypothesis is a summary of experimental facts.

(c) A theory is subject to modification in light of new experimental observations.

(d) An experiment is a well-defined, controlled procedure for obtaining facts.

1.8 Indicate whether each of the following statements is true or false.

(a) A theory is a hypothesis that has not yet been subjected to experimental testing.

(b) It is much easier to disprove a false hypothesis than it is to prove a valid one.

(c) Established theories eventually become laws.

(d) A law is an explanation of why a particular natural phenomenon occurs.

1.9 Constructively criticize the statement "You needn't take it too seriously; after all, it's only a theory."

1.10 Constructively criticize the statement "The results of the experiment do not agree with the theory. Something must be wrong with the experiment."

1.11 Assume that you have four pennies with unknown mint dates and four hypotheses concerning these dates: (1) all dates are the same, (2) two different dates are present, (3) three different dates are present, and (4) all dates are different. Which of the listed hypotheses could be eliminated by determining that

(a) two pennies have the same date?

(b) two pennies have different dates?

(c) two of three pennies have the same date?

(d) three pennies have different dates?

1.12 Assume that you have four red balls of equal size and four hypotheses concerning the masses of the balls: (1) each ball has a different mass, (2) there are balls of two masses, (3) balls of three different masses are present, and (4) all balls have the same mass. Which of the listed hypotheses could be eliminated by determining that

(a) two balls have the same mass?

(b) three balls have the same mass?

(c) there are two masses among three balls?

(d) there are two masses among four balls?

1.13 A researcher studies the behavior of a fixed amount of a gas under constant temperature conditions with the following results:

(a) At a pressure of 4.0 atmospheres the gas occupies a volume of 2.0 liters.

(b) At a pressure of 1.0 atmosphere the gas occupies a volume of 8.0 liters.

(c) At a pressure of 2.0 atmospheres the gas occupies a volume of 4.0 liters.

(d) At a pressure of 8.0 atmospheres the gas occupies a volume of 1.0 liter.

What generalization (law) concerning the relationship between volume and temperature, under the conditions of the experiments, can be obtained from these data?

1.14 A researcher studies the behavior of a gas under constant temperature and constant volume conditions with the following results:

(a) 10.0 grams of gas exerted a pressure of 4.0 atmospheres.

(b) 40.0 grams of gas exerted a pressure of 16.0 atmospheres.

(c) 5.0 grams of gas exerted a pressure of 2.0 atmospheres.

(d) 20.0 grams of gas exerted a pressure of 8.0 atmospheres.

What generalization (law) concerning the relationship between amount of gas and pressure, under the conditions of the experiments, can be obtained from these data?

1.15 What are the differences between a scientific law and a societal law?

1.16 What is the reason for repeating experiments several times before developing a law based on the experiments?

2 Numbers from Measurements

2.1 The Importance of Measurement

It would be extremely difficult for a carpenter to build cabinets without being able to use tools such as hammers, saws, and drills. They are a carpenter's "tools of the trade." Chemists also have "tools of the trade." Their most used tool is the one called *measurement*. Understanding measurement is indispensable in the study of chemistry. Questions such as "how much . . . ?," "how long . . . ?" and "how many . . . ?" simply cannot be answered without resorting to measurements.

Most of the concepts now considered to be the basic principles of chemistry had their origin in extensive tabulations of experimental data obtained by making measurements. The concepts were "discovered" as these data tabulations (measurements) were subjected to the procedures of the scientific method (Sec. 1.4).

It is the purpose of this chapter and the next to help students acquire the necessary background to deal properly with measurement. Almost all of the material of these two chapters is mathematical. An understanding of this mathematics is a necessity for students of chemistry who want their encounters with the subject to be successful. The following analogy is appropriate for the situation. Physical exertion in sports can be fun, relaxing, and challenging for those in good physical shape. But for those not in good physical condition, such exertion is not satisfying and may even be downright painful (especially the day after). Being "in good shape" mathematically has the same effect on the study of chemistry. It can cause that study to be a very satisfying and enjoyable experience. On the other hand, a lack of the necessary mathematical skills can cause "chemical exercise" to be somewhat painful. The message should be clear. The contents of this chapter (and Chapter 3) must be taken very seriously. Skimming over this material is a sure invitation to frustration and struggle with the chemical topics that follow.

2.2 ACCURACY, PRECISION, AND ERROR

It is important that measurements made by scientists be precise and accurate. Although the terms *precise* and *accurate* are used somewhat interchangeably in nonscientific discussion, they have distinctly different meanings in science. **Precision** *refers to how close multiple measurements of the same quantity are to each other.* **Accuracy** *refers to how close a measurement (or the average of multiple measurements) comes to the true or accepted value.* The activity of throwing darts at a target illustrates nicely the difference between these two terms (see Fig. 2.1). Accuracy refers to how close the darts are to the center (bull's-eye) of the target. Precision refers to how close the darts are to each other.

The precision of a measurement depends on the actual physical measuring device used. You would expect, and it is the case, that the precision of temperature readings obtained from a thermometer with a scale marked in tenths of a degree would be greater than readings obtained from a thermometer whose scale has only degree marks. A stopwatch whose dial shows tenths of a second can be read with more precision then one that shows only seconds.

In contrast to precision, accuracy depends not only on the measuring device used but also on the technical skill of the person making the measurement. How well can that person read the numerical scale of the instrument? How well can that person calibrate the instrument before its use?

Normally, high accuracy accompanies high precision. However, high precision and low accuracy are also possible. Results obtained using a high-precision, poorly calibrated instrument would give high precision but low accuracy. All measurements would be off by a constant amount as a result of the improper calibration.

Even the most accurate and precise measurements involve some amount of error. It is impossible to have a 100% accurate measurement. Flaws in measuring-device construction, improper calibration of an instrument, and the skills (or lack of skills) possessed by a person using a measuring device all contribute to error.

Errors in measurement can be classified as either random errors or systematic errors. **Random errors** *are errors originating from uncontrolled variables in an experiment.* Such errors result in experimental values that fluctuate about the true value. A variation in the angle from which a measurement scale is viewed will cause random error. Momentary changes in air currents, atmospheric pressure, or temperature near a sensitive balance for weighing would cause random errors. The net result of random errors, which can never be completely eliminated, is a decrease in the precision of measurements.

Systematic errors *are errors originating from controllable variables in an experiment.* They are "constant" errors that occur again and again. A flaw in a piece of equipment, such

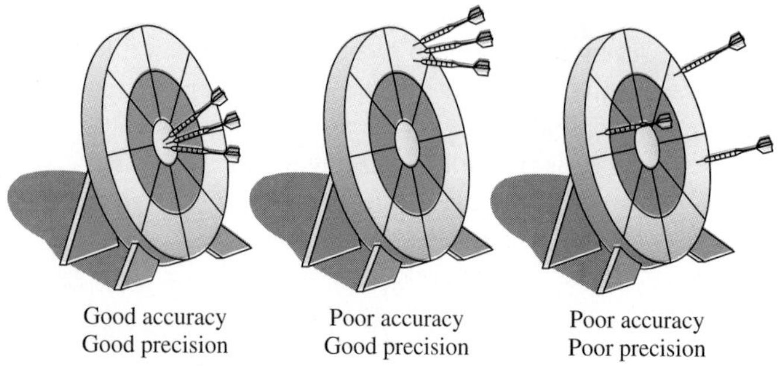

FIGURE 2.1 The difference between precision and accuracy.

as a chipped weight in a balance, would cause systematic error. All readings would be off by a specific amount because of that flaw. Systematic errors affect the accuracy of measurements. Results are consistently either too high or too low compared to the true value.

2.3 SIGNIFICANT FIGURES—A METHOD FOR HANDLING UNCERTAINTY IN MEASUREMENT

Two kinds of numbers associated with physical quantities exist—those that are *counted* or *defined* and those that are *measured*. The difference between them is that we may know the exact values of counted or defined numbers but can never know the exact values of measured numbers.

You can count the number of peaches in a bushel of peaches or the number of toes on your left foot with absolute certainty. Counting does not involve reading the scale of a measuring device, and thus counted numbers are not subject to the uncertainties inherent in a measurement.

An example of a defined number is the number of objects in a dozen—twelve. By definition, 12 and exactly 12 (not 12.01 or 12.02) objects make a dozen. There are exactly 24 hours in a day, never 24.07 hours. A square has 4 sides, never 3.75 or 3.83 sides. Thus, a defined number always has one exact value.

Another type of defined number is the numbers associated with simple fractions such as one-half ($\frac{1}{2}$), two-thirds ($\frac{2}{3}$), and three-fourths ($\frac{3}{4}$). The numbers in such fractions are always considered to be exact numbers.

In contrast to counted or defined numbers, every measured number carries with it a degree of uncertainty or error (as previously noted in Sec. 2.2). Even when very elaborate and expensive measuring devices are used, some degree of uncertainty will always be present. Let us look at the origin of this uncertainty in more detail.

Consider how two different thermometer scales, illustrated in Figure 2.2, are used to measure a given temperature. Determining the temperature involves determining the height of the mercury column in the thermometer. The scale on the left in Figure 2.2 is marked off in one-degree intervals. Using this scale we can say with certainty that the temperature is between 29 and 30 degrees. We can further say that the actual temperature is closer to 29 degrees than to 30 and estimate it to be 29.2 degrees. The scale on the right has more subdivisions, being marked off in tenths of a degree rather than in degrees. Using this scale we can definitely say that the temperature is between 29.2 and 29.3 degrees and can estimate it to be 29.25 degrees. Note how both temperature readings contain some digits (all those except the last one) that are exactly known and one digit (the last one) that is estimated. Note also that the uncertainty in the second temperature reading is less than that in the first reading—an uncertainty in the hundredths place compared to an uncertainty in the tenths place. We say that the scale on the right is *more precise* than the one on the left.

Because measurements are never exact, anytime a scientist writes down a numerical value for a measurement, two kinds of information must be conveyed: (1) the magnitude of the measurement and (2) the precision or uncertainty of the measurement. The digit values give the magnitude. Precision is indicated by the number of significant figures recorded.

Significant figures *are the digits in any measurement that are known with certainty plus one digit that is uncertain.* Only one estimated digit is ever recorded as part of a measurement. It would be incorrect for a scientist to report that the height of the mercury column in Figure 2.2, as read on the scale on the right, corresponds to a temperature of 29.247 degrees. The value 29.247 contains two estimated digits (the 4 and the 7) and would indi-

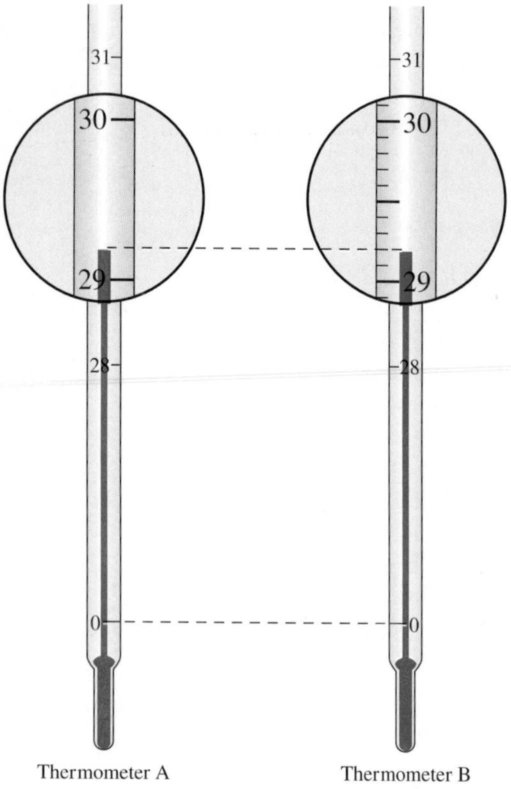

FIGURE 2.2 Measuring a temperature. A portion of the degree scale on each of the two differently scaled thermometers has been magnified.

cate a measurement of greater precision than is actually obtainable with that particular measuring device.

The magnitude of the uncertainty in the last significant digit in a measurement (the estimated digit) may be indicated using a "plus–minus" notation. The following three time measurements illustrate this notation.

$$15 \pm 1 \text{ seconds}$$
$$15.3 \pm 0.1 \text{ seconds}$$
$$15.34 \pm 0.03 \text{ seconds}$$

Most often the uncertainty in the last significant digit is one unit (as in the first two time measurements), but it may be larger (as in the third time measurement). In this text we will follow the almost universal practice of dropping the "plus–minus" notation if the magnitude of the uncertainty is one unit. Thus, in the absence of "plus–minus" notation you will be expected to assume that there is an uncertainty of one unit in the last significant digit. A measurement reported simply as 27.3 inches means 27.3 $\pm$ 0.1 inches. Only in the situation where the uncertainty is greater than one unit in the last significant digit will the amount of the uncertainty be explicitly shown.

The precision of a measurement is represented by the number of significant figures in the measurement. A measured length of 2.453 cm (centimeters) for an object is more precise than a measured length of 2.45 cm for the same object. Thus, the term *precision* refers not only to the degree of reproducibility of repeated measurements (Sec. 2.2) but also to the number of significant figures in a measurement. Example 2.1 relates measurement preciseness to actual measuring device scales.

Example 2.1

How many significant figures should be reported in each of the following volume measurements?

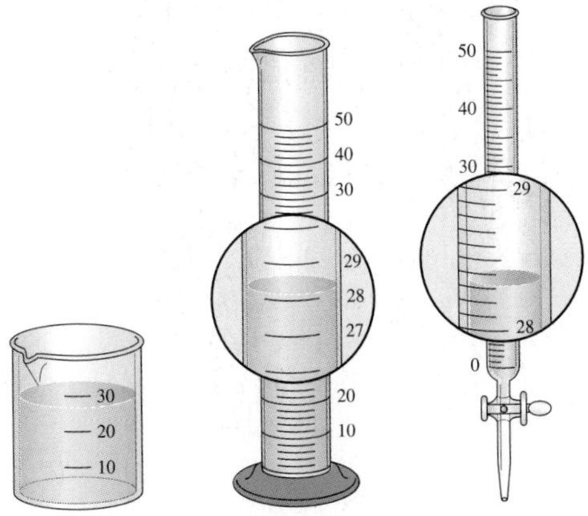

(a) Beaker calibrated in ten milliliter units

(b) Graduated cylinder calibrated in milliliters

(c) Buret calibrated in tenths of a milliliter

Solution

(a) We know definitely that the volume of liquid is between 20 and 30 milliliters. We estimate the final digit (to the closest milliliter) to be 8, giving a reading of 28. Thus, two significant figures are present.

(b) The level of the liquid is between 28 and 29 milliliters. We estimate the level to be at 28.3 milliliters. The value 28.3 has three significant figures.

(c) The buret is calibrated in tenths of a milliliter. We know for certain that the liquid level is between 28.3 and 28.4 milliliters. Adding one estimated digit (hundredths of a milliliter) gives a reading of 28.33 milliliters. This value contains four significant figures.

Practice Exercise 2.1

How many significant figures should be reported in each of the following measurements?

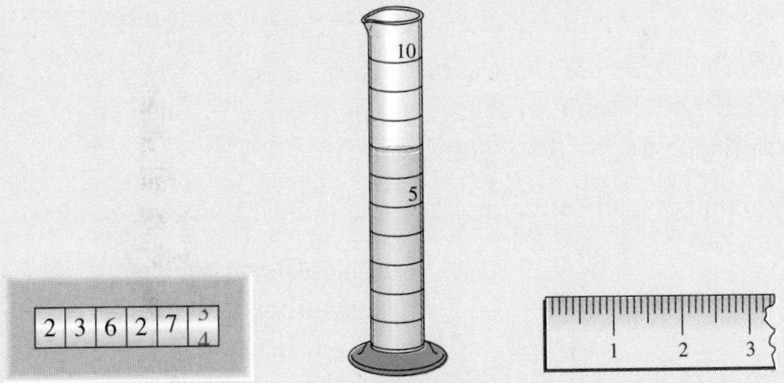

(a) Odometer calibrated in miles (the last digit is tenths of a mile)

(b) Graduated cylinder calibrated in milliliters

(c) Ruler calibrated in centimeters

Ans. (a) 7; (b) 2; (c) 3

Determining the number of significant figures in a measurement is not always as straightforward as Example 2.1 infers. In this example you knew the type of instrument used for each measurement and its limitations because you made the measurement. Quite often when someone else makes a measurement, such information is not available. All that is known is the reported final result—the numerical value of the measured quantity. In this situation questions do arise about the "significance" of various digits in the measurement. For example, consider the published value of the distance from the Earth to the sun, which is 93,000,000 miles. Intuition tells you that it is highly improbable that this distance is known to the closest mile. You suspect that this is an estimated distance. To what digit has this number been estimated? Is it to the nearest million miles, the closest hundred thousand miles, the nearest ten thousand miles, or what?

A set of guidelines has been developed to aid scientists in interpreting the significance of reported measurements or results calculated from measurements. Four rules constitute the guidelines: one rule for the digits 1 through 9 and three rules for the digit 0. A zero in a measurement may or may not be significant depending on its location in the sequence of digits forming the numerical value for the measurement. There is a rule for each of three classes of zeros—leading zeros, confined zeros, and trailing zeros.

RULE 1 The digits 1 through 9 inclusive (all of the nonzero digits) always count as significant figures.

14.232	five significant figures
3.11	three significant figures
244.6	four significant figures

RULE 2 *Leading zeros* are zeros that occur at the start of a number, that is, zeros that precede all nonzero digits. Such zeros do not count as significant figures. Their function is simply to indicate the position of the decimal point.

0.00045	two significant figures
0.0113	three significant figures
0.000000072	two significant figures

Leading zeros are always to the left to the first nonzero digit.

RULE 3 *Confined zeros* are zeros between nonzero digits. Such zeros always count as significant figures.

2.075	four significant figures
6007	four significant figures
0.03007	four significant figures

RULE 4 *Trailing zeros* are zeros at the end of a number. They are significant if (a) there is a decimal point present in the number or (b) they carry overbars. Otherwise trailing zeros are not significant.

The following numbers, all containing decimal points, illustrate condition (a) of rule 4.

62.00	four significant figures
24.70	four significant figures
0.02000	four significant figures
4300.00	six significant figures

By condition (b) trailing zeros in numbers lacking an explicitly shown decimal point become significant when marked with a bar above the zero(s).

$36{,}00\overline{0}$	five significant figures
$36{,}0\overline{0}0$	four significant figures
$36{,}\overline{0}00$	three significant figures
$10{,}02\overline{0}$	five significant figures

In cases involving trailing zeros where neither a decimal point nor overbar(s) are present the trailing zeros are not significant.

93,000,000	two significant figures
360,000	two significant figures
330,300	four significant figures
6310	three significant figures

Another method, more convenient than rule 4, for dealing with the significance of trailing zeros is to express the number in scientific notation. In this notation, to be presented in Section 2.5, only significant digits are shown.

Example 2.2

Determine the number of significant figures in the numerical value in each of the following statements.

(a) A 16 gauge wire has a diameter of 0.05082 inch.

(b) The mass of the earth is 6,600,000,000,000,000,000,000 tons.

(c) A hospital patient's blood glucose level was determined to be 4850 micrograms per milliliter of blood.

(d) Normal body temperature for a chickadee is 41.0°C.

Solution

(a) There are four significant figures. The leading zeros are not significant (rule 2) and the confined zero is significant (rule 3).

(b) There are two significant figures. The trailing zeros are not significant as no decimal point or overbar notation is present.

(c) There are three significant figures. The trailing zero is not significant (rule 4).

(d) There are three significant figures. The trailing zero is significant because a decimal point is present (rule 4).

Practice Exercise 2.2

Determine the number of significant figures in the numerical value in each of the following statements.

(a) A regular-issue U.S. postage stamp has a width of 0.021 meter.
(b) The melting point of the metal gold is 1064°C.
(c) The Earth's oceans and seas contain 330,000,000 cubic meters of seawater.
(d) The volume of a drop of water is 0.000050 liter.

Ans. (a) 2; (b) 4; (c) 2; (d) 2

Example 2.3

The number of carbon monoxide molecules in a sample of automobile exhaust is verbally reported as two hundred and five thousand. What meaning, in terms of significant figures and magnitude of uncertainty, is conveyed by each of the following written notations for this number?

(a) 205,000 **(b)** $205{,}\bar{0}00$ **(c)** 205,000. **(d)** $205{,}\overline{000}$

Solution

(a) Three significant figures are present in this number; the confined zero is significant but the trailing zeros are not. Since the last significant digit, the 5, is located in the fourth place to the left of the understood decimal point (the thousands place), the uncertainty is ±1000.

(b) This number has four significant figures. The overbar above the first of the three trailing zeros makes this zero significant. The last significant digit, the zero with the overbar, occupies the hundreds place in the number. Thus, the uncertainty is ±100.

(c) There are six significant figures present. Explicitly placing a decimal point at the end of the number makes all the trailing zeros significant. The uncertainty is ±1 since the last of the trailing zeros is in the ones position.

(d) With the overbar notation present on all trailing zeros, all six digits present are significant. The uncertainty is ±1 since the last of the trailing zeros is in the ones position.

Practice Exercise 2.3

The population of a town in southern England is verbally reported to be one hundred and thirty thousand. What meaning, in terms of significant figures and magnitude of uncertainty, is conveyed by each of the following written notations for this number?

(a) 130,000 (b) $13\bar{0}{,}000$ (c) $13\overline{0{,}00}0$ (d) 130,000.

Ans. (a) 2, ±10,000; (b) 3, ±1000; (c) 5, ±10; (d) 6, ±1

2.4 SIGNIFICANT FIGURES AND CALCULATED QUANTITIES

Most experimental measurements are not end results in themselves. Instead they function as intermediates in the calculation of other quantities. For example, we might measure in a laboratory the height, width, depth, and mass of a rectangular solid object and from this information calculate its volume and density.

In doing a calculation using experimental data, we must give major consideration to the number of significant figures in the computed result. Correct calculations never increase or decrease the precision of experimental measurements.

Concern about the number of significant figures in a calculated number is particularly critical when an electronic calculator is used to do the arithmetic of the calculation. Hand calculators now in common use are not programmed to take significant figures into account. Consequently, the digital readouts on them more often than not display more digits than are

FIGURE 2.3 The digital readout on the average electronic calculator usually shows more digits than are needed or justified. Electronic calculators are not programmed to take significant figures into account. (Science VU-Mettler/Visuals Unlimited)

needed. It is a mistake to record these extra digits, since they have no significance; that is, they are not significant figures (see Fig. 2.3).

In order to record correctly the numbers obtained through calculations, students must be able to (1) adjust (usually decrease) the number of digits in a number to give it the correct number of significant figures and (2) determine the allowable number of significant figures in the result of any mathematical operation. We will consider these skills in the order listed.

Rounding off *is the process of deleting unwanted (nonsignificant) digits from a calculated number.* Three simple rules govern the process.

RULE 1 If the first digit to be dropped is less than 5, that digit and all digits that follow it are simply dropped.

Thus, 62.312 rounded off to three significant figures becomes 62.3.

RULE 2 If the first digit to be dropped is a digit greater than 5, or a 5 followed by digits other than all zeros, the excess digits are all dropped and the last retained digit is increased in value by one unit.

Thus, 62.782 and 62.558 rounded off to three significant figures become, respectively, 62.8 and 62.6.

RULE 3 If the first digit to be dropped is a 5 not followed by any other digit or a 5 followed only by zeros, an odd–even rule applies. Drop the 5 and any zeros that follow it and then

(a) increase the last retained digit by one unit if it is *odd,* or

(b) leave the last retained digit the same if it is *even.*

Thus, 62.650 and 62.350 rounded to three significant figures become, respectively, 62.6 (even rule) and 62.4 (odd rule). The number zero as a last retained digit is always considered an even number; thus, 62.050 rounded to three significant figures becomes 62.0.

These rounding rules must be modified slightly when digits to the left of the decimal point are to be dropped. In order to maintain the inferred position of the decimal point in such situations, zeros must replace all of the dropped digits that are to the left of the inferred decimal point. Parts (c) and (f) of Example 2.4 illustrate this point.

Example 2.4

Round off each of the following numbers to two significant figures.

(a) 25.7 **(b)** 25.37 **(c)** 432,117

(d) 0.435 **(e)** 62.50 **(f)** 13,500

Solution

(a) Rule 2 applies. The last retained digit (the 5) is increased in value by one unit.

25.7 becomes 26

(b) Rule 1 applies. The last retained digit (the 5) remains the same, and all digits that follow it are simply dropped.

25.37 becomes 25

(c) Since the first digit to be dropped is a 2, rule 1 applies.

432,117 becomes 430,000

Note that to maintain the position of the inferred decimal point, zeros must replace all of the "dropped" digits. This will always be the case when digits to the left of the inferred decimal place are "dropped."

(d) Rule 3 applies. The first and only digit to be dropped is a 5. The last retained digit (the 3) is an odd number, so, using the odd–even rule, its value is increased by one unit.

0.435 becomes 0.44

(e) Rule 3 applies again. This time the last digit retained is even (the 2), so its value is not changed.

62.50 becomes 62

(f) This is a rule 3 situation again. Since an odd digit (the 3) occupies the second significant figure place, its value is increased by one.

13,500 becomes 14,000

Note again that zeros must take the place of all digits to the left of the inferred decimal place that are "dropped."

Practice Exercise 2.4

Round off each of the following numbers to three significant figures.

(a) 432.87	(b) 432.17	(c) 655,234
(d) 0.03315	(e) 352.50	(f) 162,500

Ans. (a) 433; (b) 432; (c) 655,000; (d) 0.0332; (e) 352; (f) 162,000

Calculations cannot improve the precision of measurements. Two operational rules exist to help ensure that calculations do not increase measurement precision. One rule covers the operations of multiplication and division and the other the operations of addition and subtraction.

RULE 1 *Multiplication and Division.* In multiplication and division the number of significant figures in the product or quotient is the same as in the number in the calculation that contains the fewest significant figures.

$$6.038 \times \mathbf{2.57} = 15.51766 \quad \text{(calculator answer)}$$
$$= 15.5 \quad \text{(correct answer)}$$

This number limits the answer to three significant figures.

RULE 2 *Addition and Subtraction.* In addition and subtraction the last digit retained in the sum or difference should correspond to the first doubtful (estimated) decimal position in the series of numbers.

25.657	
1.31	
13.2	This number limits the answer to the tenths decimal place.
40.167	(calculator answer)
40.2	(correct answer)

Note that for multiplication and division (rule 1) significant figures are counted and that for addition and subtraction (rule 2) decimal places are counted. The answers from either addition or subtraction can have more or fewer significant figures than any of the numbers that have been added or subtracted, as is shown in Example 2.7.

Example 2.5

Without actually doing any multiplications, indicate the number of significant figures that should be present in the answer to each of the multiplications. Assume that all numbers are measured quantities.

(a) 6.00 × 6.00 × 6.00 **(b)** 6.00 × 0.600 × 60.60

(c) 0.006 × 0.060 × 0.600 **(d)** 60,600 × 6060 × 606

Solution

(a) Each number to be multiplied contains three significant figures. Thus, the answer should also contain three significant figures.

(b) The first two input numbers contain three significant figures and the third number contains four significant figures. Only three significant figures should be present in the answer.

(c) The input numbers contain, respectively, one, two, and three significant figures. The one significant figure input number limits the answer to one significant figure.

(d) All three input numbers contain three significant figures. Thus, the answer should contain three significant figures.

Practice Exercise 2.5

Without actually doing any multiplications, indicate the number of significant figures that should be present in the answer to each of the multiplications. Assume that all numbers are measured quantities.

(a) 2.0 × 2.0 × 3.0 (b) 2.00 × 2.00 × 3.0
(c) 2.000 × 2.00 × 3.00 (d) 2.0000 × 2.00 × 3.0000

Ans. (a) 2; (b) 2; (c) 3; (d) 3

Example 2.6

Perform the following computations, all of which involve multiplication and/or division. Express your answers to the proper number of significant figures. Assume that all numbers are measured quantities.

(a) 3.751×0.42 **(b)** $\dfrac{1,810,000}{3.1453}$

(c) $\dfrac{1800.0}{6.0000}$ **(d)** $\dfrac{3.130 \times 3.140}{3.15}$

Solution

(a) The calculator answer to this problem is

$$3.751 \times 0.42 = 1.57542$$

The input number with the least number of significant figures is 0.42, which has two significant figures. Thus the calculator answer must be rounded off to two significant figures.

$$\underset{\text{(calculator answer)}}{1.57542} \quad \text{becomes} \quad \underset{\text{(correct answer)}}{1.6}$$

(b) The calculator answer to this problem is

$$\frac{1,810,000}{3.1453} = 575,461.8$$

The input number 1,810,000 contains three significant figures and the input number 3.1453 has five significant figures. Thus the correct answer is limited to three significant figures and is obtained by rounding the calculator answer to three significant figures.

$$\underset{\text{(calculator answer)}}{575,461.8} \quad \text{becomes} \quad \underset{\text{(correct answer)}}{575,000}$$

A decimal point is not explicitly shown in the number 575,000 since doing so would make the trailing zeros significant.

(c) The calculator answer to this problem is

$$\frac{1800.0}{6.0000} = 300$$

Both input numbers contain five significant figures. Thus the correct answer must also contain five significant figures.

$$\underset{\text{(calculator answer)}}{300} \quad \text{becomes} \quad \underset{\text{(correct answer)}}{300.00}$$

Note here how the calculator answer had too few significant figures. Most calculators cut off zeros after the decimal point even if they are significant. Using too few significant figures in an answer is just as wrong as using too many.

(d) This problem involves both multiplication and division. The calculator answer is

$$\frac{3.130 \times 3.140}{3.15} = 3.1200634$$

The input number with the least number of significant figures is 3.15, which contains three significant figures. Thus, the calculator answer must be rounded off to three significant figures.

3.1200634	becomes	3.12
(calculator answer)		(correct answer)

Practice Exercise 2.6

Perform the following computations, all of which involve multiplication and/or division. Express your answers to the proper number of significant figures. Assume that all numbers are measured quantities.

(a) 6.7321×0.0021

(b) $\dfrac{16{,}240}{23.42}$

(c) $\dfrac{120.0}{4.000}$

(d) $\dfrac{5.444 \times 8.670}{2.321 \times 3.27}$

Ans. (a) 0.014; (b) 693.4; (c) 30.00; (d) 6.22

Example 2.7

Perform the following computations, all of which involve addition or subtraction. Express your answers to the proper number of significant figures. Assume that all numbers are measured quantities.

(a) $13.01 + 13.001 + 13.010$

(b) $10.2 + 3.4 + 6.01$

(c) $0.6700 - 0.6644$

(d) $34.7 + 0.0007$

Solution

(a) The calculator answer to this problem is

$$13.01 + 13.001 + 13.010 = 39.021$$

Since this is an addition problem, in going from the calculator answer to the correct answer we must consider uncertainties rather than significant figures. (The multiplication–division rule is based on significant figures and the addition–subtraction rule is based on uncertainties.)

The uncertainties in the input numbers are

13.01	hundredths
13.001	thousandths
13.010	thousandths

The number 13.01 has the greatest uncertainty (hundredths) and so the last retained digit in the correct answer should reflect this uncertainty. Hence, the calculator answer is rounded off to hundredths.

39.021	becomes	39.02
(calculator answer)		(correct answer)

(b) The calculator answer to this problem is

$$10.2 + 3.4 + 6.01 = 19.61$$

The uncertainty in the first two input numbers is tenths and the third input number involves an uncertainty of hundredths. Thus, the last retained digit in the correct answer will be in the tenths place, the largest uncertainty among the input numbers.

19.61 becomes 19.6
(calculator answer) (correct answer)

Note that the input number 3.4 possesses two significant figures and yet the correct answer contains three significant figures. Why? The number of significant figures is not the determining factor in addition and subtraction (rule 2) as it is in multiplication and division (rule 1).

(c) The calculator answer to this problem is

$$0.6700 - 0.6644 = 0.0056$$

Both input numbers are known to the ten-thousandths place. Thus, the answer should also have an uncertainty involving the ten-thousandths place.

In this particular problem the calculator answer and the correct answer are the same, a situation which does not occur very often. The correct answer is 0.0056. Note that two significant figures were "lost" in the subtraction. The answer has two significant figures. The two input numbers each have four significant figures. For addition and subtraction this is allowable; for multiplication and division it would not be allowable.

(d) The calculator answer to this problem is

$$34.7 + 0.0007 = 34.7007$$

The uncertainty in the input number 34.7 is tenths and the uncertainty in the input number 0.0007 is ten-thousandths. Thus, the calculator answer must be rounded off to the tenths place.

34.7007 becomes 34.7
(calculator answer) (correct answer)

The correct answer, 34.7, is the same as one of the input numbers. The message of this is that the number 0.0007 is negligible when added to the number 34.7.

Practice Exercise 2.7

Perform the following computations, all of which involve addition or subtraction. Express your answers to the proper number of significant figures. Assume that all numbers are measured quantities.

(a) $28.7 + 7.01 + 22$
(b) $8.3 + 1.2 + 1.7$
(c) $0.4378 - 0.4367$
(d) $4200 + 14.7$

Ans. (a) 58; (b) 11.2; (c) 0.0011; (d) 4200

Not all numbers in a calculation have to originate from a measurement. Sometimes *exact numbers* (counted or defined numbers and simple fractions) are part of a calculation. Since there is no uncertainty in an exact number, it is considered to possess an infinite number of significant figures. Consequently, exact numbers, when they appear in a calculation, will never limit the number of significant figures allowable in the answer. We do not consider them when determining the number of allowable significant figures. The allowable number of significant figures is determined in the usual way, taking into account only those numbers that were experimentally determined by measurement.

Example 2.8

What would be the total combined length, in centimeters, of 7 new pencils, each of which has a length of 19.13 centimeters?

Solution

Since a pencil has a length of 19.13 centimeters, the length of 7 such pencils will be

$$7 \times 19.13 \text{ cm} = 133.91 \text{ cm} \quad \text{(calculator answer)}$$

The number 7 is an exact number (a counted number) and can be considered to have an infinite number of significant figures. Thus, the correct answer will have four significant figures, the number of significant figures in the measurement 19.13.

$$133.91 \quad \text{becomes} \quad 133.9 \quad \text{(correct answer)}$$

Practice Exercise 2.8

A nickel is found to weigh 5.0715 grams (g). What would be the mass in grams of 13 such coins?

Ans. 65.930 g

2.5 SCIENTIFIC NOTATION

In scientific work, very large and very small numbers are frequently encountered. For example, in one cup of water—an "ordinary" amount of water—there are approximately

7,910,000,000,000,000,000,000,000 molecules

(A molecule of water is the smallest possible unit of water; see Sec. 5.2.) A single water molecule has a mass of

0.0000000000000000000000299 gram

Such large and small numbers as these are difficult to use. Recording them is not only time consuming but also very prone to errors—too many or too few zeros recorded. Also such numbers are awkward to work with in calculations. Consider the problem of manually multiplying the above two numbers together. Handling all of those zeros is "mind boggling."

A method exists for expressing cumbersome many-digit numbers, such as the two just shown, in compact form. Called scientific notation, this method eliminates the need to write all the zeros. **Scientific notation** *is a system in which an ordinary decimal number is expressed as a product of a number between 1 and 10 multiplied by 10 raised to a power.* The two previously cited numbers, dealing with molecules of water, expressed in scientific notation are, respectively,

$$7.91 \times 10^{24} \text{ molecules} \qquad \text{and} \qquad 2.99 \times 10^{-23} \text{ gram}$$

Note that in scientific notation each number requires only seven digits, compared to 25 in normal arithmetic notation.

Exponents

An understanding of exponents is the key to handling scientific notation. A brief review of exponents and their use is in order before we consider the rules for converting numbers from ordinary decimal notation to scientific notation and vice versa.

In mathematics a convenient method exists for showing that a number has been multiplied by itself two or more times. It involves the use of an exponent. An **exponent** *is a number written as a superscript following another number and indicates how many times the first number, the base, is to be multiplied by itself.* The following examples illustrate the use of exponents.

$$6^2 = 6 \times 6 = 36$$
$$3^5 = 3 \times 3 \times 3 \times 3 \times 3 = 243$$
$$10^3 = 10 \times 10 \times 10 = 1000$$

Exponents are also frequently referred to as *powers* of numbers. Thus, 6^2 may be verbally read as "six to the second power," and 3^5 as "three to the fifth power." Raising a number to the second power is often called "squaring," and raising it to the third power "cubing."

Scientific notation exclusively uses powers of ten. *When ten is raised to a positive power, its decimal equivalent is the number 1 followed by as many zeros as the power.* This one-to-one correlation between power magnitude and number of zeros is shown in color in the following examples.

$10^2 = 100$ (two zeros and a power of 2)

$10^4 = 10{,}000$ (four zeros and a power of 4)

$10^6 = 1{,}000{,}000$ (six zeros and a power of 6)

The notation 10^0 is a defined quantity.

$$10^0 = 1$$

The above generalization easily explains why. Ten to the zero power is the number 1 followed by no zeros, which is simply 1.

All of the examples of exponential notation presented so far have had positive exponents. This is because each example represented a number of magnitude greater than one. Negative exponents are also possible. They are associated with numbers of magnitude less than one.

A negative sign in front of an exponent is interpreted to mean that the base and the power to which it is raised are in the denominator of a fraction in which 1 is the numerator. The following examples illustrate this interpretation.

$$10^{-1} = \frac{1}{10^1} = \frac{1}{10} = 0.1$$

$$10^{-2} = \frac{1}{10^2} = \frac{1}{10 \times 10} = \frac{1}{100} = 0.01$$

$$10^{-3} = \frac{1}{10^3} = \frac{1}{10 \times 10 \times 10} = \frac{1}{1000} = 0.001$$

When the number 10 is raised to a negative power, the absolute value of the power (the value ignoring the minus sign) is always one more than the number of zeros between the decimal point and the one. This correlation between power magnitude and number of zeros is shown in color in the following examples.

$10^{-2} = 0.01$	(one zero and a power of −2)
$10^{-4} = 0.0001$	(three zeros and power of −4)
$10^{-6} = 0.000001$	(five zeros and power of −6)

Differences in magnitude between numbers that are powers of 10 are often described with the phrase "orders of magnitude." An **order of magnitude** *is a single exponential value of the number 10.* Thus, 10^6 is four orders of magnitude larger than 10^2, and 10^7 is three orders of magnitude larger than 10^4.

Writing Numbers in Scientific Notation

A number written in scientific notation has two parts: (1) a *coefficient*, written first, which is a number between 1 and 10, and (2) an *exponential term*, which is 10 raised to a power. The coefficient part is always multiplied by the exponential term. Using the scientific notation form of the number 703 as an example, we have

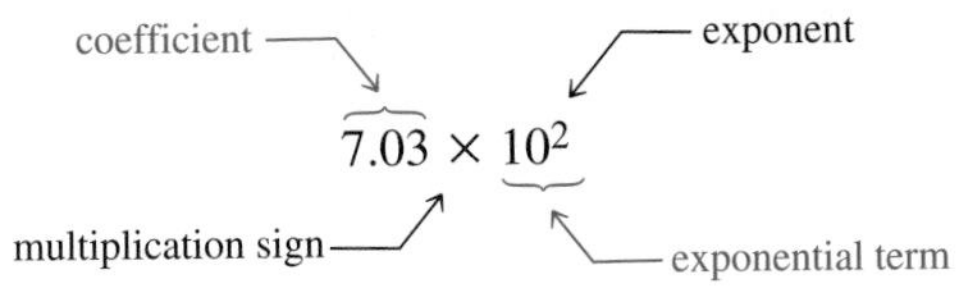

When a number is expressed in scientific notation, *only significant digits become part of the coefficient.* Because of this there is never any confusion (ambiguity) in determining the number of significant figures in a number expressed in scientific notation. There are five possible precision interpretations for the number 10,000 (1, 2, 3, 4, or 5 significant figures). In scientific notation each of these interpretations assumes a different form.

1×10^4	(10,000 with one significant figure)
1.0×10^4	($1\overline{0},000$ with two significant figures)
1.00×10^4	($10,\overline{0}00$ with three significant figures)
1.000×10^4	($10,0\overline{0}0$ with four significant figures)
1.0000×10^4	($10,00\overline{0}$ with five significant figures)

Most scientists use the preceding scientific notation method for designating significant figures in an "ambiguous" number instead of the overbar notation discussed in Section 2.3.

The overbar notation is used only in those situations where there is reason for not expressing the number in exponential notation.

Rules for converting from decimal to scientific notation are as follows.

RULE 1 The coefficient must be a number between 1 and 10 which contains the same number of significant figures as are present in the original decimal number. The coefficient is obtained by rewriting the decimal number with a decimal point after the first nonzero digit and all *nonsignificant* zeros deleted.

For 233,000 the coefficient is 2.33

For 0.00557 the coefficient is 5.57

For 0.35500 the coefficient is 3.5500

RULE 2 The value of the exponent for the power of ten is obtained by counting the number of places the decimal point in the coefficient must be moved to give back the original decimal number. If the decimal point movement is to the *right* the exponent has a *positive* value and if the decimal point movement is to the *left* the exponent has a *negative* value.

Numerous applications of these two rules are found in Examples 2.9 and 2.10

Example 2.9

Express in scientific notation the number in each of the following statements.

(a) Light travels at a speed of 186,000 miles per second.

(b) A person exhales approximately 320,000,000,000,000,000,000 molecules of carbon dioxide in one breath.

(c) The diameter of a human hair is found to be 0.0016 inch.

(d) The maximum allowable amount of chromium in drinking water (EPA standard) is 0.00000010 gram per milliliter of water.

Solution

(a) For the number 186,000 the scientific notation coefficient is 1.86. This coefficient meets the requirement that it contain the same number of significant figures as the original decimal number.

The value of the exponent in the exponential term of the scientific notation is +5 since the decimal point in the coefficient must be moved 5 places to the right to generate the original decimal number.

1.86000

5 place movement

Note that zeros are added to the coefficient as the decimal point is moved in order to obtain the original decimal point position.

Multiplying the coefficient by the exponential term 10^5 gives the scientific notation form of the number, which is

$$1.86 \times 10^5$$

(b) The coefficient for the number 320,000,000,000,000,000,000 is 3.2. It, like the original number, contains two significant figures.

The value of the exponent for the exponential term is +20 since the decimal point in the coefficient must be moved 20 places to the right to generate the original decimal number.

3.20000000000000000000

20 place movement

Movement of the decimal point in the coefficient to the *right* always results in a *positive* exponent.

The scientific notation form of the number, obtained by multiplying the coefficient and exponential term is

$$3.2 \times 10^{20}$$

(c) The scientific notation coefficient for the number 0.0016 is 1.6. The exponent for the power of ten is -3, since moving the decimal point in the coefficient three places to the left generates the original decimal number.

0.001.6

3 place movement

Movement of the decimal point in the coefficient to the *left* always means that the exponent will be *negative*. The scientific notation form of the number is, thus, 1.6×10^{-3}.

(d) The scientific notation coefficient for the number 0.00000010 is 1.0. The coefficient is 1.0 rather than 1 because the original number has two significant figures. The exponent for the power of ten is -7.

0.0000001.0

7 place movement

The number in scientific notation is, thus, 1.0×10^{-7}.

Practice Exercise 2.9

Express in scientific notation the number in each of the following statements.

(a) The Yangzi River, which flows through China, is 3915 miles long.
(b) The distance from the earth to the sun is 93,000,000 miles.
(c) The maximum allowable amount of mercury in drinking water (EPA standard) is 0.0000015 gram per milliliter of water.
(d) The mass of a single carbon monoxide molecule is 0.0000000000000000000000465 gram.

Ans. (a) 3.915×10^3; (b) 9.3×10^7; (c) 1.5×10^{-6}; (d) 4.65×10^{-23}

Scientific notation is not often used to express numbers that are more easily written in decimal form. For example, the numbers 2.4, 0.911, and 57 are simpler in their original form, and we would probably not use scientific notation for them. Although there are no fixed rules as to when scientific notation should be used, it is preferable to use it for numbers greater than 10^3 or less than 10^{-1}.

Converting from Scientific Notation to Decimal Notation

To convert a number in scientific notation, such as 6.02×10^{23}, into a regular decimal number, we start by examining the exponent. The value of the exponent tells how many places the decimal point must be moved. If the exponent is positive, movement is to the right to give a number greater than one; if it is negative, movement is to the left to give a number less than one. Zeros may have to be added to the number as the decimal point is moved.

Example 2.10

Convert the scientific notation number in each of the following statements to a decimal number.

(a) The announced attendance at a football game was 5.3127×10^4.

(b) The circumference of the Earth is 2.5×10^4 miles.

(c) The concentration of gold in seawater is 1.1×10^{-8} gram per liter.

(d) The mass of a hydrogen atom is 1.67×10^{-24} gram.

Solution

(a) The exponent +4 tells us the decimal is to be located four places to the right of where it is in 5.3127.

5.3127

decimal point shift

The decimal number is 53,127.

(b) The exponent +4 tells us the decimal is to be located four places to the right of where it is in 2.5. Trailing zeros will have to be added to accommodate the decimal point change.

2.5000 added zeros

decimal point shift

These added "trailing zeros" are not significant zeros. Thus, the number of significant digits remains at two. The decimal form of the number is

25,000

(c) The exponent −8 tells us the decimal is to be located eight places to the left of where it is in 1.1. Leading zeros will have to be added to accommodate the decimal point change.

added zeros 0.00000001.1

decimal point shift

These added "leading zeros" are not significant zeros. Thus, the number of significant digits remains at two. The decimal form of the number is

0.000000011

(d) The exponent -24 tells us the decimal is to be located twenty-four places to the left of where it is in 1.67. This will produce an extremely small number.

$$0.00000000000000000000000\underleftarrow{1\,}67$$

decimal point shift

Twenty-three leading zeros were needed to mark the new decimal place. (In numbers with negative exponents, the number of added leading zeros will always be one less than the value of the exponent.) The decimal form of this number is, thus

$$0.00000000000000000000000167$$

The number of significant figures present is three, the same number as in the original scientific notation form of the number.

Practice Example 2.10

Convert the scientific notation number in each of the following statements to a decimal number.

(a) A supersonic transport (SST) airplane consumes about 1.8×10^4 liters of fuel per hour of flight.
(b) Approximately 4.5×10^{10} aspirin tablets are consumed annually in the United States.
(c) The naked eyes can detect an object that is 1×10^{-4} meter in diameter.
(d) The diameter of an influenza virus is 1×10^{-8} meter.

2.6 SCIENTIFIC NOTATION AND MATHEMATICAL OPERATIONS

A major advantage of writing numbers in scientific notation is that it greatly simplifies the mathematical operations of multiplication and division.

Multiplication in Scientific Notation

Multiplication of two or more numbers expressed in scientific notation involves two separate operations or steps.

STEP 1 Multiply the coefficients (the decimal numbers between 1 and 10) together in the usual manner.

STEP 2 *Add* algebraically the exponents of the powers of ten to obtain a new exponent.

In general terms, we can represent the multiplication of two scientific notation numbers as follows.

$$(a \times 10^x) \times (b \times 10^y) = ab \times 10^{x+y}$$

Example 2.11

Carry out the following multiplications in scientific notation. Be sure to take into account significant figures in obtaining your final answer.

(a) $(1.113 \times 10^3) \times (7.200 \times 10^5)$ **(b)** $(2.05 \times 10^{-3}) \times (1.19 \times 10^{-7})$

(c) $(4.21 \times 10^{-9}) \times (2.107 \times 10^6)$ **(d)** $(7.92 \times 10^{10}) \times (2.3 \times 10^{-4})$

Solution

(a) Multiplying the two coefficients together gives

$$1.113 \times 7.200 = 8.0136$$

Since each of the input numbers for the multiplication has four significant figures, the answer should also have four significant figures, not five as given by the calculator. With rounding,

$$\underset{\text{(calculator answer)}}{8.0136} \quad \text{becomes} \quad \underset{\text{(correct answer)}}{8.014}$$

Multiplication of the two powers of ten to give the exponential part of the answer requires adding the exponents to give a new exponent.

$$10^3 \times 10^5 = 10^{3+5} = 10^8$$

Combining the new coefficient with the new exponential term gives the answer

$$8.014 \times 10^8$$

(b) Multiplying the two coefficients together gives

$$2.05 \times 1.19 = 2.4395$$

Since both input numbers for the multiplication contain three significant figures, the calculator answer must be rounded to three significant figures.

$$\underset{\text{(calculator answer)}}{2.4395} \quad \text{becomes} \quad \underset{\text{(correct answer)}}{2.44}$$

Next the exponents of the powers of ten are added to generate the new power of ten.

$$10^{-3} \times 10^{-7} = 10^{(-3)+(-7)} = 10^{-10}$$

(To add two numbers of the same sign, either positive or negative, just add the numbers and place the common sign in front of the sum.) Combining the coefficient and the exponential term gives the answer of

$$2.44 \times 10^{-10}$$

(c) Multiplying the two coefficients together gives

$$4.21 \times 2.107 = 8.87047$$

Because the number 4.21 contains only three significant figures, the answer is also limited to three significant figures. Thus,

$$\underset{\text{(calculator answer)}}{8.87047} \quad \text{becomes} \quad \underset{\text{(correct answer)}}{8.87}$$

In combining the exponential terms we will have to add exponents with different signs, −9 and +6. To do this, we first determine the larger number (9 is larger than 6) and then subtract the smaller number from it (9 − 6 = 3). The sign is always the sign of the larger number (minus in this case). Thus

$$(-9) + (+6) = (-3)$$

and

$$10^{-9} \times 10^{6} = 10^{(-9)+(+6)} = 10^{-3}$$

Combining the coefficient and the exponential parts gives

$$8.87 \times 10^{-3}$$

(d) Multiplying the coefficients gives

$$7.92 \times 2.3 = 18.216$$

The answer must be rounded to two significant figures because the input number 2.3 has only two significant figures.

$$\underset{\text{(calculator answer)}}{18.216} \quad \text{becomes} \quad \underset{\text{(correct answer)}}{18}$$

Again, in combining the exponential terms we have exponents of different signs. The smaller exponent (4) is subtracted from the larger exponent (10), and the sign of the larger exponent (+) is used. Thus,

$$(+10) + (-4) = (+6)$$

and

$$10^{10} \times 10^{-4} = 10^{6}$$

Combining the coefficient and the exponential term gives

$$18 \times 10^{6}$$

This answer has something wrong with it; it is not in correct scientific notation form. The coefficient is not a number between one and ten. This problem is corrected by recognizing that 18 is equal to 1.8×10^1, making this substitution for 18, and then combining exponential terms.

$$18 \times 10^{6} = 1.8 \times 10^{1} \times 10^{6} = 1.8 \times 10^{7}$$

The correct answer is 1.8×10^7.

Practice Exercise 2.11

Carry out the following multiplications in scientific notation. Be sure to take into account significant figures in obtaining your final answer.

(a) $(2.543 \times 10^3) \times (2.003 \times 10^6)$
(b) $(1.15 \times 10^{-2}) \times (4.52 \times 10^{-3})$
(c) $(4.210 \times 10^{-5}) \times (2.0 \times 10^2)$
(d) $(5.329 \times 10^{-1}) \times (3.11 \times 10^9)$

Ans. (a) 5.094×10^9; (b) 5.20×10^{-5}; (c) 8.4×10^{-3}; (d) 1.66×10^9

Division in Scientific Notation

Division of two numbers expressed in scientific notation involves two separate operations or steps.

STEP 1 Divide the coefficients (the decimal numbers between 1 and 10) in the usual manner.

STEP 2 *Subtract* algebraically the exponent in the denominator (bottom) from the exponent in the numerator (top) to give the exponent of the new power of ten.

Note in multiplication we add exponents, and in division we subtract exponents.

Example 2.12

Carry out the following divisions in scientific notation. Be sure to take significant figures into account in obtaining your final answer.

(a) $\dfrac{2.05 \times 10^5}{1.19 \times 10^3}$ **(b)** $\dfrac{7.200 \times 10^{-3}}{1.113 \times 10^{-7}}$

(c) $\dfrac{4.21 \times 10^{-9}}{2.107 \times 10^6}$ **(d)** $\dfrac{3.92 \times 10^{10}}{9.1 \times 10^{-4}}$

Solution

(a) Performing the indicated division involving the coefficients gives

$$\frac{2.05}{1.19} = 1.722689$$

Since both input numbers for the division have three significant figures, the calculator answer must be rounded off to three significant figures.

1.722689 (calculator answer) becomes 1.72 (correct answer)

Dividing exponential terms involves the algebraic subtraction of exponents.

$$\frac{10^5}{10^3} = 10^{(+5)-(+3)} = 10^2$$

Algebraic subtraction involves changing the sign of the number to be subtracted and then following the rules for addition (as outlined in Example 2.11). In this problem the number to be subtracted (+3) becomes, upon changing the sign, −3. Then we add +5 and −3. The answer is +2, as shown in the preceding equation. Combining the coefficient and the exponential term gives

$$1.72 \times 10^2$$

(b) The new coefficient is obtained by dividing 1.113 into 7.200

$$\frac{7.200}{1.113} = 6.4690026$$

The correct answer will contain four significant figures, the same number as in both input numbers.

6.4690026 (calculator answer) becomes 6.469 (correct answer)

The exponential part of the answer is obtained by subtracting −7 from −3. Changing the sign of the number to be subtracted gives +7. Adding +7 and −3 gives +4. Therefore:

$$\frac{10^{-3}}{10^{-7}} = 10^{(-3)-(-7)} = 10^{4}$$

Combining the coefficient and the exponential term gives

$$6.469 \times 10^{4}$$

(c) One input number for the coefficient division has three significant figures and the other has four significant figures. Therefore, the answer should contain three significant figures.

$$\frac{4.21}{2.107} = 1.9981015$$

1.9981015 (calculator answer) becomes 2.00 (correct answer)

The exponential term division involves subtracting +6 from −9. Changing the sign of the number to be subtracted gives a −6. Adding −6 and −9 gives −15.

$$\frac{10^{-9}}{10^{6}} = 10^{(-9)-(+6)} = 10^{-15}$$

Combining the two parts of the problem gives the number

$$2.00 \times 10^{15}$$

(d) The new coefficient, obtained by dividing 3.92 by 9.1, should contain two significant figures, the same number as in the input number 9.1

$$\frac{3.92}{9.1} = 0.43076923$$

0.43076923 (calculator answer) becomes 0.43 (correct answer)

Performing the exponential term division by subtracting the powers of the exponential terms gives

$$\frac{10^{10}}{10^{-4}} = 10^{(+10)-(-4)} = 10^{14}$$

Combining the coefficient and the exponential term gives

$$0.43 \times 10^{14}$$

which is not in correct scientific notation form because the coefficient is a number less than one. This problem is remedied by recognizing that 0.43 is equal to 4.3×10^{-1}, making this substitution for 0.43, and then combining the two exponential terms.

$$0.43 \times 10^{14} = (4.3 \times 10^{-1}) \times 10^{14} = 4.3 \times 10^{13}$$

The correct answer is, thus, 4.3×10^{13}.

Practice Exercise 2.12

Carry out the following divisions in scientific notation. Be sure to take significant figures into account in obtaining your final answer.

(a) $\dfrac{3.76 \times 10^{9}}{1.23 \times 10^{6}}$ (b) $\dfrac{9.98 \times 10^{-3}}{2.341 \times 10^{-7}}$

(c) $\dfrac{5.1 \times 10^{-10}}{8.76 \times 10^{7}}$ (d) $\dfrac{3.43 \times 10^{5}}{3.93 \times 10^{-4}}$

Ans. (a) 3.06×10^{3}; (b) 4.26×10^{4}; (c) 5.8×10^{-18}; (d) 8.73×10^{8}

Addition and Subtraction in Scientific Notation

To add or subtract numbers written in scientific notation, *the power of ten for all numbers must be the same*. More often than not, one or more exponents must be adjusted. Adjusting the exponent requires rewriting the number in a form where the coefficient is a number greater than 10 or less than 1. With exponents all the same, the *coefficients are then added or subtracted and the exponent is maintained at its now common value*. Although any of the exponents may be changed, changing the smaller exponent to a larger one will usually produce a coefficient in the answer that is a number between 1 and 10.

Example 2.13

Perform the following additions or subtractions with all numbers expressed in scientific notation. Answers will need to be checked for the correct number of significant figures and for the correct scientific notation form (coefficient is a number between 1 and 10).

(a) $(2.661 \times 10^{3}) + (3.001 \times 10^{3})$

(b) $(2.66 \times 10^{4}) - (1.03 \times 10^{3})$

(c) $(9.98 \times 10^{-3}) + (8.04 \times 10^{-5})$

Solution

(a) The exponents are the same to begin with. Therefore, we can proceed with the addition immediately

$$\begin{array}{l} 2.661 \times 10^{3} \\ \underline{3.011 \times 10^{3}} \\ 5.672 \times 10^{3} \end{array} \quad \text{(calculator and correct answer)}$$

Both input numbers have uncertainties in the thousandths place. The calculator answer has the same uncertainty. Thus, the calculator answer and correct answer are the same. Note that the exponent values are not added. They are maintained at their common value.

(b) The exponents are different, so before subtracting we must change one of the exponents. Let us change 10^{3} to 10^{4}

$$10^{3} \text{ can be written as } 10^{-1} \times 10^{4}$$

Then, by substitution, we have

$$1.03 \times \underbrace{10^3} = 1.03 \times \underbrace{10^{-1} \times 10^4}$$

The coefficient and the first exponent are then combined to give a new coefficient.

$$\underbrace{1.03 \times 10^{-1}} \times 10^4 = \underbrace{0.103} \times 10^4$$

We are now ready to make the subtraction called for in the original statement of the problem.

$$\begin{array}{rrl} 2.66 \times 10^4 = & 2.66 \times 10^4 & \text{(common exponent)} \\ -\,1.03 \times 10^3 = & -0.103 \times 10^4 & \\ \hline & 2.557 \times 10^4 & \text{(calculator answer)} \end{array}$$

The calculator answer must be adjusted for significant figures. On the common exponent basis of 10^4, the uncertainty in 2.66 lies in the hundredths place and that in 0.103 lies in the thousandths place. The correct answer, therefore, is limited to an uncertainty of hundredths. (Recall, from Example 2.6, the rules on addition and significant figures.) Thus,

$$2.557 \times 10^4 \quad \text{becomes} \quad 2.56 \times 10^4 \quad \text{(correct answer)}$$

(c) The exponents are 10^{-3} and 10^{-5}. Since -5 is smaller than -3, let us have 10^{-3} as the common exponent. (Always use the larger of the two exponents as the common exponent.)

$$10^{-5} \quad \text{can be rewritten as} \quad 10^{-2} \times 10^{-3}$$

Then by substitution we have

$$8.04 \times \underbrace{10^{-5}} = 8.04 \times \underbrace{10^{-2} \times 10^{-3}}$$

The coefficient and the first exponent are then combined to give a new coefficient.

$$\underbrace{8.04 \times 10^{-2}} \times 10^{-3} = \underbrace{0.0804} \times 10^{-3}$$

We are now ready to make the addition called for in the original statement of the problem.

$$\begin{array}{rrl} 9.98 \times 10^{-3} = & 9.98 \times 10^{-3} & \text{(common exponent)} \\ 8.04 \times 10^{-5} = & 0.0804 \times 10^{-3} & \\ \hline & 10.0604 \times 10^{-3} & \text{(calculator answer)} \end{array}$$

The calculator answer must be adjusted for significant figures and also changed into correct scientific notation since the coefficient has a value greater than 10. The significant figure adjustment rounds the answer to hundredths, giving

$$10.06 \times 10^{-3}$$

The correct scientific notation adjustment involves rewriting 10.06 as a power of ten and then simplifying the resulting expression.

$$\underbrace{10.06}\times 10^{-3} = \underbrace{1.006 \times 10^{1}}\times 10^{-3} = 1.006 \times 10^{-2} \text{ (correct answer)}$$

Practice Exercise 2.13

Perform the following additions and subtractions with all numbers expressed in scientific notation. Answers will need to be checked for the correct number of significant figures and for the correct scientific notation form (coefficient is a number between 1 and 10).

(a) $(2.723 \times 10^4) + (6.045 \times 10^4)$
(b) $(3.73 \times 10^3) - (3.73 \times 10^2)$
(c) $(9.97 \times 10^{-2}) + (5.89 \times 10^{-4})$

Ans. (a) 8.768×10^4; (b) 3.36×10^3; (c) 1.003×10^{-1}

KEY TERMS

The new terms or concepts defined in this chapter are

accuracy (Sec. 2.2) How close a measurement (or the average of multiple measurements) comes to the true or accepted value.

exponent (Sec. 2.5) A number written as a superscript following another number that indicates how many times the first number is to be multiplied by itself.

order of magnitude (Sec. 2.5) A single exponential value of the number ten.

precision (Sec. 2.2) How close multiple measurements of the same quantity come to each other.

random error (Sec. 2.2) An error originating from uncontrolled variables in an experiment.

rounding off (Sec. 2.4) The process of deleting unwanted (nonsignificant) digits from a calculated number.

scientific notation (Sec. 2.5) A system in which an ordinary decimal number is expressed as a number between 1 and 10 multiplied by 10 raised to a power.

significant figures (Sec. 2.3) Digits in any measurement that are known with certainty plus one digit that is uncertain.

systematic error (Sec. 2.2) An error originating from controlled variables in an experiment.

PRACTICE PROBLEMS

Accuracy and Precision (Sec. 2.2)

2.1 With a very precise volumetric measuring device, the volume of a liquid sample is determined to be 6.321 L (liters). Three students are asked to determine the volume of the same liquid sample using a less precise measuring device. How do you evaluate the following work of the three students with regard to precision and accuracy?

	Students		
Trials	A	B	C
1	6.35 L	6.31 L	6.36 L
2	6.31 L	6.32 L	6.36 L
3	6.38 L	6.33 L	6.35 L
4	6.32 L	6.32 L	6.36 L

2.2 With a very precise measuring device, the length of an object is determined to be 13.452 mm (millimeters). Three students are asked to determine the length of the same object using a less precise measuring device. How do you evaluate the following work of the three students with regard to precision and accuracy?

	Students		
Trials	A	B	C
1	13.6 mm	13.4 mm	13.9 mm
2	13.9 mm	13.5 mm	13.9 mm
3	13.3 mm	13.5 mm	13.3 mm
4	13.6 mm	13.4 mm	14.3 mm

Uncertainty in Measurement (Sec. 2.3)

2.3 Indicate to what decimal position readings should be recorded (nearest 0.1, nearest 0.01, etc.) for measurements made with the following devices.

(a) A thermometer with smallest scale marking of 1 degree.

(b) A cup for measuring volume with smallest scale marking of 1 fluid ounce.

(c) A volumetric device with smallest scale marking of 10 milliliters.

(d) A ruler with smallest scale marking of 1 millimeter.

2.4 Indicate to what decimal position readings should be recorded (nearest 0.1, nearest 0.01, etc.) for measurements made with the following devices.

(a) A ruler with smallest scale marking of 1 centimeter.

(b) A protracter with smallest scale marking of 1 degree.

(c) A thermometer with smallest scale marking of 10 degrees.

(d) A graduated cylinder with smallest scale marking of 0.1 milliliter.

2.5 What is the difference in meaning between the times 3.3 seconds and 3.30 seconds?

2.6 What is the difference in meaning between the lengths 0.54 inch and 0.540 inch.

2.7 Which of the following measurements are consistent with the uncertainty of the thermometers in Figure 2.2?

(a) thermometer A: 36.72°, 42.1°, 39°, 61.5°

(b) thermometer B: 35.03°, 45.1°, 62°, 47.98°

2.8 Which of the following measurements are consistent with the uncertainty of the thermometers in Figure 2.2?

(a) thermometer A: 16.7°, 16.83°, 42°, 35.4°

(b) thermometer B: 75.3°, 65.30°, 65.03°, 57°

2.9 Indicate whether each of the following statements contains an exact number or a measured (estimated) quantity within it.

(a) The beehive contains 14,000 bees.

(b) There were 97 people in attendance at the meeting.

(c) A 20-lb bag of potatos was purchased at the grocery store.

(d) There are 4 quarts in a gallon.

2.10 Indicate whether each of the following statements contains an exact number or a measured (estimated) quantity within it.

(a) Americans consume 75 acres of pizza per year.

(b) A sheet of paper is 0.0042 inch thick.

(c) There are 12 eggs in a dozen.

(d) The maximum speed of a three-toed sloth is 0.15 mile per hour.

Significant Figures (Sec. 2.3)

2.11 Determine the number of significant figures in each of the following measured values.

(a) 0.00043 **(b)** 0.0220022

(c) 0.30303030 **(d)** 0.03030303

2.12 Determine the number of significant figures in each of the following measured values.

(a) 0.111101 **(b)** 0.0000007

(c) 0.013013013 **(d)** 0.13013013

2.13 Determine the number of significant figures in each of the following measured values.

(a) 4700 (b) 37,540
(c) 67.010 (d) 3000.00

2.14 Determine the number of significant figures in each of the following measured values.

(a) 4000 (b) 4.000
(c) 67,000,100 (d) 43,200

2.15 Determine the number of significant figures in each of the following measured values.

(a) 43,657.3 (b) 3.0003
(c) 0.706050 (d) 46,000,300

2.16 Determine the number of significant figures in each of the following measured values.

(a) 113.235 (b) 2002
(c) 0.00500500 (d) 1,350,100

2.17 In the following pairs of numbers tell whether both members of the pair contain the same number of significant figures?

(a) 11.01 and 11.00 (b) 2002 and 2020
(c) 0.05700 and 0.05070 (d) 0.000066 and 660,000

2.18 In the following pairs of numbers tell whether both members of the pair contain the same number of significant figures?

(a) 2305 and 2350 (b) 0.6600 and 0.0066
(c) 23,000 and 23,001 (d) 936,000 and 0.000936

2.19 In the pairs of numbers of problem 2.17 tell whether both members of the pair have the same precision (uncertainty).

2.20 In the pairs of numbers of problem 2.18 tell whether both members of the pair have the same precision (uncertainty).

2.21 What is the magnitude of uncertainty (±10, ±0.1, etc.) of each of the following numbers?

(a) 3200 (b) 32.00
(c) $32\bar{0}0$ (d) $32,\overline{000}$

2.22 What is the magnitude of uncertainty (±10, ±0.1, etc.) of each of the following numbers?

(a) 63,000 (b) 63.000
(c) $45\bar{0}0$ (d) $11\overline{00}$

2.23 Using standard arithmetic notation and overbars (if needed), write the number twenty-three thousand in a manner such that it has the following numbers of significant figures.

(a) 2 (b) 4 (c) 6 (d) 8

2.24 Using standard arithmetic notation and overbars (if needed), write the number six hundred thousand in a manner such that it has the following numbers of significant figures.

(a) 1 (b) 3 (c) 5 (d) 7

2.25 Determine the number of significant figures in each of the following measured values.

(a) 2600 ± 10 (b) 1.375 ± 0.001
(c) 42 ± 1 (d) 73,000 ± 1

2.26 Determine the number of significant figures in each of the following measured values.

(a) 700 ± 100 (b) 700 ± 10
(c) 43.57 ± 0.01 (d) 64,000 ± 1

2.27 A balance has a precision of ±0.0001 gram. A sample that weighs about 20 grams is weighed on this balance. How many significant figures should be reported for this measurement?

2.28 A balance has a precision of ±0.01 gram. A sample that weighs about 2 grams is weighed on this balance. How many significant figures should be reported for this measurement?

Rounding Off (Sec. 2.4)

2.29 Round off the number 3.6305023 to the indicated number of significant figures.

(a) seven (b) five (c) four (d) three

2.30 Round off the number 4.7205059 to the indicated number of significant figures.

(a) seven (b) five (c) four (d) three

2.31 Round off the number 30,427.29 to the indicated number of significant figures.

(a) six (b) five (c) four (d) two

2.32 Round off the number 50,125.09 to the indicated number of significant figures.

(a) six (b) five (c) four (d) two

2.33 Using proper rounding techniques, decrease by two the number of significant figures in each of the following numbers.

(a) 0.03455 (b) 2.5003
(c) 1,456,000 (d) 100.0

2.34 Using proper rounding techniques, decrease by two the number of significant figures in each of the following numbers.

(a) 0.50505 (b) 2,000,567
(c) 2.335 (d) 1234.5

2.35 Rewrite each of the following numbers so that it contains two significant figures.

(a) 0.123 (b) 123,000
(c) 12.3 (d) 0.000123

2.36 Rewrite each of the following numbers so that it contains two significant figures.

(a) 21.000 (b) $21\overline{0,0}00$
(c) 0.0210 (d) 2.100

Significant Figures in Multiplication and Division (Sec. 2.4)

2.37 Without actually solving the problems, indicate the number of significant figures that should be present in the answers to the following multiplications and divisions. Assume that all numbers are measured quantities.

(a) $4.5 \times 4.05 \times 4.50$ **(b)** $0.100 \times 0.001 \times 0.010$

(c) $\dfrac{655{,}000}{6.5500}$ **(d)** $\dfrac{6.00}{33.000}$

2.38 Without actually solving the problems, indicate the number of significant figures that should be present in the answers to the following multiplications and divisions. Assume that all numbers are measured quantities.

(a) $3.33 \times 3.03 \times 0.0333$

(b) $300{,}003 \times 20{,}200 \times 1.33333$

(c) $\dfrac{333{,}000}{3.33000}$ **(d)** $\dfrac{0.0666}{1.3457}$

2.39 How many significant figures must the number Q possess, in each case, in order to make the following mathematical equations valid from a significant figure standpoint?

(a) $7.312 \times Q = 4.13$

(b) $7.312 \times Q = 0.0022$

(c) $7.312 \times Q = 20.44$

(d) $7.312 \times Q = 0.1100$

2.40 How many significant figures must the number Q possess, in each case, in order to make the following mathematical equations valid from a significant figure standpoint?

(a) $94{,}461 \times Q = 33{,}003$

(b $94{,}461 \times Q = 1.03$

(c) $94{,}461 \times Q = 0.6200$

(d) $94{,}461 \times Q = 233{,}620{,}000$

2.41 Carry out the following multiplications and divisions, expressing your answers to the correct number of significant figures. Assume that all numbers are measured quantities.

(a) 4.2337×0.00706 **(b)** 3700×37.00

(c) $\dfrac{5671}{4.44}$ **(d)** $\dfrac{5.01}{5.07}$

2.42 Carry out the following multiplications and divisions, expressing your answers to the correct number of significant figures. Assume that all numbers are measured quantities.

(a) 350.00×0.00072 **(b)** $620{,}000 \times 620.000$

(c) $\dfrac{3554}{2.22}$ **(d)** $\dfrac{0.000623}{0.000632}$

2.43 Carry out the following mathematical operations, expressing your answers to the correct number of significant figures. Assume that all numbers are measured quantities.

(a) $\dfrac{4.5 \times 6.3}{7.22}$ **(b)** $\dfrac{5.567 \times 3.0001}{3.45}$

(c) $\dfrac{37 \times 43}{4.2 \times 6.0}$ **(d)** $\dfrac{112 \times 20}{30 \times 63}$

2.44 Carry out the following mathematical operations, expressing your answers to the correct number of significant figures. Assume that all numbers are measured quantities.

(a) $\dfrac{2.322 \times 4.00}{3.200 \times 6.73}$ **(b)** $\dfrac{7.403}{3.220 \times 5.000}$

(c) $\dfrac{11.2 \times 11.2}{3.3 \times 6.5}$ **(d)** $\dfrac{5600 \times 300}{22 \times 97.1}$

Significant Figures in Addition and Subtraction (Sec. 2.4)

2.45 Without actually solving the problems, indicate the uncertainty (tenths, hundredths, etc.) that should be present in the answers to the following additions and subtractions. Assume that all numbers are measured quantities.

(a) $12.1 + 23.1 + 127.01$ **(b)** $43.65 - 23.7$

(c) $1237.6 + 23 + 0.12$ **(d)** $4650 + 25 + 200$

2.46 Without actually solving the problems, indicate the uncertainty (tenths, hundredths, etc.) that should be present in the answers to the following additions and subtractions. Assume that all numbers are measured quantities.

(a) $0.06 + 1.32 + 7.901$ **(b)** $4.72 - 3.908$

(c) $23.6 + 33 + 17.21$ **(d)** $46{,}230 + 325 + 45$

2.47 Perform the following additions or subtractions. Report your results to the proper number of significant figures. Assume that all numbers are measured quantities.

(a) $12 + 23 + 127$ **(b)** $3.111 + 3.11 + 3.1$

(c) $1237.6 + 23 + 0.12$ **(d)** $43.65 - 23.7$

2.48 Perform the following additions or subtractions. Report your results to the proper number of significant figures. Assume that all numbers are measured quantities.

(a) $237 + 37 + 7$ **(b)** $4.000 + 4.002 + 4.20$

(c) $235.45 + 37 + 36.4$ **(d)** $4.111 - 3.07$

2.49 Perform the following additions or subtractions. Report your results to the proper number of significant figures. Assume that all numbers are measured quantities.

(a) $999.0 + 1.7 - 43.7$ **(b)** $345 - 6.7 + 4.33$

(c) $1200 + 43 + 7$ **(d)** $132 - 0.0073$

2.50 Perform the following additions or subtractions. Report your results to the proper number of significant figures. Assume that all numbers are measured quantities.

(a) $1237.6 + 1237.4$ **(b)** $1237.6 - 1237.4$

(c) $23{,}000 + 457 + 23$ **(d)** $3.12 - 0.00007$

Significant Figures and Exact Numbers (Sec. 2.4)

2.51 A rubber heel for a man's shoe is found to have a thickness of 1.12 centimeters. What would be the height, in centimeters, of a stack of 13 such identical rubber heels?

2.52 A thumbtack is found to weigh 0.482 gram. What would be the total mass, in grams, of 125 such identical thumbtacks?

2.53 Each of the following calculations involves the numbers 4.3, 230, 20, and 13.00. The numbers 4.3 and 13.00 are measurements; 230 and 20 are exact numbers. Express each answer to the proper number of significant figures.

(a) $4.3 + 230 + 20 + 13.00$
(b) $4.3 \times 230 \times 20 \times 13.00$
(c) $4.3 + 230 - 20 - 13.00$
(d) $\dfrac{4.3 \times 230}{20 \times 13.00}$

2.54 Each of the following calculations involves the numbers 200, 17, 24, and 40. The numbers 200 and 17 are exact; 24 and 40 are measurements. Express each answer to the proper number of significant figures.

(a) $200 + 17 + 24 + 40$ **(b)** $200 \times 17 \times 24 \times 40$
(c) $200 - 17 - 24 - 40$ **(d)** $\dfrac{200 \times 17}{24 \times 40}$

Exponents and Orders of Magnitude (Sec. 2.5)

2.55 Write each of the following expressions as an exponential term.

(a) $5 \times 5 \times 5$ **(b)** $10 \times 10 \times 10 \times 10$
(c) $\dfrac{1}{3 \times 3 \times 3 \times 3}$ **(d)** $\dfrac{1}{10 \times 10 \times 10}$

2.56 Write each of the following expressions as an exponential term.

(a) $7 \times 7 \times 7 \times 7 \times 7$ **(b)** 10×10
(c) $\dfrac{1}{2 \times 2 \times 2 \times 2 \times 2}$ **(d)** $\dfrac{1}{10 \times 10 \times 10 \times 10}$

2.57 Perform the indicated changes on the following exponential terms.

(a) 10^4; increase by three orders of magnitude
(b) 10^6; decrease by seven orders of magnitude
(c) 10^{-5}; increase by two orders of magnitude
(d) 10^{-3}; decrease by four orders of magnitude

2.58 Perform the indicated changes on the following exponential terms.

(a) 10^2; increase by two orders of magnitude
(b) 10^3; decrease by four orders of magnitude
(c) 10^{-4}; increase by three orders of magnitude
(d) 10^{-1}; decrease by five orders of magnitude

Scientific Notation (Sec. 2.5)

2.59 Express the following numbers in scientific notation.

(a) 473.2 **(b)** 0.001234
(c) 231.00 **(d)** 231,000,000

2.60 Express the following numbers in scientific notation.

(a) 787.6 **(b)** 0.01798
(c) 40.0 **(d)** 675,000

2.61 How many digits will there be in the coefficient when each of the following numbers is expressed in scientific notation?

(a) 55.00 **(b)** 55,000
(c) 0.0001000 **(d)** 0.10010

2.62 How many digits will there be in the coefficient when each of the following numbers is expressed in scientific notation?

(a) 6.5000 **(b)** 672,000,000
(c) 0.10003 **(d)** 200.01

2.63 Using scientific notation, express the number sixty-seven thousand to the following number of significant figures.

(a) 1 **(b)** 3 **(c)** 5 **(d)** 7

2.64 Using scientific notation, express the number six-hundred-and-seventy-four thousand to the following number of significant figures.

(a) 2 **(b)** 4 **(c)** 6 **(d)** 8

2.65 Express the following numbers in decimal notation.

(a) 1.70×10^{-4} **(b)** 5.73×10^2
(c) 5.550×10^{-1} **(d)** 1.110×10^{10}

2.66 Express the following numbers in decimal notation.

(a) 3.57×10^{-8} **(b)** 3.500×10^{-3}
(c) 6.2134×10^3 **(d)** 2.0200×10^9

Multiplication and Division in Scientific Notation (Sec. 2.6)

2.67 Carry out the following multiplications of exponential terms.

(a) $10^5 \times 10^3$ **(b)** $10^{-5} \times 10^{-3}$
(c) $10^5 \times 10^{-3}$ **(d)** $10^{-5} \times 10^3$

2.68 Carry out the following multiplications of exponential terms.

(a) $10^7 \times 10^4$ **(b)** $10^{-7} \times 10^{-4}$
(c) $10^7 \times 10^{-4}$ **(d)** $10^{-7} \times 10^4$

2.69 Carry out the following multiplications, making sure that your answer is expressed in correct scientific notation form and to the correct number of significant figures.

(a) $(1.171 \times 10^6) \times (2.555 \times 10^2)$
(b) $(5.37 \times 10^{-3}) \times (1.7 \times 10^5)$
(c) $(9.0 \times 10^{-5}) \times (3.000 \times 10^{-5})$
(d) $(3.0 \times 10^5) \times (9.000 \times 10^5)$

2.70 Carry out the following multiplications, making sure that your answer is expressed in correct scientific notation form and to the correct number of significant figures.

(a) $(2.340 \times 10^{-3}) \times (2.60 \times 10^6)$
(b) $(1.110 \times 10^5) \times (3.333 \times 10^{-7})$
(c) $(9.8 \times 10^2) \times (7.00 \times 10^2)$
(d) $(8.77 \times 10^{-6}) \times (5.030 \times 10^2)$

2.71 Carry out the following divisions of exponential terms.

(a) $\frac{10^5}{10^3}$ **(b)** $\frac{10^5}{10^{-3}}$ **(c)** $\frac{10^{-5}}{10^3}$ **(d)** $\frac{10^{-5}}{10^{-3}}$

2.72 Carry out the following divisions of exponential terms.

(a) $\frac{10^2}{10^3}$ **(b)** $\frac{10^2}{10^{-3}}$ **(c)** $\frac{10^{-2}}{10^3}$ **(d)** $\frac{10^{-2}}{10^{-3}}$

2.73 Carry out the following divisions, making sure that your answer is expressed in correct scientific notation form and to the correct number of significant figures.

(a) $\frac{9.51167 \times 10^{-2}}{3.32 \times 10^{-3}}$ **(b)** $\frac{4.500 \times 10^{10}}{5.0005 \times 10^{-8}}$

(c) $\frac{3.32 \times 10^{-3}}{9.51167 \times 10^{-2}}$ **(d)** $\frac{5.0005 \times 10^{-8}}{4.500 \times 10^{10}}$

2.74 Carry out the following divisions, making sure that your answer is expressed in correct scientific notation form and to the correct number of significant figures.

(a) $\frac{3.5608 \times 10^3}{5.71 \times 10^5}$ **(b)** $\frac{3.300 \times 10^{-5}}{4.0003 \times 10^2}$

(c) $\frac{5.71 \times 10^5}{3.5608 \times 10^3}$ **(d)** $\frac{4.0003 \times 10^2}{3.300 \times 10^{-5}}$

2.75 Perform the following mathematical operations involving exponential terms.

(a) $\frac{10^2 \times 10^3}{10^4}$ **(b)** $\frac{10^{-2} \times 10^{-3}}{10^{-4}}$

(c) $\frac{10^6}{10^{-5} \times 10^{-9}}$ **(d)** $\frac{10^{-3} \times 10^2 \times 10^5}{10^{-6} \times 10^8}$

2.76 Perform the following mathematical operations involving exponential terms.

(a) $\frac{10^4 \times 10^5}{10^6 \times 10^3}$ **(b)** $\frac{10^{-3} \times 10^{-3} \times 10^{-3}}{10^{-6}}$

(c) $\frac{10^2 \times 10^3 \times 10^4}{10^{-2} \times 10^{-3} \times 10^{-4}}$ **(d)** $\frac{10^{-6} \times 10^4}{10^3 \times 10^{-5}}$

2.77 Perform the following mathematical operations. Be sure your answer contains the correct number of significant figures and that it is in correct scientific notation form.

(a) $\frac{(6.0 \times 10^3) \times (5.0 \times 10^3)}{2.0 \times 10^7}$

(b) $\frac{2.0 \times 10^7}{(6.0 \times 10^3) \times (5.0 \times 10^3)}$

(c) $\frac{(3.571 \times 10^{-5}) \times (4.5113 \times 10^{-9})}{(5.10 \times 10^{-6}) \times (3.71300 \times 10^{10})}$

(d) $\frac{(5 \times 10^{10}) \times (6.0 \times 10^7) \times (3.111 \times 10^{-5})}{(3 \times 10^3) \times (4.00 \times 10^{-6})}$

2.78 Perform the following mathematical operations. Be sure your answer contains the correct number of significant figures and that it is in correct scientific notation form.

(a) $\frac{(3.00 \times 10^5) \times (6.00 \times 10^3) \times (5.00 \times 10^6)}{2.00 \times 10^7}$

(b) $\frac{4.1111 \times 10^{-3}}{(3.003 \times 10^{-6}) \times (9.8760 \times 10^{-5})}$

(c) $\frac{(6 \times 10^5) \times (6 \times 10^{-5})}{(3 \times 10^2) \times (1 \times 10^{-10})}$

(d) $\frac{(3.00 \times 10^6) \times (2.7 \times 10^3) \times (8.50 \times 10^3)}{(2.22 \times 10^2) \times (8.504 \times 10^6)}$

Addition and Subtraction in Scientific Notation (Sec. 2.6)

2.79 Carry out the following additions and subtractions, expressing each answer in correct scientific notation form to the correct number of significant figures.

(a) $(3.245 \times 10^3) + (1.17 \times 10^3)$
(b) $(9.870 \times 10^{-2}) - (5.7 \times 10^{-3})$
(c) $(9.356 \times 10^5) + (3.27 \times 10^4)$
(d) $(2.030 \times 10^4) - (1.111 \times 10^3)$

2.80 Carry out the following additions and subtractions, expressing each answer in correct scientific notation form to the correct number of significant figures.

(a) $(5.405 \times 10^6) + (3.09 \times 10^5)$
(b) $(7.777 \times 10^{-1}) - (5.3 \times 10^{-1})$
(c) $(8.219 \times 10^2) - (1.901 \times 10^1)$
(d) $(3.45 \times 10^3) + (3.45 \times 10^2)$

2.81 Carry out the following subtractions, expressing each answer in correct scientific notation form to the correct number of significant figures.

(a) $(8.313 \times 10^7) - (6.00 \times 10^6)$
(b) $(8.313 \times 10^7) - (6.00 \times 10^5)$
(c) $(8.313 \times 10^7) - (6.00 \times 10^4)$
(d) $(8.313 \times 10^7) - (6.00 \times 10^2)$

2.82 Carry out the following subtractions, expressing each answer in correct scientific notation form to the correct number of significant figures.

(a) $(7.431 \times 10^8) - (4.00 \times 10^7)$
(b) $(7.431 \times 10^8) - (4.00 \times 10^6)$
(c) $(7.431 \times 10^8) - (4.00 \times 10^5)$
(d) $(7.431 \times 10^8) - (4.00 \times 10^3)$

Additional Problems

2.83 How many significant figures does each of the following numbers have?

(a) 7770 **(b)** 7.770
(c) 7.770×10^{-3} **(d)** 7.770×10^3

2.84 How many significant figures does each of the following numbers have?

(a) 0.4500 **(b)** 4.500

(c) 4.500×10^{-3} (d) 4.500×10^{3}

2.85 In the following pairs of numbers tell whether both members of the pair contain the same number of significant figures.

(a) 11.0 and 11.00 (b) 600.0 and 6.000×10^{3}
(c) 6300 and 6.3×10^{3} (d) 0.300045 and 0.345000

2.86 In the following pairs of numbers tell whether both members of the pair contain the same number of significant figures.

(a) 0.000066 and 660,000 (b) 1.500 and 1.500×10^{2}
(c) 54,000 and 5400.0 (d) 0.05700 and 0.0570

2.87 In the following pairs of numbers tell whether both members of the pair have a precision of at least ±0.01.

(a) 0.006 and 0.016 (b) 2.700 and 2.700×10^{2}
(c) 3300.00 and $33\overline{0}0$ (d) 5.750×10^{1} and 5.750×10^{-1}

2.88 In the following pairs of numbers tell whether both members of the pair have a precision of at least ±0.01.

(a) 3.71 and 3.50 (b) 4.500 and 4.500×10^{-2}
(c) 270.0 and 0.27000 (d) 4.31×10^{-2} and 4.31×10^{2}

2.89 Write each of the following numbers in scientific notation to the number of significant figures indicated in parentheses.

(a) 632,567 (4) (b) 0.312546 (3)
(c) 63,000,023 (4) (d) 0.500000 (4)

2.90 Write each of the following numbers in scientific notation to the number of significant figures indicated in parentheses.

(a) 0.00300300 (3) (b) 936,000 (2)
(c) 23.5003 (3) (d) 450,000,001 (6)

2.91 How many significant figures must the number Q possess, in each case, in order to make the following mathematical equations valid from a significant figure standpoint?

(a) $6.000 \times Q = 4.0$ (b) $\dfrac{5.000}{Q} = 3.175$

(c) $5.250 + Q = 7.03$ (d) $0.7777 - Q = 0.011$

2.92 How many significant figures must the number Q possess, in each case, in order to make the following mathematical equations valid from a significant figure standpoint?

(a) $450.0 \times Q = 3.00$ (b) $\dfrac{Q}{5.1256} = 1.703$

(c) $9.13 + Q = 10.2$ (d) $Q - 0.111 = 9.25$

2.93 Perform the following mathematical operations, expressing your answers to the correct number of significant figures. Assume all numbers are measured quantities.

(a) $4.0 \times (2.3 + 4.5)$
(b) $3.0 \times (3.7 - 3.4)$
(c) $6.0 \times (34 - 4.23)$
(d) $7.02 \times (0.0001 + 0.01)$

2.94 Perform the following mathematical operations, expressing your answers to the correct number of significant figures. Assume all numbers are measured quantities.

(a) $5.5 \times (12.3 - 3.2)$
(b) $6.0 \times (3.7 + 0.001)$
(c) $2.0 \times (3.43 - 2.415)$
(d) $3.0 \times (37 - 3.7)$

2.95 What is wrong with the statement "The number of objects is 12.00 exactly"?

2.96 What is wrong with the statement "Through counting, it was determined that the basket contained 6.70×10^{1} peaches"?

2.97 Each of the following calculations contains the numbers 4.2, 5.30, 11, and 28. The numbers 4.2 and 5.30 are measured quantities, and 11 and 28 are exact numbers. Do each calculation and express each answer to the proper number of significant figures.

(a) $(4.2 + 5.30) \times (28 + 11)$
(b) $4.2 \times 5.30 \times (28 - 11)$
(c) $\dfrac{28 - 4.2}{5.30 \times 11}$
(d) $\dfrac{28 - 4.2}{11 - 5.30}$

2.98 Each of the following calculations contains the numbers 3.111, 5.03, 100, and 33. The numbers 3.111 and 5.03 are measured quantities, and 100 and 33 are exact numbers. Do each calculation and express each answer to the proper number of significant figures.

(a) $(3.111 + 5.03) \times (100 + 33)$
(b) $3.111 \times 5.03 \times (100 + 33)$
(c) $\dfrac{3.111 + 5.03}{100 \times 33}$
(d) $\dfrac{5.03 - 3.111}{100 + 33}$

2.99 Arrange the following sets of numbers in ascending order (from smallest to largest).

(a) 2.07×10^{2}, 243, 1.03×10^{3}
(b) 0.0023, 3.04×10^{-2}, 2.11×10^{-3}
(c) 23,000, 2.30×10^{5}, 9.67×10^{4}
(d) 0.00013, 0.000014, 1.5×10^{-4}

2.100 Arrange the following sets of numbers in ascending order (from smallest to largest).

(a) 350, 3.51×10^{2}, 3.522×10^{1}
(b) 0.000234, 2.341×10^{-3}, 2.3401×10^{-4}
(c) 965,000,000, 9.76×10^{8}, 2.03×10^{8}
(d) 0.00010, 0.00023, 3.4×10^{-2}

3 Unit Systems and Dimensional Analysis

3.1 The Metric System of Units

All measurements consist of three parts: a number that tells the amount of the quantity measured, an error that affects the precision of that amount, and a unit that tells the nature of the quantity being measured. Chapter 2 dealt with the interpretation and manipulation of the number and error parts of a measurement. We now turn our attention to units.

A unit is a "label" that describes (or identifies) what is being measured (or counted). It can be almost anything: quarts, dimes, dozen frogs, bushels, inches, or pages, for example. Having units for a measurement is an absolute necessity. If you were to ask a neighbor to lend you six sugar, the immediate response would be "how much sugar?" You would then have to indicate that you wished to borrow six pounds, six ounces, six cups, six teaspoons, or whatever amount of sugar you needed.

Two formal systems of units of measurement are used in the United States today. Common measurements in commerce—in supermarkets, lumberyards, gas stations, and so on—are made in the **English system**. The units of this system include the familiar inch, foot, pound, quart, and gallon. A second system, the **metric system**, is used in scientific work. Units in this system include the gram, meter, and liter. The United States is one of only a very few countries that use different unit systems in commerce and scientific work. On a worldwide basis, almost universally, the metric system is used in both areas.

The metric system is slowly coming into use in the United States for commercial purposes. Metric units now appear on many consumer products (see Fig. 3.1). Soft drinks can now be bought in 2-liter containers, and automobile engine sizes are often given in liters. Road signs in some states display distances in both miles and kilometers. Canned and packaged goods (cereals, mixes, fruits, etc.) on grocery store shelves now have the content masses listed in grams as well as ounces or pounds.

The metric system is superior to the English system. Its superiority lies in the area of interrelationships between units of the same type (vol-

(a)

(b)

FIGURE 3.1 Metric units are becoming increasingly evident on highway signs and consumer products. (a-Tony Freeman/PhotoEdit; b-Kenlax/PH Archives)

ume, length, etc.). Metric unit interrelationships are less complicated than English unit interrelationships because the metric system is a decimal-unit system. In the metric system conversion from one unit size to another can be accomplished simply by moving the decimal point to the right or left an appropriate number of places. The metric system is no more precise than the English system, but it is more convenient.

The most recent modification of the metric system is called the International System of Units and is abbreviated SI (after the French name Système International). Since SI units are still not in universal use, we shall use the more traditional metric system in this text.

Because the SI and metric systems are very similar, switching to the SI system sometime in the future, when its use is more extensive, will present no major difficulties to the student who properly understands the traditional metric system.

Metric System Prefixes

In the metric system there is one basic unit for each type of measurement—length, volume, mass, etc. These basic units are then multiplied by appropriate powers of ten to form smaller or larger units. The names of the larger and smaller units are constructed from the basic unit name by attaching to it a prefix that tells which power of ten is involved. These prefixes are given in Table 3.1, along with their symbols or abbreviations and mathematical meanings. The prefixes in color are those most frequently used.

The use of numerical prefixes should not be new to you. Consider the use of the prefix tri- in the following words: triangle, tricycle, trio, trinity, triple. Every one of these words conveys the idea of three of something. We will use the metric system prefixes in the same way.

The meaning of a prefix always remains constant; it is independent of the base unit it modifies. For example, a kilosecond is a thousand seconds; a kilowatt, a thousand watts; and a kilocalorie, a thousand calories. The prefix kilo- will always mean a thousand.

TABLE 3.1 METRIC SYSTEM PREFIXES AND THEIR MATHEMATICAL MEANINGS

Prefix	Symbol	Mathematical Meaning	Origin	Original Meaning
Tera-	T	1,000,000,000,000 = 10^{12}	Greek	monstrous
Giga-	G	1,000,000,000 = 10^{9}	Greek	gigantic
Mega-	M	1,000,000 = 10^{6}	Greek	great
Kilo-	k	1,000 = 10^{3}	Greek	thousand
Hecto-	h	100 = 10^{2}	Greek	hundred
Deca-	da	10 = 10^{1}	Greek	ten
		1 = 10^{0}		
Deci-	d	0.1 = 10^{-1}	Latin	tenth
Centi-	c	0.01 = 10^{-2}	Latin	hundredth
Milli-	m	0.001 = 10^{-3}	Latin	thousandth
Micro-	μ[a]	0.000 001 = 10^{-6}	Greek	small
Nano-	n	0.000 000 001 = 10^{-9}	Greek	very small
Pico-[b]	p	0.000 000 000 001 = 10^{-12}	Spanish	extremely small

[a] This is the Greek letter mu (pronounced "mew," rhymes with "you").
[b] This prefix is pronounced "peek-o."

3.2 METRIC UNITS OF LENGTH

The **meter** *is the basic unit of length in the metric system.* (*Metre* has been adopted as the preferred international spelling for this unit, but *meter* is the spelling used in the United States and in this book.)

Other units of length in the metric system are derived from the meter by using the prefixes listed in Table 3.1. The kilometer (km) is 1000 meters, whereas the centimeter (cm) and millimeter (mm) are, respectively, $\frac{1}{100}$ and $\frac{1}{1000}$ of a meter. Note that the abbreviation for meter is m.

A comparison of metric lengths with the commonly used English system lengths of mile, yard, and inch gives us some size perspective. A kilometer is approximately three fifths (60%) of a mile, a meter is slightly larger (109%) than a yard (see Fig. 3.2a), and a centimeter is about two fifths (40%) of an inch (see Fig. 3.2b).

A better "feel" for lengths equivalent to a meter, centimeter, and millimeter can perhaps be obtained by relating these metric units to objects and situations we encounter in everyday life. This is done in Figure 3.3 in terms of a coin, a paper clip, some chalk, and a basketball player.

F.Y.I.
The meter was originally defined as one ten millionth of the distance from the equator to the North Pole.

F.Y.I.
Smoke particles are 0.1 to 1 micrometer in diameter.

3.3 METRIC UNITS OF MASS

Mass is a measurement of the total quantity of matter present in an object. A **gram** *is the basic unit of mass in the metric system.* The gram is relatively small compared to the com-

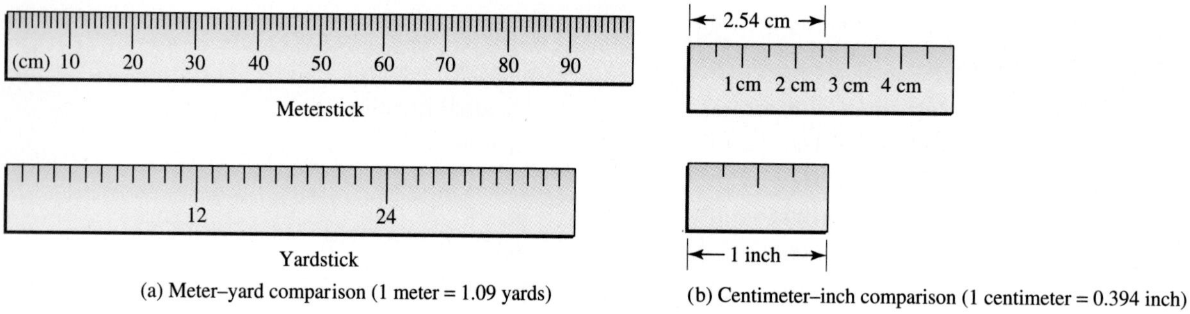

(cm) 10 20 30 40 50 60 70 80 90
Meterstick
12 24
Yardstick
(a) Meter–yard comparison (1 meter = 1.09 yards)
2.54 cm
1 cm 2 cm 3 cm 4 cm
1 inch
(b) Centimeter–inch comparison (1 centimeter = 0.394 inch)

FIGURE 3.2 A comparison of metric and English units of length.

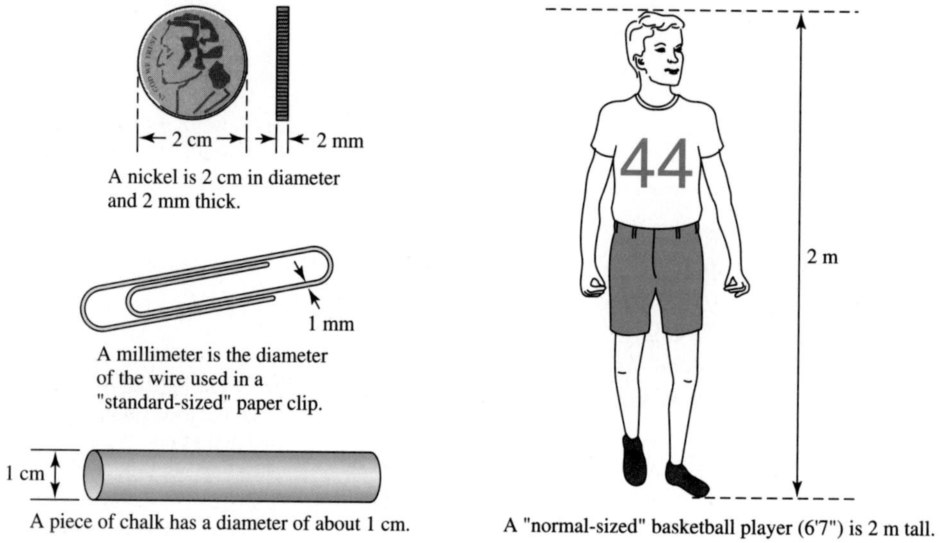

2 cm
2 mm
A nickel is 2 cm in diameter and 2 mm thick.
1 mm
A millimeter is the diameter of the wire used in a "standard-sized" paper clip.
1 cm
A piece of chalk has a diameter of about 1 cm.
44
2 m
A "normal-sized" basketball player (6'7") is 2 m tall.

FIGURE 3.3 Metric units of length and the "everyday realm."

F.Y.I.
A one karat diamond has a mass of 200 mg.

monly used English mass units of ounce and pound. It takes approximately 28 grams to equal an ounce and nearly 454 grams to equal a pound. Because of the small size of the gram, the kilogram (kg) is a very commonly used unit. (The abbreviation for gram is g.) A kilogram is equal to approximately 2.2 pounds, as shown in Figure 3.4.

To give you a "feel" for metric mass unit magnitude, Figure 3.5 relates the mass units of milligram (mg), gram, and kilogram to everyday objects. Comparisons involve a single staple, a postage stamp, two thumbtacks, a nickel, a quart of milk in a paper carton, and a "medium-sized" football player.

The terms *mass* and *weight* are frequently used interchangeably in discussions. Although in most cases this practice does no harm, technically it is incorrect to interchange the terms. Mass and weight refer to different properties of matter, and their difference in meaning should be understood.

Mass *is a measure of the total quantity of matter in an object.* **Weight** *is a measure of*

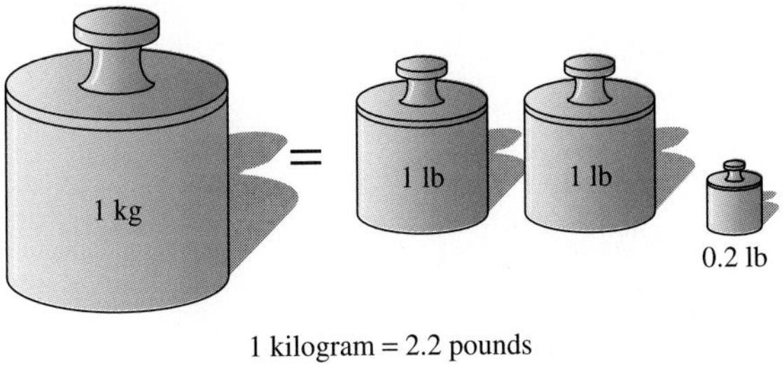

FIGURE 3.4 A comparison of the metric kilogram with the English pound.

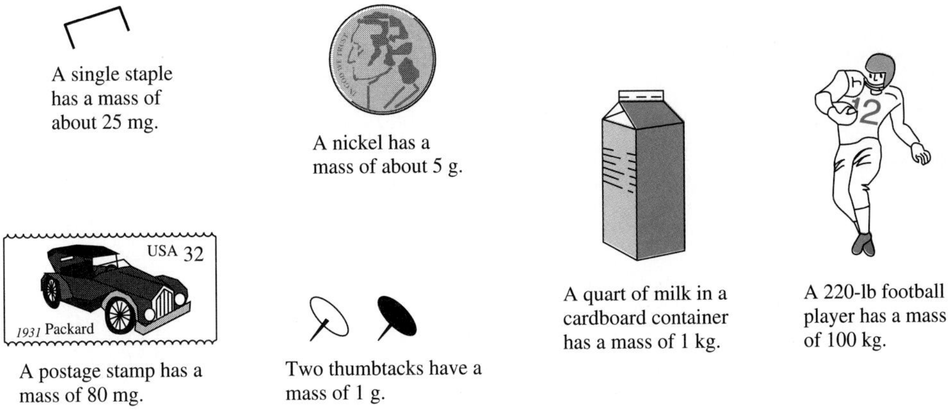
A single staple has a mass of about 25 mg.
A nickel has a mass of about 5 g.
USA 32
1931 Packard
A postage stamp has a mass of 80 mg.
Two thumbtacks have a mass of 1 g.
A quart of milk in a cardboard container has a mass of 1 kg.
A 220-lb football player has a mass of 100 kg.

FIGURE 3.5 Metric units of mass and the "everyday realm."

the force exerted on an object by the pull of gravity. The mass of a substance is a constant; the weight of an object is a variable dependent upon the geographical location of that object.

> **F.Y.I.**
> If you hit your thumb with a hammer, it is the mass of the hammer not its weight that you feel. If you hit your thumb in outer space with a weightless hammer, it would still hurt.

Matter at the equator weighs less than it would at the North Pole because the Earth is not a perfect sphere but bulges at the equator. As a result, an object at the equator is farther from the center of the Earth. It therefore weighs less because the magnitude of gravitational attraction (the measure of weight) is inversely proportional to the distance between the centers of the attracting objects; that is, the gravitational attraction is larger when the objects' centers are closer together and smaller when the objects' centers are farther apart. Gravitational attraction also depends on the masses of the attracting bodies; the greater the masses, the greater the attraction. For this reason, an object would weigh much less on the moon than on Earth because of the smaller size of the moon and the correspondingly lower gravitational attraction. Quantitatively, a 22.0-pound mass weighing 22.0 pounds at the Earth's North Pole would weigh 21.9 pounds at the Earth's equator and only 3.7 pounds on the moon. In outer space an astronaut may be weightless but never massless. In fact, he or she has the same mass in space as on Earth (see Fig. 3.6.)

A *balance* is used to determine the mass of an object. Figure 3.7 shows three common types of balances. With balances such as these, the actual mass of the object can be obtained by balancing the object against objects ("weights") of known mass. The object whose mass is to be determined is placed on the left pan of the platform or triple-beam balance (Figs. 3.7a and 3.7b), and objects ("weights") of known mass are placed on the right-hand pan and/or sliding scale to counterbalance the object. The analytical balance shown in Figure 3.7c has only one pan. The counterbalancing masses for this balance are located within the balance case and are put in place by manipulating knobs and levers. Gravitational attraction

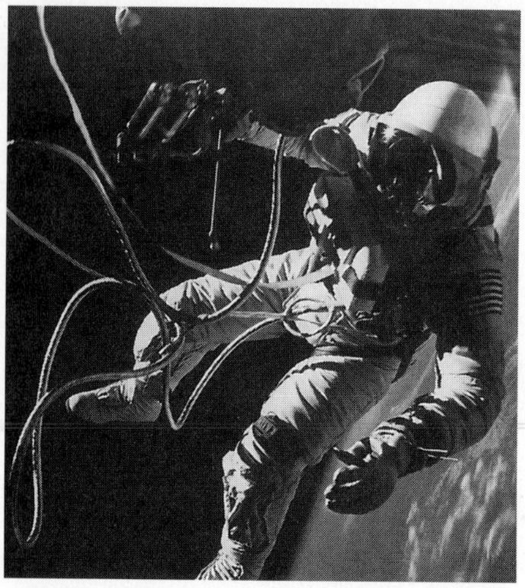

FIGURE 3.6 As astronaut on a "space walk" is weightless but not massless. He or she has exactly the same mass as on Earth. (NASA)

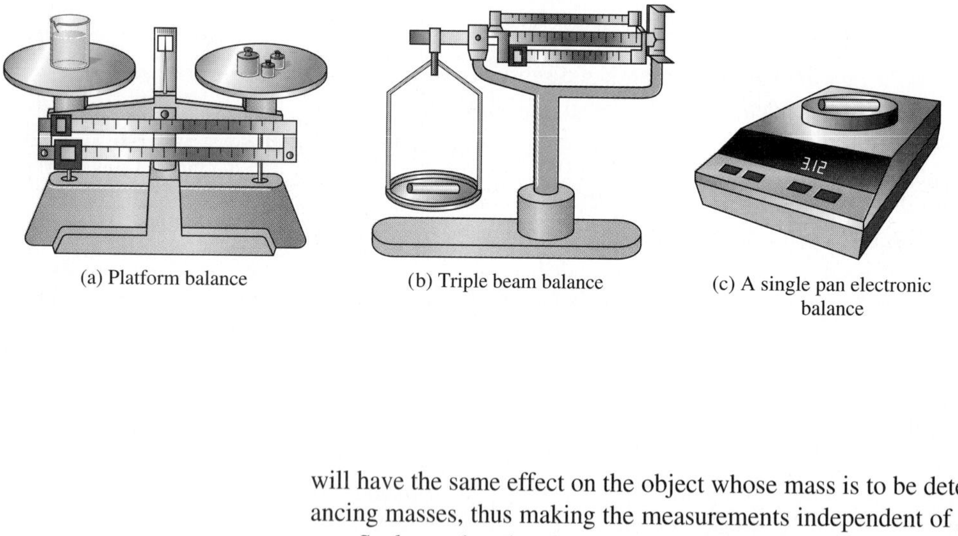

FIGURE 3.7 Common laboratory balances, used for determining mass, are (a) the platform balance (double pan balance), (b) the triple beam balance, and (c) the single pan analytical balance.

will have the same effect on the object whose mass is to be determined and the counterbalancing masses, thus making the measurements independent of geographical location.

Scales, rather than balances, are used to determine weight. Weight is found by noting the magnitude of the stretch of a calibrated spring. The extent of stretching of the spring is dependent on the gravitational attraction of the object for the Earth. Bathroom scales and grocery store scales operate on this principle.

The processes of determining mass and weight are both referred to as "weighing" since the word "massing" is not an accepted part of the English language (although it should be). This dual usage of the term weighing leads to the practice (somewhat common) of using mass and weight interchangeably. Although some textbooks follow this practice, this one will not. A mass is not a weight. A mass will be called a mass. Chemists always use balances rather than scales. Hence they always measure the mass of an object.

Square	Area = side × side $A = s \times s$ $= s^2$	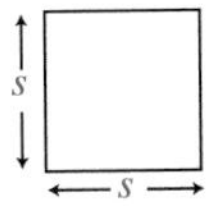
Rectangle	Area = length × width $A = l \times w$	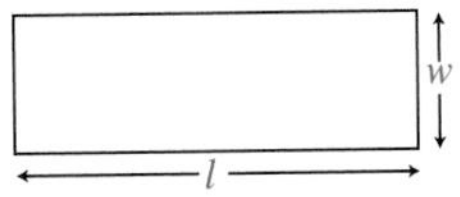
Circle	Area = π × (radius)^2 $A = \pi \times r^2$	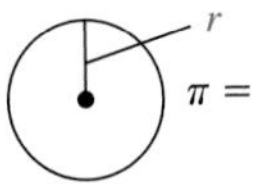 $\pi = 3.1416$
Triangle	Area = $\frac{1}{2}$ × base × height $A = \frac{1}{2} \times b \times h$	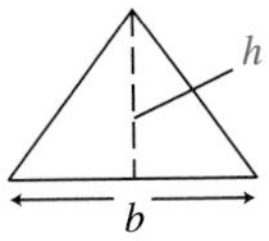

FIGURE 3.8 Formulas for calculating the areas of various geometrical figures.

3.4 METRIC UNITS OF VOLUME

Before specific metric units of volume are considered, a quick review of how the quantities area and volume are calculated is in order.

Area *is a measure of the extent of a surface.* The units for area are squared units of length. Common area units include square inches (in.^{2}), square feet (ft^2), square meters (m^2), and square centimeters (cm^2). Note that a squared unit is just that unit multiplied by itself. The unit cm^2 means centimeter × centimeter in the same way that 3^2 means 3 × 3. Figure 3.8 lists the formulas needed for determining the areas of commonly encountered geometrical shapes. Notice that in each case the formula involves taking the product of two lengths. Note also that the constant pi (π) is needed in calculating the area of a circle. Its value to three, four, and five significant figures, respectively, is 3.14, 3.142, and 3.1416.

Volume *is a measure of the amount of space occupied by an object.* It is a three-dimensional measure and thus involves units that have been cubed: in.^{3}, ft^3, m^3, cm^3, etc. Again, a cubed unit is just that unit multiplied by itself three times in the same manner that 3^3 is 3 × 3 × 3. Figure 3.9 lists volume formulas for commonly encountered three-dimensional shapes. Note particularly the manner in which the volume of a cube is calculated—side × side × side. This is the key to understanding metric units of volume whose definitions are cube-related.

A **liter** *is the basic unit of volume in the metric system.* (As with meter, we will use the United States spelling rather than the international spelling, which is litre.) A liter is abbreviated L, to avoid the confusing of lowercase l with the number 1.

A liter is a volume equal to that occupied by a perfect cube that is 10 centimeters (or 1 decimeter) on each side. Since the volume of a cube is calculated by multiplying length × width × height (which are all the same for a cube) we have

$$
\begin{aligned}
1\ \text{L} &= \text{volume of a 10-cm-edged cube} \\
&= 10\ \text{cm} \times 10\ \text{cm} \times 10\ \text{cm} \\
&= 1000\ \text{cm}^3
\end{aligned}
$$

Cube	Volume = side × side × side $V = s \times s \times s$ $= s^3$	
Rectangular solid	Volume = length × width × height $V = l \times w \times h$	
Cylinder	Volume = π × (radius)^2 × height $V = \pi \times r^2 \times h$	
Sphere	Volume = $\frac{4}{3} \times \pi \times$ (radius)^3 $V = \frac{4}{3}\pi \times r^3$	

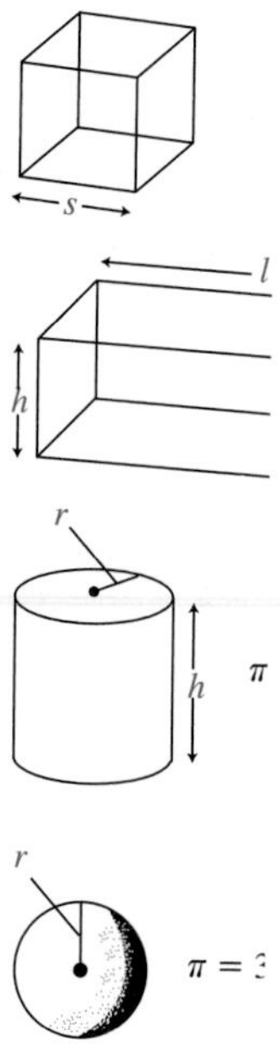
s
l
h
r
h
r

FIGURE 3.9 Formulas for calculating the volumes of various three-dimensional shapes.

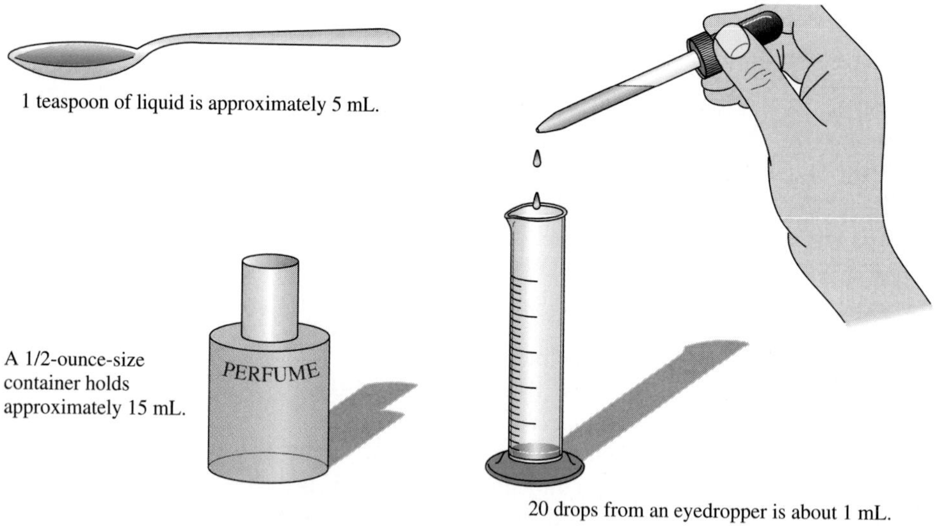
1 teaspoon of liquid is approximately 5 mL.
A 1/2-ounce-size container holds approximately 15 mL.
PERFUME
20 drops from an eyedropper is about 1 mL.

FIGURE 3.10 The metric volume unit the milliliter and the "everyday realm."

The abbreviation cc is sometimes used for the unit cubic centimeter, most commonly in medically oriented situations.

As with the units of length and mass, the basic unit of volume is modified with prefixes to represent smaller or larger units. The most commonly used prefixed volume unit is the milliliter (mL), which is $\frac{1}{1000}$ L. A milliliter is much smaller than a fluid ounce; it takes approximately 30 mL to equal 1 fluid ounce. A quart (32 fl oz.) contains 946 mL. Figure 3.10 shows some comparisons that will give you a better "feel" for the size of a milliliter.

A liter is equal to 1000 milliliters. We have also just seen that it is equal to 1000 cubic centimeters. Therefore,

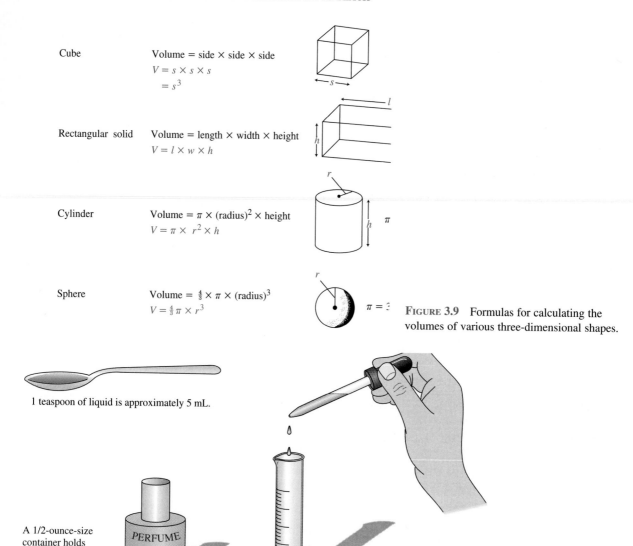

$$1000 \text{ mL} = 1000 \text{ cm}^3$$

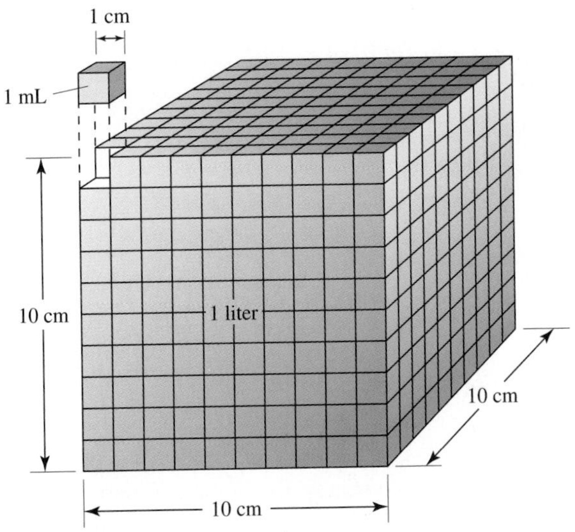

FIGURE 3.11 The relationship between the units liter and milliliter in terms of cubic centimeters.

Dividing both sides of this equation by a thousand, we find that

$$1 \text{ mL} = 1 \text{ cm}^3$$

Consequently, the units mL and cm^3 are interchangeable. In practice, mL is usually used for volumes of liquids and gases and cm^3 for volumes of solids. Figure 3.11 shows the relationship between 1 mL (1 cm^3) and its parent unit, liter, in terms of cubic measurement.

3.5 UNITS IN MATHEMATICAL OPERATIONS

Just as numbers can have exponents associated with them (3^2, 6^3, 10^2, 10^4, etc.), so also can units have exponents associated with them. In Section 3.4 in the discussion of volume, the unit cubic centimeter (cm^3) was introduced. As noted, cm^3 denotes the unit centimeter multiplied by itself three times (cm × cm × cm). Other commonly encountered units with exponents include

in.^2 (square inches)
ft^3 (cubic feet)
cm^2 (square centimeters)
m^2 (square meters)

In mathematical problems the powers on units are manipulated in the same manner as powers of ten (Sec. 2.6). Exponents are added during multiplication and subtracted during division. Just as

$$10^2 \times 10^4 = 10^6$$

so likewise

$$\text{km}^2 \times \text{km}^4 = \text{km}^6$$

Similarly,

$$\frac{10^2}{10^4} = 10^{-2}$$

and

$$\frac{km^2}{km^4} = km^{-2}$$

Example 3.1

Carry out the following multiplications and divisions of measurements and units:

(a) $2.4\text{ cm} \times 3.6\text{ cm}$ **(b)** $3.6\text{ km}^2 \times 6.0\text{ km}$

(c) $\dfrac{6.3\text{ ft}^3}{1.0\text{ ft}}$ **(d)** $\dfrac{6.0\text{ in.}^2 \times 3.0\text{ in.}}{2.0\text{ in.}}$

Solution

Each of these problems can be considered in two stages: the numerical parts of the measurements are multiplied or divided as indicated and the units are then similarly combined.

(a)
$$\begin{aligned} 2.4\text{ cm} \times 3.6\text{ cm} &= (2.4 \times 3.6) \times (\text{cm} \times \text{cm}) \\ &= 8.64 \times \text{cm}^2 \\ &= 8.64\text{ cm}^2 \quad \text{(calculator answer)} \\ &= 8.6\text{ cm}^2 \quad \text{(correct answer)} \end{aligned}$$

The *input* numbers dictate that the answer contain only two significant figures. For the units, multiplying a unit by itself gives the unit squared.

(b)
$$\begin{aligned} 3.6\text{ km}^2 \times 6.0\text{ km} &= (3.6 \times 6.0) \times (\text{km}^2 \times \text{km}) \\ &= 21.6 \times \text{km}^3 \\ &= 21.6\text{ km}^3 \quad \text{(calculator answer)} \\ &= 22\text{ km}^3 \quad \text{(correct answer)} \end{aligned}$$

Note that the exponents on the units km^2 and km^1 were added to give km^3. Only two significant figures are allowed in the answer.

(c)
$$\begin{aligned} \frac{6.3\text{ ft}^3}{1.0\text{ ft}} &= \frac{6.3}{1.0} \times \frac{\text{ft}^3}{\text{ft}} \\ &= 6.3 \times \text{ft}^2 \\ &= 6.3\text{ ft}^2 \quad \text{(calculator and correct answer)} \end{aligned}$$

In this case, according to the rule for division of exponential terms (Sec. 2.6), the unit exponents were subtracted. No adjustment for significant figures needs to be made.

(d)
$$\begin{aligned} \frac{6.0\text{ in.}^2 \times 3.0\text{ in.}}{2.0\text{ in.}} &= \frac{6.0 \times 3.0}{2.0} \times \frac{\text{in.}^2 \times \text{in.}}{\text{in.}} \\ &= 9 \times \text{in.}^2 \\ &= 9\text{ in.}^2 \quad \text{(calculator answer)} \\ &= 9.0\text{ in.}^2 \quad \text{(correct answer)} \end{aligned}$$

In this case the calculator gives too few significant figures.

Practice Exercise 3.1

Carry out the following multiplications and divisions of measurements and units. Remember that significant figures are a consideration for all measurements.

(a) $6.2\ \text{ft} \times 3.7\ \text{ft}$ (b) $3.01\ \text{mm} \times 6.007\ \text{mm}^2$

(c) $\dfrac{5.000\ \text{yd}^3}{2.00\ \text{yd}}$ (d) $\dfrac{7.0\ \text{cm}^2 \times 6.0\ \text{cm}^3}{3.0\ \text{cm}^4}$

Ans. (a) 23 ft^2; (b) 18.1 mm^3; (c) 2.50 yd^2; (d) 14 cm

3.6 CONVERSION FACTORS

Many times a need arises to change the units of a quantity or measurement to different units. In this section we deal with this topic. The new units needed may be in the same measurement system as the old ones, or they may be in a different system. With two unit systems in common use in the United States, the need to change measurements in one system to their equivalent in the other frequently occurs.

The mathematical tool we will use to accomplish the task of changing units is a general method of problem solving called *dimensional analysis*. By this method, unit conversion is accomplished by multiplying a given quantity or measurement by one or more conversion factors to obtain the desired quantity or measurement.

$$\text{Given quantity} \times \text{conversion factor(s)} = \text{desired quantity}$$

Prior to discussing the mechanics involved in using dimensional analysis to solve "unit-conversion problems," some comments concerning conversion factors are in order. A proper understanding of conversion factors is the key to being able to solve problems "comfortably" with dimensional analysis.

Formally defined, a **conversion factor** *is an equality, expressed in fractional form, that is used as a multiplier to convert a quantity in one unit into its equivalent in another unit.*

Multiplication by one or more conversion factors does not change the value of a measurement or quantity because *all conversion factors have a numerical value equal to unity (one)*. Multiplication by unity does not change the value of an expression.

A fixed relationship between two quantities gives sufficient information to construct a conversion factor. Let us construct some conversion factors to see (1) how they originate and (2) why they always have a value of unity.

The quantities "1 minute" and "60 seconds" both describe the same amount of time. We may write an equation describing this fact.

$$1\ \text{min} = 60\ \text{sec}$$

This fixed relationship can be used to construct a pair of conversion factors that relate seconds and minutes. (Conversion factors always occur in pairs, as will become obvious shortly.)

Dividing both sides of our minute–second equation by the quantity "1 minute" gives

$$\frac{1 \text{ min}}{1 \text{ min}} = \frac{60 \text{ sec}}{1 \text{ min}}$$

Since the numerator and denominator of the fraction on the left are identical, this fraction has a value of unity.

$$1 = \frac{60 \text{ sec}}{1 \text{ min}}$$

The fraction on the right side of the equation is our conversion factor. Its value is one. Note that the numerator and denominator of the conversion factor describe the same "amount" of time.

Two conversion factors are always obtainable from any given equality. For the equality we are considering (1 min = 60 sec) the second conversion factor is

$$\frac{1 \text{ min}}{60 \text{ sec}}$$

It is obtained by dividing both sides of the equality by "60 seconds" instead of "1 minute." The relationship between the two conversion factors is that of reciprocals.

$$\frac{1 \text{ min}}{60 \text{ sec}} \quad \text{and} \quad \frac{60 \text{ sec}}{1 \text{ min}}$$

In general, we will always be able to construct a set of two conversion factors, each with a value of unity, from any two terms that describe the same "amount" of whatever we are considering. The two conversion factors will always be reciprocals of each other. (Conversion factors can also be constructed from two quantities that are equivalent rather than equal. We will consider this situation in Sec. 3.8.)

For convenience in introducing the use of conversion factors we will classify them into three categories: (1) metric-to-metric, (2) English-to-English, and (3) metric-to-English or English-to-metric.

Metric-to-Metric Conversion Factors

Metric-to-metric conversion factors are used to change a measurement in one metric unit into another metric unit. Both the numerator and the denominator of such conversion factors involve metric-system units.

Conversion factors relating a prefixed metric unit to the base unit for that type of measurement are derived from the meaning of the prefix of concern. For example, the set of conversion factors involving kilogram and gram is derived from the meaning of the prefix kilo-, which is 10^3. The two conversion factors are

$$\frac{10^3 \text{ m}}{1 \text{ km}} \quad \text{and} \quad \frac{1 \text{ km}}{10^3 \text{ m}}$$

Note again the reciprocal relationship between the two conversion factors of the set. The conversion factors relating centimeter and meter involve 10^{-2}, the mathematical equivalent of centi-, and are

$$\frac{1 \text{ cm}}{10^{-2} \text{ m}} \quad \text{and} \quad \frac{10^{-2} \text{ m}}{1 \text{ cm}}$$

For conversion factors of the type now under discussion, the numerical equivalent of the prefix always goes with the base unit.

All metric-to-metric conversion factors are derived from exact definitions. Thus, they all contain an unlimited number of significant figures. This observation about significant figures will be of major importance when these conversion factors become part of an actual calculation.

English-to-English Conversion Factors

English-to-English conversion factors also contain an unlimited number of significant figures, since they also result from defined equalities. Twelve inches equals 1 foot—exactly. Three feet equal 1 yard—exactly. The majority of English-to-English conversion factors should be "second nature" to most students. You have used these relationships all your life, even though you may not consciously have thought of them as conversion factors. The following are representative of these "second-nature" conversion factors.

$$\frac{2\text{ pt}}{1\text{ qt}} \qquad \frac{36\text{ in.}}{1\text{ yd}} \qquad \frac{1\text{ mi}}{5280\text{ ft}} \qquad \frac{16\text{ oz}}{1\text{ lb}}$$

In order to avoid confusion with the word *in*, the abbreviation for inches includes a period (in.). This is the only unit abbreviation that has a period associated with it.

Metric-to-English and English-to-Metric Conversion Factors

Conversion factors that relate metric units to English units or vice versa must be established by measurement since they involve two different unit systems. One of the numbers associated with each such conversion factor must be determined experimentally. Since this number will not be *exact*, concern about the number of significant figures present arises.

To avoid confusion about the accuracy of such English-to-metric conversion factors, scientists in English-speaking countries have agreed on common "experiment-based" *definitions* for English units in terms of metric units. They have also agreed that these definitions should be taken to be *exact*. The three *exact* definitions we consider are

1 inch = 2.540005 centimeters
1 pound = 453.59237 grams
1 quart = 0.94633343 liter

For simplicity, these "exact" conversion factors are most often rounded off to four significant digits.

1 inch = 2.540 centimeters
1 pound = 453.6 grams
1 quart = 0.9463 liter

In these rounded forms, the definitions are not exact.

Table 3.2 lists some commonly encountered relationships between metric and English system units. These few factors are sufficient to solve most of the problems that we will encounter. In fact, later in this chapter we will see that only one factor of a given type (volume, mass, length) is sufficient to solve most problems.

TABLE 3.2 CONVERSION FACTORS THAT RELATE THE ENGLISH AND METRIC SYSTEMS OF MEASUREMENT TO EACH OTHER

	Factor to Convert from Metric to English Unit
Length	
Inch and centimeter	$\frac{1 \text{ in.}}{2.540 \text{ cm}}$
Inch and meter	$\frac{1 \text{ in.}}{0.02540 \text{ m}}$
Mile and kilometer	$\frac{1 \text{ mi}}{1.609 \text{ km}}$
Mass	
Pound and gram	$\frac{1 \text{ lb}}{453.6 \text{ g}}$
Pound and kilogram	$\frac{1 \text{ lb}}{0.4536 \text{ kg}}$
Ounce and gram	$\frac{1 \text{ oz}}{28.34 \text{ g}}$
Volume	
Quart and liter	$\frac{1 \text{ qt}}{0.9463 \text{ L}}$
Pint and liter	$\frac{1 \text{ pt}}{0.4732 \text{ L}}$
Fluid ounce and milliliter	$\frac{1 \text{ fl oz}}{29.57 \text{ mL}}$

3.7 DIMENSIONAL ANALYSIS

Formally defined, **dimensional analysis** *is a general problem-solving method that uses the units associated with numbers as a guide in setting up the calculation*. This method treats units in the same way as numbers, that is, they can be multiplied, divided, canceled, etc. For example, just as

$$3 \times 3 = 3^2 \quad \text{(3 squared)}$$

we have

$$\text{km} \times \text{km} = \text{km}^2 \quad \text{(km squared)}$$

Also, just as the twos cancel in the expression

$$\frac{\cancel{2} \times 3 \times 6}{\cancel{2} \times 5}$$

the inches cancel in the expression

$$\frac{\cancel{\text{inch}} \times \text{cm}}{\cancel{\text{inch}}}$$

The same units found in both the numerator and denominator of a fraction will always cancel just as like numbers do.

The steps followed in setting up a problem by dimensional analysis are as follows.

STEP 1 *Identify the known or given quantity (both a numerical value and units) and the units of the new quantity to be determined.*

This information, which serves as the starting point for setting up the problem, will always be found in the statement of the problem. Write an equation with the given quantity on the left and the units of the desired quantity on the right.

STEP 2 *Multiply the given quantity by one or more conversion factors in a manner such that the unwanted (original) units are canceled out, leaving only the new desired unit.*

The general format for the multiplication is

$$\text{Information given} \times \text{conversion factor(s)} = \text{information sought}$$

The number of conversion factors used depends on the individual problem. Except in the simplest of problems, it is a good idea to predetermine formally the sequence of unit changes to be used. This sequence will be called the unit "pathway."

STEP 3 *Perform the mathematical operations indicated by the conversion factor setup.*

In performing the calculation you need to double check that all units except the desired set have canceled out. You also need to check the numerical answer to see that it contains the proper number of significant figures.

Now let us work a number of sample problems using dimensional analysis and the steps just outlined. Our first two examples involve only metric system units and thus will involve only metric–metric conversion factors.

Example 3.2

A vitamin C tablet is found to contain 0.500 g of vitamin C. How many milligrams of vitamin C does this tablet contain?

Solution

STEP 1 The given quantity is 0.500 g, the mass of vitamin C in one tablet. The unit of the desired quantity is milligrams.

$$0.500 \text{ g} = ? \text{ mg}$$

STEP 2 Only one conversion factor will be needed to convert from grams to milligrams—one that relates grams to milligrams. Two forms of this factor exist.

$$\frac{1 \text{ mg}}{10^{-3} \text{g}} \quad \text{and} \quad \frac{10^{-3} \text{g}}{1 \text{ mg}}$$

The first factor is used because it allows for the cancellation of the gram units, leaving us with milligrams as the new units.

$$0.500 \cancel{\text{g}} \times \frac{1 \text{ mg}}{10^{-3} \cancel{\text{g}}}$$

For cancellation, a unit must appear in both the numerator and the denominator. Because the given quantity (0.500 g) has grams in the numerator, the conversion factor used must be the one with grams in the denominator.

If the other conversion factor had been used, we would have

$$0.500 \text{ g} \times \frac{10^{-3} \text{ g}}{1 \text{ mg}}$$

No unit cancellation is possible in this setup. Multiplication gives g^2/mg as the final units, which is certainly not what we want. In all cases, only one of the two conversion factors of a reciprocal pair will correctly fit into a dimensional-analysis setup.

STEP 3 Step 2 takes care of the units. All that is left is to combine numerical terms to get the final answer; we still have to do the arithmetic. Collecting the numerical terms gives

$$\frac{0.500 \times 1}{10^{-3}} \text{ mg} = 500 \text{ mg} \quad \text{(calculator answer)}$$

$$= 5.00 \times 10^2 \text{ mg} \quad \text{(correct answer)}$$

Since the conversion factor used in this problem is derived from a definition, it contains an unlimited number of significant figures and will not limit in any way the allowable number of significant figures in the answer. Therefore the answer should have three significant figures, the same number as in the given quantity

Practice Exercise 3.2

A basketball player is 192 cm tall. What is the player's height in meters?

Ans. 1.92 m

Example 3.3

The *ozone layer* is a region in the upper atmosphere, at altitudes between 25 and 35 km, where the concentration of ozone is several times higher than at ground level. Express the altitude 35 km (the upper limit of the ozone layer) in centimeter units.

Solution

STEP 1 The given quantity is 35 km, and the units of the desired quantity are centimeters.

$$35 \text{ km} = ? \text{ cm}$$

STEP 2 In dealing with metric–metric unit changes where both the original and desired units carry prefixes (which is the case in this problem), it is recommended that you always channel units through the basic unit (unprefixed unit). If you do that, you will not need to deal with any conversion factors other than those resulting from prefix definitions. Following this recommendation, the unit pathway for this problem is

$$\underset{\text{prefixed unit}}{\text{km}} \longrightarrow \underset{\text{base unit}}{\text{m}} \longrightarrow \underset{\text{prefixed unit}}{\text{cm}}$$

In the setup for this problem we will need two conversion factors, one for the kilometer-to-meter change and one for the meter-to-centimeter change.

$$35\ \cancel{\text{km}} \times \frac{10^3\ \cancel{\text{m}}}{1\ \cancel{\text{km}}} \times \frac{1\ \text{cm}}{10^{-2}\ \cancel{\text{m}}}$$

This conversion factor converts km to m (first factor); This conversion factor converts m to cm (second factor)

The units cancel except for the desired centimeters.

STEP 3 Carrying out the indicated numerical calculation gives

$$\frac{35 \times 10 \times 1}{1 \times 10^{-2}}\ \text{cm} = 3.5 \times 10^{-6} \quad \text{(calculator and correct answer)}$$

numbers from first conversion factor (10 and 1); numbers from second conversion factor (1 and 10^{-2})

The correct answer is the same as the calculator answer for this problem. The given quantity has two significant figures, and both conversion factors are exact. Thus, the correct answer should contain two significant figures.

Practice Exercise 3.3

The average diameter of a coronary artery is 0.32 cm. What is this diameter in nanometers?

Ans. 3.2×10^6 nm

Our next two worked-out example problems involve the use of English–English conversion factors. The conversion factors themselves should pose no problems for you. The examples are intended to give you further insights into dimensional analysis, particularly in the area of setting up the pathway for unit change. In addition, Example 3.5 exposes you to the complication of having to deal with "compound" units.

Example 3.4

If a person's stomach produces 87 fl oz of gastric juice in a day, what is the equivalent volume of the gastric juice in gallons?

Solution

STEP 1 The given quantity is 87 fl oz, and the unit of the desired quantity is gallons.

$$87\ \text{fl oz} = ?\ \text{gal}$$

STEP 2 The logical pathway to follow to accomplish the desired change is

$$\text{fl oz} \longrightarrow \text{qt} \longrightarrow \text{gal}$$

Always use logical steps in setting up the pathway for a unit change. It does not have to be done in one big jump. Use smaller steps for which you know the conversion factors. Big steps usually get you involved with unfamiliar conversion factors. Most people do not carry around in their head the number of fluid ounces in a gallon, but they do know that there are 32 fl oz in a quart.

The setup for this problem will require two conversion factors: fluid ounces to quarts and quarts to gallons.

$$87\ \cancel{\text{fl oz}} \times \frac{1\ \cancel{\text{qt}}}{32\ \cancel{\text{fl oz}}} \times \frac{1\ \text{gal}}{4\ \cancel{\text{qt}}}$$

$$\text{fl oz} \longrightarrow \text{qt} \longrightarrow \text{gal}$$

The units all cancel except for the desired gallons.

STEP 3 Performing the indicated multiplications, we get

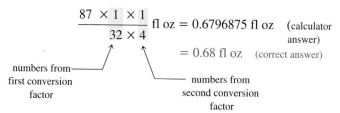

$$\frac{87 \times 1 \times 1}{32 \times 4}\ \text{fl oz} = 0.6796875\ \text{fl oz} \quad \text{(calculator answer)}$$

$$= 0.68\ \text{fl oz} \quad \text{(correct answer)}$$

numbers from first conversion factor; numbers from second conversion factor

The calculator answer contains too many significant figures. The correct answer should contain only two significant figures, the number in the given quantity. Again, both conversion factors involve exact definitions and will not enter into significant figure considerations.

Practice Exercise 3.4

The estimated amount of recoverable oil from the field at Prudhoe Bay in Alaska is 4.0×10^{11} gal. What is this amount of oil in fluid ounces (fl oz)?

Ans. 5.1×10^{13} fl oz

Example 3.5

Worldwide emissions of carbon dioxide into the atmosphere are estimated at 2×10^9 ton per year. What is this emission rate in pounds per hour?

Solution

STEP 1 The given quantity is 2×10^9 ton/year, and the units of this quantity are to be changed to lb/hr.

$$2 \times 10^9\ \frac{\text{ton}}{\text{yr}} = \frac{?\ \text{lb}}{\text{hr}}$$

STEP 2 This problem is more complex than previous examples because two different types of units are involved: mass and time. However, the approach we use to solve it is similar, other than that we must change two units instead of one. The tons must be converted to pounds and the years to hours.

The logical pathway for the mass change is the direct one-step path

$$\text{ton} \longrightarrow \text{lb}$$

For time, it is logical to make the change in two steps.

$$\text{yr} \longrightarrow \text{day} \longrightarrow \text{hr}$$

It does not matter whether time or mass is handled first in the conversion factor setup. We will arbitrarily choose to handle time first. The setup becomes

$$2 \times 10^9 \frac{\text{ton}}{\cancel{\text{yr}}} \times \frac{1\,\cancel{\text{yr}}}{365\,\cancel{\text{day}}} \times \frac{1\,\cancel{\text{day}}}{24\,\text{hr}}$$

The units at this point are ton/day (after the first conversion factor); the units at this point are ton/hr (after the second conversion factor).

Note that in the first conversion factor, years had to be in the numerator in order to cancel the years in the denominator of the given quantity.

We are not done yet. The time conversion from years to hours has been accomplished, but nothing has been done with mass. To take care of mass we do not start a new conversion factor setup. Rather, an additional conversion factor is tacked onto those we already have in place.

$$2 \times 10^9 \frac{\cancel{\text{ton}}}{\cancel{\text{yr}}} \times \frac{1\,\cancel{\text{yr}}}{365\,\cancel{\text{day}}} \times \frac{1\,\cancel{\text{day}}}{24\,\cancel{\text{hr}}} \times \frac{2000\text{ lb}}{1\,\cancel{\text{ton}}}$$

The tons in the denominator of the last factor cancel the tons in the numerator of the given quantity. With this cancellation the units now become lb/hr.

STEP 3 Collecting the numerical factors and performing the indicated math gives

$$\frac{2 \times 10^9 \times 1 \times 1 \times 2000}{365 \times 24 \times 1} \frac{\text{lb}}{\text{hr}} = 4.56621 \times 10^8 \frac{\text{lb}}{\text{hr}} \quad \text{(calculator answer)}$$

$$= 5 \times 10^8 \frac{\text{lb}}{\text{hr}} \quad \text{(correct answer)}$$

The given quantity possesses only one significant figure. Rounding the calculator answer to one significant figure produces the correct answer of 5×10^8 lb/hr. Again, none of the conversion factors plays a role in significant figure considerations since they all originated from definitions.

Practice Exercise 3.5

The average human heart pumps blood at the rate of 6.8 fl oz/sec. What is this pump rate in gallons per hour?

Ans. 190 gal/hr

We will now consider some sample problems that involve both the English and metric systems of units. As mentioned previously (Sec. 3.6), conversion factors between these two unit systems do not arise from definitions, but rather are determined experimentally. Hence they are not exact. Some of these experimentally determined conversion factors were given in Table 3.2.

Instead of trying to remember all of the conversion factors listed in Table 3.2, memorize only one factor for each type of measurement (mass, volume, length). Knowing only one factor of each type is sufficient information to work metric–English or English–metric conversion problems. The relationships that are the most useful for you to memorize are the following.

Length:	1 in. = 2.540 cm	(4 significant figures)
Mass:	1 lb = 453.6 g	(4 significant figures)
Volume:	1 qt = 0.9463 L	(4 significant figures)

These three equalities can be considered "bridge relationships" connecting English and metric system measurement units of various types. These bridge relationships are depicted in Figure 3.12.

These bridge relationships are always applicable in problem solving. For example, no matter what mass units are given or asked for in a problem, we can convert to pounds or grams (bridge units), cross the bridge with our memorized concersion factor, and then convert to the desired final unit. The only advantage that would be gained by memorizing all the factors in Table 3.2 would be that you could work some problems with fewer conversion factors, usually only one factor fewer. The reduction in the number of conversion factors used is usually not worth the added complication of keeping track of the additional conversion factors.

Exercise 3.6

Your overweight neighbor Zelda Zucclakeley weighs 244 lb. What is her mass in kilograms?

Solution

STEP 1 The given quantity is 244 lb, and the units of the desired quantity are kilograms.

$$244 \text{ lb} = ? \text{ kg}$$

STEP 2 This is an English–metric mass-unit conversion problem. The mass-unit "measurement bridge" (Fig. 3.12b) involves pounds and grams. Since pounds are the given units, we are at the bridge to start with. Pounds are converted to grams (crossing the bridge), and then the grams are converted to kilograms.

$$\text{lb} \longrightarrow \text{g} \longrightarrow \text{kg}$$

The conversion factor setup for this problem is

$$244 \cancel{\text{lb}} \times \frac{453.6 \cancel{\text{g}}}{1 \cancel{\text{lb}}} \times \frac{1 \text{ kg}}{10^3 \cancel{\text{g}}}$$

The units all cancel except for kilograms.

STEP 3 Performing the indicated arithmetic gives

$$\frac{244 \times 453.6 \times 1}{1 \times 10^3} \text{ kg} = 110.6784 \text{ kg} \quad \text{(calculator answer)}$$

$$= 111 \text{ kg} \quad \text{(correct answer)}$$

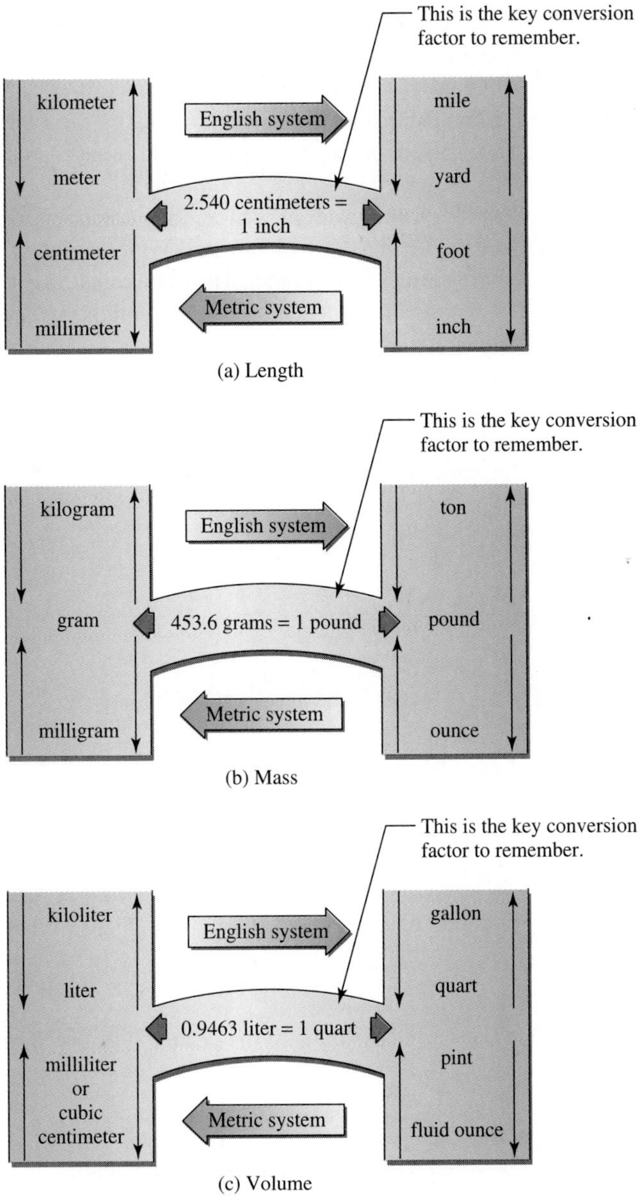

FIGURE 3.12 Measurement bridges connecting the metric and English systems of measurement for (a) length, (b) mass, and (c) volume.

The calculator answer must be rounded off to three significant figures, since the given quantity (244 lb) has only three significant figures. The first conversion factor has four significant figures, and the second one is exact.

An alternative pathway for working this problem makes use of the conversion factor in Table 3.2 involving pounds and kilograms. If we use this conversion factor, we can go directly from the given to the desired unit in one step. The setup is

$$244 \cancel{\text{lb}} \times \frac{1 \text{ kg}}{2.205 \cancel{\text{lb}}} = 110.65759 \text{ kg} \quad \text{(calculator answer)}$$

$$= 111 \text{ kg} \quad \text{(correct answer)}$$

Note that the calculator answers obtained from the two setups for this problem are different: 110.6784 and 110.65759, but each gives the same answer after rounding to three significant figures. The slight difference arises from the fact that the English-to-metric conversion factors used are not exact definitions. The two conversion factors used have different rounding-off errors in them.

As in the case here, for most problems there is more than one pathway that can be used to get the answer. When alternative pathways exist, it cannot be said that one way is more correct than another. The important concept is that you select a pathway, choose a correct set of conversion factors consistent with that pathway, and get the answer.

Although both pathways used in this problem are "correct," we prefer the first solution because it uses our mass measurement bridge. Use of the measurement bridge system cuts down on the number of conversion factors you are required to know (or look up in a table).

Practice Exercise 3.6

A piece of copper metal has a mass of 17.62 lb. What is the copper mass in kilograms?

Ans. 7.992 kg

Exercise 3.7

Typical water usage by a household washing machine, per load, is 140,000 mL. What is the water usage in gallons?

Solution

STEP 1 The given quantity is 140,000 mL, and the units of the desired quantity are gallons.

$$140{,}000 \text{ mL} = \text{? gal}$$

STEP 2 The "bridge relationship" for volume (Fig. 3.12c) involves liters and quarts. Thus, we need to convert the milliliters to liters, cross the bridge to quarts, and then convert quarts to gallons.

$$\text{mL} \longrightarrow \text{L} \longrightarrow \text{qt} \longrightarrow \text{gal}$$

Following the pathway, the setup becomes

$$140{,}000 \cancel{\text{mL}} \times \frac{10^{-3} \cancel{\text{L}}}{1 \cancel{\text{mL}}} \times \frac{1 \cancel{\text{qt}}}{0.9463 \cancel{\text{L}}} \times \frac{1 \text{ gal}}{4 \cancel{\text{qt}}}$$

STEP 3 The numerical calculation involves the following collection of numbers.

$$\frac{140{,}000 \times 10^{-3} \times 1 \times 1}{1 \times 0.9463 \times 4} \text{ gal} = 36.986156 \text{ gal} \quad \text{(calculator answer)}$$

$$= 37 \text{ gal} \quad \text{(correct answer)}$$

The calculator answer must be rounded to two significant figures because the input number 140,000 has only two significant figures.

Again, as in Example 3.6, to illustrate that there is more than one way to set up almost any unit conversion problem, let us work this problem two alternative ways using the other two volume conversion factors from Table 3.2.

To use the conversion factor based on 1 L = 2.113 pt requires a pathway of

$$\text{mL} \longrightarrow \text{L} \longrightarrow \text{pt} \longrightarrow \text{qt} \longrightarrow \text{gal}$$

The conversion factor setup is

$$140{,}000\ \text{mL} \times \frac{10^{-3}\ \cancel{\text{L}}}{1\ \cancel{\text{mL}}} \times \frac{2.113\ \cancel{\text{pt}}}{1\ \cancel{\text{L}}} \times \frac{1\ \cancel{\text{qt}}}{2\ \cancel{\text{pt}}} \times \frac{1\ \text{gal}}{4\ \cancel{\text{qt}}} = 36.9775\ \text{gal} \quad \text{(calculator answer)}$$

$$= 37\ \text{gal} \quad \text{(correct answer)}$$

The conversion factor based on 29.57 mL = 1 fl oz requires a pathway of

$$\text{mL} \longrightarrow \text{fl oz} \longrightarrow \text{qt} \longrightarrow \text{gal}$$

The conversion factor setup for this pathway is

$$140{,}000\ \cancel{\text{mL}} \times \frac{1\ \cancel{\text{fl oz}}}{29.57\ \cancel{\text{mL}}} \times \frac{1\ \cancel{\text{qt}}}{32\ \cancel{\text{fl oz}}} \times \frac{1\ \text{gal}}{4\ \cancel{\text{qt}}} = 36.988501\ \text{gal} \quad \text{(calculator answer)}$$

$$= 37\ \text{gal} \quad \text{(correct answer)}$$

Note that all three methods give the same correct answer after rounding to the correct number of significant figures. Although each of the three setups is correct, we still prefer the method of having a bridge relationship for volume, mass, and length to use in crossing the metric-to-English bridge.

Practice Exercise 3.7

A typical normal loss of water through sweating per day for a human is 450 mL. What is the volume, in pints, of this amount of water?

Ans. 0.95 pt

Example 3.8

A large flask contains 725 mL of strained (no seeds) watermelon juice. (Tests are to be run on the juice to see how much vitamin A it contains.) Specify the volume of this amount of juice in

(a) cubic centimeters **(b)** cubic feet

Solution

(a) From Section 3.4, we note that

$$1 \text{ mL} = 1 \text{ cm}^3$$

Therefore,

$$375 \text{ mL} \times \frac{1 \text{ cm}^3}{1 \text{ mL}} = 375 \text{ cm}^3 \quad \text{(calculator and correct answer)}$$

(b) We will use the answer from part (a), 375 cm^3, as the starting point for this calculation.

STEP 1 The given quantity is 375 cm^3 and the units for the desired quantity are ft^3

$$375 \text{ cm}^3 = ?\ \text{ft}^3$$

STEP 2 If this problem were a problem involving just length rather than $(\text{length})^3$, the pathway would be

$$\text{cm} \longrightarrow \text{in.} \longrightarrow \text{ft}$$

The pathway for $(\text{length})^3$ is just an adaptation of this pathway.

$$\text{cm}^3 \longrightarrow \text{in.}^3 \longrightarrow \text{ft}^3$$

The conversion factors needed for this pathway are simply those for length raised to the third power, that is,

$$375 \text{ cm}^3 \times \left(\frac{1 \text{ in.}}{2.54 \text{ cm}}\right)^3 \times \left(\frac{1 \text{ ft}}{12 \text{ in.}}\right)^3$$

Note that the *entire* conversion factor, in each case, must be cubed, not just the units on the conversion factor. The notation $\left(\frac{1 \text{ ft}}{12 \text{ in.}}\right)^3$ means that the conversion factor within the parentheses is multiplied by itself three times. Thus,

$$\left(\frac{1 \text{ ft}}{12 \text{ in.}}\right)^3 = \frac{1 \text{ ft}}{12 \text{ in.}} \times \frac{1 \text{ ft}}{12 \text{ in.}} \times \frac{1 \text{ ft}}{12 \text{ in.}}$$

$$= \frac{1 \text{ ft}^3}{1728 \text{ in.}^3}$$

The complete setup for this problem, showing the removal of parentheses and cancellation of units, is

$$375 \text{ cm}^3 \times \left(\frac{1 \text{ in.}}{2.54 \text{ cm}}\right)^3 \times \left(\frac{1 \text{ ft}}{12 \text{ in.}}\right)^3$$

$$= 375 \text{ cm}^3 \times \frac{1 \text{ in.}^3}{16.4 \text{ cm}^3} \times \frac{1 \text{ ft}^3}{1728 \text{ in.}^3}$$

Note that in going from $(2.54)^3$ to 16.4, the calculator answer was 16.387064, which, rounded to three significant figures, is 16.4.

STEP 3 The numerical setup is

$$\frac{375 \times 1 \times 1}{16.4 \times 1728} \text{ ft}^3 = 0.013232554 \text{ ft}^3 \quad \text{(calculator answer)}$$

$$= 0.0132 \text{ ft}^3 \quad \text{(correct answer)}$$

Practice Exercise 3.8

A graduated cylinder contains 37.5 mL of water. Express this water volume in

(a) cubic centimeters (b) cubic inches

Ans. (a) 37.5 cm^3; (b) 2.29 $in.^3$

Although the emphasis in this section has been on using conversion factors to change units within the English or metric systems or from one to the other, the applications of conversion factors go far beyond this type of activity. We will resort to using conversion factors time and again throughout this textbook in solving problems. What has been covered in this section is only the "tip of the iceberg" relative to dimensional analysis and conversion factors.

3.8 DENSITY

Density *is the ratio of the mass of an object to the volume occupied by that object,* that is,

$$\text{Density } (d) = \frac{\text{mass}}{\text{volume}}$$

The most frequently encountered density units in chemistry are grams per cubic centimeter (g/cm^3) for solids, grams per milliliter (g/mL) for liquids, and grams per liter (g/L) for gases. Use of these units avoids the problem of having density values that are extremely small or extremely large numbers. Table 3.3 gives density values for a number of substances.

TABLE 3.3 DENSITIES OF SELECTED SOLIDS, LIQUIDS, AND GASES

Solids	Density (g/cm^3 at 25°C)[a]	Liquids	Density (g/mL at 25°C)[a]	Gases	Density (g/L at 25°C, 1 atm)[a]
Gold	19.3	Mercury	13.55	Chlorine	3.17
Lead	11.3	Milk	1.028–1.035	Carbon dioxide	1.96
Copper	8.93	Blood plasma	1.027	Oxygen	1.42
Aluminum	2.70	Urine	1.003–1.030	Air (dry)	1.29
Table salt	2.16	Water	0.997	Nitrogen	1.25
Bone	1.7–2.0	Olive oil	0.92	Methane	0.66
Table sugar	1.59	Ethyl alcohol	0.79	Hydrogen	0.08
Wood, pine	0.30–0.50	Gasoline	0.56		

[a] Density changes with temperature. (In most cases it decreases with increasing temperature, since almost all substances expand when heated.) Consequently, the temperature must be recorded along with a density value. In addition, the pressure of gases must be specified.

People often speak of one substance being "heavier" or "lighter" than another. For example, it is said that "lead is a heavier metal than aluminum." What is actually meant by this statement is that lead has a higher density than aluminum; that is, there is more mass in a specific volume of lead than there is in the same volume of aluminum. The density of an object is a measure of how tightly the object's mass is packed into a given volume. Even though the density of lead (11.3 g/cm^3) is greater than that of aluminum (2.70 g/cm^3), 1 lb of lead weighs exactly the same as 1 lb of aluminum—1 lb is 1 lb. Because the aluminum is less dense than the lead, the mass in the 1 lb of aluminum will occupy a larger volume than the mass in the 1 lb of lead. Said another way, if equal-volume samples of lead and aluminum are weighed, the lead will have the greater mass. When we say lead is "heavier" than aluminum, we actually mean that lead is more dense than aluminum.

The densities of solids and liquids are often compared to the density of water. Anything less dense ("lighter") than water floats on it, and anything more dense ("heavier") sinks. In a similar vein, densities of gases are compared to that of air. Any gas less dense ("lighter") will rise in air, and anything more dense ("heavier") will sink in air.

To calculate an object's density, we must make two measurements; one involves determining the object's mass, and the other its volume.

Example 3.9

A student determines that the mass of a 20.0-mL sample of olive oil (to be used in oil and vinegar salad dressing) is 18.4 g.

(a) What is the density of the olive oil in grams per milliliter?

(b) Predict where the olive oil layer will be (top or bottom) in unshaken oil and vinegar salad dressing.

Solution

(a) Substituting the given mass and volume values into the formula

$$\text{Density} = \frac{\text{mass}}{\text{volume}}$$

we have

$$\text{Density} = \frac{18.4\text{ g}}{20.0\text{ mL}} = 0.92\text{ g/mL} \quad \text{(calculator answer)}$$

$$= 0.920\text{ g/mL} \quad \text{(correct answer)}$$

Since both input numbers contain three significant figures, the density is specified to three significant figures.

(b) Since vinegar is a water-based solution, its density will be slightly greater than 1.00 g/mL. The olive oil will be the top layer because its density is less than that of the vinegar.

Practice Exercise 3.9

Osmium is the densest of all metals. What is its density in grams per cubic centimeter if 50.00 g of the metal occupies a volume of 2.22 cm^3?

Ans. 22.5 g/cm^3

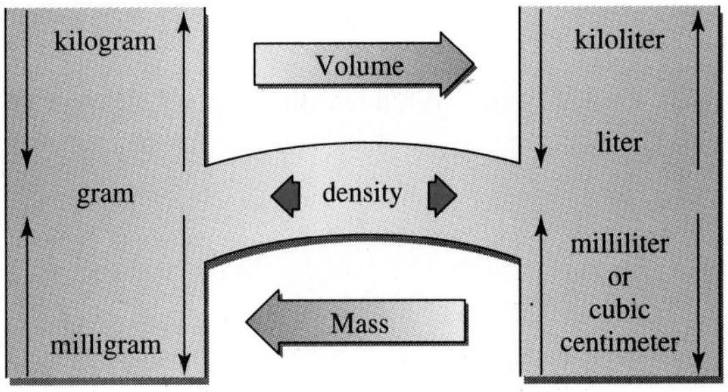

FIGURE 3.13 Density, a measurement bridge connecting mass and volume.

In a mathematical sense, density can be thought of as a conversion factor that relates the volume and mass of an object. Interpreting density in this manner enables us to calculate a substance's volume from its mass and density or its mass from its volume and density. Density is thus a "bridge" connecting mass and volume (see Fig. 3.13).

Example 3.10

Methane, a gas which contributes to the phenomenon called *global warming*, has a density of 0.714 g/L at a particular temperature and pressure. What is the mass, in grams of 10.0 L of this gas?

Solution

We will use density as a conversion factor in solving this problem by dimensional analysis.

STEP 1 The given quantity is 10.0 L of methane. The unit of the desired quantity (mass) is grams. Thus,

$$10.0 \text{ L} = ? \text{ g}$$

STEP 2 The pathway going from liters to grams involves a single step, since density, used as a conversion factor, directly relates grams and liters.

$$10.0 \not{\text{L}} \times \frac{0.714 \text{ g}}{1 \not{\text{L}}}$$

STEP 3 Doing the indicated math gives the following answer.

$$\frac{10.0 \times 0.714}{1} \text{ g} = 7.14 \text{ g} \quad \text{(calculator and correct answer)}$$

The calculator answer and correct answer turn out to be the same.

Practice Exercise 3.10

The density of octane, a component of gasoline, is 0.702 g/mL. What is the mass, in grams, of 875 mL of octane?

Ans. 614 g

Example 3.11

Common table sugar has a density of 1.587 g/cm^3. What would be the volume, in cubic centimeters, of 2.500 g of table sugar?

Solution

STEP 1 The given quantity is 2.500 g of table sugar. The unit of the desired quantity (volume) is cubic centimeters. Thus,

$$2.500 \text{ g} = ? \text{ cm}^3$$

STEP 2 The pathway going from grams to cubic centimeters involves a single step, since density, used as a conversion factor, directly relates grams and cubic centimeters.

$$2.500 \cancel{\text{g}} \times \frac{1 \text{ cm}^3}{1.587 \cancel{\text{g}}}$$

Note that the conversion factor used is actually the reciprocal of the density. Use of the inverted form is necessary in order for the gram units to cancel.

STEP 3 Doing the indicated math gives the following answer.

$$\frac{2.500 \times 1}{1.587} \text{ cm}^3 = 1.5752993 \text{ cm}^3 \quad \text{(calculator answer)}$$

$$= 1.575 \text{ cm}^3 \quad \text{(correct answer)}$$

The correct answer is the calculator answer rounded to four significant figures, the number in each input number.

Practice Exercise 3.11

The density of ethyl alcohol is 0.789 g/mL. What volume of ethyl alcohol, in milliliters, would have a mass of 25.0 g?

Ans. 31.7 mL

Density is a conversion factor of a different type than those previously used in this chapter. It is a conversion factor whose numerator and denominator are *equivalent* rather than *equal*. All previously used conversion factors have been fractions where the numerator and denominator have been the same quantity under two different names. Twelve inches and one foot are different names for the same distance. The mass and volume involved in density are not different names for the same thing, but related equivalent quantities.

A major difference between *equivalence* conversion factors and *equality* conversion factors is that the former have applicability only in the particular problem setting for which they were derived, whereas the latter are applicable in all problem-solving situations. Many different gram-to-cubic centimeter (mass-to-volume) relationships (densities) exist, but only one foot-to-inch relationship exists. Mathematically, equivalence conversion factors can be used the same way equality conversion factors are, and we use the equal sign for both relationships. Equivalence conversion factors will be used often in later chapters.

3.9 PERCENTAGE

Percent *means parts per hundred; that is, it is the number of items of a specified type in a group of 100 items*. The quantity 45% means 45 items per 100 total items.

A mathematical statement of the percent concept is

$$\text{Percent} = \frac{\text{number of items of interest}}{\text{total number of items}} \times 100$$

Example 3.12 shows a simple calculation of a percent.

Example 3.12

A professor proctoring an examination notices that 7 students out of a class of 83 students write with their left hand. What is the percent, to three significant figures, of *right-handed* students in the class?

Solution

We first calculate the number of right-handed students in the class.

$$\text{Right-handed students} = 83 - 7 = 76 \quad \text{(calculator and correct answer)}$$

The percent of right-handed students is equal to the number of right-handed students divided by the total number of students times the factor 100.

$$\text{Percent right-handed students} = \frac{76}{83} \times 100$$
$$= 91.566265 \quad \text{(calculator answer)}$$
$$= 91.6 \quad \text{(correct answer)}$$

Practice Exercise 3.12

The composition of a 14-karat gold ring is found to be 10.68 g gold and 7.62 g copper. What is the percent by mass of copper in the ring?

Ans. 41.6% copper

Percent values find use as conversion factors in problem-solving situations where dimensional analysis is employed. Let us look at the mechanics of writing a percentage as a conversion factor, paying particular attention to the units involved. The percent value in the statement "A gold alloy is found upon analysis to contain 77% gold by mass" will be our focus point for the discussion. What are the mass units associated with the value 77%? The answer is that many mass units could be appropriate. The following statements, written as fractions (conversion factors) are all consistent with our 77% gold analysis.

$$\frac{77 \text{ oz of gold}}{100 \text{ oz of gold alloy}} \qquad \frac{77 \text{ lb of gold}}{100 \text{ lb of gold alloy}}$$

$$\frac{77 \text{ g of gold}}{100 \text{ g of gold alloy}} \qquad \frac{77 \text{ kg of gold}}{100 \text{ kg of gold alloy}}$$

Thus, in writing a percent as a conversion factor, the choice of unit (for mass in this case) is arbitrary. In practice, the complete context of the problem in which the percentage is found usually makes obvious the appropriate unit choice.

In conversion factors derived from percentages, confusion about cancellation of units frequently arises. Both numerator and denominator contain the same units, for example, ounces. Yet the ounces do not cancel. Not considering the *complete unit* is what causes confusion in the minds of students. In the conversion factor

$$\frac{77 \text{ oz of gold}}{100 \text{ oz of gold alloy}}$$

the numerator and denominator dimensions are not simply "ounces." The dimensions are, respectively, "ounces of gold" and "ounces of gold alloy." Ounces cannot be canceled because they are only a part of the complete dimension. Cancellation is possible only when *complete* units are identical.

To avoid the problem of mistakenly canceling dimensions that are not the same, always write complete dimensions. Percent will always involve the same dimensional units (pounds, grams, meters) of different things. The identities of the different things, gold and gold alloy in our example, must always be included as part of the units.

Example 3.13

A sample of clean dry air is found to be 20.9% oxygen by volume. How many milliliters of oxygen are present in 375 mL of this air.

Solution

STEP 1 The given quantity is 375 mL of air, and the desired quantity is milliliters of oxygen.

$$375 \text{ mL air} = ? \text{ mL oxygen}$$

STEP 2 This is a one-conversion-factor problem with the conversion factor, obtained from the given percentage, being

$$\frac{20.9 \text{ mL oxygen}}{100 \text{ mL air}}$$

The setup for the problem is

$$375 \cancel{\text{mL air}} \times \frac{20.9 \text{ mL oxygen}}{100 \cancel{\text{mL air}}}$$

STEP 3 The numerical calculation involves the following arrangement of numbers.

$$\frac{375 \times 20.9}{100} \text{ mL oxygen} = 78.375 \text{ mL oxygen} \quad \text{(calculator answer)}$$

$$= 78.4 \text{ mL oxygen} \quad \text{(correct answer)}$$

The original percentage (20.9) and the air volume (375 mL) both have three significant figures. Therefore, the correct answer should also contain three significant figures. The number 100 in the setup is an exact number, since it originates from the definition of percent.

Practice Exercise 3.13

A 75.0-g solution containing table sugar and water is 7.0% sugar by mass. What is the mass, in grams, of sugar in the solution?

Ans. 5.2 g

Example 3.14

A study of the 184 adult males living in a small town in Weber County in the state of Utah includes the following facts: (1) 20.1% of the adult males are bald-headed; (2) 62.1% of the bald-headed adult males are left-handed; and (3) 91.3% of the left-handed bald-headed adult males are handsome. How many handsome left-handed bald-headed adult males live in the town?

Solution

STEP 1 The given quantity is 184 adult males, with the units of the desired quantity being handsome left-handed bald-headed adult males

184 adult males = ? handsome left-handed bald-headed adult males

STEP 2 When conversion factors, particularly unusual ones, are given in word form in the statement of a problem, it is suggested that you first extract them from the problem statement and write them down before starting to solve the problem. The given conversion factors in this problem are

$$\frac{20.1 \text{ bald-headed adult males}}{100 \text{ adult males}}$$

$$\frac{62.1 \text{ left-handed bald-headed adult males}}{100 \text{ bald-headed adult males}}$$

$$\frac{91.3 \text{ handsome left-handed bald-headed adult males}}{100 \text{ left-handed bald-headed adult males}}$$

The unit-conversion pathway for this problem will be

adult males ⟶ bald-headed adult males ⟶ left-handed bald-headed adult males ⟶ handsome left-handed bald-headed adult males

The dimensional analysis setup for this problem is

$$184 \text{ adult males} \times \frac{20.1 \text{ bald-headed}}{100 \text{ adult males}} \times \frac{62.1 \text{ left-handed}}{100 \text{ bald-headed}} \times \frac{91.3 \text{ handsome}}{100 \text{ left-handed}}$$

All units cancel except those in the numerator of the last conversion factor.

STEP 3 Performing the indicated arithmetic gives

$$\frac{184 \times 20.1 \times 62.1 \times 91.3}{100 \times 100 \times 100} \text{ handsome left-handed bald-headed adult males}$$

$= 20.968929$ handsome left-handed bald-headed adult males (calculator answer)
$= 21.0$ handsome left-handed bald-headed adult males (correct answer)

The calculator answer must be rounded off to three significant figures since all three of the given percentages contain only three significant figures.

Practice Exercise 3.14

A 5-lb box of chocolates contains 112 chocolates. Dark chocolates are more prevalent than light chocolates—75% versus 25%. Exactly 25% of the dark chocolates are cream-filled. How many cream-filled dark chocolates are there in the 5-lb box of chocolates?

Ans. 21 cream-filled dark chocolates

3.10 TEMPERATURE SCALES

Temperature *is a measure of the hotness or coldness of an object.* The most common instrument for measuring temperature is the mercury-in-glass thermometer, which consists of a glass bulb containing mercury sealed to a slender glass capillary tube. The higher the temperature, the farther the mercury will rise in the capillary tube. Graduations on the capillary tube indicate the height of the mercury column in terms of defined units, usually called *degrees*. A tiny superscript circle is used as the symbol for a degree.

Three different temperature scales are in common use—Celsius, Kelvin, and Fahrenheit. Both the Celsius and Kelvin scales are part of the metric measurement system, and the Fahrenheit scale belongs to the English measurement system. Different degree sizes and different reference points are what produce the various temperature scales.

The Celsius scale, named after Anders Celsius (1701–1744), a Swedish astronomer, is the scale most commonly encountered in scientific work. On this scale the boiling and freezing points of water serve as reference points, with the former having a value of 100°C (degrees Celsius) and the latter 0°C. Thus, there are 100 degree intervals between the two reference points. The Celsius scale was formerly called the centigrade scale.

The Kelvin scale is a close relative of the Celsius scale. The size of the temperature unit is the same on both scales, as is the interval between the reference points. The two scales differ only in the numerical values assigned to the reference points and the names of the units. On the Kelvin scale the boiling point of water is at 373.15 K (kelvins), and the freezing point of water at 273.15 K. The scale, proposed by the British mathematician and physi-

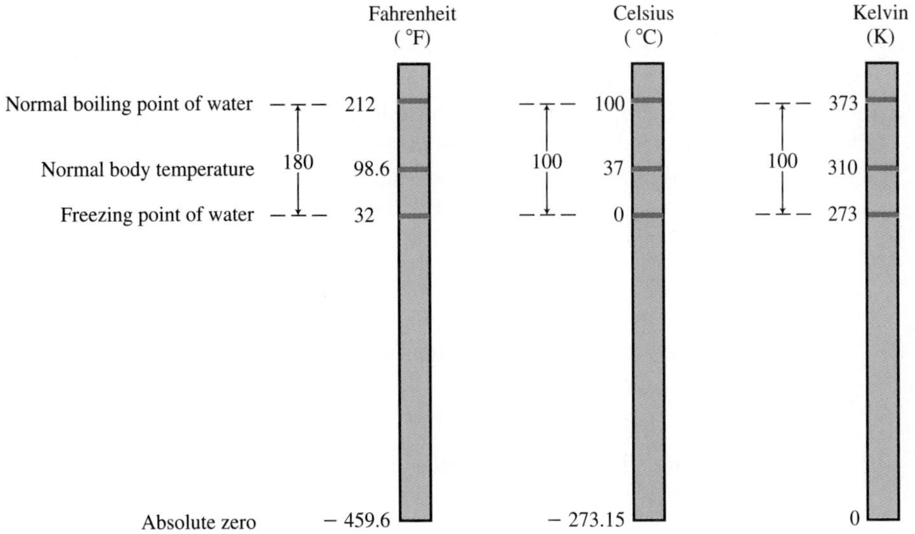

FIGURE 3.14 The relationships among the Fahrenheit, Celsius, and Kelvin temperature scales.

cist William Kelvin (1824–1907) in 1848, is particularly useful when working with relationships between temperature and pressure–volume behavior of gases (see Sec. 12.3). The degree sign is not used in specifying Kelvin temperatures (321 K instead of 321°K).

A unique feature of the Kelvin scale is that negative temperature readings never occur. The lowest possible temperature thought to be obtainable occurs at 0 on the Kelvin scale. This temperature, known as *absolute zero*, has never been produced experimentally, although scientists have come within a fraction of a degree of reaching it.

The Fahrenheit scale was designed by the German physicist Gabriel Fahrenheit (1686–1736) in the early 1700s. After proposing several scales he finally adopted a system that used a salt–ice mixture and boiling mercury as the reference points. These were the two extremes in temperature available to him at that time. A reading of 0 was assigned to the salt–ice mixture and 600 to the boiling mercury. The distance between these two points was divided into 600 equal parts or degrees. On this scale, water freezes at 32°F (degrees Fahrenheit) and boils at 212°F. Thus, there are 180 degrees between the freezing and boiling points of water on this scale as contrasted to 100 degrees on the Celsius scale and 100 kelvins on the Kelvin scale. Figure 3.14 shows a comparison of the three temperature scales.

When changing a temperature reading on one scale to its equivalent on another scale, we must take two factors into consideration: (1) the size of the unit on the two scales may differ, and (2) the zero points on the two scales do not coincide.

Difference in unit size will be a factor any time the Fahrenheit scale is involved in a conversion process. The conversion factors necessary to relate the size of the Fahrenheit degree to the size of the Celsius degree or the kelvin are obtainable from the information in Figure 3.14. From that figure we see that 180 Fahrenheit degrees are equivalent to 100 Celsius degrees or kelvins. Using this relationship and the fact that $\frac{180}{100} = \frac{9}{5}$, we obtain the following equalities.

$$5 \text{ Celsius degrees} = 9 \text{ Fahrenheit degrees}$$
$$5 \text{ kelvins} = 9 \text{ Fahrenheit degrees}$$

Conversion factors derived from these equalities will contain an infinite number of significant figures; that is, they are exact conversion factors.

F.Y.I.

How hot is hot?

100°C	Boiling point of water
800°C	Campfire
875°C	Cigarette ember
1600°C	Gas stove fire
2300°C	Filament of light bulb
7500°C	Surface of the sun
30,000°C	Typical lightning bolt

Adjustment for differing zero-point locations is carried out by considering how many degrees above or below the freezing point of water (the ice point) the original temperature is. Examples 3.14 and 3.15 show how this zero-point adjustment is carried out in addition to illustrating the use of temperature-scale conversion factors.

We should note that in all problem solving involving temperatures the precision of temperature readings is to the closest degree unless stated otherwise. Thus a temperature reading of 10°C is a two-significant-figure measurement and a reading of 100°C is a three-significant-figure measurement.

Exercise 3.15

An oven for baking pizza operates at approximately 525°F. What is the equivalent temperature on the Celsius scale?

Solution

First, we determine the number of degrees between the ice point (freezing point of water) and the given temperature on the original scale.

$$525°\text{F} - 32°\text{F} = 493 \ \ °\text{F above the ice point}$$

Second, we convert from Fahrenheit units to Celsius units.

$$493 \text{ Fahrenheit degrees} \times \frac{5 \text{ Celsius degrees}}{9 \text{ Fahrenheit degrees}} = 273.88888 \text{ Celsius degrees}$$

(calculator answer)

$$= 274 \text{ Celsius degrees}$$

(correct answer)

Third, taking into account the ice point on the new scale, we determine the new temperature. On the Celsius scale, the temperature will be 274 degrees above the ice point. Since the ice point is 0°C, the new temperature will be 274°C.

Practice Exercise 3.15

A comfortable temperature for bath water is 95°F. What is the equivalent temperature on the Celsius scale?

Ans. 35°C

Example 3.16

The ozone layer over Antarctica thins dramatically every year during spring. Antarctica's unusual winter weather conditions of extreme cold (−85°C) and total darkness are necessary prerequisites for the occurrence of the chemical reactions that lead to ozone depleton. What is the equivalent temperature on the Fahrenheit scale of the Antarctica's winter temperature of −85°C?

Solution

First, we determine how many degrees there are between the original temperature and the ice point. On the Celsius scale this will always be equal numerically to the original temperature, since the ice point is 0°C.

$$-85°C = 85 \text{ Celsius degrees below the ice point}$$

Second, we change this number of degrees from Celsius units to Fahrenheit units.

$$85 \text{ \sout{Celsius} degrees} \times \frac{9 \text{ Fahrenheit degrees}}{5 \text{ \sout{Celsius degrees}}} = 153 \text{ Fahrenheit degrees}$$

Significant figure considerations for changes in size of degree (Celsius to Fahrenheit or vice versa) represent an exception to the general rules for handling significant figures. Although size of degree changes involve a multiplication, position of uncertainty (addition–subtraction rules) is used to determine how the calculator answer should be modified to give the correct answer. If the original number of degrees is known to tenths, then the new number of degrees is also specified to tenths. Or, as in our specific problem here, if the original number of degrees is known to the closest degree (± 1), then the new number of degrees is specified to the closest degree.

Third, taking into account the ice point on the new scale, we determine the new temperature. The new temperature will be 153 Fahrenheit degrees below the ice point on the Fahrenheit scale.

$$32°F - 153 \text{ F°} = -121°F \quad \text{(calculator and correct answer)}$$

Practice Exercise 3.16

The temperature at the bottom of a blast furnace used in the production of iron (steel making) is measured at 1935°C. What is the equivalent temperature on the Fahrenheit scale?

Ans. 3515°F

Examples 3.15 and 3.16 point out that temperature-scale conversions are more complicated than the unit conversions of the last three sections. Not only is multiplication by a conversion factor required, but also addition and subtraction.

The relationship between the Kelvin and Celsius scales is very simple because the sizes of the degree unit is the same. No conversion factors are needed. All that is required is an adjustment for the differing zero points. This adjustment involves the number 273, the number of units by which the two scales are offset from each other. The adjustment factor is specifically 273, 273.2, or 273.15 depending on the precision of the temperature measurement. Since temperatures are most often stated in terms of a whole number of units (31°C, 43°C, etc.), 273 is the most used adjustment factor. However, the other two factors are needed when dealing with temperatures involving tenths or hundredths of a unit (31.5°C, 452.72°C, etc.).

To change a Celsius temperature to the Kelvin scale we add the adjustment factor 273.

$$K = °C + 273$$

To change a Kelvin temperature to Celsius we subtract this same adjustment factor.

$$°C = K - 273$$

Note that the symbol for the kelvin is K, not °K.

The relationship between the Fahrenheit scale and the Celsius scale can also be stated in an equation format.

$$°F = \frac{9}{5}(°C) + 32$$

$$°C = \frac{5}{9}(°F - 32)$$

Some students prefer to use these equations rather than the dimensional-analysis approach used in Examples 3.15 and 3.16. The use of these equations is illustrated in Example 3.17.

Example 3.17

A person suffering from heat stroke is found to have a body temperature of 41.1°C. What is this temperature on (a) the Fahrenheit scale? (b) the Kelvin scale?

Solution

(a) Substituting into the Celsius-to-Fahrenheit equation, we get

$$°F = \frac{9}{5}(°C) + 32 = \frac{9}{5}(41.1) + 32$$

$$= 74.0 + 32$$

$$= 106°F \quad \text{(calculator answer)}$$

$$= 106.0°F \quad \text{(correct answer)}$$

The multiplication $\frac{9}{5}$ (41.1) gives the calculator answer 73.98, which is rounded to 74.0, the closest tenth of a degree (see the significant figure discussion in Example 3.16). Adding 32 (an exact number by definition) to 74.0 gives 106.0 as the correct answer.

(b) Substituting into the Celsius-to-Kelvin equation, we get

$$K = °C + 273.2 = 41.1 + 273.2$$

$$= 314.3\ K \quad \text{(calculator and correct answer)}$$

Note that we used the adjustment factor 273.2 rather than simply 273 because the given temperature involved tenths of a degree.

Practice Exercise 3.17

A hospital patient with a fever has a temperature of 39.4°C. What is this temperature on (a) the Fahrenheit scale? (b) the Kelvin scale?

Ans. (a) 102.9°F; (b) 312.6 K

3.11 TYPES AND FORMS OF ENERGY

On some days you wake up feeling "very energetic." On these days you usually accomplish a great deal. By the end of the day you are tired, you have "no energy left," and you do not feel like doing any more "work." The scientific definition for energy closely parallels the ideas presented in these three sentences. **Energy** *is the capacity to do work.*

Energy can exist in any of several forms. Common forms include radiant (light) energy, chemical energy, thermal (heat) energy, electrical energy, and mechanical energy. These forms of energy are interconvertible. The heating of a home is a process that illustrates energy interconversion. As the result of burning natural gas or some other fuel, chemical energy is converted into heat energy. In large conventional power plants that are used to produce electricity, the heat energy obtained from burning coal is used to change water into stream, which can then turn a turbine (mechanical energy) to produce electricity (electrical energy).

During energy interconversions, the law of conservation of energy is always obeyed. The **law of conservation of energy** *states that in any chemical or physical change, energy can be converted from one form to another, but it is neither created nor destroyed.* Studies of energy changes in numerous systems have shown that no system acquires energy except at the expense of energy possessed by another system.

Almost all of our energy on Earth originates from the sun in the form of radiant or light energy. Green plants convert this radiant energy into chemical energy by means of a process called photosynthesis. The chemical energy is stored within the living plant. **Chemical energy** *is energy stored in a substance that can be released during a chemical change.* Energy for the human body is obtained, either directly or indirectly, from plants when they are consumed as food.

Chemical change can be used to produce other forms of energy. Chemical changes in an automobile battery produce the electrical energy needed to start a car. Chemical changes that occur in a magnesium flash bulb generate the light energy needed for photographic purposes. The burning of fuels releases both heat and light energy. The energy that "runs" our life processes—for example, breathing, muscle contraction, and blood circulation— is produced by chemical changes occurring within the cells of the body. The energy required for or generated by chemical changes will be an important point of focus in many discussions in later chapters.

In addition to the various *forms* of energy, there are two *types* of energy: potential energy and kinetic energy. The basis for determining energy types depends on whether the energy is available but not being used or is actually in use. **Potential energy** *is stored energy that results from an object's position, condition, and/or composition.* Water that is backed up behind a dam represents potential energy because of its position. When the water is released, it can be used to produce electrical energy at a hydroelectric plant. A compressed spring can spontaneously expand and do work as the result of potential energy associated with condition. Chemical energy, such as that stored in gasoline, is potential energy arising from composition. This stored energy is released when the gasoline is burned.

Kinetic energy *is energy that matter possesses because of its motion.* An object that is in motion has the capacity to do work. If it collides with another object, it will do work on that object. A hammer held in the air above an object possesses potential energy of position. As the hammer moves downward toward a nail, this potential energy becomes kinetic energy that can be used to drive the nail into a board. When water behind a dam is released and allowed to flow, its potential energy of position becomes kinetic energy. During the operation of a hydroelectric plant, some of this kinetic energy becomes mechanical and electrical energy.

Figure 3.15 summarizes the forms and types of energy discussed in the preceding paragraphs.

The concepts of potential and kinetic energy will play a part in many discussions in future chapters. For example, in Chapter 11, we will explain the differences between the solid, liquid, and gaseous states of matter in terms of relative amounts of potential and kinetic energy. The pressure that a gas exerts (Sec. 12.4) is related to kinetic energy.

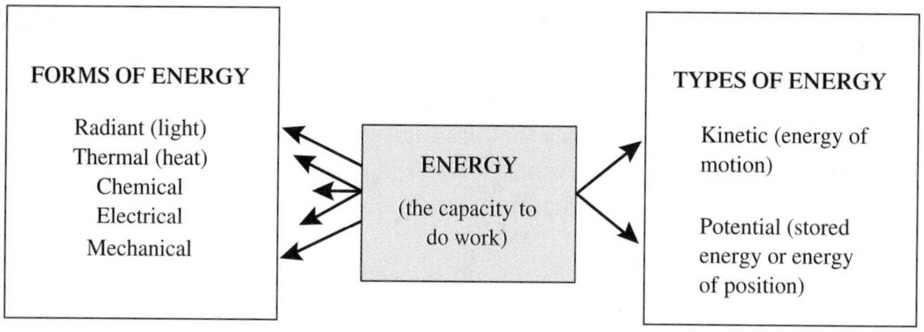

FIGURE 3.15 Energy may be classified by form and by type.

3.12 HEAT ENERGY AND SPECIFIC HEAT

The form of energy that is most often required for or released by the chemical and physical changes considered in this text is heat energy. For this reason, we will consider further particulars about this form of energy.

Units of Heat Energy

The most commonly used unit of measurement for heat energy is the joule. The joule (pronounced *jool*, rhymes with pool)—after the English physicist James Prescott Joule (1819–1889) who studied the energy concept—is a derived rather than a fundamental measurement unit. Its derivation involves the mass of an object (kg) and the square of the object's velocity (m^2/sec^2). A **joule** *is an energy unit obtained from the units kilogram, meter, and second.* Mathematically, the equation for a joule is

$$1 \text{ joule} = \frac{1 \text{ kg} \cdot \text{m}^2}{\text{sec}^2}$$

The joule unit, abbreviated as J, is suitable for measuring all types of energy, not just heat energy.

> **F.Y.I.**
> The Btu units used to rate air conditioners and heaters are the English units for heat. "Btu" stands for British thermal unit and is the amount of heat energy it takes to raise the temperature of one pound of water one degree Fahrenheit. One Btu is the equivalent of 0.818 kcal.

An additional unit for heat energy is the calorie (cal). A **calorie** *is the amount of heat energy needed to raise the temperature of one gram of water by one degree Celsius measured between 14.5 and 15.5°C.* The relationship between the calorie and the joule is

$$1 \text{ cal} = 4.184 \text{ J}$$

Both the joule and the calorie involve relatively small amounts of energy, so the kilojoule (kJ) and the kilocalorie (kcal) are often used instead. It follows from the relationship between joules and calories that

$$1 \text{ kcal} = 4.184 \text{ kJ}$$

In discussions involving nutrition, the energy content of foods, or dietary tables, the term *Calories* (spelled with a capital C) is used. The dietetic Calorie is actually 1 kilocalorie (1000 cal). The statement that an oatmeal raisin cookie contains 60 Calories means that 60 kcal (60,000 cal) of energy is released when the cookie is metabolized (undergoes chemical change) within the body.

Example 3.18

One gram of natural gas undergoes combustion (burns) to produce 55.2 kJ of heat energy. Express the energy obtained from this reaction in

(a) joules **(b)** kilocalories **(c)** calories

Solution

(a) The relationship between joules and kilojoules is

$$1000 \text{ J} = 1 \text{ kJ}$$

Therefore, using dimensional analysis, we have

$$55.2 \cancel{\text{kJ}} \times \frac{10^3 \text{ J}}{1 \cancel{\text{kJ}}} = 55{,}200 \text{ J} \quad \text{(calculator and correct answer)}$$

(b) The relationship between kilocalories and kilojoules is

$$1 \text{ kcal} = 4.184 \text{ kJ}$$

The one-step dimensional analysis setup for this problem is

$$55.2 \cancel{\text{kJ}} \times \frac{1 \text{ kcal}}{4.184 \cancel{\text{kJ}}} = 13.193116 \text{ kcal} \quad \text{(calculator answer)}$$

$$= 13.2 \text{ kcal} \quad \text{(correct answer)}$$

(c) There will be two conversion factors in this part, derived, respectively, from the relationship 1 kcal = 4.184 kJ and 1000 cal = 1 kcal. The pathway will be

$$\text{kJ} \longrightarrow \text{kcal} \longrightarrow \text{cal}$$

$$55.2 \cancel{\text{kJ}} \times \frac{1 \cancel{\text{kcal}}}{4.184 \cancel{\text{kJ}}} \times \frac{10^3 \text{ cal}}{1 \cancel{\text{kcal}}} = 13193.116 \text{ cal} \quad \text{(calculator answer)}$$

$$= 13{,}200 \text{ cal} \quad \text{(correct answer)}$$

There can be only three significant figures in the correct answer.

Practice Exercise 3.18

One gram of acetylene gas, a gas used in welding torches, undergoes combustion (burns) to produce 11,900 calories of heat energy. Express the energy obtained from this reaction in

(a) kilocalories (b) kilojoules (c) joules

Ans. (a) 11.9 kcal; (b) 49.8 kJ; (c) 49,800 J

Specific Heat

For every pure substance, in a given state (solid, liquid, or gas), we can measure a physical property called the *specific heat* of that substance. **Specific heat** *is the amount of heat needed to raise the temperature of 1 g of a substance in a specific physical state by 1°C.* The units most commonly used for specific heat, in scientific work, are joules per gram per degree Celsius [J/(g · °C)]. Specific heats for a number of substances in various states are given in Table 3.4. Note the three different entries for water in this table—one for each of the physical states. As these entries point out, the magnitude of the specific heat for a substance changes when the physical state of the substance changes. (Water is one of the few substances that we routinely encounter in all three physical states; hence the three specific-heat values in the table for this substance.)

TABLE 3.4 SPECIFIC HEATS OF SELECTED PURE SUBSTANCES

Substance	Physical State	Specific Heat [J/(g · °C)]
Aluminum	solid	0.908
Copper	solid	0.382
Ethyl alcohol	liquid	2.42
Gold	solid	0.13
Iron	solid	0.444
Nitrogen	gas	1.0
Oxygen	gas	0.92
Silver	solid	0.24
Sodium chloride	solid	0.88
Water (ice)	solid	2.1
Water	liquid	4.18
Water (steam)	gas	2.0

F.Y.I.
The warm water that flows from the tropics in ocean currents such as the Gulf Stream brings heat to the northern latitudes and makes them warmer than they would otherwise be. This is the reason Scandinavia has a much milder climate than Siberia.

F.Y.I.
Metals have low specific heats. They heat up quickly, but cool quickly.

F.Y.I.
A walk on the beach on a sunny day will demonstrate differences in specific heat. The sand, which has a relatively low specific heat, heats up rapidly in the sun. The water, with its higher specific heat, stays cool.

The higher the specific heat of a substance, the less its temperature will change when it absorbs a given amount of heat. For liquids, water has a relatively high specific heat; it is thus a very effective coolant. The moderate climates of geographical areas where large amounts of water are present—for example, the Hawaiian Islands—are related to water's ability to absorb large amounts of heat without undergoing drastic temperature changes. Desert areas, areas that lack water, are the areas where the extremes of high temperature are encountered on Earth. The temperature of a living organism remains relatively constant because of the large amounts of water present in it.

The amount of heat energy needed to cause a fixed amount of a substance to undergo a specific temperature change (within a range that causes no change of state) can easily be calculated if the substance's specific heat is known. The specific heat [in J/(g · °C)] is multiplied by the mass (in grams) and by the temperature change (in degrees Celsius) to eliminate the units of g and °C and obtain the unit calories.

$$\text{Heat absorbed} = \text{specific heat} \times \text{mass} \times \text{temperature change}$$

$$= \frac{\text{J}}{\cancel{\text{g}} \cdot \cancel{^\circ\text{C}}} \times \cancel{\text{g}} \times \cancel{^\circ\text{C}}$$

$$= \text{J}$$

The temperature change, denoted as ΔT, is always calculated as a positive number; the lower temperature is always subtracted from the higher temperature.

Example 3.19

A beaker contains 75.0 g of water at a temperature of 25°C. How many joules of heat energy must the water absorb in order for the temperature of the water to increase to 55°C?

Solution

From Table 3.4 we determine that the specific heat of liquid water is 4.18 J/(g · °C). Note that three specific heats for water are given in Table 3.4, one for each of the three physical states. In general, as is the case for water, a substance's specific heat as a solid is different from that as a liquid, which in turn is different from that as a gas.

Using this specific heat value, the given mass in grams, and the temperature change (in °C), we calculate the heat absorbed as follows.

$$\text{Heat absorbed} = \text{specific heat} \times \text{mass} \times \text{temperature change}$$

$$= \frac{4.18\ \text{J}}{\cancel{\text{g}} \cdot \cancel{^\circ\text{C}}} \times 75.0\ \cancel{\text{g}} \times 30\cancel{^\circ\text{C}}$$

$$= 9405\ \text{J} \quad \text{(calculator answer)}$$
$$= 9400\ \text{J} \quad \text{(correct answer)}$$

Practice Exercise 3.19

Calculate the number of joules of heat energy needed to increase the temperature of 50.0 g of copper metal from 21.0 to 80.0°C.

Ans. 1130 J

Example 3.20

One cup of dry roasted, salted, shelled, pistachio nuts has a caloric value of 787 Cal (78,700 cal). What would be the temperature change in a quart of water (944 g) at 10°C if the same amount of energy were added to it?

Solution

The equation

$$\text{Heat absorbed} = \text{specific heat} \times \text{mass} \times \text{temperature change}$$

is rearranged to isolate temperature change on the left side of the equation.

$$\text{Change in temperature (°C)} = \frac{\text{heat absorbed (J)}}{\text{mass (g)} \times \text{specific heat [J/(g} \cdot \text{°C)]}}$$

In this equation the heat absorbed will be the heat energy (caloric value) supplied by the pistachio nuts. Since this is given in calories, we will need to change it to joule units.

$$78{,}700\ \cancel{\text{cal}} \times \frac{4.184\ \text{J}}{1\ \cancel{\text{cal}}} = 329{,}280.8\ \text{J} \quad \text{(calculator answer)}$$
$$= 329{,}000\ \text{J} \quad \text{(correct answer)}$$

Substituting the values 329,000 J (heat absorbed), 944 g (mass of water), and 4.18 J/g · °C (specific heat of water—Table 3.4) in the previously derived equation for change in temperature (°C) gives

$$\text{Change in temperature (°C)} = \frac{329{,}000\ \cancel{\text{J}}}{944\ \cancel{\text{g}} \times 4.18\ \cancel{\text{J}}/(\cancel{\text{g}} \cdot \text{°C})}$$

$$= 83.37726\text{°C} \quad \text{(calculator answer)}$$
$$= 83.4\text{°C} \quad \text{(correct answer)}$$

The temperature of the water will increase from 10°C to 93°C.

Practice Exercise 3.20

A one-half cup serving of zucchini has a caloric value of approximately 15 Cal (15,000 cal). What would be the temperature change in a cup of water (237 g) at 20°C if the same amount of energy were added to it?

Ans. 64°C

A quantity closely related to specific heat is that of heat capacity. **Heat capacity** *is the amount of heat needed to raise the temperature of a given quantity of a substance in a specific physical state by 1°C.* The relationship between heat capacity and specific heat is

$$\text{Heat capacity} = \text{grams} \times \text{specific heat}$$

Common units for heat capacity are J/°C. Heat capacity refers to a property of a whole object (its entire mass), while specific heat refers to the heat capacity per unit mass (1 g). If either heat capacity or specific heat is known, the other quantity can always be calculated from the known quantity.

Example 3.21

A 35.0-g sample of the metal iron has a heat capacity of 15.5 J/°C. What is the specific heat, in J/g · °C, of iron?

Solution

The relationship between heat capacity and specific heat is

$$\text{Heat capacity} = \text{mass} \times \text{specific heat}$$

Rearranging this equation to isolate specific heat on one side gives

$$\begin{aligned}\text{Specific heat} &= \frac{\text{heat capacity}}{\text{mass}}\\ &= \frac{15.5\ \text{J/°C}}{35.0\ \text{g}}\\ &= 0.44285714\ \text{J/g}\cdot\text{°C} \quad \text{(calculator answer)}\\ &= 0.443\ \text{J/g}\cdot\text{°C} \quad \text{(correct answer)}\end{aligned}$$

Practice Exercise 3.21

A 80.0-g sample of the metal gold has a heat capacity of 10.5 J/°C. What is the specific heat of gold?

Ans. 0.131 J/g · °C

Another common type of heat energy calculation involves the transfer of heat from one substance to another. The following two generalizations always apply to such situations.

1. Heat always flows from the warmer body to the colder body.

2. The heat loss by the warmer body is equal to the heat gained by the colder body.

Example 3.22

A 125-g piece of a rock of unknown specific heat is heated to 93°C and then the rock is dropped into 100.0 g of water at 19°C. The temperature of the water rises to 31°C. What is the specific heat, in J/g · °C, of the rock?

Solution

The heat lost by the rock (the hotter body) is equal to the heat gained by the water (the colder body)

$$\text{Heat lost (rock)} = \text{heat gained (water)}$$

Recalling from Example 2.19, that the general equation for heat lost or heat gained is

$$\text{Heat lost or gained} = \text{specific heat} \times \text{mass} \times \text{temperature change}$$

we have for our present situation

$$\underbrace{\text{Specific heat} \times 125\text{ g} \times 62°\text{C}}_{\text{heat lost by rock}} = \underbrace{4.18\,\frac{\text{J}}{\text{g}\cdot°\text{C}} \times 100.0\text{ g} \times 12\text{ °C}}_{\text{heat gained by water}}$$

Solving this equation for the specific heat of the rock gives

$$\text{Specific heat of rock} = \frac{4.18\,\frac{\text{J}}{\text{g}\cdot°\text{C}} \times 100.0\,\cancel{\text{g}} \times 12\cancel{°\text{C}}}{125\,\cancel{\text{g}} \times 62\cancel{°\text{C}}}$$

$$= 0.6472258\text{ J/g}\cdot°\text{C} \quad \text{(calculator answer)}$$

$$= 0.65\text{ J/g}\cdot°\text{C} \quad \text{(correct answer)}$$

Practice Exercise 3.22

How many grams of copper can be heated from 20 to 30°C by the heat energy released when 245 g of aluminum cools from 80 to 50°C? Needed specific heats are found in Table 3.4.

Ans. 1700 g

KEY TERMS

The new terms or concepts defined in this chapter are

area (Sec. 3.4) A measure of the extent of a surface.

calorie (Sec. 3.12) The amount of heat energy needed to raise the temperature of 1 g of water by 1°C measured between 14.5 and 15.5°C.

chemical energy (Sec. 3.11) Energy stored in a substance that can be released during a chemical change.

conversion factor (Sec. 3.6) An equality, expressed in fractional form, that is used as a multiplier to convert a quantity in one unit into its equivalent in another unit.

density (Sec. 3.8) Ratio of the mass of an object to the volume occupied by that object.

dimensional analysis (Sec. 3.7) A general problem-solving method that uses the units associated with numbers as a guide in setting up the calculation.

energy (Sec. 3.11) The capacity to do work.

gram (Sec. 3.3) Basic unit of mass in the metric system.

heat capacity (Sec. 3.12) The amount of heat needed to raise the temperature of a given quantity of a substance in a specific physical state by 1°C.

joule (Sec. 3.12) An energy unit obtained from the units kilogram, meter, and second.

kinetic energy (Sec. 3.11) Energy that matter possesses because of its motion.

law of conservation of energy (Sec. 3.11) In any chemical or physical change, energy can be converted from one form to another, but it is neither created nor destroyed.

liter (Sec. 3.4) The basic unit of volume in the metric system.

mass (Sec. 3.3) A measure of the total quantity of matter in an object.

meter (Sec. 3.2) The basic unit of length in the metric system.

percent (Sec. 3.9) Parts per hundred, that is, the number of items of a specified type in a group of 100 items.

potential energy (Sec. 3.11) Stored energy that results from an object's position, condition, and/or composition.

specific heat (Sec. 3.12) The amount of heat needed to raise the temperature of 1 g of a substance in a specific physical state by 1°C.

temperature (Sec. 3.10) A measure of the hotness or coldness of an object.

volume (Sec. 3.4) A measure of the amount of space occupied by an object.

weight (Sec. 3.3) A measure of the force exerted on an object by the pull of gravity.

PRACTICE PROBLEMS

Metric System Units (Sec. 3.1 through 3.4)

3.1 Identify the numerical multiplier (power of ten) corresponding to each of the following metric prefixes or vice versa.

(a) 10^3 **(b)** 10^{-2} **(c)** 10^{-6}
(d) nano **(e)** mega **(f)** milli

3.2 Identify the numerical multiplier (power of ten) corresponding to each of the following metric prefixes or vice versa.

(a) 10^{-3} **(b)** 10^{-1} **(c)** 10^6
(d) pico **(e)** micro **(f)** centi

3.3 Write the symbol (abbreviation) for each of the following metric system units or vice versa.

(a) microgram **(b)** kilometer **(c)** centiliter
(d) dm **(e)** mL **(f)** pg

3.4 Write the symbol (abbreviation) for each of the following metric system units or vice versa.

(a) megagram **(b)** microliter **(c)** millimeter
(d) cL **(e)** nm **(f)** kg

3.5 Use the appropriate metric prefix abbreviation to replace the power of ten in each of the following values.

(a) 6.8×10^{-9} m **(b)** 3.2×10^{-6} L
(c) 7.23×10^3 L **(d)** 6.5×10^9 g

3.6 Use the appropriate metric prefix abbreviation to replace the power of ten in each of the following values.

(a) 4.1×10^{-3} L **(b)** 9.9×10^{-12} g
(c) 8.721×10^{-2} g **(d)** 4.4×10^6 m

3.7 For each of the pairs of units listed, indicate whether the first unit is larger or smaller than the second unit and then indicate how many times larger or smaller it is.

(a) centigram, gram **(b)** nanogram, microgram
(c) kilogram, decigram **(d)** milligram, megagram

3.8 For each of the pairs of units listed, indicate whether the first unit is larger or smaller than the second unit and then indicate how many times larger or smaller it is.

(a) milliliter and liter **(b)** kiloliter and microliter
(c) nanoliter and deciliter **(d)** centigram and megagram

3.9 What type of quantity (length, mass, area, or volume) do each of the following units represent?

(a) cm^3 **(b)** mm **(c)** ML **(d)** km^2

3.10 What type of quantity (length, mass, area, or volume) do each of the following units represent?

(a) L (b) cm^2 (c) kg (d) km^3

3.11 For each of the following choose the answer that most closely applies.

(a) Does a nickel coin weigh 5 mg or 5 g or 5 kg?

(b) Does a 220 lb football player weigh 100 g or 100 dg or 100 kg?

(c) Is the thickness of a nickel coin 2 mm or 2 cm or 2 m?

(d) Does a teaspoon hold 5 mL or 5 dL or 5 L?

3.12 For each of the following choose the answer that most closely applies.

(a) Does a thumbtack weigh $\frac{1}{2}$ mg or $\frac{1}{2}$ g or $\frac{1}{2}$ Mg?

(b) Does a $\frac{1}{2}$ fluid ounce container hold 15 μL or 15 mL or 15 L?

(c) Is the diameter of a piece of chalk 1 mm or 1 cm or 1 m?

(d) Is the length of a football field 92 cm or 92 m or 92 km?

3.13 For each of the pairs of units listed, indicate which quantity is larger.

(a) 1 centimeter, 1 inch (b) 1 meter, 1 yard

(c) 1 gram, 1 pound (d) 1 liter, 1 gallon

3.14 For each of the pairs of units listed, indicate which quantity is larger.

(a) 1 kilometer, 1 mile

(b) 1 milliliter, 1 fluid ounce

(c) 1 kilogram, 1 pound

(d) 1 liter, 1 quart

Units and Mathematical Operations (Secs. 3.4 and 3.5)

3.15 Carry out the following mathematical manipulations of units.

(a) $mm \times mm \times mm$ (b) $\frac{nm^3}{nm}$

(c) $\frac{km^3 \times km}{km^2 \times km}$ (d) $\frac{cm}{sec} \times sec$

3.16 Carry out the following mathematical manipulations of units.

(a) $cm \times cm^2$ (b) $\frac{mm^2}{mm}$

(c) $\frac{m^2 \times m}{m^3}$ (d) $\frac{km}{min} \times min$

3.17 Calculate the area of the following surfaces.

(a) a square surface whose side is 4.52 cm

(b) a rectangular surface whose width is 3.5 m and whose length is 9.2 m

(c) a circle whose radius is 4.579 mm

(d) a triangle whose height is 3.0 mm and whose base is 5.5 mm

3.18 Calculate the area of the following surfaces.

(a) a rectangular surface whose dimensions are 24.3 m and 32.1 m

(b) a circle of radius 2.7213 cm

(c) a triangular object whose base is 12.0 mm and whose height is 8.00 mm

(d) a square surface with sides of 6.7 cm

3.19 Calculate the volume of each of the following objects, each of which has a regular geometrical shape.

(a) a copper block 5.4 cm long, 0.52 cm high, and 3.4 cm wide

(b) a cylindrical piece of cheese that has a height of 7.5 cm and a radius of 2.4 cm

(c) a spherical piece of styrofoam with a radius of 87 mm

(d) a piece of gold in the shape of a cube whose edge is 7.2 cm

3.20 Calculate the volume of each of the following objects, each of which has a regular geometrical shape.

(a) a cube of steel whose edge is 3.5175 mm

(b) a spherical marble with a radius of 1.212 cm

(c) a bar of iron 6.0 m long, 0.10 m wide, and 0.20 m high

(d) a cylindrical rod of copper whose length is 62 mm and whose radius is 3.2 mm

Conversion Factors (Sec. 3.6)

3.21 Write an equation that relates the members of each of the following pairs of time units and also write the two conversion factors associated with the equation.

(a) days and hours (b) minutes and seconds

(c) decades and centuries (d) days and years

3.22 Write an equation that relates the members of each of the following pairs of time units and also write the two conversion factors associated with the equation.

(a) days and weeks (b) hours and minutes

(c) months and years (d) years and centuries

3.23 Give the two forms of the conversion factor that relates each of the following pairs of units.

(a) kL and L (b) mg and g

(c) m and cm (d) μsec and sec

3.24 Give the two forms of the conversion factor that relates each of the following pairs of units.

(a) ng and g (b) dL and L

(c) m and Mm (d) psec and sec

3.25 Indicate how each of the following conversion factors should be interpreted in terms of significant figures present.

(a) $\frac{1.609 \text{ km}}{1 \text{ mi}}$ (b) $\frac{10^{-2} \text{ m}}{1 \text{ cm}}$

(c) $\frac{28.34 \text{ g}}{1 \text{ oz}}$ (d) $\frac{12 \text{ in.}}{1 \text{ ft}}$

3.26 Indicate how each of the following conversion factors should be interpreted in terms of significant figures present.

(a) $\frac{2.540\text{ cm}}{1\text{ in.}}$ **(b)** $\frac{453.6\text{ g}}{1\text{ lb}}$

(c) $\frac{2.113\text{ pt}}{1\text{ L}}$ **(d)** $\frac{10^{-9}\text{ m}}{1\text{ nm}}$

Dimensional Analysis—Metric–Metric Unit Conversions (Sec. 3.7)

3.27 Perform the following metric system conversions using the dimensional analysis method of problem solving.

(a) 25 mg = ? g **(b)** 323 km = ? m

(c) ? dL = 25.0 L **(d)** ? pg = 0.010 g

3.28 Perform the following metric system conversions using the dimensional analysis method of problem solving.

(a) 3.50 nm = ? m **(b)** 20,000 μg = ? g

(c) ? cL = 250 L **(d)** ? Mg = 0.225 g

3.29 Perform the following metric system conversions using dimensional analysis and two conversion factors.

(a) 23 dL = ? cL **(b)** 6.00 kg = ? mg

(c) ? nL = 6×10^{-3} μL **(d)** ? nm = 25 Mm

3.30 Perform the following met-ic system conversions using dimensional analysis and two conversion factors.

(a) 3.00 km = ? μm **(b)** 35.7 cL = ? nL

(c) ? dm = 4×10^4 pm **(d)** ? mg = 5×10^{-8} Mg

3.31 Perform the following metric system conversions using dimensional analysis.

(a) 6.0 cm^2 = ? m^2 **(b)** 7.2 mm^3 = ? m^3

(c) 25 μm^2 = ? dm^2 **(d)** 0.023 km^3 = ? nm^3

3.32 Perform the following metric system conversions using dimensional analysis.

(a) 3.25 km^2 = ? m^2 **(b)** 0.30 pm^3 = ? m^3

(c) 9.552 dm^2 = ? mm^2 **(d)** 5.6 cm^3 = ? μm^3

3.33 A certain chemical process consumes water at a rate of 55 L/sec. Express this water consumption rate in the following units.

(a) L/hr **(b)** kL/sec **(c)** dL/min **(d)** mL/day

3.34 A certain petroleum refinery operation uses hydrogen gas at the rate of 5×10^4 g/min. Express this hydrogen consumption rate in the following units.

(a) dg/min **(b)** g/sec **(c)** kg/hr **(d)** ng/day

3.35 The concentration of sugar in a sugar solution is found to be 2.30 μg/L. What is the sugar concentration in the following units?

(a) mg/L **(b)** μg/mL **(c)** cg/cL **(d)** kg/m^3

3.36 The concentration of salt in a salt solution is found to be 4.5 mg/mL. What is the salt concentration in the following units?

(a) μg/L **(b)** pg/mL **(c)** g/L **(d)** kg/m^3

Dimensional Analysis—Metric–English Unit Conversions (Sec. 3.7)

3.37 Using dimensional analysis, convert the following measurements to gallons.

(a) 4.67 L **(b)** 4.670 L

(c) 4.6700 L **(d)** 4.67000 L

3.38 Using dimensional analysis, convert the following measurements to pounds.

(a) 4.67 g **(b)** 4.670 g

(c) 4.6700 g **(d)** 4.67000 g

3.39 The length of a football field, between goal lines, is 100.0 yd. Express this length in the following units.

(a) meters **(b)** centimeters

(c) kilometers **(d)** inches

3.40 The length of a football field, between goal posts, is 120.0 yd. Express this length in the following units.

(a) meters **(b)** millimeters

(c) megameters **(d)** miles

3.41 A spray steam iron has a capacity of 75 mL of water. Express this water capacity in the following units.

(a) qt **(b)** gal **(c)** fl oz **(d)** cm^3

3.42 An automobile's gasoline tank has a capacity of 64 L. Express this capacity in the following units.

(a) qt **(b)** gal **(c)** fl oz **(d)** cm^3

3.43 The mass of the Earth is estimated to be 6.6×10^{21} tons. Express the mass of the Earth in the following units.

(a) g **(b)** kg **(c)** ng **(d)** oz

3.44 A defensive lineman on a professional football team has a mass of 295 lb. Express the mass of this football player in the following units.

(a) kg **(b)** Mg **(c)** mg **(d)** ton

3.45 A regular-issue U.S. postage stamp is 2.1 cm wide and 2.5 cm long. Express the surface area of this postage stamp in **(a)** square centimeters and **(b)** square inches.

3.46 A rectangular piece of concrete has dimensions of 3.6 m and 1.2 m. Express its surface area in **(a)** square meters and **(b)** square yards.

3.47 The luggage compartment of an automobile has the dimensions 95 cm × 105 cm × 145 cm. What is the volume of this compartment in cubic feet?

3.48 A copper block is 65 cm long, 3.0 cm high, and 4.0 cm wide. What is the volume of this bar in cubic inches?

Density (Sec. 3.8)

3.49 Calculate the density, in grams per milliliter, for each of the following.

(a) 25.0 g of ethyl alcohol having a volume of 31.7 mL
(b) 25.0 g of chromium metal having a volume of 3.48 cm^3
(c) 25.0 mL of olive oil having a mass of 22.9 g
(d) 25.0 L of chloroform having a mass of 37,200 g

3.50 Calculate the density, in grams per milliliter, for each of the following.
(a) 15.0 g of sea water having a volume of 14.6 mL
(b) 15.0 g of cork having a volume of 60.0 cm^3
(c) 15.0 mL of kerosene having a mass of 12.3 g
(d) 15.0 L of helium gas having a mass of 2.67 g

3.51 Calculate the mass, in grams, for each of the following.
(a) 22.2 mL of blood plasma (d = 1.027 g/mL)
(b) 22.2 cm^3 of gold metal (d = 19.3 g/cm^3)
(c) 22.2 L of dry air (d = 1.29 g/L)
(d) 22.2 L of urine (d = 1.027 g/mL)

3.52 Calculate the mass, in grams, for each of the following.
(a) 33.3 mL of milk (d = 1.03 g/mL)
(b) 33.3 cm^3 of bone (d = 1.8 g/cm^3)
(c) 33.3 L of hydrogen gas (d = 0.087 g/L)
(d) 33.3 L of lead metal (d = 11.3 g/cm^3)

3.53 Calculate the volume, in milliliters, for each of the following.
(a) 50.0 g of acetone (d = 0.791 g/mL)
(b) 50.0 g of silver metal (d = 10.40 g/cm^3)
(c) 50.0 g of carbon monoxide gas (d = 1.25 g/L)
(d) 50.0 g of rock salt (d = 2.18 g/cm^3)

3.54 Calculate the volume, in milliliters, for each of the following.
(a) 75.0 g of gasoline (d = 0.56 g/mL)
(b) 75.0 g of sodium metal (d = 0.93 g/cm^3)
(c) 75.0 g of ammonia gas (d = 0.759 g/L)
(d) 75.0 g of mercury (d = 13.6 g/mL)

3.55 A small bottle contains 2.171 mL of a red liquid. The total mass of the bottle and liquid is 5.261 g. The empty bottle weighs 3.006 g. What is the density, in grams per milliliter, of the liquid?

3.56 A piece of metal weighing 187.6 g is placed in a graduated cylinder containing 225.2 mL of water. The combined volume of solid and liquid is 250.3 mL. From these data, calculate the density, in grams per milliliter, of the metal.

3.57 An automobile gasoline tank holds 13.0 gal when full. How many pounds of gasoline will it hold, if the gasoline has a density of 0.56 g/mL?

3.58 Liquid sodium metal has a density of 0.93 g/cm^3. How many pounds of liquid sodium are needed to fill a container whose capacity is 15.0 L?

3.59 What mass of the metal chromium (density = 7.18 g/cm^3) occupies the same volume as 100.0 g of aluminum (density = 2.70 g/cm^3)?

3.60 What volume of the metal nickel (density = 8.90 g/cm^3) has the same mass at 100.0 cm^3 of lead (density = 11.3 g/cm^3)?

Percentage (Sec. 3.9)

3.61 An assortment of coins contains 17 pennies, 5 nickels, 2 dimes, 15 quarters, and 1 half dollar. State the percent of the coins that
(a) are nickels
(b) are quarters
(c) have a face value of 10¢ or less
(d) are smaller in diameter than a nickel

3.62 An assortment of coins contains 6 pennies, 14 nickels, 9 dimes, 16 quarters, and 5 half dollars. State the percent of the coins that
(a) are dimes
(b) are pennies
(c) have a face value of 10¢ or more
(d) are larger in diameter than a dime

3.63 A 1980 U.S. penny (a zinc–copper alloy) with a mass of 3.053 g contains 2.902 g of copper. What is the mass percentage in the penny of **(a)** copper and **(b)** zinc?

3.64 A 1990 U.S. penny (zinc plated with a thin layer of copper) with a mass of 2.552 g contains 2.488 of zinc. What is the mass percentage in the penny of **(a)** copper and **(b)** zinc?

3.65 How many grams of water are contained in 65.3 g of a mixture of alcohol and water that is 34.2% water by mass?

3.66 How many grams of alcohol are contained in 467 g of a mixture of alcohol and water that is 23.0% alcohol by mass?

3.67 A solution of table salt in water contains 15.3% by mass of table salt. If 437 g of solution are evaporated to dryness, how many grams of table salt will remain?

3.68 A solution of table salt in water contains 15.3% by mass of table salt. In a 542-g sample of this solution, how many grams of water are present?

3.69 Consider the following facts about a candy mixture containing "Gummi bears" and "Gummi worms": (1) 30.9% of the 661 items present are Gummi bears; (2) 23.0% of the Gummi bears are orange; and (3) 6.4% of the orange Gummi bears have only one ear. How many one-eared orange Gummi bears are present in the candy mixture?

3.70 An analysis of the makeup of a beginning chemistry class gives the following facts: (1) 47.1% of the 87 students are female; (2) 43.9% of the female students are married; and (3) 33.3% of the married female students are sophomores.

How many students in the class are sophomore female students who are married?

Temperature Scales (Sec. 3.10)

3.71 Convert each of the following Celsius temperatures to the Fahrenheit scale.

(a) 1251°C **(b)** 23.2°C **(c)** −2°C **(d)** −87°C

3.72 Convert each of the following Celsius temperatures to the Fahrenheit scale.

(a) 950°C **(b)** 37.3°C **(c)** −9°C **(d)** −53°C

3.73 Convert each of the following Fahrenheit temperatures to the Celsius scale.

(a) 2450°F **(b)** 337°F **(c)** 11°F **(d)** −37°C

3.74 Convert each of the following Fahrenheit temperatures to the Celsius scale.

(a) 1530°F **(b)** 117°F **(c)** 2°F **(d)** −133°F

3.75 Convert each of the following temperature readings to the Kelvin scale.

(a) 231°C **(b)** 231.7°C
(c) 231.74°C **(d)** 37.3°F

3.76 Convert each of the following temperature readings to the Kelvin scale.

(a) 137°C **(b)** 137.2°C
(c) 137.23°C **(d)** 79.0°F

3.77 Carry out the following temperature scale conversions.

(a) The temperature on a hot summer day is 101°F. What is this temperature in degrees Celsius?

(b) Oxygen, the gas necessary to sustain life, freezes to a solid at −218.4°C. What is this temperature in degrees Fahrenheit?

(c) The melting point of sodium chloride (table salt) is 804°C. What is this temperature in kelvins?

(d) Liquefied nitrogen boils at 77 K. What is this temperature in degrees Fahrenheit?

3.78 Carry out the following temperature scale conversions.

(a) Mercury freezes at 234.3 K. What is this temperature in degrees Celsius?

(b) Normal body temperature for a chickadee is 41.0°C. What is this temperature in degrees Fahrenheit?

(c) A recommended temperature setting for household hot water heaters is 140°F. What is this temperature in degrees Celsius?

(d) The metal aluminum melts at 934 K. What is this temperature in degrees Fahrenheit?

3.79 Which is the higher temperature, −10°C or 10°F?

3.80 Which is the higher temperature, −15°C or 4°F?

3.81 What is the lower temperature, 223 K or −60°F?

3.82 Which is the lower temperature, 381 K or 98°C?

Types and Forms of Energy (Sec. 3.11)

3.83 List the predominant *forms* of energy produced when

(a) an electric light bulb is turned on.

(b) a log is burned in a fireplace.

(c) a green plant grows.

(d) a burner on an electric stove is turned on.

3.84 List the predominant *forms* of energy produced when each of the following processes occurs.

(a) a bicycle is pedaled

(b) a flashlight is turned on

(c) a photographer's flash bulb goes off

(d) a candle is lighted

3.85 Identify the principal *type* of energy (kinetic or potential) that is exhibited by each of the following.

(a) a car parked on a hill

(b) a car traveling 65 mi/hr on a level road

(c) an elevator stopped at the 35th floor

(d) water behind a dam

3.86 Identify the principal *type* of energy (kinetic or potential) that is exhibited by each of the following.

(a) a piece of coal

(b) a falling rock

(c) a compressed metal spring

(d) a rolling soccer ball on a level field

Heat Energy and Specific Heat (Sec. 3.12)

3.87 The energy content of a chicken sandwich is found to be 2290 kJ. Specify this amount of energy in

(a) joules **(b)** kilocalories
(c) calories **(d)** Calories

3.88 The energy content of a quarter-pound hamburger is found to be 211,000 J. Specify this amount of energy in

(a) kilojoules **(b)** kilocalories
(c) calories **(d)** Calories

3.89 Calculate the number of joules of heat energy required to heat 25.0 g of copper from 23.0°C to 34.7°C.

3.90 Calculate the number of joules of heat energy required to heat 35.0 g of aluminum from 55.0°C to 75.0°C.

3.91 Calculate the final temperature after 145 J of heat energy is added to 7.73 g of water at 43.2°C.

3.92 Calculate the final temperature after 145 J of heat energy is removed from 6.55 g of water at 34.2°C.

3.93 What is the mass, in grams, of a piece of aluminum if its temperature changes from 20.0°C to 315°C when it absorbs 422 J of heat energy?

3.94 What is the mass, in grams, of a piece of copper if its temperature changes from 20.0°C to 55.6°C when it absorbs 22.7 J of heat energy?

3.95 Calculate the specific heat of a substance given that 46.9 J of heat is required to raise the temperature of 40.0 g of the substance by 3.0°C.

3.96 Calculate the specific heat of a substance given that 221 J of heat is required to raise the temperature of 55.0 g of the substance by 4.0°C.

3.97 Which has the higher heat capacity, 40.0 g of gold or 80.0 g of copper?

3.98 Which has the higher heat capacity, 20.0 g of liquid water of 30.0 g of ethyl alcohol?

3.99 The heat capacity of 75.1 g of a metal is 28.7 J/°C. What is the specific heat, in J/g · °C, for this metal?

3.100 The heat capacity of 4.71 g of a metal is 0.60 J/°C. What is the specific heat, in J/g · °C, for this metal?

3.101 How many grams of copper can be heated from 30°C to 50°C when 1.0 g of liquid water cools from 100°C to 15°C?

3.102 How many grams of aluminum can be heated from 20°C to 45°C when 20.0 g of oxygen gas cools from 285°C to 15°C?

ADDITIONAL PROBLEMS

3.103 The heights of the starting five players on a basketball team are 20.9 dm, 2030 mm, 1.90 m, 0.00183 km, and 203 cm. What is the average height, in centimeters, of these five basketball players? (The average is the sum of the individual values divided by the number of values.)

3.104 The masses for the five heaviest defensive linemen on a football team are 141,000 g, 0.133 Mg, 1.28×10^8 mg, 126 kg, and 1.22×10^{11} μg. What is the average mass, in grams, of these defensive linemen? (The average is the sum of the individual values divided by the number of values.)

3.105 Using the dimensional analysis method of problem solving, set up and solve the following problem. "A package of 10 razor cartridges costs $5.49. One razor cartridge is good for 9 shaves. What is the cost, in cents, per shave?"

3.106 Using the dimensional analysis method of problem solving, set up and solve the following problem. "One box of envelopes contains 500 envelopes. A case of envelopes contains 8 boxes of envelopes and costs $28.49. What is the cost, in cents, of an envelope?"

3.107 If your blood has a density of 1.05 g/mL at 20°C, how many grams of blood would you lose if you donated 1.00 pint of blood?

3.108 If your urine has a density of 1.030 g/mL at 20°C, how many pounds of urine would you lose if you eliminated 0.500 pint of urine?

3.109 The concentration of carbon monoxide, a common air pollutant, is measured at 5.7×10^{-3} μg/cm^3 inside a room. How many grams of carbon monoxide are present in the room if the room's dimensions are 3.5 m × 3.0 m × 3.2 m?

3.110 If the amount of mercury in a polluted lake is 0.39 μg/mL, what is the total mass of mercury, in kilograms, present in the lake if the lake has a surface area of 125 mi^2 and an average depth of 35 ft?

3.111 Levels of blood glucose higher than 400 mg/dL are life threatening. Is either of the following laboratory-measured glucose levels life threatening?
(a) 5000 μg/mL **(b)** 0.5 g/L

3.112 Levels of blood glucose lower than 40 mg/dL are life threatening. Is either of the following laboratory-measured glucose levels life threatening?
(a) 2000 μg/L **(b)** 20,000,000 ng/cL

3.113 A square piece of aluminum foil, 4.0 in. on a side, is found to weigh 0.466 g. What is the thickness of the foil, in millimeters, if the density of the foil is 269 cg/cm^3?

3.114 The density of osmium (the densest metal) is 2260 cg/cm^3. What is the mass, in grams, of a block of osmium with dimensions 5.00 in. × 4.00 in. × 0.25 ft?

3.115 A certain brand of household disinfectant contains 2.9% by mass of active ingredient. How many grams of active ingredient are present in 200.0 mL of disinfectant that has a density of 1.09 g/mL?

3.116 Gold jewelry is usually made of 14-karat gold, a gold alloy that is 58.33% gold by mass. What would be the mass of gold, in grams, present in a 50.0 cm^3 sample of 14-karat gold of density 14.9 g/cm^3?

3.117 A scientist invented a new temperature scale called the Howard scale (°H) and assigned the boiling and freezing points of water the values 200°H, and −200°H, respectively. What is a temperature of 50°F equivalent to on the Howard scale?

3.118 A scientist invented a new temperature scale called the Stephen scale (°S) and assigned the boiling and freezing points of water the values 150°S and −50°S, respectively. What is the temperature of 50°C equivalent to on the Stephen scale?

3.119 A sample of metal weighing 366 g is heated to 95.3°C and then dropped in 342 g of water at 27.0°C. If the final temperature of the water and metal is 37.6°C, what is the specific heat of the metal? Assume that no heat is lost to the surroundings.

3.120 A sample of metal weighing 612 g is heated to 87.3°C and then dropped in 413 g of water at 35.0°C. If the final temperature of the water and metal is 47.4°C, what is the specific heat of the metal? Assume that no heat is lost to the surroundings.

3.121 The body contains approximately 5.7 L of blood. Assuming that the density of blood is 1.06 g/mL and that the specific heats of blood and water are the same, how many kilojoules of energy are required to raise the temperature of this amount of blood by 1.0°C?

3.122 Assuming that the density of blood is 1.06g/mL and that the specific heats of blood and water are the same, how many liters of blood are present in a human body if 36.0 kJ of energy raises the temperature of the blood present by 1.5°C?

3.123 Given the following information about a bag of gummi bears, calculate the number of gummi bears in the bag: (1) 30.3% of the gummi bears are orange; (2) 8.10% of the orange gummi bears have only one ear; (3) There are 3 one-eared orange gummi bears in the bag.

3.124 Given the following information about a class of students, calculate the number of students in the class: (1) 37.3% of the students have blue eyes; (2) 20.0% of the blue-eyed students are left-handed; (3) There are 5 blue-eyed left-handed students in the class.

CUMULATIVE PROBLEMS

3.125 A sample of a colorless liquid has a mass of two grams and a volume of four milliliters. Calculate the liquid's density using the following precision specifications and express your answers in scientific notation.

(a) 2.000 g and 4.000 mL.

(b) 2.00 g and 4.0 mL

(c) 2.0000 g and 4.0000 mL

(d) 2.000 g and 4.0000 mL

3.126 A one gram sample of a powdery white solid is found to have a volume of two cubic centimeters. Calculate the solid's density using the following precision specifications and express your answers in scientific notation.

(a) 1.0 g and 2.0 cm^3

(b) 1.000 g and 2.00 cm^3

(c) 1.0000 g and 2.0000 cm^3

(d) 1.000 g and 2.0000 cm^3

3.127 A small rectangular box measures 10 cm wide, 200 cm long, and 4 cm high. Calculate the volume of the box, in cubic centimeters, given that all of the dimensions are known to

(a) the closest centimeter

(b) the closest tenth of a centimeter

(c) the closest hundredth of a centimeter

(d) two significant figures

3.128 A small rectangular room has a width of 9 m and a length of 21 m. Calculate the area of this room, in square meters, given that both dimensions are known to

(a) the closest meter

(b) the closest tenth of a meter

(c) the closest hundredth of a meter

(d) three significant figures

3.129 Indicate which measurement in each of the following sets of measurements has the greatest precision.

(a) 3.256×10^3 g, 3.256×10^4 g, 3.256×10^5 g

(b) 3.34 g, 3.34 kg, 3.34 mg

(c) 4.31 g, 4.31×10^{-3} kg, 4.31×10^3 mg

(d) 325.0 cg, 3.2500 g, 0.00325 kg

3.130 Indicate which measurement in each of the following sets of measurements has the greatest precision.

(a) 2.53×10^{-3} m, 2.53×10^{-4} m, 2.53×10^{-5} m

(b) 7.612 m, 7.612 km, 7.612 cm

(c) 6.73 m, 6.73×10^2 cm, 6.73×10^6 μm

(d) 35.300 mm, 3.530 cm, 0.0353 m

4 Basic Concepts about Matter

4.1 Chemistry—The Study of Matter

Chemistry *is the branch of science that is concerned with matter and its properties*. What is matter? What is it that chemists study?

Intuitively, most people have a general feeling for the meaning of the word matter. They consider matter to be the materials of the physical universe—that is, the "stuff" from which the universe is made. Such an interpretation is a correct one.

Formally defined, **matter** *is anything that has mass and occupies space*. A substance need not be visible to the naked eye to be labeled as matter as long as it meets the two qualifications of having mass and occupying space. Wood, paper, stone, the food we eat, the air we breathe, the fluids we drink, our bodies themselves, our clothing, and our shelter are all examples of matter.

The question may be asked, "What, then, does matter not include?" Various forms of energy, such as heat, light, and electricity, are not included. Neither are wisdom, friendship, ideas, thoughts, and emotions (such as anger and love) included.

Consider the extraordinary breadth of our definition of matter. This concept encompasses all objects we know about from the largest objects in outer space to the most minute objects seen under a microscope to objects so small they cannot be seen with any known type of instrumentation. Individuals first encountering this breadth automatically suppose that the study of matter will be a very complicated subject because of the literally millions of different types of matter that exist. One purpose of this chapter is to show that all matter can be classified into a surprisingly small number of categories. The naturally occurring materials of the universe and the synthetic materials humans have fashioned from them are, indeed, much simpler in makeup than they outwardly appear.

4.2 PHYSICAL STATES OF MATTER

All matter can be classified into three physical states: solid, liquid, and gas. Such classification depends on whether the shape and volume of the sample of matter is definite or indefinite. **Solids** *have a definite shape and a definite volume*. The shape of *large* pieces of a solid is independent of its container. For example, a piece of copper wire has the same shape and volume whether it is placed in a large test tube or simply placed on a table top. For solids in powdered or granulated forms, such as sugar or salt, each individual particle has a definite shape and definite volume; however, a quantity of such a solid takes the shape of the portion of the container it occupies. **Liquids** *have an indefinite shape and a definite volume*. A liquid always takes the shape of its container to the extent that it fills it. **Gases** *have an indefinite shape and an indefinite volume*. A gas always completely fills its container, adopting both its volume and shape. Figure 4.1 summarizes the shape and volume characteristics of the three physical states of matter.

F.Y.I.

Most people think of rocks as solid, hence the sayings "solid as a rock" and "it is written in stone." But at the high temperatures and pressures within the earth, rock can turn to molten magma.

In nature we find matter in all three physical states. Rocks, sand and ice are in the solid state, water and petroleum are usually in the liquid state, and air is an example of the gaseous state.

The state of matter observed for a particular substance is always dependent on the temperature and pressure under which the observation is made. Because we live on a planet characterized by relatively narrow temperature extremes, we tend to fall into the error of believing that the commonly observed states of substances are the only states in which they occur. Under laboratory conditions, states other than the "natural" ones can be obtained for almost all substances. Oxygen, which is nearly always thought of as a gas, can be obtained in the liquid and solid states at very low temperatures. People seldom think of the metal iron as being a gas, its state at extremely high temperatures (above 3000°C). At intermediate temperatures (1535–3000°C) iron is a liquid (see Fig. 4.2). Water is one of the very few substances familiar to everyone in all three of its physical states: solid ice, liquid water, and gaseous steam.

Chapter 11 will consider in detail further properties of the different physical states of matter, changes from one state to another, and the question of why some substances decompose. Suffice it to say at present that physical state is one of the qualities by which matter can be classified.

4.3 PROPERTIES OF MATTER

How are the various kinds of matter differentiated from each other? The answer is simple—by their properties. **Properties** *are the distinguishing characteristics of a substance used in its identification and description*. Just as we recognize a friend by characteristics such as hair color, walk, tone of voice, or shape of nose, we recognize various chemical substances by how they look and behave. Each chemical substance has a unique set of properties that distinguishes it from all other substances. If two samples of matter have every property identical, they must necessarily be the same substance.

Knowledge of the properties of substances is useful in:

1. *Identifying an unknown substance.* Identifying a confiscated drug as cocaine involves comparing the properties of the drug to those of known cocaine samples.

FIGURE 4.1 Shape and volume characteristics of the three physical states of matter.

2. *Distinguishing between different substances.* A dentist can quickly tell the difference between a real tooth and a false tooth because of property differences.
3. *Characterizing a newly discovered substance.* Any new substance must have a unique set of properties different from those of any previously characterized substance.
4. *Predicting the usefulness of a substance for specific applications.* Water-soluble substances obviously should not be used in the manufacture of bathing suits.

There are two general categories of properties of matter: physical and chemical. **Physical properties** *are properties that can be observed without changing a substance into another substance.* Color, odor, taste, size, physical state, boiling point, melting point, and density are all examples of physical properties.

The physical appearance of a substance may change while a physical property is being determined, but the substance's identity will not. For example, the melting point of a solid cannot be measured without melting the solid, changing it to a liquid. Although the liquid's

FIGURE 4.2 During the making of steel, temperatures are high enough for the metal iron (the main ingredient in steel) to be in the liquid state. (Inland Steel Co.)

appearance is much different from that of the solid, the substance is still the same. Its chemical identity has not changed. Hence, melting point is a physical property.

F.Y.I.
When hydrogen peroxide solution is applied to a wound the frothy bubbles that form are oxygen escaping as a gas.

Chemical properties *are properties that matter exhibits as it undergoes changes in chemical composition.* Most often such composition changes result from the interaction (reaction) of the matter with other substances, or they may simply occur in the presence of heat or light (a process called decomposition). When copper objects are exposed to moist air for long periods of time, they turn green; this is a chemical property of copper. The green coating formed on the copper is a new substance; it results from the reaction of copper metal with the oxygen, carbon dioxide, and water in air. The properties of this green coating are very different from those of metallic copper. The substance hydrogen peroxide, in the presence of either heat or light, decomposes into the substances water and oxygen.

F.Y.I.
A chemical property of gold is that it does not react or corrode. When gold coins that have been buried at sea for centuries in shipwrecks are brought to the surface they are still as shiny as they were when ships went down.

The *failure* of a substance to undergo change in the presence of heat, light, or another substance is also considered a chemical property. Flammability and nonflammability are each a chemical property.

Most often, in describing chemical properties, conditions such as temperature and pressure are specified because they can and do influence interactions between two or more substances. For example, two substances may interact explosively at an elevated temperature and yet not interact at all at room temperature.

Selected physical and chemical properties of water are contrasted in Table 4.1. Note how the chemical properties of water cannot be described without reference to other substances. It does not make sense to say simply that a substance reacts. The substance that it interacts with must be specified because it might interact with many different substances.

4.4 CHANGES IN MATTER

Changes in matter are common and familiar occurrences. Many occur spontaneously, independent of human influence. Others must be forced to occur. Representative of the wide variety of known change processes are melting of snow, digestion of food, burning of wood, detonation of dynamite, rusting of iron, and sharpening of a pencil.

Changes in matter can be classified as physical or chemical. A **physical change** *is a*

TABLE 4.1 SELECTED PHYSICAL AND CHEMICAL PROPERTIES OF WATER

Physical Properties	Chemical Properties
1. Colorless	1. Reacts with bromine to form a mixture of two acids
2. Odorless	2. Can be decomposed by means of electricity to form hydrogen and oxygen
3. Boiling point = 100°C	3. Reacts vigorously with the metal sodium to produce hydrogen
4. Freezing point = 0°C	4. Does not react with gold even at high temperatures
5. Density = 1.000 g/mL at 4°C	5. Reacts with carbon monoxide at elevated temperatures to produce carbon dioxide and hydrogen

FIGURE 4.3 Chemical changes produce new substances that often have properties very different from the starting material. The steel from which this automobile fender was made has been converted by oxidation to powdered rust. (John Schultz, PAR/NYC)

process that does not alter the basic nature (chemical composition) of the substance under consideration. No new substance is ever formed as a result of a physical change. A **chemical change** *is a process that involves a change in the basic nature (chemical composition) of the substance.* Such changes always involve conversion of the material or materials under consideration into one or more new substances with distinctly different properties and composition from those of the original materials (see Fig. 4.3).

The comparison of two sources of light—a standard light bulb and a match—illustrates nicely the fundamental difference between physical and chemical changes. When an electric light bulb is turned on, light is emitted by a glowing filament inside the bulb. When the electricity is turned off, the filament ceases to glow. The filament almost instantaneously reacquires the characteristics (appearance) it had prior to its use in supplying light. This process can be repeated numerous times with the same results. The composition of the filament is unchanged—the characteristic of physical changes. Contrast this situation with that of a match. Prior to use, the match has a distinct red head that contains chemical substances that will ignite. After use, the red head is no longer present and a black residue is present. A chemical change has occurred. The match cannot be ignited again.

A change in physical state is the most common type of physical change. The melting of ice, the freezing of liquid water, the conversion of liquid water into steam (evaporation), the condensation of steam to water, the sublimation of ice in cold weather, and the formation of snow crystals in clouds in the winter (deposition) all represent changes of state. The terminology used in describing changes of state, with the exception of the terms sublimation and deposition, should be familiar to almost everyone. Although the processes of sublimation and deposition—going from a solid directly to the gaseous state or vice versa—are not common, they are encountered in everyday life. Dry ice sublimes, as do mothballs placed in a clothing storage area. As mentioned previously, ice or snow forming in clouds is an example of deposition. Figure 4.4 summarizes the terminology used in describing changes of state.

In any change of state, the composition of the substance undergoing change remains the same even though its physical state and outward appearance have changed. The melting of ice does not produce a new substance. The substance is water before and after the change. Similarly, the steam produced from boiling water is still water. Changes such as

F.Y.I.

Frost on a windowpane is an example of deposition.

F.Y.I.

When roasted coffee is ground, the rich aroma is due to the sublimation of coffee components.

F.Y.I.

A patch of ice on a concrete driveway, on a cold sunny day, will slowly disappear as it sublimates directly to vapor.

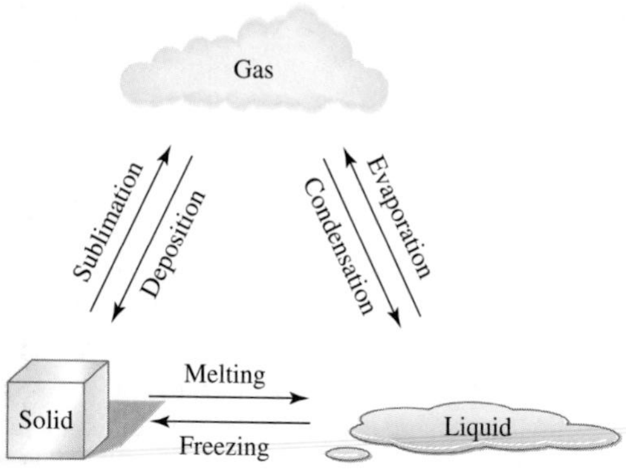

FIGURE 4.4 Terminology associated with changes of state.

these illustrate that matter can change in appearance without undergoing a change in chemical composition.

Changes in size, shape, and state of subdivision are examples of physical changes that are not changes of state. Pulverizing a lump of coal into a fine powder and tearing a piece of aluminum foil into small pieces are physical changes that involve only the solid state.

The creation of one or more new substances is always a characteristic of a chemical change. Carbon dioxide and water are two new substances produced when the chemical change associated with the burning of gasoline occurs. Ashes, carbon dioxide, and water are among the new substances produced when wood is burned. Chemical changes are often called chemical reactions. A **chemical reaction** *is a process in which at least one new substance is produced as a result of chemical change.*

Table 4.2 classifies a number of changes for matter as being either physical or chemical.

Most changes for matter can easily be classified as either physical or chemical. However, not all changes are "black" or "white." There are some "gray" areas. For example, the formation of certain solutions falls in the "gray" area. Common salt dissolves easily in water to form a solution of saltwater. The salt can easily be recovered by the physical process of evaporating the water. When gaseous hydrogen chloride is dissolved in water, again a solution results; but in this case the starting materials cannot be easily recovered by evaporation. The formation of saltwater is considered a physical change because the original components can be recovered in an unchanged form using physical methods. The second solution presents classification problems because of the possibility that a chemical reaction took place.

The changes involved in the cooking of an egg also present classification problems. The cooked egg contains the same structural units as the uncooked egg. However, some changes in structural arrangement have taken place, so, is the change physical or chemical? Despite the existence of "gray" areas, we shall continue to use the concepts of physical and chemical change because their usefulness far outweighs the problems created by a few exceptions.

The difference between the word *change* (the topic of this section) and the word *property* (Sec. 4.3) must be kept clear. A change always involves a transformation from one form to another. The new form may or may not be a new substance depending on whether the change was chemical or physical. Properties distinguish one substance from other substances.

Note that the term *physical*, when used to modify another term, as in physical property or physical change, always conveys the idea that the composition of the substances

TABLE 4.2 CLASSIFICATION OF CHANGES AS PHYSICAL OR CHEMICAL

Change	Classification
Rusting of iron	chemical
Melting of snow	physical
Sharpening a pencil	physical
Digesting food	chemical
Taking a bite of food	physical
Burning gasoline	chemical
Slicing an onion	physical
Detonation of dynamite	chemical
Souring of milk	chemical
Breaking of glass	physical

involved does not change. Similarly, the term *chemical* is always associated with the concept of change in composition.

The terms *physical* and *chemical* are commonly used to qualify the meaning of general scientific terms. For example, techniques used to accomplish physical change are called physical methods or physical means. Chemical methods and chemical means are used to bring about chemical change. A physical separation would be a separation process where none of the components experienced composition changes. Composition changes would be part of a chemical separation process. The message of the "modifiers" physical and chemical is constant: *Physical* denotes no change in composition, and *chemical* denotes change in composition.

4.5 MIXTURES AND PURE SUBSTANCES

All samples of matter can be divided into two categories: mixtures and pure substances. Most natural samples of matter are mixtures. Quite often, recognition of a substance as a mixture is easy; the distinct properties of various components are clearly visible. Other times, as we will see shortly (Sec. 4.6), the components of a mixture cannot be visually distinguished. Natural pure substances are not common; they are seldom encountered. Gold and colorless diamonds are two of the few naturally occurring pure substances. Despite the lack of their natural occurrence, many, many pure substances are known. They are obtained from natural mixtures using numerous types of separation techniques.

Formally defined, a **mixture** *is a physical combination of two or more pure substances in which the pure substances retain their identity*. The chemical identity of individual components is retained in a mixture because the components are combined physically rather than chemically. Consider, for example, a mixture of salt and pepper (see Fig. 4.5). Close examination of such a mixture will show distinct particles of salt and pepper with no obvious interaction between them. The salt particles in the mixture are identical in properties and composition to the salt particles in the salt container, and the pepper particles in the mixture are no different from those in the pepper container.

F.Y.I.

The semiconductors in computers are made from a very pure form of the substance silicon, but only work when very small amounts of impurities are added. To turn silicon into a transistor, small amounts of the substances boron and phosphorous are carefully mixed into the silicon.

F.Y.I.

A cup of coffee mixture can be made from dishwater weak to pitch black.

Once a particular mixture is made up, its composition is constant. However, mixtures of the same components with different compositions can also be made up; thus, mixtures are considered to have variable compositions. Consider the large number of salt and pepper mixtures that could be produced by varying the amounts of the two substances present.

An additional characteristic of any mixture is that its components can often be retrieved intact from the mixture by physical means, that is, without a chemical change. In many cases, the differences in properties of the various components make the separation relatively easy. For example, in our salt–pepper mixture, if the pepper grains were large enough, the two components could be separated manually by picking out all the pepper grains. Alternatively, the separation could be carried out by dissolving the salt particles in water, removing the insoluble pepper particles, and then evaporating the water to recover the salt.

Most mixture separations are not as easy as that for a salt–pepper mixture; expensive instrumentation and numerous separation steps are required. Image the logistics involved in the separation of the components of blood, a water-based mixture of varying amounts of proteins, sugar (glucose), salt (sodium chloride), oxygen, carbon dioxide, and other components.

F.Y.I.

Commercial table salt is not just the pure substance sodium chloride. It is a mixture containing small amounts of potassium iodide as a nutritional supplement, dextrose as a stabilizer, and calcium silicate as an anticaking agent (to keep it pouring when it rains) in addition to the sodium chloride.

A pure substance is exactly what its name implies—a single uncontaminated type of matter. All samples of a pure substance contain only that pure substance and nothing else. Pure water is water and nothing else. Pure table salt (sodium chloride) contains only that substance and nothing else.

A pure substance is defined in terms of its properties. A **pure substance** *is a form of matter that always has a definite and constant composition*. This constancy of composition is reflected in the properties of the pure substance. They never vary, always being the same under a given set of conditions. All samples of a pure substance, no matter what their source, must have the same properties under the same conditions. Collectively, these definite and constant physical and chemical properties of a pure substance form a set not dupli-

FIGURE 4.5 The individual particles of salt and pepper are easily recognizable in a mixture of the two substances because a mixture is a physical rather than chemical combination of substances. (Ken Lax/PH Archives)

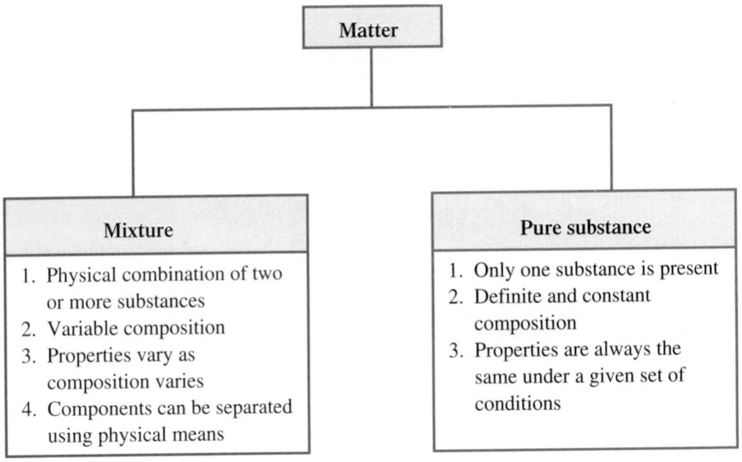

FIGURE 4.6 A comparison of the characteristics of mixtures and pure substances.

cated by any other pure substance. This unique set of properties provides the identification for the pure substance.

It is important to note that there is a significant difference between the terms *substance* and *pure substance*. Substance is a general term used to denote any variety of matter. Pure substance is a specific term that applies only to matter with those characteristics we have just noted.

Figure 4.6 summarizes the differences between mixtures and pure substances. Further considerations about mixtures are found in Section 4.6, and Section 4.7 contains further details about pure substances.

4.6 TYPES OF MIXTURES: HETEROGENEOUS AND HOMOGENEOUS

Visual identification of the components of a mixture is not always possible. Sometimes the separate ingredients of a mixture are not visible, even with the aid of a microscope. A mixture prepared by dissolving sugar in water falls in this category. Outwardly, this mixture has the same appearance as pure water.

Mixtures can be classified as being either heterogeneous or homogeneous on the basis of the visual recognizability of the components present. A **heterogeneous mixture** *contains visibly different parts, or phases, each of which has different properties*. The phases present in a heterogeneous mixture have distinct boundaries and are usually easily observed. A pepperoni and cheese pizza is obviously a heterogeneous mixture that has numerous identifiable components. Even a piece of pepperoni contains a number of phases. Common materials such as rocks and wood are also heterogeneous mixtures; different parts of these materials clearly have different properties, such as hardness and color (see Fig. 4.7).

The phases in a heterogeneous mixture may or may not be in the same physical state. Set concrete contains a number of phases, all of which are in the solid state. A mixture of sand and water contains two phases, and each is in a different state (solid and liquid). It is possible to have heterogeneous mixtures in which all components are liquids. In order for these mixtures to occur, the mixed liquids must have limited solubility in each other. When this is the case, the mixed liquids form separate layers with the least dense liquid on top. An oil-and-vinegar salad dressing is an example of such a liquid–liquid mixture (see Fig. 4.8a).

F.Y.I.

Blood appears homogeneous to the naked eye. But when looked at under a microscope, it can be seen to be a heterogeneous mixture of red and white blood cells and liquid called plasma.

F.Y.I.

Americans are accustomed to buying products like chocolate bars and peanut butter that look homogeneous. Manufacturers add substances called emulsifying agents to keep these products homogeneous. Without emulsifiers, the ingredients would slowly separate into phases and look unpalatable.

FIGURE 4.7 Common materials such as rocks and wood are heterogeneous mixtures. (a-John Schultz, PAR/NYC; b-Grant Heilman Photography)

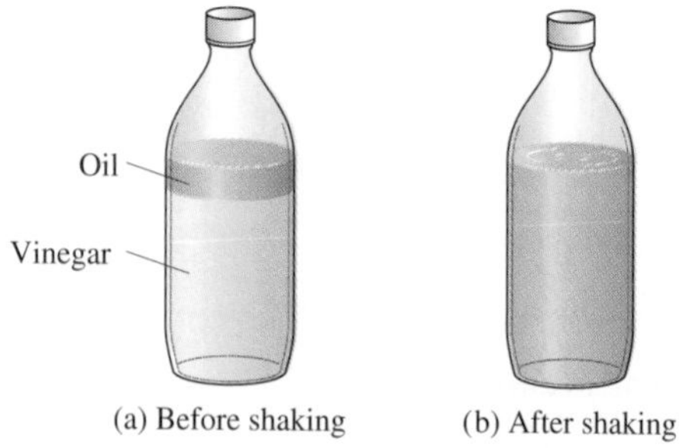

FIGURE 4.8 Oil-and-vinegar salad dressing is a two-phase heterogeneous mixture. Shaking the mixture does not change the number of phases present.

Oil-and-vinegar dressing consists of two phases (oil and vinegar) regardless of whether the mixture consists of two separate layers or of oil droplets dispersed throughout the vinegar, a condition caused by shaking the mixture (see Fig. 4.8b). All of the oil droplets together are considered to be a single phase.

A **homogeneous mixture** *contains only one visibly distinct phase, which has uniform properties throughout.* This type of mixture can have only one set of properties, which are associated with the single phase present. A spoonful of sugar water taken from the surface of a homogeneous sugar–water mixture is just as sweet to the taste as one taken from the bottom of the container. If this were not the case the mixture would not be truly homogeneous.

F.Y.I.

Modern American dimes and quarters are heterogeneous mixtures of metals. A copper phase is sandwiched between a copper-nickel phase. Older silver dimes are homogeneous.

Homogeneous mixtures are possible only when all components present are in the same physical state. Homogeneous mixtures for all three physical states are common. Air is a homogeneous mixture of gases; motor oil and gasoline are each multicomponent homogeneous mixtures of liquids; and metal alloys such as 14-karat gold (a mixture of copper and gold) are examples of solid homogeneous mixtures.

A thorough intermingling of the components in a homogeneous mixture is required in order for a single phase to exist. Sometimes this occurs almost instantaneously during the preparation of the mixture, as in the addition of alcohol to water. At other times, an extended period of mixing or stirring is required. For example, when a hard sugar cube is added to a container of water, it does not instantaneously dissolve to give a homogeneous solution. Only after much stirring does the sugar completely dissolve. Prior to that point, the mixture is heterogeneous because a solid phase (the undissolved sugar cube) is present.

4.7 TYPES OF PURE SUBSTANCES: ELEMENTS AND COMPOUNDS

There are two kinds of pure substances: elements and compounds. An **element** *is a pure substance that cannot be broken down into simpler substances by ordinary chemical means (a reaction, an electric current, a beam of light, etc.).* Elements resist all attempts to fragment them into simpler pure substances. The metals gold, silver, and copper are all elements. A **compound** *is a pure substance that can be broken down into two or more simpler substances by chemical means.* Water is a compound. By means of electric current water can be broken down into the gases hydrogen and oxygen, both of which are simpler pure substances. The properties of the simpler substances obtained from compound breakdown are always distinctly different from those of the parent compound.

Ultimately the products from the breakdown of any compound are elements. In practice the breakdown often occurs in steps, with simpler compounds resulting from the intermediate steps, as illustrated in Figure 4.9.

Presently 111 elements have been identified. These elements, which are the simplest known substances, are considered the building blocks of all other types of matter. Every object, regardless of its complexity, is a collection of substances that are made up of the 111 elements.

Compared to the total number of compounds characterized by chemists, the number of known elements is extremely small. Over 6 million different compounds are known. Each is a definite chemical combination of two or more of the known elements. Figure 4.10 summarizes the characteristics of elements and compounds.

Before a substance can be classified as an element, all possible attempts must be made chemically to subdivide it into simpler substances. If a sample of pure substance, S, is subjected to a decomposition process and two new substances, X and Y, are produced, S would be classified as a compound. If, on the other hand, a number of attempts made chemically to subdivide S proved unsuccessful, we might correctly call it an element, but until all possible reactions have proved unsuccessful, such a classification could be in error.

A heterogeneous sample of a pure substance is possible. For this situation to occur the pure substance must be present in two or more states. An ice cube floating in a container of water is such a system. The solid phase (ice) obviously has some physical properties different from those of the liquid phase (liquid water). An ice–water system, although heterogeneous, is not a mixture. Only one substance is present—water. Mixtures require the presence of two substances.

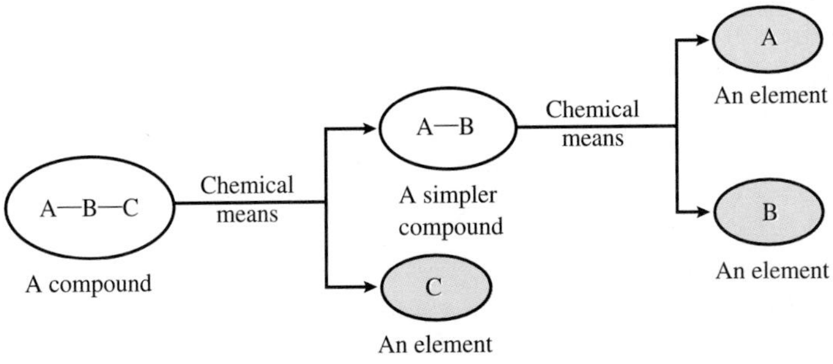

FIGURE 4.9 Stepwise breakdown of a compound containing three elements (A, B, and C) to yield the constituent elements.

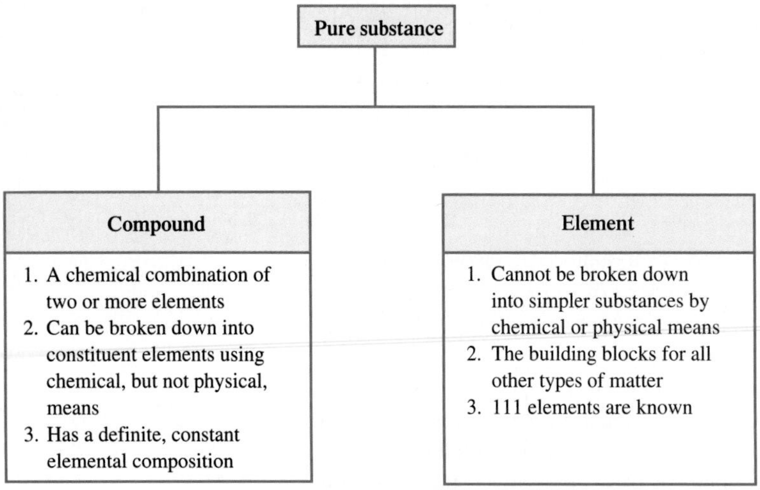

FIGURE 4.10 A comparison of the characteristics of elements and compounds.

Figure 4.11 summarizes the overall classification scheme for matter developed in Sections 4.5 through 4.7.

Students frequently have trouble with the concept that compounds are not mixtures even though two or more simple substances can be obtained from compound decomposition. It is very important that the distinction between a mixture and a compound be understood. There are three distinct areas of difference between these two classifications of matter.

1. A compound always has properties distinctly different from those of the substances (elements or compounds) used to produce the compound. This is because the combining substances are *bound* together into discrete units; the substances are *chemically* combined. The substances in a mixture retain their individual identities because they are *not bound* together; hence, their individual properties are still manifest; these substances are *physically* combined.
2. Compounds have a definite composition, a property of all pure substances. Mixtures have a variable composition.

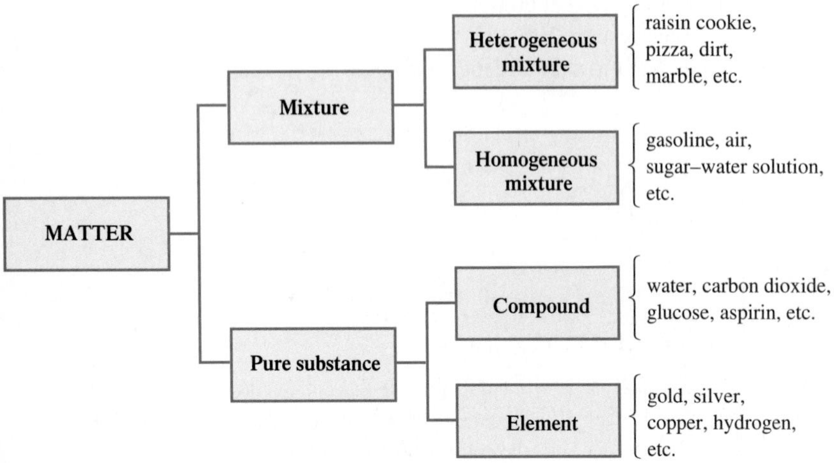

FIGURE 4.11 Categories of classification for matter.

3. The individual components of a mixture can be physically separated from each other. The elements in a compound cannot be separated from each other using physical means since they are bound together. They can only be separated by chemical means through which the compound is destroyed.

Example 4.1, which involves two comparisons involving ballpoint pens and their caps, illustrates the difference between compounds and mixtures.

Example 4.1

Consider two boxes with the following contents: the first contains 25 ballpoint pens each with its cap on; the second contains 25 ballpoint pens without caps and 25 ballpoint pen caps. Which box has contents that would be an analogy for a mixture and which box has contents that would be an analogy for a compound?

Solution

The box containing the ballpoint pens with their caps on represents the compound. Two samples withdrawn from this box will always be the same; each will be a ballpoint pen with its cap on. Each item in the box has the same "composition."

The box containing separated ballpoint pens and caps represents the mixture. Two samples withdrawn from this box need not be the same; results could be two ballpoint pens, two caps, or a cap and a ballpoint pen. All items in the box do not have the same "composition."

Practice Exercise 4.1

Consider two boxes with the following contents: the first contains 30 bolts and 30 nuts which fit the bolts; the second contains the same number of bolts and nuts with the difference that each bolt has a nut screwed on it. Which box has contents that would be an analogy for a mixture and which box has contents that would be an analogy for a compound?

Ans. first box, mixture; second box, compound

Figure 4.12 is a summary of the matter classifications presented in Sections 4.5 through 4.7. It is a visual model for the thought processes that a chemist goes through in classifying a sample of matter into one of the categories heterogeneous mixture, homogeneous mixture, element, or compound.

4.8 DISCOVERY AND ABUNDANCE OF THE ELEMENTS

The discovery and isolation of the 111 elements have taken place over several centuries. Discovery, for the most part, has occurred since 1700, and in particular during the 1800s. Table 4.3 shows how the number of known elements has increased dramatically since 1750.

Eighty-eight of the 111 elements occur naturally, and 23 are synthetic, having been produced in the laboratory from naturally occurring elements. It is generally accepted by sci-

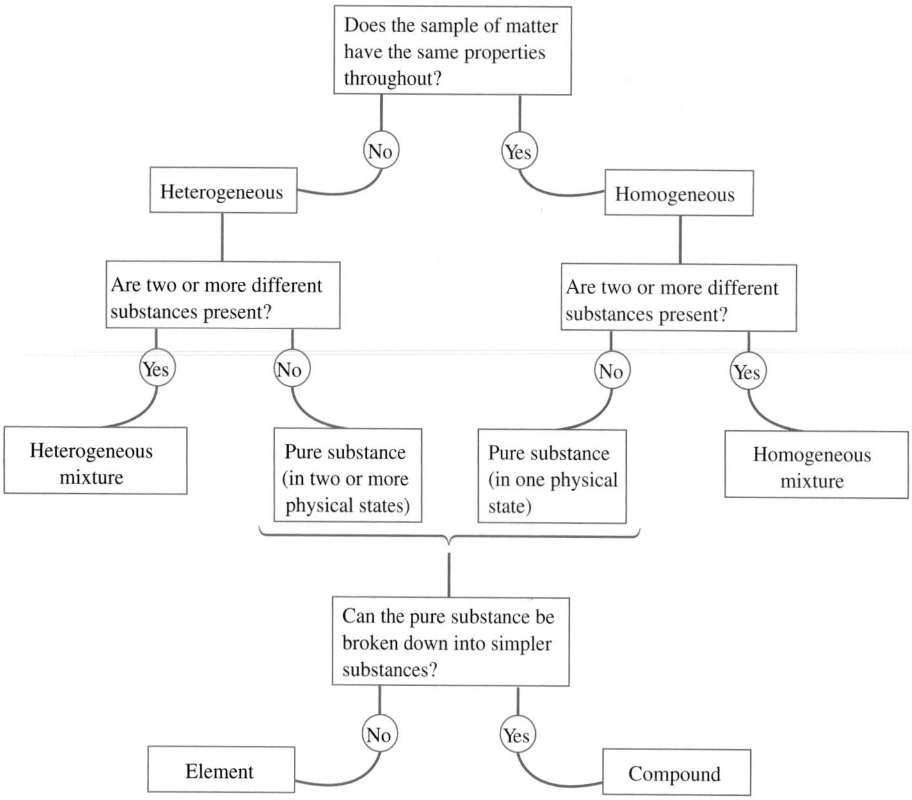

FIGURE 4.12 Classification of matter by chemical composition into various categories.

TABLE 4.3 NUMBER OF ELEMENTS DISCOVERED DURING VARIOUS TIME PERIODS

Time Period	Number of Elements Discovered During Time Period	Total Number of Elements Known at End of Time Period
Ancient–1700	13	13
1701–1750	3	16
1751–1800	18	34
1801–1850	25	59
1851–1900	23	82
1901–1950	16	98
1951–present	13	111

entists that no more naturally occurring elements will be found. The last of the naturally occurring elements was discovered in 1925. It is possible however, that additional synthetic elements will be prepared. The synthetic elements are unstable (radioactive) and most of them change rapidly into naturally occurring elements as the result of radioactive decay.

The naturally occurring elements are not evenly distributed in our world and universe. What is startling is the degree of inequality in the distribution. A very few elements com-

TABLE 4.4 ABUNDANCE OF THE ELEMENTS IN VARIOUS REALMS OF MATTER

Element	Abundance (atom %)	Element	Abundance (atom %)
Universe		*Atmosphere*	
Hydrogen	91	Nitrogen	78.3
Helium	9	Oxygen	21.0
Earth (including core)		*Hydrosphere*	
Oxygen	49.3	Hydrogen	66.4
Iron	16.5	Oxygen	33
Silicon	14.5		
Magnesium	14.2	*Human body*	
		Hydrogen	63
Earth's crust		Oxygen	25.5
Oxygen	60.1	Carbon	9.5
Silicon	20.1	Nitrogen	1.4
Aluminum	6.1		
Hydrogen	2.9	*Vegetation*	
Calcium	2.6	Hydrogen	49.8
Magnesium	2.4	Oxygen	24.9
Iron	2.2	Carbon	24.9
Sodium	2.1		

pletely dominate. In answering the question of which elements are most common, one must define the area to be considered. The abundances of the elements in the Earth's crust are considerably different from those for the Earth as a whole; they are even more different from abundances for the universe. When living organisms such as vegetation or the human body are considered, an altogether different perspective emerges. Table 4.4 gives information on element abundances. The abundance of each element is the percentage of the total elemental units (atoms; see Sec. 5.1) present that are units of that element. Only element abundances greater than 1% are listed in any given analysis.

Note from Table 4.4 that the most abundant elements in the universe are not the most abundant ones on Earth. The cosmic figures reflect the composition of stars, which are almost entirely hydrogen and helium. Significant differences also exist between the composition of the Earth as a whole and that of its crust. The figures for the Earth's crust—taken to mean the Earth's waters, atmosphere, and outer covering to a depth of 10 miles—do not take into account the composition of the Earth's core, which is mostly iron. Note also that only two elements occur in large amounts (greater than 1%) in the atmosphere and the hydrosphere and that carbon is more abundant and hydrogen less abundant in vegetation than in the human body.

4.9 NAMES AND SYMBOLS OF THE ELEMENTS

Each element has a unique name, which in most cases was selected by its discoverer. A wide variety of rationales for choosing a name is found when name origins are studied. Some ele-

TABLE 4.5 THE CHEMICAL SYMBOLS FOR THE ELEMENTS[a]

Ac	actinium	Ha	hahnium	Pr	praseodymium
Ag	**silver**[b]	**He**	**helium**	**Pt**	**platinum**
Al	**aluminum**	Hf	hafnium	Pu	plutonium
Am	americium	**Hg**	**mercury**[b]	Ra	radium
Ar	**argon**	Ho	holmium	Rb	rubidium
As	arsenic	Hs	hassium	Re	rhenium
At	astatine	**I**	**iodine**	Rf	rutherfordium
Au	**gold**[b]	In	indium	Rh	rhodium
B	**boron**	Ir	iridium	Rn	radon
Ba	**barium**	**K**	**potassium**[b]	Ru	ruthenium
Be	**beryllium**	Kr	krypton	**S**	**sulfur**
Bi	bismuth	La	lanthanum	Sb	antimony[b]
Bk	berkelium	**Li**	**lithium**	Sc	scandium
Br	**bromine**	Lu	lutetium	Se	selenium
C	**carbon**	Lr	lawrencium	Sg	seaborgium
Ca	**calcium**	Md	mendelevium	**Si**	**silicon**
Cd	cadmium	**Mg**	**magnesium**	Sm	samarium
Ce	cerium	Mn	manganese	**Sn**	**tin**[b]
Cf	californium	Mo	molybdenum	Sr	strontium
Cl	**chlorine**	Mt	meitnerium	Ta	tantalum
Cm	curium	**N**	**nitrogen**	Tb	terbium
Co	**cobalt**	**Na**	**sodium**[b]	Tc	technetium
Cr	**chromium**	Nb	niobium	Te	tellurium
Cs	cesium	Nd	neodymium	Th	thorium
Cu	**copper**[b]	**Ne**	**neon**	Ti	titanium
Dy	dysprosium	**Ni**	**nickel**	Tl	thallium
Er	erbium	No	nobelium	Tm	thulium
Es	einsteinium	Np	neptunium	**U**	**uranium**
Eu	europium	Ns	nielsbohrium	V	vanadium
F	**fluorine**	**O**	**oxygen**	W	tungsten[b]
Fe	**iron**[b]	Os	osmium	Xe	xenon
Fm	fermium	**P**	**phosphorus**	Y	yttrium
Fr	francium	Pa	protactinium	Yb	ytterbium
Ga	gallium	**Pb**	**lead**[b]	**Zn**	**zinc**
Gd	gadolinium	Pd	palladium	Zr	zirconium
Ge	germanium	Pm	promethium		
H	**hydrogen**	Po	polonium		

[a] Only 109 elements are listed in this table. Elements 110 and 111, discovered (synthesized), respectively, in November and December 1994 are yet to be named.
[b] These elements have symbols that were derived from non-English sources.

ments bear geographical names. Germanium was named after the native country of its German discoverer. The elements francium and polonium acquired names in a similar manner. The elements mercury, uranium, neptunium, and plutonium are all named for planets. Helium gets its name from the Greek word *helious* for sun, since it was first observed spec-

TABLE 4.6 ELEMENTS WHOSE SYMBOLS ARE DERIVED FROM A NON-ENGLISH NAME OF THE ELEMENT

English Name of Element	Non-English Name of Element	Symbol
	Symbols from Latin	
Antimony	stibium	Sb
Copper	cuprum	Cu
Gold	aurum	Au
Iron	ferrum	Fe
Lead	plumbum	Pb
Mercury	hydrargyrum	Hg
Potassium	kalium	K
Silver	argentum	Ag
Sodium	natrium	Na
Tin	stannum	Sn
	Symbol from German	
Tungsten	wolfram	W

troscopically in the sun's corona during an eclipse. Some elements carry names that relate to specific properties of the element or compounds containing it. Chlorine's name is derived from the Greek *chloros* denoting greenish yellow, the color of chlorine gas. Iridium gets its name from the Greek *iris* meaning rainbow because of the various colors of the compounds from which it was isolated.

In the early 1800s chemists adopted the practice of assigning chemical symbols to the elements. **Chemical symbols** *are abbreviations for the names of the elements.* These chemical symbols are used more frequently than names in referring to the elements in written communications. The system of chemical symbols now in use was first proposed in 1814 by the Swedish chemist Jöns Jakob Berzelius (1779–1848).

A list of elements and their symbols is given in Table 4.5. The symbols and names of the more frequently encountered elements are shown in color in this table. You would do well to learn the symbols of these more common elements. Learning them is a key to having a successful experience in studying chemistry.

Fourteen elements have one-letter symbols and the rest have two-letter symbols. If a symbol consists of a single letter, it is capitalized. In all two-letter symbols the first letter is always capitalized and the second letter is always lowercase. Double-letter symbols usually, but not always, start with the first letter of the element's English name. The second letter of the symbol is frequently, but not always, the second letter of the name. Consider the elements terbium, technetium, and tellurium, whose symbols are, respectively, Tb, Tc, and Te. Obviously, a variety of choices of second letters is necessary because the first two letters are the same in all three elements' names.

Eleven elements have symbols that bear no relationship to the element's English language name. In ten of these cases, the symbol is derived from the Latin name of the element; in the case of tungsten a German name is the symbol source. Most of these elements have been known for hundreds of years and date back to the time when Latin was the language of scientists. Table 4.6 shows the relationship between the symbol and the non-English name of these eleven elements.

F.Y.I.

Gold gets its symbol from the Latin "aurum," which means shiny. The liquid metal mercury gets its symbol from the Latin "hydrogyrum," which means "runs like water." Lead gets its symbol from the Latin "plumbum," from which we get the name "plumber," because pipes used to be made out of lead.

The symbols of the elements are also found on the inside front and back covers of this book. The chart of elements on the inside front cover is called a *periodic table*. More will be said about it in later chapters. Both cover listings also give other information about the elements. This additional information will be discussed in Chapter 5.

KEY TERMS

The new terms or concepts defined in this chapter are

chemical change (Sec. 4.4) A process that involves a change in the basic nature (chemical composition of a substance.

chemical properties (Sec. 4.3) Properties that matter exhibits as it undergoes changes in chemical composition.

chemical reaction (Sec. 4.4) A process in which at least one new substance is produced as a result of chemical change.

chemical symbol (Sec. 4.9) An abbreviation for the name of an element.

chemistry (Sec. 4.1) The branch of science that is concerned with matter and its properties.

compound (Sec. 4.7) A pure substance that can be broken down into two or more simpler substances by chemical means.

element (Sec. 4.7) A pure substance that cannot be broken down into simpler substances by ordinary chemical means.

gaseous state (Sec. 4.2) A state characterized by both an indefinite shape and an indefinite volume.

heterogeneous mixture (Sec. 4.6) A mixture that contains visibly different parts, or phases, each with different properties.

homogeneous mixture (Sec. 4.6) A mixture that contains only one visibly distinct phase with uniform properties throughout.

liquid state (Sec. 4.2) A state characterized by its properties of an indefinite shape and a definite volume.

matter (Sec. 4.1) Anything that has mass and occupies space.

mixture (Sec. 4.5) A physical combination of two or more pure substances in which the pure substances retain their identity.

physical change (Sec. 4.4) A process that does not alter the basic nature (chemical composition) of a substance.

physical properties (Sec. 4.3) Properties that can be observed without changing a substance into another substance.

properties (Sec. 4.3) Distinguishing characteristics of a substance used in its identification and description.

pure substance (Sec. 4.5) A form of matter that always has a definite and constant composition.

solid state (Sec. 4.2) A state characterized by its properties of a definite shape and a definite volume.

PRACTICE PROBLEMS

Physical States of Matter (Sec. 4.2)

4.1 List a characteristic that distinguishes
(a) liquids from solids **(b)** gases from liquids

4.2 List a characteristic that is the same for
(a) liquids and solids **(b)** gases and liquids

4.3 Which of the following would take the shape of its container and also have a definite volume?
(a) copper wire **(b)** granulated sugar
(c) hard sugar cube **(d)** liquid water

4.4 Which of the following would take the shape of its container and also have an indefinite volume?
(a) aluminum powder **(b)** clean air
(b) carbon dioxide gas **(d)** gasoline

4.5 Gold has a melting point of 1063°C and a boiling point of 2966°C. Specify the physical state of gold at each of the following temperatures.
(a) 500°C **(b)** 1000°C

(c) 2000°C (d) 3000°C

4.6 Oxygen has a melting point of −218°C and a boiling point of −183°C. Specify the physical state of oxygen at each of the following temperatures.

(a) −250°C (b) −200°C

(c) −100°C (d) 100°C

Properties of Matter (Sec. 4.3)

4.7 The following are properties of the metal beryllium. Classify them as physical or chemical.

(a) In powdered form, it burns brilliantly on ignition.
(b) Bulk metal does not react with steam even at red heat.
(c) It has a density of 1.85 g/cm^3 at 20°C.
(d) It is a relatively soft silvery-white metal.

4.8 The following are properties of the metal aluminum. Classify them as physical or chemical.

(a) It generates a colorless, odorless gas when added to sulfuric acid.
(b) It can easily be formed into thin foils.
(c) It is a solid at room temperature.
(d) It is a good conductor of heat.

4.9 Indicate whether each of the following statements describes a physical or chemical property.

(a) Silver compounds discolor the skin by reacting with skin protein.
(b) Hemoglobin gives blood its red color.
(c) Lithium metal is light enough to float on water.
(d) Mercury is a liquid at room temperature.

4.10 Indicate whether each of the following statements describes a physical or chemical property.

(a) Charcoal lighter fluid can be ignited with a match.
(b) Magnesium metal does not react with cold water.
(c) Carbon monoxide is a colorless gas.
(d) Sodium metal is so soft that it can be cut with a sharp knife.

Changes in Matter (Sec. 4.4)

4.11 Classify each of the following changes as physical or chemical.

(a) Grinding sugar crystals into powder.
(b) Cutting grass in your front yard.
(c) Exploding of a firecracker.
(d) Burning a log in a fireplace.

4.12 Classify each of the following changes as physical or chemical.

(a) Stretching a rubber band.
(b) Fashioning a table leg from a piece of wood.
(c) Ski goggles become fogged.
(d) Leaves turn red in autumn.

4.13 Indicate whether each of the following methods for obtaining various substances involves physical or chemical change.

(a) Sodium chloride (salt) is obtained from salt water by evaporation of the water.
(b) Nitrogen gas is obtained from air by letting the nitrogen boil off from liquid air.
(c) Oxygen gas is obtained by decomposition of the oxygen-containing compound potassium chlorate.
(d) Water is obtained by the high-temperature reaction of gaseous hydrogen with gaseous oxygen.

4.14 Indicate whether each of the following methods for obtaining various substances involves physical or chemical change.

(a) Mercury is obtained by decomposing a mercury–oxygen compound, liberating the oxygen and leaving the mercury behind.
(b) Sand is obtained from a sand–sugar mixture by adding water to the mixture and pouring off the resulting sugar–water solution.
(c) Ammonia is obtained by the high-temperature high-pressure reaction between hydrogen and nitrogen.
(d) Water is obtained from a sugar–water solution by evaporating off and then collecting the water.

4.15 Give the name of the change of state associated with each of the following processes.

(a) Water is made into ice cubes.
(b) The inside of your car window fogs up.
(c) Mothballs in the clothes closet disappear with time.
(d) Perspiration dries.

4.16 Give the name of the change of state associated with each of the following processes.

(a) Dry ice disappears without melting.
(b) Snowflakes form.
(c) Dew on the lawn disappears when the sun comes out.
(d) Ice cubes in a soft drink disappear with time.

Pure Substances and Mixtures (Secs. 4.5 and 4.6)

4.17 Classify each of the following as a heterogeneous mixture, a homogeneous mixture, or a pure substance.

(a) scrambled eggs **(b)** an apple
(c) multivitamin tablet **(d)** distilled water

4.18 Classify each of the following as a heterogeneous mixture, a homogeneous mixture, or a pure substance.

(a) copper ore **(b)** copper wire
(c) wet sand **(d)** carbonated water

4.19 Classify each of the following statements as true or false.

(a) Heterogeneous mixtures must contain three or more substances.

(b) Pure substances cannot have a variable composition.

(c) Substances maintain many of their properties in a heterogeneous mixture but not in a homogeneous mixture.

(d) Pure substances are seldom encountered in the "everyday" world.

4.20 Classify each of the following statements as true or false.

(a) Homogeneous mixtures must contain at least two substances.

(b) Heterogeneous mixtures but not homogeneous mixtures can have a variable composition.

(c) Pure substances cannot be separated into other kinds of matter using physical means.

(d) The number of known pure substances is less than one hundred thousand.

4.21 Assign each of the following descriptions of matter to one of the following categories: heterogeneous mixture, homogeneous mixture, pure substance.

(a) two substances present, two phases present

(b) two substances present, one phase present

(c) three substances present, one phase present

(d) three substances present, three phases present

4.22 Assign each of the following descriptions of matter to one of the following categories: heterogeneous mixture, homogeneous mixture, pure substance.

(a) one substance present, one phase present

(b) one substance present, two phases present

(c) one substance present, three phases present

(d) three substances present, two phases present

4.23 Classify each of the following as a heterogeneous mixture, a homogeneous mixture, or a pure substance. Also indicate how many phases are present. (In each case the substances are present in the same container.)

(a) water and dissolved salt

(b) water and dissolved sugar

(c) water and sand

(d) water and oil

4.24 Classify each of the following as a heterogeneous mixture, a homogeneous mixture, or a pure substance. Also indicate how many phases are present. (In each case the substances are present in the same container.)

(a) liquid water and ice

(b) liquid water, oil, and ice

(c) carbonated water (soda water) and ice

(d) oil, ice, saltwater solution, sugar–water solution, and pieces of copper metal

---

Elements and Compounds (Sec. 4.7)

4.25 Based on the information given, classify each of the pure substances A through D as elements or compounds, or indicate that no such classification is possible because of insufficient information.

(a) Analysis with an elaborate instrument indicates that substance A contains two elements.

(b) Substance B decomposes upon heating.

(c) Heating substance C to 1000°C causes no change in it.

(d) Heating substance D to 500°C causes it to change from a solid to a liquid.

4.26 Based on the information given, classify each of the pure substances A through D as elements or compounds, or indicate that no such classification is possible because of insufficient information.

(a) Substance A cannot be broken down into simpler substances by chemical means.

(b) Substance B cannot be broken down into simpler substances by physical means.

(c) Substance C readily dissolves in water.

(d) Substance D readily reacts with the element chlorine.

4.27 Indicate whether each of the following statements is true or false.

(a) Both elements and compounds are pure substances.

(b) A compound results from the physical combination of two or more elements.

(c) In order for matter to be heterogeneous, at least two compounds must be present.

(d) Compounds, but not elements, can have a variable composition.

4.28 Indicate whether each of the following statements is true or false.

(a) Compounds can be separated into their constituent elements using chemical means.

(b) Elements can be separated into their constituent compounds using physical means.

(c) A compound must contain at least two elements.

(d) A compound is a physical mixture of different elements.

4.29 Based on the information given in the following equations, classify each of the pure substances A through G as elements or compounds, or indicate that no such classification is possible because of insufficient information.

(a) $A + B \rightarrow C$ **(b)** $D \rightarrow E + F + G$

4.30 Based on the information given in the following equations, classify each of the pure substances A through G as elements or compounds, or indicate that no such classification is possible because of insufficient information.

(a) $A \rightarrow B + C$ **(b)** $D + E \rightarrow F + G$

4.31 Consider two boxes with the following contents: the first contains 50 individual paper clips and 50 individual rubber bands; the second contains the same number of paper clips and rubber bands with the difference that each paper clip is interlocked with a rubber band. Which box has contents that would be an analogy for a mixture and which has contents that would be an analogy for a compound?

4.32 Consider the characteristics of the two breakfast cereals "Crispy Wheat 'N Raisins" and "Crispix." The first cereal contains wheat flakes and raisins. The second cereal contains a fused two-layered flake, one side of which is rice and the other side corn. Characterize the properties of these two cereals that make one an analogy for a mixture and the other an analogy for a compound.

4.33 Based on the notation used in Figure 4.9, classify each of the following pairs of substances as (1) two elements, (2) two compounds, (3) an element and a compound, or (4) a single pure substance.

(a) (Q—X) and (Q) **(b)** (Q—X) and (X)

(c) (Q) and (X) **(d)** (Q—X) and (Q—X)

4.34 Based on the notation used in Figure 4.9, classify each of the following pairs of substances as (1) two or more elements, (2) two compounds, (3) an element and a compound, or (4) a single pure substance.

(a) (Q—X) and (C) **(b)** (Q—X) and (Q—C)

(c) (Q) and (X) and (C) **(d)** (Q—X—C) and (Q)

Discovery and Abundance of the Elements (Sec. 4.8)

4.35 What would a person have been taught about the number of known elements from our current list if he or she had taken a chemistry class in each of the following years?

(a) 1700 **(b)** 1800 **(c)** 1900

4.36 What would a person have been taught about the number of known elements from our current list if he or she had taken a chemistry class in each of the following years?

(a) 1750 **(b)** 1850 **(c)** 1950

4.37 Indicate whether each of the following statements about elemental abundances is true or false.

(a) Hydrogen is the most abundant element in the universe as a whole.

(b) Carbon is the most abundant element in vegetation but not in the human body.

(c) Oxygen and silicon are the two most abundant elements in Earth's crust.

(d) Hydrogen is the most abundant element in both Earth's atmosphere and hydrosphere.

4.38 Indicate whether each of the following statements about elemental abundances is true or false.

(a) Oxygen is the second most abundant element in the universe as a whole.

(b) Carbon is the most abundant element in the human body.

(c) Nitrogen is the most abundant element in both Earth's atmosphere and hydrosphere.

(d) Oxygen is the most abundant element in both the Earth as a whole and in Earth's crust.

Names and Symbols of the Elements (Sec. 4.9)

4.39 Name the element each of the following chemical symbols represents.

(a) Ar **(b)** Ba **(c)** Ca **(d)** F

(e) He **(f)** K **(g)** C **(h)** Cu

4.40 Name the element each of the following chemical symbols represents.

(a) N **(b)** Ni **(c)** Pb **(d)** Sn

(e) Al **(f)** B **(g)** Mg **(h)** Ne

4.41 What are the chemical symbols of the following elements?

(a) bromine **(b)** chlorine

(c) iron **(d)** mercury

(e) lithium **(f)** sodium

(g) hydrogen **(h)** iodine

4.42 What are the chemical symbols of the following elements?

(a) oxygen **(b)** sulfur

(c) zinc **(d)** silver

(e) gold **(f)** beryllium

(g) phosphorus **(h)** silicon

4.43 Each of the following names of elements is spelled incorrectly. Correct the misspelling.

(a) flourine **(b)** zink

(c) potasium **(d)** sulfer

4.44 Each of the following names of elements is spelled incorrectly. Correct the misspellings.

(a) phosphorous **(b)** murcury

(c) clorine **(d)** argone

4.45 Give the English name and symbol for each of the following elements, whose Latin name is

(a) ferrum **(b)** stannum

(c) natrium **(d)** aurum

4.46 Give the English name and symbol for each of the following elements, whose Latin name is

(a) kalium **(b)** argentum

(c) plumbum **(d)** stibium

4.47 Certain words can be viewed as sequential combinations of symbols of the elements. For example, the given name Stephen is made up of the following chemical symbol sequence: S-Te-P-He-N. Analyze each of the following given names in a similar manner

(a) Rebecca (b) Raymond
(c) Nancy (d) Bruce
(e) Sharon (f) Alice

4.48 Certain words can be viewed as sequential combinations of symbols of the elements. For example, the given name Stephen is made up of the following chemical symbol sequence: S-Te-P-He-N. Analyze each of the following given names in a similar manner

(a) Barbara (b) Eugene
(c) Heather (d) Monica
(e) Allan (f) Bryce

ADDITIONAL PROBLEMS

4.49 Carbon monoxide is a colorless, odorless gas that is toxic to humans. It combines with the metal nickel to form nickel carbonyl, a colorless liquid which boils at 43°C.

(a) List all physical properties of substances found in the preceding narrative.

(b) List all chemical properties of substances found in the preceding narrative.

4.50 A hard sugar cube is pulverized and the resulting granules are heated in air until they discolor and then finally burst into flame and burn.

(a) List all physical changes to substances found in the preceding narrative.

(b) List all chemical changes to substances found in the preceding narrative.

4.51 Assign each of the following descriptions of matter to one of the following categories: element, compound, mixture.

(a) one substance present, one phase present, substance can be decomposed by chemical means

(b) two substances present, one phase present

(c) one substance present, two elements present

(d) two elements present, composition is variable

4.52 Assign each of the following descriptions of matter to one of the following categories: element, compound, mixture.

(a) one substance present, one phase present, substance cannot be decomposed by chemical means

(b) one substance present, three elements present

(c) two substances present, two phases present

(d) two elements present, composition is definite and constant

4.53 Indicate whether each of the following samples of matter is a heterogeneous mixture, a homogeneous mixture, a compound, or an element.

(a) A colorless single-phase liquid that when boiled away (evaporated) leaves behind a solid white residue.

(b) A uniform red liquid with a boiling point of 59°C that cannot be broken down into simpler substances using chemical means.

(c) A nonuniform white crystalline substance, part of which dissolves in water and part of which does not.

(d) A colorless single-phase liquid that completely evaporates without decomposition when heated and produces a gas that can be separated into simpler components using physical means.

4.54 Indicate whether each of the following samples of matter is a heterogeneous mixture, a homogeneous mixture, a compound, or an element.

(a) A colorless gas, only part of which reacts with hot iron.

(b) A "cloudy" liquid that separates into two layers upon standing for two hours.

(c) A green solid, all of which melts at the same temperature to produce a liquid that decomposes upon further heating.

(d) A colorless gas that cannot be separated into simpler substances using physical means and that reacts with copper to produce both a copper–nitrogen compound and a copper–oxygen compound.

4.55 Does each of the elements in the following sequences have a two-letter symbol?

(a) magnesium, nitrogen, phosphorus

(b) bromine, iron, calcium

(c) aluminum, copper, chlorine

(d) boron, barium, beryllium

4.56 Does each of the elements in the following sequences have a symbol that starts with a letter not the first letter of the element's English name?

(a) silver, gold, mercury

(b) copper, helium, neon

(c) cobalt, chromium, sodium

(d) potassium, iron, lead

4.57 The chemical symbols Co and Hf when split into two capital letters produce the symbols of other elements; Co gives C and O and Hf gives H and F. Identify all the two-letter chemical symbols that when split into two separate capital letters produce the symbols of two other elements.

4.58 Reversal of the letters in the chemical symbols Ni and Ca produce the chemical symbols of other elements (In and Ac). Identify all the two-letter chemical symbols for which this reversal process produces the symbol of another element.

CUMULATIVE PROBLEMS

4.59 Specify the physical state of a pure substance at each of the following conditions or indicate that the state determination is not possible from the information given.

(a) 10°C below its freezing point
(b) 30°C above its melting point
(c) after sublimation has taken place
(d) at its boiling point

4.60 Specify the physical state of a pure substance at each of the following conditions or indicate that the state determination is not possible from the information given.

(a) 10°C below its melting point
(b) 30°C above its freezing point
(c) after decomposition has taken place
(d) after deposition has taken place

4.61 Using the information in Table 4.3, calculate the percentage (to four significant figures) of the known elements that were discovered

(a) by the year 1900.
(b) in the period 1801–1850.
(c) since the year 1800.
(d) since the year 1950.

4.62 Using the information in Table 4.3, calculate the percentage (to four significant figures) of the known elements that were discovered

(a) by the year 1750.
(b) in the period 1801–1900.
(c) since the year 1850.
(d) since the year 1900.

4.63 The following density determination data were obtained by three students for their unknowns.

Student I:	mass = 4.32 g	volume = 3.78 mL
Student II:	mass = 5.73 g	volume = 5.02 mL
Student III:	mass = 1.52 g	volume = 1.33 mL

(a) Is it likely that the students were working with different unknowns or that they were working with the same substance?
(b) Is it possible to tell from the given data whether the unknowns were elements or compounds?

4.64 The following density determination data were obtained by three students for their unknowns.

Student I:	mass = 27.2 g	volume = 23.6 mL
Student II:	mass = 30.3 g	volume = 28.3 mL
Student III:	mass = 55.6 g	volume = 42.5 mL

(a) Is it likely that the students were working with different unknowns or that they were working with the same substance?
(b) Is it possible to tell from the given data whether the unknowns were elements or compounds?

4.65 Three samples of a substance were subjected to analysis with each sample analyzed by a different technique. The results were:

Technique I:	34.1% of Q and 65.9% of X
Technique II:	34.12% of Q and 65.88% of X
Technique III:	34.12497% of Q and 65.87503% of X

Is the substance that was analyzed likely an element, a compound, or a mixture?

4.66 Three samples of a substance were subjected to analysis with each sample analyzed by a different technique. The results were:

Technique I:	34.2% of C and 65.8% of D
Technique II:	36.32% of C and 63.68% of D
Technique III:	37.2111% of C and 62.7889% of D

Is the substance that was analyzed likely an element, a compound, or a mixture?

4.67 The specific heat of substance A is 0.88 J/(g · °C) and that of substance B is 2.1 J/(g · °C). You are given an unknown that could be pure substance A, pure substance B, or a homogeneous mixture of A and B. In the laboratory you determine that it requires 59.5 J of heat energy to raise the temperature of a 35.0-g sample of the unknown by 1.0°C. What conclusions can you make about the identity of your unknown from these data.

4.68 The specific heat of substance C is 0.93 J/(g · °C) and that of substance D is 1.8 J/(g · °C). You are given an unknown that could be pure substance C, pure substance D, or a homogeneous mixture of C and D. In the laboratory you determine that it requires 23.3 J of heat energy to raise the temperature of a 25.0-g sample of the unknown by 1.0°C. What conclusions can you make about the identity of your unknown from these data.

5

Atoms, Molecules, Formulas, and Subatomic Particles

5.1 The Atom

If you took a sample of the element gold and started to break it into smaller and smaller and smaller pieces, it seems reasonable that you would eventually reach a "smallest possible piece" of gold that could not be divided further and still be called gold. This smallest possible unit of gold would be a gold atom. An **atom** *is the smallest particle of an element that can exist and still have the properties of the element.*

The concept of an atom is an old one, dating back to ancient Greece. Records indicate that around 460 B.C. Democritus, a Greek philosopher, suggested that continued subdivision of matter ultimately would yield small indivisible particles which he called atoms (from the Greek word *atomos* meaning "uncut or indivisible"). Democritus's ideas about matter were, however, lost (forgotten) during the Middle Ages, as were the ideas of many other people.

It was not until the beginning of the nineteenth century that the concept of the atom was "rediscovered." In a series of papers published in the period 1803–1807, the English chemist John Dalton (1766–1844) again proposed that the fundamental building block for all kinds of matter was an atom. This time, however, there was a firm basis for the proposal. Dalton's proposal was based on experimental observations. This is in marked contrast to the early Greek concept of atoms, which was based solely on philosophical speculation. Because of its experimental basis, Dalton's idea got wide attention and stimulated new work and thought concerning the ultimate building blocks of matter.

Additional research, carried out by many scientists, has now validated Dalton's basic conclusion that the building blocks for all types of matter are atoms. Some of the details of Dalton's original proposals have had to be modified in the light of later, more sophisticated experiments, but the basic concept of atoms remains.

Today, among scientists, the concept that atoms are the building blocks for matter is a foregone conclusion. The large accumulated amount of supporting evidence for atoms is most impressive. Key concepts about atoms, in terms of current knowledge, are found in what is known as the atomic theory of matter. The **atomic theory of matter** *is a set of five statements that summarizes modern-day scientific thought about atoms.* These five statements are

1. All matter is made up of small particles called atoms, of which 111 different "types" are known, with each "type" corresponding to a different element.
2. All atoms of a given type are similar to one another and significantly different from all other types.
3. The relative number and arrangement of different types of atoms contained in a pure substance (its composition and structure) determine its identity.
4. Chemical change is a union, separation, or rearrangement of atoms to give new substances.
5. Only whole atoms can participate in or result from any chemical change, since atoms are considered indestructible during such changes.

Just how small is an atom? Atomic dimensions and masses, although not directly measurable, are known quantities obtained by calculation. The data used for the calculations come from measurements made on macroscopic amounts of pure substances.

The diameter of an atom is on the order of 10^{-8} cm. If one were to arrange atoms of diameter 1×10^{-8} cm in a straight line, it would take 10 million of them to extend a length of 1 mm and 254 million of them to reach 1 in. (see Fig. 5.1). Indeed, atoms are very small.

The mass of an atom is also a very small quantity. The mass of a uranium atom, which is one of the heaviest of the known kinds of naturally occurring atoms, is 4×10^{-22} g or 9×10^{-25} lb. To produce a mass of 1 lb would require 1×10^{24} atoms of uranium. The number 1×10^{24} is so large it is difficult to visualize its magnitude. The following comparison "hints" at this number's magnitude. If each of the 5 billion people on Earth were made a millionaire (receiving 1 million \$1 bills), we would still need 200 million other worlds, each inhabited by the same number of millionaires, to have 1×10^{24} dollar bills in circulation.

F.Y.I.

If a 1-karat diamond were cut into 5 billion pieces and given to every person on Earth, each person would get a diamond with 2 trillion atoms. Each diamond would be too small to be seen with the naked eye.

Atoms are incredibly small particles. No one has seen or ever will see an atom with the naked eye. The question may thus be asked: "How can you be absolutely sure that something as minute as an atom really exists?" The achievements of twentieth-century scientific instrumentation have gone a long way toward removing any doubt about the existence of atoms. Electron microscopes, capable of producing magnification factors in the millions, have made it possible to photograph "images" of individual atoms. In 1976 physicists at The University of Chicago were successful in obtaining motion pictures of the movement of single atoms. One of these pictures is shown in Figure 5.2.

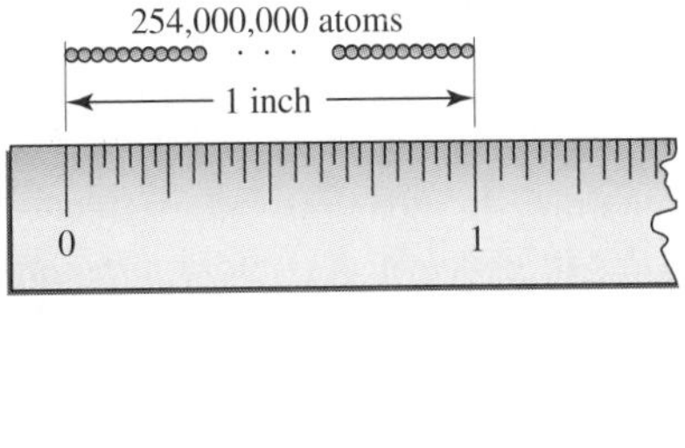

FIGURE 5.1 Comparison of atomic diameters with the common measuring unit of 1 inch.

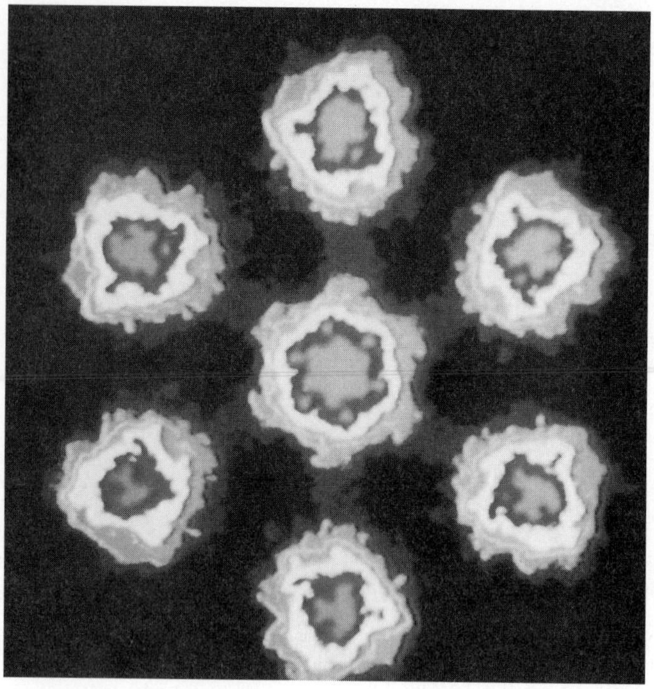

FIGURE 5.2 A uranyl acetate cluster on a very thin carbon substrate. The individual uranium atoms are the roundish spots with darker gray centers. (Courtesy of M. Isaacson, Cornell University, and M. Ohtsuki, The University of Chicago)

5.2 THE MOLECULE

Free isolated atoms are rarely encountered in nature. Instead, under normal conditions of temperature and pressure, atoms are almost always found together in aggregates or clusters ranging in size from two atoms to numbers too large to count. When the group or cluster of atoms is relatively small and bound together tightly, the resulting entity is called a molecule. Thus a **molecule** *is a group of two or more atoms that functions as a unit because the atoms are tightly bound together*. This resultant "package" of atoms behaves in many ways as a single, distinct particle would. The forces that hold the atoms of a molecule together (chemical bonds) will be discussed in Chapter 7.

A *diatomic molecule*, a molecule containing just two atoms, is the simplest type of molecule. Next in complexity are *triatomic* (three atom) and *tetraatomic* (four atom) molecules. Numerous examples of all three of these types of molecules are known. The molecule present in water, the most common of all compounds, is triatomic, containing two hydrogen atoms and one oxygen atom. The molecules present in the poisonous gas carbon monoxide are diatomic, containing one carbon atom and one oxygen atom; tetraatomic molecules are found in the common household cleaning agent called ammonia—one nitrogen atom and three hydrogen atoms are present. Substances with a much larger molecular unit than two, three, or four atoms can and do exist. Caffeine, the central nervous system stimulant present in coffee and cola drinks, has a molecule containing 24 atoms: 8 carbon, 10 hydrogen, 4 nitrogen, and 2 oxygen atoms.

The atoms contained in a molecule may all be of the same kind, or two or more kinds may be present. On the basis of this observation molecules are classified into two categories: homoatomic and heteroatomic. **Homoatomic molecules** *are molecules in which all atoms present are the same kind*. A pure substance containing homoatomic molecules is, thus, an element. **Heteroatomic molecules** *are molecules in which two or more different kinds of atoms are present*. Pure substances containing heteroatomic molecules must be

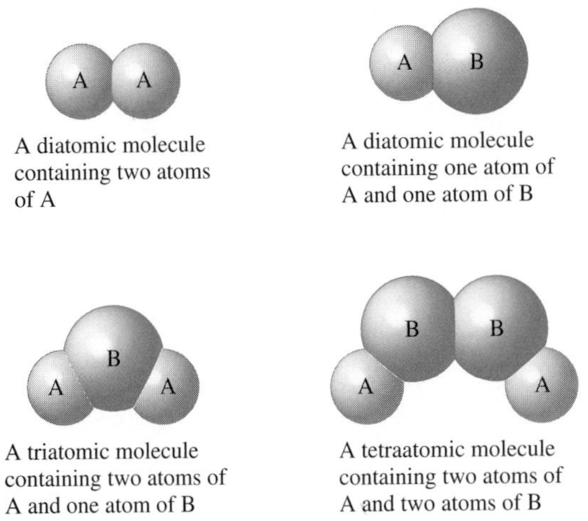
A A
A diatomic molecule containing two atoms of A
A B
A diatomic molecule containing one atom of A and one atom of B
A B A
A triatomic molecule containing two atoms of A and one atom of B
A B B A
A tetraatomic molecule containing two atoms of A and two atoms of B

FIGURE 5.3 Depictions of various simple molecules using models. Spheres of different sizes and colors are used to represent different kinds of atoms.

compounds. The previously mentioned molecules of water, carbon monoxide, ammonia, and caffeine are all heteroatomic molecules. Figure 5.3 shows general models for selected simple molecules.

The fact that homoatomic molecules exist indicates that individual atoms are not always the preferred structural unit for an element. Oxygen is an element that exists in molecular form. Almost all of the oxygen present in air is in the form of diatomic molecules. The elements hydrogen, nitrogen, and chlorine are most commonly encountered with their atoms in groups of two, that is, as diatomic molecules. Under most conditions sulfur atoms collect together in groups of eight; phosphorus atoms readily form tetraatomic molecules. Some guidelines for determining which elements have individual atoms as their basic unit and which exist in molecular form will be given in Section 7.2.

Many, but not all, compounds have heteroatomic molecules as their basic structural unit. Those compounds that do are called *molecular compounds*. Some compounds in the liquid and solid state, however, are not molecular; that is, the atoms present are not collected together into discrete heteroatomic molecules. These nonmolecular compounds still contain atoms of at least two kinds (a necessary requirement for a compound), but the form of aggregation is different. It involves an extended three-dimensional assembly of positively and negatively charged particles called *ions* (Sec. 7.4). Compounds that contain ions are called *ionic compounds*. The familiar substances sodium chloride (table salt) and calcium carbonate (limestone) are ionic compounds. The reasons some compounds have ionic rather than molecular structures are considered in Section 7.7.

For molecular compounds, the molecule is the smallest particle of the compound capable of a stable independent existence. It is the limit of physical subdivision for the compound. Consider the molecular compound sucrose (table sugar). Continued subdivision of a quantity of table sugar to yield smaller and smaller amounts would ultimately lead to the isolation of one single particle of table sugar—a molecule of table sugar. This molecule of sugar could not be broken down any further and still maintain the physical and chemical properties of table sugar. The sugar molecule could be broken down further by chemical (not physical) means to give atoms, but if that occurred we would no longer have sugar. The *molecule* is the limit of *physical* subdivision. The *atom* is the limit of *chemical* subdivision.

Every molecular compound has as its smallest characteristic unit a *unique* molecule. If two samples had the same molecule as a basic unit, both would have the same properties; thus, they would be one and the same compound. An alternative way of stating the same conclusion is: There is only one kind of molecule for any given molecular substance.

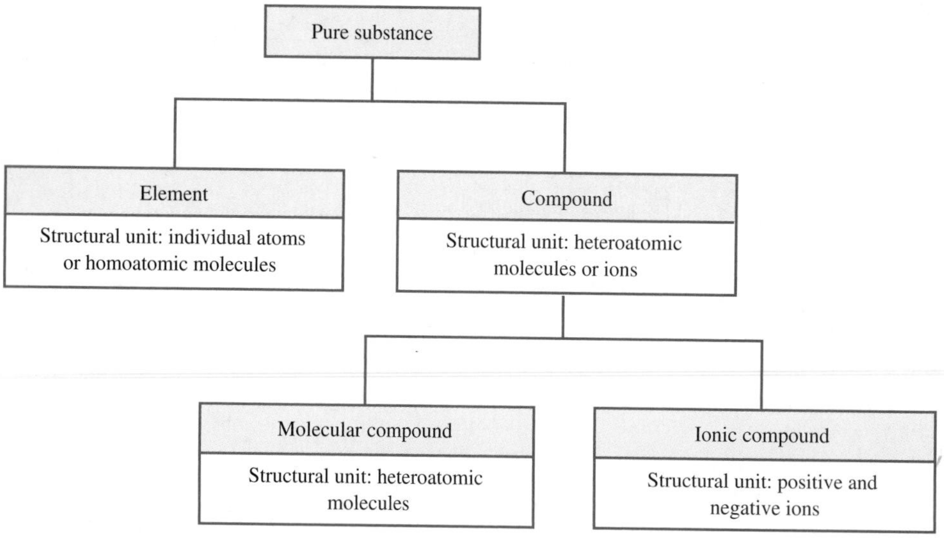

FIGURE 5.4 Structural units for various types of pure substances.

F.Y.I.
Table salt is an ionic compound made from the elements sodium and chlorine. Sodium is a soft metal that bursts into flames when it contacts water; chlorine is a very toxic greenish gas. The salt on potato chips fortunately does not have the properties of either element.

Since every molecule in a sample of a molecular compound is the same as every other molecule in the sample, it is commonly stated that molecular compounds are made up of a single kind of particle. Such terminology is correct as long as it is remembered that the particle referred to is the molecule. In a sample of a molecular compound there are at least two kinds of atoms present but only one kind of molecule.

The properties of molecules are very different from the properties of the atoms that make up the molecules. Molecules do not maintain the properties of their constituent elements. Table sugar is a white crystalline molecular compound with a sweet taste. None of the three elements present in table sugar (carbon, hydrogen, and oxygen) is a white solid or has a sweet taste. Carbon is a black solid, and hydrogen and oxygen are colorless gases.

Figure 5.4 summarizes the relationships between hetero- and homoatomic molecules and elements, compounds, and pure substances.

5.3 NATURAL AND SYNTHETIC COMPOUNDS

Approximately 6 million chemical compounds are now known, with more being characterized daily. No end appears to be in sight as to the number of compounds that can and will be prepared in the future. At present, approximately 7000 new chemical substances are registered every week with Chemical Abstracts Service, a clearing house for new information concerning chemical substances.

Many compounds, perhaps the majority now known, are not naturally occurring substances. These synthetic (laboratory-produced) compounds are legitimate compounds and should not be considered "second class" or "unimportant" simply because they lack the distinction of being natural. Many of the plastics, synthetic fibers, and prescription drugs now in common use are synthetic materials produced through controlled chemical change carried out on an industrial scale.

We have noted that chemists can produce compounds not found in nature. The reverse is also true. Nature is capable of making many compounds, especially those found in living systems, that chemists are not yet able to prepare in the laboratory.

There is a middle ground also. Many compounds that exist in nature can also be produced in the laboratory. A fallacy exists in the thinking of some people concerning these compounds that have "dual origins." A belief still persists that there is a difference between compounds prepared in the laboratory and samples of the same compounds found in nature. This is not true for pure samples of a compound. The message of the law of definite proportions is that all pure samples of a compound, regardless of their origin, have the same composition. Since compositions are the same, properties will also be the same. There is no difference, for example, between a laboratory-prepared vitamin and a "natural" vitamin if both are pure samples of the same vitamin, despite frequent claims to the contrary.

5.4 CHEMICAL FORMULAS

A most important piece of information about a compound is its composition. Chemical formulas represent a concise means of specifying compound compositions. A **chemical formula** *is a notation made up of the symbols of the elements present in a compound with numerical subscripts (located to the right of each symbol) that indicate the number of atoms of each element present in a formula unit.*

The chemical formula for the compound we call aspirin in $C_9H_8O_4$. This formula provides us with the following information about an aspirin molecule: three elements are present: carbon (C), hydrogen (H), and oxygen (O); and 21 atoms are present: 9 carbon atoms, 8 hydrogen atoms, and 4 oxygen atoms.

When only one atom of a particular element is present in a molecule of a compound, the element's symbol is written without a numerical subscript in the formula of the compound. In the formula for rubbing alcohol, C_3H_6O, for example, the subscript 1 for the element oxygen is not written.

To write formulas correctly, it is necessary to follow strictly the capitalization rules for elemental symbols (Sec. 4.9). Making the error of capitalizing the second letter of an element's symbol can dramatically alter the meaning of a chemical formula. The formulas $CoCl_2$ and $COCl_2$ illustrate this point; the symbol Co stands for the element cobalt, whereas CO stands for one atom of carbon and one atom of oxygen. The properties of the compounds $CoCl_2$ and $COCl_2$ are dramatically different. The compound $CoCl_2$ is a blue crystalline solid with a melting point of 724°C and a boiling point of 1029°C. The compound $COCl_2$ is a highly toxic colorless gas with a melting point of −118°C and a boiling point of 8°C.

For molecular compounds, chemical formulas give the composition of the molecules making up the compounds. For ionic compounds, which have no molecules, a chemical formula gives the ion ratio found in the compound. For example, the ionic compound sodium oxide contains sodium ions and oxygen ions in a two-to-one ratio—twice as many sodium ions as oxygen ions. The formula of this compound is Na_2O, which expresses the ratio between the two types of ions present. The term *formula unit* is used to describe this smallest ratio between ions. The distinction between the formula unit of an ionic compound and the molecule of a molecular compound is graphically portrayed in Figure 5.5.

Sometimes chemical formulas contain parentheses, an example being $Al_2(SO_4)_3$. The interpretation of this formula is straightforward; in a formula unit there are present two aluminum (Al) atoms and three SO_4 groups. The subscript following the parentheses always indicates the number of units in the formula of the polyatomic entity inside the parentheses. As another example, consider the compound $Pb(C_2H_5)_4$. Four units of C_2H_5 are present. In terms of atoms present, the formula $Pb(C_2H_5)_4$ represents 29 atoms: 1 lead (Pb) atom, 4 ×

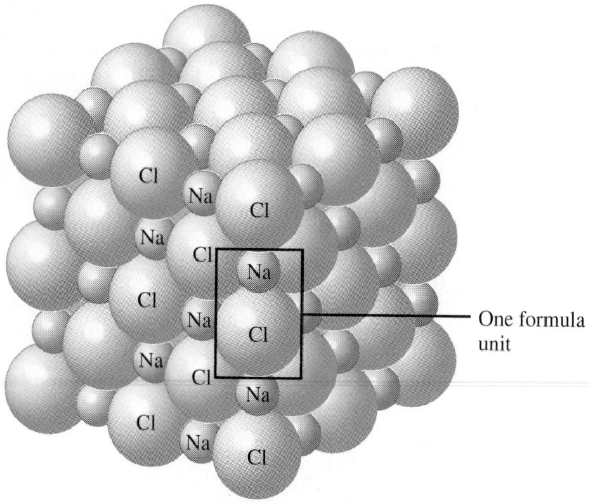

(b) A formula unit of the ionic compound sodium chloride (NaCl)

FIGURE 5.5 Comparison of a molecule of a molecular compound and a formula unit of an ionic compound. A molecule can exist as a separate unit, whereas a formula unit is simply two or more ions "plucked" from a much larger array of ions.

2 = 8 carbon (C) atoms, and 4 × 5 = 20 hydrogen (H) atoms. The formula could be (but is not) written as PbC_8H_{20}. Both versions of the formula convey the same information in terms of atoms present. However, $Pb(C_2H_5)_4$ gives the additional information that the C and H are present as C_2H_5 units and is therefore the preferred way of writing the formula. Further information concerning the use of parentheses (when and why) will be presented in Section 7.8. The important concern now is being able to interpret formulas that contain parentheses in terms of total atoms present. Example 5.1 deals with this skill in greater detail.

Example 5.1

Interpret each of the following formulas in terms of how many atoms of each element are present in one structural unit of the substance.

(a) $C_8H_9O_2N$ (acetaminophen, the active ingredient in Tylenol).

(b) $(NH_4)_2C_2O_4$ (ammonium oxalate, used in the manufacture of explosives).

(c) $Ca_{10}(PO_4)_6(OH)_2$ (hydroxyapatite, present in tooth enamel).

Solution

(a) We simply look at the subscripts following the symbols for the elements. This formula indicates that 8 carbon atoms, 9 hydrogen atoms, 2 oxygen atoms, and 1 nitrogen atom are present in one molecule of the compound.

(b) The subscript following the parenthesis, 2, indicates that two NH_4 units are present. Collectively, in these two units, we have 2 nitrogen atoms and 2 × 4 = 8 hydrogen atoms. In addition, 2 carbon atoms and 4 oxygen atoms are present.

(c) There are 10 calcium atoms. The amounts of phosphorus, hydrogen, and oxygen are affected by the subscripts outside the parentheses. There are 6 phosphorus atoms and 2 hydrogen atoms present. Oxygen atoms are present in two locations in the formula. There are a total of 26 oxygen atoms: 24 from the PO_4 subunits (6 × 4) and 2 from the OH subunits (2 × 1).

Practice Exercise 5.1

Interpret each of the following formulas in terms of how many atoms of each element are present in one structural unit of the substance.

(a) $C_8H_{10}N_4O_2$ (caffeine, an addictive central nervous system stimulant)
(b) $(NH_4)_3PO_4$ (ammonium phosphate, an ingredient in some lawn fertilizers)
(c) $Ca(NO_3)_2$ (calcium nitrate, used in fireworks to give a reddish color)

Ans. (a) 8 C, 10 H, 4 N, 2 O; (b) 3 N, 12 H, 1 P, 4 O; (c) 1 Ca, 2 N, 6 O

The phrases *atoms of an element* and *molecules of an element* are concepts often confused by students. An atom of an element is the smallest particle of that element that can combine to form a compound. A molecule of an element, if the element exists in a molecular form, is the preferred structural unit for the element as it is found in nature. It is not the preferred structural unit, however, for the element in compounds. Thus, the symbols and numbers found in chemical formulas deal with the number of *atoms* of various elements present; they have nothing to do with molecules of the elements. The formula for a molecule of oxygen is O_2. A molecule of carbon dioxide, CO_2, does not contain an oxygen molecule; it contains two oxygen atoms.

In addition to formulas, compounds have names. Naming compounds is not as simple as naming elements. Although the nomenclature of elements (Sec. 4.9) has been largely left up to the imagination of their discoverers, extensive sets of systematic rules exist for naming compounds. Rules must be used because of the large number of compounds that exist. Chapter 8 is devoted to compound nomenclature, and we will not worry about this problem until then. For the time being, our focus will be on the meaning and significance of chemical formulas; knowing how to name the compounds that the formulas represent is not a prerequisite for understanding the meaning of formulas.

5.5 SUBATOMIC PARTICLES: PROTONS, NEUTRONS, AND ELECTRONS

Until the closing decades of the nineteenth century, scientists believed that atoms were solid indivisible spheres without substructure. Today this concept is known to be incorrect. Evidence from a variety of sources, some of which will be discussed in Section 5.6, indicates that atoms themselves are made up of smaller, more fundamental particles called *subatomic particles*. **Subatomic particles,** *particles smaller than atoms, are the building blocks from which all atoms are made.*

Three major types of subatomic particles exist: the proton, the neutron, and the electron. The properties of these subatomic particles that are of most concern to us are mass and electrical charge. A **proton** *is a subatomic particle that possesses one unit of positive charge and has a mass of* 1.673×10^{-24} *g*. Protons were discovered in 1886 by the German physicist Eugene Goldstein (1850–1930). An **electron** *is a subatomic particle that possesses one unit of negative charge and has a mass of* 9.109×10^{-28} *g*. Electrons were characterized in 1897 by the English physicist Joseph John Thomson (1856–1940). A **neutron** *is a subatomic particle that is neutral (i.e., has no charge) and has a mass of* 1.675×10^{-24} *g*. The neutron, the last of the three subatomic particle types to be identified, was characterized in 1932 by the English physicist James Chadwick (1891–1974). Table 5.1 summa-

TABLE 5.1 CHARGES AND MASSES OF THE MAJOR SUBATOMIC PARTICLES

	Electron	Proton	Neutron
Charge	-1	$+1$	0
Actual mass (g)	9.109×10^{-28}	1.673×10^{-24}	1.675×10^{-24}
Relative mass (based on the electron being one unit)	1	1837	1839
Relative mass (based on the neutron being one unit)	0 (1/1839)	1	1

rizes the mass and electrical charge characteristics for subatomic particles. In this table relative masses are given in addition to actual masses.

Using the relative mass values from Table 5.1, we see that the neutron has a mass value only slightly greater than that of the proton. For most purposes their masses can be considered equal. Both neutrons and protons are very massive particles compared to an electron, being nearly 2000 times heavier. The mass of an electron is almost negligible compared to that of the other two types of subatomic particles.

Electrons and protons, the two types of *charged* subatomic particles, possess the same amount of electrical charge; the character of the charge is, however, opposite (negative versus positive). The fact that these subatomic particles are charged is most important because of the way in which charged particles interact. *Particles of opposite or unlike charge attract each other; particles of like charge repel each other.* This behavior of charged particles will be of major concern in many of the discussions in later portions of the text.

The arrangement of subatomic particles within an atom is not haphazard. As is shown in Figure 5.6, the atom can be considered to be composed of two regions: (1) a nuclear region and (2) an extranuclear region, which is everything except the nuclear region.

At the center of every atom is a nucleus. The **nucleus** *is the center region (core) of an atom and contains within it all protons and neutrons present in the atom.* Because of their presence in the nucleus, neutrons and protons are often collectively called *nucleons*. Almost all (over 99.9%) of the mass of an atom is concentrated in its nucleus; all of the heavy subatomic particles (protons and neutrons) are there. A nucleus always carries a positive charge because of the presence of the positively charged protons.

The extranuclear region contains all of the electrons. It is an extremely large region compared to the nucleus. It is mostly empty space. It is a region in which the electrons move rapidly about the nucleus. The motion of the electrons in this extranuclear region determines the volume (size) of the atom in the same way as the blades of a fan determine a volume by their motion. The volume occupied by the electrons is sometimes referred to as the *electron cloud*. Since electrons are negatively charged, the electron cloud is said to be negatively charged.

An atom as a whole is neutral. How can it be that an entity possessing positive charge (the nuclear region) and negative charge (the extranuclear region or electron cloud) can end up neutral overall? For this to occur the same amount of positive and negative charge must be present in the atom; equal amounts of positive and negative charge cancel each other. Atom neutrality thus requires that there be the same number of electrons and protons present in an atom, which is always the case for atoms.

It is important that the size relationships between the parts of an atom be correctly visualized. The nucleus is extremely small (supersmall) compared to the total atomic size. A

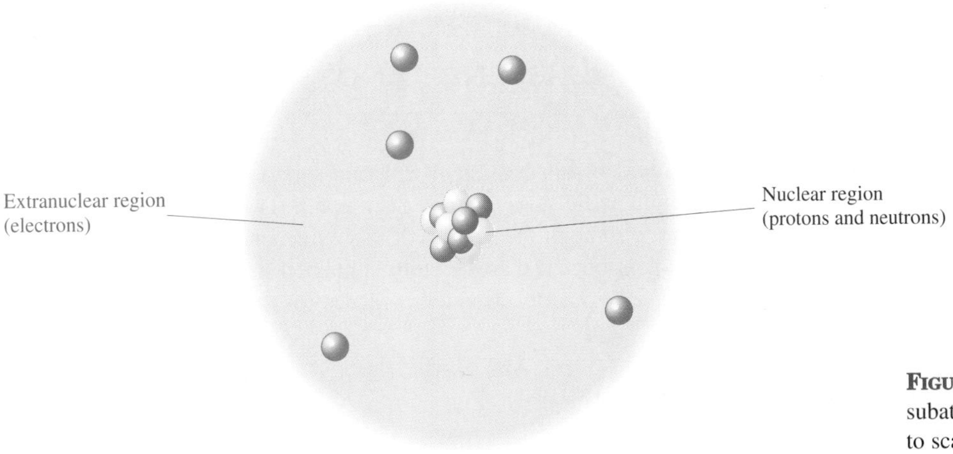

FIGURE 5.6 Arrangement of subatomic particles in an atom (not to scale).

nonmathematical conceptual model to help visualize this relationship involves clouds. Visualize a large spherical or nearly spherical fluffy cloud in the sky. Let the cloud represent the negatively charged extranuclear region of the atom. Buried deep within the cloud would be the positively charged nucleus, the size of a small pebble. As another example, consider the electron cloud to be the size of a major league baseball park. The nucleus would be no larger than a small fly located somewhere in the region behind second base. As a more quantitative example, consider enlarging (magnifying) the nucleus until it is the size of a baseball (7.4 cm in diameter). If the nucleus were this large, the whole atom would have a diameter of approximately 2.5 mi. The nuclear volume is only 1/100,000 that of the atom's total volume; that is, almost all (over 99.9%) of the volume of an atom is occupied by the electron cloud.

The concentration of almost all of the mass of an atom in the nucleus can best be illustrated by using an example. If a coin the same size as a copper penny contained copper nuclei (copper atoms stripped of their electrons) rather than copper atoms (which are mostly empty space), the coin would weigh 190,000,000 tons. Nuclei are indeed very dense matter.

Our just-completed discussion of the makeup of atoms in terms of subatomic particles is based on the existence of three types of subatomic particles: protons, neutrons, and electrons. Actually this model of the atom is an oversimplification. In recent years, as the result of research carried out by nuclear physicists, the picture of the atom has lost its simplicity. Experimental evidence now available indicates that protons and neutrons themselves are made up of even smaller particles. Numerous other particles, with names such as leptons, mesons, and baryons, have been discovered. No theory is yet available that can explain all of these new discoveries relating to the complex nature of the nucleus.

Despite the existence of other nuclear particles we will continue to use the three-subatomic-particle model of the atom. It readily explains almost all chemical observations about atoms. We will have no occasion to deal with any of the recently discovered types of subatomic (or sub-subatomic) particles. Protons, neutrons, and electrons will meet all our needs.

We will also continue to use the concept that atoms are the fundamental building blocks for all types of matter (Sec. 5.1) despite the existence of protons, neutrons, and electrons. This is because under normal conditions subatomic particles do not lead an independent existence for any appreciable length of time. The only way they gain stability is by joining together to form an atom.

5.6 EVIDENCE SUPPORTING THE EXISTENCE AND ARRANGEMENT OF SUBATOMIC PARTICLES

A significant collection of evidence is consistent with and supports the existence, nature, and arrangement of subatomic particles as given in Section 5.5. Two historically important types of experiments illustrate some of the sources of this evidence. *Discharge tube experiments* resulted in the original concept that the atom contained negatively and positively charged particles. *Metal foil experiments* provided evidence for the existence of a nucleus within the atom.

Discharge Tube Experiments

Neon signs, fluorescent lights, and television tubes are all basic ingredients of our modern technological society. The forerunner for all three of these developments was the *gas discharge tube*. Gas discharge tubes also provided some of the first evidence that an atom consisted of still smaller particles (subatomic particles).

The principle behind the operation of a gas discharge tube—that gases at low pressure conduct electricity—was discovered in 1821 by the English chemist Humphry Davy (1778–1829). Subsequently, gas discharge tube studies were carried out by many scientists.

A simplified diagram of a gas discharge tube is shown in Figure 5.7. The apparatus consists of a sealed glass tube containing two metal disks called *electrodes*. The glass tube also has a side arm for attachment to a vacuum pump. During operation, the electrodes are connected to a source of electrical power. (The electrode attached to the positive side of the electrical power source is called the *anode*; the one attached to the negative side is known as the *cathode*.) Use of the vacuum pump allows the amount of gas within the tube to be varied. The smaller the amount of gas present, the lower the pressure within the tube.

Early studies with gas discharge tubes showed that when the tube was almost evacuated (low pressure), electricity flowed from one electrode to the other and the residual gas became luminous (it glowed). Different gases in the tube gave different colors to the glow. After the pressure in the tube was reduced to still lower levels (very little gas remaining), it was found that the luminosity disappeared but the electrical conductance continued, as shown by a greenish glow given off by the tube's glass walls. This glow was the initial discovery of what became known as *cathode rays*. Their discovery marked the beginning of nearly 40 years of discharge tube experimentation that ultimately led to the characterization of both the electron and the proton.

The term *cathode rays* comes from the observation that when an obstacle is placed between the negative electrode (cathode) and the opposite glass wall, a sharp shadow the shape of the obstacle is cast on that wall. This indicates that the rays are coming from the cathode.

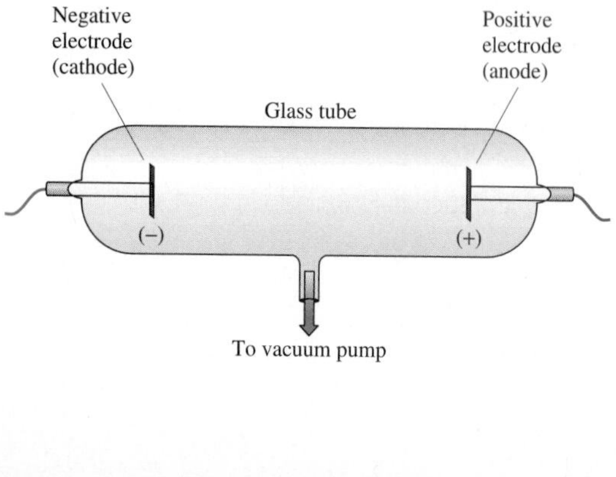

FIGURE 5.7 A simplified version of a gas discharge tube.

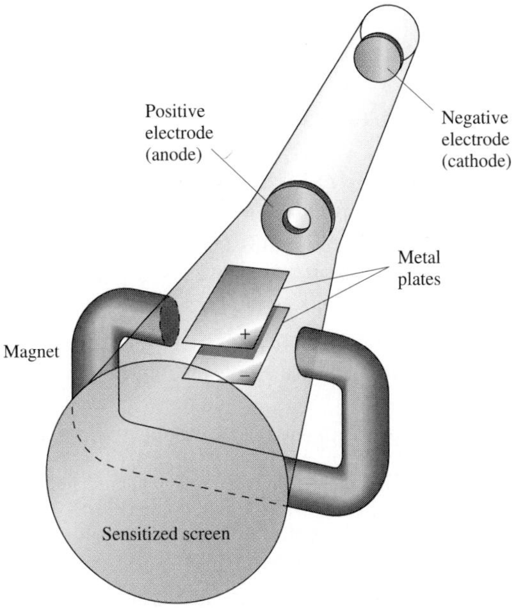

FIGURE 5.8 J. J. Thomson's cathode ray tube involved the use of both electrical and magnetic fields.

Further studies showed that these cathode rays caused certain minerals such as sphalerite (zinc sulfide) to glow. Glass plates were coated with sphalerite and observed under high magnification while being bombarded with cathode rays. The light emitted by the sphalerite coating consisted of many pinpoint flashes. This observation suggested that cathode rays were in reality a stream of extremely small particles.

F.Y.I.
Thomson's cathode ray tube experiment evolved into televisions and computer monitors (still called CRT's).

Joseph John Thomson (1856–1940), an English physicist, provided many facts about the nature of cathode rays. Using a variety of materials as cathodes, he showed that cathode ray production was a general property of matter. By using a specially designed cathode ray tube (see Fig. 5.8), he also found that cathode rays could be deflected by charged plates or a magnetic field. The rays were repelled by the north pole or negative plate and attracted to the south pole or positive plate, thus indicating that they were negatively charged. In 1897, Thomson concluded that cathode rays were streams of negatively charged particles, which today we call *electrons*. Further experiments by others proved that his conclusions were correct.

In 1886, a German physicist, Eugene Goldstein (1850–1930), showed that positive particles were also present in discharge tubes. He used a discharge tube in which the cathode was a metal plate with a large number of holes drilled in it. The usual cathode rays were observed to stream from cathode to anode. In addition, rays of light appeared to stream from each of the holes in the cathode in a direction opposite to that of the cathode rays (see Fig. 5.9). Because these rays were observed streaming through the holes or channels in the cathode, Goldstein called them *canal rays*.

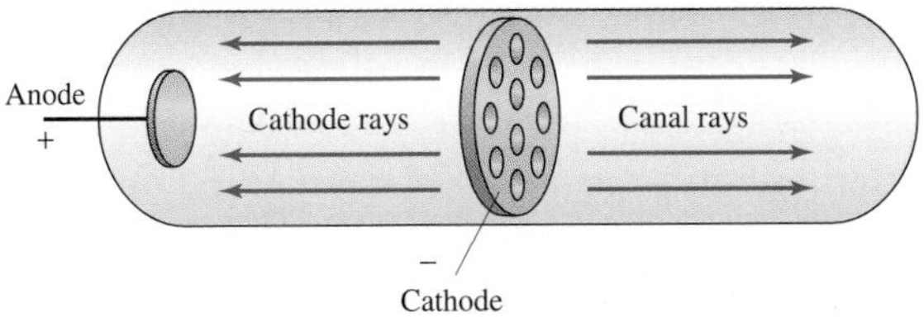

FIGURE 5.9 A gas discharge tube showing canal rays.

Further research showed that canal rays were of many different types, in contrast to cathode rays, which are only one type, and that the particles making up canal rays were much heavier than those of cathode rays. The type of canal rays produced depended upon the gas in the tube. The simplest canal rays were eventually identified as the particles now called protons.

Canal rays are now known to be gas atoms that have lost one or more electrons. Their origin and behavior in a discharge tube can be understood as follows. Electrons (cathode rays) emitted from the cathode collide with residual gas molecules (air) on the way to the anode. Some of these electrons have enough energy to knock electrons away from the gas molecules, leaving behind a positive particle (the remainder of the gas molecule). These positive particles are attracted to the cathode, and some of them pass through the holes or channels. The fact that atoms, under certain conditions, can lose electrons will be discussed further in Section 7.4.

On the basis of discharge tube experiments, Thomson proposed in 1898 that the atom was composed of a sphere of positive electricity containing most of the mass, and that small negative electrons were attached to the surface of the positive sphere. He postulated that a high voltage could pull off surface electrons to produce cathode rays. Thomson's model of the atom, sometimes referred to as the "raisin muffin" or "plum pudding" model—with the electrons as the raisins—is now known to be incorrect. Its significance is that it set the stage for an experiment, commonly called the gold foil experiment, that led to the currently accepted arrangement of protons and electrons in the atom.

Metal Foil Experiments

In 1911 Ernest Rutherford (1871–1937) designed an experiment to test the Thomson model of the atom. In this experiment thin sheets of metal foil were bombarded by alpha particles from a radioactive source. Alpha particles, which are positively charged, are ejected at high speeds from some radioactive materials. The phenomenon of radioactivity had been discovered in 1896 and gave further evidence that electrical charges existed within the atom. Gold was chosen as the target metal because it is easily hammered into very thin sheets. The

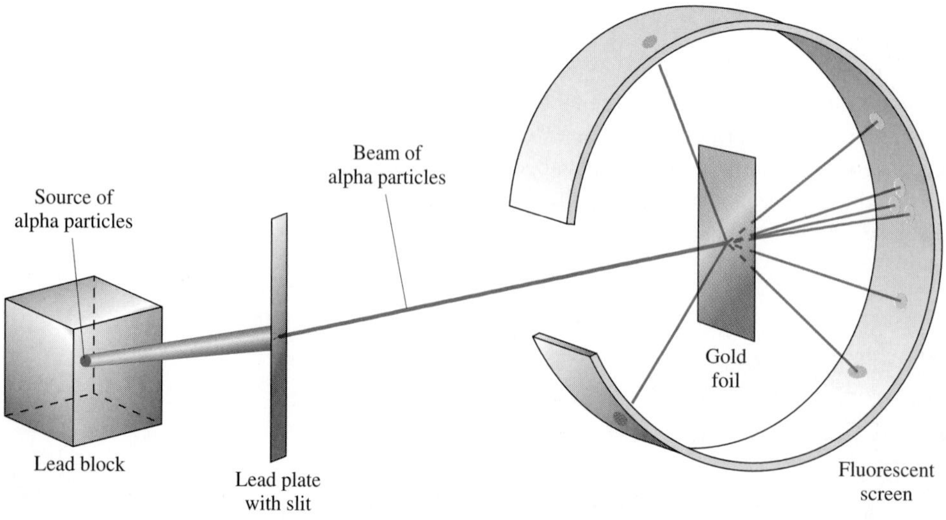

FIGURE 5.10 Rutherford's gold foil–alpha particle experiment. Most of the alpha particles went straight through the foil, but a few were deflected at large angles.

experimental setup for Rutherford's experiment is shown in Figure 5.10. Alpha particles do not appreciably penetrate lead, so a lead plate with a slit was used to produce a narrow alpha particle beam. Each time an alpha particle hit the fluorescent screen, a flash of light was produced.

Rutherford expected that all the alpha particles, since they were so energetic, would pass straight through the thin gold foil. His reasoning was based on the Thomson model, in which the mass and positive charge of the gold atoms were distributed uniformly through each atom. As each positive alpha particle neared the foil, Rutherford assumed that it would be confronted by a uniform positive charge. All particles would be affected the same way (no deflection), which would support the Thomson model of the atom.

The results from the experiment were very surprising. Most of the particles—more than 99%—went straight through as expected. A few, however, were appreciably deflected by something that had to be much heavier than the alpha particles themselves. A very few particles were deflected almost directly back toward the alpha particle source. Similar results were obtained when elements other than gold were used as targets.

Extensive study of the results of his experiments led Rutherford to propose the following explanation.

1. A very dense, small nucleus exists in the center of the atom. This nucleus contains most of the mass of the atom and all of the positive charge.
2. Electrons occupy most of the total volume of the atom and are located outside the nucleus.
3. When an alpha particle scores a direct hit on a nucleus, it is deflected back along the incoming path.
4. A near miss of a nucleus by an alpha particle results in repulsion and deflection.
5. Most of the alpha particles pass through without any interference, because most of the atomic volume is empty space.
6. Electrons have so little mass that they do not deflect the much larger alpha particles (an alpha particle is almost 8000 times heavier than an electron).

Many other experiments have since verified Rutherford's conclusion that at the center of an atom there is a nucleus that is very small and very dense.

5.7 ATOMIC NUMBER AND MASS NUMBER

What determines whether a given atom is an atom of carbon or an atom of oxygen or an atom of gold? It is the number of protons present in the nucleus of the atom. Number of protons determines atom identity. Every element has a characteristic number of protons associated with all of its atoms. For example:

All carbon atoms contain 6 protons.

All oxygen atoms contain 8 protons.

All gold atoms contain 79 protons.

If two atoms differ in the number of protons present, they must be atoms of two different elements. Conversely, if two atoms possess the same number of protons, they must be atoms of the same element.

The characteristic number of protons associated with the atoms of a particular element is called the *atomic number* of the element. An **atomic number** *is a number characteristic of an element which gives the number of protons associated with atoms of the element.* Atomic numbers are experimentally determinable quantities.

Values of atomic numbers for the various elements, along with selected additional information, are printed on the inside back cover of this book. A check of the entries in the atomic number column of the data tabulation shows that an entry exists for each of the numbers in the sequence 1 to 111. The existence of an element that corresponds to each of these numbers is an indication of the order existing in nature. Scientists also interpret this continuous sequence of atomic numbers as evidence that there are no "missing elements" yet to be discovered in nature. The highest-atomic-numbered element that is naturally occurring is element 92 (uranium); elements 93 through 111 are all synthetic (Sec. 4.8).

The atomic numbers of the elements are found inside the front cover of the text. The diagram there is a *periodic table*, a graphical presentation of selected characteristics of the elements. (The periodic table will be considered in detail in Chapter 6.) We note at this time that each box in the periodic table designates an element, with that element's symbol in the center of the box. The number above the symbol is the element's atomic number. The elements are arranged in the periodic table in order of increasing atomic number.

Previously, in Section 4.7, an element was defined as a pure substance that cannot be broken down into simpler substances by ordinary chemical means. Although this is a good historical definition for an element, we can now give a more rigorous definition using the concept of atomic number. An **element** *is a pure substance in which all atoms present have the same atomic number; that is, all atoms have the same number of protons.*

An atomic number, besides giving information about the number of protons present in an atom, also gives information about the number of electrons present. Because of charge neutrality (Sec. 5.5) an atom has the same number of protons and electrons. Thus,

$$\text{Atomic number} = \text{number of protons} = \text{number of electrons}$$

Information about the number of neutrons present in an atom is not obtainable solely from an atomic number. A second number, a *mass number*, is needed in addition to the atomic number. A **mass number** *is a number that gives the total number of protons and neutrons present in the nucleus of an atom.*

$$\text{Mass number} = \text{number of protons} + \text{number of neutrons}$$

Since only protons and neutrons are present in a nucleus, the mass number gives the total number of subatomic particles present in the nucleus. The mass of an atom is almost totally accounted for by the protons and neutrons present (Sec. 5.5); hence the designation *mass* number.

Knowing the atomic number and mass number of an atom uniquely specifies the atom's makeup in terms of subatomic particles. The following equations show the relationship between subatomic particles and the two numbers.

$$\text{Number of protons} = \text{atomic number}$$
$$\text{Number of electrons} = \text{atomic number}$$
$$\text{Number of neutrons} = \text{mass number} - \text{atomic number}$$

Note that the neutron count is obtained through subtraction of the atomic number from the mass number. (The *sum* of the mass number and atomic number also has significance; it corresponds to the *total* number of subatomic particles present in the atom.)

Like atomic numbers, mass numbers are experimentally determinable quantities. Mass numbers are not tabulated in a manner similar to atomic numbers because, as we learn in the next section, most elements lack a unique mass number.

The mass and atomic numbers of a given atom are often specified using the notation

$$^{A}_{Z}E$$

Here E represent the symbol of the element being considered. The atomic number, whose general symbol is Z, is placed as a subscript in front of the elemental symbol. The mass number, whose general symbol is A, is placed as a superscript in front of the elemental symbol. Example of such notation for actual atoms include

$$^{19}_{9}F \qquad ^{23}_{11}Na \qquad \text{and} \qquad ^{197}_{79}Au$$

The first of these notations specifies a fluorine atom that has an atomic number of 9 and a mass number of 19.

Examples 5.2 and 5.3 are sample calculations showing the interrelationships between atomic number, mass number, and the subatomic particle composition of atoms.

Example 5.2

Determine the number of protons, neutrons, and electrons present in the atom $^{27}_{13}Al$.

Solution
The atomic number of this atom is 13 and the mass number of the atom is 27.

$$\begin{aligned} \text{Number of protons} &= \text{atomic number} = 13 \\ \text{Number of electrons} &= \text{atomic number} = 13 \\ \text{Number of neutrons} &= \text{mass number} - \text{atomic number} \\ &= 27 - 13 = 14 \end{aligned}$$

Practice Example 5.2

Determine the number of protons, neutrons, and electrons present in the atom $^{197}_{79}Au$.

Ans.. 79 protons, 79 electron, 118 neutrons

Example 5.3

Write a complete symbol ($^{A}_{Z}E$) for an atom whose nucleus contains 15 protons and 16 neutrons.

Solution
Since 15 protons are present in the atom the atomic number of the atom is 15. From the atomic number the identity of the atom is determined using the information found inside the front or back covers. Using either tabulation, we find that the atomic number 15 belongs to the element phosphorus (P).

The mass number of the atom is obtained by adding the numbers of protons and neutrons. The mass number is 15 + 16 = 31.

The complete symbol of the atom is $^{31}_{15}P$. The atomic number is always the subscript and the mass number is always the superscript.

Practice Exercise 5.3

Write a complete symbol ($^{A}_{Z}E$) for an atom whose nucleus contains 53 protons and 74 neutrons.

Ans. $^{127}_{53}I$

5.8 ISOTOPES

Though all atoms of a given element have the same atomic number and therefore the same number of protons (and electrons), they need not all be identical. They can differ in the number of neutrons present. The presence of one or more additional neutrons in the tiny nucleus of an atom has essentially no effect on the way it behaves chemically. For example, all oxygen atoms have eight protons and eight electrons. Most oxygen atoms also contain eight neutrons. Some oxygen atoms exist, however, that contain nine neutrons, and a few exist that contain ten neutrons. Thus, three different kinds of oxygen atoms exist, all with the same chemical properties. Designations for these three kinds of oxygen atoms are

$$^{16}_{8}O \qquad ^{17}_{8}O \qquad \text{and} \qquad ^{18}_{8}O$$

Three *isotopes* of oxygen are said to exist. Formally defined, **isotopes** *are atoms that have the same atomic number and differing mass numbers.* Isotopes will always have the same number of protons (and electrons) and differing numbers of neutrons.

Most elements occurring naturally are mixtures of isotopes. The various isotopes of a given element are of varying abundance; usually one isotope is predominant. Typical of this situation is the element magnesium, which exists in nature in three isotopic forms: $^{24}_{12}Mg$, $^{25}_{12}Mg$, and $^{26}_{12}Mg$. The percentage abundances for these three isotopes are, respectively, 78.70%, 10.13%, and 11.17%. Percentage abundances are number percentages (number of atoms) rather than mass percentages. A sample of 10,000 magnesium atoms would contain 7870 $^{24}_{12}Mg$ atoms, 1013 $^{25}_{12}Mg$ atoms, and 1117 $^{26}_{12}Mg$ atoms. Table 5.2 gives natural isotopic abundances and isotopic masses for selected elements. The units used for specifying the mass of the various isotopes (last column of Table 5.2) will be discussed in Section 5.9.

The percentage abundances of the isotopes of an element may vary slightly in samples obtained from different locations, but such variations are ordinarily extremely small. We will assume in this text that the isotopic composition of an element is a constant.

Isotopic masses, although not whole numbers, have values that are very close to whole numbers. This fact can be verified by looking at the numbers in the last column of Table 5.2. If an isotopic mass is rounded off to the closest whole number, this value is the same as the mass number of the isotope. This statement can be verified by comparing the second and fourth columns in Table 5.2.

Twenty-three elements have only one naturally occurring form, that is, they are "monoisotopic." For these elements, all atoms found in nature are identical to each other. Of the simpler elements (atomic numbers of 20 or less) those with only one form are

$$^{9}_{4}Be \qquad ^{19}_{9}F \qquad ^{23}_{11}Na \qquad ^{27}_{13}Al \qquad ^{31}_{15}P$$

The existence of isotopes adds clarification to the wording used in some of the statements of atomic theory (Sec. 5.1). Statement 1 reads: "All matter is made up of small particles called atoms, of which 111 different 'types' are known." It should now be apparent why the word types was put in quotation marks. Because of the existence of isotopes, atoms of each type are similar, but not identical. Atoms of a given element are similar in that they

TABLE 5.2 NATURALLY OCCURRING ISOTOPIC ABUNDANCES FOR SOME COMMON ELEMENTS

Element	Isotope	Percent Natural Abundance	Isotopic Mass (amu)
Hydrogen	$^{1}_{1}H$	99.985	1.0078
	$^{2}_{1}H$	0.015	2.0141
Carbon	$^{12}_{6}C$	98.89	12.0000
	$^{13}_{6}C$	1.11	13.0033
Nitrogen	$^{14}_{7}N$	99.63	14.0031
	$^{15}_{7}N$	0.37	15.0001
Oxygen	$^{16}_{8}O$	99.759	15.9949
	$^{17}_{8}O$	0.037	16.9991
	$^{18}_{8}O$	0.204	17.9992
Sulfur	$^{32}_{16}S$	95.0	31.9721
	$^{33}_{16}S$	0.76	32.9715
	$^{34}_{16}S$	4.22	33.9679
	$^{36}_{16}S$	0.014	35.9671
Chlorine	$^{35}_{17}Cl$	75.53	34.9689
	$^{37}_{17}Cl$	24.47	36.9659
Copper	$^{63}_{29}Cu$	69.09	62.9298
	$^{65}_{29}Cu$	30.91	64.9278
Titanium	$^{46}_{22}Ti$	7.93	45.95263
	$^{47}_{22}Ti$	7.28	46.9518
	$^{48}_{22}Ti$	73.94	47.94795
	$^{49}_{22}Ti$	5.51	48.94787
	$^{50}_{22}Ti$	5.34	49.9448
Uranium	$^{234}_{92}U$	0.0057	234.0409
	$^{235}_{92}U$	0.72	235.0439
	$^{238}_{92}U$	99.27	238.0508

have the same atomic number, but not identical since they may have different mass numbers.

Statement 2 reads: "All atoms of a given type are similar to one another and significantly different from all other types." All atoms of an element are similar in chemical properties and differ significantly from atoms of other elements with different chemical properties.

It is possible for isotopes of two different elements to have the same mass number. For example, the element iron (atomic number 26) exists in nature in four isotopic forms, one of which is $^{58}_{26}Fe$. The element nickel, with an atomic number two units greater than that of iron, exists in nature in five isotopic forms, one of which is $^{58}_{28}Ni$. Thus, atoms of both iron and nickel exist with a mass number of 58. Thus, mass numbers are not unique for elements as are atomic numbers. Atoms of different elements that have the same mass number are called isobars; $^{58}_{26}Fe$ and $^{58}_{28}Ni$ are isobars. **Isobars** *are atoms that have the same mass number but different atomic numbers*. Even though atoms of two *different* elements can have the same mass number (isobars), they cannot have the same atomic number. All atoms of a given atomic number must necessarily be atoms of the same element.

Figure 5.11 gives information about number of naturally occurring isotopes and mass number identity for the simpler elements (atomic numbers 1–20). The patterns in both number of isotopes and mass number are relatively simple for the first 15 elements. For the first seven elements the number of isotopes varies in the pattern

2–2–2–1–2–2–2

For the next eight elements the pattern is

3–1–3–1–3–1–3–1

Mass numbers are sequential for the first 15 elements with the exception that mass numbers 5 and 8 are skipped. (There are no naturally occurring atoms with mass numbers 5 and 8.) Element 16 (S) is the first element with nonsequential mass numbers for its isotopes. The first isobars occur at mass number 36 (S and Ar).

The number of known naturally occurring stable isotopes, collectively for all of the elements, is approximately 270. In addition to these isotopes, over 1600 more unstable isotopes have been synthesized in the laboratory (from the naturally occurring ones) using nuclear rather than chemical reactions. These unstable synthetic isotopes all have the common characteristic of being radioactive. Radioactive isotopes eventually revert back to naturally occurring isotopes. Many of these unstable isotopes, despite their instability, have important uses in chemical and biological research as well as in medicine.

Example 5.4

Indicate whether the members of each of the following pairs are isotopes, isobars, or neither.

(a) $^{42}_{20}X$ and $^{43}_{20}Q$ **(b)** $^{40}_{19}X$ and $^{40}_{20}Q$ **(c)** $^{44}_{20}X$ and $^{45}_{21}Q$

FIRST SEVEN ELEMENTS—NUMBER OF ISOTOPES (2–2–2–1–2–2–2)

H	He		Li		Be	B	C	N
1	3	(no 5)	6	(no 8)	9	10	12	14
2	4		7			11	13	15

NEXT EIGHT ELEMENTS—NUMBER OF ISOTOPES (3–1–3–1–3–1–3–1)

O	F	Ne	Na	Mg	Al	Si	P
16	19	20	23	24	27	28	31
17		21		25		29	
18		22		26		30	

NEXT FIVE ELEMENTS—NUMBER OF ISOTOPES (irregular pattern)

S	Cl	Ar	K	Ca
32	35	36	39	40
33	37	38	40	42
34		40	41	43
36				44
				46
				48

FIGURE 5.11 Variance in mass numbers and number of stable isotopes for elements 1–20. (The mass numbers that occur for an element are listed underneath its symbol with the most abundant isotope in color.)

(d) an atom X with 20 protons and 21 neutrons and an atom Q with 19 protons and 21 neutrons

Solution

(a) These atoms are isotopes. Both atoms have the same atomic number of 20. Isotopes differ from each other in neutron count, which is the case here. Atom X has 22 neutrons, and atom Q has 23 neutrons.

(b) These atoms are isobars. They have the same mass number (40) and different atomic numbers (19 and 20).

(c) These atoms are not isotopes or isobars. Isotopes must have the same atomic number and isobars must have the same mass number. Neither is the case here.

(d) These atoms are not isotopes or isobars. They are not isotopes because a differing number of protons means differing atomic numbers. They are not isobars because the mass numbers differ; 41 for atom X and 40 for atom Q. The two atoms contain the same number of neutrons. However, the definition for isobars is based on the same mass number rather than on the same neutron count.

Practice Exercise 5.4

Indicate whether the members of each of the following pairs are isotopes, isobars, or neither.

(a) ${}^{27}_{13}X$ and ${}^{28}_{14}Q$ (b) ${}^{46}_{20}X$ and ${}^{48}_{20}Q$ (c) ${}^{36}_{16}X$ and ${}^{36}_{18}Q$

(d) an atom X with eight protons and nine neutrons and an atom Q with nine protons and ten neutrons

Ans. (a) neither; (b) isotopes; (c) isobars; (d) neither

5.9 ATOMIC MASSES

The mass of an atom of a specific element can have one of several values if the element exists in isotopic forms. For example, oxygen atoms can have any one of three masses, since three isotopes of oxygen exist. Because of the existence of isotopes, it might seem necessary to specify isotopic identity every time atomic masses must be dealt with; however, this is not the case. In practice, isotopes are seldom mentioned in discussions involving atomic masses; instead, the atoms of an element are treated as if they all had a single mass. The mass used is an average relative mass that takes into account the existence of isotopes. The use of this average relative mass concept reduces the number of masses needed for calculations from many hundreds to 111—one for each element. These average masses are called *atomic masses*.

Atomic mass values for the elements are printed inside both the front and back covers of this book. For the periodic table (Sec. 5.7) inside the front cover, the atomic mass value is the number underneath each element's symbol. (The number above each element's symbol is its atomic number—Sec. 5.7.) Typical atomic mass values include

14.0067	for the element nitrogen
32.066	for the element sulfur
126.9045	for the element iodine

Atomic mass values are not *mass numbers*. They cannot be mass numbers because mass numbers must be *whole* numbers since they are a *count* of the number of protons and neutrons in the nucleus (Sec. 5.7). Atomic masses are calculated numbers obtained from data on isotopic masses and isotopic abundances.

The starting point for understanding the origins of atomic mass values is a consideration of the formal definition for an atomic mass. An **atomic mass** *is the relative mass of an average atom of an element on a scale using the* $^{12}_{6}C$ *atom as the reference*. The meaning of two terms found within this definition, *relative mass* and *average atom*, is crucial to understanding the definition as a whole.

Relative Mass

The usual standards of mass, such as grams or pounds, are not convenient for use with atoms, because very small numbers are always encountered. For example, the mass in grams of a $^{238}_{92}U$ atom, one of the heaviest atoms known, is 3.95×10^{-22}. To avoid repeatedly encountering such small numbers scientists have chosen to work with relative rather than actual mass values.

A relative mass value for an atom is the mass of that atom relative to some standard rather than the actual mass value of the atom in grams. The term relative means "as compared to." The choice of the standard is arbitrary; this gives scientists control over the magnitude of the numbers on the relative scale, thus avoiding very small numbers.

For most purposes in chemistry, relative mass values serve just as well as actual mass values. Knowing how many times heavier one atom is than another, information obtainable from a relative mass scale, is just as useful as knowing the actual mass values of the atoms involved. Example 5.5 illustrates the procedures involved in constructing a relative mass scale and also points out some of the characteristics of such a scale.

Example 5.5

Construct a relative mass scale for the hypothetical atoms Q, X, and Z given the following information about them.

1. Atoms of Q are four times heavier than those of X.
2. Atoms of X are three times heavier than those of Z.

Solution

Atoms of Z are the lightest of the three types of atoms. We will arbitrarily assign atoms of Z a mass value of one unit. The unit name can be anything we wish, and we shall choose "snick." On this basis, one atom of Z has a mass value of 1 snick. Atoms of Z will be our scale reference point. Atoms of X will have a mass value of 3 snicks (three times as heavy as Z) and Q atoms a value of 12 snicks (four times as heavy as X).

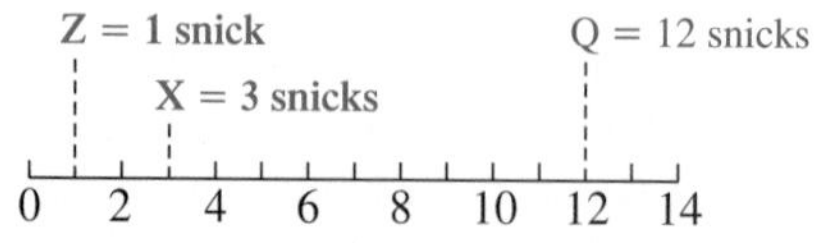

The name chosen for the mass unit was arbitrary. The assignment of the value 1 for the mass of Z, the reference point on the scale, was also arbitrary. What if we had chosen to call the unit a "smerge" and had chosen a value of 3 for the mass of an atom of Z? If this had been the case, the resulting relative scale would have appeared as

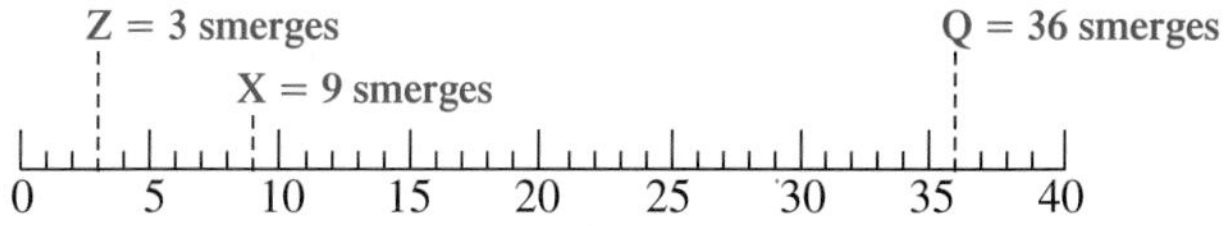

Which of the above relative scales is the "better" scale? The answer is that the scales are equivalent. The relationships between the masses of Q, X, and Z are the same on the two scales, even though the reference points and unit names differ. On the "snick" scale, for example, Q is four times heavier than X (12/3); on the "smerge" scale, Q is also four times heavier than X (36/9).

Notice that we did not need to know the actual masses of Q, X, and Z to set up either the "snick" or "smerge" scale. All that is needed to set up a relative scale is a set of interrelationships among quantities. One value—the reference point—is arbitrarily assigned, and all other values are determined by using the known interrelationships.

The information given at the start of this example is sufficient to set up an infinite number of relative mass scales. Each scale would differ from the others in choice of reference point and unit name. All the scales would, however, be equivalent to each other, and each scale would provide all of the mass relationships obtainable from an actual mass scale except for actual mass values.

Practice Exercise 5.5

Construct a relative mass scale for the hypothetical atoms Q, X, and Z given the following information about them.

(1) Atoms of Q are two times as heavy as those of X.

(2) Atoms of X are four times as heavy as those of Z.

(3) The reference point for the scale is Z = 3.00 sloops.

Ans. Z = 3.00 sloops, X = 12.00 sloops, and Q = 24.00 sloops

A relative scale of atomic masses has been set up in a manner similar to that used in Example 5.5. The unit is called the *atomic mass unit*, abbreviated amu. The arbitrary reference point involves a particular isotope of carbon, $^{12}_{6}C$. The mass of this isotope is set at 12.00000 amu. The masses of all other atoms are then determined relative to that of $^{12}_{6}C$. For example, if an atom is twice as heavy as a $^{12}_{6}C$ atom, its mass is 24.00000 amu on the scale, and if an atom weighs half as much as a $^{12}_{6}C$ atom, its scale mass is 6.00000 amu.

The masses of all atoms have been determined relative to each other experimentally. Actual values for the masses of selected isotopes on the $^{12}_{6}C$ scale are given in Figure 5.12.

Reread the formal definition of atomic mass given at the start of this section. Note how $^{12}_{6}C$ is mentioned explicitly in the definition because of the central role it plays in the setting up of the relative atomic mass scale.

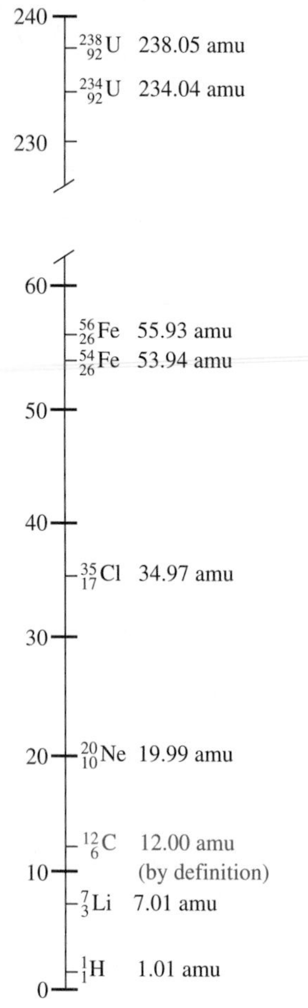

FIGURE 5.12 Relative masses of selected isotopes on the $^{12}_{6}C$ atomic mass scale.

On the basis of the values given in Figure 5.12 it is possible to state, for example, that $^{238}_{92}U$ is 4.256 times as heavy as $^{56}_{26}Fe$ (238.05 amu/55.93 amu = 4.256), and $^{56}_{26}Fe$ is 2.798 times as heavy as $^{20}_{10}Ne$ (55.93 amu/19.99 amu = 2.798). We do not need to know the actual masses of the atoms involved to make such statements; relative masses are sufficient to calculate the information.

Average Atom

Since isotopes exist, the mass of an atom of a specific element can have one of several values. For example, oxygen atoms can have any one of three masses, since three isotopes exist: $^{16}_{8}O$, $^{17}_{8}O$, and $^{18}_{8}O$. Despite mass variances among isotopes, the atoms of an element are treated as if they all had a single common mass. The common mass value used is a *weighted average mass*, which takes into account the natural abundances and atomic masses of the isotopes of an element.

The validity of the weighted average mass concept rests on two points. First, extensive studies of naturally occurring elements have shown that the percent abundance of the isotopes of a given element is generally constant. No matter where the element sample is obtained on Earth, it generally contains the same percentage of each isotope. Because of these constant isotopic ratios, the mass of an "average atom" does not vary. Second, chemical operations are always carried out with very large numbers of atoms. The tiniest piece of matter visible to the eye contains more atoms than can be counted by a person in a lifetime. The numbers are so great that any collection of atoms a chemist works with will be representative of naturally occurring isotopic ratios.

Weighted Averages

Atomic masses are weighted averages calculated from the following three pieces of information.

1. The *number* of isotopes that exist for the element.
2. The *isotopic mass* for each isotope, that is, the relative mass of each isotope on the $^{12}_{6}C$ scale.
3. The *percent abundance* of each isotope.

Table 5.2 gives these data for selected elements.

Examples 5.6 and 5.7 illustrate the operations needed to calculate weighted averages. Example 5.6 is a general exercise concerning weighted averages, and Example 5.7 illustrates the calculation of an atomic mass by the method of weighted averages.

Example 5.6

Sulfur oxides are air pollutants that arise primarily from the burning of coal. A student measures the sulfur oxide concentration in the atmosphere on five successive days with the following results:

Tuesday	0.024 ppm	(parts per million)
Wednesday	0.021 ppm	
Thursday	0.021 ppm	
Friday	0.024 ppm	
Saturday	0.015 ppm	

What is the average sulfur oxide concentration in the air over this time period based on the student's measurements?

Solution

Let us solve this problem using two different methods. The first method involves procedures familiar to you—the "normal" way of taking an average. By this method, the average is found by dividing the sum of the numbers by the number of values summed.

$$\frac{(0.024 + 0.021 + 0.021 + 0.024 + 0.015)}{5}\ \text{ppm} = 0.021\ \text{ppm}$$

(calculator and correct answer)

Now let us solve this same problem again, this time treating it as a "weighted average" problem. To do this, we organize the given information in a different way. In our list of pollutant concentrations we have three different concentration values: 0.024 ppm, 0.021 ppm, and 0.015 ppm.

Two of the five values (40.0%) are 0.024 ppm
Two of the five values (40.0%) are 0.021 ppm
One of the five values (20.0%) is 0.015 ppm

We will use the data in this "percent form" for our weighted average calculation.

To find the weighted average we multiply each distinct value (0.024, 0.021, and 0.015) by its fractional abundance, that is, by its percentage expressed in decimal form, and then we sum the products from the multiplications.

$$\begin{aligned} 0.400 \times 0.024\ \text{ppm} &= 0.0096\ \text{ppm} \\ 0.400 \times 0.021\ \text{ppm} &= 0.0084\ \text{ppm} \\ 0.200 \times 0.015\ \text{ppm} &= \underline{0.0030\ \text{ppm}} \\ &\ 0.0210\ \text{ppm} \quad \text{(same average value as before)} \end{aligned}$$

This averaging method, although it appears somewhat more involved than the "normal" method, is the one that must be used in calculating atomic masses because of the form in which data about isotopes are obtained. The percent abundances of isotopes are experimentally determinable quantities. The total number of atoms of various isotopes present in nature, a prerequisite for using the "normal" average method, is not easily determined. Therefore, we use the "percent" method.

Practice Exercise 5.6

Carbon monoxide is an air pollutant that arises primarily from the operation of automobiles. A student measures the carbon monoxide concentration in the atmosphere on five successive days with the following results:

Wednesday	11 ppm
Thursday	11 ppm
Friday	8 ppm
Saturday	8 ppm
Sunday	7 ppm

Using the "weighted average" method, calculate the average carbon monoxide concentration in the air over this time period based on the student's measurements.

Ans. 9.0 ppm

Example 5.7

Magnesium occurs in nature in three isotopic forms: ${}^{24}_{12}Mg$ (78.70% abundance), ${}^{25}_{12}Mg$ (10.13% abundance), and ${}^{26}_{12}Mg$ (11.17% abundance). The relative masses of these three isotopes, respectively, are 23.985, 24.986, and 25.983 amu. Calculate the atomic mass of magnesium from these data.

Solution

The atomic mass of an element is calculated using the weighted average method illustrated in Example 5.6. Each of the isotopic masses is multiplied by the fractional abundance associated with that mass, and then the products are summed.

$$0.7870 \times 23.985 \text{ amu} = 18.876195 \text{ amu} = 18.88 \text{ amu}$$
$$0.1013 \times 24.986 \text{ amu} = 2.5310818 \text{ amu} = 2.531 \text{ amu}$$
$$0.1117 \times 25.983 \text{ amu} = 2.9023011 \text{ amu} = 2.902 \text{ amu}$$

Note that the method for converting percentages to fractional abundances is always the same. The decimal point in the percentage is moved two places to the left. For example,

$$78.70\% \quad \text{becomes} \quad 0.7870$$

Significant figures are always an important part of an atomic mass calculation. Both isotopic masses and percent abundances are experimentally determined numbers.

Summing, to obtain the atomic mass, gives

$$(18.88 + 2.531 + 2.902) \text{ amu} = 24.313 \text{ amu} \quad \text{(calculator answer)}$$
$$= 24.31 \text{ amu} \quad \text{(correct answer)}$$

Since the number 18.88 is known only to the hundredths place, the answer can be expressed only to the hundredths place.

The above calculation involved an element that exists in three isotopic forms. An atomic mass calculation for an element having four isotopic forms would be carried out in an almost identical fashion. The only difference would be four products to calculate (instead of three) and four terms in the resulting sum.

Practice Exercise 5.7

Silicon occurs in nature in three isotopic forms: ${}^{28}_{14}Si$ (92.21% abundance), ${}^{29}_{14}Si$ (4.70% abundance), and ${}^{30}_{14}Si$ (3.09% abundance). The relative masses of these three isotopes, respectively, are 27.977, 28.976, and 29.974 amu. Calculate the atomic mass of silicon from these data.

Ans. 28.09 amu

In Example 5.7 the atomic mass of magnesium was calculated to be 24.31 amu. How many magnesium atoms have a mass of 24.31 amu? The answer is none. Magnesium atoms have a mass of 27.977, 28.976, or 29.974 amu, depending on which isotope they are. The mass 28.09 amu is the mass of an "average" magnesium atom. It is this average mass that is used in calculations even though no magnesium atoms have masses equal to this average value. Only in the case where all atoms have the same mass will the isotopic mass and the atomic mass be the same.

Atomic masses are subject to change and they do change. Every two years an updated atomic mass listing is published by the International Union of Pure and Applied Chemistry (IUPAC). This update, produced by an international committee of chemists, takes into account all new research on isotopic abundances.

New atomic mass values are almost always more precise than the older values they replace. This is a reflection of the increasingly sophisticated instrumentation available to current researchers, which enables them to make more precise measurements.

Illustrative of atomic mass changes that occur are those found in the 1991 and 1993 IUPAC reports. The atomic masses of seven elements were updated.

1991 changes:	Indium	114.818 amu instead of 114.82 amu
	Tungsten	183.84 amu instead of 183.85 amu
	Osmium	190.23 amu instead of 190.2 amu
1993 changes:	Titanium	47.867 amu instead of 47.88 amu
	Iron	55.845 amu instead of 55.847 amu
	Antimony	121.760 amu instead of 121.757 amu
	Iridium	192.217 amu instead of 192.22 amu

Atomic mass revisions such as these explain why various textbooks (and periodic tables) often differ in a few atomic mass values. The differing values come from different biannual IUPAC reports on atomic masses.

The precision of atomic mass values varies from element to element. For example, we have

B	10.811	amu
F	18.9984032	amu
Si	28.0855	amu
Pb	207.2	amu

What causes such precision variance? The key factor is the constancy of isotopic percentage abundance measurements between various samples of an element. Although all elements have an essentially constant set of isotopic percentage abundances, there is slight variation between samples obtained from different sources. All variations are small; however, for some elements they are greater than for others. When only one form of an element occurs in nature, such as F, maximum precision can be obtained for atomic mass.

An atomic mass cannot be calculated for all elements. Recall, from Section 4.8, that not all elements are naturally occurring substances. Twenty-three of the known elements are "synthetic," having been produced in the laboratory from naturally occurring elements. Obviously a weighted average atomic mass cannot be calculated for these laboratory-produced elements, since the amount of each isotope produced varies, depending on the laboratory experiment carried out.

Tabulations of atomic masses do contain entries for the synthetic elements. Such entries are the mass number of the most stable isotope of the synthetic element. (All isotopes of all synthetic elements are unstable.) Such mass numbers are always enclosed in parentheses to distinguish them from calculated atomic masses. Note the presence of such entries in both the atomic mass listing inside the back cover and the periodic table inside the front cover.

Before leaving the subject of atomic masses, we need to consider one additional ques-

F.Y.I.

If there are a dozen people in a room and six are 5′ tall and six are 6′ tall, the average height of the twelve people is 5.5′, although no particular person is that height. If, on the other hand, nine are 6′ tall and three are 5′ tall, the average would now lie closer to 6′ at 5.75′.

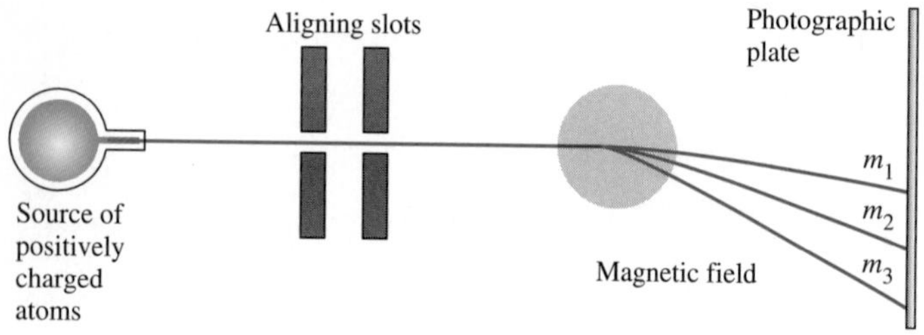

FIGURE 5.13 A simplified diagram of a mass spectrometer, an instrument used to obtain information about isotopic masses and abundances.

tion. How do scientists determine the abundances and masses of the various isotopes of an element? A device known as a *mass spectrometer* is the key to obtaining such information.

A schematic diagram of a mass spectrometer is given in Figure 5.13. The important components of this instrument are a source of charged particles of the element under investigation, an aligning system to create a narrow beam of the charged particles, a magnetic field to affect the path of the charged particles, and a means of detecting the charged particles (such as a photographic plate).

Suppose oxygen gas containing all three isotopes is admitted to the instrument. The gaseous molecules are bombarded with an energetic electron beam, producing positively charged oxygen species. The aligning system produces a narrow beam of these charged atoms, which then enters the magnetic field. The most massive particles (heaviest isotope) are not deflected by the magnetic field as much as the less massive ones, so the charged atoms are divided into separate beams that strike the photographic plate at different points depending on their masses. The more abundant isotopes will create more intense lines on the plate. The relative intensities of the lines correlates exactly with the relative abundances of the isotopes.

KEY TERMS

The new terms or concepts defined in this chapter are

atom (Sec. 5.1) The smallest particle of an element that can exist and still have the properties of the element.

atomic mass (Sec. 5.9) The relative mass of an average atom of an element on a scale using the $^{12}_{6}C$ atom as the reference.

atomic number (Sec. 5.7) The number of protons in the nucleus of an atom.

atomic theory of matter (Sec. 5.1) A set of five statements that summarize modern-day scientific thought about atoms.

chemical formula (Sec. 5.4) A notation made up of the symbols of the elements present in a compound with numerical subscripts (located to the right of each symbol) that indicate the number of atoms of each element present in a formula unit.

electron (Sec. 5.5) Subatomic particle, located outside the nucleus of an atom, that has a negative charge and a mass of 9.109×10^{-28} g.

element (Sec. 5.7) A pure substance in which all atoms present have the same atomic number; that is, all atoms have the same number of protons.

heteroatomic molecule (Sec. 5.2) Molecule in which two or more different kinds of atoms are present.

homoatomic molecule (Sec. 5.2) Molecule in which all atoms present are the same kind.

isobars (Sec. 5.8) Atoms that have the same mass number but different atomic numbers.

isotopes (Sec. 5.8) Atoms that have the same atomic number but different mass numbers.

mass number (Sec. 5.7) The number of protons plus neutrons in the nucleus of an atom, that is, the total number of nucleons present.

molecule (Sec. 5.2) A group of two or more atoms that

functions as a unit because the atoms are tightly bound together.

neutron (Sec. 5.5) Subatomic particle, located within the nucleus of an atom, that has no charge and has a mass of 1.675×10^{-24} g.

nucleus (Sec. 5.5) The center region of an atom that contains within it all protons and neutrons present in the atom.

proton (Sec. 5.5) Subatomic particle, located within the nucleus of an atom, that has a positive charge and a mass of 1.673×10^{-24} g.

subatomic particle (Sec. 5.5) A particle smaller than an atom that is a building block from which the atom is made.

PRACTICE PROBLEMS

Atoms and Molecules (Secs. 5.1 and 5.2)

5.1 Which of the following concepts are *not* consistent with the statements of modern-day atomic theory?

- **(a)** Atoms are the basic building blocks for all kinds of matter.
- **(b)** Different "types" of atoms exist.
- **(c)** All atoms of a given "type" are identical.

5.2 Which of the following concepts are *not* consistent with the statements of modern-day atomic theory?

- **(a)** Only whole atoms can participate in chemical reactions.
- **(b)** Atoms change identity during chemical change processes.
- **(c)** 113 different "types" of atoms are known.

5.3 Which of the terms *heteroatomic*, *homoatomic*, *diatomic*, *triatomic*, *element*, and *compound* apply to each of the following molecules. (More than one term may apply in a given situation.)

(a) (Q–X) **(b)** (Q–Z–X) **(c)** (X–X) **(d)** (X–Q–X)

5.4 Which of the terms *heteroatomic*, *homoatomic*, *diatomic*, *triatomic*, *element*, and *compound* apply to each of the following molecules. (More than one term may apply in a given situation.)

(a) (Q–Q–Q) **(b)** (Q–X) **(c)** (Q–X–X) **(d)** (Q–X–Z)

5.5 Indicate whether each of the following statements is *true* or *false*. If a statement is false, change it to make it true. (Such a rewriting should involve more than merely converting the statement to the negative of itself.)

- **(a)** Molecules must contain three or more atoms.
- **(b)** The atom is the limit of chemical subdivision for an element.
- **(c)** All compounds have molecules as their basic structural unit.
- **(d)** A molecule of a compound must be heteroatomic.
- **(e)** There is only one kind of molecule for any given molecular substance.

5.6 Indicate whether each of the following statements is *true* or *false*. If a statement is false, change it to make it true. (Such a rewriting should involve more than merely converting the statement to the negative of itself.)

- **(a)** A molecule of an element may be homoatomic or heteroatomic depending on which element is involved.
- **(b)** The limit of chemical subdivision for a molecular compound is a molecule.
- **(c)** Heteroatomic molecules do not maintain the properties of their constituent elements.
- **(d)** Only one kind of atom may be present in a homoatomic molecule.
- **(e)** The main difference between molecules of elements and molecules of compounds is the number of atoms they contain.

Chemical Formulas (Sec. 5.4)

5.7 What is the chemical formula for each of the following molecules?

(a) (X–Q–X) **(b)** (X–Q–Z) **(c)** (Z–Q–Z) **(d)** (Z–Z)

5.8 What is the chemical formula for each of the following molecules?

(a) (X–Q) **(b)** (X–Q–Q) **(c)** (Q–Q–Q) **(d)** (X–X–Z)

5.9 On the basis of its formula, classify each of the following substances as an element or compound.

(a) $NaClO_2$ **(b)** CO
(c) S_8 **(d)** Al

5.10 On the basis of its formula, classify each of the following substances as an element or compound.

(a) AlN **(b)** CO_2
(c) Co **(d)** O_3

5.11 Write formulas for the following substances using the information given about a molecule of each substance.

(a) A molecule of vitamin A contains 20 atoms of carbon, 30 atoms of hydrogen, and 1 atom of oxygen.

(b) A molecule of sulfuric acid contains 2 atoms of hydrogen, 1 atom of sulfur, and 4 atoms of oxygen.

(c) A molecule of elemental phosphorus is tetraatomic.

(d) A molecule of hydrogen cyanide, a compound that contains hydrogen, carbon, and nitrogen, is triatomic.

5.12 Write formulas for the following substances using the information given about a molecule of each substance.

(a) A molecule of nicotine contains 10 atoms of carbon, 14 atoms of hydrogen, and 2 atoms of nitrogen.

(b) A molecule of ethyl alcohol contains 2 atoms of carbon, 6 atoms of hydrogen, and 1 atom of oxygen.

(c) A molecule of ozone (a form of oxygen) is triatomic.

(d) A formula unit of sodium hydroxide, a compound that contains sodium, oxygen, and hydrogen, is triatomic.

5.13 In each of the following pairs of formulas indicate whether the first listed formula denotes *more total atoms,* the *same number of total atoms*, or *fewer total atoms* than the second listed formula.

(a) N_2O and NO_2

(b) KNO_3 and $Ca(OH)_2$

(c) $Ba(ClO)_2$ and $Ba(ClO_2)_2$

(d) $Al_2(SO_4)_3$ and $Ba_3(PO_4)_2$

5.14 In each of the following pairs of formulas indicate whether the first listed formula denotes *more total atoms*, the *same number of total atoms*, or *fewer total atoms* than the second listed formula.

(a) HN_3 and NH_3 (b) $NaClO_3$ and $Be(CN)_2$

(c) $CaSO_4$ and $Mg(OH)_2$ (d) $Be_3(PO_4)_2$ and $Be(C_2H_3O_2)_2$

5.15 What difference in meaning, if any, is there in the formulas Cs_2 and CS_2?

5.16 What difference in meaning, if any, is there in the formulas Hf and HF?

5.17 The following molecular formulas are incorrectly written. Rewrite each formula in the correct manner.

(a) H3PO4

(b) $SICL_4$ (a silicon–chlorine compound)

(c) NOO

(d) 2HO (two H atoms and two O atoms)

5.18 The following molecular formulas are incorrectly written. Rewrite each formula in the correct manner.

(a) H2CO3

(b) $ALBR_3$ (an aluminum–bromine compound)

(c) HSH

(d) $2NO_2$ (two N atoms and four O atoms)

Subatomic Particles (Sec. 5.5)

5.19 Match the terms *proton, neutron*, and *electron* to each of the following subatomic particle descriptions. It is possible that more than one answer may apply in a given situation.

(a) possesses a negative charge

(b) has a mass slightly less than that of a neutron

(c) can be called a nucleon

(d) is the heaviest of the three particles

5.20 Match the terms *proton, neutron*, and *electron* to each of the following subatomic particle descriptions. It is possible that more than one answer may apply in a given situation.

(a) has no charge

(b) has a charge equal to but opposite in sign to that of an electron

(c) is not found in the nucleus

(d) has a positive charge

5.21 Indicate whether each of the following statements about the nucleus of an atom is *true* or *false*.

(a) The nucleus of an atom is neutral.

(b) The nucleus of an atom contains only neutrons.

(c) The number of nucleons present in the nucleus is always equal to the number of electrons present outside the nucleus.

(d) The nucleus accounts for almost all of the mass of an atom.

5.22 Indicate whether each of the following statements about the nucleus of an atom is *true* or *false*.

(a) The nucleus accounts for almost all the volume of an atom.

(b) The nucleus can be positively or negatively charged, depending on the identity of the atom.

(c) The nucleus of an atom contains an equal number of protons, neutrons, and electrons.

(d) The nucleus of an atom is always positively charged.

Evidence Supporting the Existence of Subatomic Particles (Sec. 5.6)

5.23 Indicate whether each of the following statements about discharge tube and metal foil experiments is *true* or *false*.

(a) Information obtained from discharge tube experiments led to the characterization of both protons and electons.

(b) Canal rays are negatively charged particles produced in a discharge tube.

(c) Metal foil experiments led to the discovery of neutrons.

(d) Many different types of cathode rays are known.

5.24 Indicate whether each of the following statements about discharge tube and metal foil experiments is *true* or *false*.

(a) In metal foil experiments almost all of the bombarding particles were stopped by the metal foil.

(b) Metal foil experiments led to the concept that an atom has a nucleus.

(c) Cathode rays and canal rays move in opposite directions in a discharge tube.

(d) Many different types of canal rays have been observed but only one type of cathode ray is known.

Atomic Number and Mass Number (Sec. 5.7)

5.25 What is the value of the atomic number for each of the following elements?
(a) tin **(b)** silver **(c)** mercury **(d)** lawrencium

5.26 What is the value of the atomic number for each of the following elements?
(a) lead **(b)** beryllium **(c)** calcium **(d)** lutetium

5.27 For each of the following atoms specify the atomic number and the mass number.
(a) $^{53}_{24}Cr$ **(b)** $^{103}_{44}Ru$ **(c)** $^{256}_{101}Md$ **(d)** $^{34}_{16}S$

5.28 For each of the following atoms specify the atomic number and the mass number.
(a) $^{67}_{30}Zn$ **(b)** $^{9}_{4}Be$ **(c)** $^{40}_{20}Ca$ **(d)** $^{3}_{1}H$

5.29 What information about the subatomic makeup of an atom is given by the following?
(a) atomic number
(b) mass number − atomic number

5.30 What information about the subatomic makeup of an atom is given by the following?
(a) mass number
(b) mass number + atomic number

5.31 What is the complete symbol ($^{A}_{Z}E$) for atoms composed of the following sets of subatomic particles?
(a) 5 protons, 5 electrons, and 6 neutrons
(b) 8 protons, 8 electrons, and 8 neutrons
(c) 13 protons, 13 electrons, and 14 neutrons
(d) 18 protons, 18 electrons, and 22 neutrons

5.32 What is the complete symbol ($^{A}_{Z}E$) for atoms composed of the following sets of subatomic particles?
(a) 4 protons, 4 electrons, and 5 neutrons
(b) 7 protons, 7 electrons, and 8 neutrons
(c) 15 protons, 15 electrons, and 16 neutrons
(d) 20 protons, 20 electrons, and 28 neutrons

5.33 Determine the number of protons, electrons, and neutrons in each of the following atoms.
(a) $^{32}_{16}S$ **(b)** $^{63}_{29}Cu$ **(c)** $^{50}_{22}Ti$ **(d)** $^{238}_{92}U$

5.34 Determine the number of protons, electrons, and neutrons in each of the following atoms.
(a) $^{35}_{17}Cl$ **(b)** $^{55}_{25}Mn$ **(c)** $^{127}_{53}I$ **(d)** $^{209}_{83}Bi$

5.35 Determine the number of protons, electrons, and neutrons present in atoms with the following characteristics.
(a) atomic number = 27 and mass number = 59
(b) mass number = 103 and Z = 45
(c) Z = 69 and A = 169
(d) atomic number = 9 and A = 19

5.36 Determine the number of protons, electrons, and neutrons present in atoms with the following characteristics.
(a) A = 103 and atomic number = 44
(b) Z = 41 and A = 93
(c) mass number = 59 and atomic number = 27
(d) Z = 59 and mass number = 141

5.37 Write complete symbols ($^{A}_{Z}E$) for atoms with the following characteristics.
(a) contains 15 electrons and 16 neutrons
(b) oxygen atom with 10 neutrons
(c) chromium atom with a mass number of 54
(d) gold atom that contains 276 subatomic particles

5.38 Write complete symbols ($^{A}_{Z}E$) for atoms with the following characteristics.
(a) contains 20 electrons and 24 neutrons
(b) radon atom with a mass number of 211
(c) silver atom that contains 157 subatomic particles
(d) beryllium atom that contains 9 nucleons

5.39 Characterize each of the following pairs of atoms as containing: (1) the same number of neutrons; (2) the same number of electrons; or (3) the same total number of subatomic particles.
(a) $^{40}_{20}Ca$ and $^{41}_{19}K$ **(b)** $^{30}_{14}Si$ and $^{32}_{16}S$
(c) $^{23}_{11}Na$ and $^{24}_{12}Mg$ **(d)** $^{4}_{2}He$ and $^{6}_{3}Li$

5.40 Characterize each of the following pairs of atoms as containing: (1) the same number of neutrons; (2) the same number of electrons; or (3) the same total number of subatomic particles.
(a) $^{13}_{6}C$ and $^{14}_{7}N$ **(b)** $^{18}_{8}O$ and $^{19}_{9}F$
(c) $^{37}_{17}Cl$ and $^{36}_{18}Ar$ **(d)** $^{6}_{3}Li$ and $^{9}_{4}Be$

Isotopes and Isobars (Sec. 5.8)

5.41. Write complete symbols ($^{A}_{Z}E$) for the four naturally occurring isotopes of iron whose mass numbers are 54, 56, 57, and 58.

5.42 Write complete symbols ($^{A}_{Z}E$) for the four naturally occurring isotopes of chromium whose mass numbers are 50, 52, 53, and 54.

5.43 Five naturally occurring isotopes of the element zirconium exist. Knowing that the heaviest isotope has a mass number of 96 and that the other isotopes have, respectively, 2, 4, 5, and 6 fewer neutrons, write the complete symbol ($^{A}_{Z}E$) for each of the five isotopes.

5.44 Four naturally occurring isotopes of the element strontium exist. Knowing that the lightest isotope has a mass number of 84 and that the other isotopes have, respectively, 2, 4, and 5 more neutrons, write the complete symbol ($^{A}_{Z}E$) for each of the four isotopes.

5.45 Indicate whether the members of each of the following pairs of atoms are isotopes, isobars, or neither.

(a) $^{24}_{12}X$ and $^{26}_{12}Q$ **(b)** $^{71}_{31}X$ and $^{71}_{32}Q$

(c) $^{56}_{25}X$ and $^{54}_{24}Q$ **(d)** $^{57}_{27}X$ and $^{60}_{27}Q$

5.46 Indicate whether the members of each of the following pairs of atoms are isotopes, isobars, or neither.

(a) $^{64}_{30}X$ and $^{64}_{29}Q$ **(b)** $^{20}_{10}X$ and $^{22}_{10}Q$

(c) $^{36}_{16}X$ and $^{35}_{17}Q$ **(d)** $^{60}_{28}X$ and $^{60}_{26}Q$

5.47 Indicate whether the members of each of the following pairs of atoms, specified in terms of subatomic particle composition, are isotopes, isobars, or neither.

(a) (24p, 24e, 26n) and (24p, 24e, 28n)

(b) (24p, 24e, 28n) and (25p, 25e, 27n)

(c) (24p, 24e, 26n) and (25p, 25e, 26n)

(d) (24p, 24e, 26n) and (24p, 24e, 24n)

5.48 Indicate whether the members of each of the following pairs of atoms, specified in terms of subatomic particle composition, are isotopes, isobars, or neither.

(a) (30p, 30e, 39n) and (31p, 31e, 38n)

(b) (30p, 30e, 35n) and (29p, 29e, 36n)

(c) (30p, 30e, 34n) and (30p, 30e, 36n)

(d) (30p, 30e, 38n) and (32p, 32e, 38n)

5.49 Isobars with a mass number of 40 exist for Ar, K, and Ca. Write the complete symbol ($^{A}_{Z}E$) for each of these isobars.

5.50 Isobars with a mass number of 50 exist for Ti, V, and Cr. Write the complete symbol ($^{A}_{Z}E$) for each of these isobars.

5.51 For which elements in the atomic number sequence 1–20

(a) are all naturally occurring atoms of the element identical?

(b) are there exactly three naturally occurring stable isotopes?

(c) is the heaviest naturally occurring isotope the most abundant of the naturally occurring isotopes?

(d) do naturally occurring atoms exist that have fewer neutrons than protons?

5.52 For which elements in the atomic number sequence 1–20

(a) are there exactly two naturally occurring isotopes?

(b) are there more than three naturally occurring isotopes?

(c) is the lightest naturally occurring isotope the most abundant of the naturally occurring isotopes?

(d) do naturally occurring atoms exist that have an equal number of protons, electrons, and neutrons?

Atomic Masses (Sec. 5.9)

5.53 What is the value of the atomic mass for each of the following elements?

(a) sulfur **(b)** nitrogen **(c)** gold **(d)** iron

5.54 What is the value of the atomic mass for each of the following elements?

(a) phosphorus **(b)** iodine **(c)** silver **(d)** nickel

5.55 Construct a relative mass scale for the hypothetical atoms Q, X, and Z, given that atoms of Q are two times as heavy as those of X, atoms of X are two times as heavy as atoms of Z, and the reference point for the scale is X = 4.00 bebs.

5.56 Construct a relative mass scale for the hypothetical atoms Q, X, and Z, given that atoms of Q are three times as heavy as those of X, atoms of X are four times as heavy as atoms of Z, and the reference point for the scale is Z = 2.50 bobs.

5.57 The atoms of element Z each have an average mass three-fourths that of a $^{12}_{6}C$ atom. Another element, X, has atoms whose average mass is three times the mass of Z atoms. A third element, Q, has atoms with an average mass nine times that of $^{12}_{6}C$.

(a) Construct a relative mass scale, based on $^{12}_{6}C$ having a mass of 12 amu, for the elements Z, X, and Q.

(b) Based on atomic masses rounded off to whole numbers, what is the identity of elements Z, X, and Q?

5.58 The atoms of element Z each have an average mass one-third that of a $^{12}_{6}C$ atom. Another element, X, has atoms whose average mass is four times the mass of Z atoms. A third element, Q, has atoms whose average mass is twice the mass of X atoms.

(a) Construct a relative mass scale, based on $^{12}_{6}C$ having a mass of 12 amu, for the elements Z, X, and Q.

(b) Based on atomic masses rounded off to whole numbers, what is the identity of elements Z, X, and Q?

5.59 A football team has the following distribution of players: 22.0% are defensive linemen with an average mass of 271 lb, 19.0% are defensive backs with an average mass of 175 lb, 26.0% are offensive linemen with an average mass of 263 lb, 15.0% are offensive backs with an average mass of 182 lb, and 19.0% are specialty team members with an average mass of 191 lb. What is the average mass of a football player on this team?

5.60 The assortment of automobiles on a used car lot is categorized as follows: 18.0% are 1 year old, 10.0% are 2 years old, 33.0% are 3 years old, 4.0% are 4 years old, 31.0% are 5 years old, and 4.0% are 6 years old. What is the average age of a car on this used car lot?

5.61 Naturally occurring chlorine consists of two isotopes with the atomic masses and abundances given in Table 5.2. Using these data, calculate the atomic mass of chlorine.

5.62 Naturally occurring copper consists of two isotopes with the atomic masses and abundances given in Table 5.2. Using these data, calculate the atomic mass of copper.

5.63 Naturally occurring titanium consists of five isotopes with the atomic masses and abundances given in Table 5.2. Using these data, calculate the atomic mass of titanium.

5.64 Naturally occurring sulfur consists of four isotopes with the atomic masses and abundances given in Table 5.2. Using these data, calculate the atomic mass of sulfur.

5.65 How many times heavier, on the average, is
(a) an atom of copper than an atom of lithium?
(b) an atom of gold than an atom of beryllium?

5.66 How many times heavier, on the average, is
(a) an atom of bismuth than an atom of iodine?
(b) an atom of silver than an atom of hydrogen?

5.67 The arbitrary standard for the atomic mass scale is the exact number 12 for the mass of $^{12}_{6}C$. Why, then, is the atomic mass of carbon listed as 12.011?

5.68 The atomic mass of fluorine is 18.9984 amu and that of copper is 63.546 amu. All fluorine atoms have a mass of 18.9984 amu, and not a single copper atom has a mass of 63.546 amu. Explain.

ADDITIONAL PROBLEMS

5.69 Based on the given information, determine the numerical value of the subscript x in each of the following chemical formulas.
(a) $Na_2S_xO_3$; formula unit contains 7 atoms
(b) $Ba(ClO_x)_2$; formula unit contains 9 atoms
(c) $Na_xP_xO_{10}$; formula unit contains 16 atoms
(d) $C_xH_{2x}O_x$; formula unit contains 24 atoms

5.70 Based on the given information, determine the numerical value of the subscript x in each of the following chemical formulas.
(a) BaS_2O_x; formula unit contains 6 atoms
(b) $Al_2(SO_x)_3$; formula unit contains 17 atoms
(c) SO_xCl_x; formula unit contains 5 atoms
(d) $C_xH_{2x}Cl_x$; formula unit contains 8 atoms

5.71 How many protons are present in seven molecules of the compound $C_6H_{12}O_6$ (glucose, blood sugar)?

5.72 How many electrons are present in nine molecules of the compound $C_{12}H_{22}O_{11}$ (table sugar)?

5.73 A mixture contains the following five pure substances: O_2, N_2O, H_2O, CCl_4, and CH_2Br_2.
(a) How many different kinds of heteroatomic molecules are present in the mixture?
(b) How many different kinds of pentaatomic molecules are present in the mixture?
(c) How many different compounds are present in the mixture?
(d) How many different kinds of atoms are present in the mixture?
(e) How many total atoms are present in a mixture sample containing three molecules of each component?

5.74 A mixture contains the following five pure substances: N_2, N_2H_4, NH_3, CH_4, and CH_3Cl.
(a) How many different kinds of molecules are present in the mixture that contain four or fewer atoms?
(b) How many different kinds of homoatomic molecules are present in the mixture?
(c) How many different kinds of atoms are present in the mixture?
(d) How many total atoms are present in a mixture sample containing five molecules of each component?
(e) How many total hydrogen atoms are present in a mixture sample containing four molecules of each component?

5.75 Indicate whether each of the following statements concerning sodium isotopes is *true* or *false*.
(a) $^{23}_{11}Na$ has one more electron than does $^{24}_{11}Na$.
(b) $^{23}_{11}Na$ and $^{24}_{11}Na$ contain the same number of neutrons.
(c) $^{23}_{11}Na$ has one fewer subatomic particle than $^{24}_{11}Na$.
(d) $^{23}_{11}Na$ and $^{24}_{11}Na$ have the same atomic number.

5.76 Indicate whether each of the following statements concerning magnesium isotopes is *true* or *false*.
(a) $^{24}_{12}Mg$ has one more proton than $^{25}_{12}Mg$.
(b) $^{24}_{12}Mg$ and $^{25}_{12}Mg$ contains the same number of subatomic particles in their nucleus.
(c) $^{24}_{12}Mg$ has one fewer neutron than $^{25}_{12}Mg$.
(d) $^{24}_{12}Mg$ and $^{25}_{12}Mg$ have different mass numbers.

5.77 Arrange the five isotopes $^{42}_{20}Ca$, $^{39}_{19}K$, $^{44}_{21}Sc$, $^{37}_{18}Ar$, and $^{43}_{22}Ti$ in order of
(a) increasing number of electrons.
(b) decreasing number of neutrons.
(c) increasing number of protons.
(d) decreasing mass.

5.78 Arrange the five isotopes $^{92}_{40}Zr$, $^{89}_{39}Y$, $^{95}_{41}Nb$, $^{87}_{38}Sr$, and $^{93}_{42}Mo$ in order of
(a) decreasing number of electrons.
(b) increasing number of neutrons.
(c) increasing number of nucleons.
(d) decreasing number of subatomic particles.

5.79 Write the complete symbol for the isotope of boron with each of the following characteristics.

(a) contains two fewer neutrons than $^{10}_{5}B$

(b) contains three more subatomic particles than $^{9}_{5}B$

(c) contains the same number of neutrons as $^{14}_{7}N$

(d) contains the same number of subatomic particles as $^{14}_{7}N$

5.80 Write the complete symbol for the isotope of chromium with each of the following characteristics.

(a) contains two more neutrons than $^{55}_{24}Cr$

(b) contains two fewer subatomic particles than $^{52}_{24}Cr$

(c) contains the same number of neutrons as $^{60}_{29}Cu$

(d) contains the same number of subatomic particles as $^{60}_{29}Cu$

5.81 Copper consists of two naturally occurring isotopes with masses of 62.9298 amu and 64.9278 amu.

(a) How many protons are in the nucleus of each isotope?

(b) How many electrons are in an atom of each isotope?

(c) How many neutrons are in the nucleus of each isotope?

5.82 Silver consists of two naturally occurring isotopes with masses of 106.9041 amu and 108.9047 amu

(a) How many protons are in the nucleus of each isotope?

(b) How many electrons are in an atom of each isotope?

(c) How many neutrons are in the nucleus of each isotope?

5.83 Suppose it was decided to redefine the atomic mass scale by choosing as an arbitrary reference point a value of 20.000 amu to represent the naturally occurring mixture of nickel isotopes. What would be the atomic mass of the following elements on the new atomic mass scale?

(a) silver **(b)** gold

5.84 Suppose it was decided to redefine the atomic mass scale by choosing as an arbitrary reference point a value of 40.000 amu to represent the mass of fluorine atoms. (All fluorine atoms are identical; that is, fluorine is monoisotopic.) What would be the atomic mass of the following elements on the new atomic mass scale?

(a) sulfur **(b)** platinum

5.85 In a hypothetical molecule XZ_2 43.2% of the mass is from X and the rest from Z. If a relative mass scale were established with the mass of Z assigned a value of exactly 80, what would be the relative atomic mass of X?

5.86 In a hypothetical molecule Q_2Z, 35.1% of the mass is from Q and the rest from Z. If a relative mass scale were established with the mass of Q assigned a value of exactly 40, what would be the relative atomic mass of Z?

5.87 The following isotopic mass ratios were determined experimentally:

$^{19}_{9}F/^{12}_{6}C = 1.5832$ $^{35}_{17}Cl/^{19}_{9}F = 1.8406$ $^{81}_{35}Br/^{35}_{17}Cl = 2.3140$

Based on this information, what is the mass of $^{81}_{35}Br$ in atomic mass units?

5.88 The following isotopic mass ratios were determined experimentally:

$^{17}_{8}O/^{12}_{6}C = 1.4166$ $^{32}_{16}S/^{17}_{8}O = 1.8808$ $^{80}_{34}Se/^{32}_{16}S = 2.4996$

Based on this information, what is the mass of $^{80}_{34}Se$ in atomic mass units?

5.89 Chlorine has two isotopes with mass numbers of 35 and 37. Combination of chlorine with bromine produces BrCl. An examination of the product BrCl molecules shows the presence of only three isotopically different types of molecules with approximate masses of 114, 116, and 118 amu. Based on this information, how many isotopes of Br exist and what are the mass numbers of the isotopes?

5.90 Chlorine has two isotopes with mass numbers of 35 and 37. Combination of copper with chlorine produces CuCl. An examination of the product CuCl molecules shows the presence of only three isotopically different types of molecules with approximate masses of 98, 100, and 102 amu. Based on this information, how many isotopes of Cu exist and what are the mass numbers of the isotopes?

5.91 Three naturally occurring isotopes of potassium exist: $^{39}_{19}K$, $^{40}_{19}K$, and $^{41}_{19}K$. The atomic mass of potassium is 39.102 amu. Which of the three potassium isotopes is most abundant? Explain your answer.

5.92 Naturally occurring boron is a mixture of two isotopes: $^{10}_{5}B$ and $^{11}_{5}B$. Given that the atomic mass of boron is 10.811 amu, estimate the abundances of the two isotopes to the nearest 10%.

5.93 Using the data found in Table 5.2, calculate how many $^{17}_{8}O$ and $^{18}_{8}O$ atoms you would find in an oxygen sample containing 1 million (1.0×10^6) atoms.

5.94 Using the data found in Table 5.2, calculate how many $^{234}_{92}U$ and $^{235}_{92}U$ atoms you would find in a uranium sample containing 1 million (1.0×10^6) atoms.

Cumulaltive Problems

5.95 Assign each of the following descriptions of matter to one of the following categories: *element*, *compound*, or *mixture*.

(a) one substance present, two phases present, all molecules are heteroatomic

(b) two substances present, one phase present, all molecules are homoatomic

(c) one phase present, all molecules are homoatomic, all molecules are identical

(d) one phase present, both homoatomic and heteroatomic molecules present

5.96 Assign each of the following descriptions of matter to one of the following categories: *element*, *compound*, or *mixture*.

(a) one substance present, one phase present, one kind of homoatomic molecule present

(b) two substances present, two phases present, all molecules are heteroatomic

(c) one phase present, two kinds of homoatomic molecules present

(d) one phase present, all molecules are triatomic, all molecules are heteroatomic, all molecules are identical

---

5.97 The density of gold is 19.3 g/cm^3, and the mass of a single gold atom is 3.27×10^{-22} g. How many gold atoms are present in a piece of gold whose volume is 3.22 cm^3?

5.98 The density of copper is 8.93 g/cm^3, and the mass of a single copper atom is 1.06×10^{-22} g. How many copper atoms are present in a copper bar whose dimensions are 2.00 cm $\times$ 3.00 cm $\times$ 5.00 cm?

5.99 In 1.00 g of fluorine atoms there are 3.17×10^{22} fluorine atoms. If you lined these atoms up side by side, how many miles long would the line of fluorine atoms be? The diameter of a fluorine atom is 1.44×10^{-8} cm.

5.100 In 1.00 g of fluorine atoms there are 3.17×10^{22} fluorine atoms. If you started counting these atoms at the rate of 10 per second, how many years would it take to count all the atoms in the 1.00-g sample?

5.101 An electron has a mass of 5.5×10^{-4} amu. What percent of the total mass of a Pb atom with a mass of 207 amu is due to its electrons?

5.102 How many electrons, with a mass of 5.5×10^{-4} amu, would it take to equal the mass of one proton, which has a mass of 1.0073 amu?

5.103 The diameter of an atom is approximately 10^5 times as large as the diameter of the nucleus. If the nucleus were enlarged to the size of a Ping-Pong ball (1.5 in. in diameter), what would be the diameter of the atom, in miles?

5.104 The diameter of an atom is approximately 10^{-8} cm, and that of the nucleus is 10^{-13} cm. How many times larger is the volume of the atom than the volume of the nucleus? The formula for the volume of a sphere is $V = 0.524 \times d^3$ (where d is the diameter of the sphere).

---

6 Electronic Structure and Chemical Periodicity

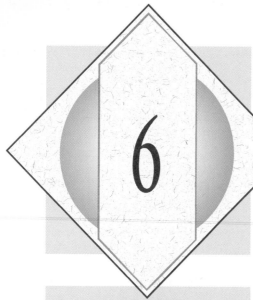

6.1 The Periodic Law

During the early part of the nineteenth century, an abundance of chemical facts became available from detailed studies of the then known elements. With the hope of providing a systematic approach to the study of chemistry, scientists began to look for some form of order in the increasing amount of chemical information. They were encouraged in their search by the unexplained but well-known fact that certain groups of elements had very similar properties. Numerous attempts were made to explain these similarities and to use them as a means for arranging or classifying the elements.

In 1869, these efforts culminated in the discovery of what is now called the *periodic law*. The **periodic law** *states that when elements are arranged in order of increasing atomic number, elements with similar properties occur at periodic (regularly recurring) intervals*. Proposed independently by both the Russian chemist Dmitri Ivanovich Mendeleev (1834–1907) and the German chemist Julius Lothar Meyer (1830–1895), the periodic law is one of the most important of all chemical laws.

The preceding statement of the periodic law is in modern-day language. It differs from the original 1869 statements in that the phrase *atomic number* has replaced "atomic mass." The use of "atomic mass" in the original statements reflected theories prevalent in 1869. According to these theories, the masses of atoms were their most important distinguishing properties—a knowledge of subatomic particles (Sec. 5.5) was still 30 years away. When the details of subatomic structure were finally discovered, it became obvious that the properties of atoms were related not to their masses but to the number and arrangement of their electrons. Thus, the periodic law was modified to reflect this new knowledge.

Figure 6.1 shows parts of the repeating pattern for chemical properties (periodic law) for the sequence of elements with atomic numbers 3 through 20. The elements within similar geometric symbols (circles and

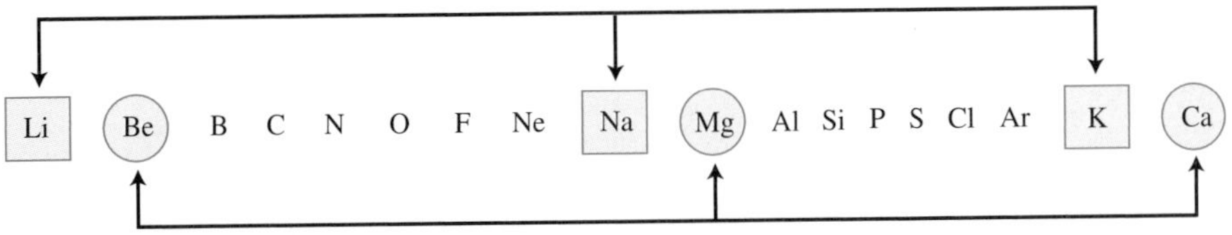

FIGURE 6.1 An illustration of the periodicity in properties that occurs when elements are arranged in order of increasing atomic number. Li, Na, and K have similar chemical properties, as do Be, Mg, and Ca.

squares) have similar chemical properties. For the sake of simplicity, only two of the periodic relationships (repeating patterns) are shown; for the elements listed, similar properties are found in every eighth element.

The concept of a periodic variation or pattern should not be new to you, as numerous everyday examples of this exist. Most, but not all, involve time. Mondays occur at regular intervals of 7 days. Office workers get paid at regular intervals, such as every two weeks. The red–green–yellow light sequence for a traffic semaphore repeats itself in a periodic fashion.

To be useful, the relationships generated by the periodic law must be easily visualized. This is the purpose of periodic tables, the subject of Section 6.2.

6.2 THE PERIODIC TABLE

A **periodic table** *is a graphical representation of the behavior described by the periodic law*. In this table, the elements are arranged according to increasing atomic number in such a way that the similarities predicted by the periodic law become readily apparent.

The most commonly used form of the periodic table is the one shown in Figure 6.2 (and also inside the front cover of this book). It is an arrangement of the elements in which vertical columns contain elements with similar chemical properties. Note how Li, Na, and K—elements with similar chemical properties (Fig. 6.1)—are all in the same vertical column in Figure 6.2. Other elements with properties similar to those of these three elements are H, Rb, Cs, and Fr.

Within the periodic table each element is represented by a rectangular box. Within each box are given the symbol, atomic number, and atomic mass of the element, as shown in Figure 6.3. It should be noted that some periodic tables, but not the ones in this text, also give the element's name within the box.

Special chemical terminology exists for specifying the position (location) of an element within the periodic table. This terminology involves the use of the words *group* and *period*. A **group** *in the periodic table is a vertical column of elements*. There are two notations in use for designating individual periodic table groups. In the first notation, which has been in use for many years, groups are designated by Roman numerals and the letters A and B. The second notation, which has been recently recommended for use by an international scientific commission, uses the Arabic numbers 1 through 18. Notice that both group notations are given at the top of each group in the periodic table of Figure 6.2. The elements with atomic numbers 8, 16, 34, 52, and 84 (O, S, Se, Te, and Po) constitute group VIA (old notation) or group 16 (new notation). Because it may be some time before the new group numbering system is widely accepted, we will use the old notation.

A **period** *in the periodic table is a horizontal row of elements*. For identification purposes, the periods are numbered sequentially, with Arabic numbers, starting at the top of the periodic table. (These period numbers are not explicitly shown on the periodic table.) Period

1 Group IA	2 Group IIA	3 Group IIIB	4 Group IVB	5 Group VB	6 Group VIB	7 Group VIIB	8 Group ←	9 Group VIIIB	10 Group →	11 Group IB	12 Group IIB	13 Group IIIA	14 Group IVA	15 Group VA	16 Group VIA	17 Group VIIA	18 Group VIIIA
1 **H** 1.00794																	2 **He** 4.002602
3 **Li** 6.941	4 **Be** 9.01218											5 **B** 10.811	6 **C** 12.011	7 **N** 14.0067	8 **O** 15.9994	9 **F** 18.998403	10 **Ne** 20.179
11 **Na** 22.98977	12 **Mg** 24.305											13 **Al** 26.98154	14 **Si** 28.0855	15 **P** 30.97376	16 **S** 32.066	17 **Cl** 35.453	18 **Ar** 39.948
19 **K** 39.0983	20 **Ca** 40.078	21 **Sc** 44.9559	22 **Ti** 47.867	23 **V** 50.9415	24 **Cr** 51.9961	25 **Mn** 54.9380	26 **Fe** 55.845	27 **Co** 58.9332	28 **Ni** 58.6934	29 **Cu** 63.546	30 **Zn** 65.38	31 **Ga** 69.723	32 **Ge** 72.59	33 **As** 74.9216	34 **Se** 78.96	35 **Br** 79.904	36 **Kr** 83.80
37 **Rb** 85.4678	38 **Sr** 87.62	39 **Y** 88.9059	40 **Zr** 91.22	41 **Nb** 92.9064	42 **Mo** 95.94	43 **Tc** (98)	44 **Ru** 101.07	45 **Rh** 102.9055	46 **Pd** 106.42	47 **Ag** 107.8682	48 **Cd** 112.41	49 **In** 114.818	50 **Sn** 118.710	51 **Sb** 121.760	52 **Te** 127.60	53 **I** 126.9045	54 **Xe** 131.29
55 **Cs** 132.9054	56 **Ba** 137.33	57 **La** 138.9055	72 **Hf** 178.49	73 **Ta** 180.9479	74 **W** 183.84	75 **Re** 186.207	76 **Os** 190.23	77 **Ir** 192.217	78 **Pt** 195.08	79 **Au** 196.9665	80 **Hg** 200.59	81 **Tl** 204.383	82 **Pb** 207.2	83 **Bi** 208.9804	84 **Po** (209)	85 **At** (210)	86 **Rn** (222)
87 **Fr** (223)	88 **Ra** 226.0254	89 **Ac** 227.0278	104 **Rf** (261)	105 **Ha** (262)	106 **Sg** (266)	107 **Ns** (262)	108 **Hs** (265)	109 **Mt** (266)	110 (271)	111 (272)							

Metals ← → Nonmetals

58 **Ce** 140.12	59 **Pr** 140.9077	60 **Nd** 144.24	61 **Pm** (145)	62 **Sm** 150.36	63 **Eu** 151.96	64 **Gd** 157.25	65 **Tb** 158.9254	66 **Dy** 162.50	67 **Ho** 164.9304	68 **Er** 167.26	69 **Tm** 168.9342	70 **Yb** 173.04	71 **Lu** 174.967
90 **Th** 232.0381	91 **Pa** 231.0359	92 **U** 238.0289	93 **Np** 237.0482	94 **Pu** (244)	95 **Am** (243)	96 **Cm** (247)	97 **Bk** (247)	98 **Cf** (251)	99 **Es** (252)	100 **Fm** (257)	101 **Md** (260)	102 **No** (259)	103 **Lr** (262)

FIGURE 6.2 The most commonly used form of the periodic table.

3 is the third row of elements, period 4 the fourth row of elements, and so forth. The elements Na, Mg, Al, Si, P, S, Cl, and Ar are all members of period 3 (see Fig. 6.2). Period 1 has only two elements—H and He.

The location of any element in the periodic table is specified by giving its group number and its period number. The element gold (Au), with an atomic number of 79, belongs to group IB and is in period 6. Nitrogen (N), with an atomic number of 7, belongs to group VA and is in period 2.

There is one area in the table where the practice of arranging the elements according to increasing atomic number seems to be violated. This is the area of elements 57 and 89, which are both in group IIIB. Shown next to them, in group IVB, are elements 72 and 104, respectively. The missing elements, 58–71 and 90–103, are located in two rows at the bottom of the periodic table. Technically, these elements should be included in the body of the table, as shown in Figure 6.4a. However, to have a more compact table, they are placed in the position shown in Figure 6.4b. This arrangement should present no problems

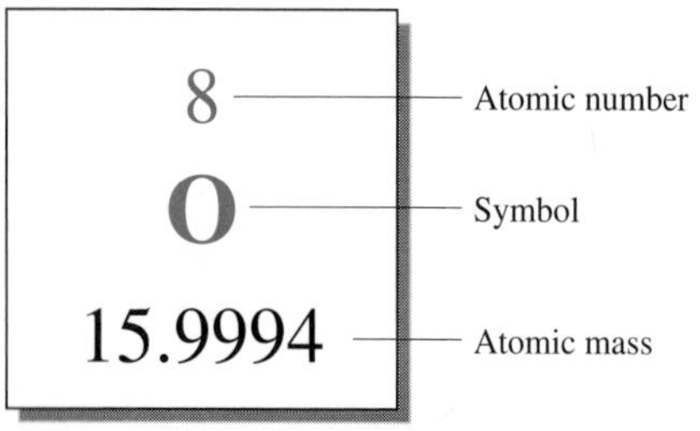

FIGURE 6.3 Arrangement of information about elements within the "boxes" of the periodic table.

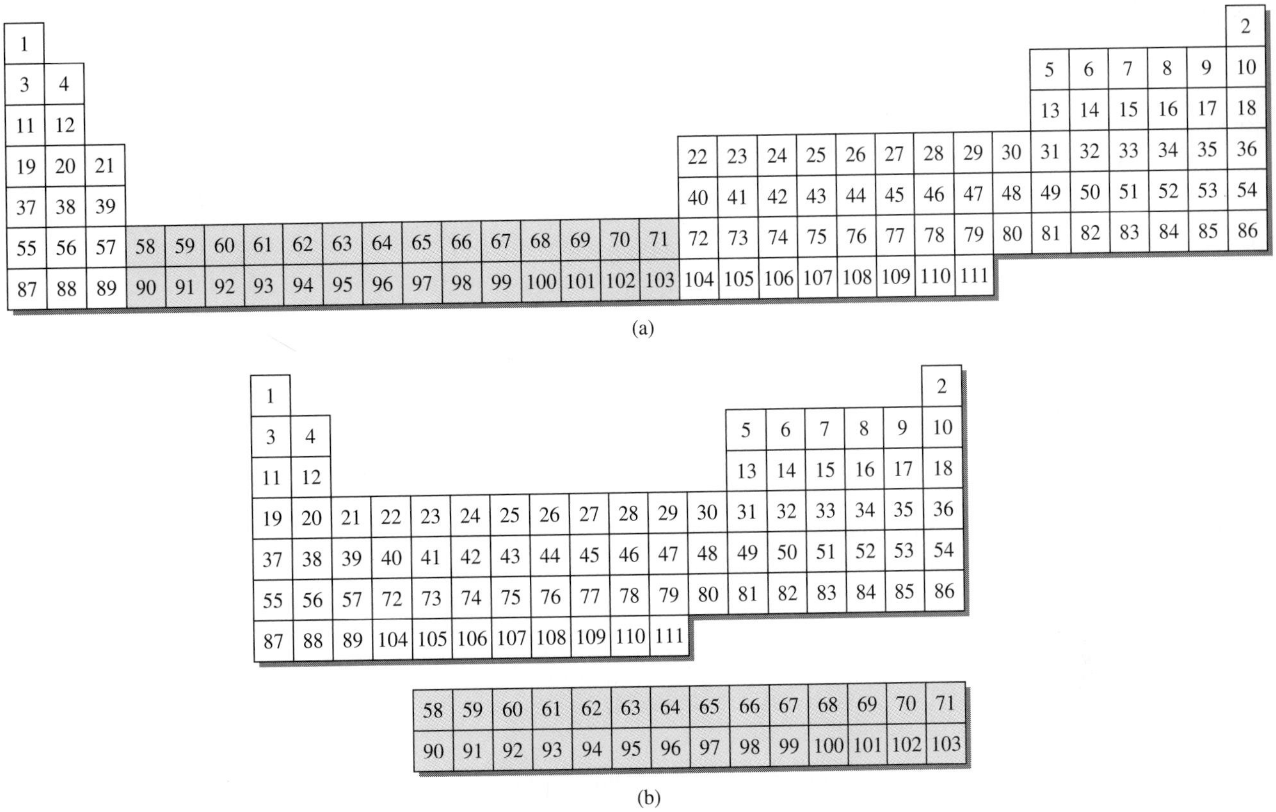

FIGURE 6.4 "Long" and "short" forms of the periodic table: (a) periodic table with elements 58–71 and 90–103 (in color) in their proper positions—the "long" form; (b) periodic table modified to conserve space by placing elements 58–71 and 90–103 (in color) below the rest of the elements—the "short" form.

to the user of the periodic table as long as it is recognized for what it is—a space-saving device.

Students often assume that a corollary of the fact that the elements are arranged in the periodic table in order of increasing atomic number is that they are also arranged in order of increasing atomic mass. This latter conclusion is correct most of the time; however, there are exceptions.

The following element sequences are exceptions to the generalization that atomic mass increases with increasing atomic number.

${}_{18}Ar$	39.95 amu	${}_{27}Co$	58.93 amu	${}_{52}Te$	127.60 amu
${}_{19}K$	39.10 amu	${}_{28}Ni$	58.69 amu	${}_{53}I$	126.90 amu
${}_{20}Ca$	40.08 amu	${}_{29}Cu$	63.55 amu	${}_{54}Xe$	131.29 amu

The inversions in atomic mass result from isotopic abundances (Sec. 5.8). Let us consider such data for the Ar–K–Ca triad.

${}^{36}Ar$	0.34%	${}^{39}K$	93.10%	${}^{40}Ca$	96.97%
${}^{38}Ar$	0.07%	${}^{40}K$	0.01%	${}^{42}Ca$	0.64%
${}^{40}Ar$	99.60%	${}^{41}K$	6.88%	${}^{43}Ca$	0.14%
				${}^{44}Ca$	2.06%
				${}^{46}Ca$	0.003%
				${}^{48}Ca$	0.18%

For argon the heaviest isotope ($A = 40$) is most abundant, and for potassium the lightest isotope ($A = 39$) is most abundant; this results in potassium having a lower atomic mass than argon. The atomic masses of argon and calcium are almost equal (39.95 and 40.08 amu, respectively); the dominant isotope for both elements has a mass number of 40.

Later in this chapter we will find that a periodic table conveys much more information about the elements than the symbols, atomic numbers, and atomic masses that are printed on it. Information about the arrangement of electrons in an atom is "coded" into the table, as is information concerning some physical and chemical property trends. Indeed, the periodic table is considered to be the single most useful study aid available for organizing information about the elements.

For many years after the formulation of the periodic law and the periodic table, both were considered to be empirical. The law worked and the table was very useful, but there was no explanation available for the law or for why the periodic table had the shape it had. It is now known that the theoretical basis for both the periodic law and the periodic table involves the arrangement of electrons in atoms. The properties of the elements repeat themselves in a periodic manner because the arrangement of electrons about the nucleus of an atom follows a periodic "pattern." Electron arrangements and an explanation of the periodic law and periodic table in terms of electronic theory are the subject matter for most of the remainder of this chapter.

6.3 THE ENERGY OF AN ELECTRON

In Section 5.5 we learned the following facts about electrons.

1. They are one of three fundamental subatomic particles.
2. They have very little mass in comparison to protons and neutrons.
3. They are located outside the nucleus of the atom.
4. They move rapidly about the nucleus in a volume that defines the size of the atom.

Much more must be known about electrons to understand the chemical behavior of the various elements, why some elements have similar chemical properties, and the theoretical basis for the periodic law.

More specific information concerning the behavior and arrangement of electrons within the extranuclear region of an atom is derived from a complex mathematical model for electron behavior called *quantum mechanics*. All of the early work on this subject was done by physicists rather than chemists.

During the early part of the twentieth century (1910–1930) a major revolution occurred in the field of physics, a revolution that profoundly affected chemistry. During this time it became clear that the "established" laws of physics, the laws that had been used for many years to predict the behavior of macroscopic objects, could not explain the behavior of extremely small objects such as atoms and electrons. The work of a number of European physicists led to this conclusion and also beyond it. These same physicists were able to formalize new laws that did apply to small objects. It is from these new laws, collectively called quantum mechanics, that information concerning the arrangement and behavior of electrons about a nucleus came.

A major force in the development of quantum mechanics was the Austrian physicist Erwin Schrödinger (1887–1961). In 1926 he showed that the laws of quantum mechanics

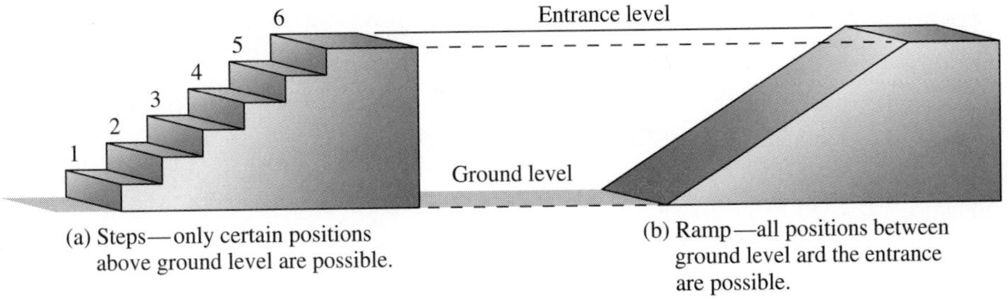
6
5
4
3
2
1
Entrance level
Ground level
(a) Steps—only certain positions above ground level are possible.
(b) Ramp—all positions between ground level ard the entrance are possible.

FIGURE 6.5 A stairway with quantized position levels versus a ramp with continuous position levels.

could be used to characterize the motion of electrons. Most of the concepts of this chapter come from solutions to equations developed by Schrödinger.

A formal discussion of quantum mechanics is beyond the scope of this course (and many other chemistry courses also) because of the rigorous mathematics involved. The answers obtained from quantum mechanics are, however, simple enough to be understood to a surprisingly large degree at the level of an introductory chemistry course. A consideration of these quantum-mechanical answers will enable us to develop a system for specifying electron arrangements around a nucleus and further to develop basic rules governing compound formation. This latter topic will occupy our attention in Chapter 7.

Present-day quantum mechanical theory describes the arrangement of an atom's electrons in terms of their energies. Indeed, the energy of an electron is the property that is most important to any consideration of its behavior about the nucleus.

The energy of an electron is manifested primarily in its velocity. The higher its energy, the higher its average velocity. The faster an electron travels, the farther it tends to move from the nucleus with which it is associated.

A most significant characteristic of an electron's energy is that it is a quantized property. A **quantized property** *is a property that can have only certain values: that is, not all values are allowed.* Since an electron's energy is quantized, an electron can have only certain specific energies.

Quantization is a phenomenon not commonly encountered in the macroscopic world. Somewhat analogous is the process of a person climbing a flight of stairs. In Figure 6.5a you see six steps between ground level and the entrance level. As a person climbs these stairs there are only six permanent positions he or she can occupy (with both feet together). Thus the person's position (height above ground level) is quantized; only certain positions are allowed. The opposite of quantized is continuousness. A person climbing a ramp up to the entrance (Fig. 6.5b) would be able to assume a continuous set of heights above ground level; all values are allowed.

The energy of an electron determines its behavior about the nucleus. Since electron energies are quantized, only certain behavior patterns are allowed. Descriptions of electron behaviors involve the use of the terms *shell*, *subshell*, and *orbital*. Sections 6.4–6.6 consider the meaning of these terms and the mathematical interrelationships between them.

F.Y.I.

The floors of a building are quantized. An elevator will stop at the 5th or the 6th floor, but not at the 5.5 floor. The frets on the neck of a guitar make it quantized because notes between the frets cannot be played. An instrument like a cello is continuous because it does not have frets; any note can be played.

6.4 ELECTRON SHELLS

It was mentioned in Section 6.3 that electrons with higher energy will tend to be found farther from the nucleus than those with lower energy. Based on energy–distance-from-the-

TABLE 6.1 IMPORTANT CHARACTERISTICS OF ELECTRON SHELLS

Shell	Number Designation (n)	Letter Designation	Electron Capacity ($2n^2$)
1st	1	K	$2 \times 1^2 = 2$
2nd	2	L	$2 \times 2^2 = 8$
3rd	3	M	$2 \times 3^2 = 18$
4th	4	N	$2 \times 4^2 = 32$
5th	5	O	$2 \times 5^2 = 50$[a]
6th	6	P	$2 \times 6^2 = 72$[a]
7th	7	Q	$2 \times 7^2 = 98$[a]

[a] The maximum number of electrons in this shell has never been attained in any element now known.

nucleus considerations, electrons can be grouped into shells or main energy levels. An **electron shell** *is an energy level in which the electrons have approximately the same energy and spend most of their time at approximately the same distance from the nucleus.*

Two different methods are used for identifying electron shells. The older method uses letters of the alphabet beginning with the letter K and then continuing sequentially. Shell K is the shell of electrons closest to the nucleus (the lowest in energy). The next closest shell is designated L; then come M, N, and so on. The more modern method designates shells by a number n, which may have the values 1, 2, 3, ... The shell with the lowest energy is assigned an n value of 1, the next higher 2, then 3, and so on. Comparing the two systems, we see that the $n = 1$ shell and the K shell are the same. Similarly, $n = 2$ and L are designations for the same shell. Only values of $n = 1$–7 (or letters K–Q) are needed at present to designate electron shells. No known atom has electrons farther from the nucleus than the seventh main energy level ($n = 7$ or letter Q).

The maximum number of electrons possible in an electron shell varies; the higher the shell energy, the more electrons the shell can accommodate. The farther electrons are from the nucleus (a higher energy shell), the greater the volume of space available for them; hence the more electrons there can be in the shell. (Conceptually, electron shells may be considered to be nested one inside another, somewhat like the layers of flavors inside a jawbreaker or similar type of candy.)

The lowest energy shell ($n = 1$) accommodates a maximum of 2 electrons. In the second, third, and fourth shells, 8, 18, and 32 electrons, respectively, are allowed. A very simple mathematical equation can be used to calculate the maximum number of electrons allowed in any given shell.

$$\text{Shell electron capacity} = 2n^2 \qquad (\text{where } n = \text{shell number})$$

For example, when $n = 4$ the value $2n^2 = 2(4^2) = 32$, which is the number previously given for the number of electrons allowed in the fourth shell. Although there is a maximum electron occupancy for each shell or main energy level, a shell may hold fewer than the allowable number of electrons in a given situation.

Table 6.1 summarizes concepts presented in this section concerning electron shells or electron main energy levels.

6.5 ELECTRON SUBSHELLS

All electrons in a shell do not have the same energy. Their energies are all close to each other in magnitude, but they are not identical. The range of energies for electrons in a shell is due to the existence of electron subshells or electron *energy sublevels*. An **electron subshell** *is an energy sublevel within an electron shell in which the electrons all have the same energy.*

The number of subshells within a shell varies. A shell contains the same number of subshells as its own shell number; that is,

$$\text{Number of subshells in a shell} = n \qquad (\text{where } n = \text{shell number})$$

Thus, each successive shell has one more subshell than the previous one. Shell 3 contains three subshells, shell 4 contains four subshells, and shell 5 contains five subshells.

Subshells are identified by a number and a letter. The number indicates the shell to which the subshell belongs. The letters are *s*, *p*, *d*, or *f* (all lowercase letters), which, in that order, denote subshells of increasing energy within a shell. The lowest energy subshell within a shell is always the *s* subshell, the next higher the *p* subshell, then the *d*, and finally the *f*. Shell 1 has only one subshell, the 1*s*. Shell 2 has two subshells, the 2*s* and 2*p*. The 3*s*, 3*p*, and 3*d* subshells are found in shell 3. Table 6.2 gives information concerning subshells for the first seven shells. Note that number-and-letter designations are not given for all the subshells in shells 5, 6, and 7; some of these subshells are not needed to describe the electron arrangements for the 111 known elements. Why they are not needed will become evident in Section 6.7.

The unusual sequence of letters (*s*, *p*, *d*, and *f*) used to denote the four subshell types has a historical origin. The early development of quantum mechanical theory (Sec. 6.3) was closely linked to the study of spectral emissions produced by atoms. The letters *s*, *p*, *d*, and *f* are derived, respectively, from old spectroscopic terminology, which describes spectral lines as *s*harp, *p*rincipal, *d*iffuse, and *f*undamental.

The maximum number of electrons that a subshell can hold varies from 2 to 14 depending on the type of subshell—*s*, *p*, *d*, or *f*. An *s* subshell can accommodate only 2 electrons. What shell the *s* subshell is located in does not affect the maximum electron occupancy figure: that is, the 1*s*, 2*s*, 3*s*, 4*s*, 5*s*, 6*s*, and 7*s* subshells all have a maximum electron occu-

TABLE 6.2 SUBSHELL ARRANGEMENTS WITHIN SHELLS

Shell Number (*n*)	Subshells						
1	1*s*						
2	2*s*	2*p*					
3	3*s*	3*p*	3*d*				
4	4*s*	4*p*	4*d*	4*f*			
5	5*s*	5*p*	5*d*	5*f*	—		
6	6*s*	6*p*	6*d*	—	—	—	
7	7*s*	—	—	—	—	—	—

TABLE 6.3 DISTRIBUTION OF ELECTRONS WITHIN SUBSHELLS

Shell	Number of Subshells within Shell	Maximum Number of Electrons within Each Subshell				Maximum Number of Electrons within Shell ($2n^2$)
		s	p	d	f	
$n = 1$ (K)	1	2				2
$n = 2$ (L)	2	2	6			8
$n = 3$ (M)	3	2	6	10		18
$n = 4$ (N)	4	2	6	10	14	32

pancy of 2. Subshells of the *p*, *d*, and *f* types can accommodate maximums of 6, 10, and 14 electrons, respectively. Again the maximum numbers of electrons in these types of subshells depend only on the subshell types and are independent of shell number. Table 6.3 summarizes the information just presented for shells 1 through 4. Notice the consistency between the numbers in columns 3 and 4 in Table 6.3. Within a shell, the sum of the subshell electron occupancies is the same as the shell electron occupancy ($2n^2$). For example, in shell 4, an *s* subshell containing 2 electrons, a *p* subshell containing 6 electrons, a *d* subshell containing 10 electrons, and an *f* subshell containing 14 electrons add up to a total of 32 electrons, which is the maximum occupancy of shell 4 as calculated by the $2n^2$ formula.

6.6 ELECTRON ORBITALS

The last and most basic of the three terms used in describing electron arrangements about nuclei is *orbital*. An **electron orbital** *is a region of space around a nucleus where an electron with a specific energy is most likely to be found.*

An analogy for the relationship between shells, subshells, and orbitals can be found in the physical layout of a high-rise condominium complex. A shell is the counterpart of a floor of the condominium in our analogy. Just as each floor will contain apartments (of different sizes), a shell contains subshells (of different sizes). Further, just as apartments contain rooms, subshells contain orbitals. An apartment is a collection of rooms; a subshell is a collection of orbitals. A floor of a condominium building is a collection of apartments; a shell is a collection of subshells.

The characteristics of orbitals, the rooms in our "electron apartment house," include

1. The number of orbitals in a subshell varies, being one for an *s* subshell, three for a *p* subshell, five for a *d* subshell, and seven for an *f* subshell.
2. The maximum number of electrons in an orbital does not vary. It is always 2.
3. The notation used to designate orbitals is the same as that used for subshells. Thus, orbitals in the 4*f* subshell (there are seven of them) are called 4*f* orbitals.

We have already noted (Sec. 6.5) that all electrons in a subshell have the same energy. Thus all electrons in orbitals of the same subshell will have the same energy. This means

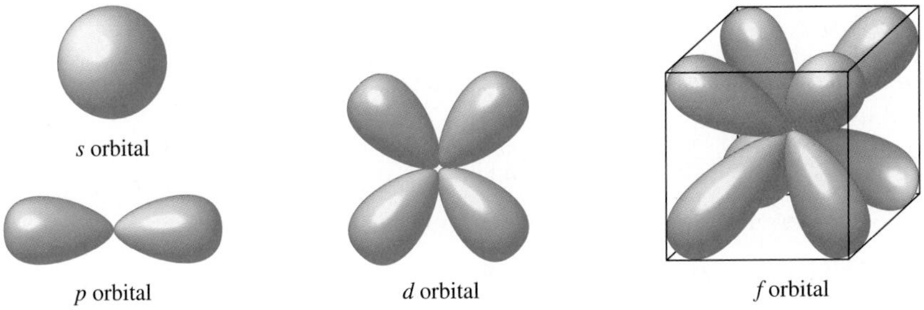

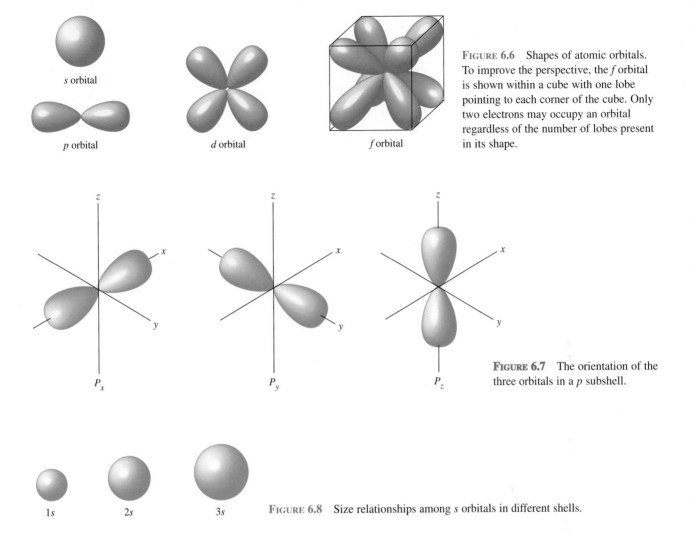

FIGURE 6.6 Shapes of atomic orbitals. To improve the perspective, the *f* orbital is shown within a cube with one lobe pointing to each corner of the cube. Only two electrons may occupy an orbital regardless of the number of lobes present in its shape.

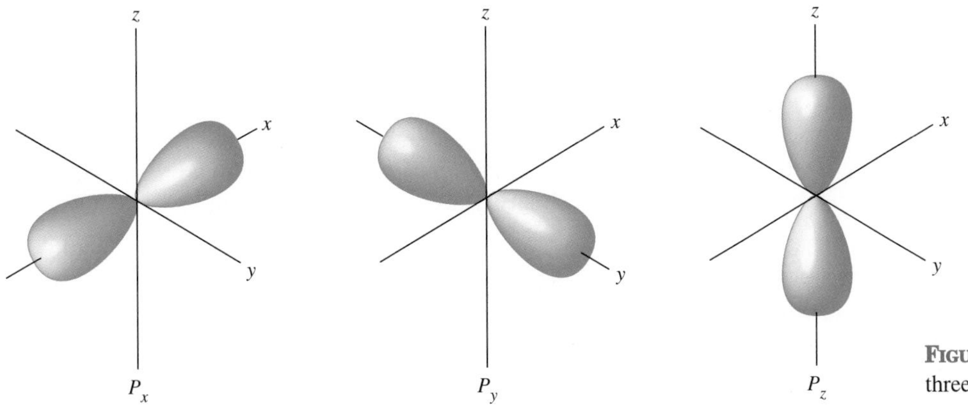

FIGURE 6.7 The orientation of the three orbitals in a *p* subshell.

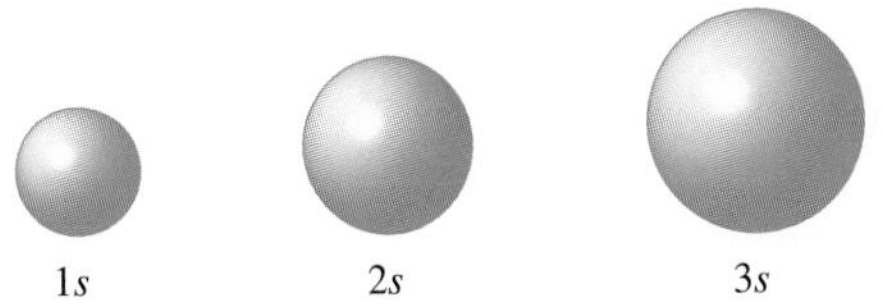

FIGURE 6.8 Size relationships among *s* orbitals in different shells.

that *shell and subshell designations are sufficient to specify the energy of an electron*. This statement will be of great importance in the discussions of Section 6.7.

Orbitals have a definite size and shape related to the type of subshell in which they are found. (Remember, an orbital is a region of space. We are not talking about the size and shape of an electron, but rather the size and shape of a region of space where an electron is found.) At any given time an electron can be at only one point in an orbital, but because of its rapid movement throughout the orbital it "occupies" the entire orbital. An analogy would be the "definite" volume occupied by rotating fan blades. Typical *s*, *p*, *d*, and *f* orbital shapes are given in Figure 6.6. Notice that the shapes increase in complexity in the order *s*, *p*, *d*, and *f*. Some of the more complex *d* and *f* orbitals have shapes related, but not identical, to those shown in Figure 6.6.

Orbitals within the same subshell differ mainly in orientation. For example, the three 2*p* orbitals look the same but are aligned in different directions—along the *x*, *y*, and *z* axes in a Cartesian coordinate system (see Fig. 6.7).

Orbitals of the same type but in different shells (for example, 1*s*, 2*s*, and 3*s*) have the same general shape but differ in size (volume) (see Fig. 6.8).

Example 6.1

Determine the following information about electron shells, electron subshells, and electron orbitals.

(a) The number of electron subshells in shell 4

(b) The number of electron orbitals in a 3*d* subshell

(c) The maximum number of electrons that could be contained in a 2*p* subshell

(d) The maximum number of electrons that could be contained in a 2*p* orbital

Solution

(a) The number of subshells in a shell is the same as the shell number. Thus, shell 4 will contain 4 subshells.

(b) The number of orbitals in a given type (*s*, *p*, *d*, *f*) subshell is independent of the shell number. Each *s* subshell (1*s*, 2*s*, 3*s*, etc.) contains one orbital; each *p* subshell contains three orbitals; each *d* subshell contains five orbitals; and each *f* subshell contains seven orbitals. Thus, a 3*d* subshell contains five orbitals.

(c) The maximum number of electrons in a subshell depends on the number of orbitals the subshell contains, with each orbital holding two electrons. In a *p* subshell there are three orbitals. Thus, a *p* subshell can accommodate 6 electrons (2 per orbital).

(d) The maximum number of electrons in an orbital is two. It does not matter what type of orbital it is (2*s*, 2*p*, 3*d*, etc.). All orbitals hold a maximum of two electrons.

Practice Exercise 6.1

Determine the following information about electron shells, electron subshells, and electron orbitals.

(a) The number of electron subshells in shell 3
(b) The number of electron orbitals in a 2*p* subshell
(c) The maximum number of electrons that could be contained in a 4*d* subshell
(d) The maximum number of electrons that could be contained in a 1*s* orbital

Ans. (a) 3; (b) 3; (c) 10; (d) 2

6.7 WRITING ELECTRON CONFIGURATIONS

An **electron configuration** *is a statement of how many electrons an atom has in each of its subshells.* Since subshells group electrons according to energy (Sec. 6.5), electron configurations indicate how many electrons an atom has of various energies.

Electron configurations are not written out in words; a shorthand system with symbols is used. Subshells containing electrons, listed in order of increasing energy, are designated using number–letter combinations (1*s*, 2*s*, 2*p*, etc.). A superscript following each subshell designation indicates the number of electrons in that subshell. The electron configuration for oxygen using this shorthand notation is

$$1s^2 2s^2 2p^4 \qquad \text{(read "one-}s\text{-two, two-}s\text{-two, two-}p\text{-four")}$$

An oxygen atom thus has an electron arrangement of two electrons in the 1*s* subshell, two electrons in the 2*s* subshell, and four electrons in the 2*p* subshell.

To determine the electron configuration for an atom, a procedure called the *Aufbau principle* (German *aufbauen*, to build) is used. The **Aufbau principle** *is that electrons normally occupy the lowest energy subshell available*. This guideline brings order to what could be a very disorganized situation. Many orbitals exist about the nucleus of any given atom. Electrons do not occupy these orbitals in a random, haphazard fashion; a very predictable pattern, governed by the Aufbau principle, exists for electron orbital occupancy. Orbitals are filled in order of increasing energy.

Use of the Aufbau principle requires knowledge concerning the electron capacities of orbitals and subshells (which we already have; see Sec. 6.6) and knowledge concerning the relative energies of subshells (which we now consider).

Figure 6.9 gives the order in which electron subshells about a nucleus acquire electrons. Note from studying this figure that the sequence of subshell filling is not as simple a pattern as might be predicted. All subshells within a given shell do not necessarily have lower energies than all subshells of higher numbered shells. Because of energy overlaps, beginning with shell 4, one or more lower energy subshells of a specific shell have energies lower than the upper subshells of a preceding shell and thus acquire electrons first. For example, the 4*s* subshell acquires electrons before the 3*d* subshell does (see Fig. 6.9). As

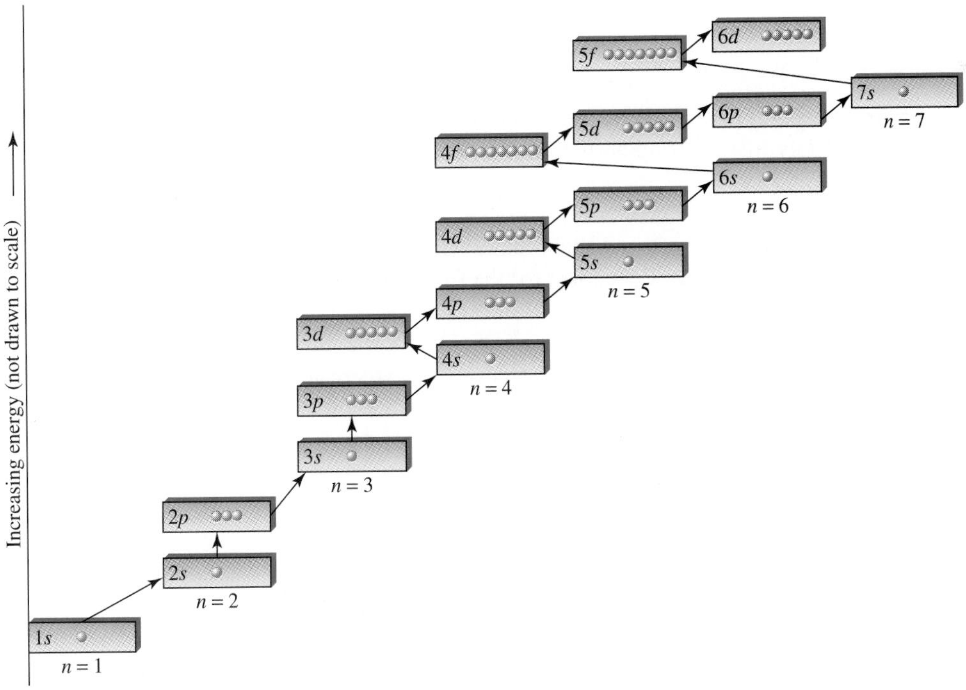

FIGURE 6.9 Relative energies and filling order for the electron subshells.

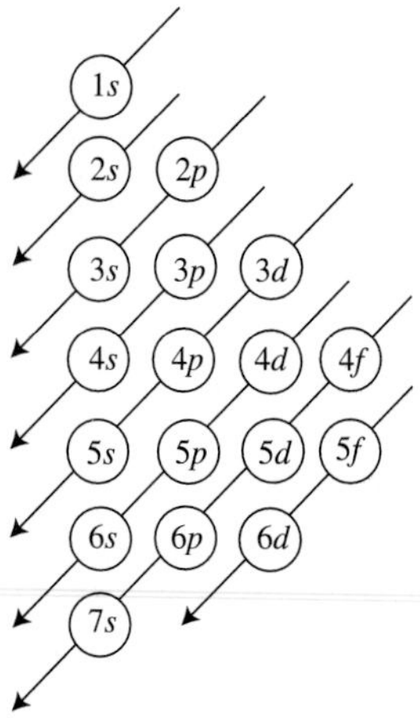

FIGURE 6.10 The Aufbau diagram—an aid to remembering subshell filling order.

another example, the *s* subshell of the sixth energy level fills before the *d* subshell of the fifth energy level or the *f* subshell of the fourth energy level (again refer to Fig. 6.9).

The sequence in which subshells acquire electrons must be learned before electron configurations can be written. A useful mnemonic (memory) device, called an *Aufbau diagram*, helps considerably with this learning process. As can be seen from Figure 6.10, an **Aufbau diagram** *is a listing of subshells in the order in which electrons occupy them.* All *s* subshells are located in column 1, all *p* subshells in column 2, and so on. Subshells belonging to the same shell are found in the same row. The order of subshell filling is given by following the diagonal arrows, starting with the top one. The 1*s* subshell fills first. The second arrow points to (goes through) the 2*s* subshell. It fills next. The third arrow points to both the 2*p* and 3*s* subshells. The 2*p* fills first, followed by the 3*s*. Any time a single arrow points to more than one subshell, start at the tail of the arrow and work to its head to determine the proper filling sequence. The 3*p* subshell fills next, and so on. An Aufbau diagram is an easy way to catalog the information given in Figure 6.9.

We are now ready to write electron configurations. Let us systematically consider electron configurations for the first few elements in the periodic table.

Hydrogen (Z = 1) has only one electron, which goes into the 1*s* subshell, which has the lowest energy of all subshells. Hydrogen's electron configuration is written as

$$1s^1$$

Helium (Z = 2) has two electrons, both of which occupy the 1*s* subshell. (Remember, an *s* subshell contains one orbital, and an orbital can accommodate two electrons.) Helium's electron configuration is

$$1s^2$$

For lithium (Z = 3), with three electrons, the third electron cannot enter the 1*s* subshell since its maximum capacity is two electrons. (All *s* subshells are completely filled with two electrons; see Sec. 6.6.) The third electron is placed in the next highest energy subshell, the 2*s*. The electron configuration for lithium is thus

$$1s^2 2s^1$$

With beryllium (Z = 4) the additional electron is placed in the 2*s* subshell, which is now completely filled, giving beryllium the electron configuration

$$1s^2 2s^2$$

In boron (Z = 5), the 2*p* subshell, the subshell of next highest energy (check Fig. 6.9 or 6.10 to be sure), becomes occupied for the first time. Boron's electron configuration is

$$1s^2 2s^2 2p^1$$

A *p* subshell can accommodate six electrons since there are three orbitals within it (Sec. 6.6). The 2*p* subshell can thus accommodate the additional electrons found in C, N, O, F, and Ne. The electron configurations of these elements are

$$\begin{aligned} \text{C (Z = 6):}\quad & 1s^2 2s^2 2p^2 \\ \text{N (Z = 7):}\quad & 1s^2 2s^2 2p^3 \\ \text{O (Z = 8):}\quad & 1s^2 2s^2 2p^4 \\ \text{F (Z = 9):}\quad & 1s^2 2s^2 2p^5 \\ \text{Ne (Z = 10):}\quad & 1s^2 2s^2 2p^6 \end{aligned}$$

With sodium (Z = 11) the 3*s* subshell acquires an electron for the first time.

$$\text{Na } (Z = 11): \quad 1s^2 2s^2 2p^6 3s^1$$

Note the pattern that is developing in the electron configurations we have written so far. Each element has an electron configuration the same as the one just before it with the addition of one electron.

Electron configurations for other elements are obtained by simply extending the principles we have just illustrated. Electrons are added to subshells, always filling one of lower energy before adding electrons to the next highest subshell, until the correct number of electrons has been accommodated.

Example 6.2

Write the electron configurations for (a) calcium ($Z = 20$) and (b) selenium ($Z = 34$).

Solution

(a) The number of electrons in a calcium atom is 20. Remember, the atomic number (Z) gives the number of electrons (Sec. 5.7). We will need to fill subshells, in order of increasing energy, until 20 electrons have been accommodated.

The $1s$, $2s$, and $2p$ subshells fill first, accommodating a total of 10 electrons among them

$$1s^2 2s^2 2p^6 \ldots$$

Next, according to Figure 6.9 or 6.10, the $3s$ fills and then the $3p$ subshell.

$$1s^2 2s^2 2p^6 3s^2 3p^6 \ldots$$

We have accommodated 18 electrons at this point. We will need to add two more electrons to get our desired number of 20.

These last two electrons are added to the $4s$ subshell

$$1s^2 2s^2 2p^6 3s^2 3p^6 4s^2$$

These last two electrons completely fill the $4s$ subshell.

(b) To write the electron configuration for selenium we continue along the same lines as in part (a), remembering that the maximum electron subshell populations are $s = 2$, $p = 6$, and $d = 10$.

The first 18 electrons, as with calcium (part a), will fill the $1s$, $2s$, $2p$, $3s$, and $3p$ subshells.

$$1s^2 2s^2 2p^6 3s^2 3p^6 \ldots$$

The $4s$ subshell fills next, accommodating 2 electrons, followed by the $3d$ subshell, which accommodates 10 electrons.

$$1s^2 2s^2 2p^6 3s^2 3p^6 4s^2 3d^{10} \ldots$$

We now have a total of 30 electrons.

Four more electrons are needed, which are added to the next higher subshell in energy, the $4p$.

$$1s^2 2s^2 2p^6 3s^2 3p^6 4s^2 3d^{10} 4p^4$$

The $4p$ subshell can accommodate 6 electrons, but we do not want it filled to capacity because that would give us too many electrons.

To double check that we have the correct number of electrons, 34, we add the superscripts in our final electron configuration.

$$2 + 2 + 6 + 2 + 6 + 2 + 10 + 4 = 34$$

The sum of the superscripts in any electron configuration should add up to the atomic number if the configuration is for a neutral atom.

Practice Exercise 6.2

Write the electron configuration for (a) chlorine ($Z = 17$) and (b) antimony ($Z = 51$).

Ans. (a) $1s^2 2s^2 2p^6 3s^2 3p^5$;
(b) $1s^2 2s^2 2p^6 3s^2 3p^6 4s^2 3d^{10} 4p^6 5s^2 4d^{10} 5p^3$

It should be noted that for a few elements in the middle of the periodic table the actual distribution of electrons within subshells differs slightly from that obtained using the Aufbau principle and Aufbau diagram. These exceptions are caused by very small energy differences between some subshells and are not important in the uses we shall make of electronic configurations.

Abbreviated electron configurations that give shell electron occupancy rather than subshell electron occupancy are sometimes written. Such configurations are simply a listing of numbers. Example 6.3 gives a comparison between a regular electron configuration (subshell occupancy) and an abbreviated electron configuration (shell occupancy).

Example 6.3

The electron configuration for calcium is

$$1s^2 2s^2 2p^6 3s^2 3p^6 4s^2$$

Write this configuration in an abbreviated form that gives only shell occupancy information.

Solution

Calcium atoms have four electron shells that contain electrons

Shell 1 contains 2 electrons: $1s^2$
Shell 2 contains 8 electrons: $2s^2 2p^6$
Shell 3 contains 8 electrons: $3s^2 3p^6$
Shell 4 contains 2 electrons: $4s^2$

The abbreviated electron configuration for calcium is

$$2, 8, 8, 2$$

In such notation the identity of each shell is not explicitly shown. It is assumed that you know that the listed numbers are ordered in terms of increasing shell number.

Calcium has 2 electrons in the first shell, 8 in the second shell, 8 in the third shell, and 2 in the fourth shell.

Practice Exercise 6.3

The electron configuration for chlorine is

$$1s^2 2s^2 2p^6 3s^2 3p^5$$

Write this configuration in an abbreviated form that gives only shell occupancy information.

Ans. 2, 8, 7

6.8 ORBITAL DIAGRAMS

The arrangement of electrons about a nucleus can be specified in terms of shell occupancy, subshell occupancy, or orbital occupancy. The notation for specifying electron arrangements in terms of shell occupancy (abbreviated electron configurations) and subshell occupancy (regular electron configurations) has been previously considered (Sec. 6.7). In this section we consider electron arrangements at the orbital level.

Two principles are needed to specify orbital occupancy for electrons.

1. The Aufbau principle.
2. Hund's rule.

The Aufbau principle was used extensively in Section 6.7 in writing electron configurations. Hund's rule has not been encountered previously. The namesake for Hund's rule is the German physicist Frederick Hund (1896–), an early worker in the field of quantum mechanics.

Hund's rule *states that when electrons are placed in a set of orbitals of equal energy (the orbitals of a subshell), the order of filling for the orbitals is such that each orbital will be occupied by one electron before any orbital receives a second electron.* Such a pattern of orbital filling minimizes repulsions between electrons. Numerous applications of Hund's rule are required in stating electron arrangements in terms of orbital occupancies, that is, in drawing orbital diagrams.

Orbital diagrams *are diagrams that show the electron occupancy of each orbital about a nucleus.* In drawing orbital diagrams, we use circles to represent orbitals and arrows with a single barb to denote electrons. The orbital diagram for hydrogen, with its one electron, is

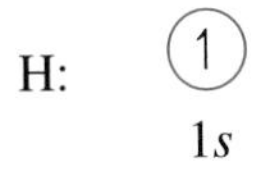

A helium atom contains two electrons, both of which occupy the 1*s* orbital. The orbital diagram for helium is

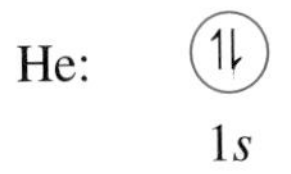

Note the notation used to denote two electrons in the same orbital—one arrow points up and the other points down. The two electrons are said to be *paired*. When only one electron is present in an orbital it is said to be *unpaired*.

Orbital diagrams for the next two elements, lithium and beryllium, are drawn according to reasoning similar to that followed for H and He. The two electrons in the 2*s* orbital of Be are paired.

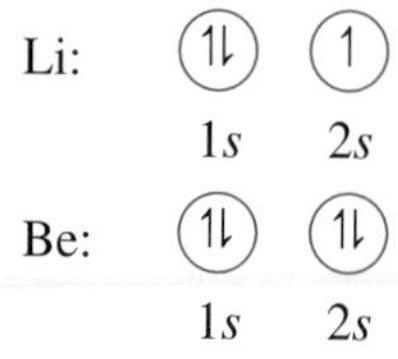

Boron has the electron configuration $1s^22s^22p^1$. The fifth electron in boron must enter a 2*p* orbital, since both the 1*s* and 2*s* orbitals are full. Boron's orbital diagram is

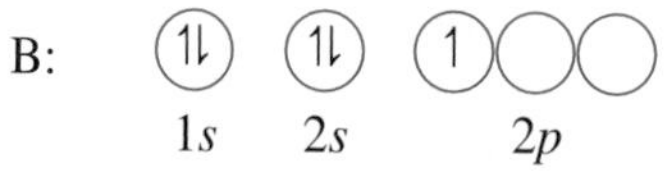

Note that it does not matter which one of the three 2*p* orbitals of boron is shown as containing an electron. All three orbitals have the same energy. Each of the following notations is correct.

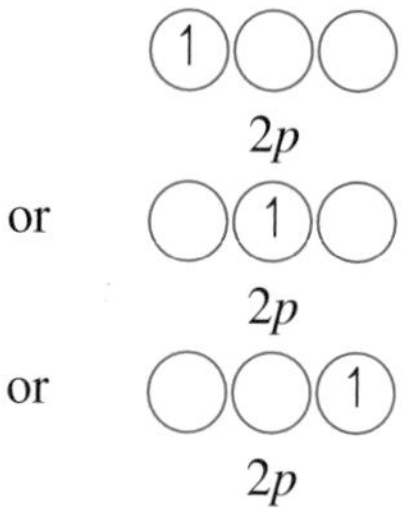

With carbon, element 6, we encounter the use of Hund's rule for the first time. Carbon has two electrons in the 2*p* subshell ($1s^22s^22p^2$). Do the two electrons go into the same orbital (paired) or do they go into separate equivalent orbitals (unpaired)? Hund's rule indicates that the latter is the case. The orbital diagram for carbon is

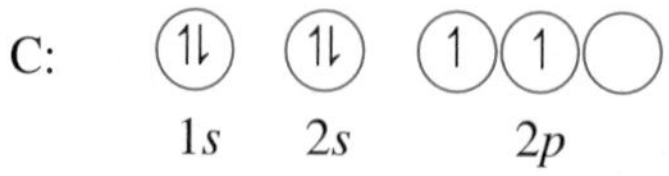

Again, it does not matter which two of the three 2*p* orbitals of carbon are shown as containing electrons since all three orbitals have the same energy.

Nitrogen, with the electron configuration $1s^22s^22p^3$, contains three unpaired electrons.

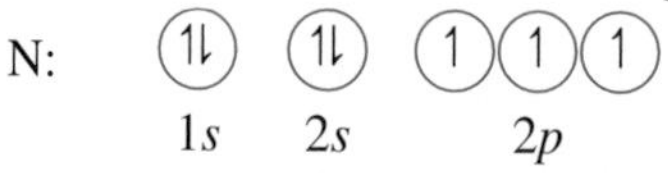

With oxygen ($1s^22s^22p^4$) two of the four 2*p* electrons must pair up, leaving two unpaired electrons. Fluorine, the next element in the periodic table, has only one unpaired electron. Finally, with neon, all of the electrons are paired up.

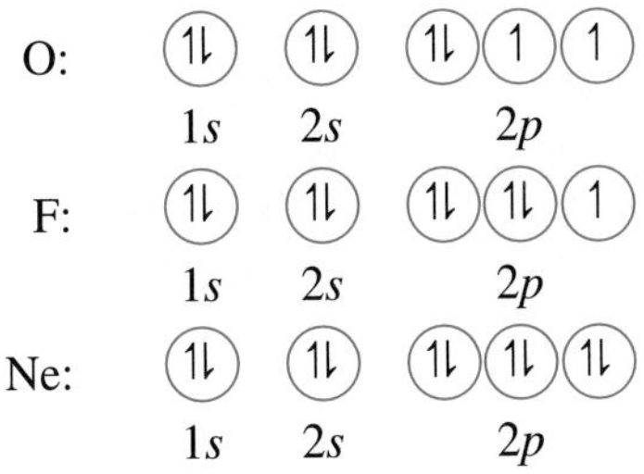

Example 6.4

Draw an orbital diagram for the electrons in the element vanadium. The electron configuration for vanadium is

$$1s^2 2s^2 2p^6 3s^2 3p^6 4s^2 3d^3$$

Solution

All of the occupied subshells of vanadium are completely filled except for the 3*d*. There are five orbitals in the 3*d* subshell, and there are three electrons present. Following Hund's rule, we will put one electron in each of three 3*d* orbitals, giving us three unpaired electrons.

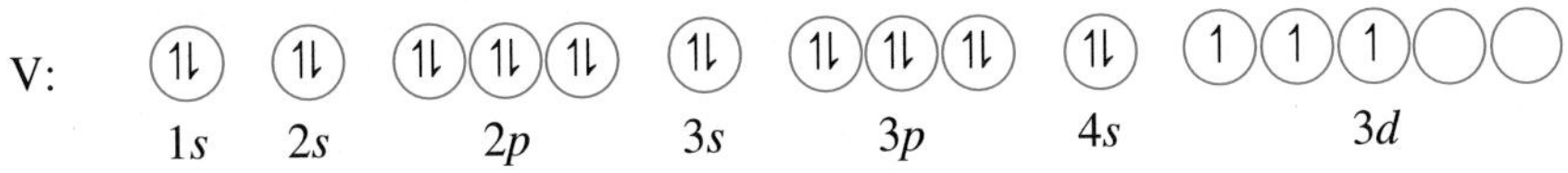

Practice Exercise 6.4

Draw an orbital diagram for the electrons in the element sulfur. The electron configuration for sulfur is

$$1s^2 2s^2 2p^6 3s^2 3p^4$$

Ans.

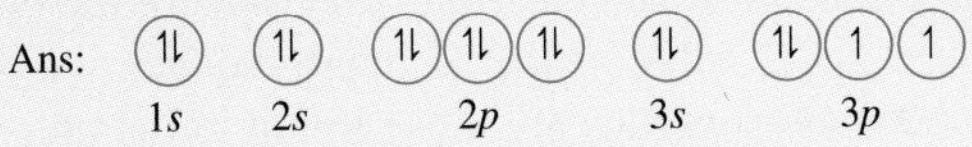

Atoms may be classified as paramagnetic or diamagnetic on the basis of unpaired electrons. A **paramagnetic atom** *has an electron arrangement containing one or more unpaired electrons*. The presence of unpaired electrons causes paramagnetic materials to be slightly attracted to a magnet. Measurement of paramagnetism provides experimental verification of the presence of unpaired electrons. A **diamagnetic atom** *has an electron arrangement in which all electrons are paired.*

6.9 ELECTRON CONFIGURATIONS AND THE PERIODIC LAW

A knowledge of electron configurations for the elements provides an explanation for the periodic law. Recall, from Section 6.1, that the periodic law points out that the properties of the elements repeat themselves in a regular manner when the elements are ordered in sequence of increasing atomic number. Those elements with similar chemical properties are placed one under another in vertical columns (groups) in a periodic table.

Groups of elements have similar chemical properties because of similarities that exist in the electron configurations of the elements of the group. *Chemical properties repeat themselves in a regular manner among the elements because electron configurations repeat themselves in a regular manner among the elements.*

To illustrate this correlation between similar chemical properties and similar electron configurations, let us look at the electron configurations of two groups of elements known to have similar chemical properties.

We begin with the elements lithium, sodium, potassium, and rubidium—all members of group IA of the periodic table. The electron configurations for these similarly propertied elements are

$$
\begin{aligned}
{}_{3}\text{Li}&: 1s^2 2s^1\\
{}_{11}\text{Na}&: 1s^2 2s^2 2p^6 3s^1\\
{}_{19}\text{K}&: 1s^2 2s^2 2p^6 3s^2 3p^6 4s^1\\
{}_{37}\text{Rb}&: 1s^2 2s^2 2p^6 3s^2 3p^6 4s^2 3d^{10} 4p^6 5s^1
\end{aligned}
$$

We see that each of these elements has one outer *s* electron (shown in color), the last *s* electron added by the Aufbau principle. It is this similarity in outer-shell electron arrangements that causes these elements to have similar chemical properties. It is found in general that elements with similar outer shell electron configurations have similar chemical properties.

Let us consider another group of elements known to have similar chemical properties: the elements fluorine, chlorine, bromine, and iodine of group VIIA of the periodic table. The electron configurations for these four elements are

$$
\begin{aligned}
{}_{9}\text{F}&: 1s^2 2s^2 2p^5\\
{}_{17}\text{Cl}&: 1s^2 2s^2 2p^6 3s^2 3p^5\\
{}_{35}\text{Br}&: 1s^2 2s^2 2p^6 3s^2 3p^6 4s^2 3d^{10} 4p^5\\
{}_{53}\text{I}&: 1s^2 2s^2 2p^6 3s^2 3p^6 4s^2 3d^{10} 4p^6 5s^2 4d^{10} 5p^5
\end{aligned}
$$

Once again similarities in electron configurations are readily apparent. This time the repeating pattern is the seven electrons (in color) in the outermost *s* and *p* subshells.

Section 7.2 will consider in depth the fact that the electrons most important in controlling chemical properties are those found in the outermost shell of an atom.

6.10 ELECTRON CONFIGURATIONS AND THE PERIODIC TABLE

One of the strongest pieces of supporting evidence for the assignment of electrons to shells, subshells, and orbitals is the periodic table itself. The basic shape and structure of this table, which was determined many years before electrons were even discovered, is consistent with and can be explained by electron configurations. Indeed, the specific location of an element in the periodic table can be used to obtain information about its electron configuration.

The concept of distinguishing electrons is the key to obtaining "electron configuration information" from the periodic table. The **distinguishing electron** *for an element is the last electron added to its electron configuration when the configuration is written according to the Aufbau principle.* This last electron added is the one that causes an element's electron configuration to differ from that of the element immediately preceding it in the periodic table; hence the term *distinguishing electron.*

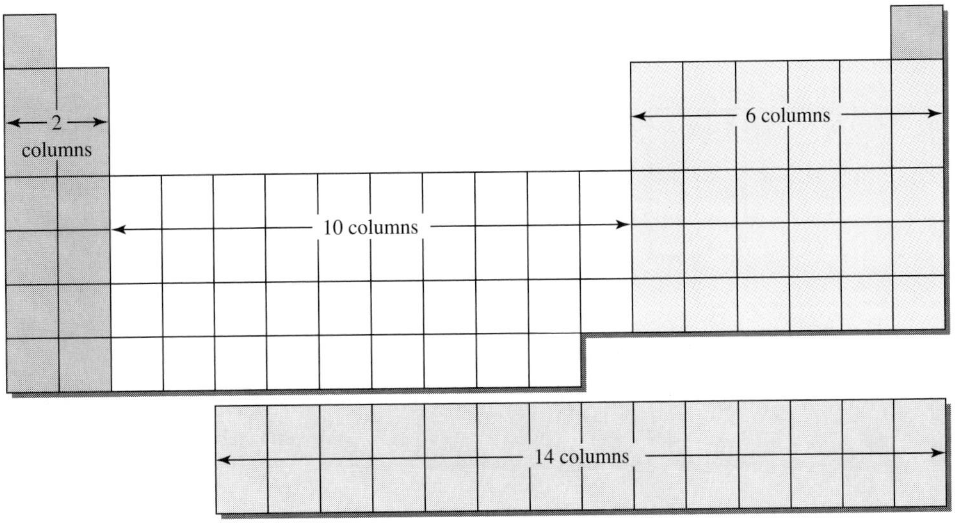

FIGURE 6.11 Structure of the periodic table in terms of columns.

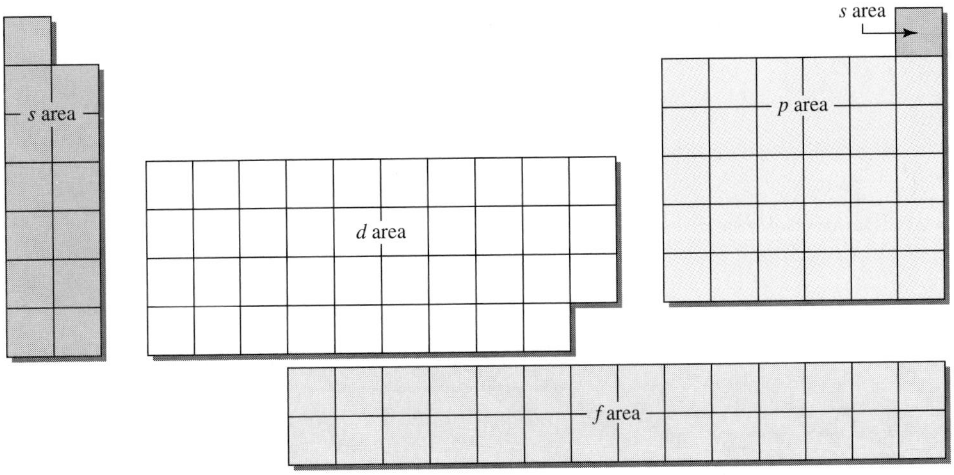

FIGURE 6.12 Areas of the periodic table.

As the first step in linking electron configurations to the periodic table, let us analyze the general shape of the periodic table in terms of columns of elements. As shown in Figure 6.11, we have on the extreme left of the table two columns of elements, in the center an area containing ten columns of elements, to the right a block of six columns of elements, and at the bottom of the table, in two rows, fourteen columns of elements. These numbers of columns of elements in the various regions of the periodic table—2, 6, 10, and 14—are the same as the maximum numbers of electrons that the various types of subshells can accommodate. We will see shortly that this is a very significant observation; the number matchup is no coincidence. The various columnar regions of the periodic table are called the *s* area (two columns), the *p* area (six columns), the *d* area (ten columns), and the *f* area (fourteen columns), as shown in Figure 6.12.

For all elements located in the *s* area of the periodic table the distinguishing electron is always found in an *s* subshell. All *p* area elements have distinguishing electrons in *p* subshells. Similarly, elements in the *d* and *f* areas of the periodic table have, respectively, distinguishing electrons located in *d* and *f* subshells. Thus, the area location of an element in

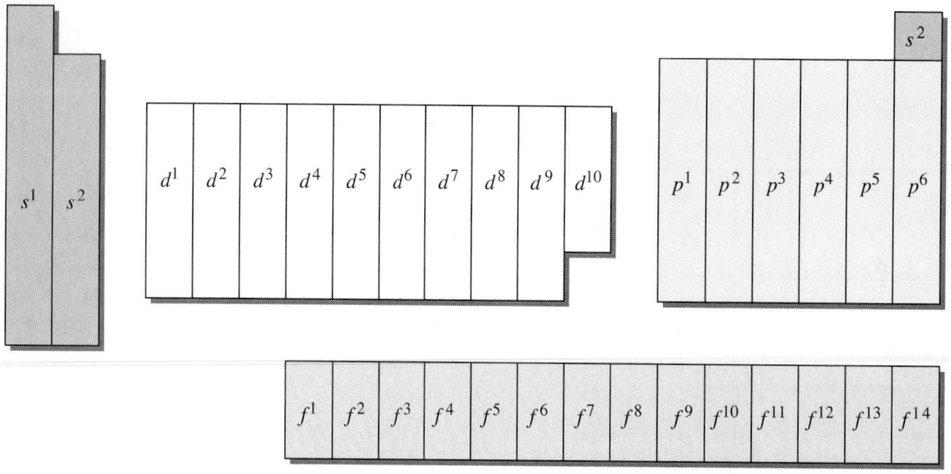

FIGURE 6.13 Extent of subshell filling as a function of periodic table location.

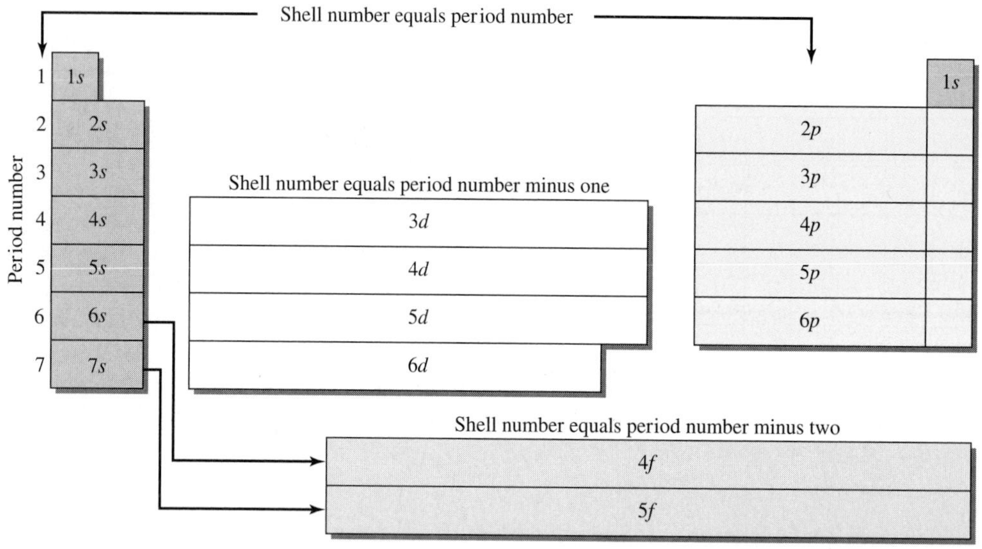

FIGURE 6.14 Relationships of period numbers to shell numbers for distinguishing electrons.

the periodic table can be used to determine the type of subshell that contains the distinguishing electron. Note that the element helium is considered to belong to the *s* rather than the *p* area of the periodic table even though its table position is on the right-hand side. (The reason for this placement of helium will be explained in Sec. 7.3.)

The extent of filling of the subshell containing an element's distinguishing electron can also be determined from the element's position in the periodic table. All elements in the first column of a specific area contain only one electron in the subshell, all elements in the second column contain two electrons in the subshell, and so on. Thus, all elements in the first column of the *p* area (group IIIA) have an electron configuration ending in p^1. Elements in the second column of the *p* area (group IVA) have electron configurations ending in p^2, and so forth. Similar relationships hold in other areas of the table, as shown in Figure 6.13. A few exceptions to these generalizations do exist in the *d* and *f* areas (because of irregular electron configurations; see Sec. 6.7), but we will not be concerned with them in this text.

We can also use the periodic table to determine the shell in which the distinguishing electron is located. The relationship used involves the number of the period in which the element is found. In the *s* and *p* areas, the period number gives the shell number directly. In the *d* area, the period number minus one is equal to the shell number. (Remember that the 3*d* subshell is filled during the fourth period.) For similar reasons, in the *f* area the period number minus two equals the shell number. Thus, the subshell that contains the distinguishing electron for elements of period 6 may be the 6*s*, 6*p*, 5*d* (period number minus one), or 4*f* (the period number minus two), depending on the location of the element in period 6. It must be remembered that even though the *f* area is located at the bottom of the table, it correctly belongs in periods 6 and 7. The complete matchup between period number and shell number for the distinguishing electron is given in the periodic table of Figure 6.14.

Example 6.5

Using the periodic table and Figures 6.12 to 6.14, determine the following for the elements calcium ($Z = 20$), manganese ($Z = 25$), and tellurium ($Z = 52$).

(a) the type of subshell in which the distinguishing electron is found

(b) the extent of filling of the subshell containing the distinguishing electron

(c) the shell in which the subshell containing the distinguishing electron is found

Solution

(a) Knowing the area of the periodic table in which an element is found is sufficient to determine the type of subshell in which the distinguishing electron is found.

Ca: Since this element is found in the *s* area of the periodic table, the distinguishing electron will be in an ***s* subshell**.

Mn: Since this element is found in the *d* area of the periodic table, the distinguishing electron will be in a ***d* subshell**.

Te: Since this element is found in the *p* area of the periodic table, the distinguishing electron will be in a ***p* subshell.**

(b) The extent of filling of the subshell containing the distinguishing electron is determined by noting the column in the area that the element occupies.

Ca: Since this element is in the second column of the *s* area, the *s* subshell involved contains **two electrons** (s^2).

Mn: Since this element is in the fifth column of the *d* area, the *d* subshell involved contains **five electrons** (d^5).

Te: Since this element is in the fourth column of the *p* area, the *p* subshell involved contains **four electrons** (p^4).

(c) The shell number of the subshell containing the distinguishing electron is obtained from the period number, sometimes directly and sometimes with modifications.

Ca: Since this element is in period 4, the *s* subshell involved is the 4*s*; therefore, the electron configuration for Ca ends in $\mathbf{4s^2}$.

Mn: Since this element is in period 4, the *d* subshell involved is the 3*d* (period number minus one); therefore, the electron configuration for Mn ends in $\mathbf{3d^5}$.

Te: Since this element is in period 5, the *p* subshell involved is the 5*p*; therefore, the electron configuration for Te ends in $5p^4$.

Practice Exercise 6.5

Using the periodic table and Figures 6.12 to 6.14, determine the following for the elements cesium (Z = 55) and silver (Z = 47).

(a) the type of subshell in which the distinguishing electron is found
(b) the extent of filling of the subshell containing the distinguishing electron
(c) the shell in which the subshell containing the distinguishing electron is found

Ans. (a) *s* for Cs, *d* for Ag; (b) s^1 for Cs, d^9 for Ag; (c) $6s^1$ for Cs, $4d^9$ for Ag

In order to write complete electron configurations, you must know the order in which the various electron subshells are filled. Up until now, we have obtained this filling order by using an Aufbau diagram (Fig. 6.10). We can also obtain this information directly from the periodic table. To obtain it, we merely follow a path of increasing atomic number through the table, noting the various subshells as we encounter them.

Figure 6.15 illustrates this way of using the periodic table. Note particularly how the *f* area of the periodic table is worked into the scheme. The results of working our way through the periodic table in terms of atomic numbers are summarized in Table 6.4.

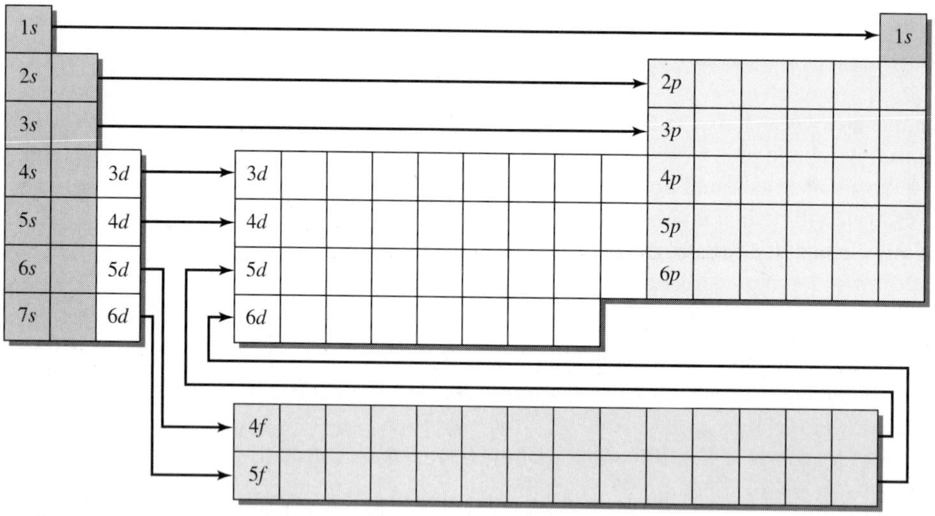

FIGURE 6.15 Using the periodic table as a guide to obtain the order in which subshells are occupied by electrons.

Example 6.6

Write the complete electron configuration of $_{86}Rn$ using the periodic table as your guide.

Solution

The needed information will be obtained by "working our way" through the periodic table. We will start at hydrogen and go from box to box, in order of increasing atomic

TABLE 6.4 SUBSHELL FILLING ORDER FROM THE PERIODIC TABLE

Atomic Numbers	Subshell Involved
1–2	1*s*
3–4	2*s*
5–10	2*p*
11–12	3*s*
13–18	3*p*
19–20	4*s*
21–30	3*d*
31–36	4*p*
37–38	5*s*
39–48	4*d*
49–54	5*p*
55–56	6*s*
57	5*d*
58–71	4*f*
72–80	5*d*
81–86	6*p*
87–88	7*s*
89	6*d*
90–103	5*f*
104–111	6*d*

number, until we arrive at position 86, radon. Every time we traverse the *s* area we will fill an *s* subshell; the *p* area, a *p* subshell; and so on. We will remember that the shell number for *s* and *p* subshells is the period number; for *d* subshells, the period number minus one; and for *f* subshells, the period number minus two.

Let us begin our journey through the periodic table. As we cross period 1 we encounter H and He, both 1*s* elements. We thus add the 1*s* electrons.

$$\text{Rn:}\quad 1s^2 \ldots$$

In traversing period 2 we pass through the *s* area (elements 3 and 4) and the *p* area (elements 5–10) in that order. We add 2*s* and 2*p* electrons.

$$\text{Rn:}\quad 1s^2 2s^2 2p^6 \ldots$$

Our trip through period 3 is very similar to that of period 2. Only *s*-area elements (11 and 12) and *p*-area elements (13–18) are encountered. The message is to add *s* and *p* electrons—3*s* and 3*p* because we are in period 3.

$$\text{Rn:}\quad 1s^2 2s^2 2p^6 3s^2 3p^6 \ldots$$

In passing through period 4 we go through the *s* area (elements 19 and 20), the *d* area (elements 21–30), and the *p* area (elements 31–36). Electrons to be added are those in the 4*s*, 3*d* (period number minus one), and 4*p* subshells, in that order.

$$\text{Rn:} \quad 1s^2 2s^2 2p^6 3s^2 3p^6 4s^2 3d^{10} 4p^6 \ldots$$

In period 5 we encounter scenery similar to that in period 4—the *s* area (elements 37 and 38), the *d* area (elements 39–48), and the *p* area (elements 49–54). Hence, the 5*s*, 4*d* (period number minus one), and 5*p* subshells are filled in order.

$$1s^2 2s^2 2p^6 3s^2 3p^6 4s^2 3d^{10} 4p^6 5s^2 4d^{10} 5p^6 \ldots$$

Our journey ends in period 6, which we completely traverse. We go through the 6*s* area (elements 55 and 56), the 4*f* area (elements 58–71), the 5*d* area (elements 57 and 72–80), and finally the 6*p* area (elements 81–86). We will completely fill the 6*p* subshell since radon is in the last column of the *p* area.

$$\text{Rn:} \quad 1s^2 2s^2 2p^6 3s^2 3p^6 4s^2 3d^{10} 4p^6 5s^2 4d^{10} 5p^6 6s^2 4f^{14} 5d^{10} 6p^6$$

Practice Exercise 6.6

Write the complete electron configuration of $_{54}Xe$ using the periodic table as a guide.

Ans. $1s^2 2s^2 2p^6 3s^2 3p^6 4s^2 3d^{10} 4p^6 5s^2 4d^{10} 5p^6$

Again, let us mention that a few slightly irregular electronic configurations are encountered in the *d* and *f* areas of the periodic table. The generalizations in this section do not address this problem. We will be working mostly with *s*- and *p*-area elements in future chapters. There are no irregularities in electron configurations for these elements.

6.11 CLASSIFICATION SYSTEMS FOR THE ELEMENTS

The elements can be classified in several ways. The two most common classification systems are

1. A system based on the electron configurations of the elements, in which elements are described as *noble gas*, *representative*, *transition*, or *inner transition* elements.
2. A system based on selected physical properties of the elements, in which elements are described as *metals* or *nonmetals*.

The classification scheme based on electron configurations of the elements is depicted in Figure 6.16. This type of classification system is used in numerous discussions in subsequent chapters.

The **noble gases** *are found in the far right column of the periodic table*. They are all gases at room temperature, and they have little tendency to form chemical compounds. With one exception, the distinguishing electron for a noble gas completes the *p* subshell. Therefore, they have electron configurations ending in p^6. The exception is helium, in which the distinguishing electron completes the first shell—a shell that has only two electrons. Helium's electron configuration is $1s^2$.

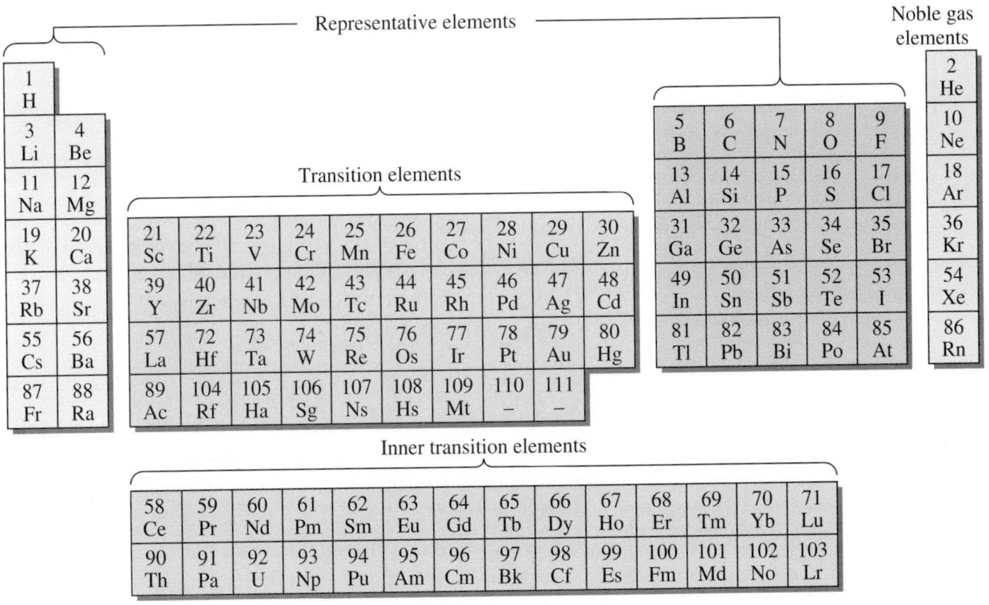

FIGURE 6.16 Elemental classification scheme based on the electron configurations of the elements

The **representative elements** *are all of the elements of the* s *and* p *areas of the periodic table with the exception of the noble gases.* The distinguishing electron in these elements partially or completely fills an *s* subshell or partially fills a *p* subshell. The representative elements include most of the more common elements.

The **transition elements** *are all of the elements of the* d *area of the periodic table.* The common feature in the electronic configurations of the transition elements is the presence of the distinguishing electron in a *d* subshell.

The **inner transition elements** *are all of the elements* of the f *area of the periodic table.* The characteristic feature of their electronic configurations is the presence of the distinguishing electron in an *f* subshell.

On the basis of selected physical properties of the elements, the second of the two classification schemes divides the elements into the categories of metals and nonmetals. A **metal** *is an element that has the characteristic properties of luster, thermal conductivity, electrical conductivity, and malleability.* With the exception of mercury, all metals are solids at room temperature (25°C). Metals are good conductors of heat and electricity. Most metals are ductile (they can be drawn into wires) and malleable (they can be rolled into sheets). Most metals have high luster (shine), high density, and high melting points. Among the more familiar metals are the elements iron, aluminum, copper, silver, and gold.

A **nonmetal** *is an element characterized by the absence of the properties of luster, thermal conductivity, electrical conductivity, and malleability.* Many of the nonmetals, such as hydrogen, oxygen, nitrogen, and the noble gases are gases at room temperature (25°C). The only nonmetal that is a liquid at room temperature is bromine. Solid nonmetals include carbon, sulfur, and phosphorus. In general, the nonmetals have lower densities and melting points than metals.

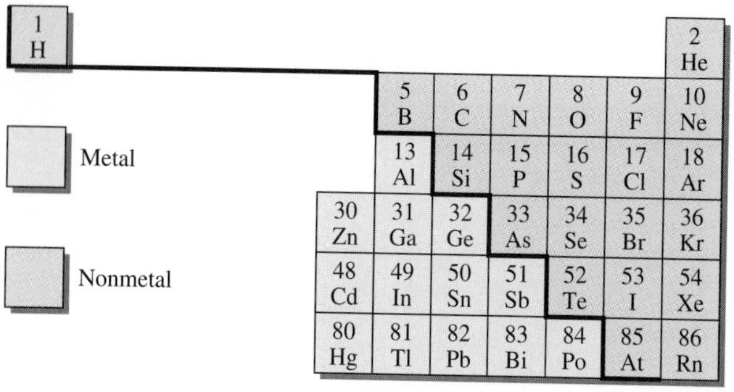

FIGURE 6.17 A portion of the periodic table showing the dividing line between metals and nonmetals. All elements that are not shown are metals.

The majority of the elements are metals; only 22 elements are nonmetals; the rest (89) are metals. It is not necessary to memorize which elements are nonmetals and which are metals. As can be seen from Figure 6.17, the location of an element in the periodic table correlates directly with its classification as a metal or nonmetal. The steplike heavy line that runs through the *p* area of the periodic table separates the metals from the nonmetals; metals are on the left and nonmetals on the right. Note that the element hydrogen is a nonmetal even though it is located on the left side of the periodic table.

The fact that the vast majority of elements are metals in no way indicates that metals are more important than nonmetals. Most nonmetals are relatively abundant and are found in many important compounds. For example, water (H_2O) is a compound involving two nonmetals. An analysis of the previously given abundances of the elements in the Earth's crust (Table 4.4) in terms of metals and nonmetals shows that three of the four most abundant elements, which account for 83% of all atoms, are nonmetals—oxygen, silicon, and hydrogen. The four most abundant elements in the human body (Table 4.4), which constitute more than 99% of all atoms in the body, are nonmetals—hydrogen, oxygen, carbon, and nitrogen.

6.12 CHEMICAL PERIODICITY

Chemical periodicity *is the variation in properties of elements with their positions in the periodic table.* In this section we consider two properties of elements that exhibit chemical periodicity—metallic–nonmetallic character and atomic size (atomic radius)—and then in Section 7.10 we consider a third property exhibiting chemical periodicity—electronegativity.

The physical properties that distinguish metals and nonmetals were considered in Section 6.11. In general, these properties are opposites for the two classes of substances.

Not all metals possess metallic properties to the same extent; they have them, but to varying degrees. For example, some metals are better conductors of electricity than other metals. Similarly, not all nonmetals possess nonmetallic properties to the same extent.

Metallic and nonmetallic character for the various elements—the extent to which they possess metallic and nonmetallic properties—can be correlated with periodic table position for the elements.

1. Metallic character increases from right to left within a given period in the periodic table.
2. Metallic character increases from top to bottom within a group in the periodic table.

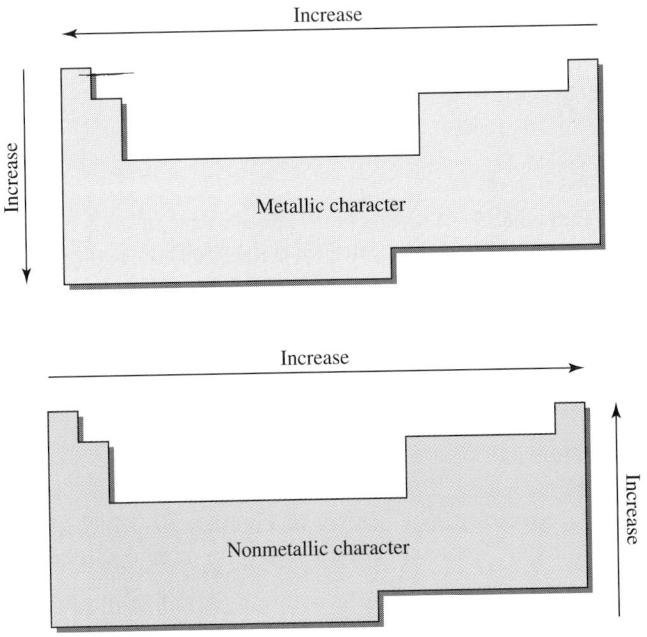

FIGURE 6.18 Chemical periodicity is associated with the intensities of metallic and nonmetallic character.

Thus, among the period-3 elements, Na is more metallic than Mg, which in turn is more metallic than Al. For the group IA elements, K is more metallic than Na, since it is farther down in group IA.

Similar, but opposite, trends exist for nonmetallic character.

1. Nonmetallic character increases from left to right within a given period in the periodic table.
2. Nonmetallic character increases from bottom to top within a group in the periodic table.

Figure 6.18 summarizes the chemical periodicity associated with the properties of metallic and nonmetallic character. The combined effect of these two trends is that the most nonmetallic elements are those in the upper right portion of the periodic table, a generalization consistent with Figure 6.17 in the preceding section of this chapter.

In Section 6.11 a "dividing line" was shown for classifying elements as metals or nonmetals (Fig. 6.17). Elements to the far left of this line have dominant metallic character. Elements to the far right of this line have dominant nonmetallic character. Most elements whose periodic table positions "touch" this metal–nonmetal dividing line actually show some properties that are characteristic of both metals and nonmetals. These elements are frequently given a classification of their own: metalloids (or semimetals). A **metalloid** *is an element with properties intermediate between those of metals and nonmetals*. Figure 6.19 identifies the metalloid elements. Several of the metalloids, including silicon, germanium, and antimony, are semiconductors. A **semiconductor** *is an element that does not conduct electrical current at room temperature but does so at higher temperatures*. Semiconductor elements are very important in the electronics industry.

The quantum mechanical model for an atom (Sec. 6.3) suggests that the shape of an atom is approximately spherical and that the size of the atom can be expressed in terms of the radius of a sphere. The most commonly used unit for expressing atomic sizes (atomic

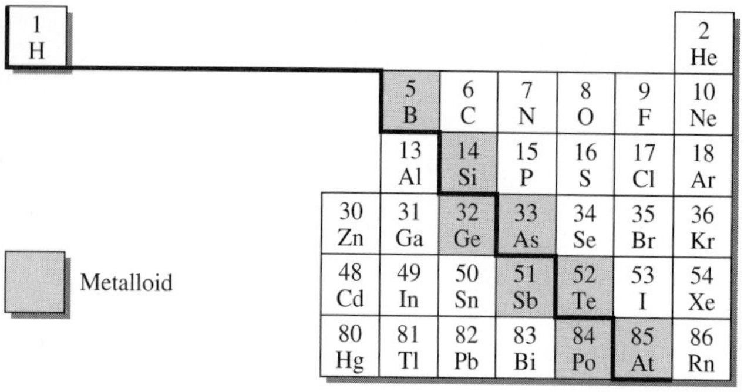
1 H
2 He
5 B 6 C 7 N 8 O 9 F 10 Ne
13 Al 14 Si 15 P 16 S 17 Cl 18 Ar
30 Zn 31 Ga 32 Ge 33 As 34 Se 35 Br 36 Kr
48 Cd 49 In 50 Sn 51 Sb 52 Te 53 I 54 Xe
80 Hg 81 Tl 82 Pb 83 Bi 84 Po 85 At 86 Rn
Metalloid

FIGURE 6.19 A portion of the periodic table showing the metalloids (shaded), elements that show some properties characteristic of metals and some other properties characteristic of nonmetals.

radii) is the picometer (Sec. 3.1). In this unit most atomic radii fall in the range of 50–200 pm.

Atomic sizes exhibit chemical periodicity, as is shown in Figure 6.20, which gives atomic radii for the representative elements. General periodic trends for atomic radii are

1. Atomic radii tend to decrease from left to right within a period of the periodic table.
2. Atomic radii tend to increase from top to bottom within a periodic table group.

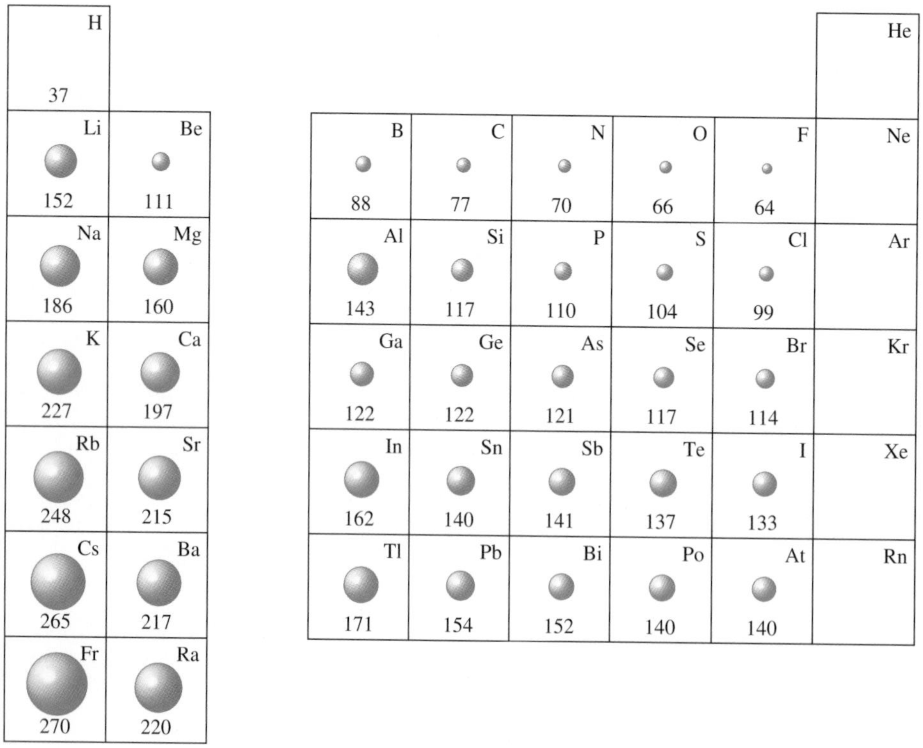
H 37
He
Li 152 Be 111
B 88 C 77 N 70 O 66 F 64 Ne
Na 186 Mg 160
Al 143 Si 117 P 110 S 104 Cl 99 Ar
K 227 Ca 197
Ga 122 Ge 122 As 121 Se 117 Br 114 Kr
Rb 248 Sr 215
In 162 Sn 140 Sb 141 Te 137 I 133 Xe
Cs 265 Ba 217
Tl 171 Pb 154 Bi 152 Po 140 At 140 Rn
Fr 270 Ra 220

FIGURE 6.20 Atomic radii, in picometers, of the representative elements.

The trends in atomic radii values can be explained on the basis of the number of electrons present in the atom and their energies. In traversing a period in the periodic table (from left to right), we note that (1) nuclear charge (atomic number) increases and (2) the added electrons enter the same shell, the outermost shell. The increased nuclear charge draws the electrons in this outermost shell closer to the nucleus, resulting in smaller atomic radii. Down a group in the periodic table, atomic radii increase because the electrons are added to shells with higher numbers. Recall, from Section 6.4, that the higher the shell number the greater the average distances between the electrons and the nucleus.

KEY TERMS

The new terms or concepts defined in this chapter are

Aufbau diagram (Sec. 6.7) A listing of subshells arranged in the order in which electrons occupy them.

Aufbau principle (Sec. 6.7) Electrons normally occupy the lowest energy subshell available.

chemical periodicity (Sec. 6.12) The variation in properties of elements with their positions in the periodic table.

diamagnetic atom (Sec. 6.8) An atom having an electron arrangement in which all electrons are paired.

distinguishing electron (Sec. 6.10) The last electron added to an element's electron configuration when the configuration is written according to the Aufbau principle.

electron configuration (Sec. 6.7) A statement of how many electrons an atom has in each of its subshells.

electron orbital (Sec. 6.6) A region of space around a nucleus where an electron with a specific energy is most likely to be found.

electron shell (Sec. 6.4) An energy level in which the electrons have approximately the same energy and spend most of their time at approximately the same distance from the nucleus.

electron subshell (Sec. 6.5) An energy sublevel within an electron shell in which the electrons all have the same energy.

group (Sec. 6.2) A vertical column of elements in the periodic table.

Hund's rule (Sec. 6.8) When electrons are placed in a set of orbitals of equal energy (the orbitals of a subshell), the order of filling for the orbitals is such that each orbital will be occupied by one electron before any orbital receives a second electron.

inner transition element (Sec. 6.11) An element of the *f* area of the periodic table.

metal (Sec. 6.11) An element that has the characteristic properties of luster, thermal conductivity, electrical conductivity, and malleability.

metalloid (Sec. 6.12) An element with properties intermediate between those of metals and nonmetals.

noble gas (Sec. 6.11) An element of the far right column of the periodic table.

nonmetal (Sec. 6.11) An element characterized by the absence of the properties of luster, thermal conductivity, electrical conductivity, and malleability.

orbital diagram (Sec. 6.8) A diagram that shows the electron occupancy of each of the orbitals about a nucleus.

paramagnetic atom (Sec. 6.8) An atom having an electron arrangement containing one or more unpaired electrons.

period (Sec. 6.2) A horizontal row of elements in the periodic table.

periodic law (Sec. 6.1) When elements are arranged in order of increasing atomic number, elements with similar properties occur at periodic (regularly recurring) intervals.

periodic table (Sec. 6.2) A graphical representation of the behavior described by the periodic law.

quantized property (Sec. 6.3) A property that can have only certain values, that is, not all values are allowed.

representative element (Sec. 6.11) An element of the *s* or *p* area of the periodic table with the exception of the noble gases.

semiconductor (Sec. 6.12) An element that does not conduct electrical current at room temperature but does so at higher temperatures.

transition element (Sec. 6.11) An element of the *d* area of the periodic table.

PRACTICE PROBLEMS

Periodic Law and Periodic Table (Secs. 6.1 and 6.2)

6.1 Give the symbol of the element that occupies each of the following positions in the periodic table.

(a) period 4, group IIIA **(b)** period 5, group IVB
(c) group IA, period 2 **(d)** group VIIA, period 3

6.2 Give the symbol of the element that occupies each of the following positions in the periodic table.

(a) period 1, group IA **(b)** period 6, group IB
(c) group IIIB, period 4 **(d)** group IVA, period 5

6.3 Which of the following sets of elements are all in the same group in the periodic table?

(a) $_{13}$Al, $_{15}$P, and $_{17}$Cl
(b) $_{14}$Si, $_{32}$Ge, and $_{82}$Pb
(c) $_{11}$Na, $_{37}$Rb, and $_{56}$Ba
(d) $_{55}$Cs, $_{66}$Dy, and $_{77}$Ir

6.4 Which of the following sets of elements are all in the same group in the periodic table?

(a) $_{21}$Sc, $_{34}$Se, and $_{39}$Y
(b) $_{12}$Mg, $_{13}$Al, and $_{18}$Ar
(c) $_{7}$N, $_{33}$As, and $_{83}$Bi
(d) $_{87}$Fr, $_{92}$U, and $_{99}$Es

6.5 Which of the sets of elements in Problem 6.3 are all in the same period in the periodic table?

6.6 Which of the sets of elements in Problem 6.4 are all in the same period in the periodic table?

6.7 The following statements either define or are closely related to the terms *periodic law*, *period*, or *group*. Match the terms to the appropriate statements.

(a) This is a vertical arrangement of elements in the periodic table.
(b) The properties of the elements repeat in a regular way as the atomic numbers increase.
(c) The chemical properties of elements 12, 20, and 38 demonstrate this principle.
(d) Elements 24 and 33 belong to this arrangement.

6.8 The following statements either define or are closely related to the terms *periodic law*, *period*, and *group*. Match the terms to the appropriate statements.

(a) This is a horizontal arrangement of elements in the periodic table.
(b) Element 19 begins this arrangement in the periodic table.
(c) The element carbon is the first member of this arrangement.
(d) Elements 10, 18, 36, and 54 belong to this arrangement.

Terminology Associated with Electron Arrangements (Secs. 6.3–6.6)

6.9 The following statements either define or are closely related to the terms *shell*, *subshell*, and *orbital*. Match the terms to the appropriate statements.

(a) In terms of electron capacity, this unit is the smallest of the three.
(b) This unit can contain a maximum of two electrons.
(c) This unit can contain as many electrons as, or more electrons than, either of the other two.
(d) This unit is designated by just a number.

6.10 The following statements either define or are closely related to the terms *shell*, *subshell*, and *orbital*. Match the terms to the appropriate statements.

(a) The term *energy sublevel* is closely associated with this unit.
(b) The formula $2n^2$ gives the maximum number of electrons that can occupy this unit.
(c) This unit is designated in the same way as the orbitals contained within it.
(d) The term *energy level* is closely associated with this unit.

6.11 Give the maximum number of electrons that can occupy each of the following electron subshells.

(a) 5*s* **(b)** 3*d* **(c)** 2*p* **(d)** 4*f*

6.12 Give the maximum number of electrons that can occupy each of the following electron subshells.

(a) 6*p* **(b)** 1*s* **(c)** 5*f* **(d)** 4*d*

6.13 Give the maximum number of electrons that can occupy each of the following electron orbitals.

(a) 3*s* **(b)** 3*p* **(c)** 4*d* **(d)** 5*f*

6.14 Give the maximum number of electrons that can occupy each of the following electron orbitals.

(a) 1*s* **(b)** 3*d* **(c)** 5*p* **(d)** 4*d*

6.15 In each of the following pairs of items identify the item that can accommodate the most electrons.

(a) 3*d* subshell, second shell
(b) shell with $n = 1$, 2*p* subshell
(c) 3*p* orbital, 3*p* subshell
(d) 4*f* subshell, third shell

6.16 In each of the following pairs of items identify the item that can accommodate the most electrons.

(a) first shell, third shell

(b) 4*f* subshell, 4*d* subshell
(c) second shell, 5*f* subshell
(d) 3*d* orbital, 3*d* subshell

6.17 Indicate whether each of the following statements is *true* or *false*.
(a) An orbital has a definite size and shape, which are related to the energy of the electrons it could contain.
(b) The *M* and $n = 3$ shells are one and the same.
(c) All of the orbitals in a subshell have the same energy.
(d) A 2*p* and a 3*p* subshell would contain the same number of orbitals.
(e) The fourth shell is made up of six subshells.

6.18 Indicate whether each of the following statements is *true* or *false*.
(a) All of the subshells in a shell have the same energy.
(b) A *d* subshell always contains five orbitals.
(c) An *s* orbital is shaped something like a four-leaf clover.
(d) The $n = 3$ shell can accommodate a maximum of 18 electrons.
(e) All subshells accommodate the same number of electrons.

6.19 Describe the general shape of each of the following orbitals.
(a) 4*s* (b) 4*p* (c) 4*d* (d) 6*s*

6.20 Describe the general shape of each of the following orbitals.
(a) 1*s* (b) 2*p* (c) 3*d* (d) 5*p*

6.21 Which of the following electron subshell and electron orbital designations is not allowed?
(a) 4*s* subshell (b) 2*d* orbital
(c) 1*p* subshell (d) 4*f* orbital

6.22 Which of the following electron subshell and electron orbital designations is not allowed?
(a) 2*d* subshell (b) 4*s* orbital
(c) 3*p* subshell (d) 2*f* orbital

Writing Electron Configurations (Sec. 6.7)

6.23 In which member of each of the following pairs of subshells would an electron have the higher energy?
(a) 2*s* or 3*s* (b) 3*p* or 3*d*
(c) 5*p* or 7*s* (d) 4*s* or 4*p*

6.24 In which member of each of the following pairs of subshells would an electron have the higher energy?
(a) 3*d* or 4*d* (b) 4*f* or 4*s*
(c) 6*s* or 6*d* (d) 5*d* or 6*p*

6.25 With the help of an Aufbau diagram, write the complete electron configuration for each of the following atoms.
(a) $_{13}$Al (b) $_{7}$N (c) $_{18}$Ar (d) $_{12}$Mg

6.26 With the help of an Aufbau diagram, write the complete electron configuration for each of the following atoms.
(a) $_{20}$Ca (b) $_{10}$Ne (c) $_{6}$C (d) $_{15}$P

6.27 With the help of an Aufbau diagram, write the complete electron configuration for each of the following atoms.
(a) $_{26}$Fe (b) $_{37}$Rb (c) $_{53}$I (d) $_{86}$Rn

6.28 With the help of an Aufbau diagram, write the complete electron configuration for each of the following atoms.
(a) $_{31}$Ga (b) $_{38}$Sr (c) $_{48}$Cd (d) $_{88}$Ra

6.29 Identify the element represented by each of the following electron configurations.
(a) $1s^22s^22p^6$
(b) $1s^22s^22p^63s^23p^64s^1$
(c) $1s^22s^22p^63s^23p^64s^23d^2$
(d) $1s^22s^22p^63s^23p^64s^23d^{10}$

6.30 Identify the element represented by each of the following electron configurations.
(a) $1s^22s^22p^2$
(b) $1s^22s^22p^63s^23p^3$
(c) $1s^22s^22p^63s^23p^64s^2$
(d) $1s^22s^22p^63s^23p^64s^23d^6$

6.31 Write abbreviated electron configurations that give only shell occupancy rather than subshell occupancy for each of the following.
(a) the electron configuration $1s^22s^22p^63s^23p^1$
(b) the electron configuration $1s^22s^22p^63s^23p^6$
(c) the element $_{14}$Si
(d) the element $_{4}$Be

6.32 Write abbreviated electron configurations that give only shell occupancy rather than subshell occupancy for each of the following.
(a) the electron configuration $1s^22s^22p^3$
(b) the electron configuration $1s^22s^22p^63s^1$
(c) the element $_{19}$K
(d) the element $_{6}$C

Orbital Diagrams (Sec. 6.8)

6.33 Draw the electron orbital diagram associated with each of the following electron configurations.
(a) $1s^22s^1$ (b) $1s^22s^22p^5$
(c) $1s^22s^22p^63s^23p^3$ (d) $1s^22s^22p^63s^23p^64s^23d^8$

6.34 Draw the electron orbital diagram associated with each of the following electron configurations.
(a) $1s^22s^22p^3$ (b) $1s^22s^22p^63s^2$
(c) $1s^22s^22p^63s^23p^5$ (d) $1s^22s^22p^63s^23p^64s^23d^7$

6.35 Draw electron orbital diagrams for the following elements.
(a) $_{6}$C (b) $_{10}$Ne (c) $_{11}$Na (d) $_{15}$P

6.36 Draw electron orbital diagrams for the following elements.

(a) $_{7}N$ **(b)** $_{9}F$ **(c)** $_{12}Mg$ **(d)** $_{16}S$

6.37 How many unpaired electrons are there in an atom of the following elements?

(a) boron **(b)** chlorine **(c)** potassium **(d)** zinc

6.38 How many unpaired electrons are there in an atom of the following elements?

(a) lithium **(b)** aluminum **(c)** calcium **(d)** bromine

6.39 Indicate whether atoms of each of the elements in Problem 6.37 are paramagnetic or diamagnetic.

6.40 Indicate whether atoms of each of the elements in Problem 6.38 are paramagnetic or diamagnetic.

Electron Configurations and the Periodic Law (Sec. 6.9)

6.41 Indicate whether or not the elements represented by the given pairs of electron configurations have similar chemical properties.

(a) $1s^22s^1$ and $1s^22s^2$

(b) $1s^22s^22p^6$ and $1s^22s^22p^63s^23p^6$

(c) $1s^22s^22p^3$ and $1s^22s^22p^63s^23p^64s^23d^3$

(d) $1s^22s^22p^63s^23p^6$ and $1s^22s^22p^63s^23p^64s^23d^{10}4p^6$

6.42 Indicate whether or not the elements represented by the given pairs of electron configurations have similar chemical properties.

(a) $1s^22s^22p^4$ and $1s^22s^22p^5$

(b) $1s^22s^2$ and $1s^22s^22p^2$

(c) $1s^22s^1$ and $1s^22s^22p^63s^23p^64s^1$

(d) $1s^22s^22p^6$ and $1s^22s^22p^63s^23p^64s^23d^6$

Electron Configurations and the Periodic Table (Sec. 6.10)

6.43 Indicate the position in the periodic table, in terms of *s*, *p*, *d*, or *f* area, of each of the following elements.

(a) $_{23}V$ **(b)** $_{34}Se$ **(c)** $_{56}Ba$ **(d)** $_{62}Sm$

6.44 Indicate the position in the periodic table, in terms of *s*, *p*, *d*, or *f* area, of each of the following elements.

(a) $_{50}Sn$ **(b)** $_{59}Pr$ **(c)** $_{78}Pt$ **(d)** $_{87}Fr$

6.45 For each of the following elements, identify the subshell (2*s*, 3*p*, 4*f*, etc.) that contains the distinguishing electron.

(a) $_{20}Ca$ **(b)** $_{31}Ga$ **(c)** $_{40}Zr$ **(d)** $_{79}Au$

6.46 For each of the following elements, identify the subshell (2*s*, 3*p*, 4*f*, etc.) that contains the distinguishing electron.

(a) $_{16}S$ **(b)** $_{38}Sr$ **(c)** $_{39}Y$ **(d)** $_{61}Pm$

6.47 Identify the element whose electron configuration, after the distinguishing electron has been added, ends in

(a) $4p^5$ **(b)** $7s^1$ **(c)** $4d^2$ **(d)** $3d^3$

6.48 Identify the element whose electron configuration, after the distinguishing electron has been added, ends in

(a) $3p^1$ **(b)** $5s^2$ **(c)** $3d^1$ **(d)** $5d^3$

6.49 Indicate the position in the periodic table where each of the following occurs by giving the symbol of the element.

(a) The 3*p* subshell begins filling.

(b) The 2*s* subshell begins filling.

(c) The 5*d* subshell begins filling.

(d) The 3*d* subshell begins filling.

6.50 Indicate the position in the periodic table where each of the following occurs by giving the symbol of the element.

(a) The 4*p* subshell begins filling.

(b) The 5*f* subshell begins filling.

(c) The 3*s* subshell begins filling.

(d) The 7*s* subshell begins filling.

6.51 Indicate the position in the periodic table where each of the following occurs by giving the symbol of the element.

(a) The 4*p* subshell becomes completely filled.

(b) The 2*s* subshell becomes half-filled.

(c) The fourth shell begins filling.

(d) The fourth shell becomes completely filled.

6.52 Indicate the position in the periodic table where each of the following occurs by giving the symbol of the element.

(a) The 3*d* subshell becomes completely filled.

(b) The 4*p* subshell becomes half-filled.

(c) The third shell becomes half-filled.

(d) The third shell becomes completely filled.

6.53 Using only the periodic table, determine the complete electron configuration for each of the following elements.

(a) $_{33}As$ **(b)** $_{51}Sb$ **(c)** $_{37}Rb$ **(d)** $_{48}Cd$

6.54 Using only the periodic table, determine the complete electron configuration for each of the following elements.

(a) $_{22}Ti$ **(b)** $_{56}Ba$ **(c)** $_{35}Br$ **(d)** $_{19}K$

6.55 Using the periodic table as a guide, indicate the number of

(a) 3*p* electrons in a $_{16}S$ atom

(b) 3*d* electrons in a $_{29}Cu$ atom

(c) 4*s* electrons in a $_{37}Rb$ atom

(d) 4*d* electrons in a $_{30}Zn$ atom

6.56 Using the periodic table as a guide, indicate the number of

(a) 3*s* electrons in a $_{12}Mg$ atom

(b) 4*p* electrons in a $_{32}Ge$ atom

(c) 3*d* electrons in a $_{47}Ag$ atom

(d) 4*p* electrons in a $_{15}P$ atom

Classification Systems for the Elements (Sec. 6.11)

6.57 Classify each of the following elements as a noble gas, representative element, transition element, or inner transition element.

(a) $_{29}Cu$ (b) $_{32}Ge$ (c) $_{36}Kr$ (d) $_{63}Eu$

6.58 Classify each of the following elements as a noble gas, representative element, transition element, or inner transition element.

(a) $_{3}Li$ (b) $_{79}Au$ (c) $_{54}Xe$ (d) $_{72}Hf$

6.59 Identify the nonmetal in each of the following sets of elements.

(a) S, Na, K (b) Cu, Li, P
(c) Be, I, Ca (d) Fe, Cl, Ga

6.60 Identify the nonmetal in each of the following sets of elements.

(a) Al, H, Mg (b) C, Sn, Pb
(c) Ti, V, F (d) Sr, Se, Sm

6.61 Identify the metal in each of the following sets of elements.

(a) H, He, Li (b) S, Cl, K
(c) N, Fe, O (d) Hg, Ne, F

6.62 Identify the metal in each of the following sets of elements.

(a) C, Br, Pb (b) Ar, Kr, Na
(c) P, Ga, Se (d) Zn, I, Xe

6.63 Identify the lowest-atomic-numbered element that is

(a) a transition metal
(b) a noble gas
(c) a representative metal
(d) an inner transition element

6.64 Identify the lowest-atomic-numbered element that is

(a) a transition element
(b) a representative nonmetal
(c) an inner transition metal
(d) a metal

Chemical Periodicity (Sec. 6.12)

6.65 Identify the metalloid in each of the following sets of elements.

(a) Ge, Ga, Zn (b) B, Al, Ga
(c) Pb, Po, P (d) Te, I, Xe

6.66 Identify the metalloid in each of the following sets of elements.

(a) Se, As, Br (b) Br, I, At
(c) Sn, Sb, Sm (d) Al, Si, P

6.67 Using the periodic table, indicate which member of each pair is more metallic.

(a) $_{12}Mg$ or $_{14}Si$ (b) $_{47}Ag$ or $_{79}Au$
(c) $_{16}S$ or $_{17}Cl$ (d) $_{4}Be$ or $_{37}Rb$

6.68 Using the periodic table, indicate which member of each pair is more metallic.

(a) $_{11}Na$ or $_{19}K$ (b) $_{22}Ti$ or $_{29}Cu$
(c) $_{32}Ge$ or $_{34}Se$ (d) $_{30}Zn$ or $_{56}Ba$

6.69 Using the periodic table, indicate which member of each pair is more nonmetallic.

(a) $_{9}F$ or $_{35}Br$ (b) $_{14}Si$ or $_{15}P$
(c) $_{25}Mn$ or $_{30}Zn$ (d) $_{17}Cl$ or $_{33}As$

6.70 Using the periodic table, indicate which member of each pair is more nonmetallic.

(a) $_{6}C$ or $_{8}O$ (b) $_{12}Mg$ or $_{20}Ca$
(c) $_{50}Sn$ or $_{52}Te$ (d) $_{53}I$ or $_{81}Tl$

6.71 Indicate which member of each of the following pairs of elements has the larger atomic radius.

(a) $_{7}N$ or $_{8}O$ (b) $_{17}Cl$ or $_{31}Ga$
(c) $_{19}K$ or $_{35}Br$ (d) $_{19}K$ or $_{37}Rb$

6.72 Indicate which member of each of the following pairs of elements has the larger atomic radius.

(a) $_{15}P$ or $_{16}S$ (b) $_{15}P$ or $_{33}As$
(c) $_{35}Br$ or $_{52}Te$ (d) $_{37}Rb$ or $_{53}I$

ADDITIONAL PROBLEMS

6.73 What is wrong with each of the following attempts to write an electron configuration?

(a) $1s^22s^3$ (b) $1s^22s^22p^23s^2$
(c) $1s^22s^23s^2$ (d) $1s^22s^22p^63s^23d^{10}$

6.74 What is wrong with each of the following attempts to write an electron configuration?

(a) $1s^21p^6$ (b) $1s^22s^4$
(c) $1s^22s^22p^43s^2$ (d) $1s^22s^22p^63s^23p^63d^{10}$

6.75 In what period and group is an element with each of the following electron configurations located?

(a) $1s^22s^22p^63s^1$
(b) $1s^22s^22p^63s^23p^1$
(c) $1s^22s^22p^63s^23p^64s^23d^1$
(d) $1s^22s^22p^63s^23p^64s^23d^{10}4p^5$

6.76 In what period and group is an element with each of the following electron configurations located?

(a) $1s^22s^22p^2$

(b) $1s^22s^22p^63s^2$
(c) $1s^22s^22p^63s^23p^64s^23d^2$
(d) $1s^22s^22p^63s^23p^64s^23d^{10}4p^65s^2$

6.77 Assign values to *x* and *y* in each of the following electron configurations.
(a) Ca: $1s^22s^22p^63s^23p^x4s^y$
(b) Al: $1s^22s^22p^63s^x3p^y$
(c) Zn: $1s^22s^22p^63s^23p^64s^x3d^y$
(d) Kr: $1s^22s^22p^63s^23p^64s^x3d^{10}4p^y$

6.78 Assign values to *x* and *y* in each of the following electron configurations.
(a) Ti: $1s^22s^22p^63s^23p^x4s^y3d^2$
(b) Ar: $1s^22s^22p^63s^x3p^y$
(c) Cl: $1s^22s^22p^x3s^23p^y$
(d) Se: $1s^22s^22p^63s^23p^64s^23d^x4p^y$

6.79 Write electron configurations for the following elements.
(a) the group IVA element in the same period as $_{15}P$
(b) the period 2 element in the same group as $_{50}Sn$
(c) the lowest-atomic-numbered nonmetal in period VA
(d) the period 2 element that has three unpaired electrons

6.80 Write electron configurations for the following elements.
(a) the group IIIA element in the same period as $_{4}Be$
(b) the period 3 element in the same group as $_{5}B$
(c) the lowest-atomic-numbered metal in group IA
(d) the period 3 element that has three unpaired electrons

6.81 Referring only to the periodic table, determine the element of lowest atomic number whose electron configuration contains the following.
(a) two completely filled orbitals
(b) two completely filled subshells
(c) two completely filled shells
(d) two completely filled *p* subshells

6.82 Referring only to the periodic table, determine the element of lowest atomic number whose electron configuration contains the following.
(a) three completely filled orbitals
(b) three completely filled subshells
(c) three completely filled shells
(d) three completely filled *s* subshells

6.83 How many electrons are in each of the following?
(a) the outermost *p* subshell of a group IVA element
(b) the outermost *p* subshell of a group VIA element
(c) the outermost *s* subshell of a group IA element
(d) the outermost *s* subshell of a group IVB element

6.84 How many electrons are in each of the following?
(a) the outermost *s* subshell of a group IIA element
(b) the outermost *d* subshell of a group IVB element
(c) the outermost *p* subshell of a group VA element
(d) the outermost *s* subshell of a group VIIA element

6.85 Determine how many elements there are whose electron configurations end in the following.
(a) ns^2
(b) ns^2np^4
(c) $ns^2\ (n-1)\ d^1$
(d) $ns^2\ (n-1)\ d^{10}np^3$

6.86 Determine how many elements there are whose electron configurations end in the following.
(a) ns^1
(b) ns^2np^6
(c) $ns^2\ (n-1)\ d^2$
(d) $ns^2\ (n-1)\ d^{10}np^2$

6.87 In what group(s) in the periodic table would you expect to find each of the following?
(a) a representative element with two unpaired electrons
(b) a transition element with three unpaired electrons
(c) an element with two unpaired *d* electrons
(d) an element with one unpaired *s* electron

6.88 In what group(s) in the periodic table would you expect to find each of the following?
(a) a transition element with no unpaired electrons
(b) a nonmetal with no unpaired electrons
(c) an element with three unpaired *p* electrons
(d) an element with one unpaired *d* electron

6.89 Indicate which element or elements have the electron characteristics below. In those cases where a series of elements have the indicated characteristics, do not write all the symbols but rather write the atomic numbers of the first and last elements in the series, for example, elements 70–83.
(a) a total of 84 electrons
(b) only four 3*d* electrons
(c) two 7*s* electrons (note that more than one element qualifies)
(d) a total of six *s* electrons (note that more than one element qualifies)

6.90 Indicate which element or elements have the electron characteristics below. In those cases where a series of elements have the indicated characteristics, do not write all the symbols but rather write the atomic numbers of the first and last elements in the series, for example, elements 70–83.
(a) only one electron in shell 7
(b) a total of fifteen *p* electrons (they are not all in the same subshell)
(c) twelve electrons in shell 3
(d) six 5*p* electrons (note that more than one element qualifies)

6.91 Referring only to the periodic table, determine the element of lowest atomic number whose electron configuration contains the following.

(a) more *p* electrons than *s* electrons

(b) more *d* electrons than *p* electrons

6.92 Referring only to the periodic table, determine the element of lowest atomic number whose electron configuration contains the following.

(a) more *f* electrons than *s* electrons

(b) more *d* electrons than *s* electrons

6.93 If an orbital could hold three electrons instead of the actual two electrons, what would be the electron configurations of the following elements? Assume that all other relationships associated with electron configurations remain unchanged.

(a) aluminum **(b)** bromine **(c)** sulfur **(d)** iron

6.94 If an orbital could hold three electrons instead of the actual two electrons, what would be the electron configurations of the following elements? Assume that all other relationships associated with electron configurations remain unchanged.

(a) chlorine **(b)** magnesium **(c)** krypton **(d)** zinc

CUMULATIVE PROBLEMS

6.95 Which combination of the six elements oxygen, lithium, helium, boron, strontium, and potassium belongs in each of the following classifications?

(a) has an atomic number less than 20

(b) is not ductile and malleable

(c) all atoms contain an equal number of protons and electrons

(d) belongs to period 3 of the periodic table

6.96 Which combination of the six elements nitrogen, beryllium, argon, aluminum, silver, and gold belongs in each of the following classifications?

(a) period and group numbers are numerically equal

(b) readily conducts electricity and heat

(c) has an atomic number less than its atomic mass

(d) all naturally occurring atoms contain the same number of neutrons, protons, and electrons

6.97 The electron configuration of the isotope ${}^{12}_{6}C$ is $1s^2 2s^2 2p^2$. What is the electron configuration for the isotope ${}^{13}_{6}C$?

6.98 The electron configuration of the isotope ${}^{16}_{8}O$ is $1s^2 2s^2 2p^4$. What is the electron configuration for the isotope ${}^{18}_{8}O$?

6.99 How many subatomic particles are present in an atom whose isotopic mass is 36.96590 amu and whose electron configuration is $1s^2 2s^2 2p^6 3s^2 3p^5$?

6.100 How many subatomic particles are present in an atom whose isotopic mass is 29.97376 amu and whose electron configuration is $1s^2 2s^2 2p^6 3s^2 3p^2$?

6.101 An element that is a member of periodic table group IIA has atoms that possess a total of eighteen *p* electrons. What is the atomic number of this element?

6.102 An element that is a member of periodic table group VIA has atoms that possess a total of ten *s* electrons. What is the atomic number of this element?

7 Chemical Bonds

7.1 Chemical Bonds

In Section 5.2 we considered the fact that chemical compounds are conveniently divided into two broad classes called *ionic compounds* and *molecular compounds*. Ionic and molecular compounds can be distinguished from each other on the basis of general physical properties. Ionic compounds tend to have high melting points (500–2000°C) and are good conductors of electricity when they are in a molten (liquid) state. Molecular compounds, on the other hand, generally have much lower melting points and tend to be gases, liquids, or low-melting solids. They do not conduct electricity in the molten state.

Ionic compounds, in contrast to molecular compounds, do not have molecules as their basic structural unit (Sec. 5.2). Instead, an extended array of positively and negatively charged particles called *ions* are present. In this chapter, we will see why some combinations of elements produce ionic compounds and why other combinations of elements produce molecular compounds.

The subject of *chemical bonding* supplies the answer to the question of what determines whether the interaction of two elements produces ions (an ionic compound) or molecules (a molecular compound). **Chemical bonds** *are the attractive forces that hold atoms together in more complex units*. Chemical bonds form as the result of interactions between electrons found in the combining atoms. Thus, chemical bond considerations are closely linked to electron configurations (Sec. 6.7).

Corresponding to the two broad categories of chemical compounds are two types of chemical attractive forces (chemical bonds): ionic bonds and covalent bonds. An **ionic bond** *results from the transfer of one or more electrons from one atom or group of atoms to another*. As suggested by its name, the ionic bond model (electron transfer) is used in describing the attractive forces in ionic compounds. A **covalent bond** *results from the sharing of one or more pairs of electrons between atoms*.

Before we consider the details of these two bonding models it is important to emphasize that the notions of ionic and covalent bonds are merely convenient concepts. Most bonds are neither 100% ionic nor 100% covalent. Instead, most bonds have at least some degree of both ionic and covalent character, that is, some degree of both the transfer and the sharing of electons. But it is easier to understand these intermediate bonds (the real bonds) by relating them to the pure or ideal bond types called ionic and covalent.

There are two fundamental concepts that are common to and necessary for understanding both the ionic and covalent bonding models. These concepts are:

1. Not all electrons in an atom are available for bonding. Those that are are called *valence electrons.*
2. Certain arrangements of electrons are more stable than other arrangements of electrons. The *octet rule* addresses this situation.

Section 7.2 deals with the concept of valence electrons and Section 7.3 discusses the octet rule.

7.2 VALENCE ELECTRONS AND ELECTRON-DOT STRUCTURES

Certain electons, called valence electrons, are particularly important in determining the bonding characteristics of a given atom. For representative elements (Sec. 6.11) **valence electrons** *are all those electrons in the outermost electron shell, that is, in the shell with the highest shell number* (n). These electrons will always be found in either *s* or *p* subshells or both. Note the restriction on the use of this definition; it applies only for representative elements. Many commonly encountered elements are representative elements; hence the definition finds much use. (We will not consider in this text the more complicated valence electron definitions for transition or inner transition elements (Sec. 6.11); the presence of incompletely filled *inner d* or *f* subshells is the complicating factor in definitions for these elements.)

Example 7.1

Determine the number of valence electrons present in atoms of each of the following elements.

(a) $_{12}Mg$ **(b)** $_{17}Cl$ **(c)** $_{34}Se$

Solution

(a) The element magnesium has two valence electrons, as can be seen by examining its electron configuration.

$$1s^22s^22p^63s^2$$

(arrow to the superscript 2 of $3s^2$: number of valence electrons; arrow to the 3 of $3s^2$: highest value of the electron shell number)

The highest value of the electron shell number is $n = 3$. Only two electrons are found in shell 3, two electrons in the 3*s* subshell.

(b) The element chlorine has seven valence electrons.

total of seven valence electrons

$$1s^22s^22p^63s^23p^5$$

highest value of the electron shell number

Electrons in two different subshells can simultaneously be valence electrons. The highest shell number is 3, and both the $3s$ and $3p$ subshells belong to shell number 3. Hence, all electrons in both subshells are valence electrons.

(c) The element selenium has six valence electrons.

total of six valance electrons

$$1s^22s^22p^63s^23p^64s^23d^{10}4p^4$$

highest value of the electron shell number

The $3d$ electrons are not counted as valence electrons because the $3d$ subshell is in shell 3 and shell 3 is not the shell with maximum n value. Shell 4 is the outermost shell, the shell with maximum n value.

Practice Exercise 7.1

Determine the number of valence electrons present in atoms of the following elements.

(a) $_{20}Ca$ (b) $_{16}S$ (c) $_{35}Br$

Ans. (a) 2; (b) 6; (c) 7

The fact that the outermost electrons of atoms are those involved in bonding seems reasonable when it is remembered that the outermost electrons will be the first to come into close proximity when atoms collide—an event necessary before atoms can combine. Also, since these electrons are located the farthest from the nucleus, they are therefore the least tightly bound (attraction to the nucleus decreases with distance) and thus the most susceptible to change (transfer or sharing).

A shorthand system for designating numbers of valence electrons, which uses electron-dot structures, has been developed. Use of this system will make it easier, later in this chapter, to picture the role that valence electrons play in chemical bonding. An **electron-dot structure** *consists of an element's symbol with one dot for each valence electron placed about the elemental symbol.* Electron-dot structures for the first 20 elements, arranged as in the periodic table, are given in Figure 7.1. Note that the location of the dots is not critical. The following all have the same meaning.

Ṁg· Mg· ·Mg ·Ṁg Ṁg ·Mg·

Electron-dot structures are also often called Lewis structures. The American chemist Gilbert Newton Lewis (1875–1946), an early contributor to chemical bonding theory, was the first to use such structures.

Three important generalizations about valence electrons can be drawn from a study of the structures in Figure 7.1.

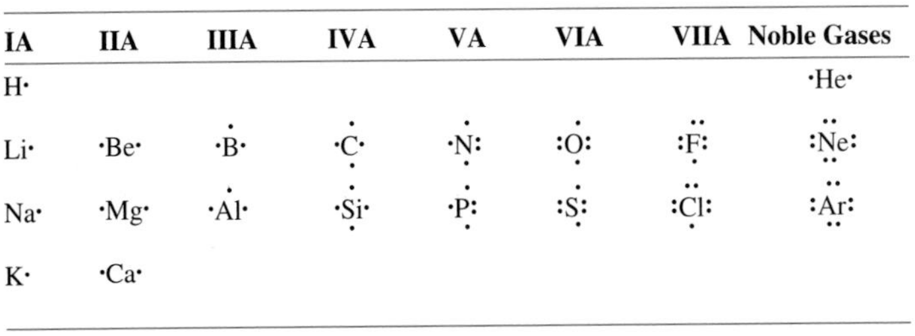
IA IIA IIIA IVA VA VIA VIIA Noble Gases
H He
Li Be B C N O F Ne
Na Mg Al Si P S Cl Ar
K Ca

FIGURE 7.1 Electron-dot structures of first 20 elements.

1. *Representative elements in the same group of the periodic table have the same number of valence electrons.* This should not be surprising to you. Elements in the same group in the periodic table have similar chemical properties as a result of having similar outer shell electron configurations (Sec. 6.9). The electrons in the outermost shell are the valence electrons.
2. *The number of valence electrons for representative elements in a group is the same as the periodic table group number.* For example, the electron-dot structures for O and S, both members of group VIA, show six dots. Similarly, the electron-dot structures of H, Li, Na, and K, all members of group IA, show one dot.
3. *The maximum number of valence electrons for any element is eight.* Only the noble gases (Sec. 6.11), beginning with Ne, have the maximum number of eight electrons. Helium, with only two valence electrons, is the exception in the noble gas family; obviously an element with a grand total of two electrons cannot have eight valence electrons. Although shells with n greater than 2 are capable of holding more than eight electrons, they do so only when they are no longer the outermost shell and thus not the valence shell. For example, selenium (Example 7.1) has 18 electrons in its third shell; however, shell 4 is the valence shell in selenium.

7.3 THE OCTET RULE

A key concept in modern bonding theory is that certain arrangements of valence electrons are more stable than others. The term *stable* as used here refers to the idea that a system (in this case an arrangement of electrons) does not easily undergo spontaneous change.

The outer shell electron configurations possessed by the noble gases (He, Ne, Ar, Kr, Xe, and Rn; see Sec. 6.11) are considered to be the *most stable of all outer shell electron configurations*. For helium, this most stable electron configuration involves two outer shell electrons. The rest of the noble gases have eight outer shell electrons.

He: $1s^2$
Ne: $1s^22s^22p^6$
Ar: $1s^22s^22p^63s^23p^6$
Kr: $1s^22s^22p^63s^23p^64s^23d^{10}4p^6$
Xe: $1s^22s^22p^63s^23p^64s^23d^{10}4p^65s^24d^{10}5p^6$
Rn: $1s^22s^22p^63s^23p^64s^23d^{10}4p^65s^24d^{10}5p^66s^24f^{14}5d^{10}6p^6$

The common feature among these noble gas electron configurations is *completely filled* outer-most s and p subshells.

The conclusion that a noble gas configuration is the most stable of all outer shell electron configurations is based on the chemical properties of the noble gases. These elements are the *most unreactive* of all the elements. They are the only elemental gases found in nature in the form of individual uncombined atoms. There are no known compounds of He, Ne, and Ar and only a very few compounds of Kr, Xe, and Rn. The noble gases appear to be "happy" the way they are. They have little or no "desire" to form bonds to other atoms.

Atoms of many elements that lack the very stable outer shell electron configuration of the noble gases tend to attain it in chemical reactions that result in compound formation. This observation has become known as the *octet rule* because of the eight outer shell electrons possessed by five of the six noble gases. A formal statement of the **octet rule** is *In forming compounds, atoms of elements lose, gain, or share electrons in such a way as to produce a noble gas electron configuration for each of the atoms involved.*

Application of the octet rule to many different systems has shown that it has value in predicting correctly the observed combining ratios of atoms. For example, it explains why two hydrogen atoms rather than some other number are bonded to one oxygen atom in the molecular compound water. It explains why the formula of the ionic compound sodium chloride is NaCl rather than $NaCl_2$, $NaCl_3$, or Na_2Cl.

There are exceptions to the octet rule, but it is still used because of the large amount of information that it is able to correlate. It is particularly effective in explaining compound formation involving only representative elements. Often complications arise with transition and inner transition elements because of the involvement of *d* and *f* electrons in the bonding.

7.4 THE IONIC BOND MODEL

Electron transfer between two or more atoms is the basic premise for the ionic-bond model. This electron transfer produces charged particles called ions. An **ion** *is an atom (or group of atoms) that is electrically charged as the result of loss or gain for electrons.* An atom is neutral only when the number of its protons (positive charges) is equal to the number of its electrons (negative charges). Loss or gain of electrons destroys this proton–electron balance and leaves a net charge on the atom.

If one or more electrons are gained by an atom, a negatively charged ion is produced; excess negative charge is present because electrons now outnumber protons. The loss of one or more electrons by an atom results in the formation of a positively charged ion; more protons than electrons are now present, resulting in excess positive charge. Note that the excess positive charge associated with a positive ion is never caused by proton gain but always by electron loss. If the number of protons remains constant and the number of electrons decreases, the result is net positive charge. The number of protons, which determines the identity of the element (Sec. 5.7), never changes during ion formation.

The charge on an ion is directly correlated with the number of electrons lost or gained. Loss of one, two, or three electrons gives ions with +1, +2, and +3 charges, respectively. Similarly, a gain of one, two, or three electrons gives ions with −1, −2, and −3 charges, respectively. (Atoms that have lost or gained more than three electrons are very seldom encountered.)

The notation for charges on ions is a superscript placed to the right of the elemental symbol. Some examples of ion symbols are

Positive ions: Na^+, K^+, Ca^{2+}, Mg^{2+}, Al^{3+}
Negative ions: Cl^-, Br^-, O^{2-}, S^{2-}, N^{3-}

Note that a single plus or minus sign is used to denote a charge of one, instead of using the notation 1+ or 1−. Also note that in multicharged ions the number precedes the charge sign; that is, the correct notation for a charge of plus two is 2+ rather than +2.

A final point about ions is that their chemical properties are very different from the neutral atoms from which they are derived. For example, water solutions containing Na^{+} ion are very stable even though the element sodium (neutral Na) reacts explosively with water.

Example 7.2

Give the symbol for each of the following ions.

(a) the ion formed when an aluminum atom loses three electrons

(b) the ion formed when a sulfur atom gains two electrons

Solution

(a) A neutral aluminum atom contains 13 protons and 13 electrons, since the atomic number of aluminum is 13 (obtained from the periodic table). The aluminum ion formed by the loss of three electrons would still contain 13 protons but would have only 10 electrons because three electrons were lost.

$$\begin{aligned} 13 \text{ protons} &= 13 + \text{ charges} \\ 10 \text{ electrons} &= \underline{10 - \text{ charges}} \\ \text{Net charge} &= 3 + \end{aligned}$$

The symbol for the aluminum ion is thus Al^{3+}.

(b) The atomic number of sulfur is 16. Thus, 16 protons and 16 electrons are present in a neutral sulfur atom. A gain of two electrons raises the electron count to 18.

$$\begin{aligned} 16 \text{ protons} &= 16 + \text{ charges} \\ 18 \text{ electrons} &= \underline{18 - \text{ charges}} \\ \text{Net charge} &= 2 - \end{aligned}$$

The symbol for the sulfur ion is thus S^{2-}.

Practice Exercise 7.2

Give the symbol for each of the following ions.

(a) the ion formed when a calcium atom loses two electrons

(b) the ion formed when a nitrogen atom gains three electrons

Ans. (a) Ca^{2+}; (b) N^{3-}

Example 7.3

Determine the number of protons and electrons present in the following ions.

(a) P^{3-} **(b)** Mg^{2+}

Solution

(a) The number of protons present is the same as in a neutral atom and is therefore given by the atomic number, which is 15 for phosphorus

$$\text{Number of protons} = \text{atomic number} = 15$$

The number of electrons in a neutral phosphorus atom is 15. The charge of −3 on the ion indicates the gain of three electrons.

$$\text{Number of electrons} = 15 + 3 = 18$$

(b) The atomic number of magnesium is 12.

$$\text{Number of protons} = \text{atomic number} = 12$$

The number of electrons in a neutral Mg atom is 12. The charge of +2 on the ion indicates the loss of two electrons.

$$\text{Number of electrons} = 12 - 2 = 10$$

Practice Exercise 7.3

Determine the number of protons and electrons present in the following ions.
(a) K^+ (b) S^{2-}

Ans. (a) 19 protons and 18 electrons; (b) 16 protons and 18 electrons.

So far our discussion about electron transfer and ion formation has focused on the loss or gain of electrons by isolated individual atoms. During ionic bond formation ion formation occurs only when atoms of two elements are present—an element that can lose electrons and an element that can gain electrons. The total number of electrons lost by atoms of the one element is the same as the total number gained by atoms of the other element. Thus, positive and negative ions must always be formed at the same time.

The mutual attraction between the positive and negative ions that results from electron transfer constitutes the force that holds the ions together as an ionic compound. This force is referred to as an ionic bond. An **ionic bond** *is the attractive force between positive and negative ions that causes them to remain together as a group.*

7.5 IONIC COMPOUND FORMATION

A simple example of ionic bonding occurs between the elements sodium and chlorine in the compound NaCl. Sodium atoms lose (transfer) one electron to chlorine atoms, producing Na^+ and Cl^- ions. These ions combine in a one-to-one ratio to give the compound NaCl.

Why do sodium atoms form Na^+ and not Na^{2+} or Na^- ions? Why do chlorine atoms form Cl^- ions rather than Cl^{2-} or Cl^+ ions? In general, what determines the specific number of electrons lost or gained in electron-transfer processes?

The octet rule (Sec. 7.3) provides very simple and straightforward answers to these questions. *Atoms tend to gain or lose electrons until they have obtained an electron configuration that is the same as that of a noble gas.*

Consider the element sodium, which has the electron configuration

$$1s^22s^22p^63s^1$$

It can attain a noble gas configuration by losing one electron (to give it the electron configuration of neon) or by gaining seven electrons (to give it the electron configuration of argon).

Na $(1s^22s^22p^63s^1)$

loss of 1 e^- → Na^+ $(1s^22s^22p^6)$
electron configuration of neon

gain of 7 e^- → Na^{7-} $(1s^22s^22p^63s^23p^6)$
electron configuration of argon

The first process, the loss of one electron, is more energetically favorable than the gain of seven electrons and is the process that occurs. The process that involves the fewer number of electrons will always be the more energetically favorable process and will be the process that occurs.

Consider the element chlorine, which has the electron configuration

$$1s^22s^22p^63s^23p^5$$

It can attain a noble gas configuration by losing seven electrons (to give it the electron configuration of neon) or by gaining one electron (to give it the electron configuration of argon). The latter occurs for the reason cited previously.

Cl $(1s^22s^22p^63s^23p^5)$

loss of 7 e^- → Cl^{7+} $(1s^22s^22p^6)$
electron configuration of neon

gain of 1 e^- → Cl^- $(1s^22s^22p^63s^23p^6)$
electron configuration of argon

The type of consideration we have just used for the elements sodium and chlorine lead to the following generalizations.

1. Metal atoms containing one, two, or three valence electrons (the metals in groups IA, IIA, and IIIA of the periodic table) tend to lose electrons to acquire a noble gas electron configuration.

   Group IA metals form +1 ions.
   Group IIA metals form +2 ions.
   Group IIIA metals form +3 ions.

   Group IA metals are all located one periodic table position past a noble gas. Thus, they will each have one more electron than the preceding noble gas. This electron must be lost if a noble gas configuration is to be obtained. Group IIA and IIIA metals are two and three periodic table positions, respectively, beyond a noble gas. Consequently, two and three electrons, respectively, must be lost for these metals to attain a noble gas electron configuration.

2. Nonmetal atoms containing five, six, or seven valence electrons (the nonmetals in groups VA, VIA, and VIIA of the periodic table) tend to gain electrons to acquire a noble gas configuration.

   Group VIIA nonmetals form − 1 ions.
   Group VIA nonmetals form − 2 ions.
   Group VA nonmetals form − 3 ions.

   The nonmetal ionic charge guidelines can be explained by reasoning similar to that for the metals, only this time the periodic table positions are those immediately preceding the noble gases. Consequently, electrons must be gained to attain noble gas configurations.

TABLE 7.1 IONS ISOELECTRONIC WITH SELECTED NOBLE GASES

Helium Structure		Neon Structure		Argon Structure	
$1s^2$		$1s^2 2s^2 2p^6$		$1s^2 2s^2 2p^6 3s^2 3p^6$	
H^-	Li^+	N^{3-}	Na^+	P^{3-}	K^+
	Be^{2+}	O^{2-}	Mg^{2+}	S^{2-}	Ca^{2+}
		F^-	Al^{3+}	Cl^-	

Elements in group IVA occupy unique positions relative to the noble gases. They are located equidistant between two noble gases. For example, the element carbon is four positions beyond helium and four positions before neon. Theoretically, ions with charges of +4 or −4 could be formed by elements in this group, but in most cases the bonding that results is more adequately described by the covalent bond model to be discussed in Section 7.9.

An ion formed in the preceding manner with an electronic configuration the same as that of a noble gas is said to be *isoelectronic* with the noble gas. **Isoelectronic species** *contain the same number of electrons.* An atom and an ion or two ions may be isoelectronic. Numerous ions that are isoelectronic with a given noble gas exist, as can be seen from the entries in Table 7.1.

It should be emphasized that an ion that is isoelectronic with a noble gas does not have the properties of the noble gas. It has not been converted into the noble gas. The number of protons in the nucleus of the isoelectronic ion is different from that in the noble gas. These points are emphasized by the comparison in Table 7.2 between Mg^{2+} and Ne, the noble gas with which Mg^{2+} is isoelectronic.

The use of electron-dot structures helps in visualizing the formation of ionic compounds through electron transfer. Let us consider, again, the reaction between sodium, which has one valence electron, and chlorine, which has seven valence electrons, to give NaCl. This reaction can be represented as follows with electron-dot structures.

$$Na\cdot + \cdot\ddot{\underset{\cdot\cdot}{Cl}}: \longrightarrow Na^+ [:\ddot{\underset{\cdot\cdot}{Cl}}:]^- \longrightarrow NaCl$$

The loss of an electron by sodium empties its valence shell. The next inner shell, which contains eight electrons (a noble gas configuration), then becomes the valence shell. After the valence shell of chlorine gains one electron, it then has the "desired" eight valence electrons.

When sodium, which has one valence electron, combines with oxygen, which has six

TABLE 7.2 A COMPARISON OF THE STRUCTURE OF A Mg^{2+} ION AND A NE ATOM, THE NOBLE GAS ATOM ISOELECTRONIC WITH THE ION

	Ne Atom	Mg^{2+} Ion
Protons (in the nucleus)	10	12
Electrons (around the nucleus)	10	10
Atomic number	10	12
Charge	0	2+

valence electrons, each oxygen atom requires two sodium atoms to meet its need of two additional electrons.

$$\begin{array}{c} \text{Na}\cdot \\ \\ \text{Na}\cdot \end{array} \!\!\!\!\!\!\! :\ddot{\text{O}}: \longrightarrow \begin{array}{c} \text{Na}^+ \\ \text{Na}^+ \end{array} [:\ddot{\text{O}}:]^{2-} \longrightarrow \text{Na}_2\text{O}$$

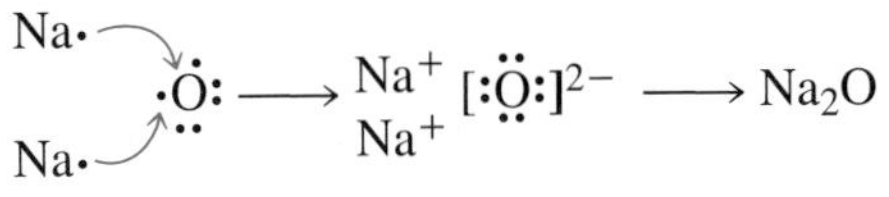

Note how oxygen's need for two additional electrons dictates that two sodium atoms are required for each oxygen atom; hence the formula is Na_2O.

An opposite situation to that for Na_2O occurs in the reaction between calcium, which has two valence electrons, and chlorine, which has seven valence electrons. Here, two chlorine atoms are required to accommodate electrons transferred from one calcium atom because a chlorine atom can accept only one electron. (It has seven valence electrons and needs only eight.)

$$\begin{array}{c} \cdot\ddot{\text{Cl}}: \\ \text{Ca}\cdot \\ \cdot\ddot{\text{Cl}}: \end{array} \longrightarrow \text{Ca}^{2+} \begin{array}{c} [:\ddot{\text{Cl}}:]^- \\ \\ [:\ddot{\text{Cl}}:]^- \end{array} \longrightarrow \text{CaCl}_2$$

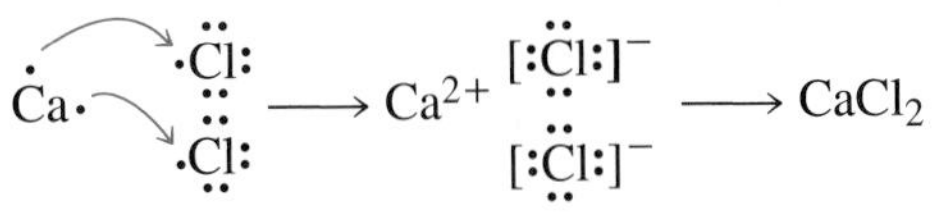

Example 7.4

Show the formation of the following ionic compounds using electron-dot structures.

(a) K_3P **(b)** NaF   **(c)** Al_2O_3

Solution

(a) Potassium (a group IA element) has one valence electron, which it would "like" to lose. Phosphorus (a group VA element) has five valence electrons and would thus "like" to acquire three more. Three potassium atoms will be required to supply enough electrons for one nitrogen atom.

$$\begin{array}{c} \text{K}\cdot \\ \text{K}\cdot \quad \cdot\ddot{\text{P}}: \\ \text{K}\cdot \end{array} \longrightarrow \begin{array}{c} \text{K}^+ \\ \text{K}^+ \; [:\ddot{\text{P}}:]^{3-} \\ \text{K}^+ \end{array} \longrightarrow \text{K}_3\text{P}$$

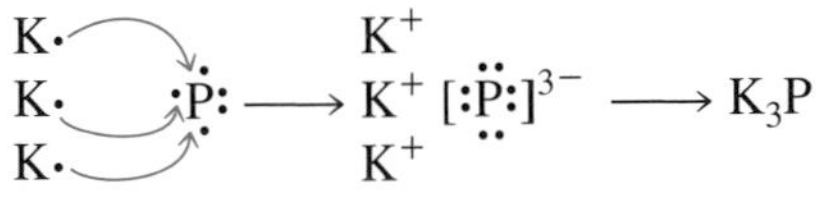

(b) Sodium (a group IA element) has one valence electron, and fluorine (a group VIIA element) has seven valence electrons. The transfer of the one sodium valence electron to a fluorine atom will result in each atom having a noble gas electron configuration. Thus, these two elements combine in a one-to-one ratio.

$$\text{Na}\cdot \quad \cdot\ddot{\text{F}}: \longrightarrow \text{Na}^+ [:\ddot{\text{F}}:]^- \longrightarrow \text{NaF}$$

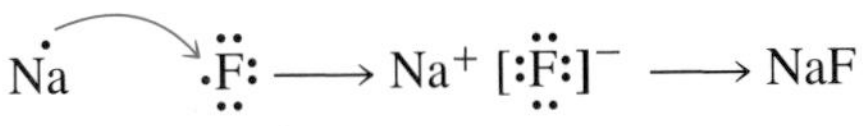

(c) Aluminum (a group IIIA element) has three valence electrons, all of which need to be lost through electron transfer. Oxygen (a group VIA element) has six valence electrons and thus needs to acquire two more. Three oxygen atoms are needed to accommodate the electrons given up by two aluminum atoms.

$$\begin{array}{c} \cdot\dot{\text{Al}}\cdot \quad \cdot\ddot{\text{O}}: \\ \\ \cdot\ddot{\text{O}}: \\ \\ \cdot\dot{\text{Al}}\cdot \quad \cdot\ddot{\text{O}}: \end{array} \longrightarrow \begin{array}{c} \text{Al}^{3+} \\ [:\ddot{\text{O}}:]^{2-} \\ \text{Al}^{3+} \; [:\ddot{\text{O}}:]^{2-} \\ [:\ddot{\text{O}}:]^{2-} \end{array} \longrightarrow \text{Al}_2\text{O}_3$$

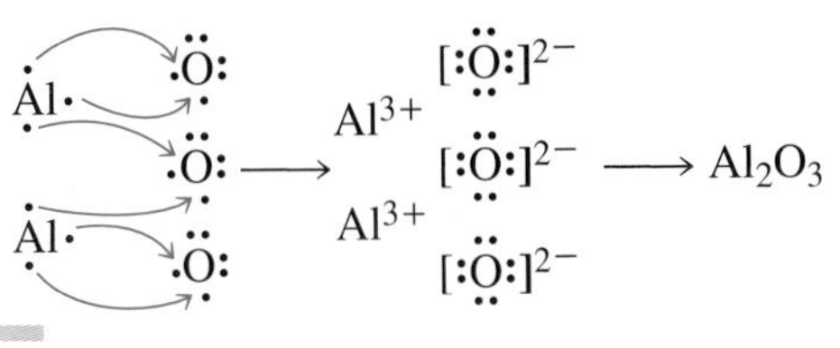

Practice Exercise 7.4

Show the formation of the following ionic compounds using electron-dot structures.
(a) KF (b) Li_2O (c) Ca_3P_2

Ans.

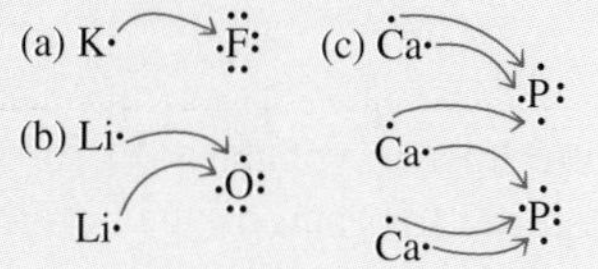

7.6 FORMULAS FOR IONIC COMPOUNDS

It is not always necessary or convenient to write electron-dot structures when determining the formula for an ionic compound. Formulas for ionic compounds can be written directly by using the charges associated with ions being combined and the fact that the total amount of positive and negative charge must add up to zero. Since total electron loss always equals total electron gain in an electron-transfer process, ionic compounds are always neutral; no net charge is present. The total positive charge on the ions that have lost electrons is always exactly counterbalanced by the total negative charge on the ions that have gained electrons. Thus, *the ratio in which positive and negative ions combine is the ratio that achieves charge neutrality for the resulting compound.*

The correct combining ratio when K^+ and S^{2-} ions combine is two to one. Two K^+ ions (each of +1 charge) will be required to balance the charge on a single S^{2-} ion.

$$\begin{aligned} 2(K^+)\text{:} &\quad (2\text{ ions}) \times (\text{charge of } +1) = +2 \\ S^{2-}\text{:} &\quad (1\text{ ion}) \times \underline{(\text{charge of } -2) = -2} \\ &\quad \text{Net charge} = 0 \end{aligned}$$

Hence, the formula is K_2S.

Example 7.5 gives further illustration of the procedures needed to determine correct combining ratios between ions and to write correct ionic formulas from the combining ratios. Correct ionic formulas are consistent with the following guidelines.

1. The symbol for the positive ion is always written first.
2. The charges on the ions that are present are *not* shown in the formula. Knowledge of charges is necessary to determine the formula, but once it is determined, the charges are not explicitly written.
3. The numbers in the formula (the subscripts) give the combining ratio for the ions.

Example 7.5

Determine the formula for the compound that is formed when each of the following types of ions interact.

(a) Ba^{2+} and Cl^- **(b)** Ba^{2+} and S^{2-} **(c)** Ba^{2+} and N^{3-}

Solution

(a) Ba^{2+} and Cl^- ions will combine in a one-to-two ratio because this combination will cause the total charge to add up to zero. One Ba^{2+} ion gives a total positive charge of 2. Two Cl^- ions give a total negative charge of 2. Thus, the formula of the compound is $BaCl_2$.

(b) The formula of this compound is simply BaS (a one-to-one ratio between ions). One Ba^{2+} ion contributes two units of positive charge, and that is counterbalanced by two units of negative charge from the S^{2-} ion.

(c) The numbers in the charges for these ions are 2 and 3. The lowest common multiple of 2 and 3 is 6 ($2 \times 3 = 6$). Thus, we will need six units of positive charge and six units of negative charge. Three Ba^{2+} ions are needed to give the six units of positive charge, and two N^{3-} ions are needed to give the six units of negative charge. The combining ratio of ions is three to two, and the formula is Ba_3N_2. The strategy used in determining this formula, finding the lowest common multiple in the charges of the ions, will always work.

Practice Exercise 7.5

Determine the formula of the compound that is formed when each of the following types of ions interact.

(a) Na^+ and P^{3-} (b) Be^{2+} and P^{3-} (c) Al^{3+} and P^{3-}

Ans. (a) Na_3P; (b) Be_3P_2; (c) AlP

Before leaving the subject of ions, ionic bonds, and formulas for ionic compounds, let us quickly review the key principles about ionic bonding that have been presented.

1. Ionic compounds usually contain both a metallic and a nonmetallic element.
2. The metallic element atoms lose electrons to produce positive ions, and the nonmetallic element atoms gain electrons to produce negative ions.
3. The electrons lost by the metal atoms are the same ones that are gained by the nonmetal atoms. Electron loss must always equal electron gain.
4. The ratio in which positive metal ions and negative nonmetal ions combine is the simplest ratio that achieves charge neutrality for the resulting compound.
5. Metals from groups IA, IIA, and IIIA of the periodic table form ions with charges of +1, +2, and +3, respectively. Nonmetals of groups VIIA, VIA, and VA of the periodic table form ions with charges of −1,−2, and −3, respectively. Table 7.3 lists, in general terms, all of the possible metal–nonmetal combinations from these periodic table groups that result in the formation of ionic compounds.

TABLE 7.3 GENERAL FORMULAS FOR IONIC COMPOUNDS AS A FUNCTION OF PERIODIC TABLE POSITION

Metal Group Number	Nonmetal Group Number	Charge on Metal (M) Ion	Charge on Nonmetal (X) Ion	Formula of Compound
IA	VIIA	+1	−1	MX
IA	VIA	+1	−2	M_2X
IA	VA	+1	−3	M_3X
IIA	VIIA	+2	−1	MX_2
IIA	VIA	+2	−2	MX
IIA	VA	+2	−3	M_3X_2
IIIA	VIIA	+3	−1	MX_3
IIIA	VIA	+3	−2	M_2X_3
IIIA	VA	+3	−3	MX

7.7 STRUCTURE OF IONIC COMPOUNDS

F.Y.I.

Looked at closely, salt grains reveal the cubical shape of the NaCl crystal. The intricate shapes of many gems such as rubies and emeralds reflect the arrangement of their microscopic ionic arrays.

The term *molecule* is not appropriate for describing the smallest unit of an ionic compound (Sec. 5.2). In the solid state, ionic compounds consist of an extended array of alternating positive and negative ions. **Ionic solids** *consist of positive and negative ions arranged in such a way that each ion is surrounded by nearest neighbors of the opposite charge.* Any given ion is bonded by electrostatic (positive–negative) attractions to all of the other ions of opposite charge immediately surrounding it. Figure 7.2 gives two three-dimensional depictions of the arrangement of ions for the ionic compound NaCl (table salt).

We can see in Figure 7.2 that discrete molecules do not exist in an ionic solid. Therefore, the formulas for these solids (Sec. 7.6) cannot represent the composition of a molecule of the substance. Instead, formulas for this type of solid represent the simplest ratio in which the atoms combine. For example, in NaCl (Fig. 7.2) there is no unique partner for a sodium ion; there are six immediate neighbors (chloride ions) that are equidistant

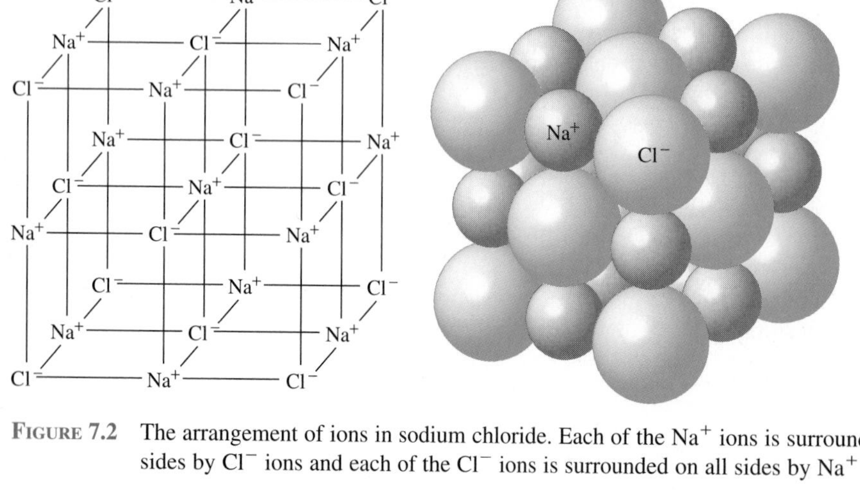

FIGURE 7.2 The arrangement of ions in sodium chloride. Each of the Na^+ ions is surrounded on all sides by Cl^- ions and each of the Cl^- ions is surrounded on all sides by Na^+ ions.

from it. A chloride ion in turn has six immediate sodium neighbors. The formula NaCl represents the fact that sodium and chloride ions are present in this solid in a one-to-one ratio.

Although the formulas for ionic solids represent only ratios, they are used in equations and chemical calculations in the same way as the formulas for molecular species. Remember, however, that they cannot be interpreted as indicating that molecules exist for these substances. They represent the simplest ratio of ions.

7.8 POLYATOMIC IONS

To this point in this chapter all references to and comments about ions have involved monoatomic ions. **Monoatomic ions** *are single atoms that have lost or gained electrons.* Such ions are very common and very important. Another large and important category of ions, called polyatomic ions, exists. Numerous ionic compounds exist in which the positive or negative ion (sometimes both) is polyatomic. Polyatomic ions are very stable species, generally maintaining their identity during chemical reactions.

An example of a polyatomic ion is the sulfate ion, SO_4^{2-}. This ion contains four oxygen atoms and one sulfur atom, and the whole group of five atoms has acquired a -2 charge. The whole sulfate group is the ion rather than any one atom within the group. Covalent bonding, discussed in Section 7.9, holds the sulfur and oxygen atoms together. A **polyatomic ion** *is a group of atoms with a net charge, held together by covalent bonds.*

Polyatomic ions are not molecules. They never occur alone as molecules do. Instead, they are always found associated with ions of opposite charge. Polyatomic ions are *pieces* of compounds, not compounds. Ionic compounds require the presence of both positive and negative ions and are neutral overall. Polyatomic ions are always charged species.

Formulas for ionic compounds containing polyatomic ions are determined in the same way as those for ionic compounds containing monoatomic ions (Sec. 7.6). The basic rule is the same: the total positive and negative charge present must add up to zero.

Two conventions not encountered previously in formula writing often arise when writing formulas with polyatomic ions. They are

1. When more than one polyatomic ion of a given kind is required in a formula, the polyatomic ion is enclosed in parentheses and a subscript is placed outside the parentheses to indicate the number of polyatomic ions needed.
2. To preserve the identity of polyatomic ions, the same elemental symbol may be used more than once in a formula.

Example 7.6 contains examples illustrating the use of both of these new conventions. Besides the sulfate ion, four other polyatomic ions are involved in this example: OH^- (hydroxide ion), NO_3^- (nitrate ion), NH_4^+ (ammonium ion), and CN^- (cyanide ion). The formulas for numerous other polyatomic ions are considered in Section 8.4.

Example 7.6

Determine the formulas for the ionic compounds containing the following pairs of ions.

(a) K^+ and SO_4^{2-} **(b)** Na^+ and NO_3^-

(c) Ca^{2+} and OH^- **(d)** NH_4^+ and CN^-

Solution

(a) In order to equalize the total positive and negative charges, we need two K^+ ions for each SO_4^{2-} ion. We indicate the presence of the two K^+ ions with the subscript 2 following the symbol of the ion. The formula of the compound is K_2SO_4. The convention that the positive ion is always written first in the formula still holds when polyatomic ions are present.

(b) Since both of these ions possess a charge of one, combining them in a one-to-one ratio will balance the charge. The formula of the compound is $NaNO_3$.

(c) Two OH^- ions are needed to balance the charge on one Ca^{2+} ion. Since more than one polyatomic ion is needed, the formula will contain parentheses: $Ca(OH)_2$. The subscript 2 outside the parentheses indicates two of what is inside the parentheses. If parentheses were not used, the formula would appear to be $CaOH_2$, which is not intended and which actually conveys false information. The formula $Ca(OH)_2$ indicates a formula unit containing one Ca atom, two O atoms, and two H atoms (Sec. 5.4); the formula $CaOH_2$ would indicate a formula unit containing one Ca atom, one O atom, and two H atoms. Verbally the correct formula, $Ca(OH)_2$, would be read as "C-A" (pause) "O-H-taken-twice."

(d) In this compound both ions are polyatomic, a perfectly legal situation. Since the ions have equal but opposite charges, they will combine in a one-to-one ratio. The formula is thus NH_4CN. No parentheses are needed because we need only one polyatomic ion of each type in a formula unit. Parentheses are used only when there are two or more polyatomic ions of a given kind in a formula unit. What is different about this formula is the appearance of the symbol for the element nitrogen (N) at two locations in the formula. This could be prevented by combining the two nitrogens, giving the formula N_2H_4C. However, combining is not done in situations like this because the identity of the polyatomic ions present is lost in the resulting combined formula. The formula N_2H_4C does not convey the message that NH_4^+ and CN^- ions are present; the formula NH_4CN does. Thus, in writing formulas that contain polyatomic ions we always maintain the identities of these ions even if it means having the same elemental symbol at more than one location in the formula.

Practice Exercise 7.6

Determine the formulas for the ionic compounds containing the following pairs of ions.

(a) K^+ and OH^- (b) Na^+ and CN^-
(c) Ca^{2+} and NO_3^- (d) NH_4^+ and SO_4^{2-}

Ans. (a) KOH; (b) NaCN; (c) $Ca(NO_3)_2$; (d) $(NH_4)_2SO_4$

Figure 7.3 gives pictorial representations of the ionic makeup of the four compounds whose formulas were determined in Example 7.6.

7.9 THE COVALENT BOND MODEL

A three-point contrast between the ionic bond model (Sec. 7.4) and the covalent bond model begins our discussion of covalent bonding and molecular compounds that result from such bonding.

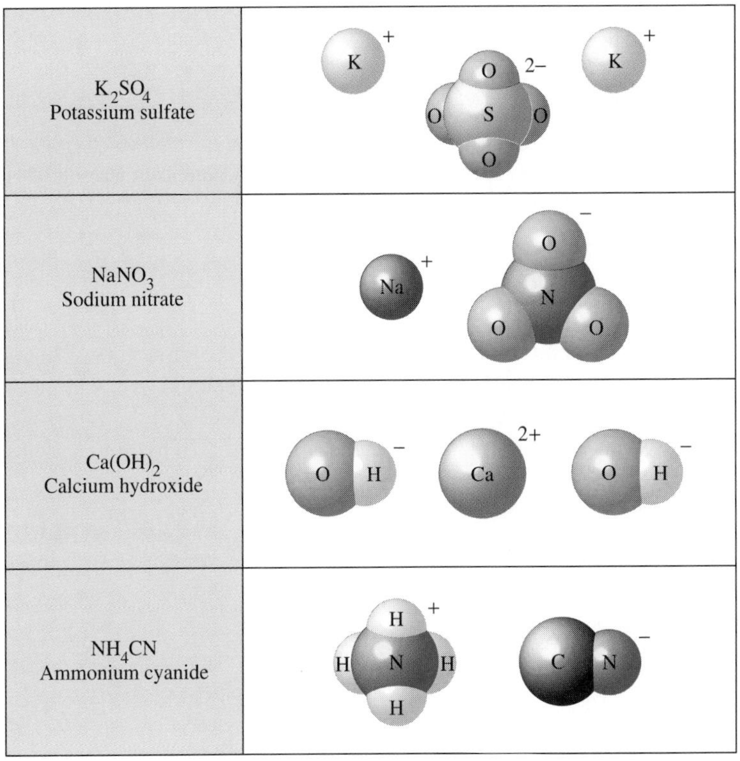

FIGURE 7.3 Ionic makeup of some compounds in which polyatomic ions are present.

1. Ionic bonds form between atoms of a metal and a nonmetal. Covalent bond formation occurs between two nonmetal atoms. The two nonmetal atoms can be identical but need not be so.
2. *Electron transfer* is the mechanism by which ionic bond formation occurs. Covalent bond formation involves *electron sharing*.
3. In an ionic compound discrete molecules do not exist since such compounds involve an extended array of alternating positive and negative ions. In covalently bonded compounds the basic structural unit is the molecule. Indeed, such compounds are called *molecular* compounds.

A **covalent bond** *is the bonding force resulting from two nuclei attracting the same shared electrons.* A consideration of the simple hydrogen (H_2) molecule provides initial insights into the nature of the covalent bond. When two hydrogen atoms, each with a single electron, are brought together, the orbitals, containing the single electrons *overlap* (as shown in Figure 7.4) to produce an orbital common to both atoms. The two electrons, one from each H atom, move throughout this new orbital and are said to be *shared* by the two nuclei.

Once two orbitals overlap, the most favorable location for the shared electrons is the "area" directly between the two nuclei. Here the two electrons can simultaneously interact with (be attracted to) both nuclei, a situation that produces increased stability. This simple analogy illustrates the "increased stability" concept. Consider the nuclei of the two hydrogen atoms in H_2 to be "old potbellied stoves" and the two electrons to be running around each of the stoves trying to keep warm. When the two nuclei are together (an H_2 molecule) the electrons have two sources of heat. In particular, in the region between the nuclei (the

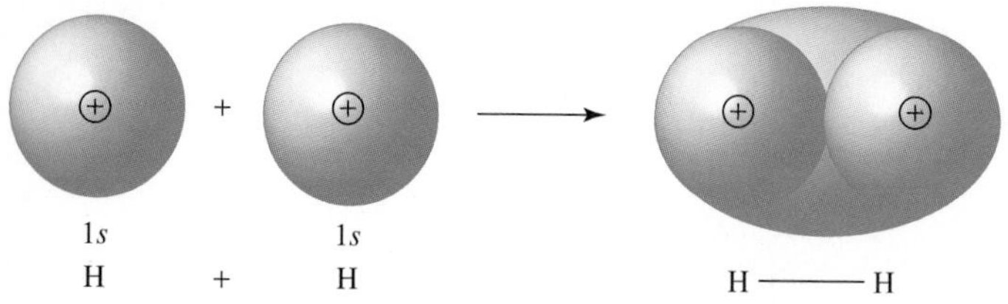

FIGURE 7.4 In the H_2 molecule, the two electrons present are shared by the two atoms as a result of 1*s* orbital overlap.

overlap region) the electrons can keep both front and back warm at the same time. This is a better situation than when each electron has only one "stove" (nucleus) as a source of heat.

In terms of electron-dot structures, this sharing of electrons by the two hydrogen atoms is diagrammed as

shared electron pair

$$\mathrm{H\cdot} \; \mathrm{\cdot H} \longrightarrow \mathrm{H{:}H}$$

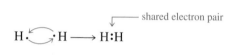

The two shared electrons do double duty, helping each of the hydrogen atoms achieve a helium noble-gas configuration.

7.10 ELECTRON-DOT STRUCTURES FOR MOLECULAR COMPOUNDS

Using the octet rule, which applies to both electron transfer and electron sharing (Sec. 7.3), and electron-dot structures (Sec. 7.2), let us now consider the formation of selected simple covalently bonded molecules containing the element chlorine. Chlorine, located in Group VIIA of the periodic table, has seven valence electrons. Its electron-dot structure is

$$\cdot\ddot{\underset{\cdot\cdot}{\mathrm{Cl}}}:$$

Chlorine needs one additional electron to achieve the octet of electrons that makes it isoelectronic with the noble gas argon. In ionic compounds, where it bonds to metals, the Cl receives the needed electron via electron transfer. When Cl combines with another nonmetal, a common situation, the octet of electrons is completed via electron sharing. Representative of the situation where chlorine obtains its eighth valence electron through an electron-sharing process are the molecules HCl, Cl_2, and BrCl, whose electron-dot structures are as follows.

$$\mathrm{H\cdot} \; \cdot\ddot{\underset{\cdot\cdot}{\mathrm{Cl}}}: \longrightarrow \mathrm{H}:\ddot{\underset{\cdot\cdot}{\mathrm{Cl}}}:$$

$$:\ddot{\underset{\cdot\cdot}{\mathrm{Cl}}}\cdot \; \cdot\ddot{\underset{\cdot\cdot}{\mathrm{Cl}}}: \longrightarrow :\ddot{\underset{\cdot\cdot}{\mathrm{Cl}}}:\ddot{\underset{\cdot\cdot}{\mathrm{Cl}}}:$$

$$:\ddot{\underset{\cdot\cdot}{\mathrm{Br}}}\cdot \; \cdot\ddot{\underset{\cdot\cdot}{\mathrm{Cl}}}: \longrightarrow :\ddot{\underset{\cdot\cdot}{\mathrm{Br}}}:\ddot{\underset{\cdot\cdot}{\mathrm{Cl}}}:$$

The HCl and BrCl molecules illustrate the point that the two atoms involved in a covalent bond need not be identical (as in H_2 or Cl_2).

A common practice when writing electon-dot structures for covalently bonded molecules is to represent the *shared* electron pairs with dashes. Using this notation, the previously discussed H_2, HCl, Cl_2, and BrCl molecules are written as

The atoms in covalently bonded molecules often possess both *bonding* and *nonbonding* electrons. **Bonding electrons** *are pairs of valence electrons that are shared between atoms in a covalent bond.* Each of the chlorine atoms in the molecules HCl, Cl_2, and BrCl possess one pair of bonding electrons. **Nonbonding electrons** *are pairs of valence electrons about an atom that are not involved in electron sharing.* Each of the chlorine atoms in HCl, Cl_2, and BrCl possesses three pairs of nonbonding electrons, as does the bromine atom in BrCl.

The number of covalent bonds that an atom forms is equal to the number of electrons it needs to achieve a noble gas configuration. Note that the chlorine atoms in HCl, Cl_2, and BrCl all formed one covalent bond. For chlorine, seven valence electrons plus one electron acquired by electron sharing (one bond) give the eight valence electrons needed for a noble gas electronic configuration. The elements oxygen, nitrogen, and carbon have, respectively, six, five, and four valence electrons. Therefore, these elements form, respectively, two, three, and four covalent bonds. The number of covalent bonds these three elements form is reflected in the formulas of their simplest hydrogen compounds H_2O, NH_3, and CH_4. Electron-dot structures for these three molecules are as follows.

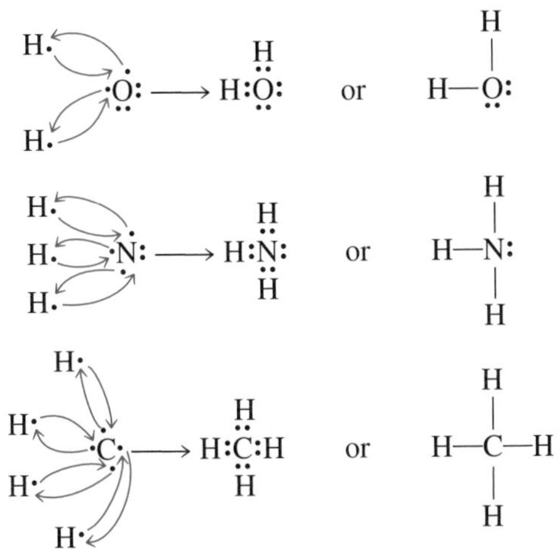

Thus, we see that just as the octet rule was useful in determining the ratio of ions in ionic compounds, we can use it to predict formulas in molecular compounds. Example 7.7 gives additional illustrations of the use of the octet rule to determine formulas for molecular compounds.

Example 7.7

Write electron-dot structures for the simplest molecular compound formed from the following pairs of nonmetals.

(a) phosphorus and hydrogen **(b)** sulfur and fluorine

(c) oxygen and chlorine

Solution

(a) Phosphorus is in group VA of the periodic table and thus has five valence electrons. It will therefore want to form three covalent bonds which through electron sharing will give it eight valence electrons (noble gas configuration). Hydrogen, in Group IA of the periodic table, has one valence electron and will want to form only one covalent bond. Remember that for H an "octet" is two electrons; the noble gas that hydrogen "mimics" is helium, which has only two valence electrons. Therefore, using electron-dot structures, we have

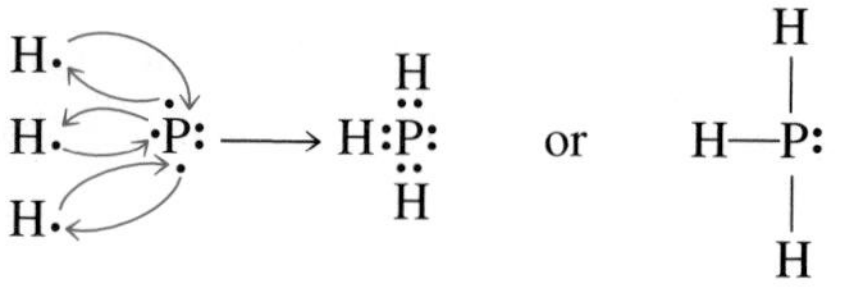

(b) Sulfur has six valence electrons and fluorine has seven valence electrons. Thus sulfur will form two covalent bonds (6 + 2 = 8), and fluorine will form one covalent bond (7 + 1 = 8).

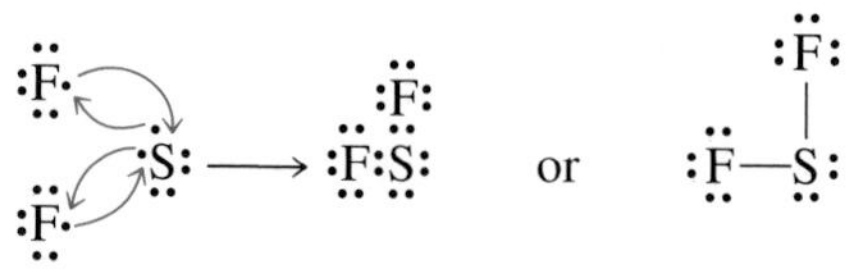

(c) Oxygen, with six valence electrons, will form two covalent bonds, and chlorine, with seven valence electrons, will form only one covalent bond. The formula of the compound is thus Cl_2O, which has the following electron-dot structure.

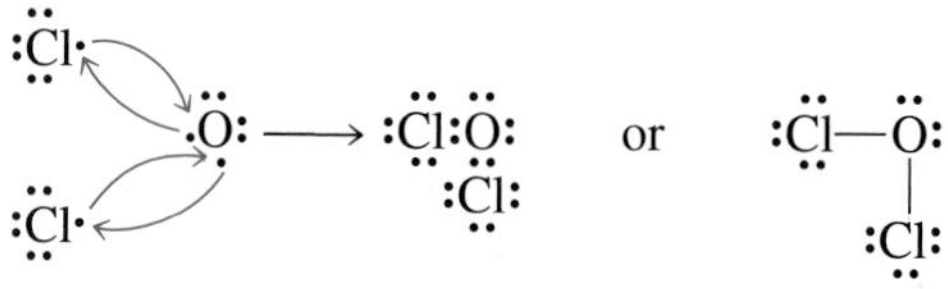

Practice Exercise 7.7

Write electron-dot structures for the simplest molecular compound formed from the following pairs of nonmetals.

(a) nitrogen and fluorine (b) carbon and chlorine
(c) sulfur and bromine

Ans.

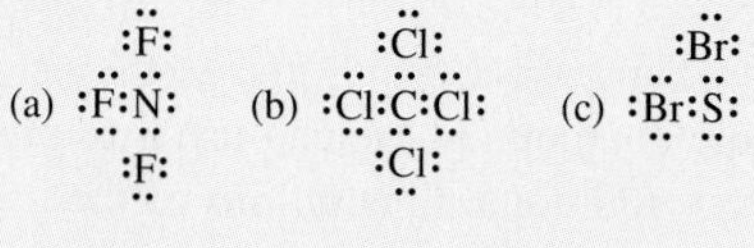

7.11 SINGLE, DOUBLE, AND TRIPLE COVALENT BONDS

A **single covalent bond** *is a chemical bond where two atoms share one pair of valence electrons.* All bonds in all of the molecules discussed in the previous section are *single* covalent bonds.

Single covalent bonds are not adequate to explain covalent bonding in all molecules. Sometimes two atoms must share two or three pairs of electrons in order to provide a complete octet of electrons for each atom involved in the bonding. Such bonds are called *double* covalent bonds and *triple* covalent bonds. A **double covalent bond** *is a chemical bond where two atoms share two pairs of valence electrons.* A double covalent bond between two atoms is approximately twice as strong as a single covalent bond between the same two atoms; that is, it takes approximately twice as much energy to break the double bond as it takes to break the single bond. A **triple covalent bond** *is a chemical bond where two atoms share three pairs of valence electrons.* A triple covalent bond is approximately three times as strong as a single covalent bond between the same two atoms. The term *multiple covalent bond* is a designation that applies collectively to both double and triple covalent bonds. This designation is often shortened to simply "multiple bond."

One of the simplest molecules possessing a multiple covalent bond is the N_2 molecule; a triple covalent bond is present. A nitrogen atom has five valence electrons and needs three additional electrons to complete its octet.

In an N_2 molecule the only sharing that can take place is between the two nitrogen atoms. They are the only atoms present. Thus, to acquire a noble gas electron configuration each nitrogen atom must share three of its electrons with the other nitrogen atom.

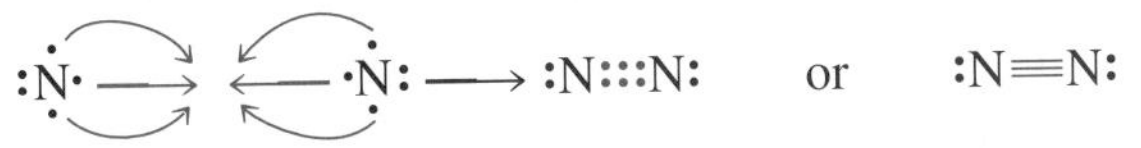

Notice how all three shared electron pairs are placed in the area between the two nitrogen atoms in the above bonding diagrams. Note also that three lines are used to denote a triple covalent bond, paralleling the use of one line to denote a single covalent bond.

In "bookkeeping" electrons in an electron-dot structure, to make sure that all atoms in the molecule have achieved their octet of electrons, *all* electrons in a multiple covalent bond are considered "to belong" to *both* of the atoms involved in that bond. The "bookkeeping" for the N_2 molecule would be

F.Y.I.

The triple bond that holds the N_2 molecule together is very stable and slow to react with other substances. The nitrogen that makes up most of our atmosphere does not react with most living things. Many manufacturers package their products in nitrogen to keep them fresh.

Each of the circles about an N atom contains eight valence electrons. Again, all of the electrons in a multiple covalent bond are considered to belong to each of the atoms in the bond. Circles are never drawn to include just some of the electrons in a multiple covalent bond.

A slightly more complicated molecule containing a triple covalent bond is the molecule C_2H_2 (acetylene). A carbon–carbon triple covalent bond is present as well as two carbon–hydrogen single covalent bonds. The arrangement of valence electrons in C_2H_2 is as follows.

The two atoms in a triple covalent bond are commonly the same element. However, they do not have to be. The molecule HCN contains a heteroatomic triple covalent bond.

H:C:::N: or H—C≡N:

Double covalent bonds are found in numerous molecules. A very common molecule that contains bonding of this type is carbon dioxide (CO_2). In fact, there are two carbon–oxygen double covalent bonds present in CO_2.

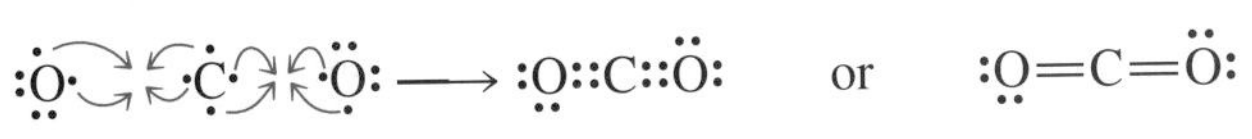

Note in the following diagram how the circles are drawn for the octet of electrons about each of the atoms in CO_2.

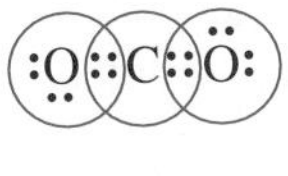

Not all elements can form multiple covalent bonds. There must be at least two vacancies in an atom's valence electron shell prior to bond formation if it is to participate in a multiple covalent bond. This requirement eliminates group VIIA elements (F, Cl, Br, I) and hydrogen from participating in such bonds. The group VIIA elements have seven valence electrons and one vacancy, and hydrogen has one valence electron and one vacancy. All bonds formed by these elements are single covalent bonds.

Double bonding becomes possible for elements needing two electrons to complete their octet, and triple bonding becomes possible when three or more electrons are needed to complete an octet. Note that the word *possible* was used twice in the previous sentence. Multiple bonding does not have to occur when an element has two or three or four vacancies in its octet; single covalent bonds can be formed instead. The "bonding behavior" of an element, when more than one behavior is possible, is determined by what other element or elements it is bonded to.

Let us consider the possible "bonding behaviors" for O (six valence electrons; two octet vacancies), N (five valence electrons; three octet vacancies), and C (four valence electrons; four octet vacancies).

To complete its octet by electron sharing, an oxygen atom can form either two single bonds or one double bond.

two single bonds   one double bond

Nitrogen is a very versatile element with respect to bonding. It can form single, double, or triple covalent bonds as dictated by the other atoms present in a molecule.

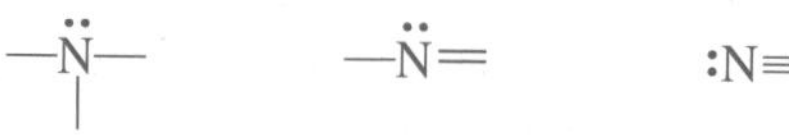

three single bonds   one single and one double bond   one triple bond

Note that in each of these bonding situations a nitrogen atom forms three bonds. A double bond counts as two bonds, and a triple bond as three bonds. Since nitrogen has only five valence electrons, it must form three covalent bonds to complete its octet.

Carbon is an even more versatile element than nitrogen with respect to variety of types of bonding as illustrated by the following possibilities for bonding.

four single bonds   two single bonds and one double bond   two double bonds   one single bond and one triple bond

7.12 COORDINATE COVALENT BONDS

In the covalent bonds considered so far (single, double, and triple), each of the participating atoms in the bond contributed an equal number of electrons to the bond. There is another *less common* way in which a covalent bond can form. It is possible for both electrons in a shared electron pair to come from the same atom; that is, one atom supplies two electrons and the other atom none. Such a covalent bond is called a *coordinate covalent bond.*

A **coordinate covalent bond** *is a bond in which both electrons of a shared pair come from one of the two atoms.* Coordinate covalent bonding allows an atom that has two (or more) vacancies in its valence shell to share a pair of nonbonding electrons located on another atom.

The ammonium ion, NH_4^+, is an example of a species containing a coordinate covalent bond. The formation of an NH_4^+ ion can be viewed as resulting from the reaction of a hydrogen ion, H^+, with an ammonia molecule, NH_3. Doing the "bookkeeping" on all of the valence electrons involved in this reaction, using × for nitrogen electrons and dots for hydrogen electrons, we get

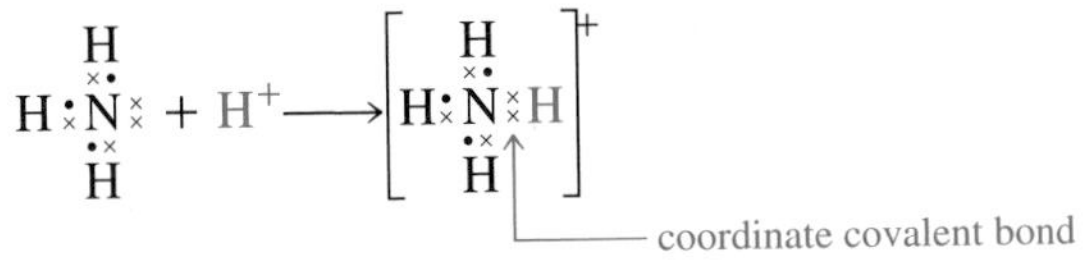

An H^+ ion has no electrons, hydrogen having lost its only electron when it became an ion. The H^+ ion has two vacancies in its valence shell; that is, it needs two electrons to become isoelectronic with the noble gas helium. The nitrogen in NH_3 possesses a pair of nonbonding electrons. These electrons are used in forming the new nitrogen–hydrogen bond. The new species formed, the NH_4^+ ion, is charged, since the H^+ was charged. The +1 charge on the NH_4^+ ion is dispersed over the *entire* molecule; it is not localized on the "new" hydrogen atom.

The element oxygen quite often forms coordinate covalent bonds. Consider the electron-dot structures of the molecules HOCl (hypochlorous acid) and $HClO_2$ (chlorous acid).

In the first structure all of the bonds are "ordinary" covalent bonds. In the second structure, which differs from the first in that a second oxygen atom is present, the "new" chlorine–oxygen bond is a coordinate covalent bond. The second oxygen atom with six valence electrons (denoted by ×'s) needs two more electrons for an octet. It shares one of the nonbonding electron pairs present on the chlorine atom. (The chlorine atom does not need any of the oxygen's electrons since it already has an octet.)

Once a coordinate covalent bond is formed, there is no way to distinguish it from any of the other covalent bonds in a molecule; all electrons are identical regardless of their source. The main use of the concept of coordinate covalency is in helping to rationalize the existence of certain molecules and ions whose bonding-electron arrangement would otherwise present problems.

Atoms participating in coordinate covalent bonds generally deviate from the common bonding pattern (Sec. 7.11) expected for that type of atom. For example, oxygen normally forms two bonds; yet in the molecules N_2O and CO, which contain coordinate covalent bonds, oxygen forms one and three bonds, respectively.

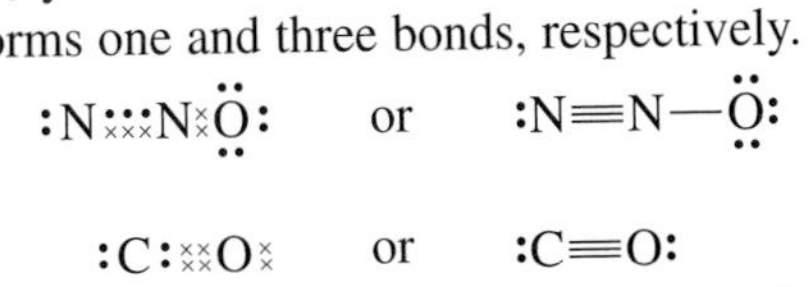

The concept of coordinate covalency is needed to rationalize the existence of such molecules.

7.13 Resonance Structures

In Section 7.11 it was noted that, in general, triple covalent bonds are stronger than double-covalent bonds, which in turn are stronger than single covalent bonds. **Bond strength *is*** *measured by the energy it takes to break a bond; that is, to separate bonded atoms to give neutral particles.* It can be determined experimentally.

Another experimentally determinable parameter of bonds is bond length. **Bond length** *is the distance between the nuclei of bonded atoms* (see Fig. 7.5). A direct relationship exists between bond strength and bond length. It is found that as bond strength increases, bond length decreases; that is, the stronger the bond, the shorter the distance between the nuclei of the atoms of the bond. Thus, in general, triple covalent bonds are shorter than double covalent bonds, which are shorter than single covalent bonds.

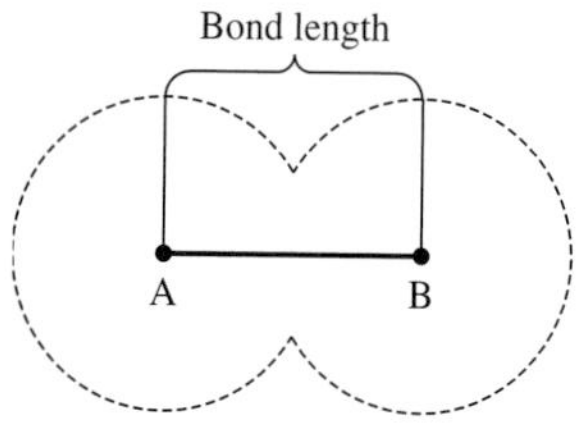

FIGURE 7.5 The length of a bond between two atoms is the distance separating the nuclei of the atoms.

Most electron-dot structures for molecules give bonding pictures, that are consistent with available experimental information on bond strength and bond length. However, there are some molecules for which no single electron-dot structure that is consistent with such information can be written.

The molecule SO_2 is an example of a situation in which a single electron-dot structure does not adequately describe bonding. A plausible electron-dot structure for SO_2, in which the octet rule is satisfied for all three atoms, is

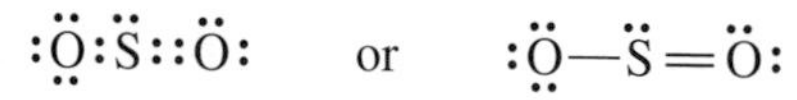

However, this structure suggests that one sulfur–oxygen bond, the double bond, should be stronger and shorter than the other sulfur–oxygen bond, the single bond. Experiment shows that this is not the case; both sulfur–oxygen bonds are equivalent, with both bond length and bond strength characteristics intermediate between those for sulfur–oxygen single and double bonds. An electron-dot structure depicting this intermediate situation cannot be written.

The solution to the phenomenon in which no single electron-dot structure adequately describes bonding involves the use of two or more electron-dot structures, known as resonance structures, to represent the bonding in the molecule. **Resonance structures** *are two or more electron-dot structures for a molecule or ion that have the same arrangement of atoms, contain the same number of electrons, and differ only in the location of the electrons.*

Two resonance structures exist for an SO_2 molecule.

A double-headed arrow is used to connect resonance structures. The only difference between the two SO_2 resonance structures is in the location of one pair of electrons. The positioning of this pair of electrons determines whether the oxygen atom on the right or the left is the oxygen atom involved in the double bond.

The actual bonding in a SO_2 molecule is said to be a *resonance hybrid* of the two contributing resonance structures. Beginning chemistry students frequently misinterpret the concept of a resonance hybrid. They incorrectly envision that a molecule—SO_2 in this case—is constantly changing (resonating) between various resonance structure forms. This is not the case. For example, SO_2 is not a mixture of two kinds of molecules, nor does a single type of molecule flip-flop back and forth between the two resonance forms. There is only one kind of SO_2 molecule, and the bonding in it is an average of those depicted by the resonance structures. SO_2 molecules exist "full time" in this average state. A mule, the offspring of a donkey and horse, can be considered a hybrid of a donkey and a horse. However, it is not a horse at one instant and a donkey at another; it is always a mule. Likewise, a molecule has only one real structure, which is different from any of the resonance structures; it has characteristics of each one but does not match any one of them exactly.

Sometimes three, four, or even more resonance structures can be drawn for a molecule. Again, such resonance structures must all contain the same number of electrons and have the same arrangement of atoms; the structures may differ only in the location of the electrons about the atoms.

7.14 A SYSTEMATIC METHOD FOR DETERMINING ELECTRON-DOT STRUCTURES

The task of constructing electron-dot structures for molecules and polyatomic ions containing many electrons or for which several resonance hybrids must be used to describe the bonding can become quite frustrating if it is approached in a nonsystematic trial-and-error manner. Use of a systematic approach to writing electron-dot structures will enable a student to avoid most of this frustration. The following guidelines make the drawing of an electron-dot structure for any molecule or polyatomic ion that obeys the octet rule, even a very complicated structure, a straightforward procedure.

STEP 1 Determine the total number of valence electrons present in the molecule, that is, the total number of dots that must appear in the electron-dot structure.

The total number of valence electrons is found by adding up the number of valence electrons each atom in the molecule or ion possesses. If the species is a polyatomic ion, add one electron for each unit of negative charge present or subtract one electron for each positive charge. Do not worry about keeping track of which electrons come from which atoms. Only their total number is important.

STEP 2 Write the symbols of the atoms in the molecule, arranged in the way in which they are bonded to each other and then place a single covalent bond (two electrons) between each pair of bonded atoms. Either a pair of dots or a dash can be used in denoting the single bond(s).

Determining which atom is the *central atom*, that is, which atom has the most other atoms bonded to it, is the key to determining the arrangement of atoms in a molecule or ion. Most other atoms present will be bonded to the central atom. For most molecular compounds containing just two elements, the molecular formula is of help in deciding the identity of the central atom. The central atom is the atom that appears only once in the formula; for example, S is the central atom in SO_3, O is the central atom in H_2O, and P is the central atom in PF_3. In a molecular compound containing hydrogen, oxygen, and an additional element, it is the additional element that is the central atom; for example, N is the central atom in HNO_3, and S is the central atom in H_2SO_4. In compounds of this type the oxygen atoms are bonded to the central atom and the hydrogen atoms are bonded to the oxygens. Carbon is the central atom in almost all carbon-containing compounds. Hydrogen and fluorine are never the central atom.

STEP 3 Add nonbonding electron pairs to the molecular structure such that each atom bonded to the central atom has an octet of electrons.

Remember that for hydrogen an "octet" of electrons is only two electrons. The noble gas electron configuration acquired by hydrogen is that of helium, and helium has only two electrons.

STEP 4 Place any remaining electrons on the central atom.

The number of remaining electrons is obtained by subtracting from the total number of valence electrons (step 1) the number of electrons used in steps 2 and 3.

STEP 5 If there are not enough electrons to give the central atom an octet, form multiple covalent bonds by shifting nonbonding electron pairs from surrounding atoms into bonding locations.

The elements C, N, and O are the elements most frequently involved in multiple bonding situations. The elements H, F, Cl, Br, and I in terminal atom positions do not participate in multiple bond formation.

The three examples that follow illustrate the use of the preceding procedures for drawing electron-dot structures. In each part of each example the outlined procedure is followed step by step so that you will become familiar with it.

Example 7.8

The acid present in the greatest amount in *acid rain*, rainfall with greater than normal acidity, is sulfuric acid, a compound with the chemical formula H_2SO_4. Draw the electron-dot structure for H_2SO_4.

Solution

STEP 1 Both sulfur and oxygen atoms have six valence electrons and hydrogen has one valence electron. The total number of valence electrons present in this molecules is 32.

1 S:	$1 \times 6 =$	6 valence electrons
4 O:	$4 \times 6 =$	24 valence electrons
2 H:	$2 \times 1 =$	2 valence electrons
		32 valence electrons

STEP 2 In compounds containing H, O, and an additional element, the additional element (S in this case) is the central atom. The oxygen atoms are attached to this central S atom, and the hydrogen atoms are attached to the oxygen atoms. Drawing this atomic arrangement with single covalent bonds (two electrons) placed between all bonded atoms gives

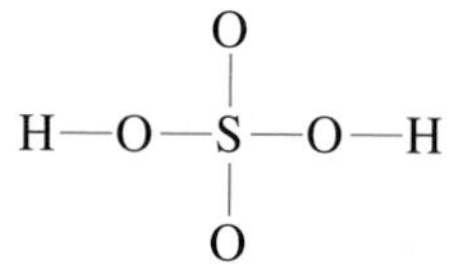

STEP 3 Adding nonbonding electrons to the structure to complete the octets of all atoms bonded to the central atom gives

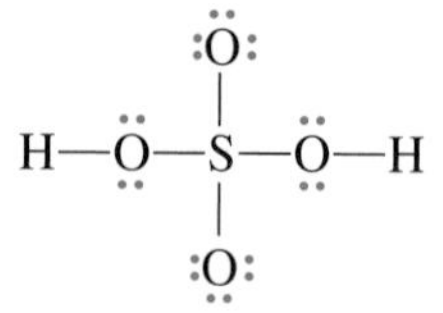

STEP 4 We started out with 32 valence electrons (step 1). Twelve were used in step 2 and twenty in step 3. This accounts for all of the available electrons. None are available to add to the central atom. This is fine because the central sulfur atom already has an octet of electrons; it is participating in four single bonds (eight electrons).

STEP 5 No double or triple bonds are needed since all atoms have an octet of electrons with only single bonds present.

Practice Exercise 7.8

Write the electron-dot structure for the molecule $HClO_3$.

Ans. 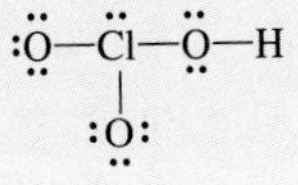

Example 7.9

The compound Na_2SO_3, which contains the polyatomic SO_3^{2-} ion (sulfite ion), is used in the manufacture of paper. Draw the electron-dot structure of the SO_3^{2-} ion.

Solution

STEP 1 The sulfur atom has 6 valence electron and each oxygen atom also has six valence electrons. (Sulfur and oxygen are in the same group in the periodic table.) There are two additional valence electrons present because of the −2 charge associated with this polyatomic ion.

1 S:	$1 \times 6 =$	6 valence electrons
3 O:	$3 \times 6 =$	18 valence electrons
charge of $-2 =$		2 valence electrons
		26 valence electrons

If the polyatomic ion had had a positive charge instead of a negative one, we would have had to subtract valence electrons from the total instead of adding. A positive charge would have denoted loss of electrons, and the electrons lost would have been valence electrons.

STEP 2 The sulfur atom is the central atom with all three oxygen atoms individually attached to it. Drawing this atomic arrangement with single covalent bonds (two electrons) placed between all bonded atoms gives

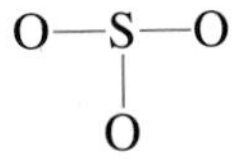

STEP 3 Adding nonbonding electrons to the structure to complete the octets of the oxygen atoms gives

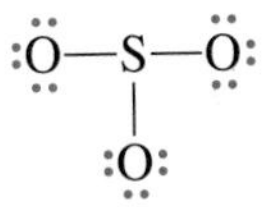

STEP 4 We started out with 26 electrons (step 1). Six electrons were used in step 2 (single bonds) and eighteen electrons in step 3 (nonbonding electron pairs). This leaves two electrons not yet used. These two remaining electrons are available for placement on the central sulfur atom, which needs two more electrons to complete its octet.

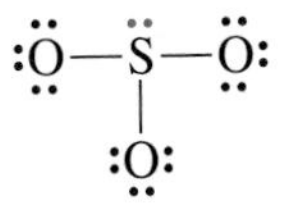

STEP 5 No double or triple bonds are needed since the action in step 4 causes the central sulfur atom to have an octet of electrons.

We must remember, however, that the electron-dot structure we have just generated is that of a polyatomic ion. We therefore need to enclose it in large brackets and place the ionic charge for it outside the right bracket in a superscript position.

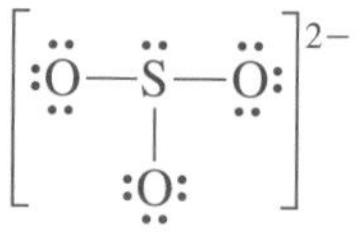

In this example we have shown the bonding within a polyatomic ion. This polyatomic ion, as well as all other polyatomic ions, is not a stable entity that exists alone. Polyatomic ions are *parts* of ionic compounds. The ion SO_3^{2-} would be found in ionic compounds such as Na_2SO_3, K_2SO_3, and $(NH_4)_2SO_3$. Ionic compounds containing polyatomic ions offer an interesting combination of both ionic and covalent bonds: covalent bonding *within* the polyatomic ion and ionic bonding *between* it and ions of opposite charge.

Practice Exercise 7.9

Draw the electron-dot structure of the phosphite ion, PO_3^{3-}.

Ans. 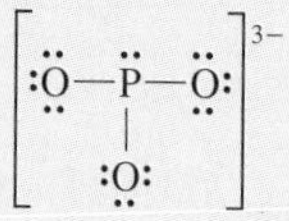

Example 7.10

Ozone, the substance present in the upper atmosphere (ozone layer) that absorbs ultraviolet light from the sun is a form of oxygen in which there are three atoms of oxygen present in a molecule (O_3). Draw the electron-dot structure of the O_3 molecule.

Solution

STEP 1 There is a total of 18 valence electrons present, six from each oxygen atom.

STEP 2 One of the oxygen atoms may be considered the central atom and the other two oxygen atoms are attached to it. Drawing this atomic arrangement with single covalent bonds (two electrons) placed between all bonded atoms gives

O—O—O

STEP 3 Adding six nonbonding electrons to each of the terminal oxygen atoms to complete their octets gives

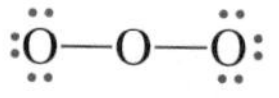

STEP 4 We started out with 18 electrons (step 1). Four electrons were used in step 2 (single bonds) and 12 in step 3 (nonbonding electron pairs). This leaves two electrons available for placement on the central oxygen atom.

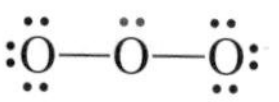

STEP 5 The addition of the two nonbonding electrons to the central oxygen atom is not sufficient to give this oxygen an octet of electrons. It still lacks two electrons. This problem is solved by moving a nonbonding pair of electrons from one of the terminal oxygen atoms into the oxygen–oxygen bonding region. This action, which produces a double bond, gives the central oxygen atom an octet of electrons.

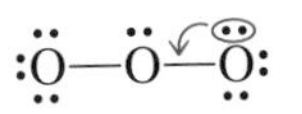

There are two choices for the terminal oxygen atom that is involved in double bond formation. The ramification of this situation is that resonance structures exist. The availability of choices for multiple bond formation (to give a central atom an octet of electrons) is always a signal for the existence of resonance structures. In this case, the resonance structures are

Practice Exercise 7.10

Draw the electron-dot structure of the SO_3 molecule.

Ans.

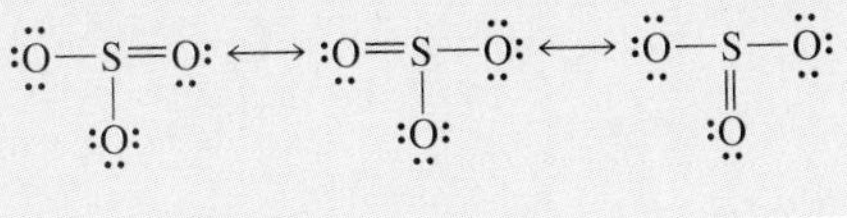

The systematic approach to drawing electron-dot structures for molecules and polyatomic ions illustrated in Examples 7.8 to 7.10 does not take into account the origin of the electrons in a chemical bond, that is, which atoms contribute which electrons to the bond. Thus, no distinction between normal covalent bonds and coordinate covalent bonds is made by the procedures of this system. This is acceptable. Each electron in a bond belongs to the bond as a whole. Electrons do not have labels of genealogy.

7.15 THE SHAPES OF MOLECULES: MOLECULAR GEOMETRY

Electron-dot structures give the number and types of bonds present in molecules. They do not, however, convey any information about molecular shape, that is, molecular geometry. **Molecular geometry** *describes the three-dimensional arrangement of atoms in a molecule.* Indeed, electron-dot structures falsely imply that all molecules have flat two-dimensional shapes.

Molecular geometry is an important factor in determining the physical and chemical properties of substances. Dramatic relationships between geometry and properties are often observed in research associated with the development of prescription drugs. A small change in overall molecular geometry, caused by addition or removal of atoms, can enhance drug effectiveness and/or decrease drug side effects. Studies also show that the human senses of taste and smell depend in part on the shapes of molecules.

For simple molecules (only a few atoms) molecular geometry can be predicted using the information present in a molecule's electron-dot structure and a procedure called valence-shell-electron-pair-replusion theory (VSEPR theory). **VSEPR theory** *is a set of* *procedures for predicting the three-dimensional shape of a molecule from the information contained in the molecule's electron-dot structure.*

The central concept of VSEPR theory is that electron pairs in the valence shell of an atom adopt an arrangement in space about the nucleus that minimizes the replusions between the like-charged (all negative) electron pairs. Minimization occurs when the electron pairs are as far away from each other as possible. Replusion-minimizing arrangements about a nucleus for two, three, and four electron pairs are as follows.

1. Two electron pairs, to be as far apart as possible from each other, will be found on opposite sides of a nucleus, that is, at 180° angles to each other. Such an electron pair arrangement is said to be *linear*.

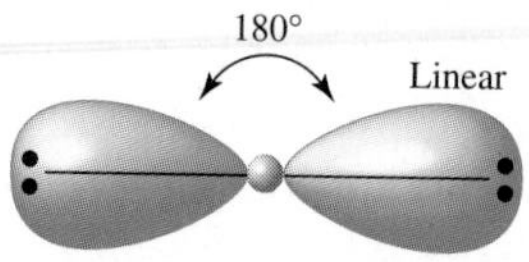

2. Three electron pairs are as far apart as possible when they are found at the corners of an equilateral triangle. In such an arrangement, they are separated by angles of 120°, giving a *trigonal planar* arrangement of the electron pairs.

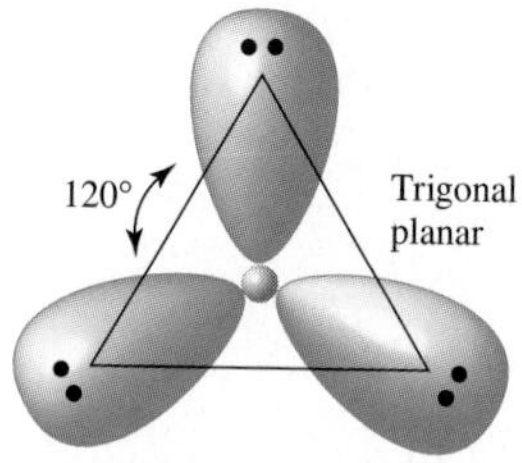

3. A *tetrahedral* arrangement of electron pairs minimizes repulsions between four sets of electron pairs. A tetrahdron is a four-sided geometrical figure, all four sides being identical equilateral triangles. The angle between any two electron pairs is 109°.

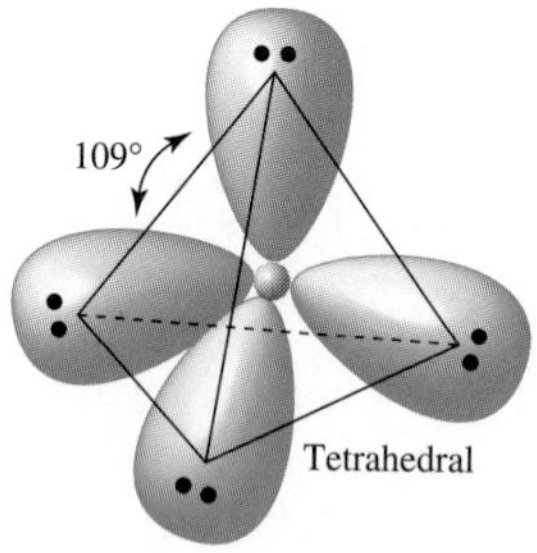

Figure 7.6 shows these three electron-pair arrangements—linear, trigonal planar, and tetrahedral—using balloons that have their ends tied together. When two, three, and four like-sized balloons are tied together they naturally assume the shapes we are talking about.

Most simple molecules have a central atom to which all other atoms are bonded. The first step in applying VSEPR theory to such molecules involves determining the number of electron pairs present on the *central atom*. This count is obtained from the molecule's electron-dot structure using the following VSEPR theory conventions.

1. Double and triple bonds between the central atom and other atoms are counted as "one pair" because they take up only one region of space.
2. No distinction is made between bonding electron pairs and nonbonding electron pairs. Each is counted in the total number of pairs.

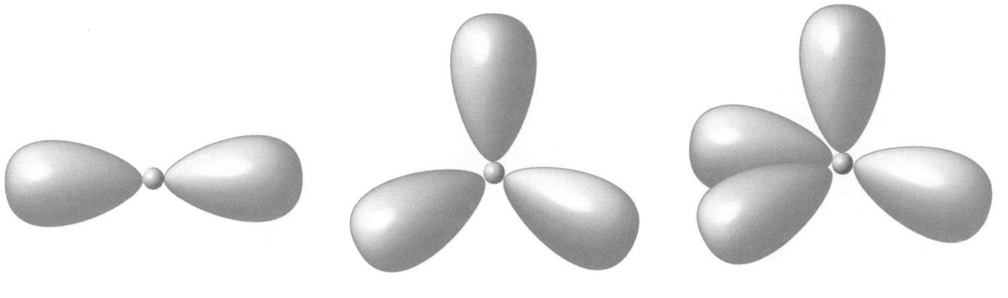

Two balloons, linear   Three balloons, trigonal planar   Four balloons, tetrahedral

FIGURE 7.6 When balloons of the same size and shape are tied together, they will assume positions in space similar to those taken by pairs of valence electrons around a central atom.

Based on these conventions, the VSEPR electron pair count about the *central atom* in molecules with the following electron-dot structures is as indicated.

$:N\equiv N-\ddot{\underset{..}{O}}:$	Central atom has 2 VSEPR electron pairs (single bond and triple bond)
$:\ddot{\underset{..}{O}}-\ddot{S}=\ddot{O}:$	Central atom has 3 VSEPR electron pairs (single bond, double bond, and nonbonding electron pair)
$H-\ddot{\underset{..}{O}}-H$	Central atom has 4 VSEPR electron pairs (2 single bonds and 2 nonbonding electron pairs)

Molecular shapes for molecules are obtained by putting together the two facets of VSEPR theory just considered: (1) the electron pair arrangements which minimize repulsions between electron pairs, and (2) the number of VSEPR electron pairs about the central atom in a molecule.

Molecules with Two VSEPR Electron Pairs

All molecules with two VSEPR electron pairs are linear. Two common molecules in this category are carbon dioxide (CO_2) and hydrogen cyanide (HCN), whose Lewis structures are

$$:\ddot{O}=C=\ddot{O}: \qquad H-C\equiv N:$$

In CO_2 the central carbon atom's two VSEPR pairs are the two double bonds. In HCN the central carbon atom's two VSEPR pairs are a single bond and a triple bond. In both molecules the VSEPR electron pairs arrange themselves on opposite sides of the carbon atom which produces a linear molecule.

Molecules with Three VSEPR Electron Pairs

Two molecular shapes are associated with molecules that have three VSEPR electron pairs: *trigonal planar* and *angular*. The former results when all three VSEPR pairs are bonding and the latter when one of the three VSEPR pairs is nonbonding. Illustrative of these two situations are the molecules H_2CO (formaldehyde) and SO_2 (sulfur dioxide), whose electron-dot structures are

trigonal planar   angular

The shape of the SO_2 molecule is described as *angular* rather than *trigonal planar* because molecular shape describes only *atom positions*. The positions of nonbonding electron pairs are "ignored" when coining words to describe molecular shape. Do not interpret this to mean that nonbonding electron pairs are unimportant in molecular shape determinations; indeed, in the case of SO_2, it is the presence of the nonbonding electron pair that makes the molecule angular rather than linear.

Molecules with Four VSEPR Electron Pairs

Three molecular shapes are possible for molecules with four VSEPR electron pairs: tetrahedral (no nonbonding electron pairs present), trigonal pyramidal (one nonbonding electron pair present), and angular (two nonbonding electron pairs present). The molecules CH_4 (methane), NH_3 (ammonia), and H_2O (water) illustrate this sequence of molecular shapes.

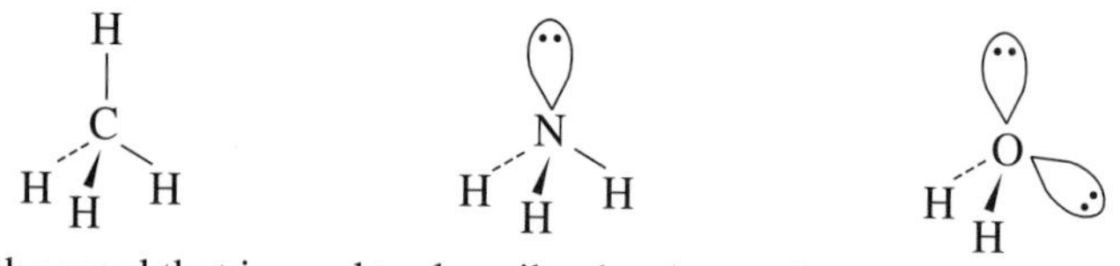

Again, note how the word that is used to describe the shape of a molecule does not take into account the positions of nonbonding electron pairs.

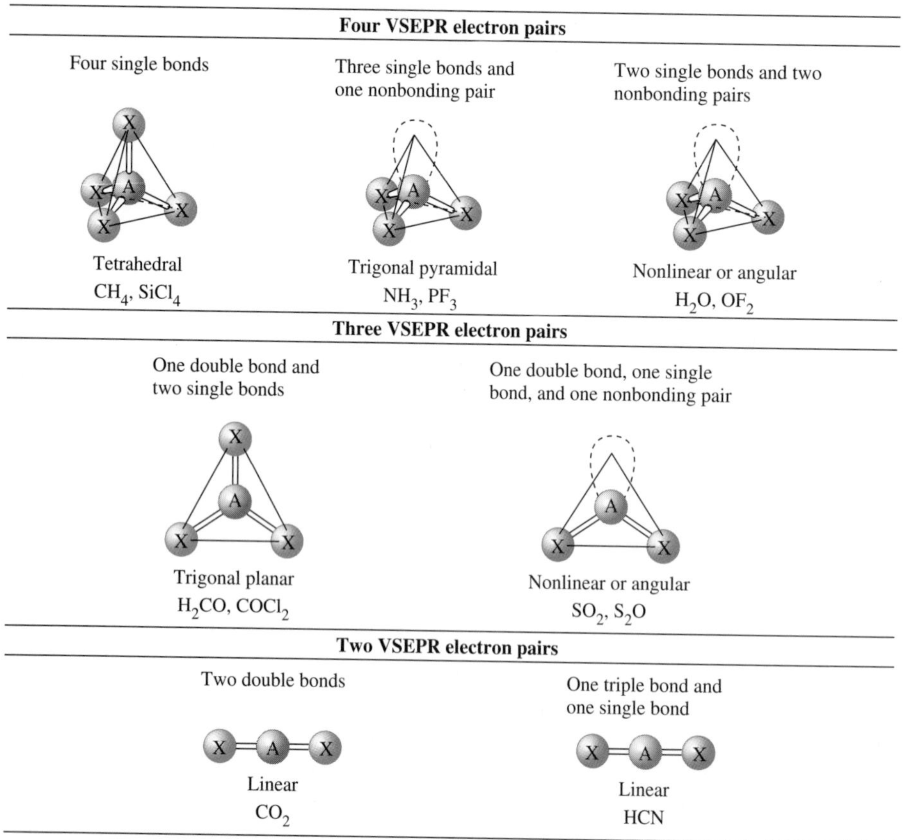

Note: The figure does not consider cases where only one bond is present — for example, one single bond and three nonbonding pairs. If only one bond is present, the molecule is diatomic. All diatomic molecules have the same geometry: The two atoms lie along a straight line.

FIGURE 7.7 Molecular geometries associated with various combinations of bonding and nonbonding electrons about a central atom that obeys the octet rule.

Molecular Formula	Electron-Dot Structure and VSEPR Electron-Pair Analysis		Molecular Geometry
C_2H_2 (acetylene)	H—C≡C—H		H—C≡C—H
	2 VSEPR electron pairs; linear C center	2 VSEPR electron pairs; linear C center	Straight chain of four atoms – linear
HN_3 (hydrogen azide)	H—N̈=N=N̈:		N̈=N=N̈: with H bonded at an angle to the first N
	3 VSEPR electron pairs; angular N center	2 VSEPR electron pairs; linear N center	Chain of four atoms with one bend
H_2O_2 (hydrogen peroxide)	H—Ö=Ö—H		O=O with H bonded at an angle to each O
	3 VSEPR electron pairs; angular O center	3 VSEPR electron pairs; angular O center	Chain of four atoms with two bends

FIGURE 7.8 Using VSEPR theory to predict molecular geometry for molecules containing more than one central atom.

Figure 7.7 summarizes the relationships among the number of VSEPR electron pairs about a central atom, molecular geometry, and terminology used to describe molecular geometry.

VSEPR theory can also be used to predict the molecular geometry of molecules that contain more than one central atom. In such situations, each central atom is considered separately and then the results of the separate analyses are combined to obtain the overall molecular geometry. Figure 7.8 shows the results of such procedures for selected molecules containing more than one central atom.

Example 7.11

Using VSEPR theory, predict the molecular geometry of each of the following molecules given their electron-dot structures.

(a) **(b)** **(c)**

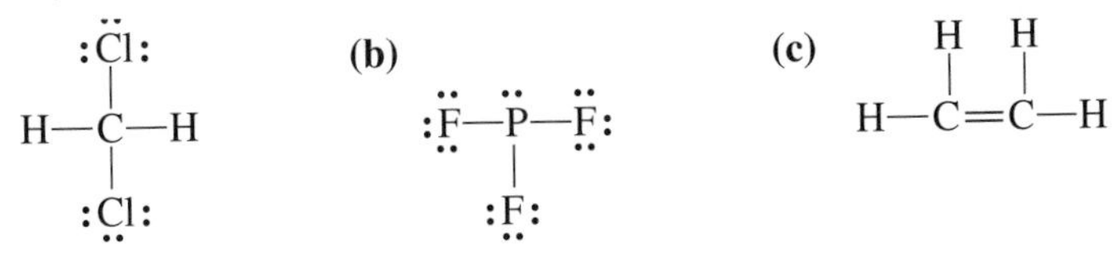

Solution

(a) The electron-dot structure for this molecule shows that the central carbon atom has four VSEPR electron pairs present (four single bonds). No nonbonding valence shell electrons are present on the carbon atom.

The arrangement of the four electron pairs about the central atom is tetrahedral.

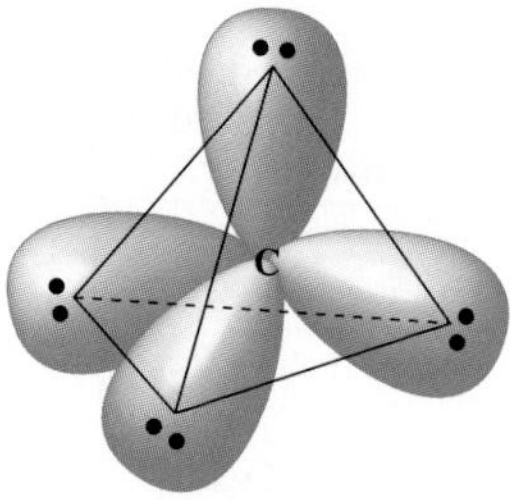

Since each electron pair is a bonding pair, the molecular shape will also be tetrahedral.

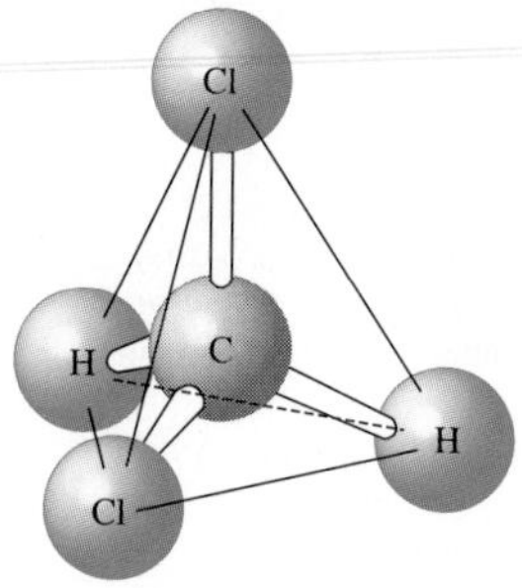

(b) Electrons are present in four locations about the central atom—three electron pairs involved in single bonds and one electron pair that is nonbonding.

The arrangement of four electron pairs about a central atom is always tetrahedral. It does not matter whether the pairs are bonding or nonbonding. Arrangement depends only on the number of pairs and not on how the pairs function.

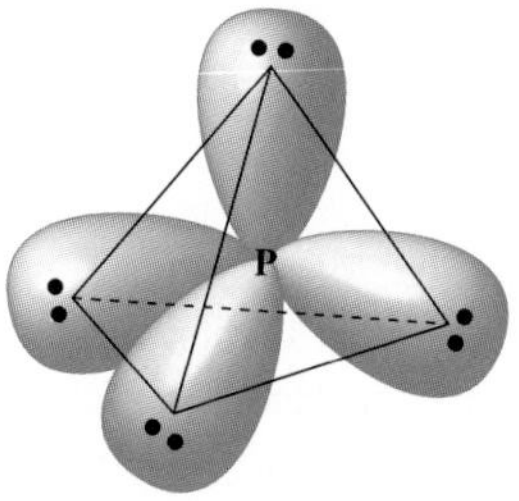

However, the molecular shape will be different from the electron-pair geometry since a nonbonding electron pair is present. The molecular shape is trigonal pyramidal.

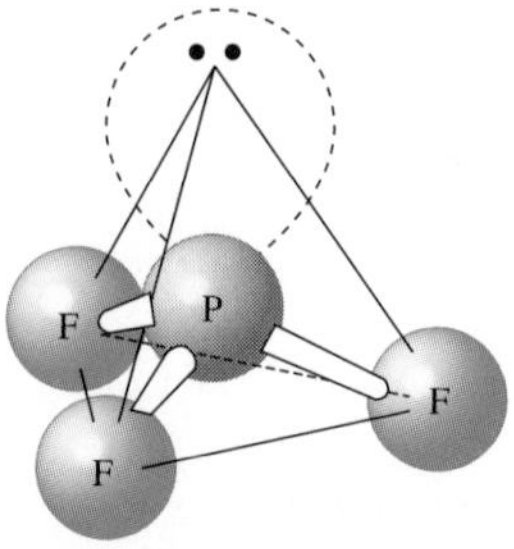

Note that the vacant corner of the tetrahedron (the corner that has a nonbonding electron pair instead of an atom) is not considered when coining a word to describe the molecular geometry. Molecular geometry describes the arrangement of *atoms only*.

(c) This molecule has two central atoms, the two carbon atoms. A VSEPR theory analysis needs to be carried out for each of the carbon atoms and then the results of the two analyses combined.

Each of the two carbon atoms has three VSEPR electron pairs about it (two single bonds and a double bond). The geometry associated with three VSEPR electron pairs is always trigonal planar.

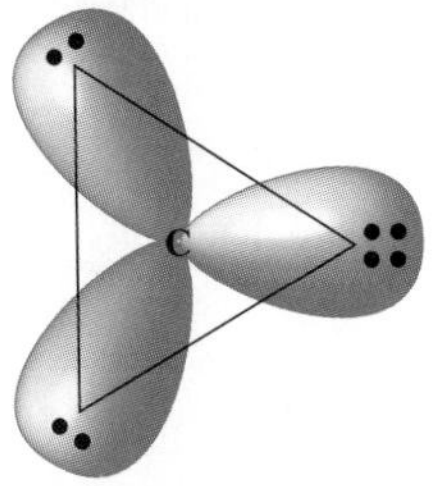

Since all positions about the carbon atoms are bonding positions (no nonbonding electron pairs are present on the carbon atoms) there will also be a trigonal planar arrangement of atoms about each carbon atom. The geometry of the molecule is therefore the following.

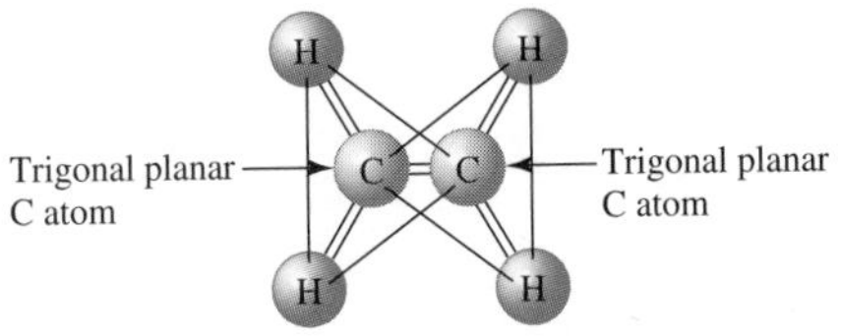

A simpler representation of the geometry of this molecule is the following

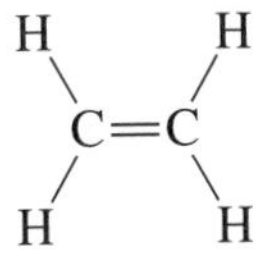

Practice Exercise 7.11

Using VSEPR theory, predict the molecular geometry of each of the following molecules given their electron-dot structures.

(a) 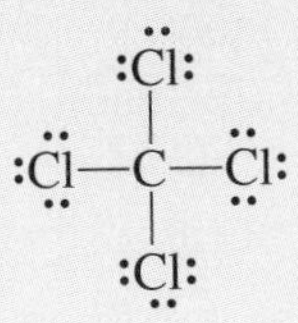 (b) 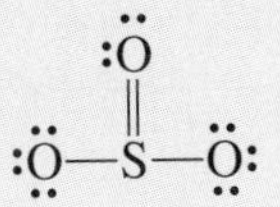 (c) 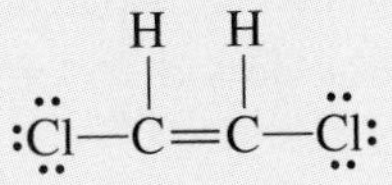

Ans. (a) tetrahedral; (b) trigonal planar; (c) trigonal planar about both carbon atoms

7.16 ELECTRONEGATIVITY

The ionic and covalent bonding models we have developed in this chapter seem to represent two very distinct forms of bonding. Actually, the two models are closely related to each other; they are the extremes of a broad continuum of bonding patterns. The close relation-

H 2.1																	He –
Li 1.0	Be 1.5											B 2.0	C 2.5	N 3.0	O 3.5	F 4.0	Ne –
Na 0.9	Mg 1.2											Al 1.5	Si 1.8	P 2.1	S 2.5	Cl 3.0	Ar –
K 0.8	Ca 1.0	Sc 1.3	Ti 1.5	V 1.6	Cr 1.6	Mn 1.5	Fe 1.8	Co 1.8	Ni 1.8	Cu 1.8	Zn 1.6	Ga 1.6	Ge 1.8	As 2.0	Se 2.4	Br 2.8	Kr –
Rb 0.8	Sr 1.0	Y 1.2	Zr 1.4	Nb 1.6	Mo 1.8	Tc 1.9	Ru 2.2	Rh 2.2	Pd 2.2	Ag 1.9	Cd 1.7	In 1.7	Sn 1.8	Sb 1.9	Te 2.1	I 2.5	Xe –
Cs 0.7	Ba 0.9	57–71 1.1–1.2	Hf 1.3	Ta 1.5	W 1.7	Re 1.9	Os 2.2	Ir 2.2	Pt 2.2	Au 2.4	Hg 1.9	Tl 1.8	Pb 1.8	Bi 1.9	Po 2.0	At 2.2	Rn –
Fr 0.7	Ra 0.9																

FIGURE 7.9 Electronegativities of the elements.

ship between the two bonding models become apparent when the concepts of *electronegativity* (discussed in this section) and *bond polarity* (discussed in the next section) are considered.

The electronegativity concept has its origins in the fact that the nuclei of various elements have differing abilities to attract shared electrons (in a bond) to themselves. Different electron-attracting abilities result from differences in size, nuclear charge, and number of nonvalence electrons present for atoms of various elements. As a result of these factors, some elements are better electron attractors than other elements.

Electronegativity *is a measure of the relative attraction that an atom has for the shared electrons in a bond.* Electronegativity values are unitless numbers on a relative scale that are obtained from bond energies and other related experimental data. Electronegativities cannot be directly measured in the laboratory. A number of electronegativity scales exist. The most widely used one, a scale developed by the American chemist Linus Pauling (1901–1994), is given in the format of a periodic table in Figure 7.9. On this scale, fluorine, the most electronegative of all elements, has arbitrarily been assigned a value of 4.0 (the maximum on the scale) and serves as the reference element. *The higher the electronegativity of an element, the greater the electron-attracting ability of atoms of that element.*

Patterns and trends in the electronegativity values given in Figure 7.9 include the following.

1. For representative elements (Sec. 6.11), electronegativity values generally increase from left to right within a period of the periodic table.
2. For the period 2 elements this left-to-right increase is regular, being 0.5 units.

Li	Be	B	C	N	O	F
1.0	1.5	2.0	2.5	3.0	3.5	4.0

3. It is important to note how the electronegativity of hydrogen (period 1) compares with that of the period 2 elements. Hydrogen's value of electronegativity, 2.1, is between that of B and C
4. For the period 3 elements the left-to-right increase is 0.3 units until the last two elements are reached where it is 0.4 and 0.5.

Na	Mg	Al	Si	P	S	Cl
0.9	1.2	1.5	1.8	2.1	2.5	3.0

5. Electronegativity values generally increase from bottom to top within a periodic table group.

The group and period trends in electronegativity values result in nonmetals generally having higher electronegativities than metals. This fact is consistent with our previous generalization (Sec. 7.5) that metals tend to lose electrons and nonmetals tend to gain electrons when an ionic bond is formed. Metals (low electronegativities, poor electron attractors) will give up electrons to nonmetals (high electronegativities, good electron attractors).

7.17 THE POLARITY OF CHEMICAL BONDS

When two atoms of equal electronegativity share one or more pairs of electrons, each atom exerts the same attraction for the electrons, which results in the electrons being *equally* shared. This type of bond is called a *nonpolar covalent bond.* A **nonpolar covalent bond** *is a covalent bond in which there is an equal sharing of electrons.*

When the two atoms involved in a covalent bond have different electronegativities, the electron-sharing situation is not equal. The atom that has the higher electronegativity will attract the electrons more strongly than the other atom; this results in an unequal sharing of electrons. This type of covalent bond is called a *polar covalent bond.* A **polar covalent bond** *is a covalent bond in which there is an unequal sharing of electrons.*

F.Y.I.

Different types of sharing are seen in everyday life. Partners in a business can own it equally, or one partner can have more than the other. It can be owned 60:40, 70:30, and so on.

The significance of unequal sharing of electrons in a covalent bond is that it creates partial (fractional) positive and negative charges on the atoms involved in a bond. Although each atom involved in a polar covalent bond is initially uncharged, the unequal sharing of electrons means the electrons spend more time near the more electronegative atom of the bond (producing a partial negative charge). The presence of such partial charges on atoms within a molecule often significantly affects molecular properties (Sec. 7.18).

The unequal sharing of electrons in a covalent bond is often indicated by a notation that uses the lowercase Greek letter δ (delta). A δ− symbol, meaning a "partial negative charge," is placed above the relatively negative atom of the bond, and a δ+ symbol, meaning a "partial positive charge," is placed above the relatively positive atom.

With "delta notation," the bond in hydrogen chloride (HCl) would be depicted as

$$\overset{\delta+}{\text{H}}\text{—}\overset{\delta-}{\text{Cl}}$$

Chlorine has an electronegativity of 3.0, and hydrogen's electronegativity is 2.1 (see Fig. 7.9). Since Cl is the more electronegative of the two elements, it dominates the electron-sharing process and draws the electrons closer to it. Hence, the Cl end of the bond has the δ− designation. Again, the δ− over the Cl atom indicates a partial negative charge; i.e., the chlorine end of the molecule is negative with respect to the hydrogen end. The meaning of the δ+ over the H is that the H end of the molecule is positive with respect to the Cl end. Partial charges are always charges less than +1 or −1. Charges of +1 and −1, full charges, would result when an electron is transferred from one atom to another. With partial charges we are talking about an intermediate charge state between 0 and 1.

The direction of polarity of a polar covalent bond, that is, which atom of the bond bears the partial negative charge, can also be designated using an arrow with a cross at its other end ($\mapsto$). The crossed end of the arrow is placed near the atom bearing the partial positive charge and the arrowhead is placed near the atom bearing the partial negative charge. Using this notation, the bond in the molecule HCl would be denoted as

$$\overset{\longmapsto}{\text{H—Cl}}$$

An extension of the reasoning used in characterizing the covalent bond in the HCl molecule as involving unequal sharing of electrons leads to the generalization that most chem-

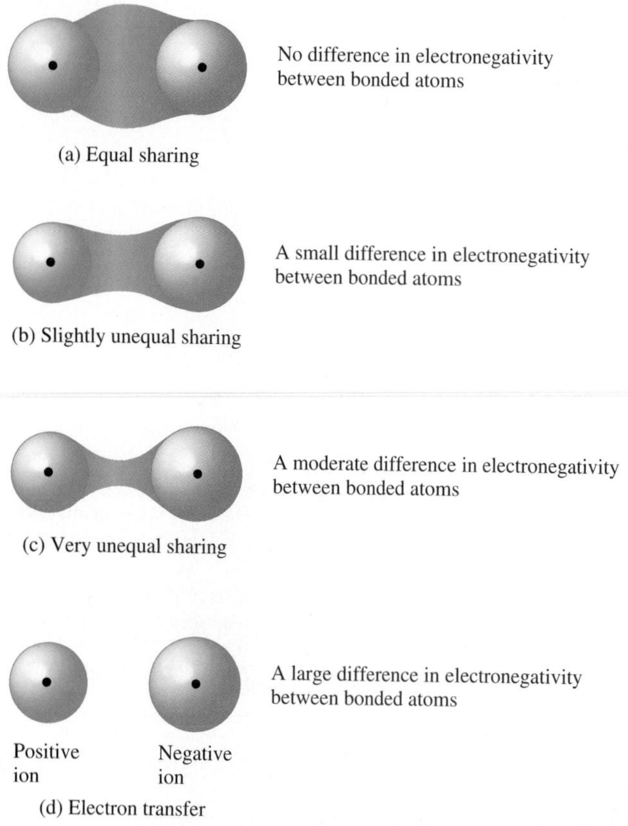

FIGURE 7.10 The continuum of bonding types includes these four situations.

ical bonds are not 100% covalent (equal sharing) or 100% ionic (electron transfer). Instead, most bonds are somewhere in between (unequal sharing). A pictorial representation of the continuum of bonding types that are possible because of the occurrence of unequal sharing of electrons is shown in Figure 7.10. Figure 7.10a shows the equal-sharing situation that results when both atoms are identical. The sharing must be equal because both atoms have the same electronegativity and must affect the bonding electrons in the same way.

Any time two nuclei differ in their ability to attract a pair of bonding electrons, unequal sharing results. Figure 7.10b shows the situation where such a difference is very small. Electron sharing will be "close" to being equal, but not exactly equal. Note that the electron distribution is no longer symmetrical. This means that the bonding electrons spend more time associated with the nucleus that has the stronger electron-attracting ability. This will be the nucleus on the right in Figure 7.10b, the side where the electron density is greater.

Figure 7.10c depicts electron density distribution when a relatively large difference in electron-attracting ability exists between nuclei. The sharing of electrons here can be described as being "very unequal."

Finally, when the electron-attracting ability difference becomes very large, one atom "wins the battle." Electron transfer occurs, and the situation depicted in Figure 7.10d results. This situation corresponds to a 100% ionic bond.

Bond polarity *is a measure of the degree of inequality in the sharing of electrons in a chemical bond.* The numerical value of the electronegativity difference between two bonded atoms is a relative gauge of the polarity of the bond. The greater the numerical difference, the greater the inequality of electron sharing and the greater the polarity of the bond.

The message of this discussion of bond polarity is that there is no natural boundary between ionic and covalent bonding. Most bonds are a mixture of pure ionic and pure covalent bonds; that is, unequal sharing of electrons occurs. Most bonds have both ionic and covalent character. Nevertheless, it is still convenient to use the terms ionic and covalent in describing chemical bonds, based on the following guidelines, which relate to electronegativity differences.

1. Bonds between identical atoms (or nonidentical atoms of equal electronegativity), where there is zero difference in electronegativity between atoms, are called *nonpolar covalent bonds.*
2. Bonds where the electronegativity difference between atoms is greater than zero but less than 2.0 are called *polar covalent bonds.*
3. Bonds where the difference in electronegativity between atoms is 2.0 or greater are called *ionic bonds.*

Example 7.12

Consider the following set of five bonds:

N—Cl, Ca—F, C—O, B—H, N—O

(a) Rank the bonds in order of increasing polarity.

(b) Determime the direction of polarity for each bond.

(c) Classify each bond as nonpolar covalent, polar covalent, or ionic.

Solution

Let us first calculate the electronegativity difference for each of the bonds using the electronegativity values of Figure 7.9

N—Cl:	$3.0 - 3.0 = 0.0$
Ca—F:	$4.0 - 1.0 = 3.0$
C—O:	$3.5 - 2.5 = 1.0$
B—H:	$2.1 - 2.0 = 0.1$
N—O:	$3.5 - 3.0 = 0.5$

(a) Bond polarity increases as electronegativity difference increases. The ranking in terms of increasing bond polarity is thus:

N—Cl	< B—H	< N—O	< C—O	< Ca—F
0.0	0.1	0.5	1.0	3.0

(b) The direction of bond polarity is from the least electronegative atom to the most electronegative atom. The more electronegative atom will bear the partial negative charge (δ^-).

	←+	+→	+→	+→
N—Cl	B—H	N—O	C—O	Ca—F

(c) Nonpolar covalent bonds require zero difference in electronegativity and ionic bonds an electronegativity difference of 2.0 or greater. The in-between region corresponds to polar covalent bonds.

nonpolar covalent: N—Cl
polar covalent: B—H, N—O, and C—O
ionic: Ca—F

Practice Exercise 7.12

Consider the following set of five bonds:

N—S, H—H, Na—F, K—Cl, and F—Cl

(a) Rank the bonds in order of increasing polarity
(b) Determine the direction of polarity for each bond
(c) Classify each bond as nonpolar covalent, polar covalent, or ionic

Ans: (a) H—H < N—S < F—Cl < K—Cl < Na—F

(b) H—H N—S (⟵+) F—Cl (⟵+) K—Cl (+⟶) Na—F (+⟶)

(c) nonpolar covalent: H—H
polar covalent: N—S, F—Cl
ionic: K—Cl, Na—F

7.18 MOLECULAR POLARITY

Molecules, as well as bonds (Sec. 7.17), can have polarity. A **polar molecule** *is a molecule in which there is an unsymmetrical distribution of electronic charge*. In a polar molecule bonding electrons are more attracted to one part of the molecule than other parts.

Molecular polarity depends on two factors: (1) bond polarities, and (2) molecular geometry (Sec. 7.15). The presence of polar bonds within a molecule does not automatically mean that the molecule as a whole is polar. In molecules that are symmetrical the effects of polar bonds may cancel each other, resulting in a molecule as a whole having no polarity.

Determining the molecular polarity of a diatomic molecule is simple because only one bond is present. If that bond is nonpolar, the molecule is nonpolar; if the bond is polar, the molecule is polar.

Molecular polarity determination in triatomic molecules is more complicated. Two different molecular geometries are possible in triatomic molecules: linear and angular. In addition, the symmetrical or unsymmetrical nature of the molecule must be considered. Let us consider the polarities of three specific triatomic molecules: CO_2 (linear), H_2O (angular), and HCN (linear).

Both bonds in the symmetrical linear CO_2 are polar. For both bonds the direction of polarity is toward oxygen since oxygen is more electronegative than carbon.

⟵+ +⟶
O=C=O

Despite the presence of the two polar bonds, CO_2 molecules are *nonpolar*. The effects of the two polar bonds cancel. The shift of electronic charge toward one oxygen atom is exactly compensated for by the shift of electronic charge toward the other oxygen atom. Thus, one end of the molecule is not negatively charged relative to the other end (a requirement for polarity) and the molecule is nonpolar.

The nonlinear (angular) triatomic H_2O molecule is polar. The bond polarities associated with the two hydrogen–oxygen bonds do not cancel each other because of the nonlinearity of the molecule.

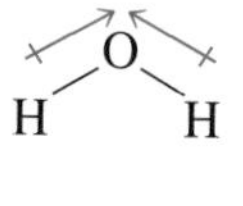

F.Y.I.

It is the polarity of the water molecule that makes cooking in microwave ovens possible. The electromagnetic field of the microwave interacts with the polar water molecules causing them to shake. The food is cooked by the heat generated by the motion of the water molecules.

Practice Exercise 7.13

Predict the polarity of the following molecules

(a) CCl_4 (tetrahedral)

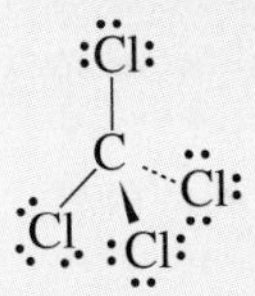

(b) OF_2 (angular)

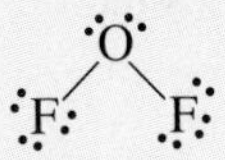

(c) H_2CO (trigonal planar)

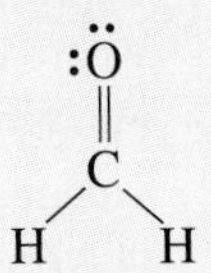

(d) HN_3 (angular)

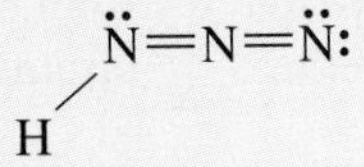

Ans. (a) nonpolar; (b) polar; (c) polar; (d) polar

KEY TERMS

The new terms or concepts defined in this chapter are

bond length (Sec. 7.13) The distance between the nuclei of bonded atoms.

bond polarity (Sec. 7.17) A measure of the degree of inequality in the sharing of electrons in a chemical bond.

bond strength (Sec. 7.13) A measure of the energy it takes to break a chemical bond.

bonding electrons (Sec. 7.10) Pairs of valence electrons that are shared between atoms in a covalent bond.

chemical bond (Sec. 7.1) Attractive forces that hold atoms or ions together in more complex aggregates.

coordinate covalent bond (Sec. 7.12) A covalent bond in which both electrons of a shared pair come from one of the two atoms.

covalent bond (Secs. 7.1 and 7.9) A bonding force resulting from two nuclei attracting the same shared electrons.

double covalent bond (Sec. 7.11) A covalent bond where two pairs of electrons are shared between the same two atoms.

electron-dot structure (Sec. 7.2) An element's symbol with one dot for each valence electron placed about the elemental symbol.

electronegativity (Sec. 7.16) A measure of the relative attraction that an atom has for the shared electrons in a bond.

ion (Sec. 7.4) An atom (or group of atoms) that is electrically charged as the result of the loss or gain of electrons.

ionic bond (Secs. 7.1 and 7.4) An attractive force resulting from the interaction of positively and negatively charged ions.

ionic solid (Sec. 7.7) An array of positive and negative ions arranged in such a way that each ion is surrounded by nearest neighbors of the opposite charge.

isoelectronic species (Sec. 7.5) Species that contain the same number of electrons.

molecular geometry (Sec. 7.15) A description of the three-dimensional arrangement of atoms within a molecule.

monoatomic ion (Sec. 7.8) A single atom that has lost or gained electrons.

nonbonding electrons (Sec. 7.10) Pairs of valence electrons about an atom that are not involved in electron sharing.

nonpolar covalent bond (Sec. 7.17) A covalent bond in which there is equal sharing of bonding electrons.

octet rule (Sec. 7.3) In forming compounds, atoms of elements lose, gain, or share electrons in such a way as to produce a noble gas electron configuration for each of the atoms involved.

polar covalent bond (Sec. 7.17) A covalent bond in which there is unequal sharing of bonding electrons.

polar molecule (Sec. 7.18) A molecule in which there is an unsymmetrical distribution of electronic charge.

polyatomic ion (Sec. 7.8) A group of atoms with a net charge, held together by covalent bonds.

resonance structures (Sec. 7.13) Two or more electron-dot structures for a molecule or ion that have the same arrangement of atoms, contain the same number of electrons, and differ only in the location of the electrons.

single covalent bond (Sec. 7.11) A covalent bond where a single pair of electrons is shared between two atoms.

triple covalent bond (Sec. 7.11) A covalent bond where three pairs of electrons are shared between two atoms.

valence electron (Sec. 7.2) An electron in the outermost shell of a representative element.

VSEPR theory (Sec. 7.15) A set of procedures for predicting the three-dimensional shape of a molecule from the information contained in the molecule's electron-dot structure.

PRACTICE PROBLEMS

Valence Electrons (Sec. 7.2)

7.1 How many valence electrons do atoms with the following electron configurations have?

(a) $1s^22s^22p^1$
(b) $1s^22s^22p^5$
(c) $1s^22s^22p^63s^1$
(d) $1s^22s^22p^63s^23p^64s^23d^{10}4p^1$

7.2 How many valence electrons do atoms with the following electron configurations have?

(a) $1s^22s^22p^3$
(b) $1s^22s^22p^63s^23p^2$
(c) $1s^22s^22p^63s^23p^64s^1$
(d) $1s^22s^22p^63s^23p^64s^23d^{10}4p^2$

7.3 How many valence electrons do atoms of each of the following elements have?

(a) beryllium **(b)** oxygen
(c) aluminum **(d)** arsenic

7.4 How many valence electrons do atoms of each of the following elements have?

(a) carbon **(b)** magnesium
(c) sulfur **(d)** iodine

7.5 Identify the element with each of the following characteristics.

(a) period 2 element with 4 valence electrons
(b) period 2 element with 7 valence electrons
(c) period 3 element with 2 valence electrons
(d) period 3 element with 5 valence electrons

7.6 Identify the element with each of the following characteristics.

(a) period 2 element with 1 valence electron
(b) period 2 element with 3 valence electrons
(c) period 3 element with 3 valence electrons
(d) period 3 element with 6 valence electrons

Electron-Dot Structures for Atoms (Sec. 7.2)

7.7 Draw electron-dot structures for atoms of the following elements.

(a) $_4Be$ **(b)** $_8O$ **(c)** $_{17}Cl$ **(d)** $_{34}Se$

7.8 Draw electron-dot structures for atoms of the following elements.

(a) $_5B$ **(b)** $_9F$ **(c)** $_{11}Na$ **(d)** $_{32}Ge$

7.9 Each of the following electron-dot structures represents a period 2 representative element. Determine the element's identity in each case.

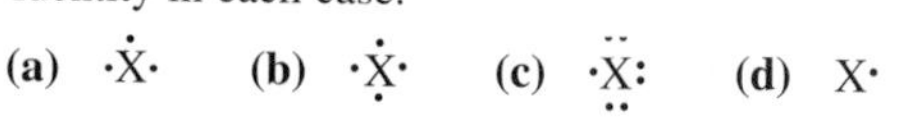

7.10 Each of the following electron-dot structures represents a period 3 representative element. Determine the element's identity in each case.

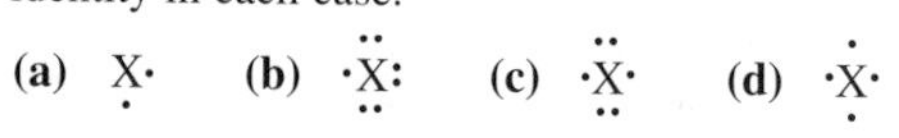

Notation for Ions (Sec. 7.4)

7.11 Give the symbol for each of the following ions.

(a) a lithium atom that has lost one electron
(b) a phosphorus atom that has gained three electrons

(c) a bromine atom that has gained one electron
(d) a barium atom that has lost two electrons

7.12 Give the symbol for each of the following ions.
(a) an iodine atom that has gained one electron
(b) a zinc atom that has lost two electrons
(c) a gallium atom that has lost three electrons
(d) a sodium atom that has lost one electron

7.13 Calculate the number of protons and electrons present in each of the following ions.
(a) Li^+ **(b)** N^{3-} **(c)** Ca^{2+} **(d)** Cl^-

7.14 Calculate the number of protons and electrons present in each of the following ions.
(a) Al^{3+} **(b)** O^{2-} **(c)** Be^{2+} **(d)** F^-

7.15 What would be the symbol for an ion with each of the following characteristics?
(a) an aluminum ion with 10 electrons
(b) an oxygen ion with 10 electrons
(c) a magnesium ion with 2 more protons than electrons
(d) a beryllium ion with 2 fewer electrons than protons

7.16 What would be the symbol for an ion with each of the following characteristics?
(a) a sodium ion with 10 electrons
(b) a fluorine ion with 10 electrons
(c) a sulfur ion with 2 fewer protons than electrons
(d) a calcium ion with 2 more protons than electrons

Ionic Charge Magnitude (Sec. 7.5)

7.17 Use the periodic table to predict the charge on the ion formed by each of the following elements.
(a) beryllium **(b)** selenium
(c) rubidium **(d)** strontium

7.18 Use the periodic table to predict the charge on the ion formed by each of the following elements.
(a) nitrogen **(b)** magnesium
(c) chlorine **(d)** iodine

7.19 Indicate the number of electrons lost or gained when each of the following atoms forms an ion.
(a) $_{12}Mg$ **(b)** $_{9}F$ **(c)** $_{7}N$ **(d)** $_{3}Li$

7.20 Indicate the number of electrons lost or gained when each of the following atoms forms an ion.
(a) $_{11}Na$ **(b)** $_{35}Br$ **(c)** $_{13}Al$ **(d)** $_{16}S$

7.21 Predict the general kind of behavior, that is, loss or gain of electrons, you would expect from atoms with the following electron configurations.
(a) $1s^2 2s^2$
(b) $1s^2 2s^2 2p^6 3s^2$
(c) $1s^2 2s^2 2p^6 3s^2 3p^1$
(d) $1s^2 2s^2 2p^6 3s^2 3p^5$

7.22 Predict the general kind of behavior, that is, loss or gain of electrons, you would expect from atoms with the following electron configurations.
(a) $1s^2 2s^2 2p^4$
(b) $1s^2 2s^2 2p^3$
(c) $1s^2 2s^1$
(d) $1s^2 2s^2 2p^6 3s^2 3p^6 4s^2 3d^{10} 4p^1$

7.23 Write the electron configurations for each of the following ions.
(a) Cl^- **(b)** Li^+ **(c)** O^{2-} **(d)** Mg^{2+}

7.24 Write the electron configurations for each of the following ions.
(a) S^{2-} **(b)** P^{3-} **(c)** Be^{2+} **(d)** Na^+

7.25 Identify the period 3 element that produces each of the following ions.
(a) X^- **(b)** X^+ **(c)** X^{3+} **(d)** X^{2-}

7.26 Identify the period 3 element that produces each of the following ions.
(a) X^{2-} **(b)** X^{2+} **(c)** X^{3-} **(d)** X^{3+}

7.27 With which noble gas is each of the following ions isoelectronic?
(a) Be^{2+} **(b)** Ba^{2+} **(c)** Br^- **(d)** I^-

7.28 With which noble gas is each of the following ions isoelectronic?
(a) Na^+ **(b)** Rb^+ **(c)** S^{2-} **(d)** Se^{2-}

7.29 Select from each of the following sets of ions the ion that is isoelectronic with the noble gas neon.
(a) Li^+, Na^+, K^+ **(b)** F^-, Cl^-, Br^-
(c) Ca^{2+}, O^{2-}, Be^{2+} **(d)** Mg^{2+}, P^{3-}, S^{2-}

7.30 Select from each of the following sets of ions the ion that is isoelectronic with the noble gas argon.
(a) Be^{2+}, Mg^{2+}, Ca^{2+} **(b)** O^{2-}, S^{2-}, Se^{2-}
(c) K^+, N^{3-}, Al^{3+} **(d)** F^-, O^{2-}, P^{3-}

7.31 In which of the following are the two members of the pair isoelectronic with each other?
(a) Ar and S^{2-} **(b)** Ne and Be^{2+}
(c) Ca^{2+} and P^{3-} **(d)** Li^+ and O^{2-}

7.32 In which of the following are the two members of the pair isoelectronic with each other?
(a) He and Li^+ **(b)** Kr and Ca^{2+}
(c) F^- and Cl^- **(d)** Al^{3+} and N^{3-}

Electron-Dot Structures for Ionic Compounds (Sec. 7.5)

7.33 Show the formation of the following ionic compounds using electron-dot structures.
(a) $CaBr_2$ **(b)** MgS **(c)** Be_3N_2 **(d)** Na_3P

7.34 Show the formation of the following ionic compounds using electron-dot structures.

(a) K_2O **(b)** MgF_2 **(c)** AlP **(d)** Na_2S

7.35 Using electron-dot structures, show how ionic compounds are formed from the following pairs of elements.

(a) lithium and nitrogen **(b)** magnesium and oxygen
(c) chlorine and barium **(d)** fluorine and potassium

7.36 Using electron-dot structures, show how ionic compounds are formed from the following pairs of elements.

(a) sodium and bromine **(b)** aluminum and sulfur
(c) phosphorus and beryllium **(d)** oxygen and calcium

Formulas for Ionic Compounds (Sec. 7.6)

7.37 Write the formula for the ionic compound formed from each of the following types of ions.

(a) Ca^{2+} and Cl^- **(b)** Be^{2+} and O^{2-}
(c) Al^{3+} and N^{3-} **(d)** K^+ and S^{2-}

7.38 Write the formula for the ionic compound formed from each of the following types of ions.

(a) Mg^{2+} and S^{2-} **(b)** Na^+ and Br^-
(c) Al^{3+} and O^{2-} **(d)** Be^{2+} and N^{3-}

7.39 Write formulas (symbol and charge) for both kinds of ions present in each of the following ionic compounds.

(a) MgS **(b)** AlN **(c)** Na_2O **(d)** Ca_3N_2

7.40 Write formulas (symbol and charge) for both kinds of ions present in each of the following ionic compounds.

(a) KCl **(b)** CaS **(c)** BeF_2 **(d)** Al_2S_3

7.41 Write the formula of the ionic compound that could form from the elements X and Z if

(a) X has 2 valence electrons and Z has 4 valence electrons
(b) X has 3 valence electrons and Z has 7 valence electrons
(c) X has 1 valence electron and Z has 5 valence electrons
(d) X has 7 valence electrons and Z has 2 valence electrons

7.42 Write the formula of the ionic compound that could form from the elements X and Z if

(a) X has 2 valence electrons and Z has 7 valence electrons
(b) X has 1 valence electron and Z has 6 valence electrons
(c) X has 3 valence electrons and Z has 5 valence electrons
(d) X has 6 valence electrons and Z has 2 valence electrons

Polyatomic-Ion-Containing Ionic Compounds (Sec. 7.8)

7.43 Determine the formula for the ionic compound formed from each of the following types of ions.

(a) Mg^{2+} and CN^- **(b)** Ca^{2+} and SO_4^{2-}
(c) Al^{3+} and OH^- **(d)** NH_4^+ and NO_3^-

7.44 Determine the formula for the ionic compound formed from each of the following types of ions.

(a) Mg^{2+} and NO_3^- **(b)** Ca^{2+} and CN^-
(c) Al^{3+} and SO_4^{2-} **(d)** NH_4^+ and OH^-

7.45 Determine the formula for the ionic compound in which the Al^{3+} ion is combined with each of the following polyatomic ions.

(a) PO_4^{3-} **(b)** CO_3^{2-}
(c) ClO_3^- **(d)** $C_2H_3O_2^-$

7.46 Determine the formula for the ionic compound in which the Mg^{2+} ion is combined with each of the following polyatomic ions.

(a) PO_4^{3-} **(b)** CO_3^{2-}
(c) ClO_3^- **(d)** $C_2H_3O_2^-$

Electron-Dot Structures for Covalent Compounds (Secs. 7.9–7.13)

7.47 Show the formation of the following covalent compounds using electron-dot structures.

(a) I_2 **(b)** ClF **(c)** H_2S **(d)** PF_3

7.48 Show the formation of the following covalent compounds using electron-dot structures.

(a) Br_2 **(b)** IF **(c)** PCl_3 **(d)** CBr_4

7.49 Write an electron-dot structure for the simplest covalent compound most likely to be formed between each of these pairs of elements.

(a) hydrogen and bromine **(b)** oxygen and fluorine
(c) nitrogen and chlorine **(d)** silicon and iodine

7.50 Write an electron-dot structure for the simplest covalent compound most likely to be formed between each of these pairs of elements.

(a) hydrogen and iodine **(b)** nitrogen and bromine
(c) oxygen and chlorine **(d)** silicon and hydrogen

7.51 How many *bonding electron pairs* and *nonbonding electron pairs* are present in each of the following electron-dot structures?

(a) :N:::N:
(b) H:C::C:H (with H under each C)
(c) :Ö=C=Ö:
(d) :N≡C—C≡N:

7.52 How many *bonding electron pairs* and *nonbonding electron pairs* are present in each of the following electron-dot structures?

(a) H:C:::C:H
(b) H:P:Cl: (with H under P)
(c) H—C—H, with :O: double-bonded (‖) to the C
(d) :F—N=N—F:

7.53 How many *single covalent bonds*, *double covalent bonds*, and *triple covalent bonds* are present in each of the molecules in Problem 7.51.

7.54 How many *single covalent bonds*, *double covalent bonds*, and *triple covalent bonds* are present in each of the molecules in Problem 7.52.

7.55 What are two differences between a single covalent bond and a double covalent bond?

7.56 What are two differences between a single covalent bond and a triple covalent bond?

7.57 Which of the following is a normally expected bonding pattern for the element shown.

(a) —Ö— **(b)** :N≡ **(c)** :C̈= **(d)** :Ö=

7.58 Which of the following is a normally expected bonding pattern for the element shown.

(a) —N̈= **(b)** :O≡ **(c)** —C≡ **(d)** =C=

7.59 What is a coordinate covalent bond?

7.60 Once formed, how (if at all) does a coordinate covalent bond differ from a regular covalent bond?

7.61 Identify the coordinate covalent bond(s) present, if any, in each of the following molecules by listing the two atoms involved in the bond. Name the atom on the left or below in the bond first.

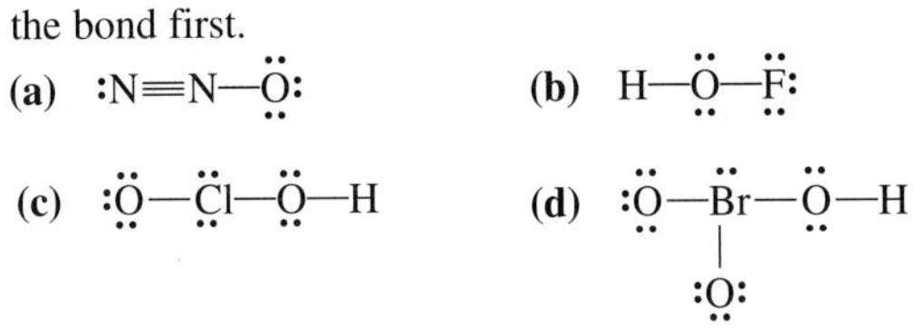

7.62 Identify the coordinate covalent bond(s) present, if any, in each of the following molecules by listing the two atoms involved in the bond. Name the atom on the left or below in the bond first.

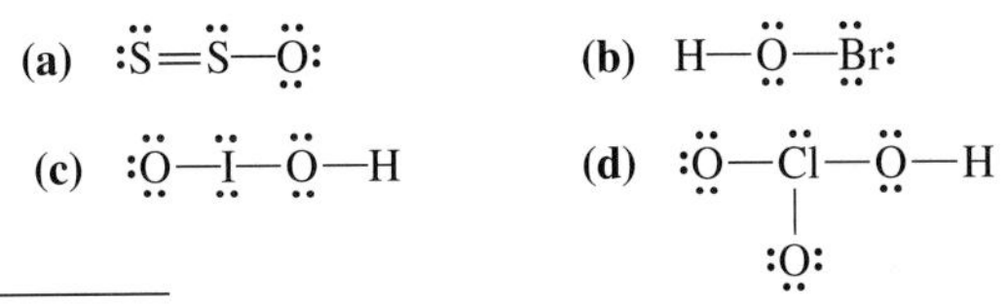

7.63 What are resonance structures?

7.64 Explain why H—C≡N: and H—N≡C: are not resonance structures.

7.65 The following is one of the three resonance structures that exist for the nitrate ion.

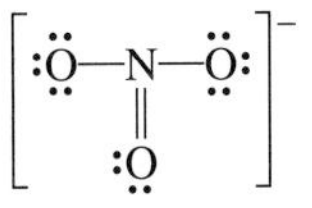

Draw the other two resonance structures.

7.66 The following is one of the three resonance structures that exist for the carbonate ion.

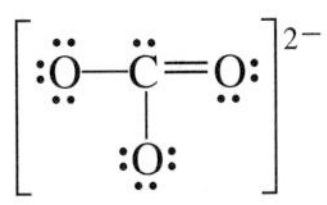

Draw the other two resonance structures.

Systematic Procedures for Determining Electron-Dot Structures (Sec. 7.14)

7.67 How many electron dots should appear in the electron-dot structures for each of the following molecules or ions?

(a) HNO_3 **(b)** PF_3 **(c)** BF_4^+ **(d)** PO_4^{3-}

7.68 How many electron dots should appear in the electron-dot structures for each of the following molecules or ions?

(a) O_2F_2 **(b)** $C_2H_2Br_2$ **(c)** S_2^{2-} **(d)** NH_4^+

7.69 Using systematic procedures, write the electron-dot structure for each of the following molecules.

(a) H_3CCH_3 (C_2H_6) **(b)** H_2NNH_2 (N_2H_4)
(c) F_2CH_2 (CH_2F_2) **(d)** H_3CCCl_3 ($C_2H_3Cl_3$)

The above formulas are written in a form that specifies atomic arrangement. When an elemental symbol carries a subscript, these atoms are directly and separately bonded to the atom immediately following or immediately preceding the subscripted symbol.

7.70 Using systematic procedures, write the electron-dot structure for each of the following molecules.

(a) H_2PPH_2 (P_2H_4) **(b)** H_3CCBr_3 ($C_2H_3Br_3$)
(c) F_2CCl_2 (CF_2Cl_2) **(d)** H_3SiSiH_3 (Si_2H_6)

The formulas are written in a way that specifies atomic arrangement, as explained in Problem 7.69.

7.71 Using systematic procedures, write the electron-dot structure for each of the following polyatomic ions.

(a) NF_4^+ **(b)** BeH_4^{2-} **(c)** ClO_3^- **(d)** IO_4^-

7.72 Using systematic procedures, write the electron-dot structure for each of the following polyatomic ions.

(a) BH_4^- **(b)** $AlCl_4^-$ **(c)** PF_4^+ **(d)** ClO_2^-

7.73 Using systematic procedures, write the electron-dot structure for each of the following molecules, each of which contains at least one multiple bond. The formulas of the molecules are written in a way that specifies atomic arrangement, as explained in Problem 7.69.

(a) Cl_2CCH_2 ($C_2H_2Cl_2$) **(b)** H_3CCN (C_2H_3N)
(c) $HCCH_3$ (C_3H_4) **(d)** FNNF (N_2F_2)

7.74 Using systematic procedures, write the electron-dot structure for each of the following molecules, each of which contains at least one multiple bond. The formulas of the molecules are written in a way that specifies atomic arrangement, as explained in Problem 7.69.

(a) H_2NCN (CH_2N_2) **(b)** Cl_2CO
(c) ClCCCl (C_2Cl_2) **(d)** NCCN (C_2N_2)

7.75 Using systematic procedures, write the electron-dot structure for each of the following molecules or polyatomic ions. Resonance structures will be needed in each case. Some formulas are written in a way that specifies atomic arrangements, as explained in Problem 7.69.

(a) H_3CNO_2 (b) NNO (c) CO_3^{2-} (d) SCN^-

7.76 Using systematic procedures, write the electron-dot structure for each of the following molecules or polyatomic ions. Resonance structures will be needed in each case. Some formulas are written in a way that specifies atomic arrangements, as explained in Problem 7.69.

(a) HONO (b) H_2NNO_2 (c) NO_3^- (d) OCN^-

7.77 How many bonding and nonbonding electron pairs are present in each of the following diatomic species?

(a) HCl (b) CO (c) OH^- (d) CN^-

7.78 How many bonding and nonbonding electron pairs are present in each of the following diatomic species?

(a) BrCl (b) N_2 (c) HS^- (d) NS^-

7.79 The electron-dot structure for the polyatomic ion ClO_4^{n-} is

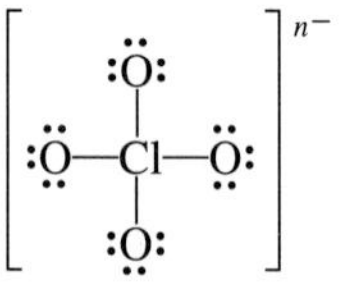

What is the value of *n*, the magnitude of the ionic charge, in the formula ClO_4^{n-}?

7.80 The electron-dot structure for the polyatomic ion $BeCl_4^{n-}$ is

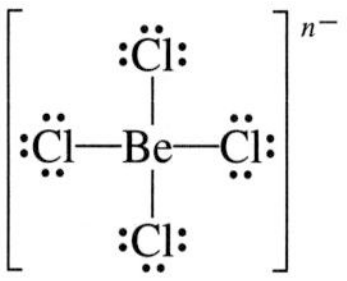

What is the value of *n*, the magnitude of the ionic charge, in the formula $BeCl_4^{n-}$?

Molecular Geometry (VSEPR Theory) (Sec. 7.16)

7.81 What is the molecular geometry that VSEPR theory predicts for a molecule whose sole central atom has the environment of

(a) 2 single bonds and 2 nonbonding electron pairs

(b) 1 single bond and 1 triple bond

(c) 1 single bond, 1 double bond, and 1 nonbonding electron pair

(d) 1 double bond and 2 nonbonding electron pairs

7.82 What is the molecular geometry that VSEPR theory predicts for a molecule whose sole central atom has the environment of

(a) 4 single bonds

(b) 2 double bonds

(c) 3 single bonds and 1 nonbonding electron pair

(d) 1 triple bond and 1 nonbonding electron pair

7.83 Using VSEPR theory predict the geometry of the following triatomic molecules or ions.

(a) H:S:H (b) :I:C:::N:

(c) [:O::N::O:]$^+$ (d) :O:O::O:

7.84 Using VSEPR theory predict the geometry of the following triatomic molecules or ions.

(a) :N::S:F: (b) :N::N::O:

(c) [H:N:H]$^-$ (d) [:S:C:::N:]$^-$

7.85 Using VSEPR theory predict the geometry of the following sulfur-containing molecules or ions.

(a) 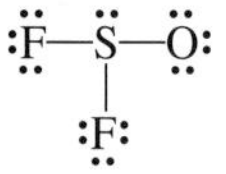 (b) 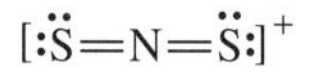

(c) 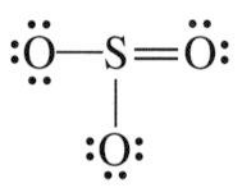 (d) 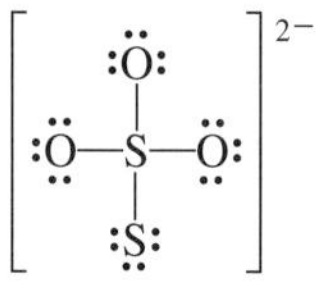

7.86 Using VSEPR theory predict the geometry of the following nitrogen-containing molecules or ions.

(a) 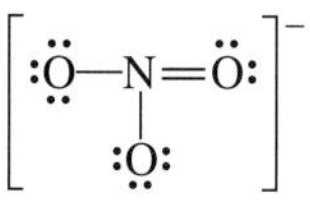 (b) 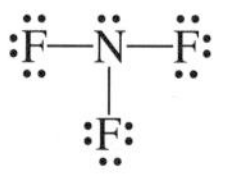

(c) 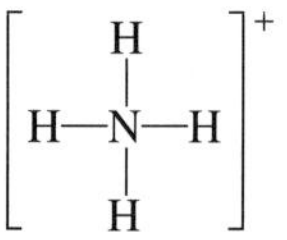 (d) 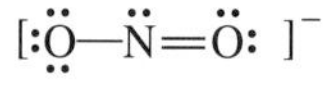

7.87 Using VSEPR theory predict the geometry of the following molecules or polyatomic ions. Some formulas are written in a way that specifies atomic arrangements as explained in Problem 7.69.

(a) $AlCl_4^-$ (b) CS_2 (c) $OSCl_2$ (d) PO_3^{3-}

7.88 Using VSEPR theory predict the geometry of the following molecules or polyatomic ions. Some formulas are written in a way that specifies atomic arrangements as explained in Problem 7.69.

(a) H_3CBr (b) $BeCl_4^{2-}$ (c) NO_2^- (d) Cl_2O

7.89 Using VSEPR theory predict the geometry of the following molecules which have more than one central atom. The formulas are written in a way that specifies atomic arrangements as explained in Problem 7.69.

(a) H_2NNH_2 (b) H_3COH

7.90 Using VSEPR theory predict the geometry of the following molecules which have more than one central atom. The formulas are written in a way that specifies atomic arrangements as explained in Problem 7.69.

(a) H_2CCCl_2 (b) H_3COCH_3

Electronegativity (Sec. 7.16)

7.91 Using chemical periodicity trends, select the more electronegative element in each of the following pairs of elements.

(a) H and O **(b)** C and O
(c) P and S **(d)** Na and Mg

7.92 Using chemical periodicity trends, select the more electronegative element in each of the following pairs of elements.

(a) N and As **(b)** Na and Al
(c) H and C **(d)** Cl and Br

7.93 Using chemical periodicity trends, arrange each of the following sets of atoms in order of increasing electronegativity.

(a) Na, Al, P, Mg **(b)** Cl, Br, I, F
(c) S, P, O, As **(d)** Ca, Ge, O, C

7.94 Using chemical periodicity trends, arrange each of the following sets of atoms in order of increasing electronegativity.

(a) Be, N, C, B **(b)** Te, S, O, Se
(c) B, Na, Al, Mg **(d)** F, Mg, Al, N

Bond Polarity (Sec. 7.17)

7.95 For which of the following bonds is the given direction of polarity *incorrect*?

(a) $\overset{\delta+\ \ \delta-}{\mathrm{O{-}F}}$ **(b)** $\overset{\delta-\ \ \delta+}{\mathrm{O{-}N}}$ **(c)** $\overset{\longleftarrow\!+}{\mathrm{H{-}O}}$ **(d)** $\overset{+\!\longrightarrow}{\mathrm{Br{-}F}}$

7.96 For which of the following bonds is the given direction of polarity *incorrect*?

(a) $\overset{\delta+\ \ \delta-}{\mathrm{N{-}C}}$ **(b)** $\overset{\delta+\ \ \delta-}{\mathrm{O{-}Se}}$ **(c)** $\overset{+\!\longrightarrow}{\mathrm{B{-}C}}$ **(d)** $\overset{\longleftarrow\!+}{\mathrm{H{-}C}}$

7.97 Identify the bond of *greatest* polarity in each of the following sets of bonds.

(a) H—Cl, H—O, H—Br
(b) O—F, O—P, O—Al
(c) H—Cl, Br—Br, B—N
(d) Al—Cl, C—N, Cl—F

7.98 Identify the bond of *greatest* polarity in each of the following sets of bonds.

(a) H—C, H—N, H—B
(b) O—N, O—S, O—Br
(c) P—N, S—O, Br—F
(d) H—F, Cl—Cl, Si—O

7.99 Classify each of the following bonds as nonpolar covalent, polar covalent, or ionic.

(a) carbon–nitrogen **(b)** beryllium–oxygen
(c) phosphorus–chlorine **(d)** silicon–silicon

7.100 Classify each of the following bonds as nonpolar covalent, polar covalent, or ionic.

(a) cesium–fluorine **(b)** potassium–chlorine
(c) hydrogen–hydrogen **(d)** silicon–phosphorus

Molecular Polarity (Sec. 7.18)

7.101 Determine the polarity of each of the following molecules.

(a) NCl_3 (trigonal pyramid)
(b) H_2S (angular)
(c) H_3CCl (tetrahedral)
(d) CS_2 (linear)

7.102 Determine the polarity of each of the following molecules.

(a) BF_3 (trigonal planar)
(b) $OSCl_2$ (trigonal pyramid)
(c) NO_2 (angular)
(d) $SiCl_4$ (tetrahedral)

7.103 For each of the following hypothetical triatomic molecules, indicate whether the *bonds* are polar or nonpolar and whether the *molecule* is polar or nonpolar. Assume that A and X have different electronegativities.

(a) X—A—X **(b)** A—X—X

(c) 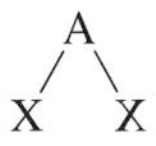 **(c)** 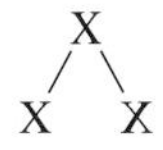

7.104 For each of the following hypothetical square planar molecules, indicate whether the *bonds* are polar or nonpolar and whether the *molecule* is polar or nonpolar. Assume that A, X, and Z have different electronegativities.

(a) 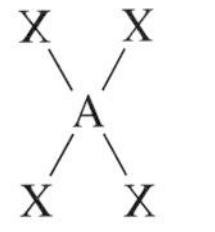 **(b)** 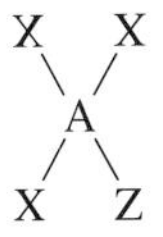

(c) 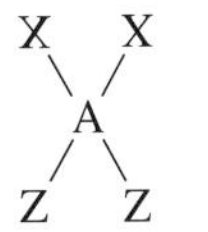 **(d)** 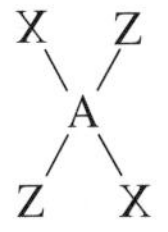

7.105 In which of the following pairs of molecules do both members of the pair have the same polarity, that is, both members are polar or both members are nonpolar?

(a) F_2 and BrF **(b)** HOCl and HCN
(c) CH_4 and CCl_4 **(d)** SO_3 and NF_3

7.106 In which of the following pairs of molecules do both members of the pair have the same polarity, that is, both members are polar or both members are nonpolar?

(a) HBr and HCl **(b)** CO_2 and SO_2
(c) SO_2 and SO_3 **(d)** CH_4 and CH_3Cl

ADDITIONAL PROBLEMS

7.107 Write the electron configurations for the following atoms and ions.

(a) Mg and Mg^{2+} **(b)** F and F^-

(c) N and N^{3-} **(d)** Ca^{2+} and S^{2-}

7.108 Write the electron configurations for the following atoms and ions.

(a) Na and Na^+ **(b)** O and O^{2-}

(c) Al and Al^{3+} **(d)** K^+ and Cl^-

7.109 What is wrong with each of the following electron-dot structures?

(a) 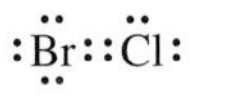 **(b)** 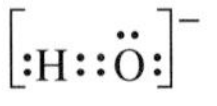

7.110 What is wrong with each of the following electron-dot structures?

(a)  **(b)** 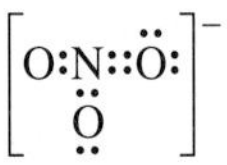

7.111 Classify the bonding in each of the following compounds as ionic or covalent.

(a) HN_3 **(b)** K_3N **(c)** CCl_4 **(d)** SF_2

7.112 Classify the bonding in each of the following compounds as ionic or covalent.

(a) NH_3 **(b)** Na_2O **(c)** SiH_4 **(d)** C_2H_4

7.113 For which of the following compounds is the basic structural unit a molecule?

(a) CO_2 **(b)** NaCl **(c)** Li_2O **(d)** HOCl

7.114 For which of the following compounds is the basic structural unit a molecule?

(a) SO_3 **(b)** KF **(c)** $Mg(OH)_2$ **(d)** BrCl

7.115 Identify the period 2 element that X represents in each of the following ionic compounds.

(a) Na_2X **(b)** Be_3X_2 **(c)** CaX **(d)** MgX_2

7.116 Identify the period 3 element that X represents in each of the following ionic compounds.

(a) Na_3X **(b)** BeX **(c)** CaX_2 **(d)** Mg_3X_2

7.117 Which of the following are *not correct* chemical formulas for ionic compounds?

(a) BeCl **(b)** Ca_2Cl_2 **(c)** Na_2S **(d)** LiS

7.118 Which of the following are *not correct* chemical formulas for ionic compounds?

(a) K_3N **(b)** MgF **(c)** $NaCl_2$ **(d)** Al_3N_3

7.119 Four hypothetical elements, A, B, C, and D, have electronegativities A = 3.8, B = 3.3, C = 2.8, and D = 1.3. These elements form the compounds BA, DA, DB, and CA. Arrange these compounds in order of increasing *ionic* bond character.

7.120 Four hypothetical elements, A, B, C, and D, have electronegativities A = 3.6, B = 3.0, C = 2.7, and D = 0.9. These elements form the compounds CA, CB, DA, and DC. Arrange these compounds in order of increasing *covalent* bond character.

7.121 In which of the following pairs of diatomic species do both members of the pair have bonds of the same multiplicity (single, double, triple)?

(a) BrCl and ClF **(b)** F_2 and N_2

(c) NO^+ and NO^- **(d)** CN^- and SN^-

7.122 In which of the following pairs of diatomic species do both members of the pair have bonds of the same multiplicity (single, double, triple)?

(a) HCl and HF **(b)** S_2 and Cl_2

(c) CO and NO^+ **(d)** OH^- and HS^-

7.123 In each of the following electron-dot structures, X represents a period 3 nonmetal. Identify the nonmetal in each case.

(a) 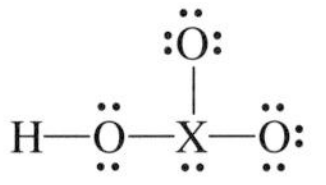 **(b)** 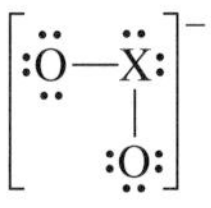

7.124 In each of the following electron-dot structures, X represents a period 3 nonmetal. Identify the nonmetal in each case.

(a) 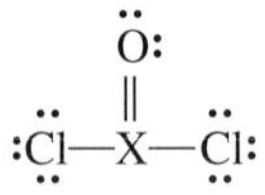 **(b)** 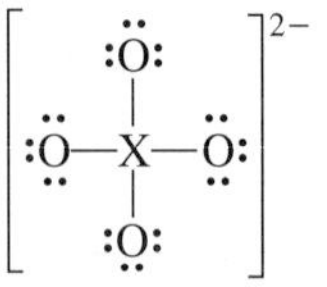

7.125 Give an electron-dot description of the bonding in each of the following compounds that takes into account the presence of both ionic and covalent bonding in each compound.

(a) $CaSO_4$ **(b)** NH_4NO_3

7.126 Give an electron-dot description of the bonding in each of the following compounds that takes into account the presence of both ionic and covalent bonding in each compound.

(a) NaOH **(b)** NH_4CN

7.127 Specify the electron-pair geometry about the central atom and the molecular geometry for each of the following species.

(a) CF_4 **(b)** NH_3 **(c)** SCN^- **(d)** PH_4^+

7.128 Specify the electron-pair geometry about the central atom and the molecular geometry for each of the following species.

(a) SiH_4 **(b)** NH_4^+ **(c)** ClNO **(d)** NO_3^-

7.129 Give an approximate value for the indicated bond angle in each of the following molecules.

(a) 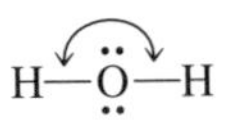

(b) 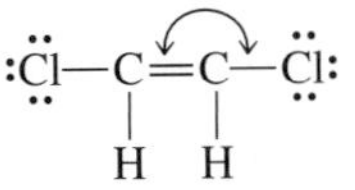

7.130 Give an approximate value for the indicated bond angle in each of the following molecules.

(a) 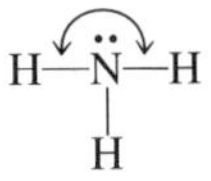

(b) 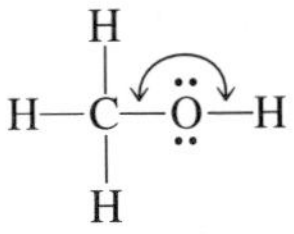

7.131 Indicate which molecule in each of the following pairs of molecules is *more* polar.

(a) HCl and HF **(b)** H_3CCl and H_3CF

(c) HCN and CO_2 **(d)** SO_2 and SO_3

7.132 Indicate which molecule in each of the following pairs of molecules is *more* polar.

(a) BrCl and BrI **(b)** CO_2 and SO_2

(c) SO_3 and NF_3 **(d)** H_3CF and Cl_3CF

CUMULATIVE PROBLEMS

7.133 Determine the identities of the elements A and D in the ionic compound AD given that

(1) the atomic number of element A is greater than that of element D.

(2) three electrons are transferred from A to D during compound formation.

(3) the sum of the atomic numbers of A and D is 20.

7.134 Determine the identities of the elements A and D in the ionic compound AD given that

(1) the atomic number of element A is less than that of element D.

(2) two electrons are transferred from A to D during compound formation.

(3) the sum of the atomic numbers of A and D is 20.

7.135 Given the following information about two elements, A and D, determine their identities and the formula of the binary ionic compound that they form.

(1) A is located in period 3 of the periodic table.

(2) The occupied electron subshell of highest energy for element D contains 2 electrons.

(3) The complete electron configuration for element D contains four fewer electrons than that for element A.

7.136 Given the following information about two elements, A and D, determine their identities and the formula of the binary ionic compound that they form.

(1) Both A and D are in period 3 of the periodic table.

(2) The occupied electron subshell of highest energy for element A contains one electron, and the one for element D contains four electrons.

(3) The atom of the element that is a nonmetal contains three more total electrons than the atom of the element that is a metal.

7.137 Determine the electron-dot structure for the covalent compound formed between the elements A and D given that

(1) both elements A and D are located in the 2*p* area of the periodic table.

(2) element A has an electronegativity greater than 3.2

(3) element D's periodic table position is next to that of a noble gas.

7.138 Determine the electron-dot structure for the covalent compound formed between the elements A and D given that

(1) both elements A and D are located in the 3*p* area of the periodic table.

(2) element A has an electronegativity greater than 2.4, and element D has an electronegativity greater than 2.0.

(3) element A's periodic table position is two positions beyond that of element D.

7.139 A covalent compound containing the elements A and D has the formula A_xD_y. Determine the identities of the elements A and D and the values for x and y in the formula for the compound given the following information about A and D.

(1) A and D are representative elements.

(2) D atoms have electron configurations ending in p^4.

(3) The isotope of D with a mass number of 16 has a percentage abundance greater than 95%.

(4) The period number for element A is one less than that for element D.

7.140 A covalent compound containing the elements A and D has the formula A_xD_y. Determine the identities of the elements A and D and the values for x and y in the formula for the compound given the following information about A and D.

(1) A and D are representative elements in the same period of the periodic table.

(2) A atoms have electron configurations ending in p^2.

(3) The isotope of A with a mass number of 12 has a percentage abundance greater than 95%.

(4) The group number for element A is three less than the group number for element D.

7.141 A polyatomic ion containing the elements A and D has the formula AD_4^{2-}. Determine the identities of the elements A and D, and then draw the electron-dot structure for the polyatomic ion given that

(1) A, the central atom in the polyatomic ion, possesses only *s* electrons, and exactly one-half of these *s* electrons are valence electrons.

(2) the D atoms present collectively possess 28 valence electrons.

(3) the atomic number of D is less than 15.

7.142 A polyatomic ion containing the elements A and D has the formula AD_4^{+}. Determine the identities of the elements A and D, and then draw the electron-dot structure for the polyatomic ion given that

(1) A, the central atom, possesses nine *p* electrons.

(2) for D atoms the period number in the periodic table and number of valence electrons are the same.

(3) for D atoms, the total number of electrons present is the same as the number of valence electrons present.

---

8 Chemical Nomenclature

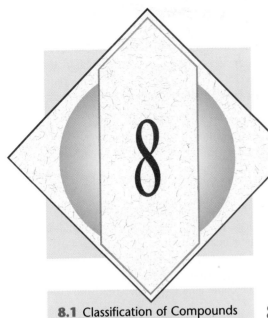

8.1 Classification of Compounds for Nomenclature Purposes

Just as it is important to be able to write formulas for compounds, it is necessary to be able to name those compounds.

Chemical nomenclature *is the system of names used to distinguish compounds from each other and the rules needed to devise these names.* In the early history of chemistry there was no system for naming compounds. Early names included quicksilver, blue vitriol, Glauber's salt, gypsum, sal ammoniac, and laughing gas. As chemistry grew, it became clear that the "anything goes" system was not acceptable. Without a system for naming compounds, coping with the multitude of known substances would be a hopeless task.

In this chapter we consider the rules now followed for naming compounds. The rules presented here follow the recommendations of the nomenclature committees of the International Union of Pure and Applied Chemistry (IUPAC). **IUPAC rules** *are a set of compound-naming rules produced by committees of the International Union of Pure and Applied Chemistry (IUPAC).* The committees of this international scientific organization meet periodically to revise and update the nomenclature rules to accommodate any newly discovered types of compounds.

For nomenclatural purposes it is convenient to classify compounds into categories based on the number of elements present in a compound. A **binary compound** *is a compound containing just two different elements.* The compounds NH_3, H_2O, CO_2, NaCl, and P_4O_{10} are all binary compounds. Any number of atoms of the two elements may be present in a molecule or formula unit of a binary compound, but only two elements may be present. A **ternary compound** *is a compound containing three different elements.* Many of the compounds that students encounter in chemistry laboratory situations are ternary compounds. Examples include nitric acid (HNO_3), sulfuric acid (H_2SO_4), and sodium hydroxide (NaOH). Compounds containing more than three kinds of elements do exist, and we will encounter a few, but will not worry about specific classification terms for such compounds.

Distinguishing between ionic and molecular compounds (Secs. 5.2 and 7.1) is an important key to becoming successful at chemical nomenclature. Although there is no sharp dividing line between ionic and covalent bonds (Sec. 7.10) and thus between ionic and molecular compounds, we will "create" such a line for nomenclatural purposes with the following simple generalizations.

1. Compounds resulting from the combination of a metal and one or more nonmetals are considered *ionic.*
2. Compounds resulting from combinations of a nonmetal with other nonmetals are considered *molecular.*

We will also treat compounds containing the positive polyatomic NH_4^+ ion (the ammonium ion) as ionic. In such ionic compounds, the NH_4^+ ion (a combination of two nonmetals) is considered to have replaced the metallic ion ordinarily present and to be functioning as a metal.

Metalloid elements (Sec. 6.12) are considered to be nonmetals for purposes of nomenclature. Thus, a compound resulting from the combination of a metalloid and nonmetals is named as a molecular compound.

8.2 TYPES OF BINARY IONIC COMPOUNDS

A **binary ionic compound** *is an ionic compound in which positively charged metallic ions and negatively charged nonmetallic ions are present.* Binary ionic compounds where both ions are monoatomic are the simplest type of ionic compound (Sec. 7.8). Such binary ionic compounds may be divided into two categories based on the metal ion that is present.

1. Type I binary ionic compounds contain a metal that can form only a *single type* of positive ion.
2. Type II binary ionic compounds contain a metal that can form *more than one type* of positive ion.

In general, all metals lose electrons when forming ions. Type I metals always exhibit the same behavior in ion formation; that is, they always lose the same number of electrons. A **type I metal** *forms only one type of positive ion, which always has the same charge magnitude*. Another name for a type I metal is *fixed-charge metal*.

Type II metals do not always lose the same number of electrons upon ion formation. For example, the type II metal iron sometimes forms an Fe^{2+} ion and at other times an Fe^{3+} ion. A **type II metal** *forms more than one type of positive ion, with the ion types differing in charge magnitude*. Another name for a type II metal is *variable-charge metal*.

Which metals are type I (fixed-charge) metals and which metals are type II (variable-charge) metals? This information is a prerequisite for naming binary ionic compounds. Type I binary ionic compounds are named in one way and type II binary ionic compounds are named in a slightly different way.

A limited number of type I metals exist. If these are learned, then all other metals will be type II metals. The type I metals are the group IA and IIA metals plus Al, Ga, Zn, Cd, and Ag. Figure 8.1 shows the periodic table positions for these fifteen type I metals. The charge magnitude for these metals can be related to their position in the periodic table. Looking at Figure 8.1, we see that all group IA elements form +1 ions. The charges for ions of elements in groups IIA and IIIA are +2 and +3, respectively. Group numbers and charge

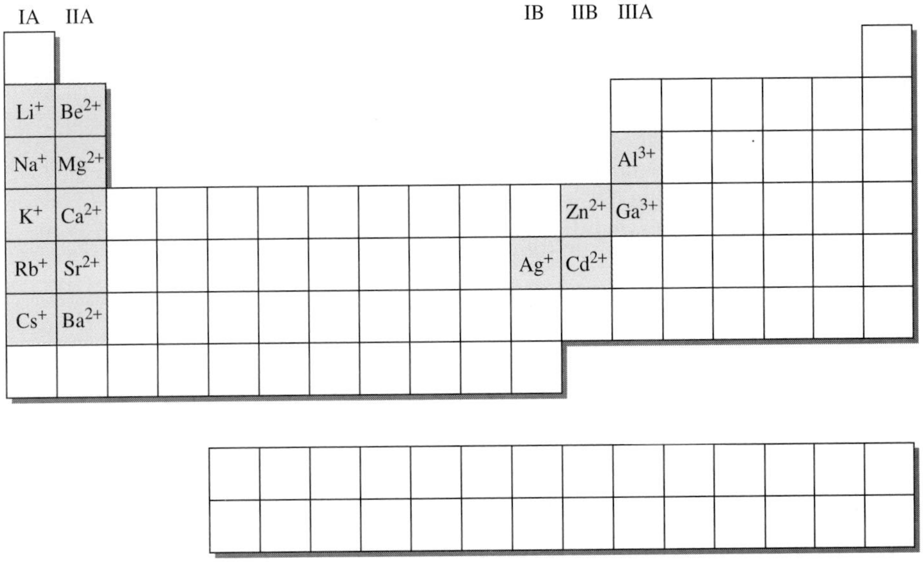

FIGURE 8.1 Periodic table showing the type I metallic ions. For these metals, ionic charge correlates with periodic table group number.

also directly correlate for Zn, Cd, and Ag, the other type I metallic ions. The reason for this charge–periodic table correlation, the octet rule, was considered in Section 7.3.

The vast majority of metals, all except the fifteen type I metals, are type II metals. Charge magnitude for ions of such metals cannot be easily related to periodic-table position. (The presence of *d* or *f* electrons in most of these metals complicates octet rule considerations in a manner beyond what we consider in this book.) Table 8.1 gives the charges for selected commonly encountered type II metal ions. Note that the charges of +2 and +3 are a common combination. However, there are also other combinations: two +2 and +4 pairings (Pb and Sn), a +1 and +2 pairing (Cu), and a +1 and +3 pairing (Au) are listed in Table 8.1.

TABLE 8.1 IONIC CHARGES ASSOCIATED WITH IONS OF THE MORE COMMON TYPE II (VARIABLE-CHARGE) METALS

Element	Ions Formed
Chromium	Cr^{2+} and Cr^{3+}
Cobalt	Co^{2+} and Co^{3+}
Copper	Cu^{+} and Cu^{2+}
Gold	Au^{+} and Au^{3+}
Iron	Fe^{2+} and Fe^{3+}
Lead	Pb^{2+} and Pb^{4+}
Manganese	Mn^{2+} and Mn^{3+}
Tin	Sn^{2+} and Sn^{4+}

8.3 NOMENCLATURE FOR BINARY IONIC COMPOUNDS

The names of binary ionic compounds are based on the names of the ions of which they are composed. Ion names are obtained as follows.

1. Type I metal ions take the full name of the element.

   Na^{+} sodium ion
   Ca^{2+} calcium ion
   Al^{3+} aluminum ion

2. Type II metal ions take the full name of the element followed by a Roman numeral (in parentheses) that gives the ionic charge.

   Fe^{2+} iron(II) ion
   Fe^{3+} iron(III) ion
   Cu^{2+} copper(II) ion

3. Nonmetal ions take the *stem* of the name of the element followed by the suffix *-ide*.

   Cl^{-} chloride ion
   S^{2-} sulfide ion
   O^{2-} oxide ion

   The stem of the name of the nonmetal is always the first few letters of the nonmetal's name, that is, the name of the nonmetal with its ending chopped off. Table 8.2 gives the stem part of the name for the more common nonmetallic ions.

Type I Binary Ionic Compounds

Names for type I binary ionic compounds are assigned using the following rule:

The full name of the metallic element is given first, followed by a separate word consisting of the stem of the nonmetallic element name and the suffix ***-ide***.

TABLE 8.2 NAMES FOR THE MORE COMMON NONMETAL IONS

Element	Stem	Name of Ion	Formula
Bromine	brom-	bromide ion	Br^{-}
Carbon	carb-	carbide ion	C^{4-}
Chlorine	chlor-	chloride ion	Cl^{-}
Fluorine	fluor-	fluoride ion	F^{-}
Hydrogen	hydr-	hydride ion	H^{-}
Iodine	iod-	iodide ion	I^{-}
Nitrogen	nitr-	nitride ion	N^{3-}
Oxygen	ox-	oxide ion	O^{2-}
Phosphorus	phosph-	phosphide ion	P^{3-}
Sulfur	sulf-	sulfide ion	S^{2-}

Thus, to name the compound NaF we start with the name of the metal (sodium), follow it with the stem of the name of the nonmetal (fluor-), and then add the suffix -ide. The name becomes *sodium fluoride*.

The name of the metal ion present is always exactly the same as the name of the metal itself. The metal's name is never shortened as are nonmetal names. Example 8.1 illustrates further the use of the preceding rule in naming type I binary ionic compounds.

Example 8.1

Name the following type I binary ionic compounds. Type 1 binary ionic compounds always contain a fixed-charge metal.

(a) KCl **(b)** MgF_2 **(c)** Na_2O **(d)** Be_3N_2

Solution

The general pattern for naming compounds of this type is

First word: name of metal
Second word: stem of name of nonmetal + -ide

(a) The metal is potassium, and the nonmetal is chlorine. Thus, the compound's name is potassium chloride (the stem of the nonmetal name is underlined).

(b) The metal is magnesium, and the nonmetal is fluorine. Hence, the name is magnesium fluoride. Note that no mention is made of the subscript 2 found after the symbol for fluorine in the formula. *The name of an ionic compound never contains any reference to formula subscript numbers.* Since magnesium is a fixed-charge metal, there is only one ratio in which magnesium and fluorine may combine. Thus, just telling the name of the elements present in the compound is adequate nomenclature.

(c) Sodium (Na) and oxygen (O) are present in the compound. Its name is sodium oxide.

(d) This compound is named beryllium nitride.

Practice Exercise 8.1

Name the following type I binary ionic compounds. Type 1 binary ionic compounds always contain a fixed-charge metal.

(a) $BaCl_2$ (b) AlP (c) K_2O (d) Ca_3N_2

Ans. (a) barium chloride; (b) aluminum phosphide; (c) potassium oxide; (d) calcium nitride

Type II Binary Ionic Compounds

Names for type II binary ionic compounds are assigned using the following rule:

The full name of the metallic element with a Roman numeral appended to it is given first, followed by a separate word consisting of the stem of the nonmetallic element name and the suffix **-ide**.

Two different iron chlorides are known; one contains Fe^{2+} ions ($FeCl_2$) and the other

F.Y.I.

The ruddy rust that forms on iron is iron(III) oxide (Fe_2O_3). This same compound gives red bricks and red clay sewer pipes their color. It also is the pigment of burnt sienna in artists' paints. Rouge, one of the most ancient of cosmetics, gets its color from iron(III) oxide as well. Barns were traditionally painted red because farmers used to paint their barns red with an iron(III) oxide paint as a way to preserve the wood. People still paint their barns red in modern times without knowing why.

contains Fe^{3+} ions ($FeCl_3$). The names of these two chlorides are, respectively, iron(II) chloride and iron(III) chloride. Without the use of Roman numerals (or some equivalent system) these two compounds would have the same name (iron chloride), an unacceptable situation.

In naming type II binary ionic compounds, the Roman numeral to be used in the name is the metal ion charge, which can be calculated using the charge on the nonmetal ion and the principle of charge neutrality. The charge neutrality principle is that the total positive charge and the total negative charge must add to zero (Sec. 7.6). Calculation of metal ion charge (Roman numeral) using this technique, as well as the use of Roman numerals in the names of type II binary ionic compounds, is illustrated in Example 8.2. As you work through Example 8.2, be sure to note that Roman numerals are *never* a part of the *formula* of a type II binary ionic compound but *always* a part of the *name* of a type II binary ionic compound.

Example 8.2

Name the following type II binary ionic compounds. Type II binary ionic compounds always contain a variable-charge metal.

(a) CuO **(b)** Cu_2O **(c)** Mn_2S_3 **(d)** $AuCl_3$

Solution

The general pattern for naming compounds of this type is

First word: Name of metal + Roman numeral
Second word: Stem of name of nonmetal + -ide

(a) The metal ion charge in this compound, a quantity needed to determine the Roman numeral to be used in the name, is easily calculated using the following procedure.

$$\text{copper charge} + \text{oxygen charge} = 0$$

The oxide ion has a -2 charge (Sec. 7.5). Therefore,

$$\text{copper charge} + (-2) = 0$$

Solving, we get

$$\text{copper charge} = +2$$

Therefore, the copper ions present are Cu^{2+}, and the name of the compound is copper(II) oxide.

(b) For charge balance in this compound we have the equation

$$2(\text{copper charge}) + \text{oxygen charge} = 0$$

Note that we have to take into account the number of copper ions present, two in this case. The oxide ion carries a -2 charge (Sec. 7.5). Therefore,

$$2(\text{copper charge}) + (-2) = 0$$
$$2(\text{copper charge}) = +2$$
$$\text{copper charge} = +1$$

Here we note that we are interested in the charge on a *single* copper ion (+1) and not in the total positive charge present (+2). Since Cu^+ ions are present, the compound is named copper(I) oxide. As is the case for all ionic compounds, the name does not contain any reference to the numerical subscripts in the compound's formula.

(c) The charge balance equation is

$$2(\text{manganese}) + 3(\text{sulfur}) = 0$$

Substituting a charge of −2 into the equation for sulfur (Sec. 7.5), we get

$$\begin{aligned} 2(\text{manganese}) + 3(-2) &= 0 \\ 2(\text{manganese}) &= +6 \\ \text{manganese} &= +3 \end{aligned}$$

The compound is thus named manganese(III) sulfide.

(d) This compound is gold(III) chloride. Chloride ions carry a −1 charge (Sec. 7.5). Since the compound contains three chloride ions (−3 total charge), the single gold ion must bear a +3 charge to counterbalance the −3 charge of the chlorides. Hence, Au^{3+} ions are present.

Practice Exercise 8.2

Name the following type II binary ionic compounds. Type II binary ionic compounds always contain a variable-charge metal.

(a) FeO (b) Fe_2O_3 (c) PbO_2 (d) AuCl

Ans. (a) iron(II) oxide; (b) iron(III) oxide; (c) lead(IV) oxide; (d) gold(I) chloride

An older method for indicating the charge on metal ions uses suffixes rather than Roman numerals. This system is more complicated and less precise than the Roman numeral system and fortunately is being abandoned. It is mentioned here because it is still encountered (especially on the labels of bottles of chemicals; see Fig. 8.2). In this system, when a metal has two common ionic charges, the suffix **-ous** is used for the ion of lower charge and the suffix **-ic** for the ion of higher charge. Table 8.3 compares the two systems for the metals where the old system is most often encountered. Note that in the old system the Latin names for metals are used (Sec. 4.9).

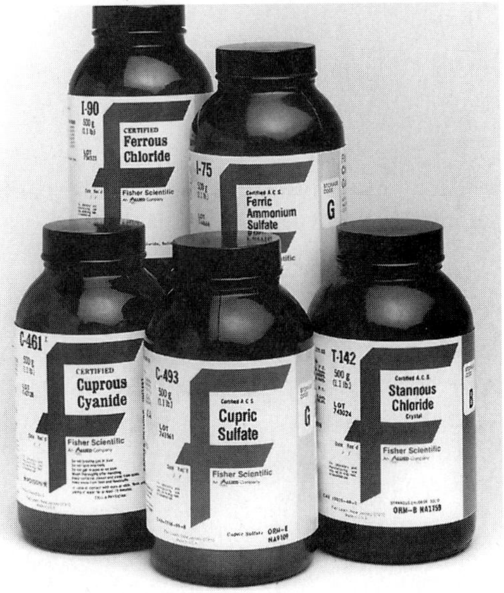

FIGURE 8.2 The older method of naming ionic compounds containing a variable-charge metal is still used on the labels of many laboratory chemicals. (Fisher Scientific)

TABLE 8.3 COMPARISON OF IUPAC AND OLD SYSTEM NAMES FOR SELECTED METAL IONS

Element	Ions	Preferred Name	Old System
Copper	Cu^{+}	copper(I) ion	cuprous ion
	Cu^{2+}	copper(II) ion	cupric ion
Iron	Fe^{2+}	iron(II) ion	ferrous ion
	Fe^{3+}	iron(III) ion	ferric ion
Tin	Sn^{2+}	tin(II) ion	stannous ion
	Sn^{4+}	tin(IV) ion	stannic ion
Lead	Pb^{2+}	lead(II) ion	plumbous ion
	Pb^{4+}	lead(IV) ion	plumbic ion
Gold	Au^{+}	gold(I) ion	aurous ion
	Au^{3+}	gold(III) ion	auric ion

Example 8.3

Two systems exist for indicating the charge on the metal ion present in type II binary ionic compounds: (1) a system that uses Roman numerals and (2) a system that uses the suffixes -ic and -ous. Change the following compound names from one system to the other system.

(a) iron(III) chloride **(b)** copper(I) fluoride

(c) stannic bromide **(d)** plumbous fluoride

Solution

The information found in Table 8.3 is needed to effect the desired name changes.

(a) The given name indicates that Fe^{3+} ion is present. The alternative name for Fe^{3+} ion, from Table 8.3, is ferric ion. The name iron(III) chloride thus becomes ferric chloride.

(b) The alternative name for the copper(I) ion, Cu^{+}, is cuprous ion. Therefore, copper(I) fluoride becomes cuprous fluoride.

(c) The stannic ion, from Table 8.3, is Sn^{4+} ion; the alternative name for this ion is tin(IV). Therefore, the changed (and preferred) name for this compound is tin(IV) bromide.

(d) Since the plumbous ion and Pb^{2+} ion are equivalent (Table 8.3), the names plumbous fluoride and lead(II) fluoride are equivalent.

Practice Exercise 8.3

Change each of the following type II binary ionic compound names from the "Roman numeral" to the "-ic, -ous" system, or vice versa.

(a) copper(II) chloride (b) tin(II) fluoride
(c) ferrous bromide (d) auric oxide

Ans. (a) cupric chloride; (b) stannous fluoride; (c) iron(II) bromide; (d) gold(III) oxide

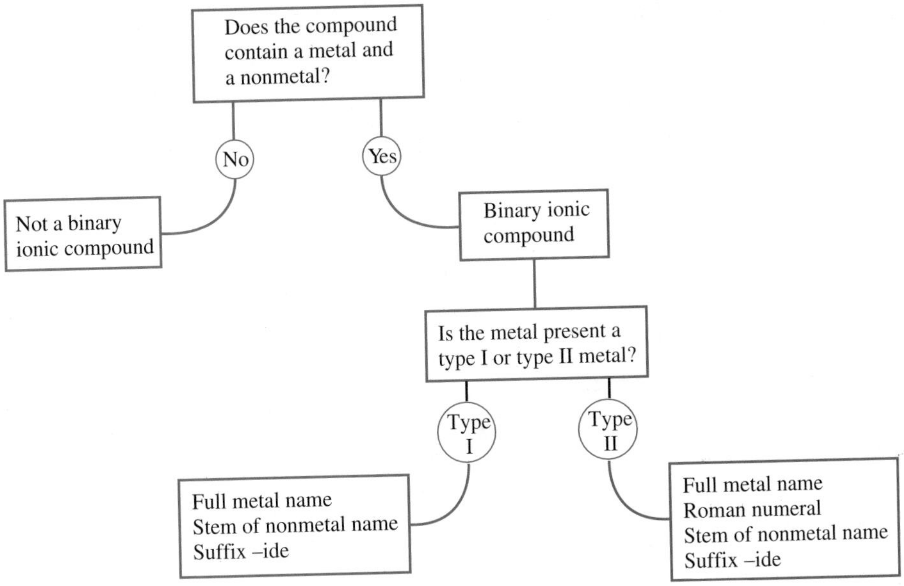

FIGURE 8.3 A "nomenclature decision tree" for naming binary ionic compounds.

The overall strategy for naming binary ionic compounds, as discussed in this section, is summarized in the "nomenclature decision tree" of Figure 8.3.

8.4 NOMENCLATURE FOR IONIC COMPOUNDS CONTAINING POLYATOMIC IONS

In Section 7.8 the topic of polyatomic ions was considered, and at that time a limited number of such ions were introduced in the context of formula writing. We now consider some additional facts about polyatomic ions.

Hundreds of different polyatomic ions exist. Table 8.4 gives the names and formulas for the most common ones. Note that almost all the polyatomic ions listed in Table 8.4 contain oxygen atoms. The names, but not necessarily the formulas, of some of these common polyatomic ions should be familiar to you. Many of these ions are found in commercial products. Examples are fertilizers (phosphates, sulfates, nitrates), baking soda and baking powder (bicarbonates), and building materials (carbonates, sulfates).

There is no easy way to learn the formulas and names for all the common polyatomic ions. Memorization is required. The charges and formulas for the various polyatomic ions cannot be related easily to the periodic table as was the case for many of the monoatomic ions. In Table 8.4 the most frequently encountered polyatomic ions are in color. Their formulas and names should definitely be memorized. Your instructor may want you to memorize others, too. The inability to recognize the presence of polyatomic ions (both by name and by formula) in a compound is a major stumbling block for many chemistry students. It requires some effort to overcome this obstacle.

Note from Table 8.4 the following facts concerning the polyatomic ions.

1. Most of the ions have a negative charge, which can vary from −1 to −3. Only two positive ions are listed in the table: NH_4^+ (ammonium) and H_3O^+ (hydronium).
2. Four of the polyatomic ions have names ending in **-ide**: OH^- (hydroxide),

F.Y.I.

Polyatomic-ion-containing compounds found in the home:

$NaHCO_3$	Baking soda
$MgSO_4$	Epsom salt
Na_3PO_4	Cleaning agent
$Al(OH)_3$	Antacid
$NaOH$	Drain cleaner
$MgCO_3$	Plaster of Paris
$NaClO$	Chlorox
$CaCO_3$	Chalk

F.Y.I.

The green corrosion on bronze statues is $Cu_2(OH)_2CO_3$. It forms from copper reacting with water, oxygen, and carbon dioxide in the air.

TABLE 8.4 FORMULAS AND NAMES OF SOME COMMON POLYATOMIC IONS

Key Element Present	Formula	Name of Ion
Nitrogen	NO_3^-	**nitrate ion**
	NO_2^-	nitrite ion
	NH_4^+	**ammonium ion**
	N_3^-	azide ion
Sulfur	SO_4^{2-}	**sulfate ion**
	HSO_4^-	hydrogen sulfate or bisulfate ion
	SO_3^{2-}	sulfite ion
	HSO_3^-	hydrogen sulfite or bisulfite ion
	$S_2O_3^{2-}$	thiosulfate ion
Phosphorus	PO_4^{3-}	**phosphate ion**
	HPO_4^{2-}	hydrogen phosphate ion
	$H_2PO_4^-$	dihydrogen phosphate ion
	PO_3^{3-}	phosphite ion
Carbon	CO_3^{2-}	**carbonate ion**
	HCO_3^-	hydrogen carbonate or bicarbonate ion
	$C_2O_4^{2-}$	oxalate ion
	$C_2H_3O_2^-$	**acetate ion**
	CN^-	**cyanide ion**
	OCN^-	cyanate ion
	SCN^-	thiocyanate ion
Chlorine	ClO_4^-	perchlorate ion
	ClO_3^-	**chlorate ion**
	ClO_2^-	chlorite ion
	ClO^-	hypochlorite ion
Oxygen	O_2^{2-}	peroxide ion
Boron	BO_3^{3-}	borate ion
Hydrogen	H_3O^+	**hydronium ion**
	OH^-	**hydroxide ion**
Metals	MnO_4^-	permanganate ion
	CrO_4^{2-}	chromate ion
	$Cr_2O_7^{2-}$	dichromate ion

CN^- (cyanide), N_3^- (azide), and O_2^{2-} (peroxide). These names present exceptions to the rule that the suffix **-ide** be reserved for use in naming monoatomic ions.

3. A number of **-ate**, **-ite** pairs of ions exist—for example, SO_4^{2-} (sulfate) and SO_3^{2-} (sulfite). The ion in the pair with the higher number of oxygens is always the *-ate* ion. The *-ite* ion always contains one fewer oxygen then the *-ate* ion.
4. A number of pairs of ions exist where one member of the pair differs from the other by having a hydrogen atom present, for example, CO_3^{2-} (carbon-

ate) and HCO_3^- (hydrogen carbonate or bicarbonate). In such pairs, the charge on the hydrogen-containing ion is always one less than the charge on the other ion.

5. Two pairs of ions exist in which the difference between pair members is that a sulfur atom has replaced an oxygen atom in one member of the pair: (SO_4^{2-}, $S_2O_3^{2-}$) and (OCN^-, SCN^-). The prefix **thio-** is used to denote this replacement of oxygen by sulfur. The names of these pairs of ions, respectively, are sulfate–thiosulfate and cyanate–thiocyanate.

The names of ionic compounds containing polyatomic ions are derived in the same way as those of binary ionic compounds (Sec. 8.3). Recall that the rule for naming binary ionic compounds is: Give the name of the metallic element first (including, when needed, a Roman numeral indicating ion charge), and then as a separate word give the stem of the nonmetallic element name to which the suffix *-ide* is appended.

For our present situation, *if the polyatomic ion is positive, its name is substituted for that of the metal. If the polyatomic ion is negative, its name is substituted for the nonmetal stem plus -ide.* In the case where both positive and negative ions are polyatomic, dual substitution occurs and the resulting name includes just the names of the polyatomic ions. Example 8.4 illustrates the use of these rules.

Example 8.4

Name the following polyatomic-ion-containing compounds.

(a) Na_3PO_4 **(b)** $Fe(NO_3)_3$ **(c)** Cu_2SO_4 **(d)** NH_4CN

Solution

(a) The positive ion present is the sodium ion (Na^+). The negative ion is the polyatomic phosphate ion (PO_4^{3-}). The name of the compound is sodium phosphate. No Roman numeral is needed in the name since sodium is a type I metal. As in naming binary ionic compounds (Sec. 8.3), subscripts in the formula are not incorporated into the name.

(b) The positive ion present is iron, and the negative ion is the nitrate ion (NO_3^-). Since iron is a type II metal, a Roman numeral must be used to indicate ionic charge. In this case, the Roman numeral is III. The fact that iron ions carry a +3 is deduced by noting that there are three nitrate ions present, each of which carries a −1 charge. The charge on the single iron ion present must be a +3 in order to counterbalance the total negative charge of −3. The name of the compound, therefore, is iron(III) nitrate.

(c) The positive ion present is Cu(I) ion. The negative ion is the polyatomic sulfate ion (SO_4^{2-}). The name of the compound is copper(I) sulfate. The determination that copper is present as copper(I) involves the following calculation dealing with charge balance.

$$\begin{aligned} 2(\text{copper charge}) + (\text{sulfate charge}) &= 0 \\ 2(\text{copper charge}) + (-2) &= 0 \\ 2(\text{copper charge}) &= +2 \\ \text{copper charge} &= +1 \end{aligned}$$

(d) Both the positive and negative ions in this compound are polyatomic—the ammonium ion (NH_4^+) and the cyanide ion (CN^-). The name of the compound is simply the combination of the names of the two polyatomic ions: ammonium cyanide.

Practice Exercise 8.4

Name the following polyatomic-ion-containing compounds.

(a) K_2CO_3 (b) $Co(NO_3)_2$ (c) $Fe_2(SO_4)_3$ (d) $(NH_4)_3PO_4$

Ans. (a) potassium carbonate; (b) cobalt(II) nitrate; (c) iron(III) sulfate; (d) ammonium phosphate

The overall strategy for naming ionic compounds—both binary (Sec. 8.3) and polyatomic-ion-containing (Sec. 8.4)—is summarized in the "nomenclature decision tree" of Figure 8.4

8.5 NOMENCLATURE FOR BINARY MOLECULAR COMPOUNDS

A **binary molecular compound** *is a covalently bonded compound in which just two nonmetallic elements are present.* Such compounds are named in the following manner. The two nonmetals present are named in the order in which they appear in the formula. [The least

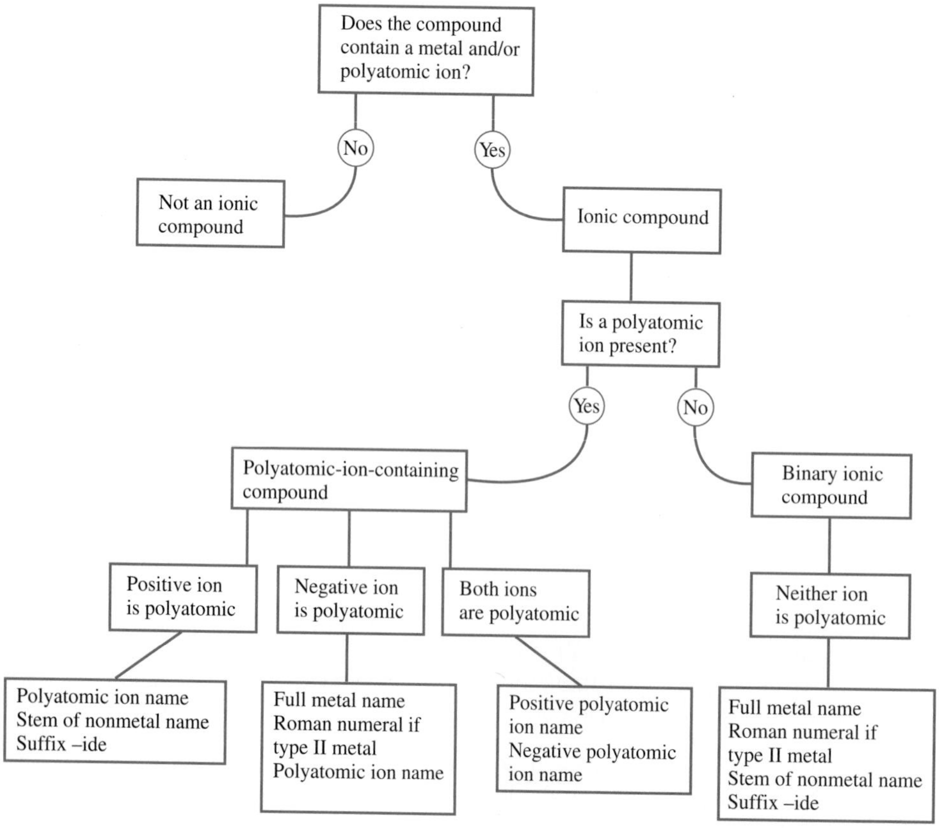

FIGURE 8.4 A "nomenclature decision tree" for naming binary and polyatomic-ion-containing ionic compounds.

TABLE 8.5 GREEK NUMERICAL PREFIXES FROM 1 TO 10

Greek Prefix	Number
Mono-	1
Di-	2
Tri-	3
Tetra-	4
Penta-	5
Hexa-	6
Hepta-	7
Octa-	8
Ennea-[a]	9
Deca-	10

[a] The prefix ennea- is preferred to the Latin nona- by IUPAC, but nona- is still frequently used.

electronegative nonmetal (Sec. 7.10) is usually written first in the formula.] *The name of the first nonmetal is used in full. The name of the second nonmetal is treated as was the nonmetal in binary ionic compounds: that is, the stem of the name is given and the suffix -ide is added.*

In addition, the number of atoms of each element present in a molecule of the compound is explicitly incorporated into the name of the compound through the use of Greek numerical prefixes. A prefix precedes the name of each nonmetal. The use of these prefixes is in direct contrast to the procedures used for naming ionic compounds. For ionic compounds, formula subscripts are not mentioned in the name.

Prefixes are needed in naming binary molecular compounds because numerous different compounds exist for many pairs of nonmetallic elements. For example, all of the following nitrogen–oxygen molecular compounds exist: NO, NO_2, N_2O, N_2O_3, N_2O_4, and N_2O_5. The prefixes used are the standard Greek numerical prefixes, which are given in Table 8.5 for the numbers 1 through 10, and which are used in words such as *mono*nucleosis, *di*chromatic, *tri*angle, *tetra*gram, *penta*gon, *hexa*pod, *hepta*thlon, *oct*ave, *ennea*hedron, and *dec*ade. Example 8.5 shows how these prefixes are used in naming binary molecular compounds.

F.Y.I.

One atom can make a big difference. Carbon dioxide (CO_2) is a nontoxic gas exhaled with every breath. Carbon monoxide (CO) is the extremely toxic gas produced when car engines idle.

Example 8.5

Name the following binary molecular compounds.

(a) N_2O_5 **(b)** PF_3 **(c)** S_4N_4 **(d)** $SiCl_4$

Solution

The name of each of these compounds will consist of two words with the following general formats.

First word: numerical prefix + full name of the first nonmetal
Second word: numerical prefix + stem of name of second nonmetal + -ide

(a) The elements present are nitrogen and oxygen. The two portions of the name before adding Greek numerical prefixes are *nitrogen* and *oxide*. Adding the prefixes gives *dinitrogen* (two nitrogen atoms are present) and *pentoxide* (five oxygen atoms are present). (When an element name begins with an *a* or *o*, the *a* or *o* at the end of the Greek prefix is dropped for ease of pronunciation—pentoxide instead of pentaoxide.) The name of this compound is dinitrogen pentoxide.

(b) When there is only one atom present of the first nonmetal present, it is standard procedure to omit the prefix *mono-* for the element. Following this guideline, we have for the name of the compound phosphorus trichloride.

(c) The prefix for four atoms is *tetra-*. This compound is, therefore, tetranitrogen tetroxide.

(d) Omitting the initial *mono-* (see part b), we name this compound silicon tetrachloride.

Practice Exercise 8.5

Name the following binary molecular compounds.

(a) S_2O (b) NCl_3 (c) P_4S_6 (d) CF_4

Ans. (a) disulfur monoxide; (b) nitrogen trichloride; (c) tetraphosphorus hexasulfide; (d) carbon tetrafluoride

There is one exception to the use of Greek numerical prefixes in naming binary molecular compounds. Binary compounds with hydrogen listed as the first element in the formula are named without prefix use. Thus, the compounds H_2S and HCl are named hydrogen sulfide and hydrogen chloride, respectively.

All procedures for naming compounds so far presented in this chapter have been based on IUPAC rules (Sec. 8.1) and have produced systematic names. A **systematic name** *for a compound is a name derived from IUPAC rules that conveys information about the composition of the compound.* A few binary molecular compounds have names completely unrelated to the IUPAC rules for naming such compounds. They have "common" or "trivial" names, which were coined before the development of the systematic rules. A **common name** *for a compound is a name not based on IUPAC rules.* Common names usually do not convey any information about the composition of compounds. At one time, in the early history of chemistry, all compounds had common names. With the advent of systematic nomenclature, most common names were discontinued. A few, however, have persisted and are now officially accepted. The most "famous" example of this is the compound H_2O, which has the systematic name hydrogen oxide. This name is never used; H_2O is known as *water*, a name that is not going to be changed. Another very common example of common nomenclature involves the compound NH_3, which is *ammonia*. Table 8.6 gives additional examples of compounds for which common names are used in preference to systematic names. Such exceptions are actually very few compared to the total number of compounds named by systematic rules.

Writing formulas for binary molecular compounds given their names is a very easy

TABLE 8.6 SOME BINARY MOLECULAR COMPOUNDS THAT HAVE COMMON NAMES

Compound Formula	Accepted Common Name
H_2O	water
H_2O_2	hydrogen peroxide
NH_3	ammonia
N_2H_4	hydrazine
CH_4	methane
C_2H_6	ethane
PH_3	phosphine
AsH_3	arsine

task. The Greek prefixes in the names of such compounds tell you exactly how many atoms of each kind are present. For example, the compounds dinitrogen pentoxide and carbon dioxide have, respectively, the formulas N_2O_5 (di- and penta-) and CO_2 (mono- and di-). Remember, the prefix mono- is always dropped at the start of a name, as in carbon dioxide.

The "nomenclature decision tree" in Figure 8.5 contrasts the strategy used in naming binary molecular compounds and binary ionic compounds.

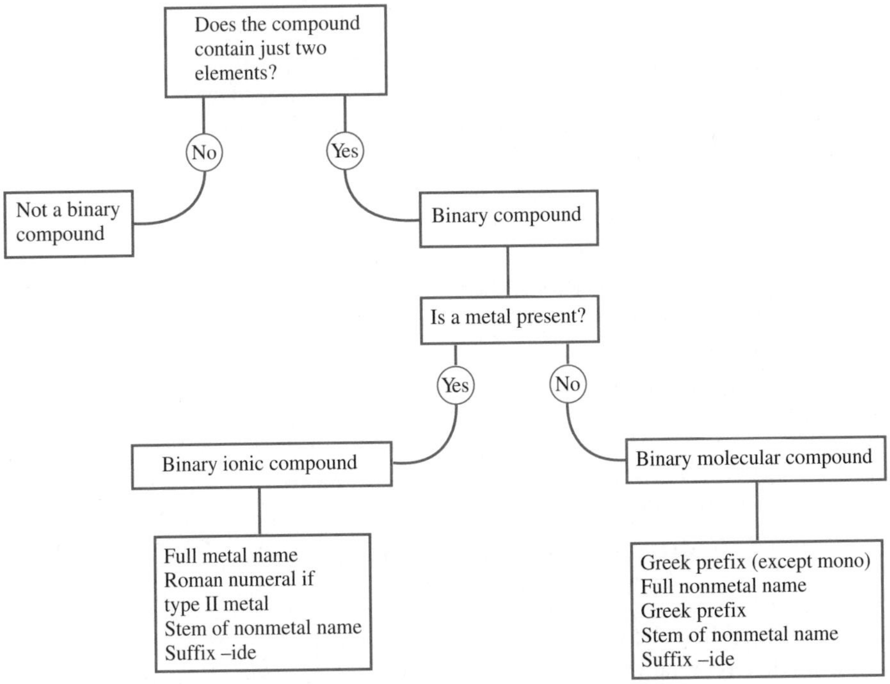

FIGURE 8.5 A "nomenclature decision tree" for naming binary molecular compounds and binary ionic compounds.

8.6 NOMENCLATURE FOR ACIDS

Many hydrogen-containing molecular compounds dissolve in water to give solutions with properties markedly different from those of the compounds that were dissolved. These solutions, which we will discuss in detail in Chapter 14, are called *acids*.

Not all hydrogen-containing molecular compounds are acids. Those that are can be recognized by their formulas, which have hydrogen written as the first element in the formula.

Acids: HCl, H_2S, H_2SO_4, HNO_3
Nonacids: NH_3, CH_4, PH_3, SiH_4

Water, which is the solvent for most acidic solutions, is generally not considered to be an acid. Despite this, H is written first in its formula (H_2O).

Because of differences in the properties of acids and the anhydrous (without water) compounds from which they are produced, the acids and the anhydrous compounds are given different names. Acid nomenclature is derived from the names of the parent anhydrous compounds (which we have learned to name in this chapter). Because of this close relationship, we will also learn to name acids at this time, even though a detailed discussion of acids and their properties does not come until Chapter 14.

Acids are molecular compounds (no ions are present). When acids are dissolved in water, the interaction with the water produces ions. The acid molecules can be considered to break apart to form ions in the presence of water. (No ion formation occurs if the water is not present.)

All acids produce the same positive ion, the H^+ ion, which is the species that gives acid's their characteristic properties. The negative ion produced varies from acid to acid. For example,

HCl, in water, produces H^+ and Cl^- ions
HNO_3, in water, produces H^+ and NO_3^- ions
HCN, in water, produces H^+ and CN^- ions

The names of acids are derived from the names of the *negative ions* produced from the acids interaction with water. There are three nomenclature rules based on whether the name of the negative ion ends in *-ide*, *-ate*, or *-ite*.

1. *-ide Rule*: When the name of the negative ion produced from the acid ends in *-ide* the acid name has four parts:

   (1) the prefix *hydro-*

   (2) the stem of the name of the negative ion

   (3) the suffix *-ic*

   (4) the word *acid*

   Examples of the application of this rule are

   HCl (chloride) is named *hydro*chlor*ic acid*
   HCN (cyanide) is named *hydro*cyan*ic acid*

2. *-ate Rule*: When the name of the negative ion produced from the acid ends in *-ate* the acid name has three parts:

   (1) the name of the negative ion less the *-ate* ending

   (2) the suffix *-ic*

(3) the word *acid*

Examples of the application of this rule are:

HNO_3 (nitrate) is named nitr*ic acid*
$HClO_4$ (perchlorate) is named perchlor*ic acid*

3. *-ite Rule*: When the name of the negative ion produced from the acid ends in *-ite* the acid name has three parts:

   (1) the name of the negative ion less the *-ite* ending
   (2) *the suffix -ous*
   (3) the word *acid*

   Examples of the application of this rule are:

   HNO_2 (nitrite) is named nitr*ous acid*
   $HClO$ (hypochlorite) is named hypochlor*ous acid*

Example 8.6 and 8.7 further illustrate the use of the three rules for naming acids.

Example 8.6

Name the following compounds as acids.

(a) H_2CO_3 **(b)** HF **(c)** $HClO_2$ **(d)** H_2SO_4

Solution

(a) The negative ion present in solutions of this acid is the carbonate ion (CO_3^{2-}). Removing the *-ate* ending from the word carbonate and replacing it with the suffix *-ic* (Rule 2) and then adding the word *acid* gives the name carbonic acid.

(b) The negative ion present in solutions of this acid is the fluoride ion (F^-). This is a rule 1 (-ide) situation. The name of the acid is formed from using the prefix *hydro-*, the suffix *-ic*, and the word *acid*. The acid name is *hydrofluoric acid.*

(c) The negative ion present in solutions of this acid is the chlorite ion (ClO_2^-). (If you are unsure of the identity of a particular negative ion you can consult Table 8.4.) This is a rule 3 (-ite) situation. Removing the *-ite* ending from the word chlorite and replacing it with the suffix *-ous* and then adding the word *acid* produces the name chlorous acid.

(d) The negative ion present in solutions of this acid is the sulfate ion (SO_4^{2-}). Changing the *-ate* ending to *-ic* (rule 2) and adding the word acid gives the name *sulfuric acid.* (For acids involving sulfur, "ur" from sulfur is reinserted into the acid name for phonetic reasons.)

Practice Exercise 8.6

Name the following compounds as acids.

(a) H_2SO_3 (b) HBr (c) H_3PO_4 (d) $HC_2H_3O_2$

Ans. (a) sulfurous acid; (b) hydrobromic acid; (c) phosphoric acid; (d) acetic acid

Example 8.7

Give the chemical formulas of each of the following acids.

(a) chlorous acid **(b)** hydrocyanic acid

Solution

In this problem we will go through a reasoning process that is the reverse of that in Example 8.6. Students often find this "reverse" process to be more challenging than the "forward" process.

(a) The *-ous* ending in the acid's name indicates that the negative ion it forms in solution is an *-ite* ion. It is the *chlorite* ion. From Table 8.4, the chlorite ion has the formula ClO_2^-. The hydrogen present in the acid, for formula-writing purposes, is considered to be H^+ ion. Combining these two ions, H^+ and ClO_2^-, in a one-to-one ratio will produce the neutral compound $HClO_2$ (chlorous acid).

(b) Since this is a hydro____ic acid, the negative ion present in acid solution will be an -ide ion. It is the cyanide ion, whose formula is CN^- (Table 8.4). Combining the H^+ and CN^- ions in the ratio that produces neutrality, one-to-one, gives the formula HCN for *hydrocyanic acid.*

Practice Exercise 8.7

Give the chemical formulas of each of the following acids.

(a) phosphorus acid (b) perchloric acid

Ans. (a) H_3PO_3; (b) $HClO_4$

The overall strategy for identifying and naming acids is summarized in the "nomenclature decision tree" of Figure 8.6.

Acids are often subclassified into the categories *nonoxyacid* and *oxyacid.* A **nonoxyacid** *is a molecular compound composed of hydrogen and one or more nonmetals other than oxygen that produces an acidic aqueous solution.* All common nonoxyacids, with one exception, HCN, are binary compounds. An **oxyacid** *is a molecular compound composed of hydrogen, oxygen, and one or more other elements that produces an acidic aqueous solution.*

Nonoxyacids exist as both pure compounds and water solutions. Conventions for distinguishing between the two situations are as follows, using HCl (a gas in the pure state) as the example.

HCl in the pure state:	HCl(g)	hydrogen chloride
HCl in aqueous solution:	HCl(aq)	hydrochloric acid

Table 8.7 gives other examples of this dual naming system for nonoxyacids.

Oxyacids are encountered only as aqueous solutions. Thus, a dual naming system is not needed here.

For many nonmetals a series of oxyacids exist. The formulas of the members of the series differ from each other only in oxygen content. For example, there are four chlorine-containing oxyacids whose formulas are

$$HClO_4, HClO_3, HClO_2, \text{ and } HClO$$

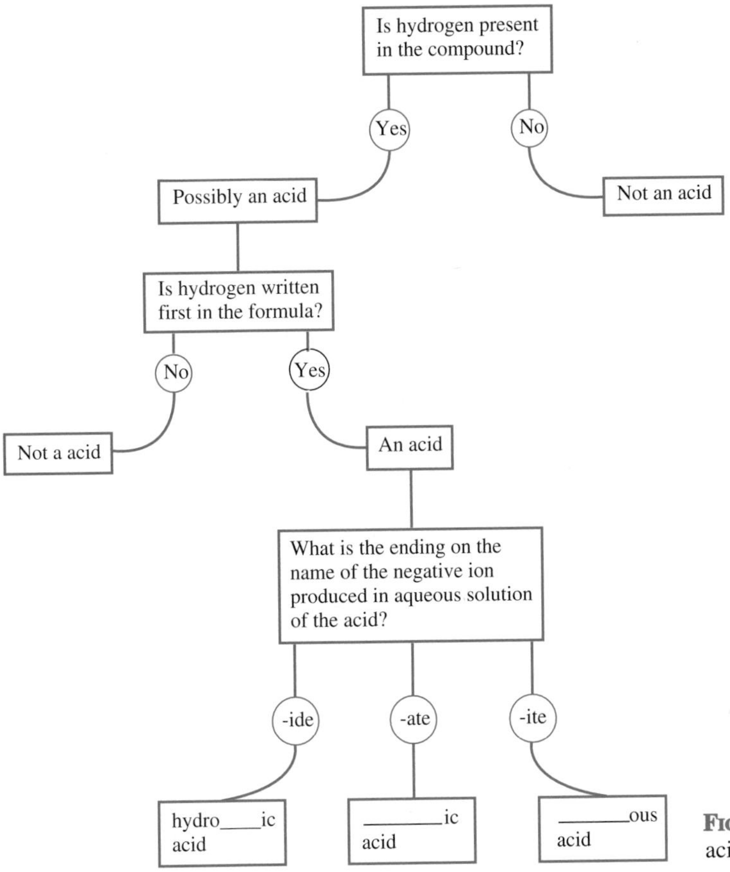

FIGURE 8.6 A "nomenclature decision tree" for naming acids.

TABLE 8.7 THE DUAL NAMING SYSTEM FOR MOLECULAR COMPOUNDS CONTAINING HYDROGEN AND A NONMETAL OTHER THAN OXYGEN

Formula	Name of Pure Compound	Name of Water Solution
HF	hydrogen fluoride	hydrofluoric acid
HBr	hydrogen bromide	hydrobromic acid
HI	hydrogen iodide	hydroiodic acid
H_2S	hydrogen sulfide	hydrosulfuric acid[a]

[a] For acids involving sulfur, "ur" from sulfur is reinserted in the acid name for pronunciation reasons.

In such a series of oxyacids, one of the acids is selected to be the "-ic" acid, the acid whose name has no prefix and ends in -ic. The names of the other acids in the series are related, in a constant manner, to the "-ic" acid by prefixes and suffixes as follows.

1. An oxyacid containing *one less oxygen atom* than the "-ic" acid is the "-ous" acid.
2. An oxyacid containing *two less oxygen atoms* than the "-ic" acid is the "hypo-____-ous" acid.

3. An oxyacid containing *one more oxygen atom* than the "-ic" acid is the "per-____-ic" acid.

These relationships between names for a series of oxyacids can be summarized as follows, where n represents the number of oxygen atoms in the "-ic" acid.

Number of Oxygen Atoms	Name
$n + 1$	per____ic acid
n	____ic acid
$n - 1$	____ous acid
$n - 2$	hypo____ous acid

Most nonmetals do not form a "complete" series of four simple oxyacids as does chlorine. Bromine is the only other nonmetal for which such a series of oxyacids exists. For both sulfur and nitrogen only the "-ic" and "-ous" acids are known. For both phosphorus and iodine there are three oxyacids; there is no $n+1$ acid (per____ic acid) for phosphorus and no $n-1$ acid (____ous acid) for iodine. Hypofluorous acid, HOF, is the only known fluorine-containing oxyacid.

8.7 NOMENCLATURE RULES—A SUMMARY

Within this chapter various sets of nomenclature rules have been presented. Students sometimes have problems deciding which set of rules to use in a given situation. This dilemma can be avoided if, when confronted with the request to name a compound, the following reasoning pattern is used.

1. Decide, first, whether the compound is ionic or molecular. If a metal or polyatomic ion is present, the compound is ionic. If not, the compound is molecular.
2. If the compound is ionic, then classify it as binary ionic or polyatomic-ion-containing ionic and use the rules appropriate for that classification.
3. If the compound is molecular, classify it as an acid or nonacid. An acid must have the element hydrogen present (written first in the chemical formula), and the compound must be in water solution. If the compound is a nonacid, name it according to the rules for binary molecular compounds.

Figure 8.7 portrays the preceding information in the format of a "nomenclature decision tree."

Example 8.8

Name the following compounds

(a) N_2O_4 **(b)** $Cu(NO_3)_2$ **(c)** $H_2S(aq)$ **(d)** Na_3N

Solution

(a) Neither a metal nor a polyatomic ion is present. Thus, the compound is not ionic. If it is not ionic, it must be molecular. Since no hydrogen is present, the compound is a nonacidic molecular compound. Such compounds are named using

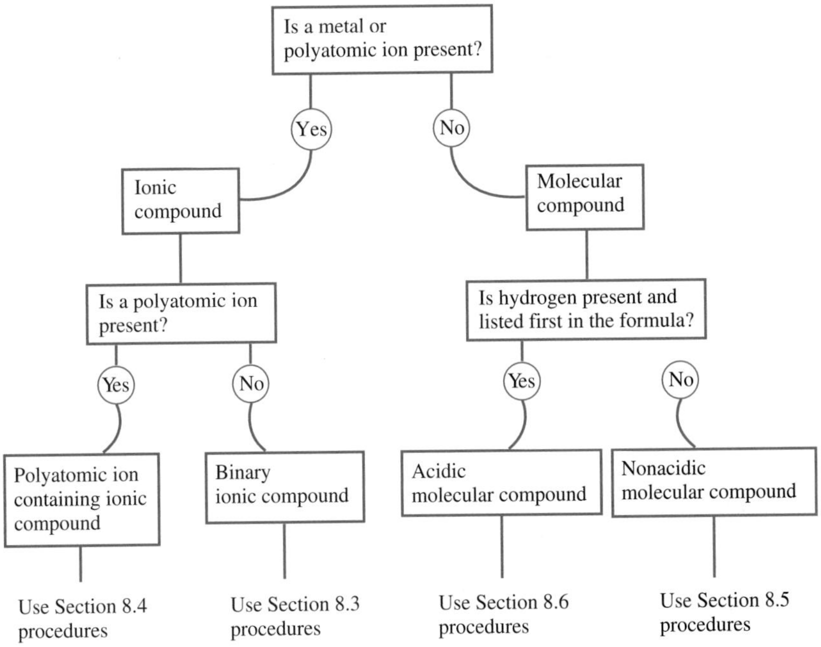

FIGURE 8.7 A "nomenclature decision tree" for deciding the classification of a compound that is to be named.

Greek numerical prefixes to denote the number of atoms present. The name of the compound is *dinitrogen tetroxide*.

(b) This compound meets both tests for an ionic compound. Both a metal and a polyatomic ion are present. The metal present is a type II metal, which means the name of the compound will have a Roman numeral in it. The metal ion present is copper(II) and the polyatomic ion is the nitrate ion. The name of the compound is *copper(II) nitrate*.

(c) This compound does not meet either test for an ionic compound. It is, thus, a molecular compound. Hydrogen is present and it is listed first in the formula. We have an acid. The (aq) means the acid is in solution rather than the pure state. The negative ion produced from the acid is sulf*ide* ion. Using the *-ide* rule for naming acids, the name of the acid becomes *hydrosulfuric acid*.

(d) A metal is present but no polyatomic ion is present. Thus, we have a binary ionic compound. The metal present is a type I metal, meaning that no Roman numeral is needed in the name. The name of the compound is *sodium nitride*. Greek numerical prefixes are never used in naming ionic compounds. The name *trisodium nitride* would be wrong.

Practice Exercise 8.8

Name the following compounds.

(a) $(NH_4)_2S$ (b) $AlCl_3$ (c) P_4O_6 (d) HNO_2

Ans. (a) ammonium sulfide; (b) aluminum chloride; (c) tetraphosphorus hexoxide; (d) nitrous acid

Example 8.8, as well as most of the other examples in this chapter, have involved naming compounds given their formulas. It is important to be able to do the reverse of this also—to write chemical formulas when given chemical names. Often students who can go "forward" without difficulty have problems with this "reverse" process. Example 8.9 involves writing chemical formulas given the names of the compounds.

Example 8.9

Write chemical formulas for the following compounds

(a) magnesium sulfide **(b)** beryllium hydroxide

(c) copper(II) carbonate **(d)** heptasulfur dioxide

Solution

(a) Magnesium is a metal and sulfur is a nonmetal. We have a binary ionic compound. The lack of a Roman numeral in the name indicates that the metal is a type I metal. Magnesium forms +2 ions (Fig. 8.1). The sulfide ion has a charge of −2 (Sec. 7.5). Combining the ions Mg^{2+} and S^{2-} in a one-to-one ratio, the ratio that will cause their charges to add to zero, gives the formula MgS.

(b) This is also an ionic compound. Both a metal (beryllium) and a polyatomic ion (hydroxide) are present. Beryllium is a type I metal, forming ions with a +2 charge (Sec. 7.5). The formula of the hydroxide ion is OH^- (Table 8.4). Combining these ions, Be^{2+} and OH^-, in the ratio that causes their charges to add to zero gives the formula $Be(OH)_2$. Remember, that any time more than one polyatomic ion is needed in a formula parentheses must be used. (Review Examples 7.5 and 7.6 if you are still having trouble determining the ratio in which ions combine.)

(c) This is another ionic compound containing a polyatomic ion. The Roman numeral associated with the word copper indicates that the copper ions present are Cu^{2+} ions. The Roman numeral gives the charge on the ions. The formula of the polyatomic carbonate ion is CO_3^{2-} (Table 8.4). Combining Cu^{2+} and CO_3^{2-} ions in a one-to-one ratio gives the formula $CuCO_3$. Note, again, that a Roman numeral in the name of a compound never becomes part of the formula of the compound. The Roman numeral's purpose is to give information about ionic charge of a type II metal.

(d) The presence of only nometals and the presence of Greek numerical prefixes both indicate that this is a molecular compound. The Greek numerical prefixes give directly the number of each type of atom present. The chemical formula is S_7O_2.

Practice Exercise 8.9

Write chemical formulas for the following compounds

(a) calcium perchlorate (b) iron(II) nitrate

(c) lithium oxide (d) sulfur trioxide

Ans. (a) $Ca(ClO_4)_2$; (b) $Fe(NO_3)_2$; (c) Li_2O; (d) SO_3

KEY TERMS

The new terms or concepts defined in this chapter are

binary compound (Sec. 8.1) A compound containing just two different elements.

binary ionic compound (Sec. 8.2) A binary compound in which positively charged metallic ions and negatively charged nonmetallic ions are present.

binary molecular compound (Sec. 8.5) A binary compound in which two nonmetallic elements are covalently bonded to each other.

chemical nomenclature (Sec. 8.1) A system of names used to distinguish compounds from each other and the rules needed to devise these names.

common name (Sec. 8.5) A name for a compound not based on IUPAC rules.

IUPAC rules (Sec. 8.1) A set of rules for naming compounds produced by committees of the International Union of Pure and Applied Chemistry (IUPAC).

nonoxyacid (Sec. 8.6) A molecular compound composed of hydrogen and one or more nonmetals other than oxygen that produces an acidic aqueous solution.

oxyacid (Sec. 8.6) A molecular compound composed of hydrogen, oxygen, and another nonmetal that produces an acidic aqueous solution.

systematic name (Sec. 8.5) A name for a compound, based on IUPAC rules, that conveys information about the composition of the compound.

ternary compound (Sec. 8.1) A compound containing three different elements.

type I metal (Sec. 8.2) A metal that forms only one type of positive ion, which always has the same charge.

type II metal (Sec. 8.2) A metal that forms more than one type of positive ion, with the ion types differing in charge magnitude.

PRACTICE PROBLEMS

Nomenclature Classifications for Compounds (Sec. 8.1)

8.1 In which of the following pairs of compounds are both members of the pair binary compounds?

(a) NO and NO_2 **(b)** H_2O and HClO
(c) HNO_3 and $HClO_3$ **(d)** N_4S_4 and P_4O_{10}

8.2 In which of the following pairs of compounds are both members of the pair ternary compounds?

(a) H_2O and H_2O_2 **(b)** $KClO_2$ and $KClO_3$
(c) HCN and KCN **(d)** NaBr and NaBrO

8.3 In which of the following pairs of compounds are both members of the pair ionic compounds?

(a) KNO_3 and K_3N **(b)** NaCl and HCl
(c) SO_2 and S_2O **(d)** K_3N and K_2O

8.4 In which of the following pairs of compounds are both members of the pair molecular compounds?

(a) BrF and BrCl **(b)** $AlBr_3$ and NaBr
(c) P_4O_6 and HNO_3 **(d)** H_2CO_3 and CO

8.5 In which of the following pairs of compounds is one member of the pair an ionic compound and the other member of the pair a molecular compound?

(a) AlN and HF **(b)** $CuCO_3$ and H_2CO_3
(c) SO_3 and H_2SO_4 **(d)** NaCl and $Mg(OH)_2$

8.6 In which of the following pairs of compounds is one member of the pair an ionic compound and the other member of the pair a molecular compound?

(a) CO and CO_2 **(b)** Be_3N_2 and Cl_2O
(c) HNO_3 and KNO_3 **(d)** MgO and CaS

Types of Binary Ionic Compounds (Sec. 8.2)

8.7 Classify each of the following metals as a type I metal or a type II metal.

(a) Mg **(b)** Ni **(c)** Ca **(d)** Au

8.8 Classify each of the following metals as a type I metal or a type II metal.

(a) Ba **(b)** Ag **(c)** Cr **(d)** Li

8.9 In which of the following pairs of type I metals do both members of the pair form ions with the same charge?

(a) Be and Mg **(b)** Ag and Zn
(c) Al and K **(d)** Na and Li

8.10 In which of the following pairs of type I metals do both members of the pair form ions with the same charge?

(a) Ag and Na (b) Al and Ga
(c) Be and Cd (d) Ca and Zn

8.11 Classify each of the following compounds as a type I binary ionic compound or a type II binary ionic compound.
(a) FeO (b) $NiCl_2$ (c) K_2O (d) Al_2S_3

8.12 Classify each of the following compounds as a type I binary ionic compound or a type II binary ionic compound.
(a) Cu_2S (b) AgCl (c) AuCl (d) Na_3N

Nomenclature of Binary Ionic Compounds (Sec. 8.2)

8.13 What is the formula of each of the following ions?
(a) copper(II) (b) iron(III) (c) lead(IV) (d) tin(II)

8.14 What is the formula of each of the following ions?
(a) nickel(II) (b) copper(I) (c) lead(II) (d) tin(IV)

8.15 What is the formula of each of the following ions?
(a) sodium (b) silver
(c) zinc (d) calcium

8.16 What is the formula of each of the following ions?
(a) lithium (b) potassium
(c) aluminum (d) magnesium

8.17 What is the formula of each of the following ions?
(a) chloride (b) oxide (c) phosphide (d) fluoride

8.18 What is the formula of each of the following ions?
(a) sulfide (b) nitride (c) iodide (d) bromide

8.19 Name each of the following type I binary ionic compounds.
(a) MgO (b) Li_2S (c) AgCl (d) $ZnBr_2$

8.20 Name each of the following type I binary ionic compounds.
(a) BeS (b) $GaCl_3$ (c) CaO (d) Cd_3P_2

8.21 Indicate whether or not a Roman numeral is required in the name of each of the following binary ionic compounds.
(a) ZnO (b) Ag_2S (c) Cu_2O (d) $FeBr_3$

8.22 Indicate whether or not a Roman numeral is required in the name of each of the following binary ionic compounds.
(a) Al_2O_3 (b) CoF_3 (c) Ag_3N (d) BaS

8.23 What is the charge on the metal ion in each of the following type II binary ionic compounds?
(a) FeO (b) $NiCl_2$ (c) Au_2O_3 (d) Co_3N_2

8.24 What is the charge on the metal ion in each of the following type II binary ionic compounds?
(a) SnO (b) PbS_2 (c) Cu_2S (d) Fe_2O_3

8.25 Name each compound in the following pairs of type II binary ionic compounds.
(a) $FeBr_2$ and $FeBr_3$ (b) Cu_2O and CuO
(c) SnS and SnS_2 (d) NiO and Ni_2O_3

8.26 Name each compound in the following pairs of type II binary ionic compounds.
(a) $SnCl_4$ and $SnCl_2$ (b) FeS and Fe_2S_3
(c) Cu_3N and Cu_3N_2 (d) NiI_2 and NiI_3

8.27 Name each of the following binary ionic compounds.
(a) $AlCl_3$ (b) $NiCl_3$ (c) ZnO (d) CoO

8.28 Name each of the following binary ionic compounds.
(a) Au_2O (b) Ag_2O (c) CuCl (d) KCl

8.29 Change each of the following compound names from the Roman numeral to the "-ic, -ous" system or vice versa.
(a) lead(IV) oxide (b) gold(III) chloride
(c) ferric iodide (d) stannous bromide

8.30 Change each of the following compound names from the Roman numeral to the "-ic, -ous" system or vice versa.
(a) cuprous chloride (b) ferrous sulfide
(c) tin(IV) nitride (d) lead(II) oxide

8.31 Write formulas for the following binary ionic compounds.
(a) iron(II) sulfide (b) tin(IV) sulfide
(c) lithium sulfide (d) zinc sulfide

8.32 Write formulas for the following binary ionic compounds.
(a) cobalt(III) sulfide (b) manganese(II) sulfide
(c) calcium sulfide (d) aluminum sulfide

8.33 In which of the following pairs of compound names do the two names in the pair denote the same compound?
(a) cupric chloride and copper(I) chloride
(b) stannic bromide and tin(IV) bromide
(c) ferrous oxide and iron(II) oxide
(d) plumbous sulfide and lead(IV) sulfide

8.34 In which of the following pairs of compound names do the two names in the pair denote the same compound?
(a) auric chloride and gold(III) chloride
(b) ferric bromide and iron(III) bromide
(c) cuprous oxide and copper(I) oxide
(d) stannous sulfide and tin(II) sulfide

Nomenclature for Ionic Compounds Containing Polyatomic Ions (Sec. 8.4)

8.35 Write the formulas, including ionic charge, of the following polyatomic ions.
(a) phosphate (b) chlorate
(c) nitrate (d) cyanide

8.36 Write the formulas, including ionic charge, of the following polyatomic ions.
(a) hydroxide (b) sulfate
(c) carbonate (d) ammonium

8.37 With the help of Table 8.4, give the names for the following polyatomic ions.

(a) O_2^{2-} (b) $S_2O_3^{2-}$ (c) $C_2O_4^{2-}$ (d) MnO_4^-

8.38 With the help of Table 8.4, give the names for the following polyatomic ions.

(a) N_3^- (b) BO_3^{3-} (c) SCN^- (d) CrO_4^{2-}

8.39 Write formulas for both ions in each of the following pairs of polyatomic ions.

(a) sulfate and sulfite
(b) phosphate and hydrogen phosphate
(c) hydroxide and peroxide
(d) chromate and dichromate

8.40 Write formulas for both ions in each of the following pairs of polyatomic ions.

(a) nitrate and nitrite
(b) chlorate and perchlorate
(c) cyanide and azide
(d) cyanate and thiocyanate

8.41 In which of the following pairs of compounds are polyatomic ions present in both members of the pair?

(a) NO_2 and KNO_2
(b) $ZnSO_4$ and NH_4Cl
(c) $Fe(NO_3)_3$ and $AlNO_3$
(d) $Ca_3(PO_4)_2$ and Ca_3N_2

8.42 In which of the following pairs of compounds are polyatomic ions present in both members of the pair?

(a) SO_3 and $CaSO_3$
(b) NH_4Br and $KClO$
(c) Cu_2CO_3 and $CuCO_3$
(d) Na_2SO_4 and Na_2S

8.43 Classify each of the following ionic compounds as a type I ionic compound or a type II ionic compound.

(a) $ZnSO_4$ (b) $Ba(OH)_2$ (c) $Fe(NO_3)_3$ (d) $CuCO_3$

8.44 Classify each of the following ionic compounds as a type I ionic compound or a type II ionic compound.

(a) $AgNO_3$ (b) $PbSO_4$ (c) $Sn(CO_3)_2$ (d) K_3PO_4

8.45 What is the charge on the type II metal ion present in each of the following polyatomic-ion-containing compounds?

(a) $FeSO_4$ (b) $Pb(SO_4)_2$ (c) $Ni_2(SO_4)_3$ (d) Cu_2SO_4

8.46 What is the charge on the type II metal ion present in each of the following polyatomic-ion-containing compounds.

(a) $PbCO_3$ (b) $Sn(CO_3)_2$ (c) $Fe_2(CO_3)_3$ (d) Cu_2CO_3

8.47 What is the name of each of the compounds in Problem 8.43?

8.48 What is the name of each of the compounds in Problem 8.44?

8.49 Name each compound in the following pairs of polyatomic-ion-containing compounds.

(a) $Fe_2(CO_3)_3$ and $FeCO_3$
(b) Au_2SO_4 and $Au_2(SO_4)_3$
(c) $Sn(OH)_2$ and $Sn(OH)_4$
(d) $Cr(C_2H_3O_2)_3$ and $Cr(C_2H_3O_2)_2$

8.50 Name each compound in the following pairs of polyatomic-ion-containing compounds.

(a) $CuNO_3$ and $Cu(NO_3)_2$
(b) $Pb_3(PO_4)_2$ and $Pb_3(PO_4)_4$
(c) $Mn(CN)_3$ and $Mn(CN)_2$
(d) $Co(ClO_3)_2$ and $Co(ClO_3)_3$

8.51 Name each of the following polyatomic-ion-containing compounds.

(a) NH_4NO_3 (b) NH_4Cl
(c) Na_3PO_4 (d) Cu_3PO_4

8.52 Name each of the following polyatomic-ion-containing compounds

(a) NH_4CN (b) $(NH_4)_2SO_4$
(c) $AgNO_3$ (d) $AuNO_3$

8.53 Write formulas for the following compounds containing polyatomic ions.

(a) silver carbonate
(b) gold(I) nitrate
(c) chromium(III) sulfate
(d) ammonium acetate

8.54 Write formulas for the following compounds containing polyatomic ions.

(a) copper(II) sulfate
(b) manganese(III) hydroxide
(c) ammonium nitrate
(d) magnesium phosphate

8.55 Write formulas for the following compounds containing polyatomic ions.

(a) ferric sulfate (b) cuprous cyanide
(c) stannic carbonate (d) plumbous hydroxide

8.56 Write formulas for the following compounds containing polyatomic ions.

(a) auric cyanide (b) ferrous nitrate
(c) plumbous phosphate (d) cupric chlorate

Nomenclature for Binary Molecular Compounds (Sec. 8.5)

8.57 Write the number that corresponds to each of the following prefixes.

(a) hepta- (b) penta- (c) tri- (d) deca-

8.58 Write the number that corresponds to each of the following prefixes.

(a) tetra- (b) octa- (c) hexa- (d) ennea-

8.59 Name the following binary molecular compounds.

(a) P_4O_{10} (b) SF_4 (c) CBr_4 (d) ClO_2

8.60 Name the following binary molecular compounds.

(a) S_4N_2 (b) SO_3 (c) IF_7 (d) N_2O_4

8.61 Write formulas for the following binary molecular compounds.

(a) iodine monochloride

(b) nitrogen trichloride

(c) sulfur hexafluoride

(d) oxygen difluoride

8.62 Write formulas for the following binary molecular compounds.

(a) disulfur monoxide

(b) tetraphosphorus hexoxide

(c) carbon dioxide

(d) silicon tetrachloride

8.63 Name the following binary molecular compounds.

(a) H_2S (b) HF (c) NH_3 (d) CH_4

8.64 Name the following binary molecular compounds.

(a) HCl (b) H_2Se (c) N_2N_4 (d) H_2O_2

8.65 Write formulas for the following binary molecular compounds.

(a) phosphine (b) hydrogen bromide

(c) ethane (d) hydrogen telluride

8.66 Write formulas for the following binary molecular compounds.

(a) hydrogen iodide (b) hydrazine

(c) hydrogen peroxide (d) arsine

Nomenclature for Acids (Sec. 8.6)

8.67 Indicate whether or not each of the following hydrogen-containing compounds forms an acid in aqueous solution.

(a) CH_4 (b) H_2S (c) HCN (d) NH_3

8.68 Indicate whether or not each of the following hydrogen-containing compounds forms an acid in aqueous solution.

(a) HCl (b) HClO (c) SiH_4 (d) CH_4

8.69 Name the acids that produce each of the following negative ions in aqueous solution.

(a) CN^- (b) SO_4^{2-} (c) NO_2^- (d) BO_3^{3-}

8.70 Name the acids that produce each of the following negative ions in aqueous solution.

(a) NO_3^- (b) I^- (c) PO_3^{3-} (d) $C_2O_4^{2-}$

8.71 What is the chemical formula for each of the acids in Problem 8.69?

8.72 What is the chemical formula for each of the acids in Problem 8.70?

8.73 Name each of the following compounds as acids.

(a) HNO_3 (b) HI (c) HClO (d) $HC_2H_3O_2$

8.74 Name each of the following compounds as acids.

(a) $HClO_3$ (b) $HClO_4$ (c) H_2S (d) HCl

8.75 Supply the missing name in each of the following pairs of name–formula combinations.

(a) H_3AsO_4 (arsenic acid); H_3AsO_3 (___?___)

(b) HIO_3 (iodic acid); HIO_4 (___?___)

(c) H_3PO_3 (phosphorus acid); H_3PO_2 (___?___)

(d) HBrO (hypobromous acid); $HBrO_2$ (___?___)

8.76 Supply the missing name in each of the following pairs of name–formula combinations.

(a) HIO_3 (iodic acid); HIO (___?___)

(b) H_2SeO_4 (selenic acid); H_2SeO_3 (___?___)

(c) HBrO (hypobromous acid); $HBrO_4$ (___?___)

(d) HNO_2 (nitrous acid); HNO_3 (___?___)

8.77 Name each of the following compounds.

(a) HBr(g) (b) HCN(aq) (c) H_2S(g) (d) HI(aq)

8.78 Name each of the following compounds.

(a) HBr(aq) (b) HCN(g) (c) H_2S(aq) (d) HI(g)

8.79 Write formulas for the following acids.

(a) chloric acid (b) nitrous acid

(c) hydrofluoric acid (d) acetic acid

8.80 Write formulas for the following acids.

(a) hypochlorous acid (b) sulfurous acid

(c) nitric acid (d) oxalic acid

ADDITIONAL PROBLEMS

8.81 In which of the following pairs of ionic compounds do both members of the pair have positive ions with the same charge?

(a) BaO and NaCl (b) Fe_2O_3 and $NiCl_3$

(c) $MgCl_2$ and BeO (d) AlN and KF

8.82 In which of the following pairs of inoic compounds do both members of the pair have positive ions with the same charge?

(a) Co_2O_3 and $CoCl_3$ (b) Cu_2O and CuO

(c) K_2O and Al_2O_3 (d) MgS and NaI

8.83 In which of the following pairs of compounds would both members of the pair have names that contain Roman numerals?

(a) $NaCl$ and $CuCl$ (b) Al_2O_3 and Fe_2O_3
(c) $Cu(NO_3)_2$ and NiO (d) Ag_2SO_4 and $CuSO_4$

8.84 In which of the following pairs of compounds would both members of the pair have names that contain Roman numerals?

(a) $AuCl_3$ and $FeCl_3$ (b) K_3N and AlN
(c) FeO and BaO (d) NiS and $Cu_3(PO_4)_2$

8.85 In which of the following pairs of compounds would both members of the pair have names that contain Greek numerical prefixes.

(a) SO_3 and N_2O (b) AlN and CO
(c) HCl and H_2S (d) PCl_3 and NF_3

8.86 In which of the following pairs of compounds would both members of the pair have names that contain Greek numerical prefixes.

(a) CO and CO_2 (b) N_2O and K_2O
(c) OF_2 and BaF_2 (d) HBr and H_2Se

8.87 How many ions are present in one formula unit of each of the following compounds?

(a) K_3N (b) KN_3 (c) Na_2O (d) Na_2O_2

8.88 How many ions are present in one formula unit of each of the following compounds?

(a) $NaCN$ (b) $NaSCN$ (c) K_2O (d) K_2O_2

8.89 Write formulas for the following compounds.

(a) calcium nitride (b) calcium nitrate
(c) calcium nitrite (d) calcium cyanide

8.90 Write formulas for the following compounds.

(a) sodium sulfide (b) sodium sulfate
(c) sodium sulfite (d) sodium thiosulfate

8.91 Write formulas for the following compounds.

(a) potassium phosphide
(b) potassium phosphate
(c) potassium hydrogen phosphate
(d) potassium dihydrogen phosphate

8.92 Write formulas for the following compounds.

(a) magnesium carbide
(b) magnesium bicarbonate
(c) magnesium carbonate
(d) magnesium hydrogen carbonate

8.93 Indicate which compounds in each of the following groups have names that contain the prefix *di-*.

(a) N_2O, Li_2O, CO_2, K_2S
(b) K_2O, K_2CO_3, NO_2, SO_2
(c) BeF_2, $BeCl_2$, SF_2, SCl_2
(d) Au_2O_3, Fe_2O_3, Al_2O_3, N_2O_3

8.94 Indicate which compounds in each of the following groups have names that contain the suffix *-ide*.

(a) CaS, CO, SO_3, Be_3N_2
(b) K_3N, KNO_3, KNO_2, $KClO$
(c) MgO, AlN, KF, KOH
(d) $NaCN$, $NaOH$, Na_2CO_3, NaF

8.95 Indicate which compounds in each of the following groups have names that contain the suffix *-ous* or *-ate*.

(a) $CaCO_3$, H_2CO_3, $Ca(NO_2)_2$, HNO_2
(b) $NaClO_4$, $NaClO_3$, $NaClO_2$, $NaClO$
(c) $HClO_4$, $HClO_3$, $HClO_2$, $HClO$
(d) $LiOH$, $LiCN$, Li_2CO_3, Li_3PO_4

8.96 Indicate which compounds in each of the following groups have names that contain the suffix *-ic* or *-ite*.

(a) $HBr(g)$, $HBr(aq)$, $HCN(g)$, $HCN(aq)$
(b) K_2SO_4, KCN, $KMnO_4$, K_3PO_3
(c) NH_4Cl, NH_4CN, $(NH_4)_2SO_4$, NH_4NO_2
(d) Li_3N, LiN_3, $LiNO_3$, $LiNO_2$

8.97 Indicate which compounds in each of the following groups possess molecules or formula units that are pentatomic.

(a) sodium cyanide, sodium thiocyanate, sodium hypochlorite, sodium nitrate
(b) aluminum sulfide, magnesium nitride, beryllium phosphide, potassium hydroxide
(c) beryllium oxide, iron(II) oxide, iron(III) oxide, sulfur dioxide
(d) gold(I) cyanide, gold(III) cyanide, gold(I) chlorate, gold(III) chlorate

8.98 Indicate which compounds in each of the following groups possess molecules or formula units that are heptatomic.

(a) magnesium cyanide, magnesium oxalate, magnesium chlorite, magnesium perchlorate
(b) aluminum oxide, calcium thiosulfate, beryllium thiocyanate, dinitrogen pentoxide
(c) iron(III) sulfide, iron(III) nitride, iron(III) sulfate, iron(III) hypochlorite
(d) dichlorine heptoxide, sulfur trioxide, sulfur hexafluoride, ammonium sulfide

8.99 The formula of a hydroxide of nickel is $Ni(OH)_3$. What are the formulas of the following nickel compounds in which nickel has the same ionic charge as in $Ni(OH)_3$?

(a) nickel sulfate (b) nickel oxide
(c) nickel oxalate (d) nickel nitrate

8.100 The formula of a phosphate of vanadium is VPO_4. What are the formulas of the following vanadium compounds in which vanadium has the same ionic charge as in VPO_4?

(a) vanadium sulfate **(b)** vanadium hydroxide

(c) vanadium oxide **(d)** vanadium nitrate

8.101 The formula of the ionic compound potassium superoxide is KO_2. The formula of the ionic compound nitronium perchlorate is NO_2ClO_4. What is the formula for the ionic compound nitronium superoxide?

8.102 The formula of the ionic compound calcium perrhenate is $Ca(ReO_4)_2$. The formula of the ionic compound nitrosonium hydrogen sulfate is $NOHSO_4$. What is the formula for the ionic compound nitrosonium perrhenate?

CUMULATIVE PROBLEMS

8.103 Give the name of the simplest binary compound that forms between elements with the following electron configurations.

(a) $1s^22s^22p^63s^23p^5$ and $1s^22s^22p^63s^2$

(b) $1s^22s^22p^4$ and $1s^22s^22p^5$

8.104 Give the name of the simplest binary compound that forms between elements with the following electron configurations.

(a) $1s^22s^22p^3$ and $1s^22s^22p^63s^23p^64s^1$

(b) $1s^22s^22p^63s^23p^4$ and $1s^22s^22p^63s^23p^5$

8.105 After determining the value of *x* in each of the following formulas, name the compounds.

(a) $SiCl_x$ **(b)** $MgCl_x$ **(c)** K_xN **(d)** NCl_x

8.106 After determining the value of *x* in each of the following formulas, name the compounds.

(a) CCl_x **(b)** BeO_x **(c)** NF_x **(d)** $AlBr_x$

8.107 A compound has the formula $M(XO_3)_2$, where M is a metal and X is a nonmetal. Assign a name to this compound given the following information.

(1) M is a metal with two valence electrons.

(2) X forms a monoatomic ion with a charge of -1.

(3) X has an electronegativity between 2.6 and 2.9.

(4) The sum of the periodic table period numbers for X and M is 6.

8.108 A compound has the formula M_3XO_3, where M is a metal and X is a nonmetal. Assign a name to this compound given the following information.

(1) M is a metal whose electron configuration "ends" in s^1.

(2) Both M and X are found in period 2 of the periodic table.

(3) There are 38 protons in one formula unit of M_3XO_3.

8.109 Determine the name for a binary ionic compound that has the following characteristics.

(1) Positive and negative ions are present in a one-to-one ratio.

(2) All ions present have the electron configuration $1s^22s^22p^6$.

(3) One of the elements present has an atomic number that is six less than the atomic number of the other element.

8.110 Determine the name for a binary ionic compound that has the following characteristics.

(1) Positive and negative ions are present in a one-to-two ratio.

(2) All ions present have the electron configuration $1s^22s^22p^63s^23p^6$.

(3) Neutral atoms of the more electronegative element present have 17 electrons.

8.111 Determine the name for a binary molecular compound that has the following characteristics.

(1) Its molecules are triatomic.

(2) There are twice as many atoms of the more electronegative element per molecule as there are of the less electronegative element.

(3) Both elements are in period 2 of the periodic table.

(4) Atoms of one element contain four valence electrons, and atoms of the other element contain six valence electrons.

8.112 Determine the name for a binary molecular compound that has the following characteristics.

(1) Its molecules are hexatomic.

(2) There are twice as many atoms of the more electronegative element per molecule as there are of the less electronegative element.

(3) The two elements occupy adjacent positions in period 2 of the periodic table.

(4) The sum of the valence electrons for an atom of each element is 11.

8.113 Determine the name for a compound that has the following characteristics.

(1) Monoatomic and polyatomic ions are present.

(2) The ratio between positive and negative ions is one to two.

(3) A group IIA element that loses one half of its total electrons upon ion formation is present.

(4) The polyatomic ion contains equal numbers of atoms of two nonmetallic elements.

(5) The sum of the atomic numbers for the two elements involved in the polyatomic ion is 13.

(6) One of the elements present in the compound forms a monoatomic ion with a charge of -3 that is isoelectronic with Ne.

8.114 Determine the name for a compound that has the following characteristics.

(1) Monoatomic and polyatomic ions are present.

(2) The ratio between positive and negative ions is one to three.

(3) A group IIIA element whose ion electron configuration is isoelectronic with Ne is present.

(4) The polyatomic ion contains equal numbers of atoms of two nonmetallic elements.

(5) The sum of the atomic numbers for the two elements involved in the polyatomic ion is 9.

9 Chemical Calculations: The Mole Concept and Chemical Formulas

9.1 The Law of Definite Proportions

This chapter is the first of two chapters dealing with "chemical arithmetic," that is, with the quantitative relationships between elements and compounds. The emphasis in this chapter will be on quantitative relationships that involve chemical formulas. Emphasis in Chapter 10 will be on quantitative relationships that involve chemical equations.

In Section 4.7 we saw that compounds are pure substances with the following characteristics.

1. They are chemical combinations of two or more elements.
2. They can be broken down into constituent elements by chemical, but not physical, means.
3. They have a definite, constant elemental composition.

In this section we consider further the fact that compounds have a definite composition.

The composition of a compound can be determined by decomposing a weighed amount of the compound into its elements and then determining the masses of the individual elements. Alternatively, determining the mass of a compound formed by the combination of known masses of elements will also allow the calculation of its composition.

Studies of composition data for many compounds have led to the conclusion that the percentage of each element present in a given compound does not vary. This conclusion has been formalized into a statement known as the **law of definite proportions**: *In a pure compound, the elements are always present in the same definite proportion by mass.* The French chemist Joseph Louis Proust (1754–1826) is responsible for the work that established this law as one of the fundamentals of chemistry.

Let us consider how composition data obtained from decomposing a compound are used to illustrate the law of definite proportions.

Samples of the compound of *known masses* are decomposed. The masses of the constituent elements present in each sample are then obtained. This experimental information (elemental masses and the original sample mass) is then used to calculate the percent composition of the samples. Within the limits of experimental error, the calculated percentages for any given element present turn out to be the same, validating the law of definite proportions. Example 9.1 gives actual decomposition data for a compound and shows how they are used to verify the law of definite proportions.

Example 9.1

Carbon dioxide is an atmospheric gas that contributes to the environmental effect called global warming. Two samples of carbon dioxide (CO_2) of differing mass and from different locations are individually decomposed to yield carbon dioxide's constituent elements (carbon and oxygen). The results of the decomposition experiments are as follows:

	Sample Mass Before Decomposition (g)	Mass of Carbon Produced (g)	Mass of Oxygen Produced (g)
Sample 1	1.537	0.419	1.118
Sample 2	3.100	0.846	2.254

Show that these data are consistent with the law of definite proportions.

Solution

Calculating the percent oxygen in each sample will be sufficient to show whether the data are consistent with the law. Both oxygen percentages should come out the same if the law is obeyed.

$$\text{Percent oxygen} = \frac{\text{mass of oxygen obtained}}{\text{total sample mass}} \times 100$$

SAMPLE 1

$$\%\ \text{O} = \frac{1.118\ \cancel{\text{g}}}{1.537\ \cancel{\text{g}}} \times 100 = 72.739102\% \quad \text{(calculator answer)}$$

$$= 72.74\% \quad \text{(correct answer)}$$

SAMPLE 2

$$\%\ \text{O} = \frac{2.254\ \cancel{\text{g}}}{3.100\ \cancel{\text{g}}} \times 100 = 72.709677\% \quad \text{(calculator answer)}$$

$$= 72.71\% \quad \text{(correct answer)}$$

Note that the percentages are close to being equal but are not identical. This is due to measuring uncertainties and rounding in the original experimental data. The two percentages are the same to three significant figures. The difference lies in the fourth significant digit. Recall from Section 2.2 that the last of the significant digits in a number (the fourth one here) has uncertainty in it. The two percentages above are considered to be the same within experimental error.

An alternative way of treating the given data to illustrate the law of definite proportions involves calculating the mass ratio between carbon and oxygen. This ratio should be the same for each sample; otherwise the two samples are not the same compound.

SAMPLE 1

$$\frac{\text{Mass of C}}{\text{Mass of O}} = \frac{0.419\ \cancel{\text{g}}}{1.118\ \cancel{\text{g}}} = 0.37477638 \quad \text{(calculator answer)}$$

$$= 0.375 \quad \text{(correct answer)}$$

SAMPLE 2

$$\frac{\text{Mass of C}}{\text{Mass of O}} = \frac{0.846\ \cancel{g}}{2.254\ \cancel{g}} = 0.37533274 \quad \text{(calculator answer)}$$

$$= 0.375 \quad \text{(correct answer)}$$

Practice Exercise 9.1

Two samples of chlorine dioxide (ClO_2) of differing mass and from different sources are individually decomposed to yield the elements chlorine and oxygen. The results of the decomposition experiments are as follows.

	Sample Mass Before Decomposition (g)	Mass of Chlorine Produced (g)	Mass of Oxygen Produced (g)
Sample 1	2.724	1.056	1.668
Sample 2	8.367	3.243	5.124

Show that these data are consistent with the law of definite proportions.

Ans. The percents by mass of chlorine in the two samples are 38.77% and 38.76%

The constancy of composition for compounds can also be examined by considering the mass ratios in which elements combine to form compounds. Let us consider the reaction between the elements calcium and sulfur to produce the compound calcium sulfide. Suppose an attempt is made to combine various masses of sulfur with a fixed mass of calcium. A set of possible experimental data for this attempt is given in the first four lines of Table 9.1. Note that, regardless of the mass of S present, only a certain amount, 44.4 g, reacts with the 55.6 g of Ca. The excess S is left over in an unreacted form. The data therefore illustrate that Ca and S will react in only one fixed mass ratio (55.6/44.4 = 1.25) to form CaS. This fact is consistent with the law of definite proportions. Note also that if the amount of Ca used is doubled (line 5 of Table 9.1), the amount of S with which it reacts also doubles (compare lines 1 and 5 of the table). Nevertheless, the ratio in which the substances react (111.2/88.8) still remains 1.25.

TABLE 9.1 DATA ILLUSTRATING THE LAW OF DEFINITE PROPORTIONS

Mass of Ca Used (g)	Mass of S Used (g)	Mass of CaS Formed (g)	Mass of Excess Unreacted Sulfur (g)	Ratio in Which Substances React
55.6	44.4	100.0	none	1.25
55.6	50.0	100.0	5.6	1.25
55.6	100.0	100.0	55.6	1.25
55.6	200.0	100.0	155.6	1.25
111.2	88.8	200.0	none	1.25

9.2 CALCULATION OF FORMULA MASSES

Formula masses play a role in almost all chemical calculations and will be used extensively in later sections of this chapter and in succeeding chapters. The **formula mass** *of a substance is the sum of the atomic masses of the atoms present in one formula unit of the substance*. Formula masses, like the atomic masses from which they are calculated, are relative masses based on the $^{12}_{6}C$ relative mass scale (Sec. 5.9). They can be calculated for compounds (both molecular and ionic) and for elements that exist in molecular form.

The term *molecular mass* is used interchangeably with *formula mass* by many chemists in referring to substances that contain discrete molecules. It is incorrect, however, to use the term *molecular mass* when talking about ionic substances, since molecules are not their basic structural unit (Sec. 7.7).

Some chemistry books use the terms *formula weight* and *molecular weight* rather than *formula mass* and *molecular mass*. This practice is not followed in this book because mass is technically more correct (Sec. 3.3).

Once the formula of a substance has been established, its formula mass is calculated by adding together the atomic masses of all the atoms in the formula. If more than one atom of any element is present, that element's atomic mass must be added as many times as there are atoms of the element present.

Example 9.2 contains sample formula mass calculations. Significant figures (Sec. 2.4) are a consideration because atomic masses are obtained from experimentally determined numbers (Sec. 5.9).

Example 9.2

Calculate the formula masses for the following compounds.

(a) N_2H_4 (hydrazine, a rocket fuel)

(b) $ClNO_3$ (chlorine nitrate, a substance involved in the Antarctic ozone hole phenomenon)

(c) CHF_2Cl (chlorodifluoromethane, a Freon replacement)

(d) $C_9H_8O_4$ (aspirin, a mild pain reliever)

Solution

Formula masses are obtained simply by adding the atomic masses of the constituent elements, counting each atomic mass as many times as the symbol for the element occurs in the formula.

(a) A molecule of N_2H_4 contains six atoms: two atoms of N and four atoms of H. The formula mass, the collective mass of these six atoms, is calculated as follows:

$$2\ \cancel{\text{atoms N}} \times \frac{14.0067\ \text{amu}}{1\ \cancel{\text{atom N}}} = 28.0134\ \text{amu}$$

$$4\ \cancel{\text{atoms H}} \times \frac{1.00794\ \text{amu}}{1\ \cancel{\text{atom H}}} = 4.03176\ \text{amu}$$

$$\begin{aligned} \text{Formula mass} &= 32.04516 \text{ amu} \quad \text{(calculator answer)} \\ &= 32.0452 \text{ amu} \quad \text{(correct answer)} \end{aligned}$$

The conversion factors in the calculation were derived from the atomic masses listed on the inside front and back covers of the text. The atomic mass, the mass of an average atom, is 14.0067 amu for N and 1.00794 amu for H. The calculator answer, 32.04516 amu, had to be rounded down one significant figure, since the atomic mass of N is known only to the fourth decimal place. (Recall the rules for significant figures in addition; Sec. 2.4.)

Often, conversion factors are not explicitly shown in a formula mass calculation as just shown; the calculation is simplified as follows:

$$\begin{aligned} \text{N:}\quad 2 \times 14.0067 \text{ amu} &= 28.0134 \text{ amu} \\ \text{H:}\quad 4 \times 1.00794 \text{ amu} &= 4.03176 \text{ amu} \\ \text{Formula mass} &= 32.04516 \text{ amu} \quad \text{(calculator answer)} \\ &= 32.0452 \text{ amu} \quad \text{(correct answer)} \end{aligned}$$

(b) Using the simplified calculation method, we calculate the formula mass for $ClNO_3$ as

$$\begin{aligned} \text{Cl:}\quad 1 \times 35.453 \text{ amu} &= 35.453 \text{ amu} \\ \text{N:}\quad 1 \times 14.0067 \text{ amu} &= 14.0067 \text{ amu} \\ \text{O:}\quad 3 \times 15.9994 \text{ amu} &= 47.9982 \text{ amu} \\ \text{Formula mass} &= 97.4579 \text{ amu} \quad \text{(calculator answer)} \\ &= 97.458 \text{ amu} \quad \text{(correct answer)} \end{aligned}$$

Note how the atomic mass of Cl limits the precision of the formula mass to thousandths. Why does the periodic table not give a more precise atomic mass for Cl, you may ask. A more fundamental question is "Why does the preciseness of atomic masses of the elements vary as much as it does?" Some elements have atomic masses known to 0.00001 amu; others are known to only 0.1 amu. A major factor in atomic mass precision is the constancy of the abundance percentages for the isotopes of an element. Abundance percentages fluctuate more for some elements than for others; the less the fluctuation, the greater the atomic mass precision. (The fluctuation is a relative matter; none of the fluctuations is large on an absolute scale.) The atomic masses given in the periodic table are stated to current precision limits.

(c) Using the simplified calculation method, we calculate the formula mass for CHF_2Cl as

$$\begin{aligned} \text{C:}\quad 1 \times 12.011 \text{ amu} &= 12.011 \text{ amu} \\ \text{H:}\quad 1 \times 1.00794 \text{ amu} &= 1.00794 \text{ amu} \\ \text{F:}\quad 2 \times 18.998403 \text{ amu} &= 37.996806 \text{ amu} \\ \text{Cl:}\quad 1 \times 35.453 \text{ amu} &= 35.453 \text{ amu} \\ \text{Formula mass} &= 86.468746 \text{ amu} \quad \text{(calculator answer)} \\ &= 86.469 \text{ amu} \quad \text{(correct answer)} \end{aligned}$$

(d) There are twenty-one atoms in a molecule of this compound. Its formula mass is

$$\begin{aligned} \text{C:}\quad 9 \times 12.011 \text{ amu} &= 108.099 \text{ amu} \\ \text{H:}\quad 8 \times 1.00794 \text{ amu} &= 8.06352 \text{ amu} \\ \text{O:}\quad 4 \times 15.9994 \text{ amu} &= 63.9976 \text{ amu} \\ \text{Formula mass} &= 180.16012 \text{ amu} \quad \text{(calculator answer)} \\ &= 180.160 \quad \text{(correct answer)} \end{aligned}$$

Practice Exercise 9.2

Calculate the formula masses for the following substances.

(a) H_2SO_4 (sulfuric acid, an industrial acid)
(b) $C_{16}H_{18}N_2O_5S$ (Penicillin V, an important antibiotic)

Ans. (a) 98.082 amu; (b) 350.395 amu

In Example 9.2, all the formula masses were calculated purposely to the maximum number of significant figures possible. This was done to emphasize the fact that the number of significant figures in "input" atomic masses determines the precision of calculated formula masses.

In most chemical calculations you will not need maximum formula mass precision. Other numbers that enter into the calculation will usually restrict the answer to three or four significant figures. Thus, in most problems, formula masses rounded to tenths or hundredths may be used. Occasionally, however, formula masses of maximum precison will be required.

Formula masses, relative masses on the $^{12}_{6}C$ relative mass scale, can be used for mass comparisons. For example, using the formula masses calculated in Example 9.2, we can make the statement that one molecule of $C_9H_8O_4$ is 1.8486 times as heavy as one molecule of $ClNO_3$ (180.160 amu/97.458 amu = 1.8486).

9.3 PERCENT COMPOSITION

A useful piece of information about a compound is its percent composition. **Percent composition** *specifies the percent by mass of each element present in a compound.* For instance, the percent composition of water is 88.81% oxygen and 11.19% hydrogen.

Percent compositions are frequently used to compare compound compositions. For example, the compounds gold(III) iodide (AuI_3), gold(III) nitrate [$Au(NO_3)_3$], and gold(I) cyanide (AuCN) contain, respectively, 34.10%, 51.43%, and 88.33% gold by mass. If you were given the choice of receiving a gift of 1 lb of one of these three gold compounds, which one would you choose?

The percent composition of a compound can be calculated from experimental decomposition data as was done in Example 9.1. Such a calculation can be carried out even if the formula or the identity of the compound is unknown. Another way to calculate percent composition for a compound is from its chemical formula. There is sufficient information in the formula for just such a calculation, as is illustrated in Example 9.3.

Example 9.3

The compound para-aminobenzoic acid (PABA) is used in sunscreen formulations to prevent sunburn from ultraviolet radiation. Calculate the percent composition (using atomic masses to two decimal places) of PABA given that its chemical formula is $C_7H_7O_2N$.

Solution

First, we calculate the formula mass of $C_7H_7O_2N$, using atomic masses rounded to the hundredths decimal place.

C:	7×12.01 amu =	84.07 amu	
H:	7×1.01 amu =	7.07 amu	
O:	2×16.00 amu =	32.00 amu	
N:	1×14.01 amu =	14.01 amu	
	Formula mass =	137.15 amu	(calculator and correct answer)

The mass percent of each element in the compound is found by dividing the mass contribution of each element, in amu, by the total mass (formula mass), in amu, and multiplying by 100.

$$\%\ \text{element} = \frac{\text{mass of element in one formula unit}}{\text{formula mass}} \times 100$$

Finding percentages, we have

$$\%\ C = \frac{84.07\ \cancel{\text{amu}}}{137.15\ \cancel{\text{amu}}} \times 100 = 61.297849\% \quad \text{(calculator answer)}$$

$$= 61.30\% \quad \text{(correct answer)}$$

$$\%\ H = \frac{7.07\ \cancel{\text{amu}}}{137.15\ \cancel{\text{amu}}} \times 100 = 5.1549398\% \quad \text{(calculator answer)}$$

$$= 5.15\% \quad \text{(correct answer)}$$

$$\%\ O = \frac{32.00\ \cancel{\text{amu}}}{137.15\ \cancel{\text{amu}}} \times 100 = 23.332118\% \quad \text{(calculator answer)}$$

$$= 23.33\% \quad \text{(correct answer)}$$

$$\%\ N = \frac{14.01\ \cancel{\text{amu}}}{137.15\ \cancel{\text{amu}}} \times 100 = 10.215092\% \quad \text{(calculator answer)}$$

$$= 10.22\% \quad \text{(correct answer)}$$

To check our work we can add the percentages of all the parts. They, of course, have to total 100. (On occasion, rounding errors may not cancel, and totals such as 99.99% or 100.01% may be obtained.)

$$61.30\% + 5.15\% + 23.33\% + 10.22\% = 100.00\%$$

Practice Exercise 9.3

What is the percent composition (using atomic masses to two decimal places) of vitamin C, a compound necessary in small amounts for the normal growth of humans, whose formula is $C_6H_8O_6$?

Ans. 40.91% C; 4.59% H; 54.50% O

Percent compositions can also be calculated from mass data obtained from compound synthesis or compound decomposition experiments. Example 9.4 shows how synthesis data are treated to yield percent composition.

Example 9.4

To produce 5.00 g of benzaldehyde, a compound responsible in part for the aroma of cherries and almonds, requires 3.960 g of C, 0.290 g of H, and 0.750 g of O. What is the percent composition of this compound?

Solution

The total mass of the compound sample is given as 5.00 g. We divide the mass of each element present by this total mass (5.00 g) and multiply by 100 to obtain percentage.

$$\% \text{ element} = \frac{\text{mass of element}}{\text{total sample mass}} \times 100$$

Finding the percentages, we have

$$\% \text{ C} = \frac{3.960 \cancel{g}}{5.00 \cancel{g}} \times 100 = 79.2\% \quad \text{(calculator and correct answer)}$$

$$\% \text{ H} = \frac{0.290 \cancel{g}}{5.00 \cancel{g}} \times 100 = 5.8\% \quad \text{(calculator answer)}$$

$$= 5.80\% \quad \text{(correct answer)}$$

$$\% \text{ O} = \frac{0.750 \cancel{g}}{5.00 \cancel{g}} \times 100 = 15\% \quad \text{(calculator answer)}$$

$$= 15.0\% \quad \text{(correct answer)}$$

Checking our work, we see that the percentages do add up correctly.

$$79.2\% + 5.80\% + 15.0\% = 100.0\%$$

Practice Exercise 9.4

The sour taste of vinegar is caused by the compound acetic acid. Calculate the percent composition of acetic acid, knowing that the synthesis of 24.03 g of this compound requires 9.61 g of C, 1.62 g of H, and 12.80 g of O.

Ans. 40.0% C; 6.74% H; 53.27% O

9.4 THE MOLE: THE CHEMIST'S COUNTING UNIT

Two common methods exist for specifying the quantity of material in a sample of a substance: (1) in terms of units of *mass* and (2) in terms of units of *amount*. We measure *mass* by using a balance (Sec. 3.3). Common mass units are gram, kilogram, and pound. For substances that consist of discrete units, we can specifiy the *amount* of substance present by indicating the number of units present—12, or 27, or 113, etc.

We all use both units of mass and units of amount on a daily basis. We work well with this dual system. Sometimes it does not matter which type of unit is used; at other times one system is preferred over the other. When buying potatoes at the grocery store we can decide on quantity in either mass units (10-lb bag, 20-lb bag, etc.) or amount units (9 potatoes, 15 potatoes, etc.). When buying eggs, amount units are used almost exclusively—12 eggs (1 dozen), 24 eggs (2 dozen), and so on. On the other hand, peanuts and grapes are almost always purchased in weighed quantities. It is impractical to count the number of grapes in a bunch. Very few people go to the store with the idea of buying 117 grapes.

In chemistry, as in everyday life, both the mass and amount methods of specifying quantity find use. Again, the specific situation dictates the method used. In laboratory work, practicality dictates working with quantities of known mass (12.3 g, 0.1365 g, etc.).

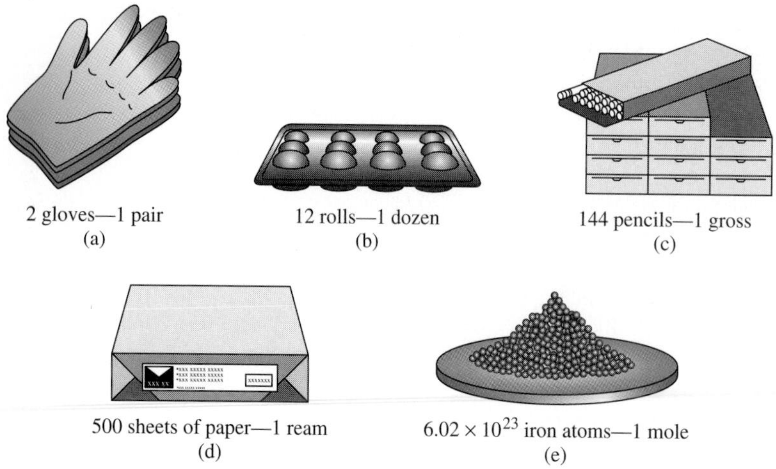

FIGURE 9.1 Some counting units used to denote quantities in terms of groups.

(Counting out a given number of atoms for a laboratory experiment is somewhat impractical, since we cannot see individual atoms.)

In performing chemical calculations, after the laboratory work has been done, it is often useful (even necessary) to think of quantities of substances present in terms of atoms or formula units. A problem exists when this is done—very, very large numbers are always encountered. Any macroscopic sample of a chemical substance contains many trillions of atoms or formula units.

In order to cope with this "large number problem" chemists have found it convenient to use a special counting unit. Employment of such a unit should not surprise you, as specialized counting units are used in many areas. The two most common counting units are *dozen* and *pair*. Other more specialized counting units exist. For example, at an office supply store, paper is sold by the *ream* (500 sheets), and pencils by the *gross* (144 pencils). (See Fig. 9.1.)

The chemist's counting unit is called a *mole*. What is unusual about the mole is its magnitude. A **mole** *is 6.02×10^{23} objects.* This extremely large number is necessitated by the extremely small size of atoms, molecules, and ions. The use of a traditional counting unit, such as a dozen, would be, at best, only a slight improvement over counting atoms singly.

$$6.02 \times 10^{23} \text{ atoms} = 5.02 \times 10^{22} \text{ dozen atoms} = 1 \text{ mole atoms}$$

Note how the use of the mole counting unit decreases very significantly the magnitude of numbers encountered. The number 1 represents 6.02×10^{23} objects, the number 2, double that number of objects. (Why the number 6.02×10^{23} was chosen as the counting unit rather than some other number will be discussed in Sec. 9.5. A more formal definition of the mole will also be presented in that section.)

The number 6.02×10^{23} also has a special name. **Avogadro's number** *is the name given to the numerical value 6.02×10^{23}, the number of particles in a mole.* This designation honors the Italian physicist Lorenzo Romano Amedeo Carlo Avogadro (1776–1856), whose pioneering work on gases later proved to be valuable in determining the number of particles present in a given volume of a substance.

In solving mathematical problems dealing with the number of atoms, molecules, or ions present in a given amount of material, Avogadro's number becomes part of the conversion factor used to relate number of particles present to moles present.

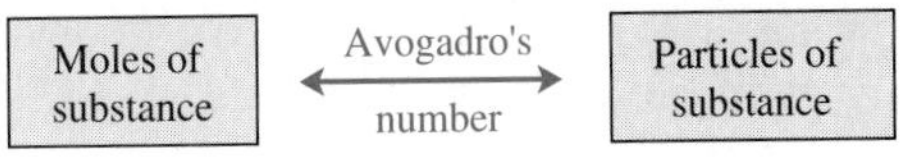

From the definition

$$1 \text{ mole} = 6.02 \times 10^{23} \text{ objects}$$

two conversion factors can be derived.

$$\frac{1 \text{ mole}}{6.02 \times 10^{23} \text{ objects}} \quad \text{and} \quad \frac{6.02 \times 10^{23} \text{ objects}}{1 \text{ mole}}$$

Example 9.5 illustrates the use of particle-to-mole conversion factors.

Example 9.5

How many objects are there in each of the following quantities?

(a) 1.20 moles of carbon monoxide (CO) molecules

(b) 2.53 moles of silver (Ag) atoms

(c) 0.025 mole of magnesium sulfate ($MgSO_4$) formula units

(d) 2.25 moles of watermelons

Solution

We will use dimensional analysis (Sec. 3.7) in solving each part of this problem. All of the parts are similar in that we are given a certain number of moles of substance and want to find the number of particles contained in the given number of moles. All parts can be classified as moles-to-particles problems, and each solution will involve the use of Avogadro's number.

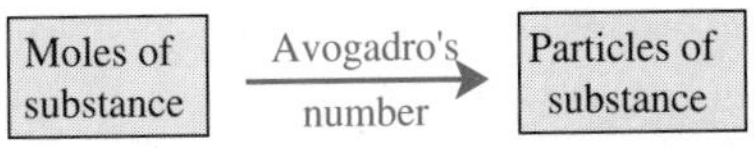

(a) The given quantity is 1.20 moles of CO molecules, and the desired quantity is number of CO molecules.

$$1.20 \text{ moles CO} = ? \text{ CO molecules}$$

The setup, by dimensional analysis, involves only one conversion factor.

$$1.20 \text{ moles CO} \times \frac{6.02 \times 10^{23} \text{ CO molecules}}{1 \text{ mole CO}}$$

$$= 7.224 \times 10^{23} \text{ CO molecules} \quad \text{(calculator answer)}$$
$$= 7.22 \times 10^{23} \text{ CO molecules} \quad \text{(correct answer)}$$

(b) The given quantity is 2.53 moles of silver atoms, and the desired quantity is the actual number of silver atoms present.

$$2.53 \text{ moles Ag} = ? \text{ Ag atoms}$$

The setup, with the same conversion factor as in part (a), is

$$2.53 \text{ moles Ag} \times \frac{6.02 \times 10^{23} \text{ Ag atoms}}{1 \text{ mole Ag}}$$

$$= 1.52306 \times 10^{24} \text{ Ag atoms} \quad \text{(calculator answer)}$$
$$= 1.52 \times 10^{24} \text{ Ag atoms} \quad \text{(correct answer)}$$

(c) The fact that we are dealing with formula units here (an ionic compound), rather than atoms or molecules, does not change the way the problem is solved.

$$0.025 \text{ mole } MgSO_4 = \text{? formula units } MgSO_4$$

The conversion factor setup is

$$0.025 \text{ mole } MgSO_4 \times \frac{6.02 \times 10^{23} \text{ formula units } MgSO_4}{1 \text{ mole } MgSO_4}$$

$$= 1.505 \times 10^{22} \; MgSO_4 \text{ formula units} \quad \text{(calculator answer)}$$
$$= 1.5 \times 10^{22} \; MgSO_4 \text{ formula units} \quad \text{(correct answer)}$$

(d) Use of the mole as a counting unit is usually found only in a chemical context. Technically, however, any type of object can be counted in units of moles. One mole denotes 6.02×10^{23} objects; it does not matter what the objects are—even watermelons. Just as we can talk about dozens of watermelons, we can talk about moles of watermelons, although the latter involves a very large watermelon patch.

$$2.25 \text{ moles watermelons} \times \frac{6.02 \times 10^{23} \text{ watermelons}}{1 \text{ mole watermelons}}$$

$$= 1.3545 \times 10^{24} \text{ watermelons} \quad \text{(calculator answer)}$$
$$= 1.35 \times 10^{24} \text{ watermelons} \quad \text{(correct answer)}$$

Practice Exercise 9.5

How many objects are there in each of the following quantities?

(a) 2.67 moles of carbon dioxide (CO_2) molecules
(b) 1.45 moles of sodium chloride (NaCl) formula units

Ans. (a) 1.61×10^{24} CO_2 molecules; (b) 8.73×10^{23} NaCl formula units

It is somewhat unfortunate, because of its similarity to the word molecule, that the name mole was selected as the name for the chemist's counting unit. Students often think that *mole* is an abbreviated form of the word *molecule*. That is not the case. The word mole comes from the Latin *moles*, which means "heap or pile." A mole is a macroscopic amount, a heap or pile of objects, that can easily be seen. A molecule is a particle too small to be seen with the naked eye.

In Example 9.5 we calculated the number of objects present in samples ranging in size from 0.025 mole to 2.53 moles. Our answers were numbers carrying the exponents 10^{22}, 10^{23}, or 10^{24}. Numbers with these exponents are inconceivably large. The magnitude of Avogadro's number itself is so large that it is almost incomprehensible. There is nothing in our experience to relate to it. (When chemists count, they really count in "big" jumps.) Many attempts have been made to create word pictures of the vast size of Avogadro's number. Such pictures, however, really only hint at its magnitude, since other large numbers must be used in the word pictures. Three such pictures are as follows.

1. Suppose a fraternity decided to throw a large pizza party and ordered 1 mole of pizza pies. Two thousand students showed up at the party. It would take these 2000 students, each downing one pizza every 3 minutes, 2×10^{15} years to finish the stack of pizzas. If everyone living on this planet attended the party (5 billion pizza eaters), it would still take 1×10^9 years to eat the pizzas—a billion years of nonstop pizza eating.
2. If each one of the 5 billion people on Earth were made a millionaire (receiving 1 million dollar bills), we would still need 120 million other worlds, each inhabited with the same number of millionaires, in order to have Avogadro's number of dollar bills in circulation. (Where would we put all of the dollar bills?) One mole of dollar bills is a lot of money, enough to pay all the expenses of the United States government for the next billion years or so (in inflated dollars).
3. It would take an ultramodern computer that can count 100 million times a second 190 million years to count 6.02×10^{23} times (Avogadro's number).

9.5 THE MASS OF A MOLE

How much does a mole weigh; that is, what is its mass? Consider a similar (but more familiar) question first: "How much does a dozen weigh?" Your response is now immediate. "A dozen what?" you reply. The mass of a dozen identical objects obviously depends on the identity of the object. For example, the mass of a dozen elephants will be "somewhat greater" than the mass of a dozen marshmallows. The mole, like the dozen, is a counting unit. Similarly, the mass of a mole of objects will depend on the identity of the object. Thus, the mass of a mole, *molar mass*, is not one set number; it varies, being different for each different chemical substance. This is in direct contrast to the *molar number*, Avogadro's number, which is the same for all chemical substances.

The **molar mass of an element**, *when the element is in atomic form, is a mass in grams that is numerically equal to the atomic mass of the element*. Thus, if we know the atomic mass of an element, we also know the mass of 1 mole of atoms of the element. The two quantities are numerically the same, differing only in units. For the elements carbon, oxygen, and sodium we can write the following mass-number relationships.

$$\text{Mass of 1 carbon atom} = 12.011\ \text{amu} \quad \text{(atomic mass)}$$
$$\text{Mass of 1 mole of carbon atoms} = 12.011\ \text{g} \quad \text{(molar mass)}$$

$$\text{Mass of 1 oxygen atom} = 15.9994\ \text{amu} \quad \text{(atomic mass)}$$
$$\text{Mass of 1 mole of oxygen atoms} = 15.9994\ \text{g} \quad \text{(molar mass)}$$

$$\text{Mass of 1 sodium atom} = 22.98977\ \text{amu} \quad \text{(atomic mass)}$$
$$\text{Mass of 1 mole of sodium atoms} = 22.98977\ \text{g} \quad \text{(molar mass)}$$

It is not a coincidence that the mass in grams of a mole of atoms of an element and the element's atomic mass are numerically equal. Avogadro's number has the value that it has in order to cause this relationship to exist. Experimentally it was determined that when 6.02×10^{23} atoms of an element are present, molar masses (in grams) and atomic masses (in amu) are numerically equal. Again, only when the chemist's counting unit has the value 6.02×10^{23} does this relationship hold. Avogadro's number is thus the "connecting link" between the chemist's microscopic mass scale (amu) and macroscopic mass scale (grams). The use of the mass unit grams in specifying the molar mass of an element is a must. The use of other mass units would require a different counting unit than 6.02×10^{23} for a numerical match between atomic mass and molar mass.

The molecular form of an element will have a different molar mass than its atomic form. Consider the element chlorine, which is found in nature in the form of diatomic molecules (Cl_2). The mass of 1 mole of chlorine atoms (Cl) is different from the mass of 1 mole of chlorine molecules (Cl_2). Since there are two atoms in each molecule of chlorine, the molar mass of molecular chlorine is twice the molar mass of atomic chlorine. The following relationships hold for chlorine, whose atomic mass is 35.453 amu.

$$6.02 \times 10^{23} \text{ Cl atoms} = 1 \text{ mole Cl atoms} = 35.453 \text{ g Cl}$$

$$6.02 \times 10^{23} \text{ Cl}_2 \text{ molecules} = 1 \text{ mole Cl}_2 \text{ molecules} = 70.906 \text{ g Cl}_2$$

Note that 1 mole of molecular chlorine contains twice as many atoms as 1 mole of atomic chlorine; however, the number of discrete particles present is the same (Avogadro's number) in both cases. For atomic chlorine, atoms are considered to be the object counted; for molecular chlorine, molecules are considered to be the discrete particle counted. There is the same number of atoms in the former case as there is of molecules in the latter case. This atomic–molecular chlorine situation is analogous to the difference between a dozen shoes and a dozen pairs of shoes. Both the mass and the actual number of shoes for the dozen pairs of shoes are double those for the dozen shoes.

The existence of some elements in molecular form becomes a source of error in chemical calculations if care is not taken to distinguish properly between atomic and molecular forms of the element. The phrase "1 mole of chlorine" is an ambiguous term. Does it mean 1 mole of chlorine atoms (Cl), or does it mean 1 mole of chlorine molecules (Cl_2)?

The **molar mass of a compound** *is a mass in grams that is numerically equal to the formula mass of the compound.* Thus, for compounds a numerical equivalence exists between molar mass and formula mass if the molar mass is specified in grams. When we add atomic masses to get the formula mass (in amu) of a compound, we are simultaneously finding the mass of 1 mole of compound (in grams). The molar mass–formula mass relationships for the compounds water (H_2O), ammonia (NH_3), and barium chloride ($BaCl_2$) are

$$\text{Mass of 1 H}_2\text{O molecule} = 18.0152 \text{ amu} \quad \text{(formula mass)}$$
$$\text{Mass of 1 mole of H}_2\text{O molecules} = 18.0152 \text{ g} \quad \text{(molar mass)}$$

$$\text{Mass of 1 NH}_3 \text{ molecule} = 17.0304 \text{ amu} \quad \text{(formula mass)}$$
$$\text{Mass of 1 mole of NH}_3 \text{ molecules} = 17.0304 \text{ g} \quad \text{(molar mass)}$$

$$\text{Mass of 1 BaCl}_2 \text{ formula unit} = 208.25 \text{ amu} \quad \text{(formula mass)}$$
$$\text{Mass of 1 mole of BaCl}_2 \text{ formula units} = 208.25 \text{ g} \quad \text{(molar mass)}$$

Figure 9.2 pictures molar quantities of a number of common substances. Note again how the mass of a mole varies in numerical value. On the other hand, all of the pictured amounts of substances contain the same number of units—6.02×10^{23} atoms, molecules, or formula units.

It should now be very evident to you why the chemist's counting unit, the mole, has the value it does—6.02×10^{23}. *Avogadro's number represents the experimentally deter-*

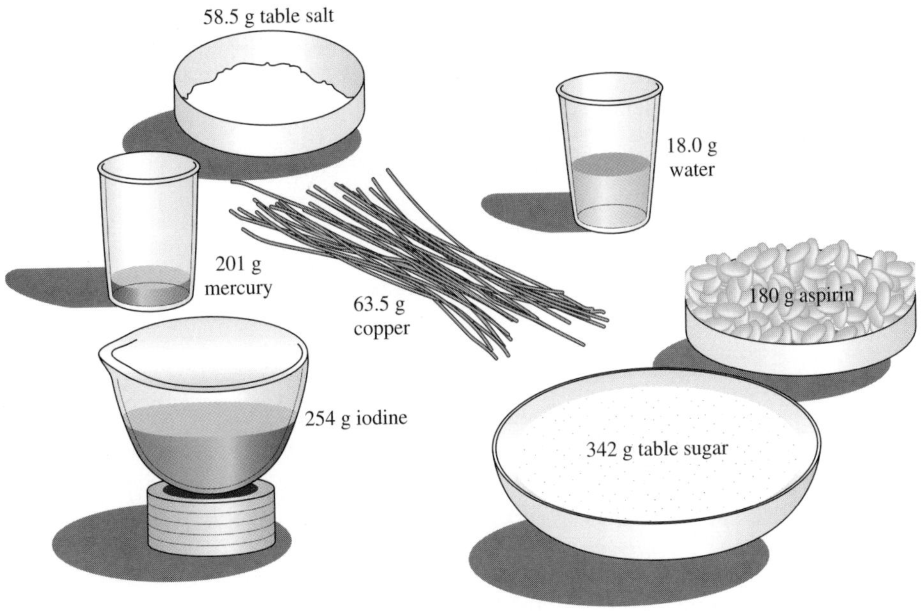

FIGURE 9.2 One mole of a substance is that amount of substance with a mass, in grams, numerically equal to the atomic mass, molecular mass, or formula mass.

mined number of atoms, molecules, or formula units contained in a sample of a pure substance with a mass in grams numerically equal to the atomic mass or formula mass of the pure substance.

The numerical match between molar mass and atomic or formula mass makes the calculation of the mass of any given number of moles of a substance a very simple procedure. In solving problems of this type, the numerical value of the molar mass becomes part of the conversion factor used to convert from moles to grams.

Moles of substance —(molar mass)→ Grams of substance

For example, for the compound CO_2, which has a formula mass of 44.0 amu, we can write the equality

$$44.0 \text{ g } CO_2 = 1 \text{ mole } CO_2$$

From this statement two conversion factors can be written.

$$\frac{44.0 \text{ g } CO_2}{1 \text{ mole } CO_2} \quad \text{and} \quad \frac{1 \text{ mole } CO_2}{44.0 \text{ g } CO_2}$$

Example 9.6 illustrates the use of gram-to-mole conversion factors like these.

Example 9.6

Calculate the mass, in grams, of each of the following quantities of matter.

(a) 1.50 moles of CH_4 molecules

(b) 2.50 moles of NaCl formula units

(c) 1.68 moles of N_2 molecules

(d) 1.68 moles of N atoms

Solution

We will use dimensional analysis to solve each of these problems. The relationship between molar mass and atomic or formula mass will serve as a conversion factor in the setup of each problem.

(a) The given quantity is 1.50 moles of CH_4 molecules, and the desired quantity is grams of CH_4. Thus,

$$1.50 \text{ moles } CH_4 = ? \text{ g } CH_4$$

The formula mass of CH_4 is calculated to be 16.0 amu.

$$\begin{aligned} \text{C:} \quad & 1 \times 12.0 \text{ amu} = 12.0 \text{ amu} \\ \text{4 H:} \quad & 4 \times \ \ 1.0 \text{ amu} = \underline{\ \ 4.0 \text{ amu}} \\ & \qquad\qquad\qquad = 16.0 \text{ amu} \end{aligned}$$

(The formula mass need contain only three significant figures because the given quantity, 1.50 moles, has only three significant figures.) Using the formula mass, we can write the equality

$$16.0 \text{ g } CH_4 = 1 \text{ mole } CH_4$$

The dimensional analysis setup for the problem with the gram-to-mole equation used as a conversion factor, is

$$1.50 \cancel{\text{moles } CH_4} \times \frac{16.0 \text{ g } CH_4}{1 \cancel{\text{mole } CH_4}} = 24 \text{ g } CH_4 \quad \text{(calculator answer)}$$

$$= 24.0 \text{ g } CH_4 \quad \text{(correct answer)}$$

(b) The given quantity is 2.50 moles of NaCl formula units and the desired quantity is grams of NaCl.

$$2.50 \text{ moles NaCl} = ? \text{ g NaCl}$$

The calculated formula mass of NaCl is 58.5 amu. Thus,

$$58.5 \text{ g NaCl} = 1 \text{ mole NaCl}$$

With this relationship as a conversion factor, the setup for the problem becomes

$$2.50 \cancel{\text{moles NaCl}} \times \frac{58.5 \text{ g NaCl}}{1 \cancel{\text{mole NaCl}}} = 146.25 \text{ g NaCl} \quad \text{(calculator answer)}$$

$$= 146 \text{ g NaCl} \quad \text{(correct answer)}$$

(c) The given quantity is 1.68 moles of N_2 molecules. The desired quantity is grams of N_2 molecules. Thus,

$$1.68 \text{ moles } N_2 = ? \text{ g } N_2$$

We are dealing here with diatomic nitrogen molecules (N_2) and not nitrogen atoms. Thus, 28.0 amu, twice the atomic mass of nitrogen, is the formula mass used in the mole-to-gram statement.

$$28.0 \text{ g } N_2 = 1 \text{ mole } N_2$$

With this relationship as a conversion factor, the setup becomes

$$1.68 \text{ moles } N_2 \times \frac{28.0 \text{ g } N_2}{1 \text{ mole } N_2} = 47.04 \text{ g } N_2 \quad \text{(calculator answer)}$$

$$= 47.0 \text{ g } N_2 \quad \text{(correct answer)}$$

(d) The given quantity is 1.68 moles of N atoms, and the desired quantity is grams of N atoms. Thus,

$$1.68 \text{ moles N} = ? \text{ g N}$$

This problem differ from the previous one in that atoms, rather than molecules, of nitrogen are being counted. The atomic mass of nitrogen is 14.0 amu, and the mole-to-gram equality statement is

$$14.0 \text{ g N} = 1 \text{ mole N}$$

With this relationship as a conversion factor, the setup becomes

$$1.68 \text{ moles N} \times \frac{14.0 \text{ g N}}{1 \text{ mole N}} = 23.52 \text{ g N} \quad \text{(calculator answer)}$$

$$= 23.5 \text{ g N} \quad \text{(correct answer)}$$

Notice that the mass of 1.68 moles of N_2 (part c) is twice the mass of 1.68 moles of N (part d). This is as expected; there are twice as many atoms of N in 1.68 moles of N_2 as in 1.68 moles of N.

Practice Exercise 9.6

Calculate the mass, in grams, of each of the following molar quantities of matter.

(a) 1.81 moles of ClF_5 molecules (b) 0.622 mole of O_3 molecules

Ans. (a) 236 g ClF_5; (b) 29.9 g O_3

The atomic mass unit (amu) and the grams (g) unit are related to each other through Avogadro's number.

$$6.02 \times 10^{23} \text{ amu} = 1.00 \text{ g}$$

That the above is the case can be deduced from the following calculation involving the element N.

$$\frac{6.02 \times 10^{23} \text{ atoms N}}{1 \text{ mole N}} \times \frac{1 \text{ mole N}}{14.0 \text{ g N}} \times \frac{14.0 \text{ amu}}{1 \text{ atom N}}$$

$$= 6.02 \times 10^{23} \text{ amu/g} \quad \text{(calculator and correct answer)}$$

Example 9.7

What is the mass, in grams, of a molecule whose mass on the amu scale is 104 amu?

Solution

This is a one-step problem based on the conversion factor

$$6.02 \times 10^{23} \text{ amu} = 1.00 \text{ g}$$

The dimensional analysis setup is

$$104 \cancel{\text{amu}} \times \frac{1.00 \text{ g}}{6.02 \times 10^{23} \cancel{\text{amu}}} = 1.7275747 \times 10^{-22} \text{ g} \quad \text{(calculator answer)}$$

$$= 1.73 \times 10^{-22} \text{ g} \quad \text{(correct answer)}$$

Practice Exercise 9.7

What is the mass in grams of a molecule whose mass on the amu scale is 169 amu?

Ans. 2.81×10^{-22} g

In Section 9.4 we defined the mole simply as

$$1 \text{ mole} = 6.02 \times 10^{23} \text{ objects}$$

Although this statement conveys correct information (the value of Avogadro's number to three significant figures is 6.02×10^{23}), it is not the officially accepted definition for Avogadro's number. The official definition, which is mass based, is: The **mole** *is the amount of substance in a system that contains as many elementary units (atoms, molecules, or formula units) as there are* $^{12}_{6}C$ *atoms in exactly 12.00000 grams of* $^{12}_{6}C$. The value of Avogadro's number is an experimentally determined quantity (the number of atoms in exactly 12.00000 g of $^{12}_{6}C$ atoms) rather than an exactly defined quantity. Its value is not even mentioned in the definition. The most up-to-date experimental value for Avogadro's number is 6.0221367×10^{23}, which is consistent with our previous definition. In calculations we will never need such a precise value as the experimentally determined one. Most often three significant figures will suffice; occasionally four significant figures (6.022×10^{23}) will be needed for a calculation. But remember, more significant figures are available if the need ever arises for their use.

9.6 COUNTING PARTICLES BY WEIGHING

In a laboratory situation it is often necessary to work with equal numbers of atoms of two different substances or twice as many atoms of one type than of another, and so forth. Atoms are so small that it is impossible to count them one by one. How does the chemist resolve this requirement of "equal or proportional numbers of atoms"? The problem is solved by means of the concept of "counting by weighing," a concept closely related to molar mass (Sec. 9.5).

To illustrate the "counting-by-weighing" concept, let us compare the masses of two kinds of atoms—oxygen and nitrogen—when varying but equal numbers of atoms are present. (The choice of elements in the comparison is arbitrary; any two could be used.) From a table of atomic masses we determine that the mass of a single oxygen atom (on the average) is 16.0 amu and that of a single nitrogen atom (on the average) is 14.0 amu. Using these single atom masses and varying the equal numbers of atoms present from 1 to 2 to 5 to 100

TABLE 9.2 MASS RATIOS FOR VARYING EQUAL NUMBERS OF OXYGEN AND NITROGEN ATOMS

Atomic Ratio	Mass Ratio of Equal Numbers of Atoms
$\frac{1 \text{ atom O}}{1 \text{ atom N}}$	$\frac{1 \times 16.00 \text{ amu}}{1 \times 14.00 \text{ amu}} = \frac{16.00}{14.00} = 1.143$
$\frac{2 \text{ atoms O}}{2 \text{ atoms N}}$	$\frac{2 \times 16.00 \text{ amu}}{2 \times 14.00 \text{ amu}} = \frac{32.00}{28.00} = 1.143$
$\frac{5 \text{ atoms O}}{5 \text{ atoms N}}$	$\frac{5 \times 16.00 \text{ amu}}{5 \times 14.00 \text{ amu}} = \frac{80.00}{70.00} = 1.143$
$\frac{100 \text{ atoms O}}{100 \text{ atoms N}}$	$\frac{100 \times 16.00 \text{ amu}}{100 \times 14.00 \text{ amu}} = \frac{1.600 \times 10^3}{1.400 \times 10^3} = 1.143$
$\frac{10^9 \text{ atoms O}}{10^9 \text{ atoms N}}$	$\frac{10^9 \times 16.00 \text{ amu}}{10^9 \times 14.00 \text{ amu}} = \frac{1.600 \times 10^{10}}{1.400 \times 10^{10}} = 1.143$
$\frac{6.022 \times 10^{23} \text{ atoms O}}{6.022 \times 10^{23} \text{ atoms N}}$	$\frac{6.022 \times 10^{23} \times 16.00 \text{ amu}}{6.022 \times 10^{23} \times 14.00 \text{ amu}} = \frac{9.635 \times 10^{24}}{8.431 \times 10^{24}} = 1.143$

to 1 billion to Avogadro's number generates the mass comparisons given in Table 9.2. Note that the mass ratio is identical in every case; also note that the numerical value of this ratio is the same as the ratio of the atomic masses of oxygen and nitrogen. What Table 9.2 establishes is that *the mass ratio of equal numbers of atoms of two elements is the same as the ratio of the atomic masses of the two elements*. This statement is true regardless of the elements involved in the comparison. The ratio between two other elements will not be 1.143 (as was the case for O and N); each pair of elements will have its own unique mass ratio determined by the atomic masses of the elements involved.

Turning the statement of the conclusion derived from Table 9.2 around (reversing it) gives an even more useful generalization (from a laboratory viewpoint): *If samples of two or more elements have mass ratios equal to their atomic mass ratios, they must contain equal numbers of atoms*. Thus, for example, since the atomic mass ratio between Au and Cu is 3.100 (197.0/63.55), any time samples of these elements are weighed out in a 3.100-to-1.000 ratio the samples will contain equal numbers of atoms. A 3.100-g sample of Au contains the same number of atoms as a 1.000-g sample of Cu. Similarly, 3.100-lb and 3.100-ton samples of Au contain, respectively, the same numbers of atoms as 1.000-lb and 1.000-ton samples of Cu. Any time, regardless of units (g, lb, kg, ton, etc.), Au and Cu are present in a 3.100-to-1.000 mass ratio, equal numbers of atoms are present.

Our comparisons (and generalizations) so far have involved atoms and elements. Parallel comparisons (and generalizations) exist for molecules (or formula units) and compounds. The generalizations for compounds differ from those for elements only in that "formula mass" has replaced atomic mass and "molecules" (or formula units) has replaced atoms. The substitution of formula masses for atomic masses is valid because both are based on the same ${}^{12}_{6}C$ scale (Sec. 9.2). Therefore, the following samples all contain the same number of molecules.

18.0 g H_2O (formula mass of H_2O = 18.0 amu)
44.0 g CO_2 (formula mass of CO_2 = 44.0 amu)
64.1 g SO_2 (formula mass of SO_2 = 64.1 amu)

We can now make the following generalization, which is applicable to both elements and compounds for **counting by weighing:** *samples of two or more pure substances (elements or compounds) found to have mass ratios equal to the ratios of their atomic or formula masses must contain identical numbers of particles (atoms, molecules, or formula units).* The following samples therefore contain the same numbers of particles.

27.0 g Al (atomic mass = 27.0 amu)
18.0 g H_2O (formula mass = 18.0 amu)

The particles for water are molecules, and the particles for aluminum are atoms.

We see, therefore, that to obtain samples of elements or compounds containing equal numbers of particles, we merely weigh out quantities (in any units) whose mass ratio is numerically equal to the ratio of the substance's atomic or formula masses.

Example 9.8

Two different hydrogen–oxygen compounds are known: water (H_2O) and hydrogen peroxide (H_2O_2). Using the concept of "counting particles by weighing," determine whether equal numbers of molecules of these two compounds are present in the following pairs of samples.

(a) 14.00 g H_2O_2 and 8.00 g H_2O

(b) 3.512 g H_2O_2 and 1.860 g H_2O

Solution

The formula mass of H_2O_2 is 34.02 amu, and that of H_2O 18.02 amu. The ratio of formula masses for these two compounds is

$$\frac{\text{Formula mass } H_2O_2}{\text{Formula mass } H_2O} = \frac{34.02 \text{ amu}}{18.02 \text{ amu}} = 1.8879023 \quad \text{(calculator answer)}$$

$$= 1.888 \quad \text{(correct answer)}$$

The ratio of masses for the two samples in each pair must equal this number, 1.888, if equal numbers of molecules are present.

(a) The ratio of the masses of the two samples is

$$\frac{\text{Mass } H_2O_2}{\text{Mass } H_2O} = \frac{14.00 \not{g}}{8.00 \not{g}} = 1.75 \quad \text{(calculator and correct answer)}$$

Since this ratio is not equal to 1.888, equal numbers of molecules are not present. The fact that the ratio is less than 1.888 indicates that there are fewer H_2O_2 molecules (the numerator of the ratio) than H_2O molecules.

(b) The ratio of the masses of the two samples is

$$\frac{\text{Mass } H_2O_2}{\text{Mass } H_2O} = \frac{3.512 \not{g}}{1.860 \not{g}} = 1.888172 \quad \text{(calculator answer)}$$

$$= 1.888 \quad \text{(correct answer)}$$

This time the formula mass ratio and the mass ratio are the same (to four significant figures). Thus, equal numbers of molecules are present (to four significant figures).

Practice Exercise 9.8

Carbon monoxide (CO) and nitrogen monoxide (NO) are air pollutants that enter the atmosphere in automobile exhaust. Using the concept of "counting particles by weighing," determine whether equal numbers of molecules of these two pollutants are present in the following pairs of samples.

(a) 10.00 g CO and 11.00 g NO (b) 7.011 g CO and 7.511 g NO

Ans. (a) different numbers of molecules present; (b) same number of molecules present

9.7 THE MOLE AND CHEMICAL FORMULAS

A chemical formula has two meanings or interpretations: (1) a microscopic level interpretation and (2) a macroscopic level interpretation.

The first of these two interpretations was discussed in Section 5.4. At a **microscopic level** *a chemical formula indicates the number of atoms of each element present in one molecule or formula unit of a substance. The subscripts in the formula are interpreted to mean the numbers of atoms of the various elements present in one unit of the substance.* The formula C_2H_6, interpreted at the microscopic level, conveys the information that two atoms of C and six atoms of H are present in one molecule of C_2H_6.

Now that the mole concept has been introduced, a macroscopic interpretation of formulas is possible. At a **macroscopic level** a *chemical formula indicates the number of moles of atoms of each element present in 1 mole of a substance. The subscripts in the formula are interpreted to mean the numbers of moles of atoms of the various elements present in 1 mole of the substance.* The designation "macroscopic" is given to this molar interpretation, since moles are "laboratory-sized" quantities of atoms. The formula C_2H_6, interpreted at the macroscopic level, conveys the information that 2 moles of C atoms and 6 moles of H atoms are present in 1 mole of C_2H_6.

It is now evident, then, that the subscripts in a formula always carry a dual meaning: "atom" at the microscopic level and "moles of atoms" at the macroscopic level.

The validity of the molar interpretation for subscripts in a formula derives from the following line of reasoning. In x molecules of C_2H_6, where x is any number, there are $2x$ atoms of C and $6x$ atoms of H. Regardless of the value of x, there must always be two times as many C atoms as molecules and six times as many H atoms as molecules; that is,

$$\begin{aligned} \text{Number of } C_2H_6 \text{ molecules} &= x \\ \text{Number of C atoms} &= 2x \\ \text{Number of H atoms} &= 6x \end{aligned}$$

Now let x equal 6.02×10^{23}, the value of Avogadro's number. With this x value, the following statements are true.

$$\text{Number of } C_2H_6 \text{ molecules} = 6.02 \times 10^{23}$$

$$\begin{aligned} \text{Number of C atoms} &= 2 \times 6.02 \times 10^{23} = 1.204 \times 10^{24} && \text{(calculator answer)} \\ &= 1.20 \times 10^{24} && \text{(correct answer)} \\ \text{Number of H atoms} &= 6 \times 6.02 \times 10^{23} = 3.612 \times 10^{24} && \text{(calculator answer)} \\ &= 3.61 \times 10^{24} && \text{(correct answer)} \end{aligned}$$

Since 6.02×10^{23} is equal to 1 mole, 1.20×10^{24} to 2 moles, and 3.61×10^{24} to 6 moles, these statements may be changed to read

$$\text{Number of } C_2H_6 \text{ molecules} = 1 \text{ mole}$$
$$\text{Number of C atoms} = 2 \text{ moles}$$
$$\text{Number of H atoms} = 6 \text{ moles}$$

Thus, the mole ratio is the same as the subscript ratio: 2 to 6.

In calculations where the moles of a particular element within a compound are asked for, the subscript of that particular element in the chemical formula of the compound becomes part of the conversion factor used to convert from moles of compound to moles of element within the compound.

[Moles of compound] —formula subscript→ [Moles of element within compound]

For example, again using C_2H_6 as our chemical formula, we can write the following conversion factors.

For C: $\dfrac{2 \text{ moles C atoms}}{1 \text{ mole } C_2H_6 \text{ molecules}}$ or $\dfrac{1 \text{ mole } C_2H_6 \text{ molecules}}{2 \text{ moles C atoms}}$

For H: $\dfrac{6 \text{ moles H atoms}}{1 \text{ mole } C_2H_6 \text{ molecules}}$ or $\dfrac{1 \text{ mole } C_2H_6 \text{ molecules}}{6 \text{ moles H atoms}}$

Example 9.9 illustrates the use of conversion factors of this type in a problem-solving context.

Example 9.9

The characteristic odor of pineapple is due to ethyl butyrate, a compound with the formula $C_6H_{12}O_2$. How many moles of each type of atom present in $C_6H_{12}O_2$ are contained in a 2.65 mole sample of this compound?

Solution

The formula $C_6H_{12}O_2$ specifies that 1 mole of this substance will contain 6 moles of carbon atoms, 12 moles of hydrogen atoms, and 2 moles of oxygen atoms. This information leads to the following conversion factors:

$$\frac{6 \text{ moles C atoms}}{1 \text{ mole } C_6H_{12}O_2} \qquad \frac{12 \text{ moles H atoms}}{1 \text{ mole } C_6H_{12}O_2} \qquad \frac{2 \text{ moles O atoms}}{1 \text{ mole } C_6H_{12}O_2}$$

Using the first of these conversion factors, the moles of C atoms present are calculated as follows:

$$2.65 \cancel{\text{moles } C_6H_{12}O_2} \times \frac{6 \text{ moles C atoms}}{1 \cancel{\text{mole } C_6H_{12}O_2}} = 15.9 \text{ moles C atoms}$$

(calculator and correct answer)

Similarly, using the second conversion factor, the moles of H atoms present are calculated.

$$2.65 \cancel{\text{moles } C_6H_{12}O_2} \times \frac{12 \text{ moles H atoms}}{1 \cancel{\text{mole } C_6H_{12}O_2}} = 31.8 \text{ moles H atoms}$$

(calculator and correct answer)

Finally, using the third conversion factor, we obtain the moles of O atoms present.

$$2.65 \text{ moles } C_6H_{12}O_2 \times \frac{2 \text{ moles O atoms}}{1 \text{ mole } C_6H_{12}O_2} = 5.3 \text{ moles O atoms} \quad \text{(calculator answer)}$$

$$= 5.30 \text{ moles O atoms} \quad \text{(correct answer)}$$

If the question "How many total moles of atoms are present?" were asked, we could obtain the answer by adding the moles of C, H, and O atoms just calculated.

$$(15.9 + 31.8 + 5.30) \text{ moles atoms} = 53 \text{ moles atoms} \quad \text{(calculator answer)}$$

$$= 53.0 \text{ moles atoms} \quad \text{(correct answer)}$$

Alternatively, by noting that there are a total of 20 moles of atoms present in 1 mole of $C_6H_{12}O_2$ (the sum of the subscripts in the formula), we could calculate the total moles of atoms present using the following setup.

$$2.65 \text{ moles } C_6H_{12}O_2 \times \frac{20 \text{ moles of atoms}}{1 \text{ mole } C_6H_{12}O_2} = 53 \text{ moles of atoms} \quad \text{(calculator answer)}$$

$$= 53.0 \text{ moles of atoms} \quad \text{(correct answer)}$$

Practice Exercise 9.9

How many moles of each type of atom are present in each of the following molar quantities?

(a) 0.753 mole of CO_2 molecules (b) 1.31 moles of P_4O_{10} molecules

Ans. (a) 0.753 mole of C atoms and 1.51 moles of O atoms; (b) 5.24 moles of P atoms and 13.1 moles of O atoms

9.8 THE MOLE AND CHEMICAL CALCULATIONS

In this section we combine the major points we've learned about moles in previous sections to produce a general approach to problem solving that is applicable to a variety of types of chemical calculations.

The three quantities most often calculated in chemical problems are

1. The number of *particles* of a substance, that is, the number of atoms, molecules, or formula units.
2. The number of *moles* of a substance.
3. The number of *grams* (mass) of a substance.

These quantities are interrelated. The conversion factors dealing with these relationships, as previously noted, involve the concepts of (1) Avogadro's number, (2) molar mass, and (3) molar interpretation of chemical formula subscripts.

1. Avogadro's number (Sec. 9.4) provides a relationship between the number of particles of a substance and the number of moles of the same substance.

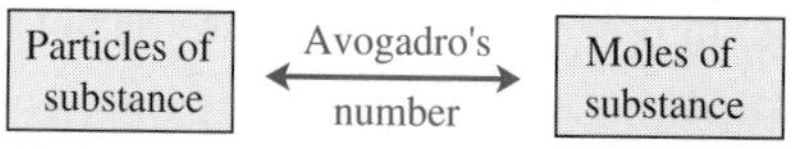

2. Molar mass (Sec. 9.5) provides a relationship between the number of grams of a substance and the number of moles of the same substance.

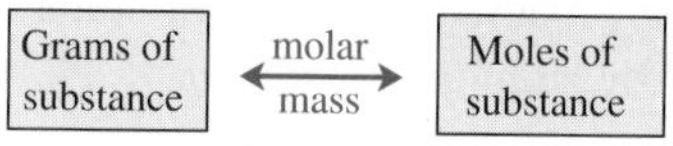

3. Molar interpretation of chemical formula subscripts (Sec. 9.7) provides a relationship between the number of moles of a substance and the number of moles of its component parts.

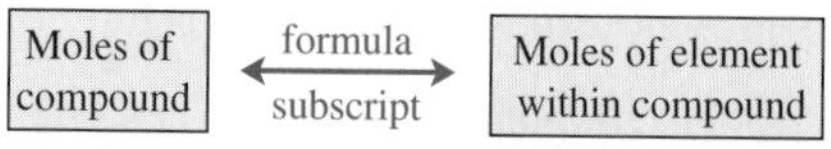

The preceding three concepts can be combined into a single diagram that is very useful in problem solving. This diagram, Figure 9.3, can be viewed as a "road map" from which conversion factor sequences (pathways) can be obtained. It gives all the needed relationships for solving two general types of problems.

1. Calculations for which information (moles, grams, particles) is given about a particular substance, and additional information (moles, grams, particles) is needed concerning the *same* substance.
2. Calculations for which information (moles, grams, particles) is given about a particular substance, and information (moles, grams, particles) is needed concerning a *component* of that same substance.

For the first type of problem, only the left side of Figure 9.3 (the A boxes) is needed. For problems of the second type, both sides of the diagram (both A and B boxes) are used.

The thinking pattern needed to use Figure 9.3 is very simple.

1. Determine which box in the diagram represents the *given* quantity in the problem.
2. Next, locate the box that represents the *desired* quantity.
3. Finally, follow the indicated pathway that takes you from the *given* quantity to the *desired* quantity. This involves simply following the arrows. There will always be only one pathway possible for the needed transition.

Examples 9.10 to 9.13 illustrate a few of the types of problems that can be solved using the relationships shown in Figure 9.3. In the first two examples we will need only the A side

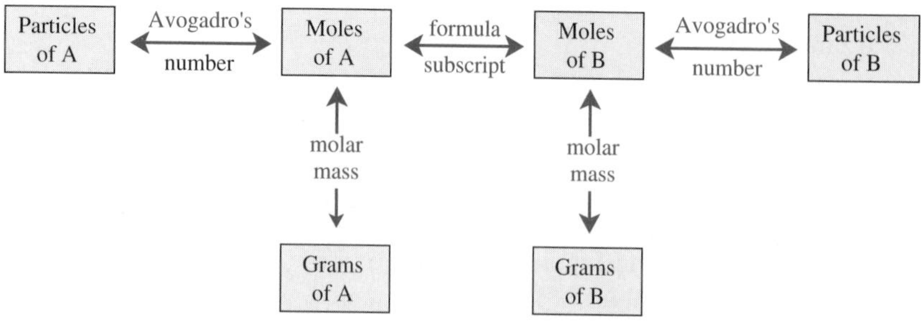

FIGURE 9.3 Useful relationships for solving chemical-formula-based problems.

of the diagram; the following two examples make use of both the A and B sides of the diagram.

Example 9.10

Nicotine, the second most widely used central nervous system stimulant in our society (caffeine is first) occurs naturally in tobacco leaves. Its chemical formula is $C_{10}H_{14}N_2$. How many nicotine molecules are present in a 0.0015-g sample of nicotine (a typical amount in a cigarette)?

Solution

We will solve this problem by using the three steps of dimensional analysis (Sec. 3.7) and Figure 9.3.

STEP 1 The given quantity is 0.0015 g of $C_{10}H_{14}N_2$, and the desired quantity is molecules of $C_{10}H_{14}N_2$.

$$0.0015 \text{ g } C_{10}H_{14}N_2 = ? \text{ molecules } C_{10}H_{14}N_2$$

In terms of Figure 9.3. this is a "grams of A" to "particles of A" problem. We are given grams of substance A and desire to find particles (molecules) of that same substance.

STEP 2 Figure 9.3 gives us the "pathway" (sequence of conversion factors) needed to work the problem. We want to start in the "grams of A" box and end up in the "particles of A" box. The pathway is

Grams of A $\xrightarrow[\text{mass}]{\text{molar}}$ Moles of A $\xrightarrow[\text{number}]{\text{Avogadro's}}$ Particles of A

Using dimensional analysis, the setup for this sequence of conversion factors is

$$0.0015 \cancel{\text{g } C_{10}H_{14}N_2} \times \frac{1 \cancel{\text{ mole } C_{10}H_{14}N_2}}{162 \cancel{\text{ g } C_{10}H_{14}N_2}} \times \frac{6.02 \times 10^{23} \text{ molecules } C_{10}H_{14}N_2}{1 \cancel{\text{ mole } C_{10}H_{14}N_2}}$$

grams A $\longrightarrow$ moles A $\longrightarrow$ particles A

The number 162 that was used in the first conversion factor is the formula mass of $C_{10}H_{14}N_2$. It is not given in the problem, but had to be calculated using atomic masses.

STEP 3 The solution to the problem, obtained by doing the arithmetic, is

$$\frac{0.0015 \times 6.02 \times 10^{23}}{162 \times 1} \text{ molecules } C_{10}H_{14}N_2$$

$$= 5.574074 \times 10^{18} \text{ molecules } C_{10}H_{14}N_2 \quad \text{(calculator answer)}$$

$$= 5.6 \times 10^{18} \text{ molecules } C_{10}H_{14}N_2 \quad \textbf{(correct answer)}$$

Practice Exercise 9.10

The koala bear feeds exclusively on eucalyptus leaves. Its digestive system detoxifies the eucalyptus oil, a poison to other animals. The predominant compound in eucalyptus oil is eucalyptol, which has the formula $C_{10}H_{18}O$. How many eucalyptol molecules are present in a 2.00-g sample of this compound?

Ans. 7.82×10^{21} molecules $C_{10}H_{18}O$

Example 9.11

The "starting material" for the production of photochemical smog is the air pollutant nitrogen dioxide (NO_2). Its interaction with ultraviolet light from the sun produces the chemical species needed for smog production. What would be the mass, in grams, of an NO_2 sample in which 100 billion (1.000×10^{11}) molecules are present?

Solution

STEP 1 The given quantity is 1.000×10^{11} NO_2 molecules. The desired quantity is grams of NO_2.

$$1.000 \times 10^{11} \text{ NO}_2 \text{ molecules} = ? \text{ g NO}_2$$

In terms of Figure 9.3, this is a "particles of A" to "grams of A" problem.

STEP 2 The pathway for this problem is the exact reverse of the one used in the previous example. We are given particles and asked to find grams of the same substance.

Particles of A —(Avogadro's number)→ Moles of A —(molar mass)→ Grams of A

Using dimensional analysis, the setup is

$$1.000 \times 10^{11} \cancel{\text{NO}_2 \text{ molecules}} \times \frac{1 \cancel{\text{mole NO}_2}}{6.022 \times 10^{23} \cancel{\text{NO}_2 \text{ molecules}}} \times \frac{46.01 \text{ g NO}_2}{1 \cancel{\text{mole NO}_2}}$$

particles A ⟶ moles A ⟶ grams A

The molar mass of NO_2, 46.01 g, which was used in the second conversion factor, was calculated from the atomic masses of nitrogen and oxygen. Avogadro's number in the first conversion factor was specified to four significant figures since the given number 1.000×10^{11} has four significant figures.

STEP 3 The final answer is obtained by doing the arithmetic.

$$\frac{1.000 \times 10^{11} \times 1 \times 46.01}{6.022 \times 10^{23} \times 1} \text{ g NO}_2$$

$= 7.6403118 \times 10^{-12}$ g NO_2 (calculator answer)
$= 7.640 \times 10^{-12}$ g NO_2 (**correct answer**)

Practice Exercise 9.11

The artifical sweetener aspartame (Nutra-Sweet), with the chemical formula $C_{14}H_{18}N_2O_5$, is the sugar substitute used in most diet soft drinks. What would be the mass in grams of a pure sample of this sweetener that contains 1 billion (1.000×10^9) molecules?

Ans. 4.887×10^{-13} g $C_{14}H_{18}N_2O_5$

Example 9.12

The industrial explosive TNT (trinitrotoluene) has the chemical formula $C_7H_5O_6N_3$. How many grams of carbon are present in a 125-g sample of TNT?

Solution

STEP 1 There is an important difference between this problem and the preceding two; here we are dealing with not one but two substances, TNT and carbon. The given quantity is grams of TNT (substance A) and we are asked to find the grams of carbon (substance B). This is a "grams of A" to "grams of B" problem.

$$125 \text{ g TNT} = ? \text{ g C}$$

STEP 2 The appropriate set of conversions for a "grams of A" to "grams of B" problem, from Figure 9.3, is

$$\boxed{\text{Grams of A}} \xrightarrow[\text{mass}]{\text{molar}} \boxed{\text{Moles of A}} \xrightarrow[\text{subscript}]{\text{formula}} \boxed{\text{Moles of B}} \xrightarrow[\text{mass}]{\text{molar}} \boxed{\text{Grams of B}}$$

The mathematical setup involving conversion factors is

$$125 \text{ g } C_7H_5O_6N_3 \times \frac{1 \text{ mole } C_7H_5O_6N_3}{227 \text{ g } C_7H_5O_6N_3} \times \frac{7 \text{ moles C}}{1 \text{ mole } C_7H_5O_6N_3} \times \frac{12.0 \text{ g C}}{1 \text{ mole C}}$$

$$\text{grams A} \longrightarrow \text{moles A} \longrightarrow \text{moles B} \longrightarrow \text{grams B}$$

The number 227 that is used in the first conversion factor is the formula mass for TNT. The conversion from "moles of A" to "moles of B" uses a conversion factor derived from the information contained in the formula for TNT. One mole of TNT contains 7 moles of C because the subscript for C in the formula is 7. The number 12.0 in the final conversion factor is the atomic mass of carbon.

STEP 3 Collecting the numbers from the various conversion factors together and doing the arithmetic gives us our answer.

$$\frac{125 \times 1 \times 7 \times 12.0}{227 \times 1 \times 1} \text{ g C} = 46.255506 \text{ g C} \quad \text{(calculator answer)}$$

$$= 46.3 \text{ g C} \quad \text{(correct answer)}$$

Practice Exercise 9.12

Glucose, $C_6H_{12}O_6$, is one of the main sources of energy used by living organisms. How many grams of oxygen are present in 10.0 g of glucose?

Ans. 5.33 g O

Example 9.13

The compound cholesterol, $C_{27}H_{46}O$, is necessary for human life. It is found in cell membranes, nerve tissue, and brain tissue. How many hydrogen atoms would be present in a 2.000 g sample of cholesterol?

Solution

STEP 1 The given quantity is 2.000 g of $C_{27}H_{46}O$ (cholesterol) and the desired quantity is atoms of hydrogen.

$$2.000 \text{ g } C_{27}H_{46}O = \text{? atoms H}$$

In the jargon of Figure 9.3, this is a "grams of A" to "particles of B" problem.

STEP 2 From Figure 9.3, the appropriate pathway for solving this problem is

Grams of A —(molar mass)→ Moles of A —(formula subscript)→ Moles of B —(Avogadro's number)→ Particles of B

Using dimensional analysis, the conversion factor setup becomes

$$2.000 \cancel{\text{ g } C_{27}H_{46}O} \times \frac{1 \cancel{\text{ mole } C_{27}H_{46}O}}{386.0 \cancel{\text{ g } C_{27}H_{46}O}} \times \frac{46 \cancel{\text{ moles H}}}{1 \cancel{\text{ mole } C_{27}H_{46}O}} \times \frac{6.022 \times 10^{23} \text{ atoms H}}{1 \cancel{\text{ mole H}}}$$

grams A ⟶ moles A ⟶ moles B ⟶ particles B

The numbers in the second conversion factor (46 and 1) were obtained from the chemical formula for cholesterol; they are, thus, exact numbers. One mole of $C_{27}H_{46}O$ contains 46 moles of hydrogen; the hydrogen subscript is 46. Other numbers in the setup must be specified to four significant figures since the given number 2.000 g contains four significant figures.

STEP 3 The solution to the problem, obtained by doing the arithmetic, is

$$\frac{2.000 \times 1 \times 46 \times 6.022 \times 10^{23}}{386.0 \times 1 \times 1} \text{ atoms H} = 1.4352953 \times 10^{23} \text{ atoms H}$$

(calculator answer)

$$= 1.435 \times 10^{23} \text{ atoms H}$$

(correct answer)

Practice Exercise 9.13

Ethylene glycol, the major ingredient in most automobile antifreeze solutions, has the formula $C_2H_6O_2$. How many atoms of oxygen are present in a 2.000-g sample of ethylene glycol?

Ans. 3.880×10^{22} atoms O

9.9 EMPIRICAL AND MOLECULAR FORMULAS

Chemical formulas provide a great deal of useful information about the substances they represent and, as seen in previous sections of this chapter, are key entities in many types of calculations. How are formulas themselves determined? They are determined by calculation from experimentally obtained information.

TABLE 9.3 A COMPARISON OF EMPIRICAL AND MOLECULAR FORMULAS FOR SELECTED COMPOUNDS

Compound	Empirical Formula	Molecular Formula	Whole Number Multiplier
Dinitrogen tetrafluoride	NF_2	N_2F_4	2
Hydrogen peroxide	HO	H_2O_2	2
Sodium chloride	NaCl	NaCl	1
Benzene	CH	C_6H_6	6

Depending on the amount of experimental information available, two types of chemical formulas may be obtained: an empirical formula or a molecular formula. The **empirical formula** *(or simplest formula) gives the* smallest *whole number ratio of atoms present in a formula unit of a compound.* In empirical formulas the subscripts in the formula cannot be reduced to a simpler set of numbers by division with a small integer. The **molecular formula** *(or true formula) gives the* actual *number of atoms present in a formula unit of a compound.*

For ionic compounds the empirical and molecular formulas are almost always the same; that is, the actual ratio of atoms present in a formula unit and the smallest ratio of atoms present are the same. For molecular compounds the two types of formulas may be the same, but frequently they are not. When they are not the same, the molecular formula is a multiple of the empirical formula.

$$\text{Molecular formula} = \text{whole number multiplier} \times \text{empirical formula}$$

Table 9.3 contrasts the differences between the two types of formulas for selected compounds.

Example 9.14

Write the empirical formula for each of the following compounds.

(a) C_2H_4 (ethene, used to make polyethylene)

(b) C_8H_{18} (octane, a component of gasoline)

(c) CH_4O (methanol, an industrial solvent)

(d) $C_6H_{12}O_6$ (glucose, often called blood sugar)

Solution

(a) Each of the formula subscripts is divided by 2 to give the empirical formula CH_2.

(b) Each of the formula subscripts is divided by 2 to give the empirical formula C_4H_9.

(c) There is no small number that will divide evenly into each of the formula subscripts other than 1. Thus, the empirical formula is the same as the given formula: CH_4O.

(d) Each of the formula subscripts is divided by 6 to give the empirical formula CH_2O.

Practice Exercise 9.14

Write the empirical formula for each of the following compounds.

(a) P_4O_{10} (b) $C_3H_{10}N_2$ (c) C_2H_6O (d) $B_3N_3H_6$

Ans. (a) P_2O_5; (b) $C_3H_{10}N_2$; (c) C_2H_6O; (d) BNH_2

9.10 DETERMINATION OF EMPIRICAL FORMULAS

The determination of a compound's chemical formula from experimental data is usually carried out in two calculational steps: (1) elemental composition data are used to determine the compound's empirical formula (simplest ratio of atoms present—Sec. 9.9), and (2) this empirical formula and molecular mass data are used to determine the compound's molecular formula (actual ratio of atoms present—Sec. 9.9).

In this section we consider the first of these two steps and then in Sec. 9.11 we learn the additional procedures needed to go from an empirical formula to a molecular formula.

Elemental composition data are sufficient for an empirical formula calculation. Such data may be in the form of percent composition or simply the mass of each element present in a known mass of compound. Example 9.15 shows the steps used in obtaining an empirical formula from percent composition data. Example 9.16 uses the mass of the elements present in a compound sample of known mass to obtain an empirical formula.

Example 9.15

The compound Freon-12 was widely used in automobile air conditioning systems until it was found that it can cause damage to the ozone layer in the upper atmosphere. Freon-12 use for this purpose is now being phased out. Percent composition data for Freon-12 is 9.90% carbon, 58.6% chlorine, and 31.5% fluorine. Based on these data, what is the empirical formula of Freon-12?

Solution

The problem of calculating the empirical formula of a compound from percent composition data can be broken down into three steps.

1. Determine the number of grams of each element in a sample of the compound.
2. Convert the grams of each element to moles of element.
3. Express the mole ratio between the elements in terms of *small whole numbers*.

STEP 1 Mass percentage values are independent of sample size; that is, they apply to samples of all sizes. In working with mass percentages in empirical formula calculations, it is convenient to assume a 100.0-g sample size. When this is done, mass percentages translate directly into gram amounts. The mass of each element in the 100.0-g sample is numerically equal to the percentage value.

C: 9.90% of 100.0 g = 9.90 g
Cl: 58.6% of 100.0 g = 58.6 g
F: 31.5% of 100.0 g = 31.5 g

STEP 2 We next convert the grams (from step 1) to moles. We need moles information in order to determine the subscripts in the formula of the compound. Formula subscripts give the ratio of the number of moles of each element present in a compound (Sec. 9.7).

$$9.90\ \cancel{\text{g C}} \times \frac{1\ \text{mole C}}{12.0\ \cancel{\text{g C}}} = 0.825\ \text{mole C} \quad \text{(calculator and correct answer)}$$

$$58.6\ \cancel{\text{g Cl}} \times \frac{1\ \text{mole Cl}}{35.5\ \cancel{\text{g Cl}}} = 1.6507042\ \text{moles Cl} \quad \text{(calculator answer)}$$

$$= 1.65\ \text{moles Cl} \quad \text{(correct answer)}$$

$$31.5\ \cancel{\text{g F}} \times \frac{1\ \text{mole F}}{19.0\ \cancel{\text{g F}}} = 1.6578947\ \text{moles F} \quad \text{(calculator answer)}$$

$$= 1.66\ \text{moles F} \quad \text{(correct answer)}$$

Thus, in 100.0 g of compound there are 0.825 mole of C, 1.65 moles of Cl, and 1.66 moles of F.

STEP 3 The subscripts in a formula are expressed as whole numbers, not as decimals. To obtain whole numbers from the decimals of step 2, each of the numbers is divided by the smallest of the numbers.

$$\text{C:} \quad \frac{0.825\ \cancel{\text{mole}}}{0.825\ \cancel{\text{mole}}} = 1 \quad \text{(calculator answer)}$$

$$= 1.00 \quad \text{(correct answer)}$$

$$\text{Cl:} \quad \frac{1.65\ \cancel{\text{moles}}}{0.825\ \cancel{\text{mole}}} = 2 \quad \text{(calculator answer)}$$

$$= 2.00 \quad \text{(correct answer)}$$

$$\text{F:} \quad \frac{1.66\ \cancel{\text{moles}}}{0.825\ \cancel{\text{mole}}} = 2.0121212 \quad \text{(calculator answer)}$$

$$= 2.01 \quad \text{(correct answer)}$$

Thus the ratio of carbon to chlorine to fluorine in this compound is 1 to 2 to 2, and the empirical formula is CCl_2F_2.

The calculated value for one of the three formula subscripts (2.01) has a nonzero digit to the right of the decimal place. This is because of experimental error in the original mass percentages. This calculated value is, however, sufficiently close to a whole number that we easily recognize what the whole number is. If experimental data having four significant figures instead of three had been used, agreement with whole numbers may have been closer.

Practice Exercise 9.15

Acetone is the chemical solvent present in fingernail polish remover. Its percent composition is 62.0% carbon, 10.4% hydrogen, and 27.5% oxygen. What is the empirical formula for the compound?

Ans. C_3H_6O

Example 9.16

Analysis of a sample of ibuprofen, the active ingredient in Advil, shows that a 10.00-g sample contains 7.56 g of carbon, 0.88 g of hydrogen, and 1.55 g of oxygen. Use these data to calculate the empirical formula of ibuprofen.

Solution

Having data given in grams instead of as percent composition (Example 9.14) simplifies the calculation of an empirical formula. The grams are changed to moles, and we are ready to find the smallest whole number ratio between elements.

STEP 1 Each of the given gram amounts is converted to moles using dimensional analysis.

$$7.56\ \cancel{\text{g C}} \times \frac{1\ \text{mole C}}{12.0\ \cancel{\text{g C}}} = 0.63\ \text{mole C} \quad \text{(calculator answer)}$$

$$= 0.630\ \text{mole C} \quad \text{(correct answer)}$$

$$0.88\ \cancel{\text{g H}} \times \frac{1\ \text{mole H}}{1.0\ \cancel{\text{g H}}} = 0.88\ \text{mole H} \quad \text{(calculator and correct answer)}$$

$$1.55\ \cancel{\text{g O}} \times \frac{1\ \text{mole O}}{16.0\ \cancel{\text{g O}}} = 0.096875\ \text{mole O} \quad \text{(calculator answer)}$$

$$= 0.0969\ \text{mole O} \quad \text{(correct answer)}$$

STEP 2 Dividing the mole quantities from step 1 by the smallest of the three numbers gives

$$\text{C:} \quad \frac{0.630\ \cancel{\text{mole}}}{0.0969\ \cancel{\text{mole}}} = 6.5015479 \quad \text{(calculator answer)}$$

$$= 6.50 \quad \text{(correct answer)}$$

$$\text{H:} \quad \frac{0.88\ \cancel{\text{mole}}}{0.0969\ \cancel{\text{mole}}} = 8.9783281 \quad \text{(calculator answer)}$$

$$= 9.0 \quad \text{(correct answer)}$$

$$\text{O:} \quad \frac{0.0969\ \cancel{\text{mole}}}{0.0969\ \cancel{\text{mole}}} = 1 \quad \text{(calculator answer)}$$

$$= 1.00 \quad \text{(correct answer)}$$

Frequently all whole numbers or near-whole numbers are obtained at this point (as was the case in Example 9.14) and the calculation is finished.

Sometimes, however, as is the case in this example, we obtain one or more numbers that are not even close to being whole numbers. The number we obtained for carbon, 6.50, is such a number. What do we do?

The number 6.50 is recognizable as being close to a simple fraction; it is $6\frac{1}{2}$. When we obtain simple fractions, we clear them by multiplying *all of the numbers* in our set of numbers by a common factor. We multiply all numbers by 2 if we are dealing with halves, 3 if we are dealing with thirds, 4 if we are dealing with fourths, and so on. Following this procedure, we obtain for the situation in this example the following:

$$\begin{aligned} \text{C:}\quad & 6\tfrac{1}{2} \times 2 = 13 \\ \text{H:}\quad & 9 \times 2 = 18 \\ \text{O:}\quad & 1 \times 2 = 2 \end{aligned}$$

We now have a whole number ratio. The empirical formula of ibuprofen is $C_{13}H_{18}O_2$.

The fractions that most commonly occur in empirical formula calculations are fourths, 0.25 ($\frac{1}{4}$) and 0.75 ($\frac{3}{4}$), which are multiplied by 4 to clear; thirds, 0.33 ($\frac{1}{3}$) and 0.67 ($\frac{2}{3}$), which are multiplied by 3 to clear; and half, 0.50 ($\frac{1}{2}$), which is multiplied by 2 to clear.

Practice Exercise 9.16

The tip of the head of a strike-anywhere match contains a phosphorus–sulfur compound that readily ignites when drawn over a rough surface. What is the empirical formula of this ignitable compound given that a 1.0000-g sample of the compound contains 0.5629 g of P and 0.4371 g of S?

Ans. P_4S_3

Examples 9.15 and 9.16 suggest that compounds are always broken down completely into their elements when they are analyzed. This is not always the case. Elemental analysis often involves changing the compound to be analyzed into other compounds that contain the elements present in the original compound. A common procedure of this type is *combustion analysis*. This procedure is used extensively for compounds containing carbon, hydrogen, and oxygen (or just carbon and hydrogen). The basic idea in combustion analysis is as follows. A sample of the compound is burned completely in pure oxygen. The products of the reaction are CO_2 and H_2O. All carbon atoms originally present end up in CO_2 molecules, and all hydrogen atoms originally present end up in H_2O molecules (see Fig. 9.4). From the

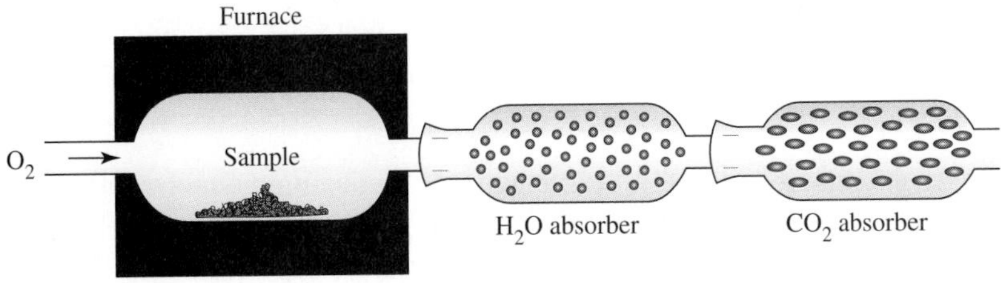

FIGURE 9.4 Combustion analysis method for determining the percentages of carbon and hydrogen in a compound. A weighed sample of the compound is burned in a stream of oxygen, producing gaseous CO_2 and H_2O. These gases then pass through a series of tubes. One tube contains a substance that absorbs H_2O; the substance in another tube absorbs CO_2. By comparing the masses of these tubes before and after the reaction, the analyst can determine the masses of hydrogen and carbon present in the compound that was burned.

mass of CO_2 and H_2O produced, it is possible to calculate the empirical formula of the original compound. Example 9.17 shows how this is done for a carbon–hydrogen compound and Example 9.18 deals with the same situation for a carbon–hydrogen–oxygen compound.

Example 9.17

Cyclopropane, a compound that contains only carbon and hydrogen, is used as an anesthetic. A sample of this compound is burned in a combustion analysis apparatus. The sample size is 1.00 g, and 3.14 g of CO_2 and 1.29 g of H_2O are produced. What is the empirical formula of cyclopropane?

Solution

Every carbon atom in the CO_2 and every hydrogen atom in the H_2O came from the cyclopropane. We will need to calculate the number of moles of carbon present in the CO_2 and then the number of moles of hydrogen present in the H_2O. Both of these calculations are "grams of A" to "moles of B" problems in terms of the jargon of Figure 9.3.

The dimensional analysis setup from which we obtain the moles of carbon is

$$3.14 \text{ g } CO_2 \times \frac{1 \text{ mole } CO_2}{44.0 \text{ g } CO_2} \times \frac{1 \text{ mole C}}{1 \text{ mole } CO_2}$$

$$= 0.071363636 \text{ mole C} \quad \text{(calculator answer)}$$
$$= 0.0714 \text{ mole C} \quad \text{(correct answer)}$$

The last conversion factor used in this calculation comes from the formula CO_2. One molecule of CO_2 contains one atom of carbon. Thus, one mole of CO_2 will contain one mole of carbon.

The dimensional analysis setup from which we obtain the moles of hydrogen is

$$1.29 \text{ g } H_2O \times \frac{1 \text{ mole } H_2O}{18.0 \text{ g } H_2O} \times \frac{2 \text{ moles H}}{1 \text{ mole } H_2O}$$

$$= 0.14333333 \text{ mole H} \quad \text{(calculator answer)}$$
$$= 0.143 \text{ mole H} \quad \text{(correct answer)}$$

The last conversion factor in this setup is obtained from the information in the formula H_2O. There are 2 moles of H in 1 mole of H_2O.

With the moles of C and of H both known, we can now calculate the whole number mole ratio by dividing each mole amount by the smallest of these amounts.

$$\text{C:} \quad \frac{0.0714 \text{ mole}}{0.0714 \text{ mole}} = 1 \quad \text{(calculator answer)}$$
$$= 1.00 \quad \text{(correct answer)}$$

$$\text{H:} \quad \frac{0.143 \text{ mole}}{0.0714 \text{ mole}} = 2.0028011 \quad \text{(calculator answer)}$$
$$= 2.00 \quad \text{(correct answer)}$$

Carbon and hydrogen are present in the compound in a one-to-two molar ratio. The empirical formula of the compounds is CH_2

Practice Exercise 9.17

A compound containing only carbon and hydrogen is subjected to combustion analysis. From a 20.0-g sample of the compound, 67.5 g of CO_2 and 13.8 g of H_2O are obtained. What is the empirical formula of the compound?

Ans. CH

Example 9.18

The compound ascorbic acid (vitamin C) is a carbon–hydrogen–oxygen compound. It cannot be stored in the body and thus must continuously be supplied by the diet. Combustion analysis of a 3.08-g ascorbic acid sample yields 6.17 g of CO_2 and 2.52 g of H_2O. What is the empirical formula for ascorbic acid?

Solution

First, we will calculate the grams of C in the original sample from the grams of CO_2 and the grams of H in the original sample from the grams of H_2O. The grams of O present in the original sample can then be calculated by difference: the difference between the original sample mass and the masses of C and H. We cannot calculate the grams of oxygen present in the original sample from the CO_2 and H_2O masses because the oxygen atoms in these compounds came from two sources: (1) the ascorbic acid, and (2) the oxygen in air.

With masses of the individual elements known, we then obtain moles of the individual elements. Dividing these molar amounts by the smallest number of moles of an element present leads to the subscripts for the empirical formula. The following flowchart outlines the overall calculational process.

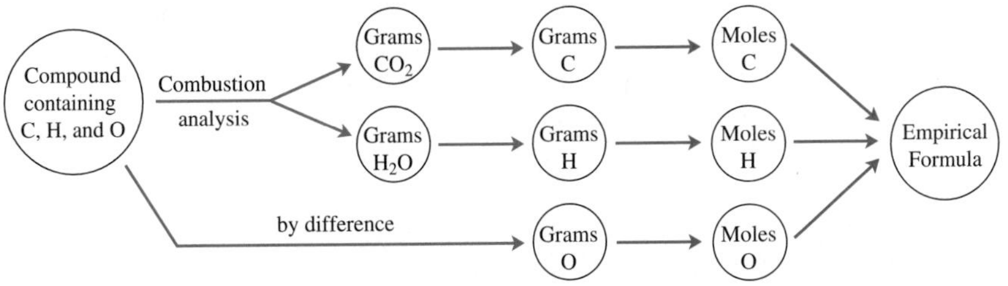

Converting the grams of CO_2 and grams of H_2O to grams of C and grams of H, respectively, involves "grams of A" to "grams of B" calculations.

$$6.17 \text{ g } \cancel{CO_2} \times \frac{1 \text{ mole } \cancel{CO_2}}{44.0 \text{ g } \cancel{CO_2}} \times \frac{1 \text{ mole } \cancel{C}}{1 \text{ mole } \cancel{CO_2}} \times \frac{12.0 \text{ g C}}{1 \text{ mole } \cancel{C}}$$

$$= 1.6827272 \text{ g C} \quad \text{(calculator answer)}$$

$$= 1.68 \text{ g C} \quad \text{(correct answer)}$$

$$2.52 \text{ g } \cancel{H_2O} \times \frac{1 \text{ mole } \cancel{H_2O}}{18.0 \text{ g } \cancel{H_2O}} \times \frac{2 \text{ moles } \cancel{H}}{1 \text{ mole } \cancel{H_2O}} \times \frac{1.01 \text{ g H}}{1 \text{ mole } \cancel{H}}$$

$$= 0.2828 \text{ g H} \quad \text{(calculator answer)}$$

$$= 0.283 \text{ g H} \quad \text{(correct answer)}$$

The next-to-last conversion factor in the CO_2 calculation comes from the formula CO_2. One CO_2 molecule contains one C atom. Likewise, the formula H_2O is the key to obtaining information about hydrogen; one H_2O molecule contains two H atoms.

The difference between the mass of the sample and the masses of C and H in the sample is the mass of O in the sample.

$$\begin{aligned}\text{grams O} &= \text{grams of sample} - \text{grams C} - \text{grams H}\\ &= 3.08\text{ g} - 1.68\text{ g} - 0.283\text{ g}\\ &= 1.117\text{ g} \quad \text{(calculator answer)}\\ &= 1.12\text{ g} \quad \text{(correct answer)}\end{aligned}$$

The next step is to go from grams of element to moles of element. These are simple one-step "grams of A" to "moles of A" calculations.

$$1.68\ \cancel{\text{g C}} \times \frac{1\text{ mole C}}{12.0\ \cancel{\text{g C}}} = 0.14\text{ mole C} \quad \text{(calculator answer)}$$

$$= 0.140\text{ mole C} \quad \text{(correct answer)}$$

$$0.283\ \cancel{\text{g H}} \times \frac{1\text{ mole H}}{1.01\ \cancel{\text{g H}}} = 0.2801980\text{ mole H} \quad \text{(calculator answer)}$$

$$= 0.280\text{ mole H} \quad \text{(correct answer)}$$

$$1.12\ \cancel{\text{g O}} \times \frac{1\text{ mole O}}{16.0\ \cancel{\text{g O}}} = 0.07\text{ mole O} \quad \text{(calculator answer)}$$

$$= 0.0700\text{ mole O} \quad \text{(correct answer)}$$

Dividing each of these molar values by the smallest among them, 0.0700 mole O, gives us the small whole numbers that lead to the empirical formula of the compound.

$$\text{C:} \quad \frac{0.140\ \cancel{\text{mole}}}{0.0700\ \cancel{\text{mole}}} = 2 \quad \text{(calculator answer)}$$

$$= 2.00 \quad \text{(correct answer)}$$

$$\text{H:} \quad \frac{0.280\ \cancel{\text{mole}}}{0.0700\ \cancel{\text{mole}}} = 4 \quad \text{(calculator answer)}$$

$$= 4.00 \quad \text{(correct answer)}$$

$$\text{O:} \quad \frac{0.0700\ \cancel{\text{mole}}}{0.0700\ \cancel{\text{mole}}} = 1 \quad \text{(calculator answer)}$$

$$= 1.00 \quad \text{(correct answer)}$$

The empirical formula of ascorbic acid is C_2H_4O

Practice Exercise 9.18

The food preservative butylated hydroxytoluene (BHT) is known to contain only the elements C, H, and O. A sample of this compound is burned in a combustion analysis apparatus. The sample size is 2.204 g, and 6.601 g of CO_2 and 2.162 g of H_2O are produced. What is the empirical formula of BHT?

Ans. $C_{15}H_{24}O$

9.11 DETERMINATION OF MOLECULAR FORMULAS

To determine the molecular formula of a compound we need additional information besides its empirical formula. That additional information is the compound's molecular mass. A number of experimental methods exist, whose details are beyond the scope of this book, for obtaining such molecular mass information.

A compound's molecular formula will always be a *whole-number multiple* of its empirical formula (Sec. 9.9).

$$\text{molecular formula} = (\text{empirical formula})_x, \text{ where } x = \text{whole number}$$

The value of x is calculated by dividing the compound's molecular mass by the compound's empirical formula mass.

$$x = \frac{\text{molecular mass (experimentally determined quantity)}}{\text{empirical formula mass (calculated from atomic masses)}}$$

If the value of x was found to be 5.00 and the compound's empirical formula was CH_2, then the compound's molecular formula would be

$$(CH_2)_5 = C_5H_{10}$$

Example 9.19 illustrates further the mechanics involved in obtaining a molecular formula from empirical formula and molecular mass data.

Example 9.19

Determine the molecular formula of each of the following compounds from the given empirical formula and molecular mass information.

(a) empirical formula = CH; molecular mass = 78.0 amu

(b) empirical formula = NH_2; molecular mass = 32.0 amu

(c) empirical formula = CO; molecular mass = 28.0 amu

(d) empirical formula = C_4H_9; molecular mass = 104 amu

Solution

In each case we will determine the whole-number multiplier (x) that relates empirical and molecular formulas to each other.

$$x = \frac{\text{molecular mass (MM)}}{\text{empirical formula mass (EFM)}}$$

(a) First we calculate the empirical formula mass of CH from atomic masses. The atomic mass of C is 12.0 amu and that of H is 1.0 amu.

$$\text{EFM of CH} = 12.0 \text{ amu} + 1.0 \text{ amu} = 13.0 \text{ amu}$$

The whole-number multiplier is

$$x = \frac{\text{MM}}{\text{EFM}} = \frac{78.0 \cancel{\text{amu}}}{13.0 \cancel{\text{amu}}} = 6.00$$

Therefore, the molecular formula is

$$(CH)_6 = C_6H_6$$

(b) The empirical formula mass of NH_2 is 16.0 amu. Nitrogen has an atomic mass of 14.0 amu and each hydrogen has an atomic mass of 1.0 amu. The whole-number multiplier, x, is

$$x = \frac{\text{MM}}{\text{EFM}} = \frac{32.0\ \cancel{\text{amu}}}{16.0\ \cancel{\text{amu}}} = 2.00$$

The molecular formula is $(NH_2)_2 = N_2H_4$.

(c) The empirical formula mass is 28.0 amu. The whole-number multiplier, x, is

$$x = \frac{\text{MM}}{\text{EFM}} = \frac{28.0\ \cancel{\text{amu}}}{28.0\ \cancel{\text{amu}}} = 1.00$$

An x value of 1.00 means that the empirical formula and the molecular formula are one and the same. Both are CO.

(d) The empirical formula mass is 57.0 amu, the value of x is 2.00, and the molecular formula is $(C_4H_9)_2 = C_8H_{18}$.

Practice Exercise 9.19

Determine the molecular formula of each of the following compounds from the given empirical formula and molecular mass information.

(a) empirical formula = HO; molecular mass = 34.0 amu
(b) empirical formula = SN; molecular mass = 184 amu

Ans. (a) H_2O_2; (b) S_4N_4

Example 9.19 showed how a molecular formula is determined when the empirical formula is already known. Example 9.20 illustrates a "complete" molecular formula determination, one in which we begin at the beginning—with percent composition data. There are two different approaches to such a calculation and both approaches are illustrated in Example 9.20. The two approaches differ in the selection of the basis (amount of compound) for the calculation. The two aproaches are:

1. *Select 100.0 g of compound as the basis for the calculation*, determine the empirical formula, and then use the empirical formula and the molecular mass to determine the molecular formula.
2. *Select 1.00 mole of compound as the basis for the calculation*, and determine directly the molecular formula.

In the first approach, the molecular mass data are used at the end of the problem. In the second approach, the molecular mass data are used at the beginning of the calculation. Either way, the answer is the same.

Example 9.20

Butane, a compound containing only carbon and hydrogen, is the fuel used in many camp stoves. Butane has a percent composition by mass of 82.63% C and 17.37% H and a molecular mass of 58.14 amu. Determine the molecular formula of butane using:

(a) 100.0 g of butane as the basis for the calculation

(b) 1.000 mole of butane as the basis for the calculation

Solution

(a) Taking 100.0 g of compound as our basis, we will have the following amounts of carbon and hydrogen present.

$$\text{C: } 82.63\% \text{ of } 100.0 \text{ g} = 82.63 \text{ g}$$
$$\text{H: } 17.37\% \text{ of } 100.0 \text{ g} = 17.37 \text{ g}$$

We next convert these gram amounts of C and H to moles of the same.

$$82.63 \cancel{\text{g C}} \times \frac{1 \text{ mole C}}{12.01 \cancel{\text{g C}}} = 6.8800999 \text{ moles C} \quad \text{(calculator answer)}$$
$$= 6.880 \text{ moles C} \quad \text{(correct answer)}$$

$$17.37 \cancel{\text{g H}} \times \frac{1 \text{ mole H}}{1.008 \cancel{\text{g H}}} = 17.232142 \text{ moles H} \quad \text{(calculator answer)}$$
$$= 17.23 \text{ moles H} \quad \text{(correct answer)}$$

Dividing each of these molar values by the smallest one (6.880) gives the following results.

$$\text{C: } \frac{6.880 \cancel{\text{mole}}}{6.880 \cancel{\text{mole}}} = 1 \quad \text{(calculator answer)}$$
$$= 1.000 \quad \text{(correct answer)}$$

$$\text{H: } \frac{17.23 \cancel{\text{mole}}}{6.880 \cancel{\text{mole}}} = 2.5043604 \quad \text{(calculator answer)}$$
$$= 2.504 \quad \text{(correct answer)}$$

Our result for hydrogen is not a whole number nor a near-whole number that can be rounded to a whole number. For hydrogen, we should recognize that the number we are dealing with is $2\frac{1}{2}$. Multiplication of both molar ratios by 2 will clear the fraction. This gives us our needed whole-number ratio.

$$\text{C: } 1 \times 2 = 2$$
$$\text{H: } 2\tfrac{1}{2} \times 2 = 5$$

The empirical formula of butane is C_2H_5.

To make the transition from empirical formula to molecular formula, we first determine the formula mass for the empirical formula.

$$\text{C: } 2 \times 12.01 \text{ amu} = 24.02 \text{ amu}$$
$$\text{H: } 5 \times 1.01 \text{ amu} = \underline{5.05 \text{ amu}}$$
$$29.07 \text{ amu}$$

The molecular mass of butane, given in the problem statement, is 58.14 amu. We next determine how many times larger the molecular mass is than the empirical formula mass. This is done by dividing the molecular mass by the empirical formula mass.

$$\frac{58.14 \cancel{\text{amu}}}{29.07 \cancel{\text{amu}}} = 2.00$$

We have just determined the multiplication factor which converts the empirical formula into a molecular formula. Each of the subscripts in the empirical formula is multiplied by 2.

$$(C_2H_5)_2 = C_4H_{10}$$

The molecular formula of butane is C_4H_{10}.

(b) We will go through this same calculation again, this time using 1.000 mole of butane as our basis. The mass of 1.000 mole of butane is 58.14 g. We know this because it was given in the problem statement; the molecular mass of butane is 58.14 amu.

We first determine the number of grams of each element present in our basis amount of 58.14 g of butane.

C: 82.63% of 58.14 g = 48.041082 g (calculator answer)
= 48.04 g (correct answer)

H: 17.37 % of 58.14 g = 10.098918 g (calculator answer)
= 10.10 g (correct answer)

We next change grams of element to moles of element. Formula subscripts are always determined from molar information.

$$48.04 \text{ g C} \times \frac{1 \text{ mole C}}{12.01 \text{ g C}} = 4 \text{ moles C} \quad \text{(calculator answer)}$$

$$= 4.000 \text{ moles C} \quad \text{(correct answer)}$$

$$10.10 \text{ g H} \times \frac{1 \text{ mole H}}{1.008 \text{ g H}} = 10.019841 \text{ moles H} \quad \text{(calculator answer)}$$

$$= 10.02 \text{ moles H} \quad \text{(correct answer)}$$

At this stage in the calculation, because the basis for the calculation is 1 mole of compound, we will always obtain whole-number molar amounts or near-whole-number molar amounts that can be rounded to whole numbers. (If we do not, we have a mistake somewhere in our calculation.) These whole-number molar amounts are the subscripts in the molecular formula. Thus, the molecular formula of butane is C_4H_{10}.

What caused this formula, C_4H_{10}, to be a molecular formula rather than an empirical formula? Again, it is the fact that the basis for the calculation was 1 mole of compound.

This second method for calculating a molecular formula can be used only when the molecular mass of the compound is known. This method is usually shorter than the other method.

Practice Exercise 9.20

The compound oxalic acid, which is used in commercial laundries to remove rust stains, has a molecular mass of 90.0 amu and a percent composition of 26.7% C, 2.2% H, and 71.1% O. Determine the molecular formula of oxalic acid using:

(a) 100.0 g of oxalic acid as the basis for the calculation
(b) 1.00 mole of oxalic acid as the basis for the calculation

Ans. (a) $H_2C_2O_4$; (b) $H_2C_2O_4$

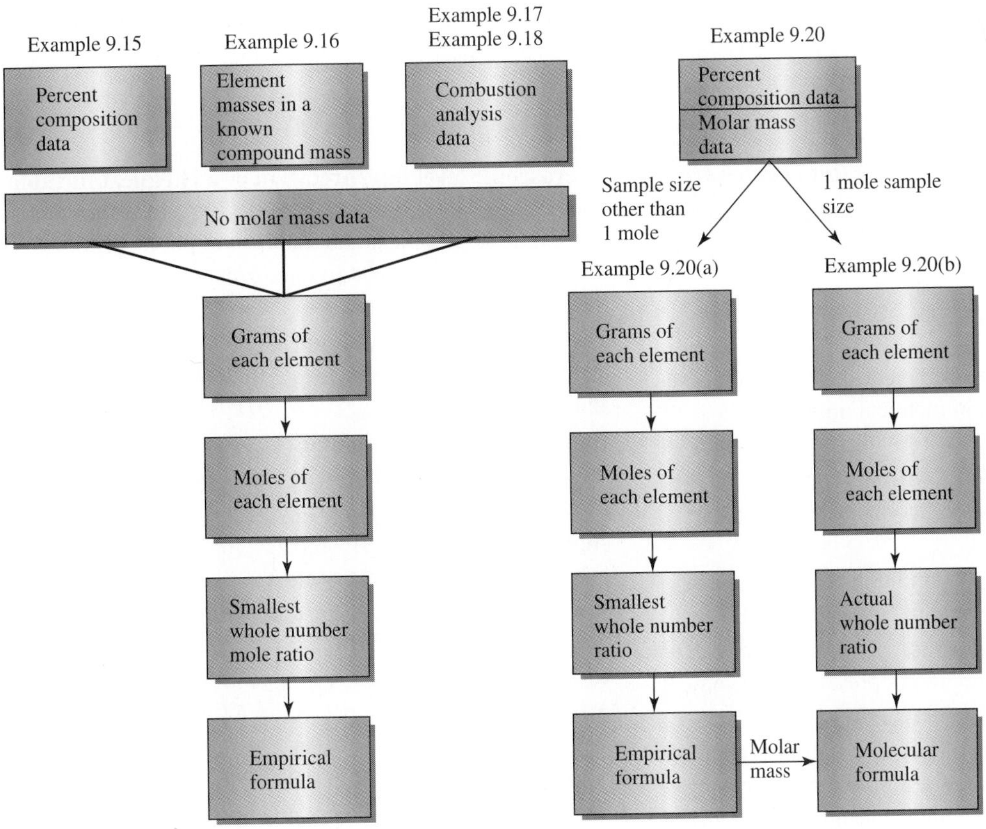

FIGURE 9.5 A summary of how empirical formulas and molecular formulas are calculated from various types of data.

Figure 9.5 summarizes the methods used in this section and the preceding one to obtain empirical and molecular formulas. It contrasts each of the example calculations we carried out in these sections.

KEY TERMS

The new terms or concepts defined in this chapter are

Avogadro's number (Sec. 9.4) The name given to the numerical value 6.02×10^{23}, the number of particles in a mole.

counting by weighing (Sec. 9.6) Samples of two or more pure substances (elements or compounds) found to have mass ratios equal to the ratios of their atomic or formula masses must contain identical numbers of particles (atoms, molecules, or formula units).

empirical formula (Sec. 9.9) A formula that gives the smallest whole number ratio of atoms present in a formula unit of a compound.

formula mass (Sec. 9.2) A number that is the sum of the atomic masses of the atoms in one formula unit.

law of definite proportions (Sec. 9.1) In a pure compound, the elements are always present in the same definite proportion by mass.

macroscopic level interpretation of a chemical formula (Sec. 9.7) The subscripts in the formula of a substance are interpreted to mean the numbers of moles of atoms of the various elements present in one mole of the substance.

microscopic level interpretation of a chemical formula (Sec. 9.7) The subscripts in the formula of a substance are interpreted to mean the numbers of atoms of the various elements present in one formula unit of the substance.

molar mass of a compound (Sec. 9.5) A mass, in grams, that is numerically equal to the formula mass of the compound.

molar mass of an element (Sec. 9.5) For an element in atomic form, a mass, in grams that is numerically equal to the atomic mass of the element.

mole (Secs. 9.4 and 9.5) 6.02×10^{23} objects or an amount of substance that contains as many elementary units (atoms, molecules, or formula units) as there are ${}^{12}_{6}C$ atoms in exactly 12.00000 g of ${}^{12}_{6}C$.

molecular formula (Sec. 9.9) A formula that gives the actual number of atoms present in a formula unit of a compound.

percent composition (Sec. 9.3) The percent by mass of each element present in a compound

PRACTICE PROBLEMS

Law of Definite Proportions (Sec. 9.1)

9.1 The air pollutant SO_2 is produced when coal is burned. A 1.00 g sample of SO_2 is found to contain 33.4% S and 66.6% O. What would be the percent compostion of a 5.00-g sample of this compound?

9.2 The air pollutant NO is a component of automobile exhaust. A 5.00-g sample of NO is found to contain 46.7% N and 53.3% O. What would be the percent composition of a 1.00-g sample of this compound?

9.3 Two different samples of a pure compound (containing elements A and D) were analyzed with the following results.

Sample I: 17.35 g of compound yielded 10.03 g of A and 7.32 g of D.

Sample II: 22.78 g of compound yielded 13.17 g of A and 9.61 g of D.

Show that these data are consistent with the law of definite proportions.

9.4 Two different samples of a pure compound (containing elements E and D) were analyzed with the following results.

Sample I: 11.21 g of compound yielded 5.965 g of E and 5.245 g of D.

Sample II: 13.65 g of compound yielded 7.263 g of E and 6.387 g of D.

Show that these data are consistent with the law of definite proportions.

9.5 It has been found experimentally that the elements X and Q react to produce two different compounds, depending on conditions. Some sample experimental data are as follows.

Experiment	Grams of X	Grams of Q	Grams of Compound
1	3.37	8.90	12.27
2	0.561	1.711	2.272
3	26.0	71.0	97.9

Which two of the three experiments produced the same compound?

9.6 It has been found experimentally that the elements R and T react to produce two different compounds, depending upon conditions. Some sample experimental data are as follows.

Experiment	Grams of R	Grams of T	Grams of Compound
1	1.926	1.074	3.000
2	3.440	1.560	5.000
3	4.494	2.506	7.000

Which two of the three experiements produced the same compound?

9.7 The elemental composition, by mass, of carbon monoxide (CO) is 42.9% C and 57.1% O. What is the maximum amount of CO that could be formed from 42.9 g of C and 80.0 g of O?

9.8 The elemental composition, by mass, of carbon dioxide (CO_2) is 27.3% C and 72.7% O. What is the maximum amount of CO_2 that could be formed from 30.0 g of C and 72.7 g of O?

Formula Masses (Sec. 9.2)

9.9 Calculate the formula mass of each of the following substances. Round off all atomic masses to the hundredths digit before using.

(a) Na_4SiO_4 (sodium silicate, a flame retardent)

(b) H_3BO_3 (boric acid, a mild antiseptic)

(c) $C_{12}H_{11}NO_2$ (Sevin, an insecticide)

(d) $C_{22}H_{30}ClNO_2$ (Darvon, a prescription pain killer)

9.10 Calculate the formula mass of each of the following substances. Round off all atomic masses to the hundredths digit before using.

(a) $NaHCO_3$ (sodium bicarbonate, baking soda)

(b) Tl_2SO_4 (thallium (I) sulfate, a rat and ant poison)

(c) $C_5H_8NO_4Na$ (MSG, a flavor enhancer used in Chinese cooking)

(d) $C_{20}H_{24}N_2O_2$ (quinine, an antimalarial drug)

9.11 Calculate the formula mass of each of the following substances. Round off all atomic masses to the tenths digit before using.

(a) $C_2H_4(OH)_2$ (ethylene glycol, automotive antifreeze component)

(b) $(H_2N)_2CO$ (urea, a component of urine)

(c) $Mg_3(Si_2O_5)_2(OH)_2$ (talc, mineral from which talcum powder comes)

(d) $Al_2Si_2O_5(OH)_2$ (kaolinite, a form of clay)

9.12 Calculate the formula mass of each of the following substances. Round off all atomic masses to the tenths digits before using.

(a) $Al(OH)_3$ (aluminum hydroxide, a water purification chemical)

(b) $C_3H_5(OH)_3$ (glycerin, a substance derived from fats)

(c) $KLi_2Al(Si_2O_5)_2(OH)_2$ (form of the mineral mica)

(d) $Ca_3Be(OH)_2(Si_3O_{10})$ (form of the mineral aminoffite)

9.13 The compound 2-butene-1-thiol, which is responsible in part for the characteristic odor of skunks, has a formula mass of 88.19 amu and the formula C_yH_8S. What number does y stand for in this formula?

9.14 The compound 1-propanethiol, which is the eye irritant that is released when fresh onions are chopped up, has a formula mass of 76.18 amu and the formula C_3H_yS. What number does y stand for in this formula?

9.15 A compound that is 51.5% oxygen by mass contains two oxygen atoms per molecule. What is the formula mass of the compound?

9.16 A compound that is 52.1% oxygen by mass contains three oxygen atoms per molecule. What is the formula mass of the compound?

Percent Composition (Sec. 9.3)

9.17 Calculate the percent composition for each of the following compounds. (Round off all atomic masses to 0.01 amu before using them.)

(a) SnF_2 (tin (II) fluoride, a toothpaste additive)

(b) $FeSO_4$ (iron (II) sulfate, used in treatment of iron deficiency)

(c) $C_{12}H_{22}O_{11}$ (sucrose, table sugar)

(d) $C_{14}H_9Cl_5$ (DDT, an insecticide)

9.18 Calculate the percent composition for each of the following compounds. (Round off all atomic masses to 0.01 amu before using them.)

(a) $C_9H_8O_4$ (naphthalene, ingredient in some mothballs)

(b) NaCN (sodium cyanide, used to extract gold from ores)

(c) C_7H_{16} (heptane, a component of gasoline)

(d) $C_{55}H_{72}MgN_4O_5$ (chlorophyll, present in all green plants)

9.19 Calculate the percent composition of each of the following compounds from the given information.

(a) 1.271 g of Cu and 0.320 g of O completely react to produce a sample of the compound.

(b) Decomposition of a 49.31-g sample of the compound yields 15.96 g of Na, 11.13 g of S, and 22.22 g of O.

(c) A 6.48-g sample of N is reacted with oxygen to give 25.00 g of compound.

(d) The reaction of 3.34 g of S with 10.00 g of O produces 10.00 g of compound and 3.34 g of unreacted (leftover) O.

9.20 Calculate the percent composition of each of the following compounds from the given information.

(a) 1.12 g of Fe and 0.48 g of O completely react to produce a sample of the compound.

(b) Decomposition of a 18.03-g sample of the compound yields 4.79 g of K, 6.38 g of Cr, and 6.86 g of O.

(c) A 5.76-g sample of P is reacted with oxygen to give 13.20 g of compound.

(d) The reaction of 4.67 g of N with 10.00 g of O produces

10.00 g of compound and 4.67 g of unreacted (leftover) O.

9.21 In which of the following pairs of compounds do both members of the pair have a mass percent oxygen that exceeds 50%?

(a) CO and CO_2 **(b)** ClO and ClO_2
(c) $LiNO_3$ and $RbNO_3$ **(d)** Na_3PO_4 and $Be_3(PO_4)_2$

9.22 In which of the following pairs of compounds do both members of the pair have a mass percent oxygen that exceeds 50%?

(a) NO and NO_2 **(b)** H_2O and H_2O_2
(c) BeO and MgO **(d)** $BaSO_4$ and $Al_2(SO_4)_3$

9.23 Calculate the percent compositions of acetylene (C_2H_2) and benzene (C_6H_6). Explain your results.

9.24 Calculate the percent compositions of propene (C_3H_6) and cyclobutane (C_4H_8). Explain your results.

9.25 A sample of the compound hydrazine, N_2H_4, contains 52.34 g of H. How many grams of N does it contain?

9.26 A sample of the compound sodium azide, NaN_3, contains 57.08 g of Na. How many grams of N does it contain?

The Mole as a Counting Unit (Sec. 9.4)

9.27 How many particles (atoms, molecules, formula units, or ions) are present in 1.00 mole of each of the following?

(a) silver (Ag) atoms
(b) water (H_2O) molecules
(c) sodium nitrate ($NaNO_3$) formula units
(d) sulfate (SO_4^{2-}) ions

9.28 How many particles (atoms, molecules, formula units, or ions) are present in 1.00 mole of each of the following?

(a) copper (Cu) atoms
(b) ammonia (NH_3) molecules
(c) potassium carbonate (K_2CO_3) formula units
(d) phosphate (PO_4^{3-}) ions

9.29 How many carbon atoms are present in each of the following molar quantities of carbon?

(a) 2.50 moles **(b)** 3.25 moles
(c) 0.23 mole **(d)** 0.3114 mole

9.30 How many boron atoms are present in each of the following molar quantities of boron?

(a) 1.50 moles **(b)** 4.21 moles
(c) 0.97 mole **(d)** 1.079 moles

9.31 Calculate the number of molecules present in each of the following samples of molecular compounds.

(a) 1.50 moles CO_2 **(b)** 0.500 mole NH_3
(c) 2.33 moles PF_3 **(d)** 1.115 moles N_2H_4

9.32 Calculate the number of molecules present in each of the following samples of molecular compounds.

(a) 4.69 moles CO **(b)** 0.433 mole SO_3
(c) 1.44 moles P_2H_4 **(d)** 2.307 moles H_2O_2

Molar Mass (Sec. 9.5)

9.33 What is the mass, in grams, of 1.00 mole of each of the following elements?

(a) Cu **(b)** Ba **(c)** Si **(d)** U

9.34 What is the mass, in grams, of 1.00 mole of each of the following elements?

(a) Ag **(b)** K **(c)** Al **(d)** I

9.35 What is the molar mass, to four significant figures, of each of the following compounds?

(a) $Al(OH)_3$ **(b)** Mg_3N_2
(c) $Cu(NO_3)_2$ **(d)** La_2O_3

9.36 What is the molar mass, to four significant figures, of each of the following compounds?

(a) NH_4Cl **(b)** $HClO_4$
(c) $Ca(ClO_4)_2$ **(d)** HfO_2

9.37 Calculate the mass, in grams, of 1.357 moles of each of the following.

(a) NaCl **(b)** Na_2S **(c)** $NaNO_3$ **(d)** Na_3PO_4

9.38 Calculate the mass, in grams, of 0.981 mole of each of the following.

(a) SO_2 **(b)** SO_2Cl_2 **(c)** S_4N_4 **(d)** $Li_2S_2O_3$

9.39 In each of the following pairs of molar-sized quantities select the quantity which has the greater mass, in grams.

(a) 2.00 moles of Cu and 2.00 moles of O
(b) 1.00 mole of Br and 5.00 moles of Be
(c) 2.00 moles of CO and 1.50 moles of N_2O
(d) 4.87 moles of B_2H_6 and 0.35 mole of U

9.40 In each of the following pairs of molar-sized quantities select the quantity which has the greater mass, in grams.

(a) 3.00 moles of Na and 3.00 moles of Al
(b) 1.00 mole of I and 9.00 moles of Li
(c) 2.00 moles of CO_2 and 1.00 mole of SO_3
(d) 9.00 moles of Be and 1.00 mole of HCl

9.41 A 0.123 mole sample of a pure substance has a mass of 5.904 g. What is the molar mass of the substance?

9.42 A 0.571 mole sample of a pure substance has a mass of 36.60 g. What is the molar mass of the substance?

9.43 What is the mass, in grams, of an atom whose mass on the amu scale is 19.0 amu?

9.44 What is the mass, in grams, of an atom whose mass on the amu scale is 23.0 amu?

9.45 Identify the element that contains atoms with an average mass of 5.143×10^{-23} g.

9.46 Identify the element that contains atoms with an average mass of 2.326×10^{-23} g.

Counting Particles by Weighing (Sec. 9.6)

9.47 What is the Ag/nonmetal mass ratio needed to obtain an equal number of atoms of Ag and each of the following nonmetals?

(a) Cl **(b)** H **(c)** O **(d)** P

9.48 What is the Cu/nonmetal mass ratio needed to obtain an equal number of atoms of Cu and each of the following nonmetals?

(a) F **(b)** He **(c)** N **(d)** S

9.49 What is the Al/Be mass ratio needed to obtain the following?

(a) an equal number of both types of atoms
(b) twice as many Al atoms as Be atoms
(c) twice as many Be atoms as Al atoms
(d) 80% Al atoms and 20% Be atoms

9.50 What is the Si/N mass ratio needed to obtain the following?

(a) an equal number of both types of atoms
(b) three times as many Si atoms as N atoms
(c) three times as many N atoms as Si atoms
(d) 75% Si atoms and 25% N atoms

9.51 Using the concept of counting by weighing, determine whether or not each of the following pairs of samples contains the same number of particles (atoms or molecules).

(a) 28.086 g of Si and 35.453 g of Cl
(b) 72.59 g of Ge and 74.72 g of Ga
(c) 1.427 g of NH_3 and 2.685 g of N_2H_4
(d) 64.14 g of S and 128.14 g of SO_2

9.52 Using the concept of counting by weighing, determine whether or not each of the following pairs of samples contains the same number of particles (atoms or molecules).

(a) 3.813 g of B and 2.314 g of Li
(b) 2.0020 g of C and 2.3347 g of N
(c) 9.01 g of H_2O and 15.78 g of H_2O_2
(d) 41.4 g of Au and 44.6 g of AuCN

9.53 Using the concept of counting by weighing, indicate whether each of the following samples contains more atoms, the same number of atoms, or fewer atoms than 23.6 g of copper (Cu).

(a) 24.3 g of Fe **(b)** 19.6 g of Ni
(c) 11.52 g of P **(d)** 11.2 g of Al

9.54 Using the concept of counting by weighing, indicate whether each of the following samples contains more atoms, the same number of atoms, or fewer atoms than 14.0 g of sodium (Na).

(a) 77.3 g of I **(b)** 126 g of Pb
(c) 20.2 g of K **(d)** 9.00 g of N

9.55 Using the concept of counting by weighing, determine the number of grams of each of the following elements that will contain three times as many atoms as 10.00 g of P.

(a) S **(b)** Be **(c)** U **(d)** Si

9.56 Using the concept of counting by weighing, determine the number of grams of each of the following elements that will contain twice as many atoms as 20.00 g of Zn.

(a) Li **(b)** Al **(c)** Cu **(d)** Ba

The Mole and Chemical Formulas (Sec. 9.7)

9.57 Write the six mole-to-mole conversion factors that can be derived from the formula Na_3PO_4.

9.58 Write the six mole-to-mole conversion factors that can be derived from the formula K_2SO_4.

9.59 In which of the following pairs of compound amounts do both members of the pair contain the same number of moles of sulfur atoms?

(a) 1.0 mole Na_2SO_4 and 0.50 mole $Na_2S_2O_3$
(b) 2.00 moles S_3Cl_2 and 1.50 moles S_2O
(c) 3.00 moles $H_2S_2O_5$ and 6.00 moles H_2SO_4
(d) 1.00 mole $Na_3Ag(S_2O_3)_2$ and 2.00 moles S_2F_{10}

9.60 In which of the following pairs of compound amounts do both members of the pair contain the same number of moles of nitrogen atoms?

(a) 0.50 mole N_2O_5 and 1.0 mole of N_2O_4
(b) 2.00 moles HNO_3 and 2.00 moles HNO_2
(c) 3.00 moles NH_3 and 1.00 mole HN_3
(d) 1.50 moles $(NH_4)_2SO_4$ and 1.00 mole $(NH_4)_3PO_4$

9.61 Which amount in each of the following pairs of amounts contains the greater number of total moles of atoms?

(a) 2.00 moles $NaAuBr_4$ and 2.00 moles Au_2Te_3
(b) 1.00 mole $C_2H_2Cl_4$ and 1.00 mole CCl_4
(c) 3.00 moles $Ba(NO_3)_2$ and 3.00 moles $BaSO_4$
(d) 1.20 moles NH_4CN and 1.30 moles NH_4Cl

9.62 Which amount in each of the following pairs of amounts contains the greater number of total moles of atoms?

(a) 3.00 moles Cl_2O and 3.00 moles Cl_2O_3
(b) 1.00 mole C_2H_6 and 1.00 mole $SOCl_2$
(c) 2.00 moles $CaSO_4$ and 2.00 moles NH_4Br
(d) 1.00 mole $(NH_4)_2CO_3$ and 3.00 moles $Ba(OH)_2$

The Mole and Chemical Calculations (Sec. 9.8)

9.63 Calculate the number of atoms present in a 7.500-g sample of each of the following elements.

(a) S **(b)** Be **(c)** Ba **(d)** Au

9.64 Calculate the number of atoms present in a 3.752-g sample of each of the following elements.

(a) Li **(b)** V **(c)** Hg **(d)** Pb

9.65 Calculate the number of molecules present in a 25.0-g sample of each of the following compounds.

(a) HF **(b)** N_2H_4 **(c)** SO_2 **(d)** H_2SO_4

9.66 Calculate the number of molecules present in a 52.0-g sample of each of the following compounds.

(a) HClO **(b)** $HClO_2$ **(c)** $HClO_3$ **(d)** $HClO_4$

9.67 What is the mass, in grams, of each of the following quantities of chemical substance?

(a) 6.022×10^{23} atoms of Ag

(b) 1.750×10^{23} atoms of N

(c) 6.5×10^{30} molecules of CO_2

(d) 2431 molecules of CO

9.68 What is the mass, in grams, of each of the following quantities of chemical substance?

(a) 6.022×10^{23} atoms of Xe

(b) 3.125×10^{22} atoms of As

(c) 3.125×10^{22} molecules of H_3AsO_4

(d) 989 molecules of H_2O

9.69 What is the mass, in grams (to four significant figures), of a single atom (for elements) or single molecule (for compounds) of the following substances?

(a) Na **(b)** Mg **(c)** C_4H_{10} **(d)** C_6H_6

9.70 What is the mass, in grams (to four significant figures), of a single atom (for elements) or single molecule (for compounds) of the following substances?

(a) Cu **(b)** Be **(c)** C_2H_6 **(d)** $C_{10}H_{20}$

9.71 Determine the number of grams of Cl present in each of the following amounts of chlorine-containing compounds.

(a) 100.0 g of NaCl **(b)** 1000.0 g of CCl_4

(c) 10.0 g of HCl **(d)** 50.0 g of $BaCl_2$

9.72 Determine the number of grams of N present in each of the following amounts of nitrogen-containing compounds.

(a) 100.0 g of $NaNO_3$ **(b)** 1000.0 g of N_2H_4

(c) 10.0 g of Na_3N **(d)** 50.0 g of KCN

9.73 Determine the number of phosphorus atoms present in 25.0 g of each of the following substances.

(a) PF_3 **(b)** Be_3P_2

(c) $POCl_3$ **(d)** $Na_5P_3O_{10}$

9.74 Determine the number of sulfur atoms present in 35.0 g of each of the following substances.

(a) CS_2 **(b)** S_4N_4

(c) SF_6 **(d)** $Al_2(SO_4)_3$

9.75 Determine the number of grams of O present in each of the following chemical samples.

(a) 2.0×10^{25} molecules of P_4O_{10} **(b)** 50.00 g of SO_3

(c) 3.50 moles of KNO_3 **(d)** 475 g of $Ca_3(PO_4)_2$

9.76 Determine the number of grams of N present in each of the following chemical samples.

(a) 6.3×10^{23} molecules of N_2O_5 **(b)** 26.00 g of HN_3

(c) 25.0 moles of NH_4Cl **(d)** 75.7 g of $Al(NO_3)_3$

9.77 What amount or mass of each of the following substances would be needed to obtain 1.000 g of S?

(a) moles of H_2S **(b)** molecules of S_8

(c) grams of $S_3N_3O_3Cl_3$ **(d)** atoms of S

9.78 What amount or mass of each of the following substances would be needed to obtain 1.000 g of Si?

(a) moles of SiH_4 **(b)** molecules of SiO_2

(c) grams of $(CH_3)_3SiCl$ **(d)** atoms of Si

9.79 Progesterone, a female hormone, has the formula $C_{21}H_{30}O_2$. In a 25.00-g sample of this compound,

(a) how many moles of atoms are present?

(b) how many atoms of C are present?

(c) how many grams of O are present?

(d) how many progesterone molecules are present?

9.80 Testosterone, a male hormone, has the formula $C_{19}H_{28}O_2$. In a 30.00-g sample of this compound,

(a) how many moles of atoms are present?

(b) how many atoms of H are present?

(c) how many grams of C are present?

(d) how many testosterone molecules are present?

Empirical and Molecular Formulas (Secs. 9.9 through 9.11)

9.81 Each of the following is a correctly written molecular formula. In each case write the empirical formula for the substance.

(a) H_2O_2 **(b)** C_8H_{14} **(c)** C_3H_8 **(d)** S_4N_4

9.82 Each of the following is a correctly written molecular formula. In each case write the empirical formula for the substance.

(a) P_4O_{10} **(b)** Pb_3S_4 **(c)** C_4H_8 **(d)** C_5H_{12}

9.83 Given the following percent compositions, determine the empirical formula.

(a) 58.91% Na and 41.09% S

(b) 24.74% K, 34.76% Mn, and 40.50% O

(c) 2.057% H, 32.69% S, and 65.25% O

(d) 19.84% C, 2.503% H, 66.08% O, and 11.57% N

9.84 Given the following percent compositions, determine the empirical formula.

(a) 47.26% Cu and 52.74% Cl

(b) 40.27% K, 26.78% Cr, and 32.96% O

(c) 40.04% Ca, 12.00% C, and 47.96% O

(d) 28.03% Na, 29.28% C, 3.690% H, and 39.01% O

9.85 Convert each of the following molar ratios between

elements in a compound to a whole-number molar ratio.

(a) 1.00-to-1.33 (b) 1.50-to-2.00
(c) 1.75-to-2.00-to-2.25 (d) 1.33-to-2.33-to-2.00

9.86 Convert each of the following molar ratios between elements in a compound to a whole-number molar ratio.

(a) 1.00-to-1.25 (b) 1.50-to-2.50
(c) 1.67-to-2.33-to-3.00 (d) 1.50-to-1.00-to-1.33

9.87 Determine the empirical formula for substances with each of the following percent compositions.

(a) 43.64% P and 56.36% O
(b) 72.24% Mg and 27.76% N
(c) 29.08% Na, 40.56% S, and 30.36% O
(d) 21.85% Mg, 27.83% P, and 50.32% O

9.88 Determine the empirical formula for substances with each of the following percent compositions.

(a) 54.88% Cr and 45.12% S
(b) 38.76% Cl and 61.24% O
(c) 59.99% C, 4.485% H, and 35.52% O
(d) 26.58% K, 35.35% Cr, and 38.06% O

9.89 Nickel forms two chlorides. One has 45.3% Ni by mass, and the other 35.5% Ni. What are the empirical formulas of these two chlorides?

9.90 Copper forms two fluorides. One has 77.0% Cu by mass, and the other 62.6% Cu. What are the empirical formulas of these two fluorides?

9.91 Ethyl mercaptan is an odorous substance added to natural gas to make leaks easily detectable. Analysis of a 15.00-g sample of ethyl mercaptan indicates that 5.798 g of C, 1.462 g of H, and 7.740 g of S are present. What is the empirical formula of this compound?

9.92 Hydroquinone is a compound used in developing photographic film. Analysis of a 25.00-g sample of hydroquinone indicates that 16.36 g of C, 1.375 g of H, and 7.265 g of O are present. What is the empirical formula of this compound?

9.93 A 2.00-g sample of beryllium metal is burned in an oxygen atmosphere to produce 5.55 g of a beryllium–oxygen compound. Determine the compound's empirical formula.

9.94 A 2.00-g sample of lithium metal is burned in an oxygen atmosphere to produce 4.31 g of a lithium–oxygen compound. Determine the compound's empirical formula.

9.95 Determine the empirical formula of the carbon–hydrogen compound that, upon combustion in a combustion analysis apparatus, generates each of the following sets of CO_2–H_2O data.

(a) 0.338 g of CO_2 and 0.277 g of H_2O
(b) 0.303 g of CO_2 and 0.0621 g of H_2O
(c) 0.225 g of CO_2 and 0.115 g of H_2O
(d) 0.314 g of CO_2 and 0.192 g of H_2O

9.96 Determine the empirical formula of the carbon–hydrogen compound that, upon combustion in a combustion analysis apparatus, generates each of the following sets of CO_2–H_2O data.

(a) 0.269 g of CO_2 and 0.221 g of H_2O
(b) 0.294 g of CO_2 and 0.120 g of H_2O
(c) 0.600 g of CO_2 and 0.184 g of H_2O
(d) 0.471 g of CO_2 and 0.0963 g of H_2O

9.97 A 3.750-g sample of the compound responsible for the odor of cloves (containing only C, H, and O) is burned in a combustion analysis apparatus. The mass of CO_2 produced is 10.05 g, and the mass of H_2O produced is 2.470 g. What is the empirical formula of the compound?

9.98 A 0.8640-g sample of the compound responsible for the pungent odor of rancid butter (containing only C, H, and O) is burned in a combustion analysis apparatus. The mass of CO_2 produced is 1.727 g, and the mass of H_2O produced is 0.7068 g. What is the empirical formula of the compound?

9.99 Determine the molecular formulas of compounds with the following empirical formulas and molecular masses.

(a) P_2O_5, 284 amu (b) SN, 184 amu
(c) $C_3H_6O_2$, 74 amu (d) BNH_2, 80.4 amu

9.100 Determine the molecular formulas of compounds with the following empirical formulas and molecular masses.

(a) NH_2, 32 amu (b) P_2O_3, 220 amu
(c) $C_5H_{10}O_2$, 102 amu (d) $SNCl_2$, 351 amu

9.101 Methyl benzoate, a compound used in the manufacture of perfumes, has a molecular mass of 136 amu, and its percent composition by mass is 70.57% C, 5.93% H, and 23.49% O. Determine the molecular formula of methyl benzoate using:

(a) 100.0 g of methyl benzoate as the basis for the calculation
(b) 1.00 mole of methyl benzoate as the basis for the calculation

9.102 Adipic acid, a compound used as a raw material for the manufacture of nylon, has a molecular mass of 146 amu, and its percent composition by mass is 49.30% C, 6.91% H, and 43.79% O. Determine the molecular formula of adipic acid using:

(a) 100.0 g of adipic acid as the basis for the calculation
(b) 1.00 mole of adipic acid as the basis for the calculation

9.103 Lactic acid, the substance that builds up in muscles and causes them to "hurt" when they are worked hard, has a molecular mass of 90.0 amu and a percent composition by mass of 40.0% C, 6.71% H, and 53.3% O. Determine the molecular formula of lactic acid using:

(a) 100.0 g of lactic acid as the basis for the calculation
(b) 1.00 mole of lactic acid as the basis for the calculation

9.104 Citric acid, a flavoring agent in many carbonated beverages, has a molecular mass of 192 amu and a percent composition by mass of 37.50% C, 4.21% H, and 58.29% O. Determine the molecular formula of citric acid using:

(a) 100.0 g of citric acid as the basis for the calculation

(b) 1.00 mole of citric acid as the basis for the calculation

ADDITIONAL PROBLEMS

9.105 How many grams of potassium and sulfur are theoretically needed to make 4.000 g of K_2S?

9.106 How many grams of beryllium and nitrogen are theoretically needed to make 3.000 g of Be_3N_2?

9.107 How many grams of B would contain the same number of atoms as there are in 3.50 moles of Xe?

9.108 How many grams of Si would contain the same number of atoms as there are in 2.10 moles of Ar?

9.109 How many grams of glucose, $C_6H_{12}O_6$, would contain the same mass of carbon as there is in 3.44 g of ethanol, C_2H_6O?

9.110 How many grams of ethanol, C_2H_6O, would contain the same mass of oxygen as there is in 7.08 g of glucose, $C_6H_{12}O_6$?

9.111 A chemist prepared a sample by mixing 20.0 g of KNO_3 with 1.00 mole of KCl. How many potassium atoms are present in the sample?

9.112 A chemist prepared a sample by mixing 15.0 g of Na_2SO_4 with 1.00 mole of Na_3PO_4. How many sodium atoms are present in the sample?

9.113 Individual samples of the three compounds Cr_2O_3, CrO_2, and CrO_3 are found to contain identical masses of Cr.

(a) Which sample has the largest total mass?

(b) Which sample contains the greatest number of oxygen atoms?

9.114 Individual samples of the three compounds V_2O_5, V_2O_3, and VO_2 are found to contain identical masses of V.

(a) Which sample has the largest total mass?

(b) Which sample contains the greatest number of oxygen atoms?

9.115 A certain alloy of Au, Cu, and Ni contains these elements in the atomic proportions 3:2:1, respectively. What is the mass, in grams, of a sample of this alloy containing a total of 1.00×10^{24} atoms?

9.116 A certain alloy of Sn, Pb, and Bi contains these elements in the atomic proportions 2:4:3, respectively. What is the mass, in grams, of a sample of this alloy containing Avogadro's number of atoms?

9.117 The U.S. recommended daily allowance of vitamin A, $C_{20}H_{30}O$, is 1.5 milligrams. What is this RDA in terms of molecules/day of vitamin A?

9.118 The U.S. recommended daily allowance of vitamin C, $C_6H_8O_6$, is 60. milligrams. What is this RDA in terms of molecules/day of vitamin C?

9.119 Identify each of the following elements.

(a) Atoms of this element have an average mass of 1.792×10^{-22} g.

(b) A sample of this element with a mass of 0.05232 g contains 4.840×10^{-3} mole of atoms.

9.120 Identify each of the following elements.

(a) Atoms of this element have an average mass of 6.647×10^{-24} g.

(b) A sample of this element with a mass of 0.2346 g contains 3.380×10^{-2} mole of atoms.

9.121 For the compound $(CH_3)_3SiCl$ calculate the

(a) mass percent of H present

(b) atom percent of H present

(c) mole percent of H present

9.122 For the compound $(CH_3)_2SiCl_2$ calculate the

(a) mass percent of H present

(b) atom percent of H present

(c) mole percent of H present

9.123 What are the empirical formulas for the compounds that contain each of the following?

(a) 9.0×10^{23} atoms of Na, 3.0×10^{23} atoms of Al, and 1.8×10^{24} atoms of F

(b) 3.2 g of S and 1.20×10^{23} atoms of O

(c) 0.36 mole of Ba, 0.36 mole of C, and 17.2 g of O

(d) 1.81×10^{23} atoms of H, 10.65 g of Cl, and 0.30 mole of O atoms

9.124 What are the empirical formulas for the compounds that contain each of the following?

(a) 3.0×10^{30} atoms of Fe, 3.0×10^{30} atoms of Cr, and 1.2×10^{31} atoms of O

(b) 0.0023 g of N and 2.0×10^{20} atoms of O

(c) 0.40 mole of Li, 6.4 g of S, and 0.80 mole of O

(d) 0.15 mole of S, 1.8×10^{23} atoms of O, and 5.7 g of F.

9.125 Calculate the number of carbon atoms in 5.25 g of a compound that contains 92.26% C and 7.74% H by mass.

9.126 Calculate the number of nitrogen atoms in 5.25 g of a compound that contains 87.39% N and 12.61% H by mass.

9.127 A compound has an empirical formula of C_2H_3O. Calculate the molecular formula of the compound for each of the following cases.

(a) The compound's molecular mass is twice the compound's empirical formula mass.

(b) Molecules of the compound contain 18 atoms.

(c) The sum of the carbon and oxygen atoms in a molecule of the compound is 18.

(d) The mass of 0.010 mole of the compound is 0.86 g.

9.128 A compound has an empirical formula of C_3H_5O. Calculate the molecular formula of the compound for each of the following cases.

(a) The compound's molecular mass and empirical formula mass differ by a factor of 2.

(b) Molecules of the compound contain 18 atoms.

(c) Molecules of the compound contain more than 20 atoms but fewer than 30 atoms.

(d) The mass of 0.010 mole of the compound is 0.57 g.

9.129 A sample of a compound containing only C and H is burned in oxygen and 13.75 g of CO_2 and 11.25 g of H_2O are obtained. What was the mass of the sample, in grams, that was burned?

9.130 A sample of a compound containing only C and H is burned in oxygen and 14.66 g of CO_2 and 9.00 g of H_2O are obtained. What was the mass of the sample, in grams, that was burned?

9.131 A sample of a compound containing only C, H, and S was burned in oxygen, and 6.601 g of CO_2, 5.406 g of H_2O, and 9.615 g of SO_2 were obtained.

(a) What is the empirical formula of the compound?

(b) What was the mass, in grams, of the sample that was burned?

9.132 A sample of a compound containing only C, H, and N was burned in oxygen, and 6.601 g of CO_2, 6.758 g of H_2O, and 4.502 g of NO were obtained.

(a) What is the empirical formula of the compound?

(b) What was the mass, in grams, of the sample that was burned?

9.133 A sample containing NaF, Na_2SO_4, and $NaNO_3$ gives the following elemental analysis by mass: 18.1% F and 6.60% N. Calculate the mass percent of each compound in the mixture.

9.134 A sample containing NaF, Na_2SO_4, and $NaNO_3$ gives the following elemental analysis by mass: 11.3% F and 5.65% S. Calculate the mass percent of each compound in the mixture.

9.135 By analysis, a compound with the formula $KClO_x$ is found to contain 28.9% chlorine by mass. What is the value of the integer x?

9.136 By analysis, a compound with the formula H_3AsO_x is found to contain 52.78% arsenic by mass. What is the value of the integer x?

9.137 A 7.503-g sample of metal is reacted with excess oxygen to yield 10.498 g of the oxide MO. Calculate the molar mass of the element M.

9.138 A 11.17-g sample of metal is reacted with excess oxygen to yield 15.97 g of the oxide M_2O_3. Calculate the molar mass of the element M.

9.139 A certain compound contains only carbon, hydrogen, and oxygen. If it contains 47.4% carbon by mass, and if there is one oxygen atom present for every four hydrogen atoms, what is its empirical formula?

9.140 A certain compound contains only lead, carbon, and hydrogen. If it contains 64.07% lead by mass, and if there are two carbon atoms present for every five hydrogen atoms, what is its empirical formula?

CUMULATIVE PROBLEMS

9.141 Calculate the molecular mass of the compound H_3AsO_4 to the following number of significant figures.

(a) three (b) four (c) five (d) six

9.142 Calculate the molecular mass of the compound H_3AsO_3 to the following number of significant figures.

(a) three (b) four (c) five (d) six

9.143 Calculate the density of the metal lead, in g/cm^3, given that 0.422 mole of lead occupies a volume of 7.74 cm^3.

9.144 Calculate the density of the metal potassium, in g/cm^3, given that 0.674 mole of potassium occupies a volume of 30.8 cm^3.

9.145 Calculate the volume of 2.50 moles of zinc, assuming the density of the metal to be 7.14 g/cm^3.

9.146 Calculate the volume of 1.25 moles of magnesium,

assuming the density of the metal to be 1.74 g/cm^3.

9.147 Calculate the molar mass, to four significant figures, of each of the following compounds.

(a) magnesium nitrate

(b) sodium azide

(c) nickel(II) fluoride

(d) ammonium perchlorate

9.148 Calculate the molar mass, to four significant figures, of each of the following compounds.

(a) sodium thiosulfate

(b) copper(II) hydroxide

(c) ammonium chlorate

(d) aluminum hydrogen phosphate

9.149 The specific heat of the metal silver is 0.24 J/(g · °C). How many joules of energy are needed to raise the temperature of a block of silver containing 2.50 moles of silver atoms by 10.0°C?

9.150 The specific heat of the metal gold is 0.13 J/(g · °C). How many joules of energy are needed to raise the temperature of a block of gold containing 1.40 moles of gold atoms by 15.0°C?

9.151 The percent natural abundance of $^{40}_{19}K$ is 0.012%. How many $^{40}_{19}K$ atoms does a person ingest by drinking one cup of whole milk containing 371 mg of K?

9.152 The percent natural abundance of $^{41}_{19}K$ is 6.88%. How many $^{41}_{19}K$ atoms does a person ingest by drinking two cups of whole milk containing 392 mg of K per cup?

9.153 A 2.33-mole sample of the ionic compound $Al_2(SO_4)_3$ contains how many of the following?

(a) $Al_2(SO_4)_3$ formula units

(b) Al^{3+} ions

(c) SO_4^{2-} ions

(d) total ions

9.154 A 3.50-mole sample of the ionic compound $(NH_4)_3PO_4$ contains how many of the following?

(a) $(NH_4)_3PO_4$ formula units

(b) NH_4^+ ions

(c) PO_4^{3-} ions

(d) total ions

9.155 A mixture consists of 42.0% NaCl and 58.0% $CaCl_2$ by mass. What is the total number of chloride ions (Cl^-) in 425 g of mixture?

9.156 A mixture consists of 22.0% $Cu(NO_3)_2$ and 78.0% $Fe(NO_3)_3$ by mass. What is the total number of nitrate ions (NO_3^-) in 25.00 g of mixture?

9.157 The mass percent composition for an alloy that has a density of 8.31 g/cm^3 is 56.0% copper, 43.0% nickel, and 1.0% manganese. How many nickel atoms are in a block of this alloy measuring 10.0 cm × 30.0 cm × 62.0 cm?

9.158 The mass percent composition for an alloy that has a density of 8.28 g/cm^3 is 64.0% iron, 12.0% cobalt, and 24.0% molybdenum. How many molybdenum atoms are in a block of this alloy measuring 2.0 cm × 1.5 cm × 6.7 cm?

9.159 What volume, in milliliters, of a NaOH solution that is 12.0% NaOH by mass contains 0.275 mole of NaOH? The density of the solution is 1.131 g/mL.

9.160 What volume, in milliliters, of a H_3PO_4 solution that is 85.5% H_3PO_4 by mass contains 0.100 mole of H_3PO_4? The density of the solution is 1.70 g/mL.

9.161 How many total atoms are present in 15.0 mL of an ethyl alcohol–water solution that is 85.0% ethanol by mass? The formula of ethyl alcohol is C_2H_6O, and the density of the solution is 0.807 g/mL.

9.162 How many total atoms are present in 500.0 mL of an ethylene glycol–water solution that is 56.0% ethylene glycol by mass? The formula for ethylene glycol is $C_2H_6O_2$, and the density of the solution is 1.072 g/mL.

9.163 When copper is selling for $0.645 per pound, how many copper atoms could you buy for 1 cent?

9.164 When gold is selling for $390.00 per ounce, how many gold atoms could you buy for 1 cent?

10 Chemical Calculations Involving Chemical Equations

10.1 The Law of Conservation of Mass

In a previous consideration of chemical change (Sec. 4.4), it was noted that chemical changes are referred to as chemical reactions. As we learned there, a *chemical reaction is a process in which at least one new substance is produced as a result of chemical change*. It is usually easy to see that a chemical reaction has occurred. Color change, emission of heat and/or light, gas evolution, and solid formation are an indication that a chemical reaction has taken place.

The starting materials for a chemical reaction are known as reactants. **Reactants** *are all substances present prior to the start of a chemical reaction*. As a chemical reaction proceeds, reactants are consumed (used up) and new materials with new chemical properties are produced. **Products** *are the substances produced as a result of a chemical reaction*.

From a molecular viewpoint, a chemical reaction (chemical change) involves the union, separation, or rearrangement of atoms to produce new substances. Figure 10.1 shows the rearrangement of atoms that occurs when methane (CH_4) reacts with oxygen (O_2) to produce carbon dioxide (CO_2) and water (H_2O). Hydrogen atoms originally associated with carbon atoms (CH_4) became associated with oxygen atoms (H_2O) as the result of the chemical reaction.

Studies of countless chemical reactions over a period of 200 years have shown that there is no detectable change in the quantity of matter present during an ordinary chemical reaction. This generalization concerning chemical reactions has been formalized into a statement known as the **law of conservation of mass**: *mass is neither created nor destroyed in any ordinary chemical reaction*. To demonstrate the validity of this law, the masses of all reactants (substances that react together) and all products (substances formed) in a chemical reaction are carefully determined. It is found that the sum of the masses of the products is

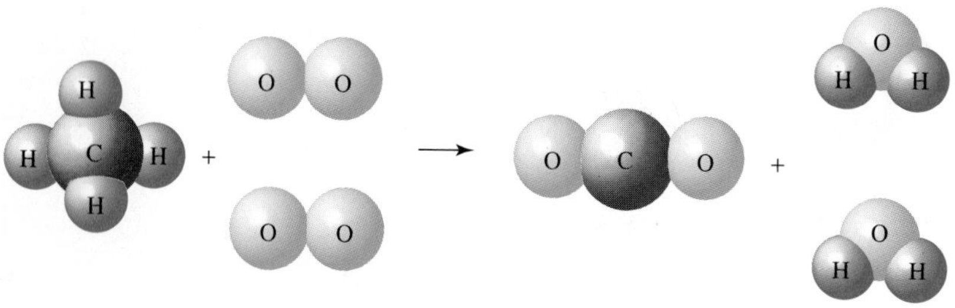

FIGURE 10.1 Rearrangement of atoms that occurs when methane (CH_4) reacts with oxygen (O_2).

always the same as the sum of the masses of the reactants. The French chemist Antoine Laurent Lavoisier (1743–1794) is given credit for being the first to state this important relationship between the reactants and products of a chemical reaction.

Consider, as an illustrative example of this law, the reaction of known masses of the elements beryllium (Be) and oxygen (O) to form the compound beryllium oxide (BeO). Experimentally it is found that 36.03 g of Be will react with *exactly* 63.97 g of O. After the reaction no Be or O remains in elemental form; the only substance present is the product BeO, combined Be and O. When this product is weighed, its mass is found to be 100.00 g, which is the sum of the masses of the reactants (36.03 g + 63.97 g = 100.00 g). It is also found that when the 100.00 g of product BeO is heated to a high temperature in the absence of air, the BeO decomposes into Be and O, producing 36.03 g of Be and 63.97 g of O. Once again, no detectable mass change is observed; the mass of the reactants is equal to the mass of the products.

The law of conservation of mass is consistent with the statements of atomic theory (Sec. 5.1). Since all reacting chemical substances are made up of atoms (statement 1), each with its unique identity (statement 2), and these atoms can be neither created nor destroyed in a chemical reaction but merely rearranged (statement 4), it follows that the total mass after the reaction must equal the total mass before the reaction. We have the same number of atoms of each kind after the reaction as we started out with. An alternative way of stating the law of conservation of mass is: *The total mass of reactants and the total mass of products in a chemical reaction are always equal.*

The law of conservation of mass applies to all ordinary chemical reactions, and there is no known case of a *measurable* change in total mass during an *ordinary* chemical reaction.* This law will be a "guiding principle" for the discussion that follows about chemical equations and their use.

* In the last half century the law of conservation of mass has had to be qualified. Certain types of reactions that involve radioactive processes have been found to deviate from this law. In these processes, there is a conversion of a small amount of matter into energy rather than into another form of matter. A more general law incorporates this apparent discrepancy—the law of conservation of mass and energy. This law takes into account the fact that matter and energy are interconvertible. Note that the statement of the law of conservation of mass as given at the start of this discussion contains the phrase "ordinary chemical reaction." Radioactive processes are not considered to be ordinary chemical reactions.

10.2 WRITING CHEMICAL EQUATIONS

A **chemical equation** *is a written statement that uses symbols and formulas instead of words to describe the changes that occur in a chemical reaction.* The following example shows the contrast between a word description of a chemical reaction and a chemical equation for the same reaction.

Word description: Magnesium oxide reacts with carbon to produce carbon monoxide and magnesium.

Chemical equation: $MgO + C \longrightarrow CO + Mg$

In the same way that chemical symbols are considered the *letters* of chemical language, and formulas the *words* of the language, chemical equations can be considered the *sentences* of chemical language.

The conventions used in writing chemical equations are

1. The correct formulas of the *reactants* are always written on the *left* side of the equation.

$$MgO + C \longrightarrow CO + Mg$$

2. The correct formulas of the *products* are always written on the *right* side of the equation.

$$MgO + C \longrightarrow CO + Mg$$

3. The reactants and products are separated by an arrow pointing toward the products.

$$MgO + C \longrightarrow CO + Mg$$

4. Plus signs are used to separate different reactants or different products from each other.

$$MgO + C \longrightarrow CO + Mg$$

In reading chemical equations, plus signs on the reactant side of the equation are taken to mean "reacts with"; the arrow, "to produce"; and plus signs on the product side, "and."

A catchy, informal way of defining a chemical equation is to say that it gives the "before and after" picture of a chemical reaction. *Before* the reaction starts, only reactants are present—the left side of the equation. *After* the reaction is completed, products are present—the right side of the equation.

In order for a chemical equation to be *valid* it must satisfy two conditions.

1. *It must be consistent with experimental facts.* Only the reactants and products actually involved in a reaction are shown in an equation. The accurate formula must be used for each of these substances. For compounds, molecular rather than empirical formulas (Sec. 9.9) are always used. Elements in the solid and liquid states are represented in equations by the chemical symbol for the element. Elements that are gases at room temperature are represented by the molecular form in which they actually occur in nature. Monoatomic, diatomic, and tetraatomic elemental gases are known.

Monoatomic:	He, Ne, Ar, Kr, Xe, Rn
Diatomic:	H_2, O_2, N_2, F_2, Cl_2, Br_2 (vapor), I_2 (vapor)
Tetraatomic:	P_4 (vapor), As_4 (vapor)*

2. *It must be consistent with the law of conservation of mass* (Sec. 10.1). There must be the same number of product atoms of each kind as there are reactant atoms of each kind, since atoms are neither created nor destroyed in an ordinary chemical reaction. Equations that satisfy the conditions of this law are said to be *balanced.* Using the four conventions previously listed for writing equations does not guarantee a balanced equation. Section 10.3 considers the steps that must be taken to ensure that an equation is balanced.

10.3 BALANCING CHEMICAL EQUATIONS

A **balanced chemical equation** *has the same number of atoms of each element involved in the reaction on each side of the equation.* It is therefore an equation consistent with the law of conservation of mass (Sec. 10.1).

An unbalanced equation is brought into balance by adding coefficients to the equation; such coefficients adjust the number of reactant and/or product molecules (or formula units) present. A **coefficient** *is a number placed to the left of the formula of a substance that denotes the amount of the substance.* In the notation 2 H_2O, the 2 on the left is a coefficient; 2 H_2O means two molecules of H_2O, and 3 H_2O means three molecules of H_2O. Coefficients tell how many formula units of a given substance are present.

The following is a balanced chemical equation with the coefficients shown in color.

$$3\ Cu + 8\ HNO_3 \longrightarrow 3\ Cu(NO_3)_2 + 2\ NO + 4\ H_2O$$

The message of this balanced equation is "three Cu atoms react with eight HNO_3 molecules to produce three $Cu(NO_3)_2$ formula units, two NO molecules, and four H_2O molecules." A coefficient of 1 in a balanced equation is not explicitly written; it is considered to be understood. Both PCl_3 and H_3PO_3 have understood coefficients of 1 in the following balanced equation.

$$PCl_3 + 3\ H_2O \longrightarrow H_3PO_3 + 3\ HCl$$

A coefficient placed in front of a formula applies to the whole formula. In contrast, subscripts, also present in formulas, affect only parts of a formula.

coefficient (affects both the H and O)

$2\ H_2O$

subscript (affects only H)

The above notation denotes two molecules of H_2O; it also denotes a total of four H atoms and two O atoms.

We now proceed to the method for determining the proper coefficients to balance a given equation. It is introduced in the context of actually balancing two equations (Examples 10.1 and 10.2). Both of these examples should be studied carefully, because each includes detailed commentary concerning the "ins and outs" of balancing equations.

* The four elements listed as vapors are not gases at room temperature but vaporize at slightly higher temperatures. The resultant vapors contain molecules with the formulas indicated. Even if these elements do not vaporize, they can still be represented with these formulas.

Example 10.1

Balance the equation $Fe_3O_4 + H_2 \longrightarrow Fe + H_2O$.

Solution

STEP 1 *Examine the equation, and pick one element to balance first.* It is often convenient to identify the most complex substance first, that is, the substance with the greatest number of atoms per formula unit. For this most complex substance, whether a reactant or product, "key in" on the ele ment within it that is present in the greatest amount (greatest number of atoms). Using this guideline, we select Fe_3O_4 and the element oxygen.

We note that there are four oxygen atoms on the left side of the equation (in Fe_3O_4) and only one oxygen atom on the right side (in H_2O). For the oxygen atoms to balance we will need four on each side. To obtain four atoms of oxygen on each side of the equation we place the coefficient 1 in front of Fe_3O_4 and the coefficient 4 in front of H_2O.

$$1\ Fe_3O_4 + H_2 \longrightarrow Fe + 4\ H_2O$$

The coefficient 1 (in front of Fe_3O_4) has been explicitly shown in the preceding equation to remind us that the Fe_3O_4 coefficient has been determined. (In the final balanced equation the 1 need not be shown.) We now have four oxygen atoms on each side of the equation.

$$1\ Fe_3O_4:\quad 1 \times 4 = 4$$
$$4\ H_2O:\quad 4 \times 1 = 4$$

STEP 2 *Now pick a second element to balance.* We will balance the element Fe next. (In this particular equation it does not matter whether we balance Fe or H second.) The number of Fe atoms on the left side of the equation is three; the coefficient 1 in front of Fe_3O_4 sets the Fe atom number at three. We will need three Fe atoms on the product side. This is accomplished by placing the coefficient 3 in front of Fe.

$$1\ Fe_3O_4 + H_2 \longrightarrow 3\ Fe + 4\ H_2O$$

Now there are three Fe atoms on each side of the equation.

STEP 3 *Now pick a third element to balance.* The only element left to balance is H. There are two H atoms on the left and eight H atoms on the right (4 H_2O involves 8 H atoms). Placing the coefficient 4 in front of H_2 on the left side gives 8 H atoms on that side.

$$1\ Fe_3O_4 + 4\ H_2 \longrightarrow 3\ Fe + 4\ H_2O$$

STEP 4 *As a final check on the correctness of the balancing procedure, count atoms on each side of the equation.* The following table can be constructed from our balanced equation.

$$Fe_3O_4 + 4\ H_2 \longrightarrow 3\ Fe + 4\ H_2O$$

Atom	Left Side	Right Side
Fe	$1 \times 3 = 3$	$3 \times 1 = 3$
O	$1 \times 4 = 4$	$4 \times 1 = 4$
H	$4 \times 2 = 8$	$4 \times 2 = 8$

All elements are in balance: three Fe atoms on each side, four O atoms on each side, and eight H atoms on each side.

Practice Exercise 10.1

Balance the equation $Fe_2O_3 + C \longrightarrow Fe + CO_2$

Ans. $2\ Fe_2O_3 + 3\ C \longrightarrow 4\ Fe + 3\ CO_2$

Example 10.2

Balance the equation $C_4H_{10} + O_2 \longrightarrow CO_2 + H_2O$.

Solution

STEP 1 *Examine the equation, and pick one element to balance first.* The formula containing the most atoms is C_4H_{10}. We will balance the element H first. We have ten H atoms on the left and two H atoms on the right. The two sides are brought into balance by placing the coefficient 5 in front of H_2O on the right side. We now have ten H atoms on each side.

$$1\ C_4H_{10}:\quad 1 \times 10 = 10$$
$$5\ H_2O:\quad 5 \times 2(1) = 10$$

Our equation now has the following appearance.

$$1\ C_4H_{10} + O_2 \longrightarrow CO_2 + 5\ H_2O$$

In setting the H balance at ten atoms we are setting the coefficient in front of C_4H_{10} at 1. The 1 has been explicitly shown in the above equation to remind us that the C_4H_{10} coefficient has been determined. (In the final balanced equation the 1 should not be shown.)

STEP 2 *Now pick a second element to balance.* We will balance C next. It is always better to balance the elements that appear in only one reactant and one product before trying to balance any elements appearing in several formulas on one side of the equation. Oxygen, our other choice for an element to balance at this stage, appears in two places on the product side of the equation. The number of carbon atoms is already set at four on the left side of the equation.

$$1\ C_4H_{10}:\quad 1 \times 4 = 4$$

We obtain a balance of two carbon atoms on each side of the equation by placing the coefficient 4 in front of CO_2.

$$1\ C_4H_{10} + O_2 \longrightarrow 4\ CO_2 + 5\ H_2O$$

STEP 3 *Now pick a third element to balance.* Only one element is left to balance—oxygen. The number of oxygen atoms on the right side of the equation is already set at thirteen: eight O atoms from the CO_2 and five O atoms from the H_2O.

$$4\ CO_2:\quad 4 \times 2 = 8$$
$$5\ H_2O:\quad 5 \times 1 = 5$$

To obtain thirteen O atoms on the left side of the equation we need a fractional coefficient, $6\frac{1}{2}$.

$$6\tfrac{1}{2}\ O_2:\quad 6\tfrac{1}{2} \times 2 = 13$$

The coefficient 6 gives 12 atoms and the coefficient 7 gives 14 atoms. The only way we can get 13 atoms is by using $6\frac{1}{2}$.

All of the coefficients in the equation have now been determined.

$$1\ C_2H_6 + 6\tfrac{1}{2}\ O_2 \longrightarrow 4\ CO_2 + 5\ H_2O$$

Equations containing fractional coefficients are not considered to be written in their most conventional form. Although such equations are "mathematically" correct, they have some problems "chemically." The above equation indicates the need for $6\frac{1}{2}$ O_2 molecules among the reactants, but half an O_2 molecule does not exist as such. Step 4 shows how to take care of this "problem."

STEP 4 *After all coefficients have been determined, clear any fractional coefficients that are present.* We can clear the fraction present in this equation, $6\frac{1}{2}$, by multiplying *each* of the coefficients in the equation by the factor 2.

$$2\ C_4H_{10} + 13\ O_2 \longrightarrow 8\ CO_2 + 10\ H_2O$$

Now we have the equation in its conventional form. Note that *all of the coefficients* had to be multiplied by 2, not just the fractional one. It will always be the case that whatever is done to a fractional coefficient to make it a whole number must also be carried out on all of the other coefficients.

If a coefficient involving $\frac{1}{3}$ had been present in the equation, we would have multiplied by 3 instead of by 2.

STEP 5 *As a final check on the correctness of the balancing procedure, count atoms on each side of the equation.* The following table can be constructed from our balanced equation.

$$2\ C_4H_{10} + 13\ O_2 \longrightarrow 8\ CO_2 + 10\ H_2O$$

Atom	Left Side	Right Side
C	$2 \times 4 = 8$	$8 \times 1 = 8$
H	$2 \times 10 = 20$	$10 \times 2 = 20$
O	$13 \times 2 = 26$	$(8 \times 2) + (10 \times 1) = 26$

All atom counts balance. We have accomplished our task.

Practice Exercise 10.2

Balance the equation $C_2H_6 + O_2 \longrightarrow CO_2 + H_2O$.

Ans. $2\ C_2H_6 + 7\ O_2 \longrightarrow 4\ CO_2 + 6\ H_2O$

Some additional comments and guidelines concerning equations in general and the process of balancing in particular are

1. The coefficients in a balanced equation are always the *smallest set of whole numbers* that will balance the equation. We mention this because more than one set of coefficients will balance an equation. Consider the following three equations.

$$2\,H_2 + O_2 \longrightarrow 2\,H_2O$$
$$4\,H_2 + 2\,O_2 \longrightarrow 4\,H_2O$$
$$8\,H_2 + 4\,O_2 \longrightarrow 8\,H_2O$$

   All three of these equations are mathematically correct; there are equal numbers of H and O atoms on each side of the equation. The first equation, however, is considered the conventional form, because the coefficients used there are the smallest set of whole numbers that will balance the equation. The coefficients in the second equation are double those in the first, and the third equation has coefficients four times those of the first equation.
2. It is helpful to consider polyatomic ions as single entities in balancing an equation, provided they maintain their identity in the chemical reaction, that is, provided they appear on both sides of the equation in the same form. For example, in an equation where sulfate units are present (SO_4^{2-}), both as reactants and products, balance them as a unit rather than trying to balance S and O separately. The reasoning would be: "We have two sulfates on this side, so we need two sulfates on that side, etc."
3. Subscripts in a formula may never be altered during the balancing process. A student might try to balance the atoms of K in KCl at two by using the notation K_2Cl_2 instead of 2 KCl. This is incorrect. The notation K_2Cl_2 denotes a formula unit containing four atoms, whereas the notation 2 KCl denotes two formula units, each containing two atoms. The experimental fact is that a formula unit of KCl contains two rather than four atoms. *The coefficient deals with the number of formula units of a substance, and the subscript deals with the composition of the substance.* Subscripts illustrate the law of definite proportions (Sec. 9.1); coefficients relate to the law of conservation of mass (Sec. 10.1).
4. You are not expected, at this point, to be able to write down the products for a chemical reaction given what the reactants are. After learning how to balance equations, students sometimes get the mistaken idea that they ought to be able to write down equations from "scratch." This is not so. At this stage, you should be able to balance simple equations given *all* of the reactants and products. In Section 10.5 guidelines are given for predicting the products for selected types of reaction.
5. Some equations are much more difficult to balance than those you encounter in this chapter's examples and problem exercises. The procedures discussed here simply are not adequate for these more difficult equations. In Chapter 15 a more systematic method for balancing equations, specifically designed for these more difficult situations, will be presented.
6. The ultimate source of any chemical equation is experimental information. The identities of the products formed in a chemical reaction are learned by experiment; we cannot discover them simply by writing an equation. The products are first identified experimentally, and then they can be represented by an equation.

7. Finally, the only way to learn to balance equations is through practice. The problem set at the end of this chapter contains numerous equations for you to practice on.

10.4 SPECIAL SYMBOLS USED IN EQUATIONS

In addition to the essential plus sign and arrow notation used in chemical equations, a number of optional symbols convey more information about a chemical reaction than just the chemical species involved. In particular, it is often useful to know the physical state of the substances involved in a chemical reaction. The optional symbols listed in Table 10.1 are used to specify the physical state of reactants and products.

The equations we balanced in Section 10.3 (Examples 10.1 and 10.2) are written as follows when the optional symbols are included.

$$Fe_3O_4(s) + 4\ H_2(g) \longrightarrow 3\ Fe(s) + 4\ H_2O(l)$$
$$2\ C_4H_{10}(g) + 13\ O_2(g) \longrightarrow 8\ CO_2(g) + 10\ H_2O(g)$$

Two more examples of the use of optional symbols are

$$NaCl(aq) + AgNO_3(aq) \longrightarrow AgCl(s) + NaNO_3(aq)$$
$$NaOH(aq) + HCl(aq) \longrightarrow NaCl(aq) + H_2O(l)$$

The optional symbols in these latter two equations indicate that both reactions take place in aqueous solution. In the first reaction, one of the products, AgCl, is insoluble, being present in the solution as a solid. In the second reaction, the product NaCl is soluble and thus remains in solution.

Also, note that in the last equation the reactant HCl is functioning as an acid and must be named as such (Sec. 8.6). The notation HCl(g) indicates hydrogen chloride in its gaseous state; the notation HCl(aq), hydrogen chloride dissolved in water, denotes hydrochloric acid.

TABLE 10.1 SYMBOLS USED IN EQUATIONS

Symbol	**Meaning**
Essential	
→	"to produce"
+	"reacts with" or "and"
Optional	
(s)	solid
(l)	liquid
(g)	gas
(aq)	aqueous solution (a substance dissolved in water)

10.5 PATTERNS IN CHEMICAL REACTIVITY

To this point in our discussion of chemical equations the focus has been on balancing them given all of the reactants and all of the products. We now consider, for selected situations, the prediction of what the products of a chemical reaction will be.

Combustion reactions are a most common type of reaction. A **combustion reaction** *involves the reaction of a substance with oxygen (usually from air) that proceeds with evolution of heat and usually also a flame.* Hydrocarbons, binary compounds of carbon and hydrogen (of which many exist), are the most common type of compound that undergoes combustion. In hydrocarbon combustion, the carbon of the hydrocarbon combines with oxygen to produce carbon dioxide (CO_2). The hydrocarbon hydrogen also interacts with oxygen of air to give water (H_2O) as a product. The relative amounts of CO_2 and H_2O produced depends on the composition of the hydrocarbon.

F.Y.I.

Combustion reactions are the basis of industrial society, whether it is burning gasoline in cars, natural gas in homes, or coal in factories. Unlike most other reactions, hydrocarbons are burned for the energy produced not the material products.

$$2\ C_2H_2(g) + 5\ O_2(g) \longrightarrow 4\ CO_2(g) + 2\ H_2O(g)$$
$$C_3H_8(g) + 5\ O_2(g) \longrightarrow 3\ CO_2(g) + 4\ H_2O(g)$$
$$C_4H_8(g) + 6\ O_2(g) \longrightarrow 4\ CO_2(g) + 4\ H_2O(g)$$

Note that the equation that was balanced in Example 10.2 was a combustion reaction.

Combustion of compounds containing oxygen as well as carbon and hydrogen (for example CH_4O or C_3H_8O) also produce CO_2 and H_2O as products.

$$2\ CH_4O(l) + 3\ O_2(g) \longrightarrow 2\ CO_2(g) + 4\ H_2O(g)$$
$$2\ C_3H_8O(l) + 9\ O_2(g) \longrightarrow 6\ CO_2(g) + 8\ H_2O(g)$$

Example 10.3 is an equation-balancing exercise involving a reaction of this type.

Example 10.3

Diethyl ether, $C_4H_{10}O$, was one of the first general anesthetics. Special precautions had to be taken when it was used because of its extreme flammability. Operating room fires are not desirable. Write the balanced chemical equation for the reaction that occurs when diethyl ether burns in air (O_2).

Solution

The C atoms in the ether will end up in product CO_2 and the H atoms of the ether in product H_2O. There are two sources for the oxygen present in the products CO_2 and H_2O. Most of it comes from the O_2 of air; however, the reactant $C_4H_{10}O$ is also an oxygen source. The unbalanced equation for this reaction is

$$C_4H_{10}O + O_2 \longrightarrow CO_2 + H_2O$$

The equation is balanced using the procedures of Section 10.3.

STEP 1 *Balancing of C atoms:* There are four C atoms on the left and only one C atom on the right. Placing the coefficient 4 in front of CO_2 balances the C atoms at four on each side.

$$1\ C_4H_{10}O + O_2 \longrightarrow 4\ CO_2 + H_2O$$

STEP 2 *Balancing of H atoms:* An effect of balancing the C atoms at four (step 1) is the setting of the H atoms on the left side of the equation at ten;

the coefficient in front of $C_4H_{10}O$ is 1. Placing the coefficient 5 in front of H_2O causes the hydrogen atoms to balance at ten on each side of the equation.

$$1\ C_4H_{10}O + O_2 \longrightarrow 4\ CO_2 + 5\ H_2O$$

STEP 3 *Balancing of O atoms:* The oxygen content of the right side of the equation is set at thirteen atoms; eight oxygen atoms from 4 CO_2 and five oxygen atoms from 5 H_2O. To obtain thirteen oxygen atoms on the left side of the equation the coefficient 6 is placed in front of O_2; 6 O_2 gives 12 oxygen atoms and there is an additional O in 1 C_2H_6O. Note that the element oxygen is present in all four formulas in the equation

$$1\ C_4H_{10}O + 6\ O_2 \longrightarrow 4\ CO_2 + 5\ H_2O$$

STEP 4 *Final check:* The equation is balanced. There are four carbon atoms, ten hydrogen atoms, and thirteen oxygen atoms on each side of the equation.

$$C_4H_{10}O + 6\ O_2 \longrightarrow 4\ CO_2 + 5\ H_2O$$

Practice Exercise 10.3

Ethyl alcohol, C_2H_6O, is a component of many oxygenated gasolines that are used during winter time. Write the balanced chemical equation for the reaction that occurs when ethyl alcohol burns in air (O_2).

Ans. $C_2H_6O + 3\ O_2 \longrightarrow 2\ CO_2 + 3\ H_2O$

Thermal decomposition of metal carbonates is another common type of reaction. Metal carbonates, when heated to a high temperature (the temperature needed varies with the metal), break down, releasing carbon dioxide gas and producing the metal oxide. Typical carbonate decomposition equations include the following.

$$Na_2CO_3(s) \longrightarrow Na_2O(s) + CO_2(g)$$
$$MgCO_3(s) \longrightarrow MgO(s) + CO_2(g)$$
$$Al_2(CO_3)_3(s) \longrightarrow Al_2O_3(s) + 3\ CO_2(g)$$

F.Y.I.
Glucose sugar ($C_6H_{12}O_6$) is "burned" in the body by the reaction
$C_6H_{12}O_6 + 6\ O_2 \longrightarrow 6\ CO_2 + 6\ H_2O$
only it is accomplished with the help of enzymes rather than an open flame.

10.6 CLASSES OF CHEMICAL REACTIONS

An almost inconceivable number of chemical reactions is possible. The problems associated with organizing our knowledge about them are diminished considerably by grouping the reactions into classes. We consider in this section an important classification system for reactions based on the *form* of the equation for the reaction. In this system four general types of reactions are recognized: synthesis, decomposition, single replacement, and double replacement.

The first of the four categories of reactions is the synthesis reaction. A **synthesis reaction** *is one in which a single product is produced from two (or more) reactants.*

F.Y.I.
Sulfuric acid (H_2SO_4), one of the major components of acid rain, is produced in the air when sulfur-contaminated fuel is burned. It happens in three synthesis reactions: First sulfur combines with oxygen to form sulfur dioxide,
$S + O_2 \longrightarrow SO_2$
Sulfur dioxide reacts with another molecule of air to form sulfur trioxide,
$2\ SO_2 + O_2 \longrightarrow 2\ SO_3$
Sulfur trioxide reacts with moisture in the air to form sulfuric acid,
$SO_3 + H_2O \longrightarrow H_2SO_4$

$$X + Y \longrightarrow XY$$

This is an extremely simple type of reaction, one in which two substances join together to form a more complicated product. The reactants X and Y may be elements or compounds or an element and a compound. The product of the reaction, XY, is always a compound.

Some representative synthesis reactions with elements as the reactants are

$$H_2 + Cl_2 \longrightarrow 2\ HCl$$
$$S + O_2 \longrightarrow SO_2$$
$$Ni + S \longrightarrow NiS$$

Some examples of synthesis reactions in which compounds are involved as reactants are

$$SO_3 + H_2O \longrightarrow H_2SO_4$$
$$K_2O + H_2O \longrightarrow 2\ KOH$$
$$2\ NO + O_2 \longrightarrow 2\ NO_2$$
$$2\ NO_2 + H_2O_2 \longrightarrow 2\ HNO_3$$

F.Y.I.
The decomposition reaction
$CaCO_3 \longrightarrow CaO + CO_2$
occurs when limestone ($CaCO_3$) is heated in cement factories to produce calcium oxide, one of the components of cement.

A **decomposition reaction** *is one in which a single reactant is converted into two or more simpler substances.* It is thus the exact opposite of a synthesis reaction. The general equation for a decomposition reaction is

$$XY \longrightarrow X + Y$$

At sufficiently high temperatures all compounds can be broken down (decomposed) into their constituent elements. Such reactions include

$$2\ CuO \longrightarrow 2\ Cu + O_2$$
$$2\ H_2O \longrightarrow 2\ H_2 + O_2$$

F.Y.I.
When hydrogen peroxide is poured on a wound, it decomposes into water and oxygen:
$2\ H_2O_2 \longrightarrow 2\ H_2O + O_2$
The bubbles that form are oxygen gas.

Examples of decomposition reactions that result in at least one compound as a product are

$$CaCO_3 \longrightarrow CaO + CO_2$$
$$2\ KClO_3 \longrightarrow 2\ KCl + 3\ O_2$$

A **single-replacement reaction** *is one in which one element within a compound is replaced by another element.* In this type of reaction there are always two reactants, an element and a compound, and two products, also an element and a compound. The general equation for a single-replacement reaction is

$$X + YZ \longrightarrow Y + XZ$$

F.Y.I.
When a can of soda pop is opened, carbonic acid decomposes to carbon dioxide and water:
$H_2CO_3 \longrightarrow CO_2 + H_2O$
The carbon dioxide escapes as bubbles. The soda goes flat when all the carbonic acid decomposes.

Both metals and nonmetals can be replaced in this manner. Such reactions usually involve aqueous solutions. Examples include

$$Zn + H_2SO_4 \longrightarrow H_2 + ZnSO_4$$
$$Ni + 2\ HCl \longrightarrow H_2 + NiCl_2$$
$$Fe + CuSO_4 \longrightarrow Cu + FeSO_4$$
$$Mg + Ni(NO_3)_2 \longrightarrow Ni + Mg(NO_3)_2$$

A **double-replacement reaction** *is one in which two compounds exchange parts with each other and form two different compounds.* The general equation for such a reaction is

$$AX + BY \longrightarrow AY + BX$$

Double-replacement reactions generally involve ionic compounds in aqueous solution. Most often the positive ion from one compound exchanges with the positive ion of the

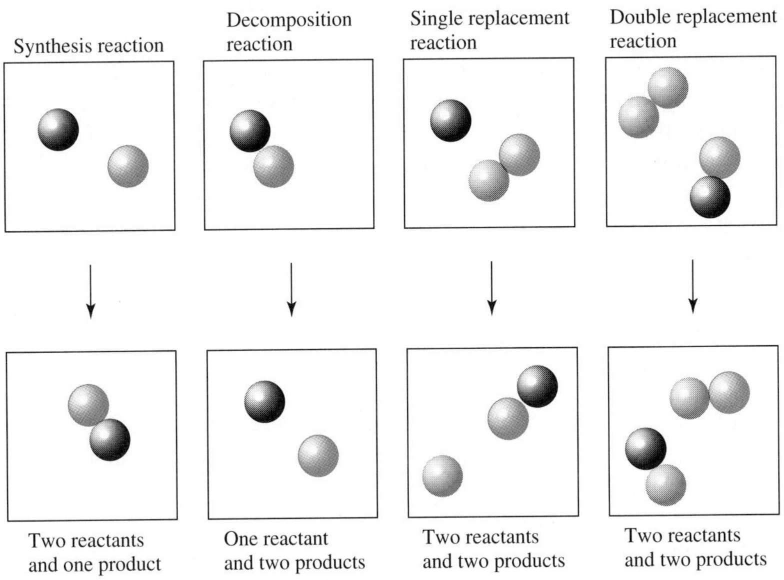

FIGURE 10.2 Diagrams of the four general types of chemical reactions.

other. The process may be thought of as "partner swapping," since each negative ion ends up paired with a new partner (positive ion). Such reactions include

$$AgNO_3 + NaCl \longrightarrow NaNO_3 + AgCl$$
$$NaF + HCl \longrightarrow NaCl + HF$$
$$AgNO_3 + HCl \longrightarrow AgCl + HNO_3$$

Figure 10.2 summarizes in pictorial form the four general types of chemical reactions.

F.Y.I.

One type of dry fire extinguisher (contains no water and can be used on electrical fires) contains finely powdered sodium bicarbonate (baking soda). When sprayed on a fire, the heat causes the sodium bicarbonate to break down into sodium carbonate, water, and carbon dioxide:
$NaHCO_3 \longrightarrow Na_2CO_3 + H_2O + CO_2$
The carbon dioxide formed smothers the flame.

Example 10.4

Classify each of the following reactions as synthesis, decomposition, single replacement, or double replacement.

(a) $CuCO_3 \longrightarrow CuO + CO_2$

(b) $Fe + Cu(NO_3)_2 \longrightarrow Cu + Fe(NO_3)_2$

(c) $3\ Mg + N_2 \longrightarrow Mg_3N_2$

(d) $NH_4Cl + AgNO_3 \longrightarrow NH_4NO_3 + AgCl$

Solution

(a) Since two substances are produced from a single substance, this reaction is a decomposition reaction.

(b) Having an element and a compound as reactants and an element and a compound as products is a characteristic of a single-replacement reaction.

(c) Two substances combine to form a single substance; hence, this reaction is classified as a synthesis reaction.

(d) This reaction is a double-replacement reaction. Ammonium ion and silver ion are changing places, that is, "swapping partners."

Practice Exercise 10.4

Classify each of the following reactions as synthesis, decomposition, single replacement, or double replacement.

(a) $2\,C + O_2 \longrightarrow 2\,CO$

(b) $2\,KNO_3 \longrightarrow 2\,KNO_2 + O_2$

(c) $Zn + 2\,AgNO_3 \longrightarrow Zn(NO_3)_2 + 2\,Ag$

(d) $Ni(NO_3)_2 + 2\,NaOH \longrightarrow Ni(OH)_2 + 2\,NaNO_3$

Ans. (a) synthesis; (b) decomposition; (c) single-replacement; (d) double-replacement

Many, but not all, reactions fall into the four categories we have just discussed. For example, combustion reactions (Sec. 10.5) do not fit any of the four general patterns. Despite the system not being all-inclusive, it is still very useful because of the many reactions it does help correlate.

We should note, before leaving this topic, that other ways of classifying chemical reactions exist. Two important additional categories of reactions are *acid–base reactions* and *oxidation–reduction reactions*. Acid–base reactions are the topic of Chapter 14 and oxidation–reduction reactions are considered in Chapter 15. Combustion reactions are examples of oxidation–reduction reactions.

10.7 CHEMICAL EQUATIONS AND THE MOLE CONCEPT

The coefficients in a balanced chemical equation, like the subscripts in a chemical formula (Sec. 9.7), have two levels of interpretation—a microscopic level of meaning and a macroscopic level of meaning.

The first of these two interpretations, the microscopic level, has been used in the previous sections of this chapter. At this level, a balanced chemical equation gives the relative numbers of formula units of the various reactants and products involved in a chemical reaction. *The coefficients in the equation give directly the numerical relationships among formula units consumed (used up) and/or produced in the chemical reaction.* Interpreted at the microscopic level, the equation

$$4\,NH_3 + 5\,O_2 \longrightarrow 4\,NO + 6\,H_2O$$

conveys the information that four molecules of NH_3 react with five molecules of O_2 to produce four molecules of NO and six molecules of H_2O.

At the macroscopic level of interpretation, chemical equations are used to relate mole-sized quantities of reactants and products to each other. At this **molar (macroscopic) level of interpretation**, *the coefficients in an equation give the fixed molar ratios between substances consumed and/or produced in the chemical reaction.* Thus the equation

$$4\,NH_3 + 5\,O_2 \longrightarrow 4\,NO + 6\,H_2O$$

conveys the information that four moles of NH_3 react with five moles of O_2 to produce four moles of NO and six moles of H_2O.

The validity of the molar interpretation of coefficients in an equation can be derived very straightforwardly from the microscopic level of interpretation. A balanced chemical equation remains valid (mathematically correct) when all of its coefficients are multiplied

F.Y.I.

To a chemist a chemical equation is a recipe. Just like the (very simple) recipe:
3 cups flour + 1 cup milk = 2 cakes
says that if 3 cups of flour are mixed with 1 cup of milk, 2 cakes can be made. The equation
$3\,H_2 + N_2 \longrightarrow 2\,NH_3$
tells a chemist that when 3 moles of H_2 are reacted with 1 mole of nitrogen, 2 moles of NH_3 are formed.

by the same number. (If molecules react in a 3-to-1 ratio, they will also react in a 6-to-2 or 9-to-3 ratio.) Multiplying the previous equation by y, where y is any number, we have

$$4y\ NH_3 + 5y\ O_2 \longrightarrow 4y\ NO + 6y\ H_2O$$

The situation where $y = 6.02 \times 10^{23}$ is of particular interest because $6.02 \times 10^{23} = 1$ mole. Using $y = 1$ mole, we have by substitution

$$4\text{ moles } NH_3 + 5\text{ moles } O_2 \longrightarrow 4\text{ moles } NO + 6\text{ moles } H_2O$$

Thus, as with the subscripts in formulas, the coefficients in equations carry a dual meaning: "number of formula units" at the microscopic level and "moles of formula units" at the macroscopic level.

In Section 10.3 it was noted that fractional equation coefficients are often obtained in the equation-balancing process. We can now further note that such fractional coefficients do have valid meaning for the macroscopic level interpretation of a chemical equation ($3\frac{1}{2}$ moles, and so on), whereas they are totally unacceptable for the microscopic level of interpretation ($3\frac{1}{2}$ molecules, and so on).

The coefficients in an equation may be used to generate conversion factors used in problem solving. Numerous conversion factors are obtainable from a single balanced equation. Consider the balanced equation

$$P_4O_{10} + 6\ H_2O \longrightarrow 4\ H_3PO_4$$

Three mole-to-mole relationships can be obtained from this equation.

1 mole of P_4O_{10} produces 4 moles of H_3PO_4
6 moles of H_2O produces 4 moles of H_3PO_4
1 mole of P_4O_{10} reacts with 6 moles of H_2O

From these three macroscopic-level relationships, six conversion factors can be written.

From the first relationship:

$$\frac{1\text{ mole } P_4O_{10}}{4\text{ moles } H_3PO_4} \quad \text{and} \quad \frac{4\text{ moles } H_3PO_4}{1\text{ mole } P_4O_{10}}$$

From the second relationship:

$$\frac{6\text{ moles } H_2O}{4\text{ moles } H_3PO_4} \quad \text{and} \quad \frac{4\text{ moles } H_3PO_4}{6\text{ moles } H_2O}$$

From the third relationship:

$$\frac{1\text{ mole } P_4O_{10}}{6\text{ moles } H_2O} \quad \text{and} \quad \frac{6\text{ moles } H_2O}{1\text{ mole } P_4O_{10}}$$

Any chemical equation can be the source of numerous conversion factors. The more reactants and products there are in the equation, the greater the number of conversion factors.

Conversion factors obtained from equations are used in many different types of calculations. Example 10.5 illustrates some very simple applications of their use. In Section 10.8 we explore their use in more complicated problem-solving situations.

Example 10.5

Two air pollutants present in automobile exhaust are carbon monoxide (CO) and nitrogen monoxide (NO). Within an automobile's catalytic converter, these two pollutants react with each other to produce carbon dioxide (CO_2) and nitrogen (N_2). The equation for the reaction is

$$2\ CO + 2\ NO \longrightarrow 2\ CO_2 + N_2$$

(a) How many moles of N_2 are produced when 3.50 moles of CO reacts?

(b) How many moles of NO are needed to react with 2.31 moles of CO?

Solution

Both parts of this problem are one-step mole-to-mole calculations. In each case the needed conversion factor is derived from the coefficients of the chemical equation.

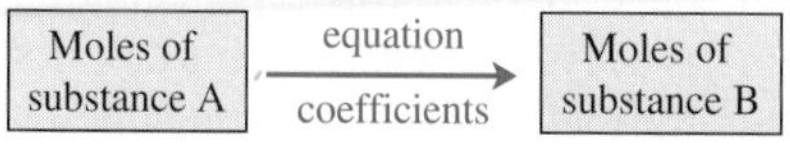

(a)

STEP 1 The given quantity is 3.50 moles of CO and the desired quantity is moles of N_2

$$3.50 \text{ moles CO} = ? \text{ moles } N_2$$

STEP 2 The conversion factor needed to convert from moles of CO to moles of N_2 is derived from the coefficients of CO and N_2 in the balanced equation. The equation tells us that 2 moles of CO produces 1 mole of N_2. From this relationship two conversion factors are obtainable

$$\frac{2 \text{ moles CO}}{1 \text{ mole } N_2} \quad \text{and} \quad \frac{1 \text{ mole } N_2}{2 \text{ moles CO}}$$

We will use the second of these conversion factors in solving the problem. The setup is

$$3.50 \cancel{\text{ moles CO}} \times \frac{1 \text{ mole } N_2}{2 \cancel{\text{ moles CO}}}$$

We used the second of the two conversion factors because it had moles of CO in the denominator, a requirement for the units moles of CO to cancel.

STEP 3 Collecting numerical terms, after cancellation of units, gives

$$\frac{3.50 \times 1}{2} \text{ moles } N_2 = 1.75 \text{ moles } N_2 \text{ (calculator and correct answer)}$$

Note that the coefficients in the equation enter directly into the numerical calculation. Having a correctly balanced equation is therefore of vital importance. Using an unbalanced or misbalanced equation as a source of a conversion factor will lead to a wrong numerical answer.

(b)

STEP 1 Both the given species, CO, and the desired species, NO, are reactants.

$$2.31 \text{ moles CO} = ? \text{ moles NO}$$

STEP 2 Molar relationships obtained from an equation are not required to always involve one reactant and one product, as was the case in part (a).

Molar relationships involving only reactants or only products are often needed and used. In this problem we will need the molar relationship between the two reactants, CO and NO, which is two to two. From this ratio the conversion factor

$$\frac{2 \text{ moles NO}}{2 \text{ moles CO}}$$

can be constructed, which is used in the setup of the problem as follows.

$$2.31 \text{ } \cancel{\text{moles CO}} \times \frac{2 \text{ moles NO}}{2 \text{ } \cancel{\text{moles CO}}}$$

STEP 3 Collecting numerical terms, after cancellation of units, gives

$$\frac{2.31 \times 2}{2} \text{ moles NO} = 2.31 \text{ moles NO (calculator and correct answer)}$$

In terms of significant figures, the numbers in conversion factors obtained from equation coefficients are considered exact numbers. Thus, since 2.31 contains three significant figures, the answer to this problem should also contain three significant figures.

Practice Exercise 10.5

The natural gas burned to provide heat in many homes is predominantly methane (CH_4). Methane burns (reacts with O_2) as shown by the equation.

$$CH_4 + 2\,O_2 \longrightarrow CO_2 + 2\,H_2O$$

(a) How many moles of H_2O are produced when 1.23 moles of CH_4 burns?
(b) How many moles of O_2 are needed to react with 3.61 moles of CH_4?

Ans. (a) 2.46 moles H_2O; (b) 7.22 moles O_2

10.8 CALCULATIONS BASED ON CHEMICAL EQUATIONS—STOICHIOMETRY

A major area of concern for chemists is the quantities of materials consumed and produced in chemical reactions. This area of study is called chemical stoichiometry. **Chemical stoichiometry** *is the study of the quantitative relationships among reactants and products in a chemical reaction.* The word *stoichiometry*, pronounced stoy-**kee**-om-eh-tree, is derived from the Greek *stoicheion* (element) and *metron* (measure). The stoichiometry of a chemical reaction always involves the *molar relationships* between reactants and products (Sec. 10.5) and thus is given by the coefficients in the balanced equation for the chemical reaction.

In a typical stoichiometric (stoy-kee-oh-met-rik) calculation, information is given about one reactant or product of a reaction (number of grams, moles, or particles), and information is requested concerning another reactant or product of the same reaction. The substances involved in such a calculation may both be reactants, may both be products, or may be one of each.

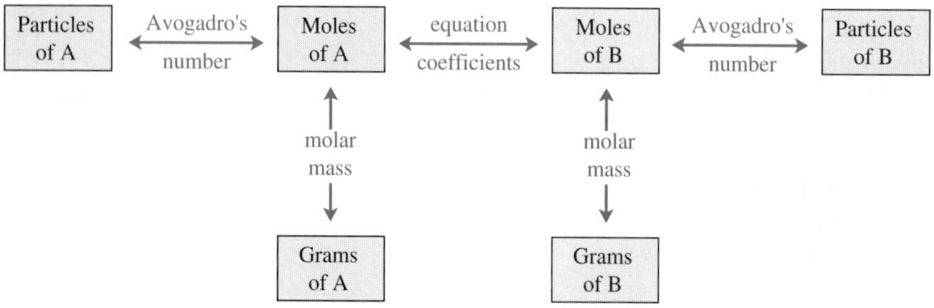

FIGURE 10.3 Conversion factor relationships needed for solving chemical-equation-based problems.

The conversion factor relationships needed to solve problems of the above general type are given in Figure 10.3. This diagram should seem very familiar to you, for it is almost identical to Figure 9.3, with which you have worked repeatedly. There is only one difference between the two. In the Chapter 9 diagram the subscripts in a chemical formula were listed as the basis for relating moles of given and desired substances to each other. In this new diagram, these same two quantities are related using the coefficients of a balanced chemical equation.

The most common type of stoichiometric calculation is a mass-to-mass (gram-to-gram) problem. In such problems the mass of one substance involved in a chemical reaction (either reactant or product) is given, and information is requested about the mass of another of the substances involved in the reaction (either a reactant or product). Situations requiring the solution of problems of this type are frequently encountered in laboratory settings. For example, a chemist has available so many grams of a certain chemical and wants to know how many grams of another substance can be produced from it, or how many grams of a third substance are needed to react with it. Examples 10.6 and 10.7 are both problem-solving situations of the gram-to-gram type.

F.Y.I.

A cement truck may not look like chemistry. But the hardening of cement is a complex chemical reaction with its own stoichiometry. If the cement mixer adds too much or too little water, the cement will not harden properly and will be too weak to support a structure or will develop cracks.

Example 10.6

A mixture of hydrazine (N_2H_4) and hydrogen peroxide (H_2O_2) is used as a fuel for rocket engines. These two substances react as shown by the equation.

$$N_2H_4(l) + 2\ H_2O_2(l) \longrightarrow N_2(g) + 4\ H_2O(g)$$

How many grams of H_2O_2 are needed to completely react with 50.0 g of N_2H_4?

Solution

STEP 1 Here we are given information about one reactant (50.0 g of N_2H_4) and asked to calculate information about the other reactant (H_2O_2)

$$50.0\ \text{g}\ N_2H_4 = ?\ \text{g}\ H_2O_2$$

STEP 2 This problem is of the "grams of A" to "grams of B" type. The pathway to be used in solving this type of problem, in terms of Figure 10.3, is

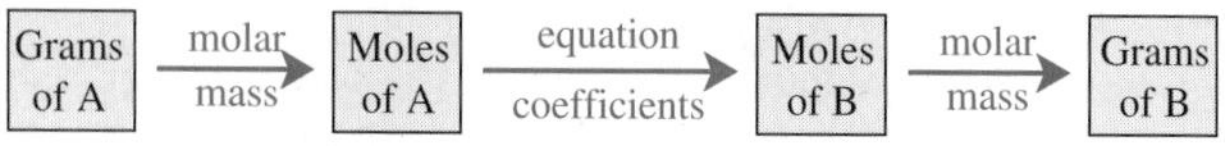

STEP 3 The dimensional analysis setup for this pathway is

$$50.0\ \cancel{\text{g}\ N_2H_4} \times \frac{1\ \cancel{\text{mole}\ N_2H_4}}{32.0\ \cancel{\text{g}\ N_2H_4}} \times \frac{2\ \cancel{\text{moles}\ H_2O_2}}{1\ \cancel{\text{mole}\ N_2H_4}} \times \frac{34.0\ \text{g}\ H_2O_2}{1\ \cancel{\text{mole}\ H_2O_2}}$$

$$\text{grams A} \longrightarrow \text{moles A} \longrightarrow \text{moles B} \longrightarrow \text{grams B}$$

The number 32.0 in the first conversion factor is the molar mass of N_2H_4; the 2 and 1 in the second conversion factors are the coefficients of H_2O_2 and N_2H_4, respectively, in the balanced chemical equation; and the number 34.0 in the last conversion factor is the molar mass of H_2O_2.

STEP 4 The solution, obtained from combining all of the numerical factors, is

$$\frac{50.0 \times 1 \times 2 \times 34.0}{32.0 \times 1 \times 1}\ \text{g } H_2O_2 = 106.25\ \text{g } H_2O_2 \text{ (calculator answer)}$$

$$= 106\ \text{g } H_2O_2 \text{ (correct answer)}$$

Practice Exercise 10.6

The active ingredient in some antacid formulations is magnesium hydroxide [$Mg(OH)_2$], which reacts with stomach acid (HCl) to produce magnesium chloride ($MgCl_2$) and water. The equation for the reaction is

$$Mg(OH)_2(s) + 2\ HCl(aq) \longrightarrow MgCl_2(aq) + 2\ H_2O(l)$$

How many grams of $Mg(OH)_2$ are needed to react with 1.00 g of HCl?

Ans. 0.799 g $Mg(OH)_2$

Example 10.7

When baking soda ($NaHCO_3$) is heated, it decomposes producing carbon dioxide gas (CO_2). This carbon dioxide is responsible for the rising of bread, donuts, and cookies. The equation for the baking soda decomposition reaction is

$$2\ NaHCO_3 \longrightarrow Na_2CO_3 + CO_2 + H_2O$$

How many grams of CO_2 are produced when 1.00 g of $NaHCO_3$ decomposes?

Solution

STEP 1 Here we are given information about the reactant (1.00 g of $NaHCO_3$) and asked to calculate information about one of the products (CO_2).

$$1.00\ \text{g } NaHCO_3 = ?\ \text{g } CO_2$$

STEP 2 This problem, like Example 10.6, is a "grams of A" to "grams of B" problem. The pathway used in solving it will be the same, which, in terms of Figure 10.3, is

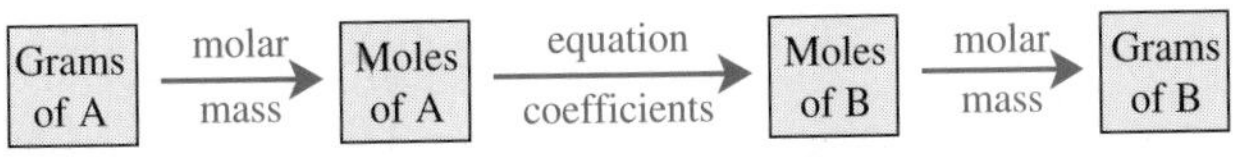

STEP 3 The dimensional analysis setup is

$$1.00\ \cancel{\text{g } NaHCO_3} \times \frac{1\ \cancel{\text{mole } NaHCO_3}}{84.0\ \cancel{\text{g } NaHCO_3}} \times \frac{1\ \cancel{\text{mole } CO_2}}{2\ \cancel{\text{moles } NaHCO_3}} \times \frac{44.0\ \text{g } CO_2}{1\ \cancel{\text{mole } CO_2}}$$

grams A ⟶ moles A ⟶ moles B ⟶ grams B

The chemical equation is the "bridge" that enables us to go from $NaHCO_3$ to CO_2. The numbers in the second conversion factor, the "bridge factor" are coefficients from this equation.

STEP 4 The solution, obtained from combining all of the numerical factors in the setup, is

$$\frac{1.00 \times 1 \times 1 \times 44.0}{84.0 \times 2 \times 1}\ \text{g } CO_2 = 0.26190476\ \text{g } CO_2 \text{ (calculator answer)}$$

$$= 0.262\ \text{g } CO_2 \text{ (correct answer)}$$

Practice Exercise 10.7

The chemical equation for the photosynthesis reaction in plants is

$$6\ CO_2 + 6\ H_2O \longrightarrow C_6H_{12}O_6 + 6\ O_2$$

How many grams of oxygen (O_2) are produced by a plant from the consumption of 25.0 g of carbon dioxide (CO_2)?

Ans. 18.2 g O_2

"Grams of A" to "grams of B" problems (Examples 10.6 and 10.7) are not the only type of problem for which the coefficients in a balanced equation can be used to relate quantities of two substances. As further examples of the use of equation coefficients in problem solving, consider Example 10.8 (a "grams of A" to "moles of B" problem) and Example 10.9 (a "particles of A" to "grams of B" problem).

Example 10.8

In the atmosphere, the air pollutant nitrogen dioxide (NO_2) reacts with water to produce nitric acid (NHO_3). Nitric acid is a component, along with sulfuric acid, of acid rain. The reaction for the formation of nitric acid is

$$3\ NO_2 + H_2O \longrightarrow 2\ HNO_3 + NO$$

How many moles of NO are produced at the same time that 5.00 g of HNO_3 is produced?

Solution

STEP 1 The given quantity is 5.00 g of HNO_3 and the desired quantity is moles of NO.

$$5.00\ \text{g } HNO_3 = ?\ \text{moles NO}$$

STEP 2 This is a "grams of A" to "moles of B" problem. The pathway used to solve such a problem is, according to Figure 10.3,

Grams of A —(molar mass)→ Moles of A —(equation coefficients)→ Moles of B

STEP 3 The dimensional analysis setup is

$$5.00\ \cancel{\text{g } HNO_3} \times \frac{1\ \cancel{\text{mole } HNO_3}}{63.0\ \cancel{\text{g } HNO_3}} \times \frac{1\ \text{mole NO}}{2\ \cancel{\text{moles } HNO_3}}$$

grams A ⟶ moles A ⟶ moles B

The number 63.0 in the first conversion factor is the molar mass of HNO_3.

STEP 4 The solution, obtained from combining all of the numbers in the manner indicated in the setup, is

$$\frac{5.00 \times 1 \times 1}{63.0 \times 2} \text{ mole NO} = 0.039682539 \text{ mole NO (calculator answer)}$$
$$= 0.0397 \text{ mole NO (correct answer)}$$

Practice Exercise 10.8

The thermal decomposition of solid potassium chlorate ($KClO_3$) serves as a convenient laboratory source of small amounts of oxygen gas. The reaction is

$$2\ KClO_3(s) \longrightarrow 2\ KCl(s) + 3\ O_2(g)$$

How many moles of $KClO_3$ must be decomposed to produce 5.00 g of O_2?

Ans. 0.104 mole $KClO_3$

Example 10.9

When silver carbonate (Ag_2CO_3) is decomposed by heating, three products are produced: metallic silver (Ag), carbon dioxide (CO_2), and oxygen (O_2). How many grams of O_2 will be produced when 100 billion (1.00×10^{11}) Ag_2CO_3 formula units decompose?

Solution

Although a calculation of this type will not have a lot of practical significance, it will test your understanding of the problem-solving relationships under discussion in this section of the text.

The specifics of the chemical reaction of concern to us in this problem were given in "word" rather than "equation" form in the problem statement. These "words" must be translated into an equation before we can proceed with the problem solving. The equation is

$$Ag_2CO_3 \longrightarrow Ag + CO_2 + O_2$$

Having a chemical equation is not enough. It must be a *balanced* chemical equation. Using the balancing procedures of Section 10.3, the preceding equation, in balanced form, becomes

$$2\ Ag_2CO_3 \longrightarrow 4\ Ag + 2\ CO_2 + O_2$$

Now we are ready to proceed with the solving of our problem.

STEP 1 We are given a certain number of particles (formula units) and asked to find the number of grams of a related substance.

$$1.00 \times 10^{11} \text{ formula units } Ag_2CO_3 = ?\ g\ O_2$$

STEP 2 This is a "particles of A" to "grams of B" problem. The pathway for this problem (see Fig. 10.3) is

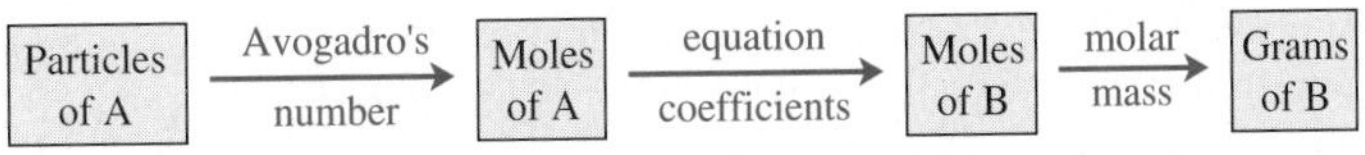

STEP 3 The dimensional analysis setup is

$$1.00 \times 10^{11} \text{ units } Ag_2CO_3 \times \frac{1 \text{ mole } Ag_2CO_3}{6.02 \times 10^{23} \text{ units } Ag_2CO_3} \times \frac{1 \text{ mole } O_2}{2 \text{ moles } Ag_2CO_3} \times \frac{32.0 \text{ g } O_2}{1 \text{ mole } O_2}$$

particles A ⟶ moles A ⟶ moles B ⟶ grams B

STEP 4 The solution, obtained by combining all of the numerical factors in the setup, is

$$\frac{1.00 \times 10^{11} \times 1 \times 1 \times 32.0}{6.02 \times 10^{23} \times 2 \times 1} \text{ g } O_2 = 2.6578073 \times 10^{-12} \text{ g } O_2 \text{ (calculator answer)}$$

$$= 2.66 \times 10^{-12} \text{ g } O_2 \text{ (correct answer)}$$

Practice Exercise 10.9

The reaction of sulfuric acid (H_2SO_4) with elemental copper (Cu) produces three products: sulfur dioxide (SO_2), water (H_2O), and copper(II) sulfate ($CuSO_4$). How many grams of water will be produced at the same time that 5 billion (5.00×10^9) sulfur dioxide molecules are produced?

Ans. 2.99×10^{-13} g H_2O

10.9 THE LIMITING REACTANT CONCEPT

When a chemical reaction is carried out in a laboratory or industrial setting, the reactants are not usually present in the exact molar ratios specified in the balanced chemical equation for the reaction. Most often, on purpose, excess quantities of one or more of the reactants are present.

Numerous reasons exist for having some reactants present in excess. Sometimes such a procedure will cause a reaction to occur more rapidly. For example, large amounts of oxygen make combustible materials burn faster. Sometimes an excess of one reactant will ensure that another reactant, perhaps a very expensive one, is completely consumed. (Reactions do not always go to completion in the way that theory predicts they should—that is, the reactants are not completely converted to products.)

When one or more reactants are present in excess, the excess will not react because there is not enough of the other reactant to react with it. The reactant *not* in excess thus limits the amount of product(s) formed and is called *the limiting reactant*. The **limiting reactant** *is the reactant in a chemical reaction that determines how much product(s) can be formed.*

The concept of a limiting reactant plays a major role in chemical calcuations of certain types. It must be thoroughly understood. Let us consider some simple but analogous nonchemical examples of a "limiting reactant" before we go on to limiting reactant calculations.

Suppose we have a vending machine that contains forty 50¢ candy bars and we have 30 quarters. In this case we can purchase only 15 candy bars. The quarters are the limiting reactant. The candy bars are present in excess. Suppose we have ten slices of cheese and eigh-

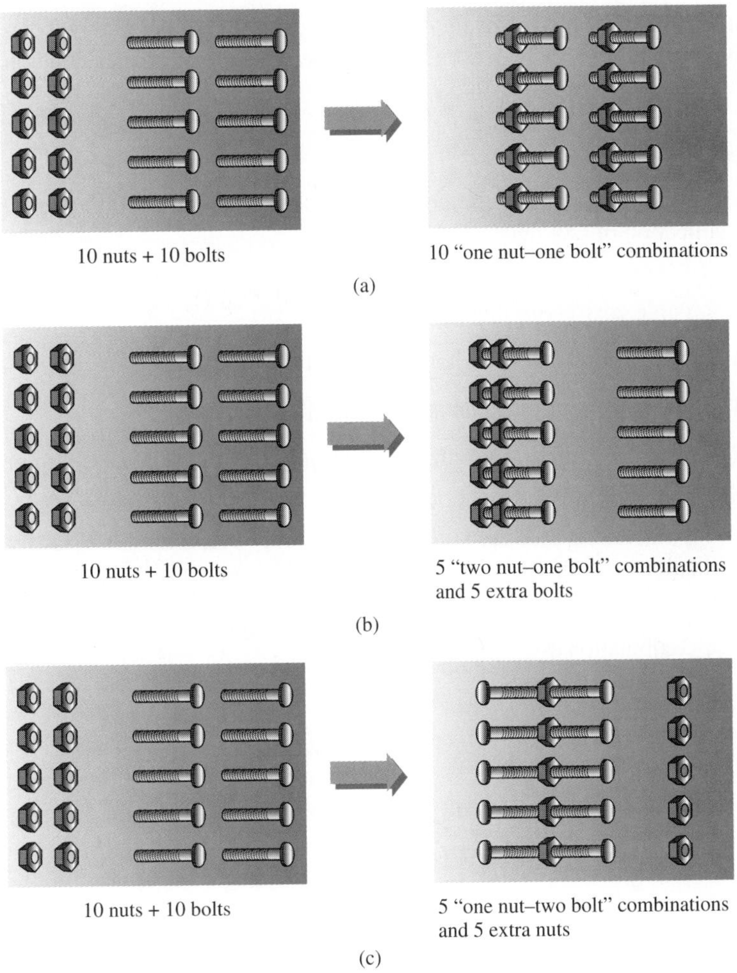

FIGURE 10.4 Starting with 10 nuts and 10 bolts, we can make (a) 10 "one nut–one bolt" combination, (b) 5 "two nut–one bolt" combinations with 5 bolts left over; the nuts are the limiting reactant, and (c) 5 "one nut–two bolt" combinations with 5 nuts left over; the bolts are the limiting reactant.

teen slices of bread and we want to make as many cheese sandwiches as possible using one slice of cheese and two slices of bread per sandwich. The eighteen slices of bread limit us to nine sandwiches; one slice of cheese is left over. The bread is the limiting reactant in this case even though initially there was more bread (18 slices) than cheese (10 slices) present. The bread is still limiting because it is used up first.

An additional limiting reactant analogy that comes closer to the realm of molecules and atoms involves nuts and bolts. Assume we have 10 identical nuts and 10 identical bolts. From this collection we can make 10 "one nut–one bolt" entities by screwing a nut on each bolt. This situation is depicted in Figure 10.4a.

Next, let us make "two nut–one bolt" entities from our same collection of nuts and bolts. This time we can make only 5 combinations and we will have 5 bolts left over, as is shown in Figure 10.4b. We run out of nuts before all of the bolts are used up. In chemical jargon, we say that the nuts are the limiting reactant.

Finally, let us consider making "one nut–two bolt" combinations. As is shown in Figure 10.4c, this time we do not have enough bolts; we can make 5 combinations and we will have 5 nuts left over. The bolts are the limiting reactant.

Now let us consider, as is presented in Example 10.10, an extension of our nut–bolt discussion to a situation that cannot easily be reasoned out in one's head. Instead a calculation must be performed.

Example 10.10

What will be the limiting reactant in the production of "two nut–three bolt" combinations from a collection of 284 nuts and 414 bolts?

Solution

There are three possible answers to this problem.

1. We will run out of bolts first; bolts are the limiting reactant.
2. We will run out of nuts first; nuts are the limiting reactant.
3. The ratio of nuts to bolts is such that we run out of both at the same time; both are limiting reactants.

We determine which of these answers is the correct one by calculating how many "nut–bolt" combinations can be made from each of the "ingredients," assuming an excess of the other.

$$284 \text{ \cancel{nuts}} \times \frac{1 \text{ combination}}{2 \text{ \cancel{nuts}}} = 142 \text{ combinations} \quad \text{(calculator and correct answer)}$$

$$414 \text{ \cancel{bolts}} \times \frac{1 \text{ combination}}{3 \text{ \cancel{bolts}}} = 138 \text{ combinations} \quad \text{(calculator and correct answer)}$$

Because fewer combinations can be made from the bolts, the bolts are the limiting reactant.

Practice Exercise 10.10

What is the limiting reactant in the production of "three nut–two bolt" combinatons from a collection of 261 nuts and 176 bolts?

Ans. Nuts are the limiting reactant

Now let us proceed to chemical calculations that involve a limiting reactant. *Whenever the quantities of two or more reactants in a chemical reaction are given, it is necessary to determine which of the given quantities is the limiting reactant.*

Determining the limiting reactant can be accomplished by the following procedure.

1. Determine the number of *moles* of each of the reactants present.
2. Calculate the number of moles of product *each* of the molar amounts of reactant would produce if it were the only reactant amount given. If more than one product is formed in the reaction, you need to do this mole calculation for only one of the products.
3. The reactant that produces the *lesser number* of moles of product is the limiting reactant.

F.Y.I.

In combustion reactions, oxygen is not usually a limiting reactant because it is so plentiful in the air. But if a jar is put over a burning candle, oxygen becomes a limiting reactant. As soon as all the oxygen in the jar is used up, the candle sputters out.

Example 10.11

Old oil paintings are darkened by PbS, which forms by the reaction of the Pb in paint (a coloring agent) with H_2S in air.

$$Pb + H_2S \longrightarrow PbS + H_2$$

If 5.00 g of Pb and 1.00 g of H_2S are present in the reaction mixture, which is the limiting reactant for the darkening reaction?

Solution

To determine the limiting reactant, we determine how many moles of product each of the reactants can form. In this particular problem there are two products: PbS and H_2. It is sufficient to calculate how many moles of either PbS or H_2 are formed. The decision as to which product to use is arbitrary; we will choose H_2.

The calculation type will be "grams of A" to "moles of B." We start with a given number of grams of reactant and desire to calculate moles of product. The pathway for the calculation, in terms of Figure 10.3, is

$$\boxed{\text{Grams of A}} \xrightarrow[\text{mass}]{\text{molar}} \boxed{\text{Moles of A}} \xrightarrow[\text{coefficients}]{\text{equation}} \boxed{\text{Moles of B}}$$

Note that we will have to go through this type of calculation twice because we have two reactants: once for Pb and once for H_2S.

For Pb:

$$5.00\ \cancel{\text{g Pb}} \times \frac{1\ \cancel{\text{mole Pb}}}{207\ \cancel{\text{g Pb}}} \times \frac{1\ \text{mole } H_2}{1\ \cancel{\text{mole Pb}}} = 0.024154589\ \text{mole } H_2 \quad \text{(calculator answer)}$$

$$= 0.0242\ \text{mole } H_2 \quad \text{(correct answer)}$$

For H_2S:

$$1.00\ \cancel{\text{g } H_2S} \times \frac{1\ \cancel{\text{mole } H_2S}}{34.1\ \cancel{\text{g } H_2S}} \times \frac{1\ \text{mole } H_2}{1\ \cancel{\text{mole } H_2S}} = 0.029325513\ \text{mole } H_2 \quad \text{(calculator answer)}$$

$$= 0.0293\ \text{mole } H_2 \quad \text{(correct answer)}$$

The limiting reactant is the reactant that will produce the fewer number of moles of H_2. Looking at the numbers just calculated we see that Pb will be the limiting reactant. Once the limiting reactant has been determined, the amount of that reactant present becomes the starting point for any further calculations about the chemical reaction under consideration. Example 10.12 illustrates this point.

Practice Exercise 10.11

Silver and silver-plated objects tarnish in the presence of hydrogen sulfide (H_2S), a gas produced from the decay of food. The reaction of Ag and H_2S is

$$4\ Ag + 2\ H_2S + O_2 \longrightarrow 2\ Ag_2S + 2\ H_2O$$

The black product Ag_2S is the tarnish. If 25.00 g of Ag, 5.00 g of H_2S, and 4.00 g of O_2 are present in a reaction mixture, which is the limiting reactant for tarnish formation?

Ans. Ag is the limiting reactant

Example 10.12

When iron(III) oxide is heated with aluminum powder, a very vigorous reaction occurs in which iron is produced from the iron(III) oxide. The equation for this reaction, which is called the thermite reaction, is

$$Fe_2O_3(s) + 2\ Al(s) \longrightarrow Al_2O_3(s) + 2\ Fe(l)$$

How many grams of Fe can be formed from a reaction mixture containing 90.0 g of Fe_2O_3 and 30.0 g of Al?

Solution

First, we must determine the limiting reactant since specific amounts of both reactants are given in the problem. This determination involves two "grams of A" to "moles of B" calculations—one for the reactant Fe_2O_3 and one for the reactant Al. The product we "key in" on is Fe because our final goal is the mass of iron produced.

Grams of A $\xrightarrow[\text{mass}]{\text{molar}}$ Moles of A $\xrightarrow[\text{coefficients}]{\text{equation}}$ Moles of B

For Fe_2O_3:

$$90.0\ \text{g } Fe_2O_3 \times \frac{1\ \text{mole } Fe_2O_3}{160.0\ \text{g } Fe_2O_3} \times \frac{2\ \text{moles Fe}}{1\ \text{mole } Fe_2O_3} = 1.125\ \text{moles Fe} \quad \text{(calculator answer)}$$

$$= 1.12\ \text{moles Fe} \quad \text{(correct answer)}$$

For Al:

$$30.0\ \text{g Al} \times \frac{1\ \text{mole Al}}{27.0\ \text{g Al}} \times \frac{2\ \text{moles Fe}}{2\ \text{moles Al}} = 1.1111111\ \text{moles Fe} \quad \text{(calculator answer)}$$

$$= 1.11\ \text{moles Fe} \quad \text{(correct answer)}$$

Thus, Al is the limiting reactant since fewer moles of Fe can be produced from it (1.11 moles) than from the Fe_2O_3 (1.12 moles).

We can now calculate the grams of Fe formed in the reaction using the 1.11 moles of Fe (formed from our limiting reactant) as our starting factor. The calculation will be a simple one-step "moles of A" to "grams of A" conversion.

Moles of Fe $\xrightarrow[\text{mass}]{\text{molar}}$ Grams of Fe

$$1.11\ \text{moles Fe} \times \frac{55.8\ \text{g Fe}}{1\ \text{mole Fe}} = 6.1938\ \text{g Fe} \quad \text{(calculator answer)}$$

$$= 6.19\ \text{g Fe} \quad \text{(correct answer)}$$

Practice Exercise 10.12

Ammonia (NH_3) reacts with oxygen as shown in the following equation.

$$4\ NH_3 + 3\ O_2 \longrightarrow 2\ N_2 + 6\ H_2O$$

How many grams of H_2O can be formed from a reaction mixture containing 35.0 g of NH_3 and 50.0 g of O_2?

Ans. 55.6 g H_2O

10.10 YIELDS: THEORETICAL, ACTUAL, AND PERCENT

When stoichiometric relationships (Secs. 10.8 and 10.9) are used to calculate the amount of a product that will be produced in a chemical reaction from given amounts of reactants, the answer obtained represents a theoretical yield. The **theoretical yield** *is the calculated amount of a product that can be obtained from given amounts of reactants in a chemical reaction that proceeds completely in the manner described by its chemical equation.* Examples 10.7 and 10.12 are theoretical yield calculations, although this fact was not noted at the time they were presented.

In most chemical reactions the amount of a given product isolated from the reaction mixture is less than that theoretically possible. Why is this so? Two major factors contribute to this situation.

1. Some product is almost always lost in the process of its isolation and purification and in such mechanical operations as transferring materials from one container to another.
2. Often a particular set of reactants undergoes two or more reactions simultaneously, forming undesired products (in small amounts) as well as the desired products. Reactants consumed in these side reactions obviously will not end up in the form of the desired products.

The net effect of these factors is that the actual quantities of product isolated—that is, the actual yield—are usually less, sometimes far less, than the theoretically possible amount. An **actual yield** *is the amount of a product actually obtained from a chemical reaction.* Actual yield is always an experimentally determined number; it cannot be calculated.

Product loss is specified in terms of percent yield. **Percent yield** *is the ratio of the actual yield to the theoretical yield multipled by 100 (to give percent).* The mathematical equation for percent yield is

$$\text{Percent yield} = \frac{\text{actual yield}}{\text{theoretical yield}} \times 100$$

If the theoretical yield of a product for a reaction is calculated to be 17.9 g and the amount of product actually obtained (the actual yield) is 15.8 g, the percent yield is 88.3%.

$$\text{Percent yield} = \frac{15.8\ \cancel{\text{g}}}{17.9\ \cancel{\text{g}}} \times 100 = 88.3\%$$

Example 10.13

The active ingredient in household laundry bleaches is sodium hypochlorite (NaClO). This bleaching agent can be prepared by the reaction

$$2\ NaOH + Cl_2 \longrightarrow NaCl + NaClO + H_2O$$

(a) What is the theoretical yield of NaClO that can be obtained from a reaction mixture containing 75.0 g of NaOH and 50.0 g of Cl_2?

(b) If the actual yield of NaClO for the reaction mixture in part (a) is 43.2 g, what is the percent yield of NaClO for the reaction?

Solution

(a) The limiting reactant must be determined before the theoretical yield can be calculated. Recalling the procedures of Examples 10.11 and 10.12 for determining the limiting reactant, we calculate the number of moles that can be produced from each individual reactant amount using the "grams of A" to "moles of B" type of calculations.

Grams of A $\xrightarrow[\text{mass}]{\text{molar}}$ Moles of A $\xrightarrow[\text{coefficients}]{\text{equation}}$ Moles of B

For NaOH:

$$75.0\ \cancel{\text{g NaOH}} \times \frac{1\ \cancel{\text{mole NaOH}}}{40.0\ \cancel{\text{g NaOH}}} \times \frac{1\ \text{mole NaClO}}{2\ \cancel{\text{moles NaOH}}} = 0.9375\ \text{mole NaClO}$$

(calculator answer)

$$= 0.938\ \text{mole NaClO}$$

(correct answer)

For Cl_2:

$$50.0\ \cancel{\text{g Cl}_2} \times \frac{1\ \cancel{\text{mole Cl}_2}}{71.0\ \cancel{\text{g Cl}_2}} \times \frac{1\ \text{mole NaClO}}{1\ \cancel{\text{mole Cl}_2}} = 0.70422535\ \text{mole NaClO}$$

(calculator answer)

$$= 0.704\ \text{mole NaClO}$$

(correct answer)

The calculations show that Cl_2 is the limiting reactant.

The maximum number of grams of NaClO obtainable from the limiting reactant, that is, the theoretical yield, can now be calculated. It is done using a one-step "moles of A" to "grams of A" setup.

Moles of NaClO $\xrightarrow[\text{mass}]{\text{molar}}$ Grams of NaClO

$$0.704\ \cancel{\text{mole NaClO}} \times \frac{74.5\ \text{g NaClO}}{1\ \cancel{\text{mole NaClO}}} = 52.448\ \text{g NaClO} \quad \text{(calculator answer)}$$

$$= 52.4\ \text{g NaClO} \quad \text{(correct answer)}$$

(b) The percent yield is obtained by dividing the actual yield by the theoretical yield and multiplying by 100.

$$\text{Percent yield} = \frac{\text{actual yield}}{\text{theoretical yield}} \times 100 = \frac{43.2\ \cancel{\text{g}}}{52.4\ \cancel{\text{g}}} \times 100 = 82.442748\%$$

(calculator answer)

$$= 82.4\%$$

(**correct answer**)

Practice Exercise 10.13

A mixture of 80.0 g of chromium(III) oxide (Cr_2O_3) and 8.00 g of carbon (C) is used to produce elemental chromium (Cr) by the reaction

$$Cr_2O_3 + 3\ C \longrightarrow 2\ Cr + 3\ CO$$

(a) What is the theoretical yield of Cr that can be obtained from the reaction mixture?

(b) The actual yield is 21.7 g Cr. What is the percent yield for the reaction?

Ans. (a) 23.1 g Cr; (b) 93.9%

10.11 SIMULTANEOUS AND CONSECUTIVE REACTIONS

The concepts presented so far in this chapter can easily be adapted to problem-solving situations that involve two or more chemical reactions. In some cases the two or more chemical reactions occur simultaneously, and in other cases they occur consecutively (one right after the other). Example 10.14 deals with a pair of simultaneous reactions, and Example 10.15 deals with three consecutive reactions.

Example 10.14

A mixture contains 47.3% by mass magnesium carbonate ($MgCO_3$) and 52.7% by mass calcium carbonate ($CaCO_3$). The mixture is heated until both carbonates completely decompose as shown by the following equations.

$$MgCO_3 \longrightarrow MgO + CO_2$$
$$CaCO_3 \longrightarrow CaO + CO_2$$

How many grams of CO_2 are produced from decomposition of 78.3 g of the mixture?

Solution

In solving this problem we will need to carry out two parallel calculations. In the one calculation we will determine the grams of CO_2 produced from the $MgCO_3$ component of the mixture and in the other the grams of CO_2 produced from the $CaCO_3$ component. Then we will add together the answers from the two parallel calculations to get our final answer, the total grams of CO_2 produced. The sequence of conversion factors for each setup is derived from the following pathway.

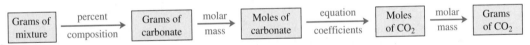

Grams of mixture —(percent composition)→ Grams of carbonate —(molar mass)→ Moles of carbonate —(equation coefficients)→ Moles of CO_2 —(molar mass)→ Grams of CO_2

The number of grams of CO_2 produced from the $MgCO_3$ is given by the following setup.

$$78.3 \text{ g mixture} \times \frac{47.3 \text{ g } MgCO_3}{100.0 \text{ g mixture}} \times \frac{1 \text{ mole } MgCO_3}{84.3 \text{ g } MgCO_3} \times \frac{1 \text{ mole } CO_2}{1 \text{ mole } MgCO_3} \times \frac{44.0 \text{ g } CO_2}{1 \text{ mole } CO_2}$$

$$= 19.330718 \text{ g } CO_2 \quad \text{(calculator answer)}$$
$$= 19.3 \text{ g } CO_2 \quad \text{(correct answer)}$$

The number of grams of CO_2 produced from the $CaCO_3$ is given by the following setup.

$$78.3 \text{ g mixture} \times \frac{52.7 \text{ g } CaCO_3}{100.0 \text{ g mixture}} \times \frac{1 \text{ mole } CaCO_3}{100.1 \text{ g } CaCO_3} \times \frac{1 \text{ mole } CO_2}{1 \text{ mole } CaCO_3} \times \frac{44.0 \text{ g } CO_2}{1 \text{ mole } CO_2}$$

$$= 18.138065 \text{ g } CO_2 \quad \text{(calculator answer)}$$
$$= 18.1 \text{ g } CO_2 \quad \text{(correct answer)}$$

The first conversion factor in each setup is derived from the given percentage of that compound in the mixture. The use of percentages as conversion factors was covered in Section 3.9

The total number of grams of CO_2 produced is the sum of the grams of CO_2 in the individual reactions.

$$(19.3 + 18.1)\text{ g } CO_2 = 37.4\text{ g } CO_2 \quad \text{(calculator and correct answer)}$$

Practice Exercise 10.14

A mixture of gaseous fuel has the composition 83.1% methane (CH_4) and 16.9% ethane (C_2H_6). The combustion of this mixture produces CO_2 and H_2O as the only products. The combustion equations are

$$CH_4 + 2\,O_2 \longrightarrow CO_2 + 2\,H_2O$$
$$2\,C_2H_6 + 7\,O_2 \longrightarrow 4\,CO_2 + 6\,H_2O$$

How many moles of CO_2 would be produced from the combustion of 72.0 g of this gaseous fuel mixture?

Ans. 4.55 moles CO_2

Example 10.15

In steelmaking, a series of three reactions is needed to convert Fe_2O_3 (the iron-containing component of iron ore) to molten iron.

$$\text{Reaction (1):}\quad 3\,Fe_2O_3 + CO \longrightarrow 2\,Fe_3O_4 + CO_2$$
$$\text{Reaction (2):}\quad Fe_3O_4 + CO \longrightarrow 3\,FeO + CO_2$$
$$\text{Reaction (3):}\quad FeO + CO \longrightarrow Fe + CO_2$$

Assuming that the reactant CO is present in excess, how many grams of Fe can be produced from 125 g of Fe_2O_3?

Solution

The key substances in this set of reactions, from a calculational point of view, are the iron-containing species: Fe_2O_3, Fe_3O_4, FeO, and Fe. Note that the iron-containing species produced in the first and second reactions (Fe_3O_4 and FeO, respectively) are the reactants for the second and third reactions, respectively.

$$Fe_2O_3 \xrightarrow{\text{reaction (1)}} Fe_3O_4 \xrightarrow{\text{reaction (2)}} FeO \xrightarrow{\text{reaction (3)}} Fe$$

We can solve this problem using a single multiple-step setup. The sequence of conversion factors needed is that for a gram-to-gram problem with two additional intermediate mole-to-mole steps added.

$$\boxed{\text{Grams of } Fe_2O_3} \xrightarrow[\text{mass}]{\text{molar}} \boxed{\text{Moles of } Fe_2O_3} \xrightarrow[\text{coeff.}]{\text{eq. (1)}} \boxed{\text{Moles of } Fe_3O_4} \xrightarrow[\text{coeff.}]{\text{eq. (2)}}$$

$$\boxed{\text{Moles of FeO}} \xrightarrow[\text{coeff.}]{\text{eq. (3)}} \boxed{\text{Moles of Fe}} \xrightarrow[\text{mass.}]{\text{molar}} \boxed{\text{Grams of Fe}}$$

The dimensional analysis setup is

$$125 \text{ g } \cancel{Fe_2O_3} \times \frac{1 \text{ mole } \cancel{Fe_2O_3}}{160.0 \text{ g } \cancel{Fe_2O_3}} \times \frac{2 \text{ moles } \cancel{Fe_3O_4}}{3 \text{ moles } \cancel{Fe_2O_3}} \times \frac{3 \text{ moles } \cancel{FeO}}{1 \text{ mole } \cancel{Fe_3O_4}} \times \frac{1 \text{ mole } \cancel{Fe}}{1 \text{ mole } \cancel{FeO}} \times \frac{55.8 \text{ g Fe}}{1 \text{ mole } \cancel{Fe}}$$

$= 87.187499$ g Fe (calculator answer)

$= 87.2$ g Fe (correct answer)

An alternative approach to solving this problem would involve setting up a separate calculation for each equation. As a first step, the number of moles of Fe_3O_4 produced in the first reaction would be calculated. In the second step, one would determine the moles of FeO obtained if all of the Fe_3O_4 produced in the first reaction entered into the second reaction. In the final step, one would determine the grams of Fe derivable from the FeO produced in the second reaction. The answer obtained from this three-setup method is the same as that obtained from the one-setup method.

Practice Exercise 10.15

An older method for preparation of nitric acid (HNO_3) involves the following sequence of three reactions.

Reaction (1): $4\ NH_3 + 5\ O_2 \longrightarrow 4\ NO + 6\ H_2O$

Reaction (2): $2\ NO + O_2 \longrightarrow 2\ NO_2$

Reaction (3): $3\ NO_2 + H_2O \longrightarrow 2\ HNO_3 + NO$

Assuming an excess of O_2 and H_2O as reactants, how many grams of HNO_3 can be produced from 244 g of NH_3?

Ans. 603 g HNO_3

KEY TERMS

The new terms or concepts defined in this chapter are

actual yield (Sec. 10.10) The amount of product actually obtained from a chemical reaction.

balanced chemical equation (Sec. 10.3) A chemical equation in which the same number of atoms of each element involved in the reaction appears on each side of the equation.

chemical equation (Sec. 10.2) A written statement that uses symbols and formulas instead of words to describe the changes that occur in a chemical reaction.

chemical stoichiometry (Sec. 10.8) The study of the quantitative relationships among reactants and products in a chemical reaction.

coefficient (Sec. 10.3) A number placed to the left of the formula of a substance that denotes the amount of the substance.

combustion reaction (Sec. 10.5) The reaction of a substance with oxygen (usually from air) that proceeds with evolution of heat and usually also a flame.

decomposition reaction (Sec. 10.6) A reaction in which a single reactant is converted into two or more simpler substances.

double-replacement reaction (Sec. 10.6) A reaction in which two compounds exchange parts with each other and form two different compounds.

law of conservation of mass (Sec. 10.1) Mass is

neither created nor destroyed in any ordinary chemical reaction.

limiting reactant (Sec. 10.9) The reactant in a chemical reaction that determines how much product(s) can be formed.

molar (macroscopic) level interpretation of coefficients (Sec. 10.7) The coefficients in a balanced chemical equation are interpreted to mean moles of reactants and moles of products.

percent yield (Sec. 10.10) The ratio of the actual yield to the theoretical yield (of a product from a chemical reaction) multiplied by 100.

products (Sec. 10.1) Substances produced as a result of a chemical reaction.

reactants (Sec. 10.1) Substances present prior to the start of a chemical reaction.

single-replacement reaction (Sec. 10.6) A reaction in which one element within a compound is replaced by another element.

synthesis reaction (Sec. 10.6) A reaction in which a single product is produced from two or more reactants.

theoretical yield (Sec. 10.10) The calculated amount of product that can be obtained from given amounts of reactants in a chemical reaction if no losses or inefficiencies of any kind occur.

PRACTICE PROBLEMS

The Law of Conservation of Mass (Sec. 10.1)

10.1 Based on the following description of a chemical reaction, calculate the numerical value of *x*.

$$(5.85\text{ g NaCl}) + (16.98\text{ g AgNO}_3) \longrightarrow (x\text{ g AgCl}) + (8.50\text{ g NaNO}_3)$$

10.2 Based on the following description of a chemical reaction, calculate the numerical value of *x*.

$$(7.62\text{ g CS}_2) + (x\text{ g O}_2) \longrightarrow (4.40\text{ g CO}_2) + (12.82\text{ g SO}_2)$$

10.3 A 4.2-g sample of sodium hydrogen carbonate is added to a solution of acetic acid weighing 10.0 g. The two substances react, releasing carbon dioxide gas to the atmosphere. After the reaction, the contents of the reaction vessel weigh 12.0 g. What is the mass of carbon dioxide given off during the reaction?

10.4 A 1.00-g sample of solid calcium carbonate is added to a reaction flask containing 10.00 g of hydrochloric acid solution. The calcium carbonate slowly dissolves in the acid solution as evidenced by the generation of carbon dioxide gas. After 5 min of reaction, 0.21 g of carbon dioxide gas has been given off. At that time, what is the mass, in grams, of the reaction flask contents?

Chemical Equation Notation (Secs. 10.2 and 10.4)

10.5 In which of the following pairs of symbols and/or formulas for gaseous elements are both members of the pair written appropriately for use in chemical equations?

(a) N_2 and O_2 **(b)** Xe and Cl
(c) H_2 and He_2 **(d)** F_2 and I_2

10.6 In which of the following pairs of symbols and/or formulas for gaseous elements are both members of the pair written appropriately for use in chemical equations?

(a) Br_2 and He **(b)** H_2 and F_2
(c) Ar and Kr **(d)** O_2 and N

10.7 What do the symbols in parentheses stand for in the following equations?

(a) $CaCO_3(s) \longrightarrow CaO(s) + CO_2(g)$
(b) $SO_2(g) + H_2O(l) \longrightarrow H_2SO_3(aq)$

10.8 What do the symbols in parentheses stand for in the following equations?

(a) $PCl_3(l) + Cl_2(g) \longrightarrow PCl_5(s)$
(b) $NaCl(aq) + AgNO_3(aq) \longrightarrow AgCl(s) + NaNO_3(aq)$

Balancing Chemical Equations (Sec. 10.3)

10.9 Classify each of the following equations as *balanced* or *unbalanced*.

(a) $SO_3 + H_2O \longrightarrow H_2SO_4$
(b) $CuO + H_2 \longrightarrow Cu + H_2O$
(c) $CS_2 + O_2 \longrightarrow CO_2 + SO_2$
(d) $AgNO_3 + KCl \longrightarrow KNO_3 + AgCl$

10.10 Classify each of the following equations as *balanced* or *unbalanced*.

(a) $BaCO_3 \longrightarrow BaO + CO_2$
(b) $KCl + O_2 \longrightarrow KClO_3$
(c) $Fe_2O_3 + CO \longrightarrow Fe + CO_2$
(d) $H_2SO_4 + CaCO_3 \longrightarrow H_2CO_3 + CaSO_4$

10.11 Balance the following equations.

(a) $Cu + O_2 \longrightarrow CuO$
(b) $Al + N_2 \longrightarrow AlN$
(c) $HgO \longrightarrow Hg + O_2$
(d) $H_2O \longrightarrow H_2 + O_2$

10.12 Balance the following equations.
(a) $Fe + O_2 \longrightarrow Fe_2O_3$
(b) $Be + N_2 \longrightarrow Be_3N_2$
(c) $NaClO_3 \longrightarrow NaCl + O_2$
(d) $NaOH \longrightarrow Na_2O + H_2O$

10.13 Balance the following equations.
(a) $BaCl_2 + Na_2S \longrightarrow BaS + NaCl$
(b) $Mg + HBr \longrightarrow MgBr_2 + H_2$
(c) $Co + HgCl_2 \longrightarrow CoCl_3 + Hg$
(d) $Na + H_2O \longrightarrow NaOH + H_2$

10.14 Balance the following equations.
(a) $Au_2S_3 + H_2 \longrightarrow H_2S + Au$
(b) $Mg_3N_2 + H_2O \longrightarrow Mg(OH)_2 + NH_3$
(c) $NH_3 + O_2 \longrightarrow N_2O + H_2O$
(d) $LiOH + CO_2 \longrightarrow Li_2CO_3 + H_2O$

10.15 Balance the following equations.
(a) $PbO + NH_3 \longrightarrow Pb + N_2 + H_2O$
(b) $NaHCO_3 + H_2SO_4 \longrightarrow Na_2SO_4 + H_2O + CO_2$
(c) $TiO_2 + C + Cl_2 \longrightarrow TiCl_4 + CO_2$
(d) $NBr_3 + NaOH \longrightarrow N_2 + NaBr + HBrO$

10.16 Balance the following equations.
(a) $NH_3 + O_2 + CH_4 \longrightarrow HCN + H_2O$
(b) $KClO_3 + HCl \longrightarrow KCl + Cl_2 + H_2O$
(c) $SO_2Cl_2 + HI \longrightarrow H_2S + H_2O + HCl + I_2$
(d) $NO + CH_4 \longrightarrow HCN + H_2O + H_2$

10.17 Balance the following equations.
(a) $Ca(OH)_2 + HNO_3 \longrightarrow Ca(NO_3)_2 + H_2O$
(b) $BaCl_2 + (NH_4)_2SO_4 \longrightarrow BaSO_4 + NH_4Cl$
(c) $Fe(OH)_3 + H_2SO_4 \longrightarrow Fe_2(SO_4)_3 + H_2O$
(d) $Na_3PO_4 + AgNO_3 \longrightarrow NaNO_3 + Ag_3PO_4$

10.18 Balance the following equations.
(a) $Al + Sn(NO_3)_2 \longrightarrow Al(NO_3)_3 + Sn$
(b) $Na_2CO_3 + Mg(NO_3)_2 \longrightarrow MgCO_3 + NaNO_3$
(c) $Al(NO_3)_3 + H_2SO_4 \longrightarrow Al_2(SO_4)_3 + HNO_3$
(d) $Ba(C_2H_3O_2)_2 + (NH_4)_3PO_4 \longrightarrow Ba_3(PO_4)_2 + NH_4C_2H_3O_2$

10.19 Each of the following *mathematically balanced* chemical equations is in a *nonconventional* form. Through coefficient adjustment, change each of these equations to conventional form without unbalancing them.
(a) $3\ AgNO_3 + 3\ KCl \longrightarrow 3\ KNO_3 + 3\ AgCl$
(b) $2\ CS_2 + 6\ O_2 \longrightarrow 2\ CO_2 + 4\ SO_2$
(c) $H_2 + \frac{1}{2}O_2 \longrightarrow H_2O$
(d) $Ag_2CO_3 \longrightarrow 2\ Ag + CO_2 + \frac{1}{2}O_2$

10.20 Each of the following *mathematically balanced* chemical equations is in a *nonconventional* form. Through coefficient adjustment, change each of these equations to conventional form without unbalancing them.
(a) $2\ Cu(NO_3)_2 + 2\ Fe \longrightarrow 2\ Cu + 2\ Fe(NO_3)_2$
(b) $2\ N_2H_4 + 4\ H_2O_2 \longrightarrow 2\ N_2 + 8\ H_2O$
(c) $Li_3N \longrightarrow 3\ Li + \frac{1}{2}N_2$
(d) $2\ HNO_3 \longrightarrow 2\ NO_2 + H_2O + \frac{1}{2}O_2$

Patterns in Chemical Reactivity (Sec. 10.5)

10.21 Write a balanced chemical equation for the combustion of each of the following hydrocarbons in air.
(a) CH_4 (b) C_6H_6 (c) C_6H_{12} (d) C_3H_4

10.22 Write a balanced chemical equation for the combustion of each of the following hydrocarbons in air.
(a) C_5H_{12} (b) C_4H_6 (c) C_7H_8 (d) C_8H_{18}

10.23 Write a balanced chemical equation for the combusion of each of the following carbon–hydrogen–oxygen compounds in air.
(a) CH_2O (b) C_3H_6O (c) CH_2O_2 (d) $C_4H_8O_2$

10.24 Write a balanced chemical equation for the combustion of each of the following carbon–hydrogen–oxygen compounds in air.
(a) C_2H_4O (b) $C_5H_{10}O$ (c) $C_2H_4O_2$ (d) $C_3H_6O_2$

10.25 Write a balanced equation for the thermal decomposition of each of the following metal carbonates to its metal oxide and carbon dioxide.
(a) K_2CO_3 (b) $CaCO_3$ (c) $NiCO_3$ (d) $Fe_2(CO_3)_3$

10.26 Write a balanced equation for the thermal decomposition of each of the following metal carbonates to its metal oxide and carbon dioxide.
(a) $BeCO_3$ (b) Li_2CO_3 (c) $ZnCO_3$ (d) Cs_2CO_3

10.27 Write a complete balanced equation for the combustion in air of each of the following compounds.
(a) C_2H_7N, where NO_2 is one of the products
(b) CH_4S, where SO_2 is one of the products

10.28 Write a complete balanced equation for the combustion in air of each of the following compounds.
(a) C_2H_6S, where SO_2 is one of the products
(b) CH_5N, where NO_2 is one of the products

Classes of Chemical Reactions (Sec. 10.6)

10.29 Classify each of the following reactions as synthesis, decomposition, single replacement, or double replacement.
(a) $SO_3 + H_2O \longrightarrow H_2SO_4$
(b) $2\ H_2 + O_2 \longrightarrow 2\ H_2O$
(c) $Na_2CO_3 + Ca(OH)_2 \longrightarrow CaCO_3 + 2\ NaOH$
(d) $Cu(NO_3)_2 + Fe \longrightarrow Cu + Fe(NO_3)_2$

10.30 Classify each of the following reactions as synthesis, decomposition, single replacement, or double replacement.
(a) $3\ CuSO_4 + 2\ Al \longrightarrow Al_2(SO_4)_3 + 3\ Cu$

(b) $K_2CO_3 \longrightarrow K_2O + CO_2$

(c) $2\ AgNO_3 + K_2SO_4 \longrightarrow Ag_2SO_4 + 2\ KNO_3$

(d) $2\ SO_2 + O_2 \longrightarrow 2\ SO_3$

10.31 Identify the products of and then write a balanced chemical equation for each of the following chemical reactions.

(a) $Zn + Cu(NO_3)_2 \longrightarrow$? + ? (single-replacement reaction)

(b) $Ca + O_2 \longrightarrow$? (synthesis reaction)

(c) $K_2SO_4 + Ba(NO_3)_2 \longrightarrow$? + ? (double-replacement reaction)

(d) $Ag_2O \longrightarrow$? + ? (decomposition reaction)

10.32 Identify the products of and then write a balanced chemical equation for each of the following chemical reactions.

(a) $AlCl_3 \longrightarrow$? + ? (decomposition reaction)

(b) $Cu(NO_3)_2 + Na_2CO_3 \longrightarrow$? + ? (double-replacement reaction)

(c) $Al + Ni(NO_3)_2 \longrightarrow$? + ? (single-replacement reaction)

(d) $Be + N_2 \longrightarrow$? (synthesis reaction)

Chemical Equations and the Mole Concept (Sec. 10.7)

10.33 Consider the general equation

$$3\ A + 2\ B \longrightarrow C + 3\ D$$

(a) How many molecules of B will react with 3 molecules of A?

(b) How many molecules of A will react with 6 molecules of B?

(c) How many molecules of D are produced when 4 molecules of B react?

(d) How many moles of B will react with 3 moles of A?

10.34 Consider the general equation

$$A + 3\ B \longrightarrow C + 2\ D$$

(a) How many molecules of A will react with 3 molecules of B?

(b) How many molecules of B will react with 3 molecules of A?

(c) How many molecules of C are produced when 2 molecules of A react?

(d) How many moles of B will react with 2 moles of A?

10.35 Write the twelve mole-to-mole conversion factors that can be derived from the balanced equation

$$4\ NH_3 + 3\ O_2 \longrightarrow 2\ N_2 + 6\ H_2O$$

10.36 Write the twelve mole-to-mole conversion factors that can be derived from the balanced equation

$$CS_2 + 3\ O_2 \longrightarrow 2\ N_2 + 2\ SO_2$$

10.37 Using each of the following equations, calculate the number of moles of the first-listed reactant that are needed to produce 3.00 moles of N_2.

(a) $2\ NaN_3 \longrightarrow 2\ Na + 3\ N_2$

(b) $3\ CO + 2\ NaCN \longrightarrow Na_2CO_3 + 4\ C + N_2$

(c) $2\ NH_2Cl + N_2H_4 \longrightarrow 2\ NH_4Cl + N_2$

(d) $4\ C_3H_5O_9N_3 \longrightarrow 12\ CO_2 + 6\ N_2 + O_2 + 10\ H_2O$

10.38 Using each of the following equations, calculate the number of moles of the first-listed reactant that are needed to produce 4.00 moles of N_2.

(a) $4\ NH_3 + 3\ O_2 \longrightarrow 2\ N_2 + 6\ H_2O$

(b) $(NH_4)_2Cr_2O_7 \longrightarrow Cr_2O_3 + N_2 + 4\ H_2O$

(c) $N_2H_4 + 2\ H_2O_2 \longrightarrow N_2 + 4\ H_2O$

(d) $2\ Li_3N \longrightarrow 6\ Li + N_2$

10.39 How many moles of the first-listed reactant in each of the following equations will completely react with 1.42 moles of the second-listed reactant?

(a) $C_7H_{16} + 11\ O_2 \longrightarrow 7\ CO_2 + 8\ H_2O$

(b) $2\ HCl + CaCO_3 \longrightarrow CaCl_2 + CO_2 + H_2O$

(c) $Na_2SO_4 + 2\ C \longrightarrow Na_2S + 2\ CO_2$

(d) $4\ Na_2CO_3 + Fe_3Br_8 \longrightarrow 8\ NaBr + 4\ CO_2 + Fe_3O_4$

10.40 How many moles of the first-listed reactant in each of the following equations will completely react with 2.03 moles of the second-listed reactant?

(a) $3\ O_2 + CS_2 \longrightarrow CO_2 + 2\ SO_2$

(b) $FeO + CO \longrightarrow Fe + CO_2$

(c) $2\ C_8H_{18} + 25\ O_2 \longrightarrow 16\ CO_2 + 18\ H_2O$

(d) $Fe_3O_4 + CO \longrightarrow 3\ FeO + CO_2$

10.41 Using each of the following equations, calculate the total number of moles of products that can be obtained from the decomposition of 1.75 moles of the reactant.

(a) $2\ NH_4NO_3 \longrightarrow 2\ N_2 + O_2 + 4\ H_2O$

(b) $2\ NaClO_3 \longrightarrow 2\ NaCl + 3\ O_2$

(c) $2\ KNO_3 \longrightarrow 2\ KNO_2 + O_2$

(d) $4\ I_4O_9 \longrightarrow 6\ I_2O_5 + 2\ I_2 + 3\ O_2$

10.42 Using each of the following equations, calculate the total number of moles of products that can be obtained from the decomposition of 2.25 moles of the reactant.

(a) $2\ Ag_2CO_3 \longrightarrow 4\ Ag + 2\ CO_2 + O_2$

(b) $2\ KClO_3 \longrightarrow 2\ KCl + 3\ O_2$

(c) $4\ HNO_3 \longrightarrow 4\ NO_2 + 2\ H_2O + O_2$

(d) $2\ H_2O_2 \longrightarrow 2\ H_2O + O_2$

10.43 For the chemical reaction

$$CH_4(g) + 4\ Cl_2(g) \longrightarrow CCl_4(l) + 4\ HCl(g)$$

(a) How many moles of Cl_2 are needed to produce 4.75 moles of CCl_4?

(b) How many moles of HCl will be produced from 0.083 mole of CH_4?

(c) How many moles of CH_4 are needed to react with 2.30 moles of Cl_2?

(d) How many moles of CCl_4 are produced at the same time that 1.23 moles of HCl are produced?

10.44 For the chemical reaction

$$4\ FeS_2(s) + 11\ O_2(g) \longrightarrow 2\ Fe_2O_3(s) + 8\ SO_2(g)$$

(a) How many moles of O_2 are needed to produce 3.50 moles of SO_2?

(b) How many moles of Fe_2O_3 will be produced from 1.02 moles of FeS_2?

(c) How many moles of FeS_2 are needed to react with 5.40 moles of O_2?

(d) How many moles of Fe_2O_3 are produced at the same time that 0.908 mole of SO_2 is produced?

Stoichiometry (Sec. 10.8)

10.45 How many grams of the second-listed reactant in each of the following reactions is needed to react completely with 1.772 g of the first-listed reactant?

(a) $SiO_2 + 3\ C \longrightarrow 2\ CO + SiC$

(b) $5\ O_2 + C_3H_8 \longrightarrow 3\ CO_2 + 4\ H_2O$

(c) $CH_4 + 4\ Cl_2 \longrightarrow 4\ HCl + CCl_4$

(d) $3\ NO_2 + H_2O \longrightarrow 2\ HNO_3 + NO$

10.46 How many grams of the second-listed reactant in each of the following reactions is needed to react completely with 12.56 g of the first-listed reactant?

(a) $H_2O_2 + H_2S \longrightarrow 2\ H_2O + S$

(b) $4\ NH_3 + 3\ O_2 \longrightarrow 2\ N_2 + 6\ H_2O$

(c) $Mg + 2\ HCl \longrightarrow MgCl_2 + H_2$

(d) $6\ HCl + 2\ Al \longrightarrow 3\ H_2 + 2\ AlCl_3$

10.47 Silicon carbide, SiC, used as an abrasive on sandpaper, is prepared using the following chemical reaction

$$SiO_2(s) + 3\ C(s) \longrightarrow SiC(s) + 2\ CO(g)$$

(a) How many grams of SiO_2 are needed to react with 1.50 moles of C?

(b) How many grams of CO are produced when 1.37 moles of SiO_2 react?

(c) How many grams of SiC are produced at the same time that 3.33 moles of CO are produced?

(d) How many grams of C must react in order to produce 0.575 mole of SiC?

10.48 The inflating gas for automobile air bags is nitrogen (N_2), generated from the decomposition of sodium azide (NaN_3). The equation for the decomposition reaction is

$$2\ NaN_3(s) \longrightarrow 2\ Na(s) + 3\ N_2(g)$$

(a) How many grams of NaN_3 must decompose to produce 3.57 moles of N_2?

(b) How many grams of NaN_3 must decompose to produce 3.57 moles of Na?

(c) How many grams of Na are produced at the same time that 5.40 moles of N_2 are produced?

(d) How many moles of NaN_3 must decompose in order to produce 10.00 g of N_2?

10.49 One way to remove gaseous carbon dioxide (CO_2) from the air in a spacecraft is to let canisters of solid lithium hydroxide (LiOH) absorb it according to the reaction

$$2\ LiOH(s) + CO_2(g) \longrightarrow Li_2CO_3(s) + H_2O(l)$$

Based on this equation, how many grams of LiOH must be used to achieve the following?

(a) absorb 4.50 moles of CO_2

(b) absorb 3.00×10^{24} molecules of CO_2

(c) produce 10.0 g of H_2O

(d) produce 10.0 g of Li_2CO_3

10.50 Tungsten (W) metal, used to make incandescent light bulb filaments, is produced by the reaction

$$WO_3(s) + 3\ H_2(g) \longrightarrow W(s) + 3\ H_2O(l)$$

Based on this equation, how many grams of WO_3 are needed to produce each of the following?

(a) 10.00 g of W

(b) 1 billion (1.00×10^9) molecules of H_2O

(c) 2.53 moles of H_2O

(d) 250,000 atoms of W

10.51 Hydrofluoric acid, HF, cannot be stored in glass bottles because it attacks silicate compounds present in the glass. For example, sodium silicate, Na_2SiO_3, reacts with HF in the following way:

$$Na_2SiO_3 + 8\ HF \longrightarrow H_2SiF_6 + 2\ NaF + 3\ H_2O$$

(a) How many moles of Na_2SiO_3 must react to produce 25.00 g of NaF?

(b) How many grams of HF must react to produce 27.00 g of H_2O?

(c) How many molecules of H_2SiF_6 are produced from the reaction of 2.000 g of Na_2SiO_3?

(d) How many grams of HF are needed to react with 50.00 g of Na_2SiO_3?

10.52 Potassium thiosulfate, $K_2S_2O_3$, is used to remove any excess chlorine from fibers and fabrics that have been bleached with that gas.

$$K_2S_2O_3 + 4\ Cl_2 + 5\ H_2O \longrightarrow 2\ KHSO_4 + 8\ HCl$$

(a) How many moles of $K_2S_2O_3$ must react to produce 2.500 g of HCl?

(b) How many grams of Cl_2 must react to produce 20.00 g of $KHSO_4$?

(c) How many molecules of HCl are produced at the same time that 2.000 g of $KHSO_4$ is produced?

(d) How many grams of H_2O are consumed as 12.50 g of Cl_2 reacts?

10.53 How many grams of sodium (Na) are needed to react completely with 16.5 g of sulfur (S) in the synthesis of Na_2S?

10.54 How many grams of beryllium (Be) are needed to react completely with 45.0 g of nitrogen (N_2) in the synthesis of Be_3N_2?

10.55 When chromium metal reacts with chlorine gas, a violet solid with the formula $CrCl_3$ is formed.

$$2\ Cr + 3\ Cl_2 \longrightarrow 2\ CrCl_3$$

How many grams of Cr and how many grams of Cl_2 are needed to produce 200.0 g of $CrCl_3$?

10.56 Black silver sulfide can be produced from the reaction of silver metal with sulfur.

$$2\ Ag + S \longrightarrow Ag_2S$$

How many grams of Ag and how many grams of S are needed to produce 150.0 g of Ag_2S?

Limiting Reactant Calculations (Sec. 10.9)

10.57 What will be the limiting reactant in the production of "three nut–four bolt" combinations from a collection of 216 nuts and 284 bolts?

10.58 What will be the limiting reactant in the production of "five nut–four bolt" combinations from a collection of 785 nuts and 660 bolts?

10.59 A model airplane kit is designed to contain two wings, one fuselage, four engines, and six wheels. How many model airplane kits can a manufacturer produce from a parts inventory of 426 wings, 224 fuselages, 860 engines, and 1578 wheels?

10.60 A model car kit is designed to contain one body, four wheels, two bumpers, and one steering wheel. How many model car kits can a manufacturer produce from a parts inventory of 137 bodies, 532 wheels, 246 bumpers, and 139 steering wheels?

10.61 At high temperatures and pressures nitrogen will react with hydrogen to produce ammonia as shown by the equation

$$N_2 + 3\ H_2 \longrightarrow 2\ NH_3$$

For each of the following combinations of reactants, decide which is the limiting reactant.

(a) 1.25 moles of N_2 and 3.65 moles of H_2
(b) 2.60 moles of N_2 and 8.00 moles of H_2
(c) 44.0 g of N_2 and 3.00 moles of H_2
(d) 55.0 g of N_2 and 15.0 g of H_2

10.62 Aluminum oxide can be prepared by the direct reaction of the elements as shown by the equation

$$4\ Al + 3\ O_2 \longrightarrow 2\ Al_2O_3$$

For each of the following combinations of reactants, decide which is the limiting reactant.

(a) 3.00 moles of Al and 4.00 moles of O_2
(b) 7.00 moles of Al and 5.40 moles of O_2
(c) 16.2 g of Al and 0.40 mole of O_2
(d) 100.0 g of Al and 100.0 g of O_2

10.63 Magnesium nitride can be prepared by the direct reaction of the elements as shown by the equation

$$3\ Mg + N_2 \longrightarrow Mg_3N_2$$

How many grams of magnesium nitride can be produced from the following amounts of reactants?

(a) 10.0 g of Mg and 10.0 g of N_2
(b) 20.0 g of Mg and 10.0 g of N_2
(c) 30.0 g of Mg and 10.0 g of N_2
(d) 40.0 g of Mg and 10.0 g of N_2

10.64 Under appropriate conditions water can be produced from the reaction of the elements hydrogen and oxygen as shown by the equation

$$2\ H_2 + O_2 \longrightarrow 2\ H_2O$$

How many grams of water can be produced from the following amounts of reactants?

(a) 10.0 g of H_2 and 40.0 g of O_2
(b) 10.0 g of H_2 and 60.0 g of O_2
(c) 10.0 g of H_2 and 80.0 g of O_2
(d) 10.0 g of H_2 and 100.0 g of O_2

10.65 Determine how many $CoCl_3$ formula units can be produced from a reaction mixture containing 525 cobalt atoms and 525 HCl molecules according to the following reaction.

$$2\ Co + 6\ HCl \longrightarrow 2\ CoCl_3 + 3\ H_2$$

10.66 Determine how many $NiCl_2$ formula units can be produced from a reaction mixture containing 782 nickel atoms and 782 HCl molecules according to the following reaction.

$$Ni + 2\ HCl \longrightarrow NiCl_2 + H_2$$

10.67 If 70.0 g of Fe_3O_4 and 12.0 g of O_2 are present in a reaction mixture, determine how many grams of each reactant will be left unreacted upon completion of the following reaction.

$$4\ Fe_3O_4 + O_2 \longrightarrow 6\ Fe_2O_3$$

10.68 If 70.0 g of $TiCl_4$ and 16.0 g of Ti are present in a reaction mixture, determine how many grams of each reactant will be left unreacted upon completion of the following reaction.

$$3\ TiCl_4 + Ti \longrightarrow 4\ TiCl_3$$

10.69 Determine the number of grams of each of the products that can be made from 8.00 g of SCl_2 and 4.00 g of NaF by the following reaction.

$$3\ SCl_2 + 4\ NaF \longrightarrow SF_4 + S_2Cl_2 + 4\ NaCl$$

10.70 Determine the number of grams of each of the products that can be made from 100.0 g of Na_2CO_3 and 300.0 g of Fe_3Br_8 by the following reaction.

$$4\ Na_2CO_3 + Fe_3Br_8 \longrightarrow 8\ NaBr + 4\ CO_2 + Fe_3O_4$$

Theoretical Yield and Percent Yield (Sec. 10.10)

10.71 Because of "sloppiness" in his procedures, a student was able to isolate only 16.0 g of a desired product from a chemical reaction rather than the 52.0 g that was theoretically possible. What was the percent yield of product that the student obtained?

10.72 The theoretical yield of product for a particular reaction is 25.31 g. A very "meticulous" student isolates 24.79 g of product when the reaction is run. What is the percent yield that this student obtained?

10.73 Aluminum and sulfur react to form aluminum sulfide by the equation.

$$2\ Al + 3\ S \longrightarrow Al_2S_3$$

In a certain experiment, 125 g of Al_2S_3 are produced from 75.0 g of Al and 300.0 g of S.

(a) What is the theoretical yield of Al_2S_3?

(b) What is the percent yield of Al_2S_3?

10.74 Aluminum and oxygen react to form aluminum oxide by the equation

$$4\ Al + 3\ O_2 \longrightarrow 2\ Al_2O_3$$

In a certain experiment, 125 g of Al_2O_3 are produced from 75.0 g of Al and 200.0 g of O_2.

(a) What is the theoretical yield of Al_2O_3?

(b) What is the percent yield of Al_2O_3?

10.75 If 74.30 g of HCl were produced from 2.130 g of H_2 and an excess of Cl_2 according to the reaction

$$H_2 + Cl_2 \longrightarrow 2\ HCl$$

what was the percent yield of HCl?

10.76 If 115.7 g of Ca_3N_2 were produced from 28.2 g of N_2 and an excess of Ca according to the reaction

$$3\ Ca + N_2 \longrightarrow Ca_3N_2$$

what was the percent yield of Ca_3N_2?

10.77 Under appropriate reaction conditions Al and S produce Al_2S_3 according to the equation

$$2\ Al + 3\ S \longrightarrow Al_2S_3$$

In a certain experiment with 55.0 g of Al and an excess of S, a percent yield of 85.6% was obtained. What was the actual yield of Al_2S_3, in grams, for this experiment?

10.78 Under appropriate reaction conditions Ag and S produce Ag_2S according to the equation

$$2\ Ag + S \longrightarrow Ag_2S$$

In a certain experiment with 75.0 g of Ag and an excess of S, a percent yield of 72.9% was obtained. What was the actual yield of Ag_2S, in grams, for this experiment?

10.79 If the percent yield for the reaction

$$2\ CO + O_2 \longrightarrow 2\ CO_2$$

were 57.8%, what mass of product, in grams, could be produced from a reactant mixture containing 35.0 g of each reactant?

10.80 If the percent yield for the reaction

$$2\ C + O_2 \longrightarrow 2\ CO$$

were 34.3%, what mass of product, in grams, could be produced from a reactant mixture containing 2.25 g of each reactant?

Simultaneous Reactions (Sec. 10.11)

10.81 A mixture of composition 60.0% ZnS and 40.0% CuS is heated in air until the sulfides are completely converted to oxides as shown by the following equations.

$$2\ ZnS + 3\ O_2 \longrightarrow 2\ ZnO + 2\ SO_2$$

$$2\ CuS + 3\ O_2 \longrightarrow 2\ CuO + 2\ SO_2$$

How many grams of SO_2 are produced from the reaction 82.5 g of the sulfide mixture?

10.82 A mixture of composition 50.0% H_2S and 50.0% CH_4 is reacted with oxygen, producing SO_2, CO_2, and H_2O. The equations for the reactions are

$$2\ H_2S + 3\ O_2 \longrightarrow 2\ SO_2 + 2\ H_2O$$

$$CH_4 + 2\ O_2 \longrightarrow CO_2 + 2\ H_2O$$

How many grams of H_2O are produced from the reaction of 65.0 g of mixture?

10.83 A mixture of composition 70.0% methane (CH_4) and 30.0% ethane (C_2H_6) by mass is burned in oxygen to produce CO_2 and H_2O. The reactions that occur are

$$CH_4 + 2\ O_2 \longrightarrow CO_2 + 2\ H_2O$$

$$2\ C_2H_6 + 7\ O_2 \longrightarrow 4\ CO_2 + 6\ H_2O$$

How many grams of O_2 are needed to react completely with 75.0 g of mixture?

10.84 A mixture of composition 64.0% Zn and 36.0% Sn by mass is dissolved in hydrochloric acid (HCl) to produce the metal chlorides and H_2. The reactions that occur are

$$Zn + 2\ HCl \longrightarrow ZnCl_2 + H_2$$

$$Sn + 2\ HCl \longrightarrow SnCl_2 + H_2$$

How many grams of HCl are needed to react completely with 50.0 g of mixture?

10.85 Consider the following two-step reaction sequence.

$$2\ NaClO_3 \longrightarrow 2\ NaCl + 3\ O_2$$

$$S + O_2 \longrightarrow SO_2$$

Assuming that all of the oxygen generated in the first step is consumed in the second step, what mass, in grams, of sodium chlorate ($NaClO_3$) is needed in the first step to produce 20.0 g of sulfur dioxide (SO_2) in the second step?

10.86 Consider the following two-step reaction sequence.

$$2\ NaCl + 2\ H_2O \longrightarrow 2\ NaOH + H_2 + Cl_2$$

$$Ni + Cl_2 \longrightarrow NiCl_2$$

Assuming that all of the chlorine generated in the first step is consumed in the second step, what mass, in grams, of sodium chloride (NaCl) is needed in the first step to produce 50.0 g of nickel(II) chloride ($NiCl_2$) in the second step?

10.87 Acid rain contains both nitric acid and sulfuric acid. The nitric acid is formed from atmospheric N_2 in a three-step process.

$$N_2 + O_2 \longrightarrow 2\ NO$$

$$2\ NO + O_2 \longrightarrow 2\ NO_2$$

$$4\ NO_2 + 2\ H_2O + O_2 \longrightarrow 4\ HNO_3$$

How many grams of HNO_3 result from the reaction of 2.00 g of N_2 in the first step?

10.88 Acid rain contains both nitric acid and sulfuric acid. The sulfuric acid is formed from S (primarily in coal) in a three-step process.

$$S + O_2 \longrightarrow SO_2$$

$$2\ SO_2 + O_2 \longrightarrow 2\ SO_3$$

$$SO_3 + H_2O \longrightarrow H_2SO_4$$

How many grams of H_2SO_4 result from the reaction of 5.00 g of S in the first step?

10.89 The following process has been used for obtaining iodine from oilfield brines.

$$NaI + AgNO_3 \longrightarrow AgI + NaNO_3$$

$$2\ AgI + Fe \longrightarrow FeI_2 + 2\ Ag$$

$$2\ FeI_2 + 3\ Cl_2 \longrightarrow 2\ FeCl_3 + 2\ I_2$$

How much $AgNO_3$, in grams, is required in the first step for every 5.00 g of I_2 produced in the third step?

10.90 Sodium bicarbonate, $NaHCO_3$, can be prepared from sodium sulfate, Na_2SO_4, using the following three-step process.

$$Na_2SO_4 + 4\ C \longrightarrow Na_2S + 4\ CO$$

$$Na_2S + CaCO_3 \longrightarrow CaS + Na_2CO_3$$

$$Na_2CO_3 + H_2O + CO_2 \longrightarrow 2\ NaHCO_3$$

How much carbon, C, in grams, is required in the first step for every 10.00 g of $NaHCO_3$ produced in the third step?

Additional Problems

10.91 Ammonium dichromate decomposes according to the following reaction.

$$(NH_4)_2Cr_2O_7 \longrightarrow N_2 + 4\ H_2O + Cr_2O_3$$

How many grams of each of the products can be formed from the decomposition of 75.0 g of $(NH_4)_2Cr_2O_7$?

10.92 Nitrous acid decomposes according to the following reaction.

$$3\ HNO_2 \longrightarrow 2\ NO + HNO_3 + H_2O$$

How many grams of each of the products can be formed from the decomposition of 63.5 g of HNO_2?

10.93 Hydrogen sulfide burns in oxygen to form sulfur dioxide and water.

$$2\ H_2S + 3\ O_2 \longrightarrow 2\ SO_2 + 2\ H_2O$$

How many grams of hydrogen sulfide must react in order to produce a total of 100.0 g of products?

10.94 Carbon disulfide burns in oxygen to form carbon dioxide and sulfur dioxide.

$$CS_2 + 3\ O_2 \longrightarrow CO_2 + 2\ SO_2$$

How many grams of carbon disulfide must react in order to produce a total of 50.0 g of products?

10.95 From each of the following pairs of ratios, determine whether reactant A or reactant B is the limiting reactant.

(a) required ratio: 3 moles A to 1 mole B
ratio actually present: 2 moles A to 0.5 mole B

(b) required ratio: 3 moles A to 2 moles B
ratio actually present: 4.8 moles A to 3.6 moles B

(c) required ratio: 1 mole A to 2 moles B
ratio actually present: 3.5 moles A to 6.9 moles B

(d) required ratio: 4 moles A to 3 moles B
ratio actually present: 1.9 moles A to 1.6 moles B

10.96 From each of the following pairs of ratios, determine whether reactant A or reactant B is the limiting reactant.

(a) required ratio: 2 moles A to 1 mole B
ratio actually present: 0.40 mole A to 0.30 mole B

(b) required ratio: 2 moles A to 3 moles B
ratio actually present: 2.4 moles A to 2.7 moles B

(c) required ratio: 1 mole A to 1 mole B
ratio actually present: 1.3 moles A to 1.5 moles B

(d) required ratio: 3 moles A to 4 moles B
ratio actually present: 3.6 moles A to 4.4 moles B

10.97 Pure, dry NO gas can be made by the following reaction.

$$3\ KNO_2 + KNO_3 + Cr_2O_3 \longrightarrow 4\ NO + 2\ K_2CrO_4$$

How many grams of NO can be produced from a reaction mixture containing 2.00 moles each of KNO_2, KNO_3, and Cr_2O_3?

10.98 Sodium cyanide, NaCN, can be made by the following reaction.

$$Na_2CO_3 + 4\ C + N_2 \longrightarrow 2\ NaCN + 3\ CO$$

How many grams of CO can be produced from a reaction mixture containing 3.00 moles each of Na_2CO_3, C, and N_2?

10.99 An *impure* sample of $CuSO_4$ weighing 7.53 g was dissolved in water. The dissolved $CuSO_4$, but not the impurities, then reacted with excess zinc.

$$CuSO_4 + Zn \longrightarrow ZnSO_4 + Cu$$

What was the mass percent $CuSO_4$ in the sample if 1.33 g of Cu were produced?

10.100 An *impure* sample of $Hg(NO_3)_2$, weighing 64.5 g was dissolved in water. The dissolved $Hg(NO_3)_2$, but not the impurities, then reacted with excess Mg metal.

$$Hg(NO_3)_2 + Mg \longrightarrow Mg(NO_3)_2 + Hg$$

What was the mass percent $Hg(NO_3)_2$ in the sample if 23.6 g of Hg were produced?

10.101 Silver oxide (Ag_2O) decomposes completely at high temperatures to produce metallic silver and oxygen gas. A 1.80-g sample of *impure* silver oxide yielded 0.115 g of O_2. Assuming that Ag_2O was the only source of O_2, what was the mass percent of Ag_2O in the sample?

10.102 Gold(III) oxide (Au_2O_3) decomposes completely at high temperatures to produce metallic gold and oxygen gas. A 2.21-g sample of *impure* gold(III) oxide yielded 0.233 g of O_2. Assuming that Au_2O_3 was the only source of O_2, what was the mass percent of Au_2O_3 in the sample?

10.103 The reaction between 113.4 g of I_2O_5 and 132.2 g of BrF_3 was found to produce 97.0 g of IF_5. The equation for the reaction is

$$6\ I_2O_5 + 20\ BrF_3 \longrightarrow 12\ IF_5 + 15\ O_2 + 10\ Br_2$$

What is the percent yield of IF_5?

10.104 The reaction between 20.0 g of NH_3 and 20.0 g of CH_4 with an excess of oxygen was found to produce 15.0 g of HCN. The equation for the reaction is

$$2\ NH_3 + 3\ O_2 + 2\ CH_4 \longrightarrow 2\ HCN + 6\ H_2O$$

What is the percent yield of HCN?

10.105 Copper metal can be recovered from an ore containing $CuCO_3$ by the decomposition reaction

$$2\ CuCO_3 \longrightarrow 2\ Cu + 2\ CO_2 + O_2$$

What mass of copper ore, in tons, is needed to produce 500.0 lb of Cu if the ore is 13.22% by mass $CuCO_3$? Assume complete decomposition of the $CuCO_3$.

10.106 Silver metal can be recovered from an ore containing Ag_2CO_3 by the decomposition reaction

$$2\ Ag_2CO_3 \longrightarrow 4\ Ag + 2\ CO_2 + O_2$$

What mass of silver ore, in tons, is needed to produce 500.0 lb of Ag if the ore is 24.21% by mass Ag_2CO_3.

10.107 A 13.20-g sample of a mixture of $CaCO_3$ and $NaHCO_3$ was heated, and the compounds decomposed as follows.

$$CaCO_3 \longrightarrow CaO + CO_2$$

$$2\ NaHCO_3 \longrightarrow Na_2CO_3 + CO_2 + H_2O$$

The decomposition of the sample yields 4.35 g of CO_2 and 0.873 g of H_2O. What percentage, by mass, of the original sample was $CaCO_3$?

10.108 A 4.00-g sample of a mixture of H_2S and CS_2 was burned in oxygen. The equations for the reactions are

$$2\ H_2S + 3\ O_2 \longrightarrow 2\ H_2O + 2\ SO_2$$

$$CS_2 + 3\ O_2 \longrightarrow CO_2 + 2\ SO_2$$

If 7.32 g of SO_2 and 0.577 g of CO_2 were produced along with some H_2O, what percentage, by mass, of the original sample was H_2S?

CUMULATIVE PROBLEMS

10.109 Write a balanced equation for each of the following chemical reactions

(a) zinc + silver nitrate ⟶ zinc nitrate + silver

(b) hydrochloric acid + sodium hydroxide ⟶ sodium chloride + water

(c) phosphorus trichloride + chlorine ⟶ phosphorus pentachloride

(d) copper + oxygen ⟶ copper(II) oxide

10.110 Write a balanced equation for each of the following chemical reactions.

(a) sodium oxide + sulfur trioxide ⟶ sodium sulfate

(b) barium carbonate ⟶ barium oxide + carbon dioxide

(c) aluminum + iron(II) oxide ⟶ iron + aluminum oxide

(d) ammonia + phosphoric acid ⟶ ammonium phosphate

10.111 After the following equation was balanced, the name of one of the reactants was substituted for its formula.

$$2\ \text{Cyclopropane} + 9\ O_2 \longrightarrow 6\ CO_2 + 6\ H_2O$$

Using only the information found within this equation, determine the molecular formula of cyclopropane.

10.112 After the following equation was balanced, the name of one of the reactants was substituted for its formula.

$$2\ \text{Butyne} + 11\ O_2 \longrightarrow 8\ CO_2 + 6\ H_2O$$

Using only the information found within this equation, determine the molecular formula of butyne.

10.113 Write a balanced chemical equation for the reaction in which Cu_2S and O_2 are reactants and SO_2 and a copper oxide containing 88.82% copper by mass are products. The molecular and empirical formulas of the copper oxide are the same.

10.114 Write a balanced chemical equation for the reaction in which NH_3 and O_2 are reactants and H_2O and a nitrogen oxide containing 46.68% nitrogen by mass are products. The molecular and empirical formulas of the nitrogen oxide are the same.

10.115 Write a balanced chemical equation for the reaction in which CO_2 and H_2O are the products, O_2 is one of the reactants, and a compound with an empirical formula of CH and a formula mass of 78.12 amu is the other.

10.116 Write a balanced chemical equation for the reaction in which CO_2 and H_2O are the products, O_2 is one of the reactants, and a compound with an empirical formula of C_3H_5 and a formula mass of 82.16 amu is the other.

10.117 Fifty (50.000) grams of Be are reacted with an excess of F_2 to produce BeF_2. The equation for the reaction is

$$Be + F_2 \longrightarrow BeF_2$$

Calculate the mass of BeF_2 produced using each of the following specifications.

(a) mass, in grams, to three significant figures
(b) mass, in kilograms, to five significant figures
(c) mass, in micrograms, to four significant figures
(d) mass, in pounds, to four significant figures

10.118 Eighty (80.000) grams of Na are reacted with an excess of P to produce Na_3P. The equation for the reaction is

$$3\ Na + P \longrightarrow Na_3P$$

Calculate the mass of Na_3P produced using each of the following specifications.

(a) mass, in grams, to four significant figures
(b) mass, in milligrams, to three significant figures
(c) mass, in pounds, to five significant figures
(d) mass, in ounces, to four significant figures

10.119 If 100.0 g of $KClO_3$ and 200.0 g of HCl are allowed to react according to the equation

$$2\ KClO_3 + 4\ HCl \longrightarrow 2\ KCl + 2\ ClO_2 + Cl_2 + 2\ H_2O$$

what is the combined total number of moles of chlorine-containing products produced?

10.120 If 50.0 g of SO_2Cl_2 and 200.0 g of HI are allowed to react according to the equation

$$SO_2Cl_2 + 8\ HI \longrightarrow H_2S + 2\ H_2O + 2\ HCl + 4\ I_2$$

what is the combined total number of moles of hydrogen-containing products produced?

10.121 The reusable booster rockets of the U.S. space shuttle employ a mixture of aluminum and ammonium perchlorate for fuel. The chemical reaction that occurs is

$$3\ Al + 3\ NH_4ClO_4 \longrightarrow Al_2O_3 + AlCl_3 + 3\ NO + 6\ H_2O$$

How many moles of electrons are present in the $AlCl_3$ produced when 10.0 g of Al react?

10.122 The fluoride in many toothpastes is tin(II) fluoride produced by the reaction of Sn metal with gaseous HF.

$$Sn + 2\ HF \longrightarrow SnF_2 + H_2$$

How many moles of electrons are present in the HF consumed when 25.0 g of H_2 are produced?

10.123 If the products produced by the reaction of 500.0 g of $CaCl_2$ with excess Na_2CO_3 according to the equation

$$CaCl_2 + Na_2CO_3 \longrightarrow 2\ NaCl + CaCO_3$$

were broken up into ions, how many positive ions would result?

10.124 If the products produced by the reaction of 220.0 g of $AgNO_3$ with excess K_3PO_4 according to the equation

$$K_3PO_4 + 3\ AgNO_3 \longrightarrow Ag_3PO_4 + 3\ KNO_3$$

were broken up into ions, how many negative ions would result?

10.125 The concentration of an aqueous NaBr solution, whose density is 1.046 g/mL, is 6.00% by mass. Determine the volume, in milliliters, of NaBr solution needed to prepare 10.0 g of AgBr by the following reaction.

$$NaBr + AgNO_3 \longrightarrow AgBr + NaNO_3$$

10.126 The concentration of an aqueous NH_3 solution, whose density is 0.979 g/mL, is 4.50% by mass. Determine the volume, in milliliters, of NH_3 solution needed to prepare 10.0 g of NH_4NO_3 by the following reaction.

$$NH_3 + HNO_3 \longrightarrow NH_4NO_3$$

10.127 A three-step process for producing nitric acid, HNO_3, from gaseous ammonia, NH_3, is

$$4\ NH_3 + 5\ O_2 \longrightarrow 4\ NO + 6\ H_2O$$

$$2\ NO + O_2 \longrightarrow 2\ NO_2$$

$$3\ NO_2 + H_2O \longrightarrow 2\ HNO_3 + NO$$

Assuming yields, respectively, of 85.2%, 82.7%, and 87.0% for the three steps, how many grams of nitric acid can be produced from 75.0 mL of ammonia with a density of 0.695 g/L?

10.128 A three-step process for producing sulfuric acid, H_2SO_4, from gaseous sulfur dioxide SO_2, is

$$2\ SO_2 + O_2 \longrightarrow 2\ SO_3$$

$$SO_3 + H_2SO_4 \longrightarrow H_2S_2O_7$$

$$H_2S_2O_7 + H_2O \longrightarrow 2\ H_2SO_4$$

Assuming yields, respectively, of 63.1%, 87.5%, and 73.8% for the three steps, how many grams of sulfuric acid can be produced from 125 mL of sulfur dioxide with a density of 0.773 g/L?

10.129 A particular coal contains 4.3% sulfur by mass as an impurity. When the coal is burned the S is converted to gaseous SO_2. The SO_2 enters the exhaust gases where it is removed by reaction with powdered CaO to produce solid $CaSO_3$. How much by-product $CaSO_3$, in tons, is produced by the burning of 1.0 ton of coal?

10.130 A particular coal contains 3.5% sulfur by mass as an impurity. When the coal is burned the S is converted to gaseous SO_2. The SO_2 enters the exhaust gases where it is removed by reaction with powdered CaO to produce solid $CaSO_3$. How much coal would have to be burned, in tons, to produce 2.0 tons of by-product $CaSO_3$?

11 States of Matter

11.1 Physical States of Matter

In this chapter we consider the physical states of matter—the solid state, the liquid state, and the gaseous state.

From everyday experience we know that the physical state of a substance is determined both by what it is—its chemical identity—and by the temperature and pressure it is under. At room temperature and pressure some substances are solids (gold, sodium chloride, etc.), others are liquids (water, mercury, etc.), and still others are gases (oxygen, carbon dioxide, etc.). Thus chemical identity must be a determining factor for physical state, since all three states are observed at room temperature and pressure. On the other hand, when the physical state of a single substance is considered, temperature and pressure are determining variables. Liquid water can be changed to a solid by lowering the temperature or to a gas by raising the temperature.

We tend to characterize a substance almost exclusively in terms of its most common physical state, that is, the state in which it is found at room temperature and pressure. Oxygen is almost always thought of as a gas, its most common state; gold is almost always thought of as a solid, its most common state. A major reason for such "single state" characterization is the narrow range of temperatures encountered on this planet. Most substances are never encountered in more than one state under "natural" conditions. We must be careful not to fall into the error of assuming that the commonly observed state of a substance is the *only* state in which it can exist. Under laboratory conditions, states other than the "natural" one can be obtained for almost all substances. Figure 11.1 shows the temperature ranges for the solid, liquid, and gaseous states of a few elements and compounds. As can be seen in this figure, the size and location of the physical-state temperature ranges vary widely among chemical substances. Extremely high temperatures are required to obtain some substances in the gaseous state; other substances are gases at temperatures below room temperature. The size of a given physical-state range also varies dramatically. For example, the elements H_2 and O_2 are

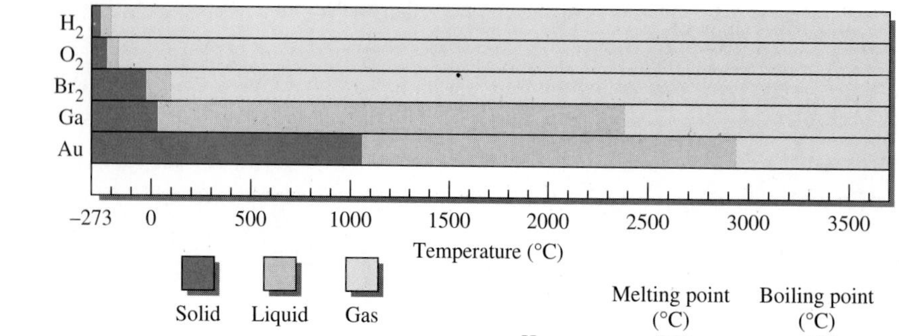

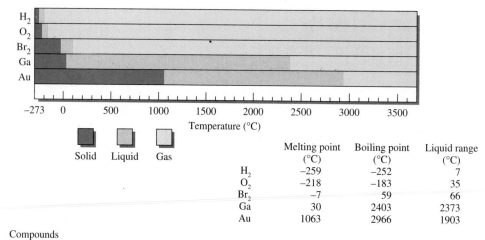

	Melting point (°C)	Boiling point (°C)	Liquid range (°C)
H_2	–259	–252	7
O_2	–218	–183	35
Br_2	–7	59	66
Ga	30	2403	2373
Au	1063	2966	1903

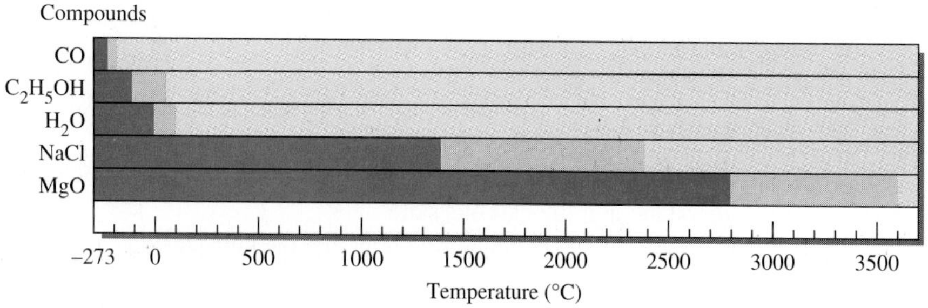

FIGURE 11.1 Solid, liquid, and gaseous temperature ranges for selected elements and compounds.

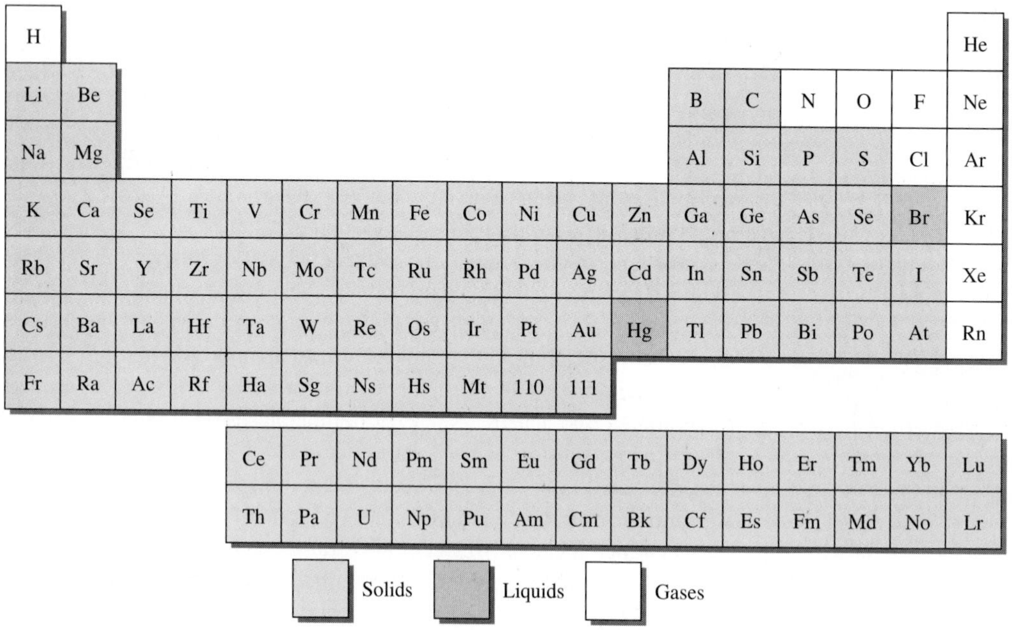

FIGURE 11.2 Physical states of the elements at room temperature and pressure.

liquids over a very narrow temperature range, whereas the elements Ga and Au remain liquids over ranges of hundreds of degrees.

Among the elements, at room temperature and pressure, the solid state is the dominant physical state. Ninety-eight of the 111 known elements are solids at room temperature, as shown in Figure 11.2. It is usually stated that only two elements, bromine and mercury, are liquids at room temperature. However, there are three other elements, cesium and francium of group IA and gallium of group IIIA, that have melting points between 25 and 30°C. Temperatures in this range are reached in a "hot" room. Thus, on "hot" days, there could be five elements that are liquids. Eleven of the elements are gases under "normal" conditions.

Explanations for the experimentally observed physical state variations among substances can be derived from the concepts of atomic structure and bonding we have considered in previous chapters. Such explanations are found in later sections of this chapter.

F.Y.I.
Atomic absorption spectroscopy, one of the techniques chemists use to measure small amounts of heavy metal pollutants such as lead in water, works by vaporizing the metals at temperatures above 2500°C.

11.2 PROPERTY DIFFERENCES OF PHYSICAL STATES

The differences among solids, liquids, and gases are so great that only a few gross distinguishing features need to be mentioned in order to differentiate them clearly. Certain obvious differences among the three states of matter are apparent to even the most casual observer—differences related to (1) volume and shape, (2) density, (3) compressibility, and (4) thermal expansion. These distinguishing properties are compared in Table 11.1 for the three states of matter. The properties of volume and density have been discussed in detail previously (Secs. 3.4 and 3.8, respectively). **Compressibility** *is a measure of the change in volume resulting from a pressure change.* **Thermal expansion** *is a measure of the volume change resulting from a temperature change.*

The contents of Table 11.1 should be studied in detail; they will serve as the starting point for further discussions about the states of matter.

F.Y.I.
Water is the only substance that is commonly encountered in all three states at temperatures normally found on earth.

F.Y.I.
Some solids, such as foam rubber, are compressible because they are full of gas.

11.3 THE KINETIC MOLECULAR THEORY

The physical characteristics of the solid, liquid, and gaseous states can be explained by kinetic molecular theory, one of the fundamental theories of chemistry. A basic idea of this theory is that the particles (atoms or molecules) present in a substance, independent of the physical state of the substance, have *motion* associated with them. The word *kinetic* comes from the Greek word *kinesis*, which means movement, hence the name *kinetic* molecular theory.

F.Y.I.
On a hot day the air molecules hit your skin with greater velocity than on a cooler day.

TABLE 11.1 DISTINGUISHING PROPERTIES OF SOLIDS, LIQUIDS, AND GASES

Property	Solid State	Liquid State	Gaseous State
Volume and shape	definite volume and definite shape	definite volume and indefinite shape; takes the shape of container to the extent it is filled	indefinite volume and indefinite shape; takes the volume and shape of container that it fills
Density	high	high, but usually lower than corresponding solid	low
Compressibility	small	small, but usually greater than corresponding solid	large
Thermal expansion	very small: about 0.01% per °C	small: about 0.10% per °C	moderate: about 0.30% per °C

The specific statements of **kinetic molecular theory** are

1. Matter is ultimately composed of tiny particles (atoms, molecules, or ions) with definite and characteristic sizes that never change.
2. The particles are in constant random motion and therefore possess kinetic energy.
3. The particles interact with each other through attractions and repulsions and therefore possess potential energy.
4. The velocity of the particles increases as the temperature is increased. The average kinetic energy of all particles in a system depends on the temperature, increasing as the temperature increases.
5. The particles in a system transfer energy from one to another during collisions in which no net energy is lost from the system. The energy of any given particle is thus continually changing.

The statements of kinetic molecular theory refer to kinetic and potential energy—the two basic types of energy. Definitions for these types of energy were previously given in Section 3.11. Recall that kinetic energy is energy associated with motion and potential energy is stored energy. We now consider additional concepts relative to these types of energy.

The amount of kinetic energy a particle possesses depends on both its mass and its velocity. The exact mathematical relationship between the kinetic energy and the mass and velocity of a particle is

$$\text{Kinetic energy} = \tfrac{1}{2}mv^2$$

where m is the mass of the particle and v is its velocity. From this expression we see that any differences in kinetic energy between particles of the same mass must be caused by differences in their velocities. Similarly, differences in kinetic energy between particles moving at the same velocity must be due to mass differences.

In any system of particles (atoms, molecules, or ions), potential energy resulting from particle interactions is present. Electrostatic interactions are the potential energy interactions of most importance for atomic-sized particles. **Electrostatic interactions** *occur between charged particles; objects of opposite electric charge (one positive and one negative) attract each other, and objects of identical charge (both positive or both negative) repel each other.* The magnitude of an electrostatic interaction depends on the sizes of the charges associated with the particles and their separation distance. Potential energy of attraction increases as the separation between particles increases, while that of repulsion increases with decreasing separation, as shown graphically in Figure 11.3. An analogy to

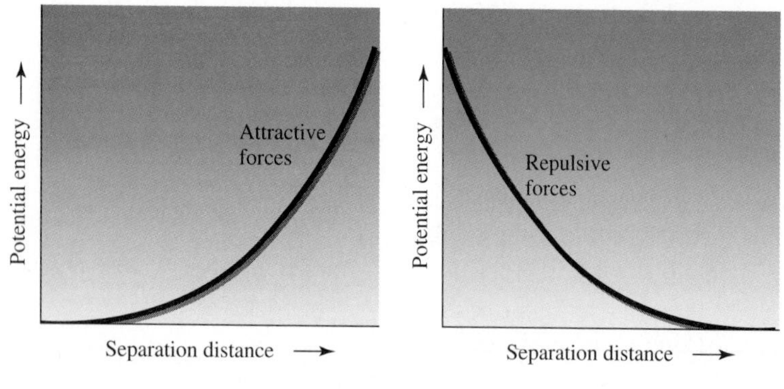

FIGURE 11.3 Effect of separation distance on potential energy.

the dependence of electrostatic potential energy on distance is found in two particles connected by a spring. When the spring is either stretched or compressed, the system has more potential energy than when the spring is in a nonextended position. The stretched spring represents attractive potential energy (the particles will come together if the spring is released), and the compressed spring represents repulsive potential energy (the particles will move apart when the spring is released). The greater the stretching of the spring (the greater the distance between particles), the greater the potential energy of attraction. The greater the compression of the spring (the smaller the distance between particles), the greater the potential energy of repulsion.

The relative influence of kinetic energy and potential energy in a chemical system is the major consideration in using kinetic molecular theory to explain the general properties of the solid, liquid, and gaseous states of matter. The important question is whether the kinetic energy or the potential energy dominates the energetics of the chemical system under study.

Kinetic energy may be considered a *disruptive force* within the chemical system, tending to make the particles of the system increasingly independent of each other. As the result of energy of motion the particles will tend to move away from each other. Potential energy may be considered a *cohesive force* tending to cause order and stability among the particles of the system.

The role that temperature plays in determining the state of a system is related to kinetic energy magnitude. Kinetic energy increases as temperature increases (statement 4 of the kinetic molecular theory). Thus, the higher the temperature the greater the magnitude of disruptive influences within a chemical system. The magnitude of potential energy is essentially independent of temperature change. Neither charge nor separation distance, the two factors on which the magnitude of potential energy depends, is affected significantly by temperature change.

Sections 11.4, 11.5, and 11.6 deal, respectively, with kinetic molecular theory explanations for the general properties of the solid, liquid, and gaseous states.

11.4 THE SOLID STATE

The solid state is characterized by a dominance of potential energy (cohesive forces) over kinetic energy (disruptive forces). The particles in a solid are drawn close together in a regular pattern by the strong cohesive forces present. Each particle occupies a fixed position about which it vibrates. An explanation of the characteristic properties of solids is obtained from this model.

1. *Definite volume and definite shape.* The strong cohesive forces hold the particles in essentially fixed positions, resulting in definite volume and definite shape.
2. *High density.* The constituent particles of solids are located as close together as possible. Therefore, large numbers of particles are contained in a unit volume, resulting in a high density.
3. *Small compressibility.* Since there is very little space between particles, increased pressure cannot push them any closer together and therefore has little effect on the solid's volume.
4. *Very small thermal expansion.* An increase in temperature increases the kinetic energy (disruptive forces), thereby causing more vibrational motion

of the particles. Each particle "occupies" a slightly larger volume. The result is a slight expansion of the solid. The strong cohesive forces prevent this effect from becoming very large.

11.5 THE LIQUID STATE

F.Y.I.
Thermometers use the thermal expansion of liquid mercury in a narrow tube. As temperature increases, the increased mercury volume raises the level in the tube.

The liquid state consists of particles randomly packed relatively close to each other. The molecules are in constant random motion, freely sliding over one another but without sufficient energy to separate from each other. *The liquid state is a situation in which neither potential energy (cohesive forces) nor kinetic energy (disruptive forces) dominates.* The fact that the particles freely slide over each other indicates the influence of disruptive forces, but the fact that the particles do not separate indicates a fairly strong influence from cohesive forces. The characteristic properties of liquids are explained by this model.

1. *Definite volume and indefinite shape.* Attractive forces are strong enough to restrict particles to movement within a definite volume. They are not strong enough, however, to prevent the particles from moving over each other in a random manner, limited only by the container walls. Thus liquids have no definite shape, with the exception that they maintain a horizontal upper surface in containers that are not completely filled.
2. *High density.* The particles in a liquid are not widely separated; they essentially touch each other. Therefore, there will be a large number of particles per unit volume and a resultant high density.
3. *Small compressibility.* Since the particles in a liquid essentially touch each other, there is very little empty space. Therefore, a pressure increase cannot squeeze the particles much closer together.
4. *Small thermal expansion.* Most of the particle movement in a liquid involves particles sliding over each other. The increased particle velocity that accompanies a temperature increase results in little change in such motion. The net effect is an increase in the effective volume a particle "occupies," which causes a slight volume increase in the liquid.

11.6 THE GASEOUS STATE

Kinetic energy (disruptive forces) completely dominates potential energy (cohesive forces) in the gaseous state. As a result, the particles of a gas are essentially independent of one another and move in a totally random manner. Under ordinary pressure, the particles are relatively far apart except, of course, when they collide with each other. Between collisions with each other or with the container walls, gas particles travel in straight lines. The particle velocities and resultant collision frequencies are extremely high; at room temperature and pressure the collisions experienced by one molecule are of the order of 10^{10} collisions per second.

The kinetic theory explanation of gaseous state properties follows the same pattern we saw earlier for solids and liquids.

1. *Indefinite volume and indefinite shape.* The attractive (cohesive) forces

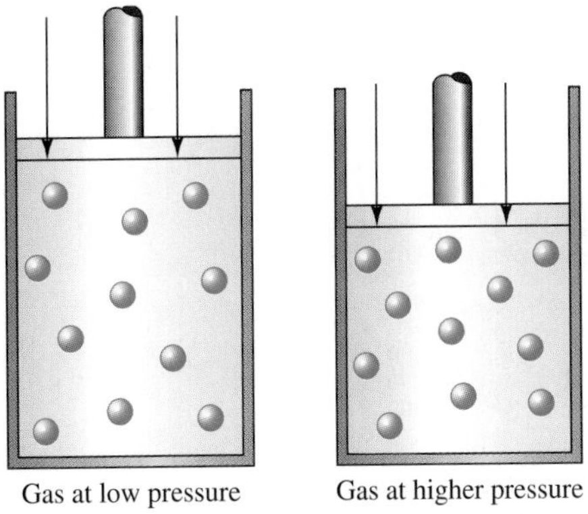

FIGURE 11.4 The compression of a gas—decreasing the amount of empty space in the container.

between particles have been overcome by kinetic energy, and the particles are free to travel in all directions. Therefore, the particles completely fill the container the gas is in and assume its shape.

2. *Low density.* The particles of a gas are widely separated. There are relatively few of them in a given volume, which means little mass per unit volume.
3. *Large compressibility.* Particles in a gas are widely separated; a gas is mostly empty space. When pressure is applied, the particles are easily pushed closer together, decreasing the amount of empty space and the volume of the gas (see Fig. 11.4).
4. *Moderate thermal expansion.* An increase in temperature means an increase in particle velocity. The increased kinetic energy of the particles enables them to push back whatever barrier is confining them into a given volume. Hence, the volume increases.

It must be understood that the size of the particles is not changed during expansion or compression of gases, solids, or liquids. The particles merely move farther apart or closer together; the space between them is what changes.

11.7 A COMPARISON OF SOLIDS, LIQUIDS, AND GASES

Two obvious conclusions about the similarities and differences between the various states of matter may be drawn from a comparison of the descriptive materials in Sections 11.4 through 11.6.

1. One of the states of matter, the gaseous state, is markedly different from the other two states.
2. Two of the states of matter, the solid and the liquid states, have many similar characteristics.

These two conclusions are illustrated diagrammatically in Figure 11.5.

The average distance between particles is only slightly different in the solid and liquid states but markedly different in the gaseous state. Roughly speaking, at ordinary temperatures and pressures, particles in a liquid are about 10% and particles in a gas about 1000%

F.Y.I.

The differences in distance between molecules in gases and liquids explains why it is easier to walk through air than through water.

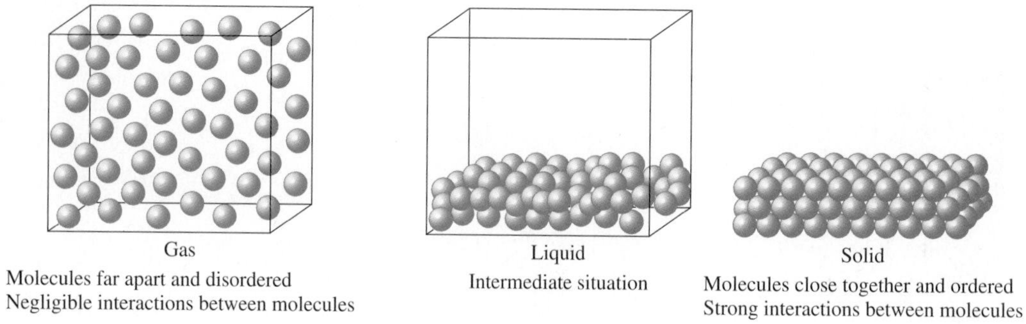

FIGURE 11.5 Similarities and differences among the states of matter.

farther apart than those in the solid state. The distance ratio between particles in the three states (solid to liquid to gas) is thus 1 to 1.1 to 10.

11.8 ENDOTHERMIC AND EXOTHERMIC CHANGES OF STATE

Physical changes of state were discussed in Section 4.4. The terminology associated with such changes—evaporation, condensation, sublimation, etc.—was introduced at that time (Fig. 4.4).

Changes of state are usually accomplished through heating or cooling a substance. (Pressure change is also a factor in some systems.) Changes of state may be classified according to whether heat (thermal energy) is given up or absorbed. An **endothermic change** *of state is a change that requires the input (absorption) of heat energy.* The endothermic changes of state are melting, sublimation, and evaporation. An **exothermic change** *of state is a change that requires heat energy to be given up (released).* Exothermic changes of state are the reverse of endothermic changes of state and include deposition, condensation, and freezing. Figure 11.6 summarizes the classification of changes of state as endothermic or exothermic.

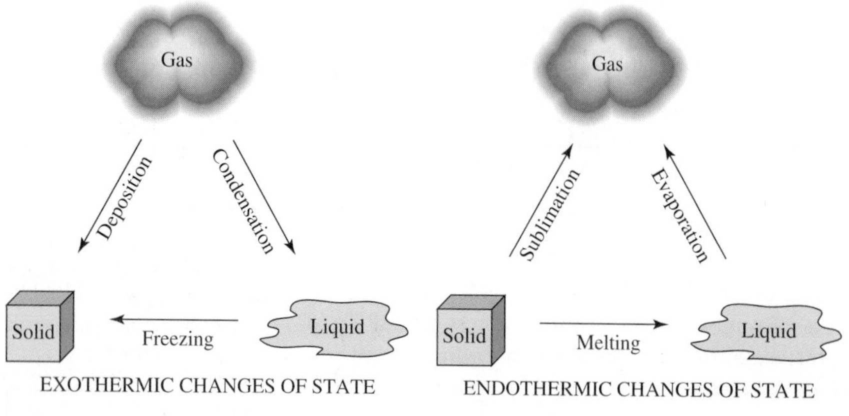

FIGURE 11.6 Endothermic and exothermic changes of state.

11.9 TEMPERATURE CHANGES AS A SUBSTANCE IS HEATED

For a given pure substance, molecules in the gaseous state contain more energy than molecules in the liquid state, which in turn contain more energy than molecules in the solid state. This fact is obvious; we know that it takes energy (heat) to melt a solid and still more energy (heat) to change the resulting liquid to a gas. Additional information concerning the relationship of energy to the states of matter can be obtained by a closer examination of what happens, step by step, to a solid (which is below its melting point) as heat is continuously supplied, causing it to melt and ultimately to change to a gas. The heating curve shown in Figure 11.7 gives the steps involved in changing a solid to a gas. As the solid is heated (region I of the graph), its temperature rises until the melting point is reached. The temperature increase indicates that the added heat causes an increase in the kinetic energy of the particles (recall Sec. 11.3, kinetic molecular theory). Once the melting point is reached, the temperature remains constant while the solid melts (region II of the graph). The constant temperature during melting indicates that the added heat has increased the potential energy of the particles without increasing their kinetic energy—the interparticle attractions are being weakened by the increase in potential energy. The addition of more heat to the system, which is now in the liquid state, increases the temperature until the boiling point is reached (region III of the graph). During this stage the molecules in the system are again gaining kinetic energy. At the boiling point another state change occurs as heat is added with the temperature remaining constant (region IV of the graph). The constant temperature again indicates an increase in potential energy. Once the system is completely vaporized, the temperature of the gas will again increase with further heating (region V of the graph).

The actual amount of energy that must be added to a system to cause it to undergo the series of changes just described depends on values of three properties of the substance. These properties are (1) specific heat, (2) heat of fusion, and (3) heat of vaporization.

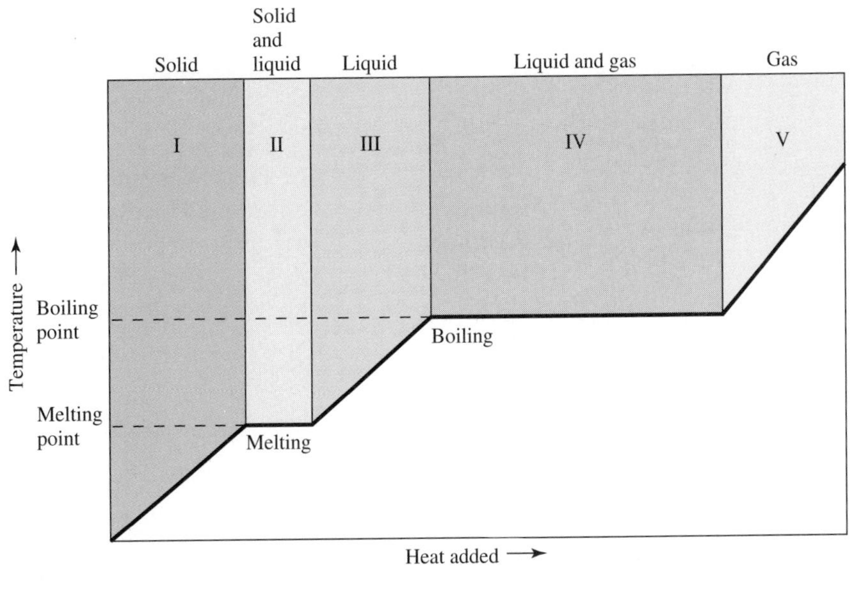

FIGURE 11.7 A heating curve depicting the addition of heat to a solid at a temperature below its melting point until it becomes a gas at a temperature above its boiling point.

Specific heat was discussed in Section 3.12. Heats of fusion and vaporization are the subject of Section 11.10.

11.10 ENERGY AND CHANGES OF STATE

F.Y.I.

Ice cubes cool drinks not only by being cold, but also by absorbing heat as they melt. Each gram of ice absorbs 334 joules from the liquid as it melts. The amount of heat it takes to melt one gram of ice can cool one gram of water from 80°C to 0°C.

F.Y.I.

Steam at 100°C is much more dangerous than water at 100°C. If steam contacts the skin, it is not only hot, it also releases its heat of condensation as it condenses on the skin. The exothermic release of energy can cause severe burns.

When heat energy is added to a solid, its temperature rises until the melting point is reached, at a rate that is governed by the specific heat (Sec. 3.12) of the solid. Once the melting point is reached, the temperature then remains constant while the solid changes to a liquid (Sec. 11.9). The "energetics" of the system during this transition from the solid state to the liquid state depend on the value of the substance's heat of fusion. (The term *fusion* means melting.)

The **heat of fusion** *is the amount of heat energy needed for the conversion of 1 g of a solid to a liquid at its melting point.* Units for heat of fusion are joules per gram (J/g). Note that these units do not involve temperature (degrees) as was the case for specific heat [J/(g · °C)]. No temperature units are needed because the temperature remains constant during a change of state.

The reverse of the fusion process is solidification (or freezing). The **heat of solidification** *is the amount of heat energy evolved in the conversion of 1 g of a liquid to a solid at its freezing point.* The heat of solidification always has the *same numerical value* as the heat of fusion. The only difference between these two entities is in the *direction* of heat flow (in or out). Heat of solidification is associated with an exothermic process, and heat of fusion with an endothermic process. The amount of heat required to melt 50.0 g of ice at its melting point is the same as the amount of heat that must be removed to freeze 50.0 g of water at its freezing point. The heats of fusion (or solidification) for selected substances are given in Table 11.2 in the units joules per gram.

Table 11.2 also gives a second heat of fusion (solidification) value for each substance, the *molar* heat of fusion (solidification), which has the units kJ/mole. In some types of calculations kJ/mole are more convenient units to use.

The magnitude of the heat of fusion for a given solid depends on the intermolecular

TABLE 11.2 HEATS OF FUSION (OR SOLIDIFICATION) FOR VARIOUS SUBSTANCES AT THEIR MELTING (OR FREEZING) POINTS

Solid	Melting (or Freezing) Point (°C)	Heat of Fusion (or Solidification) J/g	kJ/mole
Methane	−182	59	0.94
Ethyl alcohol	−117	109	5.01
Carbon tetrachloride	−23	16.3	2.51
Water	0	334	6.01
Benzene	6	126	9.87
Aluminum	658	393	10.6
Copper	1083	205	13.0

forces of attraction in the solid state. The strength of such forces is the subject of Section 11.17.

The general equation for calculating the amount of heat absorbed as a substance changes from a solid to a liquid is

$$\text{Heat absorbed (J)} = \text{heat of fusion (J/g)} \times \text{mass (g)}$$

Similarly, for the amount of heat released as a liquid freezes to a solid we have

$$\text{Heat released (J)} = \text{heat of solidification (J/g)} \times \text{mass (g)}$$

Example 11.1 illustrates the use of the first of these two equations.

Example 11.1

How much heat energy, in joules, is required to melt 25.1 g of aluminum at its melting point of 658°C?

Solution

The heat of fusion for aluminum, from Table 11.2, is 393 J/g. Therefore, the total amount of heat energy required is

$$\text{Heat absorbed} = 25.1\ \cancel{\text{g}} \times \frac{393\ \text{J}}{\cancel{\text{g}}} = 9864.3\ \text{J} \quad \text{(calculator answer)}$$
$$= 9860\ \text{J} \quad \text{(correct answer)}$$

Practice Exercise 11.1

How much heat energy, in joules, is required to melt 35.2 g of ice at its melting point of 0°C?

Ans. 11,800 J

Principles similar to those just considered for solid–liquid or liquid–solid changes apply to changes between the liquid and gaseous states. Here the specific energy quantities involved are heats of vaporization and heats of condensation. The **heat of vaporization** *is the amount of heat energy needed for the conversion of 1 g of a liquid to a gas at its boiling point*. Heats of vaporization are usually measured at the normal boiling point (Sec. 11.14) of the liquid.

The reverse of the vaporization (evaporation) process is condensation (liquefaction). The **heat of condensation** *is the amount of heat energy evolved in the conversion of 1 g of a gas to a liquid at the liquid's boiling point*. The heat of vaporzation and the heat of condensation will always have the same numerical value because these two quantities characterize processes that are the "reverse" of each other. The only difference is direction of heat flow; one process is endothermic (vaporization), and the other is exothermic (condensation). Table 11.3 gives heats of vaporization (or condensation) for selected substances.

The general equation for calculating the amount of heat absorbed as a substance changes from a liquid to a gas is

$$\text{Heat absorbed (J)} = \text{heat of vaporization (J/g)} \times \text{mass (g)}$$

Similarly, for the amount of heat released as a gas condenses to a liquid we have

$$\text{Heat released (J)} = \text{heat of condensation (J/g)} \times \text{mass (g)}$$

TABLE 11.3 HEATS OF VAPORIZATION (OR CONDENSATION) FOR VARIOUS SUBSTANCES AT THEIR NORMAL BOILING POINTS

Liquid	Normal Boiling Point (°C)	Heat of Vaporization (or Condensation) J/g	kJ/mole
Methane	−161	65.0	10.4
Ammonia	−33	1380	23.4
Diethyl ether	34.6	375	27.8
Carbon tetrachloride	77	195	30.0
Ethyl alcohol	78.3	837	38.6
Benzene	80.1	394	30.8
Water	100	2260	40.7

Example 11.2

5744 joules of heat energy are required to vaporize 15.0 g of an unknown liquid at its boiling point. What is the heat of vaporization, in joules per gram, of this unknown liquid?

Solution

Rearranging the equation

$$\text{Heat absorbed} = \text{heat of vaporization} \times \text{mass}$$

to isolate "heat of vaporization," which is our desired quantity, gives

$$\text{Heat of vaporization} = \frac{\text{heat absorbed}}{\text{mass}}$$

Substituting the given quantities into this equation gives

$$\text{Heat of vaporization} = \frac{5744\text{ J}}{15.0\text{ g}} = 382.93333\text{ J/g} \quad \text{(calculator answer)}$$

$$= 383\text{ J/g} \quad \text{(correct answer)}$$

Practice Exercise 11.2

The vaporization of 20.0 g of ethyl alcohol at its boiling point requires 16,740 J of heat energy. Using this information, calculate the heat of vaporization, in joules per gram, of ethyl alcohol.

Ans. 837 J/g

11.11 HEAT ENERGY CALCULATIONS

In doing "bookkeeping" on heat energy added to or removed from a chemical system, the following two generalizations always apply.

1. The amount of energy added to or removed from a chemical system that

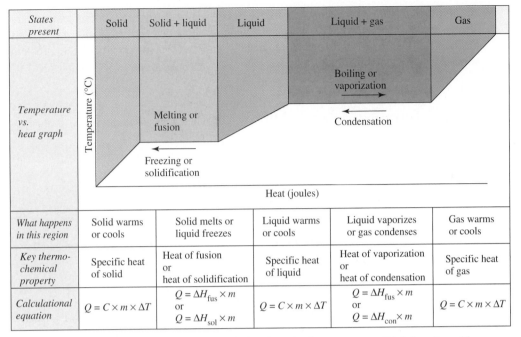

States present	Solid	Solid + liquid	Liquid	Liquid + gas	Gas
Temperature vs. heat graph	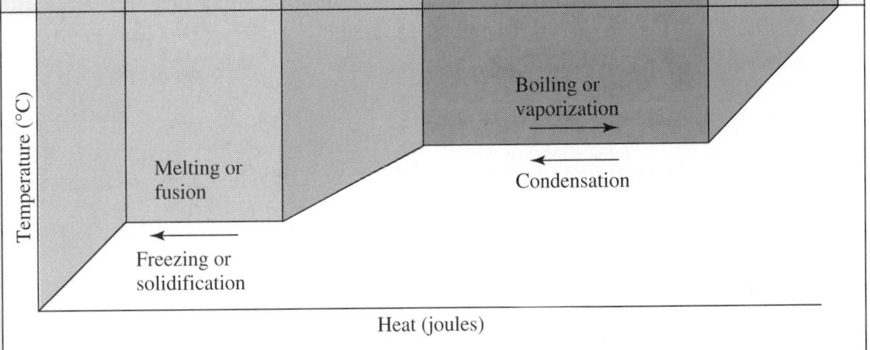				
What happens in this region	Solid warms or cools	Solid melts or liquid freezes	Liquid warms or cools	Liquid vaporizes or gas condenses	Gas warms or cools
Key thermo-chemical property	Specific heat of solid	Heat of fusion or heat of solidification	Specific heat of liquid	Heat of vaporization or heat of condensation	Specific heat of gas
Calculational equation	$Q = C \times m \times \Delta T$	$Q = \Delta H_{fus} \times m$ or $Q = \Delta H_{sol} \times m$	$Q = C \times m \times \Delta T$	$Q = \Delta H_{fus} \times m$ or $Q = \Delta H_{con} \times m$	$Q = C \times m \times \Delta T$

FIGURE 11.8 A temperature–heat-energy graph showing what happens as a solid, below its melting point, is heated to higher and higher temperatures.

undergoes a temperature change *but no change of state* is governed by the substance's specific heat. Examples 3.20 and 3.21 in Section 3.12 show how specific heat is used in calculations involving this situation.

2. The amount of energy added to or removed from a chemical system that undergoes a change of state at *constant temperature* is governed by heat of fusion or solidification (solid-to-liquid changes) and heat of vaporization or condensation (liquid-to-gas changes). Examples 11.1 and 11.2 in Section 11.10 show how these entities are used in change-of-state situations.

Figure 11.8 is a useful summary diagram of all of the possible "energy–change-of-temperature" and "energy–change-of-state" situations that occur. Also, at the bottom of this diagram are given the key equations needed for doing heat-energy calculations. The three equations that involve specific heat were originally introduced in Section 3.12, a section that it would be appropriate to review at this time. Those equations involving changes of state are the equations of Section 11.10.

The equations, as given in Figure 11.8, have new symbolism (abbreviations) associated with them. This symbolism involves the following relationships.

Q = heat absorbed or heat released
C = specific heat of the substance in a given physical state
m = mass of the substance
ΔT = change in temperature of the substance
ΔH_{fus} = heat of fusion of the substance
ΔH_{sol} = heat of solidification of the substance
ΔH_{vap} = heat of vaporization of the substance
ΔH_{con} = heat of condensation of the substance

F.Y.I.

Refrigerators work by letting a fluid (usually a Freon) evaporate in coils inside the refrigerator. The endothermic evaporation absorbs heat from inside the refrigerator. The compressor recondenses the gas back to a fluid outside of the cooling volume. (This is where most of the electricity goes.) The heat released in the condensation is radiated out of the coils in the rear. That is why it is always warm at the back of a refrigerator.

In problem-solving situations where a pure substance experiences both a temperature change and one or more changes of state (a common situation), the calculational equations of Figure 11.8 are used in an additive manner. Examples 11.3 and 11.4 illustrate problem solving in such "combination" situations.

Example 11.3

Calculate the heat released, in joules per gram, when 324 g of steam at 100°C are condensed and the resulting liquid water is cooled to 45°C.

Solution

We may consider this process to occur in two steps.

1. Condensing the steam to liquid water at 100°C.
2. Cooling the liquid water from 100°C to 45°C.

We will calculate the amount of heat released in each of the steps, then add these amounts together to get our final answer.

STEP 1 $\text{Gas}_{100°C} \longrightarrow \text{liquid}_{100°C}$

This step involves a change of state, from the gaseous to the liquid state (the fourth column of the heating curve of Figure 11.8). The heat energy equation for this change is

$$\text{Heat released} = \text{heat of condensation} \times \text{mass}$$

The heat of condensation of H_2O, from Table 11.3, is 2260 J/g. Therefore, we have

$$\text{Heat released} = \frac{2260 \text{ J}}{\cancel{\text{g}}} \times 324 \cancel{\text{g}} = 732{,}240 \text{ J} \quad \text{(calculator answer)}$$

$$= 732{,}000 \text{ J} \quad \text{(correct answer)}$$

STEP 2 $\text{Liquid}_{100°C} \longrightarrow \text{liquid}_{45°C}$

From the middle column of Figure 11.8 we note that the heat energy equation for the change in temperature of a liquid substance is

$$\text{Heat released} = \text{specific heat} \times \text{mass} \times \text{temperature change}$$

From the specific heats table in Chapter 3 (Table 3.4) we find that the specific heat of water (as a liquid) is 4.18 J/g · °C. The temperature change is 55°C and the mass is given as 324 g. Plugging these values into our equation gives

$$\text{Heat released} = \frac{4.18 \text{ J}}{\cancel{\text{g}} \cdot \cancel{°\text{C}}} \times 324 \cancel{\text{g}} \times 55 \cancel{°\text{C}} = 74{,}487.6 \text{ J} \quad \text{(calculator answer)}$$

$$= 74{,}000 \text{ J} \quad \text{(correct answer)}$$

Both the specific heat value and the termperature change limit the correct answer to two significant figures. The total heat released is the sum of that released in each step.

From step 1:	732,000 J
From step 2:	74,000 J
	806,000 J (calculator and correct answer)

Practice Exercise 11.3

Calculate the heat released when 40.0 g of steam at 115°C condenses to water at 100°C in a radiator of a steam heating system.

Ans. 91,600 J

Exercise 11.4

Calculate the amount of heat, in joules, needed to convert 23.6 g of ice at −27°C to steam at 121°C.

Solution

We may consider the change process to occur in five steps.

1. Heating the ice from −27 to 0°C, its melting point. (This change corresponds to the first column of the heating curve of Figure 11.8.)
2. Melting the ice, while the temperature remains at 0°C. (This change corresponds to the second column of the heating curve of Figure 11.8.)
3. Heating the liquid water from 0 to 100°C, its boiling point. (This change corresponds to the third column of the heating curve of Figure 11.8.)
4. Evaporating the water, while the temperature remains at 100°C. (This change corresponds to the fourth column of the heating curve of Figure 11.8.)
5. Heating the steam from 100 to 121°C. (This change corresponds to the last column of the heating curve of Figure 11.8.)

We will calculate the amount of heat required in each of the steps and then add these amounts together to get our desired answer. In doing this we will need the following heat energy values associated with water.

Step 1: specific heat of ice, 2.1 J/g · °C (Table 3.4)
Step 2: heat of fusion, 334 J/g (Table 11.2)
Step 3: specific heat of water, 4.18 J/g · °C (Table 3.4)
Step 4: heat of vaporization, 2260 J/g (Table 11.3)
Step 5: specific heat of steam, 2.0 J/g · °C (Table 3.4)

STEP 1 $\text{Solid}_{-27°\text{C}} \longrightarrow \text{solid}_{0°\text{C}}$

Heat required = specific heat × mass × temperature change

$$= \frac{2.1\ \text{J}}{\cancel{\text{g}} \cdot \cancel{°\text{C}}} \times 23.6\ \cancel{\text{g}} \times 27\cancel{°\text{C}} = 1338.12\ \text{J} \quad \text{(calculator answer)}$$

$$= 1300\ \text{J} \quad \text{(correct answer)}$$

STEP 2 $\text{Solid}_{0°\text{C}} \longrightarrow \text{liquid}_{0°\text{C}}$

Heat required = heat of fusion × mass

$$= \frac{334\ \text{J}}{\cancel{\text{g}}} \times 23.6\ \cancel{\text{g}} = 7882.4\ \text{J} \quad \text{(calculator answer)}$$

$$= 7880\ \text{J} \quad \text{(correct answer)}$$

STEP 3 $\quad \text{Liquid}_{0°\text{C}} \longrightarrow \text{liquid}_{100°\text{C}}$

Heat required = specific heat × mass × temperature change

$$= \frac{4.18\ \text{J}}{\cancel{\text{g}} \cdot \cancel{°\text{C}}} \times 23.6\ \cancel{\text{g}} \times 100\cancel{°\text{C}} = 9864.8\ \text{J} \quad \text{(calculator answer)}$$

$$= 9860\ \text{J} \quad \text{(correct answer)}$$

STEP 4 $\quad \text{Liquid}_{100°\text{C}} \longrightarrow \text{gas}_{100°\text{C}}$

Heat required = heat of vaporization × mass

$$= \frac{2260\ \text{J}}{\cancel{\text{g}}} \times 23.6\ \cancel{\text{g}} = 53{,}336\ \text{J} \quad \text{(calculator answer)}$$

$$= 53{,}300\ \text{J} \quad \text{(correct answer)}$$

STEP 5 $\quad \text{Gas}_{100°\text{C}} \longrightarrow \text{gas}_{121°\text{C}}$

Heat required = specific heat × mass × temperature change

$$= \frac{2.0\ \text{J}}{\cancel{\text{g}} \cdot \cancel{°\text{C}}} \times 23.6\ \cancel{\text{g}} \times 21\cancel{°\text{C}} = 991.2\ \text{J} \quad \text{(calculator answer)}$$

$$= 990\ \text{J} \quad \text{(correct answer)}$$

The total heat required is the sum of the results of the five steps.

$$(1300 + 7880 + 9860 + 53{,}300 + 990)\ \text{J} = 73{,}330\ \text{J} \quad \text{(calculator answer)}$$
$$= 73{,}300\ \text{J} \quad \text{(correct answer)}$$

(The answer must be rounded to the hundreds place. In adding, the numbers 1300 and 53,300 are limiting, with uncertainty in the hundreds place.)

Practice Exercise 11.4

Calculate the amount of heat, in joules, needed to convert 12.8 g of ice at −18°C to steam at 121°C.

Ans. 39,600 J

11.12 EVAPORATION OF LIQUIDS

Evaporation *is the process by which molecules escape from a liquid phase to the gas phase.* It is a familiar process. We are all aware that water left in an open container at room temperature will slowly disappear by evaporation.

The phenomenon of evaporation can readily be explained using kinetic molecular theory. Statement 5 of this theory (Sec. 11.3) indicates that the molecules in a liquid (or solid or gas) do not all possess the same kinetic energy. At any given instant some molecules will have above-average kinetic energies and others below-average kinetic energies as a result of collisions between molecules. A given molecule's energy constantly changes as a result of collisions with neighboring molecules. Molecules considerably above average in kinetic energy can overcome the attractive forces (potential energy) that are holding them in the liquid and escape if they are at the liquid surface and are moving in a favorable direction relative to the surfacae.

Note that evaporation is a surface phenomenon. Molecules within the interior of a liquid are surrounded on all sides by other molecules, making escape very improbable. Surface molecules are subject to fewer attractive forces since they are not completely surrounded by other molecules; escape is much more probable. Liquid surface area is an important factor in determining the rate at which evaporation occurs. Increased surface area results in an increased evaporation rate; a greater fraction of molecules occupy "surface" locations.

Water evaporates faster from a glass of hot water than from a glass of cold water. Why is this so? A certain minimum kinetic energy is required for molecules to escape from the attractions of neighboring molecules. As the temperature of a liquid increases, a larger fraction of the molecules present possess this needed minimum kinetic energy. Consequently, the rate of evaporation always increases as liquid temperature increases. Figure 11.9 contrasts the fraction of molecules possessing the needed minimum kinetic energy for escape at two temperatures. Note that at both the lower and higher temperatures a broad distribution of kinetic energies is present and that at each temperature some molecules possess the needed minimum kinetic energy. However, at the higher temperature a larger fraction of molecules present have the requisite kinetic energy so the rate of evaporation increases.

The escape of high-energy molecules from a liquid during evaporation affects the liquid in two ways: the amount of liquid decreases, and the liquid temperature is lowered. The temperature lowering reflects the fact that the average kinetic energy of the remaining molecules is lower than the pre-evaporation value due to the loss of the most energetic molecules. (Analogously, if all the tall people are removed from a classroom of students, the average height of the remaining students increases.) A lower average kinetic energy corresponds to a lower temperature (statement 4 of kinetic molecular theory); hence a cooling effect is produced.

Evaporative cooling is important in many processes. Our own bodies use evaporation to maintain a constant temperature. We perspire in hot weather because evaporation of the perspiration cools our skin. The cooling effect of evaporation is quite noticeable when someone first comes out of a swimming pool on a hot day (especially if a breeze is blowing). A canvas water bag keeps water cool because some of the water seeps through the canvas and evaporates, a process that removes heat from the remaining water. In medicine, the

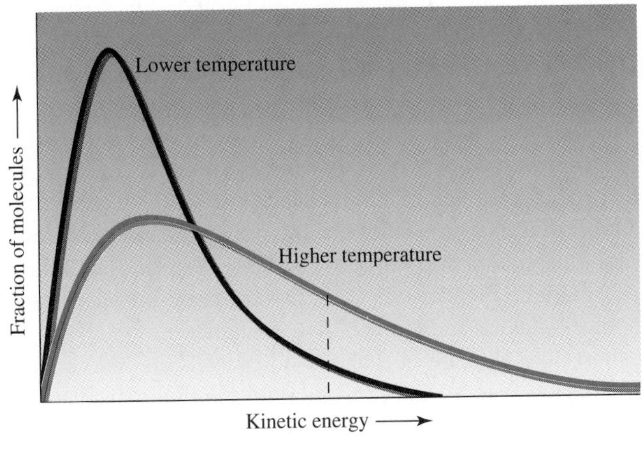

FIGURE 11.9 Kinetic energy distributions of molecules of a liquid at two different temperatures. The dashed line represents the minimum kinetic energy required for molecules of the liquid to overcome attractive forces and escape into the gas phase. Molecules in the shaded area have the necessary energy to overcome attractions.

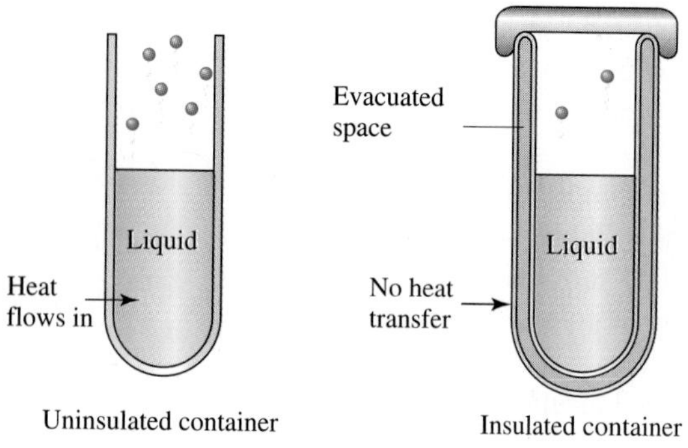

FIGURE 11.10 In order for an evaporative cooling effect to be measured, a container that minimizes heat flow into the container from its surroundings must be used.

local skin anesthetic ethyl chloride (C_2H_5Cl) exerts its effect through evaporative cooling (freezing). The minimum kinetic energy molecules of this substance need to acquire to escape (evaporate) is very low, resulting in a very rapid evaporation rate. Evaporation is so fast that the cooling "freezes" tissue near the surface of the skin, with temporary loss of feeling in the region of application. Certain alcohols also evaporate quite rapidly. An alcohol "rub" is sometimes used to reduce body temperature when a high fever is present.

For a liquid in a container, the decrease in temperature that occurs as a result of evaporation can be actually measured only if the container is an *insulated* one. When a liquid evaporates from a noninsulated container, there is sufficient heat flow from the surroundings into the container to counterbalance the loss of energy in the escaping molecules and thus prevent any cooling effect. The Thermos bottle is an insulated container that minimizes heat transfer, making it useful for maintaining liquids at a cool temperature for a short period of time. Figure 11.10 contrasts the evaporation process occurring in an uninsulated container with that which occurs within an insulated one. The temperature of liquid and surroundings is the same in an uninsulated container; in an insulated one, liquid temperature drops below that of the surroundings because of evaporative cooling.

Collectively, the molecules that escape from an evaporating liquid are often referred to as vapor rather than gas. The term **vapor** *describes the gaseous state of a substance at a temperature and pressure at which the substance is normally a liquid or solid.* For example, at room temperature and atmospheric pressure the normal state for water is the liquid state. Molecules that escape (evaporate) from liquid water at these conditions are called water vapor.

F.Y.I.

Keeping a lid on the pan allows a pot of water to boil faster. Without a lid, the heat of vaporization absorbed when water turns to steam escapes into the air. With a lid on, the water recondenses inside the lid and releases its heat of condensation back to the pot.

The evaporative behavior of a liquid in a *closed* container is quite different from that in an *open* container. In a closed container we observe that some liquid evaporation occurs, as indicated by a drop in liquid level. However, unlike the open container system, the liquid level, with time, ceases to drop (becomes constant), an indication that not all of the liquid will evaporate.

Kinetic molecular theory explains these observations in the following way. The molecules that do evaporate are unable to move completely away from the liquid as they did in the open container. They find themselves confined in a fixed space immediately above the liquid (see Fig. 11.11a). These "trapped" vapor molecules undergo many random collisions with the container walls, other vapor molecules, and the liquid surface. Molecules colliding with the liquid surface may be recaptured by the liquid. Thus, two processes—evaporation (escape) and condensation (recapture)—take place in the closed container.

In a closed container, for a short time, the rate of evaporation exceeds the rate of condensation and the liquid level drops. However, as more and more of the liquid evaporates,

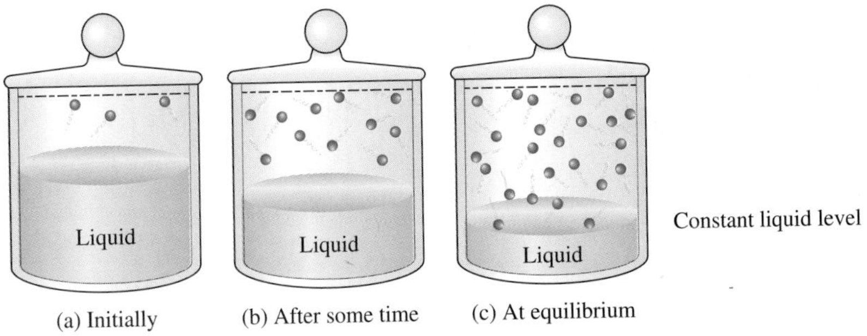

FIGURE 11.11 Evaporation of a liquid in a closed container.

the number of vapor molecules increases and the chance of their recapture through striking the liquid surface also increases. Eventually the rate of condensation becomes equal to the rate of evaporation and the liquid level stops dropping (see Fig. 11.11c). At this point the number of molecules that escape in a given time is the same as the number recaptured; a steady-state condition has been reached. The amounts of liquid and vapor in the container are not changing, even though both evaporation and condensation are still occurring.

This steady-state situation, which will continue as long as the temperature of the system remains constant, is an example of a state of equilibrium. A **state of equilibrium** *is a situation in which two opposite processes take place at equal rates.* For systems in a state of equilibrium, no net macroscopic changes can be detected. However, the system is dynamic; both forward and reverse processes are still occurring but in a manner such that they balance each other.

11.13 VAPOR PRESSURE OF LIQUIDS

For a liquid–vapor equilibrium in a closed container, the vapor in the fixed space immediately above the liquid exerts a constant pressure on the liquid surface and the walls of the container. This pressure is called the liquid's vapor pressure. **Vapor pressure** *is the pressure exerted by a vapor above a liquid when the liquid and vapor are in equilibrium.*

The magnitude of a vapor pressure depends on the nature and temperature of the liquid. Liquids with strong attractive forces between molecules will have lower vapor pressures than liquids in which only weak attractive forces exist between particles. Substances with high vapor pressures evaporate readily; that is, they are volatile. A **volatile substance** *is a substance that readily evaporates at room temperature because of a high vapor pressure.*

The vapor pressures of all liquids increase with temperature. Why? An increase in temperature results in more molecules having the minimum energy required for evaporation. Hence, at equilibrium the pressure of the vapor is greater. Table 11.4 shows the variation in vapor pressure with increasing temperature for water.

When the temperature of a liquid–vapor system in equilibrium is changed, the equilibrium is upset. The system immediately begins the process of establishing a new equilibrium. Let us consider a case where the temperature is increased. The higher temperature signifies that energy has been added to the system. More molecules will have the minimum energy needed to escape. Thus, immediately after the temperature is increased, molecules begin to escape at a rate greater than that at which they are recaptured. With time, however, the rates of escape and recapture will again become equal. The new rates will, however, be different

TABLE 11.4 VAPOR PRESSURE OF WATER AT VARIOUS TEMPERATURES

Temperature (°C)	Vapor Pressure (mm Hg)[a]	Temperature (°C)	Vapor Pressure (mm Hg)[a]
0	4.6	60	149.4
10	9.2	70	233.7
20	17.5	80	355.1
30	31.8	90	525.8
40	55.3	100	760.0
50	92.5		

[a] The units used to specify vapor pressure in this table will be discussed in detail in Sec. 12.2.

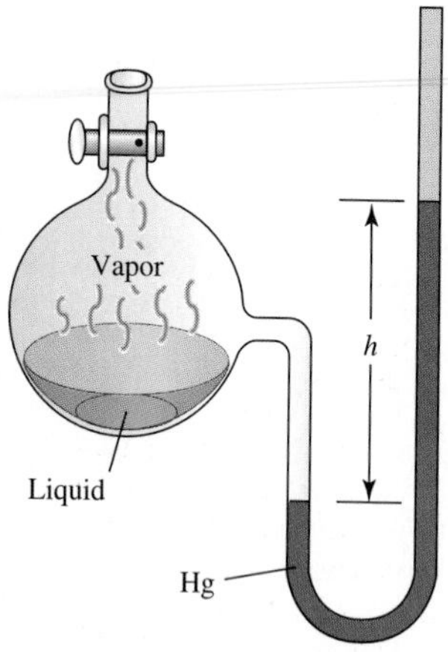

FIGURE 11.12 Apparatus for determining the vapor pressure of a liquid.

from those for the previous equilibrium. Since energy has been added to the system, the rates will be higher, resulting in a higher vapor pressure.

The size (volume) of the space that the vapor occupies does not affect the magnitude of the vapor pressure. A larger fixed space will enable more molecules to be present in the vapor at equilibrium. However, the larger number of molecules spread over a larger volume results in the same pressure as a small number of molecules in a small volume.

Vapor pressures for liquids are commonly measured in an apparatus similar to that pictured in Figure 11.12. At the moment liquid is added to the vessel, the space above the liquid is filled only with air and the levels of mercury in the U-tube are equal. With time, liquid evaporates and equilibrium is established. The vapor pressure of the liquid is proportional to the difference between the heights of the mercury columns. The larger the vapor pressure, the greater the extent to which the mercury column is "pushed up."

11.14 BOILING AND BOILING POINTS

Usually, for a molecule to escape from the liquid state, it must be on the surface of the liquid. **Boiling** *is a special form of evaporation in which conversion from the liquid to the vapor state occurs within the body of a liquid through bubble formation.* This phenomenon begins to occur when the vapor pressure of a liquid, which is steadily increasing as a liquid is heated, reaches a value equal to that of the prevailing external pressure on the liquid; for liquids in open containers this value is atmospheric pressure. When these two pressures become equal, bubbles of vapor form around any speck of dust or any rough surface of the container. Being less dense than the liquid itself, these vapor bubbles quickly rise to the surface and escape. The quick ascent of the bubbles causes the agitation associated with a boiling liquid.

Let us consider this "bubble phenomenon" in more detail. As the heating of a liquid begins, the first small bubbles that form on the bottom and sides of the container are bubbles of dissolved air (oxygen and nitrogen), which have been driven out of solution by the rising liquid temperature. (The solubilities of oxygen and nitrogen in liquids decrease with increasing temperature.)

As the liquid is heated further, larger bubbles form and begin to rise. These bubbles, usually also formed on the bottom of the container (where the heat is being applied and the liquid is hottest), are vapor bubbles rather than air bubbles. Initially, these vapor bubbles "disappear" as they rise, never reaching the liquid surface. Their disappearance is related to vapor pressure. In the hotter lower portions of the liquid, the liquid's vapor pressure is high enough to sustain bubble formation. [For a bubble to exist, the pressure within it (vapor pressure) must equal external pressure (atmospheric pressure).] In the cooler, higher portions of the liquid, where the vapor pressure is lower, the bubbles are collapsed by external pressure. Finally, with further heating, the temperature throughout the liquid becomes high enough to sustain bubble formation. Then bubbles rise all the way to the surface and escape (see Fig. 11.13). At this point we say the liquid is boiling.

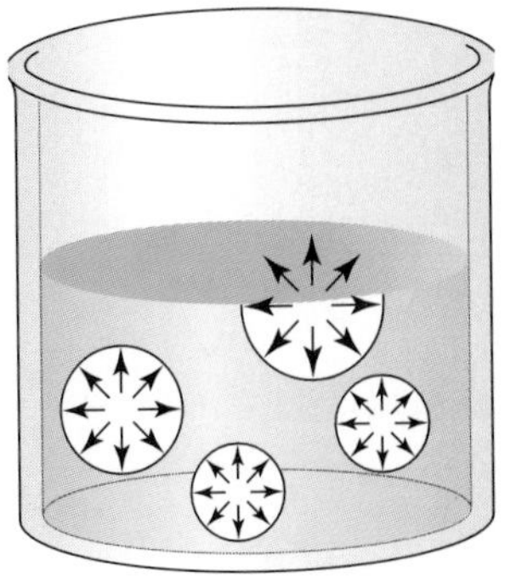

FIGURE 11.13 Bubble formation associated with a liquid that is boiling.

Like evaporation, boiling is actually a cooling process. When heat is taken away from a boiling liquid, boiling ceases almost immediately. It is the highest energy molecules that are escaping. Quickly the temperature of the remaining molecules drops below the boiling point of the liquid.

The **boiling point** *of a liquid is the temperature at which the vapor pressure of the liquid becomes equal to the external (atmospheric) pressure exerted on the liquid.* Since atmospheric pressure fluctuates from day to day, so does the boiling point of a liquid. To compare the boiling points of different liquids the external pressure must be the same. The boiling point of a liquid most often used for comparison and tabulation purposes (reference books, for example) is the normal boiling point. A liquid's **normal boiling point** *is the temperature at which a liquid boils under a pressure of 760 mm Hg.*

F.Y.I.

Anyone who has let a pan of water boil dry knows boiling is an endothermic process. As long as the water is boiling, the pan stays at a "cool" 100°C. Once all the water boils away, the pan turns red hot.

At any given location the changes in the boiling point of liquids due to *natural variation* in atmospheric pressure seldom exceed a few degrees; in the case of water the maximum is about 2°C. However, variations in boiling points *between* locations at different elevations can be quite striking, as shown by the data in Table 11.5.

The boiling point of a liquid can be increased by increasing the external pressure. Use is made of this principle in the operation of a pressure cooker. Foods cook faster in pressure cookers because the elevated pressure causes water to boil above 100°C. An increase in temperature of only 10°C will cause food to cook in approximately half the normal time. (Cooking involves chemical reactions, and the rate of a chemical reaction generally doubles with every 10°C increase in temperature.) Table 11.6 gives the boiling temperatures reached by water in normal household pressure cookers. Hospitals use the same principle in sterilizing instruments and laundry in autoclaves; sufficiently high temperatures are reached to destroy bacteria.

Liquids that have high normal boiling points or that undergo undesirable chemical reactions at boiling temperatures can be made to boil at low temperatures by reducing the external pressure. This principle is used in the preparation of numerous food products

TABLE 11.5 VARIATION OF THE BOILING POINT OF WATER WITH ELEVATION

Location	Elevation (ft above sea level)	Boiling Point of Water (°C)
San Francisco, CA	0	100.0
Salt Lake City, UT	4,390	95.6
Denver, CO	5,280	95.0
La Paz, Bolivia	12,795	91.4
Mount Everest	28,028	76.5

TABLE 11.6 BOILING POINT OF WATER IN A PRESSURE COOKER

Pressure above Atmospheric		Boiling Point
lb/in.^{2}	mm Hg	of Water (°C)
5	259	108
10	517	116
15	776	121

including frozen fruit juice concentrates. At a reduced presssure some of the water in a fruit juice is boiled away, concentrating the juice without having to heat it to a high temperature. Heating to a high temperature would cause changes that would spoil the taste of the juice and/or reduce its nutritional value.

11.15 INTERMOLECULAR FORCES IN LIQUIDS

In order for a liquid in an open container to boil, its vapor pressure must reach atmospheric pressure. For some substances this occurs at temperatures well below zero; for example, oxygen has a boiling point of −183°C. Other substances do not boil until the temperature is much higher. Mercury, for example, has a boiling point of 357°C, which is 540°C higher than that of oxygen. An explanation for this variation involves a consideration of the nature of the intermolecular forces that must be overcome in order for molecules to escape from the liquid state into the vapor state. **Intermolecular forces** *are forces that act between a molecule and another molecule.*

Intermolecular forces are similar in one way to the previously discussed *intra*molecular forces (*within* molecules) involved in covalent bonding (Sec. 7.9). They are electrostatic in origin. A major difference between inter- and intramolecular forces is their magnitude; the former are much weaker. However, intermolecular forces, despite their relative weakness, are sufficiently strong to influence the behavior of liquids, often in a very dramatic way. There are three principal types of intermolecular forces: dipole–dipole interactions, hydrogen bonds, and London forces.

Dipole–dipole interactions *are electrostatic attractions between polar molecules.* Polar molecules (which are often called dipoles), it should be recalled, are electrically unsymmetrical (Sec. 7.10). Therefore, when polar molecules approach each other, they tend to line up so that the relatively positive end of one molecule is directed toward the relatively negative end of the other molecule. As a result, there is an electrostatic attraction between the molecules. The greater the polarity of the molecules, the greater the strength of the dipole–dipole interaction. Figure 11.14 shows the many dipole–dipole interactions possible for a random arrangement of polar ClF molecules.

A **hydrogen bond** *is a special type of dipole–dipole interaction that occurs between polar molecules when one molecule contains hydrogen bonded to a very electronegative ele-*

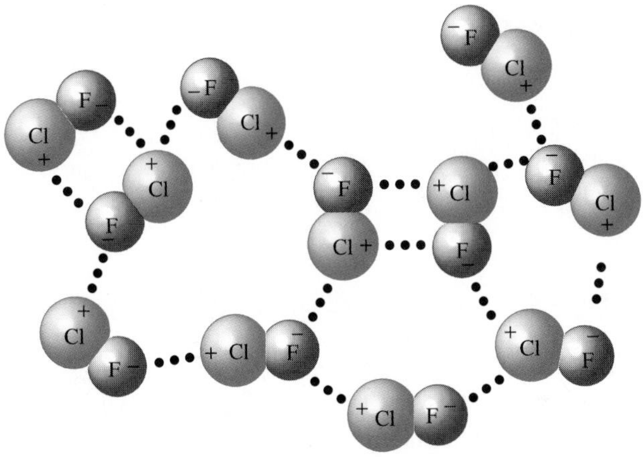

FIGURE 11.14 Dipole–dipole interactions between randomly arranged ClF molecules.

ment. When hydrogen is bonded to a very electronegative element (F, O, and N are the main elements involved), the electronegativity difference is sufficient essentially to strip the H atom of its electron, leaving a nearly exposed nucleus. (Remember, hydrogen has only one electron.) The small size of the hydrogen nucleus allows it to approach the F, O, or N atoms of other molecules very closely, resulting in a much stronger than normal dipole–dipole interaction. It is significant that hydrogen bonding appears to be limited to compounds containing the three elements F, O, and N, all of which are very small in addition to being very electronegative. The larger Cl and S atoms, even though Cl has an electronegativity similar to that of N, show little tendency to hydrogen bond. The strongest hydrogen bond is about one-tenth as strong as a covalent bond and is the strongest of all intermolecular forces. Figure 11.15 shows hydrogen bonding in liquid water.

It is not an overstatement to say that hydrogen bonding makes life possible. Were it not for hydrogen bonding, water, the most abundant compound in the human body, would be a gas at room temperature. Life as we know it could not exist under such a condition. Section

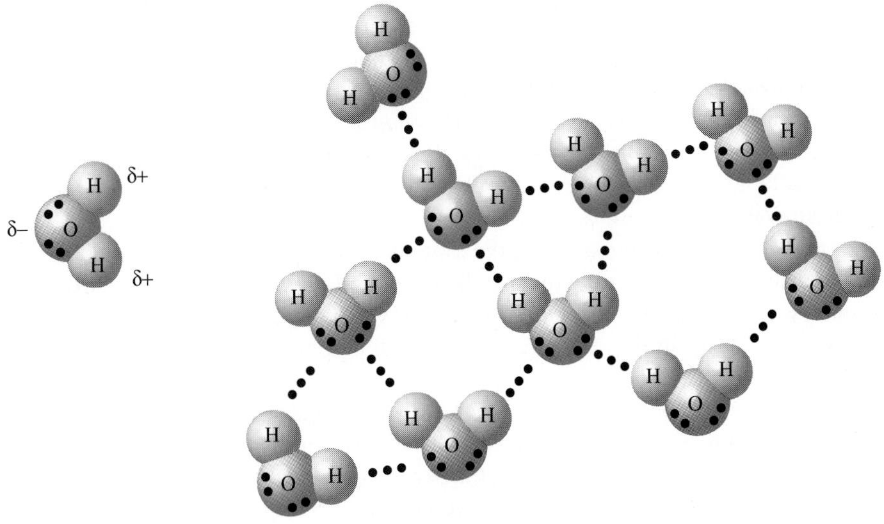

FIGURE 11.15 Hydrogen bonds, indicated by dotted lines, between water molecules in liquid water.

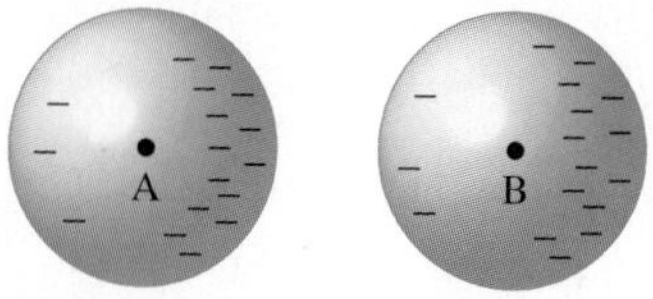

FIGURE 11.16 A London force. The instantaneous (temporary) polarity present in atom A results in induced polarity in atom B.

F.Y.I.
The change in texture of an egg as it cooks is due to the disruption of hydrogen bonds in protein.

11.16 will consider in detail the effects that hydrogen bonding have on the properties of water. Many other molecules of biological importance, such as DNA and proteins, contain O—H and N—H bonds, and hydrogen bonding plays a role in the behavior of these substances. Certain bonds in such compounds must be capable of breaking and reforming with relative ease. Only hydrogen bonds have just the right energies to permit this.

The third type of intermolecular force, and the weakest, is the London force, named after the German physicist Fritz London (1900–1954), who first postulated its existence. **London forces** *are instantaneous dipole–dipole interactions that exist between all atoms and molecules, nonpolar as well as polar.* The origin of London forces is more difficult to visualize than that of dipole–dipole interactions.

London forces result from momentary (temporary) uneven electron distributions in molecules. Most of the time the electrons can be visualized as being distributed in a molecule in a definite pattern determined by their energies and the electronegativities of the atoms. However, there is a small statistical chance (probability) that the electrons will deviate from their normal pattern. For example, in the case of a nonpolar diatomic molecule, more electron density may temporarily be located on one side of the molecule than on the other. This condition causes the molecule to become polar for an instant. The negative side of this instantaneously polar molecule will tend to repel electrons of adjoining molecules and cause these molecules to also become polar (an *induced polarity*). The original (statistical) polar molecule and all of the molecules with induced polarity are then attracted to each other. This happens many, many times per second throughout the liquid, resulting in a net attractive force. Figure 11.16 depicts the situation present when London forces exist.

The strength of London forces depends on the ease with which an electron distribution in a molecule can be distorted (polarized) by the polarity present in another molecule. In large diameter molecules the outermost electrons are necessarily located farther from the nucleus than are the outermost electrons in small molecules. The farther electrons are located from the nucleus, the more freedom they have and the more susceptible they are to polarization. This leads to the observation that for *related* molecules boiling points increase with molecular mass, which usually parallels size. This trend is reflected in the boiling points given in Table 11.7 for two series of related substances: the noble gases and the elements of group VIIA.

If two molecules have approximately the same molecular mass and polarity but are of different diameter, the smaller one will have the lower boiling point, since its electrons are less susceptible to polarization. The nonpolar SF_6 and $C_{10}H_{22}$ molecules (molecular masses 146 and 142 amu, respectively) reflect this trend with boiling points of −64 and 174°C, respectively.

F.Y.I.
Gasoline is a liquid at room temperature because of London forces.

Of the three types of intermolecular forces (dipole–dipole interactions, hydrogen bonds, and London forces), the London forces are the most common and, in the majority of cases, the most prevalent. They are the only attractive forces present between *nonpolar* molecules, and it is estimated that London forces contribute 85% of the total intermolecular force in the *polar* HCl molecule. Only where hydrogen bonding is involved do London

TABLE 11.7 BOILING POINT TRENDS FOR RELATED SERIES OF NONPOLAR MOLECULES

Substance	Molecular Mass (amu)	Boiling Point (°C)
Noble Gases		
He	4.0	−269
Ne	20.2	−246
Ar	39.9	−186
Kr	83.8	−153
Xe	131.3	−107
Rn	222	−62
Group VIIA Elements		
F_2	39.0	−187
Cl_2	70.9	−35
Br_2	159.8	+59
I_2	253.8	+184

forces play a minor role. In water, for example, about 80% of the intermolecular attraction is attributed to hydrogen bonding and only 20% to London forces.

In this section it has been assumed that the particles making up the liquids are molecules or atoms. This is a valid assumption for liquids at normal temperatures. Only at extremely high temperatures are liquids encountered where this is not the case. Liquids obtained by melting ionic compounds, which always requires an extremely high temperature, are ionic in the liquid state. The attractive forces in such liquids are those that result from the attraction of positive and negative ions.

11.16 WATER—A MOST UNUSUAL (UNIQUE) SUBSTANCE

Water is the most abundant and most essential compound known to human beings. This substance covers approximately 75% of the Earth's surface and constitutes 60–70% of the mass of a human body.

The structure of a water molecule is very simple. Two hydrogen atoms are attached to a central oxygen atom to give a V-shaped (angular) molecule, which is polar. Despite this molecular simplicity, the behavior and properties of water are both complex and unusual. The unusual behavior of water makes it the key substance that it is. Most of water's unusual behavior is a consequence of the extensive hydrogen bonding (Sec. 11.15) that occurs between water molecules, in both the liquid and solid states.

We will consider here four unusual hydrogen-bonding-caused properties of water.

1. A lower than expected vapor pressure.
2. Higher than expected thermal properties.

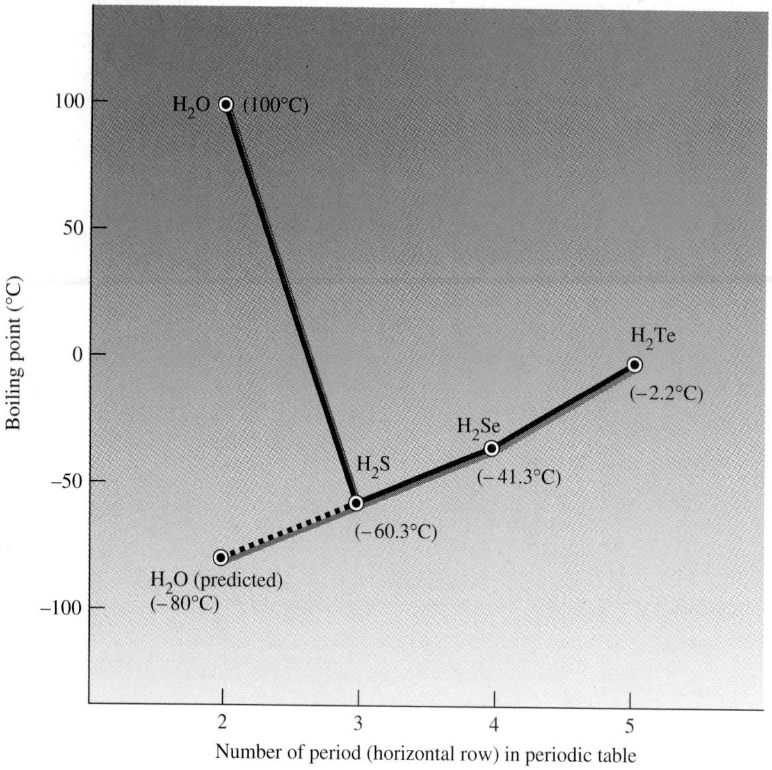

FIGURE 11.17 Boiling points of the hydrogen compounds of group VIA elements. Water is the only one of the four compounds in which significant hydrogen bonding occurs.

3. A very uncommon temperature–density relationship.
4. A higher than expected surface tension.

Vapor Pressure

Liquids in which significant hydrogen bonding occurs have vapor pressures that are significantly lower than those of similar liquids where little or no hydrogen bonding occurs. The presence of hydrogen bonds makes it more difficult for molecules to escape from the condensed state; additional energy is needed to overcome the hydrogen bonds. The greater the hydrogen bond strength, the lower the vapor pressure at any given temperature.

Since the boiling point of a liquid depends upon vapor pressure (Sec. 11.14), liquids with low vapor pressures will have to be heated to higher temperatures to bring their vapor pressures up to the point where boiling occurs (atmospheric pressure). Hence, boiling points are much higher for liquids in which hydrogen bonding occurs.

The effect that hydrogen bonding has on water's boiling point can be seen by comparing it with the boiling points of other hydrogen compounds of group VIA elements—H_2S, H_2Se, and H_2Te. In this series of compounds—H_2O, H_2S, H_2Se, and H_2Te—water is the only one in which significant hydrogen bonding occurs (Sec. 11.15). Normally, the boiling points of a series of compounds containing elements in the same periodic table group increase with increasing molar mass. Thus, in the hydrogen–group VIA element series, we would expect that H_2Te, the heaviest member of the series, would have the highest boiling point and that water, the compound of lowest molar mass, would have the lowest boiling point. Contrary to expectation, H_2O has the highest boiling point, as can be seen from the boiling-point data shown in Figure 11.17. The data in Figure 11.17 indicate that water "should have" a boiling point of approximately −80°C, a value obtained by extrapolation of the line connecting the three heavier compounds. The actual boiling point of water, 100°C, is nearly 200°C higher than predicted. Indeed, in the absence of hydrogen bonding, water would be a gas at room temperature, as mentioned in Section 11.15.

A higher than expected freezing point is also characteristic of water, and a plot of freezing points for similar compounds would have the same general shape as that in Figure 11.17.

Thermal Properties

The presence of extensive hydrogen bonding between water molecules increases significantly the ability of water to absorb heat energy when evaporating (an endothermic process) and to release heat energy when freezing (an exothermic process). Water's thermal properties, together with its abundance, largely account for its widespread use as a coolant.

Large bodies of water exert a temperature-moderating effect on their surroundings, primarily because of water's ability to absorb and release large amounts of heat energy without significant change in temperature. In the heat of a summer day, extensive water evaporation occurs, and in the process energy is absorbed from the surroundings. The net effect of the evaporation is a lowering of the temperature of the surroundings. In the cool of evening, some of this water vapor condenses back to the liquid state, releasing heat that raises the temperature of the surroundings. In this manner the temperature variation between night and day is reduced. In the winter a similar process occurs; water freezes on cold days and releases heat energy to the surroundings. The hottest and coldest regions on Earth are all inland regions, those distant from the moderating effects of large bodies of water.

The large release of heat from freezing water is the basis for the practice of spraying orange trees with water when freezing temperatures are expected. The freezing water liberates enough heat energy to keep the temperature of the air higher than the freezing point of the fruits.

Water's thermal properties are a factor in the cooling of the human body. The coolant in this case is perspiration, which evaporates, absorbs heat in the process, and lowers skin temperature. For similar reasons, a swimmer upon emerging from the water on a warm but windy day will feel cold and will shiver as excess water rapidly evaporates from his or her body. Additionally, water's ability to absorb large amounts of heat helps keep water loss to a minimum, thus making it easier for humans, animals, and plants to exist in environments where water is scarce.

Density

A very striking and unusual behavior pattern occurs in the variation of the density of water with temperature. For most liquids, density increases with decreasing temperature, reaching a maximum for the liquid at its freezing point. The density pattern for water is different. Maximum density is reached not at its freezing point, but at a temperature a few degrees higher than the freezing point. As shown in Figure 11.18, the maximum density for liquid water occurs at 4°C. This abnormality, that water at its freezing point is less dense than water at slightly higher temperatures, has tremendous ecological significance. Furthermore, at 0°C, solid water (ice), is significantly less dense than liquid water—0.9170 g/mL versus 0.9998 g/mL. All of this "strange" density behavior of water is directly related to hydrogen bonding.

Of primary importance is the fact that hydrogen bonding between water molecules is directional. Hydrogen bonds can form only at certain angles between molecules because of the angular geometry of the water molecule. The net result is that water molecules that are hydrogen bonded are farther apart than those that are not.

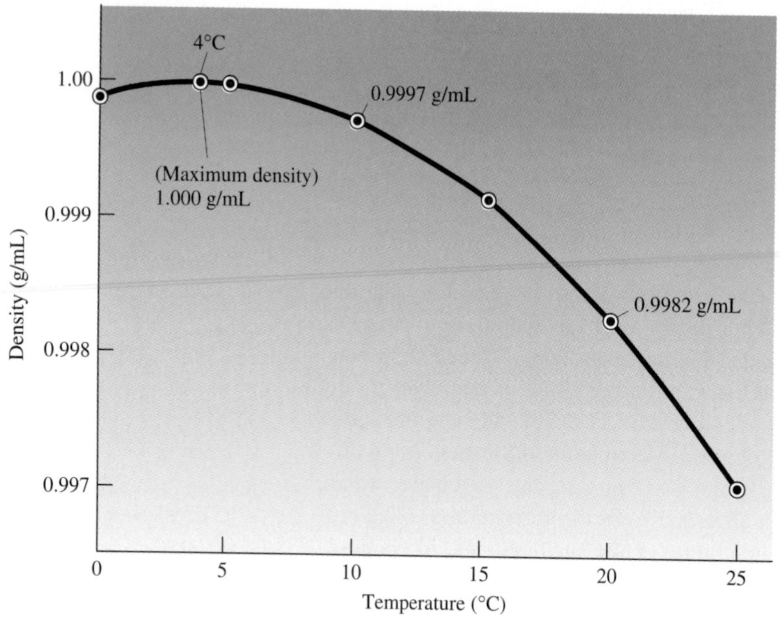

FIGURE 11.18 A plot of density versus temperature for liquid water.

On a molecular scale, let us now consider what happens to water molecules as the temperature of water is lowered. At high temperatures, such as 80°C, the kinetic energy of the water molecules is sufficiently great to prevent hydrogen bonding from having much of an orientation effect on the molecules. Hydrogen bonds are rapidly and continually being formed and broken. As the temperature is lowered, the accompanying decrease in kinetic energy decreases molecular motion and the molecules move closer together. This results in an increase in density. The kinetic energy is still sufficient to negate most of the orientation effects from hydrogen bonding. When the temperature is lowered still further, the kinetic energy finally becomes insufficient to prevent hydrogen bonding from orienting molecules into definite patterns that require "open spaces" between molecules. At 4°C, the temperature at which water has its maximum density, the trade-off between random motion from kinetic energy and orientation from hydrogen bonding is such that the molecules are as close together as they will ever be. Cooling below 4°C causes a decrease in density as hydrogen bonding causes more and more "open spaces" to be present in the liquid. Density decrease continues down to the freezing point, at which temperature the hydrogen bonding causes molecular orientation to the maximum degree, and the solid crystal lattice of ice is formed. This solid crystal lattice of ice (Fig. 11.19) is an open structure, with the result that solid ice has a lower density than liquid water. Water is one of only a few substances known whose solid phase is less dense than the liquid phase at the freezing point.

Two important consequences of ice being less dense than water are that ice floats in liquid water and liquid water expands upon freezing.

When ice floats in liquid water, approximately 8% of its volume is above water. In the case of icebergs found in far northern locations, 92% of the volume of an iceberg is located below the liquid surface. The common expression "it is only the tip of the iceberg" is literally true; what is seen is only a very small part of what is actually there.

If a container is filled with water and sealed, the force generated from the expansion of the water upon freezing will break the container. Antifreeze is added to car radiators in the

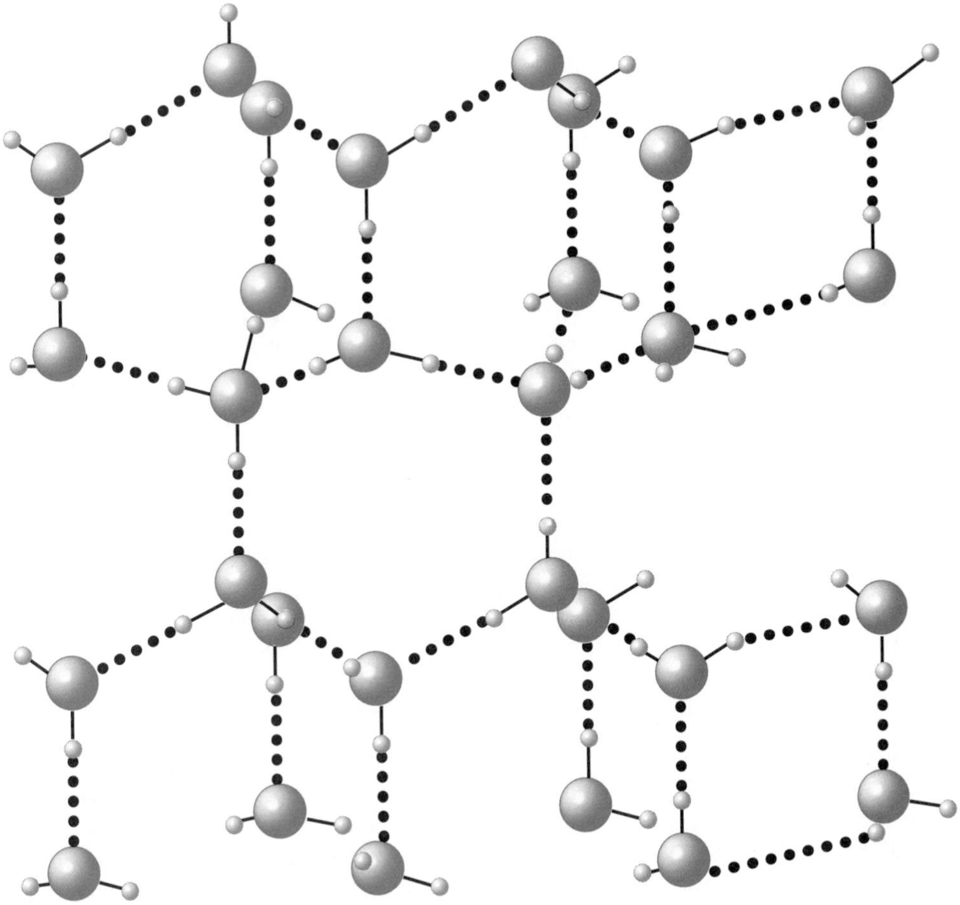

FIGURE 11.19 The crystalline structure of ice. Normal covalent bonds between oxygen and hydrogen, which hold water molecules together, are shown by solid short lines. The weaker hydrogen bonds are shown by dotted lines.

winter to prevent the water present from freezing and cracking the engine block; sufficient force is generatead from water expansion upon freezing to burst even iron or copper parts. During the winter season, also, the weathering of rocks and concrete and the formation of "potholes" in the streets are hastened by the expansion of water upon freezing.

Water's density pattern also explains why lakes freeze from top to bottom and not vice versa, and why aquatic life can continue to exist for extended periods of time in bodies of water that are "frozen over." In the fall of the year, surface water is cooled through contact with cold air. This water becomes denser than the warmer water underneath and sinks. In this way, cool water is circulated from the top of a lake to the bottom until the entire lake has reached the temperature of water's maximum density, 4°C. During the circulation process, oxygen and nutrients are distributed throughout the water. Upon further cooling, below 4°C, a new behavior pattern emerges. Surface water, upon cooling, no longer sinks, since it is less dense than the water underneath. Eventually a thin layer of surface water is cooled to the freezing point and changed to ice, which floats because of its still lower density. Even in the coldest winters, lakes will usually not freeze to a depth of more than a few feet because the ice forms an insulating layer over the water. Thus, aquatic life can live throughout the winter, under the ice, in water that is "thermally insulated" and contains nutrients. If water behaved as "normal" substances do, freezing would occur from the bottom up, and most, if not all, aquatic life would be destroyed.

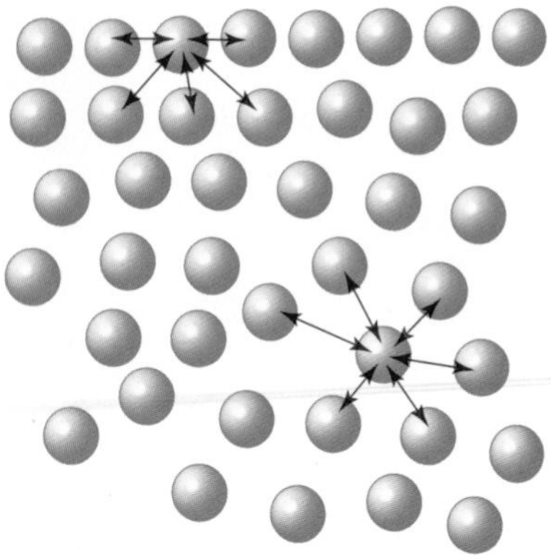

FIGURE 11.20 Molecules on the surface of a liquid experience an imbalance in intermolecular forces. Molecules within the liquid are uniformly attracted in all directions.

For a few weeks in the early spring, circulation again occurs in a body of water during the melting and warming of the surface water. This process stops when the surface water is warmed about 4°C. Above this temperature surface water is less dense and forms a layer over the colder water of higher density. This causes a thermal stratification that persists through the hot summer months. In the fall, circulation again begins, and the cycle is repeated.

Surface Tension

Surface tension is a property of liquids that is directly related to intermolecular forces. Molecules *within* a liquid, because they are completely surrounded by other molecules, experience equal intermolecular attractions in all directions. Molecules *on the surface* of a liquid, because there are no molecules above them, experience an imbalance in intermolecular attractions that pulls them toward the interior of the liquid. Figure 11.20 illustrates this imbalance of force. **Surface tension** *is a measure of the inward force on the surface of a liquid caused by unbalanced intermolecular forces.*

The strong attraction of water molecules for each other as a result of hydrogen bonding gives water a very high surface tension. This high surface tension causes water surfaces to seem to have a "membrane" or "skin" covering them. A steel needle or a razor blade will float on the surface of water because of surface tension even though each has a density greater than that of water. Water bugs are able to skitter across water because of surface tension.

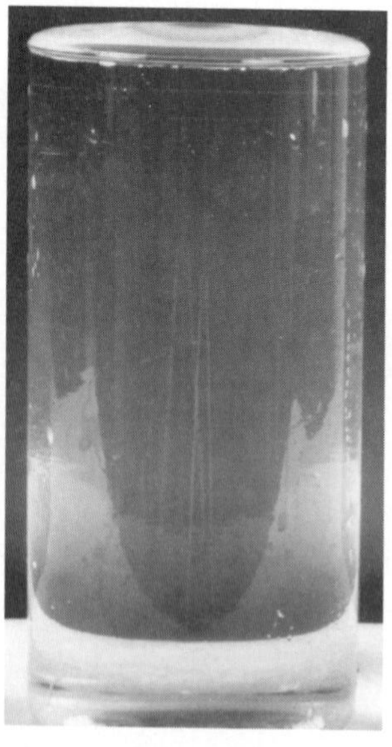

FIGURE 11.21 Surface tension enables one to fill a glass with water above the rim. (John Schultz, PAR/NYC)

Because of surface tension it is possible to fill a container with water slightly above the rim (see Fig. 11.21). Surface tension prevents the water from running over. It is also surface tension that causes water to form droplets in many situations. Almost everyone has observed water droplets that have formed on a car windshield or on greasy or waxed surfaces. Surface tension causes water to resist increasing its surface area. For a given volume of any liquid, the geometrical shape having the minimum surface area is a sphere; hence, water tends to form spherical droplets.

The property of surface tension helps water to rise in the narrow vessels in plant stems and roots. High surface tension also helps to hold water in the small spaces between soil particles.

The surface tension of a liquid can be "broken" by dissolved substances. Soaps and detergents are used for this purpose. Water bugs will sink in soapy water. Soapy water spreads out over a glass surface instead of forming droplets as pure water would.

11.17 TYPES OF SOLIDS

Solids may be classified into two categories: crystalline solids and amorphous solids. Most solids are crystalline. **Crystalline solids** *are solids characterized by a regular three-dimensional arrangement of the atoms, ions, or molecules present.* This regular arrangement of particles is manifested in the outward appearance of the solid; crystalline features are discernible. These features are often very obvious; at other times, however, the crystals may be so tiny as to be visible only with the aid of a microscope. Many times the crystals are imperfect, making it difficult to discern them, but they are there.

Amorphous solids *are solids characterized by a random, nonrepetitive three-dimensional arrangement of the atoms, ions, or molecules present.* Only a few such solids exist. The word *amorphous* comes from the Greek *amorphos* meaning "without form." Examples of amorphous solids include glass, tar, rubber, and many plastics. In some textbooks amorphous solids are referred to as *supercooled liquids*, since they have the molecular disorder associated with liquids. However, they differ from liquids and resemble solids in being rigid and immobile.

Most amorphous solids consist of very long chainlike molecules that are randomly intertwined. To form a crystalline material from such a solid, the intertwined molecules would have to become regularly arranged in the melted state. When such a liquid is cooled to a solid, molecular motion decreases to the point where "solidification" takes place before the untwining can occur, resulting in a disordered structure.

The remainder of this section deals with the more commonly encountered crystalline solid. Indeed, in this text the word *solid* without a modifier will always mean a *crystalline solid.*

The highly ordered pattern of particles found in a crystalline solid is called a *crystal lattice*. The positions occupied by the particles in the lattice are called *crystal lattice sites*. Crystalline solids may be classified into five groups based on the type of particles at the crystal lattice sites and the forces that hold the particles together. A consideration of these classifications provides some insight into the reasons some solids have very low and others very high melting points and some solids are more volatile than others. The five classes of crystalline solids are ionic, polar molecular, nonpolar molecular, macromolecular, and metallic.

Ionic solids *consist of positive and negative ions arranged in such a way that each ion is surrounded by nearest neighbors of the opposite charge (as was previously discussed in Sec. 7.7).* Any given ion is bonded by electrostatic attractions to all the ions of opposite

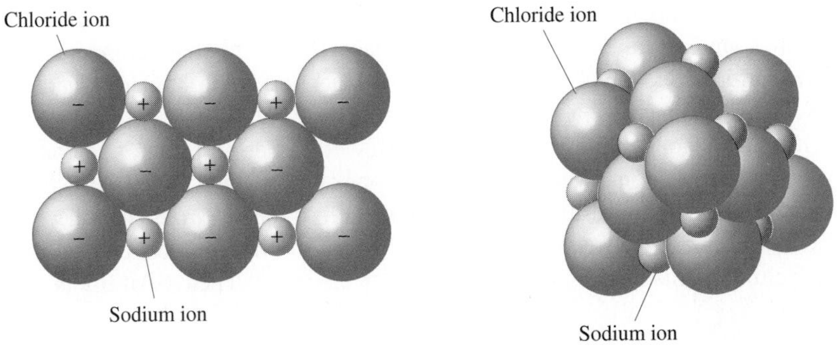

FIGURE 11.22 Two-dimensional cross section and three-dimensional view of the ionic compound NaCl.

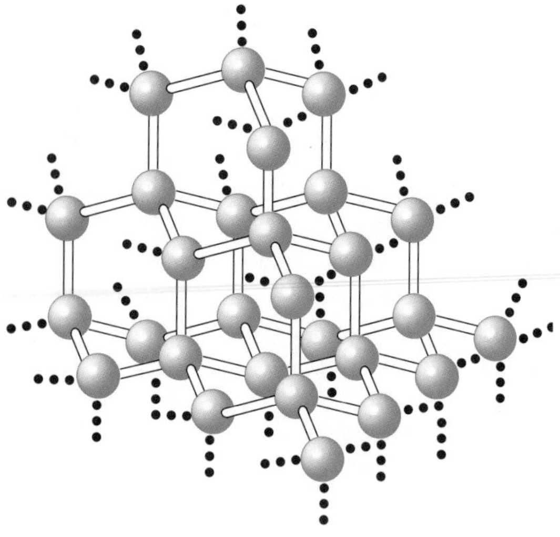

FIGURE 11.23 The diamond crystal lattice—a macromolecular solid.

charge immediately surrounding it. Since these interionic attractions are relatively strong, ionic solids have high melting points and negligible vapor pressures. Figure 11.22 shows a two-dimensional cross section and also a three-dimensional view of the arrangement of ions for the compound NaCl.

Polar molecular solids *have polar molecules at the crystal lattice sites.* The molecules are oriented so that the positive end of each molecule is near the negative end of an adjacent molecule in the lattice. Dipole–dipole interactions and London forces (Sec. 11.15) hold the molecules in the lattice positions. Since these forces are relatively weak compared to other types of electrostatic forces (ionic and covalent bonds), the resulting solids have moderate melting points and moderate vapor pressures.

F.Y.I.

Jewelry components

Diamond	
C	macromolecular
Amethyst	
SiO_2	macromolecular
Ruby	
Al_2O_3	ionic
Sapphire	
Al_2O_3	ionic
Pearl	
$CaCO_3$	ionic
Gold	
Au	metallic
Opal	
SiO_2	macromolecular

Nonpolar molecular solids *have nonpolar molecules (or atoms in the case of the noble gases) at the crystal lattice sites.* They are held togther solely by London forces, the weakest of electrostatic forces. Consequently, these solids have low melting points and are volatile.

Macromolecular solids *have atoms that are bonded to their nearest neighbors by covalent bonds at the crystal lattice sites.* A crystal of such a solid can be visualized as a gigantic molecule. Such a molecule, for the form of carbon called diamond, is shown in Figure 11.23. Each lattice site in diamond is occupied by a carbon atom. Macromolecular solids are not common, but they are noteworthy because of their extremely high melting points and their low volatility. Covalent bonds must be broken to cause physical change in such solids. The atoms at the lattice sites in a macromolecular solid need not all be the same kind, as was the case for diamond. Examples of macromolecular solids where different kinds of atoms are found at the lattice sites include quartz (SiO_2, the main component of sand) and silicon carbide (SiC, an abrasive used in sandpaper and other similar materials).

Metallic solids *have metal atoms occupying the crystal lattice sites.* The nature of bonding in such solids is not completely understood. However, the solids are often considered to have a lattice of metal ions, with the outer electrons of each metal atom free to move about in the lattice. The bonding results from the interaction of the electrons with the various nuclei. The movement of the free electrons also accounts for the electrical conductivity of metals. The melting points of metallic solids show a wide range. Their volatility is generally low.

TABLE 11.8 CHARACTERISTICS AND EXAMPLES OF THE VARIOUS TYPES OF CRYSTALLINE SOLIDS

Type	Particles Occupying Lattice Sites	Forces Between Particles	Properties of Solids	Examples
Ionic	positive and negative ions	electrostatic attraction between oppositely charged ions	high melting points; nonvolatile	NaCl $MgSO_4$ KBr
Polar molecular	polar molecules	dipole–dipole attractions and London forces	moderate melting points; moderate volatility	H_2O NH_3 SO_2
Nonpolar molecular	nonpolar molecules (or atoms)	London forces	low melting points; volatile	CO_2 CH_4 I_2 O_2 Ar
Macromolecular	atoms	covalent bonds between atoms	extremely high melting points; nonvolatile	C (diamond) SiO_2 (sand) SiC
Metallic	metal atoms	attraction between outer electrons and positive atomic centers	variable melting points; low volatility	Cu Ag Au Fe Al

Table 11.8 gives a summary of the various types of crystalline solids and their distinguishing characteristics along with examples of each type.

KEY TERMS

The new terms or concepts defined in this chapter are

amorphous solid (Sec. 11.17) A solid characterized by a random, nonrepetitive three-dimensional arrangement of atoms or molecules.

boiling (Sec. 11.14) A special form of evaporation in which conversion from the liquid to the vapor state occurs within the body of a liquid through bubble formation.

boiling point (Sec. 11.14) The temperature at which the vapor pressure of a liquid equals the external (atmospheric) pressure exerted on the liquid.

compressibility (Sec. 11.2) A measure of the change in volume resulting from a pressure change.

crystalline solid (Sec. 11.17) A solid characterized by a regular three-dimensional arrangement of atoms, ions, or molecules.

dipole–dipole interaction (Sec. 11.15) Electrostatic attraction between polar molecules.

electrostatic interaction (Sec. 11.13) Interaction between charged particles; objects of opposite electric charge attract each other, and objects of the same electric charge repel each other.

endothermic change (Sec. 11.8) A change that requires the input (absorption) of heat energy.

evaporation (Sec. 11.12) Process in which molecules escape from a liquid phase to a gaseous phase.

exothermic change (Sec. 11.8) A change that requires heat energy to be given up (released).

heat of condensation (Sec. 11.10) Amount of heat energy evolved in the conversion of 1 g of a gas to a liquid at the liquid's boiling point.

heat of fusion (Sec. 11.10) Amount of heat energy needed for the conversion of 1 g of a solid to a liquid at its melting point.

heat of solidification (Sec. 11.10) Amount of heat energy evolved in the conversion of 1 g of a liquid to a solid at its freezing point.

heat of vaporization (Sec. 11.10) Amount of heat energy needed for the conversion of 1 g of a liquid to a gas at its boiling point.

hydrogen bond (Sec. 11.15) A special type of dipole–dipole interaction that occurs between polar molecules when one molecule contains hydrogen bonded to a very electronegative element.

intermolecular force (Sec. 11.15) A force that acts between a molecule and another molecule.

ionic solid (Sec. 11.17) A solid consisting of positive and negative ions arranged in such a way that each ion is surrounded by nearest neighbors of the opposite charge.

kinetic molecular theory (Sec. 11.3) A series of statements useful in explaining the physical characteristics of the solid, liquid, and gaseous states.

London force (Sec. 11.15) Instantaneous dipole–dipole interaction that exists between all atoms and molecules, nonpolar as well as polar.

macromolecular solid (Sec. 11.17) A solid in which atoms are bonded to their nearest neighbors by covalent bonds.

metallic solid (Sec. 11.17) A solid that has metal atoms occupying the crystal lattice sites.

nonpolar molecular solid (Sec. 11.17) A solid in which nonpolar molecules (or atoms in the case of the noble gases) are at the lattice sites.

normal boiling point (Sec. 11.14) The temperature at which a liquid boils under a pressure of 760 mm Hg.

polar molecular solid (Sec. 11.17) A solid in which polar molecules are at the crystal lattice points.

state of equilibrium (Sec. 11.12) A situation in which two opposite processes take place at equal rates.

surface tension (Sec. 11.16) A measure of the inward force on the surface of a liquid caused by unbalanced intermolecular forces.

thermal expansion (Sec. 11.2) A measure of the volume change resulting from a temperature change.

vapor (Sec. 11.12) The gaseous state of a substance at a temperature and pressure at which the substance is normally a liquid or solid.

vapor pressure (Sec. 11.13) The pressure exerted by a vapor above a liquid when the liquid and vapor are in equilibrium.

volatile substance (Sec. 11.13) A substance that readily evaporates at room temperature because of a high vapor pressure.

PRACTICE PROBLEMS

States of Matter (Secs. 11.1 and 11.2)

11.1 The following statements relate to the terms *solid state*, *liquid state*, and *gaseous state*. Match the terms to the appropriate statements.

(a) This state is characterized by the lowest density of the three.

(b) This state is characterized by an indefinite shape and a definite volume.

(c) Temperature changes influence the volume of this state significantly.

(d) In this state constituent particles are more free to move about than in other states.

11.2 The following statements relate to the terms *solid state*, *liquid state*, and *gaseous state*. Match the terms to the appropriate statements.

(a) This state is characterized by an indefinite shape and high density.

(b) Pressure changes influence the volume of this state more than that of the other two.

(c) In this state constituent particles are less free to move about than in other states.

(d) This state is characterized by a definite shape and a definite volume.

11.3 For which of the following pairs of elements do both members of the pair have the same physical state (solid, liquid, gas) at room temperature and pressure?

(a) chlorine and oxygen
(b) mercury and silver
(c) krypton and neon
(d) phosphorus and aluminum

11.4 For which of the following pairs of elements do both members of the pair have the same physical state (solid, liquid, gas) at room temperature and pressure

(a) fluorine and iodine
(b) nitrogen and hydrogen
(c) sodium and sulfur
(d) bromine and mercury

Kinetic Molecular Theory (Secs. 11.3 through 11.6)

11.5 Using kinetic molecular theory concepts, answer the following questions.

(a) What is the relationship between temperature and the average velocity at which particles move?
(b) What type of energy is related to cohesive forces?
(c) What effect does temperature have on the magnitude of disruptive forces?
(d) In which of the three states of matter are disruptive forces present?

11.6 Using kinetic molecular theory concepts, answer the following questions.

(a) How do molecules transfer energy from one to another?
(b) What type of energy is related to disruptive forces?
(c) What effect does temperature have on the magnitude of cohesive forces?
(d) In which of the three states of matter are cohesive forces present?

11.7 Classify each of the following as a description of the solid, liquid, or gaseous state.

(a) cohesive forces dominate over disruptive forces
(b) neither potential energy nor kinetic energy dominates
(c) potential energy dominates over kinetic energy
(d) very small thermal expansion and small compressibility

11.8 Classify each of the following as a description of the solid, liquid, or gaseous state.

(a) disruptive forces dominate over cohesive forces
(b) kinetic energy dominates over potential energy
(c) neither disruptive forces nor cohesive forces dominate
(d) large compressibility and moderate thermal expansion

11.9 Explain each of the following observations using the kinetic molecular theory.

(a) Particles in a solid occupy essentially fixed positions.
(b) The compressibility of gases is much greater than that of liquids and solids.
(c) Liquids show little change in volume with changes in temperature.
(d) In gases particles are free to move about in all directions.

11.10 Explain each of the following observations using the kinetic molecular theory.

(a) A gas always exerts a pressure on the object or container with which it is in contact.
(b) The particles in both liquids and solids essentially touch each other.
(c) An aerosol can heated in an open fire may explode.
(d) Solids maintain characteristic shapes.

Physical Changes of State (Sec. 11.8)

11.11 In which of the following pairs of state changes are both members of the pair of the same thermicity (both exothermic or both endothermic)?

(a) freezing and melting
(b) sublimation and evaporation
(c) condensation and freezing
(d) deposition and sublimation

11.12 In which of the following pairs of state changes are both members of the pair of the same thermicity (both exothermic or both endothermic)?

(a) condensation and evaporation
(b) deposition and freezing
(c) sublimation and melting
(d) condensation and melting

11.13 In which of the pairs of state changes in Problem 11.11 is the final state (solid, liquid, gas) the same for both members of the pair?

11.14 In which of the pairs of state changes in Problem 11.12 is the final state (solid, liquid, gas) the same for both members of the pair?

11.15 In which of the pairs of state changes in Problem 11.11 are the two members of the pair opposite changes?

11.16 In which of the pairs of state changes in Problem 11.12 are the two members of the pair opposite changes?

Energy and Changes of State (Sec. 11.10)

11.17 Which one of the quantities *heat of fusion*, *heat of solidification*, *heat of vaporization*, or *heat of condensation* is needed to calculate how much energy is absorbed or released during each of the following changes? (Do not actually carry out the energy calculation.)

(a) changing of 50.0 g of molten aluminum at 658°C (its melting point) to solid aluminum at the same temperature
(b) changing of 50.0 g of steam at 100°C to liquid water at 100°C

(c) changing of 50.0 g of solid copper at its melting point (1083°C) to molten copper at the same temperature

(d) changing of 50.0 g of liquid water at 100°C to steam at 100°C

11.18 Which one of the quantities *heat of fusion, heat of solidification, heat of vaporization*, or *heat of condensation* is needed to calculate how much energy is absorbed or released during each of the following changes? (Do not actually carry out the energy calculation.)

(a) changing of 35.0 g of liquid water at its boiling point to steam at the same temperature

(b) changing of 35.0 g of molten copper at 1083°C (its melting point) to solid copper at 1083°C

(c) changing of 35.0 g of steam at 100°C to liquid water at the same temperature

(d) changing of 35.0 g of solid aluminum at its melting point (658°C) to molten aluminum at 658°C

11.19 Calculate how much heat energy, in joules, would be absorbed or evolved in each of the changes listed in Problem 11.17.

11.20 Calculate how much heat energy, in joules, would be absorbed or evolved in each of the changes listed in Problem 11.18.

11.21 If 6680 J of energy are absorbed in changing 20.0 g of ice at its melting point to the liquid state at the same temperature, how much energy, in joules, is released when 20.0 g of liquid water at its freezing point are changed to ice at the same temperature?

11.22 If 33,900 J of energy are evolved in changing 15.0 g of steam at 100°C to the liquid state at the same temperature, how much energy, in joules, is needed to change 15.0 g of liquid water at 100°C to steam at the same temperature?

11.23 The heats of fusion, respectively, of sodium hydroxide and sodium sulfate are 15.79 kJ/mole and 80.93 kJ/mole. How many times greater is the energy input needed to melt 1.00 mole of sodium sulfate than 1.00 mole of sodium hydroxide at their melting points?

11.24 The heats of fusion, respectively, of sodium thiosulfate and sodium carbonate are 49.75 kJ/mole and 71.88 kJ/mole. How many times greater is the energy input needed to melt 1.00 mole of sodium carbonate than 1.00 mole of sodium thiosulfate at their melting points?

11.25 Based on the data in Table 11.3, calculate how many moles of gaseous carbon tetrachloride at its boiling point must be condensed to the liquid state at the same temperature in order to release 115 kJ of heat energy.

11.26 Based on the data in Table 11.3, calculate how many moles of gaseous ethyl alcohol at its boiling point must be condensed to the liquid state at the same temperature in order to release 87.3 kJ of heat energy.

11.27 Experiments involving small samples of two metals, each at its melting point temperature, are carried out with the following results.

(a) 1016 J of heat energy are required to melt 3.25 g of metal A.

(b) 983 J of heat energy are required to melt 3.20 g of metal B.

Which metal has the higher heat of fusion? By how many joules per gram does this metal's heat of fusion exceed that of the other metal?

11.28 Experiments involving small samples of two liquids, each at its boiling point temperature, are carried out with the following results.

(a) 2238 J of heat energy are required to vaporize 1.79 g of liquid A.

(b) 1933 J of heat energy are required to vaporize 1.66 g of liquid B.

Which liquid has the higher heat of vaporization? By how many joules per gram does this liquid's heat of vaporization exceed that of the other liquid?

11.29 The heat of fusion of Na at its melting point is 2.40 kJ/mole. How much heat, in joules, must be absorbed by 7.00 g of solid Na at its melting point to convert it to molten Na?

11.30 The heat of vaporization of Hg at its boiling point is 58.6 kJ/mole. How much heat, in joules, must be absorbed by 13.0 g of liquid Hg at its boiling point to convert it to gaseous Hg?

Heat Energy Calculations (Sec. 11.11)

11.31 Draw the general shape of the temperature–energy graph (heating curve; Figure 11.7) for carbon tetrachloride from −100°C to 100°C. The melting and boiling points of carbon tetrachloride are, respectively, −23°C and 77°C.

11.32 Draw the general shape of the temperature–energy graph (heating curve, Figure 11.7) for benzene from −50°C to 150°C. The melting and boiling points of benzene are, respectively, 6°C and 80°C.

11.33 Calculate the amount of heat needed, in joules, to convert 75.0 g of ice at −20°C to each of the following.

(a) ice at −5°C **(b)** water at 25°C

(c) steam at 100°C **(d)** steam at 120°C

11.34 Calculate the amount of heat needed, in joules, to convert 85.0 g of ice at −15°C to each of the following.

(a) ice at 0°C **(b)** water at 0°C

(c) water at 85°C **(d)** steam at 110°C

11.35 The melting point of iron is 1530°C, its solid state specific heat is 7.78 J/(g · °C), and its heat of fusion is 63.8 J/g. How much heat energy, in joules, is required to melt 35.2 g of iron that is at each of the following initial temperatures?

(a) 20.0°C below its melting point

(b) 855°C below its melting point

(c) its melting point

(d) 1235°C

11.36 The melting point of calcium is 851°C, its solid state specific heat is 11.4 J/(g · °C), and its heat of fusion is 233 J/g. How much heat energy, in joules, is required to melt 52.0 g of calcium that is at each of the following initial temperatures?

(a) 45.0°C below its melting point

(b) 685°C below its melting point

(c) its melting point

(d) 35.0°C

11.37 Calculate the heat required to convert 15.0 g of ethyl alcohol, C_2H_6O, from a solid at −135°C into the gaseous state at 95°C. The normal melting and boiling points of this substance are −117 and 78°C, respectively. The heat of fusion is 109 J/g, and the heat of vaporization is 837 J/g. The specific heats of the solid, liquid, and gaseous states are 0.97, 2.3, and 0.95 J/(g · °C).

11.38 Calculate the heat required to convert 25.0 g of propyl alcohol, C_3H_8O, from a solid at −150°C into the gaseous state at 115°C. The normal melting and boiling points of this substance are −127 and 97°C, respectively. The heat of fusion is 86.2 J/g, and the heat of vaporization is 694 J/g. The specific heats of the solid, liquid, and gaseous states are 2.36, 2.83, and 1.76 J/g · °C).

Properties of Liquids (Secs. 11.12 through 11.14)

11.39 Match the following statements to the appropriate term: *vapor, vapor pressure, volatile, boiling, boiling point.*

(a) This is a temperature at which the liquid vapor pressure is equal to the external pressure on a liquid.

(b) This property can be measured by allowing a liquid to evaporate in a closed container.

(c) In this process bubbles of vapor form within a liquid.

(d) This temperature changes with changes in atmospheric pressure.

11.40 Match the following statements to the appropriate term: *vapor, vapor pressure, volatile, boiling, boiling point.*

(a) This state involves gaseous molecules of a substance at a temperature and pressure where we would ordinarily expect the substance to be a liquid.

(b) A substance that readily evaporates at room temperature because of a high vapor pressure has this property.

(c) This process is a special form of evaporation.

(d) This property always increases in magnitude with increasing temperature.

11.41 Offer a concise clear explanation for each of the following observations.

(a) Increasing the temperature of a liquid increases its vapor pressure.

(b) It takes more time to cook an egg in boiling water on a mountain top than at sea level.

(c) Food cooks faster in a pressure cooker than in an open pan.

(d) Evaporation is a cooling process.

11.42 Offer a concise clear explanation for each of the following observations.

(a) All liquids do not have the same vapor pressure at a given temperature.

(b) The boiling point of a liquid varies with atmospheric pressure.

(c) A person emerging from an outdoor swimming pool on a breezy warm day gets the shivers.

(d) Food will cook just as fast in boiling water with the stove set at low heat as in boiling water at high heat.

11.43 What effect (increase, decrease, or no change) will each of the following changes have on the *rate of evaporation* of a liquid in an open cylindrical container?

(a) increasing the temperature of the liquid by 10°C at constant pressure

(b) moving the container to a higher elevation (altitude) at constant temperature

(c) transferring the liquid, at constant temperature and pressure, to a new container whose dimensions are double those of the old container.

(d) doubling the amount of liquid in the container at constant temperature and pressure

11.44 What effect (increase, decrease, or no change) will each of the following changes have on the *rate of evaporation* of a liquid in an open cylindrical container?

(a) decreasing the external pressure on the surface of the liquid at constant temperature

(b) transferring the liquid, at constant temperature and pressure, to a new container which doubles the surface area for the liquid

(c) decreasing, at constant temperature and pressure, the amount of liquid in the container

(d) decreasing the temperature of the liquid by 20°C at constant pressure.

11.45 What effect (increase, decrease, or no change) will each of the changes in Problem 11.43 have on the *magnitude of the boiling point* of a liquid in an open container?

11.46 What effect (increase, decrease, or no change) will each of the changes in Problem 11.44 have on the *magnitude of the boiling point* of a liquid in an open container?

11.47 What effect (increase, decrease, or no change) will each of the changes in Problem 11.43 have on the *magnitude of the* *vapor pressure* of a liquid in a closed rigid container?

11.48 What effect (increase, decrease, or no change) will each of the changes in Problem 11.44 have on the *magnitude of the vapor pressure* of a liquid in a closed rigid container?

11.49 Identical amounts of liquids A and B are placed in identical open containers on a tabletop. Liquid B evaporates at a faster rate than liquid A even though both liquids are at the same temperature. Explain why this could be so?

11.50 Identical amounts of liquid A are placed in different open containers on a tabletop. The liquid in one container evaporates faster than the liquid in the other container even though both liquids are at the same temperature. Explain why this could be so?

11.51 Given that the vapor pressures, at 25°C, of CS_2 and CCl_4 are, respectively, 309 and 107 mm Hg, which substance is more volatile? Explain your answer.

11.52 Given that the vapor pressures at 25°C, of CS_2 and CCl_4 are, respectively, 309 and 107 mm Hg, which substance would you predict to have the lower boiling point? Explain your answer.

Intermolecular Forces in Liquids (Sec. 11.15)

11.53 Describe the molecular conditions necessary for the existence of a dipole–dipole interaction.

11.54 Describe the molecular conditions necessary for the existence of a London force.

11.55 In liquids, what is the relationship between boiling point and the strength of intermolecular forces?

11.56 In liquids, what is the relationship between vapor pressure magnitude and the strength of intermolecular forces?

11.57 For the following diatomic molecules, classify the intermolecular forces present as London forces, dipole–dipole interactions, hydrogen bonds, or combinations of these (specify the combination).

(a) H_2 **(b)** HF **(c)** CO **(d)** F_2

11.58 For the following diatomic molecules, classify the intermolecular forces present as London forces, dipole–dipole interactions, hydrogen bonds, or combinations of these (specify the combination).

(a) O_2 **(b)** HCl **(c)** Cl_2 **(d)** BrCl

11.59 In which of the following hydrogen-containing substances, in the pure liquid state, would hydrogen bonding occur?

(a) 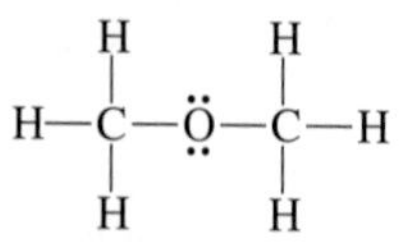

(b) 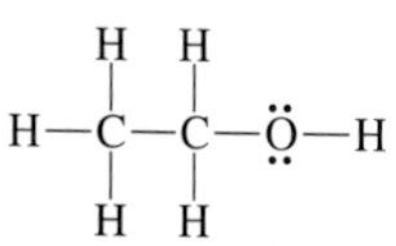

(c) 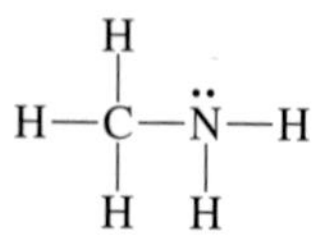

(d) 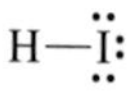

11.60 In which of the following hydrogen-containing substances, in the pure liquid state, would hydrogen bonding occur?

(a) 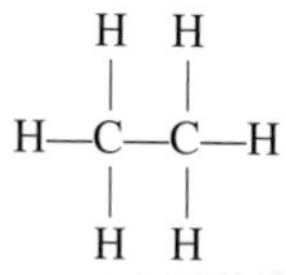

(b) 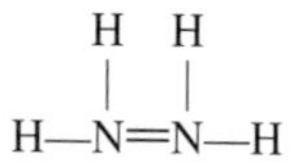

(c) 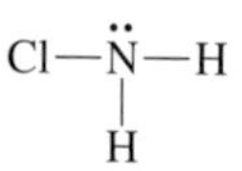

(d) 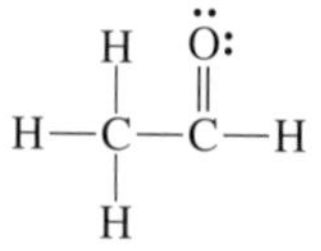

11.61 In each of the following pairs of molecules, predict which member of the pair would be expected to have the higher boiling point.

(a) F_2 and Cl_2 **(b)** HF and HBr
(c) N_2 and NO **(d)** C_2H_6 and O_2

11.62 In each of the following pairs of molecules, predict which member of the pair would be expected to have the higher boiling point.

(a) Cl_2 and Br_2 **(b)** H_2O and H_2S
(c) O_2 and CO **(d)** C_3H_8 and CO_2

Properties of Water (Sec. 11.16)

11.63 Offer a concise clear explanation for each of the following observations that involve the substance water.

(a) Large bodies of water have a moderating effect on the climate of the surrounding area.

(b) If water were a "normal" compound, it would be a gas at room temperature.

(c) During cold winters, lakes freeze from the top to bottom rather than from bottom to top.

(d) Perspiring is one of the human body's mechanisms for cooling itself.

11.64 Offer a concise clear explanation for each of the following observations that involve the substance water.

(a) The formation of potholes in streets is hastened during the winter season.

(b) Fruit trees are often sprayed with water to protect blossoms from early spring frosts.

(c) The temperature variation between night and day is smaller in locations where large bodies of water exist.

(d) An ice cube will float in a glass of water.

11.65 In the liquid state, how many hydrogen bonds can form between a single water molecule and other water molecules?

11.66 In the solid state, how many hydrogen bonds can form between a single water molecule and other water molecules?

Types of Solids (Sec. 11.17)

11.67 Indicate whether each of the following statements about various types of crystalline solids is *true* or *false*.

(a) In macromolecular solids, covalent bonds link the particles present at the lattice sites.

(b) Nonpolar molecular solids have moderately high melting points.

(c) Both ionic and macromolecular solids are nonvolatile substances.

(d) Polar molecular, nonpolar molecular, and macromolecular solids all have molecules present at the lattice sites.

11.68 Indicate whether each of the following statements about various types of crystalline solids is *true* or *false*.

(a) Metallic solids have metal atoms occupying the lattice sites.

(b) Nonpolar molecular solids usually have high volatility.

(c) Ionic molecular solids generally have low melting points because of the presence of ions at the lattice sites.

(d) In polar molecular solids, dipole–dipole interactions are important forces between lattice site particles.

11.69 For each of the following pairs of substances, predict which member of the pair will have the higher melting point and indicate why.

(a) NaCl and Cl_2 **(b)** SiO_2 and CO_2
(c) Cu and CO **(d)** NaF and MgO

11.70 For each of the following pairs of substances, predict which member of the pair will have the higher melting point and indicate why.

(a) HCl and HBr **(b)** Ar and NO
(c) Na_2O and SiC **(d)** CaO and AlN

Additional Problems

11.71 The vapor pressure of $SnCl_4$ reaches 400 mm Hg at 92°C. The vapor pressure of SnI_4 reaches 400 mm Hg at 315°C.

(a) At 100°C which substance should evaporate at the faster rate?

(b) Which substance should have the higher normal boiling point?

(c) Which substance should have the stronger intermolecular forces?

(d) At 80°C which substance should have the lower vapor pressure?

11.72 The vapor pressure of PBr_3 reaches 400 mm Hg at 150°C. The vapor pressure of PCl_3 reaches 400 mm Hg at 57°C.

(a) At 100°C which substance should evaporate at the faster rate?

(b) Which substance should have the lower normal boiling point?

(c) Which substance should have the weaker intermolecular forces?

(d) At 50°C which substance should have the higher vapor pressure?

11.73 If heat is supplied at an identical, constant rate in each case, which would take longer, heating 50.0 g of water from 0.0 to 80.0°C or vaporizing 50.0 g of water at 100.0°C?

11.74 If heat is supplied at an identical, constant rate in each case, which would take longer, heating 80.0 g of water from 20.0 to 100.0°C or melting 80.0 g of ice at 0.0°C?

11.75 A quantity of ice at 0.0°C was added to 40.0 g of water at 19.0°C in an insulated container. All of the ice melted, and the water temperature decreased to 0.0°C. How many grams of ice were added?

11.76 A quantity of ice at 0.0°C was added to 55.0 g of water at 25.3°C in an insulated container. All of the ice melted, and the water temperature decreased to 0.0°C. How many grams of ice were added?

11.77 If 10.0 g of ice at −10.0°C and 30.0 g of liquid water at 80.0°C are mixed in an insulated container, what will the final temperature of the liquid be?

11.78 If 10.0 g of ice at −20.0°C and 60.0 g of liquid water at 70.0°C are mixed in an insulated container, what will the final temperature of the liquid be?

11.79 A 500.0-g piece of metal at 50.0°C is placed in 100.0 g of water at 10.0°C in an insulated container. The metal and water come to the same temperature of 23.4°C. What is the specific heat, in joules per gram per degree Celsius, of the metal?

11.80 A 100.0-g piece of metal at 80.0°C is placed in 200.0 g of water at 20.0°C in an insulated container. The metal and water come to the same temperature of 33.6°C. What is the specific heat, in joules per gram per degree Celsius, of the metal?

11.81 Heat is added to a 25.3-g sample of a metal at the constant rate of 136 J/sec. After the sample reaches the metal's melting point temperature, the temperature remains constant for 4.3 min. Calculate the heat of fusion for this metal in joules per gram.

11.82 Heat is added to a 35.2-g sample of a metal at the constant rate of 189 J/sec. After the sample reaches the metal's melting point temperature, the temperature remains constant for 5.2 min. Calculate the heat of fusion for this metal in joules per gram.

CUMULATIVE PROBLEMS

11.83 To solidify a 10.0-g sample of a liquid at its freezing point, 1485 J of heat must be removed. To solidify a 10.0-mL sample of this same liquid at its freezing point, 1151 J of heat must be removed. What is the density, in grams per milliliter, of this liquid?

11.84 To solidify a 10.0-mL sample of a liquid at its freezing point, 1063 J of heat must be removed. To solidify a 15.0-g sample of this same liquid at its freezing point, 1276 J of heat must be removed. What is the density, in grams per milliliter, of this liquid?

11.85 The heat of vaporization for a substance is 18.1 kJ/mole. Determine the molecular mass of this substance given that 1.00 g, at its boiling point, releases 602 J of heat as it condenses.

11.86 The heat of vaporization for a substance is 36.6 kJ/mole. Determine the molecular mass of this substance given that 1.00 g, at its boiling point, releases 314 J of heat as it condenses.

11.87 How many kilojoules of heat energy are needed to melt completely at its melting point a pure sample of copper that was obtained from the complete decomposition of 52.0 g of Cu_2O?

11.88 How many kilojoules of heat energy are needed to melt completely at its melting point a pure sample of aluminum that was obtained from the complete decomposition of 75.0 g of Al_2O_3?

11.89 How many joules of heat energy are needed to heat a sample of liquid benzene (C_6H_6) from a temperature 10°C below its boiling point to its boiling point, given that the sample contains 6.32×10^{24} hydrogen atoms and that the specific heat of liquid benzene is 1.74 J/(g · °C)?

11.90 How many joules of heat energy must be removed from a sample of liquid benzene (C_6H_6) to lower its temperature from 10°C above its melting point to its melting point, given that the sample contains 4.03×10^{23} carbon atoms and the specific heat of liquid benzene is 1.74 J/(g · °C).

11.91 A nuclear power plant produces 3.3×10^6 kJ of waste heat energy every second. If this heat is to be dissipated by evaporating 25°C water (in cooling towers), how many gallons of water per day would be needed? The density of water is 0.997 g/mL at 25°C. Atmospheric pressure is 1.00 atm. The heat of vaporization of water at 25°C is 2420J/g.

11.92 A nuclear power plant produces 3.3×10^6 kJ of waste heat energy every second. If this waste heat is dumped into a river that is 42 ft wide and has an average depth of 5.1 ft and a current of 3.0 mi/hr, what is the temperature difference between the water upstream and the water downstream? Assume that the density of water is 1.0 g/mL.

12 Gas Laws

12.1 Properties of Some Common Gases

The word *gas* is used to refer to a substance that is normally in the gaseous state at ordinary temperatures and pressures. The word *vapor* (Sec. 11.12) describes the gaseous form of any substance that is a liquid or solid at normal temperatures and pressures. Thus, we speak of oxygen gas and water vapor.

The normal state for eleven of the elements is that of a gas (Sec. 11.1). A listing of these elements along with some of their properties is given in Table 12.1. This table also lists some common compounds that are gases at room temperature and pressure.

The three most commonly encountered elemental gases—hydrogen, oxygen, and nitrogen—are colorless and odorless. So also are all of the noble gases (Sec. 6.11). From these observations one should *not* conclude that all gases, or even all elemental gases, are colorless and odorless. Note from Table 12.1 the pale yellow and greenish yellow colors associated with elemental fluorine and chlorine and the irritating odors of both of them. Many gaseous compounds have pungent odors.

Table 12.1 also brings to your attention again (recall Sec. 10.2) that all of the elemental gases, except for the noble gases, exist in the form of diatomic molecules (H_2, O_2, N_2, F_2, and Cl_2).

Many of the gaseous compounds listed in Table 12.1 are colorless, have odors, and are toxic. Note that odor and toxicity do not have to go together. Carbon monoxide, a "deadly" poison, is odorless but toxic.

12.2 Gas Law Variables

The behavior of a gas can be described reasonably well by *simple* quantitative relationships called *gas laws*. **Gas laws** *are generalizations that describe in mathematical terms the relationships among the* pressure, temperature, *and* volume *of a specific quantity of a gas.*

It is only the gaseous state that is describable by simple mathematical relationships. Laws describing liquid and solid state behavior are

TABLE 12.1 COLOR, ODOR, AND TOXICITY OF ELEMENTS AND COMMON COMPOUNDS THAT ARE GASES AT ORDINARY TEMPERATURES AND PRESSURES

Element		Properties
H_2	hydrogen	colorless, odorless
O_2	oxygen	colorless, odorless
N_2	nitrogen	colorless, odorless
Cl_2	chlorine	greenish yellow, choking odor, toxic
F_2	fluorine	pale yellow, pungent odor, toxic
He	helium	colorless, odorless
Ne	neon	colorless, odorless
Ar	argon	colorless, odorless
Kr	krypton	colorless, odorless
Xe	xenon	colorless, odorless
Rn	radon	colorless, odorless
Compound		**Properties**
CO_2	carbon dioxide	colorless, faintly pungent odor
CO	carbon monoxide	colorless, odorless, toxic
NH_3	ammonia	colorless, pungent odor, toxic
CH_4	methane	colorless, odorless
SO_2	sulfur dioxide	colorless, pungent choking odor, toxic
H_2S	hydrogen sulfide	colorless, rotten egg odor, toxic
HCl	hydrogen chloride	colorless, choking odor, toxic
NO_2	nitrogen dioxide	reddish brown, irritating odor, toxic

F.Y.I.

A sharp knife cuts better than a dull knife. Because the pushing force acts on a smaller area in a sharp knife, the pressure is greater and the knife cuts better with less effort.

F.Y.I.

The popping feeling in the ears encountered driving up a steep mountainous road is caused by the decrease in pressure with altitude.

mathematically more complex. Consequently, quantitative treatments of these latter behaviors will not be given in this text.

Before we discuss the mathematical form of the various gas laws, some comments concerning the major variables involved in gas law calculations—volume, temperature, and pressure—are in order. Two of these three variables, volume and temperature, have been discussed previously (Sec. 3.4 and 3.10, respectively). The units of *liter* or *milliliter* are usually used in specifying gas volume. Only one of the three temperature scales discussed in Section 3.10, *the Kelvin scale*, can be used in gas law calculations if the results are to be valid. Therefore, you should be thoroughly familiar with the conversion of Celsius and Fahrenheit scale readings to kelvins (Sec. 3.6). We have not yet discussed pressure, the third variable. Comments concerning pressure will occupy the remainder of this section.

Pressure *is defined as the force applied per unit area, that is, the total force on a surface divided by the area of that surface.*

$$P \text{ (pressure)} = \frac{F \text{ (force)}}{A \text{ (area)}}$$

Note that pressure and force are not the same. Identical forces give rise to different pressures if they are acting on areas of different size. For areas of the same size, the larger the force, the greater the pressure.

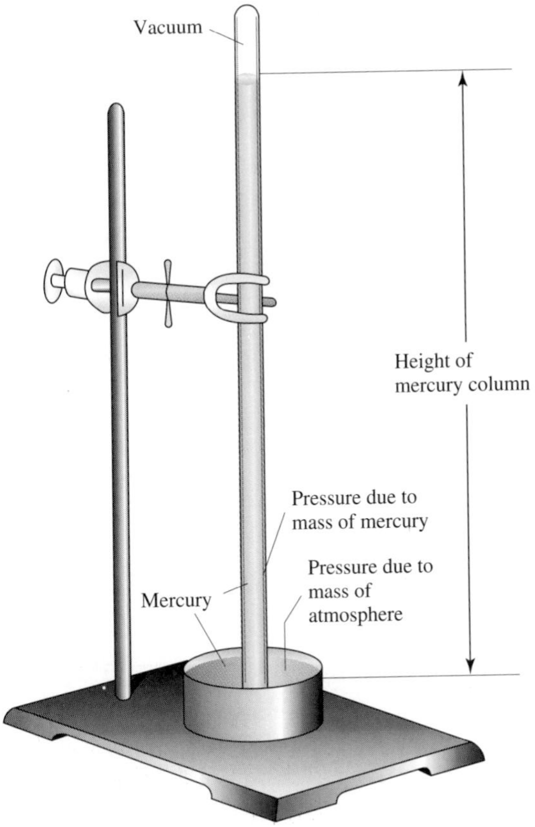

FIGURE 12.1 The essential components of a mercury baromater.

Barometers, manometers, and gauges are the instruments commonly used by chemists to measure gas pressure. Barometers and manometers measure pressure in terms of the height of a column of mercury, whereas gauges are usually calibrated in terms of force per area, for example, in pounds per square inch.

The air that surrounds the Earth exerts a pressure on all objects with which it has contact. The **barometer** *is the most commonly used device for measuring atmospheric pressure.* It was invented by the Italian physicist Evangelista Torricelli (1608–1647) in 1643. The essential components of a barometer are shown in Figure 12.1. A barometer can be constructed by filling a long glass tube, sealed at one end, all the way to the top with mercury and then inverting the tube (without letting any air in) into a dish of mercury. The mercury in the tube falls until the pressure from the mass of the mercury in the tube is just balanced by the pressure of the atmosphere on the mercury in the dish. The pressure of the atmosphere is then expressed in terms of the height of the supported column of mercury.

Mercury is the liquid of choice in a barometer for two reasons: (1) it is a very dense liquid, and therefore only a short glass tube is needed; and (2) it has a very low vapor pressure, so the pressure reading does not have to be corrected for vapor pressure.

The height of the mercury column is most often expressed in millimeters or inches. Millimeters of mercury (mm Hg) are used in laboratory work. The most common use of inches of mercury (in. Hg) is in weather reporting.

The pressure of the atmosphere varies with altitude, decreasing at the rate of approximately 25 mm Hg per 1000 ft increase in altitude. It also fluctuates with weather conditions.

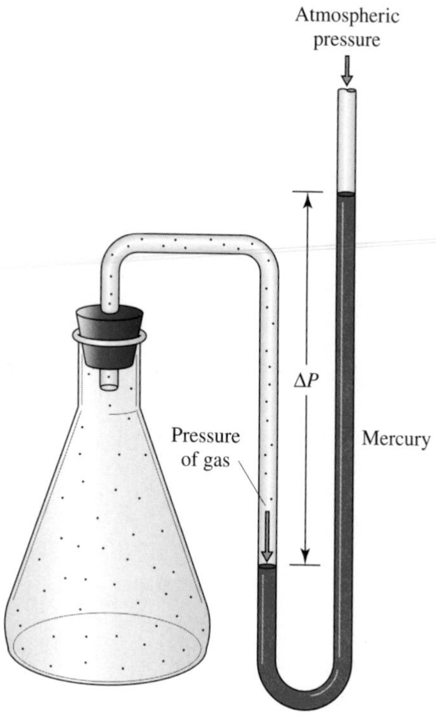

FIGURE 12.2 Pressure measurement by means of a manometer.

Recall the terminology used in a weather report: high pressure front, low pressure front, and so on. At sea level the height of a column of mercury in a barometer fluctuates with weather conditions between about 740 and 770 mm Hg and averages about 760 mm Hg. Another pressure unit, the atmosphere (atm), is defined in terms of this average sea-level pressure.

$$1 \text{ atm} = 760 \text{ mm Hg}$$

Because of its size, 760 times larger than 1 mm Hg, the atmosphere unit is frequently used to express the high pressures encountered in many industrial processes and in some experimental work.

It is impractical to measure the pressure of gases other than air with a barometer. One cannot usually introduce a mercury barometer directly into a container of a gas. A **manometer** *is the device used to measure gas pressures in a laboratory*. It is a U-tube filled with mercury. One side of the U-tube is connected to the container in which the pressure is to be measured, and the other side is connected to a region of known pressure. Such an arrangement is depicted in Figure 12.2. The gas in the container exerts a pressure on the Hg on that side of the tube, while atmospheric pressure pushes on the open end. A difference in the heights of the Hg columns in the two arms of the manometer indicates a pressure difference between the gas and the atmosphere. In Figure 12.2 the gas pressure in the container exceeds atmospheric pressure by an amount, ΔP, that is equal to the difference in the heights of the Hg columns in the two arms.

$$P_{\text{container}} = P_{\text{atm}} + \Delta P$$

Pressures of contained gas samples can also be measured by special gas pressure gauges attached to their containers. Such gauges are commonly found on tanks (cylinders)

TABLE 12.2 UNITS OF PRESSURE AND THEIR RELATIONSHIP TO THE UNIT ATMOSPHERE

Unit	Relationship to Atmosphere	Area of Use
Atmosphere	—	gas law calculations
Millimeters of mercury	760 mm Hg = 1 atm	gas law calculations
Inches of mercury	29.92 in. Hg = 1 atm	weather reports
Pound per square inch	14.68 psi = 1 atm	stored or bottled gases
Pascal	1.013×10^5 Pa = 1 atm	calculations requiring SI units

of gas purchased commercially. These gauges are most often calibrated in terms of pounds per square inch (lb/in.^{2} or psi). The relationship between psi and atmospheres is

$$14.68 \text{ psi} = 1 \text{ atm}$$

F.Y.I.
Automobile tire pressures are specified in psi units.

Table 12.2 summarizes the relationships among the various pressure units we have discussed and also lists one additional pressure unit, the *pascal*. The pascal is the SI unit (Sec. 3.1) of pressure.

The values 1 atm and 760 mm Hg are exact numbers since they arise from a definition. Consequently, these two values have an infinite number of significant figures. All other values in Table 12.2 are not exact numbers; they are given to four significant figures in the table.

The equalities of Table 12.2 can be used to construct conversion factors for problem solving in the usual way (Sec. 3.6). Example 12.1 illustrates the use of such conversion factors in the context of change in pressure unit problems.

Example 12.1

At a certain altitude, a weather balloon recorded a barometric pressure of 367 mm Hg. Express this barometric pressure in

(a) atmospheres **(b)** inches Hg **(c)** pounds per square inch

Solution

(a) The given quantity is 367 mm Hg, and the unit of the desired quantity is atmospheres

$$367 \text{ mm Hg} = ? \text{ atm}$$

The conversion factor relating these two units, found in Table 12.1, is

$$\frac{1 \text{ atm}}{760 \text{ mm Hg}}$$

The result of using this conversion factor is

$$367 \cancel{\text{mm Hg}} \times \frac{1 \text{ atm}}{760 \cancel{\text{mm Hg}}} = 0.48289473 \text{ atm} \quad \text{(calculator answer)}$$

$$= 0.483 \text{ atm} \quad \text{(correct answer)}$$

(b) This time, the unit for the desired quantity is in. Hg.

$$367 \text{ mm Hg} = ? \text{ in. Hg}$$

A direct conversion factor between mm Hg and in. Hg is not given in Table 12.1. However, the table does give the relationships of both inches and millimeters of mercury to atmospheres. Hence, we can use atmospheres as an intermediate step in the unit conversion sequence.

$$\text{mm Hg} \longrightarrow \text{atm} \longrightarrow \text{in. Hg}$$

The dimensional analysis setup is

$$367 \, \cancel{\text{mm Hg}} \times \frac{1 \, \cancel{\text{atm}}}{760 \, \cancel{\text{mm Hg}}} \times \frac{29.92 \text{ in. Hg}}{1 \, \cancel{\text{atm}}}$$

$$= 14.44821 \text{ in. Hg} \quad \text{(calculator answer)}$$
$$= 14.4 \text{ in. Hg} \quad \text{(correct answer)}$$

(c) The problem to be solved here is

$$367 \text{ mm Hg} = ? \text{ psi}$$

The unit conversion sequence will be

$$\text{mm Hg} \longrightarrow \text{atm} \longrightarrow \text{psi}$$

The dimensional analysis sequence of conversion factors, for this pathway, is

$$367 \, \cancel{\text{mm Hg}} \times \frac{1 \, \cancel{\text{atm}}}{760 \, \cancel{\text{mm Hg}}} \times \frac{14.68 \text{ psi}}{1 \, \cancel{\text{atm}}} = 7.0888947 \text{ psi} \quad \text{(calculator answer)}$$
$$= 7.09 \text{ psi} \quad \text{(correct answer)}$$

Practice Exercise 12.1

A weather reporter gives the day's barometric pressure as 30.12 in. of mercury. Express this pressure in the units of

(a) atmospheres (b) millimeters of mercury (c) pounds per square inch

Ans. (a) 1.007 atm; (b) 765.1 mm Hg; (c) 14.78 psi

12.3 BOYLE'S LAW: A PRESSURE–VOLUME RELATIONSHIP

Of the several relationships that exist between gas-law variables, the first to be discovered was the one that relates gas pressure to gas volume. It was formulated over 300 years ago, in 1662, by the British chemist and physicist Robert Boyle (1627–1691) and is known as Boyle's law. **Boyle's law** *states that the volume of a sample of gas is* inversely proportional *to the pressure applied to the gas if the temperature is kept constant*. This means that if the pressure on the gas increases, the volume decreases proportionally; and conversely, if the pressure is decreased, the volume will increase. Doubling the pressure cuts the volume in half; tripling the pressure cuts the volume to one-third its original value; quadrupling the pressure cuts the volume to one-fourth; and so on. Any time two quantities are *inversely proportional*, as pressure and volume are (Boyle's law), one increases as the other decreases. Data illustrating Boyle's law are given in Figure 12.3.

Boyle's law can be illustrated physically quite simply with the J-tube apparatus shown in Figure 12.4. The pressure on the trapped gas is increased by adding mercury to the J-tube. The volume of the trapped gas decreases as the pressure is raised.

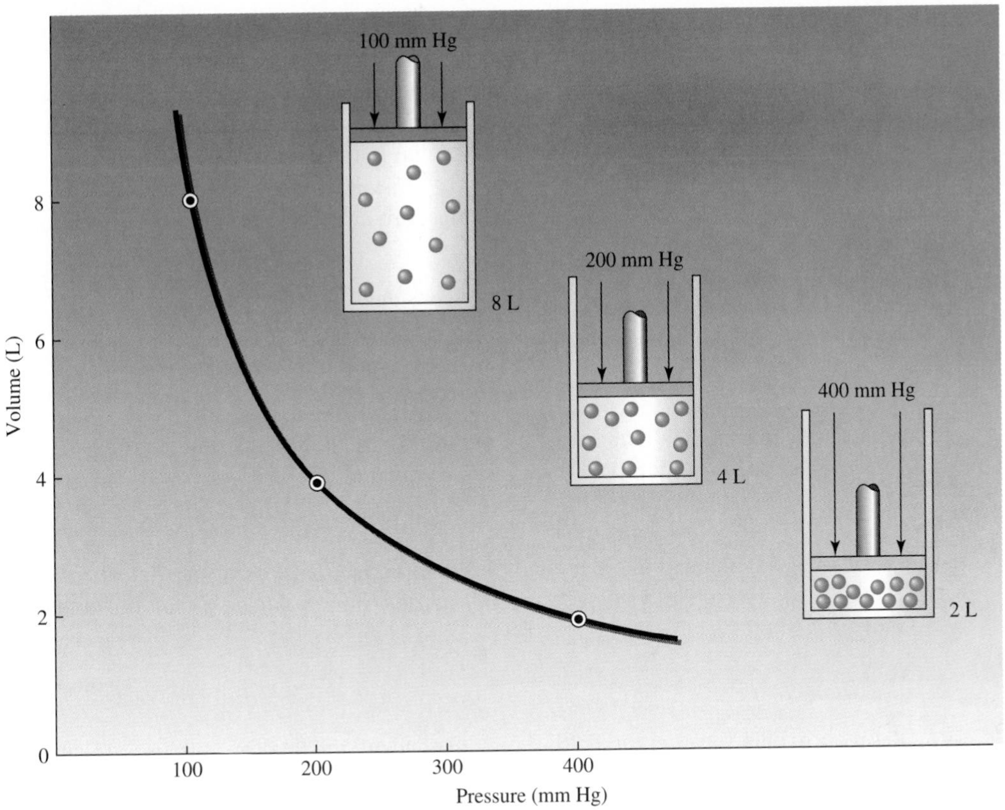

FIGURE 12.3 Data illustrating the "inverse proportionality" associated with Boyle's law.

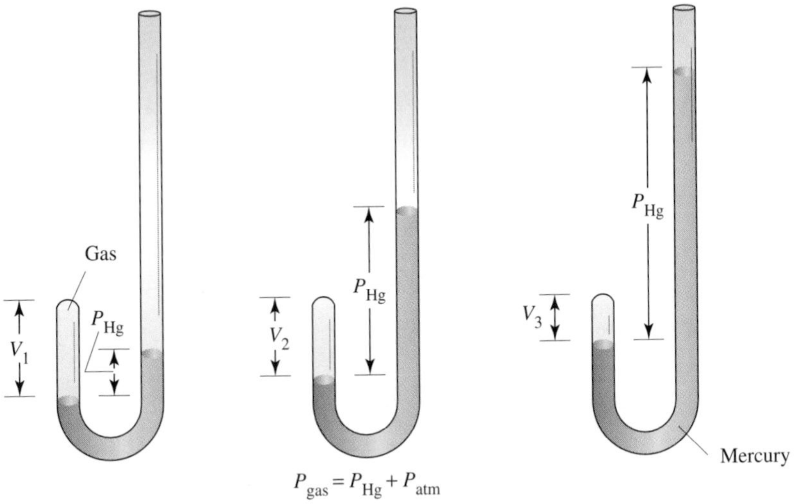

FIGURE 12.4 Boyle's law apparatus. The volume of the trapped gas in the closed end of the tube (V_1, V_2, and V_3) decreases as mercury is added through the open end of the tube.

Boyle's law can be stated mathematically as

$$P \times V = \text{constant}$$

In this expression V is the volume of the gas at a given temperature and P is the pressure. This expression thus indicates that at constant temperature the product of the pressure times the volume is always the same (or constant). (Note that Boyle's law is valid only if the temperature of the gas does not change.)

F.Y.I.

Car engines utilize Boyle's law. As the piston moves up in the cylinder, the volume in the cylinder gets smaller and the pressure of the fuel and air mixture increases. When the pressure is at a maximum, the spark plug ignites the mixture.

An alternative, and more useful, mathematical form of Boyle's law can be derived by considering the following situation. Suppose a gas is at an initial pressure P_1 and has a volume V_1. (We will use the subscript 1 to indicate the initial conditions.) Now imagine that the pressure is changed to some final pressure P_2. The volume will also change, and we will call the final volume V_2. (We will use the subscript 2 to indicate the final conditions.) According to Boyle's law,

$$P_1 \times V_1 = \text{constant}$$

After the change in pressure and volume, we have

$$P_2 \times V_2 = \text{constant}$$

The constant is the same in both cases; we are dealing with the same sample of gas. Thus, we can combine the two PV products to give the equation

$$P_1 \times V_1 = P_2 \times V_2$$

When we know any three of the four quantities in this equation, we can calculate the fourth, which will usually be the final pressure, P_2, or the final volume, V_2, as illustrated in Examples 12.2 and 12.3.

Example 12.2

An air bubble forms at the bottom of a lake, where the total pressure is 2.45 atm. At this pressure, the bubble has a volume of 3.1 mL. What volume, in milliliters, will it have when it rises to the surface, where the pressure is 0.93 atm? Assume that the temperature remains constant, as does the amount of gas in the bubble.

Solution

A suggested first step in working all gas-law problems involving two sets of conditions is to analyze the given data in terms of initial and final conditions. Doing this, we find that

$$P_1 = 2.45 \text{ atm} \qquad P_2 = 0.93 \text{ atm}$$
$$V_1 = 3.1 \text{ mL} \qquad V_2 = ? \text{ mL}$$

Next, we arrange Boyle's law to isolate V_2 (the quantity to be calculated) on one side of the equation. This is accomplished by dividing both sides of the Boyle's law equation by P_2.

$$P_1V_1 = P_2V_2 \quad \text{(Boyle's law)}$$

$$\frac{P_1V_1}{P_2} = \frac{\cancel{P_2}V_2}{\cancel{P_2}} \quad \text{(division of each side of the equation by } P_2\text{)}$$

$$V_2 = V_1 \times \frac{P_1}{P_2}$$

Substituting the given data in the rearranged equation and doing the arithmetic gives

$$V_2 = 3.1 \text{ mL} \times \frac{2.45 \cancel{\text{atm}}}{0.93 \cancel{\text{atm}}} = 8.1666666 \text{ mL} \quad \text{(calculator answer)}$$
$$= 8.2 \text{ mL} \quad \text{(correct answer)}$$

In gas law problems, when possible, it is always a good idea to check qualitatively the reasonableness of your final answer. This is a double check against two types of errors frequently made: improper substitution of variables into the equation and improper rearrangement of the equation. Decreasing the pressure on a fixed amount of gas at constant temperature, which is the case in this problem, should result in a volume increase for the gas. Our answer is consistent with this conclusion.

Practice Exercise 12.2

An inflated air mattress has a volume of 114 L at a pressure of 0.925 atm. The mattress, still inflated, is taken to a new location where the pressure is 0.975 atm. What will be the air-mattress volume, in liters, at this higher pressure, assuming that temperature remains constant?

Ans. 108 L

Exercise 12.3

A sample of O_2 gas occupies a volume of 1.51 L at a pressure of 1.27 atm and a temperature of 25°C. What volume, in liters, will it occupy if the pressure is reduced to 742 mm Hg with no change in temperature?

Solution

Analyzing the given data in terms of initial and final conditions gives

$$P_1 = 1.27 \text{ atm} \qquad P_2 = 742 \text{ mm Hg}$$
$$V_1 = 1.51 \text{ L} \qquad V_2 = ?\text{ L}$$

The temperature is constant; it will not enter into the calculation.

A slight complication exists with the pressure values; they are not given in the same units. We must make the units the same before proceeding with the calculation. It does not matter whether they are both millimeters of mercury or both atmospheres. The same answer is obtained with either unit. Let us arbitrarily decide to change atmospheres to mm Hg.

$$1.27 \cancel{\text{atm}} \times \frac{760 \text{ mm Hg}}{1 \cancel{\text{atm}}} = 965.2 \text{ mm Hg} \quad \text{(calculator answer)}$$
$$= 965 \text{ mm Hg} \quad \text{(correct answer)}$$

Our given conditions are now

$$P_1 = 965 \text{ mm Hg} \qquad P_2 = 742 \text{ mm Hg}$$
$$V_1 = 1.51 \text{ L} \qquad V_2 = ?\text{ L}$$

Boyle's law with V_2 isolated on the left side has the form

$$V_2 = V_1 \times \frac{P_1}{P_2}$$

Plugging the given quantities into this equation and doing the arithmetic gives

$$V_2 = 1.51 \text{ L} \times \frac{965 \cancel{\text{mm Hg}}}{742 \cancel{\text{mm Hg}}} = 1.963814 \text{ L} \quad \text{(calculator answer)}$$
$$= 1.96 \text{ L} \quad \text{(correct answer)}$$

The answer is reasonable. Decreased pressure, at constant temperature, should produce a volume increase.

In solving this problem we arbitrarily chose to use mm Hg pressure units. If we had used atmosphere pressure units instead, the answer obtained would still be the same, as illustrated below.

$$742 \cancel{\text{mm Hg}} \times \frac{1 \text{ atm}}{760 \cancel{\text{mm Hg}}} = 0.97631578 \text{ atm} \quad \text{(calculator answer)}$$
$$= 0.976 \text{ atm} \quad \text{(correct answer)}$$

Using $P_1 = 1.27$ atm and $P_2 = 0.976$ atm, we get

$$V_2 = 1.51 \text{ L} \times \frac{1.27 \cancel{\text{atm}}}{0.976 \cancel{\text{atm}}} = 1.9648565 \text{ L} \quad \text{(calculator answer)}$$
$$= 1.96 \text{ L} \quad \text{(correct answer)}$$

Practice Exercise 12.3

A sample of N_2 gas occupies a volume of 29.3 mL at a pressure of 745 mm Hg and a temperature of 25°C. What volume, in milliliters, will it occupy if the pressure is reduced to 0.849 atm with no change in temperature?

Ans. 33.8 mL

Boyle's law is consistent with kinetic molecular theory (Sec. 11.3). The theory states that the pressure a gas exerts results from collisions of the gas molecules with the sides of the container. The pressure of the gas at a given temperature is proportional to the number of collisions within a given area on the container wall in a given time. If the volume of a container holding a specific number of gas molecules is increased, the total wall area of the container will also increase and the number of collisions in a given area (the pressure) will decrease due to the greater wall area. Conversely, if the volume of the container is decreased, the wall area will be smaller and there will be more collisions in a given wall area. This means an increase in pressure. Figure 12.5 illustrates this idea.

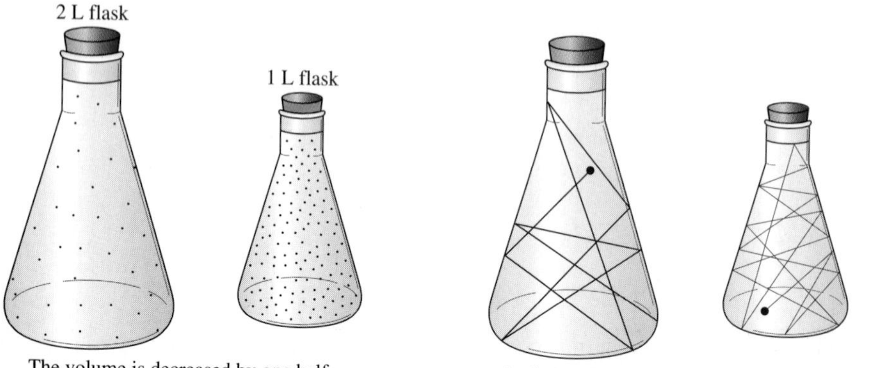

The volume is decreased by one half.

(a)

A given molecule hits container walls twice as often.

(b)

FIGURE 12.5 The pressure exerted by a gas is doubled when the volume of the gas, at constant temperature, is cut by one half.

The phenomenon described by Boyle's law has practical importance. Helium-filled research balloons, used to study the upper atmosphere, are only half-filled with helium when launched. As the balloon ascends, it encounters lower and lower pressures. As the pressure decreases, the balloon expands until it reaches full inflation. A balloon launched at full inflation would burst in the upper atmosphere because of the reduced external pressure.

Breathing is an example of Boyle's law in action, as is the operation of a respirator, a machine designed to help patients with respiration difficulties to breathe. A respirator contains a movable diaphragm that works in opposition to the patient's lungs. When the diaphragm is moved out so that the volume inside the respirator increases, the lower pressure in the respirator allows air to expand out of the patient's lungs. When the diaphragm is moved in the opposite direction, the higher pressure inside the respirator compresses the air into the lungs and causes them to increase in volume.

F.Y.I.

Air bubbles in an aquarium expand as they rise. Just as in the atmosphere, the pressure is higher at the bottom of the tank than nearer the surface.

12.4 CHARLES'S LAW: A TEMPERATURE–VOLUME RELATIONSHIP

How does the volume of a fixed quantity of gas respond to a temperature change at constant pressure? In 1787 the French scientist Jacques Alexander Cesar Charles (1746–1823) showed that a simple mathematical relationship exists between the volume and temperature of a gas, at constant pressure, *provided the temperature is expressed in kelvins.* **Charles's law** *states that the volume of a sample of gas is* directly proportional *to its Kelvin temperature if the pressure is kept constant.* Contained within the wording of this law is the phrase *directly proportional*; this contrasts with Boyle's law, which contains the phrase *inversely proportional.* Any time a *direct* proportion exists between two quantities, one increases when the other increases, and one decreases when the other decreases. Thus a direct proportion and an inverse proportion portray "opposite" behaviors. The direct proportion relationship of Charles's law means that if the temperature increases the volume will also increase and if the temperature decreases the volume will also decrease. Data illustrating Charles's law are given in Figure 12.6.

Charles's law can be qualitatively illustrated by a balloon filled with air. If the balloon is placed near a heat source, such as a light bulb that has been on for some time, the heat will cause the balloon to increase in size (volume). The change in volume is usually apparent. Putting the same balloon in the refrigerator will cause it to shrink.

Charles's law can be stated mathematically as

$$\frac{V}{T} = \text{constant}$$

In this expression V is the volume of a gas at a given pressure and T is the temperature, expressed on the Kelvin temperature scale. A consideration of two sets of temperature–volume conditions for a gas, in a manner similar to that done for Boyle's law, leads to the following useful form of the law.

$$\frac{V_1}{T_1} = \frac{V_2}{T_2}$$

Again, note that in any of the mathematical expressions of Charles's laws, or any other gas law, the symbol T is understood to mean the Kelvin temperature.

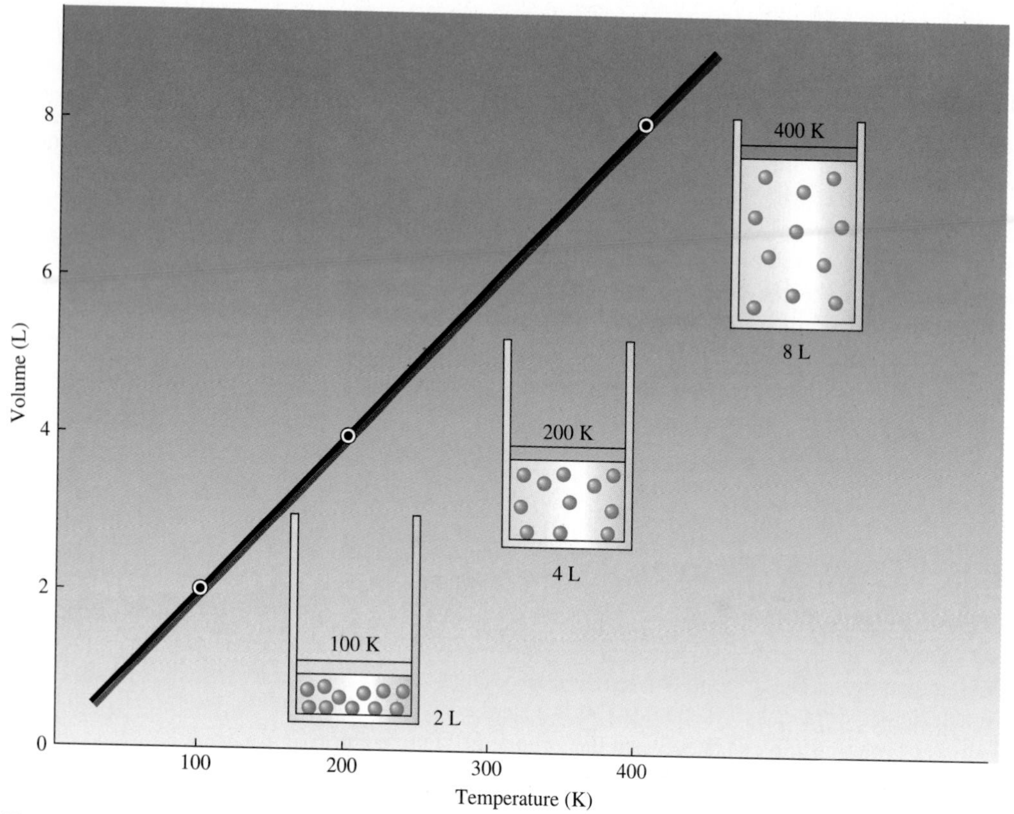

FIGURE 12.6 Data illustrating the "direct proportionality" associated with Charles's law

Example 12.4

A helium-filled heart-shaped Valentine's balloon has a volume of 247 mL at a temperature of 24°C (room temperature). The balloon is placed in the refrigerator overnight where the temperature drops to 2°C. What is the balloon's volume, in milliliters, when it is removed from the refrigerator in the morning (assuming no helium has escaped)?

Solution

Writing all the given data in the form of initial and final conditions, we have

$$V_1 = 247 \text{ mL} \qquad V_2 = ? \text{ mL}$$
$$T_1 = 24°\text{C} = 297 \text{ K} \qquad T_2 = 2°\text{C} = 275 \text{ K}$$

Note that both of the given temperatures have been converted to Kelvin scale readings by adding 273 to them.

Rearranging Charles's law to isolate V_2, the desired quantity, on one side of the equation is accomplished by multiplying each side of the equation by T_2.

$$\frac{V_1}{T_1} = \frac{V_2}{T_2} \qquad \text{(Charles's law)}$$

$$\frac{V_1 T_2}{T_1} = \frac{V_2 \cancel{T_2}}{\cancel{T_2}} \qquad \text{(multiplication of each side by } T_2\text{)}$$

$$V_2 = V_1 \times \frac{T_2}{T_1}$$

Substituting the given data into the equation and doing the arithmetic gives

$$V_2 = 247 \text{ mL} \times \frac{275 \cancel{\text{K}}}{297 \cancel{\text{K}}} = 228.7037 \text{ mL} \quad \text{(calculator answer)}$$

$$= 229 \text{ mL} \quad \text{(correct answer)}$$

Our answer is consistent with what reasoning says it should be. Decreasing the temperature of a gas, at constant pressure, should result in a volume decrease.

Practice Exercise 12.4

A helium-filled birthday balloon has a volume of 223 mL at a temperature of 24°C (room temperature). The balloon is placed in a car overnight during the winter where the temperature drops to −13°C. What is the balloon's volume, in milliliters, when it is removed from the car in the morning (assuming no helium has escaped)?

Ans. 195 mL

Figure 12.7 is a graph showing four sets of volume–temperature data for a gas, each data set being at a different constant pressure. The plot of each data set gives a straight line. These four straight lines can be used to show the basis for the use of the Kelvin temperature scale in gas-law calculations. If each of these lines is extrapolated (extended) to lower temperatures (the dashed portions of the lines in Fig. 12.7), we find that they all intersect at a common point on the temperature axis. This point of intersection, corresponding to a temperature value of −273°C, is the point at which the volume of the gas "would become" zero. Notice that the words "would become" in the last sentence are in quotation marks. In reality, the volume of a gas never reaches zero; all gases would liquefy before they reach a tem-

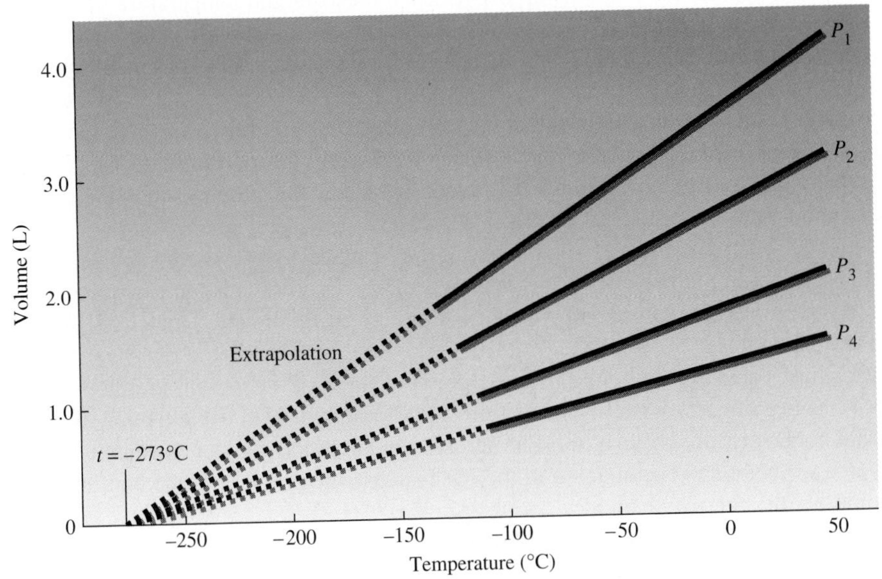

FIGURE 12.7 A graph showing the variation of the volume of a gas with temperature at a constant pressure. Each line in the graph represents a different fixed pressure for the gas. The pressures increase from P_1 to P_4. The solid portion of each line represents the temperature region before the gas condenses to a liquid. The extrapolated (dashed) portion of each line represents the hypothetical volume of the gas if it had not condensed. All extrapolated lines intersect at the same point, a point corresponding to zero volume and a temperature of −273°C.

F.Y.I.
Hawks can be seen riding the thermals of heated air that is warmed by the ground.

F.Y.I.
Thunder can be related to Charles's law. A typical lightning bolt heats the air around it to 30,000°C in a few microseconds. The heated air expands rapidly and pushes against the cooler surrounding air. This creates the shock wave of sound known as thunder.

perature of −273°C and Charles's law would no longer reply. Thus, the extrapolated portions of the four straight lines portray a hypothetical situation. It is not, however, a situation without significance.

The Scottish mathematician and physicist William Thomson (1824–1907), better known as Lord Kelvin, was the first to recognize the importance of the "zero volume" temperature value of −273°C. It is the lowest temperature that is theoretically attainable, a temperature now referred to as *absolute zero*. It is this temperature that is now used as the starting point for the temperature scale called the Kelvin temperature scale. When temperatures are specified in kelvins, Charles's law assumes the simple mathematical form in which it was previously given in this section.

Charles's law is readily understood in terms of kinetic molecular theory. The theory states that when the temperature of a gas increases, the velocity of the gas particles increases. The speedier particles hit the container walls harder and more often. In order for the pressure of the gas to remain constant, it is necessary for the container volume to increase. Moving in a larger volume, the particles will hit the container walls less often, and thus the pressure can remain the same. A similar argument applies if the temperature of the gas is lowered, only this time the velocity of the molecules decreases and the wall area (volume) must decrease in order to increase the number of collisions in a given area in a given time.

Charles's law is the principle used in the operation of a convection heater. When air comes in contact with the heating element it expands (its density becomes less). The hot, less dense air rises, causing continuous circulation of warm air. This same principle has ramifications in closed rooms in which there is not effective air circulation. The warmer and less dense air stays near the top of the room. This is desirable in the summer but not in the winter.

12.5 GAY-LUSSAC'S LAW: A TEMPERATURE–PRESSURE RELATIONSHIP

If a gas is placed in a rigid container, one that cannot expand, the volume of the gas must remain constant. What is the relationship between the pressure and temperature of a fixed amount of gas in such a situation? This question was answered in 1802 by the French scientist Joseph Louis Gay-Lussac (1778–1850). Performing a number of experiments similar to those performed by Charles, he discovered the relationship between pressure and the temperature of a gas when it is held at constant volume. **Gay-Lussac's law** *states that the pressure of a sample of gas, at constant volume, is* directly proportional *to its Kelvin temperature*. Thus, as the temperature of the gas increases, the pressure increases; conversely, as temperature is decreased, pressure decreases.

The kinetic molecular theory explanation for Gay-Lussac's law is very simple. As the temperature increases, the velocity of the gas molecules increases. This increases the number of times the molecules hit the container walls in a given time, which translates into an increase in pressure.

Gay-Lussac's law explains the observed pressure increase inside an automobile tire after a car has been driven for a period of time on a hot day. It also explains why aerosol cans should not be disposed of in a fire. The aerosol can keeps the gas at a constant volume. The elevated temperatures encountered in the fire increase the pressure of the confined gas to the point that the can may explode.

Mathematical forms of Gay-Lussac's law are

$$\frac{P}{T} = \text{constant} \qquad \text{and} \qquad \frac{P_1}{T_1} = \frac{P_2}{T_2}$$

Again, note that the symbol T is understood to mean the Kelvin temperature. Any presure unit is acceptable; P_1 and P_2 must, however, be in the same units.

Example 12.5

A pressurized can of shaving cream contains a gas under an internal pressure of 1.065 atm at 24°C (room temperature). The can itself is able to withstand an internal presure of 3.00 atm before it explodes. What temperature, in degrees Celsius, is needed to elevate the internal pressure within the can to its maximum value?

Solution

Writing all the given data in the form of initial and final conditions, we have

$$P_1 = 1.065 \text{ atm} \qquad P_2 = 3.00 \text{ atm}$$
$$T_1 = 24°\text{C} = 297 \text{ K} \qquad T_2 = ?\text{ K}$$

Rearrangement of Gay-Lussac's law to isolate T_2, the quantity we desire, on the left side of the equation gives

$$T_2 = T_1 \times \frac{P_2}{P_1}$$

Substituting the given data into the equation and doing the arithmetic gives

$$T_2 = 297 \text{ K} \times \frac{3.00 \cancel{\text{atm}}}{1.065 \cancel{\text{atm}}} = 836.61971 \text{ K} \quad \text{(calculator answer)}$$

$$= 837 \text{ K} \quad \text{(correct answer)}$$

Our answer is in kelvins. We were asked for the temperature in degrees Celsius. Subtracting 273 from the Kelvin temperature will give us the Celsius temperature.

$$837 \text{ K} - 273 = 564°\text{C}$$

The final answer is reasonable. Increased temperature, at constant volume, should produce a pressure increase.

Practice Exercise 12.5

An aerosol spray can contains gas under a pressure of 2.00 atm at 27°C. The can itself is only able to withstand a pressure of 3.00 atm. To what temperature, in degrees Celsius, could the can and its contents be heated before the can explodes?

Ans. 177°C

12.6 THE COMBINED GAS LAW

The **combined gas law** *is the expression obtained from mathematically combining Boyle's, Charles's, and Gay-Lussac's laws.* Its mathematical form is

$$\frac{P_1V_1}{T_1} = \frac{P_2V_2}{T_2}$$

This combined gas law is a much more versatile equation than the individual gas laws. With it, a change in any one of the three gas law variables brought about by changes in *both* of the other two variables can be calculated. Each of the individual gas laws requires that one of the three variables be held constant.

The combined gas law reduces (simplifies) to each of the equations for the individual gas laws when the appropriate variable is held constant. These reduction relationships are given in Table 12.3. Because of the ease with which they can be derived from the combined gas law, you need not memorize the mathematical forms for the individual gas laws if you know the mathematical form for the combined gas law.

The three most used forms of the combined gas law are those that isolate V_2, P_2, and T_2 on the left side of the equation.

$$V_2 = V_1 \times \frac{P_1}{P_2} \times \frac{T_2}{T_1}$$

$$P_2 = P_1 \times \frac{V_1}{V_2} \times \frac{T_2}{T_1}$$

$$T_2 = T_1 \times \frac{P_2}{P_1} \times \frac{V_2}{V_1}$$

Students frequently have questions about the algebra involved in accomplishing these rearrangements. Example 12.6 should clear up any such questions. Examples 12.7 and 12.8 illustrate the use of the combined gas law equation.

TABLE 12.3 RELATIONSHIP OF THE INDIVIDUAL GAS LAWS TO THE COMBINED GAS LAW

Law	Constancy Requirement	Mathmatical Form of the Law
Combined gas law	none	$\frac{P_1V_1}{T_1} = \frac{P_2V_2}{T_2}$
Boyle's law	$T_1 = T_2$	Since T_1 and T_2 are equal, substitute T_1 for T_2 in the combined gas law and cancel $\frac{P_1V_1}{\cancel{T_1}} = \frac{P_2V_2}{\cancel{T_1}}$ or $P_1V_1 = P_2V_2$
Charles's law	$P_1 = P_2$	Since P_1 and P_2 are equal, substitute P_1 for P_2 in the combined gas law and cancel. $\frac{\cancel{P_1}V_1}{T_1} = \frac{\cancel{P_1}V_2}{T_2}$ or $\frac{V_1}{T_1} = \frac{V_2}{T_2}$
Gay-Lussac's law	$V_1 = V_2$	Since V_1 and V_2 are equal, substitute V_1 for V_2 in the combined gas law and cancel. $\frac{P_1\cancel{V_1}}{T_1} = \frac{P_2\cancel{V_1}}{T_2}$ or $\frac{P_1}{T_1} = \frac{P_2}{T_2}$

Example 12.6

Rearrange the standard form of the combined gas law equation such that the variable P_2 is by itself on the left side of the equation.

Solution

In rearranging the standard form of the combined gas law into various formats, the following rule from algebra is useful.

If two fractions are equal,

$$\frac{a}{b} = \frac{c}{d}$$

then the numerator of the first fraction (a) times the denominator of the second fraction (d) is equal to the numerator of the second fraction (c) times the denominator of the first fraction (b).

$$\text{If} \quad \frac{a}{b} = \frac{c}{d}, \quad \text{then} \quad a \times d = c \times b.$$

Applying this rule to the standard form of the combined gas law gives

$$\frac{P_1V_1}{T_1} = \frac{P_2V_2}{T_2} \longrightarrow P_1V_1T_2 = P_2V_2T_1$$

With the combined gas law in the form $P_1V_1T_2 = P_2V_2T_1$, any of the six variables can be isolated by a simple division. To isolate P_2, we divide both sides of the equation by V_2T_1 (the other quantities on the same side of the equation as P_2).

$$\frac{P_1V_1T_2}{V_2T_1} = \frac{P_2\cancel{V_2T_1}}{\cancel{V_2T_1}}$$

$$P_2 = \frac{P_1V_1T_2}{V_2T_1} \quad \text{or} \quad P_2 = P_1 \times \frac{V_1}{V_2} \times \frac{T_2}{T_1}$$

Practice Exercise 12.6

Rearrange the standard form of the combined gas law equation such that the variable V_2 is by itself on the left side of the equation.

Ans. $V_2 = V_1 \times \frac{P_1}{P_2} \times \frac{T_2}{T_1}$

Example 12.7

Sulfur dioxide (SO_2) is an air pollutant whose primary source is combustion of coal. A sample of SO_2 gas is found to occupy a volume of 1.23 L at 755 mm Hg and 0°C. What volume, in liters, will this same gas sample occupy at 735 mm Hg pressure and a temperature of 50°C?

Solution

Writing all the given data in the form of initial and final conditions, we have

$P_1 = 755$ mm Hg	$P_2 = 735$ mm Hg
$V_1 = 1.23$ L	$V_2 = ?$ L
$T_1 = 0°C = 273$ K	$T_2 = 50°C = 323$ K

Rearrangement of the combined gas law expression to isolate V_2 on the left side gives

$$V_2 = V_1 \times \frac{P_1}{P_2} \times \frac{T_2}{T_1}$$

Substituting numerical values into this equation and doing the arithmetic gives

$$V_2 = 1.23 \text{ L} \times \frac{755 \text{ mm Hg}}{735 \text{ mm Hg}} \times \frac{323 \text{ K}}{273 \text{ K}}$$

$$= 1.494874 \text{ L (calculator answer)}$$
$$= 1.49 \text{ L (correct answer)}$$

Practice Exercise 12.7

On a winter day, a person takes in a breath of 425 mL of cold air (−12°C). The atmospheric pressure is 682 mm Hg. What is the volume of this air, in milliliters, in the lungs where the temperature is 37°C and the pressure is 684 mm Hg?

Ans. 503 mL

Example 12.8

A sample of helium gas [the gas used in lighter-than-air airships (blimps)] occupies a volume of 180.0 mL at a pressure of 0.800 atm and a temperature of 29°C. What will be the temperature of the gas, in degrees Celsius, if the volume of the container is deceased to 90.0 mL and the pressure is increased to 1.60 atm?

Solution

Writing all the given data in the form of intial and final conditions we have

$P_1 = 0.800$ atm	$P_2 = 1.60$ atm
$V_1 = 180.0$ mL	$V_2 = 90.0$ mL
$T_1 = 29°C = 302$ K	$T_2 = ?°C$

Rearrangement of the combined gas law to isolate T_2 on the left side of the equation gives

$$T_2 = T_1 \times \frac{P_2}{P_1} \times \frac{V_2}{V_1}$$

Substituting the given data into this equation and doing the arithmetic gives

$$T_2 = 302 \text{ K} \times \frac{1.60 \text{ atm}}{0.800 \text{ atm}} \times \frac{90.0 \text{ mL}}{180.0 \text{ mL}}$$

$$= 302 \text{ K (calculator and correct answer in kelvins)}$$

Converting the temperature to the Celsius scale by subtracting 273 gives 29°C as the final answer.

$$302 \text{ K} - 273 = 29°\text{C}$$

The temperature did not change! The pressure correction factor, considered by itself, would cause the temperature to increase by a factor of 2; the pressure was doubled. The volume correction factor, considered by itself, would cause the temperature to decrease by a factor of 2; the volume was halved. Considered together, the effects of the two factors cancel each other and result in the temperature not changing.

Practice Exercise 12.8

A sample of argon gas (the gas used in electric light bulbs) occupies a volume of 80.0 mL at a pressure of 1.10 atm and a temperature of 29°C. What will be the temperature of the gas, in degrees Celsius, if the volume of the container is decreased to 40.0 mL and the pressure is increased to 2.20 atm?

Ans. 29°C

12.7 STANDARD CONDITIONS FOR TEMPERATURE AND PRESSURE

The volumes of liquids and solids change only slightly with temperature and pressure changes (Sec. 11.2). This is not the case for volumes of gases. As the gas-law discussions of previous sections pointedly show, the volume of a gas can change markedly with changes in temperature and pressure.

Gas volumes can be compared only if the gases are at the same temperature and pressure. It is convenient to specify a particular temperature and pressure as standards for comparison purposes. **Standard temperature** *is 0°C (273 K).* **Standard pressure** *is 1 atm (760 mm Hg).* **STP conditions** *are those of standard temperature and standard pressure.*

Example 12.9 is a problem involving STP conditions.

Example 12.9

A sample of neon gas [the gas employed in luminescent lighting (neon signs)] occupies a volume of 23.4 L at STP. What volume, in liters, will it occupy at a pressure of 0.750 atm and a temperature of −20°C?

Solution

STP conditions denote a temperature of 0°C and a pressure of 1.00 atm. Writing all the given data in the form of initial and final conditions, we have

$$\begin{aligned} P_1 &= 1.00 \text{ atm} & P_2 &= 0.750 \text{ atm} \\ V_1 &= 23.4 \text{ L} & V_2 &= ?\text{ L} \\ T_1 &= 0°\text{C} = 273 \text{ K} & T_2 &= -20°\text{C} = 253 \text{ K} \end{aligned}$$

Rearrangement of the combined gas law expression to isolate V_2 on the left side gives

$$V_2 = V_1 \times \frac{P_1}{P_2} \times \frac{T_2}{T_1}$$

Substituting numerical values into the equation gives

$$V_2 = 23.4\ \text{L} \times \frac{1.00\ \cancel{\text{atm}}}{0.750\ \cancel{\text{atm}}} \times \frac{253\ \cancel{\text{K}}}{273\ \cancel{\text{K}}}$$

$$= 28.914285\ \text{L} \quad \text{(calculator answer)}$$

$$= 28.9\ \text{L} \quad \text{(correct answer)}$$

In this problem the pressure correction factor (P_1/P_2) and the temperature correction factor (T_2/T_1) work in opposite directions. The decreased pressure should cause the volume to increase. The decreased temperature should cause the volume to decrease. The fact that, overall, the volume increases indicates that the pressure change is the larger change.

Practice Exercise 12.9

A sample of H_2 gas occupies a volume of 1.37 L at STP. What volume, in liters, will it occupy at a pressure of 4.00 atm and a temperature of 340°C?

Ans. 0.769 L

12.8 GAY-LUSSAC'S LAW OF COMBINING VOLUMES

Gases are involved as reactants or products in many chemical reactions. In such reactions it is usually easier to determine the volumes rather than the masses of the gases involved.

A very simple relationship exists between the volumes of different gases consumed or produced in chemical reactions, provided the volumes are all determined at the same temperature and pressure. This relationship, known as Gay-Lussac's law of combining volumes, was first formulated in 1808 by Joseph Louis Gay-Lussac (the same Gay-Lussac discussed in Sec. 12.5). **Gay-Lussac's law of combining volumes** *states that the volumes of different gases involved in a reaction, measured at the same temperature and pressure, are in the same ratio as the coefficients for these gases in the balanced equation for the reaction.* For example, 1 volume of nitrogen and 3 volumes of hydrogen react to give 2 volumes of ammonia (NH_3).

$$\underset{\text{1 volume}}{N_2(g)} + \underset{\text{3 volumes}}{3\,H_2(g)} \longrightarrow \underset{\text{2 volumes}}{2\,NH_3(g)}$$

"Volume" is used here in the general sense of relative volume in any units. It could, for example, be 1 L of N_2, 3 L of H_2, and 2 L of NH_3 or 0.1 mL of N_2, 0.3 mL of H_2, and 0.2 mL of NH_3. Note that in making volume comparisons the units must always be the same; if the volume of N_2 is measured in milliliters, the volumes of H_2 and NH_3 must also be in milliliters.

It is important to remember that these volume relationships apply only to gases, and then only when all gaseous volumes are measured at the same temperature and pressure. The volumes of solids and liquids involved in reactions cannot be treated this way.

The previously given statement of the law of combining volumes is a modern version of the law. At the time the law was first formulated by Gay-Lussac, chemists were still struggling with the difference between atoms and molecules. Equations as we now write them were unknown; formulas for substances were still to be determined. The original statement of the law simply noted that the ratio of volumes was always a ratio of small whole numbers. Gay-Lussac's work was based solely on volume measurements and had nothing to do with chemical equations. The explanation for the small whole numbers and the linkup with equations came later.

In Section 10.5 we learned that coefficients in a balanced equation can be interpreted in terms of moles. For example, for the equation

$$4\ NH_3(g) + 3\ O_2(g) \longrightarrow 2\ N_2(g) + 6\ H_2O(g)$$

it is correct to say

$$4 \text{ moles } NH_3(g) + 3 \text{ moles } O_2(g) \longrightarrow 2 \text{ moles } N_2(g) + 6 \text{ moles } H_2O(g)$$

As a result of Gay-Lussac's law of combining volumes, all of the mole designations in this equation can be replaced with volume designations.

$$4 \text{ volumes } NH_3(g) + 3 \text{ volumes } O_2(g) \longrightarrow 2 \text{ volumes } N_2(g) + 6 \text{ volumes } H_2O(g)$$

Again, all of the volumes must be measured at the same temperature and pressure for the above relationship to be valid.

In calculations involving chemical reactions, where two or more gases are participants, this volume interpretation of coefficients can be used to generate conversion factors useful in problem solving. Consider the balanced equation

$$2\ CO(g) + O_2(g) \longrightarrow 2\ CO_2(g)$$

At constant temperature and pressure, three volume–volume relationships are obtainable from this equation.

2 volumes CO	produce	2 volumes CO_2
1 volume O_2	produces	2 volumes CO_2
2 volumes CO	react with	1 volume O_2

From these three relationships, six conversion factors can be written.
From the first relationship:

$$\frac{2 \text{ volumes CO}}{2 \text{ volumes } CO_2} \quad \text{and} \quad \frac{2 \text{ volumes } CO_2}{2 \text{ volumes CO}}$$

From the second relationship:

$$\frac{1 \text{ volume } O_2}{2 \text{ volumes } CO_2} \quad \text{and} \quad \frac{2 \text{ volumes } CO_2}{1 \text{ volume } O_2}$$

From the third relationship:

$$\frac{2 \text{ volumes CO}}{1 \text{ volume } O_2} \quad \text{and} \quad \frac{1 \text{ volume } O_2}{2 \text{ volumes CO}}$$

The more gaseous reactants and products there are in a chemical reaction, the greater the number of volume–volume conversion factors obtainable from the equation for the chemical reaction. The use of volume–volume conversion factors is illustrated in Example 12.10.

Example 12.10

Nitrogen (N_2) reacts with hydrogen (H_2) to produce ammonia (NH_3) as shown by the equation

$$N_2(g) + 3\ H_2(g) \longrightarrow 2\ NH_3(g)$$

What volume of H_2, in liters, at 750 mm Hg and 25°C is required to produce 1.75 L of NH_3 at the same temperature and pressure?

Solution

STEP 1 The given quantity is 1.75 L of NH_3 and the desired quantity is liters of H_2.

$$1.75\ \text{L}\ NH_3 = ?\ \text{L}\ H_2$$

STEP 2 This is a volume-to-volume problem. The conversion factor needed for this one-step problem is derived from the coefficients of H_2 and NH_3 in the equation for the chemical reaction. The equation tells us that at constant temperature and pressure three volumes of H_2 are needed to prepare two volumes of NH_3. From this relationship the conversion factor

$$\frac{3\ \text{L}\ H_2}{2\ \text{L}\ NH_3}$$

is obtained.

STEP 3 The dimensional analysis setup for this problem is

$$1.75\ \cancel{\text{L}\ NH_3} \times \frac{3\ \text{L}\ H_2}{2\ \cancel{\text{L}\ NH_3}}$$

Note that it makes no difference what the temperature and pressure are, as long as they are the same for the two gases involved in the calculation.

STEP 4 Doing the arithmetic, after cancellation of units, gives

$$\frac{1.75 \times 3}{2}\ \text{L}\ H_2 = 2.625\ \text{L}\ H_2 \quad \text{(calculator answer)}$$

$$= 2.62\ \text{L}\ H_2 \quad \text{(correct answer)}$$

Practice Exercise 12.10

Hydrogen, H_2, reacts with acetylene, C_2H_2, according to the equation

$$2\ H_2(g) + C_2H_2(g) \longrightarrow C_2H_6(g)$$

What volume of H_2, in liters, is required to react with 13.7 L of C_2H_2? Assume that both gases are at the same conditions of temperature and pressure.

Ans. 27.4 L H_2

12.9 AVOGADRO'S LAW: A VOLUME–QUANTITY RELATIONSHIP

In the year 1811, Amedeo Avogadro, the Avogadro involved in Avogadro's number (Sec. 9.4), published work in which he proposed an explanation for Gay-Lussac's observations

(Sec. 12.8). His proposal, a hypothesis at the time it was first published, is now known as Avogadro's law. Its validity has been demonstrated in a number of ways.

Avogadro's law *states that equal volumes of different gases, measured at the same temperature and pressure, contain equal numbers of molecules.* Although Avogadro's law seems very simple in terms of today's scientific knowledge, it was a very astute conclusion at the time it was first proposed. At that time scientists were still struggling with the differences between atoms and molecules, and Avogadro's reasoning eventually led to the realization that many of the common gaseous elements, such as hydrogen, oxygen, nitrogen, and chlorine, occur naturally as diatomic molecules (H_2, O_2, N_2, Cl_2) rather than single atoms.

Equal volumes of gases at the same temperature and pressure, because they contain equal numbers of molecules, must also contain equal numbers of moles of molecules. Thus, Avogadro's law can also be stated in terms of volumes and moles rather than volumes and molecules.

Avogadro's law can also be stated in another alternative form, which uses terminology similar to that used in Boyle's, Charles's, and Gay-Lussac's laws. This alternative form deals with two samples of the same gas rather than two samples of different gases. *The volume of a gas, at constant temperature and pressure, is directly proportional to the number of moles of gas present.* When the original number of moles of gas present is doubled, the volume of the gas increases twofold. Halving the number of moles present halves the volume.

Experimentally, it is easy to show the direct relationship between volume and amount of gas present at constant temperature and pressure. All that is needed is an inflated balloon. Adding more gas to the balloon increases the volume of the balloon. The more gas added to the balloon, the greater the volume increase. Conversely, letting gas out of the balloon decreases its volume. It should be noted that this balloon demonstration is an approximate rather than exact demonstration for the law, since the tension of the rubber material from which the balloon is made exerts an effect on the proportionality between volume and moles.

Mathematical statements of Avogadro's law, where n represents the number of moles, are

$$\frac{V}{n} = \text{constant} \qquad \text{and} \qquad \frac{V_1}{n_1} = \frac{V_2}{n_2}$$

Note the similarity in mathematical form between this law and the laws of Charles and Gay-Lussac. The similarity results from all three laws being direct proportionality relationships between two variables.

Avogadro's law and the combined gas law can be combined to give the expression

$$\frac{P_1V_1}{n_1T_1} = \frac{P_2V_2}{n_2T_2}$$

This equation covers the situation where none of the four variables, P, T, V, and n, is constant. With it, a change in any one of the four variables brought about by changes in the other three variables can be calculated.

Example 12.11

A balloon containing 2.00 moles of He has a volume of 0.500 L at a given temperature and pressure. If 0.48 mole of He gas is removed from the balloon without changing temperature and pressure conditions, what will be the new volume, in liters, of the balloon?

Solution
We will use the equation

$$\frac{V_1}{n_1} = \frac{V_2}{n_2}$$

in solving the problem.

Writing the given data in terms of initial and final conditions, we have

$$V_1 = 0.500 \text{ L} \qquad V_2 = ?\text{ L}$$
$$n_1 = 2.00 \text{ moles} \qquad n_2 = (2.00 - 0.48) \text{ moles} = 1.52 \text{ moles}$$

Rearrangement of the Avogadro's law expression is isolate V_2 on the left side gives

$$V_2 = V_1 \times \frac{n_2}{n_1}$$

Substituting numerical values into the equation gives

$$V_2 = 0.500 \text{ L} \times \frac{1.52 \text{ }\cancel{\text{moles}}}{2.00 \text{ }\cancel{\text{moles}}} = 0.38 \text{ L} \quad \text{(calculator answer)}$$

$$= 0.380 \text{ L} \quad \text{(correct answer)}$$

Our answer is consistent with reasoning. Decreasing the number of moles of gas present in the balloon at constant temperature and pressure should decrease the volume of the balloon.

Practice Exercise 12.11

A balloon containing 1.83 moles of He has a volume of 0.673 L at a given temperature and pressure. If an additional 0.50 mole of He gas is introduced into the balloon without changing temperature and pressure conditions, what will be the new volume, in liters, of the balloon?

Ans. 0.857 L

12.10 MOLAR VOLUME OF A GAS

F.Y.I.
22.4 L is a little over half the volume of a ten-gallon aquarium.

The **molar volume** *of a gas is the volume occupied by 1 mole of the gas at STP conditions.* It follows from Avogadro's law (Sec. 12.9) that all gases will have the same molar volume. Experimentally, it is found that the molar volume of a gas is 22.4 L. To visualize a volume this size, 22.4 L. think of standard-sized basketballs. The volume occupied by three standard-sized basketballs is very close to 22.4 L.

The fact that all gases have the same molar volume at STP (or any other temperature–pressure combination) is a property unique to the gaseous state. Similar statements cannot be made about the liquid and solid states. The reason for the difference can be understood by considering the relationship between the volume the molecules occupy in the given physical state and the volume of the molecules themselves. In the gaseous state, because the molecules are so far apart—a gas is mostly empty space—the volume of the gas molecules

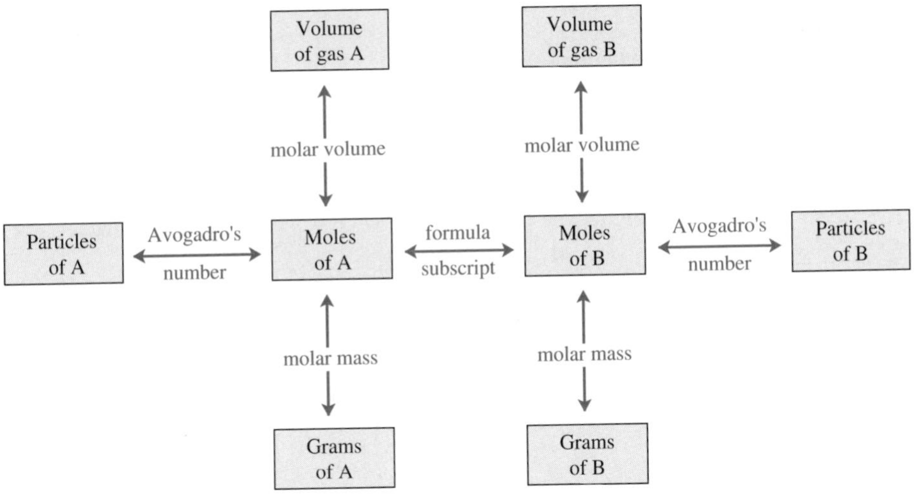

FIGURE 12.8 Quantitative relationships needed for solving mass-to-volume and volume-to-mass chemical-formula-based problems.

themselves is negligible compared to the total volume. This is not the case in the liquid and solid states, where the molecules are in close contact with each other.

It is informative to contrast the properties of molar volume and molar mass, the latter having been discussed in Section 9.5. The *molar mass* is a constant for a given substance, and is independent of temperature and pressure. Each substance has its own unique molar mass. The *molar volume* of a gas has a temperature and pressure specification (STP conditions). All gases have the same molar volume at the same temperature and pressure (STP conditions).

The molar volume concept is very useful in a variety of types of calculations. When used in calculations, the concept "translates" into a conversion factor having either of two forms

$$\frac{1 \text{ mole gas}}{22.4 \text{ L gas}} \quad \text{or} \quad \frac{22.4 \text{ L gas}}{1 \text{ mole gas}}$$

A most common type of problem involving these conversion factors is one where the volume of a gas at STP is known and you are asked to calculate from it the moles, grams, or particles of gas present, or vice versa. Figure 12.8 summarizes the relationships needed in performing calculations of this general type.

Perhaps you recognize Figure 12.8 as being very similar to a diagram you have already encountered many times in problem solving, Figure 9.3. This "new diagram" differs from Figure 9.3 in only one way; volume boxes have been added at the top of the diagram. The diagram in Figure 12.8 is used in the same way as the earlier one. The given and desired quantities are determined, and the arrows of the diagram are used to "map out" the pathway to be used in going from the given quantity to the desired quantity. Examples 12.12 through 12.14 illustrate the usefulness of Figure 12.8.

Example 12.12

Hydrogen sulfide, H_2S, is a colorless gas with a strong foul odor. This gas is responsible for the smell of rotten eggs. What would be the volume, in liters, of a 25.0 g sample of this gas at STP conditions?

Solution

STEP 1 The given quantity is 25.0 g of H_2S, and the unit of the desired quantity is liters of H_2S.

$$25.0 \text{ g } H_2S = ? \text{ L } H_2S$$

STEP 2 In terms of Figure 12.8, this problem is a "grams of A" to "volume of gas A" problem. The pathway is

Grams of A $\xrightarrow[\text{mass}]{\text{molar}}$ Moles of A $\xrightarrow[\text{volume}]{\text{molar}}$ Volume of gas A

Using dimensional analysis, we set up this sequence of conversion factors as

$$25.0 \cancel{\text{g } H_2S} \times \frac{1 \cancel{\text{mole } H_2S}}{34.1 \cancel{\text{g } H_2S}} \times \frac{22.4 \text{ L } H_2S}{1 \cancel{\text{mole } H_2S}}$$

$$\text{grams A} \longrightarrow \text{moles A} \longrightarrow \text{volume A}$$

Note how the molar volume relationship that 1 mole of gas is equal to 22.4 L is used as the last conversion factor.

STEP 3 The solution, obtained by doing the arithmetic, is

$$\frac{25.0 \times 1 \times 22.4}{34.1 \times 1} \text{ L } H_2S = 16.422287 \text{ L } H_2S \quad \text{(calculator answer)}$$

$$= 16.4 \text{ L } H_2S \quad \text{(correct answer)}$$

Practice Exercise 12.12

Fluorine gas, F_2, is one of the most reactive of all gases. What would be the volume, in liters of a 15.0-g sample of this gas at STP conditions?

Ans. 8.84 L

Equal volumes of different gases at STP do not have equal masses. This is because different kinds of molecules have different masses. Example 12.13 illustrates the validity of this concept.

Example 12.13

What is the mass, in grams, of 2.50 L of each of the following gases at STP conditions?

(a) carbon monoxide (CO) **(b)** carbon dioxide (CO_2)

Solution

(a)

STEP 1 The given quantity is 2.50 L of CO at STP, and the desired quantity is grams of CO.

$$2.50 \text{ L CO} = \text{g CO}$$

STEP 2 This is a "volume of gas A" to "grams of A" problem. The pathway appropriate for solving this problem as indicated in Figure 12.8, is

$$\boxed{\text{Volume of A}} \xrightarrow[\text{volume}]{\text{molar}} \boxed{\text{Moles of A}} \xrightarrow[\text{mass}]{\text{molar}} \boxed{\text{Grams of gas A}}$$

The setup, from dimensional analysis, is

$$2.50\ \cancel{\text{L CO}} \times \frac{1\ \cancel{\text{mole CO}}}{22.4\ \cancel{\text{L CO}}} \times \frac{28.0\text{ g CO}}{1\ \cancel{\text{mole CO}}}$$

STEP 3 Collecting the numerical terms after cancellation of units and doing the arithmetic gives

$$\frac{2.50 \times 1 \times 28.0}{22.4 \times 1}\text{ g CO} = 3.125\text{ g CO} \quad \text{(calculator answer)}$$

$$= 3.12\text{ g CO} \quad \text{(correct answer)}$$

(b) The analysis and setup for this part are identical to those in (a) except that CO_2 and its molecular mass replace CO and its molecular mass. Thus, the setup is

$$2.50\ \cancel{\text{L CO}_2} \times \frac{1\ \cancel{\text{mole CO}_2}}{22.4\ \cancel{\text{L CO}_2}} \times \frac{44.0\text{ g CO}_2}{1\ \cancel{\text{mole CO}_2}} = 4.9107142\text{ g CO}_2 \quad \text{(calculator answer)}$$

$$= 4.91\text{ g CO}_2 \quad \text{(correct answer)}$$

Equal volumes of different gases at STP do not have equal masses; 2.50 L of CO has a mass of 3.12 g, and 2.50 L of CO_2 has a mass of 4.91 g.

Practice Exercise 12.13

What is the mass, in grams, of 4.00 L of each of the following gases at STP conditions?

(a) sulfur dioxide (SO_2) (b) sulfur trioxide (SO_3)

Ans. (a) 11.4 g; (b) 14.3 g

The density of a gas at STP conditions can be calculated by dividing the molar mass of the substance by its molar volume.

$$\text{Density (STP)} = \frac{\text{molar mass}}{\text{molar volume (STP)}}$$

The units of density, g/L, are, indeed, obtained from this division.

$$\text{Density (STP)} = \frac{\dfrac{\text{grams}}{\text{mole}}}{\dfrac{\text{liters}}{\text{mole}}} = \frac{\text{grams} \times \cancel{\text{mole}}}{\cancel{\text{mole}} \times \text{liters}} = \frac{\text{grams}}{\text{liters}}$$

Rearrangement of this density equation enables one to calculate molar mass

$$\text{Molar mass} = \text{density (STP)} \times \text{molar volume (STP)}$$

Since molar volume is the same for all gases, the density of a gas at a given temperature and pressure is a function of (depends only on) the molar mass.

Example 12.14

Gaseous uranium hexafluoride, UF_6, is one of the heaviest gases known. Calculate the density of UF_6 gas at STP conditions, in grams per liter.

Solution

The molar mass and molar volume relationships needed to solve this problem are

$$\text{Molar mass} = \frac{352\text{ g } UF_6}{1\text{ mole } UF_6} \qquad \text{Molar volume} = \frac{22.4\text{ L } UF_6}{1\text{ mole } UF_6}$$

Substituting into the density equation gives

$$\text{Density} = \frac{\text{molar mass}}{\text{molar volume}} = \frac{\dfrac{352\text{ g } UF_6}{1\text{ mole } UF_6}}{\dfrac{22.4\text{ L } UF_6}{1\text{ mole } UF_6}}$$

$$= \frac{352\text{ g } UF_6 \times 1\ \cancel{\text{mole } UF_6}}{1\ \cancel{\text{mole } UF_6} \times 22.4\text{ L } UF_6}$$

$$= \frac{352 \times 1}{1 \times 22.4}\text{ g/L } UF_6$$

$$= 15.714285\text{ g/L } UF_6 \quad \text{(calculator answer)}$$
$$= 15.7\text{ g/L } UF_6 \quad \text{(correct answer)}$$

Practice Exercise 12.14

Gaseous hydrogen, H_2, is the lighest of all gases. Calculate the density of H_2 gas at STP conditions, in grams per liter.

Ans. 0.0902 g/L H_2

Example 12.15 differs from previous examples in that it involves both the A and B sides of Figure 12.8.

Example 12.15

If the Xe atoms present in 25.0 g of solid XeO_3 are converted to Xe gas, what volume, in liters, will the Xe gas occupy at STP conditions?

Solution

STEP 1 The given quantity is 25.0 g of XeO_3, and the desired quantity is liters of Xe at STP. In the jargon of Figure 12.8, this is a "grams of A" to "volume of gas B" problem.

STEP 2 The sequence of conversion factors needed for this problem, using Figure 12.8 as a guide, follows the pathway

Grams of A —(molar mass)→ Moles of A —(formula subscript)→ Moles of B —(molar volume)→ Volume of gas B

Translating this pathway into a dimensional analysis setup gives

$$25.0\ \cancel{\text{g XeO}_3} \times \frac{1\ \cancel{\text{mole XeO}_3}}{179\ \cancel{\text{g XeO}_3}} \times \frac{1\ \cancel{\text{mole Xe}}}{1\ \cancel{\text{mole XeO}_3}} \times \frac{22.4\ \text{L Xe}}{1\ \cancel{\text{mole Xe}}}$$

STEP 3 Collecting numerical terms after cancellation of units and doing the arithmetic gives

$$\frac{25.0 \times 1 \times 1 \times 22.4}{179 \times 1 \times 1}\ \text{L Xe} = 3.1284916\ \text{L Xe} \quad \text{(calculator answer)}$$

$$= 3.13\ \text{L Xe} \quad \text{(correct answer)}$$

Practice Exercise 12.15

If the Xe atoms present in 14.5 g of gaseous XeF_2 are converted to Xe gas, what volume, in liters, will the Xe gas occupy at STP conditions?

Ans. 1.92 L Xe

12.11 THE IDEAL GAS LAW

The **ideal gas law** *describes the relationships among the four variables temperature* (T), *pressure* (P), *volume* (V), *and moles of gas* (n) *for gaseous substances.*

In previous sections we have discussed three independent relationships dealing with the volume of a gas. They are

Boyle's law:	$V = k \times \frac{1}{P}$	(n and T constant)
Charles's law:	$V = kT$	(n and P constant)
Avogadro's law:	$V = kn$	(P and T constant)

We can combine these three equations into a single expression

$$V = \frac{kTn}{P}$$

since we know (from mathematics) that if a quantity is independently porportional to two or more quantities, it is also proportional to their product. This combined equation is a mathematical statement of the ideal gas law. Any gas that obeys the individual laws of Boyle, Charles, and Avogadro will also obey the ideal gas law.

$$PV = nRT$$

This form of the idea gas law, besides being rearranged, differs from the previous form in that the proportionality constant k has been given the symbol R. The constant R is called the *ideal gas constant*. In order to use the ideal gas equation we must know the value of R. This can be determined by substituting STP values for P, V, T, and n into the idea gas equation and solving for R. For 1 mole of gas at STP, $P = 1$ atm, $V = 22.4$ L, $T = 273$ K, and $n = 1$ mole. Substituting these values into the ideal gas equation arranged to have R isolated on the left side gives

$$R = \frac{PV}{nT} = \frac{(1\text{ atm})\,(22.4\text{ L})}{(1\text{ mole})\,(273\text{ K})} = 0.0821\,\frac{\text{atm}\cdot\text{L}}{\text{mole}\cdot\text{K}}$$

Notice the complex units associated with R—the four variables temperature, pressure, volume, and moles are all involved. This will always be the case.

The value of R is dependent on the units used to express pressure and volume. If we use 760 mm Hg instead of 1 atm as standard pressure, then R would have the value 62.4 and the units (mm Hg · L)/(mole · K).

$$R = \frac{PV}{nT} = \frac{(760\text{ mm Hg})\,(22.4\text{ L})}{(1\text{ mole})\,(273\text{ K})} = 62.4\,\frac{\text{mm Hg}\cdot\text{L}}{\text{mole}\cdot\text{K}}$$

These two values of R should be memorized. When pressure units other than millimeters of mercury or atmosphere and volume units other than liters are encountered, convert them to these units and then use one of these two known R values.

If three of the four variables in the ideal gas equation are known, then the fourth can be calculated by the equation. The ideal gas equation is used in calculations when *one* set of conditions is given with one missing variable. The combined gas law (Sec. 12.6) is used when *two* sets of conditions are given with one missing variable. Examples 12.16 and 12.17 illustrate the use of the ideal gas equation.

Example 12.16

Ozone, a form of oxygen with the formula O_3, is naturally found in the upper atmosphere where it screens out 95–99% of the ultraviolet solar radiation headed for earth. Calculate the volume, in liters, occupied by 1.52 moles of ozone at 0.992 atm pressure and a temperature of 65°C.

Solution

This problem deals with only one set of conditions, a situation where the ideal gas equation is applicable. Three of the four variables in the ideal gas equation (P, n, and T) are given, and the fourth (V) is to be calculated.

$$P = 0.992\text{ atm} \qquad n = 1.52\text{ moles}$$
$$V = ?\text{ L} \qquad T = 65°\text{C} = 338\text{ K}$$

Rearranging the ideal gas equation to isolate V on the left side of the equation gives

$$V = \frac{nRT}{P}$$

Since the pressure is given in atmospheres and the volume unit is liters, the appropriate R value is

$$R = 0.0821\,\frac{\text{atm}\cdot\text{L}}{\text{mole}\cdot\text{K}}$$

Substituting the given numerical values into the equation and canceling units gives

$$V = \frac{(1.52\ \cancel{\text{moles}})\left(0.0821\ \dfrac{\cancel{\text{atm}} \cdot \text{L}}{\cancel{\text{mole}} \cdot \cancel{\text{K}}}\right)(338\ \cancel{\text{K}})}{0.992\ \cancel{\text{atm}}}$$

Note how all of the parts of the ideal gas constant unit cancel except for one, the volume part.

Doing the arithmetic, we get as an answer 42.5 L O_3

$$V = \frac{1.52 \times 0.0821 \times 338}{0.992}\ \text{L}\ O_3$$

$$= 42.519854\ \text{L}\ O_3 \quad \text{(calculator answer)}$$
$$= 42.5\ \text{L}\ O_3 \quad \text{(correct answer)}$$

Practice Exercise 12.16

Carbon monoxide, a colorless, odorless, tasteless gas with the formula CO, is formed in small amounts any time a material that contains the element carbon is burned. Calculate the volume, in liters, occupied by 4.45 moles of CO gas at 1.54 atm pressure and a temperature of 213°C.

Ans. 111 L

Example 12.17

Decomposition processes in wet locations where oxygen is not available, such as swamps and natural wetlands, produce the gas methane, CH_4. Calculate the temperature, in degrees Celsius, of a 1.53 mole sample of CH_4 gas under a pressure of 5.00 atm in a 7.00-L container.

Solution

Three of the four variables in the ideal gas equation (P, V, and n) are given, and the fourth (T) is to be calculated.

$$P = 5.00\ \text{atm} \qquad n = 1.53\ \text{moles}$$
$$V = 7.00\ \text{L} \qquad T = ?\ \text{K}$$

Rearranging the ideal gas equation to isolate T on the left side gives

$$T = \frac{PV}{nR}$$

Since the pressure is given in atmospheres and the volume in liters, the value of R to be used is

$$R = 0.0821\ \frac{\text{atm} \cdot \text{L}}{\text{mole} \cdot \text{K}}$$

Substituting numerical values into the equation gives

$$T = \frac{(5.00\ \cancel{\text{atm}})(7.00\ \cancel{\text{L}})}{(1.53\ \cancel{\text{moles}})\left(0.0821\ \dfrac{\cancel{\text{atm}} \cdot \cancel{\text{L}}}{\cancel{\text{mole}} \cdot \text{K}}\right)}$$

Notice again how the gas constant units, except for K, cancel. After cancellation, the expression 1/(1/K) remains. This expression is equivalent to K. That this is the case can be easily shown. All we need to do is multiply both the numerator and denominator of the fraction by K.

$$\frac{1 \times K}{\frac{1}{K} \times K} = K$$

Doing the arithmetic, we get as an answer 279 K for the temperature of the CH_4 gas.

$$T = \frac{(5.00)(7.00)}{(1.53)(0.0821)}\ K = 278.63358\ K \quad \text{(calculator answer)}$$

$$= 279\ K \quad \text{(correct answer)}$$

The calculated temperature is in kelvins. To convert to degrees Celsius, the unit specified in the problem statement, we subtract 273 from the Kelvin temperature.

$$T(°C) = 279\ K - 273 = 6°C$$

Practice Exercise 12.17

Dinitrogen monoxide (nitrous oxide), N_2O, is naturally present in trace amounts in the atmosphere. Its source is decomposition reactions occurring in soils. Calculate the temperature, in degrees Celsius, of a 1.76-mole sample of N_2O gas under a pressure of 3.15 atm in a 7.37-L container.

Ans. −112°C

A word about the phrase "*ideal gas*" in the name ideal gas law is in order. An **ideal gas** *is a gas that obeys exactly all of the statements of kinetic molecular theory (Sec. 11.3) and obeys exactly the ideal gas law*. Real gases are not ideal gases; that is, real gases do not obey exactly the ideal gas equation. Nonetheless, for real gases under ordinary conditions of temperature and pressure, deviations from ideal gas behavior are small, and the idea gas law (as well as the other laws discussed in this chapter) gives accurate information about gas behavior. It is only at low temperatures and/or high pressures (near liquefaction conditions) that ideal gas behavior breaks down. At low temperatures, because of the slower motion of the molecules, attractive forces between molecules begin to become important. At high pressures, where the molecules are forced closer together, attractive forces again begin to cause deviation from behavior predicted by the ideal gas law.

12.12 EQUATIONS DERIVED FROM THE IDEAL GAS LAW

Some of the most useful calculations involving the ideal gas equation are those in which the mass, molar mass, or density of a gas is determined. Such calculations are performed by using modified forms of the ideal gas equation.

The number of moles of any substance is equal to the mass (m) of the substance in grams divided by the substance's molar mass (MM).

$$n = \frac{m}{(MM)}$$

Replacing n in the ideal gas equation with this equivalent expression gives

$$PV = \frac{m}{(MM)} RT$$

This equation, in the rearranged form

$$m = \frac{PV(MM)}{RT}$$

is used to calculate the mass, in grams, of a gas. This same equation, in the rearranged form

$$(MM) = \frac{mRT}{PV}$$

is used to calculate the molar mass of a gas.

The density, d, of a gas has the units of mass (grams) per unit volume (liters).

$$d = \frac{m}{V}$$

Solving the modified ideal gas equation for m/V gives

$$\frac{m}{V} = \frac{P(MM)}{RT} \quad \text{or} \quad d = \frac{P\,(MM)}{RT}$$

Examples 12.18 and 12.19 illustrate the use of these new ideal-gas-equation relationships.

Example 12.18

A 4.25-g sample of the refrigerant Freon-11 is vaporized and found to occupy a volume of 1.20 L at a temperature of 50°C and a pressure of 521 mm Hg. What is the molar mass of Freon-11?

Solution

The molar mass of a gas is calculated by the ideal gas equation in the modified form

$$(MM) = \frac{mRT}{PV}$$

All of the quantities on the right side of the equation are known.

$$m = 4.25 \text{ g} \qquad R = 62.4\,\frac{\text{mm Hg} \cdot \text{L}}{\text{mole} \cdot \text{K}} \qquad T = 50°\text{C} = 323 \text{ K}$$

$$P = 521 \text{ mm Hg} \qquad V = 1.20 \text{ L}$$

Substitution of these values into the equation gives

$$(MM) = \frac{(4.25 \text{ g})\left(62.4\,\frac{\cancel{\text{mm Hg}} \cdot \cancel{\text{L}}}{\text{mole} \cdot \cancel{\text{K}}}\right)(323\ \cancel{\text{K}})}{(521\ \cancel{\text{mm Hg}})(1.20\ \cancel{\text{L}})}$$

All units cancel except for g/mole, the units of molar mass. Recall that the molar mass of a substance is the mass in grams of 1 mole of the substance.

Doing the arithmetic, we obtain a value of 137 g/mole for the molar mass of Freon-11.

$$(MM) = \frac{4.25 \times 62.4 \times 323}{521 \times 1.20} \text{ g/mole} = 137.01151 \text{ g/mole} \quad \text{(calculator answer)}$$

$$= 137 \text{ g/mole} \quad \text{(correct answer)}$$

Freon-11 is the compound $CFCl_3$. The molar mass of $CFCl_3$, using a table of atomic masses is 137 g/mole. Thus, the experimental molar mass value and the molar mass obtained from a table of atomic masses are in agreement.

Practice Exercise 12.18

A 4.25-g sample of the refrigerant Freon-12 is vaporized and found to occupy a volume of 1.36 L at a temperature of 50°C and a pressure of 521 mm Hg. What is the molar mass of Freon-12?

Ans. 121 g/mole (CF_2Cl_2)

Example 12.19

In gas-law calculations, air is often considered to be a single gas with a molar mass of 29 g/mole. On this basis, calculate the density of air, in grams/liter, on a hot summer day (41°C) when the atmospheric pressure is 0.91 atm.

Solution

The ideal gas equation in the modified form

$$d = \frac{m}{V} = \frac{P(MM)}{RT}$$

is used to calculate the density of a gas.

All of the quantities on the right side of this equation are known.

$$P = 0.91 \text{ atm} \qquad (MM) = 29 \text{ g/mole}$$

$$R = 0.0821 \frac{\text{atm} \cdot \text{L}}{\text{mole} \cdot \text{K}} \qquad T = 41°\text{C} = 214 \text{ K}$$

Substitution of these values into the equation gives

$$d = \frac{(0.91 \cancel{\text{atm}})\left(29 \frac{\text{g}}{\cancel{\text{mole}}}\right)}{\left(0.0821 \frac{\cancel{\text{atm}} \cdot \text{L}}{\cancel{\text{mole}} \cdot \cancel{\text{K}}}\right)(214 \cancel{\text{K}})}$$

All units cancel except for the desired ones, grams per liter.

Doing the arithmetic, we obtain a value of 1.5 g/L for the density of gas at the specified temperature and pressure.

$$d = \frac{0.91 \times 29}{0.0821 \times 214} \text{ g/L} = 1.5020433 \text{ g/L} \quad \text{(calculator answer)}$$

$$= 1.5 \text{ g/L} \quad \text{(correct answer)}$$

Practice Exercise 12.19

Calculate the density of carbon dioxide gas, CO_2, in grams per liter, at 1.21 atm pressure and a temperature of 35°C.

Ans. 2.11 g/L

12.13 GAS LAWS AND CHEMICAL EQUATIONS

In Section 10.8 we learned to calculate the mass of any component (reactant or product) of a chemical reaction, given the balanced equation for the reaction and mass of any other component—a mass-to-mass (or gram-to-gram) problem. A simple extension of the procedures involved in solving such problems will enable us to do mass-to-volume and volume-to-mass calculations for reactions where at least one gas is involved.

If you are given the number of liters of any gaseous component of a reaction and are asked to calculate the mass of any other component (gaseous or otherwise)—a *volume-to-mass problem*—first convert the given volume of gas to moles of gas, using the principles in this chapter. If the reaction is at STP conditions, use molar volume as the conversion factor to go from volume to moles. At other temperatures and pressures, use the ideal gas law (or the combined gas law and molar volume) to make the conversion. Once you obtain moles of gas the rest of the calculation is a standard mole-to-gram conversion (Sec. 10.8).

F.Y.I.

Compounds such as TNT are explosive because they produce large amounts of hot, gaseous products that instantaneously expand.

If you are given the number of grams of any component (gaseous or otherwise) in a chemical reaction and are asked to calculate the volume of any gaseous component (reactant or product)—*a mass-to-volume problem*—use standard procedures to calculate the moles of gas and then the new procedures of this chapter to go from moles to volume.

Figure 12.9 summarizes the relationships between the conversion factors needed to solve mass-to-volume and volume-to-mass types of chemical-equation-based problems.

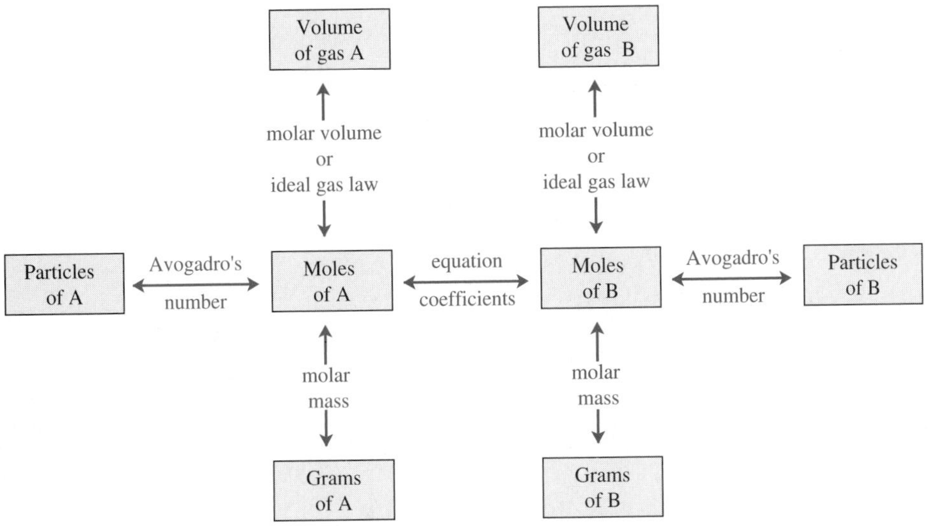

FIGURE 12.9 Quantitative relationships needed for solving mass-to-volume and volume-to-mass chemical-equation-based problems.

Example 12.20

Aluminum and oxygen react to produce aluminum oxide as shown by the equation

$$4\ Al(s) + 3\ O_2\ (g) \longrightarrow 2\ Al_2O_3(s)$$

What volume, in liters, of O_2 at STP conditions will completely react with 75.0 g of Al?

Solution

STEP 1 The given quantity is 75.0 of Al, and the desired quantity is liters of O_2 at STP conditions.

$$75.0\ g\ Al = ?\ L\ O_2$$

STEP 2 This is a "grams of A" to "volume of gas B" type of problem. The pathway used in solving it, in terms of Figure 12.9, is

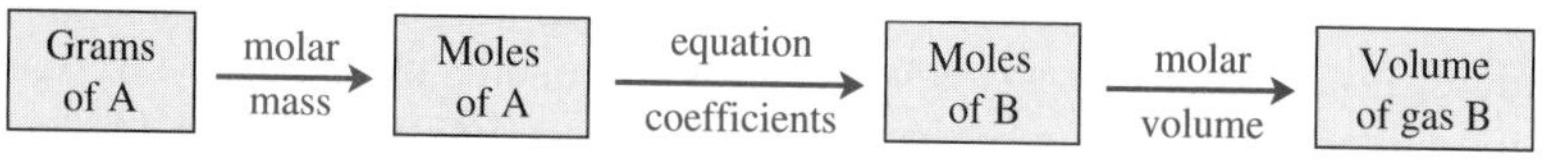

The dimensional analysis setup for the calculation is

$$75.0\ \cancel{g\ Al} \times \frac{1\ \cancel{mole\ Al}}{27.0\ \cancel{g\ Al}} \times \frac{3\ \cancel{moles\ O_2}}{4\ \cancel{moles\ Al}} \times \frac{22.4\ L\ O_2}{1\ \cancel{mole\ O_2}}$$

STEP 3 The solution, obtained from combining all the numerical factors, is

$$\frac{75.0 \times 1 \times 3 \times 22.4}{27.0 \times 4 \times 1}\ L\ O_2 = 46.666666\ L\ O_2 \quad \text{(calculator answer)}$$

$$= 46.7\ L\ O_2 \quad \text{(correct answer)}$$

Practice Exercise 12.20

Magnesium and nitrogen react to produce magnesium nitride as shown by the equation

$$3\ Mg(s) + N_2(g) \longrightarrow Mg_3N_2(s)$$

What volume, in liters, of N_2 at STP conditions will completely react with 63.0 g of Mg?

Ans. 19.4 L N_2

Example 12.21

Oxygen gas can be generated by heating $KClO_3$ to a high temperature.

$$2\ KClO_3(s) \longrightarrow 2\ KCl(s) + 3\ O_2(g)$$

How much $KClO_3$, in grams, is needed to generate 7.50 L of O_2 at a pressure of 1.00 atm and a temperature of 37°C?

Solution

STEP 1 The given quantity is 7.50 L of O_2, and the desired quantity is grams of $KClO_3$

$$7.50\ L\ O_2 = ?\ g\ KClO_3$$

STEP 2 This is a "volume of gas A" to "grams of B" problem. The pathway, in terms of Figure 12.9, is

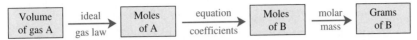

Volume of gas A —(ideal gas law)→ Moles of A —(equation coefficients)→ Moles of B —(molar mass)→ Grams of B

The first conversion step, from volume of O_2 to moles of O_2, can be made by applying the ideal gas law.

$$n = \frac{PV}{RT} = \frac{(1.00\ \cancel{\text{atm}})\ (7.50\ \cancel{\text{L}})}{\left(0.0821\ \frac{\cancel{\text{atm}}\cdot\cancel{\text{L}}}{\text{mole}\cdot\cancel{\text{K}}}\right)(310\ \cancel{\text{K}})}$$

$$= 0.2946839\ \text{mole } O_2 \quad \text{(calculator answer)}$$

$$= 0.295\ \text{mole } O_2 \quad \text{(correct answer)}$$

The dimensional analysis setup for the remaining conversion steps is

$$0.295\ \cancel{\text{mole } O_2} \times \frac{2\ \cancel{\text{moles } KClO_3}}{3\ \cancel{\text{moles } O_2}} \times \frac{123\ \text{g } KClO_3}{1\ \cancel{\text{mole } KClO_3}}$$

STEP 3 The solution, obtained from combining all the numerical factors, is

$$\frac{0.295 \times 2 \times 123}{3 \times 1}\ \text{g } KClO_3 = 24.19\ \text{g } KClO_3 \quad \text{(calculator answer)}$$

$$= 24.2\ \text{g } KClO_3 \quad \text{(correct answer)}$$

An alternative method can be used for converting the liters of O_2 to moles of O_2. The combined gas law can be used to calculate the STP volume of oxygen, and then the molar volume relationship can be used to calculate the moles of oxygen.

The combined gas law, rearranged to isolate V_2 on the left side, is

$$V_2 = V_1 \times \frac{P_1}{P_2} \times \frac{T_2}{T_1}$$

The given values for the variables are

$$P_1 = 1.00\ \text{atm} \qquad P_2 = 1.00\ \text{atm (STP)}$$
$$V_1 = 7.50\ \text{L} \qquad V_2 = ?\ \text{L}$$
$$T_1 = 37°\text{C} = 310\ \text{K} \qquad T_2 = 0°\text{C} = 273\ \text{K (STP)}$$

Substituting these values into the combined gas law equation we get

$$V_2 = 7.50\ \text{L} \times \frac{1.00\ \cancel{\text{atm}}}{1.00\ \cancel{\text{atm}}} \times \frac{273\ \cancel{\text{K}}}{310\ \cancel{\text{K}}}$$

$$= 6.6048387\ \text{L} \quad \text{(calculator answer)}$$

$$= 6.60\ \text{L} \quad \text{(correct answer)}$$

Now, changing the STP volume to moles by the molar volume relationship, we get

$$6.60\ \cancel{\text{L } O_2} \times \frac{1\ \text{mole } O_2}{22.4\ \cancel{\text{L } O_2}} = 0.29464285\ \text{mole } O_2 \quad \text{(calculator answer)}$$

$$= 0.295\ \text{mole } O_2 \quad \text{(correct answer)}$$

This is the same value for moles of O_2 we previously obtained with the ideal gas law.

Practice Exercise 12.21

Lead(IV) chloride ($PbCl_4$) can be produced from its constituent elements as shown by the equation

$$Pb(s) + 2\ Cl_2(g) \longrightarrow PbCl_4(s)$$

How many grams of $PbCl_4$ can be produced from the reaction of 5.00 L of Cl_2 at 1.33 atm pressure and a temperature of 7°C with an excess of Pb metal?

Ans. 50.4 g $PbCl_4$

12.14 DALTON'S LAW OF PARTIAL PRESSURES

In a mixture of gases that do not react with each other, each type of molecule moves about in the container as if the other kinds were not there. This type of behavior is possible because attractions between molecules in the gaseous state are negligible at most temperatures and pressures and because a gas is mostly empty space (Sec. 11.6). Each gas in the mixture occupies the entire volume of the container, that is, it distributes itself uniformly throughout the container. The molecules of each type strike the walls of the container as frequently and with the same energy as if they were the only gas in the mixture. Consequently, the pressure exerted by a gas in a mixture is the same as it would be if the gas were alone in the same container under the same conditions.

John Dalton—the same John Dalton discussed in Section 5.1—was the first to notice this independent behavior of gases in mixtures. In 1803 he published a summary statement concerning such behavior, which is now known as **Dalton's law of partial pressures:** *The total pressure exerted by a mixture of gases is the sum of the partial pressures of the individual gases.* A new term, *partial pressure*, is used in stating Dalton's law. A **partial pressure** *is the pressure that a gas in mixture would exert if it were the only gas present under the same conditions.*

Expressed mathematically, Dalton's law states that

$$P_T = P_1 + P_2 + P_3 + \cdots$$

where P_T is the total pressure of a gaseous mixture and P_1, P_2, P_3, and so on are the partial pressures of the individual gaseous components of the mixture. (When the identity of a gas is known, its molecular formula is used as a subscript in the partial-pressure notation; for example, P_{CO_2} is the partial pressure of carbon dioxide in a mixture.)

To illustrate Dalton's law, consider the four identical gas containers shown in Figure 12.10. Suppose we place amounts of three different gases (represented by A, B, and C) into three of the containers and measure the pressure exerted by each sample. We then place all three samples in the fourth container and measure the pressure exerted by this mixture of gases. It is found that

$$P_{\text{total}} = P_A + P_B + P_C$$

Using the pressures given in Figure 12.10, we see that

$$P_{\text{total}} = 1 + 3 + 2 = 6$$

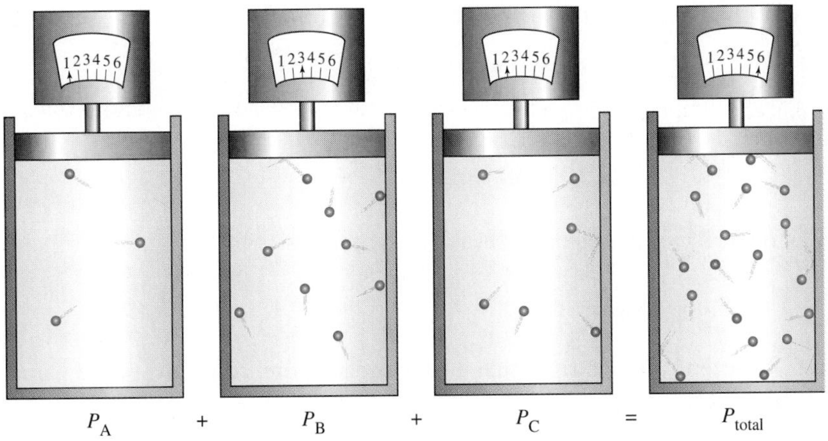

FIGURE 12.10 Dalton's law of partial pressures.

Example 12.22

An unknown quantity of the noble gas xenon (Xe) is added to a cylinder already containing a mixture of the noble gases helium (He) and argon (Ar) at partial pressures, respectively, of 3.00 atm and 1.00 atm. After the Xe addition, the total pressure in the cylinder is 5.80 atm. What is the partial pressure, in atmospheres, of the Xe gas?

Solution

The partial pressures of the He and Ar in the mixture will not be affected by the addition of the Xe (Dalton's law). Thus, the partial pressures of the He and Ar remain at 3.00 atm and 1.00 atm, respectively.

The sum of the partial pressures of He and Ar is 4.00 atm.

$$3.00 \text{ atm} + 1.00 \text{ atm} = 4.00 \text{ atm}$$

The difference between this pressure and the total pressure in the cylinder is caused by the Xe present. Thus, the partial pressure of the Xe is

$$\underset{P_{total}}{5.80 \text{ atm}} - \underset{(P_{He} + P_{Ar})}{4.00 \text{ atm}} = \underset{P_{Xe}}{1.80 \text{ atm}}$$

Practice Exercise 12.22

A gaseous mixture contains the three noble gases helium, argon, and krypton. The total pressure exerted by the mixture is 1.57 atm, and the partial pressures of the helium and argon are 0.33 and 0.39 atm, respectively. What is the partial pressure of the krypton present in the mixture?

Ans. 0.85 atm

The validity of Dalton's law of partial pressures is easily demonstrated by using the ideal gas law (Sec. 12.11). For a mixture of three gases (A, B, and C) the total pressure is given by the expression

$$P_{total} = n_{total}\frac{RT}{V}$$

The total number of moles present is the sum of the moles of A, B, and C; that is,

$$n_{\text{total}} = n_A + n_B + n_C$$

Substituting this equation into the previous one gives

$$P_{\text{total}} = (n_A + n_B + n_C)\frac{RT}{V}$$

Expanding the right side of this equation results in the expression

$$P_{\text{total}} = n_A\frac{RT}{V} + n_B\frac{RT}{V} + n_C\frac{RT}{V}$$

The three individual terms on the right side of this equation are, respectively, the partial pressure of A, B, and C.

$$P_A = n_A\frac{RT}{V} \qquad P_B = n_B\frac{RT}{V} \qquad P_C = n_C\frac{RT}{V}$$

Substitution of this information into the previous equation gives

$$P_{\text{total}} = P_A + P_B + P_C$$

which is a statement of Dalton's law of partial pressures for a mixture of three gases.

In the preceding derivation two of the expressions that are encountered are

$$P_A = n_A\frac{RT}{V} \quad \text{and} \quad P_{\text{total}} = n_{\text{total}}\frac{RT}{V}$$

If we divide the first of these expressions by the second, we obtain

$$\frac{P_A}{P_{\text{total}}} = \frac{n_A\frac{\cancel{RT}}{\cancel{V}}}{n_{\text{total}}\frac{\cancel{RT}}{\cancel{V}}} = \frac{n_A}{n_{\text{total}}} \quad \text{or} \quad \frac{P_A}{P_{\text{total}}} = \frac{n_A}{n_{\text{total}}}$$

Rearrangement of this equation gives

$$P_A = P_{\text{total}} \times \frac{n_A}{n_{\text{total}}}$$

The fraction n_A/n_{total} in this equation is called the *mole fraction* of A in the mixture. It is the fraction of the total moles that is accounted for by gas A. A **mole fraction** *is a dimensionless quantity that gives the ratio of the number of moles of a component in a mixture to the number of moles of all components present.* The symbol X is used to denote a mole fraction.

$$X_A = \frac{n_A}{n_{\text{total}}}$$

From mole fractions and total pressure, we can calculate the partial pressures of the individual components of a gaseous mixture. The partial pressure of a gas in a mixture is equal to its mole fraction multiplied by the total pressure.

$$P_A = X_A \times P_{\text{total}}$$

Example 12.23 shows how mole fractions are calculated and Example 12.24 shows how mole fractions are used in obtaining partial pressures.

Example 12.23

A gaseous mixture contains 10.0 g each of the gases N_2, O_2, and Ar. What is the mole fraction of each gas in the mixture?

Solution

We first calculate the number of moles of each gas present.

$$N_2: \quad 10.0\ \cancel{g\ N_2} \times \frac{1\ \text{mole}\ N_2}{28.0\ \cancel{g\ N_2}} = 0.35714285\ \text{mole}\ N_2 \quad \text{(calculator answer)}$$

$$= 0.357\ \text{mole}\ N_2 \quad \text{(correct answer)}$$

$$O_2: \quad 10.0\ \cancel{g\ O_2} \times \frac{1\ \text{mole}\ O_2}{32.0\ \cancel{g\ O_2}} = 0.3125\ \text{mole}\ O_2 \quad \text{(calculator answer)}$$

$$= 0.312\ \text{mole}\ O_2 \quad \text{(correct answer)}$$

$$\text{Ar}: \quad 10.0\ \cancel{g\ Ar} \times \frac{1\ \text{mole Ar}}{39.9\ \cancel{g\ Ar}} = 0.25062656\ \text{mole Ar} \quad \text{(calculator answer)}$$

$$= 0.251\ \text{mole Ar} \quad \text{(correct answer)}$$

The total number of moles of gas present is

$$\begin{aligned} n_{\text{total}} &= n_{N_2} + n_{O_2} + n_{\text{Ar}} \\ &= (0.357 + 0.312 + 0.251)\ \text{mole} \\ &= 0.92\ \text{mole} \quad \text{(calculator answer)} \\ &= 0.920\ \text{mole} \quad \text{(correct answer)} \end{aligned}$$

Mole fractions are calculated as ratios of individual component moles to total moles.

$$X_{N_2} = \frac{0.357\ \cancel{\text{mole}}}{0.920\ \cancel{\text{mole}}} = 0.38804347 \quad \text{(calculator answer)}$$

$$= 0.388 \quad \text{(correct answer)}$$

$$X_{O_2} = \frac{0.312\ \cancel{\text{mole}}}{0.920\ \cancel{\text{mole}}} = 0.33913043 \quad \text{(calculator answer)}$$

$$= 0.339 \quad \text{(correct answer)}$$

$$X_{\text{Ar}} = \frac{0.251\ \cancel{\text{mole}}}{0.920\ \cancel{\text{mole}}} = 0.27282608 \quad \text{(calculator answer)}$$

$$= 0.273 \quad \text{(correct answer)}$$

The sum of all mole fractions should always add to one.

$$\begin{aligned} 0.388 + 0.339 + 0.273 &= 1 \quad \text{(calculator answer)} \\ &= 1.000 \quad \text{(correct answer)} \end{aligned}$$

Practice Exercise 12.23

A gaseous mixture contains 5.00 g each of the gases Ar, Kr, and Xe. What is the mole fraction of each gas in the mixture?

Ans. $X_{\text{Ar}} = 0.561$; $X_{\text{Kr}} = 0.268$; $X_{\text{Xe}} = 0.171$

Example 12.24

A mixture of gases contains 4.23 moles of neon (Ne), 0.93 mole of argon (Ar), and 7.65 moles of hydrogen (H_2). Calculate the partial pressures of the gases if the total pressure is 5.00 atm at a certain temperature.

Solution

We first calculate the mole fraction of each gas.

$$X_{Ne} = \frac{4.23 \text{ moles}}{(4.23 + 0.93 + 7.65) \text{ moles}} = 0.33021077 \quad \text{(calculator answer)}$$
$$= 0.330 \quad \text{(correct answer)}$$

$$X_{Ar} = \frac{0.93 \text{ mole}}{(4.23 + 0.93 + 7.65) \text{ moles}} = 0.072599531 \quad \text{(calculator answer)}$$
$$= 0.073 \quad \text{(correct answer)}$$

$$X_{H_2} = \frac{7.65 \text{ moles}}{(4.23 + 0.93 + 7.65) \text{ moles}} = 0.59718969 \quad \text{(calculator answer)}$$
$$= 0.597 \quad \text{(correct answer)}$$

To calculate partial pressures, we rearrange the equation

$$\frac{P_A}{P_{total}} = X_A$$

to isolate the partial pressure on a side by itself.

$$P_A = X_A \times P_{total}$$

Substituting known quantities into this equation gives the partial pressures.

$$P_{Ne} = 0.330 \times 5.00 \text{ atm} = 1.65 \text{ atm} \quad \text{(calculator and correct answer)}$$
$$P_{Ar} = 0.073 \times 5.00 \text{ atm} = 0.365 \text{ atm} \quad \text{(calculator answer)}$$
$$= 0.36 \text{ atm} \quad \text{(correct answer)}$$
$$P_{H_2} = 0.597 \times 5.00 \text{ atm} = 2.985 \text{ atm} \quad \text{(calculator answer)}$$
$$= 2.98 \text{ atm} \quad \text{(correct answer)}$$

Practice Exercise 12.24

A sample of natural gas contains 6.20 moles of methane (CH_4), 0.317 mole of ethane (C_2H_6), and 0.0872 mole of propane (C_3H_8). If the total pressure of the gases is 2.43 atm, what are the partial pressures, in atmospheres, of the gases?

Ans. $P_{CH_4} = 2.28$ atm; $P_{C_2H_6} = 0.117$ atm; $P_{C_3H_8} = 0.0321$ atm

The air we breathe is a most important mixture of gases. The composition of clean air from which all water vapor has been removed (dry air) is found to be virtually constant over the entire Earth. Table 12.4 gives the composition of clean dry air in terms of mole fractions. All components that have a mole fraction of at least 1×10^{-5} (0.001%) are listed.

TABLE 12.4 THE MAJOR COMPONENTS OF CLEAN, DRY AIR

Gaseous Component	Formula	Mole Fraction	Partial Pressure (mm Hg) when total pressure is 760.0 mm Hg
Nitrogen	N_2	0.78084	593.4
Oxygen	O_2	0.20948	159.2
Argon	Ar	9.34×10^{-3}	7.1
Carbon dioxide	CO_2	3.1×10^{-4}	0.2
Neon	Ne	2×10^{-5}	—
Helium	He	1×10^{-5}	—

Atmospheric pressure is the sum of the partial pressures of the gaseous components present in air. Table 12.4 also gives the partial pressure of each component of air in a situation where total atmospheric pressure is 760 mm Hg.

The composition of air is not absolutely constant. The variability in composition is caused predominantly by the presence of water vapor, a substance not listed in Table 12.4 because those statistics are for *dry* air. The amount of water vapor in air varies between almost zero and a mole fraction of 0.05-0.06, depending on weather and temperature.

A common application of Dalton's law of partial pressures is encountered in the laboratory preparation of gases. Such gases are often collected by displacement of water. Figure 12.11 shows O_2, prepared from the decomposition of $KClO_3$, being collected by water displacement. A gas collected by water displacement is never pure. It always contains some water vapor. The total pressure exerted by the gaseous mixture is the sum of the partial pressures of the gas being collected and the water vapor.

$$P_{total} = P_{gas} + P_{H_2O}$$

The pressure exerted by the water vapor in the mixture will be constant at any given temperature if sufficient time has been allowed to establish equilibrium conditions.

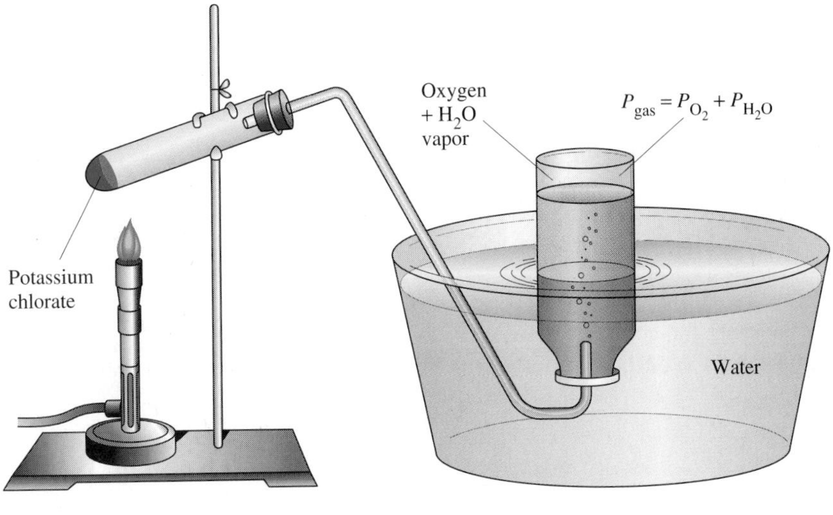

FIGURE 12.11 Collection of oxygen gas by water displacement. Potassium chlorate ($KClO_3$) decomposes to form oxygen (O_2), which is collected over water.

TABLE 12.5 VAPOR PRESSURE OF WATER AT VARIOUS TEMPERATURES

T (°C)	Vapor Pressure (mm Hg)	T (°C)	Vapor Pressure (mm Hg)	T (°C)	Vapor Pressure (mm Hg)
15	12.8	22	19.8	29	30.0
16	13.6	23	21.1	30	31.8
17	14.5	24	22.4	31	33.7
18	15.5	25	23.8	32	35.7
19	16.5	26	25.2	33	37.7
20	17.5	27	26.7	34	39.9
21	18.7	28	28.3	35	42.2

For gases collected by water displacement, the partial pressure of the water vapor can be obtained from a table showing the variation of water vapor pressure with temperature (see Table 12.5). Thus the partial pressure of the collected gas is easily determined.

$$P_{\text{gas}} = P_{\text{atm}} - P_{H_2O}$$

Example 12.25

What is the partial pressure of oxygen collected over water at 17°C on a day when the barometric pressure is 743 mm Hg?

Solution

From Table 12.5 we determine that water has a vapor pressure of 14.5 mm Hg at 17°C.

Using the equation

$$P_{O_2} = P_{\text{atm}} - P_{H_2O}$$

we find the partial pressure of the oxygen to be 728 mm Hg.

$$P_{O_2} = (743 - 14.5)\text{ mm Hg} = 728.5\text{ mm Hg} \quad \text{(calculator answer)}$$
$$= 728\text{ mm Hg} \quad \text{(correct answer)}$$

Practice Exercise 12.25

What is the partial pressure of nitrogen collected over water at 33°C on a day when the barometric pressure is 657 mm Hg?

Ans. 619 mm Hg

When considering the components of a gaseous mixture, a useful set of equalities for any given component (A) of the gaseous mixture, at a given temperature and pressure, is

$$\text{mole percent A} = \text{pressure percent A} = \text{volume percent A}$$

where

$$\text{mole \% A} = \frac{n_A}{n_{\text{total}}} \times 100$$

$$\text{pressure \% A} = \frac{P_A}{P_{total}} \times 100$$

$$\text{volume \% A} = \frac{V_A}{V_{total}} \times 100$$

Example 12.26 is a calculation showing the equality of these three quantities.

Example 12.26

A 17.92-L flask, at STP, contains 0.200 mole of O_2, 0.300 mole of N_2, and 0.300 mole of Ar. For this gaseous mixture, calculate the

(a) mole percent O_2 present

(b) pressure percent O_2 present

(c) volume percent O_2 present

Solution

(a) The total number of moles of gas present is 0.800 mole (0.200 mole + 0.300 mole + 0.300 mole). The mole percent O_2 is

$$\text{mole \% A} = \frac{n_A}{n_{total}} \times 100 = \frac{0.200\ \cancel{\text{mole}}}{0.800\ \cancel{\text{mole}}} \times 100 = 25\% \quad \text{(calculator answer)}$$

$$= 25.0\% \quad \text{(correct answer)}$$

(b) Since conditions are specified as STP, the total pressure in the flask is 1.00 atm. The partial pressure of the O_2 is

$$P_{O_2} = X_{O_2} \times P_{total}$$

$$= \frac{0.200\ \cancel{\text{mole}}}{0.800\ \cancel{\text{mole}}} \times 1.00\ \text{atm}$$

$$= 0.25\ \text{atm} \quad \text{(calculator answer)}$$

$$= 0.250\ \text{atm} \quad \text{(correct answer)}$$

The pressure percent O_2 is

$$\text{pressure \% } O_2 = \frac{P_{O_2}}{P_{total}} \times 100 = \frac{0.250\ \cancel{\text{atm}}}{1.00\ \cancel{\text{atm}}} \times 100$$

$$= 25\% \quad \text{(calculator answer)}$$

$$= 25.0\% \quad \text{(correct answer)}$$

(c) The total volume of the flask is given as 17.92 L. The volume of the oxygen present, if it were alone at STP conditions, in a different container, is

$$0.200\ \cancel{\text{mole } O_2} \times \frac{22.4\ \text{L } O_2}{1\ \cancel{\text{mole } O_2}} = 4.48\ \text{L } O_2 \quad \text{(calculator and correct answer)}$$

The volume percent O_2 is

$$\text{volume \% } O_2 = \frac{VO_2}{V_{total}} \times 100 = \frac{4.48\ \cancel{\text{L}}}{22.4\ \cancel{\text{L}}} \times 100$$

$$= 25\% \quad \text{(calculator answer)}$$

$$= 25.0\% \quad \text{(correct answer)}$$

Note, from the answers to parts (a), (b), and (c), that for O_2 the

$$\text{mole percent} = \text{pressure percent} = \text{volume percent}$$

Practice Exercise 12.26

A 20.16-L flask, at STP, contains 0.200 mole of O_2, 0.300 mole of N_2, and 0.400 mole of Ar. For this gaseous mixture, calculate the
(a) mole percent of N_2 present
(b) pressure percent of N_2 present
(c) volume percent of N_2 present

Ans. (a) 33.3%; (b) 33.3%; (c) 33.3%

KEY TERMS

The new terms or concepts defined in this chapter are

Avogadro's law (Sec. 12.9) Equal volumes of different gases, measured at the same temperature and pressure, contain equal numbers of molecules.

barometer (Sec. 12.2) Commonly used device for measuring atmospheric pressure.

Boyle's law (Sec. 12.3) The volume of a sample of gas is inversely proportional to the pressure applied to the gas if the temperature is kept constant.

Charles's law (Sec. 12.4) The volume of a sample of gas is directly proportional to its Kelvin temperature if the pressure is kept constant.

combined gas law (Sec. 12.6) The expression obtained from mathematically combining Boyle's, Charles's, and Gay-Lussac's laws.

Dalton's law of partial pressures (Sec. 12.14) The total pressure exerted by a mixture of gases is the sum of the partial pressures of the individual gases.

gas laws (Sec. 12.2) Generalizations that describe in mathematical terms the relationships among the pressure, temperature, and volume of a specific quantity of a gas.

Gay-Lussac's law (Sec. 12.5) The pressure of a sample of gas, at constant volume, is directly proportional to its Kelvin temperature.

Gay-Lussac's law of combining volumes (Sec. 12.8) The volumes of different gases involved in a reaction, if measured at the same temperature and pressure, are in the same ratio as the coefficients for these gases in the balanced equation for the reaction.

ideal gas (Sec. 12.11) A gas that obeys exactly all of the statements of kinetic molecular theory.

ideal gas law (Sec. 12.11) An expression that describes the relationships among the four variables temperature (T), pressure (P), volume (V), and moles of gas (n) for gaseous substances.

manometer (Sec. 12.2) A device used to measure gas pressures in the laboratory.

molar volume (Sec. 12.10) The volume occupied by 1 mole of a gas at STP conditions.

mole fraction (Sec. 12.14) A dimensionless quantity that gives the ratio of the number of moles of a component in a mixture to the number of moles of all components present.

partial pressure (Sec. 12.14) The pressure that a gas in a mixture would exert if it were present alone under the same conditions.

pressure (Sec. 12.2) The force applied per unit area, that is, the total force on a surface divided by the area of that surface.

standard pressure (Sec. 12.7) A pressure of 1 atm (760 mm Hg).

standard temperature (Sec. 12.7) A temperature of 0 °C (273 K).

STP conditions (Sec. 12.7) The conditions of standard temperature and standard pressure.

PRACTICE PROBLEMS

Measurement of Pressure (Sec. 12.2)

12.1 If the helium gas in a steel cylinder is at a pressure of 6.20 atm, what is the pressure in each of the following pressure units?

(a) millimeters of mercury **(b)** inches of mercury
(c) pounds per square inch **(d)** centimeters of mercury

12.2 If the oxygen gas in a steel cylinder is at a pressure of 9570 millimeters of mercury, what is the pressure in each of the following pressure units?

(a) inches of mercury **(b)** pounds per square inch
(c) atmospheres **(d)** centimeters of mercury

12.3 For each of the following pairs of pressure measurements, indicate whether the first listed measurement is larger than, equal to, or smaller than the second listed measurement.

(a) 457 mm Hg and 1.07 atm
(b) 14.68 lb/in.^{2} and 760 mm Hg
(c) 29.9 in. Hg and 585 mm Hg
(d) 4.639 atm and 68.10 psi

12.4 For each of the following pairs of pressure measurements, indicate whether the first listed measurement is larger than, equal to, or smaller than the second listed measurement.

(a) 0.998 atm and 762 mm Hg
(b) 29.92 psi and 29.92 in. Hg
(c) 17.8 psi and 545 mm Hg
(d) 1.34 atm and 40.1 in. Hg

12.5 The mercury level in the arm of a manometer (see Fig. 12.2) that is open to the atmosphere is found to be 237 mm higher than the mercury level in the arm of the manometer connected to the container of gas. Measured barometric pressure is 762 mm Hg. What is the pressure, in millimeters of mercury, of the gas in the container?

12.6 The mercury level in the arm of a manometer (see Fig. 12.2) that is open to the atmosphere is found to be 35 mm lower than the mercury level in the arm of the manometer connected to the container of gas. Measured barometric pressure is 743 mm Hg. What is the pressure, in millimeters of mercury, of the gas in the container?

Boyle's Law (Sec. 12.3)

12.7 A sample of O_2 gas occupies a volume of 2.00 L at 27°C and 2.00 atm pressure. What volume, in liters, will this O_2 sample occupy at the same temperature but at each of the following pressures?

(a) 3.13 atm **(b)** 0.723 atm
(c) 762 mm Hg **(d)** 37.2 mm Hg

12.8 A sample of N_2 gas occupies a volume of 3.00 L at 37°C and 3.00 atm pressure. What volume, in liters, will this N_2 sample occupy at the same temperature but at each of the following pressures?

(a) 2.98 atm **(b)** 10.5 atm
(c) 453 mm Hg **(d)** 54.2 mm Hg

12.9 At constant temperature, the pressure on a sample of H_2 gas is decreased from 4.0 atm to 2.5 atm. What was the original volume, in milliliters, of the gas sample if this action increases the sample volume to each of the following amounts?

(a) 425 mL **(b)** 25.4 mL
(c) 1.08 L **(d)** 4.68 L

12.10 At constant temperature, the pressure on a sample of H_2 gas is increased from 2.5 atm to 4.0 atm. What was the original volume, in milliliters, of the gas sample if this action decreases the sample volume to each of the following amounts?

(a) 322 mL **(b)** 15.0 mL
(c) 2.24 L **(d)** 0.88 L

12.11 A sample of O_2 gas under a pressure of 1.0 atm occupies 231 mL in a piston–cylinder arrangement before compression. If the gas is compressed to 0.15 its original volume, what must the new pressure, in atmospheres, be? Assume that the temperature remains constant.

12.12 A sample of N_2 gas under a pressure of 15.4 atm occupies 47.3 mL in a piston–cylinder arrangement before compression. If the gas is compressed to 0.88 its original volume, what must the new pressure, in atmospheres, be? Assume that the temperature remains constant.

12.13 A sample of Cl_2 gas at a pressure of 645 mm Hg is transferred to a new container having a volume one-third that of the original container. What pressure, in millimeters of mercury, does the Cl_2 exert in the new container? Assume that the temperature does not change.

12.14 A sample of F_2 gas at a pressure of 1.03 atm is transferred to a new container having a volume 2.50 times that of the original container. What pressure, in atmospheres, does the F_2 exert in the new container? Assume that the temperature does not change.

Charles's Law (Sec. 12.4)

12.15 A sample of carbon dioxide gas, CO_2, has a volume of 5.00 L at 35°C. What volume, in liters, will this CO_2 gas occupy at each of the following temperatures if the pressure is held constant?

(a) 125°C **(b)** 5°C **(c)** −5°C **(d)** 985°C

12.16 A sample of nitrogen dioxide gas, NO_2, has a volume of 1.50 L at 23°C. What volume, in liters, will this NO_2 gas occupy at each of the following temperatures if the pressure is held constant?

(a) 123°C **(b)** 223°C **(c)** −25°C **(d)** 883°C

12.17 At constant pressure, the temperature of a sample of He gas is decreased from 73°C to 0°C. What was the original volume of the gas sample, in milliliters, if this action decreases the sample volume to each of the following amounts?

(a) 15.2 mL **(b)** 879 mL **(c)** 1.20 L **(d)** 10.7 L

12.18 At constant pressure, the temperature of a sample of Ne gas is increased from 0°C to 73°C. What was the original volume of the gas sample, in milliliters, if this action increases the sample volume to each of the following amounts?

(a) 17.5 mL **(b)** 742 mL **(c)** 3.42 L **(d)** 0.90 L

12.19 At constant pressure, at what temperature, in degrees Celsius, will a gas have exactly one-half the volume that it has at room temperature (24°C)?

12.20 At constant pressure, at what temperature, in degrees Celsius, will a gas have exactly double the volume that it has at room temperature (24°C)?

12.21 It is desired to increase the volume of 64.0 mL of Ar gas by 20.0% while holding the pressure constant. To what temperature, in degrees Celsius, must the gas be heated if the initial temperature is 35°C?

12.22 It is desired to decrease the volume of 125 mL of Ne gas by 30.0% while holding the pressure constant. To what temperature, in degrees Celsius, must the gas be cooled if the initial temperature is 22°C?

Gay-Lussac's Law (Sec. 12.5)

12.23 A sample of air exerts a pressure of 1.00 atm at 22°C. If the sample volume remains constant, what is the new pressure, in atmospheres, exerted by the air when it is heated to each of the following temperatures?

(a) 122°C **(b)** 222°C **(c)** 422°C **(d)** 722°C

12.24 A sample of air exerts a pressure of 2.00 atm at 37°C. If the sample volume remains constant, what is the new pressure, in atmospheres, exerted by the air when it is cooled to each of the following temperatures?

(a) 31°C **(b)** 6°C **(c)** −37°C **(d)** −137°C

12.25 At constant volume, the temperature of a sample of sulfur dioxide gas, SO_2, is decreased from 97°C to 27°C. What was the original pressure of the gas, in atmospheres, if this action decreases the sample pressure to the following?

(a) 3.00 atm **(b)** 1.00 atm
(c) 1375 mm Hg **(d)** 375 mm Hg

12.26 At constant volume, the temperature of a sample of carbon monoxide gas, CO, is decreased from 122°C to 22°C. What was the original pressure of the gas, in millimeters of mercury, if this action decreases the sample pressure to the following?

(a) 762 mm Hg **(b)** 662 mm Hg
(c) 1.05 atm **(d)** 25.0 atm

12.27 At constant volume, at what temperature, in degrees Celsius, will a gas exert exactly one-half the pressure that it did at room temperature (24°C)?

12.28 At constant volume, at what temperature, in degrees Celsius, will a gas exert exactly double the pressure that it did at room temperature (24°C)?

12.29 A spray can is empty except for the propellant gas, which exerts a pressure of 1.2 atm at 24°C. If the can is thrown into a fire (485°C), what will be the pressure, in atmospheres, inside the hot can?

12.30 A spray can is empty except for the propellant gas, which exerts a pressure of 1.2 atm at 24°C. If the can is placed in a refrigerator (3°C), what will be the pressure, in atmospheres, inside the cold can?

The Combined Gas Law (Sec. 12.6)

12.31 Rearrange the standard form of the combined gas law to result in the following.

(a) the variable T_2 is isolated on the left side of the equation
(b) the quantity V_2/P_1 is isolated on the left side of the equation

12.32 Rearrange the standard form of the combined gas law to result in the following.

(a) the variable T_1 is isolated on the left side of the equation
(b) the quantity P_2/V_1 is isolated on the left side of the equation

12.33 What is the new volume, in milliliters, of a 3.00-mL sample of air at 0.980 atm and 230°C that is compressed and cooled to each of the following sets of conditions?

(a) 185°C and 1.50 atm **(b)** 35°C and 2.00 atm
(c) −35°C and 4.00 atm **(d)** −125°C and 5.67 atm

12.34 What is the new volume, in liters, of a 25.0-L sample of air at 1.11 atm and 152°C that is compressed and cooled to each of the following sets of conditions?

(a) 25°C and 2.00 atm **(b)** −25°C and 3.00 atm
(c) −75°C and 5.00 atm **(d)** −125°C and 7.75 atm

12.35 A sample of CO_2 gas has a volume of 15.2 L at a pressure of 1.35 atm and a temperature of 33°C. Determine the following for this gas sample.

(a) volume, in liters, at T = 35°C and P = 3.50 atm

(b) volume, in milliliters, at $T = 97°C$ and $P = 6.70$ atm

(c) pressure, in atmospheres, at $T = 42°C$ and $V = 10.0$ L

(d) temperature, in degrees Celsius, at $P = 7.00$ atm and $V = 0.973$ L

12.36 A sample of NO_2 gas has a volume of 37.3 mL at a pressure of 621 mm Hg and a temperature of 52°C. Determine the following for this gas sample.

(a) volume, in milliliters, at $T = 35°C$ and $P = 650$ mm Hg

(b) volume, in liters, at $T = 43°C$ and $P = 1.11$ atm

(c) pressure, in millimeters of mercury, at $T = 125°C$ and $V = 52.4$ mL

(d) temperature, in degrees Celsius, at $P = 775$ mm Hg and $V = 23.0$ mL

12.37 A sample of ammonia gas, NH_3, in a 375-mL container at a pressure of 1.03 atm and a temperature of 27°C is transferred to a container with a volume of 1.25 L.

(a) What is the new pressure, in millimeters of mercury, if no change in temperature occurs?

(b) What is the new temperature, in degrees Celsius, if no change in pressure occurs?

12.38 A sample of nitrous oxide gas, N_2O, in a 475-mL container at a pressure of 676 mm Hg and a temperature of 22°C is transferred to a container with a volume of 5.00 L.

(a) What is the new pressure, in atmospheres, if no change in temperature occurs?

(b) What is the new temperature, in degrees Celsius, if no change in pressure occurs?

12.39 A sample of nitrous oxide gas, N_2O, in a nonrigid container, at a temperature of 33°C, occupies a certain volume at a certain pressure. What will be its temperature, in degrees Celsius, in each of the following situations?

(a) Both pressure and volume are tripled.

(b) Both pressure and volume are cut in half.

(c) The pressure is tripled and the volume is cut in half.

(d) The pressure is cut in half and the volume is doubled.

12.40 A sample of nitric oxide gas, NO, in a nonrigid container, occupies a volume of 2.50 L at a certain temperature and pressure. What will be its volume, in liters, in each of the following situations?

(a) Both pressure and Kelvin temperature are doubled.

(b) Both pressure and Kelvin temperature are cut by one-third.

(c) The pressure is doubled, and the Kelvin temperature is cut by one-third.

(d) The pressure is cut in half, and the Kelvin temperature is tripled.

STP Conditions (Sec. 12.7)

12.41 What is the volume, in liters, at STP of 3.50 L of methane gas, CH_4, at each of the following initial conditions?

(a) 30°C and 1.00 atm **(b)** 0°C and 3.00 atm

(c) 55°C and 1.25 atm **(d)** 135°C and 852 mm Hg

12.42 What is the volume, in liters, at STP of 4.25 L of acetylene gas, C_2H_2, at each of the following initial conditions?

(a) 52°C and 760 mm Hg **(b)** 0°C and 452 mm Hg

(c) −8°C and 683 mm Hg **(d)** 125°C and 1.25 atm

12.43 A quantity of air has a volume of 1.00 L at STP. What volume, in liters, will the air occupy if the pressure and temperature are changed to the following values?

(a) 1.08 atm and 8°C

(b) 6.20 atm and 875°C

(c) 0.500 atm and −15°C

(d) 680 mm Hg and −30°C

12.44 A quantity of air has a volume of 856 mL at STP. What volume, in milliliters, will the air occupy if the pressure and temperature are changed to the following values?

(a) 589 mm Hg and 11°C

(b) 11 mm Hg and 589°C

(c) 1575 mm Hg and −25°C

(d) 1.01 atm and −1°C

Gay-Lussac's Law of Combining Volumes (Sec. 12.8)

12.45 Ammonia reacts with oxygen to form nitrogen (N_2) and water:

$$4\ NH_3(g) + 3\ O_2(g) \longrightarrow 2\ N_2(g) + 6\ H_2O(g)$$

It can also react with oxygen to form nitric oxide (NO) and water:

$$4\ NH_3(g) + 5\ O_2(g) \longrightarrow 4\ NO(g) + 6\ H_2O(g)$$

Which of these two reactions occurred if 1.60 liters of NH_3 are found by experiment to react with 2.00 liters of O_2? Assume that both gas volumes are measured at the same temperature and pressure.

12.46 Methane, CH_4, reacts with steam to form hydrogen and carbon monoxide:

$$CH_4(g) + H_2O(g) \longrightarrow CO(g) + 3\ H_2(g)$$

It can also react with steam to form hydrogen and carbon dioxide:

$$CH_4(g) + 2\ H_2O(g) \longrightarrow CO_2(g) + 4\ H_2(g)$$

Which of these two reactions occurred if 2.33 L of methane are found by experiment to react with 2.33 L of steam? Assume that both gas volumes are measured at the same temperature and pressure.

12.47 The equation for the combustion of the fuel propane, C_3H_8, is

$$C_3H_8(g) + 5\ O_2(g) \longrightarrow 3\ CO_2(g) + 4\ H_2O(g)$$

(a) How many liters of C_3H_8 must be burned to produce 1.30 L of CO_2 if both volumes are measured at STP?

(b) How many liters of C_3H_8 must be burned to produce 1.30 L of H_2O if both volumes are measured at 2.00 atm and 56°C?

12.48 The equation for the combustion of the fuel butane, C_4H_{10}, is

$$2\ C_4H_{10}(g) + 13\ O_2(g) \longrightarrow 8\ CO_2(g) + 10\ H_2O(g)$$

(a) How many liters of C_4H_{10} must be burned to produce 2.60 L of CO_2 if both volumes are measured at STP?

(b) How many liters of C_4H_{10} must be burned to produce 2.60 L of H_2O if both volumes are measured at 1.75 atm and 43°C?

Avogadro's Law (Sec. 12.9)

12.49 Consider two samples of Ar gas at the same temperature and pressure. The first sample contains 0.573 mole Ar and has a volume of 37.2 mL. How many moles of Ar are present in the second sample if it has a volume of 50.7 mL?

12.50 Consider two samples of Xe gas at the same temperature and pressure. The first sample contains 1.00 mole of Xe and has a volume of 123 mL. How many moles of Xe are present in the second sample if it has a volume of 57 mL?

12.51 A balloon containing 1.83 moles of He has a volume of 0.673 L at a certain temperature and pressure. How many *grams* of He would have to be added to the balloon in order for the volume to increase to 0.811 L at the same temperature and pressure?

12.52 A balloon containing 1.83 moles of He has a volume of 0.673 L at a certain temperature and pressure. How many *grams* of He would have to be removed from the balloon in order for the volume to decrease to 0.455 L at the same temperature and pressure?

12.53 A 0.625-mole sample of N_2 gas at 1.50 atm and 45°C occupies a volume of 1.33 L. What volume, in liters, would a 0.625-mole sample of N_2O gas occupy at the same temperature and pressure?

12.54 A 1.34-mole sample of Cl_2 gas at 2.00 atm and 32°C occupies a volume of 85.0 mL. What volume, in liters, would a 1.34-mole sample of Cl_2O gas occupy at the same temperature and pressure?

12.55 In each of the following pairs of gas samples, select the pair member that would have the larger volume at 27°C and 1.00 atm.

(a) 0.400 mole N_2 and 0.450 mole O_2

(b) 2.32 moles CH_4 and 2.00 moles C_2H_6

(c) 100.0 g NO_2 and 100.0 g N_2O

(d) 100.0 g CO and 100.0 g CO_2

12.56 In each of the following pairs of gas samples, select the pair member that would have the smaller volume at 33°C and 2.00 atm.

(a) 0.300 mole He and 0.200 mole H_2

(b) 1.05 moles SO_2 and 2.10 moles SO_3

(c) 50.0 g F_2 and 50.0 g OF_2

(d) 50.0 g NH_3 and 50.0 g PH_3

12.57 A gaseous nitrogen–fluorine compound decomposes to give nitrogen and fluorine gas. Using Avogadro's law, determine the molecular formula of this compound given that 2.88 L of the compound decompose to form 1.44 L of N_2 and 4.32 L of F_2, all volumes being measured at the same temperature and pressure.

12.58 A gaseous nitrogen–oxygen compound decomposes to give nitrogen and oxygen gas. Using Avogadro's law, determine the molecular formula of this compound given that 1.24 L of the compound decompose to form 1.24 L of N_2 and 2.48 L of O_2, all volumes being measured at the same temperature and pressure.

Molar Volume (Sec. 12.10)

12.59 What is the volume, in liters, at STP occupied by 1.25 moles of each of the following gases?

(a) N_2 (b) NH_3 (c) CO_2 (d) Cl_2

12.60 What is the volume, in liters, at STP occupied by 1.75 moles of each of the following gases?

(a) O_3 (b) O_2 (c) NO_2 (d) C_2H_6

12.61 In each of the following pairs of gas samples, select the pair member that occupies the larger volume at STP.

(a) 24.5 g N_2 and 24.5 g NH_3

(b) 30.0 g O_2 and 30.0 g O_3

(c) 10.0 g SO_2 and 20.0 g NO_2

(d) 15.0 g N_2O and 20.0 g NO

12.62 In each of the following pairs of gas samples, select the pair member that occupies the larger volume at STP.

(a) 10.0 g H_2 and 10.0 g He

(b) 25.0 g F_2 and 25.0 g Cl_2

(c) 15.0 g CH_4 and 5.00 g PH_3

(d) 100.0 g HCN and 2.00 g UF_6

12.63 Under STP conditions, what is the mass, in grams, of 23.7-L samples of each of the following gases?

(a) Ar (b) N_2O (c) SO_3 (d) PH_3

12.64 Under STP conditions, what is the mass, in grams, of 35.2-L samples of each of the following gases?

(a) Kr (b) S_2O (c) O_3 (d) NH_3

12.65 Calculate the density at STP conditions, in grams per liter, of each of the following gases.

(a) N_2O (b) N_2O_4 (c) NO_2 (d) NO

12.66 Calculate the density at STP conditions, in grams per liter, of each of the following gases.

(a) CH_4 (b) C_2H_2 (c) C_2H_4 (d) C_2H_6

12.67 Which gas in each of the following pairs of gases will have the greater density at STP?

(a) O_2 and O_3 (b) NH_3 and PH_3
(c) CO and CO_2 (d) F_2 and CH_4

12.68 Which gas in each of the following pairs of gases will have the greater density at STP?

(a) SO_2 and S_2O (b) F_2 and OF_2
(c) NO and CO (d) SF_6 and UF_6

12.69 Calculate molar masses for gases with the following densities at STP.

(a) 1.97 g/L (b) 1.25 g/L
(c) 0.714 g/L (d) 3.17 g/L

12.70 Calculate molar masses for gases with the following densities at STP.

(a) 1.70 g/L (b) 0.897 g/L
(c) 1.16 g/L (d) 0.759 g/L

12.71 If the Xe atoms present in 22.5 g of XeO_2F_2 are converted to gaseous Xe, what volume, in liters, will they occupy at STP conditions?

12.72 If the Xe atoms present in 235 g of $XeOF_4$ are converted to gaseous Xe, what volume, in liters, will they occupy at STP conditions?

The Ideal Gas Law (Sec. 12.11)

12.73 Using the ideal gas law, calculate the volume, in liters, of 1.20 moles of Cl_2 gas at each of the following sets of conditions.

(a) STP (b) 73°C and 1.54 atm
(c) 525°C and 15.0 atm (d) −23°C and 765 mm Hg

12.74 Using the ideal gas law, calculate the volume, in liters, of 1.15 moles of F_2 gas at each of the following sets of conditions.

(a) STP (b) 315°C and 456 mm Hg
(c) −45°C and 4.50 atm (d) 23°C and 762 mm Hg

12.75 How many moles of H_2 gas does it take to fill a 6.00-L container to a pressure of 3.67 atm at 25°C?

12.76 How many moles of He gas does it take to fill a 5.00-L container to a pressure of 2.50 atm at 35°C?

12.77 If 0.332 mole of He gas has a volume of 1275 mL and a pressure of 5.78 atm, what is its temperature in degrees Celsius?

12.78 If 0.504 mole of Ar gas has a volume of 1975 mL and a pressure of 4.60 atm, what is its temperature in degrees Celsius?

12.79 What is the pressure, in atmospheres, inside a 6.00-L container that contains the following amounts of N_2 gas at 25°C?

(a) 0.30 mole (b) 1.20 moles (c) 0.30 g (d) 1.20 g

12.80 What is the pressure, in millimeters of mercury, inside a 4.00-L container that contains the following amounts of O_2 gas at 40°C?

(a) 0.72 mole (b) 4.5 moles (c) 0.72 g (d) 4.5 g

12.81 1.14 moles of the noble gas argon occupy a volume of 3.00 L at a temperature of 152°C and a pressure of 13.3 atm. Use this information to calculate the value of the ideal gas constant R in the units of atm · L/mole · K.

12.82 0.317 mole of the noble gas helium occupies a volume of 8.67 L at a temperature of 25°C and a pressure of 679 mm Hg. Use this information to calculate the value of the ideal gas constant R in the units of mm Hg · L/mole · K.

12.83 A 1.00-mole sample of liquid water is placed in a flexible sealed container and allowed to evaporate. After complete evaporation, what will be the container volume, in liters, at 127°C and 0.908 atm pressure?

12.84 A 1.00-mole sample of dry ice (solid CO_2) is placed in a flexible sealed container and allowed to sublime. After all the CO_2 has changed from solid to gas, what will be the container volume, in liters, at 23°C and 0.983 atm pressure?

Equations Derived from the Ideal Gas Law (Sec. 12.12)

12.85 Calculate the mass, in grams, of each of the following quantities of gas.

(a) 25.0 L of C_2H_6 at 0.972 atm and 29°C
(b) 2.22 L of HCl at 854 mm Hg and 75°C
(c) 5.50 L of SO_2 at STP
(d) 783 mL of N_2O at 359 mm Hg and 273°C

12.86 Calculate the mass, in grams, of each of the following quantities of gas.

(a) 3.00 L of NO at 1.23 atm and 35°C
(b) 1.37 L of CO_2 at 498 mm Hg and 285°C
(c) 3.50 L of HF at STP
(d) 1780 mL of C_2H_2 at 3.00 atm and 585°C

12.87 A 30.0-L cylinder contains 100.0 g of Cl_2 at 23°C. How many grams of Cl_2 must be added to the container to increase the pressure in the cylinder to 1.65 atm? Assume that the temperature remains constant.

12.88 A 50.0-L cylinder contains 100.0 g of F_2 at 30°C. How many grams of F_2 must be removed from the container to decrease the pressure in the cylinder to 1.00 atm? Assume that the temperature remains constant.

12.89 A gas is either CO or CO_2. Based on the following information about a sample of the gas, what is its identity? A 0.902-L sample of the gas exerts a pressure of 1.65 atm, has a mass of 1.61 g, and is at a temperature of 20°C.

12.90 A gas is either CH_4 or HF. Based on the following information about a sample of the gas, what is its identity? A

1.98-L sample of the gas exerts a pressure of 2.00 atm, has a mass of 3.20 g, and is at a temperature of 30°C.

12.91 Calculate the density of H_2S gas, in grams per liter, at each of the following temperature–pressure conditions.
(a) 24°C and 675 mm Hg
(b) 24°C and 1.20 atm
(c) 370°C and 1.30 atm
(d) −25°C and 452 mm Hg

12.92 Calculate the density of SCl_2 gas, in grams per liter, at each of the following temperature–pressure conditions.
(a) 27°C and 1.20 atm
(b) 37°C and 5.00 atm
(c) 227°C and 4.00 atm
(d) 550°C and 1.00 atm

12.93 What pressure, in atmospheres, is required to cause each of the following gases to have a density of 1.00 g/L at 47°C?
(a) N_2 **(b)** Xe **(c)** ClF **(d)** N_2O

12.94 What pressure, in atmospheres, is required to cause each of the following gases to have a density of 1.00 g/L at 53°C?
(a) O_2 **(b)** O_3 **(c)** SO_2 **(d)** CH_4

12.95 A gas is either NO or CO. Based on the following information about a sample of the gas, what is its identity? A sample of the gas at a temperature of 27°C and a pressure of 2.00 atm has a density of 2.27 g/L.

12.96 A gas is either H_2S or HF. Based on the following information about a sample of the gas, what is its identity? A sample of the gas at a temperature of 127°C and a pressure of 3.00 atm has a density of 3.12 g/L.

Gas Laws and Chemical Equations (Sec. 12.13)

12.97 A mixture of 25.0 g of NO and an excess of O_2 reacts according to the balanced equation:

$$2\ NO(g) + O_2(g) \longrightarrow 2\ NO_2(g)$$

How many liters of NO_2, at STP, are produced?

12.98 A mixture of 25.0 g of H_2 and an excess of N_2 reacts according to the balanced equation:

$$3\ H_2(g) + N_2(g) \longrightarrow 2\ NH_3(g)$$

How many liters of NH_3, at STP, are produced?

12.99 A sample of O_2 with a volume of 25.0 L at 27°C and 1.00 atm is reacted with excess N_2 to produce NO. The equation for the reaction is

$$O_2(g) + N_2(g) \longrightarrow 2\ NO(g)$$

How many grams of NO are produced?

12.100 A sample of H_2 with a volume of 35.0 L at 35°C and 1.35 atm is reacted with excess O_2 to produce H_2O. The equation for the reaction is

$$2\ H_2(g) + O_2(g) \longrightarrow 2\ H_2O(g)$$

How many grams of H_2O are produced?

12.101 Hydrogen gas can be produced in the laboratory through reaction of magnesium metal with hydrochloric acid:

$$Mg(s) + 2\ HCl(aq) \longrightarrow MgCl_2(aq) + H_2(g)$$

What volume, in liters of H_2 at 23°C and 0.980 atm pressure can be produced from the reaction of 12.0 g of Mg with an excess of HCl?

12.102 A common laboratory preparation for O_2 gas involves the thermal decomposition of potassium nitrate:

$$2\ KNO_3(s) \longrightarrow 2\ KNO_2(s) + O_2(g)$$

What volume in liters, of O_2 at 35°C and 1.31 atm pressure can be produced from the decomposition of 35.0 g of KNO_3?

12.103 Ammonium nitrate, NH_4NO_3, can decompose explosively when heated to a high temperature, according to the equation:

$$2\ NH_4NO_3(s) \longrightarrow 2\ N_2(g) + 4\ H_2O(g) + O_2(g)$$

If a 100.0-g sample of ammonium nitrate decomposes at 450°C, how many liters of gaseous products would be formed? Assume that atmospheric pressure is 1.00 atm.

12.104 The industrial explosive nitroglycerin, $C_3H_5N_3O_9$, detonates according to the equation:

$$4\ C_3H_5N_3O_9(s) \longrightarrow 6\ N_2(g) + O_2(g) + 12\ CO_2(g) + 10\ H_2O(g)$$

At a detonation temperature of 1950°C, how many liters of gaseous products would be formed from 100.0 g of nitroglycerin? Assume that atmospheric pressure is 1.00 atm.

12.105 How many liters of NO_2 gas at 21°C and 2.31 atm must be consumed in producing 75.0 L of NO gas at 38°C and 645 mm Hg according to the following reaction?

$$3\ NO_2(g) + H_2O(l) \longrightarrow 2\ HNO_3(aq) + NO(g)$$

12.106 How many liters of $Cl_2(g)$ at 25°C and 1.50 atm are needed to react completely with 3.42 L of NH_3 gas at 50°C and 2.50 atm according to the following reaction?

$$2\ NH_3(g) + 3\ Cl_2(g) \longrightarrow N_2(g) + 6\ HCl(g)$$

Dalton's Law of Partial Pressures (Sec. 12.14)

12.107 A mixture of H_2, N_2, and Ar gases is present in a steel cylinder. The total pressure within the cylinder is 675 mm Hg and the partial pressures of N_2 and Ar are, respectively, 354 mm Hg and 235 mm Hg. If CO_2 gas is added to the mixture, at constant temperature, until the total pressure reaches 842 mm Hg, what is the partial pressure, in millimeters of Hg, of the following?

(a) CO_2 **(b)** N_2 **(c)** Ar **(d)** H_2

12.108 A mixture of O_2, He, and Ne gases is present in a steel cylinder. The total pressure within the cylinder is 652 mm Hg and the partial pressures of He and Ne are, respectively, 251 mm Hg and 152 mm Hg. If CO_2 gas is added to the mixture, at constant temperature, until the total pressure reaches 704 mm Hg, what is the partial pressure, in millimeters of Hg, of the following?

(a) CO_2 **(b)** He **(c)** Ne **(d)** O_2

12.109 A gaseous mixture contains 25.0 g of each of the gases CO, CO_2, and H_2S. Assuming that the gases do not react with each other, calculate the following.

(a) the mole fraction of each gas present in the mixture

(b) the partial pressure of each gas present given that the total pressure exerted by the gaseous mixture is 1.72 atm.

12.110 A gaseous mixture contains 15.0 g of each of the gases HCl, H_2S, and Xe. Assuming that the gases do not react with each other, calculate the following.

(a) the mole fraction of each gas present in the mixture

(b) the partial pressure of each gas present given that the total pressure exerted by the gaseous mixture is 2.24 atm.

12.111 Calculate the partial pressure of O_2, in atmospheres, in a gaseous mixture with a volume of 2.50 L at 20°C given that the mixture composition is

(a) 0.50 mole O_2 and 0.50 mole N_2

(b) 0.50 mole O_2 and 0.75 mole N_2

(c) 0.50 mole O_2, 0.75 mole N_2, and 0.75 mole Ar

(d) 0.50 g O_2 and 0.75 g N_2

12.112 Calculate the partial pressure of Xe, in atmospheres, in a gaseous mixture with a volume of 1.20 L at 32°C given that the mixture composition is

(a) 0.40 mole Xe and 0.40 mole Ne

(b) 0.40 mole Xe and 0.60 mole Ne

(c) 0.40 mole Xe, 0.60 mole Ne, and 1.25 moles O_2

(d) 0.40 g Xe and 0.60 g Ne

12.113 What is the partial pressure of O_2 in a gaseous mixture whose total pressure is 1.20 atm given the following mixture compositions?

(a) 0.40 mole O_2 and 0.40 mole Ne

(b) 0.40 mole O_2 and 0.80 mole Ne

(c) an equal number of moles of O_2, N_2, and H_2

(d) an equal number of molecules of O_2, N_2, and H_2

12.114 What is the partial pressure of Xe in a gaseous mixture whose total pressure is 1.55 atm given the following mixture compositions?

(a) 0.50 mole Xe and 0.50 mole Ne

(b) 0.50 mole Xe and 1.00 mole Ne

(c) an equal number of moles of Xe, Ne, and He

(d) an equal number of atoms of Xe, Ne, and He

12.115 What is the total pressure in a flask that contains 4.0 moles He, 2.0 moles Ne, and 0.50 mole Ar, and in which the partial pressure of Ar is 0.40 atm?

12.116 What is the total pressure in a flask that contains 2.0 moles H_2, 6.0 moles O_2, and 0.50 mole N_2, and in which the partial pressure of H_2 is 0.80 atm?

12.117 A sample of ammonia, NH_3, is *completely* decomposed to its constituent elements.

$$2\,NH_3(g) \longrightarrow N_2(g) + 3\,H_2(g)$$

If the total pressure of the N_2 and H_2 produced is 852 mm Hg, calculate the partial pressures, in millimeters of mercury, of N_2 and H_2.

12.118 A sample of steam, H_2O, is *completely* decomposed to its constituent elements.

$$2\,H_2O(g) \longrightarrow 2\,H_2(g) + O_2(g)$$

If the total pressure of the H_2 and O_2 produced is 1.35 atm, calculate the partial pressures, in atmospheres, of H_2 and O_2.

12.119 What would be the partial pressure, in millimeters of mercury, of O_2, collected over water at the following conditions of temperature and atmospheric pressure?

(a) 19°C and 743 mm Hg

(b) 28°C and 645 mm Hg

(c) 34°C and 762 mm Hg

(d) 21°C and 0.933 atm

12.120 What would be the partial pressure, in millimeters of mercury, of O_2 collected over water at the following conditions of temperature and atmospheric pressure?

(a) 15°C and 632 mm Hg

(b) 35°C and 749 mm Hg

(c) 31°C and 682 mm Hg

(d) 26°C and 0.975 atm

12.121 A 24.64-L flask, at STP, contains 0.100 mole of He, 0.200 mole of Ne and 0.800 mole of Ar. For this gaseous mixture, calculate the

(a) mole fraction He

(b) mole percent Ar

(c) pressure percent Ne

(d) volume percent He

12.122 A 15.68-L flask, at STP, contains 0.200 mole of Ar, 0.400 mole of Kr and 0.100 mole of Xe. For this gaseous mixture, calculate the

(a) mole fraction Ar

(b) mole percent Kr

(c) pressure percent Xe

(d) volume percent Ar

12.123 Three containers of gases are combined into a single large container: 2.0 L of O_2 at STP, 3.0 L of Ar at STP, and 3.0 L of Ne at STP are put into an 8.0-L container at STP. Calculate the following items pertaining to the gaseous mixture.

(a) volume percent O_2

(b) mole percent Ar

(c) pressure percent Ne

(d) partial pressure O_2

12.124 Three containers of gases are combined into a single large container: 1.0 L of N_2 at STP, 4.0 L of He at STP, and 1.0 L of Xe at STP are put into a 6.0-L container at STP. Calculate the following items pertaining to the gaseous mixture.

(a) volume percent He

(b) mole percent Xe

(c) pressure percent He

(d) partial pressure N_2

ADDITIONAL PROBLEMS

12.125 How many molecules of carbon dioxide (CO_2) gas are contained in 1.00 L of CO_2 at STP?

12.126 How many molecules of hydrogen sulfide (H_2S) gas are contained in 2.00 L of H_2S at STP?

12.127 A near-vacuum pressure of 0.0010 mm Hg is readily obtained in a laboratory by means of a vacuum pump. Calculate the number of molecules in 1.00 mL of O_2 gas at this pressure and 23°C.

12.128 A near-vacuum pressure of 0.0010 mm Hg is readily obtained in a laboratory by means of a vacuum pump. Calculate the number of atoms in 1.00 mL of Xe gas at this pressure and 27°C.

12.129 At a particular temperature and pressure, 8.00 g of N_2 gas occupy 6.00 L. What would be the volume, in liters, occupied by 1.00×10^{23} molecules of NH_3 at the same temperature and pressure?

12.130 At a particular temperature and pressure, 2.00×10^{23} molecules of N_2 gas occupy 5.00 L. What would be the volume, in liters, occupied by 25.7 g of SO_2 at the same temperature and pressure?

12.131 A piece of Al is placed in a 1.00-L container with pure O_2. The O_2 is at a pressure of 1.00 atm and a temperature of 25°C. One hour later the pressure has dropped to 0.880 atm and the temperature has dropped to 22°C. Calculate the number of grams of O_2 that reacted with the Al.

12.132 A piece of Ca is placed in a 1.00-L container with pure N_2. The N_2 is at a pressure of 1.12 atm and a temperature of 26°C. One hour later the pressure has dropped to 0.924 atm and the temperature has dropped to 24°C. Calculate the number of grams of N_2 that reacted with the Ca.

12.133 At constant temperature, the pressure on a sample of H_2 gas is increased from 1.50 atm to 3.50 atm. This action decreases the sample volume *by* 25.0 mL. What was the original volume, in milliliters, of the gas sample?

12.134 At constant temperature, the pressure on a sample of HCl gas is decreased from 3.50 atm to 1.50 atm. This action increases the sample volume *by* 75.0 mL. What was the original volume, in milliliters, of the gas sample?

12.135 A large flask (of unknown volume) is filled with air until the pressure reaches 3.6 atm. The flask is then attached to a second evacuated flask of known volume, and the air from the first flask is allowed to expand into the second flask. The final pressure of the air (in both flasks) is 2.6 atm, and the volume of the second flask is 5.21 L. Calculate the volume, in liters, of the first flask.

12.136 A large 4.2-L flask is filled with air to a pressure that is unknown. This flask is then attached to a second evacuated flask of known volume, and the air from the first flask is allowed to expand into the second flask. The final pressure of the air (in both flasks) is 2.6 atm, and the volume of the second flask is 5.21 L. Calculate the original pressure, in atmospheres, in the first flask.

12.137 The volume of a fixed quantity of gas, at constant temperature, is decreased by 20.0%. What is the resulting percentage increase in the pressure of the gas?

12.138 The volume of a fixed quantity of gas, at constant temperature, is increased by 30.0%. What is the resulting percentage decrease in the pressure of the gas?

12.139 The volume of a fixed quantity of gas, at constant pressure, is increased by 50.0%. What is the resulting percentage increase in the Kelvin temperature of the gas?

12.140 The volume of a fixed quantity of gas, at constant pressure, is decreased by 10.0%. What is the resulting percentage decrease in the Kelvin temperature of the gas?

12.141 Calculate the ratio of the densities of O_2 and N_2 at

(a) STP conditions

(b) 1.25 atm and 25°C

12.142 Calculate the ratio of the densities of He and Ne at

(a) STP conditions

(b) 2.30 atm and 57°C

12.143 At constant temperature, a 2.000-L mixture of gases is produced by combining 1.000 L of N_2 at 350.0 mm Hg, 6.000 L of O_2 at 300.0 mm Hg, and 1.000 L of H_2 at 250.0 mm Hg. What is the pressure, in millimeters of mercury, of the mixture? Assume that no chemical reactions occur.

12.144 At constant temperature, a 2.000-L mixture of gases is produced by combining 2.000 L of N_2 at 250.0 mm Hg, 4.000 L of O_2 at 250.0 mm Hg, and 1.000 L of H_2 at 600.0 mm Hg. What is the pressure, in millimeters of mercury, of the mixture? Assume that no chemical reactions occur.

12.145 Suppose 532 mL of Ne gas at 20°C and 1.04 atm and 376 mL of SF_6 gas at 20°C and 0.97 atm are put into a 275-mL flask. Assuming that the temperature remains constant, calculate the partial pressure, in atmospheres, of the SF_6 gas in the mixture.

12.146 Suppose 734 mL of Ar gas at 18°C and 1.25 atm and 252 mL of HF gas at 18°C and 2.25 atm are put into a 545-mL flask. Assuming that the temperature remains constant, calculate the partial pressure, in atmospheres, of the HF gas in the mixture.

12.147 A mixture of 15.0 g of Ar and 15.0 g of CH_4 occupies a 4.0-L container at 8.80 atm and 54°C. What is the partial pressure, in atmospheres, of Ar in the mixture?

12.148 A mixture of 30.0 g of Ar and 15.0 g of CH_4 occupies a 4.0-L container at 11.9 atm and 27°C. What is the partial pressure, in atmospheres, of CH_4 in the mixture?

12.149 Suppose 30.0 mL of N_2 gas at 27°C and 645 mm Hg pressure are added to a 40.0-mL container that already contains He at 37°C and 765 mm Hg. If the resulting mixture is brought to 32°C, what is the total pressure in millimeters of mercury, of the mixture?

12.150 Suppose 50.0 mL of Xe gas at 45°C and 0.998 atm pressure are added to a 100.0-mL container that already contains He at 37°C and 765 mm Hg. If the resulting mixture is warmed to 75°C, what is the total pressure, in millimeters of mercury, of the mixture?

CUMULATIVE PROBLEMS

12.151 Which sample contains more molecules: 20.0 L of steam at 135°C and 1.02 atm pressure or 10.5 mL of ice with a density of 0.917 g/mL at 0°C and 1.02 atm pressure?

12.152 Which sample contains more molecules: 30.0 L of steam at 135°C and 0.879 atm pressure or 10.5 mL of liquid water with a density of 0.998 g/mL at 20°C and 0.879 atm pressure?

12.153 A 2.24-L sample of a gaseous compoound has a mass of 7.5 g at STP. What is the molecular mass of a molecule of this compound in atomic mass units?

12.154 A 4.48-L sample of a gaseous compound has a mass of 20.0 g at STP. What is the molecular mass of a molecule of this compound in atomic mass units?

12.155 What is the value of x in the formula PH_x if the density of the PH_x gas is 1.517 g/L at 0°C and 1.00 atm?

12.156 What is the value of x in the formula P_2H_x if the density of P_2H_x gas is 2.944 g/L at 0°C and 1.00 atm?

12.157 The composition of a gaseous mixture in terms of mass percent is 75.0% HCl, 5.00% H_2, and 20.0% He. For this mixture, calculate the following.

(a) the mole fraction of each component

(b) the partial pressure of each component, given that the total pressure is 1.20 atm

(c) a weighted average molar mass for the mixture as a whole

(d) the density of the mixture, at STP, based on the weighted average molar mass

12.158 The composition of a gaseous mixture in terms of mass percent is 43.0% Ar, 15.0% He, and 42.0% H_2S. For this mixture, calculate the following.

(a) the mole fraction of each component

(b) the partial pressure of each component, given that the total pressure is 3.50 atm

(c) a weighted average molar mass for the mixture as a whole

(d) the density of the mixture, at STP, based on the weighted average molar mass

12.159 A 0.581-g sample of a gaseous compound containing only carbon and hydrogen contains 0.480 g of carbon and 0.101 g of hydrogen. At STP, 33.6 mL of the gas have a mass of 0.0869 g. What is the molecular formula for the compound?

12.160 A 6.01-g sample of a gaseous compound containing

only carbon and hydrogen contains 4.80 g of carbon and 1.21 g of hydrogen. At STP, 762 mL of the gas have a mass of 1.02 g. What is the molecular formula for the compound?

12.161 The elemental analysis of a certain compound is 24.3% C, 4.1% H, and 71.6% Cl by mass. If 0.132 g of compound occupies 41.4 mL at 741 mm Hg pressure and 96°C, what are the molar mass and molecular formula of the compound?

12.162 The elemental analysis of a certain compound is 88.82% C and 11.18% H by mass. A 20.87-mg sample of compound vapor occupies 11.63 mL at 1.016 atm pressure and 100°C. What are the molar mass and the molecular formula of the compound?

12.163 Ammonia, NH_3, burns in oxygen to form nitric oxide, NO, and water.

$$4\ NH_3(g) + 5\ O_2(g) \longrightarrow 4\ NO(g) + 6\ H_2O(g)$$

Calculate the total volume, in liters, of products formed when 60.0 L of NH_3 burn in the presence of 60.0 L of O_2. Assume that all volume measurements are made at the same temperature and pressure.

12.164 Methane, CH_4, burns in oxygen to form carbon dioxide, CO_2, and water.

$$CH_4(g) + 2\ O_2(g) \longrightarrow CO_2(g) + 2\ H_2O(g)$$

Calculate the total volume, in liters, of products formed when 30.0 L of CH_4 burn in the presence of 50.0 L of O_2. Assume that all volume measurements are made at the same temperature and pressure.

12.165 A 1.75-L sample of H_2S, measured at 25.0°C and 625 mm Hg, is mixed with 5.75 L of O_2, measured at 10.0°C and 715 mm Hg, and the mixture is allowed to react.

$$2\ H_2S(g) + 3\ O_2(g) \longrightarrow 2\ SO_2(g) + 2\ H_2O(g)$$

How much H_2O, in grams, is produced?

12.166 A 2.00-L sample of N_2, measured at 30.0°C and 1.08 atm, is mixed with 4.00 L of O_2, measured at 25.0°C and 0.118 atm, and the mixture is allowed to react.

$$N_2(g) + O_2(g) \longrightarrow 2\ NO(g)$$

How much NO, in grams, is produced?

12.167 A 22.0-g sample of NH_3 reacts with an excess of Cl_2 gas according to the equation

$$2\ NH_3(g) + 3\ Cl_2(g) \longrightarrow N_2(g) + 6\ HCl(g)$$

What volume, in cubic meters at STP, of HCl gas is produced?

12.168 A 43.0-g sample of NO reacts with an excess of O_2 gas according to the equation

$$2\ NO(g) + O_2(g) \longrightarrow 2\ NO_2(g)$$

What volume, in cubic meters at STP, of NO_2 gas is produced?

12.169 Ammonia gas reacts with hydrogen chloride gas according to the equation

$$NH_3(g) + HCl(g) \longrightarrow NH_4Cl(s)$$

If 7.00 g of NH_3 are reacted with 12.0 g of HCl in a 1.00-L container at 25°C, what will be the final pressure, in atmospheres, in the reaction container?

12.170 At elevated temperatures phosphorus reacts with oxygen according to the equation

$$4\ P(g) + 5\ O_2(g) \longrightarrow P_4O_{10}(s)$$

If 50.0 g of P are reacted with 25.0 g of O_2 in a 8.00-L container at 175°C, what will be the final pressure, in atmospheres, in the reaction container?

12.171 If the price of nitrogen gas (N_2) is \$5.20 per thousand cubic feet at 0°C and 1.00 atm pressure, what is the price of the gas per gram of nitrogen?

12.172 If the price of oxygen gas (O_2) is \$5.40 per thousand cubic feet at 0°C and 1.00 atm pressure, what is the price of the gas per gram of oxygen?

13 Solutions

13.1 Types of Solutions

A **solution** *is a homogeneous (uniform) mixture of two or more substances.* To achieve a homogeneous mixture, the intermingling of components must be on the molecular level; that is, the particles present must be of atomic and molecular size.

In discussing solutions it is often convenient to categorize the components of a solution as *solvent* and *solute(s)*. The **solvent** *is the component of the solution present in the greatest amount.* The solvent may be thought of as the medium in which the other substances present are *dissolved.* A **solute** *is a solution component present in a small amount relative to that of solvent.* More than one solute may be present in the same solution. For example, both sugar and salt (two solutes) may be dissolved in water (solvent).

In most situations we will encounter, the solutes present in a solution will be of more interest to us than the solvent. The solutes are the "active ingredients" in the solution. They are the substances that may react when solutions are mixed.

Solutions used in the laboratory are usually liquids, and the solvent is almost always water. However, as we shall see shortly, gaseous solutions and solid solutions of numerous types do exist.

A solution, since it is homogeneous, will have the same properties throughout. No matter from where we take a sample in a solution it will have the same composition as that of any other sample from the solution. The composition of a solution can be varied, usually within certain limits, by changing the relative amounts of solvent and solute present. (If the composition limits are violated, a heterogeneous mixture is formed.)

Two-component solutions can be classified into nine types according to the physical states of the solvent and solute before mixing. These types, along with an example of each, are listed in Table 13.1. Solutions in which the final state of the solution components is liquid are the most common and are the type that are emphasized in this book.

F.Y.I.
Fish intake through their gills the small amount of oxygen gas dissolved in water.

TABLE 13.1 EXAMPLES OF VARIOUS TYPES OF SOLUTIONS

Solution Type (solute listed first)	Example
Gaseous Solutions	
Gas dissolved in gas	Dry air (oxygen and other gases dissolved in nitrogen)
Liquid dissolved in gas[a]	Wet air (water vapor in air)
Solid dissolved in gas[a]	Moth repellent (or moth balls) sublimed into air
Liquid Solutions	
Gas dissolved in liquid	Carbonated beverage (carbon dioxide in water)
Liquid dissolved in liquid	Vinegar (acetic acid dissolved in water)
Solid dissolved in liquid	Saltwater
Solid Solutions	
Gas dissolved in solid	Hydrogen in platinum
Liquid dissolved in solid	Dental filling (mercury dissolved in silver)
Solid dissolved in solid	Sterling silver (copper dissolved in silver)

[a] An alternative viewpoint is that liquid-in-gas and solid-in-gas solutions do not actually exist as true solutions. From this viewpoint water vapor or moth repellent in air is considered to be a gas-in-gas solution since the water or moth repellent must evaporate or sublime first in order to enter the air.

The physical state of the solute becomes that of the solvent when a solution is formed. For example, solid naphthalene (moth repellent) must be sublimed (Sec. 4.4) in order for it to dissolve in air. Pulverizing a solid to a fine powder and dispersing it in air does not produce a solution. (Dust particles in air would be an example of this.) The particles of the solid must be subdivided to the molecular level; the solid must sublime. Similarly, fog is a suspension of water droplets in air; the droplets are large enough to reflect light, a fact that becomes evident when we drive an automobile on a foggy night. Thus, fog is not a solution. Water vapor, however, is present in solution form in air. When hydrogen gas dissolves in platinum metal (a gas-in-solid solution), the gas molecules take up fixed positions in the metal lattice. The gas is "solidified" as a result.

13.2 TERMINOLOGY USED IN DESCRIBING SOLUTIONS

In addition to *solvent* and *solute*, several other terms are useful in describing characteristics of solutions.

The **solubility** *of a solute is the maximum amount of solute that will dissolve in a given amount of solvent.* Numerous factors affect the numerical value of a solute's solubility in a given solvent, including the nature of the solvent itself, the temperature, and in some cases the pressure and the presence of other solutes.

Common units for expressing solubility are grams of solute per 100 g of solvent. The temperature of the solvent must also be specified. Table 13.2 gives the solubilities of selected solutes in the solvent water at three different temperatures.

TABLE 13.2 SOLUBILITIES OF VARIOUS COMPOUNDS IN WATER AT 0°C, 50°C, AND 100°C

	Solubility (g solute/100 g H_2O)		
Solute	**0°C**	**50°C**	**100°C**
Lead(II) bromide ($PbBr_2$)	0.455	1.94	4.75
Silver sulfate (Ag_2SO_4)	0.573	1.08	1.41
Copper(II) sulfate ($CuSO_4$)	14.3	33.3	75.4
Sodium chloride (NaCl)	35.7	37.0	39.8
Silver nitrate ($AgNO_3$)	122	455	952
Cesium chloride (CsCl)	161.4	218.5	270.5

The use of specific units for solubility, as in Table 13.2, allows us to compare solubilities quite precisely. Such precision is often unnecessary, and instead *qualitative* statements about solubilities are made by using terms such as *very soluble*, *slightly soluble*, and so forth. The guidelines for the use of such terms are given in Table 13.3.

A **saturated solution** *contains the maximum amount of solute that can be dissolved under the conditions at which the solution exists.* A saturated solution together with excess undissolved solute is an equilibrium situation where the rate of dissolution of undissolved solute is equal to the rate of crystallization of dissolved solute. Consider the process of adding table sugar (sucrose) to a container of water. Initially the sugar dissolves as the solution is stirred. Finally, as we add sufficient sugar, a point is reached where no amount of stirring will cause the added sugar to dissolve. Sugar remains as a solid on the bottom of the container; the solution is saturated. Although it appears to the eye that nothing is happening once the saturation point is reached, on the molecular level this is not the case. Solid sugar from the bottom of the container is continuously dissolving in the water, and an equal amount of sugar is coming out of solution. Accordingly, the net number of sugar molecules in the liquid remains the same, and outwardly it appears that the dissolution process has stopped. This equilibrium situation in the saturated solution is somewhat similar to the previously discussed evaporation of a liquid in a closed container (Sec. 11.12). Figure 13.1 illustrates the dynamic equilibrium process occurring in a saturated solution in the presence of undissolved excess solute.

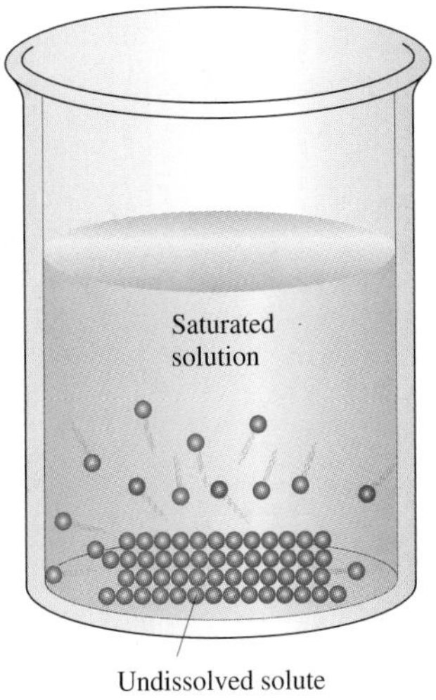

FIGURE 13.1 The dynamic equilibrium process occurring in a saturated solution that contains undissolved excess solute.

An **unsaturated solution** *is a solution where less solute than the maximum amount possible is dissolved in the solution.*

TABLE 13.3 QUALITATIVE SOLUBILITY TERMS

Solute Solubility (g solute/100 g solvent)	Qualitative Solubility Description
Less than 0.1	insoluble
0.1–1	slightly soluble
1–10	soluble
Greater than 10	very soluble

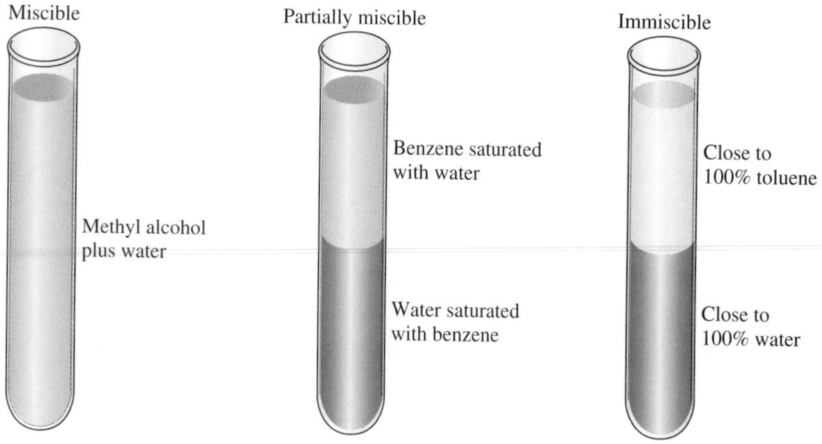

FIGURE 13.2 Miscibility of selected liquids with each other.

The terms *dilute* and *concentrated* are also used to convey qualitative information about the degree of saturation of a solution. A **dilute solution** *contains a small amount of solute in solution relative to the amount that could dissolve.* On the other hand, a **concentrated solution** *contains a large amount of solute relative to the amount that could dissolve.* A concentrated solution need not be a saturated solution.

In dealing with liquid-in-liquid solutions, the terms *miscible*, *partially miscible*, and *immiscible* are frequently used to describe solubility characteristics associated with the liquids. **Miscible substances** *dissolve in any amount in each other.* For example, methyl alcohol (CH_3OH) and water are miscible—they completely mix with each other in any and all proportions. Always, after these two liquids are mixed, only one phase is present. **Partially miscible substances** *have limited solubility in each other.* Benzene (C_6H_6) and water are partially miscible. If benzene is added slowly to water, a small amount of benzene initially dissolves; a single phase results. However, as soon as the benzene solubility limit is reached, the excess benzene (saturated with water) forms a separate layer on top of the water (on top because it is less dense). **Immiscible substances** *do not dissolve in each other.* When such substances are mixed, two layers (phases) immediately form. Very few liquids are totally immiscible in each other; toluene (C_7H_8) and water approach this limiting case. Figure 13.2 illustrates the results obtained by mixing liquids of various miscibilities with each other.

Another term commonly encountered in discussions of solutions is *aqueous solution.* An **aqueous solution** *is a solution in which water is the solvent.*

13.3 SOLUTION FORMATION

In a solution, solute particles are uniformly dispersed throughout the solvent. Considering what happens at the molecular level during the solution process will help us to understand how this is achieved.

For a solute to dissolve in a solvent, two types of interparticle attractions must be overcome: (1) attractions between solute particles (solute–solute attractions) and (2) attractions between solvent particles (solvent–solvent attractions). Only when these attractions are overcome can particles in both pure solute and pure solvent begin to intermingle. A new type of interaction, which exists only in solutions, arises as the result of the mixing of solute and solvent. This new interaction is the attraction between solute and solvent particles (solute–solvent attractions). These new attractions are the primary driving force for solution formation. The extent to which a substance dissolves depends on the degree to which the newly formed solute–solvent attractions are able to compensate for the energy needed to overcome the solute–solute and solvent–solvent interactions. A solute will not dissolve in a solvent if either solute–solute or solvent–solvent interactions are too strong to be offset by the formation of the new solute–solvent interactions.

A most important type of solution process is the dissolution of an ionic solid in water. Let us consider in detail the process of dissolving sodium chloride, a typical ionic solid, in water. We will consider the process to occur in steps. The fact that water molecules are polar (Sec. 7.15) is important in our considerations.

Figure 13.3 shows what is thought to happen when sodium chloride is placed in water. The polar water molecules become oriented so that the negative oxygen portion points toward positive sodium ions and the positive hydrogen portion points toward negative chloride ions. As the polar water molecules begin to surround ions on the crystal surface, they exert sufficient attraction to cause these ions to break away from the crystal surface. After leaving the crystal, the ion retains its surrounding group of water molecules; it has become a *hydrated ion*. As each hydrated ion leaves the surface, other ions are exposed to the water, and the crystal is picked apart ion by ion. Once in solution, the hydrated ions are uniformly distributed by stirring or by random collisions with other molecules or ions.

The random motion of solute ions in solution causes them to collide with each other, with solvent molecules, and occasionally with the surface of the undissolved solute. Ions undergoing this last type of collision occasionally stick to the solid surface and thus leave the solution. When the number of ions in solution is low, the chances for collision with the undissolved solute are low. However, as the number of ions in solution increases, so do the chances for such collisions, and more ions are recaptured by the undissolved solute.

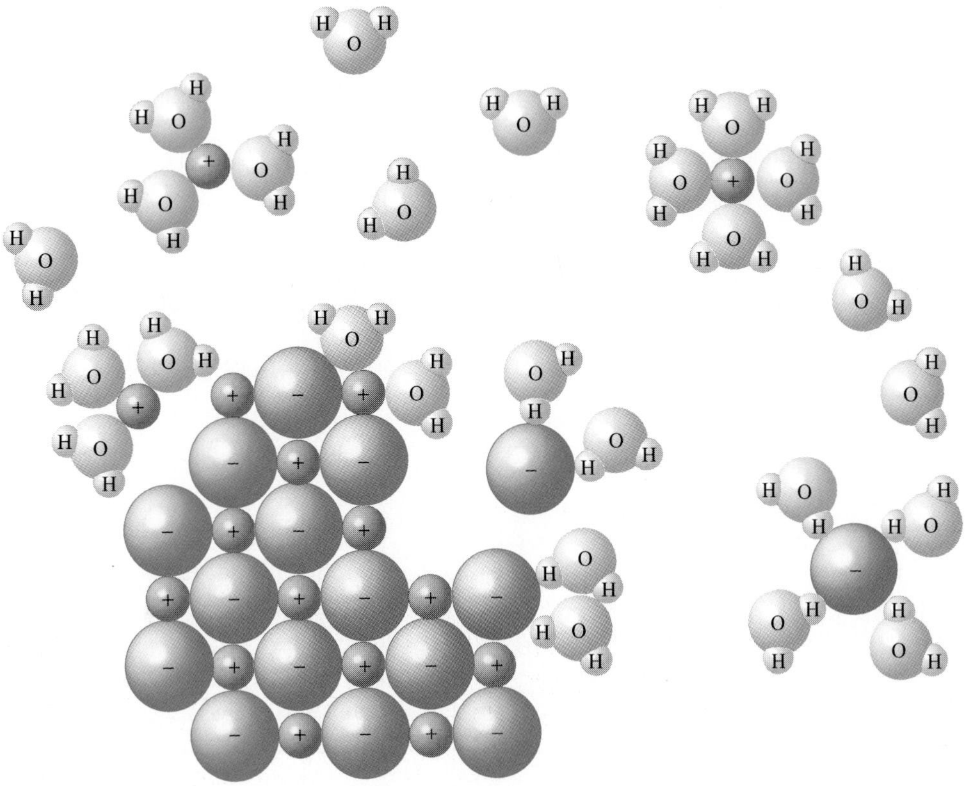

FIGURE 13.3 The solution process for an ionic solid in water.

Eventually, the number of ions in solution reaches such a level that ions return to the undissolved solute at the same rate as other ions leave. At this point the solution is saturated, and the equilibrium process discussed in the last section is in operation.

13.4 SOLUBILITY RULES

F.Y.I.

The principle of "like dissolves like" is used in offset lithography printing. Most books, including this one, are printed with this process

1. The printing plate is treated with a nonpolar, greasy lacquer in places where the image is to be printed on the paper.
2. The plate is moistened. The lacquer does not dissolve the polar water. The rest of the plate does.
3. The plate is inked. The ink is nonpolar and dissolves in the lacquer but not in the moistened areas.
4. Paper is pressed against the plate.
5. The ink transfers from the lacquer to the paper.

In this section we present some rules for qualitatively predicting solubilities. These rules summarize in a concise form the results of thousands of experimental solute–solvent solubility determinations.

A very useful generalization that relates polarity to solubility is *substances of like polarity tend to be more soluble in each other than substances that differ in polarity*. This conclusion is often expressed as the simple phrase *"like dissolves like."* Polar substances, in general, are good solvents for other polar substances but not for nonpolar substances. Similarly, nonpolar substances exhibit greater solubility in nonpolar solvents than they do in polar solvents.

The generalization "like dissolves like" is a useful tool for predicting solubility behavior in many, but not all, solute–solvent situations. Results that agree with the generalization are almost always obtained in the cases of gas-in-liquid and liquid-in-liquid solutions and for solid-in-liquid solutions in which the solute is not an ionic compound. For example, NH_3 gas (a polar gas) is much more soluble in H_2O (a polar liquid) than is O_2 gas (a nonpolar gas). (The actual solubilities of NH_3 and O_2 in water at 20°C are, respectively, 51.8 g/100 g H_2O and 0.0043 g/100 g H_2O.)

For those solid-in-liquid solutions in which the solute is an ionic compound—a very common situation—the rule "like dissolves like" is not adequate. One would predict that since all ionic compounds are polar, they all would dissolve in polar solvents such as water. This is not the case. The failure of the generalization here is related to the complexity of the factors involved in determining the magnitude of the solute–solute (ion–ion) and solute–solvent (ion–polar solvent molecule) interactions. Among other things, both the charge on and size of the ions in the solute must be considered. Changes in these factors affect both types of interactions, but not to the same extent.

Some guidelines concerning the solubility of ionic compounds in water, which should be used in place of "like dissolves like," are given in Table 13.4.

TABLE 13.4 SOLUBILITY GUIDELINES FOR IONIC COMPOUNDS IN WATER

If ____ Ion Is Present	Compound Is ____	Unless ____ Ion Also Is Present
Group IA (Li^+, Na^+, K^+, etc.)	soluble	
Ammonium (NH_4^+)	soluble	
Acetate ($C_2H_3O_2^-$)	soluble	
Nitrate (NO_3^-)	soluble	
Chloride (Cl^-), bromide (Br^-), and iodide (I^-)	soluble	Ag^+, Pb^{2+}, Hg_2^{2+}
Sulfate (SO_4^{2-})	soluble	Ca^{2+}, Sr^{2+}, Ba^{2+}, Pb^{2+}
Carbonate (CO_3^{2-})	insoluble	group IA and NH_4^+
Phosphate (PO_4^{3-})	insoluble	group IA and NH_4^+
Sulfide (S^{2-})	insoluble	groups IA and IIA and NH_4^+
Hydroxide (OH^-)	insoluble	group IA, Ba^{2+}, Sr^{2+}, Ca^{2+}

All ionic compounds, even the most insoluble ones, dissolve to a slight extent (see Table 13.2) in water. The insoluble classification used in Table 13.4 thus really means ionic compounds that have very limited solubility in water.

You should become thoroughly familiar with the rules in Table 13.4. They find extensive use in chemical discussions. We will next encounter them in Section 14.7 when the topic of net ionic equations is presented.

Example 13.1

Predict the solubility of each solute-in-solvent combination.

(a) acetone (a polar liquid) in water

(b) ammonia (a polar gas) in benzene (a nonpolar liquid)

(c) NaBr (an ionic solid) in water

(d) $CaCO_3$ (an ionic solid) in water

(e) AgCl (an ionic solid) in water

Solution

(a) Soluble. Acetone is polar, as is water. Like dissolves like.

(b) Insoluble. Since the two substances are of unlike polarity they should be relatively insoluble in each other.

(c) Soluble. Table 13.4 indicates that all Na^+ salts are soluble.

(d) Insoluble. Table 13.4 indicates that all carbonates are insoluble except those of group IA ions and NH_4^+ ion.

(e) Insoluble. Table 13.4 indicates that silver is an exception to the rule that all chlorides are soluble.

Practice Exercise 13.1

Predict the solubility of each solute-in-solvent combination.

(a) SO_2 (a polar gas) in water
(b) paraffin wax (a nonpolar solid) in CCl_4 (a nonpolar liquid)
(c) $AgNO_3$ (an ionic solid) in water
(d) $BaSO_4$ (an ionic solid) in water
(e) NH_4Cl (an ionic solid) in water

Ans. (a) soluble; (b) soluble; (c) soluble; (d) insoluble; (e) soluble

13.5 SOLUTION CONCENTRATIONS

In Section 13.2 we learned that, in general, there is a limit to the amount of solute that can be dissolved in a specified amount of solvent and also that a solution is said to be saturated when this maximum amount of solute has been dissolved. The amount of dissolved solute in a saturated solution is given by the solute's solubility.

Most solutions chemists deal with are *unsaturated* rather than saturated solutions. The amount of solute present in an unsaturated solution is specified by stating the concentration of the solution. The **concentration** *of a solution is the amount of solute present in a specified amount of solvent or a specified amount of solution*. Thus, concentration is a ratio of two quantities, being either the ratio

$$\frac{\text{Amount of solute}}{\text{Amount of solvent}} \quad \text{or} \quad \frac{\text{Amount of solute}}{\text{Amount of solution}}$$

In specifying a concentration, what are the units used to indicate the amounts of solute and solvent or solution present? In practice, a number of different unit combinations are used, with the choice of units depending on the use to be made of the concentration units. In each of the next four sections we shall discuss a commonly encountered set of units used to express solution concentration. The concentration expressions to be discussed are (1) percentage of solute (Sec. 13.6), (2) parts per million and parts per billion (Sec. 13.7), (3) molarity (Sec. 13.8), and (4) molality (Sec. 13.9). A fifth concentration unit, normality, which is used extensively in situations that involve the reactions of acids and bases, will be discussed in the next chapter (Sec. 14.17) where the topic of acids and bases is considered.

13.6 CONCENTRATION: PERCENTAGE OF SOLUTE

F.Y.I.

Normal saline used in hypodermic solutions is 0.9% NaCl (m/v).

F.Y.I.

D5W solutions used in hospital IVs are 5% glucose sugar (m/v).

F.Y.I.

The karat system for gold solutions (alloys) is
24 karat is 100% gold,
18 karat is 75% gold (m/m), and
14 karat is 58.5% gold (m/m).

The concentration of a solution is often specified in terms of the percentage of solute in the total amount of solution. Since the amounts of solute and solution present can be stated in terms of either mass or volume, different types of percent units exist. The three most common are

1. Percent by mass (or mass–mass percent).
2. Percent by volume (or volume–volume percent).
3. Mass–volume percent.

The percent unit most frequently used by chemists is percent by mass (or mass–mass percent). **Percent by mass** *is equal to the mass of solute divided by the total mass of solution multiplied by 100 (to put the value in terms of percentage)*. (Percentage is always part of the whole divided by the whole times 100; see Sec. 3.9.)

$$\text{Percent by mass} = \frac{\text{mass of solute}}{\text{mass of solution}} \times 100$$

The solute and solution masses must be in the same units but any units are allowed. The mass of solution is equal to the mass of *solute* plus the mass of *solvent*.

$$\text{Percent by mass} = \frac{\text{mass of solute}}{\text{mass of solute} + \text{mass of solvent}} \times 100$$

A solution of 5.0% by mass concentration would contain 5.0 g of solute in 100.0 g of solution (5.0 g of solute and 95.0 g of solvent). Thus, percent by mass gives directly the number of grams of solute in 100 g of solution. The abbreviation for percent by mass is % (m/m).

Example 13.2

A solution of hydrogen sulfide, H_2S, in water is prepared by bubbling H_2S gas into water. Calculate the solution concentration, as percent by mass, given that 0.290 g of H_2S dissolves in 75.00 g of water.

Solution

To calculate percent by mass we need both mass of solute and mass of solution.

$$\text{Percent by mass} = \frac{\text{mass of solute}}{\text{mass of solution}} \times 100$$

The mass of solute is given (0.290 g) and the mass of solution is calculated by adding together mass of solute and mass of solvent.

$$\text{Mass of solution} = 0.290 \text{ g} + 75.00 \text{ g} = 75.29 \text{ g} \quad \text{(calculator and correct answer)}$$

Substituting known values into the defining equation for percent by mass gives

$$\text{Percent by mass} = \frac{0.290 \cancel{\text{g}}}{75.29 \cancel{\text{g}}} \times 100 = 0.38517731\% \quad \text{(calculator answer)}$$

$$= 0.385\% \quad \text{(correct answer)}$$

Practice Exercise 13.2

What is the percent by mass, % (m/m), concentration of sucrose (table sugar) in a solution made by dissolving 5.4 g of sucrose in enough water to give 87.3 g of solution?

Ans. 6.2% (m/m)

Example 13.3

Lactose, $C_{12}H_{22}O_{11}$, is a naturally occurring sugar found in mammalian milk. How many grams of lactose must be added to 25.0 g of water to prepare a 5.00% (m/m) aqueous solution of lactose?

Solution

Often, when a solution concentration is given as part of a problem statement, the concentration information is used in the form of a conversion factor in solving the problem. That will be the case in this problem.

The given quantity is 25.0 g of H_2O (grams of solvent), and the desired quantity is grams of lactose (grams of solute).

$$25.0 \text{ g } H_2O = ? \text{ g lactose}$$

The conversion factor relating these two quantities (solvent and solute) is obtained from the given concentration. In a 5.00% (m/m) lactose solution there are 5.00 g of lactose for every 95.00 g of H_2O.

$$100.00 \text{ g solution} - 5.00 \text{ g lactose} = 95.00 \text{ g } H_2O$$

This relationship between grams of solute and grams of solvent (5.00 to 95.00) gives us the needed conversion factor.

$$\frac{5.00 \text{ g lactose}}{95.00 \text{ g } H_2O}$$

Dimensional analysis gives the problem setup, which is solved in the following manner.

$$25.0 \cancel{\text{g } H_2O} \times \frac{5.00 \text{ g lactose}}{95.00 \cancel{\text{g } H_2O}} = 1.3157894 \text{ g lactose} \quad \text{(calculator answer)}$$

$$= 1.32 \text{ g lactose} \quad \text{(correct answer)}$$

Practice Exercise 13.3

How many grams of lithium nitrate, $LiNO_3$, must be added to 15.0 g of water to prepare a 7.50% (m/m) solution of lithium nitrate?

Ans. 1.22 g $LiNO_3$

Percent by volume (or volume–volume percent) finds use as a concentration unit when both the solute and solvent are liquids or gases. In such cases it is often more convenient to measure volumes than masses. **Percent by volume** *is equal to the volume of solute divided by the total volume of solution multiplied by 100.*

$$\text{Percent by volume} = \frac{\text{volume of solute}}{\text{volume of solution}} \times 100$$

F.Y.I.

The proof system used for alcoholic beverages is twice the volume/volume percent. 40 proof is 20% alcohol (v/v), 100 proof is 50% (v/v) alcohol.

Solute and solution volumes must always be expressed in the same units when this expression is used. The abbreviation for percent by volume is % (v/v).

The numerical value of a concentration expressed as a percent by volume gives directly the number of milliliters of solute in 100 mL of solution. Thus, a 100-mL sample of a 5.0% alcohol-in-water solution contains 5.0 mL of alcohol dissolved in enough water to give 100 mL of solution. Note that such a 5.0% by volume solution could not be made by adding 5 mL of alcohol to 95 mL of water, since volumes of liquids are not usually additive. Differences in the way molecules are packed as well as in the distances between molecules almost always result in the volume of a solution being different from the sum of the volumes of solute and solvent. For example, the final volume resulting from the addition of 50.0 mL of ethyl alcohol to 50.0 mL of water is 96.5 mL of solution.

Example 13.4

A windshield washer solution is made by mixing 37.8 mL of methanol with 56.2 mL of water to produce 80.0 mL of solution. What is the concentration of methanol in the solution expressed as percent by volume methanol?

Solution

To calculate a percent by volume, the volumes of methanol and solution are needed. Both are given in this problem.

$$\text{Methanol volume} = 37.8 \text{ mL}$$
$$\text{Solution volume} = 80.0 \text{ mL}$$

Note that the solution volume is not the sum of the solute and solvent volumes. As previously mentioned, liquid volumes are generally not additive.

Substituting the given values into the equation

$$\text{Percent by volume} = \frac{\text{volume of methanol}}{\text{volume of solution}} \times 100$$

gives

$$\text{Percent by volume} = \frac{37.8 \text{ mL}}{80.0 \text{ mL}} \times 100 = 47.25\% \quad \text{(calculator answer)}$$
$$= 47.2\% \quad \text{(correct answer)}$$

Practice Exercise 3.4

The final volume of a solution resulting from the addition of 50.0 mL of ethyl alcohol to 50.0 mL of water is 96.5 mL. What is the volume percent of ethyl alcohol in the solution?

Ans. 51.8% (v/v)

The third type of percentage unit in common use is mass–volume percentage. This unit, which is often encountered in hospital and industrial settings, is particularly convenient to use when working with a solid solute (which is easily weighed) and a liquid solvent. Concentrations are specified using this unit when dealing with physiological fluids such as blood and urine. **Mass–volume percent** *is equal to the mass of solute (in grams) divided by the total volume of solution (milliliters) multiplied by 100.*

$$\text{Mass–volume percent} = \frac{\text{mass of solute (g)}}{\text{volume of solution (mL)}} \times 100$$

Note that specific mass and volume units are given in the definition of mass–volume percent. This is necessary because the units do not cancel as was the case with mass percent and volume percent. The abbreviation for mass–volume percent is % (m/v).

Example 13.5

Vinegar is a 5.0% (m/v) aqueous solution of acetic acid ($HC_2H_3O_2$). How much acetic acid, in grams, is present in one teaspoon (5.0 mL) of vinegar?

Solution

The given quantity is 5.0 mL of vinegar, and the desired quantity is grams of acetic acid.

$$5.0 \text{ mL vinegar} = ? \text{ g acetic acid}$$

The given concentration of 5.0% (m/v), which means 5.0 g acetic acid per 100 mL vinegar, can be used as a conversion factor to go from milliliters of vinegar to grams of acetic acid. The setup for the conversion is

$$5.0\ \cancel{\text{mL vinegar}} \times \frac{5.0\ \text{g acetic acid}}{100\ \cancel{\text{mL vinegar}}}$$

Doing the arithmetic, after cancellation of units, gives

$$\frac{5.0 \times 5.0}{100}\ \text{g acetic acid} = 0.25\ \text{g acetic acid (calculator and correct answer)}$$

Practice Exercise 13.5

Saline solution, a 0.92% (m/v) sodium chloride (NaCl) solution, is often administered intravenously to hospital patients. How many grams of sodium chloride are required to prepare 345 mL of saline solution?

Ans. 3.2 g NaCl

13.7 CONCENTRATION: PARTS PER MILLION AND PARTS PER BILLION

The concentration units parts per million (ppm) and parts per billion (ppb) find use when dealing with extremely dilute solutions. Environmental chemists frequently use such units in specifying the concentrations of the minute amounts of trace pollutants or toxic chemicals in air and water samples.

F.Y.I.
A ppm is the equivalent of one second in eleven days and twelve hours.

F.Y.I.
A ppb is the equivalent of one second in thirty-one years and eight months.

F.Y.I.
The maximum concentration of the pollutant vinyl chloride allowed by the EPA in drinking water is 2 ppb (m/v).

F.Y.I.
Most municipalities that fluorinate their water add fluoride at 1 ppm (m/v).

Parts per million and parts per billion units are closely related to percentage concentration units. Not only are the defining equations very similar, but also various forms of the units exist. Because amounts of solute and solution present may be stated in terms of either mass or volume, there are three different forms for each unit: mass–mass (m/m), volume–volume (v/v), and mass–volume (m/v).

A **part per million** *(ppm) is one part of solute per million parts of solution.* In terms of defining equations, we can write

$$\text{ppm (m/m)} = \frac{\text{mass of solute}}{\text{mass of solution}} \times 10^6$$

$$\text{ppm (v/v)} = \frac{\text{volume of solute}}{\text{volume of solution}} \times 10^6$$

$$\text{ppm (m/v)} = \frac{\text{mass of solute (g)}}{\text{volume of solution (mL)}} \times 10^6$$

Note that the units of grams and milliliters are specified in the last of the three defining equations, but that no units are given in the first two equations. For the first two equations, the only unit restriction is that the units be the same for both numerator and denominator.

A **part per billion** *(ppb) is one part of solute per billion parts of solution.* The mathematical defining equations for the three types of part per billion units are identical to those just shown for parts per million except that a multiplicative factor of 10^9 instead of 10^6 is used.

The use of parts per million and parts per billion in specifying concentration often avoids the very small numbers that result when other concentration units are used. For example, a pollutant in water might be present at a level of 0.0013 g per 100 mL of solution. In terms of mass–volume percent, this concentration is 0.0013%. In parts per million, however, the concentration is 13.

$$\text{ppm (m/v)} = \frac{0.0013\text{ g}}{100\text{ mL}} \times 10^6 = 13$$

The only difference in the ways in which percent concentrations and parts per million or billion are calculated is in the multiplicative factor used. For percentages it is 10^2, for parts per million 10^6, and for parts per billion 10^9. An alternative name for percentage concentration units would be *parts per hundred*.

Example 13.6

The concentration of sodium fluoride, NaF, in a town's fluoridated tap water is found to be 32.2 mg of NaF per 20.0 kg of tap water. Express this NaF concentration in : **(a)** ppm (m/m) and **(b)** ppb (m/m).

Solution

(a) The defining equation for ppm (m/m) is

$$\text{ppm (m/m)} = \frac{\text{mass of solute}}{\text{mass of solution}} \times 10^6$$

The two masses in this equation must be in the same units. Let us use grams as our mass unit. Expressing the given quantities in terms of grams, we have

$$32.3\text{ mg NaF} = 3.23 \times 10^{-2}\text{ g NaF}$$
$$20.0\text{ kg tap water} = 2.00 \times 10^4\text{ g tap water}$$

Substituting these gram quantities into the defining equation for ppm (m/m) gives

$$\text{ppm (m/m)} = \frac{3.23 \times 10^{-2}\,\cancel{\text{g}}}{2.00 \times 10^4\,\cancel{\text{g}}} \times 10^6$$

$$= 1.615 \quad \text{(calculator answer)}$$
$$= 1.62 \quad \text{(correct answer)}$$

(b) For parts per billion we have

$$\text{ppb (m/m)} = \frac{3.23 \times 10^{-2}\,\cancel{\text{g}}}{2.00 \times 10^4\,\cancel{\text{g}}} \times 10^9$$

$$= 1615 \quad \text{(calculator answer)}$$
$$= 1620 \quad \text{(correct answer)}$$

Any time a concentration is expressed in both parts per million and parts per billion, the parts per billion value will be 1000 times larger than the parts per million value.

Practice Exercise 13.6

A 500.0-mg aspirin tablet is found to contain 19 μg of a nontoxic contaminant. What is the concentration of the contaminant in (a) ppm (m/m) and (b) ppb (m/m)?

Ans. (a) 38 ppm (m/m); (b) 38,000 ppb (m/m)

Example 13.7

Agent Orange, a defoliant used on a large scale during the Vietnam war, contains about 2 ppm (m/v) of dioxin, a very toxic substance, as an impurity. How many milliliters of Agent Orange would have to be used in order to place 0.01 g of dioxin in the environment?

Solution

The given quantity is 0.01 g of dioxin, and the desired quantity is milliliters of Agent Orange

$$0.01 \text{ g dioxin} = ? \text{ mL Agent Orange}$$

The given concentration of 2ppm (m/v), which means 2 grams of dioxin per 10^6 mL of Agent Orange solution can be used as a conversion factor to go from grams of dioxin to milliliters of Agent Orange. The setup for the conversion is

$$0.01 \cancel{\text{g dioxin}} \times \frac{10^6 \text{ mL Agent Orange}}{2 \cancel{\text{g dioxin}}}$$

Doing the arithmetic, after cancellation of units, gives

$$\frac{0.01 \times 10^6}{2} \text{ mL Agent Orange} = 5000 \text{ mL Agent Orange} \quad \text{(calculator and correct answer)}$$

Practice Exercise 13.7

The carbon monoxide, CO, content of the tobacco smoke that reaches a smoker's lungs is estimated to be $2\bar{0}0$ ppm (v/v). At this concentration, how much CO, in milliliters, would be present in a sample of air the size of a standard-sized basketball (7.5 L)?

Ans. 1.5 mL CO

13.8 CONCENTRATION: MOLARITY

The **molarity** *of a solution, abbreviated* M, *is a ratio giving the number of moles of solute per liter of solution.*

$$\text{Molarity (M)} = \frac{\text{moles of solute}}{\text{liters of solution}}$$

A solution containing 1 mole of KBr in 1 L of solution has a molarity of 1 and is said to be a 1-M (1 *molar*) solution.

When a solution is to be used for a chemical reaction, concentration is almost always expressed in units of molarity. A major reason for this is the fact that the amount of solute is expressed in moles, a most convenient unit for dealing with stoichiometry in chemical reactions. Because chemical reactions occur between molecules and atoms, a unit that counts particles, as the mole does, is desirable.

To find the molarity of a solution we need to know the number of moles of solute present and the solution volume in liters and then take the ratio of the two quantities. An alternative to knowing the number of moles of solute is knowledge about the grams of solute present and the solute's molar mass.

Example 13.8

Determine the molarities of the following solutions.

(a) 1.45 moles of KCl dissolved in enough water to give 875 mL of solution

(b) 57.2 g of NH_4Br dissolved in enough water to give 2.15 L of solution

Solution

(a) The number of moles of solute is given in the problem statement.

$$\text{Moles of solute} = 1.45 \text{ moles KCl}$$

The volume of the solution is also given in the problem statement, but not in the right units. Molarity requires liters for volume units. Making the unit change gives

$$875 \text{ mL} \times \frac{10^{-3}\text{ L}}{1 \text{ mL}} = 0.875 \text{ L} \quad \text{(calculator and correct answer)}$$

The molarity of the solution is obtained by substituting the known quantities into the equation

$$\text{Molarity} = \frac{\text{moles of solute}}{\text{L of solution}}$$

which gives

$$\text{M} = \frac{1.45 \text{ moles KCl}}{0.875 \text{ L solution}} = 1.6571428 \frac{\text{moles KCl}}{\text{L solution}} \quad \text{(calculator answer)}$$

$$= 1.66 \frac{\text{moles KCl}}{\text{L solution}} \quad \text{(correct answer)}$$

Note that the units of molarity are always moles per liter.

(b) This time the volume of solution is given in the right units, liters.

$$\text{Volume of solution} = 2.15 \text{ L}$$

The moles of solute must be calculated from the grams of solute (given) and the solute's formula mass, which is 97.9 amu (calculated from a table of atomic masses).

$$57.2 \text{ g NH}_4\text{Br} \times \frac{1 \text{ mole NH}_4\text{Br}}{97.9 \text{ g NH}_4\text{Br}} = 0.58426966 \text{ mole NH}_4\text{Br} \quad \text{(calculator answer)}$$

$$= 0.584 \text{ mole NH}_4\text{Br} \quad \text{(correct answer)}$$

Substituting the known quantities into the defining equation for molarity gives

$$\text{M} = \frac{0.584 \text{ mole NH}_4\text{Br}}{2.15 \text{ L solution}} = 0.2716279 \frac{\text{mole NH}_4\text{Br}}{\text{L solution}} \quad \text{(calculator answer)}$$

$$= 0.272 \frac{\text{mole NH}_4\text{Br}}{\text{L solution}} \quad \text{(correct answer)}$$

Practice Exercise 13.8

What is the molarity of a solution prepared by dissolving 25.0 g of NaOH in enough water to give 2.50 L of solution?

Ans. 0.250 M NaOH

As the previous example indicates, when you perform a molarity calculation the chemical formula of the solute is always needed. You cannot calculate moles of solute without knowing the chemical formula of the solute. In contrast, when you perform percent concentration calculations (or parts per million or billion)—Sections 13.6 and 13.7—the chemical formula of the solute is not used in the calculation.

The moles of solute present in a known volume of solution is an easily calculated quantity if the molarity of the solution is known. In doing such a calculation, molarity serves as a conversion factor relating liters of solution to moles of solute.

Volume of solution (liters)	←molarity→	Moles of solute

Example 13.9

Citric acid, $C_6H_8O_7$, is the substance that gives lemon juice and other citrus fruit juices a sour taste. How many grams of citric acid are present in 125 mL of a 0.400 M citric acid solution?

Solution

The given quantity is 125 mL of solution, and the desired quantity is grams of $C_6H_8O_7$.

$$125 \text{ mL solution} = ? \text{ g } C_6H_8O_7$$

The pathway to be used in solving this problem is

$$\text{mL solution} \longrightarrow \text{L solution} \longrightarrow \text{moles } C_6H_8O_7 \longrightarrow \text{g } C_6H_8O_7$$

The given molarity (0.400 M) will serve as the conversion factor for the second unit change; the molecular mass of citric acid (which must be calculated as it is not given) is used in accomplishing the third unit charge.

The dimensional analysis setup from this pathway is

$$125 \cancel{\text{mL solution}} \times \frac{10^{-3} \cancel{\text{L solution}}}{1 \cancel{\text{mL solution}}} \times \frac{0.400 \cancel{\text{mole } C_6H_8O_7}}{1 \cancel{\text{L solution}}} \times \frac{192 \text{ g } C_6H_8O_7}{1 \cancel{\text{mole } C_6H_8O_7}}$$

Canceling units and doing the arithmetic gives

$$\frac{125 \times 10^{-3} \times 0.400 \times 192}{1 \times 1 \times 1} \text{ g } C_6H_8O_7 = 9.6 \text{ g } C_6H_8O_7 \quad \text{(calculator answer)}$$

$$= 9.60 \text{ g } C_6H_8O_7 \quad \text{(correct answer)}$$

Practice Example 13.9

How many grams of ascorbic acid (vitamin C), $C_6H_8O_6$, are present in 125 mL of a 0.400-M vitamin C solution?

Ans. 8.80 g $C_6H_8O_6$

Example 13.10

The chemical formula for sucrose (table sugar) is $C_{12}H_{22}O_{11}$. How many liters of 2.00 M aqueous sucrose solution can be prepared from 25.0 g of sucrose?

Solution

The given quantity is 25.0 g of $C_{12}H_{22}O_{11}$, and the desired quantity is liters of solution.

$$25.0\text{ g } C_{12}H_{22}O_{11} = ?\text{ L } C_{12}H_{22}O_{11}\text{ solution}$$

The pathway to be used in solving this problem will involve the following steps.

$$\text{g } C_{12}H_{22}O_{11} \longrightarrow \text{moles } C_{12}H_{22}O_{11} \longrightarrow \text{L } C_{12}H_{22}O_{11}\text{ solution}$$

The first unit conversion will be accomplished by using the molar mass of $C_{12}H_{22}O_{11}$ (which must be calculated since it is not given) as a conversion factor. The second unit conversion involves the use of the given molarity as a conversion factor.

$$25.0\ \cancel{\text{g } C_{12}H_{22}O_{11}} \times \frac{1\ \cancel{\text{mole } C_{12}H_{22}O_{11}}}{342\ \cancel{\text{g } C_{12}H_{22}O_{11}}} \times \frac{1\text{ L } C_{12}H_{22}O_{11}\text{ solution}}{2.00\ \cancel{\text{moles } C_{12}H_{22}O_{11}}}$$

Canceling units and doing the arithmetic gives

$$\frac{25.0 \times 1 \times 1}{342 \times 2.00}\text{ L } C_{12}H_{22}O_{11}\text{ solution} = 0.036549707\text{ L } C_{12}H_{22}O_{11}\text{ solution}$$

(calculator answer)

$$= 0.365\text{ L } C_{12}H_{22}O_{11}\text{ solution}$$ (correct answer)

Practice Exercise 13.10

How many liters of 0.100 M aqueous sodium hydroxide (NaOH) solution can be prepared from 10.0 g of sodium hydroxide?

Ans. 2.50 L solution

Molarity and mass percent are probably the two most commonly used concentration units. The need to convert from one to the other often arises. Such a conversion can easily be done provided the density of the solution is known. Figure 13.4 shows schematically the steps involved in converting one of these concentration units to the other.

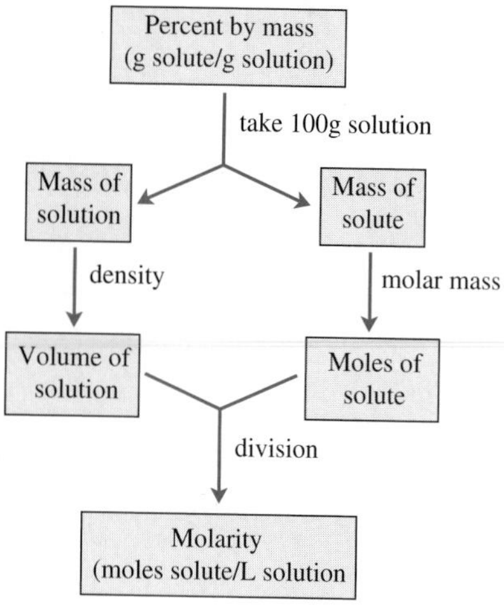

FIGURE 13.4 A "road-map" diagram showing the steps involved in converting from percent by mass to molarity or vice versa. (For the reverse process, reverse the direction of the arrows in the diagram.)

Example 13.11

The skin irritation that accompanies insect bites is often caused by formic acid (H_2CO_2). A 40.00 percent by mass aqueous solution of formic acid has a density of 1.098 g/mL. What is the molarity of this solution?

Solution

Calculate the moles of solute and liters of solution present in a sample of this solution. Since solution concentration is independent of sample size, any size sample can be the basis for the calculation. To simplify the math take a 100.0-g sample of solution.

STEP 1 *Moles of solute.* The given quantity is 100.0 g of solution, and the desired quantity is moles of H_2CO_2.

$$100.0 \text{ g solution} = ? \text{ moles } H_2CO_2$$

The pathway to be used in solving this problem is

$$\text{g solution} \longrightarrow \text{g solute} \longrightarrow \text{moles solute}$$

The known mass percent concentration will be the basis for the conversion factor that takes us from grams of solution to grams of solute.

$$100.0 \cancel{\text{ g solution}} \times \frac{40.00 \cancel{\text{ g } H_2CO_2}}{100 \cancel{\text{ g solution}}} \times \frac{1 \text{ mole } H_2CO_2}{46.03 \cancel{\text{ g } H_2CO_2}}$$

$$= 0.86899847 \text{ mole } H_2CO_2 \quad \text{(calculator answer)}$$
$$= 0.8690 \text{ mole } H_2CO_2 \quad \text{(correct answer)}$$

STEP 2 *Liters of solution.* The density of the solution is used as a conversion factor in obtaining the volume of solution. The pathway for the calculation is

$$\text{g solution} \longrightarrow \text{mL solution} \longrightarrow \text{L solution}$$

The set up is

$$100.0 \cancel{\text{ g solution}} \times \frac{1 \cancel{\text{ mL solution}}}{1.098 \cancel{\text{ g solution}}} \times \frac{10^{-3} \text{ L solution}}{1 \cancel{\text{ mL solution}}}$$

$$= 0.091074681 \text{ L solution} \quad \text{(calculator answer)}$$
$$= 0.09107 \text{ L solution} \quad \text{(correct answer)}$$

STEP 3 *Molarity.* With both moles of solute and liters of solution known, the molarity is obtained by substitution into the defining equation for molarity:

$$M = \frac{\text{moles } H_2CO_2}{\text{L solution}} = \frac{0.8690 \text{ mole } H_2CO_2}{0.09107 \text{ L solution}}$$

$$= \frac{9.5421104 \text{ moles } H_2CO_2}{\text{L solution}} \quad \text{(calculator answer)}$$

$$= \frac{9.542 \text{ moles } H_2CO_2}{\text{L solution}} \quad \text{(correct answer)}$$

Practice Exercise 13.11

A 15.00 percent by mass aqueous solution of silver nitrate ($AgNO_3$) has a density of 1.141 g/mL. What is the molarity of this solution?

Ans. 1.007 M

Molar concentrations do not give information about the amount of *solvent* present. All that is known is that enough solvent is present to give a specific volume of *solution*. The amount of solvent present in a solution of a known molarity can be calculated if the density of the solution is known. Without the density it cannot be calculated.

Sample 13.12

Large amounts of sulfuric acid (H_2SO_4) are used in the production of phosphate fertilizers. A 2.324 M H_2SO_4 solution has a density of 1.142 g/mL. How many grams of solvent (water) are present in 25.0 mL of this solution?

Solution

To find the grams of solvent present we must first find the grams of solute (H_2SO_4) and the grams of solution. The grams of solvent present is then obtained by calculating the difference.

$$\text{g solvent} = \text{g solution} - \text{g solute}$$

STEP 1 *Grams of solution.* The volume of solution is given. Density, used as a conversion factor, will enable us to convert this volume to grams of solution.

$$25.0 \text{ mL solution} \times \frac{1.142 \text{ g solution}}{1 \text{ mL solution}} = 28.55 \text{ g solution} \quad \text{(calculator answer)}$$

$$= 28.6 \text{ g solution} \quad \text{(correct answer)}$$

STEP 2 *Grams of solute.* We will use the molarity of the solution as a conversion factor in obtaining the grams of solute. The setup for this calculation is similar to that in Example 13.9.

$$25.0 \text{ mL solution} \times \frac{10^{-3} \text{ L solution}}{1 \text{ mL solution}} \times \frac{2.324 \text{ moles } H_2SO_4}{1 \text{ L solution}} \times \frac{98.1 \text{ g } H_2SO_4}{1 \text{ mole } H_2SO_4}$$

$= 5.69961 \text{ g } H_2SO_4$ (calculator answer)
$= 5.70 \text{ g } H_2SO_4$ (correct answer)

STEP 3 *Grams of solvent.* The grams of solvent will be the difference in mass between the grams of solution and the grams of solute.

28.6 g solution − 5.70 g solute = 22.9 g solvent (calculator and correct answer)

Practice Exercise 13.12

A 0.750-M acetic acid ($HC_2H_3O_2$) solution has a density of 1.01 g/mL. How many grams of solvent are present in 125 mL of this solution?

Ans. $12\bar{0}$ g solvent

13.9 CONCENTRATION: MOLALITY

Molality is a concentration unit based on a fixed amount of *solvent* and is used in areas where this is a concern. Despite this unit having a name very similar to molarity, molality differs distinctly from molarity; molarity is a unit based on a fixed amount of *solution* rather than a fixed amount of *solvent*. *The* **molality** *of a solution, abbreviated m, is a ratio giving the number of moles of solute per kilogram of solvent.*

$$\text{Molality } (m) = \frac{\text{moles of solute}}{\text{kilograms of solvent}}$$

Molality also finds use, in preference to molarity, in experimental situations where changes in temperature are of concern. Molality is a temperature-independent concentration unit; molarity is not. To be temperature independent, a concentration unit cannot involve a volume measurement. Volumes of solutions change (expand or contract) with changes in temperature. A change in temperature thus means a change in concentration, even though the amount of solute remains constant, if a concentration unit has a volume dependency. Volume changes caused by temperature change are usually very, very small; consequently, temperature independence or dependence is a factor in only the most precise experimental measurements.

Careful note should be taken of the fact that the same letter of the alphabet is used as an abbreviation for both molality and molarity—a lowercase, italic m for molality (m) and a capitalized M for molarity (M).

In dilute aqueous solutions molarity and molality are practically identical in numerical value. This results from dilute aqueous solution having a density of 1.0 g/L. Molarity and molality have significantly different values when the solvent has a density that is not equal to unity or when the solution is concentrated.

Example 13.13

A typical dose of iron(II) sulfate used in the treatment of iron-deficiency anemia is 0.35 g. What would be the molality of a solution made by dissolving 0.35 g of $FeSO_4$ in 225 g of H_2O?

Solution

To calculate molality the number of moles of solute and the solvent mass in kilograms must be known.

In this problem the solvent mass is given, but in grams rather than kilograms. We, thus, need to change the grams unit to kilograms.

$$225 \text{ g H}_2\text{O} \times \frac{1 \text{ kg H}_2\text{O}}{10^3 \text{ g H}_2\text{O}} = 0.225 \text{ kg H}_2\text{O} \quad \text{(calculator and correct answer)}$$

Information about the solute is given in terms of grams. We can calculate moles of solute from the given information by using molar mass (which is not given and must be calculated) as a conversion factor.

$$0.35 \text{ g FeSO}_4 \times \frac{1 \text{ mole FeSO}_4}{152 \text{ g FeSO}_4} = 2.3026315 \times 10^{-3} \text{ mole FeSO}_4 \quad \text{(calculator answer)}$$

$$= 2.3 \times 10^{-3} \text{ mole FeSO}_4 \quad \text{(correct answer)}$$

Substituting moles of solute and kilograms of solvent into the defining equation for molality gives

$$m = \frac{\text{moles solute}}{\text{kg solvent}} = \frac{2.3 \times 10^{-3} \text{ mole FeSO}_4}{0.225 \text{ kg H}_2\text{O}}$$

$$= 0.010222222 \frac{\text{mole FeSO}_4}{\text{kg H}_2\text{O}} \quad \text{(calculator answer)}$$

$$= 0.010 \frac{\text{mole FeSO}_4}{\text{kg H}_2\text{O}} \quad \text{(correct answer)}$$

Practice Exercise 13.13

Calculate the molality of a solution made by dissolving 25.0 g of potassium carbonate (K_2CO_3) in 725 g of H_2O.

Ans. 0.250 m

Example 13.14

Calculate the number of grams of isopropyl alcohol, C_3H_8O, which must be added to 275 g of water to prepare a 2.00 *m* solution of isopropyl alcohol.

Solution

The given quantity is 275 g of H_2O and the desired quantity is grams of C_3H_8O.

$$275 \text{ g H}_2\text{O} = \text{g C}_3\text{H}_8\text{O}$$

The pathway to be used in solving this problem is

g solvent ⟶ kg solvent ⟶ moles solute ⟶ g solute

The molality of the solution, which is given, will serve as a conversion factor to effect the change from kilograms of solvent to moles of solute.

The dimensional analysis setup for the problem is

$$275\ \text{g}\ \cancel{H_2O} \times \frac{1\ \cancel{\text{kg}\ H_2O}}{10^3\ \cancel{\text{g}\ H_2O}} \times \frac{2.00\ \cancel{\text{moles}\ C_3H_8O}}{1\ \cancel{\text{kg}\ H_2O}} \times \frac{60.1\ \text{g}\ C_3H_8O}{1\ \cancel{\text{mole}\ C_3H_8O}}$$

The second conversion factor involves the numerical value of the molality and the third conversion factor is based on the molar mass of C_3H_8O.

Canceling units and then doing the arithmetic gives

$$\frac{275 \times 1 \times 2.00 \times 60.1}{10^3 \times 1 \times 1}\ \text{g}\ C_3H_8O = 33.055\ \text{g}\ C_3H_8O \quad \text{(calculator answer)}$$

$$= 33.1\ \text{g}\ C_3H_8O \quad \text{(correct answer)}$$

Practice Exercise 13.14

Calculate the number of grams of potassium hydroxide, KOH, that must be added to 25.0 g of water to prepare a 0.0100 *m* solution.

Ans. 0.0140 g KOH

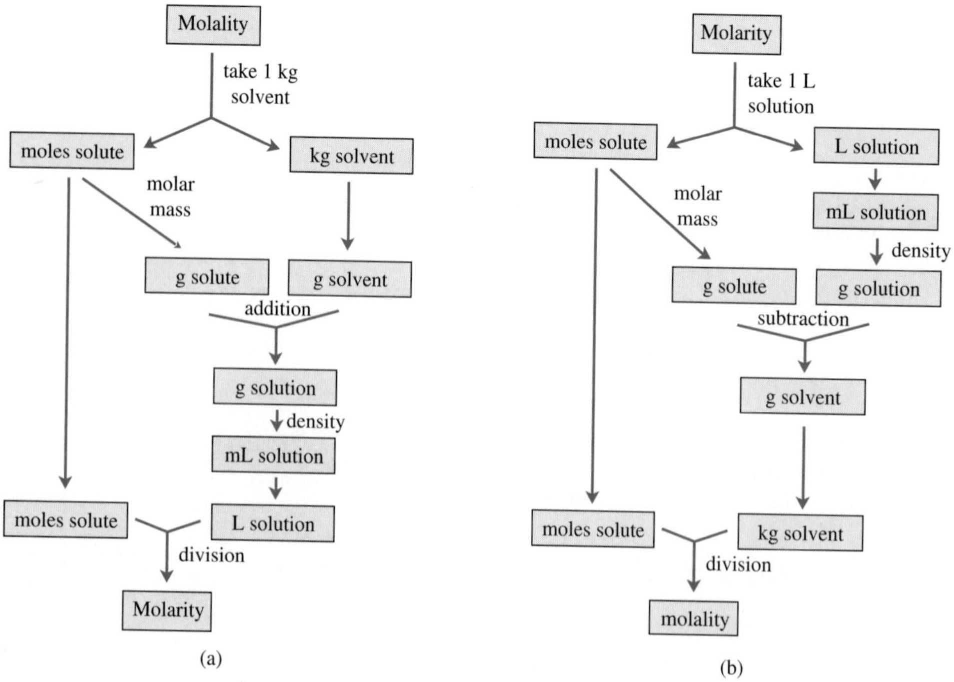

FIGURE 13.5 "Road-map" diagrams showing the steps involved in converting from molality to molarity concentration units and vice versa. (a) Molality to molality; (b) molarity to molarity.

Interconversion between molarity and molality concentration units requires knowledge of solution density. Figure 13.5 shows schematically the steps involved in such interconversions. Example 13.15 is a sample molality-to-molarity interconversion.

Example 13.15

Calculate the molarity of an 8.92 *m* (molal) ethyl alcohol (C_2H_6O) solution whose density is 0.927 g/mL.

Solution

The defining equation for molarity involves moles of solute (numerator) and liters of solution (denominator). The defining equation for molality involves moles of solute (numerator) and kilograms of solvent (denominator). The numerators of the two defining equations are the same and the denominators are different. The essence of this problem is, thus, the conversion of kilograms of solvent to liters of solution.

$$\text{kg solvent} \longrightarrow \text{L solution}$$

The "road-map" in Figure 13.5a will be our guide in effecting this change. The general pathway is

$$\text{kg solvent} \longrightarrow \text{g solvent} \longrightarrow \text{g solution} \longrightarrow \text{mL solution} \longrightarrow \text{L solution}$$

The basis for our calculation is an amount of solution containing a kilogram (1.000×10^3 g) of water. From the given molality, we know that there are 8.92 moles of ethyl alcohol per kilogram of water. The mass, in grams, of this amount of alcohol is

$$8.92 \text{ moles } C_2H_6O \times \frac{46.1 \text{ g } C_2H_6O}{1 \text{ mole } C_2H_6O} = 411.212 \text{ g } C_2H_6O \quad \text{(calculator answer)}$$

$$= 411 \text{ g } C_2H_6O \quad \text{(correct answer)}$$

The total mass of solution, in grams, equals the mass of water plus this mass of ethyl alcohol.

$$\underbrace{1.000 \times 10^3 \text{ g}}_{\text{water}} + \underbrace{411 \text{ g}}_{\text{ethyl alcohol}} = \underbrace{1411 \text{ g}}_{\text{solution}} \quad \text{(calculator and correct answer)}$$

The volume of this amount of solution is obtained by using density as a conversion factor.

$$1411 \text{ g solution} \times \frac{1 \text{ mL solution}}{0.927 \text{ g solution}} \times \frac{10^{-3} \text{ L solution}}{1 \text{ mL solution}}$$

$$= 1.5221143 \text{ L solution} \quad \text{(calculator answer)}$$
$$= 1.52 \text{ L solution} \quad \text{(correct answer)}$$

The molarity of the solution is

$$M = \frac{\text{moles solute}}{\text{L solution}} = \frac{8.92 \text{ moles solute}}{1.52 \text{ L solution}}$$

$$= 5.868421 \frac{\text{mole solute}}{\text{L solution}} \quad \text{(calculator answer)}$$

$$= 5.87 \frac{\text{moles solute}}{\text{L solution}} \quad \text{(correct answer)}$$

Solution
Three of the four variables in the equation

$$M_s \times V_s = M_d \times V_d$$

are known.

$$M_s = 0.95\ M \qquad M_d = ?\ M$$
$$V_s = 65\ mL \qquad V_d = 135\ mL$$

Rearranging the equation to isolate M_d on the left side and substituting the known variables into it gives

$$M_d = M_s \times \frac{V_s}{V_d}$$
$$= 0.95\ M \times \frac{65\ \cancel{mL}}{135\ \cancel{mL}} = 0.4574074\ M \quad \text{(calculator answer)}$$
$$= 0.46\ M \quad \text{(correct answer)}$$

Thus, the diluted solution's concentration is 0.46 M.

Practice Exercise 13.16

What is the molarity of the solution prepared by diluting 75 mL of 1.50 M silver nitrate ($AgNO_3$) solution to a final volume of 225 mL?

Ans. 0.50 M

Example 13.17

How much solvent must be added to 200.0 mL of a 1.25 M sodium chloride (NaCl) solution to decrease its concentration to 0.770 M?

Solution
The volume of solvent added is equal to the difference between the final and initial volumes. The initial volume is known. The final volume can be calculated using the equation

$$M_s \times V_s = M_d \times V_d$$

Once the final volume is known, the difference between the two volumes can be obtained.

Substituting the known quantities into the dilution equation, rearranged to isolate V_d on the left side, gives

$$V_d = V_s \times \frac{M_s}{M_d}$$
$$= 200.0\ mL \times \frac{1.25\ \cancel{M}}{0.770\ \cancel{M}} = 324.67532\ mL \quad \text{(calculator answer)}$$
$$= 325\ mL \quad \text{(correct answer)}$$

The solvent added is

$$V_d - V_s = (325 - 200.0)\ mL = 125\ mL \quad \text{(calculator and correct answer)}$$

Practice Exercise 13.17

How much solvent must be added to 50.0 mL of 2.20 M potassium chloride (KCl) solution to decrease its concentration to 0.0113 M?

Ans. 9680 mL

When two "like" solutions—that is, solutions that contain the same solute and the same solvent—of differing known molarities and volumes are mixed together, the molarity of the newly formed solution can be calculated by the same principles that apply in a simple dilution problem.

Again, the key concept involves the amount of solute present; it is constant. The sum of the amounts of solute present in the individual solutions prior to mixing is the same as the total amount of solute present in the solution after mixing. No solute is lost or gained in the mixing process. Thus, we can write

$$\text{Moles solute}_{\text{first solution}} + \text{moles solute}_{\text{second solution}} = \text{moles solute}_{\text{combined solution}}$$

Substituting the expression (M × V) for moles solute in this equation gives

$$(M_1 \times V_1) + (M_2 \times V_2) = M_3 \times V_3$$

where the subscripts 1 and 2 denote the solutions to be mixed and the subscript 3 is the solution resulting from the mixing. Again, this expression is valid only when the solutions that are mixed are "like" solutions.

Example 13.18

What is the molarity of the solution obtained by mixing 50.0 mL of 2.25 M hydrochloric acid (HCl) solution with 160.0 mL of 1.25 M hydrochloric acid solution?

Solution

Five of the six variables in the equation

$$(M_1 \times V_1) + (M_2 \times V_2) = M_3 \times V_3$$

are known:

$$\begin{aligned} M_1 &= 2.25\ \text{M} & V_1 &= 50.0\ \text{mL} \\ M_2 &= 1.25\ \text{M} & V_2 &= 160.0\ \text{mL} \\ M_3 &= ?\ \text{M} & V_3 &= 210.0\ \text{mL} \end{aligned}$$

Note that in the mixing process we consider the volumes of the solution to be additive; that is,

$$V_3 = V_1 + V_2$$

This is a valid assumption for "like" solutions.

Solving our equation for M_3 and then substituting the known quantities into it gives

$$M_3 = \frac{(M_1 \times V_1) + (M_2 \times V_2)}{V_3} = \frac{(2.25\ \text{M} \times 50.0\ \cancel{\text{mL}}) + (1.25\ \text{M} \times 160.0\ \cancel{\text{mL}})}{(210.0\ \cancel{\text{mL}})}$$

= 1.4904761 M (calculator answer)
= 1.49 M (correct answer)

Practice Exercise 13.18

What is the molarity of the solution obtained by mixing 50.0 mL of 1.25 M ammonium chloride (NH_4Cl) solution with 175 mL of 0.125 M ammonium chloride solution?

Ans. 0.375 M

In the solution of Example 13.18 the given liquid volumes were considered additive. In Section 13.6, when discussing volume percent, it was stressed that volumes were not additive. Why the difference? Volumes of different liquids (Sec. 13.6) are not additive; volumes of the same liquid (Example 13.18) are additive.

13.11 MOLARITY AND CHEMICAL EQUATIONS

Section 10.8 introduced a general problem-solving procedure for setting up problems that involve chemical equations. With this procedure, if information is given about one reactant or product in a chemical reaction (number of grams, moles, or particles), similar information can easily be obtained for any other reactant or product.

In Section 12.13 this procedure was refined to allow us to do mass-to-volume or volume-to-mass calculations for reactions when at least one reactant or product is a gas.

This section further refines our problem-solving procedure in order to deal efficiently with reactions that occur in aqueous solution. Of primary importance in this new area of problem solving will be *solution volume*. In most situations, solution volume is more conveniently determined than solution mass.

When solution concentrations are expressed in terms of molarity, a direct relationship exists between solution volume (in liters) and moles of solute present. The definition of molarity itself gives the relationship; molarity is the ratio of moles of solute to volume (in liters) of solution. Thus, molarity is the connection that links volume of solution to the other common problem-solving parameters, such as moles and grams. Figure 13.6 shows diagrammatically the place that volume of solution occupies, relative to other parameters, in the overall scheme of chemical-equation-based problem solving. This diagram is a simple modification of Figure 12.9; "volume of solution" boxes have replaced "particles" boxes. It is used in the same way as Figure 12.9 was.

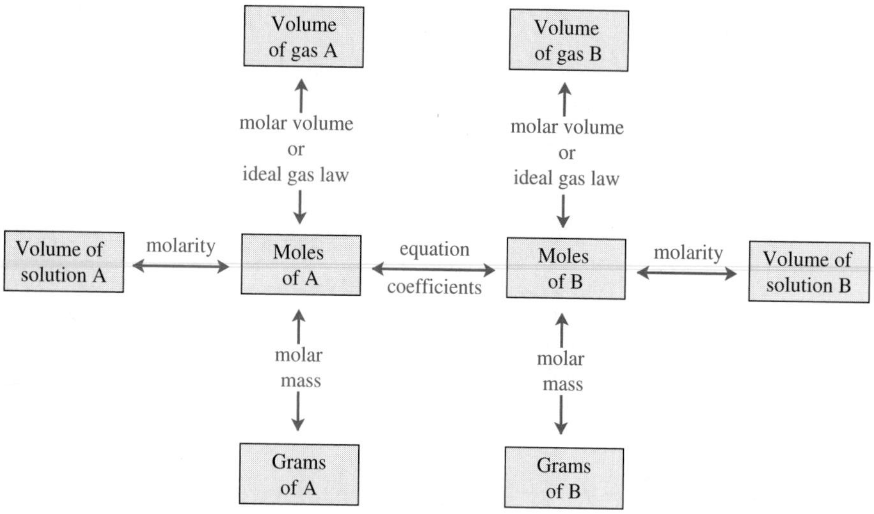

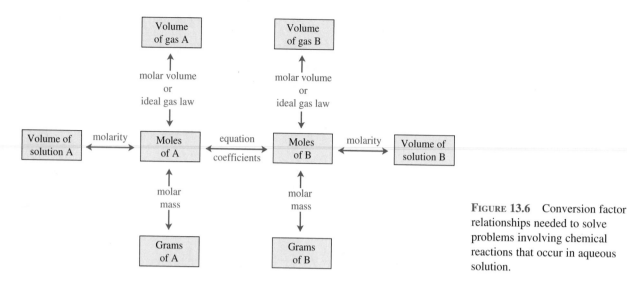

FIGURE 13.6 Conversion factor relationships needed to solve problems involving chemical reactions that occur in aqueous solution.

Example 13.19

The fizz produced when an Alka-Seltzer tablet is dissolved in water is due to the reaction between sodium bicarbonate, $NaHCO_3$, and citric acid, $C_6H_8O_7$.

$$3\ NaHCO_3(aq) + C_6H_8O_7(aq) \longrightarrow 3\ CO_2\ (g) + 3\ H_2O(l) + Na_3C_6H_8O_7(aq)$$

If this reaction were run in a laboratory, what volume, in liters, of 2.50 M $NaHCO_3$ solution is needed to react completely with 0.025 L of 3.50 M $C_6H_8O_7$ solution?

Solution

STEP 1 The given quantity is 0.025 L of $C_6H_8O_7$ solution, and the desired quantity is liters of $NaHCO_3$ solution.

$$0.025\ L\ C_6H_8O_7 = ?\ L\ NaHCO_3$$

STEP 2 This problem is a "volume of solution A" to "volume of solution B" problem. The pathway used in solving it, in terms of Figure 13.6, is

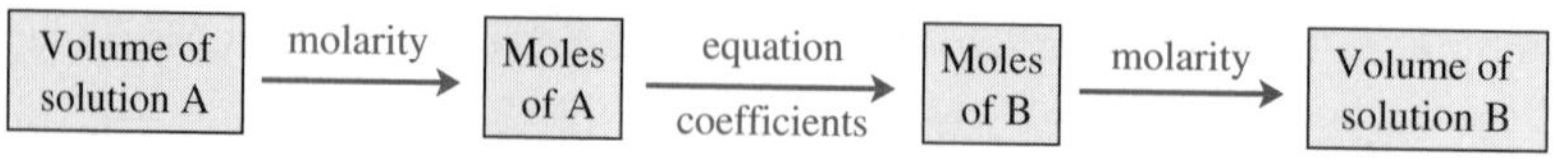

STEP 3 The dimensional analysis setup for the calculation is

$$0.025\ \cancel{L\ C_6H_8O_7} \times \frac{3.50\ \cancel{moles\ C_6H_8O_7}}{1\ \cancel{L\ C_6H_8O_7}} \times \frac{3\ \cancel{moles\ NaHCO_3}}{1\ \cancel{mole\ C_6H_8O_7}} \times \frac{1\ L\ NaHCO_3}{2.50\ \cancel{moles\ NaHCO_3}}$$

STEP 4 Combining all of the numerical factors gives

$$\frac{0.025 \times 3.50 \times 3 \times 1}{1 \times 1 \times 2.50}\ L\ NaHCO_3 = 0.105\ L\ NaHCO_3 \quad \text{(calculator answer)}$$

$$= 0.10\ L\ NaHCO_3 \quad \text{(correct answer)}$$

Practice Exercise 13.19

What volume, in liters, of a 3.40 M potassium hydroxide (KOH) solution is needed to react completely with 0.100 L of a 6.72-M sulfuric acid (H_2SO_4) solution according to the following equation?

$$2\ KOH(aq) + H_2SO_4(aq) \longrightarrow K_2SO_4(aq) + 2\ H_2O(l)$$

Ans. 0.395 L KOH

Example 13.20

How many grams of lead(II) chloride can be produced from the reaction of 1.05 L of 0.470 M potassium chloride (KCl) solution with an excess of 4.00 M lead(II) nitrate [(Pb $(NO_3)_2$] solution according to the following equation.

$$2\ KCl(aq) + Pb\ (NO_3)_2(aq) \longrightarrow PbCl_2(s) + 2\ KNO_3(aq)$$

Solution

STEP 1 The given quantity is 1.05 L of KCl solution, and the desired quantity is grams of $PbCl_2$.

$$1.05\ L\ KCl = ?\ g\ PbCl_2$$

STEP 2 This is a "volume of solution A" to "grams of B" problem. The pathway, in terms of Figure 13.6, is

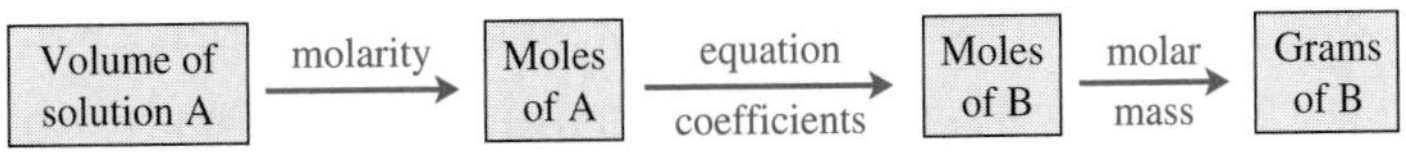

STEP 3 The dimensional analysis setup for the calculation is

$$1.05\ \cancel{L\ KCl} \times \frac{0.470\ \cancel{mole\ KCl}}{1\ \cancel{L\ KCl}} \times \frac{1\ \cancel{mole\ PbCl_2}}{2\ \cancel{mole\ KCl}} \times \frac{278\ g\ PbCl_2}{1\ \cancel{mole\ PbCl_2}}$$

STEP 4 The answer, obtained from combining all of the numerical factors, is

$$\frac{1.05 \times 0.470 \times 1 \times 278}{1 \times 2 \times 1}\ g\ PbCl_2 = 68.5965\ g\ PbCl_2 \quad \text{(calculator answer)}$$

$$= 68.6\ g\ PbCl_2 \quad \text{(correct answer)}$$

Note that the concentration of $Pb(NO_3)_2$ solution, given as 4.00 M in the problem statement, did not enter into the calculation. This is because the $Pb(NO_3)_2$ solution is present in excess; we know that we have enough of it. If a specific volume of $Pb(NO_3)_2$ solution had been given in the problem statement, we would have had to determine the limiting reactant ($Pb(NO_3)_2$ or KCl) as the first step in working the problem. The concept of a limiting reactant was discussed in Section 10.9.

Practice Exercise 13.20

How many grams of $BaCrO_4$ can be produced from the reaction of 1.05 L of 0.470 M $BaCl_2$, solution with an excess of 2.00 M K_2CrO_4 solution according to the following equation?

$$BaCl_2(aq) + K_2CrO_4(aq) \longrightarrow BaCrO_4(s) + 2\ KCl(aq)$$

Ans. 125 g $BaCrO_4$

Example 13.21

What volume, in liters of nitric oxide gas, NO, measured at STP can be produced from 1.75 L of 0.550 M nitric acid, HNO_3, and an excess of 0.650 M hydrosulfuric acid, H_2S, according to the following reaction?

$$2\ HNO_3(aq) + 3\ H_2S(aq) \longrightarrow 2\ NO(g) + 3\ S(s) + 4\ H_2O(l)$$

Solution

STEP 1 The given quantity is 1.75 L of HNO_3 solution, and the desired quantity is liters of NO gas at STP.

$$1.75\ \text{L}\ HNO_3 = ?\ \text{L NO (at STP)}$$

STEP 2 This is a "volume of solution A" to "volume of gas B" problem. From Figure 13.6 the pathway for solving the problem is

[Volume of solution A] —molarity→ [Moles of A] —equation coefficients→ [Moles of B] —molar volume→ [Volume of gas B]

STEP 3 The dimensional analysis setup for the calculation is

$$1.75\ \cancel{\text{L}\ HNO_3} \times \frac{0.550\ \cancel{\text{mole}\ HNO_3}}{1\ \cancel{\text{L}\ HNO_3}} \times \frac{2\ \cancel{\text{moles NO}}}{2\ \cancel{\text{moles}\ HNO_3}} \times \frac{22.4\ \text{L NO}}{1\ \cancel{\text{mole NO}}}$$

The last conversion factor is derived from the fact that 1 mole of any gas occupies 22.4 L at STP conditions (Sec. 12.10).

STEP 4 The result, obtained by combining all of the numerical factors, is

$$\frac{1.75 \times 0.550 \times 2 \times 22.4}{1 \times 2 \times 1}\ \text{L NO} = 21.56\ \text{L NO} \quad \text{(calculator answer)}$$

$$= 21.6\ \text{L NO} \quad \text{(correct answer)}$$

Practice Exercise 13.21

What volume, in liters, of H_2 gas measured at STP can be produced from 0.575 L of 1.22 M HBr solution and an excess of Zn according to the following equation?

$$2\ HBr(aq) + Zn(s) \longrightarrow ZnBr_2(aq) + H_2(g)$$

Ans. 7.86 L H_2

Key Terms

The new terms or concepts defined in this chapter are

aqueous solution (Sec. 13.2) A solution in which water is the solvent.

concentrated solution (Sec. 13.2) A solution that contains a large amount of solute relative to the amount that could dissolve.

concentration (Sec. 13.5) The amount of solute present in a specified amount of solvent or solution.

dilute solution (Sec. 13.2) A solution that contains a small amount of solute relative to the amount that could dissolve.

dilution (Sec. 13.10) Process in which more solvent is added to a solution in order to lower its concentration.

immiscible substances (Sec. 13.2) Liquid substances that do not dissolve in each other.

mass–volume percent (Sec. 13.6) Mass of solute (in grams) divided by the total volume of solution (in milliliters) multiplied by 100.

miscible substances (Sec. 13.2) Liquid substances that dissolve in any amount in each other.

molality (Sec. 13.9) Concentration of a solution in terms of moles of solute per kilogram of solvent.

molarity (Sec. 13.8) Concentration of a solution in terms of moles of solute per liter of solution.

partially miscible substances (Sec. 13.2) Liquid substances that have limited solubility in each other.

parts per billion (Sec. 13.7) Number of parts of solute per billion parts of solution.

parts per million (Sec. 13.7) Number of parts of solute per million parts of solution.

percent by mass (Sec. 13.6) Mass of solute divided by the total mass of solution multiplied by 100.

percent by volume (Sec. 13.6) Volume of solute divided by the total volume of solution multiplied by 100.

saturated solution (Sec. 13.2) A solution that contains the maximum amount of solute that can be dissolved under the conditions at which the solution exists.

solubility (Sec. 13.2) The maximum amount of solute that will dissolve in a given amount of solvent under a fixed set of conditions.

solute (Sec. 13.1) Component of a solution present in a small amount relative to that of the solvent.

solution (Sec. 13.1) Homogeneous mixture of two or more substances.

solvent (Sec. 13.1) Component of a solution present in the greatest amount.

unsaturated solution (Sec. 13.2) A solution where less solute than the maximum amount possible is dissolved in the solution.

Practice Problems

Solution Terminology (Sec. 13.2)

13.1 Indicate which substance is the solvent in each of the following solutions.

(a) a solution containing 10.0 g of glucose ($C_6H_{12}O_6$) and 500.0 mL of water

(b) a solution containing 60.0 mL of ethyl alcohol and 30.0 mL of methyl alcohol

13.2 Indicate which substance is the solvent in each of the following solutions.

(a) a solution containing 1.50 g of potassium chloride (KCl) and 35.0 mL of water

(b) a solution containing 30.0 mL of ethyl alcohol and 60.0 mL of methyl alcohol

---

13.3 Use the terms *soluble*, *insoluble*, *miscible*, or *immiscible* to describe the behavior of the following pairs of substances. More than one term may apply to a given case.

(a) 75 mL of water and 1 g of table sugar are shaken together. The resulting solution is clear and colorless.

(b) 25 mL of water and 10 mL of mineral oil are shaken together. The mixture is cloudy and gradually separates into two layers.

13.4 Use the terms *soluble*, *insoluble*, *miscible*, or *immiscible* to describe the behavior of the following pairs of substances. More than one term may apply to a given case.

(a) 10 mL of water and 0.1 g of fine white sand are shaken together. The resulting mixture is very cloudy, and the sand rapidly settles to the bottom of the container.

(b) 100 mL of water and 100 mL of ethyl alcohol are shaken together. A clear colorless solution results that with time shows no separation into layers.

13.5 Use Table 13.2 to determine whether each of the following silver nitrate ($AgNO_3$) solutions is saturated or unsaturated.

(a) 455 g of $AgNO_3$ in 100 g of H_2O at 100°C

(b) 455 g of $AgNO_3$ in 100 g of H_2O at 50°C

(c) 910 g of $AgNO_3$ in 200 g of H_2O at 50°C

(d) 55 g of $AgNO_3$ in 50 g of H_2O at 0°C

13.6 Use Table 13.2 to determine whether each of the following silver sulfate (Ag_2SO_4) solutions is saturated or unsaturated.

(a) 1.08 g of Ag_2SO_4 in 100 g of H_2O at 100°C

(b) 0.54 g of Ag_2SO_4 in 50 g of H_2O at 50°C

(c) 1.08 g of Ag_2SO_4 in 200 g of H_2O at 50°C

(d) 0.25 g of Ag_2SO_4 in 50 g of H_2O at 0°C

13.7 Using the solubilities given in Table 13.2, characterize each of the following solids as insoluble, slightly soluble, soluble, or very soluble in water at the indicated temperature.

(a) lead(II) bromide at 50°C

(b) cesium chloride at 0°C

(c) silver nitrate at 100°C

(d) silver sulfate at 0°C

13.8 Using the solubilities given in Table 13.2, characterize each of the following solids as insoluble, slightly soluble, soluble, or very soluble in water at the indicated temperature.

(a) copper(II) sulfate at 100°C

(b) lead(II) bromide at 0°C

(c) silver nitrate at 50°C

(d) sodium chloride at 0°C

13.9 Based on the solubilities in Table 13.2, characterize each of the following sodium chloride (NaCl) solutions as dilute or concentrated.

(a) 35.0 g of NaCl in 100 g of H_2O at 100°C

(b) 35.0 g of NaCl in 100 g of H_2O at 50°C

(c) 3.50 g of NaCl in 10 g of H_2O at 50°C

(d) 3.50 g of NaCl in 100 g of H_2O at 50°C

13.10 Based on the solubilities in Table 13.2, characterize each of the following copper(II) sulfate solutions as dilute or concentrated.

(a) 14.3 g of $CuSO_4$ in 100 g of H_2O at 100°C

(b) 14.3 g of $CuSO_4$ in 100 g of H_2O at 0°C

(c) 14.3 g of $CuSO_4$ in 50 g of H_2O at 50°C

(d) 1.43 g of $CuSO_4$ in 10 g of H_2O at 0°C

Solubility Rules (Sec. 13.4)

13.11 Predict whether the following solutes are very soluble or slightly soluble in water.

(a) SO_3 (a polar gas)

(b) CCl_4 (a nonpolar liquid)

(c) CH_4 (a nonpolar gas)

(d) $C_6H_{12}O_6$ (a polar nonionic solid)

13.12 Predict whether the following solutes are very soluble or slightly soluble in water.

(a) F_2 (a nonpolar gas)

(b) CO_2 (a nonpolar gas)

(c) $C_{12}H_{22}O_{11}$ (a polar nonionic solid)

(d) C_2H_5OH (a polar liquid)

13.13 In which of the following pairs of ionic compounds do both members of the pair have like solubility (both soluble or both insoluble) in water?

(a) $NaNO_3$ and $Pb(NO_3)_2$ **(b)** KCl and AgCl

(c) $BeSO_4$ and $MgSO_4$ **(d)** $FePO_4$ and $Ca_3(PO_4)_2$

13.14 In which of the following pairs of ionic compounds do both members of the pair have like solubility (both soluble or both insoluble) in water?

(a) $NaC_2H_3O_2$ and $Mg(C_2H_3O_2)_2$ **(b)** NH_4F and NH_4Cl

(c) KOH and $Ba(OH)_2$ **(d)** Al_2S_3 and CuS

13.15 Which of the following ions would react with both Mg^{2+} and Ca^{2+} ions to form water-insoluble compounds?

(a) sulfate **(b)** phosphate

(c) sulfide **(d)** nitrate

13.16 Which of the following ions would react with both Cu^{2+} and Ba^{2+} ions to form water-insoluble compounds?

(a) chloride **(b)** carbonate

(c) hydroxide **(d)** acetate

Mass Percent (Sec. 13.6)

13.17 Calculate the mass percent of sodium iodide (NaI) in each of the following solutions.

(a) 6.43 g of NaI dissolved in 85.0 g of H_2O

(b) 3.23 g of NaI dissolved in 175.00 g of H_2O

(c) 10.3 g of NaI dissolved in 53.0 g of solution

(d) 0.030 mole of NaI dissolved in 100.0 g of H_2O

13.18 Calculate the mass percent of potassium hydroxide (KOH) in each of the following solutions.

(a) 10.2 g of KOH dissolved in 135 g of H_2O

(b) 3.14 g of KOH dissolved in 53.14 g of H_2O

(c) 1.33 g of KOH dissolved in 23.50 g of solution

(d) 0.500 mole of KOH dissolved in 1375 g of H_2O

13.19 How many grams of solute are dissolved in the following amounts of solution?

(a) 35.0 g of 2.00% (m/m) NaCl solution

(b) 125 g of 3.50% (m/m) $AgNO_3$ solution

(c) 1355 g of 10.00% (m/m) K_2SO_4 solution

(d) 43.3 g of 8.25% (m/m) HCl solution

13.20 How many grams of solute are dissolved in the following amounts of solution?

(a) 134 g of 3.00% (m/m) KNO_3 solution

(b) 75.02 g of 9.735% (m/m) NaOH solution

(c) 1576 g of 0.800% (m/m) HI solution

(d) 1.23 g of 12.0% (m/m) NH_4Cl solution

13.21 What mass of water, in grams, is needed to prepare each of the following calcium chloride ($CaCl_2$) solutions?

(a) 5.75 g of 10.00% (m/m) $CaCl_2$ solution

(b) 57.5 g of 10.00% (m/m) $CaCl_2$ solution

(c) 57.5 g of 1.00% (m/m) $CaCl_2$ solution

(d) 2.3 g of 0.80% (m/m) $CaCl_2$ solution

13.22 What mass of water, in grams, is needed to prepare each of the following lithium nitrate ($LiNO_3$) solutions?

(a) 34.7 g of 5.00% (m/m) $LiNO_3$ solution

(b) 3.47 g of 5.00% (m/m) $LiNO_3$ solution

(c) 235 g of 12.75% (m/m) $LiNO_3$ solution

(d) 1352 g of 0.0032% (m/m) $LiNO_3$ solution

13.23 How many grams of water must be added to 50.0 g of each of the following solutes in order to prepare a 5.00% (m/m) solution?

(a) NaCl **(b)** KCl

(c) Na_2SO_4 **(d)** $LiNO_3$

13.24 How many grams of water must be added to 20.0 g of each of the following solutes in order to prepare a 2.00% (m/m) solution?

(a) NaOH **(b)** LiBr

(c) Li_2SO_4 **(d)** $Ca(NO_3)_2$

Volume Percent (Sec. 13.6)

13.25 What is the volume percent ethyl alcohol in a solution containing 257 mL of ethyl alcohol and enough water to give the following amounts of solution?

(a) 325 mL **(b)** 675 mL

(c) 1.23 L **(d)** 5.000 L

13.26 What is the volume percent water in a solution containing 35.0 mL of water and enough ethyl alcohol to give the following amounts of solution?

(a) 45.0 mL **(b)** 675 mL **(c)** 1.08 L **(d)** 4.500 L

13.27 The final volume of a solution made by adding 360.6 mL of methyl alcohol to 667.2 mL of water is 1000.0 mL. Determine the volume percent of the following.

(a) methyl alcohol in the solution

(b) water in the solution

13.28 The final volume of a solution made by adding 678.2 mL of methyl alcohol to 358.4 mL of water is 1000.0 mL. Determine the volume percent of the following.

(a) methyl alcohol in the solution

(b) water in the solution

13.29 How much hydrogen peroxide (H_2O_2), in milliliters, is needed to prepare 375 mL of a 3.00% (v/v) solution of hydrogen peroxide in water?

13.30 How much hydrogen peroxide (H_2O_2), in milliliters, is needed to prepare 525 mL of a 2.00% (v/v) solution of hydrogen peroxide in water?

13.31 What volume of water, in gallons, is contained in 4.00 gal of a 35.0% (v/v) solution of water in acetone?

13.32 What volume of water, in quarts, is contained in 3.50 qt of a 2.00% (v/v) solution of water in acetone?

Mass–Volume Percent (Sec. 13.6)

13.33 Calculate the concentration, as mass–volume percent, for potassium iodide (KI) solutions with the following characteristics.

(a) 2.00 g solute, 75.0 mL solution

(b) 15.00 g solute, 1.25 L solution

(c) 2.00 moles solute, 10.00 L solution

(d) 0.0020 mole solute, 5.00 mL solution

13.34 Calculate the concentration, as mass-volume percent, for magnesium fluoride (MgF_2) solutions with the following characteristics.

(a) 5.00 g solute, 125 mL solution

(b) 25.0 g solute, 2.20 L solution

(c) 0.150 mole solute, 105 mL solution

(d) 0.32 mole solute, 0.100 L solution

13.35 How many milliliters of a 6.0% (m/v) sodium nitrate ($NaNO_3$) solution would fulfill each of the following requirements?

(a) contain 45.0 g of $NaNO_3$

(b) supply 2.00 g of $NaNO_3$

13.36 How many milliliters of a 3.5% (m/v) potassium acetate ($KC_2H_3O_2$) solution would fulfill each of the following requirements?

(a) contain 6.60 g of $KC_2H_3O_2$

(b) supply 25.00 g of $KC_2H_3O_2$

13.37 Determine how many grams of sodium phosphate (Na_3PO_4) would

(a) be needed to prepare 455 mL of a 2.50% (m/v) Na_3PO_4 solution.

(b) be present in 50.0 L of a 7.50% (m/v) Na_3PO_4 solution.

13.38 Determine how many grams of potassium carbonate (K_2CO_3) would

(a) be needed to prepare 4.55 mL of a 15.00% (m/v) K_2CO_3 solution.

(b) be present in 1.06 L of a 0.800% (m/v) K_2CO_3 solution.

13.39 Calculate the concentration, as mass–volume percent cesium chloride (CsCl), for a solution prepared by adding 5.0 g of CsCl to 20.0 g of H_2O to give a solution with a density of 1.18 g/mL.

13.40 Calculate the concentration, as mass–volume percent ammonium sulfate [$(NH_4)_2SO_4$], for a solution prepared by adding 3.0 g of $(NH_4)_2SO_4$ to 17.0 g of H_2O to give a solution with a density of 1.09 g/mL.

Parts per Million and Parts per Billion (Sec. 13.7)

13.41 What is the concentration of sodium chloride (NaCl), in ppm (m/m), in each of the following NaCl solutions?

(a) 37.5 mg of NaCl in 21.0 kg of water

(b) 2.12 cg of NaCl in 125 g of water

(c) 1.00 μg of NaCl in 32.0 dg of water

(d) 35.7 mg of NaCl in 15.7 g of water

13.42 What is the concentration of sodium bromide (NaBr), in ppm (m/m), in each of the following NaBr solutions?

(a) 37.5 cg of NaBr in 33.0 kg of water

(b) 2.12 mg of NaBr in 375 dg of water

(c) 3.00 μg of NaBr in 45.0 g of water

(d) 125 dg of NaBr in 255 kg of water

13.43 What is the concentration of each of the solutions in Problem 13.41 in parts per billion (mass/mass)?

13.44 What is the concentration of each of the solutions in Problem 13.42 in parts per billion (mass/mass)?

13.45 Fish generally need an oxygen concentration in water of at least 5 ppm (m/v) for survival. Will river water that contains 7 mg of O_2 per liter contain sufficient O_2 to sustain fish life?

13.46 A carbon dioxide concentration in water of 200 ppm (m/v) or higher is lethal to fish. Will river water that contains 0.62 g of dissolved CO_2 per 2.0 L be toxic to fish?

13.47 A typical concentration of the air pollutant sulfur dioxide (SO_2) in urban atmospheres is 0.087 ppm (v/v). At this concentration, how many milliliters of SO_2 are present in 5.000 L of air?

13.48 A typical concentration of the air pollutant nitrogen dioxide (NO_2) in urban atmospheres is 0.30 ppm (v/v). At this concentration, how many liters of air would be needed to extract 1.00 mL of NO_2?

13.49 Determine how much ammonia (NH_3), in grams, must be present in a 725-mL sample of air to give the following NH_3 concentrations.

(a) 3.6 ppm (m/v) **(b)** 7.5 ppb (m/v) **(c)** 1.2% (m/v)

13.50 Determine how much hydrogen sulfide (H_2S), in grams, must be present in a 475-mL sample of air to give the following H_2S concentrations.

(a) 2.0 ppm (m/v) **(b)** 9.7 ppb (m/v) **(c)** 5.2% (m/v)

Molarity (Sec. 13.8)

13.51 Calculate the molarity of each of the following aqueous sodium hydroxide (NaOH) solutions.

(a) 2.0 moles NaOH in 0.50 L of solution

(b) 13.7 g NaOH in 90.0 mL of solution

(c) 53.0 g NaOH in 1.255 L of solution

(d) 0.0020 mole NaOH in 5.00 mL of solution

13.52 Calculate the molarity of each of the following aqueous potassium chloride (KCl) solutions.

(a) 1.45 moles KCl in 2.50 L of solution

(b) 12.5 g KCl in 85.0 mL of solution

(c) 27.0 g KCl in 1.055 L of solution

(d) 0.0500 mole KCl in 12.0 mL of solution

13.53 Calculate the number of grams of solute in each of the following nitric acid (HNO_3) solutions.

(a) 35.0 mL of a 6.00-M solution

(b) 10.0 mL of a 0.600-M solution

(c) 375 L of a 1.00 M-solution

(d) 375 g of a 7.91 M-solution with a density of 1.25 g/mL

13.54 Calculate the number of grams of solute in each of the following sulfuric acid (H_2SO_4) solutions.

(a) 27.0 mL of a 3.00-M solution

(b) 20.0 mL of a 6.00-M solution

(c) 125 L of a 0.100-M solution

(d) 125 g of a 7.50-M solution with a density of 1.42 g/mL

13.55 Calculate the volume, in milliliters, of the following sodium thiosulfate ($Na_2S_2O_3$) solutions needed to provide the indicated amounts of solute.

(a) 2.50 g of $Na_2S_2O_3$ from a 0.468-M solution

(b) 125 g of $Na_2S_2O_3$ from a 3.50-M solution

(c) 4.50 moles of $Na_2S_2O_3$ from a 2.50-M solution

(d) 0.0015 mole of $Na_2S_2O_3$ from a 0.990-M solution

13.56 Calculate the volume, in milliliters, of the following sodium sulfate (Na_2SO_4) solutions needed to provide the indicated amounts of solute.

(a) 2.50 g of Na_2SO_4 from a 0.468-M solution

(b) 125 g of Na_2SO_4 from a 3.50-M solution

(c) 4.50 moles of Na_2SO_4 from a 2.50-M solution

(d) 0.0015 mole of Na_2SO_4 from a 0.990-M solution

13.57 How many liters of 0.775 M solution can be prepared from 55.0 g of each of the following solutes?

(a) HBr **(b)** NH_4Br **(c)** $MgBr_2$ **(d)** CsBr

13.58 How many liters of 1.30 M solution can be prepared from 34.3 g of each of the following solutes?

(a) HI **(b)** NH_4I **(c)** BaI_2 **(d)** LiI

13.59 The density of a 88.00% (m/m) methanol (CH_3OH) solution is 0.8274 g/mL. What is the molarity of the solution?

13.60 The density of a 60.00% (m/m) ethanol (C_2H_5OH) solution is 0.8937 g/mL. What is the molarity of the solution?

13.61 The density of a 2.019-M sodium bromide (NaBr) solution is 1.157 g/mL. What is the concentration of this solution expressed as % (m/m) NaBr?

13.62 The density of a 2.687-M sodium acetate ($NaC_2H_3O_2$) solution is 1.104 g/mL. What is the concentration of this solution expressed as % (m/m) $NaC_2H_3O_2$?

13.63 What is the molarity of a 15.0% (m/v) sodium hydroxide (NaOH) solution?

13.64 What is the molarity of a 15.0% (m/v) nitric acid (HNO_3) solution?

Molality (Sec. 13.9)

13.65 Calculate the molality of each of the following sucrose ($C_{12}H_{22}O_{11}$) solutions

(a) 16.5 g of sucrose in 1.35 kg of water

(b) 3.15 moles of sucrose in 455 g of water

(c) 0.0356 g of sucrose in 13.0 g of water

(d) 45.0 g of sucrose in enough water to give 318 mL of solution with a density of 1.06 g/mL

13.66 Calculate the molality of each of the following glucose ($C_6H_{12}O_6$) solutions.

(a) 23.0 g of glucose in 2.40 kg of water

(b) 2.00 moles of glucose in 975 g of water

(c) 0.230 g of glucose in 22.0 g of water

(d) 30.0 g of glucose in enough water to give 312 mL of solution with a density of 1.04 g/mL

13.67 Calculate the number of grams of each solute that must be added to 125 g of water to prepare a 0.400-*m* solution of

(a) $Al(NO_3)_3$ **(b)** $MgCl_2$ **(c)** Na_3PO_4 **(d)** K_2SO_4

13.68 Calculate the number of grams of each solute that must be added to 235 g of water to prepare a 0.600-*m* solution of

(a) $Be(NO_3)_2$ **(b)** $CaBr_2$ **(c)** Na_2CO_3 **(d)** $Al_2(SO_4)_3$

13.69 How many grams of water must be added to 80.0 g of sodium chloride (NaCl) to prepare the following molal solutions?

(a) 0.050 *m* **(b)** 0.23 *m* **(c)** 1.345 *m* **(d)** 2.8 *m*

13.70 How many grams of water must be added to 70.0 g of potassium chloride (KCl) to prepare the following molal solutions?

(a) 0.010 *m* **(b)** 0.45 *m* **(c)** 1.23 *m* **(d)** 2.45 *m*

13.71 An aqueous solution of oxalic acid ($H_2C_2O_4$) is 0.568 M and has a density of 1.022 g/mL. What is the molality of the solution?

13.72 An aqueous solution of citric acid ($H_3C_6H_5O_7$) is 0.655 M and has a density of 1.049 g/mL. What is the molality of the solution?

13.73 An aqueous solution of acetic acid ($HC_2H_3O_2$) is 0.796 *m* and has a density of 1.004 g/mL. What is the molarity of the solution?

13.74 An aqueous solution of tartaric acid ($H_2C_4H_4O_6$) is 0.278 *m* and has a density of 1.006 g/mL. What is the molarity of the solution?

13.75 Calculate the molality of a 14.0% by mass nitric acid (HNO_3) solution.

13.76 Calculate the molality of a 23.0% by mass acetic acid ($HC_2H_3O_2$) solution.

Dilution (Sec. 13.10)

13.77 What is the molarity of a solution prepared by diluting 25.0 mL of 0.400 M potassium hydroxide (KOH) to each of the following volumes?

(a) 50.0 mL **(b)** 83.0 mL **(c)** 375 mL **(d)** 2.67 L

13.78 What is the molarity of a solution prepared by diluting 50.0 mL of 0.300 M sodium nitrate ($NaNO_3$) to each of the following volumes?

(a) 60.0 mL **(b)** 97.0 mL **(c)** 452 mL **(d)** 8.75 L

13.79 What is the molarity of the solution prepared by concentrating, by evaporation of solvent, 1353 mL of 0.500 M ammonium chloride (NH_4Cl) solution to each of the following final volumes?

(a) 1125 mL **(b)** 1.06 L **(c)** 975 mL **(d)** 297.5 mL

13.80 What is the molarity of the solution prepared by concentrating, by evaporation of solvent, 2212 mL of 0.400 M potassium sulfate (K_2SO_4) solution to each of the following final volumes?

(a) 1875 mL **(b)** 1.25 L **(c)** 853 mL **(d)** 553 mL

13.81 How many milliliters of 6.0 M potassium hydroxide (KOH) solution is required to produce, using dilution, the following KOH solutions?

(a) 30.0 mL of 5.0 M solution

(b) 6.5 L of 1.0 M solution

(c) 275 mL of 5.9 M solution

(d) 3.0 mL of 0.10 M solution

13.82 How many milliliters of 3.0 M potassium chloride (KCl) solution is required to produce, using dilution, the following KCl solutions?

(a) 20.0 mL of 2.5 M solution

(b) 3.5 L of 1.0 M solution

(c) 352 mL of 2.9 M solution

(d) 4.2 mL of 0.25 M solution

13.83 In each of the following silver nitrate ($AgNO_3$) solutions, how many milliliters of water should be added to obtain a solution that has a concentration of 0.100 M?

(a) 20.0 mL of a 2.00-M solution

(b) 20.0 mL of a 0.250-M solution

(c) 358 mL of a 0.950-M solution

(d) 2.3 L of a 6.00 M-solution

13.84 In each of the following silver acetate ($AgC_2H_3O_2$) solutions, how many milliliters of water should be added to obtain a solution that has a concentration of 0.200 M?

(a) 30.0 mL of a 4.00-M solution

(b) 30.0 mL of a 0.400-M solution

(c) 785 mL of a 0.230-M solution

(d) 1.25 L of a 1.50-M solution

13.85 What will be the final concentration of each of the following solutions if the volume of the solution is increased by 20.0 mL by adding water?

(a) 25.0 mL of 6.0 M Na_2SO_4

(b) 100.0 mL of 3.0 M K_2SO_4

(c) 0.155 L of 10.0 M CsCl

(d) 2.00 mL of 0.100 M $MgCl_2$

13.86 What will be the final concentration of each of the following solutions if the volume of the solution is increased by 20.0 mL by adding water?

(a) 50.0 mL of 2.0 M KNO_3

(b) 50.0 mL of 3.0 M $AgNO_3$

(c) 1.0000 L of 1.2131 M $NaNO_3$

(d) 1.0000 mL of 1.000 M $LiNO_3$

13.87 What would be the molarity of a solution obtained when 275 mL of 6.00 M sodium hydroxide (NaOH) solution is mixed with each of the following?

(a) 3.254 L of H_2O

(b) 125 mL of 6.00 M NaOH solution

(c) 125 mL of 2.00 M NaOH solution

(d) 27 mL of 5.80 M NaOH solution

13.88 What would be the molarity of a solution obtained when 352 mL of 4.00 M sodium bromide (NaBr) solution is mixed with each of the following?

(a) 425 mL of water

(b) 225 mL of 4.00 M NaBr solution

(c) 225 mL of 2.00 M NaBr solution

(d) 15 mL of 4.20 M NaBr solution

Molarity and Chemical Equations (Sec. 13.11)

13.89 What volume, in liters, of 1.00 M $Pb(NO_3)_2$ is needed to react completely with 0.500 L of 4.00 M NaCl according to the following equation?

$$Pb(NO_3)_2(aq) + 2\ NaCl(aq) \longrightarrow PbCl_2(s) + 2\ NaNO_3(aq)$$

13.90 What volume, in milliliters, of 0.300 M $CaCl_2$ is needed to react completely with 40.0 mL of 0.200 M H_3PO_4 according to the following equation?

$$3\ CaCl_2(aq) + 2\ H_3PO_4(aq) \longrightarrow Ca_3(PO_4)_2(s) + 6\ HCl(aq)$$

13.91 How many grams of S can be produced from the reaction of 30.0 mL of 12.0 M HNO_3 with an excess of 0.035 M H_2S solution according to the following equation?

$$2\ HNO_3(aq) + 3\ H_2S(aq) \longrightarrow 2\ NO(g) + 3\ S(s) + 4\ H_2O(l)$$

13.92 How many grams of Ag_3PO_4 can be produced from the reaction of 2.50 L of 0.200 M $AgNO_3$ with an excess of 0.750 M K_3PO_4 solution according to the following equation?

$$3\ AgNO_3(aq) + K_3PO_4(aq) \longrightarrow Ag_3PO_4(s) + 3\ KNO_3(aq)$$

13.93 What volume, in milliliters, of 0.50 M H_2SO_4 is required to react with 18.0 g of nickel according to the following equation?

$$Ni(s) + H_2SO_4(aq) \longrightarrow NiSO_4(aq) + H_2(g)$$

13.94 What volume, in milliliters, of 1.50 M HNO_3 is required to react with 100.0 g of tin according to the following equation?

$$Sn(s) + 2\ HNO_3(aq) \longrightarrow 2\ Sn(NO_3)_2(aq) + H_2(g)$$

13.95 What is the molarity of a 37.5 mL sample of HNO_3 solution that will completely react with 23.7 mL of 0.100 M NaOH according to the following equation?

$$HNO_3(aq) + NaOH(aq) \longrightarrow NaNO_3(aq) + H_2O(l)$$

13.96 What is the molarity of a 50.0-mL sample of H_2SO_4 solution that will completely react with 40.0 mL of 0.200 M $Mg(OH)_2$ according to the following equation?

$$H_2SO_4(aq) + Mg(OH)_2(aq) \longrightarrow MgSO_4(aq) + 2\ H_2O(l)$$

13.97 What volume, in liters, of NO gas measured at STP can be produced from 50.0 mL of 6.0 M HNO_3 solution and an excess of Cu metal according to the following reaction?

$$8\ HNO_3(aq) + 3\ Cu(s) \longrightarrow 3\ Cu(NO_3)_2(aq) + 2\ NO(g) + 4\ H_2O(l)$$

13.98 What volume, in liters of H_2 gas measured at STP can be produced from 50.0 mL of 3.0 M HBr solution and an excess of Zn metal according to the following reaction?

$$2\ HBr(aq) + Zn(s) \longrightarrow ZnBr_2(aq) + H_2(g)$$

13.99 What is the molarity of a 1.75-L $Ca(OH)_2$ solution that would completely react with 2.00 L of CO_2 gas measured at STP according to the following reaction?

$$CO_2(g) + Ca(OH)_2(aq) \longrightarrow CaCO_3(s) + H_2O(l)$$

13.100 What is the molarity of a 5.00-L NaOH solution that would completely react with 4.00 L of CO_2 gas measured at STP according to the following reaction?

$$CO_2(g) + 2\ NaOH(aq) \longrightarrow Na_2CO_3(aq) + H_2O(l)$$

ADDITIONAL PROBLEMS

13.101 In each of the following sets of ionic compounds, identify the members of the set that are soluble in water.

(a) $Be_3(PO_4)_2$, $AlPO_4$, $FePO_4$, $(NH_4)_3PO_4$

(b) $Cu(OH)_2$, $Be(OH)_2$, $Ca(OH)_2$, $Zn(OH)_2$

(c) Ag_3PO_4, $AgNO_3$, $AgCl$, $AgBr$

(d) CaS, $Ca(NO_3)_2$, $CaSO_4$, $Ca(C_2H_3O_2)_2$

13.102 In each of the following sets of ionic compounds, identify the members of the set that are soluble in water.

(a) K_2CO_3, $MgCO_3$, $NiCO_3$, $Al_2(CO_3)_3$

(b) MgS, Rb_2S, Al_2S_3, CaS

(c) $Pb(OH)_2$, $PbCl_2$, $Pb(NO_3)_2$, $PbSO_4$

(d) BaS, $BaCl_2$, $BaSO_4$, $Ba(OH)_2$

13.103 The solubility of $CuSO_4$ in water at 50°C is 33.3 g/100 g H_2O. If 400.0 g of a 75% saturated $CuSO_4$ solution at 50°C is heated to evaporate completely the water, how much solid $CuSO_4$ should be recovered?

13.104 The solubility of NaCl in water at 50°C is 37.0 g/100 g H_2O. If 300.0 g of a 85% saturated NaCl solution at 50°C is heated to evaporate completely the water, how much solid NaCl should be recovered?

13.105 After all of the water is evaporated from 254 mL of a $AgNO_3$ solution, 45.2 g of $AgNO_3$ remain. Express the original concentration of the $AgNO_3$ solution in each of the following units.

(a) mass–volume percent **(b)** molarity

13.106 After all of the water is evaporated from 10.0 mL of a CsCl solution, 3.75 g of CsCl remains. Express the original concentration of CsCl solution in each of the following units.

(a) mass–volume percent **(b)** molarity

13.107 What mass, in grams, of Na_2SO_4 would be required to prepare 425 mL of a 1.55% (m/m) Na_2SO_4 solution whose density is 1.02 g/mL?

13.108 What mass, in grams, of NaCl would be required to prepare 275 mL of a 30.0% (m/m) NaCl solution whose density is 1.18 g/mL?

13.109 A 3.000-M $NaNO_3$ solution has a density of 1.161 g/mL at 20°C. How many grams of solvent are present in 1.375 L of this solution?

13.110 A 0.157-M NaCl solution has a density of 1.09 g/mL at 20°C. How many grams of solvent are present in 80.0 mL of this solution?

13.111 Calculate the volume, in milliliters, of 0.125 M Na_2SO_4 solution needed to provide each of the following.

(a) 10.0 g of Na_2SO_4

(b) 2.5 g of Na^+ ion

(c) 0.567 mole of Na_2SO_4

(d) 0.112 mole of SO_4^{2-} ion

13.112 Calculate the volume, in milliliters, of 1.25 M $Mg(NO_3)_2$ solution needed to provide each of the following.

(a) 15.7 g of $Mg(NO_3)_2$

(b) 3.57 g of Mg^{2+} ion

(c) 1.2 moles of $Mg(NO_3)_2$

(d) 0.57 mole of NO_3^- ion

13.113 A solution is made by diluting 225 mL of a 0.245 M aluminum nitrate ($Al(NO_3)_3$) solution with water to a final volume of 0.750 L. Calculate the following.

(a) the molarities of Al^{3+} ion, and NO_3^- ion in the original solution.

(b) the molarities of $Al(NO_3)_3$, Al^{3+} ion, and NO_3^- ion in the diluted solution

13.114 A solution is made by diluting 315 mL of a 0.115 M potassium phosphate (K_3PO_4) solution with water to a final volume of 0.650 L. Calculate the following.

(a) the molarities of K^+ ion, and PO_4^{3-} ion in the original solution

(b) the molarities of K_3PO_4, K^+ ion, and PO_4^{3-} ion in the diluted solution

13.115 A solution is made by mixing 175 mL of 0.100 M K_3PO_4 with 27 mL of 0.200 M KCl. Assuming that the volumes are additive, what are the molar concentrations of the following ions in the new solution?

(a) K^+ ion **(b)** Cl^- ion **(c)** PO_4^{3-} ion

13.116 A solution is made by mixing 50.0 mL of 0.300 M Na_2SO_4 with 30.0 mL of 0.900 M K_2SO_4. Assuming that the volumes are additive, what are the molar concentrations of the following ions in the new solution?

(a) Na^+ ion **(b)** K^+ ion **(c)** SO_4^{2-} ion

13.117 A solute concentration is 3.74 ppm (m/m). What would this concentration be in the units of mg of solute per kg of solution?

13.118 A solute concentration is 5.14 ppm (m/m). What would this concentration be in the units of μg of solute per mg of solution?

13.119 A solution is prepared by dissolving 1.00 g of NaCl in enough water to make 10.00 mL of solution. A 1.00-mL portion of this solution is then diluted to a final volume of 10.00 mL. What is the molarity of the final NaCl solution?

13.120 A solution is prepared by dissolving 30.0 g of Na_2SO_4 in enough water to make 750.0 mL of solution. A 10.00-mL portion of this solution is then diluted to a final volume of 100.0 mL. What is the molarity of the final Na_2SO_4 solution?

13.121 How many milliliters of 38.0% (m/m) HCl (density 1.19 g/mL) are needed to make, using a dilution procedure, 1.00 L of 0.100 M HCl?

13.122 How many milliliters of 20.0% (m/m) NaCl (density 1.15 g/mL) are needed to make, using a dilution procedure, 3.50 L of 0.150 M NaCl?

13.123 How many grams of water should you add to a 1.23-*m* NaCl solution containing 1.50 kg H_2O to reduce the molality to 1.00 *m*?

13.124 How many grams of water should you add to a 0.0883-*m* NaCl solution containing 0.650 kg H_2O to reduce the molality to 0.0100 *m*?

13.125 Calculate the total mass, in grams, and the total volume, in milliliters, of a 2.16-*m* H_3PO_4 solution containing 52.0 g of solute. The density of the solution is 1.12 g/mL.

13.126 Calculate the total mass, in grams, and the total volume, in milliliters, of a 0.710 *m* $H_3C_6H_5O_7$ (citric acid) solution containing 23.0 g of solute. The density of the solution is 1.05 g/mL.

13.127 An aqueous solution having a density of 0.980 g/mL is prepared by dissolving 11.3 mL of CH_3OH (density of 0.793 g/mL) in enough water to produce 75.0 mL of solution. Express the percent CH_3OH in this solution as

(a) % (m/v) **(b)** % (m/m) **(c)** % (v/v)

13.128 An aqueous solution having a density of 0.993 g/mL is prepared by dissolving 20.0 mL of C_2H_5OH (density of 0.789 g/mL) in enough water to produce 85.0 mL of solution. Express the percent C_2H_5OH in this solution as

(a) % (m/v) **(b)** % (m/m) **(c)** % (v/v)

13.129 The concentration of a KCl solution is 0.273 molal, *m*, and 0.271 molar, M. What is the density of the solution, in grams per milliliter?

13.130 The concentration of a $Pb(NO_3)_2$ solution is 0.953 molal, *m*, and 0.907 molar, M. What is the density of the solution, in grams per milliliter?

CUMULATIVE PROBLEMS

13.131 Identify the insoluble substance(s) formed when each of the following pairs of soluble substances react in aqueous solution through a double-replacement reaction.

(a) NaCl and $AgNO_3$

(b) $Ba(C_2H_3O_2)_2$ and K_3PO_4

(c) $Pb(NO_3)_2$ and Ag_2SO_4

(d) $CuSO_4$ and BaS

13.132 Identify the insoluble substance(s) formed when each of the following pairs of soluble substances react in aqueous solution through a double-replacement reaction.

(a) $MgCl_2$ and $Ba(OH)_2$

(b) NH_4Cl and $Pb(NO_3)_2$

(c) MgS and Na_2CO_3

(d) $SrCl_2$ and Ag_2SO_4

13.133 How many liters of NH_3 gas at 25°C and 1.46 atm pressure are required to prepare 2.00 L of a 3.50-M solution of NH_3?

13.134 How many liters of HCl gas at 35°C and 1.05 atm pressure are required to prepare 4.00 L of a 0.500-M solution of HCl?

13.135 Calculate the theoretical yield, in grams, of AgCl formed from the reaction of 6.41 g of $ZnCl_2$ with 40.0 mL of a 0.404-M $AgNO_3$ solution according to the reaction.

$$ZnCl_2(s) + 2\,AgNO_3(aq) \longrightarrow Zn(NO_3)_2(aq) + 2\,AgCl(s)$$

13.136 Calculate the theoretical yield, in grams, of AgCl formed from the reaction of 10.00 g of KCl with 100.0 mL of a 2.50 M $AgC_2H_3O_2$ solution according to the reaction.

$$KCl(s) + AgC_2H_3O_2(aq) \longrightarrow KC_2H_3O_2(aq) + AgCl(s)$$

13.137 What mass, in grams, of $BaCrO_4$ would be produced by mixing $35\bar{0}$ mL of a 3.25 M $BaCl_2$ solution with $45\bar{0}$ mL of a 4.50-M K_2CrO_4 solution? The two solutions react according to the equation

$$BaCl_2(aq) + K_2CrO_4(aq) \longrightarrow BaCrO_4(s) + 2\,KCl(aq)$$

13.138 What mass, in grams, of $BaSO_4$ would be produced by mixing 1.53 L of a 4.50-M Na_2SO_4 solution with 3.20 L of a 2.50-M $Ba(NO_3)_2$ solution? The two solutions react according to the equation

$$Na_2SO_4(aq) + Ba(NO_3)_2(aq) \longrightarrow 2\,NaNO_3(aq) + BaSO_4(s)$$

13.139 A 1.25-g sample of *impure* Na_2CO_3 is found to react completely with 70.0 mL of 0.125 M HCl. The equation for the reaction is

$$Na_2CO_3(s) + 2\,HCl(aq) \longrightarrow 2\,NaCl(aq) + CO_2(g) + H_2O(l)$$

What is the mass percent Na_2CO_3 in the impure sample?

13.140 A 5.00-g sample of *impure* $CaCO_3$ is found to react completely with 100.0 mL of 0.100 M H_2SO_4. The equation for the reaction is

$$CaCO_3(s) + H_2SO_4(aq) \longrightarrow CaSO_4(s) + CO_2(g) + H_2O(l)$$

What is the mass percent $CaCO_3$ in the impure sample?

13.141 Magnesium, calcium, and zinc all react with hydrochloric acid as follows (where M represents any of these metals).

$$M(s) + 2\ HCl(aq) \longrightarrow MCl_2(aq) + H_2(g)$$

A sample of one of these metals reacts completely with the acid in 27.9 mL of 2.48 M HCl, and the resulting solution is evaporated to dryness. The residue MCl_2 has a mass of 4.72 g. What is the identity of the metal used?

13.142 Iron, nickel, and tin all react with hydrochloric acid as follows (where M represents any of these metals).

$$M(s) + 2\ HCl(aq) \longrightarrow MCl_2(aq) + H_2(g)$$

A sample of one of these metals reacts completely with the acid in 34.2 mL of 4.00 M HCl, and the resulting solution is evaporated to dryness. The residue MCl_2 has a mass of 8.87 g. What is the identity of the metal used?

13.143 A quantity of sodium peroxide (Na_2O_2) is added to water and the following reaction occurs.

$$2\ Na_2O_2(s) + 2\ H_2O(l) \longrightarrow 4\ NaOH(aq) + O_2(g)$$

If 70.0 mL of O_2 gas (at STP) and 150 mL of NaOH solution are produced, what is the molarity of the NaOH solution?

13.144 A quantity of lithium nitride (Li_3N) is added to water, and the following reaction occurs.

$$Li_3N(s) + 3\ H_2O(l) \longrightarrow 3\ LiOH(aq) + NH_3(g)$$

If 100.0 mL of NH_3 gas (at STP) and 255 mL of LiOH solution are produced, what is the molarity of the LiOH solution?

14 Acids, Bases, and Salts

14.1 Arrhenius Acid–Base Theory

Acids and bases are among the most common and important compounds known. Aqueous solutions of acids and bases are key materials in both biological systems and chemical industrial processes.

Historically, as early as the seventeenth century, acids and bases were recognized as important groups of compounds. Such early recognition was based on what the substances did rather than on their chemical composition.

Early known facts about acids include

1. Acids, when dissolved in water, have a sour taste. (The name acid comes from the Latin word *acidus*, which means "sour.")
2. Acids cause the dye litmus to change from blue to red. (Litmus is a naturally occurring vegetable dye obtained from lichens.)
3. When certain metals, such as zinc and iron, are placed in acids, they dissolve, liberating hydrogen gas.

Early known characteristics of bases include

1. Water solutions of bases feel slippery or soapy to the touch and have a bitter taste.
2. Bases cause the dye litmus to change from red to blue.
3. When fatty substances are placed in a base solution, they dissolve.

It was not until 1884 that acids and bases were defined in terms of chemical composition. In that year, the Swedish chemist Svante August Arrhenius (1859–1927) proposed that acids and bases be defined in terms of the species they form upon dissolution in water. His definitions are the simplest and most commonly used today. An **Arrhenius acid** *is a hydrogen-containing compound that, in water, produces hydrogen ions*

(H^+). The acidic species in Arrhenius theory is, thus, the hydrogen ion. An **Arrhenius base** *is a hydroxide-containing compound that, in water, produces hydroxide ions (OH^-)*. The basic species in Arrhenius theory is, thus, the hydroxide ion.

Two common examples of acids, according to the Arrhenius definition, are the substances HNO_3 and HCl.

$$HNO_3(l) \xrightarrow{H_2O} H^+(aq) + NO_3^-(aq)$$

$$HCl(g) \xrightarrow{H_2O} H^+(aq) + Cl^-(aq)$$

Arrhenius acids in the pure state (not in solution) are covalent compounds; that is, they do not contain H^+ ions. The ions are formed, through a chemical reaction, when the acid is mixed with water. This chemical reaction, between water and the acid molecules, results in removal of H^+ ions from acid molecules.

Two common examples of Arrhenius bases are NaOH and KOH.

$$NaOH(s) \xrightarrow{H_2O} Na^+(aq) + OH^-(aq)$$

$$KOH(s) \xrightarrow{H_2O} K^+(aq) + OH^-(aq)$$

Arrhenius bases are usually ionic compounds in the pure state, in direct contrast to acids. When such compounds dissolve in water, the ions separate to yield the OH^- ions.

14.2 BRØNSTED–LOWRY ACID–BASE THEORY

Although widely used, Arrhenius acid–base theory has some shortcomings. Two disadvantages are that it is restricted to aqueous solution and it cannot explain why compounds like ammonia (NH_3), which do not contain hydroxide ion, produce a basic water solution.

In 1923, Johannes Nicolaus Brønsted (1879–1947), a Danish chemist, and Thomas Martin Lowry (1874–1936), a British chemist, independently and almost simultaneously proposed broadened definitions for acids and bases that applied in both aqueous and non-aqueous solution and that also explained how some non–hydroxide-containing substances, when added to water, produce basic solutions.

A **Brønsted–Lowry acid** *is any substance that can donate a proton (H^+) to some other substance*. A **Brønsted–Lowry base** *is any substance that can accept a proton (H^+) from some other substance*. In simpler terms, a Brønsted–Lowry *acid* is a *proton donor* (or hydrogen ion donor) and a Brønsted–Lowry *base* is a *proton acceptor* (or hydrogen ion acceptor). Note that in these definitions the terms *proton* and *hydrogen ion* are used interchangeably. This is acceptable because a H^+ ion is a hydrogen atom (proton plus electron) that has lost its electron; hence, it is a proton.

Three important additional concepts associated with Brønsted–Lowry acid–base theory are

1. Any chemical reaction involving a Brønsted–Lowry acid must also involve a Brønsted–Lowry base. You cannot have one without the other. Proton donation (from an acid) cannot occur unless an acceptor (a base) is present.
2. All the acids and bases included in the Arrhenius theory (Sec. 14.1) are also acids and bases according to the Brønsted–Lowry theory. However, the converse is not true; some substances not considered Arrhenius bases are Brønsted–Lowry bases.

3. The identity of the acidic species in *aqueous solution* is not the Arrhenius H^+ ion but rather the H_3O^+ ion. Hydrogen ions in solution react with water. The attraction between a hydrogen ion and a water molecule is sufficiently strong to bond the hydrogen ion to the water molecule to form a hydronium ion (H_3O^+). The bond between them is a coordinate covalent bond (Sec. 7.12) because both electrons are furnished by the oxygen atom.

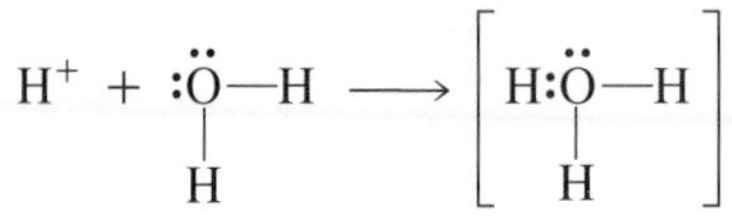

hydronium ion

The Brønsted–Lowry acid–base definitions can best be illustrated by example. Consider the formation reaction for hydrochloric acid, which involves the dissolving of hydrogen chloride gas in water.

$$\underset{\text{acid}}{HCl(g)} + \underset{\text{base}}{H_2O(l)} \longrightarrow H_3O^+(aq) + Cl^-(aq)$$

The HCl behaves as a Brønsted–Lowry acid by donating a proton to a water molecule. Note that a hydronium ion is formed as a result. The base in this reaction is water because it has accepted a proton; no hydroxide ions are involved. The Brønsted–Lowry definition of a base includes all species that accept a proton; hydroxide ions can do this, but so can many other substances.

It is not necessary that a water molecule be one of the reactants in a Brønsted–Lowry acid–base reaction or that the reaction take place in the liquid state. An important application of Brønsted–Lowry acid–base theory is to gas-phase reactions. The white solid haze that often covers glassware in a chemistry laboratory results from the gas-phase reaction between HCl and NH_3.

$$\underset{\text{acid}}{HCl(g)} + \underset{\text{base}}{NH_3(g)} \longrightarrow NH_4^+(g) + Cl^-(g) \longrightarrow NH_4Cl(s)$$

This is a Brønsted–Lowry acid–base reaction, because the HCl molecules donate protons to the NH_3, forming NH_4^+ and Cl^- ions. These ions instantaneously combine to form the white solid NH_4Cl.

Another example of a Brønsted–Lowry acid–base reaction involves the dissolving of ammonia (a nonhydroxide base) in water. In the following equation, note how hydroxide ion is produced as the result of the transfer of a proton from water (written as HOH) to the ammonia

$$\underset{\text{base}}{NH_3(g)} + \underset{\text{acid}}{HOH(l)} \longrightarrow NH_4^+(aq) + OH^-(aq)$$

As an ionic solid dissolves in water to produce an aqueous solution, the solid ionic lattice breaks up, producing individual ions that are free to move about in the solution (Sec. 13.3). Ions so formed can function as Brønsted–Lowry acids or bases, as is illustrated in the following two equations.

$$\underset{\text{acid}}{HCO_3^-(aq)} + \underset{\text{base}}{H_2O(l)} \longrightarrow CO_3^{2-}(aq) + H_3O^+(aq)$$

$$H_2PO_4^-(aq) + OH^-(aq) \longrightarrow HPO_4^{2-}(aq) + H_2O(l)$$

acid base

14.3 CONJUGATE ACIDS AND BASES

For most Brønsted–Lowry acid–base reactions, a 100% proton transfer does not occur. Instead, a state of equilibrium (Sec. 11.12) is reached in which forward and reverse reactions are occurring at an equal rate.

The equilibrium mixture of a Brønsted–Lowry acid–base reaction always has *two* acids and *two* bases present. To illustrate this, consider the acid–base reaction involving hydrogen fluoride and water.

$$HF(aq) + H_2O(l) \rightleftharpoons H_3O^+(aq) + F^-(aq)$$

(The double arrows in this equation indicate a state of equilibrium—both a forward and a reverse reaction are occurring.) For the forward reaction, the HF molecules donate protons to water molecules. Thus, the HF is functioning as an acid and the H_2O is functioning as a base.

$$HF(aq) + H_2O(l) \longrightarrow H_3O^+(aq) + F^-(aq)$$

acid base

For the reverse reaction, the one going from right to left, a different picture emerges. Here, H_3O^+ is functioning as an acid (by donating a proton) and F^- behaves as a base (by accepting the proton).

$$H_3O^+(aq) + F^-(aq) \longrightarrow HF(aq) + H_2O(l)$$

acid base

The two acids and two bases involved in a Brønsted–Lowry equilibrium situation can be grouped into two conjugate acid–base pairs. A **conjugate acid–base pair** *is two species that differ from each other by one proton.* The two conjugate acid–base pairs in our example are

conjugate pair

$$HF(aq) + H_2O(l) \rightleftharpoons H_3O^+(aq) + F^-(aq)$$

acid base acid base

conjugate pair

The **conjugate base** *of an acid is the species that remains when an acid loses a proton.* The conjugate base of HF is F^-. The **conjugate acid** *of a base is the species formed when a base accepts a proton.* The H_3O^+ is the conjugate acid of H_2O. Every acid has a conjugate base, and every base has a conjugate acid. In general terms, these relationships can be diagrammed as follows.

$$HA + B \rightleftharpoons HB^+ + A^-$$

acid base conjugate acid conjugate base

Example 14.1

Identify the conjugate acid–base pairs in the following reaction.

$$HBr(aq) + H_2O(l) \longrightarrow H_3O^+(aq) + Br^-(aq)$$

Solution

To determine the conjugate acid–base pairs, we look for formulas that differ only by one H^+ ion. For this reaction, one pair must be HBr and Br^-, and the other pair must be H_2O and H_3O^+. In each pair, the acid is the substance with one more hydrogen atom, so the two *acids* are HBr and H_3O^+, and the two *bases* are Br^- and H_2O.

conjugate pair

$$HBr(aq) + H_2O(l) \rightleftharpoons H_3O^+(aq) + Br^-(aq)$$

conjugate pair

Practice Exercise 14.1

Identify the conjugate acid–base pairs in the following reaction:

$$HCN(aq) + H_2O(l) \rightleftharpoons H_3O^+(aq) + CN^-(aq)$$

Ans. HCN, CN^-; H_3O^+, H_2O

Example 14.2

Write formulas for the following.

(a) the conjugate base of HCO_3^- **(b)** the conjugate acid of PO_4^{3-}

Solution

(a) A conjugate base can always be found by removing one H^+ ion from a given acid. Removing one H^+ (both the atom and the charge) from HCO_3^- leaves CO_3^{2-}. Thus, CO_3^{2-} is the *conjugate base* of HCO_3^-.

(b) A conjugate acid can always be found by adding one H^+ ion to a given base. Adding one H^+ (both the atom and the charge) to PO_4^{3-} produces HPO_4^{2-}. Thus, HPO_4^{2-} is the *conjugate acid* of PO_4^{3-}.

Practice Exercise 14.2

Write formulas for the following.
(a) the conjugate base of HSO_4^- (b) the conjugate acid of HPO_4^{2-}

Ans. (a) SO_4^{2-}; (b) $H_2PO_4^-$

Some molecules or ions are able to function as either an acid or a base, depending on the kind of substance with which they react. Such molecules are said to be amphoteric. An **amphoteric substance** *can either lose or accept a proton and thus can function as either* *an acid or a base*. (The term *amphoteric* comes from the Greek *amphoteres*, meaning "partly one and partly the other.") Just as an *amphibian* is an animal that lives partly on land and partly in the water, an amphoteric substance is sometimes an acid and sometimes a base.

Water is the most common example of an amphoteric substance. In the first of the following two reactions, water functions as a base and in the second it functions as an acid.

$$\underset{\text{acid}}{HNO_3(l)} + \underset{\text{base}}{H_2O(l)} \rightleftharpoons H_3O^+(aq) + NO_3^-(aq)$$

$$\underset{\text{base}}{NH_3(g)} + \underset{\text{acid}}{H_2O(l)} \rightleftharpoons NH_4^+(aq) + OH^-(aq)$$

Another example of an amphoteric substance is the hydrogen carbonate ion.

$$\underset{\text{acid}}{HCO_3^-(aq)} + \underset{\text{base}}{OH^-(aq)} \rightleftharpoons CO_3^{2-}(aq) + H_2O(l)$$

$$\underset{\text{base}}{HCO_3^-(aq)} + \underset{\text{acid}}{H_3O^+(aq)} \rightleftharpoons H_2CO_3(aq) + H_2O(l)$$

14.4 POLYPROTIC ACIDS

Acids can be classified according to the number of hydrogen ions (protons) they can transfer per molecule during an acid–base reaction. A **monoprotic acid** *is an acid that can transfer only one H^+ ion (proton) per molecule during an acid–base reaction.* Hydrochloric acid (HCl) and nitric acid (HNO_3) are both monoprotic acids.

A **diprotic acid** *is an acid that can transfer two H^+ ions (two protons) per molecule during an acid–base reaction.* Sulfric acid (H_2SO_4) and carbonic acid (H_2CO_3) are examples of diprotic acids. The transfer of protons for a diprotic acid always occurs in steps. For H_2SO_4, the two steps are as follows.

$$H_2SO_4(aq) + H_2O(l) \longrightarrow H_3O^+(aq) + HSO_4^-(aq)$$
$$HSO_4^-(aq) + H_2O(l) \longrightarrow H_3O^+(aq) + SO_4^{2-}(aq)$$

A few triprotic acids exist. A **triprotic acid** *is an acid that can transfer three H^+ ions (three protons) per molecule during an acid–base reaction.* Phosphoric acid, H_3PO_4, is the most common triprotic acid. The three-proton-transfer steps for this acid are as follows.

$$H_3PO_4(aq) + H_2O(l) \longrightarrow H_3O^+(aq) + H_2PO_4^-(aq)$$
$$H_2PO_4^-(aq) + H_2O(l) \longrightarrow H_3O^+(aq) + HPO_4^{2-}(aq)$$
$$HPO_4^{2-}(aq) + H_2O(l) \longrightarrow H_3O^+(aq) + PO_4^{3-}(aq)$$

The general term **polyprotic acid** *describes acids that can transfer two or more H^+ ions (protons) per molecule during an acid–base reaction.*

The number of hydrogen atoms present in one molecule of an acid cannot always be used to classify the acid as mono-, di-, or triprotic. For example, a molecule of acetic acid contains four hydrogen atoms and yet it is a monoprotic acid. Only one of the hydrogen atoms in acetic acid is acidic. An **acidic hydrogen atom** *is a hydrogen atom in an acid molecule that can be transferred to a base during an acid–base reaction.*

Whether or not a hydrogen atom is acidic is related to its location in a molecule, that is, to which other atom it is bonded. Let us consider our previously mentioned acetic acid example in more detail by looking at the structure of this acid. A *structural* equation for the acidic behavior of acetic acid is

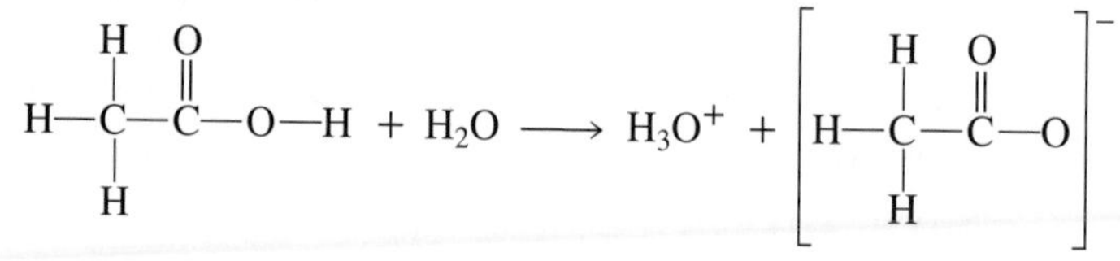

Note the structure of the acetic acid molecule (reactant side of the equation): one hydrogen atom is bonded to an oxygen atom and the other three hydrogen atoms are each bonded to a carbon atom. It is only the hydrogen atom bonded to the oxygen atom that is acidic. The hydrogen atoms bonded to the carbon atom are too tightly held to be removed by reaction with water molecules. Water has very little effect on a carbon–hydrogen bond because it is essentially nonpolar (Sec. 7.8). On the other hand, the hydrogen bonded to oxygen is involved in a very polar bond because of oxygen's large electronegativity (Sec. 7.6). Water, which is a polar molecule, readily attacks polar bonds but has very little effect on nonpolar bonds.

We now see why the formula for acetic acid is usually written as $HC_2H_3O_2$ rather than as $C_2H_4O_2$. In the situation where some hydrogens are easily removed (acidic) and others are not (nonacidic), it is accepted procedure to write the acidic hydrogens first, separated from the other hydrogens in the formula. Citric acid, the principal acid in citrus fruits, is another example of an acid that contains both acidic and nonacidic hydrogens. Its formula, $H_3C_6H_5O_7$, indicates that three of the eight hydrogen atoms present in a molecule are acidic. Table 14.1 gives the formulas, classifications, and common occurrences of selected mono-, di-, and triprotic acids, many of which contain nonacidic hydrogen atoms.

TABLE 14.1 SELECTED COMMON MONO-, DI-, AND TRIPROTIC ACIDS

Name	Formula	Classification	Number of Nonacidic Hydrogen Atoms	Common Occurrence
Acetic acid	$HC_2H_3O_2$	monoprotic	three	vinegar
Lactic acid	$HC_3H_5O_3$	monoprotic	five	sour milk, cheese; produced during muscle contraction
Salicylic acid	$HC_7H_5O_3$	monoprotic	five	present in chemically combined form in aspirin
Hydrochloric acid	HCl	monoprotic	zero	constituent of gastric juice; industrial cleaning agent
Nitric acid	HNO_3	monoprotic	zero	used in urinalysis test for protein; used in manufacture of dyes and explosives
Tartaric acid	$H_2C_4H_4O_6$	diprotic	four	grapes
Carbonic acid	H_2CO_3	diprotic	zero	carbonated beverages; produced in the body from carbon dioxide
Sulfuric acid	H_2SO_4	diprotic	zero	storage batteries; manufacture of fertilizer
Citric acid	$H_3C_6H_5O_7$	triprotic	five	citrus fruits
Boric acid	H_3BO_3	triprotic	zero	antiseptic eyewash
Phosphoric acid	H_3PO_4	triprotic	zero	found in dissociated form (HPO_4^{2-}, $H_2PO_4^-$) in intracellular fluid; component of DNA

We have focused our attention on acids in the preceding discussion. It should be noted that similar concepts can be applied to bases. From an Arrhenius standpoint, bases can release more than one hydroxide ion; for example, $Ca(OH)_2$ is a base that produces two OH^- ions per molecule. From a Brønsted–Lowry viewpoint, bases exist that can accept more than one proton, in a stepwise manner; for example, the PO_4^{3-} ion is a Brønsted–Lowry base that can ultimately accept three protons through reaction with three H_3O^+ ions:

$$PO_4^{3-} \xrightarrow{+H_3O^+} HPO_4^{2-} \xrightarrow{+H_3O^+} H_2PO_4^- \xrightarrow{+H_3O^+} H_3PO_4$$

14.5 STRENGTHS OF ACIDS AND BASES

Brønsted–Lowry acids vary in their ability to transfer protons and produce hydronium ions in aqueous solution. Such acids are classified as strong or weak on the basis of the extent that proton transfer occurs in aqueous solution. A **strong acid** *is a substance that transfers 100%, or very nearly 100%, of its acidic hydrogen atoms to water*. Thus, if an acid is strong, almost all of the acid molecules present give up protons to water. This extensive transfer of protons produces many hydronium ions (the acidic species) within the solution. A **weak acid** *is a substance that transfers only a small percentage of its acidic hydrogen atoms to water*. The extent of proton transfer for weak acids is usually less than 5%. The actual percentage of molecules involved in proton transfer to water depends on the molecular structure of the acid; molecular polarity and the strength and polarity of individual bonds are important factors in determining whether an acid is strong or weak.

A graphical representation of the differences between strong and weak acids, in terms of species present in solution, is given in Figure 14.1. The formula HA represents the acid and H_3O^+ and A^- are the products from the proton transfer to H_2O.

A 0.1-M solution of nitric acid (HNO_3) or sulfuric acid (H_2SO_4) when spilled on your clothes and not immediately washed off will "eat" holes in your clothing. If 0.1-M solutions of either acetic acid ($HC_2H_3O_2$) or carbonic acid (H_2CO_3) were spilled on your clothes, the previously noted "corrosive" effects would not be observed. Why? All four acid solutions are of equal concentration; all are 0.1-M solutions. The difference in behavior relates to the *strength* of the acids; nitric and sulfuric acids are *strong* acids, whereas acetic and carbonic acids are *weak* acids. The number of H_3O^+ ions (the active species) present in the strong

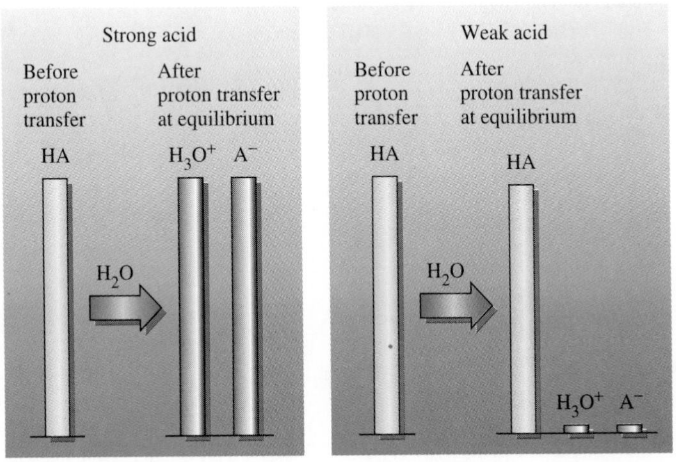

FIGURE 14.1 Many H_3O^+ ions are present in a solution of a strong acid, but only a few such ions are present in the solution of a weak acid.

TABLE 14.2 COMMONLY ENCOUNTERED STRONG ACIDS

Name[a]	Molecular Formula	Molecular Structure
Nitric Acid	HNO_3	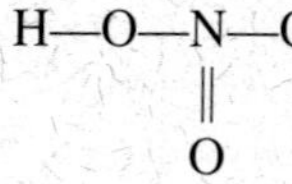
Sulfuric acid	H_2SO_4	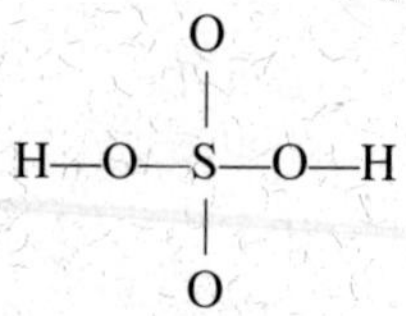
Perchloric acid	$HClO_4$	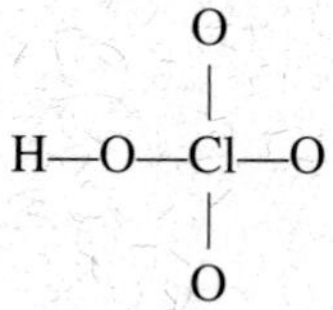
Chloric acid	$HClO_3$	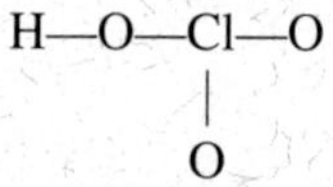
Hydrochloric acid	HCl	H—Cl
Hydrobromic acid	HBr	H—Br
Hydroiodic acid	HI	H—I

[a] Nomenclature for acids was discussed in Section 8.6.

acid solutions is many times greater than for the weak acid solutions even though all the solutions had the same number of acid molecules present (before reaction with water).

There are very few strong acids; the formulas and structures of the seven most commonly encountered strong acids are given in Table 14.2. You should know the identity of

TABLE 14.3 PERCENT PROTON-TRANSFER VALUES FOR 1.0 M SOLUTIONS (AT 25°C) OF SELECTED WEAK ACIDS

Name of Acid	Formula	Percent Proton Transfer
Phosphoric acid	H_3PO_4	8.3
Nitrous acid	HNO_2	2.7
Hydrofluoric acid	HF	2.5
Acetic acid	$HC_2H_3O_2$	0.42
Carbonic acid	H_2CO_3	0.065
Dihydrogen phosphate ion	$H_2PO_4^-$	0.025
Hydrocyanic acid	HCN	0.0020
Hydrogen carbonate ion	HCO_3^-	0.00075
Hydrogen phosphate ion	HPO_4^{2-}	0.000047

these seven strong acids; you will need such knowledge to write net ionic equations, the topic of Section 14.7.

The vast majority of acids that exist are weak acids. Familiar weak acids include acetic acid ($HC_2H_3O_2$), the acidic component of vinegar; boric acid (H_3BO_3), a common ingredient for eyewashes; and carbonic acid (H_2CO_3), found in carbonated beverages. Weak acids are not all equally weak; proton transfer occurs to a greater extent for some weak acids than for others. Table 14.3 gives percent proton-transfer values for selected weak acids. The calculational techniques needed to determine percent proton-transfer values, such as those in Table 14.3, will not be considered in this text.

For polyprotic acids the stepwise proton-transfer sequence that occurs (Sec. 14.4) can be used to determine relative acid strengths for the related acidic species. Consider the two-step proton-transfer process for carbonic acid.

$$H_2CO_3(aq) + H_2O(l) \longrightarrow H_3O^+(aq) + HCO_3^-(aq)$$
$$HCO_3^-(aq) + H_2O(l) \longrightarrow H_3O^+(aq) + CO_3^{2-}(aq)$$

The second proton is not as easily removed as the first because it must be pulled away from a negatively charged particle, HCO_3^-. Accordingly, HCO_3^- is a weaker acid than H_2CO_3. In general, each successive step in a stepwise proton-transfer process occurs to a lesser extent than the previous step. Thus, for triprotic H_3PO_4, the parent H_3PO_4 species (first-step reactant) is a stronger acid than $H_2PO_4^-$ (second-step reactant), which in turn is a stronger acid than HPO_4^{2-} (third-step reactant). This ordering of the phosphoric acid-derived species is reflected in the values given in Table 14.3.

Just as there are strong acids and weak acids, there are also strong bases and weak bases. As with acids, there are only a few strong bases. Strong bases are limited to the hydroxides of groups IA and IIA of the periodic table listed in Table 14.4. Of the strong bases, only NaOH and KOH are commonly used in the chemical laboratory. The low solubility of the group IIA hydroxides in water limits their use. However, despite this low solubility, these hydroxides are still considered to be strong bases because whatever dissolves dissociates into ions 100%.

Only one weak base will be considered in this text: aqueous ammonia. It furnishes small amounts of OH^- ions through reaction with water molecules.

$$NH_3(g) + H_2O(l) \longrightarrow NH_4^+(aq) + OH^-(aq)$$

A solution of ammonia in water is most properly called *aqueous ammonia*, although it is also commonly called ammonium hydroxide, its commercial name. Aqueous ammonia is the preferred designation, since most of the NH_3 present is in molecular form. Only a very few NH_3 molecules have reacted with the water to give ammonium (NH_4^+) and hydroxide (OH^-) ions.

It is important to remember that the terms *strong* and *weak* apply to the extent of dis-

TABLE 14.4 COMMON STRONG BASES

Group IA Hydroxides	Group IIA Hydroxides
LiOH	
NaOH	
KOH	$Ca(OH)_2$
RbOH	$Sr(OH)_2$
CsOH	$Ba(OH)_2$

sociation and not to the concentrations of acid or base. For example, stomach acid (gastric juice) is a *dilute* (not weak) solution of a strong acid; it is a 5% (m/m) solution of hydrochloric acid. On the other hand, a 35% (m/m) solution of hydrochloric acid would be considered to be a *concentrated* (not strong) solution of a strong acid.

14.6 SALTS

The title of this chapter is "Acids, Bases, and Salts." In preceding sections, we have discussed acids and bases, but not salts. What is a salt? To a nonscientist the word *salt* connotes a white granular substance used as a seasoning for food. To the chemist it has a much broader meaning. Sodium chloride, or table salt, is only one of thousands of salts known to a chemist. "Pass the salt" is a very ambiguous request to a chemist.

From a chemical viewpoint, a **salt** *is an ionic compound containing a metal or polyatomic ion as the positive ion and a nonmetal or polyatomic ion (except hydroxide) as the negative ion.* (Ionic compounds containing hydroxide ion are bases rather than salts.)

Many salts occur in nature, and numerous others have been prepared in the laboratory. The wide variety of uses found for salts can be seen from Table 14.5, a listing of selected salts and their uses.

Much information concerning salts has been presented in previous chapters, although the term *salt* was not explicitly used in these discussions. Formula writing and nomenclature for binary ionic compounds (salts) was covered in Sections 7.5 and 7.6. Many salts, as shown in Table 14.5, contain polyatomic ions such as nitrate and sulfate. Such ions were discussed in Sections 7.8 and 8.4. The solubility of ionic compounds (salts) in water was the topic of Section 13.4.

In solution all common salts are dissociated into ions (Sec. 13.3). Even if a salt is only slightly soluble, the small amount that does dissolve completely dissociates. Thus, the terms weak and strong, used to denote qualitatively the percent dissociation of acids and bases, are not applicable to common salts. We do not use the terms *strong salt* and *weak salt*.

Acids, bases, and salts are related in that a salt is one of the products resulting from the reaction of an acid with a hydroxide base. This particular type of reaction, called neutralization, will be discussed in Section 14.8.

TABLE 14.5 SOME COMMON SALTS AND THEIR USES

Name	Formula	Uses
Ammonium nitrate	NH_4NO_3	fertilizer; explosive
Barium sulfate	$BaSO_4$	X-rays of gastrointestinal tract
Calcium carbonate	$CaCO_3$	chalk; limestone
Calcium chloride	$CaCl_2$	drying agent for removal of small amounts of water
Iron(II) sulfate	$FeSO_4$	treatment for anemia
Potassium chloride	KCl	"salt" substitute for low sodium diets
Sodium chloride	$NaCl$	table salt; used as a deicer (to melt ice)
Sodium bicarbonate	$NaHCO_3$	ingredient in baking powder
Sodium hypochlorite	$NaClO$	bleaching agent
Silver bromide	$AgBr$	light-sensitive material in photographic film
Tin(II) fluoride	SnF_2	toothpaste additive

14.7 IONIC AND NET IONIC EQUATIONS

Soluble strong acids, soluble strong bases, and soluble salts all dissociate 100% (or nearly 100%) in aqueous solution to produce ions (Secs. 14.5 and 14.6). It is extremely useful to discuss the reactions of such acids, bases, and salts in aqueous solution in terms of the ions present. This is most easily done using a new type of equation—a net ionic equation.

Up to this point in the text, most equations we have used have been *molecular equations*, equations where the complete formulas of all reactants and products are shown. From molecular equations, ionic equations may be written. An **ionic equation** *is an equation in which the formulas of the predominant form of each compound in aqueous solution are used; dissociated compounds are written as ions, and undissociated compounds are written in molecular form.* Net ionic equations are derived from ionic equations. A **net ionic equation** *is an ionic equation from which nonparticipating (spectator) species have been eliminated.*

The differences between molecular, ionic, and net ionic equations can best be illustrated by examples. Let us consider the chemical reaction that results when a solution of potassium chloride (KCl) is added to a solution of silver nitrate ($AgNO_3$). An insoluble salt, silver chloride (AgCl), is produced as a result of the mixing. The *molecular equation* for this reaction is

$$AgNO_3(aq) + KCl(aq) \longrightarrow KNO_3(aq) + AgCl(s)$$

Three of the four substances involved in this reaction—$AgNO_3$, KCl, and KNO_3—are soluble salts and thus exist in solution in ionic form. This is shown by writing the *ionic equation* for the reaction.

$$\underbrace{Ag^+(aq) + NO_3^-(aq)}_{AgNO_3 \text{ in ionic form}} + \underbrace{K^+(aq) + Cl^-(aq)}_{KCl \text{ in ionic form}} \longrightarrow \underbrace{K^+(aq) + NO_3^-(aq)}_{KNO_3 \text{ in ionic form}} + AgCl(s)$$

In this equation each of the three soluble salts is shown in dissociated (ionic) form rather than in undissociated form. A close look at this ionic equation shows that the potassium ions (K^+) and nitrate ions (NO_3^-) appear on both sides of the equation, indicating that they did not undergo any chemical change. In other words, they are *spectator ions*; they did not participate in the reaction. The *net ionic equation* for this reaction is written by dropping (canceling) all spectator ions from the ionic equation. In our case, the net ionic equation becomes

$$Ag^+(aq) + Cl^-(aq) \longrightarrow AgCl(s)$$

This net ionic equation indicates that the product AgCl was formed by the reaction of silver ions (Ag^+) with chloride ions (Cl^-). It totally ignores the presence of those ions that are not taking part in the reaction. Thus, a net ionic equation focuses on only those species in a solution actually involved in a chemical reaction. It does not give all species present in the solution.

If you can write equations in molecular form, you will find it a straightforward process to convert these equations to net ionic form. Follow these three steps.

1. Check the given molecular equation to make sure that it is balanced.
2. Expand the molecular equation into an ionic equation.

3. Convert the ionic equation into a net ionic equation by eliminating spectator ions.

In expanding a molecular equation into an ionic equation (step 2) you must decide whether to write each reactant and product in dissociated (ionic) form or undissociated (molecular) form. The following rules serve as guidelines in making such decisions.

1. Soluble compounds that completely dissociate in aqueous solution are written in ionic form. They include
   (a) all soluble salts (see Sec. 13.4 for solubility rules).
   (b) all strong acids (see Table 14.3)
   (c) all strong bases (see Table 14.4)
2. Soluble weak acids and weak bases are written in molecular form, since they are incompletely ionized in solution and thus exist predominantly in the undissociated form. The following are considered to be weak acids or weak bases.
   (a) All acids not listed in Table 14.3 as strong acids. Common examples are HNO_2, HF, H_2S, $HC_2H_3O_2$, H_2CO_3, and H_3PO_4.
   (b) All bases not listed in Table 14.4 as strong bases. Aqueous ammonia (NH_3) is the most common weak base.
3. All insoluble substances (solids, liquids, and gases), whether ionic or covalent, exist as molecules or neutral ionic units and are written as such.
4. All soluble covalent substances, for example, carbon dioxide (CO_2) or sucrose ($C_{12}H_{22}O_{11}$), are written in molecular form.
5. If water, the solvent, appears in the equation, it is written in molecular form.

Now, we apply these guidelines by writing some net ionic equations.

Example 14.3

Write the net ionic equation for the following aqueous solution reaction.

$$MgCl_2 + AgNO_3 \longrightarrow Mg(NO_3)_2 + AgCl \quad \text{(unbalanced equation)}$$

Solution

STEP 1 To balance the given molecular equation, the coefficient 2 must be placed in front of both $AgNO_3$ and AgCl.

$$MgCl_2 + 2\,AgNO_3 \longrightarrow Mg(NO_3)_2 + 2\,AgCl$$

STEP 2 A decision must be made to write each reactant and product in ionic or molecular form. Let us consider them one by one.

$MgCl_2$: This compound is a salt. The solubility rules indicate that it is soluble. Thus, $MgCl_2$ will be written in ionic form: $Mg^{2+} + 2\,Cl^-$. Note that three ions (one Mg^{2+} ion and two Cl^- ions) are produced from the dissociation of one $MgCl_2$ unit.

$AgNO_3$: This compound is also a soluble salt; it will be written in ionic form. Each $AgNO_3$ unit (there are two) produces one Ag^+ ion and one $NO_3{}^-$ ion.

$Mg(NO_3)_2$: All nitrate salts are soluble. Thus, $Mg(NO_3)_2$ will be written in ionic form. Three ions are produced upon dissociation of one $Mg(NO_3)_2$ unit: one Mg^{2+} and two NO_3^-.

AgCl: The solubility rules indicate that this compound is an insoluble salt. Thus, it will be written in molecular form in the ionic equation.

The ionic equation will have the form

$$Mg^{2+} + 2\ Cl^- + 2\ Ag^+ + 2\ NO_3^- \longrightarrow Mg^{2+} + 2\ NO_3^- + 2\ AgCl$$

Note how the coefficient 2 in front of $AgNO_3$ and AgCl in the molecular equation affects the ionic equation. The dissociation of two $AgNO_3$ units produces two Ag^+ ions and two NO_3^- ions. Similarly, two AgCl units are present.

STEP 3 Inspection of the ionic equation shows that Mg^{2+} ion and NO_3^- ions (two of each of them) are spectator ions. Cancellation of these ions from the equation will give the net ionic equation.

$$\cancel{Mg^{2+}} + 2\ Cl^- + 2\ Ag^+ + \cancel{2\ NO_3^-} \longrightarrow \cancel{Mg^{2+}} + \cancel{2\ NO_3^-} + 2\ AgCl$$
$$2\ Ag^+ + 2\ Cl^- \longrightarrow 2\ AgCl$$

The coefficients in the net ionic equation should be the smallest set of numbers that correctly balance the equation. In this case, all the coefficients are divisible by 2. Dividing by 2, we get

$$Ag^+ + Cl^- \longrightarrow AgCl$$

Practice Exercise 14.3

Write the net ionic equation for the following aqueous solution reaction.

$$KCl + AgC_2H_3O_2 \longrightarrow KC_2H_3O_2 + AgCl$$

Ans. $Ag^+ + Cl^- \longrightarrow AgCl$

Example 14.4

Write the net ionic equation for the following aqueous solution reaction.

$$H_2S + AlI_3 \longrightarrow Al_2S_3 + HI \quad \text{(unbalanced equation)}$$

Solution

STEP 1 Balancing the molecular equation, we get

$$3\ H_2S + 2\ AlI_3 \longrightarrow Al_2S_3 + 6\ HI$$

STEP 2 The expansion of the molecular equation into an ionic equation is accomplished by the following analysis.

H_2S: This is a weak acid. All weak acids are written in **molecular form** in ionic equations.

AlI_3: This is a soluble salt. All iodide salts are soluble, with three exceptions; this is not one of the exceptions. Soluble salts are written in **ionic form** in ionic equations.

Al_2S_3: This is an insoluble salt. All sulfides are insoluble except for groups IA and IIA and NH_4^+. Thus, Al_2S_3 will remain in **molecular form** in the ionic equation.

HI: This compound is an acid. It is one of the seven strong acids listed in Table 14.3. Strong acids are written in **ionic form**.

The ionic equation for the reaction is

$$3\ H_2S + 2\ Al^{3+} + 6I^- \longrightarrow Al_2S_3 + 6\ H^+ + 6\ I^-$$

Note again that the coefficients present in the balanced molecular equation must be taken into consideration when determining the total number of ions produced from dissociation. On dissociation an AlI_3 unit produces four ions: one Al^{3+} ion and three I^- ions. This number must be doubled for the ionic equation because AlI_3 carries the coefficient 2 in the balanced molecular equation. Similar considerations apply to HI in this equation.

STEP 3 Inspection of the ionic equation shows that only I^- ions (six of them) are spectator ions. Cancellation of these ions from the equation will give the net ionic equation.

$$3\ H_2S + 2\ Al^{3+} + \cancel{6I^-} \longrightarrow Al_2S_3 + 6\ H^+ + \cancel{6I^-}$$

$$3\ H_2S + 2\ Al^{3+} \longrightarrow Al_2S_3 + 6\ H^+$$

Practice Exercise 14.4

Write the net ionic equation for the following aqueous solution reaction.

$$H_2CO_3 + MgCl_2 \longrightarrow MgCO_3 + HCl \quad \text{(unbalanced equation)}$$

Ans. $H_2CO_3 + Mg^{2+} \longrightarrow MgCO_3 + 2\ H^+$

Example 14.5

Write the net ionic equation for the following aqueous solution reaction.

$$HNO_3 + LiOH \longrightarrow LiNO_3 + H_2O$$

Solution

STEP 1 All coefficients in this equation are 1; the equation is balanced as written

$$HNO_3 + LiOH \longrightarrow LiNO_3 + H_2O$$

STEP 2 The expansion of the molecular equation into an ionic equation is based on the following analysis.

HNO_3: This compound is an acid. It is one of the seven strong acids listed in Table 14.3. Strong acids are written in **ionic form** in ionic equations.

LiOH: This compound is a base. It is one of the strong bases listed in Table 14.4. Strong bases are written in **ionic form** in ionic equations.

$LiNO_3$: This compound is a soluble salt. All nitrate salts are soluble. Thus, $LiNO_3$ is written in **ionic form**.

H_2O: This compound is a covalent compound; two nonmetals are present. Covalent compounds are always written in **molecular form**.

The ionic equation for the reaction, from the above information, is

$$H^+ + NO_3^- + Li^+ + OH^- \longrightarrow Li^+ + NO_3^- + H_2O$$

STEP 3 Inspection of the ionic equation shows that NO_3^- ions and Li^+ ions are spectator ions. Cancellation of these ions from the equation gives the net ionic equation.

$$H^+ + \cancel{NO_3^-} + \cancel{Li^+} + OH^- \longrightarrow \cancel{Li^+} + \cancel{NO_3^-} + H_2O$$
$$H^+ + OH^- \longrightarrow H_2O$$

Practice Exercise 14.5

Write the net ionic equation for the following aqueous solution reaction.

$$HNO_3 + Ba(OH)_2 \longrightarrow Ba(NO_3)_2 + H_2O \quad \text{(unbalanced equation)}$$

Ans. $H^+ + OH^- \longrightarrow H_2O$

14.8 REACTIONS OF ACIDS

All acids have some unique properties that adapt them for use in specific situations. In addition, all acids have certain chemical properties in common, properties related to the presence of H_3O^+ ions in aqueous solution. In this section we consider three types of chemical reactions that acids characteristically undergo.

1. Acids react with active metals to produce hydrogen gas and a salt.
2. Acids react with bases to produce a salt and water.
3. Acids react with carbonates and bicarbonates to produce carbon dioxide, a salt, and water.

Reaction with Metals

Acids react with many, but not all, metals. When they do react, the metal dissolves and hydrogen gas (H_2) is liberated. In the reaction the metal atoms lose electrons and become metal ions. The lost electrons are taken up by the hydrogen ions (protons) of the acid; the hydrogen ions become electrically neutral, combine into molecules, and emerge from the reaction mixture as hydrogen gas. Illustrative of the reaction of an acid and a metal is the reaction between zinc and sulfuric acid.

Molecular equation: $Zn + H_2SO_4 \longrightarrow ZnSO_4 + H_2$
Net ionic equation: $Zn + 2\,H^+ \longrightarrow Zn^{2+} + H_2$

In terms of the reaction types discussed in Section 10.6 the reaction of an acid with a metal to produce hydrogen gas is a *single-replacement reaction*; the metal replaces the hydrogen from the acid. Recall, from Section 10.6, that a single-replacement reaction has the general form

TABLE 14.6 ACTIVITY SERIES FOR COMMON METALS

Increasing tendency to react	Group	Metal	Symbol	Remarks
↑	React with H^+ ions to liberate hydrogen gas	Potassium	K	react violently with cold water
		Sodium	Na	(react violently with cold water)
		Calcium	Ca	reacts slowly with cold water
		Magnesium	Mg	react slowly with hot water (steam)
		Aluminum	Al	(react slowly with hot water (steam))
		Zinc	Zn	(react slowly with hot water (steam))
		Chromium	Cr	(react slowly with hot water (steam))
		Iron	Fe	
		Nickel	Ni	
		Tin	Sn	
		Lead	Pb	
		Hydrogen	H	
	Do not react with H^+ ions	Copper	Cu	
		Mercury	Hg	
		Silver	Ag	
		Platinum	Pt	
		Gold	Au	

F.Y.I.
Because they do not react with acids on the skin, gold, platinum, and silver are good metals for jewelry. Jewelry made with these metals will not tarnish like jewelry made from cheaper metals.

F.Y.I.
The reaction that occurs when stomach acid reacts with a typical antacid is $2\ HCl + Mg(OH)_2 \longrightarrow MgCl_2 + 2\ H_2O$

$$X + YZ \longrightarrow Y + XZ$$

Metals can be arranged in a reactivity order based on their ability to react with acids. Such an ordering for the more common metals is given in Table 14.6. Any metal above hydrogen in the activity series will dissolve in an acid solution and form H_2. The closer a metal is to the top of the series, the more rapid the reaction. Those metals below hydrogen in the series do not dissolve in an acid to form H_2.

As noted in Table 14.6, the most active metals (those nearest the top in the activity series) also react with water. Again, hydrogen gas is produced. In the cases of potassium and sodium, the reaction is sometimes violent enough to cause explosions as the result of H_2 ignition. The equation for the reaction of potassium with water, which is also a single-replacement reaction, is

$$\text{Molecular equation:}\quad 2\ K + 2\ H_2O \longrightarrow 2\ KOH + H_2$$
$$\text{Net ionic equation:}\quad 2\ K + 2\ H_2O \longrightarrow 2\ K^+ + 2\ OH^- + H_2$$

Note that the resulting solution is basic when a metal reacts with water; hydroxide ions are present.

Reaction with Bases

When Arrhenius acids and bases are mixed, they react with each other; their acidic and basic properties disappear, and we say that they have *neutralized* each other.

Neutralization *is the reaction between an acid and a base to form a salt and water.* The hydrogen ions from the acid combine with the hydroxide ions from the base to form water. The salt formed contains the negative ion from the acid and the positive ion from the base. Neutralization is a *double-replacement* reaction (Sec. 10.6).

$$AX + BY \longrightarrow AY + BX$$

$$\underset{\text{acid}}{HCl} + \underset{\text{base}}{KOH} \longrightarrow \underset{\text{water}}{HOH} + \underset{\text{salt}}{KCl}$$

Any time an acid is completely reacted with a base, neutralization occurs. It does not matter whether the acid and base are strong or weak. Sodium hydroxide (a strong base) and nitric acid (a strong acid) react as follows.

Molecular equation: $HNO_3 + NaOH \longrightarrow NaNO_3 + H_2O$
Net ionic equation: $H^+ + OH^- \longrightarrow H_2O$

The equations for the reaction of potassium hydroxide (a strong base) with hydrocyanic acid (a weak acid) are

Molecular equation: $HCN + KOH \longrightarrow KCN + H_2O$
Net ionic equation: $HCN + OH^- \longrightarrow CN^- + H_2O$

Note that in each case the products are a salt ($NaNO_3$ in the first reaction, KCN in the second) and water. Note also that the net ionic equations for the two neutralization reactions are different. In the second set of equations the acid must remain written in molecular form, since it is a weak acid.

F.Y.I.
The reaction equation for the "volcanoes" that children enjoy making by mixing vinegar (5% acetic acid (v/v)) and baking soda is $HC_2H_3O_2 + NaHCO_3 \longrightarrow CO_2 + H_2O + NaC_2H_3O_2$ The carbon dioxide gas makes the foamy "volcano."

Reaction with Carbonates and Bicarbonates

Carbon dioxide gas (CO_2), water, and a salt are always the products of the reaction of acids with carbonates or bicarbonates, as illustrated by the following equations.

Molecular equation: $2\ HCl + Na_2CO_3 \longrightarrow 2\ NaCl + CO_2 + H_2O$
Molecular equation: $HCl + NaHCO_3 \longrightarrow NaCl + CO_2 + H_2O$

Baking powder is a mixture of a bicarbonate and an acid-forming solid. The addition of water to this mixture generates the acid that then reacts with the bicarbonate to release carbon dioxide into the batter. It is the generated carbon dioxide that causes the batter to rise. Baking soda is pure $NaHCO_3$. To cause it to release carbon dioxide, an acid-containing substance, such as buttermilk, sour milk, or fruit juices, must be added to it.

F.Y.I.
Alka-Seltzer contains citric acid and baking soda. When the tablets are dropped into water, the two react and carbon dioxide bubbles are produced.

14.9 REACTIONS OF BASES

The most important characteristic reaction of bases is their reaction with acids (neutralization), discussed in the preceding section. Another characteristic reaction, that of bases with certain salts, is discussed in Section 14.10.

Bases react with fats and oils and convert them into smaller, soluble molecules. For this reason most household cleaning products contain basic substances. Lye (impure NaOH) is an active ingredient in numerous drain cleaners. Also, many advertisements for liquid household cleaners emphasize the fact that aqueous ammonia (a weak base) is present in the product.

In Section 14.1 we noted that one of the general properties of bases is a "slippery or soapy" feeling to the touch. The bases themselves are not slippery; the slipperiness results as the bases react with fats and oils in the skin to form "slippery or soapy" compounds.

14.10 REACTIONS OF SALTS

Dissolved salts will react with metals, acids, bases, and other salts under specific conditions.

1. Salts react with some metals to convert the metallic ion of the salt to free metal and the free metal to its salt.
2. Salts react with some acid solutions to form other acids and salts.
3. Salts react with some base solutions to form other bases and salts.
4. Salts react with some solutions of other salts to form new salts.

The tendency for salts to react with metals is related to the relative positions of the two involved metals in the activity series (Table 14.6). In order for salts to react with acids, bases, or other salts, one of the reaction products must be (1) an insoluble salt, (2) a gas that is evolved from the solution, or (3) an undissociated soluble species, such as a weak acid or a weak base. The formation of any of these products serves as the driving force to cause the reaction to occur.

Reaction with Metals

If an iron nail is placed in a solution of copper sulfate ($CuSO_4$), metallic copper will be deposited on the nail and some of the iron will dissolve.

Molecular equation: $Fe(s) + CuSO_4(aq) \longrightarrow Cu(s) + FeSO_4(aq)$
Net ionic equation: $Fe + Cu^{2+} \longrightarrow Cu + Fe^{2+}$

One metal has replaced the other; a single-replacement reaction (Sec. 10.6) has occurred. This type of reaction will occur only if the metal going into solution is above the replaced metal in the activity series. Iron is above copper in the activity series and can replace it. If a strip of copper were placed in a solution of $FeSO_4$—just the opposite situation to what we have been discussing—no reaction would occur since copper is below iron in the activity series.

Reaction with Acids

In order for a salt to react with an acid, a new weaker acid, a new insoluble salt, or a gaseous compound must be one of the products.

An example of a reaction in which the formation of an *insoluble salt* is the driving force for the reaction to occur is

$$AgNO_3(aq) + HCl(aq) \longrightarrow AgCl(s) + HNO_3(aq)$$

This is a double-replacement reaction (Sec. 10.6); the silver and hydrogen have traded partners.

The conclusion that this reaction will occur comes from a consideration of the possible recombinations of the reacting species. In a solution made by mixing silver nitrate ($AgNO_3$) and hydrochloric acid (HCl), four kinds of ions are present initially (before any reaction occurs): Ag^+ and NO_3^- (since $AgNO_3$ is a soluble salt) and H^+ and Cl^- (since HCl is a strong acid). The question is whether these ions can get together in new appropriate combinations. The possible new combinations of oppositely charged ions are $H^+NO_3^-$ and Ag^+Cl^-. The first of these combinations would result in the formation of the strong acid HNO_3. Strong acids in solution exist in dissociated form; therefore, these ions will not combine. The second combination does occur because AgCl is an insoluble salt. Thus, the overall reaction takes place as a result of the formation of this insoluble salt. The net result of the reaction is that the original ions exchange partners.

The double-replacement reaction of sodium fluoride (a soluble salt) with hydrochloric acid (a strong acid) illustrates the case where formation of a *new weaker acid* is the driving force for the reaction.

$$NaF(aq) + HCl(aq) \longrightarrow NaCl(aq) + HF(aq)$$

Using an analysis pattern similar to that in the previous example, we find that four types of ions are present initially: Na^+ and F^- (from the soluble salt) and H^+ and Cl^- (from the strong acid). Possible new combinations are Na^+Cl^- and H^+F^-. Sodium chloride, the result of the first combination, will not form because this salt is soluble. The combination of H^+ ion with F^- ion does occur because it yields the weak acid HF. In solution weak acids exist predominantly in molecular form. In all reactions of this general type, the acid formed in the reaction must be weaker than the reactant acid. If the reactant acid is strong, as in this example, such a determination is obvious. If both the reactant and product acids are weak, information such as that given in Table 14.3 would be needed to predict which of the two acids is the weaker.

The most common type of reaction in which the driving force is the *evolution of a gas* involves a carbonate or bicarbonate. This type of reaction was discussed in Section 14.8.

Note that a reaction does not always occur when acid and salt solutions are mixed. Consider the possible reaction of NaCl and HNO_3 solutions. Initially, four types of ions are present: Na^+ and Cl^- (from the soluble salt) and H^+ and NO_3^- (from the strong acid). The new combinations, if a reaction did occur, would be $Na^+NO_3^-$ (a soluble salt) and H^+Cl^- (a strong acid). Since both of the products would exist in dissociated form in solution, no recombination of ions occurs; hence, no reaction occurs.

Reaction with Bases

The criteria for the reaction of bases with salts are similar to those for acid–salt reactions, except that weaker base formation replaces weaker acid formation as one of the three driving forces. An example of a base–salt reaction involving the formation of an *insoluble salt* is

$$Ba(OH)_2(aq) + Na_2SO_4(aq) \longrightarrow BaSO_4(s) + 2\ NaOH(aq)$$

The most common situation in which *gas evolution* is the driving force for base–salt reactions is where ammonium salts are involved. In such cases, ammonia gas is given off, as illustrated by the reaction of NH_4Cl and KOH.

$$NH_4Cl(aq) + KOH(aq) \longrightarrow KCl(aq) + NH_3(g) + H_2O(l)$$

Reaction of Salts with Each Other

Two different salt solutions will react when mixed, in a double-replacement reaction, only if an *insoluble salt* can be formed. Consider the following possible reactions.

$$AgNO_3(aq) + NaCl(aq) \longrightarrow AgCl(s) + NaNO_3(aq)$$
$$KNO_3(aq) + NaCl(aq) \longrightarrow KCl(aq) + NaNO_3(aq)$$

The first reaction occurs because AgCl is an insoluble salt. The second reaction does not occur since both of the possible products are soluble salts, which means there is no driving force for the reaction.

Example 14.6

Write molecular, ionic, and net ionic equations for the reaction that occurs, if any, when 0.1 M solutions of the following substances are mixed.

(a) $Fe(NO_3)_2$ and K_2S **(b)** $CaCl_2$ and H_2SO_4 **(c)** HNO_3 and $NaC_2H_3O_2$

Solution

(a) Both of the reactants are soluble salts. Two different salt solutions will react when mixed only if an insoluble salt can be formed.

In a solution made by mixing $Fe(NO_3)_2$ and K_2S, four kinds of ions are present initially (before any reaction occurs): Fe^{2+} and NO_3^- [from the $Fe(NO_3)_2$] and K^+ and S^{2-} (from the K_2S). The possible new combinations of oppositely charged ions are Fe^{2+} with S^{2-} and K^+ with NO_3^-

Original Ion Combinations *Possible New Combinations*

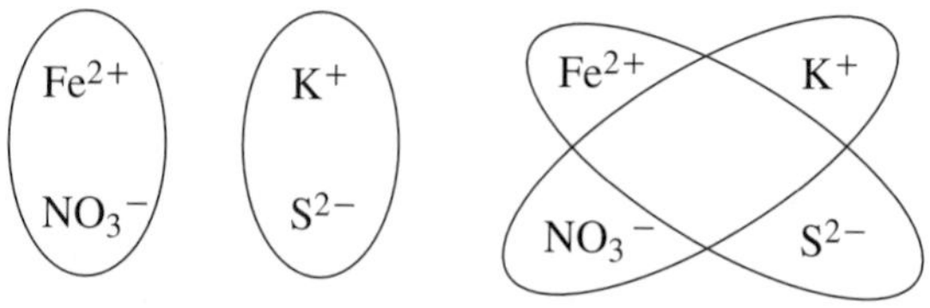

The first one of these new combinations, the formation of FeS, is the one that will be the driving force for the reaction to occur. FeS is an insoluble salt. The second new combination, the formation of KNO_3, does not occur because KNO_3 is a soluble salt and soluble salts exist in dissociated form in solution. The equations for the reaction are

Molecular: $Fe(NO_3)_2 + K_2S \longrightarrow FeS + 2\ KNO_3$
Ionic: $Fe^{2+} + \cancel{2\ NO_3^-} + \cancel{2\ K^+} + S^{2-} \longrightarrow FeS + \cancel{2\ K^+} + \cancel{2\ NO_3^-}$
Net ionic: $Fe^{2+} + S^{2-} \longrightarrow FeS$

(b) One of the reactants, $CaCl_2$, is a soluble salt and the other reactant, H_2SO_4 is a strong acid. Both reactants exist in solution in dissociated form; thus, four types of ions are present in the mixed solution (before any reaction occurs): Ca^{2+}, Cl^-, H^+, and SO_4^{2-}. The conclusion that a reaction will occur comes from a consideration of the possible new combinations of the reacting species.

Original Ion Combinations *Possible New Combinations*

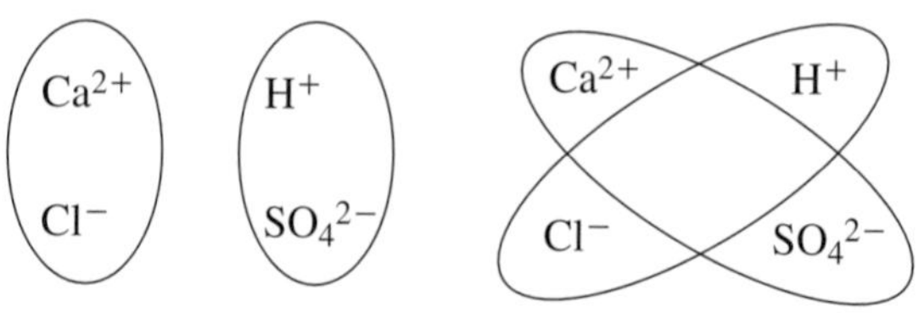

The first one of these new combinations, the formation of $CaSO_4$ is the driving force for the reaction to occur; $CaSO_4$ is an insoluble salt. The other new combination, the formation of HCl, does not occur because HCl is a strong acid and will exist in solution in dissociated form. The equations for the reaction are

Molecular:	$CaCl_2 + H_2SO_4 \longrightarrow CaSO_4 + 2\ HCl$
Ionic:	$Ca^{2+} + \cancel{2\ Cl^-} + \cancel{2\ H^+} + SO_4^{2-} \longrightarrow CaSO_4 + \cancel{2\ H^+} + \cancel{2\ Cl^-}$
Net ionic:	$Ca^{2+} + SO_4^{2-} \longrightarrow CaSO_4$

(c) The reactants are a strong acid (HNO_3) and a soluble salt ($NaC_2H_3O_2$). Both are dissociated in solution; hence, H^+, NO_3^-, Na^+, and $C_2H_3O_2^-$ ions are present in the reaction mixture (before any reaction occurs). Possible new combinations of the reacting species are

Original Ion Combinations *Possible New Combinations*

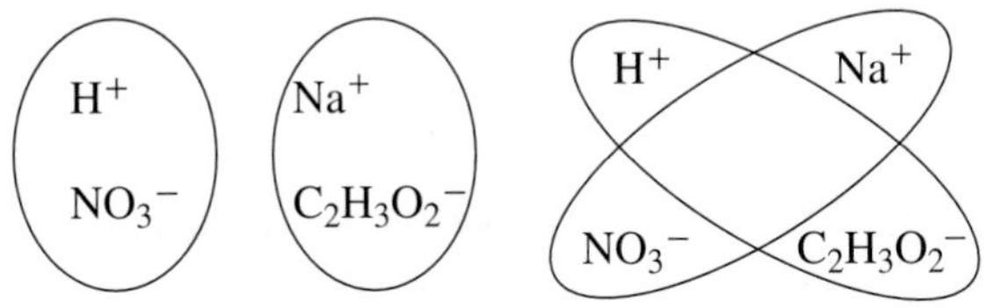

Weak acid formation is the driving force for the reaction; acetic acid ($HC_2H_3O_2$) forms from the combination of H^+ and $C_2H_3O_2^-$ ions. The Na^+ and NO_3^- will not combine because a soluble salt would be the product. The equations for the reaction are

Molecular:	$HNO_3 + NaC_2H_3O_2 \longrightarrow HC_2H_3O_2 + NaNO_3$
Ionic:	$H^+ + \cancel{NO_3^-} + \cancel{Na^+} + C_2H_3O_2^- \longrightarrow HC_2H_3O_2 + \cancel{Na^+} + \cancel{NO_3^-}$
Net ionic:	$H^+ + C_2H_3O_2^- \longrightarrow HC_2H_3O_2$

Practice Exercise 14.6

Write molecular, ionic, and net ionic equations for the reaction that occurs, if any, when 0.1 M solutions of the following substances are mixed.

(a) $Fe(NO_3)_3$ and Na_3PO_4 (b) $CaCl_2$ and HNO_3 (c) HCl and Na_2S

Ans. (a)	Molecular:	$Fe(NO_3)_3 + Na_3PO_4 \longrightarrow FePO_4 + 3\ NaNO_3$
	Ionic:	$Fe^{3+} + 3\ NO_3^- + 3\ Na^+ + PO_4^{3-} \longrightarrow FePO_4 + 3\ Na^+ + 3\ NO_3^-$
	Net ionic:	$Fe^{3+} + PO_4^{3-} \longrightarrow FePO_4$
(b)	Molecular:	$CaCl_2 + 2\ HNO_3 \longrightarrow Ca(NO_3)_2 + 2\ HCl$
	Ionic:	$Ca^{2+} + 2\ Cl^- + 2\ H^+ + 2\ NO_3^- \longrightarrow Ca^{2+} + 2\ NO_3^- + 2\ H^+ + 2\ Cl^-$
	Net ionic:	all species cancel (no reaction occurs)
(c)	Molecular:	$2\ HCl + Na_2S \longrightarrow H_2S + 2\ NaCl$
	Ionic:	$2\ H^+ + 2\ Cl^- + 2\ Na^+ + S^{2-} \longrightarrow H_2S + 2\ Na^+ + 2\ Cl^-$
	Net ionic:	$2\ H^+ + S^{2-} \longrightarrow H_2S$

14.11 DISSOCIATION OF WATER

Normally, we think of water as a covalent, nondissociating substance. Experiments show, however, that in a sample of pure water a very small percentage of the water molecules have undergone dissociation to produce ions. The dissociation reaction may be thought of as involving the transfer of a proton from one water molecule to another (Brønsted–Lowry theory; Sec. 14.2)

$$H_2O + H_2O \longrightarrow H_3O^+ + OH^-$$

or simply as the dissociation of a single water molecule (Arrhenius theory; Sec. 14.1).

$$H_2O\ (HOH) \longrightarrow H^+ + OH^-$$

From either viewpoint, the net result is the formation of *equal amounts* of hydronium (hydrogen) ion and hydroxide ion.

The dissociation of water molecules is part of an equilibrium situation. Individual water molecules are continually dissociating. This process is balanced by hydroxide and hydronium ions recombining to form water at the same rate. At equilibrium, at 25°C, the H_3O^+ and OH^- ion concentrations are each 1.00×10^{-7} M (0.000000100 M). This very, very small concentration is equivalent to there being one H_3O^+ and one OH^- ion present for every 550,000,000 undissociated water molecules. Even though the H_3O^+ and OH^- ion concentrations are very minute, they are important, as we shall shortly see.

Experimentally it is found that, at any given temperature, the product of the concentrations of H_3O^+ ion and OH^- ion in water is a constant. We can calculate the value of this constant at 25°C, since we know that the concentration of each ion is 1.00×10^{-7} M at this temperature. The brackets [ ] specifically denote ion concentration in moles per liter.

$$[H_3O^+] \times [OH^-] = \text{constant}$$
$$(1.00 \times 10^{-7}) \times (1.00 \times 10^{-7}) = 1.00 \times 10^{-14}$$

Ion product for water *is the name given to the numerical value (1.00×10^{-14}) associated with the product of the H_3O^+ ion and OH^- ion molar concentrations in water.* Note that the ion concentrations must be expressed in moles per liter (M) in order to obtain the value 1.00×10^{-14} for the ion product for water. In the general expression for the ion product for water.

$$[H_3O^+] \times [OH^-] = \text{ion product for water}$$

The ion product expression for water is valid not only in pure water but also when solutes are present in the water. At all times, the product of the hydronium and hydroxide ion molarities in an aqueous solution, at 25°C, must equal 1.00×10^{-14}. Thus if the $[H_3O^+]$ is increased by the addition of an acidic solute, the $[OH^-]$ must decrease until the expression.

$$[H_3O^+] \times [OH^-] = 1.00 \times 10^{-14}$$

is satisfied. Similarly, if OH^- ions are added to the water, the $[H_3O^+]$ must correspondingly decrease. The extent of the decrease in $[H_3O^+]$ or $[OH^-]$, as the result of the addition of a quantity of the other ion, is easily calculated by the ion product expression.

Example 14.7

Sufficient acidic solute is added to a quantity of water to produce $[H_3O^+] = 7.50 \times 10^{-5}$. What is the $[OH^-]$ in this solution?

Solution

The $[OH^-]$ can be calculated using the ion product expression for water. Solving this expression for $[OH^-]$ gives

$$[OH^-] = \frac{1.00 \times 10^{-14}}{[H_3O^+]}$$

Substituting into this expression the known $[H_3O^+]$ and doing the arithmetic gives

$$[OH^-] = \frac{1.00 \times 10^{-14}}{7.50 \times 10^{-5}} = 1.3333333 \times 10^{-10} \quad \text{(calculator answer)}$$

$$= 1.33 \times 10^{-10} \quad \text{(correct answer)}$$

Practice Exercise 14.7

Sufficient acidic solute is added to a quantity of water to produce $[H_3O^+] = 4.50 \times 10^{-2}$. What is the $[OH^-]$ in this solution?

Ans. 2.22×10^{-13} M

In Section 14.2 we learned that the acidic species in aqueous solution is the H_3O^+ ion. Now that we have noted that a small number of H_3O^+ ions are present in all aqueous solutions, even basic ones, a refining of our concepts of acids and bases is in order. An acid is a substance that increases the H_3O^+ ion concentration in water. **Acidic solutions** *have a higher* $[H_3O^+]$ *than* $[OH^-]$. In a similar manner, a base is a substance that increases the OH^- ion concentration in water. **Basic solutions** *have a higher* $[OH^-]$ *than* $[H_3O^+]$. In a **neutral solution** *the concentrations of both* H_3O^+ *ions and* OH^- *ions are equal.* Table 14.7 summarizes the relationships between $[H_3O^+]$ and $[OH^-]$ that we have just considered.

TABLE 14.7 RELATIONSHIPS BETWEEN $[H_3O^+]$ and $[OH^-]$ in Aqueous Solutions at 25°C

Neutral solution:	$[H_3O^+] = [OH^-] = 1.00 \times 10^{-7}$
Acidic solution:	$[H_3O^+]$ is greater than 1.00×10^{-7}
	$[OH^-]$ is less than 1.00×10^{-7}
Basic solution:	$[H_3O^+]$ is less than 1.00×10^{-7}
	$[OH^-]$ is greater than 1.00×10^{-7}

14.12 THE pH SCALE

Hydronium ion concentrations in aqueous solution range from relatively high values (10 M) to extremely small ones (10^{-14} M). It is inconvenient to work with numbers that extend over such a wide range; a hydronium ion concentration of 10 M is 1000 trillion times larger than a hydronium ion concentration of 10^{-14} M. The pH scale, proposed by the Danish chemist Sören Peter Lauritz Sörensen (1868–1939) in 1909, is a more practical way to handle such a wide range of numbers.

The **pH** *of a solution is defined as the negative logarithm of the molar hydronium ion concentration.* Expressed mathematically, the pH definition is

$$pH = -\log [H_3O^+]$$

TABLE 14.8 LOGARITHM VALUES FOR SELECTED NUMBERS

Number	Number Expressed as a Power of 10	Common Logarithm
10,000	1×10^4	4.0
1,000	1×10^3	3.0
100	1×10^2	2.0
10	1×10^1	1.0
1	1×10^0	0.0
0.1	1×10^{-1}	−1.0
0.01	1×10^{-2}	−2.0
0.001	1×10^{-3}	−3.0
0.0001	1×10^{-4}	−4.0

Logarithms are simply exponents. The *common logarithm*, abbreviated log, which is the type of logarithm used in the definition of pH, is based on powers of 10. *For a number expressed in scientific notation that has a coefficient of 1, the log of that number is the value of the exponent.* For instance, the log of 1×10^{-8} is −8.0, and the log of 1×10^{6} is 6.0. Table 14.8 gives more examples of the relationship between powers of 10 and logarithmic values for numbers in scientific notation whose coefficients are 1. Note from this table that log values may be either positive or negative depending on the sign of the exponent. (How we determine significant figures in logarithmic calculations is discussed later in this section.)

Integral pH Values

It is easy to calculate the pH value for a solution when the molar hydronium ion concentration is an exact power of 10, for example, 1.0×10^{-4}. In this situation the pH is given directly by the negative of the exponent value on the power of 10.

$$[H_3O^+] = 1.0 \times 10^{-x}$$
$$pH = x$$

Thus, if the hydronium ion concentration is 1.0×10^{-9}, the pH will be 9.00.

This simple relationship between pH and power of 10 is obtained from the formal definition of pH as follows.

$$\begin{aligned} pH &= -\log [H_3O^+] \\ &= -\log [1.0 \times 10^{-x}] \\ &= -(-x) \\ &= x \end{aligned}$$

Again, it should be noted that this simple relationship is valid only when the coefficient in the exponential expression for the hydronium ion concentration is 1.0. How the pH is calculated when the coefficient is not 1.0 will be covered later in this section.

Example 14.8

Calculate the pH for each of the following solutions.

(a) $[H_3O^+] = 1 \times 10^{-3}$ **(b)** $[H_3O^+] = 1 \times 10^{-9}$

(c) $[OH^-] = 1 \times 10^{-5}$

Solution

(a) Let us use the formal definition of pH in obtaining this first pH value.

$$pH = -\log [H_3O^+]$$

This expression indicates that to obtain a pH we must first take the logarithm of the molar hydronium ion concentration and then change the sign of that logarithm.

The logarithm of 1×10^{-3} is a -3.0. Thus, we have

$$\begin{aligned} pH &= -\log(1 \times 10^{-3}) \\ &= -(-3.0) \\ &= 3.0 \end{aligned}$$

(b) Let us use the shorter, more direct way for obtaining pH this time, the method based on the relationship

$$\begin{aligned} [H_3O^+] &= 1 \times 10^{-x} \\ pH &= x \end{aligned}$$

Since the power of 10 is -9 in this case, the pH will be 9.0.

(c) The given quantity involves hydroxide ion rather than hydronium ion. Thus, we must first calculate the hydronium ion concentration, and then the pH.

$$[H_3O^+] = \frac{1.00 \times 10^{-14}}{[OH^-]} = \frac{1.00 \times 10^{-14}}{1 \times 10^{-5}}$$

$$= 1 \times 10^{-9} \quad \text{(calculator and correct answer)}$$

A solution with a hydronium ion concentration of 1×10^{-9} M will have a pH of 9.0.

Practice Exercise 14.8

Calculate the pH for each of the following solutions.

(a) $[H_3O^+] = 1 \times 10^{-6}$ (b) $[H_3O^+] = 1 \times 10^{-12}$ (c) $[OH^-] = 1 \times 10^{-3}$

Ans. (a) 6.0; (b) 12.0; (c) 11.0

Since pH is simply another way of expressing hydronium ion concentration, acidic, neutral, and basic solutions can be identified by their pH values.

A neutral solution ($[H_3O^+] = 1.0 \times 10^{-7}$) has a pH of 7.00. Values of pH less than 7 correspond to acidic solutions. The lower the pH value, the greater the acidity. Values of pH greater than 7 represent basic solutions. The higher the pH value, the greater the basicity. The relationships between $[H_3O^+]$, $[OH^-]$, and pH are summarized in Table 14.9. Note that a change of *one* unit in pH corresponds to a *tenfold* increase or decrease in $[H_3O^+]$. Also note that *lowering* the pH always corresponds to *increasing* the H_3O^+ ion concentration.

Table 14.10 lists the pH values of a number of common substances. Except for gastric juices, most human body fluids have pH values within a couple of units of neutrality. Almost all foods are acidic. Tart taste is associated with food of low pH.

TABLE 14.9 THE pH SCALE

pH	$[H_3O^+]$	$[OH^-]$	
0.0	1	10^{-14}	↑
1.0	10^{-1}	10^{-13}	
2.0	10^{-2}	10^{-12}	
3.0	10^{-3}	10^{-11}	Acidic
4.0	10^{-4}	10^{-10}	
5.0	10^{-5}	10^{-9}	
6.0	10^{-6}	10^{-8}	
7.0	10^{-7}	10^{-7}	Neutral
8.0	10^{-8}	10^{-6}	
9.0	10^{-9}	10^{-5}	
10.0	10^{-10}	10^{-4}	
11.0	10^{-11}	10^{-3}	Basic
12.0	10^{-12}	10^{-2}	
13.0	10^{-13}	10^{-1}	
14.0	10^{-14}	1	

TABLE 14.10 APPROXIMATE pH VALUES OF SOME COMMON SUBSTANCES

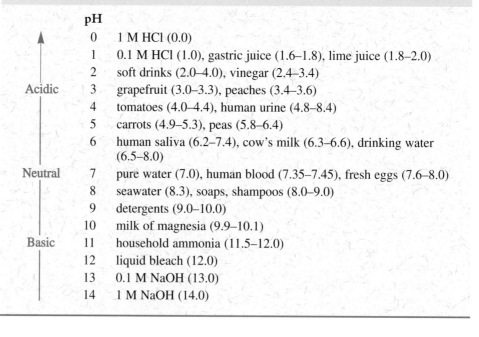

	pH	
↑	0	1 M HCl (0.0)
	1	0.1 M HCl (1.0), gastric juice (1.6–1.8), lime juice (1.8–2.0)
	2	soft drinks (2.0–4.0), vinegar (2.4–3.4)
Acidic	3	grapefruit (3.0–3.3), peaches (3.4–3.6)
	4	tomatoes (4.0–4.4), human urine (4.8–8.4)
	5	carrots (4.9–5.3), peas (5.8–6.4)
	6	human saliva (6.2–7.4), cow's milk (6.3–6.6), drinking water (6.5–8.0)
Neutral	7	pure water (7.0), human blood (7.35–7.45), fresh eggs (7.6–8.0)
	8	seawater (8.3), soaps, shampoos (8.0–9.0)
	9	detergents (9.0–10.0)
	10	milk of magnesia (9.9–10.1)
Basic	11	household ammonia (11.5–12.0)
	12	liquid bleach (12.0)
	13	0.1 M NaOH (13.0)
	14	1 M NaOH (14.0)

Nonintegral pH Values

Note that some of the pH values in Table 14.10 are nonintegral, that is, not whole numbers. Nonintegral pH values result from molar hydronium ion concentrations where the coefficient in the exponential expression for concentration has a value other than 1. For example, consider the following matchups between hydronium ion concentration and pH.

$$[H_3O^+] = 6.3 \times 10^{-5} \qquad pH = 4.20$$
$$[H_3O^+] = 4.0 \times 10^{-5} \qquad pH = 4.40$$
$$[H_3O^+] = 2.0 \times 10^{-5} \qquad pH = 4.70$$

Obtaining nonintegral pH values like these from hydronium ion concentrations requires an electronic calculator that allows for the input of exponential numbers and has a base 10 logarithm key (LOG).

In using an electronic calculator, depending on the model you have, you can obtain logarithm values simply by pressing the LOG key after having entered the number whose log value is desired or vice versa. For pH, you must remember that after obtaining the log value you must change its sign because of the negative sign in the defining equation for pH.

Significant figure considerations for log values involve a concept not previously encountered. It can best be illustrated by considering some actual log values. Consider the following five related numbers and their log values.

Number	**Logarithm**
2.43×10^0	0.38560627
2.43×10^2	2.38560627
2.43×10^4	4.38560627
2.43×10^7	7.38560627
2.43×10^{11}	11.38560627

This tabulation shows that

1. The number to the left of the decimal point in each logarithm (called the *characteristic*) is related only to the exponent of 10 in the number whose logarithm was taken.
2. The number to the right of the decimal point in each logarithm (called the *mantissa*) is related only to the coefficient in the exponential notation form of the number. Since the coefficient is 2.43 in each case, the mantissas are all the same (0.38560627).

Combining generalizations (1) and (2) gives us the significant figure rule for logarithms. *In a logarithm the digits to the left of the decimal point are not counted as significant figures.* These digits relate to the placement of the decimal point in the number. From our previous significant figure work (Sec. 2.3) they are somewhat analogous to the leading zeros (which are not significant) in a number such as 0.0000243.

Thus, the *coefficient* of the number whose logarithm has been taken and the *mantissa* of the logarithm must have the same number of digits. Rewriting the previous tabulation of logarithms to the correct number of significant figures gives

2.43×10^0	0.386
2.43×10^2	2.386
2.43×10^4	4.386
2.43×10^7	7.386
2.43×10^{11}	11.386
3 digits	3 digits

Now we can understand the following number–logarithm relationships that were given, without explanation, earlier in this section.

$$\log 1 \times 10^{-4} = -4.0$$
$$\log 1.0 \times 10^{-9} = -9.00$$

The first exponential number has one significant figure, and the second one has two significant figures.

Example 14.9

Calculate the pH of a solution with $[H_3O^+] = 3.9 \times 10^{-5}$.

Solution

With an electronic calculator, we first enter the number 3.9×10^{-5} into the calculator. We then use the LOG key to obtain the logarithm value, −4.4089353. (With some calculators, the LOG key is pressed before entering the number.)

$$\begin{aligned} pH &= -\log (3.9 \times 10^{-5}) \\ &= -(-4.4089353) \\ &= 4.4089353 \quad \text{(calculator answer)} \\ &= 4.41 \quad \text{(correct answer)} \end{aligned}$$

The given hydronium ion concentration has two significant figures. Therefore, the logarithm should have two significant figures, as 4.41 does. In a logarithm only the digits to the right of the decimal place are considered significant.

Practice Exercise 14.9

Calculate the pH of a solution with $[H_3O^+] = 7.9 \times 10^{-11}$.

Ans. 10.10

It is frequently necessary to calculate the hydronium ion concentration for a solution from its pH value. This type of calculation, which is the reverse of that just illustrated, is shown in Example 14.10.

Example 14.10

The pH of a solution is 5.70. What is the molar hydronium ion concentration for this solution?

Solution

Because the pH is between 5 and 6, we know immediately that $[H_3O^+]$ will be between 10^{-5} and 10^{-6} M. From the defining equation for pH, we have

$$\text{pH} = -\log [H_3O^+] = 5.70$$
$$\log [H_3O^+] = -5.70$$

To find $[H_3O^+]$ we need to determine the *antilog* of −5.70.

How an antilog is obtained using a calculator depends on the type of calculator you have. Many calculators have an antilog function (sometimes labed INV log) that performs this operation. If this key is present, then

1. Enter the number −5.70. Note that it is the *negative* of the pH that is entered into the calculator.

2. Press the INV log key (or an inverse key and then a log key). The result is the desired hydronium ion concentration.

$$\log [H_3O^+] = -5.70$$
$$\text{antilog } [H_3O^+] = 1.9952623 \times 10^{-6} \quad \text{(calculator answer)}$$
$$[H_3O^+] = 2.0 \times 10^{-6} \quad \text{(correct answer)}$$

(With some calculators, step 2 and step 1 reversed.)
Remember, that the original pH value was a two-significant-figure pH.

Some calculators use a 10^x key to perform the antilog operation. Use of this key is based on the mathematical identity

$$\text{antilog } X = 10^x$$

For our case this means

$$\text{antilog } -5.70 = 10^{-5.70}$$

If the 10^x key is present, then

1. Enter the number −5.70 (the negative of the pH).

2. Press the function key 10^x. The result is the desired hydronium ion concentration.

$$[H_3O^+] = 10^{-5.70} = 1.9952623 \times 10^{-6} \quad \text{(calculator answer)}$$
$$= 2.0 \times 10^{-6} \quad \text{(correct answer)}$$

Practice Exercise 14.10

The pH of a solution is 8.40. What is the molar hydronium ion concentration for this solution?

Ans. 4.0×10^{-9} M

14.13 HYDROLYSIS OF SALTS

The addition of an acid to water produces an acidic solution. The addition of a base to water produces a basic solution. What type of solution is produced when a salt is added to water? Since salts are the products of acid–base neutralizations, a logical supposition would be that salts dissolve in water to produce neutral (pH = 7) solutions. Such is the case for a *few* salts. Aqueous solutions of *most* salts, however, are either acidic or basic rather than neutral. Let us consider why this is so.

When a salt is dissolved in water it completely ionizes, that is, it completely breaks up into the ions of which it is composed (Sec. 13.3). For many salts, one or more of the ions so produced is reactive toward water. The ensuing reaction, which is called *hydrolysis*, causes the solution to have a nonneutral pH. **Hydrolysis** *is the reaction of a substance with water to produce hydronium ion or hydroxide ion or both.*

Not all salts hydrolyze. Which ones do and which ones don't? Of those salts that do hydrolyze, which ones produce acidic solutions and which ones produce basic solutions? The following guidelines, based on the neutralization "parentage" of a salt, that is, on the acid and base that will produce the salt through neutralization, can be used to answer these questions.

1. The salt of a *strong acid* and a *strong base* does not hydrolyze and therefore the salt solution is neutral.
2. The salt of a *strong acid* and a *weak base* hydrolyzes to produce an acidic solution.
3. The salt of a *weak acid* and a *strong base* hydrolyzes to produce a basic solution.
4. The salt of a *weak acid* and a *weak base* hydrolyzes to produce a slightly acidic, neutral, or slightly basic solution depending on the relative weaknesses of the acid and base.

The first prerequisite for using these guidelines is the ability to classify a salt into one of the four categories mentioned in the guidelines. This classification is accomplished by writing the neutralization equation (Sec. 14.8) that produces the salt and then specifying the strength (strong or weak) of the involved acid and base. The "parent" acid and base for the salt are identified by pairing the negative ion of the salt with H^+ (to form the acid) and pairing the positive ion of the salt with OH^- (to form the base). The following two equations illustrate the overall procedure.

$$\boxed{Na}OH + H\boxed{Cl} \longrightarrow H_2O + \boxed{NaCl}$$

strong base strong acid strong acid–strong base salt

$$\boxed{K}OH + H\boxed{CN} \longrightarrow H_2O + \boxed{KCN}$$

strong base weak acid weak acid–strong base salt

Note that knowledge of which acids and bases are strong and which are weak (Sec. 14.5) is a necessary part of the classification process. Once the salt classification has been determined, the guideline that is appropriate for the situation is easily selected.

Table 14.11 summarizes the concepts of this section to this point in our discussion.

TABLE 14.11 NEUTRALIZATION "PARENTAGE" OF SALTS AND THE NATURE OF THE AQUEOUS SOLUTIONS THEY FORM

Type of Salt	Nature of Aqueous Solution	Examples
Weak base–strong acid	Acidic	NH_4Cl, NH_4NO_3
Strong base–weak acid	Basic	$NaC_2H_3O_2$, K_2CO_3
Weak base–weak acid	Depends on the salt	$NH_4C_2H_3O_2$, NH_4NO_2
Strong base–strong acid	Neutral	NaCl, KBr

Example 14.11

Determine the acid–base "parentage" of each of the following salts and then use this information to predict whether each salt's aqueous solution is acidic, basic, or neutral.

(a) Sodium acetate, $NaC_2H_3O_2$ **(b)** Ammonium chloride, NH_4Cl

(c) Potassium chloride, KCl **(d)** Ammonium fluoride, NH_4F

Solution

(a) The ions present are Na^+ and $C_2H_3O_2^-$. The "parent" base of Na^+ is NaOH, a strong base. The "parent" acid of $C_2H_3O_2^-$ is $HC_2H_3O_2$, a weak acid. Thus, the acid–base neutralization that produces this salt is

$$\underset{\text{strong base}}{NaOH} + \underset{\text{weak acid}}{HC_2H_3O_2} \longrightarrow H_2O + \underset{\text{weak acid–strong base salt}}{NaC_2H_3O_2}$$

The solution of a weak acid–strong base salt (guideline 3) produces a basic solution.

(b) The ions present are NH_4^+ and Cl^-. The "parent" base of NH_4^+ is NH_3, a weak base. The "parent" acid of Cl^- is HCl, a strong acid. This "parentage" will produce a strong acid–weak base salt through neutralization. Such a salt gives an acidic solution upon hydrolysis (guideline 2).

(c) The ions present are K^+ and Cl^-. The "parent" base is KOH (a strong base) and the "parent" acid is HCl (a strong acid). The salt produced from neutralization involving this acid–base pair will be a strong acid–strong base salt. Such salts do not hydrolyze. The aqueous solution is neutral (guideline 1).

(d) The ions present are NH_4^+ and F^-. Both ions are of weak "parentage"; NH_3 is a weak base and HF is a weak acid. Thus, NH_4F is a weak acid–weak base salt. This is a guideline 4 situation. In this situation you cannot predict the effect of hydrolysis unless you know the relative strengths of the weak acid and weak base (which is the weaker of the two). Guidelines to determine such information are not given in this text and thus we cannot predict the final acidity of the solution.

Practice Exercise 14.11

Predict whether an aqueous solution of each of the following salts will be acidic, basic, or neutral.

(a) sodium bromide, NaBr (b) potassium cyanide, KCN
(c) ammonium iodide, NH_4I (d) barium chloride, $BaCl_2$

Ans. (a) neutral; (b) basic; (c) acidic; (d) neutral

Salt hydrolysis reactions are Brønsted–Lowry acid–base (proton transfer) reactions (Sec. 14.2). Such reactions are of the following two general types.

1. *Basic hydrolysis*: The reaction of the ***negative ion*** from a salt with water to produce the ion's conjugate acid and hydroxide ion. Examples of such reactions are:

conjugate acid–base pair (CN^- and HCN)

$$CN^- + H_2O \longrightarrow HCN + OH^-$$

CN^-: proton acceptor; H_2O: proton donor; HCN: weak acid; OH^-: makes solution basic

conjugate acid–base pair (F^- and HF)

$$F^- + H_2O \longrightarrow HF + OH^-$$

F^-: proton acceptor; H_2O: proton donor; HF: weak acid; OH^-: makes solution basic

The only negative ions that undergo hydrolysis are those of "weak acid parentage." The driving force for the reaction is the formation of the weak acid "parent."

2. *Acidic hydrolysis*: The reaction of the ***positive ion*** from a salt with water to produce the ion's conjugate base and hydronium ion. The most common ion to undergo this type of reaction is the NH_4^+ ion.

conjugate acid–base pair (NH_4^+ and NH_3)

$$NH_4^+ + H_2O \longrightarrow NH_3 + H_3O^+$$

NH_4^+: proton donor; H_2O: proton acceptor; NH_3: weak base; H_3O^+: makes solution acidic

The only positive ions that undergo hydrolysis are those of "weak base parentage." The driving force for the reaction is the formation of the weak base "parent."

Example 14.12

For each of the following salts identify the ion or ions present that will hydrolyze and then write net ionic equations for hydrolysis reactions that occur.

(a) sodium fluoride, NaF **(b)** potassium bromide, KBr

(c) ammonium nitrate, NH_4NO_3 **(d)** ammonium cyanide, NH_4CN

Solution

(a) The ions produced when NaF dissolves are Na^+ and F^-. The Na^+ ion will not hydrolyze since its "parent" base, NaOH, is *strong*. The F^- ion will hydrolyze since its "parent" acid, HF, is *weak*. The equation for the hydrolysis reaction is

$$F^- + H_2O \longrightarrow HF + OH^-$$

The product OH^- causes the solution to be basic.

(b) Dissolution of KBr in water produces K^+ and Br^- ions. Neither of these ions will hydrolyze. The "parent" base of K^+ is KOH (strong) and the "parent" acid of Br^- is HBr (strong).

(c) This salt ionizes to produce $NH_4{}^+$ and $NO_3{}^-$ ions. The $NH_4{}^+$ ion is associated with the *weak* base NH_3 and the $NO_3{}^-$ ion is associated with the *strong* acid HNO_3. The former will hydrolyze, the latter will not. The hydrolysis reaction, which produces an acidic solution (H_3O^+) is

$$NH_4{}^+ + H_2O \longrightarrow NH_3 + H_3O^+$$

(d) Both the $NH_4{}^+$ ion (from the weak base NH_3) and the CN^- ion (from the weak acid HCN) will hydrolyze.

$$NH_4{}^+ + H_2O \longrightarrow NH_3 + H_3O^+$$
$$CN^- + H_2O \longrightarrow HCN + OH^-$$

The pH of the solution will be determined by the reaction that occurs to the greater extent. If the first reaction occurs to the greater extent the solution will be acidic; conversely, if the second reaction is dominant, a basic solution results. (In this course you are not expected to be able to make such a determination, which involves comparison of the relative acid and base strengths of HCN and NH_3. For the record, the CN^- hydrolysis dominates and the solution is basic.)

Practice Exercise 14.12

For each of the following salts, identify the ion or ions present that will hydrolyze and then write net ionic equations for the hydrolysis reactions that occur.
(a) potassium chloride, KCl (b) ammonium sulfate, $(NH_4)_2SO_4$

Ans. (a) neither ion hydrolyzes

(b) $NH_4{}^+$; $NH_4{}^+ + H_2O \longrightarrow NH_3 + H_3O^+$

14.14 BUFFER SOLUTIONS

As we discussed in the last section, certain salts that hydrolyze can change the pH of water. In this section, we will consider a second phenomenon involving salts and their effect on solution pH. Certain *combinations* of compounds can protect the pH of a solution from change. These combinations, which always involve at least one salt, are called *buffers*, and the solutions containing them are called *buffer solutions*. A **buffer solution** *is a solution that resists a change in pH when small amounts of acid or base are added to it*. A **buffer** *is the solute (or solutes) present in a buffer solution that causes it to be resistant to a change in pH.*

Buffers contain two species: (1) a substance to react with and remove added base, and (2) a substance to react with and remove added acid. The "chemical combination" encountered most commonly that meets these requirements is a weak acid and a salt of its conjugate base or a weak base and salt of its conjugate acid. Thus, most common buffers involve conjugate acid–base pairs (Sec. 14.3). Such pairs employed as buffers include $HC_2H_3O_2/C_2H_3O_2^-$, NH_4^+/NH_3, $H_2PO_4^-/HPO_4^{2-}$, and H_2CO_3/HCO_3^-.

Consider a buffer solution containing approximately equal concentrations of acetic acid (a weak acid) and sodium acetate (a salt of this weak acid). This solution resists pH change by the following mechanisms.

1. When a small amount of a strong acid such as HCl is added to the solution, the added H_3O^+ ions react with the acetate ions from the sodium acetate to give acetic acid.

$$H_3O^+ + C_2H_3O_2^- \longrightarrow HC_2H_3O_2 + H_2O$$

Most of the added H_3O^+ ions are tied up in acetic acid molecules, and the pH changes very little.

2. When a small amount of a strong base such as NaOH is added to the solution, the added OH^- ions react with the acetic acid (neutralization) to give acetate ions and water.

$$OH^- + HC_2H_3O_2 \longrightarrow C_2H_3O_2^- + H_2O$$

Most of the added OH^- ions are converted to water, and the pH changes only slightly.

The reactions that are responsible for the buffering action in the acetic acid–acetate ion system can be summarized as follows.

$$HC_2H_3O_2 \underset{H_3O^+}{\overset{OH^-}{\rightleftharpoons}} C_2H_3O_2^-$$

Note that one member of the buffer pair removes excess H_3O^+ ion and the other removes excess OH^- ion. The buffering action always results in the active species being converted to its partner species.

Buffer systems have their limits. If large amounts of H_3O^+ or OH^- are added to a buffer, the buffering capacity can be exceeded; then the buffer system is overwhelmed, and the pH changes. For example, if large amounts of H_3O^+ were added to the acetate–acetic acid buffer just discussed, the H_3O^+ ion would react with acetate ion until the acetate was depleted. Then the pH would begin to drop as free H_3O^+ ions accumulate in the solution.

14.15 BUFFERS IN THE HUMAN BODY

Buffer solutions play an important role in the functioning of the human body. All body fluids have definite pH values that must be maintained within very narrow ranges because living cells are extremely sensitive to even slight changes in pH. The protection against pH change is provided by buffers, which can be referred to as "chemical shock absorbers" or "chemical sponges" because of their key protective role.

Blood is a vital buffer solution. Even small departures from the normal pH range of blood (7.35–7.45) can cause serious illness, and death can result from variations that exceed a few tenths of a pH unit. This situation results because many of the key reactions that take place in blood are enzyme catalyzed and reach optimum conditions only within the narrow pH range. Altering the pH slows down or stops the action of the enzymes.

The major buffer system in blood is composed of carbonic acid (H_2CO_3) and bicarbonate salts such as sodium bicarbonate ($NaHCO_3$). Certain proteins and, to a small extent, hydrogen phosphate ions also help buffer blood. The carbonic acid–bicarbonate buffering system in blood operates in the following manner. Any H_3O^+ formed in the blood reacts with bicarbonate ion to give carbonic acid.

$$H_3O^+ + HCO_3^- \longrightarrow H_2CO_3 + H_2O$$

Carbonic acid is an unstable acid that readily decomposes to give carbon dioxide and water.

$$H_2CO_3 \longrightarrow H_2O + CO_2$$

Excess carbon dioxide in the blood, which is formed from this decomposition, is removed from the blood in the lungs and is exhaled.

The presence of OH^- in the blood is not a common occurrence. If it does occur, the carbonic acid–bicarbonate buffer adjusts for its presence through the reaction

$$H_2CO_3 + OH^- \longrightarrow HCO_3^- + H_2O$$

Excess HCO_3^- ions can be eliminated from the body through the kidneys. Figure 14.2 summarizes the workings of the carbonic acid–bicarbonate ion buffer system in the blood. The capacity of the carbonic acid–bicarbonate buffer system to handle increases in OH^- ion is not as great as its capacity to handle increases in H_3O^+ ion. This is because the ratio of HCO_3^- ion to H_2CO_3 in blood is about 20 to 1.

A very important buffer system within cells is the dihydrogen phosphate ($H_2PO_4^-$)–hydrogen phosphate (HPO_4^{2-}) system. The control of pH within cellular fluids is maintained by the reaction of OH^- ion with $H_2PO_4^-$ and the reaction of H_3O^+ with HPO_4^{2-}.

$$H_2PO_4^- + OH^- \longrightarrow HPO_4^{2-} + H_2O$$
$$HPO_4^{2-} + H_3O^+ \longrightarrow H_2PO_4^- + H_2O$$

The overall dihydrogen phosphate–hydrogen phosphate buffering action can be summarized as

$$H_2PO_4^- \underset{H_3O^+}{\overset{OH^-}{\rightleftharpoons}} HPO_4^{2-}$$

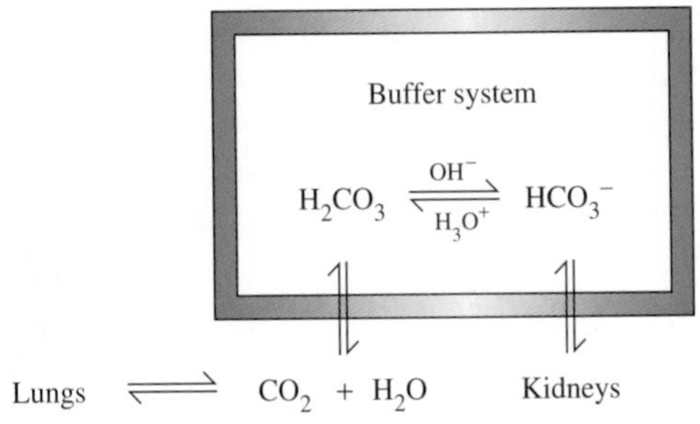

FIGURE 14.2 The carbonic acid–bicarbonate ion buffer system in human blood.

The normal hydrogen phosphate–dihydrogen phosphate ion ratio in cellular fluids is about 4 to 1. Thus, the phosphate buffer system is better equipped to handle influxes of acid than influxes of base. Significant amounts of acids (up to 10 moles/day) are produced in a human body as the result of normal metabolic reactions. For example, lactic acid ($HC_3H_5O_3$) is produced in muscle tissue during exercise.

Under normal conditions, the body's carbonate and phosphate buffer systems are adequate to maintain pH in the normal range. However, under certain stress conditions, these buffer systems can be temporarily overwhelmed. When this happens, compensatory mechanisms involving the lungs and kidneys help to return the pH to normal. Both the lungs and kidneys play a role in pH control at all times, but this role is more important during periods of stress.

Acidosis *is a body condition in which the pH of blood drops from its normal value of 7.4 to 7.1–7.2.* Various factors can cause acidosis, including emphysema-caused hypoventilation (breathing too little), congestive heart failure, diabetes mellitus, excess loss of bicarbonate ion in severe diarrhea, or decreased excretion of hydrogen ions in kidney failure. A temporary condition of acidosis can result from prolonged, intensive exercise. The body reacts to alleviate this acidosis condition in two ways. Excess carbon dioxide (formed from the decomposition of carbonic acid) is expelled by increasing the rate of respiration. Also, kidney system changes occur that increase excretion of H_3O^+ and retention of HCO_3^-; this results in acidic urine.

Alkalosis *is a body condition in which the pH of blood increases from its normal value of 7.4 to a value of 7.5.* Alkalosis can result from hyperventilation (excess breathing) caused by anxiety or hysteria or from extreme fever, severe vomiting, or exposure to high altitude (altitude sickness). The body's responses to alkalosis include a decrease in respiration rate (less expulsion of carbon dioxide by the lungs) and an increase in HCO_3^- excretion by the kidneys, resulting in alkaline urine. Alkalosis is not as common as acidosis.

14.16 ACID–BASE TITRATIONS

Determining the concentration of acid or base in a solution is a regular activity in many laboratories. The concentration of an acid or base in a solution and the solution's pH are two different entities. The pH of a solution gives information about the concentration of hydrogen (hydronium) ions in solution. Only dissociated molecules influence the pH value. The concentration of an acid or base solution gives information about the *total number* of acid or base molecules present; both dissociated and undissociated molecules are counted.

The procedure most frequently used to determine the concentration of an acidic or basic solution is that of titration. **Titration** *is the gradual adding of one solution to another until the solute in the first solution has reacted completely with the solute in the second solution.*

Suppose we want to determine the concentration of an acid solution by titration. We would first measure out a *known volume* of the acid solution into a flask. We would then slowly add a solution of base of *known concentration* to the flask by means of a buret (see Fig. 14.3). Base addition continues until all the acid has completely reacted with added base. The *volume of base* needed to reach this point is obtained from the buret readings. Knowing the original volume of acid, the concentration of the base, and the volume of added base, we can calculate the concentration of the acid (Sec. 14.17).

In order to complete a titration successfully, we must be able to detect when the reaction between acid and base is complete. One way to do this is to add an indicator to the solution being titrated. An **indicator** *is a compound that exhibits different colors depending on*

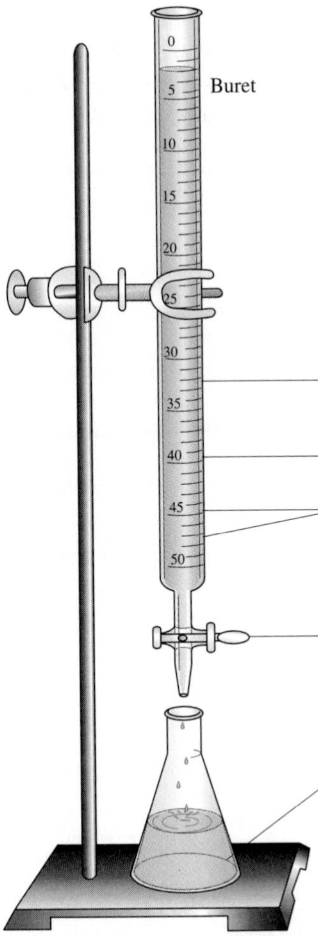

FIGURE 14.3 Use of a buret in a titration procedure.

the pH of its surroundings. Typically, an indicator is a weak acid or weak base whose conjugate base or acid is a different color. An indicator is selected that will change color at a pH corresponding as nearly as possible to the pH of the solution when the titration is complete. This pH can be calculated ahead of time based on the identities of the acid and base involved in the titration.

14.17 ACID–BASE TITRATION CALCULATIONS USING NORMALITY

The concentration unit *normality* is the most convenient unit to work with when doing acid–base calculations. This is a concentration unit that is very closely related to molarity (Sec. 13.8). The difference between the two units involves the expression of the amount of solute present. Moles of solute are specified for molarity and *equivalents* of solute are specified for normality. The **normality** *of a solution, which is designated by N, is a ratio giving the number of equivalents of solute per liter of solution.*

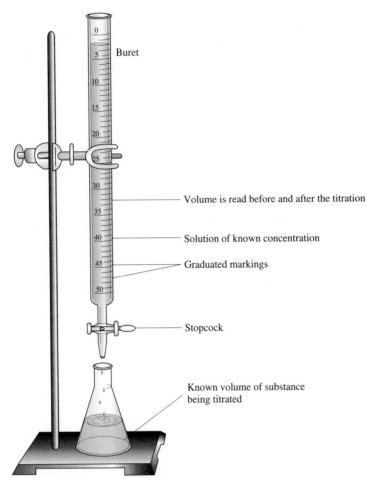

$$\text{Normality} = \frac{\text{equivalents of solute}}{\text{liters of solution}}$$

Normality takes into account the fact that all acids and bases do not yield the same number of hydronium or hydroxide ions per molecule on dissociation. Even though their molar concentrations are equal, 0.10 M H_2SO_4 (which can supply two H_3O^+ ions per molecule) will neutralize twice as much base as will the same volume of 0.10 M HNO_3 (which can supply only one H_3O^+ ion per molecule). However, equal volumes of 0.10 N H_2SO_4 and 0.10 N HNO_3 will neutralize exactly the same amount of base.

The key to comprehending normality concentration units is an understanding of the concept of equivalents. Definitions of equivalent for use in acid–base work are as follows. An **equivalent of acid** *is the quantity of acid that will supply 1 mole of* H_3O^+ *ion during an acid–base reaction.* Similarly, an **equivalent of base** *is the quantity of base that will react with 1 mole of* H_3O^+ *ion during an acid–base reaction.* Examples 14.13 and 14.14 illustrate the use of these two definitions for an equivalent in problem-solving contexts.

Example 14.13

Determine the number of equivalents of acid or base in each of the following samples.

(a) 1 mole of H_2CO_3 **(b)** 2 moles of $Ca(OH)_2$ **(c)** 3 moles of HNO_3

Solution

(a) The equation for the complete dissociation of H_2CO_3 is

$$H_2CO_3 + 2\ H_2O \longrightarrow 2\ H_3O^+ + CO_3^{2-}$$

Thus, 1 mole of H_2CO_3 produces 2 moles of H_3O^+ ion upon complete reaction. Two moles of H_3O^+ ion is equal to 2 equiv of H_3O^+ ion by definition. Therefore,

$$1\ \cancel{\text{mole } H_2CO_3} \times \frac{2\text{ equiv } H_2CO_3}{1\ \cancel{\text{mole } H_2CO_3}} = 2\text{ equiv } H_2CO_3$$

(b) The equation for the complete dissociation of $Ca(OH)_2$ is

$$Ca(OH)_2 \longrightarrow Ca^{2+} + 2\ OH^-$$

Thus, 1 mole of $Ca(OH)_2$ yields 2 moles of OH^- ion (which can react with 2 moles of H_3O^+). Therefore

$$2\ \cancel{\text{moles } Ca(OH)_2} \times \frac{2\text{ equiv } Ca(OH)_2}{1\ \cancel{\text{mole } Ca(OH)_2}} = 4\text{ equiv } Ca(OH)_2$$

(c) The equation for the complete dissociation of HNO_3 is

$$HNO_3 + H_2O \longrightarrow H_3O^+ + NO_3^-$$

One mole of acid yields only 1 mole (or equiv) of H_3O^+ ion. Therefore,

$$3\ \cancel{\text{moles } HNO_3} \times \frac{1\text{ equiv } HNO_3}{1\ \cancel{\text{mole } HNO_3}} = 3\text{ equiv } HNO_3$$

Practice Exercise 14.13

Determine the number of equivalents of acid or base in each of the following samples.
(a) 1 mole of H_3PO_4 (b) 1 mole of KOH (c) 2 moles of $HC_2H_3O_2$

Ans. (a) 3 equiv H_3PO_4; (b) 1 equiv KOH; (c) 2 equiv $HC_2H_3O_2$

Example 14.14

Calculate the normality of a solution made by dissolving 35.0 g of LiOH in enough water to give 600.0 mL of solution.

Solution

To calculate normality, we need to know the number of equivalents of solute and solution volume (in liters).

For LiOH,

$$1 \text{ equiv} = 1 \text{ mole}$$

since 1 mole of LiOH yields 1 mole of OH^- ion,

$$LiOH \longrightarrow Li^+ + OH^-$$

and 1 mole of OH^- reacts with 1 mole of H_3O^+ ion. (Hydronium and hydroxide ions always react in a one-to-one ratio.)

The number of equivalents of LiOH in 35.0 g of LiOH is obtained by converting the grams of LiOH to moles of LiOH and then using the above mole–equivalent relationship as a conversion factor to obtain equivalents.

$$35.0 \cancel{\text{g LiOH}} \times \frac{1 \cancel{\text{mole LiOH}}}{23.9 \cancel{\text{g LiOH}}} \times \frac{1 \text{ equiv LiOH}}{1 \cancel{\text{mole LiOH}}}$$

$$= 1.4644351 \text{ equiv LiOH} \quad \text{(Calculator answer)}$$

$$= 1.46 \text{ equiv LiOH} \quad \text{(correct answer)}$$

The solution volume of 600.0 mL, changed to liter units, is 0.6000 L.

Both quantities called for in the defining equation for normality, equivalents of solute and liters of solution, are now known. Therefore, substituting into the defining equation, we get

$$N = \frac{1.46 \text{ equiv LiOH}}{0.6000 \text{ L solution}} = 2.4333333 \frac{\text{equiv LiOH}}{\text{L solution}} \quad \text{(calculator answer)}$$

$$= 2.43 \frac{\text{equiv LiOH}}{\text{L solution}} \quad \text{(correct answer)}$$

Practice Exercise 14.14

Calculate the normality of a solution made by dissolving 5.00 g of $Sr(OH)_2$ in enough water to give 750.0 mL of solution.

Ans. 0.109 N $Sr(OH)_2$

In the laboratory, solution concentrations are specified in terms of molarity rather than normality. To convert acid or base molarities to normalities (or vice versa) is a very simple operation. Only the conversion factor relating moles to equivalents is needed.

Example 14.15

Express the following molar concentrations as normalities.

(a) 0.030 M H_2CO_3 **(b)** 0.10 M $Ca(OH)_2$ **(c)** 2.0 M HNO_3

Solution

(a) A 0.030-M H_2CO_3 solution contains 0.030 mole of H_2CO_3 per liter of solution. Using this fact and the fact (from Example 14.13) that

$$1 \text{ mole } H_2CO_3 = 2 \text{ equiv } H_2CO_3$$

we calculate the normality as follows.

$$\frac{0.030 \text{ mole } H_2CO_3}{1 \text{ L solution}} \times \frac{2 \text{ equiv } H_2CO_3}{1 \text{ mole } H_2CO_3} = \frac{0.06 \text{ equiv } H_2CO_3}{\text{L solution}} \quad \text{(calculator answer)}$$

$$= 0.060 \text{ N } H_2CO_3 \quad \text{(correct answer)}$$

Therefore, the solution is 0.060 N.

(b) A 0.10-M $Ca(OH)_2$ solution contains 0.10 mole of $Ca(OH)_2$ per liter of solution. This information, coupled with the fact (from Example 14.13) that

$$1 \text{ mole } Ca(OH)_2 = 2 \text{ equiv } Ca(OH)_2$$

enables us to calculate the normality as follows

$$\frac{0.10 \text{ mole } Ca(OH)_2}{1 \text{ L solution}} \times \frac{2 \text{ equiv } Ca(OH)_2}{1 \text{ mole } Ca(OH)_2} = \frac{0.2 \text{ equiv } Ca(OH)_2}{\text{L solution}} \quad \text{(calculator answer)}$$

$$= 0.20 \text{ N } Ca(OH)_2 \quad \text{(correct answer)}$$

(c) For 2.0 M HNO_3, we calculate the normality, using reasoning similar to that in parts (a) and (b), as follows.

$$\frac{2.0 \text{ moles } HNO_3}{1 \text{ L solution}} \times \frac{1 \text{ equiv } HNO_3}{1 \text{ mole } HNO_3} = \frac{2 \text{ equiv } HNO_3}{\text{L solution}} \quad \text{(calculator answer)}$$

$$= 2.0 \text{ N } HNO_3 \quad \text{(correct answer)}$$

Practice Exercise 14.15

Express the following molar concentrations as normalities.

(a) 0.050 M H_3PO_4 (b) 1.50 M KOH (c) 0.500 M $HC_2H_3O_2$

Ans. (a) 0.15 N H_3PO_4; (b) 1.50 N KOH; (c) 0.500 N $HC_2H_3O_2$

Molarities can also be converted to normalities, without use of dimensional analysis, by the following equations.

$$N_{\text{acid}} = M_{\text{acid}} \times \text{number of acidic hydrogens in acid}$$
$$N_{\text{base}} = M_{\text{base}} \times \text{number of hydroxide ions in base}$$

Both of these equations are based on the Arrhenius system for defining acids and bases. It is not recommended that you use these "shortcut" equations until you thoroughly understand the dimensional analysis approach to obtaining normalities (Example 14.15).

In an acid–base titration, the chemical reaction that occurs is that of neutralization (Sec. 14.8). The H_3O^+ ions from the acid react with the OH^- ions from the base to produce water.

$$H_3O^+ + OH^- \longrightarrow 2\ H_2O$$

These two ions always react in a one-to-one ratio; thus, 1 mole of H_3O^+ ions always reacts with 1 mole of OH^- ions. Since 1 mole of each of these ions is an equivalent, it follows that at the endpoint of an acid–base titration equal numbers of equivalents of acid and base have been consumed; that is,

$$\text{Equiv of reacted acid} = \text{equiv of reacted base}$$

A simple equation, whose derivation is based on the fact that equal numbers of equivalents of acid and base are consumed during an acid–base titration, is employed in titration calculations. It is

$$N_{\text{acid}} \times V_{\text{acid}} = N_{\text{base}} \times V_{\text{base}}$$

The derivation of this equation is as follows. The defining equations for the normality of the acid and base are

$$N_{\text{acid}} = \frac{E_{\text{acid}}}{V_{\text{acid}}} \quad \text{and} \quad N_{\text{base}} = \frac{E_{\text{base}}}{V_{\text{base}}}$$

where E is the equivalents of solute and V is the volume of solution (in liters). Each of these equations can be rearranged as follows.

$$E_{\text{acid}} = N_{\text{acid}} \times V_{\text{acid}}$$
$$E_{\text{base}} = N_{\text{base}} \times V_{\text{base}}$$

At the endpoint of the titration, where $E_{\text{acid}} = E_{\text{base}}$, we can therefore write

$$N_{\text{acid}} \times V_{\text{acid}} = N_{\text{base}} \times V_{\text{base}}$$

The volumes V_{acid} and V_{base} can be expressed in liters or milliliters, provided the same unit is used for both. Since burets are calibrated in milliliters, the milliliter is the unit most often used in titration calculations.

The data obtained from an acid–base titration (Sec. 14.12) are (1) the volume of base used, (2) the volume of acid used, and (3) the concentration of the acid or base used to titrate the solution of unknown concentration. These quantities are three of the four variables in the equation $N_{\text{acid}} \times V_{\text{acid}} = N_{\text{base}} \times V_{\text{base}}$. Hence, the fourth variable, the concentration of the acid or base that was titrated, can easily be calculated.

Example 14.16

What volume of 0.400 M H_2SO_4, in milliliters, would be needed to neutralize completely 375 mL of 0.225 M $Ba(OH)_2$ solution?

Solution

The concentrations of the two solutions must be converted from molarities to normalities. Since H_2SO_4 yields two H_3O^+ ions per acid molecule, 1 mole of H_2SO_4 contains 2 equiv of acid; hence, the normality of the H_2SO_4 solution is two times the molarity.

$$\frac{0.400\ \cancel{\text{mole } H_2SO_4}}{1\ \text{L solution}} \times \frac{2\ \text{equiv } H_2SO_4}{1\ \cancel{\text{mole } H_2SO_4}} = \frac{0.8\ \text{equiv } H_2SO_4}{\text{L solution}} \quad \text{(calculator answer)}$$

$$= 0.800\ \text{N } H_2SO_4 \quad \text{(correct answer)}$$

For $Ba(OH)_2$, the normality will be double the molarity, since $Ba(OH)_2$ yields two OH^- ions per molecule on dissociation.

$$\frac{0.225\ \cancel{\text{mole Ba(OH)}_2}}{1\ \text{L solution}} \times \frac{2\ \text{equiv Ba(OH)}_2}{1\ \cancel{\text{mole Ba(OH)}_2}} = \frac{0.45\ \text{equiv Ba(OH)}_2}{\text{L solution}} \quad \text{(calculator answer)}$$

$$= 0.450\ \text{N Ba (OH)}_2 \quad \text{(correct answer)}$$

Substituting the known values in the equation

$$V_{acid} = V_{base} \times \frac{N_{base}}{N_{acid}}$$

gives

$$V_{acid} = 375\ \text{mL} \times \frac{0.450\ \cancel{N}}{0.800\ \cancel{N}} = 210.9375\ \text{mL} \quad \text{(calculator answer)}$$

$$= 211\ \text{mL} \quad \text{(correct answer)}$$

Practice Exercise 14.16

What volume of 0.500 M HNO_3, in milliliters, would be needed to completely neutralize 375 mL of 0.515 M $Ba(OH)_2$ solution?

Ans. 772 mL HNO_3

Example 14.17

In an acid–base titration, 32.7 mL of 0.105 M KOH is required to neutralize completely 50.0 mL of H_3PO_4. What are the normality and the molarity of the H_3PO_4?

Solution

We will use the equation

$$N_{acid} \times V_{acid} = N_{base} \times V_{base}$$

to solve this problem. Three of the four quantities in this equation are known. The volumes of acid and base are, respectively, 50.0 and 32.7 mL. The normality of the base is not directly given, but can easily be calculated from the molarity of the base. Since 1 mole of KOH contains 1 equiv of OH^- ions, the normality of the base will be the same as the molarity; that is, the base is not only 0.105 M but also 0.105 N.

Substitution of these known values into the equation, rearranged so that N_{acid} is isolated on the left side, gives

$$N_{acid} = N_{base} \times \frac{V_{base}}{V_{acid}}$$

$$= 0.105\ \text{N} \times \frac{32.7\ \cancel{\text{mL}}}{50.0\ \cancel{\text{mL}}} = 0.06867\ \text{N} \quad \text{(calculator answer)}$$

$$= 0.0687\ \text{N} \quad \text{(correct answer)}$$

The molarity of H_3PO_4, which was also asked for in this problem, will be one-third the normality since H_3PO_4 is a triprotic acid. Thus, the H_3PO_4 is 0.0229 M.

Practice Exercise 14.17

In an acid–base titration, 50.2 mL of 0.252 M NaOH is required to neutralize 32.7 mL of H_2SO_4. What are the normality and the molarity of the H_2SO_4 solution?

Ans. 0.387 N and 0.194 M H_2SO_4

14.18 ACID AND BASE STOCK SOLUTIONS

Acids and bases are used so often in most laboratories that stock solutions (Sec. 13.10) of the most common ones are made readily available at each work space. The concentrations of such solutions are traditionally the same from laboratory to laboratory and are given in Table 14.12. Ordinarily the concentrations of the stock solutions are not given on their containers. Only the name of the acid or base and the term dilute (dil) or concentrated (conc) are found. It is assumed that students or researchers know what is implied by the designations dil and conc in each specific case; that is, that they know the information found in Table 14.12. Note from Table 14.12 that the designations dil and conc do not have a constant meaning in terms of molarity. For example, concentrated solutions of sulfuric acid, nitric acid, and hydrochloric acid have molarities of 18, 16, and 12, respectively. There is not as much variation in the meaning of the term dilute; except for sulfuric acid, all of the listed dilute solutions are 6 M. Note that all the listed dilute solutions, including sulfuric acid, are 6 N.

TABLE 14.12 CONCENTRATIONS OF COMMON LABORATORY STOCK SOLUTIONS OF ACIDS AND BASES

Label Designation	Chemical Formula	Concentration	
		Molarity	Normality
Acids			
Dilute hydrochloric acid	HCl	6	6
Concentrated hydrochloric acid	HCl	12	12
Dilute nitric acid	HNO_3	6	6
Concentrated nitric acid	HNO_3	16	16
Dilute sulfuric acid	H_2SO_4	3	6
Concentrated sulfuric acid	H_2SO_4	18	36
Dilute acetic acid	$HC_2H_3O_2$	6	6
Concentrated acetic acid[a]	$HC_2H_3O_2$	18	18
Bases			
Dilute aqueous ammonia[b]	$NH_3(aq)$	6	6
Concentrated aqueous ammonia[b]	$NH_3(aq)$	15	15
Dilute sodium hydroxide[c]	$NaOH$	6	6

[a] Often labeled "glacial acetic acid."
[b] Often labeled "ammonium hydroxide."
[c] Often labeled simply "sodium hydroxide."

KEY TERMS

The new terms or concepts defined in this chapter are

acidic hydrogen atom (Sec. 14.4) A hydrogen atom in an acid molecule that can be transferred to a base during an acid–base reaction.

acidic solution (Sec. 14.11) A solution that has a higher H_3O^+ concentration than OH^- concentration.

acidosis (Sec. 14.15) A body condition in which the pH of blood drops from its normal value of 7.4 to 7.1–7.2.

alkalosis (Sec. 14.15) A body condition in which the pH of blood increases from its normal value of 7.4 to a value of 7.5.

amphoteric substance (Sec. 14.3) A substance that can either lose or accept protons and thus can function as either an acid or a base.

Arrhenius acid (Sec. 14.1) A hydrogen-containing compound that, in water, produces hydrogen ions (H^+).

Arrhenius base (Sec. 14.1) A hydroxide-containing compound that, in water, produces hydroxide ions (OH^-).

basic solution (Sec. 14.11) A solution that has a higher OH^- concentration than H_3O^+ concentration.

Brønsted–Lowry acid (Sec. 14.2) Any substance that can donate a proton (H^+) to some other substance.

Brønsted–Lowry base (Sec. 14.2) Any substance that can accept a proton (H^+) from some other substance.

buffer (Sec. 14.14) The solute (or solutes) present in a buffer solution that cause it to be resistant to a change in pH.

buffer solution (Sec. 14.14) A solution that resists a change in pH when small amounts of acid or base are added to it.

conjugate acid (Sec. 14.3) The species formed when a Brønsted–Lowry base accepts a proton.

conjugate acid–base pair (Sec. 14.3) Two species that differ from each other by one proton (H^+).

conjugate base (Sec. 14.3) The species that remains when a Brønsted–Lowry acid loses a proton.

diprotic acid (Sec. 14.4) An acid that can transfer two H^+ ions (two protons) per molecule during an acid–base reaction.

equivalent of acid (Sec. 14.17) Quantity of an acid that will supply 1 mole of H_3O^+ ion during an acid–base reaction.

equivalent of base (Sec. 14.17) Quantity of a base that will react with 1 mole of H_3O^+ ion during an acid–base reaction.

hydrolysis (Sec. 14.13) The reaction of a substance with water to produce hydronium ion or hydroxide ion or both.

indicator (Sec. 14.16) A compound that exhibits different colors depending on the pH of its surroundings.

ionic equation (Sec. 14.7) An equation in which the formulas of dissociated compounds are written in terms of ions and the formulas of undissociated compounds are written in terms of molecules.

ion product for water (Sec. 14.11) Name given to the numerical value 1.00×10^{-14} which is the product of the H_3O^+ ion and OH^- ion molar concentrations in water.

monoprotic acid (Sec. 14.4) An acid that can transfer only one H^+ ion (proton) per molecule during an acid–base reaction.

net ionic equation (Sec. 14.7) An ionic equation from which nonparticipating (spectator) species have been eliminated.

neutralization (Sec. 14.8) Reaction between an acid and a base to form a salt and water.

neutral solution (Sec. 14.11) A solution in which the concentrations of H_3O^+ ion and OH^- ion are equal.

normality (Sec. 14.17) Concentration of a solution in terms of equivalents of solute per liter of solution.

pH (Sec. 14.12) The negative logarithm of the molar hydronium ion concentration.

polyprotic acid (Sec. 14.4) An acid that can transfer two or more H^+ ions (protons) per molecule during an acid–base reaction.

salt (Sec. 14.6) An ionic compound containing a metal or polyatomic ion as the positive ion and a nonmetal or polyatomic ion (except hydroxide) as the negative ion.

strong acid (Sec. 14.5) An acid that transfers 100%, or very nearly 100%, of its acidic hydrogen atoms to water.

titration (Sec. 14.16) A procedure in which one solution is gradually added to another until the solute in the first solution has reacted completely with the solute in the second solution.

triprotic acid (Sec. 14.4) An acid that can transfer three H^+ ions (three protons) per molecule during an acid–base reaction.

weak acid (Sec. 14.5) An acid that transfers only a small percentage of its acidic hydrogen atoms to water.

PRACTICE PROBLEMS

Acid–Base Definitions (Secs. 14.1 and 14.2)

14.1 Write equations for the dissociation of the following Arrhenius acids and bases in water.

(a) HBr (hydrobromic acid)
(b) $HClO_2$ (chlorous acid)
(c) LiOH (lithium hydroxide)
(d) $Ba(OH)_2$ (barium hydroxide)

14.2 Write equations for the dissociation of the following Arrhenius acids and bases in water.

(a) $HClO_3$ (chloric acid)
(b) HI (hydroiodic acid)
(c) H_2SO_4 (sulfuric acid)
(d) CsOH (cesium hydroxide)

14.3 In each of the following reactions, decide whether the first listed species is a Brønsted–Lowry acid or base.

(a) $NH_4^+ + OH^- \longrightarrow H_2O + NH_3$
(b) $HS^- + H_2O \longrightarrow H_2S + OH^-$
(c) $HClO_4 + NO_2^- \longrightarrow HNO_2 + ClO_4^-$
(d) $H_2O + HC_2O_4^- \longrightarrow H_3O^+ + C_2O_4^{2-}$

14.4 In each of the following reactions, decide whether the first listed species is a Brønsted–Lowry acid or base.

(a) $HClO_4 + H_2O \longrightarrow H_3O^+ + ClO_4^-$
(b) $HClO + NH_3 \longrightarrow NH_4^+ + ClO^-$
(c) $H_2O + CN^- \longrightarrow HCN + OH^-$
(d) $NH_3 + H_3O^+ \longrightarrow NH_4^+ + H_2O$

14.5 Write equations to illustrate the acid–base reactions that can take place between the following Brønsted–Lowry acids and bases.

(a) acid, HBr; base, H_2O
(b) acid, H_2O; base, N_3^-
(c) acid, H_2S; base, H_2O
(d) acid, $HClO_4$; base, NO_2^-

14.6 Write equations to illustrate the acid–base reactions that can take place between the following Brønsted–Lowry acids and bases.

(a) acid, H_3PO_4; base, NH_3
(b) acid, H_3O^+; base, OH^-
(c) acid, HSO_4^-; base, H_2O
(d) acid, H_2O; base, S^{2-}

Conjugate Acids and Bases (Sec. 14.3)

14.7 Write the formula of each of the following conjugate acids or bases.

(a) conjugate base of H_2SO_3
(b) conjugate acid of CN^-
(c) conjugate acid of S^{2-}
(d) conjugate base of HClO

14.8 Write the formula of each of the following conjugate acids or bases.

(a) conjugate acid of HCO_3^-
(b) conjugate base of NH_3
(c) conjugate acid of $H_2PO_4^-$
(d) conjugate base of H_3O^+

14.9 Identify the two conjugate acid–base pairs involved in each of the following reactions.

(a) $H_2C_2O_4 + ClO^- \rightleftharpoons HC_2O_4^- + HClO$
(b) $HSO_4^- + H_2O \rightleftharpoons H_3O^+ + SO_4^{2-}$
(c) $HPO_4^{2-} + NH_4^+ \rightleftharpoons NH_3 + H_2PO_4^-$
(d) $HCO_3^- + H_2O \rightleftharpoons OH^- + H_2CO_3$

14.10 Identify the two conjugate acid–base pairs involved in each of the following reactions.

(a) $SO_4^{2-} + H_2O \rightleftharpoons HSO_4^- + OH^-$
(b) $CN^- + H_2O \rightleftharpoons HCN + OH^-$
(c) $HSO_4^- + HCO_3^- \rightleftharpoons SO_4^{2-} + H_2CO_3$
(d) $H_3PO_4 + PO_4^{3-} \rightleftharpoons H_2PO_4^- + HPO_4^{2-}$

14.11 In which of the following pairs of substances do the two members of the pair constitute a conjugate acid–base pair?

(a) HCN and CN^- **(b)** H_3PO_4 and PO_4^{3-}
(c) HCO_3^- and HSO_4^- **(d)** NH_4^+ and NH_3

14.12 In which of the following pairs of substances do the two members of the pair constitute a conjugate acid–base pair?

(a) HN_3 and N_3^- **(b)** H_2SO_4 and SO_4^{2-}
(c) H_2CO_3 and $HClO_3$ **(d)** NH_3 and NH_2^-

14.13 For each of the following amphoteric substances, write the two equations needed to describe its behavior in aqueous solution.

(a) HS^- **(b)** HPO_4^{2-} **(c)** HCO_3^- **(d)** $H_2PO_3^-$

14.14 For each of the following amphoteric substances, write the two equations needed to describe its behavior in aqueous solution.

(a) $H_2PO_4^-$ **(b)** HSO_3^- **(c)** $HC_2O_4^-$ **(d)** PH_3

Polyprotic Acids (Sec. 14.4)

14.15 Classify each of the following acids as monoprotic, diprotic, or triprotic.

(a) $HClO_3$ (chloric acid) **(b)** H_3BO_3 (boric acid)
(c) $HC_3H_3O_3$ (pyruvic acid) **(d)** $H_2C_3H_2O_4$ (malonic acid)

14.16 Classify each of the following acids as monoprotic, diprotic, or triprotic.

(a) HNO_3 (nitric acid) **(b)** H_2SeO_4 (selenic acid)
(c) $HC_3H_5O_3$ (lactic acid) **(d)** $H_2C_4H_4O_4$ (succinic acid)

14.17 Write equations for the stepwise proton-transfer process that occurs in aqueous solution for each of the following acids.

(a) $H_2C_4H_4O_4$ (succinic acid)
(b) H_3BO_3 (boric acid)

14.18 Write equations for the stepwise proton-transfer process that occurs in aqueous solution for each of the following acids.

(a) $H_2C_3H_2O_4$ (malonic acid)
(b) $H_3C_6H_5O_7$ (citric acid)

14.19 The formula for salicylic acid is preferably written as $HC_7H_5O_3$ rather than $C_7H_6O_3$. Explain why this is so.

14.20 The formula for tartaric acid is preferably written as $H_2C_4H_4O_6$ rather than $C_4H_6O_6$. Explain why this is so.

14.21 Pyruvic acid, which is produced in metabolic reactions within the human body, has the following structure.

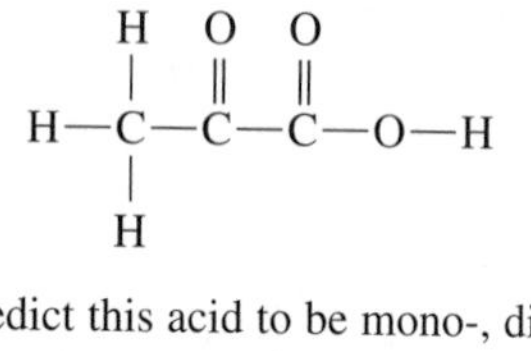

Would you predict this acid to be mono-, di-, tri-, or tetraprotic? Give your reasoning for your answer.

14.22 Succinic acid, a biologically important substance, has the following structure.

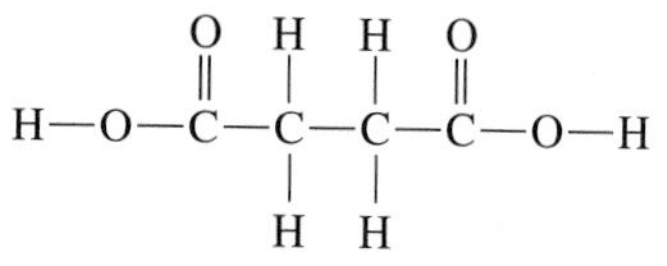

How many acidic hydrogen atoms are present in the structure? Give your reasoning for your answer.

Strength of Acids and Bases (Sec. 14.5)

14.23 Classify each of the acids in Problem 14.15 as a strong acid or a weak acid.

14.24 Classify each of the acids in Problem 14.16 as a strong acid or a weak acid.

14.25 For which of the following pairs of acids are both members of the pair of "like strength," that is, both strong or both weak?

(a) H_2SO_4 and H_2SO_3 **(b)** $HClO_4$ and HCN
(c) HF and HI **(d)** HNO_2 and $HClO_2$

14.26 For which of the following pairs of acids are both members of the pair of "like strength," that is, both strong or both weak?

(a) HNO_3 and HNO_2 **(b)** HCl and HBr
(c) H_3PO_4 and H_3BO_3 **(d)** H_2CO_3 and $H_2C_2O_4$

14.27 In which of the following acid–base pairs are both a strong acid and a strong base present?

(a) H_2SO_4 and KOH **(b)** HNO_3 and NaOH
(c) H_3PO_4 and LiOH **(d)** HF and $Ba(OH)_2$

14.28 In which of the following acid–base pairs are both a strong acid and a strong base present?

(a) HCl and NaOH **(b)** H_2CO_3 and $Ca(OH)_2$
(c) $HC_2H_3O_2$ and KOH **(d)** $HClO_4$ and $Sr(OH)_2$

14.29 With the help of Table 14.3 when necessary, indicate which acid in each of the following pairs of acids is the stronger.

(a) HNO_3 and HNO_2 **(b)** $HClO_4$ and $HC_2H_3O_2$
(c) H_3PO_4 and HCN **(d)** H_2CO_3 and HF

14.30 With the help of Table 14.3 when necessary, indicate which acid in each of the following pairs of acids is the stronger.

(a) H_2SO_4 and H_2SO_3 **(b)** HCl and HF
(c) HCN and HNO_2 **(d)** $HC_2H_3O_2$ and H_3PO_4

Salts (Sec. 14.6)

14.31 Identify each of the following substances as an acid, a base, or a salt.

(a) NH_4NO_3 **(b)** KOH **(c)** Na_2SO_4 **(d)** HCN

14.32 Identify each of the following substances as an acid, a base, or a salt.

(a) $CaCO_3$ **(b)** HF **(c)** KF **(d)** LiOH

14.33 Give the formula and name of the positive and negative ions present in each of the following salts.

(a) Na_3PO_4 **(b)** $LiNO_3$ **(c)** NH_4Cl **(d)** KCN

14.34 Give the formula and name of the positive and negative ions present in each of the following salts.
(a) K_2SO_4 (b) $MgCO_3$ (c) K_2S (d) NH_4NO_3

14.35 Indicate whether each of the salts in Problem 14.33 is soluble or insoluble in water (Sec. 13.4).

14.36 Indicate whether each of the salts in Problem 14.34 is soluble or insoluble in water (Sec. 13.4).

14.37 Write a balanced equation for the dissociation in water of each of the following soluble salts into ions.
(a) NaI (b) BaS (c) Li_2SO_4 (d) $Al(NO_3)_3$

14.38 Write a balanced equation for the dissociation in water of each of the following soluble salts into ions.
(a) $NaC_2H_3O_2$ (b) $Be(NO_3)_2$
(c) $MgCl_2$ (d) $(NH_4)_2CO_3$

Ionic and Net Ionic Equations (Sec. 14.7)

14.39 Classify the following equations for reactions occurring in aqueous solution as molecular, ionic, or net ionic.
(a) $MgCO_3 + 2\ HBr \longrightarrow MgBr_2 + H_2O + CO_2$
(b) $2\ OH^- + H_2CO_3 \longrightarrow CO_3^{2-} + 2\ H_2O$
(c) $K^+ + OH^- + H^+ + I^- \longrightarrow K^+ + I^- + H_2O$
(d) $Ag^+ + Cl^- \longrightarrow AgCl$

14.40 Classify the following equations for reactions occurring in aqueous solution as molecular, ionic, or net ionic.
(a) $CaCO_3 + 2\ H^+ + 2\ NO_3^- \longrightarrow Ca^{2+} + 2\ NO_3^- + H_2O + CO_2$
(b) $Ni + Cu^{2+} \longrightarrow Cu + Ni^{2+}$
(c) $NaCl + AgNO_3 \longrightarrow NaNO_3 + AgCl$
(d) $Ca^{2+} + 2\ OH^- \longrightarrow Ca(OH)_2$

14.41 Write a balanced net ionic equation for each of the following reactions, each of which occurs in aqueous solution.
(a) $2\ NaBr + Pb(NO_3)_2 \longrightarrow 2\ NaNO_3 + PbBr_2$
(b) $FeCl_3 + 3\ NaOH \longrightarrow Fe(OH)_3 + 3\ NaCl$
(c) $Zn + 2\ HCl \longrightarrow ZnCl_2 + H_2$
(d) $H_2S + 2\ KOH \longrightarrow K_2S + 2\ H_2O$

14.42 Write a balanced net ionic equation for each of the following reactions, each of which occurs in aqueous solution.
(a) $CaCl_2 + CuSO_4 \longrightarrow CaSO_4 + CuCl_2$
(b) $Ca(NO_3)_2 + K_2CO_3 \longrightarrow CaCO_3 + 2\ KNO_3$
(c) $Mg + 2\ HBr \longrightarrow MgBr_2 + H_2$
(d) $H_3PO_4 + 3\ NaOH \longrightarrow Na_3PO_4 + 3\ H_2O$

14.43 Write a balanced net ionic equation for each of the following reactions, each of which occurs in aqueous solution.
(a) $Pb + 2\ AgNO_3 \longrightarrow 2\ Ag + Pb(NO_3)_2$
(b) $Cl_2 + 2\ NaBr \longrightarrow 2\ NaCl + Br_2$
(c) $2\ Al(NO_3)_3 + 3\ Na_2S \longrightarrow Al_2S_3 + 6\ NaNO_3$
(d) $NaC_2H_3O_2 + NH_4Cl \longrightarrow NH_4C_2H_3O_2 + NaCl$

14.44 Write a balanced net ionic equation for each of the following reactions, each of which occurs in aqueous solution.
(a) $Ni + Cu(NO_3)_2 \longrightarrow Ni(NO_3)_2 + Cu$
(b) $Br_2 + 2\ NaI \longrightarrow 2\ NaBr + I_2$
(c) $Hg(NO_3)_2 + K_2S \longrightarrow 2\ KNO_3 + HgS$
(d) $(NH_4)_2SO_4 + 2\ NaBr \longrightarrow 2\ NH_4Br + Na_2SO_4$

Reactions of Acids and Bases (Secs. 14.8 and 14.9)

14.45 On the basis of the activity series (Table 14.6) predict whether a reaction takes place when
(a) iron metal is added to hydrochloric acid
(b) potassium metal is added to cold water
(c) gold metal is added to hydrochloric acid
(d) aluminum metal is added to hot water (steam)

14.46 On the basis of the activity series (Table 14.6) predict whether a reaction takes place when
(a) chromium metal is added to hydrochloric acid
(b) silver metal is added to hot water (steam)
(c) sodium metal is added to cold water
(d) copper metal is added to hydrochloric acid

14.47 Write the balanced molecular equation for each of the following hydrogen-gas-producing reactions.
(a) nickel metal is added to hydrochloric acid giving $NiCl_2$
(b) calcium metal is added to cold water
(c) magnesium metal is added to hydrochloric acid
(d) zinc metal is added to hot water (steam)

14.48 Write the balanced molecular equation for each of the following hydrogen-gas-producing reactions.
(a) magnesium metal is added to hot water (steam)
(b) calcium metal is added to hydrochloric acid
(c) tin metal is added to hydrochloric acid giving $SnCl_2$
(d) lead metal is added to hydrochloric acid giving $PbCl_2$

14.49 Write balanced net ionic equations showing the total neutralization between the following acids and bases.
(a) HBr and $Sr(OH)_2$ (b) $HC_2H_3O_2$ and LiOH
(c) H_2SO_4 and $Mg(OH)_2$ (d) H_3PO_4 and KOH

14.50 Write balanced net ionic equations showing the total neutralization between the following acids and bases.
(a) HCl and NaOH (b) HNO_3 and KOH
(c) H_2S and $Ba(OH)_2$ (d) $HClO_4$ and LiOH

14.51 Give the formulas of the acid and base needed to prepare the following salts by neutralization reactions.
(a) Na_3PO_4 (b) KCN
(c) $BeCl_2$ (d) $Ca(C_2H_3O_2)_2$

14.52 Give the formulas of the acid and base needed to prepare the following salts by neutralization reactions.

(a) $LiNO_3$ (b) K_2S (c) $Al_2(SO_4)_3$ (d) NaBr

14.53 Write a molecular equation for the action of HCl on each of the following. (Zn metal forms a +2 ion in solution.)

(a) Zn (b) NaOH
(c) Na_2CO_3 (d) $NaHCO_3$

14.54 Write a molecular equation for the action of HNO_3 on each of the following. (Ni metal forms a +2 ion in solution.)

(a) Ni (b) KOH
(c) Li_2CO_3 (d) $LiHCO_3$

Reactions of Salts (Sec. 14.10)

14.55 On the basis of the activity series (Table 14.6) predict whether or not a reaction occurs when each of the following metals is added to a nickel(II) nitrate solution.

(a) Cu (b) Zn (c) Au (d) Fe

14.56 On the basis of the activity series (Table 14.6) predict whether or not a reaction occurs when each of the following metals is added to an iron(II) nitrate solution.

(a) Ag (b) Ni (c) Cr (d) Pb

14.57 Write a net ionic equation for each of the following metal-replacement reactions. Assume that metals going into solution form +2 ions.

(a) iron is added to a $CuSO_4$ solution
(b) tin is added to a $AgNO_3$ solution
(c) zinc is added to $NiCl_2$ solution
(d) chromium is added to $Pb(C_2H_3O_2)_2$ solution

14.58 Write a net ionic equation for each of the following metal-replacement reactions. Assume that metals going into solution form +2 ions.

(a) lead is added to Cu_2SO_4 solution
(b) mercury is added to $Au(NO_3)_3$ solution
(c) chromium is added to $FeCl_2$ solution
(d) iron is added to $Ni(C_2H_3O_2)_2$ solution

14.59 Indicate the driving force (condition) that causes each of the following acid–salt reactions to occur.

(a) $Ba(NO_3)_2 + H_2SO_4 \longrightarrow BaSO_4 + 2\ HNO_3$
(b) $3\ CaCl_2 + 2\ H_3PO_4 \longrightarrow Ca_3(PO_4)_2 + 6\ HCl$
(c) $AgC_2H_3O_2 + HCl \longrightarrow AgCl + HC_2H_3O_2$
(d) $K_2CO_3 + 2\ HNO_3 \longrightarrow 2\ KNO_3 + CO_2 + H_2O$

14.60 Indicate the driving force (condition) that causes each of the following acid–salt reactions to occur.

(a) $NaCN + HCl \longrightarrow NaCl + HCN$
(b) $3\ MgSO_4 + 2\ H_3PO_4 \longrightarrow Mg_3(PO_4)_2 + 3\ H_2SO_4$
(c) $Li_2CO_3 + 2\ HBr \longrightarrow 2\ LiBr + CO_2 + H_2O$
(d) $Na_3PO_4 + 3\ HCl \longrightarrow 3\ NaCl + H_3PO_4$

14.61 Complete and balance a net ionic equation for the reaction, if any, between each of the following pairs of aqueous solutions. If no reaction occurs, write "no reaction."

(a) $Al(NO_3)_3$ and $(NH_4)_2S$ (b) HCl and $Ba(OH)_2$
(c) Na_2SO_4 and HNO_3 (d) KNO_3 and $HC_2H_3O_2$

14.62 Complete and balance a net ionic equation for the reaction, if any, between each of the following pairs of aqueous solutions. If no reaction occurs, write "no reaction."

(a) H_2CO_3 and KOH (b) K_3PO_4 and HCl
(c) $CaCl_2$ and HNO_3 (d) $Fe(NO_3)_2$ and $MgSO_4$

Hydronium Ion and Hydroxide Ion Concentrations (Sec. 14.11)

14.63 What is the molar H_3O^+ ion concentration in solutions with the following OH^- ion concentrations?

(a) 2.0×10^{-4} M (b) 7.3×10^{-7} M
(c) 3.0×10^{-10} M (d) 2.5×10^{-8} M

14.64 What is the molar H_3O^+ ion concentration in solutions with the following OH^- ion concentrations?

(a) 9.7×10^{-6} M (b) 3.8×10^{-2} M
(c) 3.3×10^{-7} M (d) 9.3×10^{-11} M

14.65 Indicate whether each of the solutions in Problem 14.63 is acidic, basic, or neutral.

14.66 Indicate whether each of the solutions in Problem 14.64 is acidic, basic, or neutral.

14.67 What is the molar OH^- ion concentration in solutions with the following H_3O^+ ion concentrations?

(a) 2.7×10^{-4} M (b) 7.5×10^{-9} M
(c) 1.0×10^{-7} M (d) 5.0×10^{-8} M

14.68 What is the molar OH^- ion concentration in solutions with the following H_3O^+ ion concentrations?

(a) 3.3×10^{-2} M (b) 6.9×10^{-12} M
(c) 1.5×10^{-6} M (d) 4.7×10^{-8} M

14.69 Indicate whether each of the solutions in Problem 14.67 is acidic, basic, or neutral.

14.70 Indicate whether each of the solutions in Problem 14.68 is acidic, basic, or netural.

The pH Scale (Sec. 14.12)

14.71 Calculate the pH of solutions with the following hydronium ion concentrations.

(a) 1×10^{-4} M (b) 1×10^{-9} M
(c) 0.00001 M (d) 0.000000001 M

14.72 Calculate the pH of solutions with the following hydronium ion concentrations.

(a) 1×10^{-3} M (b) 1×10^{-11} M
(c) 0.0001 M (d) 0.0000001 M

14.73 Calculate the pH of solutions with the following hydronium ion concentrations.

(a) 4×10^{-2} M (b) 7×10^{-4} M
(c) 8×10^{-10} M (d) 5×10^{-7} M

14.74 Calculate the pH of solutions with the following hydronium ion concentrations.

(a) 6×10^{-4} M **(b)** 6×10^{-8} M
(c) 9×10^{-3} M **(d)** 7×10^{-6} M

14.75 Calculate the pH of solutions with the following hydronium ion concentrations.

(a) 3×10^{-3} M **(b)** 3.0×10^{-3} M
(c) 3.00×10^{-3} M **(d)** 3.000×10^{-3} M

14.76 Calculate the pH of solutions with the following hydronium ion concentrations.

(a) 7×10^{-6} M **(b)** 7.0×10^{-6} M
(c) 7.00×10^{-6} M **(d)** 7.000×10^{-6} M

14.77 In which of the following pairs of pH values do both values represent acidic solution conditions?

(a) 5.31 and 6.31 **(b)** 6.31 and 7.31
(c) 7.31 and 8.31 **(d)** 6.90 and 7.00

14.78 In which of the following pairs of pH values do both values represent basic solution conditions?

(a) 5.92 and 6.92 **(b)** 6.92 and 7.92
(c) 7.92 and 8.92 **(d)** 7.01 and 7.10

14.79 What is the molar hydronium ion concentration associated with each of the following pH values?

(a) 3.0 **(b)** 5.0 **(c)** 5.7 **(d)** 6.3

14.80 What is the molar hydronium ion concentration associated with each of the following pH values?

(a) 4.0 **(b)** 8.0 **(c)** 8.2 **(d)** 10.1

14.81 What is the molar hydronium ion concentration associated with each of the following pH values?

(a) 2.43 **(b)** 3.43 **(c)** 7.43 **(d)** 7.45

14.82 What is the molar hydronium ion concentration associated with each of the following pH values?

(a) 4.05 **(b)** 5.05 **(c)** 8.05 **(d)** 8.15

14.83 Solution A has $[OH^-] = 4.3 \times 10^{-4}$. Solution B has $[H_3O^+] = 7.3 \times 10^{-10}$.

(a) Which solution is more basic?
(b) Which solution has the lower pH?

14.84 Solution A has $[H_3O^+] = 2.7 \times 10^{-6}$. Solution B has $\{OH^-] = 4.5 \times 10^{-8}$.

(a) Which solution is more acidic?
(b) Which solution has the higher pH?

14.85 A solution has a pH of 4.500. What will be the pH of this solution if the hydronium ion concentration is

(a) doubled
(b) quadrupled
(c) increased by a factor of 10
(d) increased by a factor of 1000

14.86 A solution has a pH of 3.699. What will be the pH of this solution if the hydronium ion concentration is

(a) tripled
(b) cut in half
(c) increased by a factor of 100
(d) decreased by a factor of 10^4

14.87 Calculate the pH of each of the following solutions.

(a) 6.3×10^{-3} M HNO_3 **(b)** 0.20 M HCl
(c) 0.000021 M H_2SO_4 **(d)** 2.3×10^{-4} M NaOH

14.88 Calculate the pH of each of the following solutions.

(a) 4.02×10^{-5} M HCl **(b)** 4.02×10^{-5} M H_2SO_4
(c) 0.00035 M HNO_3 **(d)** 5.7×10^{-3} M KOH

Hydrolysis of Salts (Sec. 14.13)

14.89 Identify the ion (or ions) present, if any, that will undergo hydrolysis in aqueous solution in each of the following salts.

(a) Na_3PO_4 **(b)** NaCN
(c) NH_4Cl **(d)** LiCl

14.90 Identify the ion (or ions) present, if any, that will undergo hydrolysis in aqueous solution in each of the following salts.

(a) $KC_2H_3O_2$ **(b)** NH_4F
(c) $Ca(CN)_2$ **(d)** NaBr

14.91 Predict whether each of the following aqueous salt solutions will be acidic, basic, or neutral.

(a) Na_2SO_4 **(b)** LiCN
(c) NH_4Br **(d)** KI

14.92 Predict whether each of the following aqueous salt solutions will be acidic, basic, or neutral.

(a) KNO_3 **(b)** NaCN
(c) $LiC_2H_3O_2$ **(d)** $(NH_4)_2SO_4$

14.93 Write a net ionic equation for the hydrolysis of each of the following salts in aqueous solution.

(a) NH_4Cl **(b)** $NaC_2H_3O_2$
(c) KF **(d)** LiCN

14.94 Write a net ionic equation for the hydrolysis of each of the following salts in aqueous solution.

(a) NaF **(b)** KCN
(c) $LiNO_2$ **(d)** NH_4I

Buffer Solutions (Sec. 14.14)

14.95 Predict whether or not each of the following pairs of substances could function as a buffer system in aqueous solution.

(a) HNO_3 and $NaNO_3$ **(b)** NaCl and NaCN
(c) HF and NaF **(d)** NH_4Cl and NH_3

14.96 Predict whether or not each of the following pairs of substances could function as a buffer system in aqueous solution.

(a) HNO_3 and HCl (b) HNO_2 and KNO_2
(c) HNO_3 and KNO_3 (d) $NaC_2H_3O_2$ and $KC_2H_3O_2$

14.97 Write a net ionic equation for each of the following buffering actions.

(a) The response of a HF/F^- buffer to the addition of H_3O^+ ions.

(b) The response of a H_2CO_3/HCO_3^- buffer to the addition of OH^- ions.

(c) The response of a NH_4^+/NH_3 buffer to the addition of H_3O^+ ions.

(d) The response of a $H_3PO_4/H_2PO_4^-$ buffer to the addition of OH^- ions.

14.98 Write a net ionic equation for each of the following buffering actions.

(a) The response of a HPO_4^{2-}/PO_4^{3-} buffer to the addition of OH^- ions.

(b) The response of a HF/F^- buffer to the addition of OH^- ions.

(c) The response of a HCN/CN^- buffer to the addition of H_3O^+ ions.

(d) The response of a $H_3PO_4/H_2PO_4^-$ buffer to the addition of H_3O^+ ions.

Normality of Acids and Bases (Sec. 14.17)

14.99 How many equivalents of acid or base are present in each of the following samples of acid or base?

(a) 2.50 moles of KOH (b) 2.50 moles of H_2SO_4
(c) 1.50 moles of $HC_2H_3O_2$ (d) 37.2 g of HNO_3

14.100 How many equivalents of acid or base are present in each of the following samples of acid or base?

(a) 0.250 mole of NaOH (b) 1.50 moles of H_3PO_4
(c) 2.00 moles of $H_2C_2O_4$ (d) 25.0 g of HCl

14.101 Calculate the normality of each of the following solutions.

(a) 250.0 mL of solution containing 5.00 g of NaOH

(b) 750.0 mL of solution containing 10.0 g of H_3PO_4

(c) 100.0 mL of solution containing 0.250 mole of H_2SO_4

(d) 2.00 L of solution containing 0.400 equivalent of $H_2C_2O_4$

14.02 Calculate the normality of each of the following solutions.

(a) 100.0 mL of solution containing 1.000 g of $Ba(OH)_2$

(b) 250.0 mL of solution containing 5.00 g of HNO_3

(c) 57.0 mL of solution containing 0.100 mole of HCl

(d) 3.75 L of solution containing 7.00 equivalents of H_2SO_4

14.103 Express the following molarities as normalities.

(a) 0.130 M H_3PO_4 (b) 2.30 M KOH
(c) 0.045 M H_2CO_3 (d) 1.00 M $HC_2H_3O_2$

14.104 Express the following molarities as normalities.

(a) 7.20 M HCl (b) 0.400 M $Ca(OH)_2$
(c) 3.2 M $H_2C_2O_4$ (d) 3.00 M H_2SO_4

14.105 Express the following normalities as molarities.

(a) 6.00 N H_2SO_4 (b) 4.12 N $Ba(OH)_2$
(c) 1.00 N HCl (d) 3.00 N H_3PO_4

14.106 Express the following normalities as molarities.

(a) 3.00 N HNO_3 (b) 3.00 N H_3PO_3
(c) 9.00 N H_2CO_3 (d) 2.00 N KOH

14.107 How many grams of oxalic acid ($H_2C_2O_4$) are needed to produce 400.0 mL of 0.0500 N $H_2C_2O_4$ solution?

14.108 How many grams of lactic acid ($HC_3H_5O_3$) are needed to produce 1.20 L of 1.000 N $HC_3H_5O_3$ solution?

Acid–Base Titration Calculations (Sec. 14.17)

14.109 What volume, in milliliters, of a 0.100-N NaOH solution would be needed to neutralize each of the following acid samples?

(a) 10.00 mL of 0.350 N H_2SO_4

(b) 50.00 mL of 1.500 N H_3PO_4

(c) 5.00 mL of 0.500 N HNO_3

(d) 75.00 mL of 0.00030 N HCl

14.110 What volume, in milliliters, of a 2.00-N KOH solution would be needed to neutralize each of the following acid samples?

(a) 50.00 mL of 6.50 N HCl

(b) 5.00 mL of 0.0500 N H_2SO_4

(c) 250.0 mL of 1.00 N H_3PO_4

(d) 75.00 mL of 2.50 N HNO_3

14.111 What volume, in milliliters, of 2.50 M aqueous ammonia [$NH_3(aq)$] would be needed to neutralize each of the following acid samples?

(a) 10.00 mL of 2.00 M HNO_3

(b) 20.00 mL of 6.00 M H_2SO_4

(c) 30.00 mL of 7.50 M H_3PO_4

(d) 100.0 mL of 0.100 M HCl

14.112 What volume, in milliliters, of a 1.30-M LiOH solution would be needed to neutralize each of the following acid samples?

(a) 7.50 mL of 0.100 M H_3PO_4

(b) 75.00 mL of 0.100 M H_2SO_4

(c) 150.0 mL of 7.50 M HNO_3

(d) 100.0 mL of 12.0 M HCl

14.113 It requires 34.5 mL of 0.102 N NaOH to neutralize each of the following acid solution samples. What are the normality and molarity of each of the acid samples?

(a) 25.0 mL of H_2SO_4

(b) 20.0 mL of HClO
(c) 20.0 mL of H_3PO_4
(d) 10.0 mL of HNO_3

14.114 It requires 21.4 mL of 0.198 N NaOH to neutralize each of the following acid solution samples. What are the normality and molarity of each of the acid samples?

(a) 30.0 mL of H_2CO_3
(b) 25.0 mL of $H_2C_2O_4$
(c) 25.0 mL of $HC_2H_3O_2$
(d) 20.0 mL of HCl

Acid and Base Stock Solutions (Sec. 14.18)

14.115 What is the molarity of each of the following stock solutions?

(a) dilute H_2SO_4 (b) dilute HCl
(c) concentrated HNO_3 (d) dilute NaOH

14.116 What is the molarity of each of the following stock solutions?

(a) dilute HNO_3 (b) concentrated NH_3
(c) concentrated H_2SO_4 (d) dilute $HC_2H_3O_2$

ADDITIONAL PROBLEMS

14.117 Identify each of the following species as the conjugate base of a strong acid or the conjugate base of a weak acid.

(a) Br^- (b) CN^- (c) $H_2PO_4^-$ (d) NO_3^-

14.118 Identify each of the following species as the conjugate base of a strong acid or the conjugate base of a weak acid.

(a) $C_2H_3O_2^-$ (b) Cl^- (c) F^- (d) HPO_4^{2-}

14.119 A solution has a pH of 2.2. Another solution has a pH of 4.5. How many times greater is the $[H_3O^+]$ in the first solution than the second one?

14.120 A solution has a pH of 3.4. Another solution has a pH of 6.7. How many times greater is the $[H_3O^+]$ in the first solution than the second one?

14.121 In which of the following pairs of solutions does the first listed solution have a lower pH than the second listed solution?

(a) 0.1 M HCl and 0.2 M HCl
(b) 0.1 M HCl and 0.1 M H_2SO_4
(c) 0.20 M H_2SO_4 and 0.25 M HNO_3
(d) 0.2 M H_2SO_4 and 0.2 M H_2CO_3

14.122 In which of the following pairs of solutions does the first listed solution have a lower pH than the second listed solution?

(a) 0.2 M HNO_3 and 0.3 M HNO_3
(b) 0.1 M HNO_3 and 0.1 M H_2SO_4
(c) 0.1 M H_2SO_4 and 0.12 M HCl
(d) 0.2 M H_2SO_4 and 0.2 M $H_2C_2O_4$

14.123 What would be the pH of a solution that contains 0.1 mole of each of the solutes NaCl, HNO_3, HCl, and NaOH in enough water to give 3.00 L of solution?

14.124 What would be the pH of a solution that contains 0.1 mole of each of the solutes NaBr, HBr, KOH, and NaOH in enough water to give 5.00 L of solution?

14.125 Arrange the following 0.1-M aqueous solutions in order of decreasing pH: NH_4Br, $Ba(OH)_2$, $HClO_4$, K_2SO_4, and LiCN.

14.126 Arrange the following 0.1-M aqueous solutions in order of increasing pH: HBr, $HC_2H_3O_2$, $NaC_2H_3O_2$, KOH, and $Ca(NO_3)_2$.

14.127 Identify the buffer systems (conjugate acid–base pairs) present in a solution containing equal molar amounts of $HC_2H_3O_2$, $NaHCO_3$, $KC_2H_3O_2$, and Na_2CO_3.

14.128 Identify the buffer systems (conjugate acid–base pairs) present in a solution containing equal molar amounts of HCN, Na_2HPO_4, KH_2PO_4, and KCN.

14.129 A 20.00-mL sample of a solution of citric acid ($H_3C_6H_5O_7$) was titrated with a 0.250-N NaOH solution. A total of 27.86 mL of NaOH was required.

(a) What was the normality of the $H_3C_6H_5O_7$ solution?
(b) How many equivalents of $H_3C_6H_5O_7$ were present in the 20.0-mL sample?
(c) How many moles of $H_3C_6H_5O_7$ were present in the 20.0-mL sample?
(d) What was the molarity of the $H_3C_6H_5O_7$ solution?

14.130 A 20.00-mL sample of a solution of lactic acid ($HC_3H_5O_3$) was titrated with a 0.100-N $Ba(OH)_2$ solution. A total of 17.03 mL of $Ba(OH)_2$ was required.

(a) What was the normality of the $HC_3H_5O_3$ solution?
(b) How many equivalents of $HC_3H_5O_3$ were present in the 20.0-mL sample?
(c) How many moles of $HC_3H_5O_3$ were present in the 20.0-mL sample?
(d) What was the molarity of the $HC_3H_5O_3$ solution?

CUMULATIVE PROBLEMS

14.131 Name each of the following species.

(a) conjugate base of hydroiodic acid
(b) conjugate base of the dihydrogen phosphate ion
(c) conjugate acid of the oxide ion
(d) conjugate acid of water

14.132 Name each of the following species.

(a) conjugate base of perchloric acid
(b) conjugate base of the bicarbonate ion
(c) conjugate acid of the hydroxide ion
(d) conjugate acid of ammonia

14.133 Write electron-dot structures for hydrocyanic acid and its conjugate base.

14.134 Write electron-dot structures for hypochlorous acid and its conjugate base.

14.135 What is the pH of a solution obtained by each of the following operations?

(a) dissolving 4.8 g of HCl in enough water to obtain 0.40 L of solution
(b) dissolving 12.5 g of LiOH in enough water to obtain 255 mL of solution
(c) diluting 75 mL of 0.10 M HCl to a volume of 125 mL
(d) mixing equal volumes of 0.20 M HCl and 0.50 M HNO_3

14.136 What is the pH of a solution obtained by each of the following operations?

(a) dissolving 4.8 g of HBr in enough water to obtain 0.30 L of solution
(b) dissolving 3.50 g of NaOH in enough water to obtain 45 mL of solution
(c) diluting 25 mL of 0.10 HNO_3 to a volume of 375 mL
(d) mixing equal volumes of 0.20 M HCl and 0.20 M HNO_3

14.137 The pH of a hydrochloric acid solution with a density of 1.10 g/mL is 3.40. Calculate the concentration of the acid in

(a) molarity units **(b)** mass percent units

14.138 The pH of a nitric acid solution with a density of 1.20 g/mL is 2.70. Calculate the concentration of the acid in

(a) molarity units **(b)** mass percent units

14.139 How many hydronium ions are present in a 10.0-mL sample of hydrochloric acid that has a pH of 5.42?

14.140 How many hydronium ions are present in a 50.0-mL sample of nitric acid that has a pH of 1.32?

14.141 How many ions are present in a 236-mL sample of a HNO_3 solution with a pH of 2.37 to which 0.100 mole of Na_2SO_4 has been added and dissolved?

14.142 How many ions are present in a 435-mL sample of a HCl solution with a pH of 1.54 to which 0.050 mole of $Mg(NO_3)_2$ has been added and dissolved?

14.143 An impure 1.00-g sample of the monoprotic acid potassium hydrogen phthalate ($KHC_8H_4O_4$) is dissolved in water and titrated with 32.3 mL of 0.1000 N NaOH solution. Calculate the mass percent $KHC_8H_4O_4$ in the impure sample.

14.144 An impure 1.00-g sample of the diprotic oxalic acid ($H_2C_2O_4$) is dissolved in water and titrated with 17.6 mL of 0.200 N NaOH solution. Calculate the mass percent $H_2C_2O_4$ in the impure sample.

15

Oxidation and Reduction

15.1 Oxidation–Reduction Terminology

Oxidation–reduction reactions are a very important class of chemical reactions. They occur all around us and even within us. The bulk of the energy needed for the functioning of all living organisms, including humans, is obtained from food via oxidation–reduction processes. Such diverse phenomena as the electricity obtained from a battery to start a car, the use of natural gas to heat a home, iron rusting, and the functioning of antiseptic agents to kill or prevent the growth of bacteria all involve oxidation–reduction reactions. In short, knowledge of this type of reaction is fundamental to understanding many biological and technological processes.

The terms oxidation and reduction, like the terms acid and base (Sec. 14.1), have several definitions. Historically, the word *oxidation* was first used to describe the reaction of a substance with oxygen. According to this historical definition, each of the following reactions involves oxidation.

$$4\,Fe + 3\,O_2 \longrightarrow 2\,Fe_2O_3$$
$$S + O_2 \longrightarrow SO_2$$
$$CH_4 + 2\,O_2 \longrightarrow CO_2 + 2\,H_2O$$

The substance on the far left in each of these equations is said to have been *oxidized*.

Originally, the term *reduction* referred to processes where oxygen was removed from a compound. A particularly common type of reduction reaction, according to this original definition, is the removal of oxygen from a metal oxide to produce the free metal.

$$CuO + H_2 \longrightarrow Cu + H_2O$$
$$2\,Fe_2O_3 + 3\,C \longrightarrow 4\,Fe + 3\,CO_2$$

The word reduction comes from the reduction in mass of the metal-containing species; the metal has a mass less than that of the metal oxide.

Today the words oxidation and reduction are used in a much

F.Y.I.

The equation
$4\ Fe + 3\ O_2 \longrightarrow 2\ Fe_2O_3$
describes the oxidation or rusting of iron.

F.Y.I.

The equation
$2\ Fe_2O_3 + 3\ C \longrightarrow 4\ Fe + 3\ CO_2$
describes the reduction process used to produce iron from iron ore.

broader sense. Current definitions include the previous examples but also much more. It is now recognized that the same changes brought about in a substance from reaction with oxygen can be caused by reaction with numerous non–oxygen-containing substances. For example, consider the following reactions.

$$2\ Mg + O_2 \longrightarrow 2\ MgO$$
$$Mg + S \longrightarrow MgS$$
$$Mg + F_2 \longrightarrow MgF_2$$
$$3\ Mg + N_2 \longrightarrow Mg_3N_2$$

In each of these reactions magnesium metal is converted to a magnesium compound that contains Mg^{2+} ions. The process is the same—the changing of magnesium atoms, through the loss of two electrons, to magnesium ions; the only difference is the identity of the substance that causes magnesium to undergo the change. All of these reactions are considered to involve oxidation by the current definition. **Oxidation** *is the process whereby a substance in a chemical reaction loses one or more electrons.* The current definition for reduction involves the use of similar terminology. **Reduction** *is the process whereby a substance in a chemical reaction gains one or more electrons.*

Oxidation and reduction are complementary processes rather than isolated phenomena. They *always* occur together, you cannot have one without the other. If electrons are lost by one species, they cannot just disappear; they must be gained by another species. Electron transfer, then, is the basis for oxidation and reduction. The collective term **oxidation–reduction reaction** *is used to describe any reaction involving the transfer of electrons between reactants.* This designation is often shortened to simply **redox reaction**.

There are two different ways of looking at the reactants in a redox reaction. First, the reactants can be viewed as being acted upon. From this viewpoint one reactant is *oxidized* (the one that loses electrons) and one is *reduced* (the one that gains electrons). Second, the reactants can be looked at as bringing about the reaction. In this approach the terms *oxidizing agent* and *reducing agent* are used. An **oxidizing agent** *causes oxidation by accepting electrons from the other reactant.* Such acceptance, the gain of electrons, means that the oxidizing agent itself is reduced. Similarly, the **reducing agent** *causes reduction by providing electrons for the other reactant to accept.* As a result of providing electrons, the reducing agent itself becomes oxidized. Note, then, that the reducing agent and substance oxidized are one and the same, as are the oxidizing agent and substance reduced.

The terms oxidizing agent and reducing agent sometimes cause confusion because the oxidizing agent is not oxidized (it is reduced) and the reducing agent is not reduced (it is oxidized). By simple analogy, a travel agent is not the one who takes a trip—he or she is the one who causes the trip to be taken.

Table 15.1 summarizes the terms presented in this section.

TABLE 15.1 OXIDATION–REDUCTION TERMINOLOGY IN TERMS OF LOSS AND GAIN OF ELECTRONS

Terms Associated with the Loss of Electrons	Terms Associataed with the Gain of Electrons
Process of oxidation	Process of reduction
Substance oxidized	Substance reduced
Reducing agent	Oxidizing agent

15.2 OXIDATION NUMBERS

Oxidation numbers are used to help determine whether oxidation or reduction has occurred in a reaction, and if such is the case, the identity of the oxidizing and reducing agents. Formally defined, an **oxidation number** *is the charge that an atom appears to have when the electrons in each bond it is participating in are assigned to the more electronegative of the two atoms involved in the bond.**

Consider an HCl molecule, a molecule in which there is one bond involving two shared electrons.

According to the definition for oxidation number, the electrons in this bond are assigned to the chlorine atom (the more electronegative atom; Sec. 7.10). This results in the chlorine atom having one more electron than a neutral Cl atom; hence, the oxidation number of chlorine is -1 (one extra electron). At the same time, the H atom in the HCl molecule has one fewer electron than a neutral H atom; its electron was given to the chlorine. This electron deficiency of one results in an oxidation number of $+1$ for hydrogen.

As a second example, consider the molecule CF_4.

Fluorine is more electronegative than carbon. Hence, the two shared electrons in each of the four carbon–fluorine bonds are assigned to the fluorine atom. Each F atom thus gains an extra electron, resulting in F having a -1 oxidation number. The carbon atom loses a total of four electrons, one to each F atom, as a result of the electron "assignments." Hence, its oxidation number is $+4$, indicating the loss of the four electrons.

As a third example, consider the N_2 molecule where like atoms are involved in a triple bond.

Since the identical atoms are of equal electronegativity, the shared electrons are "divided" equally between the two atoms; each N receives three of the bonding electrons to count as its own. This results in each N atom having five valence electrons (three from the triple bond and two nonbonding electrons), the same number of valence electrons as in a neutral N atom. Hence, the oxidation number of N in N_2 is zero.

Before going any further in our discussion of oxidation numbers, it should be noted that *calculated* oxidation numbers are *not* actual charges on atoms. This is why the phrase "appears to have" is found in the definition of oxidation number given at the start of this section. In assigning oxidation numbers, we assume when we give the bonding electrons to the more electronegative element that each bond is ionic (complete transfer of electrons). We know that this is not always the case. Sometimes it is a good approximation, sometimes it is not. Why, then, do we do this when we know that it does not always correspond to reality? Oxidation numbers, as we shall see shortly, serve as a very convenient device for "keeping track" of electron transfer in redox reactions. Even though they do not always correspond to physical reality, they are very, very useful entitites.

*In some textbooks the term *oxidation state* is used in place of oxidation number. In other textbooks the two terms are used interchangeably. We will use oxidation number.

In principle, the procedures used to determine oxidation numbers for the atoms in the molecules HCl, CF_4, and N_2 can be used to determine oxidation numbers in all molecules. However, the procedures become very laborious in many cases, especially when complicated electron-dot structures are involved. In practice, an alternative, much simpler procedure that does not require electron-dot structures is used to obtain oxidation numbers. This alternative procedure is based on a set of operational rules that are consistent with and derivable from the general definition for oxidation numbers. The operational rules are

RULE 1 The oxidation number of an atom in its elemental state is zero.

For example, the oxidation number of Cu is zero, and the oxidation number of Cl in Cl_2 is zero.

RULE 2 The oxidation number of any monoatomic ion is equal to the charge on the ion.

For example, the Na^+ ion has an oxidation number of +1, and the S^{2-} ion has an oxidation number of −2.

RULE 3 The oxidation numbers of groups IA and IIA elements are always +1 and +2, respectively.

RULE 4 The oxidation number of fluorine is always −1 and that of the other group VIIA elements (Cl, Br, and I) is usually −1.

The exception for these latter elements is when they are bonded to more electronegative elements. In which case they are assigned positive oxidation numbers.

RULE 5 The usual oxidation number for oxygen is −2.

The exceptions occur when oxygen is bonded to the more electronegative fluorine (O then is assigned a positive oxidation number) or found in compounds containing oxygen–oxygen bonds (peroxides). In peroxides the oxidation number −1 is assigned to oxygen. Peroxides form between oxygen and hydrogen (H_2O_2), group IA elements (Na_2O_2, etc.), and group IIA elements (BaO_2, etc.).

RULE 6 The usual oxidation number for hydrogen is +1.

The exception occurs in hydrides, compounds where hydrogen is bonded to a metal of lower electronegativity. In such compounds hydrogen is assigned an oxidation number of −1. Examples of hydrides are NaH, CaH_2, and LiH.

RULE 7 The algebraic sum of the oxidation numbers of all atoms in a neutral molecule must be zero.

RULE 8 The algebraic sum of the oxidation numbers of all atoms in a polyatomic ion is equal to the charge on the ion.

The use of these rules is illustrated in Example 15.1.

Example 15.1

Assign oxidation numbers to each element in the following chemical species.

(a) SO_3 **(b)** N_2H_4 **(c)** $KMnO_4$ **(d)** ClO_4^-

Solution

(a) Oxygen has an oxidation number of −2 (rule 5). The oxidation number of S can be calculated by rule 7. Letting x equal the oxidation number of S, we have

$$\begin{aligned} &\text{S:} \quad 1 \text{ atom} \times (x) &&= x \\ &\text{O:} \quad 3 \text{ atoms} \times (-2) &&= \underline{-6} \\ & && \text{sum} = 0 \quad \text{(rule 7)} \end{aligned}$$

Solving for x algebraically, we get

$$\begin{aligned} x + (-6) &= 0 \\ x &= +6 \end{aligned}$$

Consequently, the oxidation number of sulfur is $+6$ in the compound SO_3.

(b) Hydrogen has an oxidation number of $+1$ (rule 6). Rule 7 will allow us to calculate the oxidation number of N; the sum of the oxidation numbers must be zero. Letting x equal the oxidation number of N, we have

$$\begin{aligned} &\text{H:} \quad 4 \text{ atoms} \times (+1) &&= +4 \\ &\text{N:} \quad 2 \text{ atoms} \times (x) &&= \underline{2x} \\ & && \text{sum} = 0 \quad \text{(rule 7)} \end{aligned}$$

Solving for x algebraically, we get

$$\begin{aligned} 2x + (+4) &= 0 \\ x &= -2 \end{aligned}$$

Thus, the oxidation number of nitrogen in N_2H_4 is -2. Note that the oxidation number of N is not -4 (the calculated charge associated with two N atoms). Oxidation number is always specified on a *per atom* basis.

(c) Potassium has an oxidation number of $+1$ (rule 3), and oxygen has an oxidation number of -2 (rule 5). Letting x equal the oxidation number of manganese and using rule 7, we get

$$\begin{aligned} &\text{K:} \quad 1 \text{ atom} \times (+1) &&= +1 \\ &\text{Mn:} \quad 1 \text{ atom} \times (x) &&= x \\ &\text{O:} \quad 4 \text{ atoms} \times (-2) &&= \underline{-8} \\ & && \text{sum} = 0 \quad \text{(rule 7)} \end{aligned}$$

Solving for x algebraically, we get

$$\begin{aligned} (+1) + x + (-8) &= 0 \\ x &= +7 \end{aligned}$$

Thus, the oxidation number of manganese in $KMnO_4$ is $+7$.

(d) According to rule 8, the sum of the oxidation numbers must equal -1, the charge on this polyatomic ion. The oxidation number of oxygen is -2 (rule 5). Chlorine will have a positive oxidation number, since it is bonded to a more electronegative element (rule 4). Letting x equal the oxidation number of chlorine, we have

$$\begin{aligned} &\text{Cl:} \quad 1 \text{ atom} \times (x) &&= x \\ &\text{O:} \quad 4 \text{ atoms} \times (-2) &&= \underline{-8} \\ & && \text{sum} = -1 \quad \text{(rule 8)} \end{aligned}$$

Solving for x algebraically, we get

$$\begin{aligned} x + (-8) &= -1 \\ x &= +7 \end{aligned}$$

Thus, chlorine has an oxidation number of $+7$ in this ion.

Practice Exercise 15.1

Assign oxidation numbers to each element in the following chemical species.

(a) SO_2 (b) NH_4^+ (c) $NaNO_3$ (d) N_2F_4

Ans. (a) +4 for S, −2 for O; (b) −3 for N, +1 for H; (c) +1 for Na, +5 for N, −2 for O; (d) +2 for N, −1 for F

Many elements display a range of oxidation numbers in their various compounds. For example, nitrogen exhibits oxidation numbers ranging from −3 to +5 in various compounds. Selected examples are

NH_3	N_2H_4	N_2O	NO	N_2O_3	NO_2	HNO_3
−3	−2	+1	+2	+3	+4	+5

As shown in this listing of nitrogen-containing compounds, the oxidation number of an atom is written *underneath* the atom in the formula. This convention is used to avoid confusion with the charge on an ion.

Although not common, nonintegral oxidation numbers are possible. For example, the oxidation number of iron in the compound Fe_3O_4 is +2.67. The oxidation numbers of the oxygens in the compound add up to −8. Therefore, the iron atoms must have an oxidation number sum of +8. Dividing +8 by 3 (the number of iron atoms) gives +2.67.

Oxidizing and reducing agents can be defined in terms of changes in oxidation numbers. The **oxidizing agent** *in a redox reaction is the substance that* contains *the atom that shows an increase in oxidation number*. Since the oxidizing agent is the substance reduced in a reaction, reduction involves a decrease in oxidation number; the oxidation number is reduced (decreased) in a reduction. The **reducing agent** *in a redox reaction is the substance that contains the atom that shows an increase in oxidation number.* Since the reducing agent is the substance oxidized in a reaction, oxidation involves an increase in oxidation number.

Table 15.2 summarizes the relationships between oxidation–reduction terms and oxidation number changes. A comparison of Table 15.2 with Table 15.1 shows that the loss of electrons and oxidation number increases are synonymous as are the gain of electrons and oxidation number decreases. The fact that the oxidation number becomes more positive (increases) as electrons are lost is consistent with our understanding that electrons are negatively charged.

Example 15.2

Determine oxidation numbers for each atom in the following reactions, and identify the oxidizing and reducing agents.

(a) $2\,NO + O_2 \longrightarrow 2\,NO_2$ **(b)** $Zn + 2\,HCl \longrightarrow ZnCl_2 + H_2$

TABLE 15.2 OXIDATION–REDUCTION TERMINOLOGY IN TERMS OF OXIDATION NUMBER CHANGE

Terms Associated with an Increase in Oxidation Number	Terms Associated with a Decrease in Oxidation Number
Process of oxidation	Process of reduction
Substance oxidized	Substance reduced
Reducing agent	Oxidizing agent

(c) $Cl_2 + 2\,I^- \longrightarrow I_2 + 2\,Cl^-$

Solution

The oxidation numbers are calculated by the methods illustrated in Example 15.1.

(a)

2 NO	+	O_2	$\longrightarrow$	$2\,NO_2$
+2 −2		0		+4 −2
rules 5, 7		rule 1		rules 5, 7

The oxidation number of N has increased from +2 to +4. Therefore, the substance that contains N, NO, has been oxidized and is the reducing agent.

The oxidation number of the O in O_2 has decreased from 0 to −2. Therefore, the O_2 has been reduced and is the oxidizing agent.

(b)

Zn	+	2 HCl	$\longrightarrow$	$ZnCl_2$	+	H_2
0		+1 −1		+2 −1		0
rule 1		rules 6, 7		rules 4, 7		rule 1

The oxidation number of Zn has increased from 0 to +2. An increase in oxidation number is associated with oxidation. Therefore, the element Zn, has been oxidized and is the reducing agent.

The oxidation number of H has decreased from +1 to 0. A decrease in oxidation number is associated with reduction. Therefore, the HCl, the hydrogen-containing compound, is the oxidizing agent.

(c)

Cl_2	+	$2\,I^-$	$\longrightarrow$	I_2	+	Cl^-
0		−1		0		−1
rule 1		rule 2		rule 1		rule 2

The oxidation number of I has increased from −1 to 0. Thus, I^-, the iodine-containing reactant, has been oxidized and is the reducing agent.

The oxidation number of Cl has decreased from 0 to −1. Thus, Cl_2, the chlorine-containing reactant, has been reduced and is the oxidizing agent.

Practice Exercise 15.2

Determine oxidation numbers for each atom in the following reactions, and identify the oxidizing and reducing agents.

(a) $2\,SO_2 + O_2 \longrightarrow 2\,SO_3$
(b) $2\,Fe_2O_3 + 3\,C \longrightarrow 4\,Fe + 3\,CO_2$
(c) $Pb + Cu^{2+} \longrightarrow Cu + Pb^{2+}$

Ans. (a) $2\,SO_2 + O_2 \longrightarrow 2\,SO_3$
SO_2: +4 −2; O_2: 0; SO_3: +6 −2
reducing agent: SO_2; oxidizing agent: O_2

(b) $2\,Fe_2O_3 + 3\,C \longrightarrow 4\,Fe + 3\,CO_2$
Fe_2O_3: +3 −2; C: 0; Fe: 0; CO_2: +4 −2
reducing agent: C; oxidizing agent: Fe_2O_3

(c) $Pb + Cu^{2+} \longrightarrow Cu + Pb^{2+}$
Pb: 0; Cu^{2+}: +2; Cu: 0; Pb^{2+}: +2
reducing agent: Pb; oxidizing agent: Cu^{2+}

15.3 TYPES OF CHEMICAL REACTIONS

Two classification systems for chemical reactions are in common use. We have now encountered both of them.

The first system, presented initially in Section 10.6, recognized four types of reactions.

1. Synthesis ($X + Y \longrightarrow XY$)
2. Decomposition ($XY \longrightarrow X + Y$)
3. Single-replacement ($X + YZ \longrightarrow Y + XZ$)
4. Double-replacement ($AX + BY \longrightarrow AY + BX$)

The second system involves two reaction types.

1. Oxidation–reduction (or redox)
2. Non-oxidation–reduction (or nonredox)

As we have just learned (Sec. 15.2), reactions in which oxidation numbers change are called oxidation–reduction reactions. When there are no changes in oxidation numbers we have a nonredox reaction.

These two classification systems are not mutually exclusive and are commonly used together. For example, a particular reaction may be characterized as a single-replacement redox reaction.

Synthesis reactions with only elements as reactants are always oxidation–reduction reactions. Oxidation number changes must occur because all elements (the reactants) have an oxidation number of zero and all of the constituent elements of a compound *cannot* have oxidation numbers of zero. Synthesis reactions in which compounds are the reactants may or may not be redox reactions.

$$S + O_2 \longrightarrow SO_2 \quad \text{(redox synthesis)}$$
$$K_2O + H_2O \longrightarrow 2\ KOH \quad \text{(nonredox synthesis)}$$
$$2\ NO + O_2 \longrightarrow 2\ NO_2 \quad \text{(redox synthesis)}$$

Both redox and nonredox decomposition reactions are common. At sufficiently high temperatures all compounds can be broken down (decomposed) into their constituent elements. Such reactions, where only elements are the products, are always redox reactions. Decomposition reactions where compounds are the products are most often nonredox reactions.

$$2\ CuO \longrightarrow 2\ Cu + O_2 \quad \text{(redox decomposition)}$$
$$2\ KClO_3 \longrightarrow 2\ KCl + O_2 \quad \text{(redox decomposition)}$$
$$CaCO_3 \longrightarrow CaO + CO_2 \quad \text{(nonredox decomposition)}$$

Single-replacement reactions are always redox reactions. By definition, an element and a compound are reactants and an element and a compound are products. The elements always undergo oxidation number change. Two of the reaction types studied in Chapter 14 are redox single-replacement reactions—the reaction between an acid and an active metal (Sec. 14.8) and the reaction between a metal and an aqueous salt solution (Sec. 14.10).

Double-replacement reactions generally involve acids, bases, and salts in aqueous solution. In such reactions ions, which maintain their identity, are generally trading places.

Such reactions will always be nonredox reactions. All acid–base neutralization reactions (Sec. 14.8) are nonredox double-replacement reactions.

Combustion reactions (Sec. 10.5) are always redox reactions. However, as mentioned in Section 10.5, they do not fit any of the four general reaction patterns of synthesis, decomposition, single-replacement, and double-replacement.

Example 15.3

Classify the following reactions as redox or nonredox. Further classify them, when possible, as synthesis, decomposition, single-replacement, or double-replacement.

(a) $Ni + F_2 \longrightarrow NiF_2$

(b) $Fe_2O_3 + 3\ C \longrightarrow 2\ Fe + 3\ CO$

(c) $C_4H_8 + 6\ O_2 \longrightarrow 4\ CO_2 + 4\ H_2O$

(d) $H_2SO_4 + 2\ NaOH \longrightarrow Na_2SO_4 + 2\ H_2O$

Solution

The oxidation numbers are calculated by the method illustrated in Example 15.1.

(a)

Ni	+	F_2	$\longrightarrow$	NiF_2
0		0		+2 −1
rule 1		rule 1		rules 4, 7

This is a redox reaction; the oxidations numbers of both Ni and F change. Since one substance is produced from two substances, it is also a synthesis reaction. We thus have a redox synthesis reaction.

(b)

Fe_2O_3	+	3 C	$\longrightarrow$	2 Fe	+	3 CO
+3 −2		0		0		+2 −2
rules 5, 7		rule 1		rule 1		rules 5, 7

This is a redox reaction; carbon is oxidized, iron is reduced. Having an element and a compound as reactants and an element and compound as products is a characteristic of a single-replacement reaction. That is the type of reaction we have here: iron and carbon are exchanging places. We thus have a redox single-replacement reaction.

(c)

C_4H_8	+	$6\ O_2$	$\longrightarrow$	$4\ CO_2$	+	$4\ H_2O$
−2 +1		0		+4 −2		+1 −2
rules 6, 7		rule 1		rules 5, 7		rules 5, 7

This is a redox reaction; the oxidation numbers of both carbon and oxygen change. This reaction does not fit any of the four reaction patterns of synthesis, decomposition, single replacement, and double replacement. (It is a combustion reaction.)

(d)

H_2SO_4	+	2 NaOH	$\longrightarrow$	Na_2SO_4	+	$2\ H_2O$
+1 +6 −2		+1 −2 +1		+1 +6 −2		+1 −2
rules 5, 6, 7		rules 3, 5, 6		rules 3, 5, 7		rules 5, 6

This is a nonredox reaction; there are no oxidation number changes. The reaction is also a double-replacement reaction; hydrogen and sodium are changing places, that is, "swapping partners." Thus we have a nonredox double-replacement reaction.

Practice Exercise 15.3

Classify the following reactions as redox or nonredox. Further classify them, when possible, as synthesis, decomposition, single replacement, or double replacement.

(a) $2\ KNO_3 \longrightarrow 2\ KNO_2 + O_2$
(b) $Zn + CuBr_2 \longrightarrow ZnBr_2 + Cu$
(c) $CH_4 + 2\ O_2 \longrightarrow CO_2 + 2\ H_2O$
(d) $NiCl_2 + 2\ NaOH \longrightarrow Ni(OH)_2 + 2\ NaCl$

Ans. (a) redox decomposition; (b) redox single-replacement; (c) redox; (d) nonredox double-replacement

15.4 BALANCING OXIDATION–REDUCTION EQUATIONS

Balancing an equation is not a new topic to us. In Section 10.3 we learned how to balance equations by the *inspection method*. With that method, we start with the most complicated compound within the equation and balance one of the elements in it. Then we balance the atoms of a second element, then a third, and so on until all elements are balanced. This inspection procedure is a useful method for balancing simple equations with small coefficients. However, it breaks down when applied to complicated equations.

Equations for redox reactions are often quite complicated and contain numerous reactants and products and large coefficients. Trying to balance redox equations such as

$$PH_3 + CrO_4^{2-} + H_2O \longrightarrow P_4 + Cr(OH)_4^- + OH^-$$

or

$$As_4O_6 + MnO_4^- + H_2O \longrightarrow AsO_4^{3-} + H^+ + Mn^{2+}$$

by inspection is a tedious, time-consuming, frustrating experience. Balancing such equations is, however, easily accomplished by systematic equation-balancing procedures that use oxidation numbers and focus on the fact that the numbers of electrons lost and gained in a redox reaction must be equal.

Two distinctly different approaches for systematically balancing redox equations are in common use; the oxidation-number method and the half-reaction method. Each method has advantages and disadvantages. We will consider both methods.

15.5 OXIDATION-NUMBER METHOD FOR BALANCING REDOX EQUATIONS

A useful feature of oxidation numbers is that they provide a rather easy method for recognizing and balancing redox equations. The steps involved in their use in this balancing process are as follows.

STEP 1 Assign oxidation numbers to all atoms in the equation and determine which atoms are undergoing a change in oxidation number.

STEP 2 Determine the magnitude of the change in oxidation number *per atom* for the elements undergoing a change in oxidation number.

Draw a large bracket from the element in the reactant to the element in the product, and write the increase or decrease in oxidation number at the middle of the bracket. (See the examples that follow.)

STEP 3 When more than one atom of an element that changes oxidation number is present in a formula unit (of either reactant or product), determine the change in oxidation number per *formula unit*.

Indicate this change per formula unit by multiplying the oxidation number change per atom, already written on the brackets, by an appropriate factor.

STEP 4 Determine multiplying factors that make the total increase in oxidation number equal to the total decrease in oxidation number.

Place them on the bracket also.

STEP 5 Place in front of the oxidizing and reducing agents and their products in the equation coefficients that are consistent with the total number of atoms of the elements undergoing oxidation-number change.

STEP 6 Balance all other atoms in the equation except those of hydrogen and oxygen.

In doing this, do not alter the coefficients determined in the previous step.

STEP 7 Balance the charge (the sum of all the ionic charges) so that it is the same on both sides of the equation by adding H^+ or OH^- ions.

This step is necessary only when dealing with net ionic equations describing aqueous solution reactions. If the reaction takes place in acidic solution, add H^+ ion to the side deficient in positive charge. If the reaction takes place in basic solution, add OH^- ions to the side deficient in negative charge.

STEP 8 Balance the hydrogen atoms.

For net ionic equations, H_2O must usually be added to an appropriate side of the equation to achieve hydrogen balance. Water is, of course, present in all aqueous solutions and can be either a reactant or a product.

STEP 9 Balance the oxygen atoms.

The oxygens should automatically be balanced. If oxygens do not balance, there is a mistake in a previous step. Check your work.

Now let us consider some examples where these rules are applied. The first two examples will involve molecular equations. The third example involves a net ionic equation. In balancing net ionic equations, any H_2O, H^+, or OH^- present is usually left out of the unbalanced equation that we start with and then added as needed during the balancing process.

Example 15.4

Balance the following molecular redox equation by the oxidation-number method of balancing.

$$\mathrm{Cr + O_2 + HBr \longrightarrow CrBr_3 + H_2O}$$

Solution

STEP 1 We identify the elements being oxidized and reduced by assigning oxidation numbers.

$$\underset{0}{\mathrm{Cr}} + \underset{0}{\mathrm{O_2}} + \underset{+1\ -1}{\mathrm{HBr}} \longrightarrow \underset{+3\ -1}{\mathrm{CrBr_3}} + \underset{+1\ -2}{\mathrm{H_2O}}$$

Chromium (Cr) and oxygen (O) are the elements that undergo oxidation-number change.

STEP 2 The change in oxidation number *per atom* is shown by drawing brackets connecting the oxidizing and reducing agents to their products and indicating the change at the middle of the bracket.

0 (+3) +3

$$Cr + O_2 + HBr \longrightarrow CrBr_3 + H_2O$$

0 (−2) −2

change in oxidation number per atom

STEP 3 For Cr the change in oxidation number per formula unit is the same as the change per atom, since both Cr and $CrBr_3$, the two Cr-containing species, contain only one Cr atom. For O the change in oxidation number per formula unit will be double the change per atom since O_2 contains two atoms. The change per formula unit is indicated by multiplying the per atom change by an appropriate numerical factor, which is 2 in this case.

(+3)

$$Cr + O_2 + HBr \longrightarrow CrBr_3 + H_2O$$

2(−2)

change in oxidation number per formula unit

STEP 4 For Cr, the total increase in oxidation number per formula unit is +3. For oxygen, the total decrease in oxidation number per formula unit is −4. To make the increase equal to the decrease, we must multiply the oxidation-number change for the element oxidized (Cr) by 4 and the oxidation-number change for the element reduced (O) by 3. This will make the increase and decrease both numerically equal to 12.

4(+3)

$$Cr + O_2 + HBr \longrightarrow CrBr_3 + H_2O$$

3[2(−2)]

oxidation-number increase equals oxidation-number decrease

STEP 5 We are now ready to place coefficients in the equation in front of the oxidizing and reducing agents and their products. The bracket notation indicates that four Cr atoms undergo an oxidation-number change. Place the coefficient 4 in front of both Cr and $CrBr_3$. The bracket notation also indicates that six O atoms (3 × 2) undergo an oxidation-number decrease of two units. Thus, we need six oxygen atoms on each side. Place the coefficient 3 in front of O_2 (6 atoms of O), and the coefficient 6 in front of H_2O (6 atoms of O).

$$4\ Cr + 3\ O_2 + HBr \longrightarrow 4\ CrBr_3 + 6\ H_2O$$

The equation is only partially balanced at this point; only Cr and O atoms are balanced.

STEP 6 We next balance the element Br (by inspection). There are twelve Br atoms on the right side. Thus, to obtain twelve Br atoms on the left side we place the coefficient 12 in front of HBr.

$$4\ Cr + 3\ O_2 + 12\ HBr \longrightarrow 4\ CrBr_3 + 6\ H_2O$$

STEP 7 This step is not needed when the equation is a molecular equation.

STEP 8 In this particular equation the H atoms are already balanced. There are 12 hydrogen atoms on each side of the equation.

STEP 9 If all of the previous procedures (steps) have been carried out correctly, the O atoms should automatically balance. They do. There are six O atoms on each side of the equation. The balanced equation is thus

$$4\,Cr + 3\,O_2 + 12\,HBr \longrightarrow 4\,CrBr_3 + 6\,H_2O$$

Practice Example 15.4

Balance the following molecule redox equation by the oxidation-number method.

$$Zn + O_2 + HCl \longrightarrow ZnCl_2 + H_2O$$

Ans. $2\,Zn + O_2 + 4\,HCl \longrightarrow 2\,ZnCl_2 + 2\,H_2O$

Example 15.5

Balance the following molecular redox equation by the oxidation-number method of balancing.

$$HNO_2 + HI \longrightarrow NO + I_2 + H_2O$$

Solution

STEP 1 $HNO_2 + HI \longrightarrow NO + I_2 + H_2O$

$+1{+3}{-2}\;\; +1{-1} \qquad +2{-2}\quad 0 \quad +1{-2}$

The two elements undergoing oxidation-number change are N and I.

STEP 2

$$\overset{+3 \qquad (-1) \qquad +2}{\overbrace{HNO_2 + HI \longrightarrow NO}} + I_2 + H_2O$$

$$HNO_2 + \underbrace{HI \longrightarrow NO + I_2}_{-1 \qquad (+1) \qquad 0} + H_2O$$

change in oxidation number per atom

STEP 3

$$\overbrace{HNO_2 + HI \longrightarrow NO}^{(-1)} + I_2 + H_2O$$

$$HNO_2 + \underbrace{HI \longrightarrow NO + I_2}_{2(+1)} + H_2O$$

change in oxidation number per formula unit

The iodine oxidation-number change per atom had to be multiplied by 2 since there are two iodine atoms per molecule in I_2. Thus, a minimum of two I atoms must undergo an oxidation-number increase. This illustrates that *both* reactant and product formulas must be considered when determining the change in oxidation number per formula unit.

STEP 4

$$\overbrace{HNO_2 + HI \longrightarrow NO}^{2(-1)} + I_2 + H_2O$$

$$HNO_2 + \underbrace{HI \longrightarrow NO + I_2}_{2(+1)} + H_2O$$

oxidation-number increase equals oxidation-number decrease

By multiplying the N per formula unit oxidation-number decrease of -1 by 2 we make the oxidation-number increase and decrease equal; both are at two units.

STEP 5 The coefficients for HNO_2, HI, NO, and I_2 in the equation are determined from the bracket information, which indicates that two N atoms undergo an oxidation number change for every two I atoms that change.

$$2\ HNO_2 + 2\ HI \longrightarrow 2\ NO + 1\ I_2 + H_2O$$

STEP 6 The only atoms left to balance are H and O

STEP 7 This step is not needed for a molecular equation.

STEP 8 We balance the hydrogen by placing the coefficient 2 in front of the H_2O on the right side.

$$2\ HNO_2 + 2\ HI \longrightarrow 2\ NO + 1\ I_2 + 2\ H_2O$$

STEP 9 If all of the procedures in previous steps have been carried out correctly, the O atoms should automatically balance. They do. There are four O atoms on each side of the equation.

$$2\ HNO_2 + 2\ HI \longrightarrow 2\ NO + I_2 + 2\ H_2O$$

Practice Exercise 15.5

Balance the following molecular redox equation by the oxidation-number method of balancing.

$$NF_3 + AlCl_3 \longrightarrow N_2 + Cl_2 + AlF_3$$

Ans. $2\ NF_3 + 2\ AlCl_3 \longrightarrow N_2 + 3\ Cl_2 + 2\ AlF_3$

Example 15.6

Balance the following net ionic redox equation by the oxidation-number method of balancing.

$$Cu + NO_3^- \longrightarrow Cu^{2+} + NO_2$$

This reaction occurs in acidic solution.

Solution

STEP 1 $Cu + NO_3^- \longrightarrow Cu^{2+} + NO_2$

0 +5 −2 +2 +4 −2

STEP 2 The two elements undergoing oxidation-number change are Cu and N.

0 (+2) +2

$$Cu + NO_3^- \longrightarrow Cu^{2+} + NO_2$$

+5 (−1) +4

change in oxidation number per atom

STEP 3 For both Cu and N the oxidation-number change per formula unit is the same as per atom. Both Cu and Cu^{2+} contain only one Cu atom; similarly, both NO_3^- and NO_2 contain only one N atom.

STEP 4 By multiplying the N oxidation-number decrease by 2 we make the oxidation-number increase and decrease per formula unit the same—two units.

$$\overbrace{Cu + NO_3^- \longrightarrow Cu^{2+}}^{(+2)} + NO_2$$

$$Cu + \underbrace{NO_3^- \longrightarrow Cu^{2+} + NO_2}_{2(-1)}$$

oxidation-number increase equals oxidation-number decrease

STEP 5 The bracket notation indicates that two N atoms and one Cu atom undergo an oxidation-number change. Translating this information into coefficients, we get

$$1\,Cu + 2\,NO_3^- \longrightarrow 1\,Cu^{2+} + 2\,NO_2$$

STEP 6 The only atoms left to balance are H and O.

STEP 7 Since this is a net ionic equation, the charges must balance; that is, the sum of the ionic charges of all species on each side of the equation must be equal. (They do not have to add up to zero; they just have to be equal.) In acidic solution, which is the case in this example, charge balance is accomplished by adding H^+ ion.

As the equation now stands, we have a charge of -2 on the left side (two nitrate ions each with a -1 charge) and a charge of $+2$ on the right side (one copper ion). By adding four H^+ ions to the left side we balance the charge at $+2$.

$$-2 + (+4) = +2$$

The equation at this point becomes

$$1\,Cu + 2\,NO_3^- + 4\,H^+ \longrightarrow 1\,Cu^{2+} + 2\,NO_2$$

STEP 8 The hydrogen atoms are balanced through the addition of H_2O molecules. There are four H atoms on the left side (4 H^+ ions) and none on the right side. Addition of two H_2O molecules to the right side will balance the H atoms at four per side.

$$1\,Cu + 2\,NO_3^- + 4\,H^+ \longrightarrow 1\,Cu^{2+} + 2\,NO_2 + 2\,H_2O$$

:STEP 9 The O atoms automatically balance at six atoms on each side. This is our double check that previous steps have been correctly carried out. The balanced net ionic equation is thus

$$Cu + 2\,NO_3^- + 4\,H^+ \longrightarrow Cu^{2+} + 2\,NO_2 + 2\,H_2O$$

Practice Exercise 15.6

Balance the following net ionic redox equation by the oxidation-number method of balancing.

$$UO_2^+ + Cr_2O_7^{2-} \longrightarrow UO_2^{2+} + Cr^{3+}$$

This reaction occurs in acidic solution.

Ans. $6\,UO_2^+ + Cr_2O_7^{2-} + 14\,H^+ \longrightarrow 6\,UO_2^{2+} + 2\,Cr^{3+} + 7\,H_2O$

15.6 HALF-REACTION METHOD FOR BALANCING REDOX EQUATIONS

The basis for the half-reaction method for balancing redox equations is the separation of the unbalanced redox equation into two half-reactions, one for oxidation and one for reduction. A **half-reaction** *is a part of a redox equation that shows only the oxidation process or only the reduction process*. The half-reactions, once obtained, are then balanced separately. The two balanced half-reactions are added together to generate the overall balanced equation.

The division of the original unbalanced redox equation into two parts (two half-reactions) is artificial. One half-reaction does not really take place independently of the other; we cannot have oxidation without reduction. Nevertheless, this method of balancing redox equations is preferred in certain areas of redox chemistry. In particular, it leads to an increased understanding of the reactions that take place in electrochemical cells such as batteries. Electrochemical cells are the topic of Section 15.8.

As we did with the oxidation-number method for balancing redox equations, we will break the half-reaction balancing process into a series of steps.

STEP 1 Using oxidation numbers, determine which atoms are oxidized and reduced. Based on this information, split the redox equation into two skeletal half-reaction equations.

- **(a)** an *oxidation* half-reaction equation, which involves the formula of the substance containing the element oxidized along with other species associated with it.
- **(b)** a *reduction* half-reaction equation, which involves the formula of the substance containing the element reduced along with other species associated with it.

STEP 2 Balance each of the half-reactions.

- **(a)** First, balance the element oxidized or reduced and then balance any other elements present in the skeletal equation other than oxygen or hydrogen.
- **(b)** Next, show the number of electrons lost or gained in the oxidation or reduction. Use the change in oxidation number and the number of atoms oxidized or reduced to determine the number of electrons lost or gained. Electrons lost (oxidation) are shown on the product side of the equation and electrons gained (reduction) on the reactant side of the equation.
- **(c)** Balance the ionic charge by adding H^+ ions (acidic solution) or OH^- ions (basic solution) as reactant or product. (Remember that the electrons previously added must be considered in balancing charge.)
- **(d)** Balance the hydrogen atoms by adding H_2O molecules as reactant or product.
- **(e)** Verify that the oxygen atoms are balanced. (If they do not balance, a mistake has been made in a previous step.)

STEP 3 Multiply each balanced half-reaction by appropriate integers to make the total number of electrons lost equal the total number of electrons gained.

STEP 4 Add the two half-reactions together and cancel identical species, including electrons, on each side of the equation. See if the coefficients obtained can be simplified.

Examples 15.7 through 15.9 illustrate how the preceding guidelines for balancing redox equations are applied.

Example 15.7

Balance the following net ionic equation that occurs in acidic solution using the half-reaction method of balancing.

$$S^{2-} + NO_3^- \longrightarrow S + NO \quad \text{(acidic solution)}$$

Solution

STEP 1 *Determine the oxidation and reduction skeletal half-reactions.*
Assigning oxidation numbers, we get

$$\underset{-2}{S^{2-}} + \underset{+5\,-2}{NO_3^-} \longrightarrow \underset{0}{S} + \underset{+2\,-2}{NO}$$

Sulfur is oxidized, increasing in oxidation number from −2 to 0.
Nitrogen is reduced, decreasing in oxidation number from +5 to +2.
The skeletal half-reactions for oxidation and reduction are

Oxidation: $S^{2-} \longrightarrow S$
Reduction: $NO_3^- \longrightarrow NO$

STEP 2 *Balance the individual half-reactions.*

(a) In both half-reactions, the element being oxidized or reduced is already balanced—one atom of S on both sides in the first half-reaction and one atom of N on both sides in the second half-reaction. There are no other elements present except oxygen.

Oxidation: $S^{2-} \longrightarrow S$
Reduction: $NO_3^- \longrightarrow NO$

(b) The oxidation number increase for S is +2. This is caused by the loss of two electrons, which are shown on the product side of the oxidation half-reaction.

Oxidation: $S^{2-} \longrightarrow S + 2e^-$

The oxidation number decrease for N is −3. This results from the gain of three electrons, which are shown on the reactant side of the reduction half-reaction.

Reduction: $NO_3^- + 3\,e^- \longrightarrow NO$

(c) Since this is an acidic solution reaction, charge balance is achieved by adding H^+ ions. In the oxidation half-reaction there is a charge of −2 on each side of the equation. No H^+ ions are needed, since the charge is already in balance.

Oxidation: $S^{2-} \longrightarrow S + 2\,e^-$

In the reduction half-reaction there is a charge of −4 on the left side of the equation (−1 from the NO_3^- ion and −3 from the three electrons). There is no charge on the right side of the equation.

Charge balance is achieved by adding four H^+ ions to the left side of the equation. Each side of the equation will now have zero charge.

$$\text{Reduction:}\quad NO_3^- + 3\,e^- + 4\,H^+ \longrightarrow NO$$

(d) Water molecules are used to achieve hydrogen balance. Since no hydrogen is present in the oxidation half-reaction, no water molecules are needed.

$$\text{Oxidation:}\quad S^{2-} \longrightarrow S + 2\,e^-$$

In the reduction half-reaction, two water molecules are added to the right side of the equation. We now have four hydrogen atoms of each side of the equation.

$$\text{Reduction:}\quad NO_3^- + 3\,e^- + 4\,H^+ \longrightarrow NO + 2\,H_2O$$

(e) There are no oxygen atoms present in the oxidation half-reaction. In the reduction half-reaction, the oxygen balances at three atoms of each side of the equation. The two balanced half-reactions are

$$\text{Oxidation:}\quad S^{2-} \longrightarrow S + 2\,e^-$$
$$\text{Reduction:}\quad NO_3^- + 3\,e^- + 4\,H^+ \longrightarrow NO + 2\,H_2O$$

STEP 3 *Equalize the electron loss and electron gain.*
Two electrons are produced in the oxidation half-reduction and three electrons are gained in the reduction half-reaction. To equalize electron loss and electron gain, we multiply the oxidation half-reaction by three and the reduction half-reaction by two. We have then an electron loss of 6 and an electron gain of 6.

$$\text{Oxidation:}\quad 3\,(S^{2-} \longrightarrow S + 2\,e^-)$$
$$\text{Reduction:}\quad 2\,(NO_3^- + 3\,e^- + 4\,H^+ \longrightarrow NO + 2\,H_2O)$$

STEP 4 *Add the half-reactions and cancel the identical species.*
Adding the two half-reactions together, we get

$$\text{Oxidation:}\quad 3\,S^{2-} \longrightarrow 3\,S + \cancel{6\,e^-}$$
$$\text{Reduction:}\quad 2\,NO_3^- + \cancel{6\,e^-} + 8\,H^+ \longrightarrow 2\,NO + 4\,H_2O$$
$$3\,S^{2-} + 2\,NO_3^- + 8\,H^+ \longrightarrow 3\,S + 2\,NO + 4\,H_2O$$

There are no species to cancel other than the electrons. The electrons always cancel. If they do not, we have made a mistake in step 3.

Practice Exercise 15.7

Balance the following net ionic equation that occurs in acidic solution using the half-reaction method of balancing.

$$Fe^{2+} + Cr_2O_7^{2-} \longrightarrow Fe^{3+} + Cr^{3+} \quad \text{(acidic solution)}$$

Ans. $6\,Fe^{2+} + Cr_2O_7^{2-} + 14\,H^+ \longrightarrow 6\,Fe^{3+} + 2\,Cr^{3+} + 7\,H_2O$

Example 15.8

Balance the following net ionic equation that occurs in basic solution using the half-reaction method of balancing.

$$S^{2-} + Cl_2 \longrightarrow SO_4^{2-} + Cl^- \quad \text{(basic solution)}$$

Solution

STEP 1 *Determine the oxidation and reduction skeletal half-reactions.*

Assigning oxidation numbers, we get

$$\underset{-2}{S^{2-}} + \underset{0}{Cl_2} \longrightarrow \underset{+6\ -2}{SO_4^{2-}} + \underset{-1}{Cl^-}$$

Sulfur is oxidized, increasing in oxidation number from -2 to $+6$. Chlorine is reduced, decreasing in oxidation number from 0 to -1. The skeletal half-reactions for oxidation and reduction are

$$\text{Oxidation:} \quad S^{2-} \longrightarrow SO_4^{2-}$$
$$\text{Reduction:} \quad Cl_2 \longrightarrow Cl^-$$

STEP 2 *Balance the individual half-reactions.*

(a) In the oxidation half-reaction the S is already balanced.

$$\text{Oxidation:} \quad S^{2-} \longrightarrow SO_4^{2-}$$

To balance the Cl in the reduction half-reaction the coefficient 2 must be added on the right side.

$$\text{Reduction:} \quad Cl_2 \longrightarrow 2\ Cl^-$$

(b) The oxidation number increase for S is $+8$, which corresponds to the loss of 8 electrons.

$$\text{Oxidation:} \quad S^{2-} \longrightarrow SO_4^{2-} + 8\ e^-$$

The oxidation number decrease for Cl is -1, which corresponds to a gain of one electron. Since there are two Cl atoms changing, the total electron gain is two electrons.

$$\text{Reduction:} \quad Cl_2 + 2\ e^- \longrightarrow 2\ Cl^-$$

(c) Since this reaction occurs in basic solution, charge balance is achieved by adding OH^- ions. In the oxidation half-reaction there is a charge of -2 on the left side and a charge of -10 (one sulfate and eight electrons) on the right side. The charge is brought into balance, at a -10, by adding 8 OH^- ions to the left side of the equation.

$$\text{Oxidation:} \quad S^{2-} + 8\ OH^- \longrightarrow SO_4^{2-} + 8\ e^-$$

In the reduction half-reaction, the charge is already balanced at a -2 on each side. No OH^- ions are needed.

$$\text{Reduction:} \quad Cl_2 + 2\ e^- \longrightarrow 2\ Cl^-$$

(d) Hydrogen balance is achieved in the oxidation half-reaction by adding four H_2O molecules to the right side of the equation.

$$\text{Oxidation:}\quad S^{2-} + 8\,OH^- \longrightarrow SO_4^{2-} + 8\,e^- + 4\,H_2O$$

Hydrogen balance is not needed in the reduction half-reaction since no hydrogen is present.

$$\text{Reduction:}\quad Cl_2 + 2\,e^- \longrightarrow 2\,Cl^-$$

(e) Oxygen balances at 8 atoms on each side of the equation in the oxidation half-reaction. Oxygen is not present in the reduction half-reaction. The two balanced half-reactions are

$$\text{Oxidation:}\quad S^{2-} + 8\,OH^- \longrightarrow SO_4^{2-} + 8\,e^- + 4\,H_2O$$
$$\text{Reduction:}\quad Cl_2 + 2\,e^- \longrightarrow 2\,Cl^-$$

STEP 3 *Equalize the electron loss and electron gain.*
Eight electrons are produced in the oxidation half-reaction and two electrons are gained in the reduction half-reaction. Multiplying the reduction half-reaction by 4 will cause electron loss and electgron gain to be equal at eight electrons.

$$\text{Oxidation:}\quad S^{2-} + 8\,OH^- \longrightarrow SO_4^{2-} + 8\,e^- + 4\,H_2O$$
$$\text{Reduction:}\quad 4\,(Cl_2 + 2\,e^- \longrightarrow 2\,Cl^-)$$

STEP 4 *Add the half-reactions and cancel the identical species.*
Adding the two half-reactions together, we get

$$\text{Oxidation:}\quad S^{2-} + 8\,OH^- \longrightarrow SO_4^{2-} + \cancel{8\,e^-} + 4\,H_2O$$
$$\text{Reduction:}\quad 4\,Cl_2 + \cancel{8\,e^-} \longrightarrow 8\,Cl^-$$
$$S^{2-} + 8\,OH^- + 4\,Cl_2 \longrightarrow SO_4^{2-} + 4\,H_2O + 8\,Cl^-$$

There are no species to cancel other than the electrons.

Practice Exercise 15.8

Balance the following net ionic equation that occurs in basic solution using the half-reacation method of balancing.

$$Zn + MnO_4^- \longrightarrow Zn(OH)_2 + MnO_2 \quad \text{(basic solution)}$$

Ans. $3\,Zn + 2\,MnO_4^- + 4\,H_2O \longrightarrow 3\,Zn(OH)_2 + 2\,MnO_2 + 2\,OH^-$

Example 15.9

Balance the following net ionic equation that occurs in acidic solution using the half-reaction method of balancing.

$$H_3AsO_3 + MnO_4^- \longrightarrow H_3AsO_4 + Mn^{2+} \quad \text{(acidic solution)}$$

Solution

STEP 1 *Determine the oxidation and reduction skeletal half-reactions.*
Assigning oxidation numbers, we get

$$\underset{+1\;+3\;-2}{H_3AsO_3} + \underset{+7\;-2}{MnO_4^-} \longrightarrow \underset{+1\;+5\;-2}{H_3AsO_4} + \underset{+2}{Mn^{2+}}$$

Arsenic is oxidized, increasing in oxidation number from +3 to +5.

Manganese is reduced, decreasing in oxidation number from +7 to +2. The skeletal half-reactions for oxidation and reduction are

$$\text{Oxidation:}\quad H_3AsO_3 \longrightarrow H_3AsO_4$$
$$\text{Reduction:}\quad MnO_4^- \longrightarrow Mn^{2+}$$

STEP 2 *Balance the individual half-reactions.*

(a) In both half-reactions, the element being oxidized or reduced is already balanced—one atom of As on both sides in the oxidation half-reaction and one atom of Mn on both sides in the reduction half-reaction.

$$\text{Oxidation:}\quad H_3AsO_3 \longrightarrow H_3AsO_4$$
$$\text{Reduction:}\quad MnO_4^- \longrightarrow Mn^{2+}$$

(b) The oxidation number increase for As is +2, which corresponds to the loss of two electrons.

$$\text{Oxidation:}\quad H_3AsO_3 \longrightarrow H_3AsO_4 + 2\,e^-$$

The oxidation number decrease for Mn is −5, which corresponds to the gain of five electrons.

$$\text{Reduction:}\quad MnO_4^- + 5\,e^- \longrightarrow Mn^{2+}$$

(c) Since this reaction occurs in acidic solution, charge balance is achieved by adding H^+ ions. In the oxidation half-reaction there is a charge of zero on the left side and a charge of −2 (two electrons) on the right side. Charge balance, at zero, is achieved by adding two H^+ ions to the right side of the equation.

$$\text{Oxidation:}\quad H_3AsO_3 \longrightarrow H_3AsO_4 + 2\,e^- + 2\,H^+$$

In the reduction half-reaction there is a charge of −6 on the left side (one MnO_4^- ion and 5 electrons) and a charge of +2 on the right side. Charge balance, at a +2, is achieved by adding 8 H^+ ions to the left side of the equation.

$$\text{Reduction:}\quad MnO_4^- + 5\,e^- + 8\,H^+ \longrightarrow Mn^{2+}$$

(d) Hydrogen balance is obtained in the oxidation half-reaction by adding one H_2O molecule to the left side of the equation.

$$\text{Oxidation:}\quad H_3AsO_3 + H_2O \longrightarrow H_3AsO_4 + 2\,e^- + 2\,H^+$$

Hydrogen balance is obtained in the reduction half-reaction by adding four H_2O molecules to the right side of the equation.

$$\text{Reduction:}\quad MnO_4^- + 5\,e^- + 8\,H^+ \longrightarrow Mn^{2+} + 4\,H_2O$$

(e) Oxygen balances at 4 atoms on each side in both the oxidation and reduction half-reactions. The two balanced half-reactions are

$$\text{Oxidation:}\quad H_3AsO_3 + H_2O \longrightarrow H_3AsO_4 + 2\,e^- + 2\,H^+$$
$$\text{Reduction:}\quad MnO_4^- + 5\,e^- + 8\,H^+ \longrightarrow Mn^{2+} + 4\,H_2O$$

STEP 3 *Equalize the electron loss and electron gain.*

The lowest common multiple for an electron loss of 2 and an electron gain of 5 is 10 electrons. Thus, we multiply the oxidation half-reaction by 5 and the reduction half-reaction by two.

$$\text{Oxidation:} \quad 5\,(H_3AsO_3 + H_2O \longrightarrow H_3AsO_4 + 2\,e^- + 2\,H^+)$$
$$\text{Reduction:} \quad 2\,(MnO_4^- + 5\,e^- + 8\,H^+ \longrightarrow Mn^{2+} + 4\,H_2O)$$

STEP 4 *Add the half-reactions and cancel the identical species.*

Adding the two half-reactions together, we get

$$\text{Oxidation:} \quad 5\,H_3AsO_3 + 5\,H_2O \longrightarrow 5\,H_3AsO_4 + \cancel{10\,e^-} + 10\,H^+$$
$$\text{Reduction:} \quad 2\,MnO_4^- + \cancel{10\,e^-} + 16\,H^+ \longrightarrow 2\,Mn^{2+} + 8\,H_2O$$
$$5\,H_3AsO_3 + 5\,H_2O + 2\,MnO_4^- + 16\,H^+ \longrightarrow 5\,H_3AsO_4 + 10\,H^+ + 2\,Mn^{2+} + 8\,H_2O$$

Both H^+ ion and H_2O are on both sides of the equation. We can cancel 5 H_2O molecules from each side and 10 H^+ ions from each side. The final balanced equation becomes

$$5\,H_3AsO_3 + 2\,MnO_4^- + 6\,H^+ \longrightarrow 5\,H_3AsO_4 + 2\,Mn^{2+} + 3\,H_2O$$

Practice Example 15.9

Balance the following net ionic equation which occurs in acidic solution using the half-reaction method of balancing.

$$HNO_2 + Cr_2O_7^{2-} \longrightarrow Cr^{3+} + NO_3^- \quad \text{(acidic solution)}$$

Ans. $3\,HNO_2 + Cr_2O_7^{2-} + 5\,H^+ \longrightarrow 3\,NO_3^- + 2\,Cr^{3+} + 4\,H_2O$

In each of the three examples we have just considered, the oxidation and reduction half-reactions were simultaneously balanced. This approach was used in the examples to enable us to make comparisons. In practice, particularly when you are thoroughly familiar with the balancing procedure, one half-reaction is usually completely balanced before work begins on balancing the other half-reaction. Usually, it is better to work on just one reaction at a time.

A comparison of the two methods for balancing redox equations is in order. Basic to each method is being able to recognize the elements involved in the actual oxidation–reduction process. The oxidation–number method works on the principle that the increase in oxidation number must equal the decrease in oxidation number. The half-reaction method involves equalizing the number of electrons lost by the substance oxidized with the number of electrons gained by the substance reduced.

The oxidation–number method is usually faster, particularly for simple equations. This potential speed is considered the major advantage of the oxidation–number method. The half-reaction method's focus on electron transfer is its major advantage. The feature becomes particularly important in electrochemistry (Sec. 15.8). In this field it is most useful to discuss chemical reactions in terms of half-reactions occurring at different locations (electrodes) in an electrochemical cell.

15.7 DISPROPORTIONATION REACTIONS

A disproportionation reaction is a special type of oxidation–reduction reaction. A **disproportionation reaction** *is a reaction in which some atoms of a single element in a reactant are oxidized and others are reduced.* For such reactant behavior to be possible, the reactant must contain an element that is capable of having at least three oxidation numbers: its original number plus one higher and one lower oxidation number. Note that any given atom is not both oxidized and reduced. Some atoms are oxidized, and other atoms of the same element are reduced.

An example of a disproportionation reaction is

$$3\ Br_2 + 3\ H_2O \longrightarrow HBrO_3 + 5\ HBr$$

Note that two bromine-containing products have been produced from one bromine-containing reactant. The reactant bromine atoms have an oxidation number of zero. Bromine in $HBrO_3$ has a $+5$ oxidation number (it has been oxidized), and bromine in HBr has a -1 oxidation number (it has been reduced).

$$\underset{0}{3\ Br_2} + 3\ H_2O \longrightarrow \underset{+5}{HBrO_3} + \underset{-1}{5\ HBr}$$

Thus, some of the reactant bromine atoms have been oxidized, while others have been reduced. A disproportionation reaction has taken place.

Example 15.10 shows how the procedures for balancing redox equations (Sec. 15.5 and 15.6) are slightly modified to balance disproportionation reaction equations.

Example 15.10

Balance the following disproportionation redox reaction using (a) the oxidation-number method and (b) the half-reaction method.

$$NO_2 \longrightarrow NO_3^- + NO \quad \text{(acidic solution)}$$

Solution

(a) *Oxidation-Number Method*

STEP 1 In assigning oxidation numbers we immediately become aware that this is a disproportionation reaction. Nitrogen is the only element for which an oxidation-number change occurs.

$$\underset{+4\ -2}{NO_2} \longrightarrow \underset{+5\ -2}{NO_3^-} + \underset{+2\ -2}{NO}$$

STEP 2 Since the species NO_2 is undergoing both oxidation and reduction, for balancing purposes we will write it twice on the reactant side of the equation. With the NO_2 in two places, brackets can then be drawn in the "normal" manner to connect the substances involved in oxidation and reduction.

$$\overset{+4 \quad (+1) \quad +5}{\overbrace{NO_2 + NO_2 \longrightarrow NO_3^-}} + NO \quad \text{change in oxidation number per atom}$$

$$NO_2 + \underbrace{NO_2 \longrightarrow NO_3^- + NO}_{+4 \quad (-2) \quad +2}$$

(The NO_2 molecules will be recombined later into one location.)

STEP 3 The change in oxidation number per formula unit in both cases is the same as per atom.

STEP 4 By multiplying the oxidation-number increase by 2 we equalize the oxidation-number increase and decrease per formula unit.

$$\overbrace{NO_2 + NO_2 \longrightarrow NO_3^-}^{2(+1)} + NO$$

$$\underbrace{NO_2 \longrightarrow NO_3^- + NO}_{(-2)}$$

oxidation-number increase equals oxidation-number decrease

STEP 5 The bracket notation indicates that two N atoms undergo an increase in oxidation number for every one that undergoes a decrease in oxidation number. Translating this information into equation coefficients gives

$$2\ NO_2 + 1\ NO_2 \longrightarrow 2\ NO_3^- + 1\ NO$$

Now that the equation coefficients for the substance involved in oxidation and reduction, NO_2, have been determined, we can combine the NO_2 into one location, reversing the process carried out in step 2.

$$3\ NO_2 \longrightarrow 2\ NO_3^- + 1\ NO$$

STEP 6 The only atoms left to balance are oxygen atoms.

STEP 7 Since this is a net ionic equation, charge must be balanced. In acidic solution, which is the case here, we balance the charge by adding H^+ ion. As the equation now stands, we have a charge of -2 on the right side (two NO_3^- ions). By adding 2 H^+ ions to the right side of the equation we balance the charge at zero

$$3\ NO_2 \longrightarrow 2\ NO_3^- + 1\ NO + 2\ H^+$$

STEP 8 Hydrogen atom balance is achieved through the addition of H_2O molecules. There are no H atoms on the left side and 2 H atoms on the right side. Addition of one H_2O molecule to the left side will balance the H atoms at 2 on each side.

$$1\ H_2O + 3\ NO_2 \longrightarrow 2\ NO_3^- + 1\ NO + 2\ H^+$$

STEP 9 The oxygen atoms should automatically balance. They do, at 7 atoms on each side.

$$H_2O + 3\ NO_2 \longrightarrow 2\ NO_3^- + NO + 2\ H^+$$

(b) *Half-reaction Method*

STEP 1 *Determine the oxidation and reduction skeletal half-reactions.* Assignment of oxidation numbers is the same as in part (a).

$$\underset{+4\ -2}{NO_2} \longrightarrow \underset{+5\ -2}{NO_3^-} + \underset{+2\ -2}{NO}$$

Nitrogen is undergoing both oxidation and reduction. The skeleton half-reactions for oxidation and reduction are

$$\text{Oxidation:}\quad NO_2 \longrightarrow NO_3^-$$
$$\text{Reduction:}\quad NO_2 \longrightarrow NO$$

Note how disproportionation is handled at this point. The substance undergoing disproportionation appears as a reactant in both the oxidation and reduction half-reaction.

STEP 2 *Balance the individual half-reactions.*

(a) In both half-reactions, the element being oxidized or reduced is already balanced—one atom of N in both cases.

$$\text{Oxidation:}\quad NO_2 \longrightarrow NO_3^-$$
$$\text{Reduction:}\quad NO_2 \longrightarrow NO$$

(b) The oxidation number increase for the oxidized N is +1, which corresponds to the loss of one electron.

$$\text{Oxidation:}\quad NO_2 \longrightarrow NO_3^- + 1\ e^-$$

The oxidation number decrease for the reduced N is −2, which corresponds to the gain of two electrons.

$$\text{Reduction:}\quad NO_2 + 2\ e^- \longrightarrow NO$$

(c) Since this reaction occurs in acidic solution, charge balance is achieved by adding H^+ ions. In the oxidation half-reaction there is no charge on the left side of the equation and a charge of −2 on the right side. Adding two H^+ ions to the right side of the equation will balance the charge at zero on both sides.

$$\text{Oxidation:}\quad NO_2 \longrightarrow NO_3^- + 1\ e^- + 2\ H^+$$

In the reduction half-reaction there is a charge of −2 on the left side of the equation and no charge on the right side. Charge balance is achieved by adding two H^+ ions to the left side of the equation.

$$\text{Reduction:}\quad NO_2 + 2\ e^- + 2\ H^+ \longrightarrow NO$$

(d) Hydrogen balance is obtained in the oxidation half-reaction by adding one H_2O to the left side of the equation.

$$\text{Oxidation:}\quad NO_2 + H_2O \longrightarrow NO_3^- + 1\ e^- + 2\ H^+$$

Hydrogen balance is obtained in the reduction half-reaction by adding one H_2O to the right side of the equation.

$$\text{Reduction:}\quad NO_2 + 2\ e^- + 2\ H^+ \longrightarrow NO + H_2O$$

(e) Oxygen balances at 2 atoms on each side in both the oxidation and reduction half-reactions. The two balanced half-reactions are

$$\text{Oxidation:}\quad NO_2 + H_2O \longrightarrow NO_3^- + 1\ e^- + 2\ H^+$$
$$\text{Reduction:}\quad NO_2 + 2\ e^- + 2\ H^+ \longrightarrow NO + H_2O$$

STEP 3 *Equalize the electron loss and electron gain.*

The oxidation half-reaction involves the loss of one electron. The reduction half-reaction involves the gain of two electrons. Multiplying the oxidation half-reaction by a factor of 2 will cause electron loss and gain to be equal at two electrons.

$$\text{Oxidation:}\quad 2\ (NO_2 + H_2O \longrightarrow NO_3^- + 1\ e^- + 2\ H^+)$$
$$\text{Reduction:}\quad NO_2 + 2\ e^- + 2\ H^+ \longrightarrow NO + H_2O$$

STEP 4 *Add the half-reactions and cancel the identical species.*
Adding the two half-reactions together, we get

$$\begin{array}{ll} \text{Oxidation:} & 2\ NO_2 + 2\ H_2O \longrightarrow 2\ NO_3^- + \cancel{2e^-} + 4\ H^+ \\ \text{Reduction:} & NO_2 + \cancel{2e^-} + 2\ H^+ \longrightarrow NO + H_2O \\ \hline & 2\ NO_2 + NO_2 + 2\ H_2O + 2\ H^+ \longrightarrow 2\ NO_3^- + 4\ H^+ + NO + H_2O \end{array}$$

Both H_2O and H^+ are on both sides of the equation and some of each can be canceled. Also NO_2 appears in two places on the left side of the equation and needs to be combined. The final balanced equation is

$$3\ NO_2 + H_2O \longrightarrow 2\ NO_3^- + NO + 2\ H^+$$

Practice Exercise 15.10

Balance the following disproportionation redox reaction using: (a) the oxidation-number method and (b) the half-reaction method.

$$MnO_4^{2-} \longrightarrow MnO_2 + MnO_4^- \quad \text{(acidic solution)}$$

Ans. $3\ MnO_4^{2-} + 4\ H^+ \longrightarrow 2\ MnO_4^- + MnO_2 + 2\ H_2O$

15.8 SOME IMPORTANT OXIDATION–REDUCTION REACTIONS

In this section we consider two important applications of redox reactions.

1. A *spontaneous* oxidation–reduction reaction can be used to convert chemical energy into electrical energy. For this to occur, a redox reaction must be carried out in a specially designed apparatus called a *galvanic cell.*
2. A *nonspontaneous* oxidation–reduction reaction can be caused to occur by using electrical energy to produce chemical energy. Such a process is called *electrolysis*, and the apparatus in which the reaction is carried out is called an *electrolytic cell.*

Galvanic Cells

When a strip of zinc metal is placed in a solution of copper(II) sulfate, a source of Cu^{2+} ion, a coating of copper metal forms on the zinc strip (see Fig. 15.1). At the same time this occurs, some of the zinc dissolves to give Zn^{2+} ions in solution. The reaction occurring is

$$Zn(s) + Cu^{2+}(aq) \longrightarrow Zn^{2+}(aq) + Cu(s)$$

The sulfate ions (SO_4^{2-}) present in the copper(II) sulfate solution remain unaffected by this change. This reaction occurs because Zn is more active than Cu and will replace it (activity series; Sec. 14.10).

This reaction is an example of a spontaneous oxidation–reduction reaction. When zinc metal and a solution of Cu^{2+} ions come into contact with each other, the Zn metal is spontaneously oxidized and the Cu^{2+} ions are spontaneously reduced; a direct transfer of electrons from the zinc atoms to the Cu^{2+} ions occurs. The products of this reaction are copper

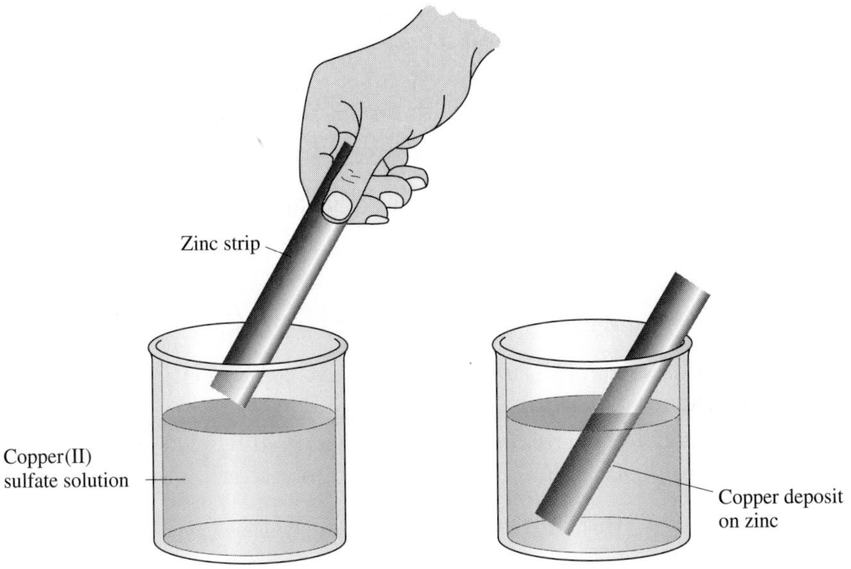

FIGURE 15.1 A spontaneous redox reaction occurs when zinc metal is placed in a solution of Cu^{2+} ions. A coating of copper metal quickly deposits on the zinc.

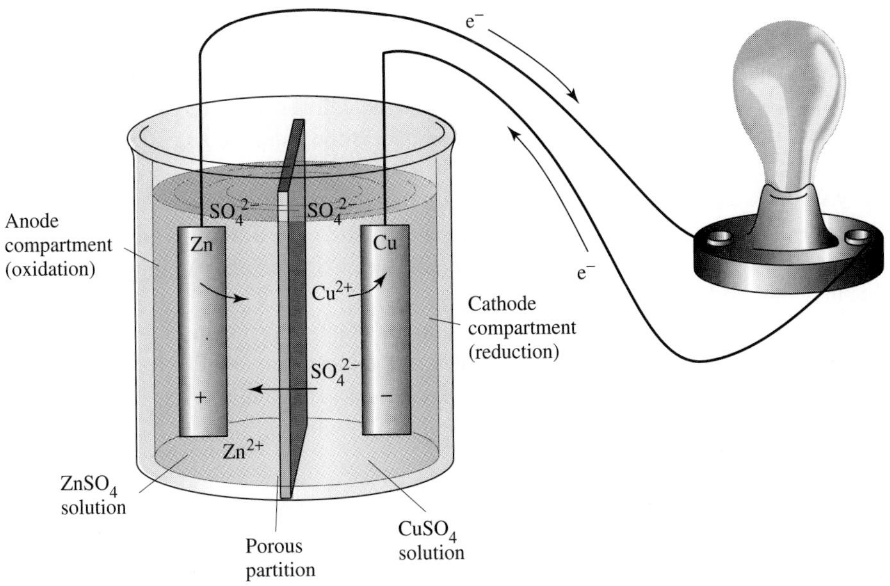

FIGURE 15.2 A zinc–copper galvanic cell. The electrical energy produced by this cell is generated by the spontaneous redox reaction $Zn(s) + Cu^{2+}(aq) \longrightarrow Zn^{2+}(aq) + Cu(s)$

atoms (Cu^{2+} ions that have gained electrons) and Zn^{2+} ions (zinc atoms that have lost electrons). Heat is liberated as the reaction proceeds, as evidenced by a slight warming of the solution.

By rearranging the reactants for this spontaneous redox reaction, one can obtain energy in the form of electricity rather than heat. The desired arrangement, a simple galvanic cell, is shown in Figure 15.2. A **galvanic cell** *is an appartus in which a spontaneous redox reaction is used to convert chemical energy to electrical energy.*

The galvanic cell of Figure 15.2 has two compartments separated by a porous partition. One compartment contains a strip of zinc metal immersed in a solution of zinc sulfate ($ZnSO_4$), and the other contains a strip of copper metal immersed in a solution of copper(II) sulfate ($CuSO_4$). The porous partition prevents the solutions from mixing freely but it does allow for passage of ions from one compartment to the other, a necessity for proper operation of the cell. The two strips of metal, called *electrodes*, are connected by a wire. This wire allows electrons to be transferred from one electrode to the other. As the spontaneous reaction between Zn and Cu^{2+} ions occurs, the flow of electrons through the wire can be demonstrated by placing a light bulb in the external circuit (see Fig. 15.2). The light bulb glows.

What is actually happening in the cell to cause the light bulb to glow?

1. Electrons are produced at the zinc electrode through the process of oxidation.

$$Zn(s) \longrightarrow Zn^{2+}(aq) + 2\ e^-$$

2. The electrons pass from the zinc electrode to the copper electrode through the external circuit (the wire).
3. Electrons enter the copper electrode and are accepted by Cu^{2+} ions in solution adjacent to the electrode. This is a reduction reaction.

$$Cu^{2+}(aq) + 2\ e^- \longrightarrow Cu(s)$$

4. To complete the circuit, ions (both positive and negative) move through the solution, passing through the porous membrane as needed.

Special names are given to the two electrodes in a galvanic cell; one is called the *cathode* and the other is the *anode*. The **cathode** *is the electrode at which reduction takes place*. Here electrons enter the galvanic cell from the external circuit. The **anode** *is the electrode at which oxidation takes place*. Here electrons leave the cell for the external circuit. In the cell now under discussion, the copper electrode is the cathode and the zinc electrode is the anode.

Many students have a hard time remembering the relationship between cathode–anode and oxidation–reduction. A mnemonic device can be helpful here. The two words that begin with vowels (anode and oxidation) go together, and the two words beginning with consonants (cathode and reduction) go together.

In principle, any spontaneous oxidation–reduction reaction can be used to build a galvanic cell, and many such cells have been studied in the laboratory. A selected few of such galvanic cells are now used commercially. We will discuss two that are very common: (1) the dry cell and (2) the lead storage battery.

The Dry Cell

The *dry cell* is widely used in flashlights, portable radios and tape recorders, and battery-powered toys; it is often referred to as a flashlight battery. Two versions of the dry cell are marketed: an acidic version and an alkaline version.

The *acidic version* of the dry cell contains a zinc outer surface (covered with cardboard or paint for protection) that functions as the anode and a carbon (graphite) rod, in contact with a moist paste, that serves as the cathode (see Fig. 15.3). The paste (the cell is not truly dry) is a mixture of solid MnO_2, solid NH_4Cl, and graphite powder (C) moistened with water. In the operation of the cell, Zn is oxidized to Zn^{2+} ion, and MnO_2 (a paste component) is reduced to Mn_2O_3. The reduction takes place at the interface between the carbon cathode and the paste; the inert carbon electrode conducts the electrons to the external circuit. The electrode reactions are

Anode (oxidation):

$$Zn(s) \longrightarrow Zn^{2+}(aq) + 2\ e^-$$

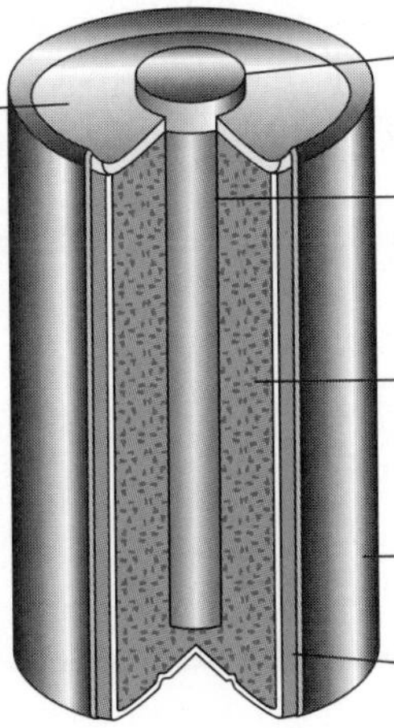

FIGURE 15.3 The acidic Zn–MnO_2 dry cell is commonly known as a flashlight battery.

Cathode (reduction):

$$2\ MnO_2(s) + 2\ NH_4^+(aq) + 2\ e^- \longrightarrow Mn_2O_3(s) + 2\ NH_3(aq) + H_2O$$

The useful life of acidic dry cells can be shortened if the slightly acidic paste corrodes the zinc. A protective paper is inserted between the paste and zinc to minimize this problem.

In the *alkaline version* of the dry cell, the solid NH_4Cl is replaced with KOH, and a steel rod rather than a graphite one is the cathode. The anode reaction still involves oxidation of zinc, but the zinc is present as a powder in a gel formulation. The cathode reaction also still involves the reduction of MnO_2. Equations for the electrode reactions are

Anode (oxidation):

$$Zn(s) + 2\ OH^-(aq) \longrightarrow ZnO(s) + H_2O(l) + 2\ e^-$$

Cathode (reduction):

$$2\ MnO_2(s) + H_2O(l) + 2\ e^- \longrightarrow Mn_2O_3(s) + 2\ OH^-(aq)$$

The alkaline dry cell costs roughly three times as much to produce as the acidic version. A major cost factor is the more elaborate internal construction needed to prevent leakage of the KOH solution. These cells provide up to 50% more total energy than the less expensive acidic model because they maintain usable voltage over a larger fraction of the lifetime of the cathode and anode materials. Miniature alkaline cells find extensive use in calculators, watches, and camera exposure controls.

Lead Storage Battery

The lead storage battery provides the starting power for automobiles. A 12-volt lead storage battery, the standard size, consists of six galvanic cells connected together (see Fig. 15.4). Each cell generates 2 volts.

Both electrodes in a lead storage battery involve the element lead. Lead serves as the anode, and lead coated with lead dioxide serves as the cathode. Glass fiber spacers separate electrodes to prevent them from touching each other. The electrodes are immersed in a 38% (m/m) sulfuric acid (H_2SO_4) solution. Details of one of the six cells of a lead storage battery are shown in Figure 15.5.

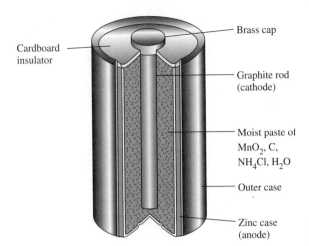

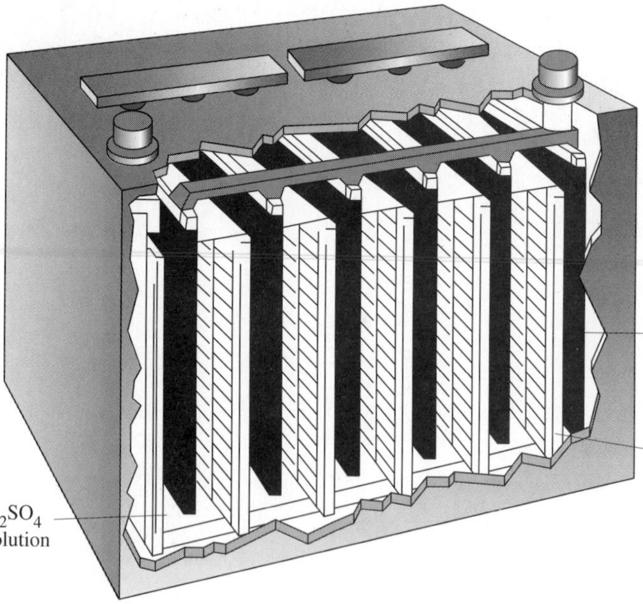

FIGURE 15.4 The six galvanic cells in a 12-volt automobile battery.

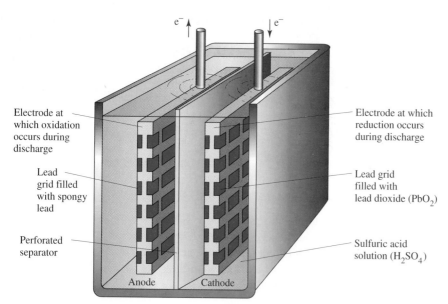

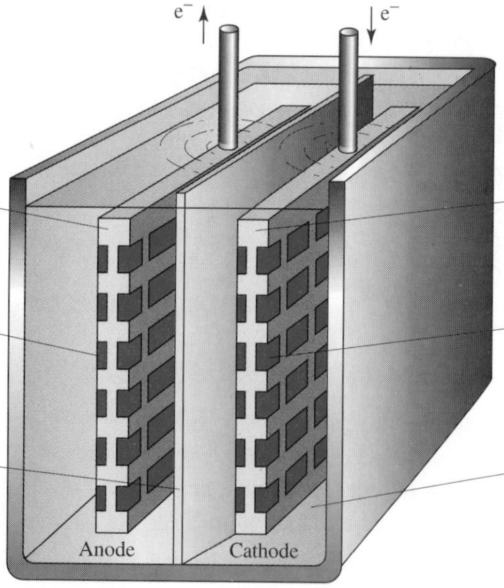

FIGURE 15.5 A single lead storage battery cell.

The electrode reactions when a lead battery is used to supply power (discharging) are

Anode (oxidation):

$$Pb(s) + SO_4^{2-}(aq) \longrightarrow PbSO_4(s) + 2\,e^-$$

Cathode (reduction):

$$PbO_2(s) + 4\,H^+(aq) + SO_4^{2-}(aq) + 2\,e^- \longrightarrow PbSO_4(s) + 2\,H_2O(l)$$

Battery discharge results in a buildup of $PbSO_4$ on the electrodes and a decrease in the density of the sulfuric acid. (The state of charge of a lead storage battery can thus be checked by a service station attendant by measuring the density (Sec. 3.8) of the sulfuric acid.)

Unlike dry cells, a lead storage battery can be recharged. This reverse process, which is nonspontaneous, uses an external source of electrical energy; in the automobile the external energy source is an alternator driven by the automobile engine.

In recharging, the $PbSO_4$ on the electrodes (formed during discharge) is converted back to Pb at one electrode and to PbO_2 at the other and H_2SO_4 is also produced. The electrode reactions are the reverse of what occurs during discharge.

Theoretically a lead storage battery should be rechargeable indefinitely. In practice, such batteries have a lifetime of 3–5 yr because small amounts of lead sulfate continually fall from the electrodes (to the bottom of the cell) as a result of "road shock" and chemical side reactions. Eventually the electrodes lose so much lead sulfate that the recharging process is no longer effective.

In "standard" lead storage batteries, water must be added to the individual cells on a regular basis. Recharging the battery, besides converting $PbSO_4$ back to Pb and PbO_2, also decomposes small amounts of water to give H_2 and O_2; hence, the H_2O must be replenished. Because of the possible presence of H_2 gas in a lead storage battery, a person should wear glasses (for eye protection) when releasing the cap of a battery since escaping gas can force sulfuric acid out. In addition, a person should not smoke while doing this, since hydrogen gas is flammable and forms explosive mixtures with oxygen. Newer automobile batteries have electrodes made of an alloy of calcium and lead. The presence of the calcium minimizes the decomposition of water during recharging. Thus, batteries with these alloy electrodes can be sealed; there is no need to add water.

Electrolytic Cells

The application of electrical energy from an external power source can be used to cause a nonspontaneous redox reaction to occur. The charging of the lead storage battery previously discussed is an example of this. The general term for such a process is *electrolysis*. **Electrolysis** *is the process in which electrical energy is used to cause a nonspontaneous redox reaction to occur.* During recharging the lead storage battery is functioning as an

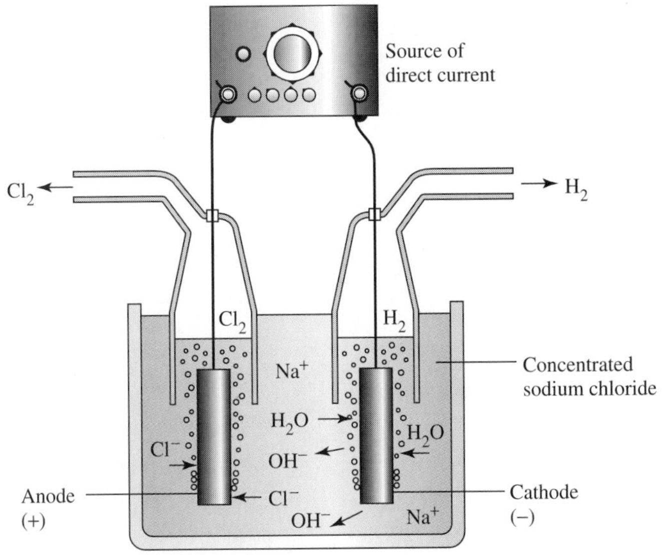

FIGURE 15.6 The electrolysis of aqueous sodium chloride solution (saltwater brine) produces hydrogen gas, chlorine gas, and sodium hydroxide solution.

electrolytic cell. An **electrolytic cell** *is an apparatus in which chemical change is caused to occur through the application of electrical energy.*

Electrolytic cells have a number of important commercial applications, including (1) production of important industrial chemicals, (2) electrorefining and purification of metals, and (3) electroplating. Let us consider examples in all three of these areas.

Chlorine gas, hydrogen gas, and sodium hydroxide—three important industrial chemicals—can be produced simultaneously from the electrolysis of a concentrated aqueous sodium chloride solution (saltwater or brine solution). The type of electrolytic cell needed is shown in Figure 15.6.

In this cell the two electrodes are inert; that is, they do not participate in the redox reactions themselves but serve as surfaces on which the redox reactions occur. As with galvanic cells, reduction occurs at the cathode and oxidation at the anode.

The negative ions in the solution, the Cl^- ions, react at the anode, where they give up electrons and are reduced to produce Cl_2 gas.

$$\text{Anode (oxidation):}\quad 2\,Cl^-(aq) \longrightarrow Cl_2(g) + 2\,e^-$$

The positive ions in the solution, the Na^+ ions, do not react at the cathode to produce Na metal atoms. Instead, water, the solvent in the solution, reacts at the cathode; it is more easily reduced than Na^+ ion. The cathode reaction is

$$\text{Cathode (reduction):}\quad 2\,H_2O(l) + 2\,e^- \longrightarrow H_2(g) + 2\,OH^-(aq)$$

This reduction yields H_2 gas as well as OH^- ions. The OH^- ions remain in solution, producing a solution that now contains Na^+ ions and OH^- ions; our solution of sodium chloride has been changed to one of sodium hydroxide. Thus, three substances—Cl_2 gas, H_2 gas, and NaOH solution—result from the electrolysis of a concentrated NaCl solution.

Metal purification frequently depends on electrolysis. All copper used in electrical wire is electrolytically purified; impurities decrease the electrical conductance of the wire. In an electrolytic cell used to purify copper, large slabs of impure copper (obtained from the reduction of copper ores) serve as anodes, and thin sheets of very pure copper serve as cathodes. The solution in which the electrodes are immersed is an acidic copper(II) sulfate solution. As the cell is operated, the anodes (the impure copper) decrease in size and the cathodes (the pure copper) increase in size (see Fig. 15.7). What is happening? At the anodes, oxidation of Cu causes it to dissolve. The reaction is

$$\text{Anode (oxidation):}\quad Cu(s) \longrightarrow Cu^{2+}(aq) + 2\,e^-$$

As the anodic copper dissolves, the impurities present also go into solution (some are also oxidized) or fall to the bottom of the cell. At the cathode, reduction causes copper to come out of solution.

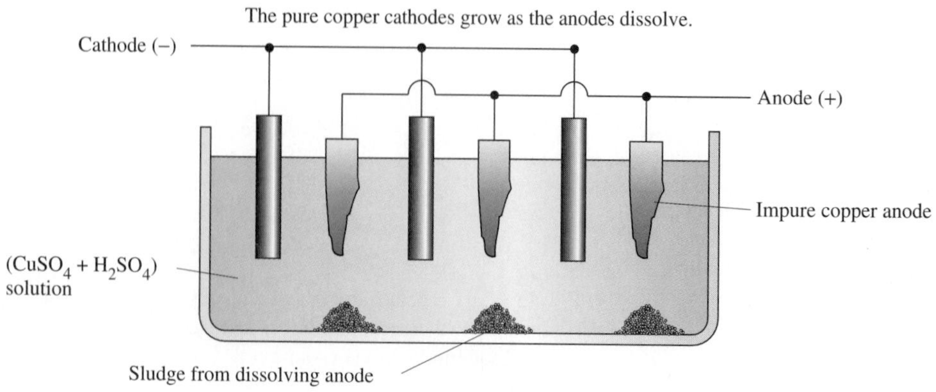

FIGURE 15.7 Cross section of an electrolytic cell for purifying copper.

$$\text{Cathode (reduction):} \quad Cu^{2+}(aq) + 2\,e^- \longrightarrow 2\,Cu(s)$$

The net effect of this oxidation–reduction process is that copper is transferred via solution from one electrode (the impure one) to the other (the pure one). The electrical voltage supplied to the cell is set at a value that allows copper ions to be reduced but is not sufficient to reduce any dissolved impurities. Thus, only copper, and not impurities, deposits on the cathode.

Electroplating *is the deposition of a thin layer of a metal on an object through the process of electrolysis*. Electroplated objects are common in our society. Jewelry is plated with silver and gold. Tableware is often plated with silver. Gold-plated electrical contacts are used extensively. "Tin cans" are actually steel cans with a thin coating of tin. Chromium-plated steel automobile bumpers have been used for many years. The thin metallic layer deposited during electroplating is generally only 0.001–0.002 in. thick.

Figure 15.8 shows a typical apparatus used for electroplating items with silver. The object to be plated is made the cathode in a solution containing Ag^+ ions. The anode is a bar of the plating metal, silver in this case. At the cathode, Ag^+ ions from solution are deposited as metallic silver.

$$\text{Cathode (reduction):} \quad Ag^+(aq) + e^- \longrightarrow Ag(s)$$

At the anode, silver from that electrode is oxidized to give Ag^+ ions in solution; this replenishes the supply of Ag^+ ions in solution needed for the plating process.

$$\text{Anode (oxidation):} \quad Ag(s) \longrightarrow Ag^+(aq) + e^-$$

The electroplating bath, the solution around the electrodes, usually contains other chemicals besides the plating metal. For example, silver plating is usually done from a solution containing both AgCN and KCN.

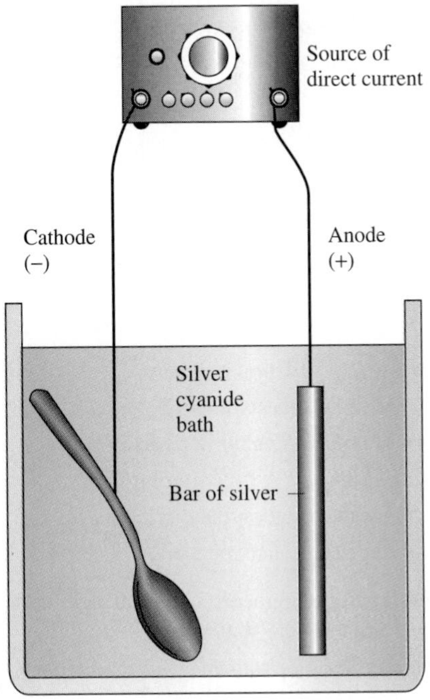

FIGURE 15.8 An apparatus for electroplating silver.

KEY TERMS

The new terms or concepts defined in this chapter are

anode (Sec. 15.8) The electrode in an electrochemical cell at which oxidation takes place.

cathode (Sec. 15.8) The electrode in an electrochemical cell at which reduction takes place.

disproportionation reaction (Sec. 15.7) A redox reaction in which some atoms of a single element in a reactant are oxidized and others are reduced.

electrolysis (Sec. 15.8) The process in which electrical energy is used to cause a nonspontaneous redox reaction to occur.

electrolytic cell (Sec. 15.8) An apparatus in which chemical change is caused to occur through the application of electrical energy.

electroplating (Sec. 15.8) The deposition of a thin layer of metal on an object through the process of electrolysis.

galvanic cell (Sec. 15.8) An apparatus in which a spontaneous redox reaction is used to convert chemical energy to electrical energy.

half-reaction (Sec. 15.6) A part of a redox equation that shows only the oxidation process or only the reduction process.

half-reaction method for balancing redox equations (Sec. 15.6) A method for balancing redox equations that employs two partial equations, an oxidation half-reaction and a reduction half-reaction.

oxidation (Sec. 15.1) Process whereby a substance in a chemical reaction loses one or more electrons.

oxidation number (Sec. 15.2) Charge that an atom appears to have when the electrons in each bond it is participating in are assigned to the more electronegative of the two atoms involved in the bond.

oxidation-number method for balancing redox equations (Sec. 15.5) A method for balancing redox equations in which oxidation numbers are assigned to each element present and oxidation-number increase is made to equal oxidation-number decrease.

oxidation–reduction reaction (Sec. 15.1) A reaction in which there is a transfer of electrons between reactants.

oxidizing agent (Secs. 15.1 and 15.2) The substance that accepts electrons in a redox reaction or the substance that contains the atom that shows a decrease in oxidation number in a redox reaction.

redox reaction (Sec. 15.1) A shortened designation for an oxidation–reduction reaction.

reducing agent (Secs. 15.1 and 15.2) The substance that provides electrons in a redox reaction or the substance that contains the atom that shows an increase in oxidation number in a redox reaction.

reduction (Sec. 15.1) The process whereby a substance in a chemical reaction gains one or more electrons.

PRACTICE PROBLEMS

Oxidation–Reduction Terminology (Secs. 15.1 and 15.2)

15.1 Give definitions of *oxidation* in terms of
- **(a)** loss or gain of electrons
- **(b)** increase or decrease in oxidation number

15.2 Give definitions of *reduction* in terms of
- **(a)** loss or gain of electrons
- **(b)** increase or decrease in oxidation number

15.3 Give definitions of *oxidizing agent* in terms of
- **(a)** loss or gain of electrons
- **(b)** increase or decrease in oxidation number
- **(c)** substance oxidized or substance reduced

15.4 Give definitions of *reducing agent* in terms of
- **(a)** loss or gain of electrons
- **(b)** increase or decrease in oxidation number
- **(c)** substance oxidized or substance reduced

15.5 In each of the following statements, choose the word in parentheses that best completes the statement.
- **(a)** An element that has lost electrons in a redox reaction is

said to have been (oxidized, reduced).

(b) Reduction always results in an (increase, decrease) in the oxidation number.

(c) The substance oxidized in a redox reaction is the (oxidizing, reducing) agent.

(d) The reducing agent (gains, loses) electrons during a redox reaction.

15.6 In each of the following statements, choose the word in parentheses that best completes the statement.

(a) The reducing agent causes an (increase, decrease) in the oxidation number of the oxidizing agent in a redox reaction.

(b) The oxidizing agent (gains, loses) electrons during a redox reaction.

(c) Oxidation always results in an (increase, decrease) in the oxidation number.

(d) An element that has gained electrons in a redox reaction is said to have been (oxidized, reduced).

Assignment of Oxidation Numbers (Sec. 15.2)

15.7 Assign oxidation numbers to the atoms in each of the following compounds.

(a) NH_3 **(b)** H_2SO_3 **(c)** HNO_2 **(d)** Na_3PO_4

15.8 Assign oxidation numbers to the atoms in each of the following compounds.

(a) NO_2 **(b)** H_3PO_4 **(c)** AlF_3 **(d)** $KClO_4$

15.9 Assign oxidation numbers to the atoms in each of the following ions.

(a) P^{3-} **(b)** Mg^{2+} **(c)** NH_2^- **(d)** PO_4^{3-}

15.10 Assign oxidation numbers to the atoms in each of the following ions.

(a) Se^{2-} **(b)** Al^{3+} **(c)** SO_4^{2-} **(d)** ClF_4^-

15.11 What is the oxidation number of carbon in each of the following carbon-containing compounds?

(a) H_2CO_3 **(b)** $H_2C_2O_4$
(c) C_2H_4 **(d)** C_3H_8

15.12 What is the oxidation number of carbon in each of the following carbon-containing compounds?

(a) $HC_2H_3O_2$ **(b)** H_2CO
(c) C_3H_6 **(d)** C_4H_{10}

15.13 What is the oxidation number of the metal present in each of the following compounds or ions?

(a) $Ni(NO_3)_2$ **(b)** $FeSO_4$
(c) $Rh_2(CO_3)_3$ **(d)** $Zn(CN)_4^{2-}$

15.14 What is the oxidation number of the metal present in each of the following compounds or ions?

(a) $Ni_2(SO_4)_3$ **(b)** $Fe(NO_3)_3$
(c) $Rh_3(PO_4)_2$ **(d)** $Ag(CN)_2^-$

15.15 Indicate whether oxygen has a −2, −1, or positive oxidation number in each of the following oxygen-containing species.

(a) Na_2O **(b)** OF_2 **(c)** Na_2O_2 **(d)** BaO

15.16 Indicate whether oxygen has a −2, −1, or positive oxidation number in each of the following oxygen-containing species.

(a) O_2F_2 **(b)** SO_3 **(c)** BaO_2 **(d)** BeO

15.17 Indicate whether hydrogen has a +1 or −1 oxidation number in each of the following hydrogen-containing species.

(a) NaH **(b)** CH_4 **(c)** HCl **(d)** CaH_2

15.18 Indicate whether hydrogen has a +1 or −1 oxidation number in each of the following hydrogen-containing species.

(a) H_2Se **(b)** N_2H_2 **(c)** KH **(d)** MgH_2

Characteristics of Oxidation–Reduction Reactions (Sec. 15.2)

15.19 Identify which substance is oxidized and which substance is reduced in each of the following redox reactions.

(a) $N_2 + 3\ H_2 \longrightarrow 2\ NH_3$

(b) $Cl_2 + 2\ KI \longrightarrow 2\ KCl + I_2$

(c) $Sb_2O_3 + 3\ Fe \longrightarrow 2\ Sb + 3\ FeO$

(d) $3\ H_2SO_3 + 2\ HNO_3 \longrightarrow 2\ NO + H_2O + 3\ H_2SO_4$

15.20 Identify which substance is oxidized and which substance is reduced in each of the following redox reactions.

(a) $2\ Al + 3\ Cl_2 \longrightarrow 2\ AlCl_3$

(b) $Zn + CuCl_2 \longrightarrow ZnCl_2 + Cu$

(c) $2\ NiS + 3\ O_2 \longrightarrow 2\ NiO + 2\ SO_2$

(d) $3\ H_2S + 2\ HNO_3 \longrightarrow 3\ S + 2\ NO + 4\ H_2O$

15.21 Identify which substance is the oxidizing agent and which substance is the reducing agent in each of the redox reactions in Problem 15.19.

15.22 Identify which substance is the oxidizing agent and which substance is the reducing agent in each of the redox reactions in Problem 15.20.

15.23 Identify the following species for the redox reaction

$$2\ HNO_3 + SO_2 \longrightarrow H_2SO_4 + 2\ NO_2$$

(a) substance that is oxidized

(b) oxidizing agent

(c) substance that contains the element that decreases in oxidation number

(d) substance that contains the element that loses electrons during the oxidation–reduction reaction

15.24 Identify the following species for the redox reaction

$$PH_3 + 2\ NO_2 \longrightarrow H_3PO_4 + N_2$$

(a) substance that is reduced

(b) reducing agent

(c) substance that contains the element that increases in oxidation number

(d) substance that contains the element that gains electrons during the oxidation–reduction process

Types of Chemical Reactions (Sec. 15.3)

15.25 Characterize each of the following reactions using one selection from the choices *redox* and *nonredox* combined with one selection from the choices *synthesis, decomposition, single replacement*, and *double replacement.*

(a) $H_2 + Cl_2 \longrightarrow 2\ HCl$

(b) $2\ HBr + Mg \longrightarrow MgBr_2 + H_2$

(c) $MgCO_3 \longrightarrow MgO + CO_2$

(d) $2\ KOH + H_2SO_4 \longrightarrow K_2SO_4 + 2\ H_2O$

15.26 Characterize each of the following reactions using one selection from the choices *redox* and *nonredox* combined with one selection from the choices *synthesis, decomposition, single replacement*, and *double replacement.*

(a) $Zn + Cu(NO_3)_2 \longrightarrow Zn(NO_3)_2 + Cu$

(b) $2\ SO_2 + O_2 \longrightarrow 2\ SO_3$

(c) $2\ CuO \longrightarrow 2\ Cu + O_2$

(d) $NaCl + AgNO_3 \longrightarrow AgCl + NaNO_3$

15.27 Characterize each of the following reactions as (1) a redox reaction, (2) a nonredox reaction, or (3) "can't classify" because of insufficient information.

(a) a synthesis reaction in which both reactants are elements

(b) a combustion reaction

(c) a decomposition reaction in which the products are all compounds

(d) a decomposition reaction in which an element and a compound are products

15.28 Characterize each of the following reactions as (1) a redox reaction, (2) a nonredox reaction, or (3) "can't classify" because of insufficient information.

(a) a synthesis reaction in which one reactant is an element and the other is a compound

(b) an acid–base neutralization reaction

(c) a decomposition reaction in which the products are all elements

(d) a single-replacement reaction involving an active metal and an acid

Balancing Redox Equations: Oxidation-Number Method (Sec. 15.5)

15.29 Balance the following equations by the oxidation-number method.

(a) $Cr + HCl \longrightarrow CrCl_3 + H_2$

(b) $Cr_2O_3 + C \longrightarrow Cr + CO_2$

(c) $SO_2 + NO_2 \longrightarrow SO_3 + NO$

(d) $BaSO_4 + C \longrightarrow BaS + CO$

15.30 Balance the following equations by the oxidation-number method.

(a) $Fe_2O_3 + CO \longrightarrow Fe + CO_2$

(b) $Al + MnO_2 \longrightarrow Al_2O_3 + Mn$

(c) $I_2O_5 + CO \longrightarrow I_2 + CO_2$

(d) $N_2H_4 + O_2 \longrightarrow N_2 + H_2O$

15.31 Balance the following equations by the oxidation-number method.

(a) $Br_2 + H_2O + SO_2 \longrightarrow HBr + H_2SO_4$

(b) $H_2S + HNO_3 \longrightarrow S + NO + H_2O$

(c) $SnSO_4 + FeSO_4 \longrightarrow Sn + Fe_2(SO_4)_3$

(d) $Na_2TeO_3 + NaI + HCl \longrightarrow NaCl + Te + H_2O + I_2$

15.32 Balance the following equations by the oxidation-number method.

(a) $HNO_3 + I_2 \longrightarrow NO_2 + H_2O + HIO_3$

(b) $As_4O_6 + Cl_2 + H_2O \longrightarrow H_3AsO_4 + HCl$

(c) $HI + HNO_3 \longrightarrow I_2 + NO + H_2O$

(d) $PbO_2 + Sb + NaOH \longrightarrow PbO + NaSbO_2 + H_2O$

15.33 Balance the following equations by the oxidation-number method. All reactions occur in acidic solution.

(a) $I_2 + Cl_2 \longrightarrow HIO_3 + Cl^-$

(b) $MnO_4^- + AsH_3 \longrightarrow H_3AsO_4 + Mn^{2+}$

(c) $Br^- + SO_4^{2-} \longrightarrow Br_2 + SO_2$

(d) $Au + Cl^- + NO_3^- \longrightarrow AuCl_4^- + NO_2$

15.34 Balance the following equations by the oxidation-number method. All reactions occur in acidic solution.

(a) $I^- + SO_4^{2-} \longrightarrow H_2S + I_2$

(b) $Mn^{2+} + BiO_3^- \longrightarrow MnO_4^- + Bi^{3+}$

(c) $Fe^{2+} + ClO_3^- \longrightarrow Fe^{3+} + Cl^-$

(d) $Pt + Cl^- + NO_3^- \longrightarrow PtCl_6^{2-} + NO_2$

15.35 Balance the following equations by the oxidation-number method. All reactions occur in basic solution.

(a) $S^{2-} + Cl_2 \longrightarrow SO_4^{2-} + Cl^-$

(b) $SO_3^{2-} + CrO_4^{2-} \longrightarrow Cr(OH)_4^- + SO_4^{2-}$

(c) $MnO_4^- + IO_3^- \longrightarrow MnO_2 + IO_4^-$

(d) $I_2 + Cl_2 \longrightarrow H_3IO_6^{2-} + Cl^-$

15.36 Balance the following equations by the oxidation-number method. All reactions occur in basic solution.

(a) $Zn + MnO_4^- \longrightarrow Zn(OH)_2 + MnO_2$

(b) $NO_2^- + Al \longrightarrow NH_3 + AlO_2^-$

(c) $NO_2^- + MnO_4^- \longrightarrow NO_3^- + MnO_2$

(d) $Al + NO_3^- \longrightarrow Al(OH)_4^- + NH_3$

Balancing Redox Equations: Half-Reaction Method (Sec. 15.6)

15.37 Balance the following half-reactions occurring in acidic solution.

(a) $MnO_2 \longrightarrow Mn^{3+}$

(b) $H_3MnO_4 \longrightarrow Mn$

(c) $MnO_4^- \longrightarrow Mn^{2+}$

(d) $MnO_4^- \longrightarrow MnO_2$

15.38 Balance the following half-reactions occurring in acidic solution.

(a) $V^{2+} \longrightarrow VO_2^+$
(b) $V^{3+} \longrightarrow VO^{2+}$
(c) $VO^{2+} \longrightarrow VO_2^+$
(d) $V \longrightarrow VO_2^+$

15.39 Balance the following half-reactions occurring in basic solution.

(a) $SeO_4^{2-} \longrightarrow Se$
(b) $Se^{2-} \longrightarrow SeO_3^{2-}$
(c) $SeO_4^{2-} \longrightarrow SeO_3^{2-}$
(d) $Se \longrightarrow SeO_3^{2-}$

15.40 Balance the following half-reactions occurring in basic solution.

(a) $H_3IO_6^{2-} \longrightarrow I_2$
(b) $IO_3^- \longrightarrow IO^-$
(c) $I^- \longrightarrow IO^-$
(d) $IO^- \longrightarrow H_3IO_6^{2-}$

15.41 Balance each of the following redox reactions by the half-reaction method. Each reaction occurs in acidic solution.

(a) $Zn + Cu^{2+} \longrightarrow Cu + Zn^{2+}$
(b) $Br_2 + I^- \longrightarrow Br^- + I_2$
(c) $S_2O_3^{2-} + Cl_2 \longrightarrow HSO_4^- + Cl^-$
(d) $Zn + As_2O_3 \longrightarrow AsH_3 + Zn^{2+}$

15.42 Balance each of the following redox reactions by the half-reaction method. Each reaction occurs in acidic solution.

(a) $Fe + Ag^+ \longrightarrow Fe^{3+} + Ag$
(b) $Cl_2 + Br^- \longrightarrow Cl^- + Br_2$
(c) $S_2O_3^{2-} + Cu^{2+} \longrightarrow S_4O_6^{2-} + Cu$
(d) $C_2O_4^{2-} + MnO_4^- \longrightarrow CO_2 + Mn^{2+}$

15.43 Balance each of the equations in Problem 15.33 using the half-reaction method for balancing.

15.44 Balance each of the equations in Problem 15.34 using the half-reaction method for balancing.

15.45 Balance each of the following redox reactions by the half-reaction method. Each reaction occurs in basic solution.

(a) $NH_3 + ClO^- \longrightarrow N_2H_4 + Cl^-$
(b) $Cr(OH)_2 + BrO^- \longrightarrow CrO_4^{2-} + Br^-$
(c) $CrO_2^- + H_2O_2 \longrightarrow CrO_4^{2-} + OH^-$
(d) $Bi(OH)_3 + Sn(OH)_3^- \longrightarrow Sn(OH)_6^{2-} + Bi$

15.46 Balance each of the following redox reactions by the half-reaction method. Each reaction occurs in basic solution.

(a) $Cr_2O_3 + ClO^- \longrightarrow CrO_4^{2-} + Cl^-$
(b) $NO + MnO_4^- \longrightarrow NO_3^- + MnO_2$
(c) $Al + PO_3^- \longrightarrow PH_3 + AlO_2^-$
(d) $Bi(OH)_3 + SnO_2^{2-} \longrightarrow SnO_3^{2-} + Bi$

15.47 Balance each of the equations in Problem 15.35 using the half-reaction method for balancing.

15.48 Balance each of the equations in Problem 15.36 using the half-reaction method for balancing.

Balancing Redox Equations: Disproportionation Reactions (Sec. 15.7)

15.49 Balance each of the following redox reactions by the oxidation-number method.

(a) $HNO_2 \longrightarrow NO + NO_3^-$ (acidic solution)
(b) $ClO^- + Cl^- \longrightarrow Cl_2$ (acidic solution)
(c) $S \longrightarrow S^{2-} + SO_3^{2-}$ (basic solution)
(d) $Br_2 \longrightarrow BrO_3^- + Br^-$ (basic solution)

15.50 Balance each of the following redox reactions by the oxidation-number method.

(a) $NO + NO_3^- \longrightarrow N_2O_4$ (acidic solution)
(b) $H_5IO_6 + I^- \longrightarrow I_2$ (acidic solution)
(c) $P_4 \longrightarrow HPO_3^{2-} + PH_3$ (basic solution)
(d) $HClO_2 \longrightarrow ClO_2 + Cl^-$ (basic solution)

15.51 Balance each of the redox reactions in Problem 15.49 using the half-reaction method.

15.52 Balance each of the redox reactions in Problem 15.50 using the half-reaction method.

Important Oxidation–Reduction Processes (Sec. 15.8)

15.53 The spontaneous redox reaction between lead metal and copper(II) nitrate solution occurs according to the following ionic equation.

$$Pb(s) + Cu^{2+}(aq) \longrightarrow Cu(s) + Pb^{2+}(aq)$$

Assume the reactants are separated into two compartments. A lead electrode is immersed in 1.00 M $Pb(NO_3)_2$ and a copper electrode in 1.00 M $Cu(NO_3)_2$. For this galvanic cell, indicate each of the following.

(a) equation for the oxidation half reaction
(b) equation for the reduction half reaction
(c) identity of anode and cathode
(d) direction of electron flow

15.54 The spontaneous redox reaction between nickel metal and cadmium nitrate solution occurs according to the following ionic equation.

$$Ni(s) + Cd^{2+}(aq) \longrightarrow Cd(s) + Ni^{2+}(aq)$$

Assume the reactants are separated into two compartments. A nickel electrode is immersed in 1.00 M $Ni(NO_3)_2$ and a cadmium electrode in 1.00 M $Cd(NO_3)_2$. For this galvanic cell, indicate each of the following.

(a) equation for the oxidation half reaction
(b) equation for the reduction half reaction
(c) identity of anode and cathode

(d) direction of electron flow

15.55 What are the anode and cathode reactions during the operation of an acidic dry cell?

15.56 What are the anode and cathode reactions during the operation of an alkaline dry cell?

15.57 What are the anode and cathode reactions during the discharging of a lead storage battery?

15.58 What are the anode and cathode reactions during the charging of a lead storage battery?

15.59 Why does the density of the H_2SO_4 in a lead storage battery decrease as the cell discharges?

15.60 Why is it not possible to recharge a lead storage battery an infinite number of times?

15.61 Write an equation for what happens at the anode during the electrolysis of a concentrated aqueous NaCl solution.

15.62 Write an equation for what happens at the cathode during the electrolysis of a concentrated aqueous NaCl solution.

ADDITIONAL PROBLEMS

15.63 Nitrogen forms a number of oxides including NO_2, N_2O_3, NO, N_2O, and N_2O_5. Arrange these oxides in order of increasing oxidation number of nitrogen.

15.64 Sulfur forms a number of oxides including S_2O, S_7O_2, SO_2, SO_3, and S_6O. Arrange these oxides in order of increasing oxidation number of sulfur.

15.65 In which of the following pairs of ionic compounds is the oxidation number of the metal the same in both members of the pair?

(a) $CuSO_4$ and Cu_2SO_4 **(b)** $Fe(NO_3)_3$ and $FePO_4$
(c) AuCl and $AgNO_3$ **(d)** $AlPO_4$ and GaN

15.66 In which of the following pairs of ionic compounds is the oxidation number of the metal the same in both members of the pair?

(a) CuO and $CuCl_2$ **(b)** Ni_2O_3 and NiN
(c) PbO_2 and $SnCl_4$ **(d)** Be_3N_2 and MgO

15.67 Classify each of the following pairs of balanced half-reactions as *two reduction half-reactions, two oxidation half-reactions*, or *one reduction and one oxidation half-reaction.*

(a) $Fe^{3+} + e^- \longrightarrow Fe^{2+}$ and $Fe^{2+} + 2\,e^- \longrightarrow Fe$
(b) $Ni^{3+} + e^- \longrightarrow Ni^{2+}$ and $Ni \longrightarrow Ni^{2+} + 2\,e^-$
(c) $Cu \longrightarrow Cu^+ + e^-$ and $Cu \longrightarrow Cu^{2+} + 2\,e^-$
(d) $Au \longrightarrow Au^{3+} + 3\,e^-$ and $Au^{3+} + 3\,e^- \longrightarrow Au$

15.68 Classify each of the following pairs of balanced half-reactions as *two reduction half-reactions, two oxidation half-reactions*, or *one reduction and one oxidation half-reaction.*

(a) $Sn^{2+} \longrightarrow Sn^{4+} + 2\,e^-$ and $Sn \longrightarrow Sn^{2+} + 2\,e^-$
(b) $Pb^{2+} \longrightarrow Pb^{4+} + 2\,e^-$ and $Pb^{2+} + 2\,e^- \longrightarrow Pb$
(c) $Co^{2+} + 2\,e^- \longrightarrow Co$ and $Co^{3+} + 3\,e^- \longrightarrow Co$
(d) $Ag \longrightarrow Ag^+ + e^-$ and $Ag^+ + e^- \longrightarrow Ag$

15.69 Write balanced equations for all possible redox reactions obtainable by combining the following balanced half-reactions in sets of two. The half-reactions must be used as written; they cannot be reversed in direction.

$$2\,H_2O + PH_3 \longrightarrow H_3PO_2 + 4\,H^+ + 4\,e^-$$
$$3\,H_2O + As \longrightarrow H_3AsO_3 + 3\,H^+ + 3\,e^-$$
$$MnO_4^- + 8\,H^+ + 5\,e^- \longrightarrow Mn^{2+} + 4\,H_2O$$
$$SO_4^{2-} + 4\,H^+ + 2\,e^- \longrightarrow SO_2 + 2\,H_2O$$

15.70 Write balanced equations for all possible redox reactions obtainable by combining the following balanced half-reactions in sets of two. The half-reactions must be used as written; they cannot be reversed in direction.

$$4\,OH^- + ClO_2^- \longrightarrow ClO_4^- + 2\,H_2O + 4\,e^-$$
$$2\,H_2O + MnO_4^- + 3\,e^- \longrightarrow MnO_2 + 4\,OH^-$$
$$6\,H_2O + NO_3^- + 8\,e^- \longrightarrow NH_3 + 9\,OH^-$$
$$4\,OH^- + Al \longrightarrow AlO_2^- + 2\,H_2O + 3\,e^-$$

15.71 Write the two balanced half-reactions associated with the following redox reaction.

$$4\,Zn + 10\,HNO_3 \longrightarrow 4\,Zn(NO_3)_2 + NH_4NO_3 + 3\,H_2O$$

15.72 Write the two balanced half-reactions associated with the following redox reaction.

$$14\,HNO_3 + 3\,Cu_2O \longrightarrow 6\,Cu(NO_3)_2 + 2\,NO + 7\,H_2O$$

CUMULATIVE PROBLEMS

15.73 Classify each of the following water-producing reactions as an acid–base reaction or as an oxidation–reduction reaction. If it is an acid–base reaction, identify the acid; if it is an oxidation–reduction reaction, identify the oxidizing agent.

(a) $2\ HNO_3 + 3\ H_2S \longrightarrow 3\ S + 2\ NO + 4\ H_2O$

(b) $2\ KOH + H_2S \longrightarrow K_2S + 2\ H_2O$

(c) $2\ HI + H_2O_2 \longrightarrow I_2 + 2\ H_2O$

(d) $H_2SO_4 + 2\ NaOH \longrightarrow Na_2SO_4 + 2\ H_2O$

15.74 Classify each of the following water-producing reactions as an acid–base reaction or as an oxidation–reduction reaction. If it is an acid–base reaction, identify the acid; if it is an oxidation–reduction reaction, identify the oxidizing agent.

(a) $7\ HI + H_5IO_6 \longrightarrow 4\ I_2 + 6\ H_2O$

(b) $H_2CO_3 + 2\ KOH \longrightarrow K_2CO_3 + 2\ H_2O$

(c) $5\ HClO_2 + NaOH \longrightarrow 4\ ClO_2 + 3\ H_2O + NaCl$

(d) $14\ HNO_3 + 3\ Cu_2O \longrightarrow 6\ Cu(NO_3)_2 + 2\ NO + 7H_2O$

15.75 Convert each of the following balanced molecular redox equations to balanced net ionic redox equations.

(a) $SnSO_4(aq) + 2\ FeSO_4(aq) \longrightarrow Sn(s) + Fe_2(SO_4)_3(aq)$

(b) $PH_3(g) + 2\ NO_2(g) \longrightarrow H_3PO_4(aq) + N_2(g)$

(c) $S(s) + 3\ H_2O\ (l) + 2\ Pb(NO_3)_2 \longrightarrow 2\ Pb(s) + H_2SO_3(aq) + 4\ HNO_3(aq)$

(d) $4\ Zn(s) + 10\ HNO_3(aq) \longrightarrow 4\ Zn(NO_3)_2(aq) + NH_4NO_3(aq) + 3\ H_2O(l)$

15.76 Convert each of the following balanced molecular redox equations to balanced net ionic redox equations.

(a) $NaNO_3(aq) + Pb(s) \longrightarrow NaNO_2(aq) + PbO(s)$

(b) $3\ H_2S(aq) + 2\ HNO_3(aq) \longrightarrow 3\ S(s) + 2\ NO(g) + 4\ H_2O(l)$

(c) $2\ HNO_3(aq) + SO_2(g) \longrightarrow H_2SO_4(aq) + 2\ NO_2(g)$

(d) $10\ FeSO_4(aq) + 2\ KMnO_4(aq) + 8\ H_2SO_4(aq) \longrightarrow 5\ Fe_2(SO_4)_3(aq) + 2\ MnSO_4(aq) + K_2SO_4(aq) + 8\ H_2O(l)$

15.77 Balance each of the following ionic redox reactions.

(a) hydrosulfuric acid plus dichromate ion produces chromium(III) ion plus sulfur (acidic solution)

(b) chlorate ion plus iodine produces iodate ion plus chloride ion (acidic solution)

(c) sulfide ion plus bromine produces sulfate ion plus bromide ion (basic solution)

(d) nitrogen dioxide disproportionates to produce nitrate ion plus nitrite ion (basic solution)

15.78 Balance each of the following ionic redox reactions.

(a) iron(II) ion plus permanganate ion produces iron(III) ion plus manganese(II) ion (acidic solution)

(b) iodine plus sulfur dioxide produces iodide ion plus sulfate ion (acidic solution)

(c) manganese(II) hydroxide plus nickel(IV) oxide produces manganese(III) oxide plus nickel(II) hydroxide (basic solution)

(d) chlorine disproportionates to produce chlorate ion and chloride ion (basic solution)

15.79 The amount of $I_3^-(aq)$ in a solution can be determined by reacting it with a solution containing $S_2O_3^{2-}(aq)$

$$I_3^-(aq) + 2\ S_2O_3^{2-}(aq) \longrightarrow 3\ I^-(aq) + S_4O_6^{2-}(aq)$$

Calculate the molarity of I_3^- in a solution given that 43.2 mL of 0.300 M $S_2O_3^{2-}$ solution reacts with a 20.0-mL sample of the I_3^- solution.

15.80 Oxalic acid, $H_2C_2O_4$, reacts with dichromate ion, $Cr_2O_7^{2-}$, in acidic solution as follows.

$$3\ H_2C_2O_4(aq) + Cr_2O_7^{2-}(aq) + 8\ H^+(aq) \longrightarrow 6\ CO_2(g) + 2\ Cr^{3+}(aq) + 7\ H_2O(l)$$

If 35.0 mL of an oxalic acid solution reacts completely with 25.0 mL of a 0.0500-M $Cr_2O_7^{2-}$ solution, what is the molarity of the oxalic acid solution?

15.81 The amount of ozone, O_3, in polluted air can be determined in a two-step process. First the ozone is reacted with an acidic solution containing iodide ion.

$$O_3(g) + 2\ I^-(aq) + 2\ H^+(aq) \longrightarrow O_2(g) + I_2(s) + H_2O(l)$$

The iodine so produced is then reacted with thiosulfate, $S_2O_3^{2-}$, solution

$$2\ S_2O_3^{2-}(aq) + I_2(s) \longrightarrow S_4O_6^{2-}(aq) + 2\ I^-(aq)$$

If 18.03 mL of a 0.00200-M $S_2O_3^{2-}$ solution completely reacts with the I_2 produced by a 28.09-g sample of polluted air, calculate the O_3 concentration, in ppm (m/m), in the sample of air.

15.82 The active ingredient in household bleach is the hypochlorite ion, ClO^-. The concentration of this ion in bleach can be determined using a two-step process. First, the bleach is reacted with an $I^-(aq)$ solution.

$$ClO^-(aq) + 2\,I^-(aq) + 2\,H^+(aq) \longrightarrow I_2(s) + Cl^-(aq) + H_2O(l)$$

The iodine so produced is then reacted with thiosulfate, $S_2O_3^{2-}$, solution.

$$I_2(s) + 2\,S_2O_3^{2-}(aq) \longrightarrow 2\,I^-(aq) + S_4O_6^{2-}(aq)$$

A 50.00-g sample of a certain household bleach is found to react completely with 42.5 mL of a 0.0150 M $S_2O_3^{2-}$ solution. Calculate the concentration, in mass percent, of ClO^- ion in the bleach.

16 Reaction Rates and Chemical Equilibrium

16.1 Theory of Reaction Rates

In Chapter 10 we learned how to write (and balance) chemical equations to represent chemical reactions and then use these equations to calculate amounts of products produced and reactants consumed in such reactions. We now concern ourselves with another important topic relative to chemical reactions: the rate at which they occur.

Formally defined, the **rate of a chemical reaction** *is the rate or speed at which reactants are consumed or products are produced.* A number of variables affect the rate of a reaction, some of which we encounter routinely. One variable is temperature. Food is stored in a refrigerator to reduce the rate at which spoiling (a chemical reaction) occurs. State of subdivision of solids is another reaction rate variable. Sawdust and kindling wood burn (a chemical reaction) much faster than do large logs. To understand why these variables and others affect reaction rates we will need first to examine the conditions that are necessary for a reaction to take place.

Collison theory *is a set of statements that specifies the conditions necessary for a reaction to occur.* Developed from the study of reaction-rate information for many different reactions, collision theory contains three fundamental statements.

1. Reactant particles must collide with each other in order for a reaction to occur.
2. Colliding particles must collide with a certain minimum total amount of energy if the collision is to result in a reaction.
3. In some cases reactants must be oriented in a specific way upon collision if a reaction is to occur.

Let us consider each of these three statements separately.

Molecular Collisions

When reactions involve two or more reactants, collision theory assumes that the reactant molecules, ions, or atoms must come into contact (collide) with each other in order for a reaction to occur (statement 1). The validity of this assumption is fairly obvious. Reactants cannot react with each other if they are miles apart.

Most reactions are carried out in liquid solution or in the gaseous phase. The reason for this is simple. In these situations reacting particles are more free to move about, and thus it is easier for the reactants to come into contact with each other. Reactions of solids usually take place only on the solid surface and therefore include only a small fraction of the total particles present in the solid. As the reaction proceeds and products dissolve, diffuse, or fall from the surface, fresh solid is exposed. In this way, the reaction eventually can consume all of the solid. The rusting of iron is an example of this type of process.

Activation Energy

Not all collisions between reactant particles result in the formation of reaction products. Sometimes reactant particles rebound from a collision unchanged. Statement 2 of collision theory indicates that for a reaction to occur, colliding particles must impact with a certain minimum energy; that is, the sum of the kinetic energies of the colliding particles must add to a certain minimum value. **Activation energy** *is the minimum combined kinetic energy reactant particles must possess in order for their collision to result in a reaction.* Each chemical reaction has a different activation energy.

In a slow reaction, the activation energy is far above the average energy content of the reacting particles. Only a few particles, those with well above average energy, will undergo collisions that result in a reaction—hence the slowness of the reaction.

It is *sometimes* possible to start a reaction by providing activation energy and then have it continue on its own. Once the reaction is started, enough energy is released to activate other molecules and keep the reaction going. The striking of a kitchen match is an example of such a situation. Activation energy is initially provided by rubbing the match head against a rough surface; heat is generated by friction. Once the reaction is started, the match continues to burn.

Collision Orientation

Even when activation energy requirements are met, some collisions between reactant particles still do not result in product formation. How can this be? Statement 3 of collision theory, which deals with the orientation of colliding particles at the moment of collision, relates to this situation. For nonspherical molecules and polyatomic ions, their orientation relative to each other at the moment of collision is a factor in determining whether a collision is effective.

Consider the following hypothetical reaction with the diatomic molecules AB and CD as reactants.

$$AB + CD \longrightarrow AC + BD$$

In this reaction B and C exchange places. The most favorable orientation during reactant molecule collisions would be one that simultaneously puts A and C in close proximity to each other (to form the molecule AC) and B and D near each other (to form the molecule BD). A possible orientation in which this is the situation is shown in Figure 16.1a. The possibility of a reaction resulting from this orientation is much greater than if the molecules were to collide while oriented as shown in Figure 16.1b or c. In (b), A is not near C, nor is

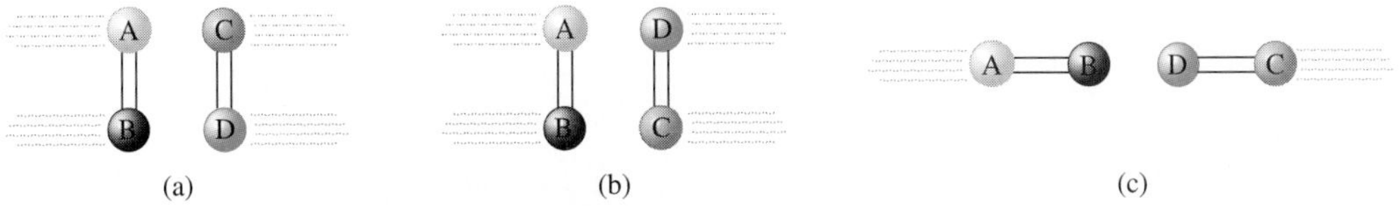

FIGURE 16.1 Different collision orientations for the reacting molecules AB and CD.

B near D. In (c), B is near D, but A and C are far removed from each other. Thus, certain collision orientations are preferred over others. The undesirable collision orientations of Figures 16.1b and c, however, could still result in a reaction if the molecules collided with abnormally high energies.

16.2 POTENTIAL ENERGY DIAGRAMS FOR CHEMICAL REACTIONS

A **potential energy diagram** *shows graphically the relationship between the activation energy of a chemical reaction and the total potential energy (Sec. 11.3) of the reactants and products*. Two such diagrams, one representing an exothermic chemical reaction (Sec. 11.8) and the other an endothermic chemical reaction (Sec. 11.8), are shown in Figure 16.2. Note that in each of these diagrams the "hill" or "hump" corresponds to the activation energy. It is the barrier that must be overcome in order for the reaction to proceed. Note also that activation energy is needed regardless of whether a given reaction is exothermic or endothermic.

Whether a reaction is exothermic or endothermic is determined by how the total potential energy of the products of a chemical reaction compares with the total potential energy of the reactants. As shown in Figure 16.2a, an exothermic reaction is one in which the prod-

FIGURE 16.2 Potential energy diagrams for (a) an exothermic chemical reaction and (b) an endothermic reaction.

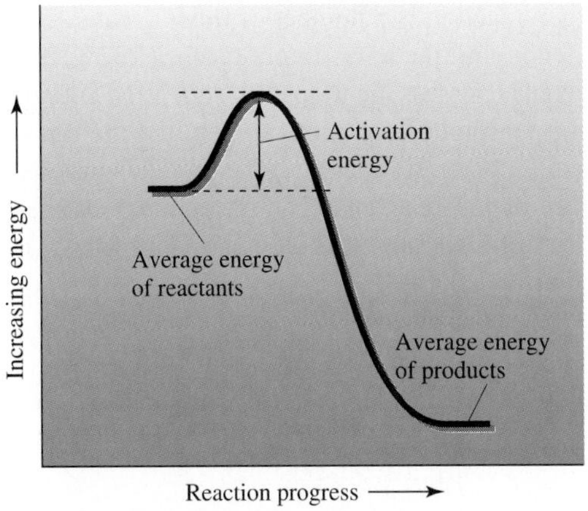

(a) Exothermic reaction. The average energy of the reactants is *higher* than that of the products, indicating energy has been *released* in the reaction.

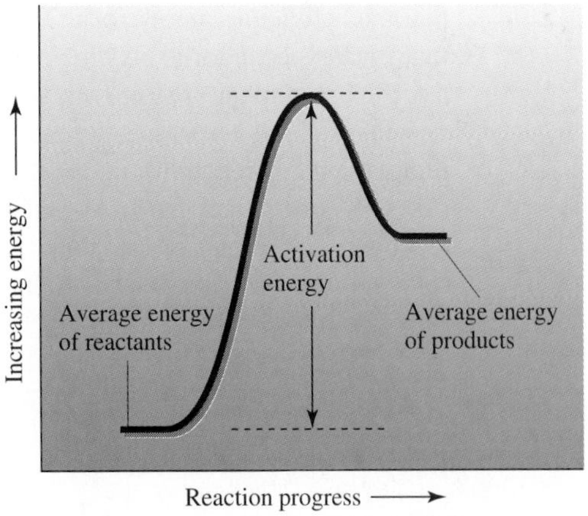

(b) Endothermic reaction. The average energy of the reactants is *less* than that of the products, indicating energy has been *absorbed* in the reaction.

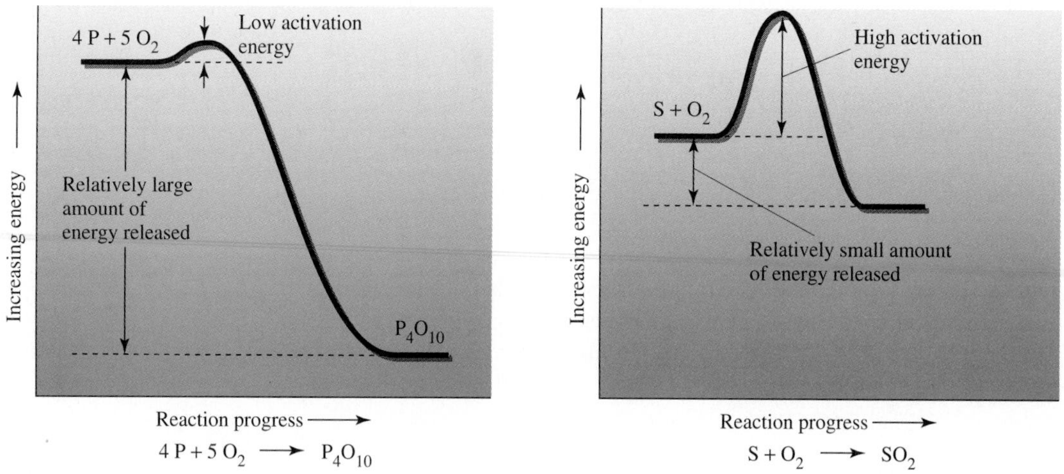

FIGURE 16.3 Potential energy diagrams for the reactions of the nonmetals phosphorus and sulfur with oxygen.

ucts are at a lower potential energy than the reactants; energy has been lost (released) in such a reaction. Conversely, in an endothermic reaction the products are at a higher potential energy than the reactants; thus, an input (absorption) of energy into the system is required.

Further insights about the energy relationships between reactants and products in a chemical reaction can be obtained from a study of Figure 16.3, which gives potential energy diagrams for two specific reactions: the reaction of phosphorus with oxygen and the reaction of sulfur with oxygen.

White phosphorus, a form of elemental phosphorus, spontaneously reacts with oxygen (bursts into flame) at a temperature of 34°C. Sulfur will also ignite spontaneously in the presence of oxygen, but not until it is heated to a temperature of 232°C. The difference in behavior is readily explained in terms of activation energies. Looking at Figure 16.3 we see that the activation energy for the sulfur–oxygen reaction is a number of times greater than that for the phosphorus–oxygen reaction. It is not until the temperature reaches 232°C that a sufficient number of sulfur and oxygen molecules possess the necessary activation energy.

Figure 16.3 also gives us the information that both the sulfur–oxygen and the phosphorus–oxygen reactions are exothermic; the total potential energy of the products, in both cases, is lower than the total potential energy of the reactants. The phosphorus–oxygen reaction is more exothermic than the sulfur–oxygen reaction; that is, much more energy is released in the former than in the latter reaction.

16.3 FACTORS THAT INFLUENCE REACTION RATES

Reaction rates are influenced by a number of factors. Four that affect the rate of all reactions are: (1) the physical nature of the reactants, (2) reactant concentrations, (3) reactant temperature, and (4) the presence of catalysts.

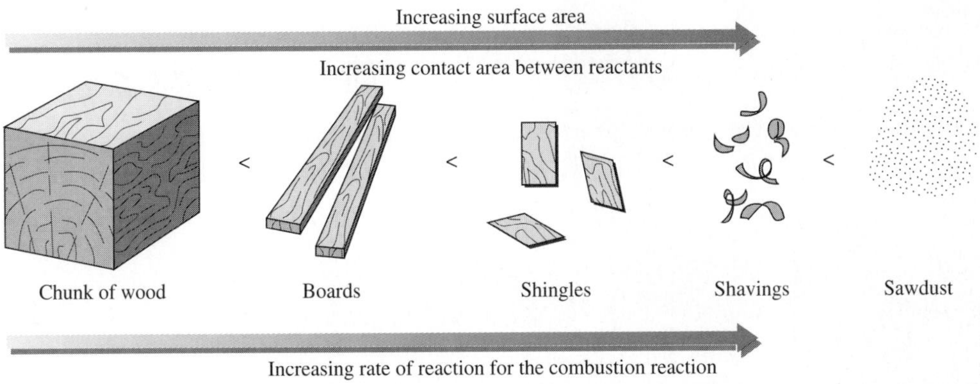

FIGURE 16.4 Greater surface area increases the reaction rate.

Physical Nature of Reactants

The physical nature of reactants refers not only to the physical state of each reactant (solid, liquid, or gas) but also to the state of subdivision, that is, particle size. In reactions where the reactants are all in the same physical state, reaction rate is generally faster between liquid reactants than between solid reactants and faster still between gaseous reactants. Of the three states of matter the gaseous state is the one where there is the most freedom of movement; hence in this state there is a greater frequency of collision (reaction) between reactants.

In reactions involving solids and heterogeneous liquid mixtures, reaction occurs at the boundary surface between reactants. The greater the amount of boundary surface area, the greater the reaction rate. Subdividing a solid into smaller particles will increase surface area and thus increase reaction rate. For example, large pieces of wood are difficult to ignite, smaller pieces burn more rapidly, and wood shavings ignite instantaneously (see Fig. 16.4).

When particle size becomes extremely small, reaction rates can be so fast that an explosion results. A lump of coal is difficult to ignite; coal dust ignites explosively. The spontaneous ignition of coal dust is a real threat to underground coal-mining operations. Grain dust (very finely divided grain particles) is a problem in grain-storage elevators; explosive ignition of the dust from an accidential spark is always a possibility. Figure 16.5 shows the destruction that can result from the accidental ignition of grain dust in a storage elevator.

Reactant Concentration

An increase in the concentration of a reactant causes an increase in the rate of the reaction. Combustible substances burn much more rapidly in pure oxygen than they do in air (21% oxygen). Increasing the concentration of a reactant means that there are more molecules of that reactant present in the reaction mixture and therefore there is a greater possibility for collisions between this reactant and other reactant particles. An analogy to the reaction-rate–reactant-concentration relationship can be drawn from the game of billiards. The more billiard balls there are on the table, the greater the probability of a moving cue ball striking one of them.

The actual quantitative change in reaction rate as the concentration of reactants is increased varies with the specific reaction. The rate always increases, but not to the same extent in all cases. Simply looking at the balanced equation for a reaction will not enable you to determine how changes in concentration will affect the reaction rate. This must be determined by actual experimentation. In some reactions the rate doubles with a doubling of concentration; however, this is not always the case.

FIGURE 16.5 Extremely rapid combustion of grain dust produced the explosive effect that destroyed this grain elevator. (AP/Wide World Photos)

Reaction Temperature

The effect of temperature on reaction rates can also be explained by using the molecular-collision concept. An increase in the temperature of a system results in an increase in the average kinetic energy of the reacting molecules. The increased molecular speed causes more collisions to take place in a given time. Also, since the average kinetic energy of the colliding molecules is greater, a larger fraction of the collisions will have sufficient kinetic energy to equal or exceed the activation energy.

As a rough rule of thumb it has been found that for many common reactions the rate of a chemical reaction doubles for every 10°C increase in temperature in the temperature range we normally encounter. The chemical reaction of cooking takes place faster in a pressure cooker because of a higher cooking temperature (Sec. 11.14). On the other hand, foods are cooled or frozen to slow down the chemical reactions that result in the spoiling of food, the souring of milk, and the ripening of fruit.

Presence of Catalysts

A **catalyst** *is a substance that, when added to a reaction mixture, increases the rate of the reaction but is itself unchanged after the reaction is completed.* Catalysts can be classified into two categories: (1) homogeneous catalysts and (2) heterogeneous catalysts. **Homogeneous catalysts** *exist in the same phase as the reactants.* They are usually dispersed uniformly throughout the reaction mixture. **Heterogeneous catalysts** *exist as a separate phase from the reactants.* Such catalysts are usually solids.

Catalysts increase reaction rates by providing alternative reaction pathways with lower activation energies than the original uncatalyzed pathway. This lowering of activation energy effect is illustrated in Figure 16.6.

In homogeneous catalysis, it is thought, the alternative pathway involves the formation of an intermediate "complex" that contains the catalyst. This catalyst-containing intermediate then breaks up to give the final products and regenerate the catalyst. The following equations illustrate this concept.

Uncatalyzed reaction: $A + B \longrightarrow A\text{—}B$

Catalyzed reaction: $A + B + \text{catalyst} \longrightarrow \underset{\text{catalyst}}{A\text{—}B} \longrightarrow A\text{—}B + \text{catalyst}$

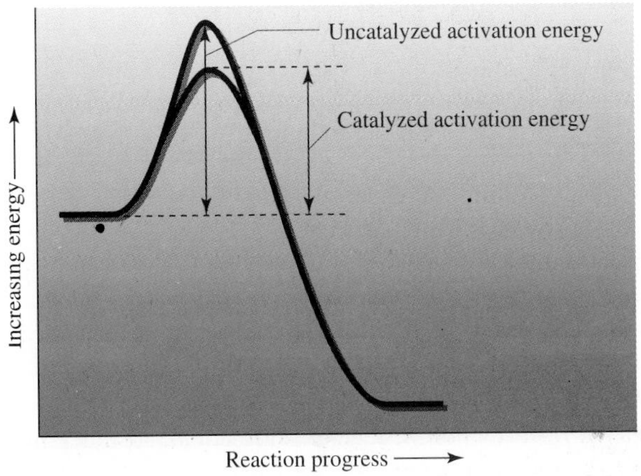

FIGURE 16.6 The effect of a catalyst on activation energy.

Catalysts that are solids are thought to provide a surface to which impacting reactant molecules are physically attracted and on which they are held with a particular orientation. Reactants so held are sufficiently close and favorably oriented toward each other to allow the reaction to take place. The products of the reaction then leave the surface and make it available to catalyze other reactants.

Catalysts are used extensively in the chemical industry. Usually, very specific catalysts are used that accelerate one chemical reaction without influencing other possible competitive reactions. The small amounts of catalysts required, coupled with the fact that they are not used up, make the use of catalysts economically desirable in industrial processes. Often a catalyst is what makes a particular process economically feasible. For example, a catalyst often makes it possible to avoid the high temperatures (costly) that would otherwise be necessary to cause a reaction with high activation energy to proceed.

Catalysts are a key element in the functioning of automobile-emission control systems. In such systems, heterogeneous catalysts speed up reactions that convert air pollutants in the exhaust to less harmful products (see Fig. 16.7). For example, carbon monoxide is converted to carbon dioxide through reaction with the oxygen in air.

Catalysts are of extreme importance for the proper functioning of the human body and other biological systems. In the human body, catalysts called *enzymes*, which are proteins, cause many reactions to take place rapidly at body temperature and under mild conditions. These same reactions, uncatalyzed, proceed very slowly and then only under harsher conditions.

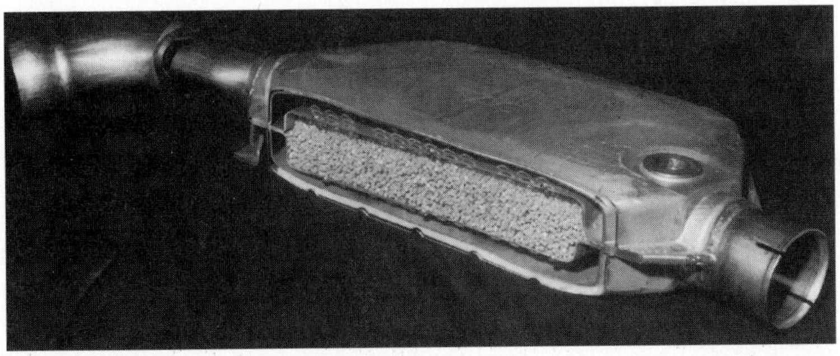

FIGURE 16.7 A cross-sectional view of a catalytic converter used in automobiles. The beaded materials are the catalyst. They contain the metals platinum, palladium, and rhodium. Air is injected into the chamber between the catalytic beads. (General Motors Corporation)

16.4 CHEMICAL EQUILIBRIUM

In our discussions of chemical reactions, up to this point, we have assumed that chemical reactions go to completion, that is, that reactions continue until one or more of the reactants is used up. Strictly speaking, this is not true. Experiments show that in most chemical reactions the complete conversion of reactants to products does not occur regardless of the time allowed for the reactions to take place. The reason for this is that product molecules (provided they are not allowed to escape from the reaction mixture) begin to react with each other to form again the reactants. With time, a steady-state situation results where the rate of formation of products and the rate of re-formation of reactants are equal. At this point the concentrations of all reactants and all products remain constant; the reaction has reached a stage of chemical equilibrium. **Chemical equilibrium** *is a condition in which two opposing chemical reactions occur simultaneously at the same rate*. We have discussed equilibrium situations in previous chapters—see Sections 11.13 (vapor pressure), 13.2 (saturated solutions), and 14.3 (conjugate acids and bases). The first two of these previous equilibrium situations involved physical equilibrium (no chemical reaction) rather than chemical equilibrium. Conjugate acid–base relationships involve chemical equilibrium, a topic we now consider in detail.

The conditions that exist in a system in a state of chemical equilibrium can best be visualized by considering an actual chemical reaction. Suppose equal molar amounts of gaseous H_2 and I_2 are mixed together in a closed container and allowed to react.

$$H_2 + I_2 \longrightarrow 2\,HI$$

Initially, no HI is present, so the only possible reaction that can occur is the one between H_2 and I_2. However, with time, as the HI concentration increases, some HI molecules collide with each other in a way that causes the reverse reaction to occur.

$$2\,HI \longrightarrow H_2 + I_2$$

The initial low concentration of HI makes this reverse action slow at first, but as the concentration of HI increases, so does the reaction rate. At the same time the reverse reaction rate is increasing, the forward reaction rate (production of HI) is decreasing as reactants are used up. Eventually, the concentrations of H_2, I_2, and HI in the reaction mixture reach a level at which the rates of the forward and reverse reactions become equal. At this point a state of chemical equilibrium has been reached.

Figure 16.8 shows graphically the behavior of reaction rates and reaction concentrations with time for both the forward and reverse reactions in the H_2–I_2–HI system. Figure 16.8a shows that the forward and reverse reaction rates become equal as a result of the forward reaction rate decreasing (as reactants are used up) and the reverse reaction rate increasing (as product concentration increases). Figure 16.8b shows the important point that reactant and product concentrations are usually not equal at the point at which equilibrium is reached. Rates are equal, but concentrations are not. For the H_2–I_2–HI system, much more product HI is present than reactants H_2 and I_2 at equilibrium. In Figure 16.8b note that the point at which equilibrium is established is the point where the two curves become straight lines.

The equilibrium involving H_2, I_2, and HI could have been established just as easily by starting with pure HI and allowing it to change into H_2 and I_2 (the reverse reaction). The

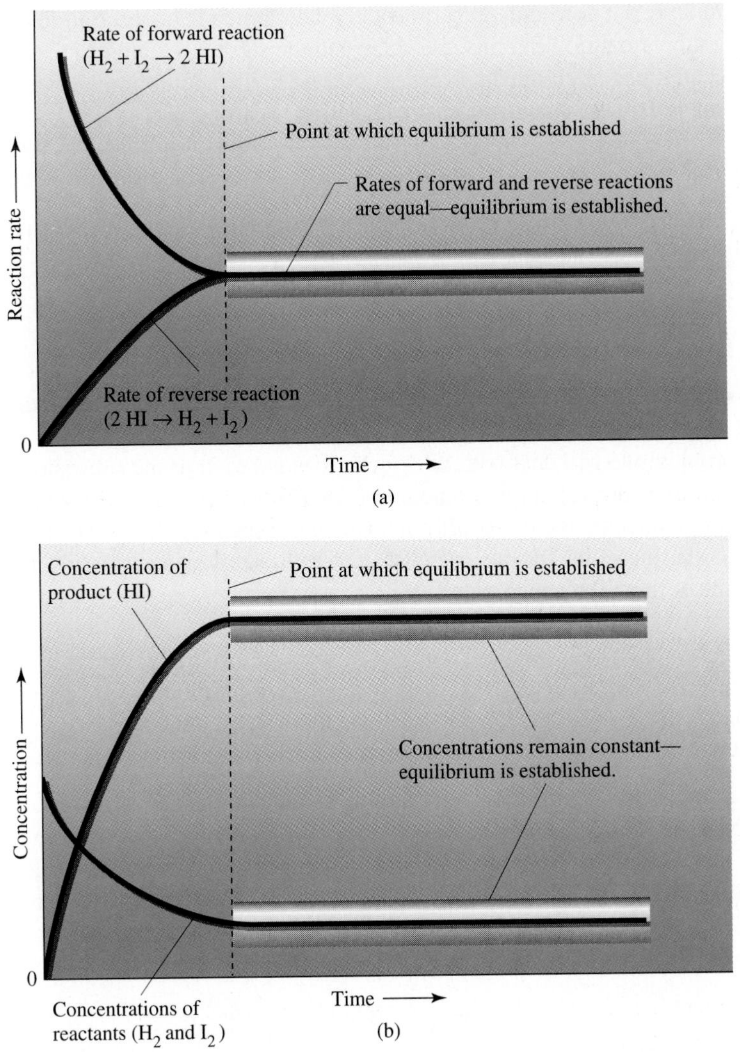

FIGURE 16.8 Variation of (a) reaction rates and (b) reactant concentrations with time in the chemical system H_2–I_2–HI.

final position of equilibrium does not depend on the direction from which equilibrium is approached.

Instead of separate equations for the forward and reverse reactions for a system at equilibrium, it is normal procedure to represent the equilibrium by using a single equation and double arrows. Thus, the reactions between H_2 and I_2 and between 2 HI, at equilibrium, are written as

$$H_2 + I_2 \rightleftharpoons 2\ HI$$

16.5 EQUILIBRIUM MIXTURE STOICHIOMETRY

Suppose that known amounts of reactants are placed in a reaction vessel and the system is allowed to reach chemical equilibrium (Sec. 16.4). To determine the composition of the resulting equilibrium mixture we need only experimentally determine the concentration of one of the substances in the mixture. With this one value and the substance amounts originally present, the concentrations of all other substances present in the equilibrium mixture can be calculated. Example 16.1 shows how such a calculation is carried out.

Example 16.1

0.0930 mole of NO and 0.0652 mole of Br_2 are placed in a container and allowed to react until equilibrium is established.

$$2\ NO(g) + Br_2(g) \rightleftharpoons 2\ NOBr(g)$$

At equilibrium 0.0612 mole of NOBr is present. What is the composition of the equilibrium mixture in terms of moles of each substance present?

Solution

In solving this problem we will deal with three quantities for each of the substances involved in the equilibrium: (1) starting amount of each substance, (2) amount that changes (undergoes reaction), and (3) equilibrium amount of each substance. The following table, the starting point for our calculation, summarizes the known (given) quantities in terms of these three parameters.

	2 NO(g)	+	$Br_2(g)$	$\rightleftharpoons$	2 NOBr(g)
Start	0.0930 mole		0.0652 mole		0 mole
Change	—		—		—
Equilibrium	—		—		0.0612 mole

Four of the nine "blanks" in the table have numbers in them. The key observation is that two of the three "blanks" for NOBr are known. In a problem of this type, any time two of the three key items (start, change, and equilibrium) are known for a substance, the third can be calculated by addition or subtraction. For NOBr, we started with zero amount and ended up with 0.0612 mole at equilibrium. Obviously, the amount of change for NOBr is +0.0612 mole, the amount of NOBr formed.

	2 NO(g)	+	$Br_2(g)$	$\rightleftharpoons$	2 NOBr(g)
Start	0.0930 mole		0.0652 mole		0 mole
Change	—		—		+0.0612 mole
Equilibrium	—		—		0.0612 mole

Once one of the change values is known, all other change quantities can quickly be calculated. The molar-change values are related to each other in the same manner as the coefficients in the equation are related to each other. Thus, we know that

(1) The molar amount of NO that reacts is the same as the molar amount of NOBr produced, since these two substances have the same coefficients in the equation, and
(2) The molar amount of Br_2 that changes (reacts) is one-half the molar amount of NOBr produced since the Br_2/NOBr coefficient ratio is 1 to 2.

Placing this information into the "table" gives

	2 NO(g)	+	$Br_2(g)$	$\rightleftharpoons$	2 NOBr(g)
Start	0.0930 mole		0.0652 mole		0 mole
Change	−0.0612 mole		−0.0306 mole		+0.0612 mole
Equilibrium	—		—		0.0612 mole

Note the minus signs placed in front of the NO and Br_2 change amounts. This is because these amounts are consumed (used up in the reaction). The plus sign in front of the NOBr change amount indicates a gain in the amount of this substance.

The last two blanks in the table are now easily determined through subtraction. For NO, 0.0930 mole (start) − 0.0612 mole (change) = 0.0318 mole (equilibrium). Similarly, for Br_2 we have 0.0652 mole − 0.0306 mole = 0.0346 mole. Our completed table is

	$2\ NO(g)$ +	$Br_2(g)$ ⇌	$2\ NOBr(g)$
Start	0.0930 mole	0.0652 mole	0 mole
Change	−0.0612 mole	−0.0306 mole	+0.0612 mole
Equilibrium	0.0318 mole	0.0346 mole	0.0612 mole

The equilibrium mixture composition, which is the bottom line of the table, is

0.0318 mole NO; 0.0346 mole Br_2; 0.0612 mole NOBr

Practice Exercise 16.1

Sulfur dioxide and oxygen react according to the following equation.

$$2\,SO_2(g) + O_2(g) \rightleftharpoons 2\,SO_3(g)$$

When 4.00 moles of SO_2 and 2.00 moles of O_2 are placed in an appropriate container and allowed to react until equilibrium is established, it is found that 2.96 moles of SO_3 have been formed. What is the composition of the equilibrium mixture in terms of moles of each substance present?

Ans. 1.04 moles SO_2; 0.52 mole O_2; 2.96 moles SO_3

16.6 EQUILIBRIUM CONSTANTS

The concentrations of reactants and products, at the time chemical equilibrium is reached in a chemical reaction, can be experimentally determined through analysis of samples of the equilibrium mixture. By using these equilibrium concentrations and the balanced chemical equation for the reaction, an *equilibrium constant* can be calculated for the reaction. This equilibrium constant, which relates the concentrations of reactants and products, gives information about the extent to which the reaction has occured at the point that equilibrium is reached.

To illustrate the calculation of an equilibrium constant, let us consider a general gas phase reaction in which a moles of A and b moles of B react to produce c moles of C and d moles of D.

$$aA(g) + bB(g) \rightleftharpoons cC(g) + dD(g)$$

The equilibrium constant for this reaction is

$$K_{eq} = \frac{[C]^c[D]^d}{[A]^a[B]^b}$$

Formally defined, an **equilibrium constant** *is the product of the molar concentrations of* *the products for a chemical reaction, each raised to the power of its respective coefficient in the equation, divided by the product of the molar concentrations of the reactants, each raised to the power of its respective coefficient in the equation.*

Note that this definition of an equilibrium constant implies the following about equilibrium constants.

1. Concentrations are always expressed in moles per liter.
2. Product concentrations are always placed in the numerator of the equilibrium-constant expression.
3. Reactant concentrations are always placed in the denominator of the equilibrium-constant expression.
4. The powers to which concentrations are raised are always determined by the coefficients in the balanced equation for the reaction.

An additional convention in writing equilibrium constants, not apparent from the equilibrium constant definition, is: Only *concentrations of gases and substances in solution* are written in an equilibrium constant expression. The reason for this convention is that other substances (pure solids and pure liquids) have constant concentrations. These constant concentrations are incorporated into the equilibrium constant itself. For example, pure water in the liquid state has a concentration of 55.5 moles/L. It does not matter whether we have 1.00, 50.0, or 750 mL of liquid water, the concentration will be the same. In the liquid state, pure water is pure water, and it has only one concentration. Similar reasoning applies to other pure liquids and pure solids. All such substances have constant concentrations.

Example 16.2

Write the equilibrium constant expression for each of the following reactions.

(a) $4\ NH_3(g) + 3\ O_2(g) \rightleftharpoons 2\ N_2(g) + 6\ H_2O(g)$

(b) $6\ Ca(s) + 2\ NH_3(g) \rightleftharpoons 3\ CaH_2(s) + Ca_3N_2(s)$

(c) $2\ Ag_2CO_3(s) \rightleftharpoons 4\ Ag(s) + 2\ CO_2(g) + O_2(g)$

(d) $NaCl(aq) + AgNO_3(aq) \rightleftharpoons AgCl(s) + NaNO_3(aq)$

Solution

(a) All of the substances involved in this reaction are gases. Therefore, each reactant and product will appear in the equilibrium constant expression.

The product concentrations, each raised to the power of its coefficient in the balanced equation, are placed in the numerator.

equation coefficients

$$K_{eq} = \frac{[N_2]^2[H_2O]^6}{-}$$

The reactant concentrations, each raised to the power of its coefficient in the balanced equation, are placed in the denominator.

$$K_{eq} = \frac{[N_2]^2[H_2O]^6}{[NH_3]^4[O_2]^3}$$

Note that H_2O as a gas (water vapor or steam) is included in an equilibrium constant expression. The concentration of a gas can vary. Water, as a liquid, is never included in equilibrium constant expressions.

(b) Three of the four substances involved in this reaction are solids and thus will not appear in the equilibrium expression. Since both products are solids, the numerator of the equilibrium expression is 1. The concentration of NH_3 raised to the second power is the only factor in the denominator since the other reactant is a solid.

$$K_{eq} = \frac{1}{[NH_3]^2}$$

(c) The reactant Ag_2CO_3 is a solid and thus will not appear in the equilibrium expression. Since Ag_2CO_3 is the only reactant, this means there will be no denominator in the equilibrium expression. Two of the three products are gases and will appear in the numerator of the equilibrium expression.

$$K_{eq} = [CO_2]^2[O_2]$$

(d) All of the powers in this equilibrium expression are "1" because all of the coefficients in the balanced equation are ones.

$$K_{eq} = \frac{[NaNO_3]}{[NaCl][AgNO_3]}$$

AgCl is not included in the equilibrium expression as it is a solid.

Practice Exercise 16.2

Write the equilibrium constant expression for each of the following reactions.

(a) $2\ NO_2(g) + 7\ H_2(g) \rightleftharpoons 2\ NH_3(g) + 4\ H_2O(g)$
(b) $C(s) + H_2O(g) \rightleftharpoons CO(g) + H_2(g)$
(c) $NH_4Cl(s) \rightleftharpoons NH_3(g) + HCl(g)$
(d) $Fe_2(SO_4)_3(s) \rightleftharpoons 2\ Fe^{3+}(aq) + 3\ SO_4^{2-}(aq)$

Ans. (a) $K_{eq} = \frac{[NH_3]^2[H_2O]^4}{[NO_2]^2[H_2]^7}$; (b) $K_{eq} = \frac{[CO][H_2]}{[H_2O]}$;

(c) $K_{eq} = [NH_3][HCl]$; (d) $K_{eq} = [Fe^{3+}]^2\ [SO_4^{2-}]^3$

Equilibrium constant values vary with temperature changes. A change in temperature changes molecular energies, and molecular energies have a direct effect on the relative amounts of reactants and products present in an equilibrium mixture.

At a given temperature, the numerical value of the equilibrium constant for a reaction is obtained by substituting the experimentally determined equilibrium concentrations at that temperature into the equilibrium constant expression for the reaction.

Example 16.3

At a temperature of 927°C, the equilibrium molar concentrations for the reaction

$$CO(g) + 3\ H_2(g) \rightleftharpoons CH_4(g) + H_2O(g)$$

are

$[CO] = 0.613$, $[H_2] = 1.839$, $[CH_4] = 0.387$, $[H_2O] = 0.387$

Calculate the value of the equilibrium constant for this reaction at 927°C.

Solution

The general expression for the equilibrium constant is

$$K_{eq} = \frac{[CH_4][H_2O]}{[CO][H_2]^3}$$

Substituting the known equilibrium concentrations into this expression gives

$$K_{eq} = \frac{[0.387][0.387]}{[0.613][1.839]^3}$$

$$= 0.039284051 \quad \text{(calculator answer)}$$

$$= 0.0393 \quad \textbf{(correct answer)}$$

Practice Exercise 16.3

At a temperature of 350°C, the equilibrium concentrations for the reaction

$$N_2(g) + 3\ H_2(g) \rightleftharpoons 2\ NH_3(g)$$

are

$$[N_2] = 0.885, \qquad [H_2] = 0.665, \qquad \text{and} \qquad [NH_3] = 1.230$$

Calculate the value of the equilibrium constant for this reaction at 350°C.

Ans. 5.81

The term **position of equilibrium** *specifies qualitatively, the extent to which a chemical reaction, at equilibrium, has proceeded toward completion.* In equilibrium situations where concentrations of products are greater than those of reactants, the equilibrium position is said to lie *to the right,* that is, toward the product side of the equation. If, at equilibrium, product concentrations are less than reactant concentrations, the equilibrium position lies *to the left*.

The magnitude of the equilibrium constant for a reaction gives information about how far a reaction proceeds toward completion, that is, about where the equilibrium position lies. A large value of K_{eq} (greater than 10^3) means that the numerical value of the numerator is significantly greater than that of the denominator. In terms of reactants and products, this means that the concentrations of the products are greater than those of the reactants. The position of the equilibrium lies to the right.

Conversely, if the equilibrium constant is small (less than 10^{-3}), we have a situation where reactants will predominate over products in the reaction mixture. The equilibrium position is said to lie to the left in this situation.

For equilibrium conditions where K_{eq} has a value close between 10^3 and 10^{-3} appreciable concentrations of both products and reactants are present. The reaction described in Example 16.3 falls into this category.

Table 16.1 summarizes the relationship between equilibrium constant magnitude and the extent to which a reaction proceeds toward completion.

TABLE 16.1 COMPARISON OF K_{eq} Values and Reactant and Product Concentrations at Equilibrium

Value of K_{eq}	Relative Amounts of Products and Reactant	Description of Equilibrium Position
Very large (10^{30})	essentially all products	far to the right
Large (10^3)	more products than reactants	to the right
Near unity (between 10^3 and 10^{-3})	significant amounts of both reactants and products	neither to right nor to left
Small (10^{-3})	more reactants than products	to the left
Very small (10^{-30})	essentially all reactants	far to the left

Example 16.4

Describe qualitatively the position of equilibrium for each of the following reactions.

(a) $2\ NO(g) + O_2(g) \rightleftharpoons 2\ NO(g)$ K_{eq} (at 25°C) = 2×10^{12}

(b) $2\ HF(g) \rightleftharpoons H_2(g) + F_2(g)$ K_{eq} (at 25°C) = 1×10^{-95}

(c) $2\ SO_2(g) + O_2(g) \rightleftharpoons 2\ SO_3(g)$ K_{eq} (at 25°C) = 8×10^{35}

(d) $CO(g) + 3\ H_2(g) \rightleftharpoons CH_4(g) + H_2O(g)$ K_{eq} (at 927°C) = 4

Solution

The guidelines for qualitatively specifying equilibrium position are given in Table 16.1.

(a) The position of equilibrium will lie to the right; that is, more products are present than reactants.

(b) The position of equilibrium lies far to the left. Essentially, the reaction has not occurred. Only a minute amount of product is in the equilibrium mixture.

(c) The position of equilibrium is far to the right. For all intents and purposes the reaction has gone to completion. With such a large K_{eq}, only traces of the reactants would be present in the equilibrium mixture.

(d) The position of equilibrium lies neither to the right nor left. Significant amounts of both products and reactants are present.

Practice Exercise 16.4

Describe qualitatively the position of equilibrium for each of the following reactions.

(a) $I_2(g) + Cl_2(g) \rightleftharpoons 2\ ICl(g)$ K_{eq} (at 25°C) = 2×10^5
(b) $N_2(g) + O_2(g) \rightleftharpoons 2\ NO(g)$ K_{eq} (at 25°C) = 1×10^{-30}
(c) $Si(s) + O_2(g) \rightleftharpoons SiO_2(s)$ K_{eq} (at 25°C) = 2×10^{142}
(d) $Ag_2CrO_4(s) \rightleftharpoons 2\ Ag^+(aq) + CrO_4^{2-}(aq)$ K_{eq} (at 25°C) = 9×10^{-12}

Ans. (a) to the right; (b) far to the left; (c) far to the right; (d) to the left

16.7 LE CHÂTELIER'S PRINCIPLE

A chemical system in a state of equilibrium remains in that state until it is disturbed by some change of condition. Disturbing an equilibrium has one of two results: Either the forward reaction speeds up (to produce additional products) or the reverse reaction speeds up (to produce additional reactants). Then with time, the forward and reverse reactions again become equal and a new equilibrium, not identical to the previous one, is established. If more products have been produced as a result of the disruption, the equilibrium is said to have *shifted to the right*. Similarly, when the disruption causes more reactants to form, the equilibrium has *shifted to the left*.

Qualitative predictions about the direction in which chemical equilibria shift can be made using a guideline (principle) introduced in 1888 by the French chemist Henri Louis Le Châtelier* (1850–1936). **Le Châtelier's principle** *states that if a change of conditions (a stress) is applied to a system in equilibrium, the position of the equilibrium will shift in a direction that best reduces the stress and a new equilibrium position will be reached.* We will use this principle in considering how four types of changes affect equilibrium position. The changes are: (1) concentration changes, (2) temperature changes, (3) pressure changes, and (4) addition of a catalyst.

Concentration Changes

Adding or removing a *gaseous* reactant or *gaseous* product from a reaction mixture at equilibrium will always upset the equilibrium. Le Châtelier's principle predicts that the reaction will shift in the direction that will minimize the change in concentration caused by the addition or removal. If an additional amount of any *gaseous* reactant or product has been *added* to the system, the stress is relieved by shifting the equilibrium in the direction that *consumes* (uses up) some of the added reactant or product. Conversely, if a *gaseous* reactant or product is *removed* from an equilibrium system, the equilibrium will shift in a direction that will *produce* more of the substance that was removed.

Let us consider the effect that selected concentration changes will have on the gaseous equilibrium.

$$N_2(g) + 3\,H_2(g) \rightleftharpoons 2\,NH_3(g)$$

Suppose some additional H_2 is added to the equilibrium mixture. The equilibrium will shift to the right; that is, the forward reaction rate will increase in order to use up additional H_2. Eventually a new equilibrium position will be reached. At this new position the H_2 concentration will still be higher than it was before the addition; that is, not all of the added H_2 is consumed. In addition, the N_2 concentration will have decreased (some N_2 had to react with the H_2) and the NH_3 concentration will have increased as the product of the H_2 and N_2 reacting.

Removal of some NH_3 from this newly established equilibrium position will cause an additional shift to the right. The concentration of H_2 and N_2 will decrease as the system attempts to replace the NH_3 that was removed by producing more of it. Again, not all of the removed NH_3 will be replaced. When the new equilibrium position is achieved, the NH_3 concentration will be less than it was before the NH_3 removal.

Figure 16.9 shows graphically the effects that the H_2 addition and NH_3 removal just discussed have on the concentration of all substances present in the N_2–H_2–NH_3 equilibrium mixture.

*Pronounced "le-SHOT-lee-ay."

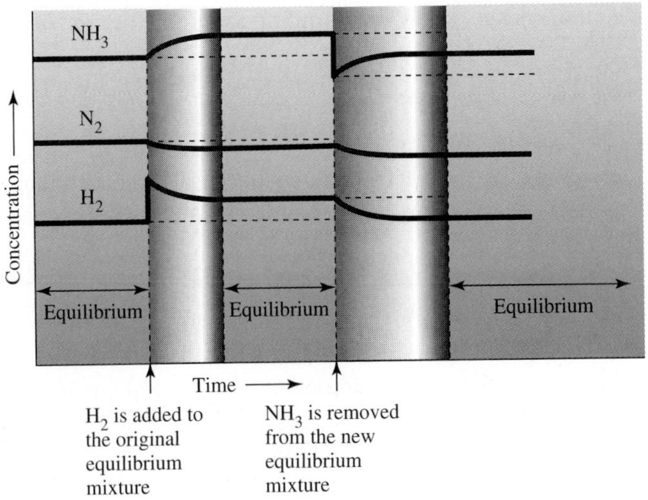

FIGURE 16.9 Concentration changes that result when H_2 is added to, and then NH_3 is removed from, an equilibrium mixture involving the reaction $N_2(g) + 3\ H_2(g) \rightleftharpoons 2\ NH_3(g)$.

Throughout this discussion of the effect of concentration changes on an equilibrium system we have referred to the effect of adding or removing a *gaseous* species. If we add a pure solid or liquid to a gas phase equilibrium system there will be no shift in the equilibrium position; nothing happens. This is because there will be no change in concentration. The solid or liquid was 100% pure before the addition and it is still "100%" after the addition.

More specifically, consider the equilibrium system

$$CaCO_3(s) \rightleftharpoons CaO(s) + CO_2(g)$$

Adding or removing $CaCO_3$ or CaO from this system will cause no change in the equilibrium position. In general, *adding or removing a species disturbs an equilibrium system only if the concentration of that species appears in the expression for the equilibrium constant* (Sec. 16.6).

The concentrations of solids and liquids do not appear in equilibrium constants. The concentrations of gases and species in *aqueous solution* do appear in equilibrium constants. The changing of the concentration of one of the two ions in the equilibrium

$$Pb^{2+}(aq) + 2\ Cl^-(aq) \rightleftharpoons PbCl_2(s)$$

would affect the equilibrium position.

Temperature Changes

Le Châtelier's principle can be used to predict the influence of temperature changes on an equilibrium provided it is known whether the reaction of concern is endothermic or exothermic.

Consider the following general *exothermic* reaction.

$$A + B \rightleftharpoons C + D + \text{heat}$$

Heat is produced when the reaction proceeds to the right. Thus, if we add heat to an exothermic system at equilibrium (by raising the temperature), the system will shift to the left in an attempt to use up the added heat. When equilibrium is reestablished, the concentrations of A and B will be higher and the concentrations of C and D will have decreased. Lowering the temperature of an exothermic system at equilibrium will cause the reaction to shift to the right as the system attempts to replace the lost heat.

The behavior, with temperature change, of an equilibrium reaction mixture involving an *endothermic* reaction

$$E + F + \text{heat} \rightleftharpoons G + H$$

is just the opposite of that of an exothermic reaction, since a shift to the left (rather than the right) produces heat. Consequently, an increase in temperature will cause the equilibrium to shift to the right (to use up the added heat), and a decrease in temperature will produce a shift to the left (to generate more heat).

Pressure Changes

Changes in pressure do not significantly affect the concentrations of solids and liquids (Sec. 11.2) but do alter significantly the concentrations of gases. Pressure changes affect systems at equilibrium only when gases are involved, and then only in cases where the chemical reaction is such that a change occurs in the total number of moles in the gaseous state. This latter point can be illustrated by considering the following two gas-phase reactions.

$$2\,H_2(g) + O_2(g) \rightleftharpoons 2\,H_2O(g)$$
$$H_2(g) + Cl_2(g) \rightleftharpoons 2\,HCl(g)$$

In the first reaction the total number of moles of gaseous reactants and products decreases as the reaction proceeds to the right, since 3 moles of reactants combine to give only 2 moles of products. In the second reaction there is no change in the total number of moles of gaseous substances present as the reaction proceeds, since 2 moles of reactants combine to give 2 moles of products. Thus, a pressure change will shift the position of equilibrium in the first reaction but not in the second reaction.

Pressure changes are usually brought about through volume changes. A pressure increase results from a volume decrease, and a pressure decrease from a volume increase (Sec. 12.3). The use of Le Châtelier's principle correctly predicts the direction of the equilibrium position shift resulting from a pressure change only when the pressure change is due to a volume change. It does not apply to pressure increases caused by the addition of a nonreactive (inert) gas to the reaction mixture. Such an addition has no effect on the equilibrium position. The partial pressure (Sec. 12.14) of each of the gases involved in the reaction remains the same.

According to the Le Châtelier's principle, the stress of increased pressure is relieved by decreasing the number of moles of gaseous substances in the system. This is accomplished by the reaction shifting in the direction of the smaller number of moles, that is, to the side of the equation that contains the smaller number of moles of gaseous substances. For the reaction

$$2\,NO_2(g) + 7\,H_2(g) \rightleftharpoons 2\,NH_3(g) + 4\,H_2O(g)$$

an increase in pressure would shift the equilibrium position to the right, since there are 9 moles of gaseous reactants and only 6 moles of gaseous products. A stress of decreased pressure will result in an equilibrium system reacting in such a way as to produce more moles of gaseous substances.

Addition of a Catalyst

Catalysts do not change the position of equilibrium. This fact becomes clear when we remember that a catalyst functions by lowering the activation energy for a reaction (Sec.

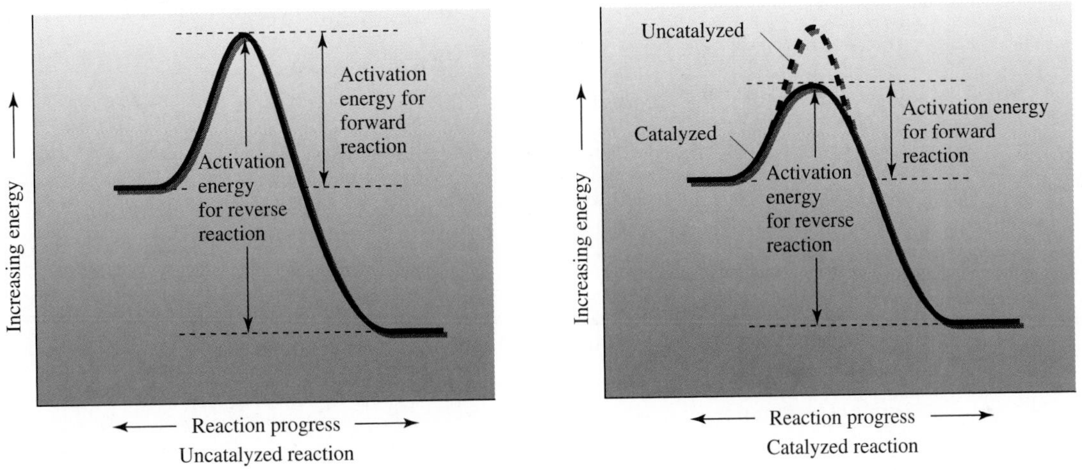

FIGURE 16.10 Influence of a catalyst on the activation energies of forward and reverse reactions.

16.3). As shown in Figure 16.10, as the activation energy of the forward reaction is lowered, so is the activation energy of the reverse reaction. Hence, a catalyst speeds up both the forward and reverse reactions and has no effect on the position of equilibrium. However, the lowered activation energy allows equilibrium to be established more quickly than if the catalyst were absent.

Example 16.5

How will the gas-phase equilibrium

$$PCl_3(g) + Cl_2(g) \rightleftharpoons PCl_5(g) + \text{heat}$$

be affected by each of the following?

(a) removal of $PCl_5(g)$
(b) addition of $Cl_2(g)$
(c) temperature decrease
(d) an increase in the volume of the container (pressure decrease)

Solution

(a) The equilibrium will shift to the right, according to Le Châtelier's principle, in an attempt to replenish the PCl_5 removed.
(b) The equilibrium will shift to the right in an attempt to use up the extra Cl_2 that has been placed in the system.
(c) Lowering the temperatures means that heat energy has been removed. The position of equilibrium will shift to the right in order to produce more heat to take the place of that removed.
(d) The system will shift to the left in an attempt to produce more moles of gaseous reactants; this will increase the pressure. In going to the left the reaction produces 2 moles of gaseous reactants for every 1 mole of gaseous product consumed.

Practice Exercise 16.5

How will the gas-phase equilibrium

$$CH_4(g) + 2\ H_2S(g) + \text{heat} \rightleftharpoons CS_2(g) + 4\ H_2(g)$$

be affected by each of the following?

(a) removal of $H_2(g)$
(b) addition of $CS_2(g)$
(c) temperature increase
(d) an increase in the volume of the container (pressure decrease)

Ans. (a) shift to the right; (b) shift to the left; (c) shift to the right; (d) shift to the right

16.8 FORCING REACTIONS TO COMPLETION

Reactions that would ordinarily reach a state of equilibrium can be forced to completion by using experimental conditions that place a "continual stress" on the potential equilibrium condition. Let us consider a few ways in which this "forcing" is done.

Continuous removal of one or more products of a reaction will force the reaction to completion. To reach equilibrium both reactants and products must be present. The removal of a product continually shifts the reaction to the right, that is, toward completion, according to Le Châtelier's principle. Eventually, one or more of the reactants is depleted, the sign of a completed reaction.

Product removal is very easy to arrange in situations that involve gaseous products. If the reaction is run in an open container, the gaseous products automatically escape to the atmosphere as fast as they are produced. Such a reaction will never reach equilibrium and will continue until the limiting reactant is used up.

Sometimes another chemical reaction is used to remove a product. Consider the situation of a saturated solution of NaCl.

$$NaCl(s) \rightleftharpoons Na^+(aq) + Cl^-(aq)$$

In such a solution, as much NaCl is dissolved as is possible. Adding $AgNO_3$ to the saturated solution will cause more NaCl to dissolve. The Ag^+ ions from the $AgNO_3$ react with the Cl^- ions in the saturated solution to form insoluble AgCl.

$$Ag^+(aq) + Cl^-(aq) \rightleftharpoons AgCl(s)$$

This removes Cl^- (one of the products for the original equilibrium) from solution, thus upsetting the equilibrium. More NaCl will dissolve to compensate for the loss of the chloride ions (Le Châtelier's principle). Continued addition of $AgNO_3$ will eventually cause all of the NaCl to dissolve.

It is also possible to drive a reaction to completion by ensuring that an excess of one of the reactants is always present. The system will continually shift to the right (Le Châtelier's principle) to remove the stress caused by the excess reactant. Eventually other reactants will be depleted and the reaction will be completed. A procedure such as this is useful in a situation where one reactant is very expensive and others are much cheaper. To ensure that none of the expensive reactant goes unreacted, an excess of one of the less expensive reactants is used.

KEY TERMS

The new terms or concepts defined in this chapter are

activation energy (Sec. 16.1) The minimum combined kinetic energy reactant particles must possess in order for their collision to result in a reaction.

catalyst (Sec. 16.3) A substance that, when added to a reaction mixture, increases the rate of the reaction but is itself unchanged after the reaction is completed.

chemical equilibrium (Sec. 16.4) A condition in which two opposing chemical reactions occur simultaneously at the same rate.

collision theory (Sec. 16.1) A set of statements that specifies the conditions necessary for a reaction to occur.

equilibrium constant (Sec. 16.6) The product of the molar concentrations of the products for a chemical reaction, each raised to the power of its respective coefficient in the equation, divided by the product of the molar concentrations of the reactants, each raised to the power of its respective coefficient in the equation.

heterogeneous catalyst (Sec. 16.3) A catalyst that exists as a separate phase from the reactants.

homogeneous catalyst (Sec. 16.3) A catalyst that exists in the same phase as the reactants.

Le Châtelier's principle (Sec. 16.7) If a change of conditions (a stress) is applied to a system in equilibrium, the position of equilibrium will shift in a direction that reduces the stress and a new equilibrium position will be reached.

position of equilibrium (Sec. 16.6) A qualitative specification of the extent to which a chemical reaction, at equilibrium, has proceeded toward completion.

potential energy diagram (Sec. 16.2) A graphical representation of the relationship between the activation energy for a chemical reaction and the total potential energy of the reactants and products.

rate of a chemical reaction (Sec. 16.1) The rate or speed at which reactants are consumed or products are produced in a chemical reaction.

PRACTICE PROBLEMS

Theory of Reaction Rates (Sec. 16.1)

16.1 Why are reactions between substances in solution usually faster than reactions between solid-state reactants?

16.2 Why are gas-phase reactions usually faster than solid-phase reactions?

16.3 Under similar concentration and temperature conditions, would a reaction with an activation energy of 65 kJ/mole or one with an activation energy of 45 kJ/mole proceed at a faster rate? Explain your answer.

16.4 What is the relationship between activation energy and the minimum combined kinetic energy reactant particles must possess in order for their collision to result in a reaction?

16.5 What two factors determine whether a collision between two reactant molecules will result in a reaction?

16.6 What happens to the reactants in an ineffective molecular collison?

16.7 For the reaction

$$H_2 + Cl_2 \longrightarrow 2\ HCl$$

draw a sketch of a molecular orientation that is highly favorable for an effective collision.

16.8 For the reaction

$$H_2 + Cl_2 \longrightarrow 2\ HCl$$

draw a sketch of a molecular orientation that is unfavorable for an effective collision.

Potential Energy Diagrams for Chemical Reactions (Sec. 16.2)

16.9 Sketch a potential energy diagram for a hypothetical chemical reaction that is exothermic by 35 kJ/mole and has an activation energy of 75 kJ/mole. Label the following on the diagram.

(a) average energy of the reactants

(b) average energy of the products

(c) activation energy

(d) amount of energy liberated during the reaction

16.10 Sketch a potential energy diagram for a hypothetical chemical reaction that is endothermic by 75 kJ/mole and has an activation energy of 35 kJ/mole. Label the following on the diagram.

(a) average energy of the reactants

(b) average energy of the products

(c) activation energy

(d) amount of energy absorbed during the reaction

16.11 Reaction A occurs at room temperature and liberates 200 kJ of energy per mole of reactant. Reaction B does not occur until a temperature of 150°C is reached; it also liberates 200 kJ of energy per mole of reactant. Draw a potential energy diagram for each reaction, and indicate the similarities and differences between the two diagrams.

16.12 Reaction C occurs at room temperature and liberates 200 kJ of energy per mole of reactant. Reaction D also occurs at room temperature but absorbs 200 kJ of energy per mole of reactant. Draw a potential energy diagram for each reaction, and indicate the similarities and differences between the two diagrams.

Factors That Influence Reaction Rates (Sec. 16.3)

16.13 Give two reasons why reactions occur more rapidly at higher temperatures.

16.14 Why does increasing reactant concentrations generally increase the rate of a reaction?

16.15 What effect does increasing surface area have on a reaction rate involving a solid reactant? Explain.

16.16 Describe how a catalyst affects activation energy and reaction rate.

16.17 Why will a spark cause coal dust in a mine to explode and yet not cause an explosion with charcoal in a barbeque?

16.18 Milk will sour in a couple of days when left at room temperature yet can remain unspoiled for 2 weeks when refrigerated. Explain why.

16.19 The characteristics of four reactions, each of which involves only two reactants, are as follows.

Reaction	*Activation Energy*	*Temperature*	*Concentration of Reactants*
1	low	low	1 mole/L of each
2	high	low	1 mole/L of each
3	low	high	1 mole/L of each
4	low	low	1mole/L of 1st reactant; 4 moles/L of 2nd reactant

For each of the following pairs of the preceding reactions, indicate which reaction is faster. The rates to be compared are the rates when the two reactants are first mixed. Explain each of your answers.

(a) 1 and 2 **(b)** 1 and 3

(c) 1 and 4 **(d)** 2 and 3

16.20 The characteristics of four reactions, each of which involves only two reactants, are as follows.

Reaction	*Activation Energy*	*Temperature*	*Concentration of Reactants*
1	high	low	1 mole/L of each
2	high	high	1 mole/L of each
3	low	low	1 mole/L of 1st reactant; 4 mole/L of 2nd reactant
4	low	low	4 moles/L of each

For each of the following pairs of the preceding reactions, indicate which reaction is faster. The rates to be compared are the rates when the two reactants are first mixed. Explain each of your answers.

(a) 1 and 2 **(b)** 1 and 3

(c) 1 and 4 **(d)** 3 and 4

16.21 Draw a potential energy diagram for an endothermic reaction where no catalyst is present. Then draw an energy diagram for the same reaction when a catalyst is present. Indicate the similarities and differences between the two diagrams.

16.22 Draw a potential energy diagram for an exothermic reaction where no catalyst is present. Then draw an energy diagram for the same reaction when a catalyst is present. Indicate the similarities and differences between the two diagrams.

Chemical Equilibrium (Sec. 16.4)

16.23 Compare the magnitudes of the rates of the forward and reverse reactions for a chemical system that is in a state of equilibrium.

16.24 For a system in a state of chemical equilibrium, concentrations of reactants and products are constant. Explain why.

16.25 What is the difference between a physical equilibrium and a chemical equilibrium?

16.26 What is the difference between *equal* product and reactant concentrations and *constant* product and reactant concentrations?

Equilibrium Mixture Stoichiometry (Sec. 16.5)

16.27 A 0.0200-mole sample of SO_3 is placed in a reaction container and allowed to decompose until equilibrium is established.

$$2\,SO_3(g) \rightleftharpoons 2\,SO_2(g) + O_2(g)$$

At equilibrium 0.0029 mole of O_2 is present. What is the composition of the equilibrium mixture in terms of moles of each substance present?

16.28 A mixture of 0.100 mole of SO_2 and 0.100 mole of O_2 is placed in a reaction container and allowed to react until equilibrium is established.

$$2\,SO_2(g) + O_2(g) \rightleftharpoons 2\,SO_3(g)$$

At equilibrium 0.0916 mole of SO_3 is present. What is the composition of the equilibrium mixture in terms of moles of each substance present?

16.29 A mixture of 0.296 mole of NH_3, 0.170 mole of N_2, and 0.095 mole of H_2 is allowed to reach equilibrium.

$$2\ NH_3(g) \rightleftharpoons N_2(g) + 3\ H_2(g)$$

At equilibrium, it is found that 0.268 mole of NH_3 is present. What is the composition of the equilibrium mixture in terms of moles of each substance present?

16.30 A mixture of 0.520 mole of NOCl, 0.010 mole of NO, and 0.053 mole of Cl_2 is allowed to reach equilibrium.

$$2\ NOCl(g) \rightleftharpoons 2\ NO(g) + Cl_2(g)$$

At equilibrium, it is found that 0.022 mole of NO is present. What is the composition of the equilibrium mixture in terms of moles of each substance present?

Equilibrium Constants (Sec. 16.6)

16.31 Write the expression for the equilibrium constant for each of the following reactions.

(a) $SO_2(g) + Cl_2(g) \rightleftharpoons SO_2Cl_2(g)$

(b) $2\ NO_2(g) \rightleftharpoons N_2(g) + 2\ O_2(g)$

(c) $2\ SO_3(g) + CO_2(g) \rightleftharpoons CS_2(g) + 4\ O_2(g)$

(d) $4\ H_2(g) + CS_2(g) \rightleftharpoons CH_4(g) + 2\ H_2S(g)$

16.32 Write the expression for the equilibrium constant for each of the following reactions.

(a) $PCl_5(g) \rightleftharpoons PCl_3(g) + Cl_2(g)$

(b) $2\ NO(g) \rightleftharpoons N_2(g) + O_2(g)$

(c) $4\ NH_3(g) + 7\ O_2(g) \rightleftharpoons 4\ NO_2(g) + 6\ H_2O(g)$

(d) $CO(g) + 3\ H_2(g) \rightleftharpoons CH_4(g) + H_2O(g)$

16.33 Write the expression for the equilibrium constant for each of the following reactions.

(a) $2\ Pb(NO_3)_2(s) \rightleftharpoons 2\ PbO(s) + 4\ NO_2(g) + O_2(g)$

(b) $2\ KClO_3(s) \rightleftharpoons 2\ KCl(s) + 3\ O_2(g)$

(c) $2\ Ag(s) + Cl_2(g) \rightleftharpoons 2\ AgCl(s)$

(d) $PCl_5(s) \rightleftharpoons PCl_3(l) + Cl_2(g)$

16.34 Write the expression for the equilibrium constant for each of the following reactions.

(a) $H_2SO_4(l) \rightleftharpoons SO_3(g) + H_2O(l)$

(b) $2\ FeBr_3(s) \rightleftharpoons 2\ FeBr_2(s) + Br_2(g)$

(c) $BaCl_2(aq) + Na_2SO_4(aq) \rightleftharpoons 2\ NaCl(aq) + BaSO_4(s)$

(d) $2\ Na_2O(s) \rightleftharpoons 4\ Na(l) + O_2(g)$

16.35 At a particular temperature, a hypothetical chemical system has the following equilibrium molar concentrations: A = 3.00, B = 2.00, and C = 5.00. Calculate the value of the equilibrium constant for the system if the reaction occurring were each of the following.

(a) $A(g) \rightleftharpoons 2\ B(g) + C(g)$

(b) $A(g) + 3\ B(g) \rightleftharpoons 2\ C(g)$

(c) $2\ B(g) \rightleftharpoons A(g) + C(g)$

(d) $4\ C(g) + B(g) \rightleftharpoons 3\ A(g)$

16.36 At a particular temperature, a hypothetical chemical system has the following equilibrium molar concentrations: A = 2.00, B = 4.00, and C = 3.00. Calculate the value of the equilibrium constant for the system if the reaction occurring were each of the following.

(a) $A(g) + 2\ B(g) \rightleftharpoons C(g)$

(b) $A(g) \rightleftharpoons B(g) + 3\ C(g)$

(c) $2\ C(g) + B(g) \rightleftharpoons 2\ A(g)$

(d) $3\ A(g) + 2\ B(g) \rightleftharpoons 4\ C(g)$

16.37 The equilibrium constant for the reaction

$$CS_2(g) + 4\ H_2(g) \rightleftharpoons CH_4(g) + 2\ H_2S(g)$$

is 0.0280 at a particular temperature. The system at equilibrium has $[H_2S] = 1.43$, $[H_2] = 0.100$, and $[CH_4] = 0.00100$. What is $[CS_2]$?

16.38 The equilibrium constant for the reaction

$$CH_4(g) + 2\ H_2S(g) \rightleftharpoons CS_2(g) + 4\ H_2(g)$$

is 3.30×10^4 at a particular temperature. The system at equilibrium has $[CH_4] = 0.709$, $[H_2S] = 0.0100$, and $[H_2] = 2.34$. What is $[CS_2]$?

16.39 A 6.00-L vessel contained 0.0222 mole of PCl_3, 0.0189 mole of PCl_5, and 0.1044 mole of Cl_2 at 230°C in an equilibrium mixture. Calculate the value of K_{eq} for the reaction

$$PCl_3(g) + Cl_2(g) \rightleftharpoons PCl_5(g)$$

16.40 An 8.00-L vessel contained 0.650 mole of H_2, 0.275 mole of I_2, and 2.86 moles of HI at 491°C in an equilibrium mixture. Calculate the value of K_{eq} for the reaction

$$2\ HI(g) \rightleftharpoons H_2(g) + I_2(g)$$

16.41 Describe qualitatively the position of equilibrium for reactions with each of the following equilibrium constants.

(a) 10^{-10} at 25°C **(b)** 10^{30} at 25°C

(c) 10^{9} at 127°C **(d)** 10^{2} at 327°C

16.42 Describe qualitatively the position of equilibrium for reactions with each of the following equilibrium constants.

(a) 10^{25} at 25°C **(b)** 10^{-19} at 25°C

(c) 10^{-11} at 235°C **(d)** 10^{-1} at 1235°C

Le Châtelier's Principle (Sec. 16.7)

16.43 For the reaction

$$CO(g) + 3\ H_2(g) \rightleftharpoons CH_4(g) + H_2O(g)$$

determine the direction that the equilibrium will be shifted by each of the following changes.

(a) increase in CO concentration

(b) increase in CH_4 concentration

(c) decrease in H_2 concentration

(d) decrease in H_2O concentration

16.44 For the reaction

$$CH_4(g) + 2\ O_2(g) \rightleftharpoons CO_2(g) + 2\ H_2O(g)$$

determine the direction that the equilibrium will be shifted by each of the following changes.

(a) increase in O_2 concentration
(b) increase in CO_2 concentration
(c) decrease in CH_4 concentration
(d) decrease in H_2O concentration

16.45 For the reaction

$$2\ C_2H_2(g) + 5\ O_2(g) \rightleftharpoons 4\ CO_2(g) + 2\ H_2O(g) + \text{heat}$$

determine the direction that the equilibrium will be shifted by each of the following changes.

(a) increasing the concentration of C_2H_2
(b) decreasing the concentration of O_2
(c) increasing the temperature
(d) increasing the pressure by decreasing the volume of the container.

16.46 For the reaction

$$C_3H_8(g) + 5\ O_2(g) \rightleftharpoons 3\ CO_2(g) + 4\ H_2O(g) + \text{heat}$$

determine the direction that the equilibrium will be shifted by each of the following changes.

(a) increasing the concentration of CO_2
(b) decreasing the concentration of C_3H_8
(c) decreasing the temperature
(d) increasing the pressure by decreasing the volume of the container

16.47 Consider the following chemical system at equilibrium.

$$2\ H_2O(g) + 2\ Cl_2(g) + \text{heat} \rightleftharpoons 4\ HCl(g) + O_2(g)$$

For each of the following adjustments of conditions, indicate the effect on the position of equilibrium: shifts left, shifts right, no effect.

(a) heating the equilibrium mixture
(b) adding O_2 to the mixture
(c) increasing the pressure on the equilibrium mixture by adding a noble gas
(d) increasing the size of the reaction container

16.48 Consider the following chemical system at equilibrium

$$2\ N_2(g) + 6\ H_2O(g) + \text{heat} \rightleftharpoons 4\ NH_3(g) + O_2(g)$$

For each of the following adjustments of conditions, indicate the effect on the position of equilibrium: shifts left, shifts right, no effect.

(a) adding N_2 to the mixture
(b) decreasing the size of the container holding the mixture
(c) adding a catalyst to the mixture
(d) refrigerating the warm equilibrium mixture

16.49 For which of the following reactions is product formation favored by high temperature?

(a) $N_2(g) + O_2(g) \rightleftharpoons 2\ NO(g) + \text{heat}$
(b) $N_2(g) + 3\ H_2(g) \rightleftharpoons 2\ NH_3(g) + \text{heat}$
(c) $CO(g) + 3\ H_2(g) + \text{heat} \rightleftharpoons CH_4(g) + H_2O(g)$
(d) $2\ H_2O(g) \rightleftharpoons 2\ H_2(g) + O_2(g) + \text{heat}$

16.50 For which of the reactions in Problem 16.49 is product formation favored by high pressure?

ADDITIONAL PROBLEMS

16.51 Write a balanced chemical equation for a totally gaseous equilibrium system that would lead to the following expressions for the equilibrium constant.

(a) $\dfrac{[NH_3]^2}{[N_2][H_2]^3}$ **(b)** $\dfrac{[N_2]^2[H_2O]^6}{[NH_3]^4[O_2]^3}$

(c) $\dfrac{[N_2][O_2]}{[NO]^2}$ **(d)** $\dfrac{[NO]^2}{[N_2][O_2]}$

16.52 Write a balanced chemical equation for a totally gaseous equilibrium system that would lead to the following expressions for the equilibrium constant.

(a) $\dfrac{[HCN]^2}{[H_2][C_2N_2]}$ **(b)** $\dfrac{[CH_4][H_2S]^2}{[CS_2][H_2]^4}$

(c) $\dfrac{[NOBr]^2}{[NO]^2[Br_2]}$ **(d)** $\dfrac{[NO]^2[Br_2]}{[NOBr]^2}$

16.53 The following reaction at a certain temperature has an equilibrium-constant value of 25.9.

$$2\ CO_2(g) \rightleftharpoons 2\ CO(g) + O_2(g)$$

For each of the following compositions, decide whether the reaction mixture is at equilibrium. If it is not, decide which direction the reaction shifts to reach equilibrium.

(a) $[CO_2] = 0.0300$, $[CO] = 0.350$, $[O_2] = 0.190$
(b) $[CO_2] = 0.0600$, $[CO] = 0.700$, $[O_2] = 0.380$
(c) $[CO_2] = 0.0280$, $[CO] = 0.356$, $[O_2] = 0.160$
(d) $[CO_2] = 0.0100$, $[CO] = 0.330$, $[O_2] = 0.180$

16.54 The following reaction at a certain temperature has an equilibrium constant value of 0.016.

$$2\ HI(g) \rightleftharpoons H_2(g) + I_2(g)$$

For each of the following compositions, decide whether the reaction mixture is at equilibrium. If it is not, decide which direction the reaction shifts to reach equilibrium.

(a) [HI] = 0.080, $[H_2]$ = 0.010, $[I_2]$ = 0.010

(b) [HI] = 0.084, $[H_2]$ = 0.012, $[I_2]$ = 0.012

(c) [HI] = 0.076, $[H_2]$ = 0.012, $[I_2]$ = 0.012

(d) [HI] = 0.140, $[H_2]$ = 0.031, $[I_2]$ = 0.010

16.55 At a given temperature, the equilibrium constant for the reaction

$$2\ NO(g) + Br_2(g) \rightleftharpoons 2\ NOBr(g)$$

is 2×10^3. What is the equilibrium constant, at the same temperature, for the following reaction?

$$2\ NOBr(g) \rightleftharpoons 2\ NO(g) + Br_2(g)$$

16.56 At a given temperature, the equilibrium constant for the reaction

$$CO(g) + H_2O(g) \rightleftharpoons CO_2(g) + H_2(g)$$

is 0.034. What is the equilibrium constant, at the same temperature for the following reaction?

$$CO_2(g) + H_2(g) \rightleftharpoons CO(g) + H_2O(g)$$

16.57 Which of the following changes would change the *value* of a system's equilibrium constant?

(a) addition of a reactant or product

(b) increase in the total pressure

(c) decrease in the temperature.

(d) addition of an inert gas

16.58 Which of the following changes would change the *value* of a system's equilibrium constant?

(a) removal of a reactant or product

(b) decrease in the total pressure

(c) increase in the temperature

(d) addition of a catalyst

16.59 Predict the direction in which each of the following equilibria will shift if the pressure on the system is decreased by expansion.

(a) $2\ SO_3(g) \rightleftharpoons 2\ SO_2(g) + O_2(g)$

(b) $2\ HI(g) \rightleftharpoons H_2(g) + I_2(g)$

(c) $ClF_5(g) \rightleftharpoons ClF_3(g) + Cl_2(g)$

(d) $C(s) + CO_2(g) \rightleftharpoons 2\ CO(g)$

16.60 Predict the direction in which each of the following equilibria will shift if the pressure on the system is increased by compression.

(a) $H_2(g) + C_2N_2(g) \rightleftharpoons 2\ HCN(g)$

(b) $CO(g) + Br_2(g) \rightleftharpoons COBr_2(g)$

(c) $CS_2(g) + 4\ H_2(g) \rightleftharpoons CH_4(g) + 2\ H_2S(g)$

(d) $Ni(s) + 4\ CO(g) \rightleftharpoons Ni(CO)_4(g)$

CUMULATIVE PROBLEMS

16.61 Given the following descriptions of reversible reactions, write the equilibrium constant expression for each.

(a) In a decomposition reaction sulfur trioxide gas produces sulfur dioxide gas and oxygen gas.

(b) Hydrogen gas reduces nitrogen dioxide gas to produce ammonia gas and steam.

(c) Solid iron(II) oxide and carbon monoxide gas react to produce solid iron and carbon dioxide gas.

(d) Solid magnesium carbonate decomposes to produce solid magnesium oxide and carbon dioxide gas.

16.62 Given the following descriptions of reversible reactions, write the equilibrium constant expression for each.

(a) In a synthesis reaction hydrogen and gaseous bromine react to produce gaseous hydrogen bromide.

(b) In a redox reaction carbon disulfide gas reacts with hydrogen gas to produce methane gas and hydrogen sulfide gas.

(c) Solid sodium carbonate reacts with gaseous sulfur dioxide and oxygen gas to produce solid sodium sulfate and carbon dioxide gas.

(d) Chlorine gas reacts with liquid carbon disulfide to produce the liquids carbon tetrachloride and disulfur dichloride.

16.63 An equilibrium mixture, at 900°C in a 1725-mL container, involving the chemical system

$$CH_4(g) + 2\ H_2S(g) \rightleftharpoons CS_2(g) + 4\ H_2(g)$$

is found to contain 17.6 g CH_4, 50.8 g H_2S, 83.8 g CS_2, and 8.10 g H_2. Calculate the equilibrium constant for this reaction at the given temperature.

16.64 An equilibrium mixture, at 472°C in a 1325-mL container, involving the chemical system

$$N_2(g) + 3\ H_2(g) \rightleftharpoons 2\ NH_3(g)$$

is found to contain 4.23 g N_2, 0.915 g H_2, and 0.496 g NH_3. Calculate the equilibrium constant for this reaction at the given temperature.

16.65 For the chemical system

$$SbCl_5(g) \rightleftharpoons SbCl_3(g) + Cl_2(g)$$

it is found that an equilibrium mixture in a 1.00-L flask contains 2.48×10^{20} molecules of $SbCl_5$, 0.723 g of $SbCl_3$, and 0.00317 mole of Cl_2. Calculate the equilibrium constant for the reaction.

16.66 For the chemical system

$$PCl_5(g) \rightleftharpoons PCl_3(g) + Cl_2(g)$$

it is found that an equilibrium mixture in a 1.00-L flask contains 2.95×10^{20} molecules of PCl_5, 0.00451 mole of PCl_3, and 0.320 g of Cl_2. Calculate the equilibrium constant for the reaction.

16.67 Consider the following equilibrium situation, at constant temperature.

$$N_2O_4(g) \rightleftharpoons 2\ NO_2(g)$$

An empty container is charged with pure $N_2O_4(g)$ until the pressure reaches 1.50 atm. The $N_2O_4(g)$ is then allowed to reach equilibrium with $NO_2(g)$, at which time the partial pressure of $N_2O_4(g)$ is 0.80 atm. What is the total pressure in atmospheres in the flask?

16.68 Consider the following equilibrium situation, at constant temperature.

$$2\ O_3(g) \rightleftharpoons 3\ O_2(g)$$

An empty container is charged with pure $O_3(g)$ until the pressure reaches 2.25 atm. The $O_3(g)$ is then allowed to reach equilibrium with $O_2(g)$, at which time the partial pressure of $O_3(g)$ is 0.11 atm. What is the total pressure, in atmospheres, in the flask.

16.69 At 750°C and 1.000 atm pressure, a gaseous mixture of carbon monoxide and carbon dioxide is in equilibrium with solid carbon. The gaseous mixture is 87.43% CO by mass.

$$C(s) + CO_2(g) \rightleftharpoons 2\ CO(g)$$

Calculate the equilibrium constant for this reaction from the given information.

16.70 At 35°C and 1.00 atm pressure, a gaseous mixture of dinitrogen tetroxide and nitrogen dioxide at equilibrium is found to contain 32.7% by mass of N_2O_4

$$N_2O_4(g) \rightleftharpoons 2\ NO_2(g)$$

Calculate the equilibrium constant for this reaction from the given information.

17 Nuclear Chemistry

17.1 Atomic Nuclei

Early in the text (Sec. 5.5) we considered the structure of the atom. An atom is composed of two regions: (1) a nuclear region where all protons and neutrons are found and (2) an extranuclear region where the electrons are found. Since that initial discussion of atomic structure, we have focused primarily on the behavior of electrons in atoms, ions, and molecules, because the arrangement of electrons determines the physical and chemical properties of substances. Little has been said about the nucleus because it remains unchanged in ordinary chemical reactions and is important only insofar as it influences the electrons.

In this chapter we examine a group of processes known as nuclear reactions. A **nuclear reaction** *is a reaction in which changes occur in the nucleus of an atom.* Nuclear reactions are not considered to be ordinary chemical reactions. It is in the field of nuclear reactions that we encounter the terms *radioactivity*, *nuclear power plant*, *A-bomb*, and *H-bomb*. All of these terms are now part of our everyday vocabulary.

A brief review of previously discussed concepts about atomic nuclei as well as some new material concerning them will serve as the starting point for the discussion of nuclear reactions.

Atomic nuclei are the very dense positively charged centers of atoms about which the electrons move. All nuclei of atoms of a given element contain the same number of protons. It is this characteristic number of protons that determines the identity of the element. The *atomic number* for an atom gives the number of protons in the nucleus. The number of neutrons associated with the nuclei of a given element may vary within a limited range. Atoms of a given element that differ in the number of neutrons in the nucleus are called *isotopes*. The *mass number* of an atom is equal to the total number of protons and neutrons present in the nucleus. Isotopes of an element have different mass numbers but the same atomic number.

A term new to us, used when talking about a single type of atom (or nucleus), is *nuclide*. A **nuclide** *is an atom with specific numbers of pro-*

tons and neutrons in its nucleus. All atoms of a given nuclide must have the same number of protons and the same number of neutrons. The term *isotope* refers to different forms of the same element; the term *nuclide* is used in describing atomic forms of different elements. The species $^{12}_{6}C$ and $^{13}_{6}C$ are isotopes of the element carbon. The species $^{12}_{6}C$ and $^{16}_{8}O$ are nuclides of different elements.

In order to uniquely identify a nucleus, or an atom for that matter, both the atomic number and mass number must be specified. Two notation systems exist for doing this. Consider a nuclide of nitrogen with seven protons and eight neutrons. This nuclide can be denoted as $^{15}_{7}N$ or nitrogen-15. In the first notation the superscript is the mass number and the subscript is the atomic number. In the second notation the mass number is appended to the name of the element with a hyphen. An advantage of the first notation is that the atomic number is shown; a disadvantage is the need for superscripts and subscripts. Both types of notation will be used in this chapter.

Nuclides may be divided into two categories based on nuclear stability: stable and unstable. A **stable nuclide** *has a nucleus that is stable*. An **unstable nuclide** *has a nucleus that is unstable*. As a mechanism for achieving stability, all unstable nuclei spontaneously emit energy (radiation). Such radiation is called *radioactivity*. **Radioactivity** *is the radiation spontaneously emitted from the nucleus of an unstable nuclide*. Only a few or the naturally occurring elements, approximately twenty, have at least one isotope that has an unstable nucleus, that is, is a radioactive nuclide. A **radioactive nuclide** *is an atom with an unstable nucleus that spontaneously emits energy (radiation)*. (The term *radioactive nuclide* is often shortened to *radionuclide*.*)

Radioactive nuclides are known for *all* elements even though relatively few such species exist in nature. This is because laboratory procedures have been discovered by which scientists can convert nonradioactive nuclides into radioactive nuclides. Further details concerning this "conversion process" are given in a later section of this chapter (Sec. 17.6).

17.2 DISCOVERY OF RADIOACTIVITY

The fact that certain naturally occurring nuclides are radioactive was unexpectedly (accidentally) discovered by the French physicist and engineer Antoine Henri Becquerel (1852–1908) while he was studying certain minerals called phosphors, which glow in the dark (phosphoresce) after exposure to radiation such as sunlight or ultraviolet light. The rays emitted by these phosphorescing minerals, like visible light, darken a photographic plate. One day, in 1896, while working with a uranium ore sample that phosphoresced in the normal manner, he was interrupted. As he left, he inadvertently placed the uranium ore sample on top of an unexposed photographic plate packaged to protect it from light. Later it was determined that this photographic plate had been exposed by the uranium ore despite its being protectively wrapped. Becquerel correctly concluded that this plate exposure was due to radiation emitted by the uranium ore without external stimulus. As a result of this incident Becquerel is credited with having discovered the phenomenon we now call radioactivity. Further studies by Becquerel showed that this "radioactivity" was not unique to the one uranium ore he had used, but was a characteristic of all uranium-containing substances independently of whether they phosphoresced or not. (It seems strange now that this phenomenon was not detected earlier, since the element uranium had been isolated more than 100 years before Becquerel's discovery.)

* In some textbooks the term *radioactive isotope* (*or radioisotope*) is used in place of radioactive nuclide (radionuclide). In other textbooks the two terms are used interchangeably. We will use nuclide rather than isotope.

Becquerel chose not to continue to work in this new field of study. Instead, he suggested to two of his colleagues, Pierre and Marie Curie (who had just recently been married), that they continue the project and find out what other substances besides uranium possessed "radioactive" properties.

Pierre Curie (1859–1906), a physicist, and Marie Sklodowska Curie (1859–1934), a chemist, conducted a systematic search of the then-known elements to see how widespread this phenomenon was. The only other radioactive element they found was thorium. In further investigations, the Curies discovered a uranium ore sample that exhibited four times the radioactivity of a similar quantity of pure uranium or thorium, thus indicating the presence of a new substance more radioactive than either of these elements. From this ore the Curies were able to isolate, in 1898, two new radioactive elements: polonium, 400 times as radioactive as uranium, and radium, over a million times as radioactive as uranium. It was the Curies who coined the word *radioactive* to describe elements that spontaneously emit radiation.

The Curies spent the rest of their lifetimes studying radioactive phenomena. Notwithstanding the accidental death of her husband in 1906 (he was killed by a heavy horse-drawn wagon), Marie continued on tirelessly. For her monumental efforts, Madame Curie became one of the most celebrated scientists of all time.

17.3 NATURE OF NATURAL RADIOACTIVE EMISSIONS

The first information concerning the nature of the radiation emanating from naturally radioactive materials was obtained by Ernest Rutherford in the years 1898–1899. Using an apparatus similar to that shown in Figure 17.1, he found that if a beam of radiation is passed between electric plates it is split into three components, indicating the presence of three different types of emissions from radioactive materials. A closer analysis of Rutherford's experiment reveals that one radiation component is positively charged (it is attracted to the negative plate), a second component is negatively charged (it is attracted to the positive plate), and the third component carries no charge (it is unaffected by either charged plate). Rutherford chose to call the three radiation components alpha rays (α rays) (the positive component), beta rays (β rays) (the negative component), and gamma rays (γ rays) (the uncharged component). (Alpha, beta, and gamma are the first three letters of the Greek alphabet.) We mention Rutherford's nomenclature system because it "stuck"; we still use

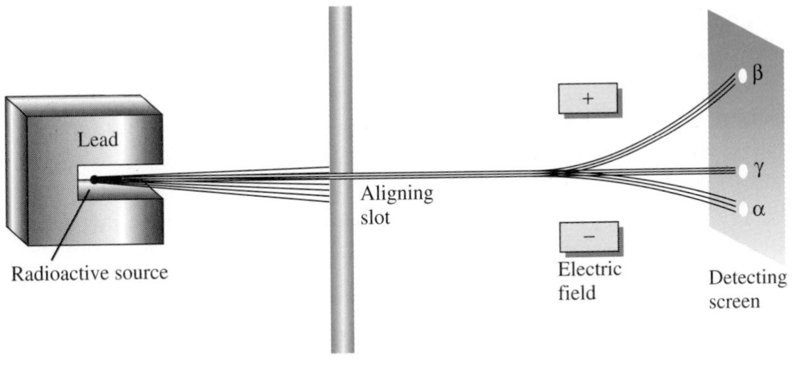

FIGURE 17.1 Effect of an electric field on radiation emanating from a naturally radioactive substance. Gamma rays are unaffected. The lighter beta rays are deflected considerably more than the heavier alpha rays.

these names today for these radiation types even though we know much, much more about their identity.

Additional research has substantiated Rutherford's conclusion that three distinct types of radiation are present in the emissions from naturally radioactive substances. This research has also supplied the necessary information for their complete characterization. Complete identification required many years. Early researchers in the field were hampered by the fact that many of the details concerning atomic structure were not yet known. (Recall that the neutron was not identified until 1932; Sec. 5.5.)

Alpha rays *consist of a stream of positively charged particles (alpha particles), each of which is made up of two protons and two neutrons.* The notation used to represent an alpha particle is ${}^{4}_{2}\alpha$. The numerical subscript indicates that the charge on the particle is +2 (from the two protons). The numerical superscript indicates a mass of 4 amu. On the atomic mass scale (Sec. 5.9) protons and neutrons both have masses equal to 1.0 amu. Thus, the total mass of an alpha particle (two protons and two neutrons) is about 4 amu. Alpha particles are identical with the nuclei of helium-4 (${}^{4}_{2}He$) atoms.

Beta rays *consist of a stream of negatively charged particles (beta particles) whose charge and mass are identical to those of an electron.* However, beta particles are not extranuclear electrons; they are particles that have been produced inside the nucleus and then ejected. More concerning this process will be given in Section 17.4. The symbol used to represent a beta particle is ${}^{0}_{-1}\beta$. The numerical subscript indicates that the charge on the beta particle is −1, that of an electron. The use of the superscript zero for the mass of a beta particle is not to be interpreted as meaning that a beta particle has no mass, but rather that its mass number (protons + neutrons) is zero. The actual mass of a beta particle on the atomic mass scale is 0.00055 amu.

Gamma rays *are not considered to be particles; rather, they are high energy radiation without mass or charge.* They are very high-energy radiation somewhat like X rays. The symbol for gamma rays is ${}^{0}_{0}\gamma$.

17.4 EQUATIONS FOR NUCLEAR REACTIONS

Alpha, beta, and gamma emissions come from the nucleus of an atom. These spontaneous emissions alter nuclei; obviously, if a nucleus loses an alpha particle (two protons and two neutrons), it will not be the same as it was before the departure of the particle. In the case of alpha and beta emissions the nuclear alteration causes the identity of the atom to change; that is, a new element is formed. Nuclear reactions thus differ dramatically from ordinary chemical reactions. In the latter the identity of the elements is always maintained. This is not the case in nuclear reactions.

The term *decay* or *disintegration* is used in describing a nuclear process where an element changes into another element as a result of radiation emission. **Radioactive decay** *is the process whereby a radionuclide is transformed into a nuclide of another element as a result of the emission of radiation.*

Radioactive decay occurs only in certain ways, called *modes of decay*. For naturally occurring radioactive substances the modes of decay are: (1) alpha particle decay and (2) beta particle decay. Separate consideration of these two modes of decay provides further insights into the nature of nuclear reactions and also illustrates how equations for nuclear reactions are written.

Alpha-Particle Decay

Alpha-particle decay, the emission of an alpha particle from a nucleus, always results in the formation of a nuclide of a different element. The product nucleus of such decay has an atomic number that is 2 less than that of the original nucleus and a mass number that is 4 less. We can represent alpha-particle decay in general terms by the equation

$$^{A}_{Z}\text{X} \longrightarrow {}^{4}_{2}\alpha + {}^{(A-4)}_{(Z-2)}\text{Y}$$

where X is the symbol for the nucleus of the original element undergoing decay and Y is the symbol of the element formed as a result of the decay.

As an introduction to nuclear equations, let us write such equations for two actual processes. Both $^{211}_{83}\text{Bi}$ and $^{238}_{92}\text{U}$ are alpha emitters; that is, they are radionuclides that undergo alpha-particle decay. The nuclear equations for these two decay processes are

$$^{211}_{83}\text{Bi} \longrightarrow {}^{4}_{2}\alpha + {}^{207}_{81}\text{Tl}$$

$$^{238}_{92}\text{U} \longrightarrow {}^{4}_{2}\alpha + {}^{234}_{90}\text{Th}$$

How do these two equations differ from ordinary chemical equations? First of all, nuclear equations convey a different type of information from that found in ordinary chemical equations. The symbols in nuclear equations stand for nuclei rather than atoms. (We do not worry about electrons when writing nuclear equations.) Second, mass numbers and atomic numbers (nuclear charge) are always used in conjunction with elemental symbols in nuclear equations. Third, the elemental symbols on both sides of the equation frequently are not the same in nuclear equations.

The procedures for balancing nuclear equations are different from those used for ordinary chemical equations. In a **balanced nuclear equation** *the sums of the subscripts (atomic number or particle charge) on both sides of the equation are equal and the sums of the superscripts (mass number) on both sides of the equation are equal.* Both of our example equations are balanced. In the alpha decay of $^{211}_{83}\text{Bi}$, the subscripts on both sides total 83 and the superscripts total 211. For the decay of $^{238}_{92}\text{U}$, the subscripts total 92 on both sides and the superscripts total 238 on both sides.

The terms *parent nuclide* and *daughter nuclide* are often used in describing radioactive decay processes. The **parent nuclide** *is the nuclide undergoing decay.* The **daughter nuclide** *is the product nuclide resulting from the decay.* In our two previous equations, thallium-207 and thorium-234 are the daughter nuclides.

Beta-Particle Decay

Beta-particle decay also always results in the formation of a nuclide of a different element. The mass number of the new nuclide is the same as that of the original atom. The atomic number, however, has increased by one unit. The general equation for beta decay is

$$^{A}_{Z}\text{X} \longrightarrow {}^{0}_{-1}\beta + {}^{A}_{(Z+1)}\text{Y}$$

Specific examples of beta particle decay are

$$^{10}_{4}\text{Be} \longrightarrow {}^{0}_{-1}\beta + {}^{10}_{5}\text{B}$$

$$^{234}_{90}\text{Th} \longrightarrow {}^{0}_{-1}\beta + {}^{234}_{91}\text{Pa}$$

Both of these nuclear equations are balanced; superscripts and subscripts add to the same sums on both sides of the equation.

It is not immediately apparent how a nucleus, composed only of neutrons and protons, ejects a negative particle (beta particle) when no such particle is present in the nucleus. The accepted explanation is that through a complex series of steps a neutron in the nucleus is transformed into a proton and a beta particle; that is,

$$^{1}_{0}\text{n} \longrightarrow {}^{1}_{1}\text{p} + {}^{0}_{-1}\beta$$

Once formed within the nucleus, the beta particle is ejected with a high velocity. The net result of beta-particle formation is an increase by one in the number of protons present in the nucleus and a decrease by one in the number of neutrons present in the nucleus. Note in our two examples of beta emission that the daughter nuclide has one more proton than the parent as evidenced by the atomic number of the daughter being greater than that of the parent by one unit. Subtraction of the atomic number of the daughter nuclide from its mass number in each case—to get the number of neutrons—will reveal that the daughter nuclide contains one fewer neutron than the parent. An explanation of why and when beta-particle formation occurs is considered in Section 17.8.

Gamma-Ray Emission

For naturally occurring radionuclides, gamma-ray emission always occurs in conjunction with an alpha- or beta-decay process; it never occurs independently. Such gamma rays are most often not included in the nuclear equation, since they do not affect the balancing of the equation or the identity of the decay product. Gamma rays are to nuclear reactions what "heat" is to chemical reactions.

Just because gamma rays are usually left out of nuclear equations, it should not be assumed that they are not important. On the contrary, gamma rays are more important than alpha and beta particles when the effects of external radiation exposure on living organisms are considered (Sec. 17.11).

Example 17.1

Write a balanced nuclear equation for the decay of each of the following radioactive nuclides. The mode of decay is indicated in parentheses.

(a) $^{138}_{54}\text{Xe}$ (beta emission) **(c)** $^{190}_{78}\text{Pt}$ (alpha emission)

(b) $^{142}_{58}\text{Ce}$ (alpha emission) **(d)** $^{82}_{35}\text{Br}$ (beta emission)

Solution

In each case the atomic and mass numbers of the daughter nucleus are obtained by first writing the symbols of the parent nucleus and the particule emitted by the nucleus (alpha or beta particle) and then balancing the equation.

(a) Let X represent the product of the radioactive decay, that is, the daughter nuclide. Then

$$^{138}_{54}\text{Xe} \longrightarrow {}^{0}_{-1}\beta + \text{X}$$

Since the sums of the superscripts on both sides of the equation must be equal, the superscript for X must be 138. In order for the sums of the subscripts on both sides of the equation to be equal, the subscript for X must be 55. Then $54 = (-1) + (55)$. As soon as the subscript of X is determined, the identity of X can be determined from a periodic table. The element with an atomic number of 55 is cesium (Cs). Therefore,

$$^{138}_{54}\text{Xe} \longrightarrow {}^{0}_{-1}\beta + {}^{138}_{55}\text{Cs}$$

(b) Similarly, letting X represent the product of the radioactive decay, we have for the alpha decay of $^{142}_{58}Ce$

$$^{142}_{58}Ce \longrightarrow {}^{4}_{2}\alpha + X$$

Balancing the equation, making the superscripts on the right side of the equation total 142 and the subscripts total 58, we get

$$^{142}_{58}Ce \longrightarrow {}^{4}_{2}\alpha + {}^{138}_{56}Ba$$

(c) Similarly, we write

$$^{190}_{78}Pt \longrightarrow {}^{4}_{2}\alpha + X$$

Balancing superscripts and subscripts, we get

$$^{190}_{78}Pt \longrightarrow {}^{4}_{2}\alpha + {}^{186}_{76}Os$$

(d) Finally, we write

$$^{82}_{35}Br \longrightarrow {}^{0}_{-1}\beta + X$$

In beta emission the atomic number of the daughter nuclide is always greater by one and the mass number does not change from that of the parent. The balancing procedure gives us this result.

$$^{82}_{35}Br \longrightarrow {}^{0}_{-1}\beta + {}^{82}_{36}Kr$$

Practice Exercise 17.1

Write a balanced nuclear equation for the decay of each of the following radioactive nuclides. The mode of decay is indicated in parentheses.

(a) $^{147}_{62}Sm$ (alpha emission) (b) $^{117}_{48}Cd$ (beta emission)

Ans. (a) $^{147}_{62}Sm \longrightarrow {}^{4}_{2}\alpha + {}^{143}_{60}Nd$; (b) $^{117}_{48}Cd \longrightarrow {}^{0}_{-1}\beta + {}^{117}_{49}In$

17.5 RATE OF RADIOACTIVE DECAY

Each type of radioactive nuclide does not decay at the same rate as all other types. Some decay very rapidly; others undergo disintegration at extremely slow rates. This indicates that radionuclides are not all equally unstable. The faster the decay rate, the lower the stability.

The concept of half-life is used to express quantitatively nuclear stability. The **half-life** *is the time required for one half of any given quantity of a radioactive substance to undergo decay.* For example, if a radionuclide's half-life is 12 days and you have a 4.00-g sample of it, then after 12 days (one half-life) only 2.00 g of the sample (half the original amount) will remain undecayed; the other half will have decayed into some other substance.

TABLE 17.1 RANGE OF HALF-LIVES FOUND FOR NATURALLY OCCURRING RADIONUCLIDES

Element	Half-life
Vanadium-50	6×10^{15} yr
Platinum-190	6.9×10^{11} yr
Uranium-238	4.5×10^{9} yr
Uranium-235	7.1×10^{8} yr
Thorium-230	7.5×10^{4} yr
Lead-210	22 yr
Bismuth-214	19.7 min
Polonium-212	3.0×10^{-7} sec

Half-lives of billions of years and as short as a fraction of a second have been determined. Table 17.1 contains examples of the wide range of half-life values.

Most naturally occurring radionuclides have long half-lives. Some radionuclides with *short* half-lives, however, are also found in nature. Such short-lived species, since they decay rapidly, must be continually produced in order to be present. Processes that result in their production are: (1) the decay of naturally occurring long-lived nuclides, (2) the decay of short-lived nuclides (daughter nuclides) that have been produced in the previous manner, and (3) bombardment reactions involving cosmic rays, which take place naturally in the upper atmosphere. Examples of the second method of producing short-lived nuclides are presented in Section 17.8.

The decay rate (half-life) of a radionuclide is constant. It is independent of outward conditions such as temperature, pressure, and state of chemical combination. It is dependent only on the identity of the radionuclide. For example, radioactive sodium-24, whether incorporated into NaCl, NaBr, Na_2SO_4, or $NaC_2H_3O_2$, decays at the same rate. If a nuclide is radioactive, nothing will stop it from decaying and nothing will increase or decrease its decay rate.

Figure 17.2 shows graphically the meaning of half-life. After one half-life has passed, one half of the original atoms have decayed, so half remain. During the next half-life, one-half of the remaining half will decay, and one fourth of the original atoms remain undecayed. After three half-lives, $\frac{1}{2} \times \frac{1}{2} \times \frac{1}{2} = \frac{1}{8}$, of the original atoms remain undecayed, and so on. Note from Figure 17.2 that only a very small amount of original material (less than 1%) remains after seven half-lives have elapsed.

Calculations involving amounts of radioactive material decayed, amounts remaining undecayed, and time elapsed can be carried out by using the following equation.

$$\begin{matrix}\text{Amount of radionuclide} \\ \text{undecayed after } n \text{ half-lives}\end{matrix} = \begin{matrix}\text{original amount} \\ \text{of radionuclide}\end{matrix} \times \frac{1}{2^n}$$

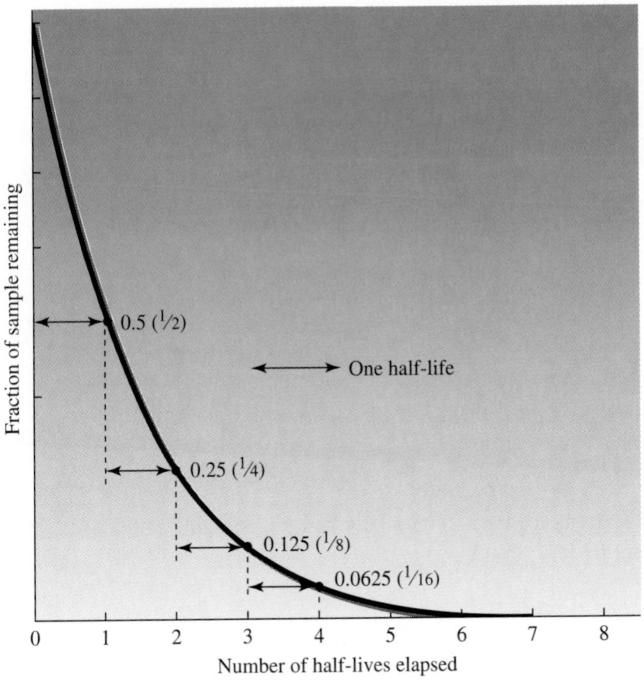

FIGURE 17.2 A general half-life decay curve for a radionuclide.

Example 17.2

The half-life of cobalt-60 is 5.3 yr. If 2.0 g of cobalt-60 is allowed to decay for a period of 15.9 yr, how many grams of cobalt-60 remain?

Solution

First, we must determine the number of half-lives that have elapsed.

$$15.9 \cancel{\text{yr}} \times \frac{1 \text{ half-life}}{5.3 \cancel{\text{yr}}} = 3.0 \text{ half-lives}$$

Knowing the number of elapsed half-lives and the original amount of radioactive cobalt present, we can use the equation

$$\begin{matrix}\text{Amount of radionuclide} \\ \text{undecayed after } n \text{ half-lives}\end{matrix} = \begin{matrix}\text{original amount} \\ \text{of radionuclide}\end{matrix} \times \frac{1}{2^n}$$

to get our answer.

$$\begin{matrix}\text{Amount of radionuclide} \\ \text{undecayed after } n \text{ half-lives}\end{matrix} = 2.0 \text{ g} \times \frac{1}{2^3} \leftarrow \text{three half-lives}$$

$$= 2.0 \text{ g} \times \frac{1}{8}$$

$$= 0.25 \text{ g (calculator and correct answer)}$$

Practice Exercise 17.2

The half-life of iodine-131 is 8.0 days. If 12 g of iodine-131 is allowed to decay for a period of 32 days, how many grams of iodine-131 remain?

Ans. 0.75 g

Example 17.3

Iodine-135 is a nuclide found in radioactive fallout from nuclear weapon explosions. Its half-life is 6.70 hr. How long will it take for 93.75% (15/16) of the iodine-135 atoms in a "fallout" sample to undergo decay?

Solution

If 15/16 of the sample has decayed, then 1/16 of the sample remains undecayed. In terms of $1/2^n$, 1/16 is equal to $1/2^4$; that is,

$$\frac{1}{2} \times \frac{1}{2} \times \frac{1}{2} \times \frac{1}{2} = \frac{1}{2^4} = \frac{1}{16}$$

Thus four half-lives have elapsed in reducing the amount of iodine-135 to 1/16 of its original amount.

Since the half-life of iodine-135 is 6.70 hr, the total time elapsed will be

$$4 \cancel{\text{half lives}} \times \frac{6.70 \text{ hr}}{1 \cancel{\text{half life}}} = 26.8 \text{ hr} \quad \text{(calculator and \textbf{correct answer})}$$

Practice Exercise 17.3

Strontium-90 is a nuclide found in radioactive fallout from nuclear weapon explosions. Its half-life is 28.0 yr. How long will it take for 75.0% (3/4) of the strontium-90 atoms in a "fallout" sample to undergo decay?

Ans. 56.0 yr

In both Examples 17.2 and 17.3 the time elapsed was equivalent to a whole number of half-lives. In order to work problems involving a fractional number of half-lives, more complicated equations with logarithms must be used. Such equations will not be presented in this text; hence only problems that involve a whole number of half-lives will be considered.

The half-life concept can be used to determine the age of objects that contain a radionuclide. The best-known radiochemical dating technique is radiocarbon dating. Radiocarbon dating involves measuring the amount of carbon-14 present in an object. Many archeological artifacts can be dated by carbon-14 techniques, since they were made from or contain once-living carbon-containing materials.

The isotope carbon-14, the only naturally occurring radionuclide of carbon, has a half-life of 5730 yr. It is continually produced in the upper atmosphere as a result of cosmic ray bombardment.

$$^{14}_{7}N + ^{1}_{0}n \longrightarrow ^{14}_{6}C + ^{1}_{1}p$$

The steady-state (equilibrium) concentration of carbon-14, which reflects both its rate of formation and its rate of decay, is one carbon-14 atom for every 10^{12} nonradioactive carbon atoms. This trace amount of carbon-14 in the atmosphere reacts with oxygen to give carbon dioxide in the same manner that nonradioactive carbon does. Thus, approximately 1 out of every 10^{12} carbon dioxide molecules is radioactive. This radioactive carbon is incorporated into the structure of plants through photosynthesis and into animals and human beings through the food chain. A steady-state concentration of carbon-14, equal to that found in the atmosphere, is thus found in all living organisms. Upon the death of an organism, the intake of carbon-14 ceases and the natural level of radioacative carbon present within the structure begins to decrease as the result of carbon-14 decay.

$$^{14}_{6}C \longrightarrow ^{0}_{-1}\beta + ^{14}_{7}N$$

In carbon-14 dating the ratio of carbon-14 to total carbon in an object that contains once-living material (parchment, cloth, charcoal, etc.) is compared to the ratio for living matter. A wooden object with a ratio of carbon-14 to total carbon one fourth that of a living tree would be approximately 11,400 years (two half-lives) old. An important assumption in the carbon-14 dating method is that the flow of carbon-14 into the biosphere is constant with time. There is some evidence, such as the carbon-14 content of the growth rings in old trees, to indicate that this is approximately true.

17.6 INDUCED RADIOACTIVITY: BOMBARDMENT REACTIONS

A **transmutation reaction** *is a nuclear reaction in which one nuclide is changed into a nuclide of another element.* Radioactive decay, discussed in Section 17.4, is an example of a natural transmutation process. It is also possible to cause transmutation to occur in a laboratory setting through use of bombardment reactions. A **bombardment reaction** *is a nuclear reaction in which small particles traveling at very high speeds are made to collide with target nuclei and cause them to undergo nuclear change.*

The first successful bombardment reaction was carried out in 1919, twenty-five years after the discovery of radioactive decay, by Ernest Rutherford, the same Rutherford who earlier had investigated the nature of alpha, beta, and gamma rays (Sec. 17.3). Rutherford's initial successful bombardment experiment consisted of letting alpha particles from a natural source (radium) bombard nitrogen gas. In this process he found that a new stable nuclide was formed: oxygen-17. The nuclear equation for this transmutation is

$$^{14}_{7}N + ^{4}_{2}\alpha \longrightarrow ^{17}_{8}O + ^{1}_{1}p$$

Since this initial successful reaction, further research carried out by many investigators has shown that numerous nuclei experience change under the stress of small particle bombardment. In most cases the new nuclide produced as a result of the transmutation is radioactive (unstable) rather than stable as was the case with oxygen-17.

The radioactivity associated with bombardment-reaction produced unstable nuclei is called *induced* radioactivity. **Induced radioactivity** *is the radioactivity associated with* *radionuclides produced from nonradioactive nuclides through bombardment reactions.* Radionuclides possessing induced radioactivity undergo radioactive decay, like naturally occurring radionuclides. In many cases, the previously discussed alpha- and beta-particle modes of decay (Sec. 17.4) occur. Additional modes of decay, to be discussed in Section 17.7, are also encountered.

In the 75 plus years since Rutherford's discovery of "bombardment," more than 1600 radionuclides that do not occur naturally have been produced. Included in this total is at least one radionuclide of every naturally occurring element. In addition, nuclides of 23 elements that do not occur in nature have been produced in small quantities as the result of bombardment reactions. Four of these "synthetic" elements, produced between 1937 and 1941, filled gaps in the periodic table for which no naturally occurring element had been found. These four elements are technetium (Tc, element 43), an element with numerous uses in nuclear medicine (Sec. 17.14); promethium (Pm, element 61); astatine (At, element 85); and francium (Fr, element 87). The remainder of the synthetic elements, elements 93 to 111, are called the *transuranium elements* because of their occurrence immediately following uranium in the periodic table. (Uranium is the highest atomic-numbered, naturally occurring element.) All isotopes of all of the transuranium elements are radioactive. Table 17.2 gives information about the stability of the transuranium elements. Note the extremely short half-lives of the more recently produced elements.

There are approximately five times as many laboratory-produced nuclides (1600) as naturally occurring nuclides (330). Significant uses exist for some "synthetic" radionu-

TABLE 17.2 STABILITY OF TRANSURANIUM ELEMENTS

Name	Symbol	Atomic Number	Mass Number of Most Stable Isotope	Half-life of Most Stable Isotope	Year of Discovery
Neptunium	Np	93	237	2.14×10^6 yr	1940
Plutonium	Pu	94	244	7.6×10^7 yr	1940
Americium	Am	95	243	8.0×10^3 yr	1944
Curium	Cm	96	247	1.6×10^7 yr	1944
Berkelium	Bk	97	247	1400 yr	1950
Californium	Cf	98	251	900 yr	1950
Einsteinium	Es	99	252	472 days	1952
Fermium	Fm	100	257	100 days	1953
Mendelevium	Md	101	258	52 days	1955
Nobelium	No	102	259	58 min	1958
Lawrencium	Lr	103	262	3.6 hr	1961
Rutherfordium	Rf	104	261	65 sec	1969
Hahnium	Ha	105	262	34 sec	1970
Seaborgium	Sg	106	266	20 sec	1974
Nielsbohrium	Ns	107	264	1.4 sec	1980
Hassium	Hs	108	267	0.074 sec	1984
Meitnerium	Mt	109	268	0.072 sec	1982
Element 110	—	110	271	0.006 sec	1994
Element 111	—	111	272	0.020 sec	1994

clides, particularly in the field of medicine. For example, the synthetic radionuclides cobalt-60, yttrium-90, iodine-131, and gold-198 find use in radiotherapy treatments for cancer.

Many of the early bombardment reactions were carried out with alpha particles ejected from naturally radioactive materials. Today, many other types of particles, generated in the laboratory by particle accelerators, are available to bombard nuclei. They include protons (hydrogen-1 nuclei), neutrons, and deuterons (hydrogen-2 nuclei). Gamma rays have also been used successfully to produce artificially radioactive nuclides through bombardment. Two examples of bombardment reactions now carried out in laboratories are

$$^{44}_{20}\text{Ca} + {}^{1}_{1}\text{H} \longrightarrow {}^{44}_{21}\text{Sc} + {}^{1}_{0}\text{n}$$
$$^{23}_{11}\text{Na} + {}^{2}_{1}\text{H} \longrightarrow {}^{21}_{10}\text{Ne} + {}^{4}_{2}\alpha$$

Gold can be produced from platinum by the bombardment technique. However, the process is astronomically expensive compared to the worth of the gold so produced. Platinum-196 is bombarded with deuterons (${}^{2}_{1}\text{H}$) to produce platinum-197.

$$^{196}_{78}\text{Pt} + {}^{2}_{1}\text{H} \longrightarrow {}^{197}_{78}\text{Pt} + {}^{1}_{1}\text{H}$$

The platinum-197 decays through beta-particle emission to produce gold-197.

$$^{197}_{78}\text{Pt} \longrightarrow {}^{0}_{-1}\beta + {}^{197}_{79}\text{Au}$$

17.7 POSITRON EMISSION AND ELECTRON CAPTURE

Laboratory-produced radionuclides (induced radioactivity) undergo radioactive decay just as do radionuclides from nature. Four modes of decay are encountered. They are alpha particle emission and beta particle emission (the same as for naturally occurring radionuclides—Sec. 17.4) and two new modes not found for naturally radioactive substances: positron emission and electron capture.

The particle involved in positron emission, the positron, is a particle we have not previously discussed. A **positron**, *designated by the symbol* ${}^{0}_{1}\beta$, *is identical to an electron or a beta particle except that it has a positive charge.* Its production in the nucleus is due to the conversion within the nucleus of a proton to a neutron

$$^{1}_{1}\text{p} \longrightarrow {}^{1}_{0}\text{n} + {}^{0}_{1}\beta$$

This process is just the opposite of that occurring during beta-particle emission (Sec. 17.4). The net effect of positron emission is thus to decrease the atomic number (number of protons), while the mass number remains constant. An example of a radioactive decay process involving positron emission is

$$^{30}_{15}\text{P} \longrightarrow {}^{0}_{1}\beta + {}^{30}_{14}\text{Si}$$

In **electron capture** *an electron in a low-energy orbital, such as the 1*s *orbital, is pulled into the nucleus, converting a proton to a neutron.*

$$^{0}_{-1}\text{e} + {}^{1}_{1}\text{p} \longrightarrow {}^{1}_{0}\text{n}$$

An example of such a process is the reaction

$$^{87}_{37}\text{Rb} + {}^{0}_{-1}\text{e} \longrightarrow {}^{87}_{36}\text{Kr}$$

Why and when positron emission and electron capture occur is considered in Section 17.8.

Example 17.4

Write a balanced nuclear equation for the decay of each of the following radioactive nuclides. The mode of decay is indicated in parentheses.

(a) $^{62}_{29}Cu$ (positron emission) **(c)** $^{105}_{47}Ag$ (electron capture)

(b) $^{118}_{52}Te$ (electron capture) **(d)** $^{82}_{37}Rb$ (positron emission)

Solution

In each case the atomic number and mass number of the daughter nucleus are obtained by first writing the symbols of the parent nucleus and the particle emitted (positron) or absorbed (electron) and then balancing the equation.

(a) Let X represent the product of the radioactive decay, that is, the daughter nuclide. Then

$$^{62}_{29}Cu \longrightarrow {}^{0}_{1}\beta + X$$

Note that the Greek letter β is used to denote not only a beta particle but also a positron. The difference between the two particles is that the former is negatively charged ($^{0}_{-1}\beta$) and the latter is positively charged ($^{0}_{1}\beta$).

Since the sum of the superscripts on each side of the equation must be equal, the superscript for X must be 62. In order for the sums of the subscripts on both sides of the equation to be equal, at 29, the subscript for X must be 28. As soon as the subscript of X is determined, the identity of X is known. Looking at a periodic table we determine that the element with an atomic number of 28 is nickel (Ni). Therefore,

$$^{62}_{29}Cu \longrightarrow {}^{0}_{1}\beta + {}^{62}_{28}Ni$$

(b) Similarly, letting X represent the product daughter nuclide of the radioactive decay, we have for $^{118}_{52}Te$ decaying by the electron capture mechanism

$$^{118}_{52}Te + {}^{0}_{-1}e \longrightarrow X$$

Note that in electron capture the electron appears on the reactant side of the equation. This makes equations for electron capture different from those for alpha, beta, and positron emissions where in each case the small particle involved is placed on the product side of the equation.

Balancing the above equation, making the superscripts on each side of the equation total 118 and the subscripts total 51, we get

$$^{118}_{52}Te + {}^{0}_{-1}e \longrightarrow {}^{118}_{51}Sb$$

(c) Similarly, for this electron capture we write

$$^{105}_{47}Ag + {}^{0}_{-1}e \longrightarrow X$$

Balancing superscripts and subscripts, we get

$$^{105}_{47}Ag + {}^{0}_{-1}e \longrightarrow {}^{105}_{46}Pd$$

(d) Finally, we write

$$^{82}_{37}Rb \longrightarrow {}^{0}_{1}\beta + X$$

In both positron emission and electron capture the atomic number of the daughter nuclide decreases by one and the mass number does not change from that of the parent. The balancing process gives results consistent with this generalization in this part as well as in the previous three parts of this problem.

$$^{82}_{37}Rb \longrightarrow ^{0}_{1}\beta + ^{82}_{36}Kr$$

Practice Exercise 17.4

Write a balanced nuclear equation for the decay of each of the following radioactive nuclides. The mode of decay is indicated in parentheses.

(a) $^{21}_{11}Na$ (positron emission) (b) $^{7}_{4}Be$ (electron capture)

Ans. (a) $^{21}_{11}Na \longrightarrow ^{0}_{1}\beta + ^{21}_{10}Ne$ (b) $^{7}_{4}Be + ^{0}_{-1}e \longrightarrow ^{7}_{3}Li$

17.8 NUCLEAR STABILITY

No simple rule exists that allows us to predict whether a particular nucleus is radioactive and how it might decay. However, a consideration of some observations about those nuclei that are stable is helpful in understanding why some nuclei are stable and others are not.

Two generalizations readily apparent from a study of the properties of the stable nuclei in nature are

1. *There is a correlation between nuclear stability and the total number of nucleons found in a nucleus.* All nuclei with 84 or more protons are unstable. The largest stable nucleus known is that of $^{209}_{83}Bi$, a nucleus that contains 209 nucleons. It thus appears that there is a limit to the number of nucleons that can be packed into a stable nucleus.
2. *There is a correlation between nuclear stability and neutron-to-proton ratio in a nucleus.* The number of neutrons necessary to create a stable nucleus increases as the number of protons increases. This pattern of increasing neutron-to-proton ratio with increasing atomic number is shown in Figure 17.3. For elements of low atomic number, neutron-to-proton ratios of stable nuclides are very close to 1. For heavier elements, stable nuclides have higher neutron-to-proton ratios, with the ratio reaching approximately 1.5 for the heaviest stable elements.

The fact that stable nuclei fall into a rather narrow zone of stability defined by neutron-to-proton ratios strongly suggests that neutrons are at least partially responsible for the stability of the nucleus. It should be remembered that like charges repel each other and that most nuclei contain many protons (with identical positive charges) squeezed together into a

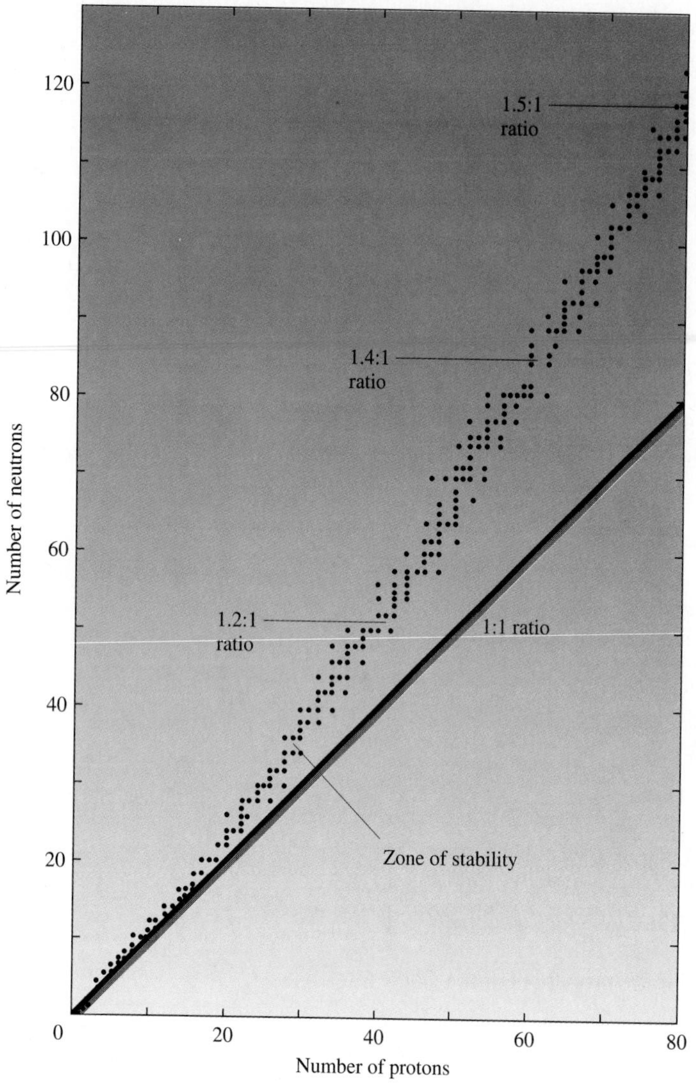

FIGURE 17.3 A graph of the number of neutrons versus the number of protons for stable nuclei.

very small volume. As the number of protons increases, the electrostatic forces of repulsion between protons increase sharply. Therefore a greater number of neutrons is required to counteract the increased repulsive forces. Finally, at element 84, the repulsive forces become great enough that the nuclei are unstable regardless of the number of neutrons present.

The mode by which a radionuclide decays depends to a large extent upon how the neutron-to-proton ratio for the radionuclide compares with that of stable nuclei containing approximately the same number of nucleons. The driving force for radioactive decay is the tendency of unstable nuclei to adjust neutron-to-proton ratios in such a way that stability is achieved.

Unstable nuclei can be classed into three categories on the basis of their position on the graph in Figure 17.3 relative to where the stable nuclei are found (zone of stability). These categories are

1. *Unstable nuclei in which the neutron-to-proton ratio is too high.* Such nuclides, which can be considered proton-poor, lie to the left of the zone of stability in Figure 17.3.

2. *Unstable nuclei in which the neutron-to-proton ratio is too low.* Such nuclides, which can be considered proton-rich, lie to the right of the zone of stability in Figure 17.3.
3. *Unstable nuclei in which the total number of nucleons exceeds 209—the limit for a stable nucleus.* Such nuclei lie beyond the zone of stability in Figure 17.3.

Nuclei in each of these categories have a particular mode of decay that predominates over others.

Beta emission is the predominant decay mode for nuclides having neutron-to-proton ratios that are *too high* for stability (category 1). In almost all cases, nuclei in this category have mass numbers greater than the atomic mass of the element. In the following example of beta emission, note that the mass number of the radioactive manganese nuclide (56) is greater than the atomic mass of manganese (54.9 amu).

$$^{56}_{25}\text{Mn} \longrightarrow ^{56}_{26}\text{Fe} + ^{0}_{-1}\beta$$

As discussed previously (Sec. 17.4), beta emission involves the transformation of a neutron into a proton. This increases the number of protons, decreases the number of neutrons, and causes a decrease in the neutron-to-proton ratio—the desired result. Note that in the above reaction for the beta decay of manganese-56, the neutron-to-proton ratio of the parent nuclide is 1.24 and that of the daughter iron-56 is 1.15.

Radionuclides lying to the right of the zone of stability, that is, those with *too low* a neutron-to-proton ratio (category 2), decay by converting a proton into a neutron—just the opposite of the process that occurs in beta-particle emission. Radionuclei in this category generally have mass numbers that are lower than the atomic mass of the element. The conversion of a proton into a neutron can be accomplished by either of two ways: (1) by positron emission or (2) by electron capture. These decay modes were described in Section 17.7. The process of electron capture seems to be preferred over positron emission for nuclides of high atomic number. For lighter nuclides numerous examples of both types of decay processes are known.

Alpha-particle emission is found primarily among elements in which the total number of nucleons exceeds 209 (category 3). These are the nuclei that lie *beyond* the region of stability (Fig. 17.3). Alpha-particle emission is not, however, the only mode of decay for elements in this region; beta-particle emission is also common.

Another characteristic of category 3 radionuclides is that the decay process for reaching stability involves more than one step. This results in the formation of a *decay series*, the topic of Section 17.9.

Before we leave this section, we should note that the preceding guidelines work in most instances; however, there are exceptions. For example, both $^{146}_{60}\text{Nd}$ and $^{148}_{60}\text{Nd}$ are stable and lie in the region of stability in Figure 17.3, but $^{147}_{60}\text{Nd}$, which is radioactive, also lies in the region of stability (between the two stable nuclides).

17.9 RADIOACTIVE DECAY SERIES

Radioactive nuclides with high atomic numbers cannot attain nuclear stability with a single emission; they are too far away from the zone of stability (Fig. 17.3) for this to occur. Instead, a series of decay steps is required for such nuclei to reach stability. A large radionuclide undergoes decay to produce a product nucleus that is also radioactive. This product in turn produces a third radionuclide; this in turn decays to produce a fourth nuclide, and so

TABLE 17.3 THE FOUR KNOWN RADIOACTIVE DECAY SERIES

Parent	Number of Decay Steps	Final Product of Series
Uranium-238	14	lead-206
Thorium-232	10	lead-208
Uranium-235	11	lead-207
Plutonium-241	13	bismuth-209

forth, until ultimately a stable nucleus is produced. Such a sequence of decay products is called a radioactive decay series. A **radioactive decay series** *is a sequence of nuclear reactions in which one radioactive nuclide decays to a second, which then decays to a third, and so forth, until a stable nuclide is finally produced.*

Three naturally occurring decay series are known. Each starts with a long-lived radionuclide and ends with a stable nuclide. A fourth decay series was discovered after the synthesis of certain elements not found in nature (transuranium elements—Sec. 17.6). Because the parent of this fourth series, plutonium-241, does not occur in nature to a measurable extent, the series is not classified as naturally occurring. General characteristics of the four decay series are given in Table 17.3.

Figure 17.4 shows all of the members of the uranium-238 decay series. It is represen-

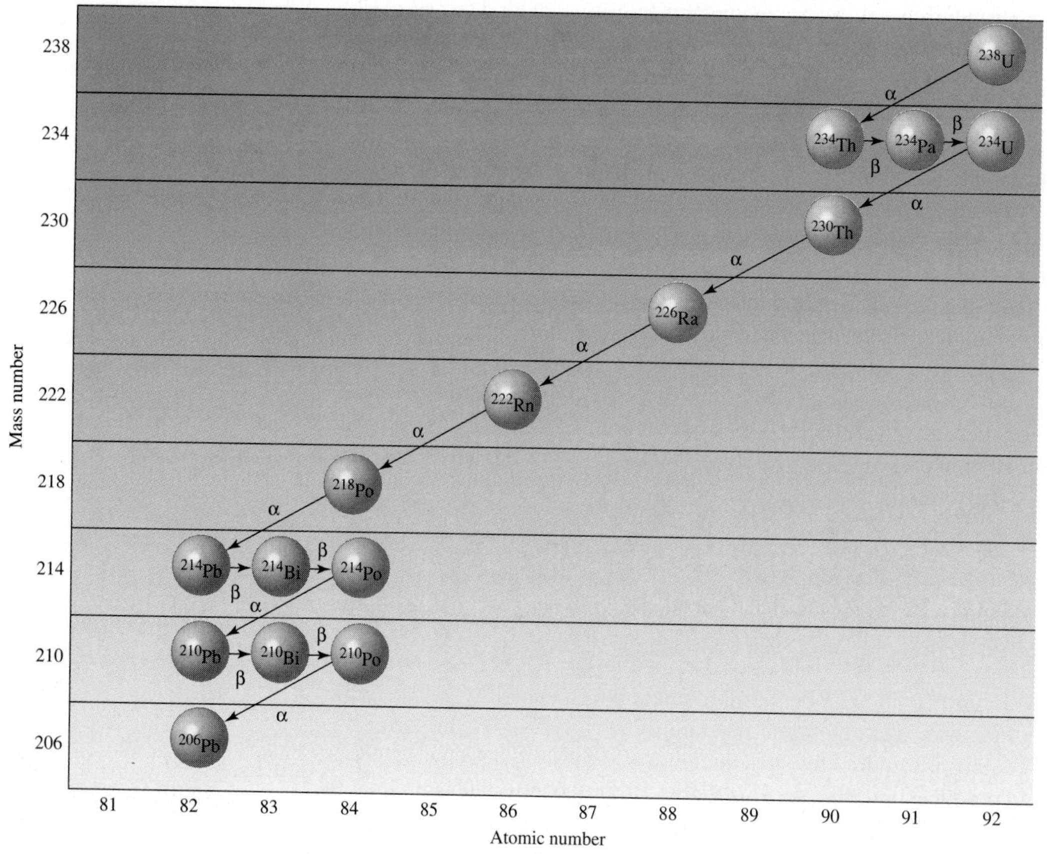

FIGURE 17.4 In this uranium-238 decay series, each nuclide, except lead-206 (the final product), is radioactive; the successive transformations continue until this stable product is obtained.

tative of the other three series. Note, from Figure 17.4, that both alpha and beta particles are part of the decay sequence and that there is no pattern relative to which is emitted when.

One of the intermediate products in the uranium-238 decay series is the radionuclide radon-222 (see Fig. 17.4). This substance is a gas at normal temperatures, and is therefore a very mobile species. Its presence has been detected in both aqueous and atmospheric environments. For the typical American, radon-222 is a major exposure source (see Sec. 17.13).

17.10 IONIZING EFFECTS OF RADIATION

The alpha, beta, and gamma radiations produced from radioactive decay travel outward from their nuclear sources into the material surrounding the radioactive substance. There, they interact with the atoms and molecules of the material and lose their energy. Let us consider in closer detail these interactions between radiation and atoms and molecules.

Because the nucleus of an atom occupies such a small portion of the total volume of an atom (Sec. 5.5), it is not surprising that in the great majority of radiation–atom interactions the extranuclear electrons of an atom are more directly involved than the nucleus is. In many cases, energy transfer during radiation–atom interaction is sufficient to knock electrons away from atoms; that is, ionization occurs and ion pairs are formed. An **ion pair** *is the electron and positive ion that are produced during an ionization collision between an atom or molecule and radiation*. This ionization process is not the voluntary transfer of electrons that occurs during ionic compound formation, but rather a nonchemical, involuntary removal of electrons from atoms to form ions. Figure 17.5 diagrammatically shows ion pair formation. Many ion pairs are produced by a single "particle" of radiation because such a particle must

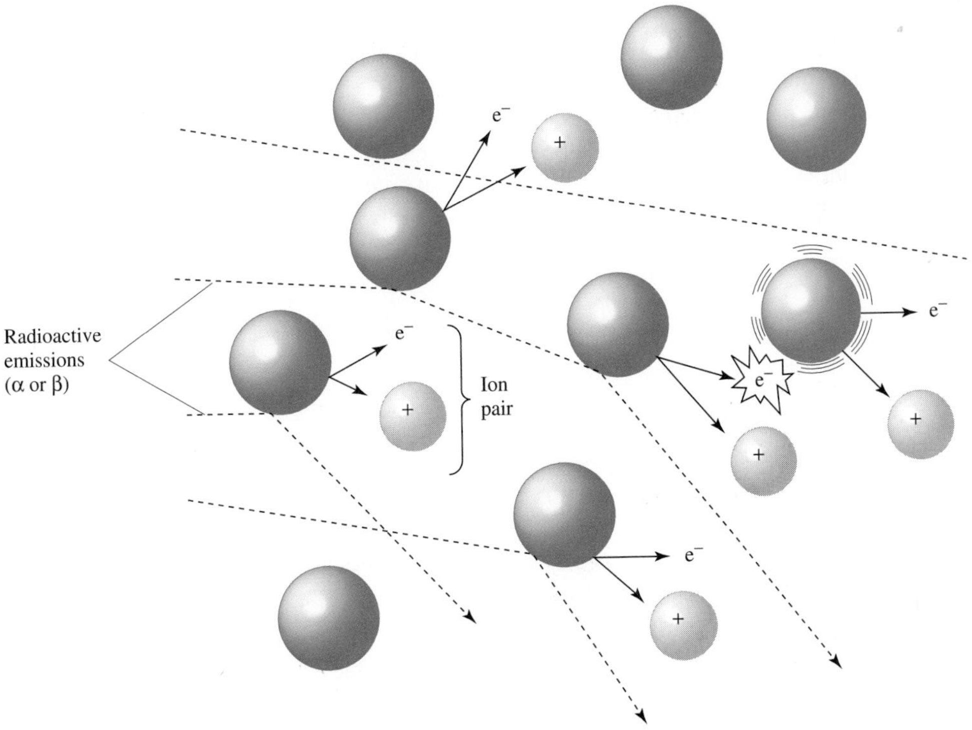

FIGURE 17.5 The interaction of radiation with atoms to form ion pairs.

undergo many collisions before its energy is reduced to the level of the surrounding material. The electrons ejected from an atom frequently have enough energy to bombard neighboring molecules and cause additional ionization.

Free-radical formation is another effect caused by alpha, beta, and gamma radiation. A **free radical** *is a highly reactive uncharged molecule fragment, that is, an* uncharged *piece of a molecule*. A free radical is a very reactive entity because it contains an unpaired electron. (Recall from Sec. 7.9 that electrons occur in pairs in normal bonding situations.)

Ion-pair formation and free-radical formation are the factors that makes radiation harmful to both living matter and inert materials. In living matter, the formation of ions and free radicals disrupts cellular function.

17.11 BIOLOGICAL EFFECTS OF VARIOUS TYPES OF RADIATION

The three types of naturally occurring radioactive emissions—alpha particles, beta particles, and gamma rays—differ in their ability to penetrate matter and to cause ionization. Consequently, the extent of the biological effects of radiation depends on the type of radiation involved.

Alpha particles are the most massive and also the slowest moving particles involved in natural radioactive decay; they are emitted from nuclei at a velocity of about one tenth of the speed of light. They have low penetrating power and cannot penetrate the body's outer layers of skin. The major danger from alpha radiation occurs when alpha-emitting radionuclides are ingested, for example, in contaminated food, or inhaled in air. There are no protective layers of skin within the body.

Beta particles are emitted from nuclei at speeds of up to nine-tenths that of light. With their greater velocity, they can penetrate much deeper than alpha particles and can cause severe skin burns if their source remains in contact with the skin for an appreciable time. They do not ionize molecules as readily as alpha particles do because of their much smaller size. An alpha particle is approximately 8000 times heavier than a beta particle. It is estimated that a typical alpha particle travels about 6 cm in air and produces 40,000 ion pairs and a typical beta particle travels 1000 cm in air and produces about 2000 ion pairs. Internal exposure to beta radiation is also serious.

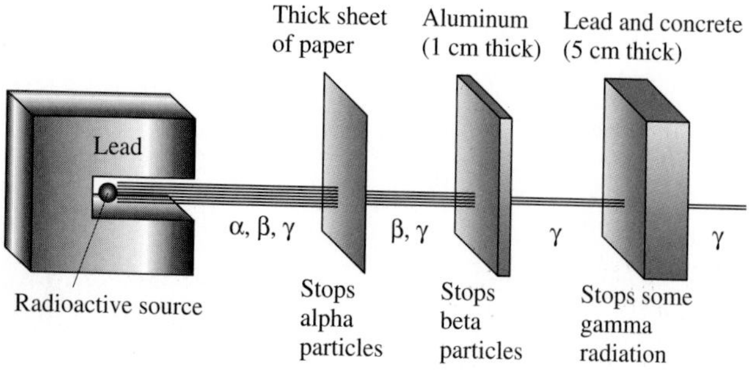

FIGURE 17.6 Penetrating abilities of alpha, beta, and gamma radiation.

TABLE 17.4 THE EFFECTS OF SHORT-TERM WHOLE-BODY RADIATION EXPOSURE ON HUMANS

Dose (rems)[a]	Effects
0–25	no detectable clinical effects
25–100	slight, short-term reduction in number of some blood cells; disabling sickness not common
100–200	nausea and fatigue, vomiting if dose is greater than 125 rems; longer term reduction in number of some blood cells
200–300	nausea and vomiting first day of exposure; up to 2-week latent period followed by appetite loss, general malaise, sore throat, pallor, diarrhea, and moderate emaciation; recovery in about 3 months unless complicated by infection or injury
300–600	nausea, vomiting, and diarrhea in first hours; up to a 1-week latent period followed by loss of appetite, fever, and general malaise in the second week, followed by hemorrhage, inflammation of mouth and throat, diarrhea, and emaciation; some deaths in 2–6 weeks, eventual death for 50% if exposure is above 450 rems; others recover in about 6 months
600 or more	nausea, vomiting, and diarrhea in first few hours; rapid emaciation and death as early as second week; eventual death of nearly 100%

[a] A rem is the quantity of ionizing radiation that must be absorbed by a human to produce the same biological effect as 1 roentgen of high-penetration X rays. A roentgen is the quantity of high-penetration X rays that produces approximately 2×10^9 ion pairs per cubic centimeter in dry air at 0°C and 1 atm.

Gamma radiation travels at the speed of light. Gamma rays readily penetrate deeply into organs, bones, and tissues.

Figure 17.6 contrasts the abilities of alpha, beta, and gamma radiations to penetrate paper, aluminum foil, and a thin layer of a lead–concrete mixture.

The minimum radiation dosage that causes human injury is unknown. However, the effects of larger doses have been studied and are listed in Table 17.4. As you can see, very serious damage or death can result from large doses of ionizing radiation. The doses causing the various effects listed in Table 17.4 are given in terms of the radiation unit called a *rem*, which is defined in the footnote to the table.

17.12 DETECTION OF RADIATION

You cannot hear, feel, taste, see, or smell low levels of radiation. However, there are numerous methods for detecting the presence of radiation. Becquerel's initial discovery of radioactivity (Sec. 17.2) was the result of the effect of radiation on photographic plates. Radiation affects photographic film in the same way as ordinary light; the film is exposed. Technicians

FIGURE 17.7 Film badges are worn by technicians and others whose work requires them to be around radiation. The extent of exposure of all personnel is carefully logged to help prevent overexposure. (Tech/Ops Landauer, Inc.—PAR/NYC photo)

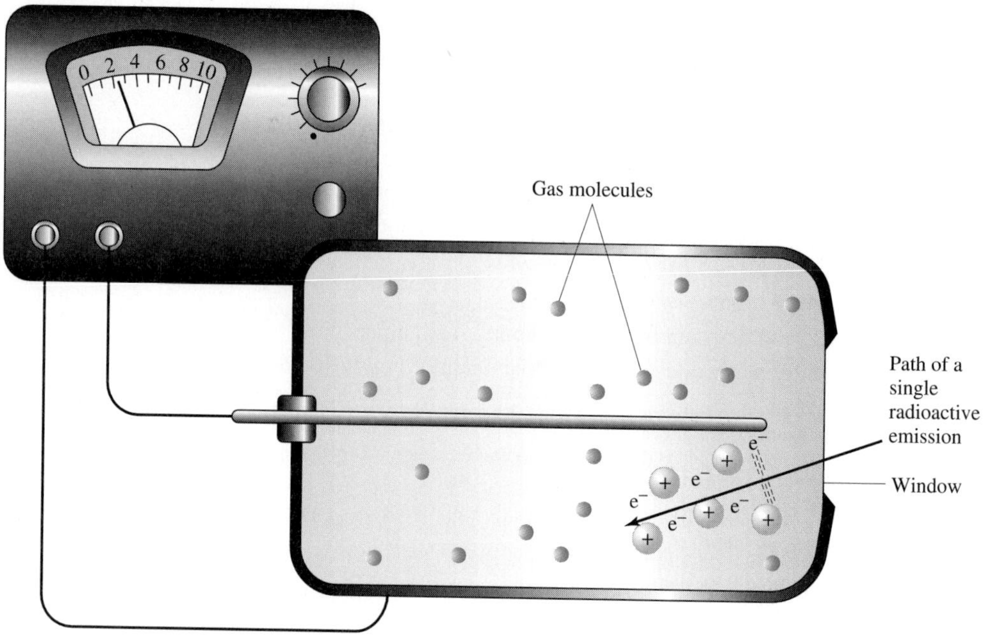

FIGURE 17.8 The principle of operation of a Geiger counter. When radiation enters through the window, it ionizes one or more gas atoms, producing ion pairs. The electrons from the ion pairs are attracted to the central wire, and the positive ions are drawn to the metal tube. This constitutes a pulse of electric current, which is amplified and displayed on the meter or other readout.

and others who work around radiation usually wear film badges (see Fig. 17.7) to record the extent of their exposure to radiation. When the film from the badge is developed, the degree of darkening of the film indicates the extent of radiation exposure. By using different filters, various parts of the film register exposures to the types of radiation (alpha, beta, gamma, and X rays).

Radiation can also be detected by making use of the fact that radiation ionizes atoms and molecules (Sec. 17.10). The Geiger counter operates on this principle. The basic components of a Geiger counter are shown in Figure 17.8. The detection part of such a counter is a metal tube filled with a gas (usually argon). The tube has a thin-walled window made of a material that can be penetrated by the radiation. In the center of the tube is a wire attached to the positive terminal of an electric power source. The shell of the metal tube is attached to the negative terminal of the same source. Radiation entering the tube ionizes the gas, which allows a pulse of electricity to flow. This pulse is amplified and displayed on a meter or some other type of readout display.

17.13 SOURCES OF RADIATION EXPOSURE

Most of us will never come into contact with the radiation dosage necessary to cause the effects listed in Table 17.4. Nevertheless, *low level* exposure to ionizing radiation is something we constantly encounter. In fact, there is no way we can totally avoid this low level exposure because much of it results from naturally occurring materials in the environment. Estimates of per capita radiation exposure from various sources are given in Figure 17.9.

A comparison of the values of Figure 17.9, which are in *millirem* units, with those of Table 17.4, which are in *rem* units, shows that the current dosage levels experienced by the general population are very small compared to those known to cause serious radiation sickness. Nevertheless, it is important to monitor carefully the amount of radioactive materials in the environment in order to ensure that radiation levels remain low.

With low-level radiation exposure, cell damage rather than cell death frequently occurs. If the damaged cells repair themselves improperly, which often occurs, new, abnormal cells are produced when the cells replicate. Much still needs to be learned about the long-term effects of cell damage caused by low-level radiation exposure.

Cells that reproduce at a rapid rate, such as those in bone marrow, lymph nodes, and embryonic tissue, are the most sensitive to radiation damage. One of the first signs of overexposure to radiation is a drop in red blood cell count. This directly relates to the sensitivity of bone marrow, the site of red cell formation, to radiation. The sensitivity of embryonic tissue to radiation damage is the reason that pregnant women need to be protected from radiation exposure.

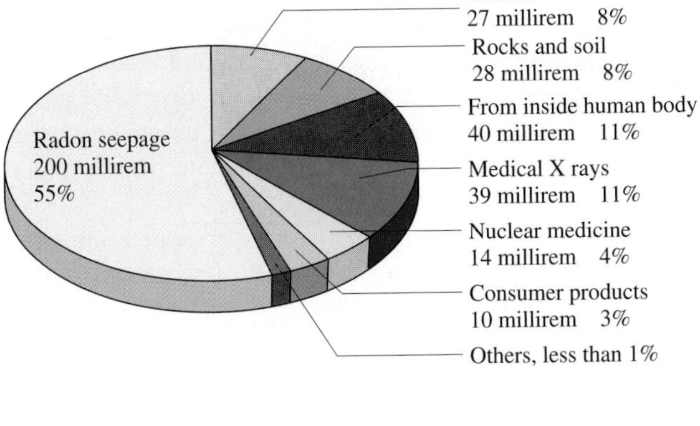

FIGURE 17.9 Estimated annual radiation exposure, in millirems, of an average American. Individual exposures vary widely. Shaded area is radiation from human activities.

17.14 NUCLEAR MEDICINE

Radionuclides are used both diagnostically and therapeutically in medicine. In diagnostic applications, technicians use small amounts of radionuclides whose progress through the body or localization in specific organs can be followed. Larger quantities of radionuclides are used in therapeutic applications.

The fundamental chemical principle behind the use of radionuclides in diagnostic medical work is the fact that a radioactive isotope of an element has the same chemical properties as a nonradioactive isotope of the element. (Radioactive and nonradioactive isotopes of an element differ in nuclear properties but not in chemical properties.) Thus, body chemistry is not upset by the presence of a small amount of a radioactive substance that is already present in the body in a nonradioactive form.

The criteria used in selecting radionuclides for diagnostic procedures include the following.

1. At low concentrations (to minimize radiation damage), the radionuclide must be detectable by instrumentation placed outside the body. Almost all diagnostic radionuclides are gamma emitters because the penetrating power of alpha and beta particles is too low.
2. It must have a short half-life so that the intensity of the radiation is sufficiently great to be detected. A short half-life also limits the time period of radiation exposure.
3. It must have a known mechanism for elimination from the body so that the material does not remain in the body indefinitely.
4. Its chemical properties must be such that it is compatible with normal body chemistry. It must be able to be selectively transmitted to the part or system of the body that is under study.

The circulation of blood in the body can be followed by using radioactive sodium-24. A small amount of this isotope is injected into the bloodstream in the form of a sodium chloride solution. The movement of this radionuclide through the circulation system can be followed easily with a radiation detector. If the nuclide takes longer than normal to move through a particular part of the body, this is an indication that the circulation is impaired at that spot. Sodium-24 can also be used to locate blood clots because the amount of radiation will be high on one side and low on the opposite side of the clot.

Radiologists evaluate the functioning of the thyroid gland by administering iodine-131 to a patient, usually in the form of a sodium iodide (NaI) solution. The radioactive iodine behaves in the same manner as ordinary iodine and is absorbed by the thyroid at a rate related to the activity of the gland. If a hypothyroid condition exists, the amount accumulated is less than normal, and if a hyperthyroid condition exists, a greater than average amount accumulates.

The size and shape of organs, as well as the presence of tumors, can be determined in some situations by scanning the organ in which a radionuclide tends to concentrate. Iodine-131 and technetium-99, in the form of a polyatomic ion (TcO_4^-), concentrate in brain tumors more than in normal brain tissue; this helps radiologists determine the presence, size, and location of brain tumors.

Table 17.5 lists a number of radionuclides that are used in diagnostic procedures. The

TABLE 17.5 SELECTED RADIONUCLIDES USED IN DIAGNOSTIC PROCEDURES

Isotope	Half-life	Part of Body	Use in Diagnosis
Barium-131	11.6 days	bone	detection of bone tumors
Chromium-51	27.8 days	blood	determination of blood volume and red blood cell lifetime
		kidney	assessment of kidney activity
Iodine-131	8.05 days	brain	detection of fluid buildup in the brain
		kidney	location of cysts
		lung	location of blood clots
		thyroid	assessment of iodine uptake by thyroid
Iron-59	45 days	blood	evaluation of iron metabolism in blood
Phosphorus-32	14.3 days	blood	blood studies
		breast	assessment of breast carcinoma
Potassium-42	12.4 hr	tissue	determination of intercellular spaces in fluids
Sodium-24	15.0 hr	blood	detection of circulatory problem; assessment of peripheral vascular disease
Technetium-99	6.0 hr	brain	detection of brain tumors, hemorrhages, or blood clots
		spleen	measurement of size and shape of spleen
		thyroid	measurement of size and shape of thyroid
		lung	location of blood clots

half-life of the radionuclide, the parts of the body it concentrates in, and its diagnostic value are also given.

When radionuclides are used for therapeutic purposes, the objectives are entirely different from those for diagnostic procedures. The main objective for radionuclides in therapeutic use is to destroy *selectively* abnormal (usually cancerous) cells. The radionuclide is often, but not always, placed within the body. There is no need to monitor the radiation produced with an external detector. Therapeutic radionuclides implanted in the body are usually alpha or beta emitters because an intense dose of radiation is needed in a small localized area.

A commonly used implantation radionuclide that is effective in the localized treatment of tumors is yttrium-90, a beta emitter with a half-life of 64 hr. Yttrium-90 salts are implanted by using small hollow needles that are inserted into the tumor.

External, high-energy beams of gamma radiation are also used extensively in the treatment of certain cancers. Cobalt-60 is frequently used for this purpose; a beam of radiation is focused on the small area of the body where the tumor is located. This therapy usually causes some radiation sickness because normal cells are also affected, although to a lesser extent. The operating principle here is that the more rapidly dividing abnormal cells are more susceptible to radiation damage than normal cells are (Sec. 17.13). Radiation sickness

TABLE 17.6 SELECTED RADIONUCLIDES USED IN RADIATION THERAPY

Isotope	Half-life	Type of Emitter	Use in Therapy
Cobalt-60	5.3 yr	beta, gamma	external source of radiation for treatment of cancer
Iodine-131	8 days	beta, gamma	cancer of thryoid
Phosphorus-32	14.3 days	beta, gamma	treatment of some types of leukemia and widespread carcinomas
Radium-226	1620 yr	alpha, gamma	used in implantation cancer therapy
Radon-222	3.8 days	alpha, gamma	used in treatment of uterine, cervical, oral, and bladder cancers
Yttrium-90	64 hr	beta, gamma	implantation therapy

is the price paid for the destruction of abnormal cells. Table 17.6 lists selected radionuclides that are used in therapy.

17.15 NUCLEAR FISSION AND NUCLEAR FUSION

Our glimpse into the world of nuclear chemistry would not be complete without a brief mention of two additional types of nuclear reactions: nuclear fission and nuclear fusion. **Nuclear fission** *is the process in which the nucleus of a heavy element splits into two or more lighter nuclei as the result of nuclear bombardment.*

The first fissionable nucleus to be discovered, which remains the most important one, was uranium-235. Bombardment of this nucleus with neutrons causes it to split into two fragments. Characteristics of the uranium-235 fission reaction include the following.

1. There is no unique way in which the uranium-235 nucleus splits. The following are examples of ways in which this fission process may proceed.

$$^{235}_{92}U + ^{1}_{0}n \longrightarrow ^{135}_{53}I + ^{97}_{39}Y + 4\,^{1}_{0}n$$
$$^{235}_{92}U + ^{1}_{0}n \longrightarrow ^{139}_{56}Ba + ^{94}_{36}Kr + 3\,^{1}_{0}n$$
$$^{235}_{92}U + ^{1}_{0}n \longrightarrow ^{131}_{50}Sn + ^{103}_{42}Mo + 2\,^{1}_{0}n$$
$$^{235}_{92}U + ^{1}_{0}n \longrightarrow ^{139}_{54}Xe + ^{95}_{38}Sr + 2\,^{1}_{0}n$$

2. Very large amounts of energy, which are many times greater than that released by ordinary radioactive decay, are emitted during the fission process. It is this large release of energy that makes nuclear fission of uranium-235 the important process that it is. In general, the term *nuclear energy* is used to refer to the energy released during the nuclear fission process. An older term for this energy is *atomic energy.*
3. Neutrons, which are reactants in the fission process, are also produced as products. The number of neutrons produced per fission depends on the way

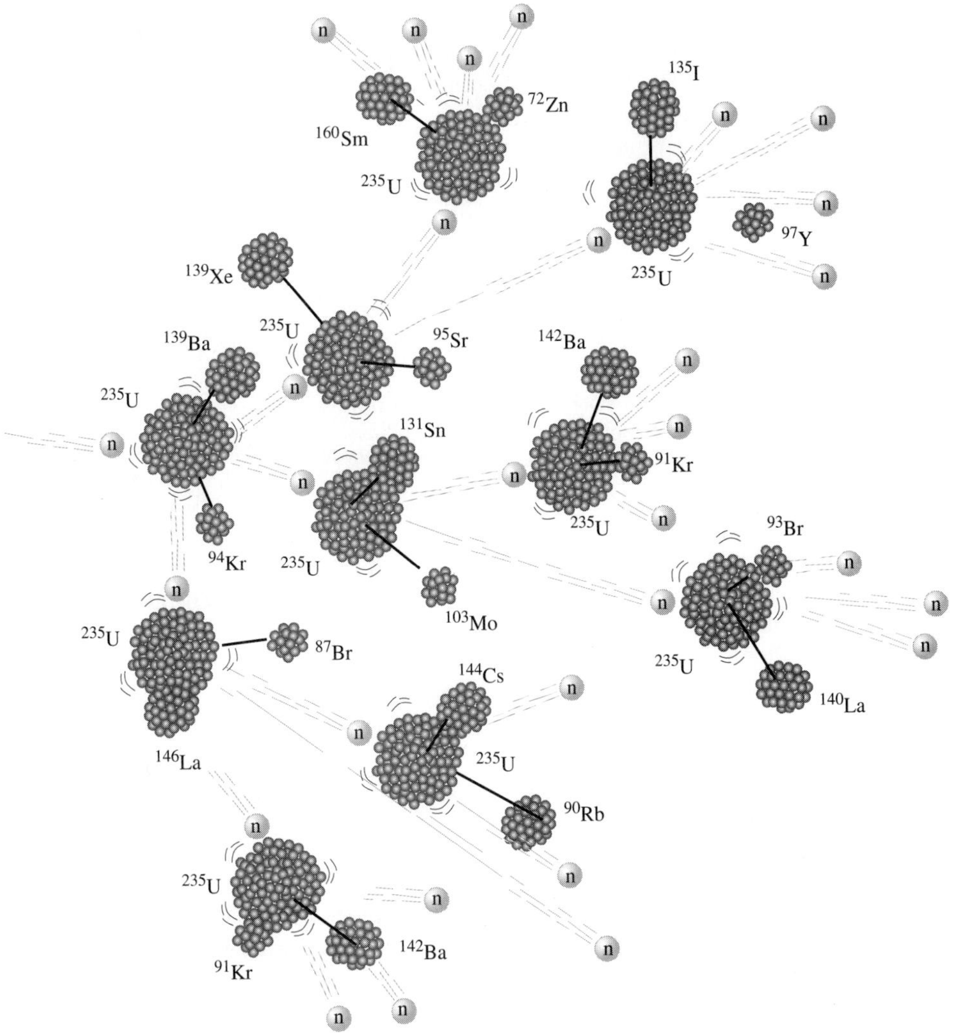

FIGURE 17.10 A fission chain reaction caused by further reaction of the neutrons produced during fission. (The "free" neutrons are shown here proportionally larger than those contained within the nucleus.)

in which the nucleus splits and ranges from two to four (as can be seen from the fission equations just given). On the average, 2.4 neutrons are produced per fission. The significance of the produced neutrons is that they can cause the fission process to continue by colliding with other uranium-235 nuclei. Figure 17.10 shows diagrammatically the chain reaction that can occur once the fission process is started.

The process of nuclear fission, or "splitting the atom" as it is called in popularized science, can be carried out in both an uncontrolled manner and a controlled manner. The key question is "What happens to the neutrons produced during fission?" Do they react further, causing additional fission, or do they escape into the surroundings? If the majority of the neutrons produced react further (Fig. 17.10), an uncontrolled nuclear reaction (an atomic bomb) results. When only a few neutrons react further (on the average, one per fission), the fission reaction self-propagates in a controlled manner.

The process of nuclear fission is the basis for the operation of nuclear power plants to produce electricity. In this case, the fission process is carried out in a controlled manner.

The reaction is controlled with rods that absorb excess neutrons (so that they cannot cause unwanted fissions) or with moderating substances that decrease the speed of the neutrons. The energy produced during the fission process, which appears as heat, is used to operate steam-powered, electricity-generating equipment.

There is another type of nuclear reaction that produces even more energy than nuclear fission. **Nuclear fusion** *is the process in which small nuclei are put together to make larger ones.* This process is essentially the opposite of nuclear fission. Fusion requires a very high temperature—several million degrees.

One place that is hot enough for nuclear fusion to occur is the interior of the sun. Nuclear fusion is the process by which the sun generates its energy. Within the sun, in a three-step reaction, hydrogen-1 nuclei are converted to helium-4 nuclei with the release of extraordinarily large amounts of energy.

The use of nuclear fusion on Earth might seem impossible because of the high temperatures required. It has, however, been accomplished in a hydrogen bomb. In such a bomb, a *fission* device (an atomic bomb) is used to achieve the high temperatures needed to start the fusion reaction.

$$^{3}_{1}H + ^{2}_{1}H \longrightarrow ^{4}_{2}He + ^{1}_{0}n$$

At the high temperature of fusion reactions, electrons completely separate from nuclei. Neutral atoms cannot exist. The high temperature, gaslike mixture of nuclei and electrons that results is called a *plasma* and is considered by some scientists to represent a fourth state of matter (in addition to solids, liquids, and gases).

17.16 A COMPARISON OF NUCLEAR AND CHEMICAL REACTIONS

As can be seen from the discussion of the previous sections of this chapter, nuclear chemistry is quite different from ordinary chemistry. Many of the laws of chemistry must be modified when we consider nuclear reactions. The major differences between nuclear reactions and ordinary chemical reactions are listed in Table 17.7. This table serves as a summary of many of the concepts presented in this chapter.

TABLE 17.7 DIFFERENCES BETWEEN NUCLEAR AND CHEMICAL REACTIONS

Chemical Reaction	Nuclear Reaction
1. Different isotopes of an element have practically identical chemical properties.	1. Different isotopes of an element have different properties in nuclear processes.
2. The chemical reactivity of an element depends on the element's state of combination (free element, compound etc.).	2. The nuclear reactivity of an element is independent of the state of chemical combination.
3. Elements retain their identity in chemical reactions.	3. Elements may be changed into other elements during nuclear reactions.
4. Energy changes that accompany chemical reactions are relatively small.	4. Nuclear reactions involve energy changes a number of orders of magnitude larger than those in chemical reactions.

KEY TERMS

The new terms or concepts defined in this chapter are

alpha rays (Sec. 17.3) A stream of positively charged particles (alpha particles), each of which is made up of two protons and two neutrons.

balanced nuclear equation (Sec. 17.4) A nuclear equation in which the sums of the subscripts (atomic number or particle charge) on both sides of the equation are equal and the sums of the superscripts (mass number) on both sides of the equation are equal.

beta rays (Sec. 17.3) A stream of negatively charged particles (beta particles) whose charge and mass are identical to those of an electron.

bombardment reaction (Sec. 17.6) A nuclear reaction in which small particles traveling at very high speeds are made to collide with target nuclei and cause them to undergo nuclear change.

daughter nuclide (Sec. 17.4) The product nuclide in a radioactive decay process.

electron capture (Sec. 17.7) Process in which an electron in a low-energy orbital is pulled into the nucleus, converting a proton to a neutron.

free radical (Sec. 17.10) A highly reactive uncharged molecular fragment, that is, an uncharged piece of a molecule.

gamma rays (Sec. 17.3) High energy radiation without mass or charge.

half-life (Sec. 17.5) Time required for one half of any given quantity of a radioactive substance to undergo decay.

induced radioactivity (Sec. 17.6) The radioactivity associated with radionuclides produced from nonradioactive nuclides through bombardment reactions.

ion pair (Sec. 17.10) The electron and positive ion that are produced during an ionization collison between an atom or molecule and radiation.

nuclear fission (Sec. 17.15) The process in which the nucleus of a heavy element splits into two or more lighter nuclei as the result of nuclear bombardment.

nuclear fusion (Sec. 17.15) The process in which small nuclei are put together to make larger ones.

nuclear reaction (Sec. 17.1) A reaction in which changes in the nucleus occur.

nuclide (Sec. 17.1) An atom with specific numbers of protons and neutrons in its nucleus.

parent nuclide (Sec. 17.4) The nuclide undergoing decay in a nuclear reaction.

positron (Sec. 17.7) A particle identical to an electron or beta particle except that it has a positive charge.

radioactive decay (Sec. 17.4) Process whereby a radionuclide is transformed into another nuclide as a result of emission of radiation from its nucleus.

radioactive decay series (Sec. 17.9) A sequence of nuclear reactions in which one radioactive nuclide decays to a second, which then decays to a third, and so on, until a stable nuclide is finally produced.

radioactive nuclide (Sec. 17.1) A nuclide with an unstable nucleus that spontaneously emits energy (radiation).

radioactivity (Sec. 17.1) The radiation spontaneously emitted from the nucleus of an unstable nuclide.

radionuclide (Sec. 17.1) A shortened form of the term *radioactive nuclide*.

stable nuclide (Sec. 17.1) A nuclide that has a nucleus that is stable.

transmutation reaction (Sec. 17.6) A nuclear reaction in which one nuclide is changed into a nuclide of another element.

unstable nuclide (Sec. 17.1) A nuclide that has a nucleus that is unstable.

PRACTICE PROBLEMS

Atomic Nuclei (Sec. 17.1)

17.1 What is the difference in meaning between the terms *radioactive nuclide* and *nonradioactive nuclide*?

17.2 What is the difference in meaning between the terms *nuclide* and *isotope*?

17.3 Use two different notations to denote each of the following nuclides.
(a) contains 5 protons, 4 neutrons, and 5 electrons
(b) contains 19 protons, 25 neutrons, and 19 electrons
(c) contains 45 protons, 51 neutrons, and 45 electrons
(d) contains 73 protons, 109 neutrons, and 73 electrons

17.4 Use two different notations to denote each of the following nuclides.
(a) contains 7 protons, 9 neutrons, and 7 electrons
(b) contains 31 protons, 45 neutrons, and 31 electrons
(c) contains 56 protons, 84 neutrons, and 56 electrons
(d) contains 75 protons, 105 neutrons, and 75 electrons.

The Nature of Naturally Occurring Radioactive Emissions (Sec. 17.3)

17.5 Supply a complete symbol, with superscript and subscript, for each of the following types of radiation.
(a) alpha particle **(b)** beta particle
(c) gamma ray

17.6 Give the charge and mass (in amu) of each of the following types of radiation.
(a) alpha particle **(b)** beta particle
(c) gamma ray

17.7 State the composition of an alpha particle in terms of protons and neutrons.

17.8 What is the relationship between a beta particle and an electron?

Alpha-Particle Decay and Beta-Particle Decay (Sec. 17.4)

17.9 Write balanced nuclear equations for the alpha decay of the following radionuclides.
(a) ${}^{192}_{78}Pt$ **(b)** radon-217
(c) ${}^{212}_{85}At$ **(d)** curium-244

17.10 Write balanced nuclear equations for the alpha decay of the following radionuclides.
(a) ${}^{144}_{60}Nd$ **(b)** polonium-194
(c) ${}^{147}_{62}Sm$ **(d)** radium-224

17.11 Write balanced nuclear equations for the beta decay of the following radionuclides.
(a) ${}^{48}_{21}Sc$ **(b)** silver-117
(c) ${}^{92}_{36}Kr$ **(d)** cesium-138

17.12 Write balanced nuclear equations for the beta decay of the following radionuclides.
(a) ${}^{25}_{11}Na$ **(b)** rhodium-107
(c) ${}^{67}_{29}Cu$ **(d)** rhenium-190

17.13 How does alpha particle decay affect the mass number and atomic number of the parent nuclide undergoing decay?

17.14 How does beta particle decay affect the mass number and atomic number of the parent nuclide undergoing decay?

17.15 Identify X in each of the following radioactive-decay processes.
(a) ${}^{12}_{5}B \longrightarrow {}^{12}_{6}C + X$ **(b)** ${}^{125}_{51}Sb \longrightarrow {}^{0}_{-1}\beta^{-} + X$
(c) ${}^{251}_{98}Cf \longrightarrow {}^{247}_{96}Cm + X$ **(d)** ${}^{233}_{92}U \longrightarrow {}^{4}_{2}\alpha + X$

17.16 Identify X in each of the following radioactive-decay processes.
(a) ${}^{195}_{84}Po \longrightarrow {}^{4}_{2}\alpha + X$ **(b)** ${}^{75}_{31}Ga \longrightarrow {}^{75}_{32}Ge + X$
(c) $X \longrightarrow {}^{0}_{-1}\beta + {}^{88}_{36}Kr$ **(d)** ${}^{222}_{89}Ac \longrightarrow {}^{218}_{87}Fr + X$

17.17 Write nuclear equations for each of the following radioactive-decay processes.
(a) mercury-199 is formed by beta emission
(b) cadmium-120 undergoes beta emission
(c) terbium-148 is formed by alpha emission
(d) radium-226 undergoes alpha emission

17.18 Write nuclear equations for each of the following radioactive-decay processes.
(a) thallium-206 is formed by beta emission
(b) palladium-109 undergoes beta emission
(c) plutonium-241 is formed by alpha emission
(d) fermium-249 undergoes alpha emission

Half-life (Sec. 17.5)

17.19 Palladium-114 has a half-life of 2.4 minutes. What fraction of palladium-114 atoms in a sample will remain in an undecayed state after the following periods of time?
(a) 12 minutes **(b)** 4 half-lives
(c) 7.2 minutes **(d)** 8 half-lives

17.20 Nickel-66 has a half-life of 6.4 days. What fraction of nickel-66 atoms in a sample will remain in an undecayed state after the following periods of time?
(a) 25.6 days **(b)** 3 half-lives
(c) 32 days **(d)** 7 half-lives

17.21 Determine the half-life of a radionuclide if after 12 years the fraction of undecayed nuclides present is
(a) 1/4 **(b)** 1/32 **(c)** 1/256 **(d)** 1/1024

17.22 Determine the half-life of a radionuclide if after 5.2 days the fraction of undecayed nuclides present is
(a) 1/16 **(b)** 1/64 **(c)** 1/128 **(d)** 1/512

17.23 The half-life of copper-60 is 2.4 minutes. How many grams of copper-60 in a 8.0-g sample will remain undecayed after the following periods of time?
(a) 4.8 min **(b)** 9.6 min **(c)** 16.8 min **(d)** 24.0 min

17.24 The half-life of arsenic-69 is 15 minutes. How many grams of arsenic-69 in a 6.0-g sample will remain undecayed after the following periods of time?
(a) 45 min **(b)** 75 min **(c)** 105 min **(d)** 165 min

17.25 The half-life of silver-112 is 3.2 hours. How many grams of this nuclide in a 10.0-g sample will have decayed after the following periods of time?

(a) 6.4 hr **(b)** 19.2 hr **(c)** 28.8 hr **(d)** 48 hr

17.26 The half-life of barium-129 is 2.5 hours. How many grams of this nuclide in a 4.00-g sample will have decayed after the following periods of time?

(a) 5.0 hr **(b)** 15 hr **(c)** 17.5 hr **(d)** 32.5 hr

17.27 Iron-52 has a half-life of 8.0 hours. How long will it take, in hours, for the following fractions of nuclides in an iron-52 sample to decay?

(a) 3/4 **(b)** 15/16 **(c)** 31/32 **(d)** 127/128

17.28 Cobalt-55 has a half-life of 18 hours. How long will it take, in hours, for the following fractions of nuclides in a cobalt-55 sample to decay?

(a) 7/8 **(b)** 31/32 **(c)** 63/64 **(d)** 127/128

Bombardment Reactions (Sec. 17.6)

17.29 Identify X in each of the following bombardment reactions.

(a) ${}^{15}_{7}N + {}^{1}_{1}H \longrightarrow X + {}^{12}_{6}C$

(b) ${}^{27}_{13}Al + X \longrightarrow {}^{4}_{2}\alpha + {}^{25}_{12}Mg$

(c) ${}^{80}_{34}Se + {}^{2}_{1}H \longrightarrow X + {}^{1}_{1}H$

(d) $X + {}^{2}_{1}H \longrightarrow {}^{9}_{3}Li + 2\ {}^{1}_{1}H$

17.30 Identify X in each of the following bombardment reactions.

(a) ${}^{14}_{7}N + X \longrightarrow {}^{14}_{6}C + {}^{1}_{1}H$

(b) ${}^{27}_{13}Al + {}^{4}_{2}\alpha \longrightarrow X + {}^{1}_{0}n$

(c) ${}^{56}_{26}Fe + {}^{2}_{1}H \longrightarrow {}^{54}_{25}Mn + X$

(d) $X + {}^{4}_{2}\alpha \longrightarrow {}^{109}_{47}Ag + {}^{1}_{1}H$

17.31 Write equations for the following nuclear bombardment processes.

(a) Beryllium-9 is bombarded with an alpha particle and emits a neutron.

(b) Nickel-58 is bombarded with a proton, and an alpha particle is emitted.

(c) Bombardment of cadmium-113 produces cadmium-114 and a gamma ray.

(d) Bombardment of a nuclide with an alpha particle produces phosphorus-30 and a neutron.

17.32 Write equations for the following nuclear bombardment processes.

(a) Bombardment of a radionuclide with an alpha particle produces curium-242 and one neutron.

(b) Bombardment of curium-246 with a small particle produces nobelium-254 and four neutrons.

(c) Aluminum-27 is bombarded with an alpha particle and produces a neutron.

(d) Bombardment of sodium-23 with hydrogen-2 produces neon-21.

17.33 Identify the starting material in each of the following transuranium element preparation reactions:

(a) $X + {}^{4}_{2}\alpha \longrightarrow {}^{245}_{98}Cf + {}^{1}_{0}n$

(b) $X + {}^{14}_{7}N \longrightarrow {}^{247}_{99}Es + 5\ {}^{1}_{0}n$

(c) $X + {}^{10}_{5}B \longrightarrow {}^{257}_{103}Lr + 5\ {}^{1}_{0}n$

(d) $X + {}^{58}_{26}Fe \longrightarrow {}^{266}_{109}Mt + {}^{1}_{0}n$

17.34 Identify the starting material in each of the following transuranium element preparation reactions.

(a) $X + {}^{4}_{2}\alpha \longrightarrow {}^{239}_{94}Pu + 3\ {}^{1}_{0}n$

(b) $X + {}^{12}_{6}C \longrightarrow {}^{254}_{102}No + 4\ {}^{1}_{0}n$

(c) $X + {}^{15}_{7}N \longrightarrow {}^{260}_{105}Ha + 4\ {}^{1}_{0}n$

(d) $X + {}^{58}_{26}Fe \longrightarrow {}^{265}_{108}Hs + {}^{1}_{0}n$

17.35 Using Table 17.2 as your source of information, for how many of the transuranium elements does the most stable isotope have a half-life greater than 1 month?

17.36 Using Table 17.2 as your source of information, for how many of the transuranium elements does the most stable isotope have a half-life less than 1 day?

17.37 Approximately how many "synthetic" radionuclides are known.

17.38 How does the number of laboratory-produced radionuclides compare to the number of naturally occurring nuclides?

Positron Emission and Electron Capture (Sec. 17.7)

17.39 How does positron emission affect the mass number and atomic number of the parent nuclide undergoing decay?

17.40 How does electron capture affect the mass number and atomic number of a parent nuclide that undergoes electron capture?

17.41 What mode or modes of decay are associated with the conversion within the nucleus of a neutron to a proton?

17.42 What mode or modes of decay are associated with the conversion within the nucleus of a proton to a neutron?

17.43 Write balanced nuclear equations for the positron decay of the following radionuclides.

(a) ${}^{29}_{15}P$ **(b)** antimony-112

(c) ${}^{46}_{23}V$ **(d)** cerium-132

17.44 Write balanced nuclear equations for the positron decay of the following radionuclides.

(a) ${}^{33}_{17}Cl$ **(b)** selenium-70

(c) ${}^{40}_{21}Sc$ **(d)** tin-109

17.45 Write balanced nuclear equations for the electron capture decay of the following radionuclides.

(a) ${}^{76}_{36}Kr$ **(b)** xenon-122

(c) ${}^{100}_{46}Pd$ **(d)** tantalum-175

17.46 Write balanced nuclear equations for the electron capture decay of the following radionuclides.

(a) ${}^{80}_{38}Sr$ **(b)** rhenium-181

(c) $^{88}_{40}Zr$ **(d)** lead-196

17.47 Identify X in each of the following radioactive decay processes.

(a) $^{17}_{9}F \longrightarrow X + ^{17}_{8}O$

(b) $^{83}_{37}Rb + X \longrightarrow ^{83}_{36}Kr$

(c) $X \longrightarrow ^{0}_{1}\beta + ^{103}_{46}Pd$

(d) $^{133}_{56}Ba + ^{0}_{-1}e \longrightarrow X$

17.48 Identify X in each of the following radioactive decay processes.

(a) $^{191}_{80}Hg + X \longrightarrow ^{191}_{79}Au$

(b) $X \longrightarrow ^{0}_{1}\beta + ^{172}_{72}Hf$

(c) $X + ^{0}_{-1}e \longrightarrow ^{108}_{49}In$

(d) $^{47}_{24}Cr \longrightarrow X + ^{47}_{23}V$

17.49 Most zinc-63 atoms (93%) decay through electron capture. Some (7%), however, decay through positron emission. Write a nuclear equation for each of these modes of decay for zinc-63.

17.50 Most gallium-68 atoms (86%) decay through positron emission. Some (14%), however, decay through electron capture. Write a nuclear equation for each of these modes of decay for gallium-68.

Nuclear Stability (Sec. 17.8)

17.51 What is the dominant mode of decay for a radionuclide in which the neutron-to-proton ratio is too high for stability?

17.52 What is the dominant mode of decay for a radionuclide in which the neutron-to-proton ratio is too low for stability?

17.53 Calculate the neutron-to-proton ratio before and after each of the following processes takes place.

(a) beta decay of $^{65}_{28}Ni$

(b) alpha decay of $^{192}_{78}Pt$

(c) electron capture by $^{165}_{69}Tm$

(d) positron decay of $^{107}_{49}In$

17.54 Calculate the neutron-to-proton ratio before and after each of the following processes takes place.

(a) electron capture by $^{75}_{34}Se$

(b) beta decay of $^{78}_{33}As$

(c) positron decay of $^{37}_{19}K$

(d) alpha decay of $^{253}_{102}No$

17.55 One member of each of the following pairs of radionuclides decays by beta-particle emission and the other by positron emission. Which is which? Explain your reasoning.

(a) $^{74}_{36}Kr$ and $^{87}_{36}Kr$ **(b)** $^{68}_{33}As$ and $^{84}_{34}Se$

(c) $^{74}_{31}Ga$ and $^{64}_{31}Ga$ **(d)** $^{99}_{41}Nb$ and $^{99}_{46}Pd$

17.56 One member of each of the following pairs of radionuclides decays by beta-particle emission and the other by positron emission. Which is which? Explain your reasoning.

(a) $^{82}_{39}Y$ and $^{92}_{39}Y$ **(b)** $^{53}_{26}Fe$ and $^{68}_{29}Cu$

(c) $^{99}_{46}Pd$ and $^{115}_{46}Pd$ **(d)** $^{50}_{25}Mn$ and $^{50}_{22}Sc$

Radioactive Decay Series (Sec. 17.9)

17.57 The plutonium-241 decay series terminates with bismuth-209. Would you expect bismuth-209 to be a stable or unstable nuclide? Explain your answer.

17.58 The thorium-232 decay series terminates with lead-208. Would you expect lead-208 to be a stable or unstable nuclide? Explain your answer.

17.59 In the thorium-232 natural decay series, the intermediate radon-220 undergoes alpha decay, the resulting daughter also undergoes alpha decay, and the succeeding two daughters both emit beta particles. Write four nuclear equations to represent these four steps of the thorium-232 natural decay series.

17.60 In the uranium-235 natural decay series, the intermediate thorium-231 undergoes beta decay, the resulting daughter undergoes alpha decay, and the two succeeding daughters emit a beta particle and an alpha particle in that order. Write four nuclear equations to represent these four steps of the uranium-235 natural decay series.

Effects of Radiation (Secs. 17.10 and 17.11)

17.61 What are ion pairs, and how are they produced?

17.62 What is a free radical, and how is it produced?

17.63 Contrast the abilities of alpha, beta, and gamma radiations to penetrate a thick sheet of paper.

17.64 Contrast the abilities of alpha, beta, and gamma radiations to penetrate a 1-cm-thick piece of aluminum.

17.65 Contrast the speeds with which alpha, beta, and gamma radiations are emitted by nuclei.

17.66 Contrast the ionizing ability of alpha and beta radiations.

Nuclear Medicine (Sec. 17.14)

17.67 The radionuclides used for diagnostic procedures are almost always gamma emitters. Why is this so?

17.68 How do radionuclides used for therapeutic purposes differ from those used for diagnostic purposes?

17.69 Explain how each of the following radionuclides is used in nuclear medicine.

(a) radium-226 **(b)** potassium-42

(c) cobalt-60 **(d)** chromium-51

17.70 Explain how each of the following radionuclides is used in nuclear medicine.

(a) technetium-99 **(b)** iron-59

(c) phosphorus-32 **(d)** radon-222

Nuclear Fission and Nuclear Fusion (Sec. 17.15)

17.71 Tell whether the following characteristics apply to the fission process, to the fusion process, or to both processes.

(a) An extremely high temperature is required to start the process.

(b) Large amounts of energy are released in the process.

(c) Transmutation of elements occurs.

(d) A fourth state of matter called a *plasma* is encountered in this process.

17.72 Tell whether the following characteristics apply to the fission process, to the fusion process, or to both processes.

(a) An example of this process occurs on the sun.

(b) Neutrons are needed to start the process.

(c) Energy released in the process is called *nuclear energy*.

(d) The process is now used to generate some electrical power in the United States.

17.73 Identify each of the following as a fission reaction, a fusion reaction, or as neither a fission nor fusion reaction.

(a) ${}^{1}_{1}H + {}^{1}_{1}H \longrightarrow {}^{2}_{1}H + {}^{0}_{1}\beta$

(b) ${}^{234}_{92}U \longrightarrow {}^{230}_{90}Th + {}^{4}_{2}\alpha$

(c) ${}^{242}_{96}Cm + {}^{4}_{2}\alpha \longrightarrow {}^{245}_{98}Cf + {}^{0}_{1}n$

(d) ${}^{235}_{92}U + {}^{1}_{0}n \longrightarrow {}^{139}_{56}Ba + {}^{94}_{36}Kr + 3\ {}^{1}_{0}n$

17.74 Identify each of the following as a fission reaction, a fusion reaction, or as neither a fission nor fusion reaction.

(a) ${}^{239}_{94}Pu + {}^{1}_{0}n \longrightarrow {}^{132}_{52}Te + {}^{105}_{42}Mo + 3\ {}^{1}_{0}n$

(b) ${}^{3}_{1}H + {}^{2}_{1}H \longrightarrow {}^{4}_{2}He + {}^{1}_{0}n$

(c) ${}^{208}_{82}Pb + {}^{58}_{26}Fe \longrightarrow {}^{265}_{108}Hs + {}^{1}_{0}n$

(d) ${}^{241}_{95}Am + {}^{1}_{0}n \longrightarrow {}^{144}_{57}La + {}^{95}_{38}Sr + 3\ {}^{1}_{0}n$

ADDITIONAL PROBLEMS

17.75 Fill in the blanks in the following radioactive decay series.

(a) ${}^{232}Th \xrightarrow{\alpha} \_\_\_ \xrightarrow{\beta^-} \_\_\_ \xrightarrow{\beta^-} {}^{228}Th$

(b) $\_\_\_ \xrightarrow{\beta^-} \_\_\_ \xrightarrow{\alpha} {}^{224}Ra \xrightarrow{\alpha} \_\_\_$

17.76 Fill in the blanks in the following radioactive decay series

(a) ${}^{223}Fr \xrightarrow{\beta^-} \_\_\_ \xrightarrow{\alpha} \_\_\_ \xrightarrow{\alpha} {}^{215}Po$

(b) $\_\_\_ \xrightarrow{\beta^-} \_\_\_ \xrightarrow{\beta^-} {}^{210}Po \xrightarrow{\alpha} \_\_\_$

17.77 The radioactive-decay series that starts with thorium-232 stops with the formation of lead-208. The decays proceed through a series of alpha- and beta-particle emissions. How many of each of these types of emissions are involved in the series?

17.78 The radioactive-decay series that starts with plutonium-241 stops with the formation of bismuth-209. The decays proceed through a series of alpha- and beta-particle emissions. How many of each of these types of emissions are involved in the series?

17.79 Phosphorus-31 is the only stable isotope of phosphorus. Predict how phosphorus-28 will decay and how phosphorus-34 will decay.

17.80 Fluorine-19 is the only stable istope of fluorine. Predict how fluorine-16 will decay and how fluorine-21 will decay.

17.81 Consider the decay series

$$A \longrightarrow B \longrightarrow C \longrightarrow D$$

where A, B, and C are radioactive, with half-lives of 3.2 min, 25 days, and 9.0 sec, respectively, and D is nonradioactive. Starting with 1.00 mole of A, and none of B, C, and D, calculate the numbers of moles of A, B, C, and D present after 50 days.

17.82 Consider the decay series

$$E \longrightarrow F \longrightarrow G \longrightarrow H$$

where E, F, and G are radioactive, with half-lives of 5.0 min, 2.3 min, and 35 days, respectively, and D is nonradioactive. Starting with 1.00 mole of E, and none of F, G, and H, calculate the number of moles of E, F, G, and H present after 105 days.

17.83 Two radionuclides are compared. Nuclide A requires 16 hr for its decay rate to fall to 1/64 of its initial value. Nuclide B has a half-life that is 45% shorter than that of nuclide A. How long does it take for the decay rate of nuclide B to fall to 1/16 of its initial value?

17.84 Two radionuclides are compared. Nuclide C requires 8.0 hr for its decay rate to fall to 1/64 of its initial value. Nuclide D has a half-life that is 35% longer than that of nuclide C. How long does it take for the decay rate of nuclide D to fall to 1/32 of its initial value?

17.85 A sample of radioactive ${}^{210}X$ initially weighed 4.000 g. After 35 days, 0.125 g of ${}^{210}X$ remained, the rest having decayed to the stable ${}^{206}Q$. Calculate

(a) the half-life of ${}^{210}X$

(b) the mass of ${}^{206}Q$ formed

17.86 A sample of radioactive ${}^{213}X$ initially weighed 2.000 g. After 15 days, 0.250 g of ${}^{213}X$ remained, the rest having decayed to the stable ${}^{209}Q$. Calculate

(a) the half-life of ${}^{213}X$

(b) the mass of ${}^{209}Q$ formed

CUMULATIVE PROBLEMS

17.87 For the following nuclear processes, characterize the relationship between parent and daughter nuclides as that of isobars, isotopes, or neither.

(a) positron emission

(b) alpha emission

(c) neutron absorption

(d) alpha decay followed by two beta decays

17.88 For the following nuclear processes, characterize the relationship between parent and daughter nuclides as that of isobars, isotopes, or neither.

(a) electron capture

(b) beta emission

(c) neutron emission

(d) beta emission followed by electron capture

17.89 An isotope of a particular element has a half-life of 4.7 days. The emissions from a sample of the isotope was measured as 7.2×10^5 disintegrations per second (dps). (The number of disintegrations from a radioactive sample is proportional to the mass of the sample, so dps values can be treated the same as mass values.) At a later time the decay rate (dps) is measured at 4.5×10^4 dps. How much time has elapsed, in days, between the measurements?

17.90 An isotope of a particular element has a half-life of 10.2 days. The emissions from a sample of the isotope was measured as 6.4×10^4 dps. (See problem 17.89 for an explanation of dps.) At a later time the decay rate (dps) is measured at 2.0×10^3 dps. How much time has elapsed, in days, between the measurements?

17.91 If the uranium present in pure UF_6 and UBr_4 has a normal natural distribution of isotopes, which of the following will produce the greater amount of radioactivity?

(a) 50.0 g of UF_6 or 50.0 g of UBr_4

(b) 0.250 mole of UF_6 or 0.200 mole of UBr_4

17.92 If the thorium present in pure ThO_2 and ThF_4 has a normal natural distribution of isotopes, which of the following will produce the greater amount of radioactivity?

(a) 25.0 g of ThO_2 or 25.0 g ThF_4

(b) 0.300 mole of ThO_2 or 0.350 mole of ThF_4

17.93 Polonium-210, with a half-life of 138 days, decays by alpha emission to stable lead-206. The alpha particles so emitted, once they lose their energy, pick up electrons from their surroundings and become helium atoms. Suppose the helium gas originating from all of the alpha particles is collected. What volume of helium gas, in liters, at 25°C and 0.800 atm pressure, would be obtained from a 5.00-g sample of polonium-210 left to decay for 138 days?

17.94 Polonium-212, with a half-life of 0.30 microsecond, decays by alpha emission to stable lead-208. The alpha particles so emitted become helium gas as explained in problem 17.93. What volume of helium gas, in liters, at 25°C and 1.03 atm pressure, would be obtained from a 30.0-g sample of polonium-212 left to decay for 1.0 day?

17.95 The energy released when 1.00 kg of uranium-235 undergoes the fission process

$$^{235}_{92}\text{U} + ^{1}_{0}\text{n} \longrightarrow ^{136}_{53}\text{I} + ^{96}_{39}\text{Y} + 4\,^{1}_{0}\text{n}$$

is 1.01×10^{11} kJ. How many kilograms of carbon (graphite) would have to be burned to CO_2 to release the same amount of energy? The heat of combustion of graphite is 394 kJ/mole.

17.96 The energy released when 1.00 kg of hydrogen-1 undergoes the fusion process

$$4\,^{1}_{1}\text{H} \longrightarrow ^{4}_{2}\text{He} + 2\,^{0}_{1}\beta$$

is 5.92×10^{11} kJ. How many kilograms of carbon (graphite) would have to be burned to CO_2 to release the same amount of energy? The heat of combustion of graphite is 394 kJ/mole.

18 Hydrocarbons and Hydrocarbon Derivatives

18.1 Organic Chemistry—A Historical Perspective

In 1675, Nicholas Lemery (1645–1715), a French scientist and philosopher and an early writer of chemistry textbooks, published a book entitled *Cours de Chymie*. Its significance in the context of this chapter is that Lemery chose to distinguish between substances derived from plant or animal sources and those obtained from mineral constituents.

During the 1700s, this classification scheme became almost universally accepted by scientists. The term *organic* was introduced to refer to substances obtained from living matter (*organisms*), and the term *inorganic* to denote materials that originated from nonliving (mineral) matter. In addition, the belief developed during this same time period that living organisms contained some mysterious "vital force" that was necessary to produce organic substances. This belief arose because no scientists of the time were able to synthesize any known organic material from inorganic reactants. With each additional failure at such a synthesis, the "vital force theory" became more firmly entrenched. By 1800, it was universally accepted.

The vital force theory is now known to be incorrect. Its demise began in 1828 as a result of an experiment performed by the German chemist Friedrich Wöhler (1800–1882). While heating two inorganic salts in an attempt to produce a new inorganic salt, Wöhler found that, instead, he had produced urea, a compound found in urine and very well known as an organic compound. Wöhler's successful synthesis was the stimulus for renewed efforts by many scientists to synthesize organic substances from inorganic materials. This time, after a century of negative results, numerous successful reactions were discovered. By 1850 the vital force theory was laid to rest.

The significance of this portion of chemical history, in terms of modern chemistry, is that the terms *organic* and *inorganic*, whose origins lie in the vital force theory, are still in use. The original definitions, however, are not.

Today, **organic chemistry** *is defined as the study of hydrocarbons (binary compounds of hydrogen and carbon) and their derivatives*. Interestingly, almost all compounds found in living organisms still fall in the field of organic chemistry when this modern definition is applied. In addition, many compounds synthesized in the laboratory, which have never been found in nature or in living organisms, are considered to be organic compounds.

In a less rigorous manner, organic chemistry is often defined as the study of carbon-containing compounds. It is true that almost all carbon-containing compounds qualify as organic compounds. There are, however, some exceptions. The oxides of carbon, carbonates, cyanides, and metallic carbides are all considered to be inorganic compounds rather than organic compounds. The field of *inorganic chemistry* encompasses the study of all noncarbon-containing compounds (the other 110 elements) plus the few carbon-containing compounds just mentioned.

In essence, organic chemistry is the study of one element (carbon), and inorganic chemistry the study of 110 elements. Why, relative to their study, is there such an unequal partitioning of the elements? The answer is simple. The chemistry of carbon is so much more extensive than that of the other elements that there is justification in making its study a field by itself. Approximately 6 million organic compounds are known. Fewer than 250,000 inorganic compounds exist. This is an approximate 25-to-1 ratio between organic and inorganic compounds.

Why does carbon form 24 times as many compounds as all the other elements combined? The reason is that carbon atoms possess the unique ability to bond to other carbon atoms to form long chains, rings, and complex combinations of both. Chains and rings of all lengths are possible. All such chains and rings may contain carbon atom side chains as well.

The number of possible arrangements for carbon atoms bonded to each other is literally limitless. It has been calculated that there are 366,319 different ways of arranging 20 carbon atoms based on a chain of atoms and allowing for side chains.

Figure 18.1 illustrates some of the possible ways of arranging carbon atoms to form organic molecules. Each of the carbon atoms in the structures of Figure 18.1 will also be bonded to additional atoms (not shown). Most often the additional bonds will be to hydrogen atoms.

Organic compounds are the chemical basis for life itself as well as the basis for our current high standard of living. Not only are proteins, carbohydrates, enzymes, and hormones

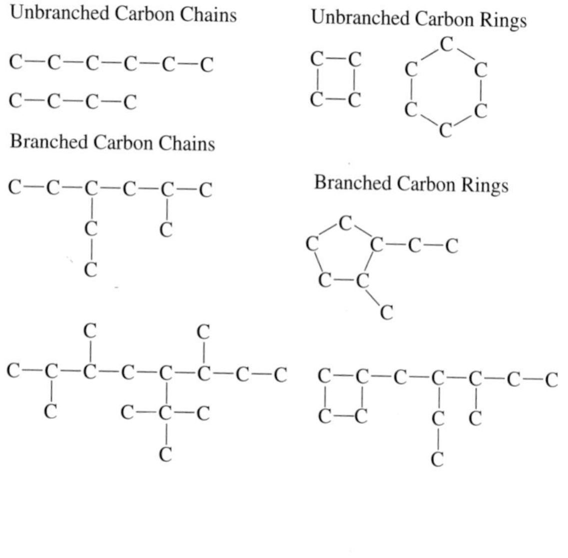

FIGURE 18.1 Simple and complex chains and rings of carbon atoms.

organic molecules, but so are natural gas, petroleum, coal, gasoline, and many synthetic materials such as plastics, dyes, and fibers (rayon, nylon, polyester, etc.).

18.2 HYDROCARBONS

The formal definition of organic chemistry given in Section 18.1 suggests a logical way for organizing our study of the many organic compounds that are known. First, we will consider *hydrocarbons* and then *derivatives of hydrocarbons*. This section, and those that immediately follow it, deal with hydrocarbons, the simplest type of organic compound. Beginning in Section 18.10, hydrocarbon derivatives will occupy our attention.

As the name implies, **hydrocarbons** *are compounds that contain only the two elements hydrogen and carbon*. Several different series of hydrocarbons are known. These include alkanes, alkenes, alkynes, and aromatic hydrocarbons, each of which is discussed in this chapter.

18.3 ALKANES

Alkanes *are hydrocarbons in which all chemical bonds are single bonds*. Straight-chain, branched-chain, and cyclic structures are possible for such compounds. Since an alkane molecule contains only single bonds, each carbon atom in such a molecule is bonded to the maximum number of atoms possible: four. Alkanes are also referred to as *saturated* hydrocarbons. The term **saturated compound** *describes any organic compound that contains only single bonds between carbon atoms*.

All noncyclic alkanes have molecular formulas that fit the general formula C_nH_{2n+2}, where n is the number of carbon atoms present. Cyclic alkanes always contain two fewer hydrogen atoms than their noncyclic counterparts and hence have molecular formulas that fit the general formula C_nH_{2n}.

The first member of the noncyclic alkane series is *methane*, which contains one carbon atom ($n = 1$) and therefore has the formula CH_4. The methane molecule has a tetrahedral structure. The carbon atom is found at the center of the tetrahedron, and the four hydrogen atoms bonded to the carbon are at the corners of the tetrahedron. Different ways of representing this tetrahedral structure for methane are shown in Figure 18.2.

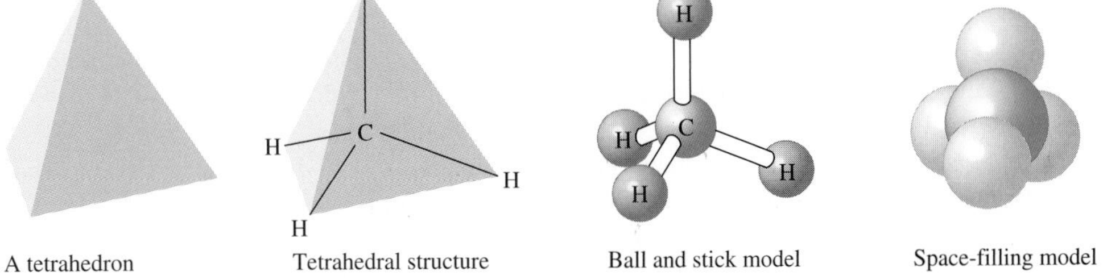

FIGURE 18.2 Different ways of showing the tetrahedral arrangement of atoms in the methane (CH_4) molecule.

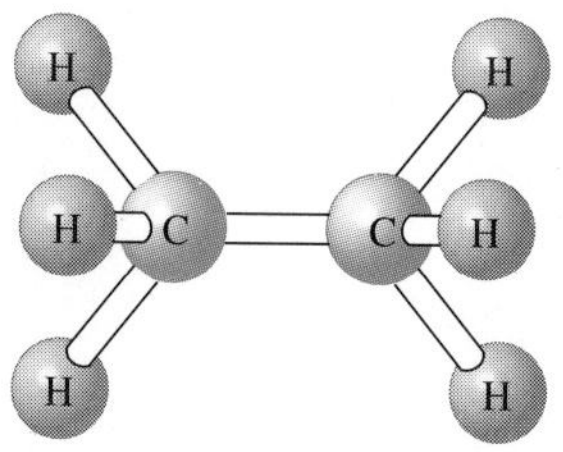

Ball and stick model

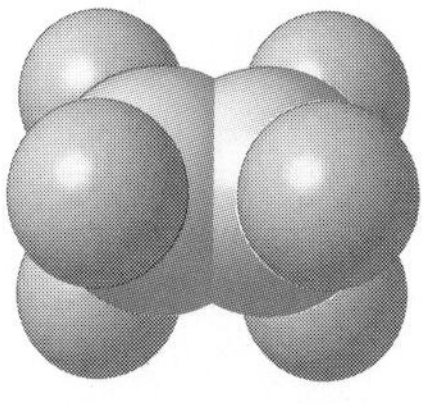

Space-filling model

FIGURE 18.3 Ball and stick and space-filling models of the hydrocarbon ethane (C_2H_6).

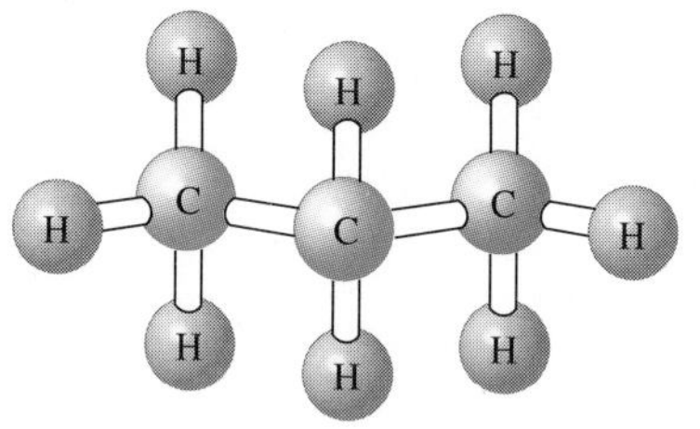

Ball and stick model

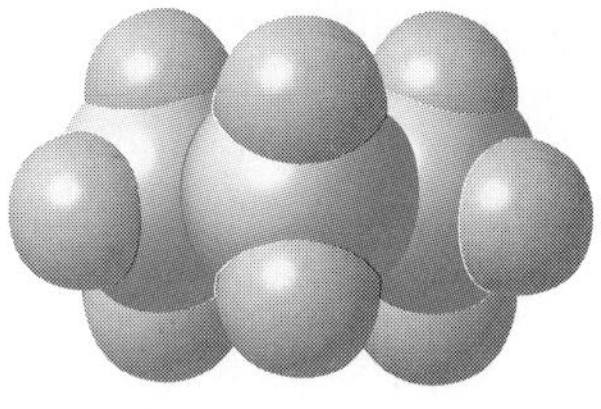

Space-filling model

FIGURE 18.4 Ball and stick and space-filling models of the hydrocarbon propane (C_3H_8).

The next member of the noncyclic alkane series, with $n = 2$ and a molecular formula of C_2H_6, is *ethane*. This molecule may be thought of as a methane molecule with one hydrogen atom removed and a $—CH_3$ group put in its place. Perspective drawings of the ethane molecule, given in Figure 18.3, illustrate that the carbon bonds still have a tetrahedral geometry. Carbon atoms in alkanes, and also in other kinds of organic molecules, always form four bonds. Since carbon is in group IVA of the periodic table and thus has four valence electrons, it must form a total of four bonds in order to achieve an octet of outer shell electrons (Sec. 7.9).

Propane, the third alkane, has the molecular formula C_3H_8 ($n = 3$). Once again, we can produce this formula by removing a hydrogen atom from the preceding compound (ethane) and substituting a $—CH_3$ group in its place. All six hydrogen atoms of ethane are equivalent, so it makes no difference which one we choose to replace. Both ball and stick and space-filling models of the propane molecule are given in Figure 18.4.

The noncyclic alkanes methane, ethane, and propane are the first three members of a homologous series of compounds. In a **homologous series** *of compounds, each compound in the series differs from the previous one in the series by a constant amount*. For the noncyclic alkanes this factor by which consecutive series members differ is CH_2.

18.4 STRUCTURAL ISOMERISM OF HYDROCARBONS

It should be apparent that the procedures outlined in the last section in establishing the structures of the ethane and propane molecules (replacement of a hydrogen with a $—CH_3$ group) can be used to generate other members of the homologous noncyclic alkane series. However, a complication arises when four or more carbon atoms are present: different structures may be obtained depending on which hydrogen atom is replaced.

Two different structural arrangements of four carbons can be produced by removing a hydrogen atom from propane and replacing it with a $—CH_3$ group because not all the hydrogens in propane are geometrically equivalent. The two hydrogens attached to the central carbon atom in propane (Fig. 18.4) are equivalent to each other but distinct from the six hydrogens associated with the end carbons, which in turn are all equivalent to each other. Replacement of a hydrogen on an end carbon gives *butane*, the compound shown in Figure 18.5a. The compound in Figure 18.5b, *isobutane*, is the result of a $—CH_3$ group replacing a hydrogen on the central carbon atom.

Butane and isobutane, although they both have the same molecular formula of C_4H_{10}, are two different compounds with different properties. The melting point of butane is $-138.3°C$, and that of isobutane is $-160°C$. The boiling points of the two compounds are, respectively, $-0.5°C$ and $-12°C$. Their densities as gases at 20°C also differ: 0.579 g/mL for butane and 0.557 g/mL for isobutane.

Compounds such as butane and isobutane are called *structural isomers*. **Structural isomers** *are compounds that have the same molecular formula but different structural formulas, that is, different arrangements of atoms within the molecule.* Structural isomers, as we shall see, are not rare in organic chemistry; in fact, they are the rule rather than the exception. The phenomenon of structural isomerism is one of the major reasons why there are so many organic compounds.

In Figure 18.5, three-dimensional representations were used to illustrate the structures of butane and isobutane. Such representations give both the arrangement and spatial orientation of the atoms in a molecule. However, it is often difficult to draw such structures, especially if artistic talent is lacking, and it is also time consuming. Because of these drawbacks, an easier system for indicating structure has been developed. This alternative system involves structural formulas. **Structural formulas** *are two-dimensional (planar) representations of the arrangement of the atoms in a molecule.* Such formulas give complete information about the arrangement of the atoms in a molecule, but nothing about the spatial orientation of the atoms. The structural formulas for butane and isobutane are

$$\underset{\text{butane}}{CH_3—CH_2—CH_2—CH_3} \qquad \underset{\text{isobutane}}{\begin{array}{c} CH_3—\underset{\displaystyle CH_3}{\underset{|}{CH}}—CH_3 \end{array}}$$

Note that in writing structural formulas each carbon atom is followed by its attached hydrogens. The number of attached hydrogens is determined by the number of other carbon atoms to which a given carbon atom is bonded. Since each carbon atom must have four bonds (Sec. 18.3), carbon atoms attached to only one other carbon will need three hydrogen atoms (CH_3). Carbons bonded to two other carbons will need two hydrogens (CH_2), those bonded to three other carbons need only one hydrogen (CH), and those bonded to four other carbons need no hydrogens.

When we consider noncyclic alkanes with five carbon atoms, we find that three structural isomers exist.

$$\underset{\text{pentane}}{CH_3—CH_2—CH_2—CH_2—CH_3} \qquad \underset{\text{isopentane}}{CH_3—\underset{\displaystyle CH_3}{\underset{|}{CH}}—CH_2—CH_3} \qquad \underset{\text{neopentane}}{CH_3—\overset{\displaystyle CH_3}{\overset{|}{\underset{\displaystyle CH_3}{\underset{|}{C}}}}—CH_3}$$

These three C_5H_{12} isomers like the two C_4H_{10} isomers, are distinctly different compounds with different properties.

The number of possible structural isomers increases rapidly with the number of carbon atoms in an alkane. There are five noncyclic alkane isomers with six carbons (all C_6H_{14})

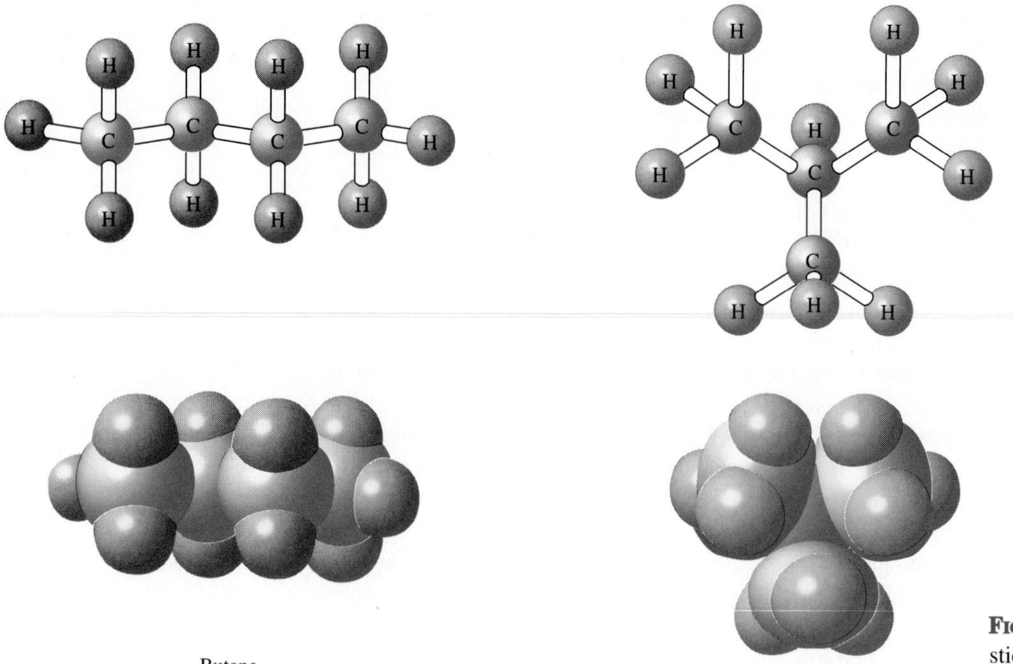

Butane

(a)

Isobutane

(b)

FIGURE 18.5 Ball and stick and space-filling models of (a) butane and (b) isobutane.

and nine isomers with seven carbons (all C_7H_{16}). A listing of the number of possible noncyclic alkane isomers, as a function of the number of carbon atoms present, is given in Table 18.1. Obviously, no one has prepared all the isomers for compounds that contain a

TABLE 18.1 POSSIBLE NUMBER OF ISOMERS FOR SELECTED NONCYCLIC ALKANES

Molecular Formula	Possible Number of Isomers
CH_4	1
C_2H_6	1
C_3H_8	1
C_4H_{10}	2
C_5H_{12}	3
C_6H_{14}	5
C_7H_{16}	9
C_8H_{18}	18
C_9H_{20}	35
$C_{10}H_{22}$	75
$C_{15}H_{32}$	4,347
$C_{20}H_{42}$	336,319
$C_{30}H_{62}$	4,111,846,763

large number of carbon atoms. However, methods are available for their synthesis if the need arises.

The three isomers of C_5H_{12} are called pentane, isopentane, and neopentane. What prefixes do we use to name the five isomers of C_6H_{14}? What about names for the 9 C_7H_{16} isomers and the 75 $C_{10}H_{22}$ isomers? Obviously, using prefixes to distinguish between various isomers is not a workable system when the number of isomers becomes large. We would rapidly run out of prefixes to use. Even if we had a sufficient number of prefixes, remembering the exact meaning of each prefix would be a problem. It would be an almost impossible task. There must be a better way, and there is. By use of a nomenclature system called the IUPAC system (for the International Union of Pure and Applied Chemistry) organic compounds are assigned names that relate directly to their structures. This system keeps prefix use to a minimum. The logic of the system also minimizes the amount of rote memorization required. The IUPAC ("eye-you-pack") rules for naming noncyclic alkanes (both branched and unbranched) are given in Section 18.5.

18.5 IUPAC NOMENCLATURE FOR NONCYCLIC ALKANES

When only a relatively few organic compounds were known, chemists named them by what are today called common names. Such names were selected arbitrarily. Isopentane and neopentane (Sec. 18.4) are examples of such names. As more and more compounds became known, it became obvious that a systematic method for naming organic compounds was needed. The IUPAC nomenclature system meets that need.

For nomenclatural purposes we will classify noncyclic alkanes into two categories: normal chain and branched chain. In **normal-chain noncyclic hydrocarbons,** *all carbon atoms are connected in a continuous nonbranching chain.* In **branched-chain noncyclic hydrocarbons,** *one or more side chains of carbon atoms are attached at some point to a continuous chain of carbon atoms.* The two four-carbon noncyclic alkane isomers (Sec. 18.4) illustrate this classification system.

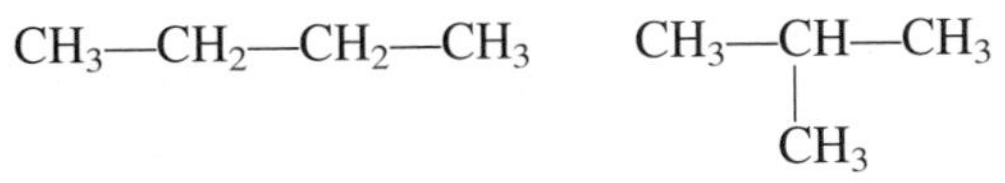

normal chain   branched chain

In the branched-chain compound the longest continuous chain of carbon atoms has three carbons, with a —CH_3 group (the branch) attached to the middle carbon.

Let us first consider IUPAC names for normal-chain noncyclic alkanes, compounds that are structurally the simplest of all alkanes. IUPAC names for such alkanes with one through ten carbons are given in Table 18.2. Note that all of the names end in *-ane*, the characteristic ending for all alkanes. The first four names do not indicate the number of carbon atoms in the molecule, but beginning with five carbon atoms Greek numerical prefixes are used that give directly the number of carbon atoms present. It is important to memorize the names for these normal-chain alkanes because they are the basis for the entire IUPAC nomenclature system. (The IUPAC system also includes names for normal-chain alkanes with more than ten carbon atoms, but we will not consider them. The names given in Table 18.2 will be sufficient for our purposes.)

TABLE 18.2 IUPAC NAMES FOR NORMAL-CHAIN NONCYCLIC ALKANES THROUGH DECANE

CH_4	methane
C_2H_6	ethane
C_3H_8	propane
C_4H_{10}	butane
C_5H_{12}	pentane
C_6H_{14}	hexane
C_7H_{16}	heptane
C_8H_{18}	octane
C_9H_{20}	nonane
$C_{10}H_{22}$	decane

Branched-chain noncyclic alkanes always contain alkyl groups. The alkyl groups are the branches. Formally defined, an **alkyl group** *is the fragment produced by the removal of one hydrogen atom from an alkane.* The general formula for an alkyl group is C_nH_{2n+1}. Alkyl groups do not lead a stable independent existence, but rather are always found attached to another group of carbon atoms.

The two most commonly encountered alkyl groups are the two simplest ones, the methyl group and the ethyl group.

$$—CH_3 \qquad —CH_2—CH_3$$

methyl group   ethyl group

(The bond on the left in the alkyl group structures denotes the point of attachment to the carbon chain.) Note how alkyl groups are named: the stem of the name of the parent alkane plus the ending *-yl*.

Two different three-carbon alkyl groups exist. They differ from each other in the point of attachment to the main carbon chain.

$$—CH_2—CH_2—CH_3 \qquad \begin{array}{l} —CH—CH_3 \\ \quad\;\; | \\ \quad\; CH_3 \end{array}$$

propyl group   isopropyl group

A propyl group is attached to a carbon chain through an end carbon atom, whereas an isopropyl group is attached via the middle carbon of the three carbons of the propyl group. The isopropyl group is an example of a branched-chain alkyl group, that is, a branched branch. Other branched alkyl groups, containing more carbon atoms, also exist. Rules for naming these more complex groups will not be considered in this text.

To name branched-chain noncyclic alkanes—that is, alkanes where the carbon atoms are not arranged in a continuous chain—the following rules are used.

RULE 1 Select the longest continuous carbon atom chain in the molecule as the base for the name. This longest carbon atom chain is named as in Table 18.2.

Example:

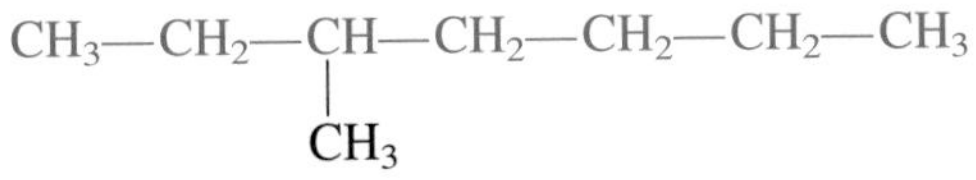

In this example the longest continuous carbon atom chain (shown in color) is seven carbon atoms. Therefore, the base name (but not the complete name) for this compound is **heptane**.

Example:

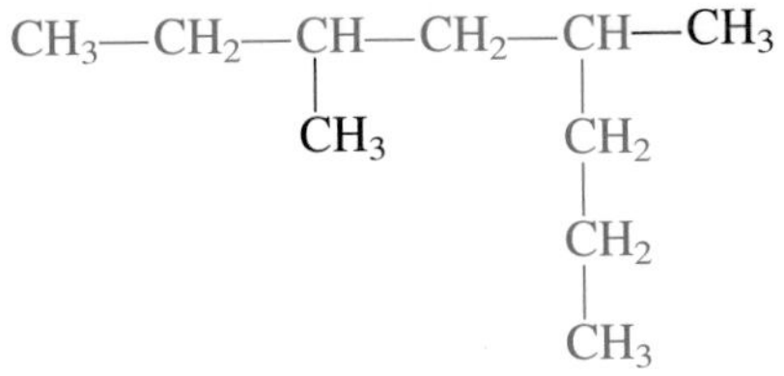

In this example the longest continuous carbon chain (shown in color) possesses eight carbon atoms. Note that the carbon atoms in the longest continuous chain do not necessarily have to lie in a straight line. The base name (but not the complete name) for this alkane will be **octane**.

RULE 2 The carbon atoms in the longest continuous chain of carbon atoms are numbered consecutively from the end nearest a branch (alkyl group).

Since a chain always has two ends, there are always two ways to number the chain; either from left to right or right to left. The effect of this rule is to give the first-encountered alkyl group the lowest number possible.

Example:

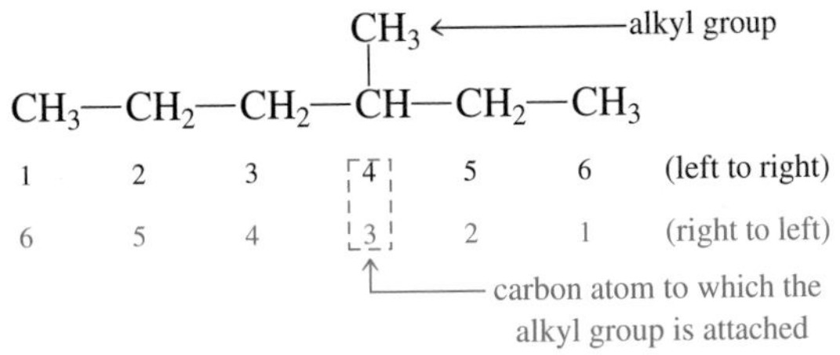

In this example, the right-to-left numbering system is used because the alkyl group is nearer the right end of the chain than the left end.

If two or more alkyl groups are present on a carbon chain and both numbering systems (left–right and right–left) give the same number for the first-encountered alkyl group, then the second-encountered alkyl group is used to distinguish between numbering systems.

Example:

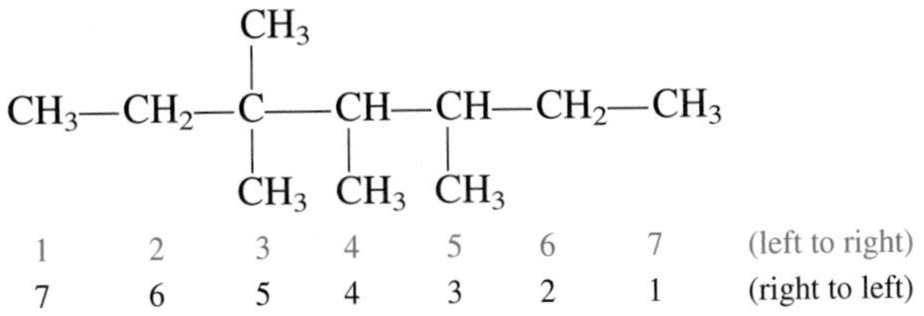

In this example, there are four alkyl groups (side chains) attached to the main chain. Both numbering systems give carbon 3 as the position of the first alkyl group. If we consider the first two encountered alkyl groups, the left–right numbering system gives 3 and 3 for the alkyl group locations, and the right–left numbering system gives 3 and 4. (Note that when a carbon atom carries two alkyl groups, that carbon's number must be repeated for each group.) Thus, the left–right numbering system (3,3) is used instead of the right–left numbering system (3,4).

RULE 3 The complete name for a noncyclic alkane contains the location by number and the name of each alkyl group and the name of the longest carbon chain.

The alkyl group names with their locations always precede the name of the base chain of carbon atoms.

Examples:

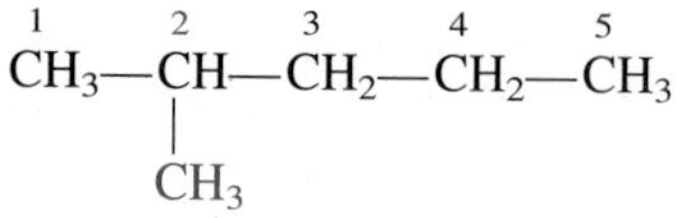 is 2-methylpentane

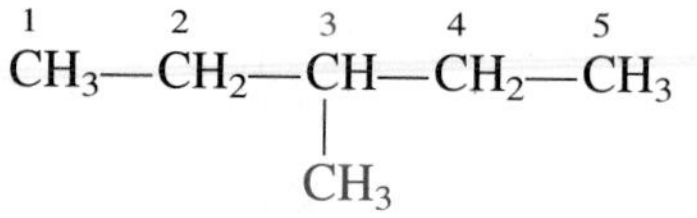 is 3-methylpentane

Note that the names are written as one word with a hyphen between the number (location) and the name of the alkyl group.

RULE 4 If two or more of the same kind of alkyl group (two methyl groups or two ethyl groups, and so forth) are present in a molecule, the number of them is indicated by the prefixes *di-*, *tri-*, *tetra-*, *penta-*, and so on, and the location of each is again indicated by a number.

These position numbers, separated by commas, are put just before the numerical prefix, with hyphens before and after the numbers when necessary.

Examples:

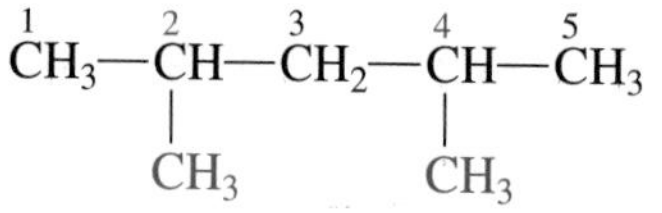 is 2,4-dimethylpentane

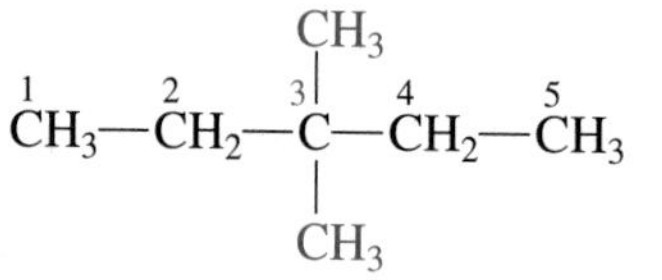 is 3,3-dimethylpentane

Note that the prefix di- must always be accompanied by two numbers, tri- by three, and so on, even if the same number must be written twice, as in 3,3-dimethylpentane.

RULE 5 When two different kinds of alkyl groups are present on the same carbon chain, each group is separately numbered with the names of the alkyl groups listed in alphabetical order.

Example:

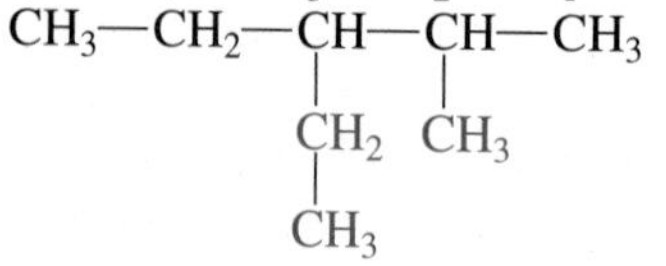 is 3-ethyl-2-methylpentane

Note that ethyl is named first in accordance with the alphabetical rule determining the order in which alkyl groups are listed.

Example:

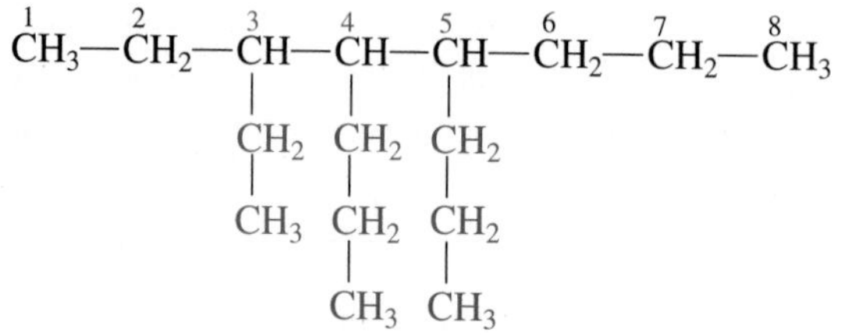 is 3-ethyl-4,5-dipropyloctane

Note that the prefix di- does not affect the alphabetical order for alkyl groups; “e” from ethyl is compared to “p” from propyl.

There are additional rules and conventions used in naming very complicated alkanes, but the five fundamental rules given will suffice for the compounds we are likely to encounter. In using the five given rules always use the correct “punctuation.” Remember to

1. *always* use commas between numbers.
2. *always* use hyphens between numbers and words.
3. *never* leave spaces in the name.

Example 18.1

What is the IUPAC name for each of the following hydrocarbons?

(a) 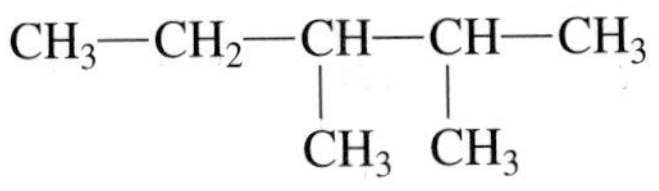

(b) 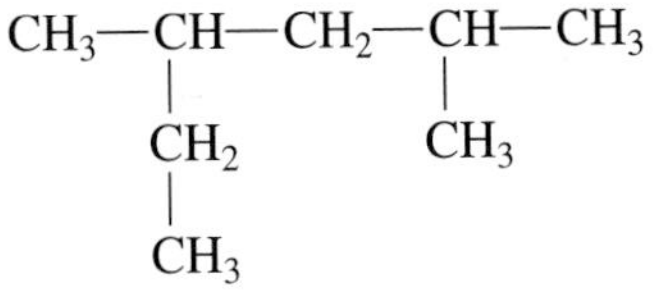

Solution

(a) The longest carbon chain possesses five carbon atoms. Thus, the parent chain name is pentane.

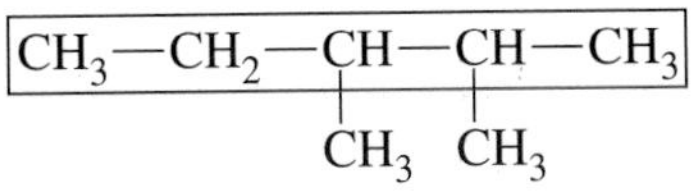

This parent chain is numbered from right-to-left, since an alkyl group (methyl group) is closer to the right end of the chain than to the left end.

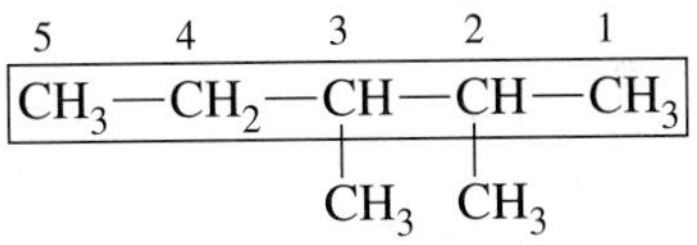

One methyl group is on carbon 2 and the other is on carbon 3. The IUPAC name of the compound is **2,3-dimethylpentane**.

(b) The carbon atoms in the longest carbon chain, which is six, are not horizontally arranged.

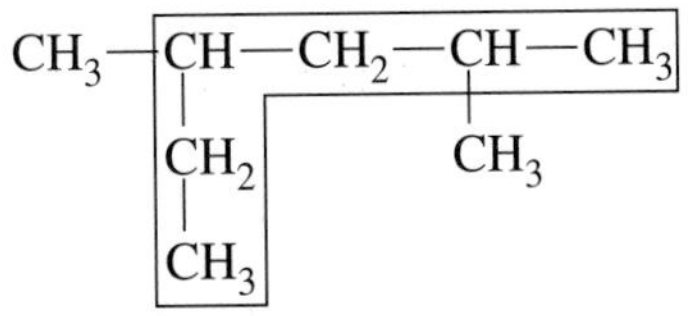

Again, a right-to-left numbering system is used since there is a methyl group closer to the right end of the chain than to the left end.

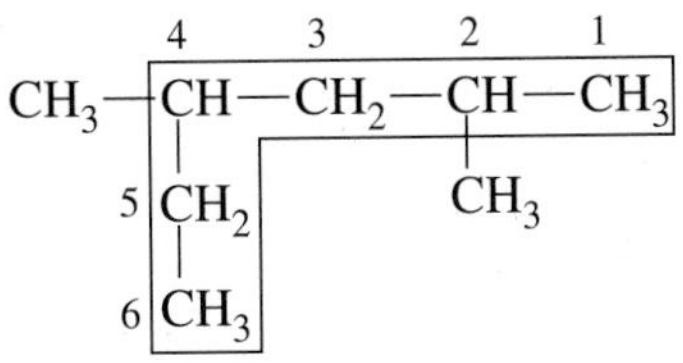

The IUPAC name for the compound is **2,4-dimethylhexane**

Practice Exercise 18.1

What is the IUPAC name for each of the following compounds?

(a) 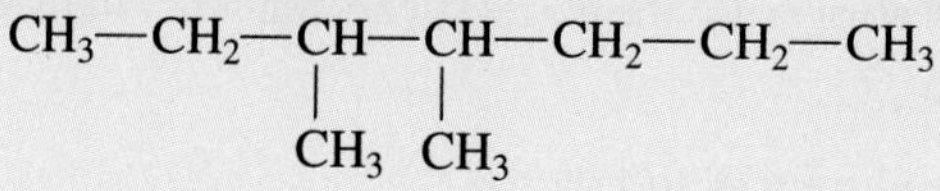

(b) 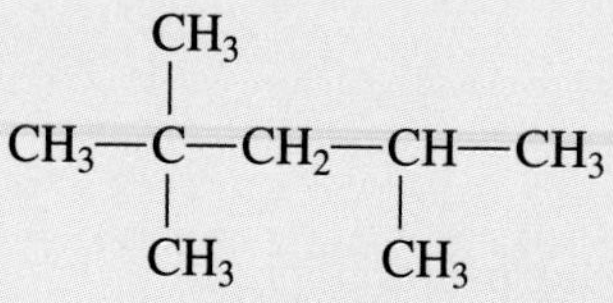

Ans. (a) 3,4-dimethylheptane; (b) 2,2,4-trimethylpentane

Structural formulas can easily be obtained from correct IUPAC names, since all the information necessary to draw a structure is contained within the IUPAC name. Example 18.2 illustrates the process of going from an IUPAC name to a structural formula.

Example 18.2

Draw the structural formula for the alkane whose IUPAC name is 4,4-diethyl-2-methylheptane.

Solution

STEP 1 The IUPAC name indicates that the base chain contains seven carbon atoms (heptane). Draw a heptane skeleton (no hydrogens) and number it.

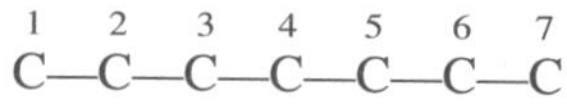

STEP 2 Place two ethyl groups on carbon number 4.

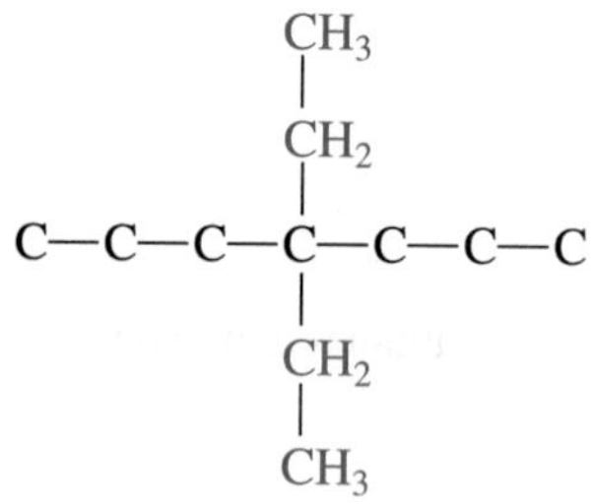

STEP 3 Place a methyl group on carbon 2.

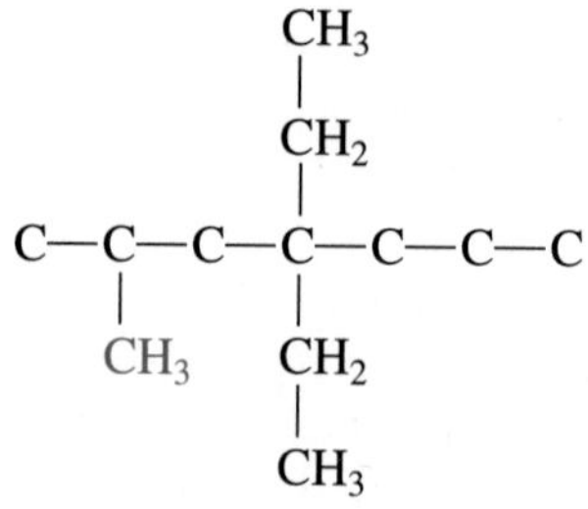

STEP 4 Add necessary hydrogen atoms to the carbon base chain so that each carbon atom has four bonds.

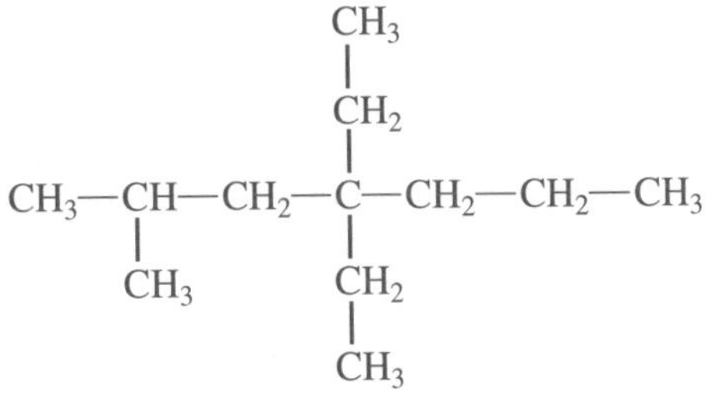

This structure has 12 carbon atoms. It should therefore have 26 hydrogen atoms (C_nH_{2n+2}). Counting the number of hydrogen atoms enables you to check the correctness of the structure.

Practice Exercise 18.2

Draw the structural formula of the alkane whose IUPAC name is 4-ethyl-2,4,6-trimethyldecane.

Ans.

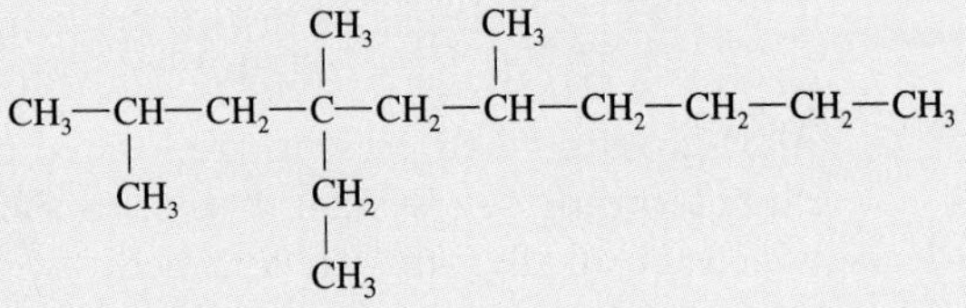

18.6 STRUCTURE AND NOMENCLATURE OF CYCLOALKANES

In addition to normal-chain and branched-chain noncyclic alkanes, a third type of alkane exists: the cycloalkanes. A **cycloalkane** *contains a ring of carbon atoms and has the general formula* C_nH_{2n}. Cycloalkanes thus contain two fewer hydrogen atoms than the other types of alkanes. The reason for this deficiency can be visualized by considering that cycloalkanes arise from the removal of one hydrogen atom from each of the terminal carbons in a linear alkane. The two end carbons, which now need to form one more bond, then join together to give a cyclic structure.

The simplest cycloalkane possible contains three carbon atoms. Figure 18.6 shows ball and stick models of cycloalkanes containing three, four, and five carbons.

For convenience, geometric figures are often used to represent cycloalkanes: a triangle for a three-carbon ring, a square for a four-carbon ring, and so forth. When such figures are used, it is assumed that each corner of the figure represents a carbon atom together with the number of hydrogens needed to give the carbon four bonds. Examples of this geometric-figure type of notation along with IUPAC names for the example compounds are

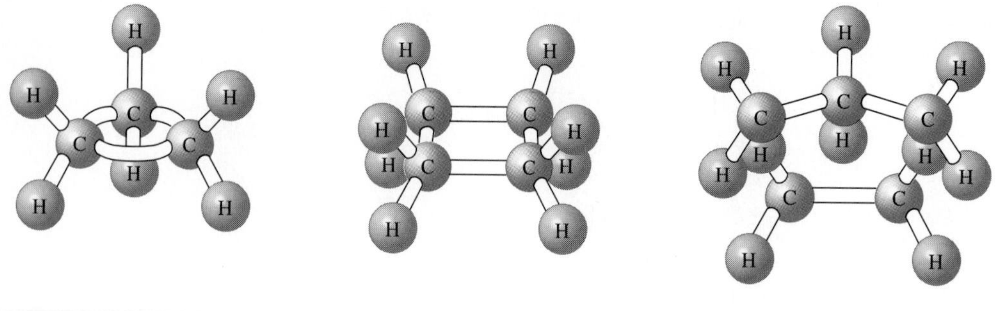

FIGURE 18.6 Ball and stick models of simple cycloalkanes.

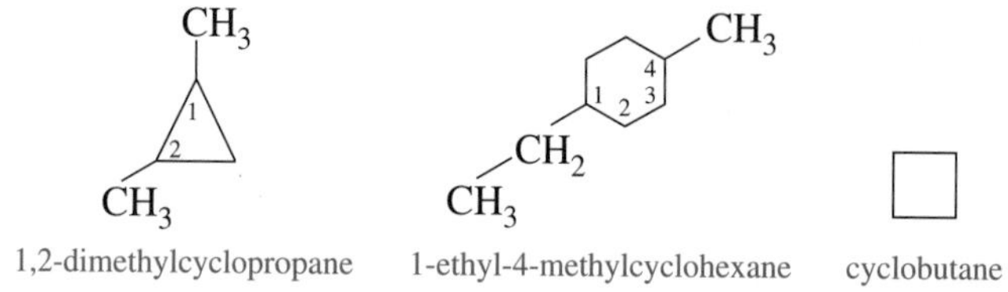

As can be seen from the above examples, IUPAC nomenclature for cycloalkanes is very similar to that for noncyclic alkanes. The only modifications to the noncyclic alkane rules are

1. The prefix *cyclo-* is placed before the name that corresponds to the noncyclic chain that has the same number of carbon atoms as the ring.
2. Alkyl groups, when present, are located by numbering the carbons in the ring according to a system that yields the lowest numbers for the carbons at which the alkyl groups are attached.

18.7 STRUCTURE AND NOMENCLATURE OF ALKENES AND ALKYNES

An **alkene** *is a hydrocarbon in which there is one carbon–carbon* double bond. Numerous noncyclic and cyclic alkenes are known. An **alkyne** *is a hydrocarbon in which there is one carbon–carbon* triple bond. Both noncyclic and cyclic alkynes are known, although cyclic alkynes are not common.

Alkenes and alkynes are classified as unsaturated hydrocarbons, in contrast to alkanes, which are saturated hydrocarbons. An **unsaturated compound** *is an organic compound that contains fewer hydrogen atoms than the maximum possible owing to the presence of one or more multiple (double or triple) carbon–carbon bonds.*

The simplest alkene is ethene (common name, ethylene), which has the formula C_2H_4. Obviously a one-carbon alkene cannot exist because two carbon atoms are required for a carbon–carbon double bond to exist. The simplest alkyne is also a two-carbon species with the formula C_2H_2. The name of this compound is ethyne (common name, acetylene). Ball and stick models as well as space-filling models of ethene and ethyne are shown in Figure 18.7.

The general formula for a noncyclic alkene is C_nH_{2n}, and for a noncyclic alkyne C_nH_{2n-2}. The general formula for a noncyclic alkene is thus identical to that for a cycloalkane (Sec. 18.6). This observation shows that compounds belonging to different

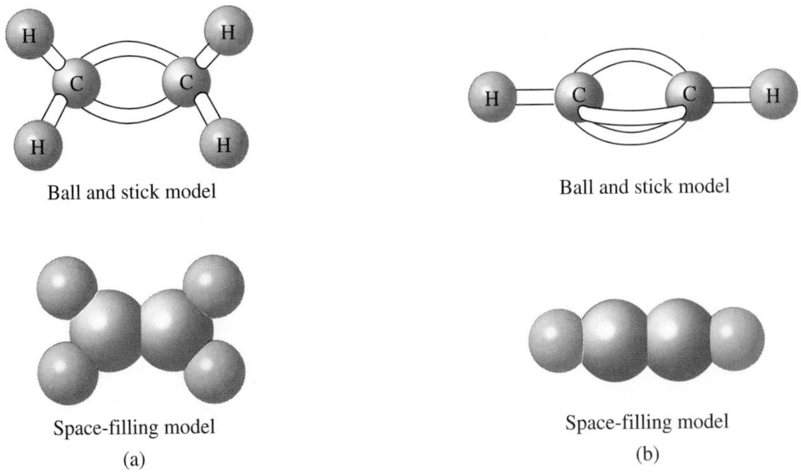

FIGURE 18.7 Ball and stick and space-filling models of (a) ethene, an unsaturated hydrocarbon with one double bond, and (b) ethyne, an unsaturated hydrocarbon with one triple bond.

hydrocarbon classes may be isomeric with each other. Both alkenes and cycloalkanes contain two fewer hydrogen atoms than the maximum number possible for a hydrocarbon; noncyclic alkanes contain the maximum number. In alkenes, the lower number of hydrogen atoms is caused by the presence of the double bond. In cycloalkanes, the cyclization process brings about the loss of two hydrogens.

The rules we have previously developed for naming alkanes, with slight modifications, can be used to name alkenes and alkynes. The modifications are

1. The *-ane* ending characteristic of alkanes is changed to *-ene* for alkenes and to *-yne* for alkynes.
2. For noncyclic molecules containing more than three carbon atoms, the position of the multiple bond must be specified, since there is more than one position it may occupy. The position is given by a single number (the number of the lower-numbered carbon involved in the multiple bond), which is placed immediately in front of the base chain name.

Examples:

$$\overset{1}{CH_2}{=}\overset{2}{CH}{-}\overset{3}{CH_2}{-}\overset{4}{CH_3} \quad \text{is} \quad \text{1-butene}$$

$$\overset{1}{CH_3}{-}\overset{2}{CH}{=}\overset{3}{CH}{-}\overset{4}{CH_3} \quad \text{is} \quad \text{2-butene}$$

The compounds 1-butene and 2-butene are isomers. Note that in the case of 1-butene the double bond involves carbons 1 and 2, but we write only the lower of the two numbers in the name.

Example:

$$\overset{4}{CH_3}{-}\underset{\displaystyle CH_3}{\overset{3}{CH}}{-}\overset{2}{CH}{=}\overset{1}{CH_2} \quad \text{is} \quad \text{3-methyl-1-butene}$$

The chain is always numbered in such a way as to give the multiple bond the lowest number possible, even if this means that alkyl groups must get higher numbers; that is, multiple bonds always take precedence over alkyl groups when deciding upon a numbering system.

Example 18.3

What is the IUPAC name for each of the following unsaturated hydrocarbons?

(a) $CH_2{=}CH{-}CH{-}CH_3$ (with CH_3 on the third carbon)

(b) $CH_3{-}CH{-}C{\equiv}CH$ (with CH_3 on the second carbon)

(c)

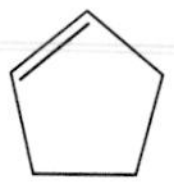

(d)

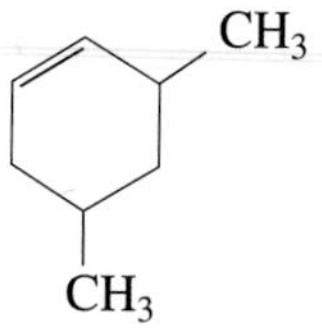

Solution

(a)

$$\overset{1}{C}H_2{=}\overset{2}{C}H{-}\overset{3}{C}H{-}\overset{4}{C}H_3$$

(with CH_3 on carbon 3)

The longest continuous chain containing the double bond has four carbons; therefore, the base name for the compounds is *butene*. We number from the left (the chain end closest to the double bond) to the right. Thus, the double bond position is 1, and the location of the methyl group 3. The IUPAC name is **3-methyl-1-butene.**

(b)

$$\overset{4}{C}H_3{-}\overset{3}{C}H{-}\overset{2}{C}{\equiv}\overset{1}{C}H$$

(with CH_3 on carbon 3)

The base name is *butyne* since the longest carbon chain containing the triple bond has four carbons in it. Numbering from right to left, we get the number 1 for the position of the triple bond and the number 3 for the location of the methyl group. The IUPAC name is thus 3-methyl-1-butyne.

(c) This compound is simply cyclopentene. In cyclic structures containing only one multiple bond no number is needed to specify bond position. It is understood that the multiple bond involves carbons 1 and 2.

(d)

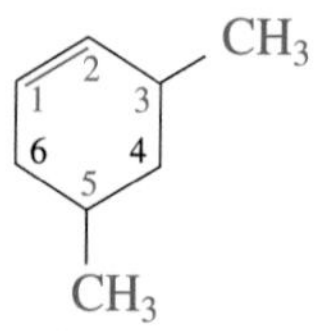

The compound is a *cyclohexene*. The double-bonded carbons become carbons 1 and 2. We number in the clockwise direction in order for the methyl groups to be on the lowest-numbered carbons possible (3 and 5). (Numbering in the counterclockwise direction would put the methyl groups on carbons 4 and 6.) The IUPAC name of the compound is 3,5-dimethylcyclohexene.

Practice Exercise 18.3

Using the IUPAC system, name the following unsaturated hydrocarbons.

(a) $CH_3{-}CH{=}CH{-}CH_2{-}CH_3$

(b)
$$\begin{array}{l} CH{\equiv}C{-}\underset{\displaystyle CH_3}{\underset{|}{CH}}{-}\underset{\displaystyle CH_3}{\underset{|}{CH}}{-}CH_3 \end{array}$$

(c) 

(d) 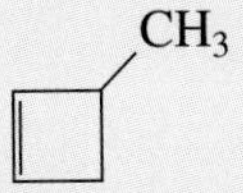

Ans. (a) 2-pentene; (b) 3,4-dimethyl-1-pentyne; (c) cyclopropene; (d) 3-methylcyclobutene

Numerous unsaturated hydrocarbons contain more than one multiple bond—for example, two double bonds or two triple bonds. Compounds containing two double bonds are properly referred to as *alkadienes* and those with two triple bonds as *alkadiynes*. If three double bonds are present, we have an *alkatriene*.

The general family names just mentioned give us the key to naming compounds containing more than one multiple bond. A prefix is added to the base chain ending to indicate the number of double bonds: *di*ene for two double bonds, *tri*ene for three double bonds, *di*yne for two triple bonds, and so on. In addition, a number is used to locate each multiple bond. If there are three multiple bonds, three numbers are needed.

Example 18.4

Using the IUPAC system, name the following hydrocarbons, each of which contains more than one carbon–carbon multiple bond.

(a) $CH_2{=}CH{-}CH_2{-}CH{=}CH_2$

(b)
$$CH_2{=}\underset{\displaystyle CH_3}{\underset{|}{C}}{-}CH{=}CH_2$$

Solution

(a) The base name for this hydrocarbon is pentadiene, since there are two double bonds (diene) present in a chain of five carbon atoms. It does not matter which way we number the chain because the molecule is symmetrical. Either way, the double bond positions are 1 and 4. Therefore, the name of the compound is 1,4-pentadiene.

(b)
$$\overset{1}{C}H_2{=}\underset{\displaystyle CH_3}{\underset{|}{\overset{2}{C}}}{-}\overset{3}{C}H{=}\overset{4}{C}H_2$$

The molecule is a butadiene with a methyl group on it. It does not matter which end we number from relative to the double bonds because of their symmetrical positioning within the molecule. It does matter, however, where we start relative to the methyl group. Numbering from the left gives a lower position number for the methyl group. Using the left-to-right numbering systems, the IUPAC name for the compound becomes 2-methyl-1,3-butadiene.

Practice Exercise 18.4

Using the IUPAC system, name the following compounds, each of which contains more than one multiple bond.

(a) $CH_2{=}CH{-}CH{=}CH{-}CH_3$ (b) $\underset{\displaystyle CH_3}{\underset{|}{CH_2{=}C}}{-}\underset{\displaystyle CH_3}{\underset{|}{CH}}{=}CH{-}CH_2{-}CH_3$

Ans. (a) 1,3-pentadiene; (b) 2,3-dimethyl-1,3-hexadiene

Unsaturated *cyclic* hydrocarbons may also contain more than one multiple bond. Introducing a second bond into the molecule cyclohexene

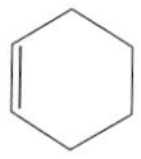

would produce either

1,3-cyclohexadiene

or

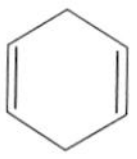

1,4-cyclohexadiene

depending on where the additional double bond is positioned relative to the first one. These two compounds are isomers, differing only in the positions of the double bonds (1,3- versus 1,4-).

The introduction of a third double bond into 1,3-cyclohexadiene produces the compound

Using the nomenclature rules of this section, we would expect to call this compound 1,3,5-cyclohexatriene. However, the compound is not called that. Instead, its name is *benzene*. The reason for this "rule violation" is that benzene does not possess the chemical properties normally associated with cyclic hydrocarbons containing multiple bonds. Its properties are different enough from those of these other compounds that it is considered to be the first member of a new series of hydrocarbons called aromatic hydrocarbons. The reason for benzene's different behavior is the subject of Section 18.8. Benzene is more than just a cycloalkatriene.

18.8 AROMATIC HYDROCARBONS

The cyclic hydrocarbon benzene (C_6H_6) is the parent molecule for a large family of hydrocarbons called *aromatic hydrocarbons*. The structural feature in aromatic hydrocarbons that is responsible for their distinctive properties is the *benzene ring*. **Aromatic hydrocarbons** *are hydrocarbons that contain the characteristic benzene ring*. An understanding of the uniqueness of the benzene ring is the key to understanding aromatic hydrocarbon chemistry.

At the end of Section 18.7, the structure of benzene was given as

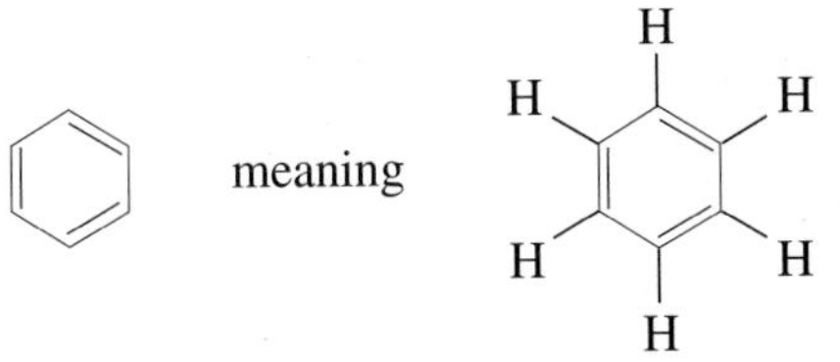

It was noted at that time that this compound was not named cyclohexatriene. What is the reason for this?

The cyclohexatriene interpretation of the benzene structure implies that alternating single and double carbon–carbon bonds are present about the carbon ring. With this interpretation, we would predict that not all carbon–carbon bonds in the ring would be of the same length, since carbon–carbon double bonds are known to be shorter in length than carbon–carbon single bonds. This prediction is contrary to the experimental information available on bond lengths in the benzene molecule. Experimental studies indicate that all carbon–carbon bonds in the benzene molecule are equal in length. Thus, the bonding present in the carbon ring must be something different from alternating single and double bonds.

If all of the carbon–carbon bonds in benzene were of the same kind, that is, all single bonds or all double bonds, we would have a molecule whose carbon–carbon bond lengths were consistent with experimental bond-length data. However, neither of these structures is acceptable; in both structures the carbon atoms violate the octet rule (Sec. 7.3). The all-single-bond suggestion leaves each carbon atom with only three bonds instead of the needed four bonds. In an all-double-bond molecule, each carbon atom would have too many bonds: five per carbon atom.

To obtain a structure for benzene that is consistent with both the octet rule and experimental bond length information, a new type of bonding concept must be invoked, that of delocalized bonding. In **delocalized bonding** *three or more atoms share the same valence electrons*. (Up to now all bonding discussions in the text have involved localized bonds, that is, bonds in which electrons are shared between *two* atoms.) Let us now consider how the delocalized-bond concept helps explain the characteristics of benzene and other aromatic compounds.

A carbon atom has four valence electrons for use in bonding. Three of the four electrons on each carbon atom in benzene will be used in forming the "structural framework" of a benzene molecule, as shown in the following diagram.

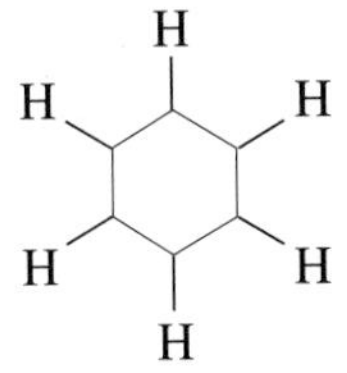

In this structure each carbon atom has formed three bonds. Thus, each carbon atom still has one more electron available for bonding, which is shown as a dot in the following structure.

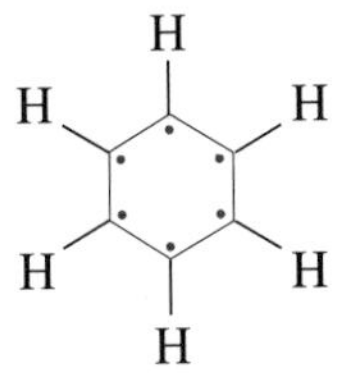

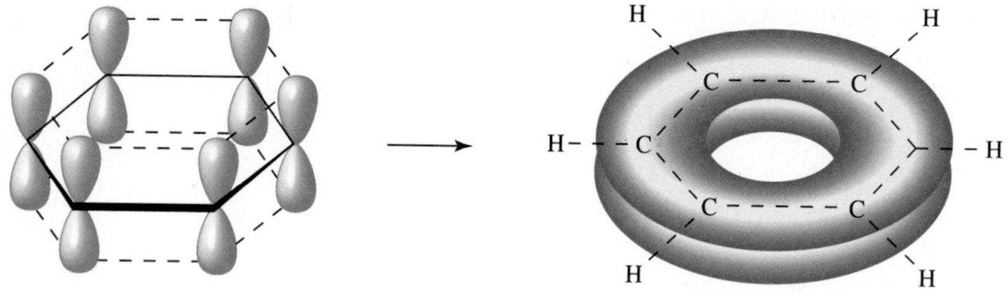

FIGURE 18.8 The six p orbitals of benzene interact to form a delocalized bond "running" completely around the ring.

This additional electron, in each case, is in a p orbital (Sec. 6.6) whose orientation is such that it can interact simultaneously with two other p orbitals—the p orbitals on the carbon atom on each side of it in the carbon ring. This results in a continuous interaction around the ring and a delocalized bond that "runs" completely around the ring, as shown in Figure 18.8. In this delocalized bond, the six electrons (one from each carbon atom) are considered to be shared equally by all six carbon atoms.

A circle drawn inside the benzene ring is used to denote the delocalized bond present in this molecule.

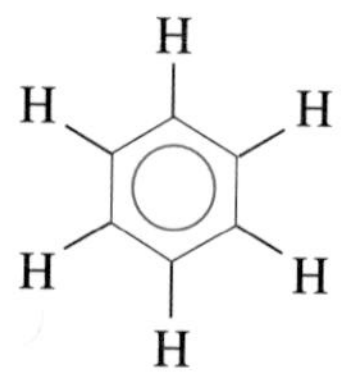

Note that in this structure each carbon atom has four bonds: one bond to a H atom, two localized bonds to other C atoms, and one bond that is part of the delocalized system.

Delocalized bonding is possible only for specific orientations of atomic orbitals. Aromatic hydrocarbons are the only class of hydrocarbons where proper orbital orientation is present for delocalized bonding to occur.

The presence of a delocalized bond in a molecule gives it extra stability. It is this extra stability that causes benzene and the other aromatic hydrocarbons to have chemical properties that are different from those of other hydrocarbons.

The hydrogen atoms on benzene can be replaced by alkyl groups without destroying the delocalized bonding in the carbon ring. Replacements of this type give rise to numerous other members of the aromatic hydrocarbon family. The simplest aromatic hydrocarbon (next to benzene) is methylbenzene (common name, toluene), a compound in which one hydrogen on the ring has been replaced by a methyl group.

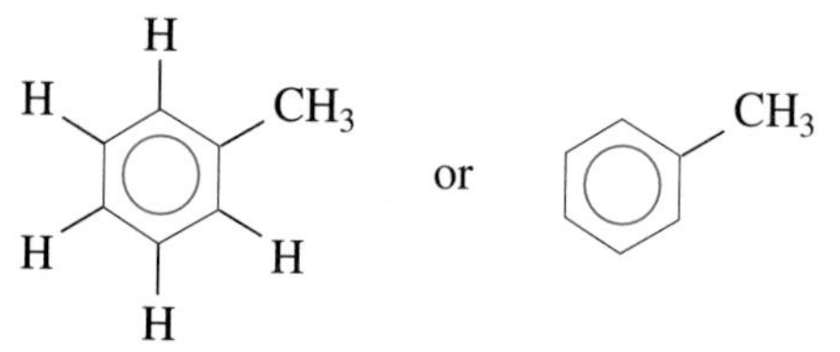

methylbenzene (toluene)

Three isomeric compounds are possible when two methyl groups are placed on a benzene ring. IUPAC names for these compounds are 1,2-, 1,3-, and 1,4-dimethyl-benzene. Their structures are

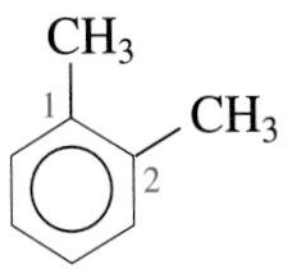

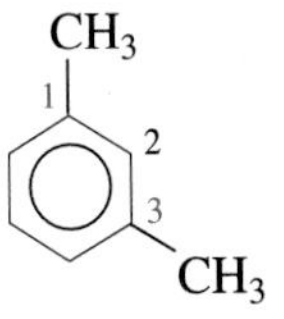

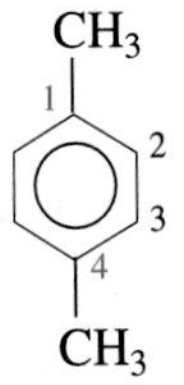

1,2-dimethylbenzene 1,3-dimethylbenzene 1,4-dimethylbenzene

A commonly used alternative method of indicating positioning on the carbon ring in disubstituted benzenes uses the prefixes *ortho-* (substituents that are adjacent to each other), *meta-* (substituents that are one carbon removed from each other), and *para* (substituents that are two carbons removed from each other). The prefixes are often abbreviated as *o*-, *m*-, and *p*-. Thus, we have for the three dimethylbenzenes (common name, xylene) the following additional acceptable names.

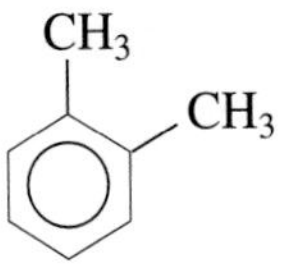

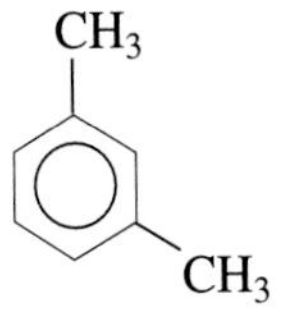

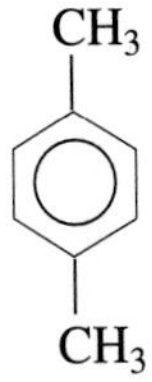

ortho-dimethylbenzene *meta*-dimethylbenzene *para*-dimethylbenzene
ortho-xylene *meta*-xylene *para*-xylene
o-xylene *m*-xylene *p*-xylene

When three or more groups are attached to a benzene ring, the carbon atoms must be numbered to indicate the positions of the groups. The number system that gives the lowest possible sum is the system selected.

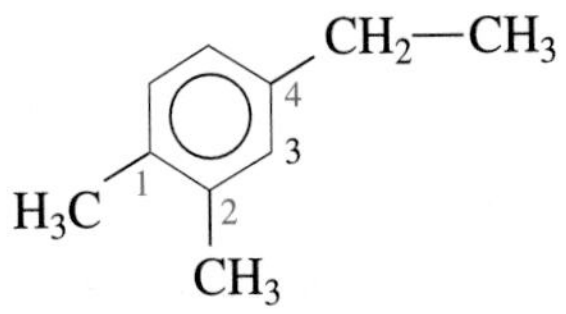

4-ethyl-1,2-dimethylbenzene

Note that the attachments to the benzene ring need not be identical.

Another type of aromatic hydrocarbon contains *fused* benzene rings. The simplest example of such a compound is naphthalene, which is used as a moth repellent.

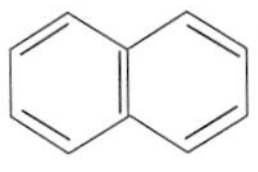

naphthalene

Another fused-ring aromatic hydrocarbon is benzypyrene, which has been identified as one of the cancer-causing agents in tobacco smoke.

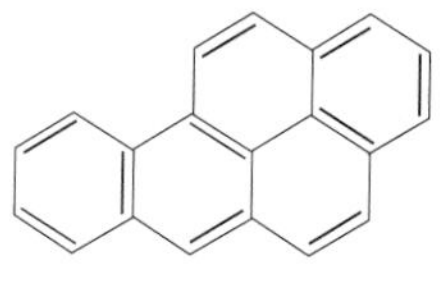

benzpyrene

18.9 SOURCES AND USES OF HYDROCARBONS

There are three major natural sources of hydrocarbons: natural gas, crude oil (petroleum), and coal.

Unprocessed natural gas contains methane (50–90%) and ethane (1–10%) with smaller amounts of propane and butanes. Processed natural gas, the "natural gas" used in homes for cooking and as a heating fuel, is mainly methane.

Crude petroleum is an extremely complex mixture of hydrocarbons. Some crudes consist chiefly of noncyclic alkanes; others contain as much as 40% of other types of hydrocarbons—cycloalkanes and aromatics.

In its natural state, crude petroleum has very few uses, but this complex hydrocarbon mixture can be separated into various useful fractions through refining. The resulting fractions are still hydrocarbon mixtures, but each one is simpler (fewer compounds are present). During refining, the physical separation of the crude into component fractions is accomplished by *fractional distillation*, a process that takes advantage of the different boiling points of the various components of the crude (see Fig. 18.9). Each fraction contains hydrocarbons within a specific boiling point range. Table 18.3 characterizes some of these specific fractions in terms of size of hydrocarbon molecule, boiling point range, and uses. Note the many important and familiar products obtained from petroleum. All of these products are mixtures of hydrocarbons.

Although petroleum is by far the leading natural source of hydrocarbons, coal is

FIGURE 18.9 These tall towers, part of a modern petroleum refinery, are the distillation towers in which the crude oil is separated into fractions. (Courtesy American Petroleum Institute Photographic and Film Services)

TABLE 18.3 HYDROCARBON FRACTIONS OBTAINED FROM CRUDE OIL BY FRACTIONAL DISTILLATION

Fraction Name	Size of Hydrocarbon Molecule	Boiling Point Range (°C)	Uses
Natural gas	1–4 carbon atoms	below room temperature	home heating, cooking fuel
Petroleum ether	5–7 carbon atoms	20–60	solvents, dry cleaning
Gasoline	6–12 carbon atoms	50–200	automobile fuel
Kerosene	12–16 carbon atoms	175–275	jet fuel, diesel fuel
Fuel oil	15–18 carbon atoms	250–400	furnace oil
Lubricating oils	16–20 carbon atoms	350 and above	lubricants
Greases	more than 18 carbon atoms	semisolid	lubricants
Asphalt	high number of carbon atoms	solid	paving, roofing

another significant source. When coal is heated in the absence of air, a process called *destructive distillation*, the coal breaks down into three components: (1) coal gas, (2) coke, and (3) coal tar (a liquid). Coal gas is mainly methane and hydrogen and finds use as a fuel. Coke, the principal product of destructive distillation, is essentially pure carbon and is a vital raw material for the production of steel from iron ore. Coal tar is a mixture of numerous hydrocarbons, many of which are aromatic. Coal tar is a source of aromatic hydrocarbons.

There are no large natural sources of unsaturated hydrocarbons (alkenes and alkynes), although small amounts of alkenes are present in some crude petroleums. Most alkenes and alkynes are produced industrially from alkanes.

When hydrocarbons are burned as fuel, their major use, the products of combustion (reaction with oxygen) are carbon dioxide, water, and heat energy. Representative combustion reactions for methane (natural gas) and octane (a component of gasoline) are

$$CH_4 + 2\,O_2 \longrightarrow CO_2 + 2\,H_2O + \text{energy}$$

$$2\,C_8H_{18} + 25\,O_2 \longrightarrow 16\,CO_2 + 18\,H_2O + \text{energy}$$

A small but significant percentage of hydrocarbons find use as chemical feed stock for the petrochemical industry. In this industry many of the consumer products that are considered to be necessities today are manufactured. Examples of materials produced from hydrocarbon starting materials are plastics, synthetic fibers, adhesives, dyes, and many pharmaceuticals.

18.10 DERIVATIVES OF HYDROCARBONS

Hydrocarbons, as numerous as they are, make up only a small fraction of known organic compounds. The majority of organic compounds are hydrocarbon derivatives. A **hydrocarbon derivative** *is a hydrocarbon molecule in which one or more of the hydrogen atoms has*

TABLE 18.4 CLASSES OF HYDROCARBON DERIVATIVES

Class Name	Functional Group	General Formula for Class[a]	Example
Halide	—X (X = F, Cl, Br, or I)	R—X	CH_3—Br
Alcohol	—OH	R—OH	CH_3—OH
Ether	—O—	R—O—R′	CH_3—O—CH_3
Aldehyde	$-\overset{O}{\overset{\|}{C}}-H$	$R-\overset{O}{\overset{\|}{C}}-H$	$CH_3-\overset{O}{\overset{\|}{C}}-H$
Ketone	$-\overset{O}{\overset{\|}{C}}-$	$R-\overset{O}{\overset{\|}{C}}-R'$	$CH_3-\overset{O}{\overset{\|}{C}}-CH_3$
Carboxylic acid	$-\overset{O}{\overset{\|}{C}}-OH$	$R-\overset{O}{\overset{\|}{C}}-OH$	$CH_3-\overset{O}{\overset{\|}{C}}-OH$
Ester	$-\overset{O}{\overset{\|}{C}}-O-$	$R-\overset{O}{\overset{\|}{C}}-O-R'$	$CH_3-\overset{O}{\overset{\|}{C}}-O-CH_3$
Amine	$-NH_2$	$R-NH_2$	CH_3-NH_2
Amide	$-\overset{O}{\overset{\|}{C}}-NH_2$	$R-\overset{O}{\overset{\|}{C}}-NH_2$	$CH_3-\overset{O}{\overset{\|}{C}}-NH_2$

[a] The symbol R′ represents a hydrocarbon group that may or may not be the same as R.

been replaced with a new atom or group of atoms called a functional group. This newly attached part of the hydrocarbon, which always contains an element or elements other than carbon and hydrogen, serves as a basis for characterizing the molecule. A **functional group** *is that part of a hydrocarbon derivative that contains the elements other than carbon and hydrogen.* The name functional group is very appropriate because it is at the functional group site that most chemical reactions of organic compounds occur. The functional group literally determines how the hydrocarbon derivative functions in chemical reactions.

The use of the functional-group concept greatly simplifies the study of hyrocarbon derivatives. All compounds containing the same functional group are studied as a class because of the similarities of their chemistry. Table 18.4 lists the classes of hydrocarbon derivatives most commonly encountered in organic chemistry as well as the functional group that characterizes each class.

Note that the element oxygen is a particularly common component of functional groups. The element nitrogen is the next most frequently encountered functional group constituent. The symbol R, used in the general formulas for a class of hydrocarbon derivatives (third column of Table 18.4), is a designation for the hydrocarbon part of any hydrocarbon derivative. For example, the general formula for a carboxylic acid

$$R-\overset{O}{\overset{\|}{C}}-OH$$

collectively designates the following compounds

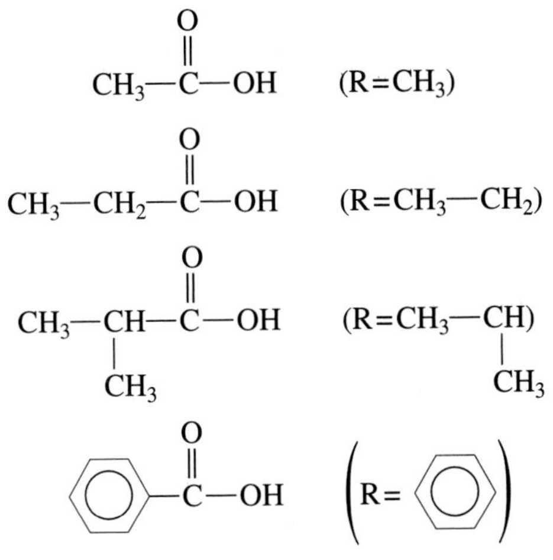

$$\underset{CH_3}{\overset{O}{\underset{\|}{C}}}\text{—OH} \qquad (R=CH_3)$$

$$CH_3\text{—}CH_2\text{—}\overset{O}{\overset{\|}{C}}\text{—OH} \qquad (R=CH_3\text{—}CH_2)$$

$$CH_3\text{—}\underset{\underset{CH_3}{|}}{CH}\text{—}\overset{O}{\overset{\|}{C}}\text{—OH} \qquad (R=CH_3\text{—}\underset{\underset{CH_3}{|}}{CH})$$

as well as hundreds of other similar compounds that differ from each other only in the identity of the hydrocarbon part of the molecule.

We shall not try to discuss all of the functional groups listed in Table 18.4. As representative of the field of hydrocarbon derivatives, we will consider briefly five classes of such derivatives: halides, alcohols, ethers, carboxylic acids, and esters.

18.11 HALOGENATED HYDROCARBONS (ORGANIC HALIDES)

Halogenated hydrocarbons (*or* organic halides) *are hydrocarbon derivatives in which one or more halogen atoms have replaced hydrogen atoms in the parent hydrocarbon.* (The group name for the elements in group VIIA of the periodic table—fluorine, chlorine, bromine, and iodine—is *halogen.*)

In the IUPAC system for naming organic halides, the prefixes fluoro-, chloro-, bromo-, and iodo-, are used to designate the various halogen atoms. These prefixes (and location numbers for the halogen atoms if isomers are possible) are attached to the name of the longest continuous hydrocarbon chain. Thus,

$$CH_3\text{—}\underset{\underset{Cl}{|}}{CH}\text{—}CH_2\text{—}CH_3$$

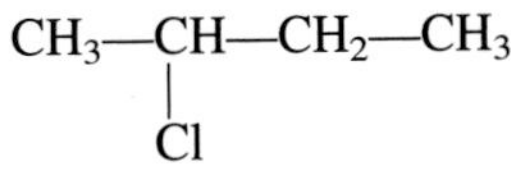

would be named 2-chlorobutane.

Many halogenated methanes, ethanes, and ethenes have important commercial uses. Information concerning selected members of these families of compounds is given in Table 18.5.

Freons, chlorofluoromethanes, are used as refrigerants (air conditioning, refrigeration, and so on) and prior to 1979 as propellants in aerosol cans. Aerosal-can use for these compounds was discontinued when studies indicated that they could possibly harm the ozone

TABLE 18.5 SOME COMMERCIALLY IMPORTANT HALOGENATED HYDROCARBONS

Formula	Common Name	IUPAC Name	Uses
CH_3Cl	methyl chloride	chloromethane	local anesthetic
CH_2Cl_2	methylene chloride	dichloromethane	laboratory solvent
$CHCl_3$	chloroform	trichloromethane	laboratory solvent
CCl_4	carbon tetrachloride	tetrachloromethane	stain remover
CCl_3F	Freon-11	trichlorofluoromethane	refrigerant
CCl_2F_2	Freon-12	dichlorodifluoromethane	refrigerant
CH_3CH_2Cl	ethyl chloride	chloroethane	local anesthetic
$Cl_2C{=}CHCl$	trichloroethylene	trichloroethene	dry-cleaning solvent
$Cl_2C{=}CCl_2$	tetrachloroethylene	tetrachloroethene	dry-cleaning solvent

layer in the Earth's upper atmosphere. (The function of the ozone layer is to filter out most of the ultraviolet radiation given off from the sun.)

Chlorinated ethenes (trichloroethene and tetrachloroethene) are very important solvents in the dry-cleaning industry. Carbon tetrachloride was used as a stain remover because of its ability to dissolve greases. The use of this compound as a general dry-cleaning solvent was discontinued because of concerns about its toxicity.

Chloromethane and chloroethane, both gases at room temperature, are used in a pressurized liquid form as local anesthetics to alleviate pain from muscular injuries and bruises. Sprayed onto the skin, these compounds rapidly evaporate, causing an intense cooling effect. This cooling minimizes swelling and reduces pain by deadening nerves.

As with hydrocarbons, isomers are possible for hydrocarbon derivatives. Many isomeric halogenated hydrocarbons exist. For example, four isomeric dichloropropanes are possible.

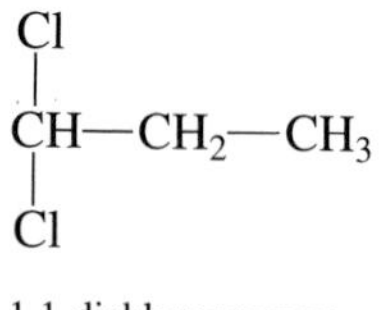

1,1-dichloropropane

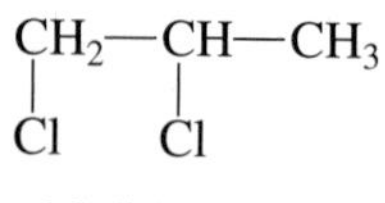

1,2-dichloropropane

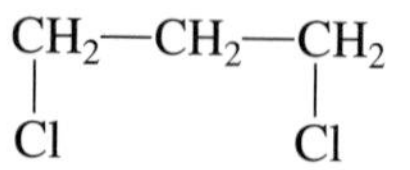

1,3-dichloropropane

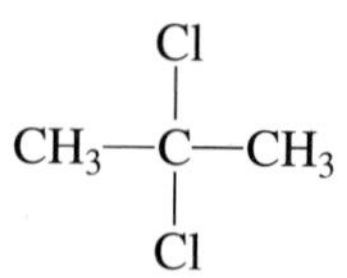

2,2-dichloropropane

The plastics polyvinyl chloride (PVC) and Teflon are made from the halogenated hydrocarbon starting materials chloroethene and tetrafluoroethene, respectively.

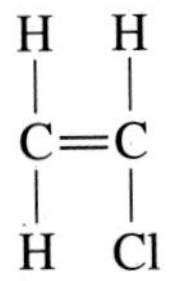

chloroethene
(vinyl chloride)

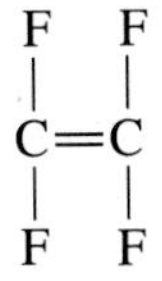

tetrafluoroethene

Both PVC and Teflon are polymeric materials. A **polymer** *is a giant molecule that contains thousands of small molecules (monomers) linked together in long chains.* Chloroethene and tetrafluoroethene are the monomers for the two plastics we have mentioned. In the process of linking the monomers together to generate the polymers, the double bonds revert to single bonds. Thus, the structures of PVC and Teflon are

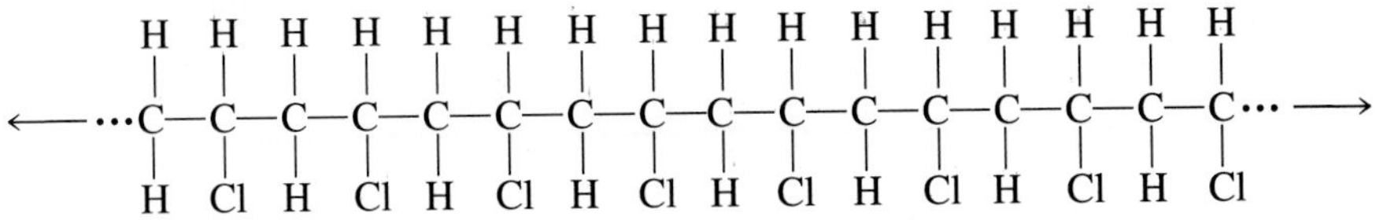

poly(vinyl chloride) (PVC)

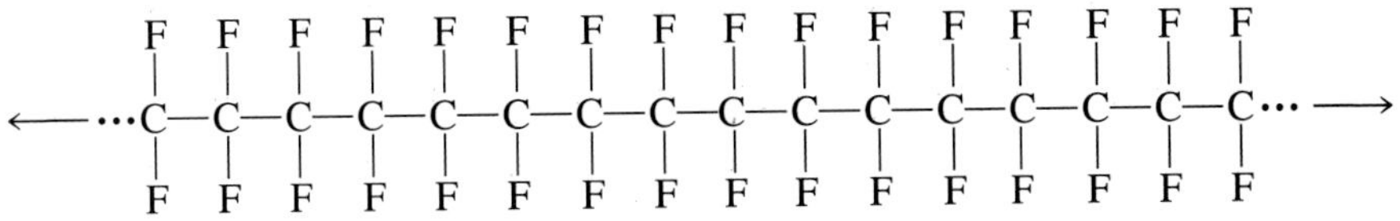

teflon

The most used of all plastics is polyethylene. Its structure is similar (but simpler) to that of PVC and Teflon. It is not a halogenated hydrocarbon but simply a hydrocarbon; the monomer (starting material) is $CH_2{=}CH_2$ (ethene).

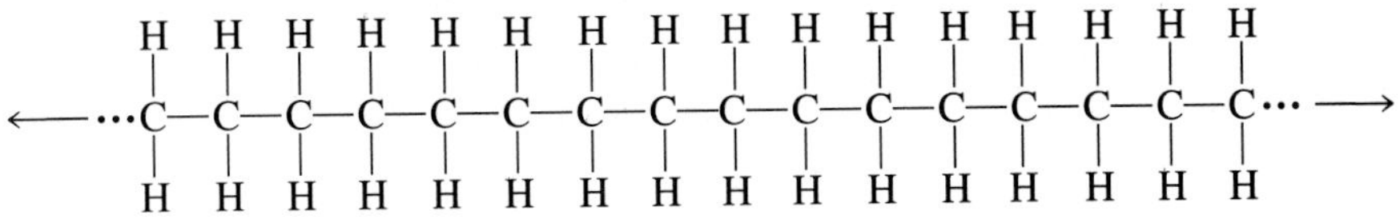

polyethylene

18.12 ALCOHOLS

Alcohols *are hydrocarbon derivatives in which one or more hydroxyl groups (—OH groups) have replaced hydrogen atoms in the parent hydrocarbon.* Besides being hydrocarbon derivatives, alcohols can also be thought of as derivatives of water. If one of the hydrogen atoms in water is replaced with an R group, an alcohol results.

H—O—H　　R—O—H

water　　an alcohol

The IUPAC name of an alcohol containing only one hydroxyl group is obtained from the name of the parent alkane by replacing the *-e* ending with *-ol*. When necessary the position of the —OH functional group is specified by a number. The simpler alcohols are also often known by common names, where the alkyl group name is followed by the word alcohol.

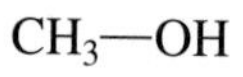

methanol
or
methyl alcohol

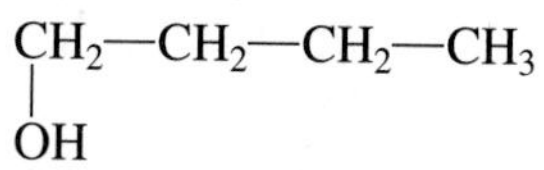

1-butanol
or
butyl alcohol

Alcohols that contain more than one hydroxyl group are called *polyhydric alcohols*. Dihydric alcohols (two —OH groups) are called *glycols* or *diols*. Trihydric alcohols (three —OH groups) are called *triols*.

Five of the most common alcohols are methanol, ethanol, 2-propanol, 1,2-ethanediol, and 1,2,3-propanetriol. Structural formulas and common names (which are used more often then the IUPAC names) for these alcohols are given in Table 18.6. From Table 18.6 note particularly how IUPAC names take into account the presence of more than one hydroxyl group in an alcohol.

Further insights concerning IUPAC alcohol nomenclature is obtained by considering the IUPAC names for the following four isometric four-carbon alcohols.

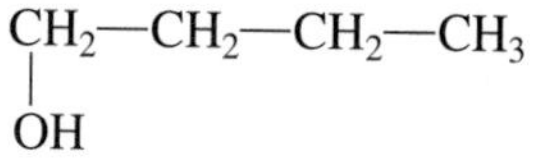

1-butanol

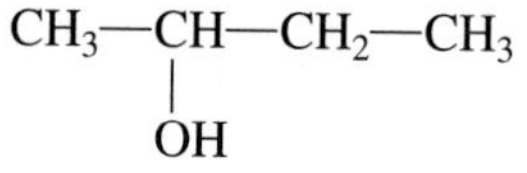

2-butanol

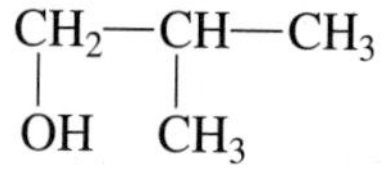

2-methyl-1-propanol

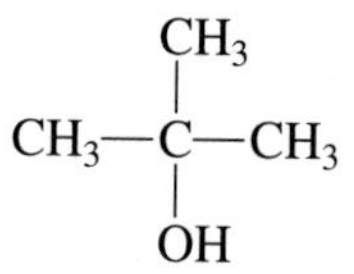

2-methyl-2-propanol

The longest carbon chain to which the —OH group is attached is the basis for the name, and all attachments to the chain, including the —OH group(s), are located by number. The numbering system for the chain is such that the —OH group(s) receive the lowest numbers possible.

TABLE 18.6 SOME IMPORTANT COMMON ALCOHOLS

IUPAC Name	Structural Formula	Common Names
Methanol	$CH_3—OH$	methyl alcohol wood alcohol
Ethanol	$CH_3—CH_2—OH$	ethyl alcohol grain alcohol drinking alcohol
2-Propanol	$CH_3—CH—CH_3$ (OH on the middle carbon)	isopropyl alcohol rubbing alcohol
1,2-Ethanediol	$CH_2—CH_2$ (OH on each carbon)	ethylene glycol
1,2,3-Propanetriol	$CH_2—CH—CH_2$ (OH on each carbon)	glycerol glycerin

Methanol, which is the simplest of all alcohols, is a colorless liquid with a characteristic odor. Its common name, wood alcohol, comes from the fact that its principal source for many years was the destructive distillation of wood. Today, it is synthetically produced in large quantities from hydrogen and carbon monoxide gases.

Methanol is used as a chemical raw material in the production of some plastics, as a solvent for shellacs and varnishes, and as a fuel for motor vehicles. It is a highly toxic substance and should never be used for medicinal purposes. If taken internally, even small amounts can cause permanent blindness and paralysis, and large amounts can be fatal.

Ethanol is the compound commonly referred to as simply "alcohol" by the layperson. Besides being the physiologically active ingredient in alcoholic beverages, it is also used in the pharmaceutical industry as a solvent (tinctures are ethanol solutions) and medicinal ingredient (cough syrups often contain as much as 20% ethanol), and as an industrial solvent. A 70% ethanol solution is an excellent antiseptic. Of all the monohydroxy alcohols, only ethyl alcohol has toxic effects mild enough to render it "safe" for human consumption. Long-term, excessive use, however, is known to cause undesirable effects such as cirrhosis of the liver.

Gasohol, a mixture of 10% ethanol and gasoline, is a fuel for automobiles that was developed in response to the energy crisis of the early 1970s. The ethanol is prepared by fermenting cereal grains.

Ethanol destined for human consumption is carefully controlled and heavily taxed. Ethanol for use in industry is not taxed, and substances are usually added to it to make it unfit to drink, thus removing any temptation. Such alcohol is called *denatured alcohol*. Common denaturants are wood alcohol and formaldehyde. The small amounts added do not interfere with the industrial uses made of ethanol, but definitely make it unfit to drink.

Isopropyl alcohol is the alcohol that is commonly called "rubbing alcohol." It is also used as an astringent, a medication used externally to contract tissue and decrease the size of blood vessels.

Ethylene glycol is the most important and also the simplest of the dihydric alcohols. It is used extensively as the basic ingredient in permanent antifreeze for automobile radiators. It is less volatile than water, completely miscible with water, noncorrosive, and relatively inexpensive to produce; hence its use as antifreeze. Ethylene glycol is also used to make some polymers, the most common of which is the synthetic fiber Dacron.

The most important trihydric alcohol is glycerin or glycerol. It has an affinity for water and finds uses as a moistening agent in tobacco and many food products such as candy and shredded coconut. Florists use glycerin on cut flowers to retain water and maintain freshness. Glycerin is also used for its soothing qualities in shaving and toilet soaps and in many cosmetics.

Alcohols in general are very important biologically, since the hydroxyl group occurs in a variety of compounds associated with living systems. Sugars contain several hydroxyl groups, and starch and cellulose contain thousands of hydroxyl groups; both starch and cellulose are polymeric materials.

18.13 ETHERS

Ethers *are compounds that contain the —O— functional group*. The R groups attached to this functional group may or may not be the same. When both R groups are the same, the compound is called a simple ether. When the R groups are different, a mixed ether results.

$$\underset{\text{a simple ether}}{CH_3{-}O{-}CH_3} \qquad \underset{\text{a mixed ether}}{CH_3{-}CH_2{-}O{-}CH_3}$$

Common names for ethers are obtained by first naming the two hydrocarbon groups attached to the oxygen and then adding the word *ether*. If the two hydrocarbon groups are identical, the prefix di- is often omitted.

$$\underset{\substack{\text{dimethyl ether}\\ \text{(methyl ether)}}}{CH_3\text{—}O\text{—}CH_3} \qquad \underset{\text{ethyl methyl ether}}{CH_3\text{—}CH_2\text{—}O\text{—}CH_3}$$

IUPAC names are obtained by calling an —O—R group an alkoxy group. Thus, $\text{—}O\text{—}CH_3$ is called a methoxy group. The alkoxy group is then treated as a substituent on a parent hydrocarbon molecule. In mixed ethers, the smaller alkyl group becomes the alkoxy group.

$$\underset{\text{methoxyethane}}{CH_3\text{—}O\text{—}CH_2\text{—}CH_3}$$

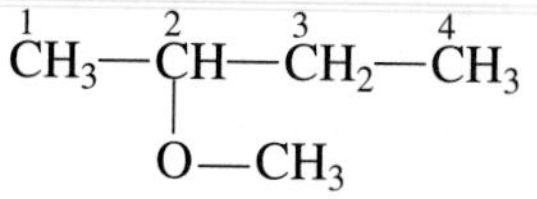

2-methoxybutane

Ethers can be considered to be organic derivatives of water in which both hydrogens have been replaced with R groups. In addition, they can be treated as derivatives of alcohols in which the hydrogen of the hydroxyl group has been replaced by an R group

$$\underset{\text{water}}{H\text{—}O\text{—}H} \xrightarrow[\text{hydrogen}]{\text{replace one}} \underset{\text{an alcohol}}{R\text{—}O\text{—}H} \xrightarrow[\text{hydrogen}]{\text{replace second}} \underset{\text{an ether}}{R\text{—}O\text{—}R'}$$

Many ethers have been found to have general anesthetic properties. Diethyl ether, which is often called simply "ether," has in the past been used extensively as an anesthetic. It was easy to administer and caused excellent relaxation of the muscles. Its drawbacks were that it caused a slight respiratory-passage irritation and some post-anesthetic nausea. In addition, it is a very flammable compound; hence, extreme care had to be taken in administering it. Other anesthetics that do not have these disadvantages have now taken its place. Many of the newer anesthetics are halogenated ethers. It has been found that introduction of

TABLE 18.7 SOME COMMON INHALATION ANESTHETICS THAT ARE ETHERS

Structure	Chemical Name	Trade Name
$CH_3\text{—}O\text{—}CH_2\text{—}CH_2\text{—}CH_3$	methyl propyl ether	Neothyl
$CH_2\text{=}CH\text{—}O\text{—}CH\text{=}CH_2$	divinyl ether	Vinethene
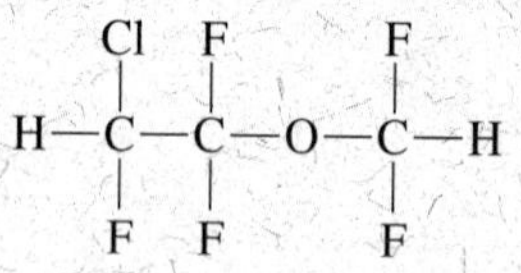	2-chloro-1,1,2-trifluoroethyl difluoromethyl ether	Enflurane or Ethrane
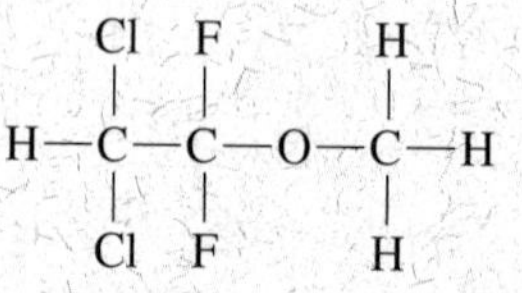	2,2-dichloro-1,1-difluoroethyl methyl ether	Methoxyglurane or Penthrane
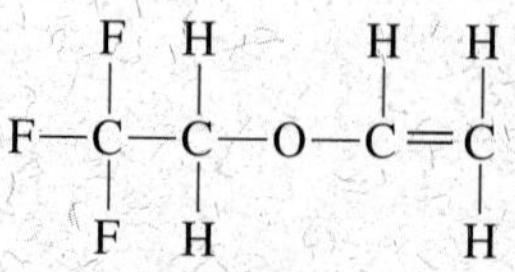	2,2,2-trifluoroethyl vinyl ether	Fluoxene or Fluoromar

halogen atoms into an ether in many cases cuts down on the flammability of the compound. Table 18.7 gives the structures and names of some of the ethers now used in producing anesthesia.

A monohydroxyl alcohol contains one oxygen atom that is involved in two single bonds: one bond to a carbon atom and one bond to a hydrogen atom. Ethers also possess a single oxygen atom that forms two single bonds. This similarity involving oxygen translates in the situation that alcohols and ethers can be isomeric with each other. For example, the following ether and alcohol have the same molecular formula ($C_4H_{10}O$) and thus are isomers.

$$CH_3{-}CH_2{-}CH_2{-}O{-}CH_3 \qquad CH_3{-}CH_2{-}CH_2{-}CH_2{-}OH$$

methoxypropane (methyl propyl ether) — 1-butanol (butyl alcohol)

There are also additional ethers and additional alcohols with the formula $C_4H_{10}O$ that are isomeric with these two compounds.

18.14 CARBOXYLIC ACIDS

Carboxylic acids *are compounds that contain the* $-\overset{\overset{\displaystyle O}{\|}}{C}-OH$ *functional group, that is, the carboxyl group.* They were some of the first organic compounds studied in detail because of their wide distribution and abundance in natural products. Many of the tart or sour tastes we encounter in foods are due to the presence of carboxylic acids. Some of the unpleasant odors associated with spoiled or spoiling foods are also due to carboxylic acids.

IUPAC names for carboxylic acids are derived by replacing the *-e* of the name of the longest carbon chain containing the functional group with *-oic* and then adding the word *acid.* The functional-group carbon atom is counted as part of the longest chain and is assigned the number 1 when a numbering system is needed. The IUPAC names and structural formulas for the two simplest carboxylic acids are

$$H{-}\overset{\overset{\displaystyle O}{\|}}{C}{-}OH \qquad CH_3{-}\overset{\overset{\displaystyle O}{\|}}{C}{-}OH$$

methanoic acid — ethanoic acid

Since many carboxylic acids were isolated from natural sources long before systematic nomenclature was established, we still often refer to them by their common names, which usually indicate their early sources. The common names for methanoic and ethanoic acid are, respectively, formic acid and acetic acid. Formic acid is an active irritant in both ant and bee stings. (The Latin word for ant is *formica.*) Vinegar is a dilute solution of acetic acid, and the pungent odor of this substance is due to the acetic acid present. (The Latin word for vinegar is *acetum.*) Butyric acid, a four-carbon acid, which occurs in rancid butter, gets its common name from the Latin *butyrum* for butter.

Some carboxylic acids contain more than one carboxyl group per molecule. Such acids are known almost exclusively by their common names. The tart (sour) taste associated with citrus fruits is due to the presence of citric acid, a carboxylic acid containing three carboxyl groups. Oxalic acid, a poisonous material found in the leaves of the rhubarb plant, is a dicarboxylic acid.

TABLE 18.8 SELECTED EXAMPLES OF CARBOXYLIC ACIDS

Common Name	IUPAC Name	Structural Formula	Characteristics and Typical Uses
Formic acid	methanoic acid	$H{-}C({=}O){-}OH$	stinging agent of red ants, bees, and nettles
Acetic acid	ethanoic acid	$CH_3{-}C({=}O){-}OH$	active ingredient in vinegar
Propionic acid	propanoic acid	$CH_3{-}CH_2{-}C({=}O){-}OH$	salts of this acid are used as mold inhibitors in breads and cereals
Butyric acid	butanoic acid	$CH_3{-}(CH_2)_2{-}C({=}O){-}OH$	odor-causing agent in rancid butter; present in human perspiration
Caproic acid	hexanoic acid	$CH_3{-}(CH_2)_4{-}C({=}O){-}OH$	characteristic odor of Limburger cheese
Oxalic acid	ethanedioic acid	$HO{-}C({=}O){-}C({=}O){-}OH$	poisonous material in leaves of some plants such as rhubarb; used as cleaning agent for rust stains on fabric and porcelain
Citric acid	3-hydroxy-3-carboxy-pentanedioic acid	$HO{-}C({=}O){-}CH_2{-}C(OH)(C({=}O){-}OH){-}CH_2{-}C({=}O){-}OH$	present in citrus fruits; used as a flavoring agent in foods
Lactic acid	2-hydroxypropanoic acid	$CH_3{-}CH(OH){-}C({=}O){-}OH$	found in sour milk and sauerkraut; formed in muscles during exercise

The structures of all the previously mentioned acids, as well as those of selected other commonly encountered acids, are given in Table 18.8. The table also includes both common and IUPAC names for each acid. Note how dicarboxylic acids are named by the IUPAC system.

18.15 ESTERS

Esters *are compounds that contain the* $-\overset{\displaystyle O}{\overset{\|}{C}}-O-R$ *functional group.* Thus, esters are structurally very closely related to carboxylic acids. The ester functional group may be visualized as a carboxyl group in which an R group has been substituted for the hydrogen atom in the carboxyl group.

Names of esters have a direct relationship to the names of the acids from which they are derived. The names of the ester is formed by changing the *-ic* ending of the acid name (either common or IUPAC name) to *-ate* and preceding this name with the name of the R group attached to the oxygen atom as a separate word. The relatively simple ester

$$CH_3-\overset{\displaystyle O}{\overset{\|}{C}}-O-CH_2-CH_3$$

has the IUPAC name ethyl ethanoate (from ethanoic acid).

Esters are found widely distributed in nature. Many of the fragrances of flowers and fruits as well as the flavors of many fruits are due to the presence of esters. Esters are also the chemical components of many of the artificial fruit flavors that are used in cakes, candies, ice cream, and soft drinks. Table 18.9 lists some common ester flavoring agents and the odors or tastes we associate with them. Note from the table entries how a relatively small change in the size of the R group of the ester functional group significantly alters our perception of flavor. A five-carbon R group (pentyl acetate) is perceived by us to be banana flavor, whereas an eight-carbon R group (octyl acetate) registers as orange flavor. The difference between apple and pineapple flavor is a methyl group versus an ethyl group.

A number of esters find use in the field of medicine. The most used of all medicinal esters is acetylsalicylic acid, more commonly known as aspirin. The structure of aspirin is

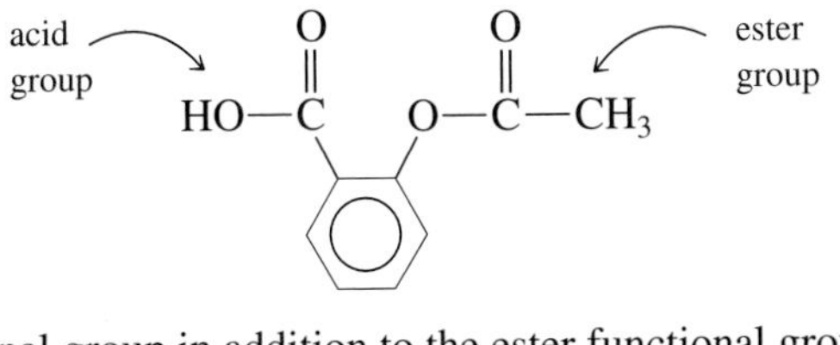

Note that an acid functional group in addition to the ester functional group is present in this molecule.

The environment of the oxygen atoms in simple esters and simple carboxylic acids is similar. Each type of compound has an oxygen atom involved in a double bond to a carbon atom and an oxygen atom involved in two single bonds. The result is that esters and carboxylic acids with the same number of carbon atoms can be isomers. The following four compounds (two acids and two esters) are isomeric; they all have the formula $C_4H_8O_2$.

$$CH_3-CH_2-\overset{\displaystyle O}{\overset{\|}{C}}-O-CH_3 \qquad CH_3-\overset{\displaystyle O}{\overset{\|}{C}}-O-CH_2-CH_3$$

methyl propanoate — ethyl ethanoate

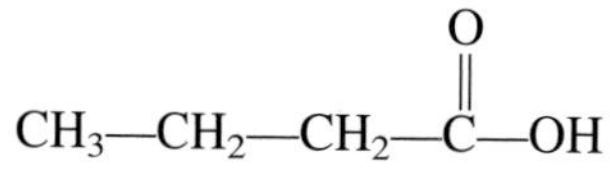

butanoic acid

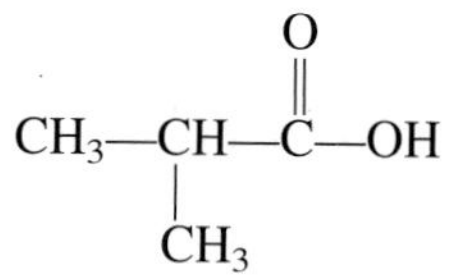

2-methylpropanoic acid

TABLE 18.9 FRUIT ODOR OR FLAVOR ASSOCIATED WITH SELECTED ESTER FLAVORING AGENTS

Name	**Structural Formula**	**Characteristic Flavor and Odor**
Isobutyl methanoate (isobutyl formate)	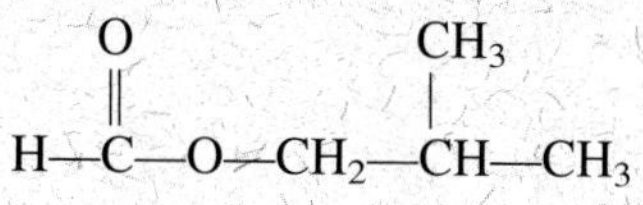	raspberry
Pentyl ethanoate (pentyl acetate)	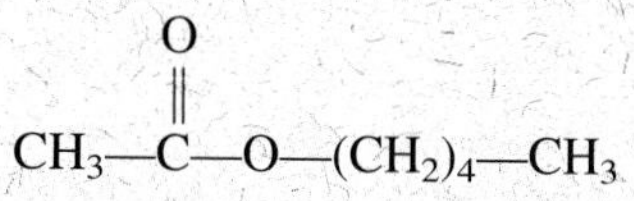	banana
Octyl ethanoate (octyl acetate)	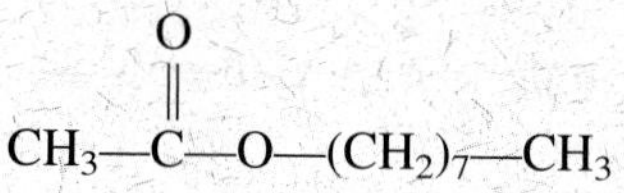	orange
Pentyl propanoate (pentyl propionate)	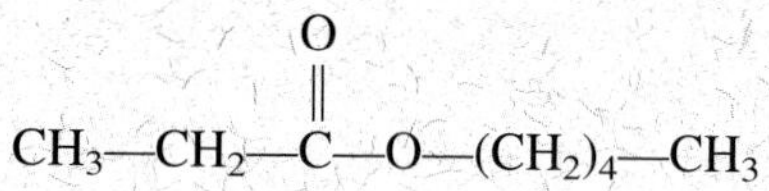	apricot
Methyl butanoate (methyl butyrate)	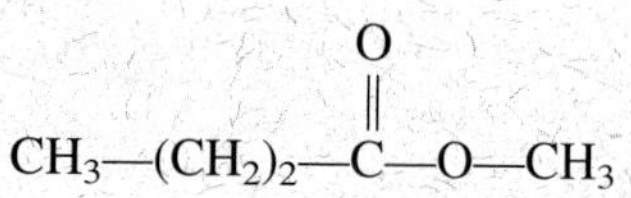	apple
Ethyl butanoate (ethyl butyrate)	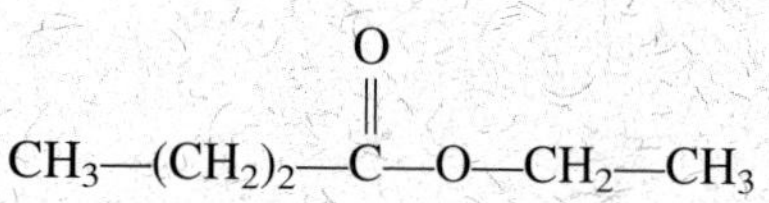	pineapple
Methyl 2-aminobenzoate (methyl anthranilate)	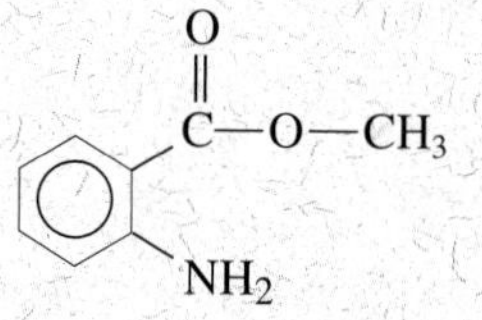	grape

KEY TERMS

The new terms or concepts defined in this chapter are

alcohol (Sec. 18.12) A hydrocarbon derivative in which one or more hydroxyl groups (—OH) have replaced hydrogen atoms in the parent hydrocarbon.

alkane (Sec. 18.3) A hydrocarbon in which all chemical bonds are single bonds.

alkene (Sec. 18.7) A hydrocarbon in which there is one carbon–carbon double bond.

alkyl group (Sec. 18.5) The fragment produced by the removal of one hydrogen atom from an alkane.

alkyne (Sec. 18.7) A hydrocarbon in which there is one carbon–carbon triple bond.

aromatic hydrocarbon (Sec. 18.8) A hydrocarbon that contains a benzene ring.

branched-chain noncyclic hydrocarbon (Sec. 18.5) A hydrocarbon in which one or more side chains of carbon atoms are attached at some point to a continuous chain of carbon atoms.

carboxylic acid (Sec. 18.14) A hydrocarbon derivative that contains the carboxyl functional group

O
||
(—C—OH)

cycloalkane (Sec. 18.6) A hydrocarbon that contains a ring of carbon atoms and has the general formula C_nH_{2n}.

delocalized bonding (Sec. 18.8) Bonding in which three or more atoms share the same valence electrons.

ester (Sec. 18.15) A hydrocarbon derivative that contains an ester functional group

O
||
(—C—O—R)

ether (Sec. 18.13) A hydrocarbon derivative that contains an ether functional group (—O—).

functional group (Sec. 18.10) That part of a hydrocarbon derivative that contains the elements other than carbon and hydrogen.

halogenated hydrocarbon (Sec. 18.11) A hydrocarbon derivative in which one or more halogen atoms have replaced hydrogen atoms in the parent hydrocarbon.

homologous series (Sec. 18.3) A series of compounds in which each compound in the series differs from the previous one in the series by a constant amount.

hydrocarbon (Sec. 18.2) A compound that contains only the two elements hydrogen and carbon.

hydrocarbon derivative (Sec. 18.10) A hydrocarbon molecule in which one or more of the hydrogen atoms have been replaced with a new atom or group of atoms called a functional group.

normal chain noncyclic hydrocarbon (Sec. 18.5) A hydrocarbon in which all carbon atoms are connected in a continuous nonbranching chain.

organic chemistry (Sec. 18.1) The study of hydrocarbons and their derivatives.

polymer (Sec. 18.11) A giant molecule that contains thousands of small molecules (monomers) linked together in long chains.

saturated compound (Sec. 18.3) An organic compound that contains only single bonds between carbon atoms.

structural formula (Sec. 18.4) Two-dimensional (planar) representation of the arrangement of the atoms in a molecule.

structural isomers (Sec. 18.4) Compounds that have the same molecular formulas but different structural formulas.

unsaturated compound (Sec. 18.7) An organic compound that contains fewer hydrogen atoms than the maximum possible owing to the presence of one or more multiple (double or triple) carbon–carbon bonds.

PRACTICE PROBLEMS

Alkanes (Secs. 18.3–18.5)

18.1 How many hydrogen atoms would be present in noncyclic alkanes with the following numbers of carbon atoms?

(a) 3 **(b)** 8 **(c)** 11 **(d)** 18

18.2 How many carbon atoms would be present in noncyclic alkanes with the following numbers of hydrogen atoms?

(a) 8 **(b)** 12 **(c)** 26 **(d)** 40

18.3 The following structural formulas for noncyclic alkanes are incomplete in that the hydrogen atoms attached to each carbon atom are not shown. Complete each of these formulas by adding the correct number of hydrogen atoms to each carbon atom.

(a) 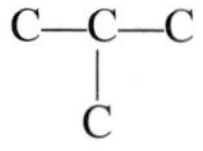

(b) 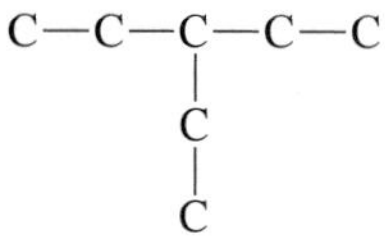

(c) 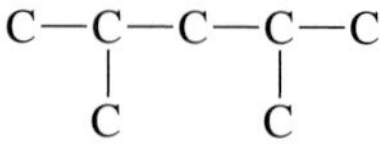

(d) 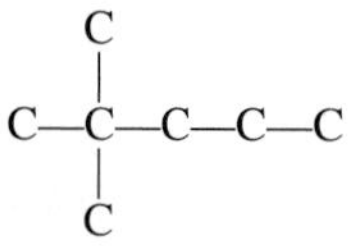

18.4 The following structural formulas for noncyclic alkanes are incomplete in that the hydrogen atoms attach to each carbon atom are not shown. Complete each of these formulas by adding the correct number of hydrogen atoms to each carbon atom.

(a) C—C—C—C—C

(b) 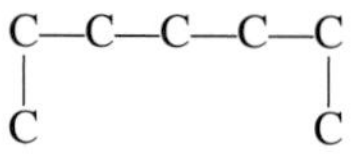

(c) 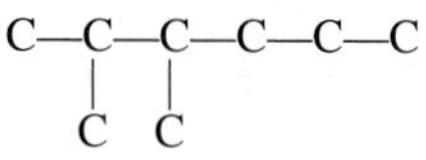

(d) 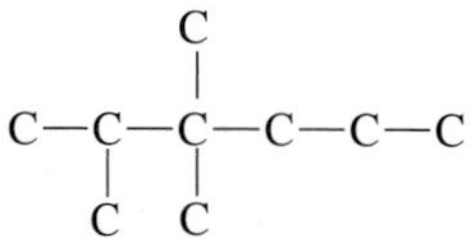

18.5 The first step in naming an alkane is to identify the longest continuous carbon-atom chain. Give the number of carbon atoms in the longest continuous chain in each of the following carbon-atom arrangements.

(a) 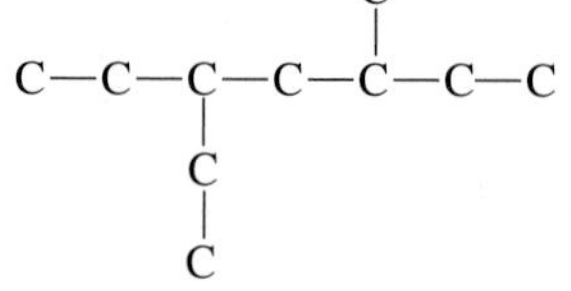

(b) 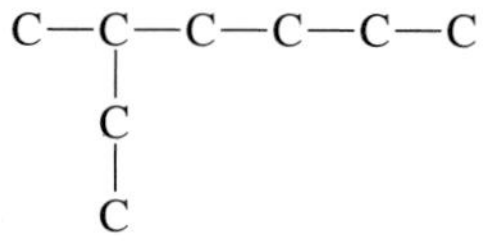

(c) 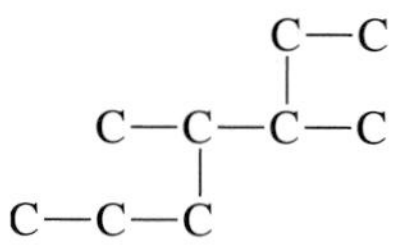

(d) 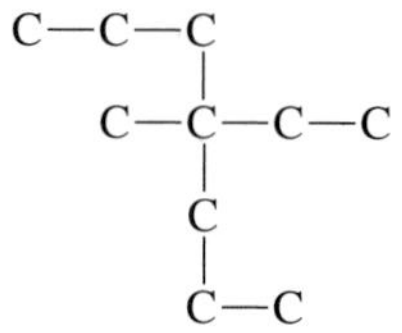

18.6 The first step in naming an alkane is to identify the longest continuous carbon-atom chain. Give the number of carbon atoms in the longest continuous chain in each of the following carbon-atom arrangements.

(a) 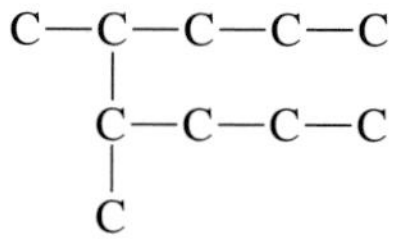

(b) 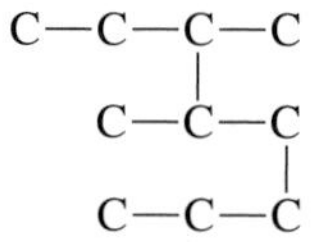

(c) 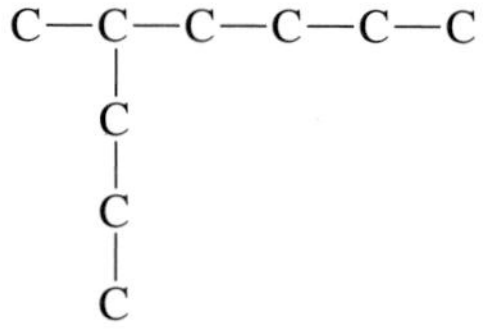

(d) 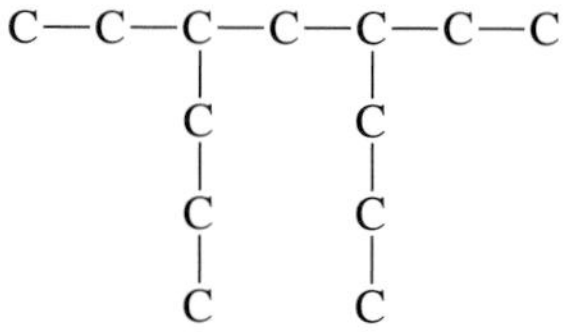

18.7 Using the IUPAC system, name the following alkanes.

(a) $CH_3—CH_2—CH_2—CH_2—CH_2—CH_3$

(b) 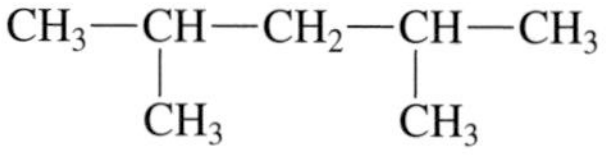

(c) 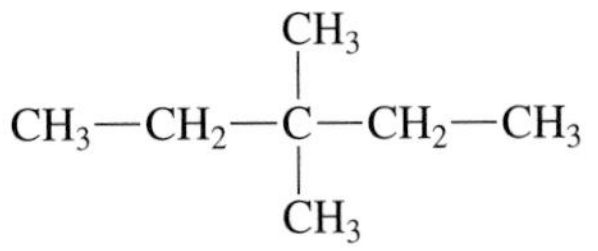

(d) 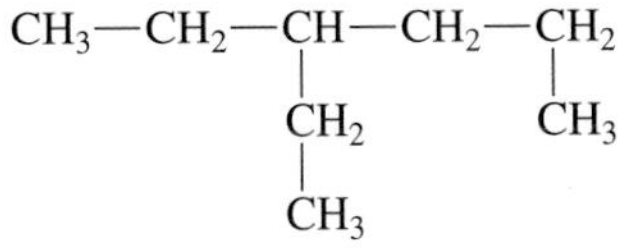

18.8 Using the IUPAC system, name the following alkanes.

(a) $CH_3—CH_2—CH_3$

(b) 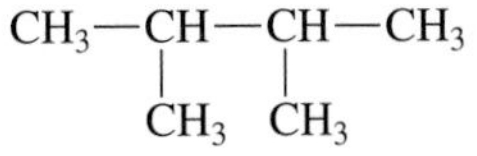

(c) 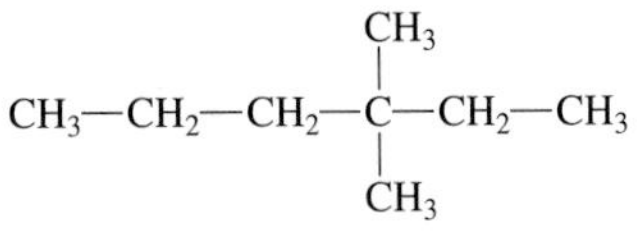

(d) 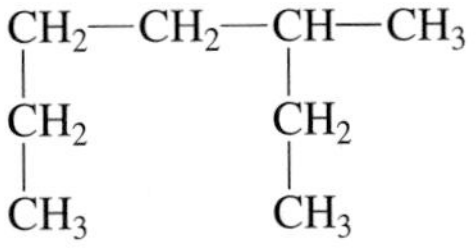

18.9 Indicate whether each of the alkanes in Problem 18.7 is a branched-chain alkane or a normal-chain alkane.

18.10 Indicate whether each of the alkanes in Problem 18.8 is a branched-chain alkane or a normal-chain alkane.

18.11 Indicate the number of alkyl groups present in each of the alkanes in Problem 18.7.

18.12 Indicate the number of alkyl groups present in each of the alkanes in Problem 18.8.

18.13 Draw the structural formula for each of the following alkanes.

(a) pentane
(b) 2,3-dimethylpentane
(c) 2,4-dimethylpentane
(d) 2,2,4,4-tetramethylpentane

18.14 Draw the structural formula for each of the following alkanes.

(a) hexane
(b) 2-methylhexane
(c) 2,2,3-trimethylhexane
(d) 2,3,4-trimethylhexane

18.15 Draw the structural formula for each of the following alkanes.

(a) 4-ethylheptane
(b) 4-ethyl-3-methylheptane
(c) 4,5-diethyloctane
(d) 4-isopropyl-5-propyldecane

18.16 Draw the structural formula for each of the following alkanes.

(a) 4-ethyloctane
(b) 4,4-diethyloctane
(c) 4-propylheptane
(d) 4-isopropyldecane

18.17 The following names, although incorrect according to IUPAC rules, contain sufficient information to enable you to draw structural formulas. Draw the structural formulas and then explain why each name is incorrect. Write the correct IUPAC name for each alkane.

(a) 4-methylpentane
(b) 3,3-dimethylbutane
(c) 4-ethyl-5-methylhexane
(d) 2,5-dimethyl-2-propylhexane

18.18 The following names, although incorrect according to IUPAC rules, contain sufficient information to enable you to draw structural formulas. Draw the structural formulas and then explain why each name is incorrect. Write the correct IUPAC name for each alkane.

(a) 4,4-dimethylhexane
(b) 2-ethyl-2-methylpropane
(c) 1,2-dimethylpropane
(d) 4-ethylpentane

18.19 Write structural formulas (showing only carbon atoms) for the five alkane isomers with the molecular formula C_6H_{14}. Also assign an IUPAC name to each isomer.

18.20 Write structural formulas (showing only carbon atoms) for the nine alkane isomers with the molecular formula C_7H_{16}. Also assign an IUPAC name to each isomer.

18.21 Indicate whether or not the following pairs of alkanes are structural isomers.

(a) 2-methylhexane and 3-methylhexane
(b) 2-methylhexane and 2-methylheptane
(c) 2-methylpentane and 2,2-dimethylpentane
(d) 2-methylpentane and 2,3-dimethylbutane

18.22 Indicate whether or not the following pairs of alkanes are structural isomers.

(a) 2-methylbutane and 2-methylpentane
(b) 2,2-dimethylbutane and 2-methylpentane
(c) 4-ethyl-4-methylheptane and 4-propylheptane
(d) 5-isopropyldecane and 5-propyldecane

Cycloalkanes (Sec. 18.6)

18.23 How many hydrogen atoms are present in each of the following cycloalkane molecules?

(a) 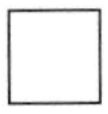

(b) 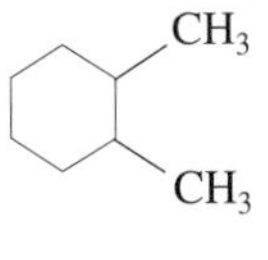

(c) 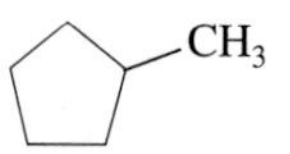 **(d)** 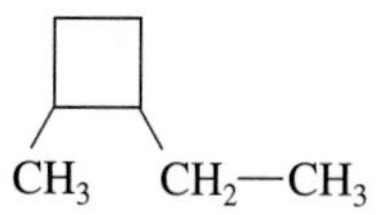

18.24 How many hydrogen atoms are present in each of the following cycloalkane molecules?

(a)  **(b)** 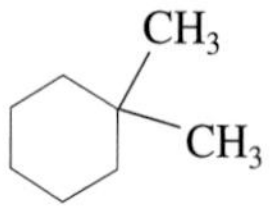

(c) 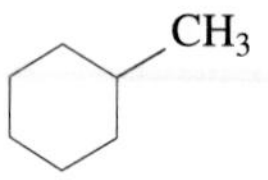 **(d)** 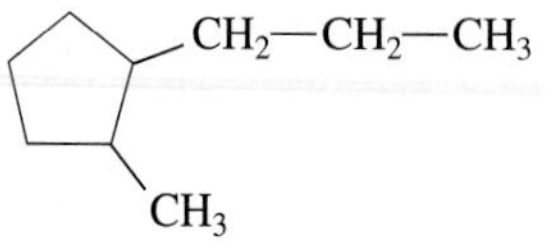

18.25 Assign IUPAC names to each of the compounds in Problem 18.23.

18.26 Assign IUPAC names to each of the compounds in Problem 18.24.

18.27 Draw structural formulas for each of the following cycloalkanes, using a geometrical figure to denote the ring system.

(a) 1,3-dimethylcyclopentane
(b) ethylcyclopropane
(c) 1,2-dimethylcyclobutane
(d) 1,3,5-trimethylcyclohexane

18.28 Draw structural formulas for each of the following cycloalkanes, using a geometrical figure to denote the ring system.

(a) 1,1-dimethylcyclobutane
(b) propylcyclohexane
(c) 1,3-diethylcyclopentane
(d) 1-isopropyl-2-methylcyclohexane

18.29 Draw and name the three possible dimethylcyclobutane isomers.

18.30 Draw and name the three possible dimethylcyclopentane isomers.

Alkenes and Alkynes (Sec. 18.7)

18.31 Classify each of the following hydrocarbons as unsaturated or saturated.

(a) $CH_3—CH=CH—CH_3$ **(b)** $CH≡C—CH_2—\underset{\displaystyle CH_3}{\underset{|}{CH}}—CH_3$

(c) 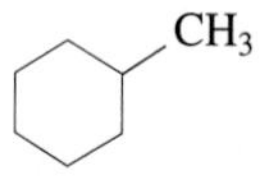 **(d)** 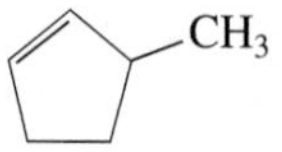

18.32 Classify each of the following hydrocarbons as unsaturated or saturated.

(a) $CH_3—CH_2—CH_2—\underset{\displaystyle CH_3}{\underset{|}{CH}}—CH_3$

(b) $CH_3—CH=CH—CH_3$

(c) 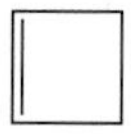 **(d)** 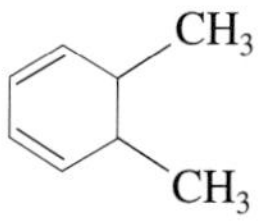

18.33 Using the IUPAC system, name each of the following unsaturated hydrocarbons.

(a) $CH_3—CH=CH—CH_3$ **(b)** $CH_3—C≡C—CH_3$

(c) 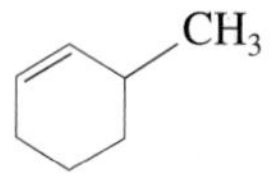 **(d)** 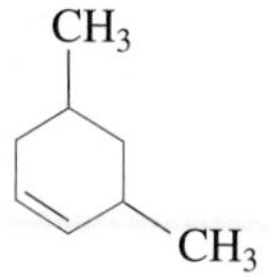

18.34 Using the IUPAC system, name each of the following unsaturated hydrocarbons.

(a) $CH_3—\underset{\displaystyle CH_3}{\underset{|}{CH}}—CH=CH_2$

(b) $CH_3—C≡C—CH_2—CH_3$

(c) 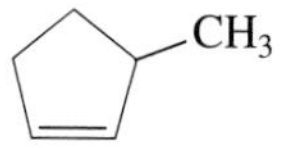 **(d)** 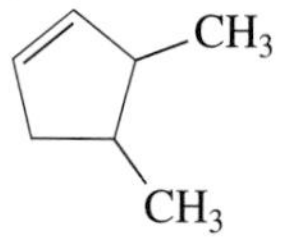

18.35 Using the IUPAC system, name each of the following unsaturated hydrocarbons.

(a) $CH_3—CH=CH—CH=CH_2$

(b) $CH_2=\underset{\displaystyle \underset{\displaystyle CH_3}{\underset{|}{CH_2}}}{\underset{|}{C}}—CH=CH_2$

(c)  **(d)** 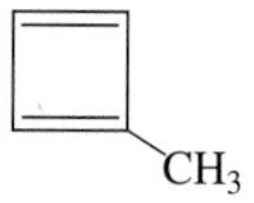

18.36 Using the IUPAC system, name each of the following unsaturated hydrocarbons.

(a) $CH_2=CH—CH_2—CH=CH_2$

(b) $CH_2=\underset{\displaystyle \underset{\displaystyle CH_3}{\underset{|}{CH_2}}}{\underset{|}{C}}—CH_3$

(c)  **(d)** 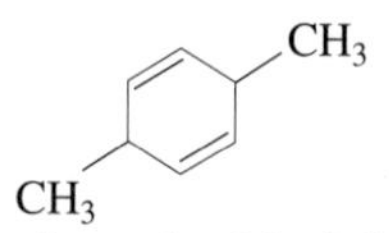

18.37 Draw structural formulas for each of the following unsaturated hydrocarbons.

(a) 4-methyl-2-hexyne
(b) 3-methyl-2-pentene
(c) 2,3-dimethylcyclohexene
(d) 2-methyl-1,4-pentadiene

18.38 Draw structural formulas for each of the following unsaturated hydrocarbons.

(a) 2-methyl-1-butene

(b) 4-methyl-2-pentyne
(c) 1,2-dimethylcyclohexene
(d) 2,3-dimethyl-1,3-butadiene

18.39 Draw the structural formulas (showing only carbon atoms) and give the IUPAC names of the five isomeric alkenes with the formula C_5H_{10}.

18.40 Draw the structural formulas (showing only carbon atoms) and give the IUPAC names of the thirteen isomeric alkenes with the formula C_6H_{12}. (Three of the isomers are hexenes, six are methylpentenes, three are dimethyl butenes, and one is an ethylbutene.)

Aromatic Hydrocarbons (Sec. 18.8)

18.41 Classify each of the following cyclic hydrocarbons as aromatic or nonaromatic.

(a) 
(b) 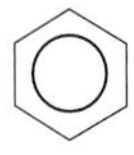
(c) 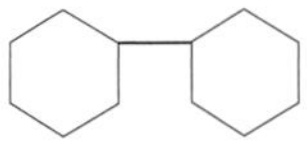
(d) 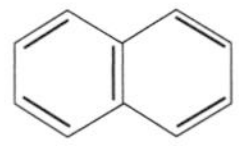

18.42 Classify each of the following cyclic hydrocarbons as aromatic or nonaromatic.

(a) 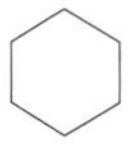
(b) 
(c) 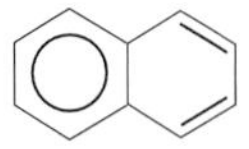
(d) 

18.43 How many hydrogen atoms are present in each of the molecules in Problem 18.41?

18.44 How many hydrogen atoms are present in each of the molecules in Problem 18.42?

18.45 Give the IUPAC name for each of the following aromatic hydrocarbons using the word benzene in the name and using numbers to locate any alkyl groups present.

(a) 
(b) 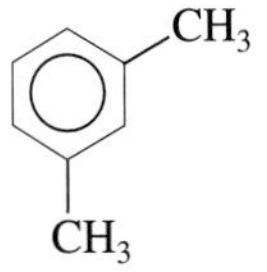

(c) 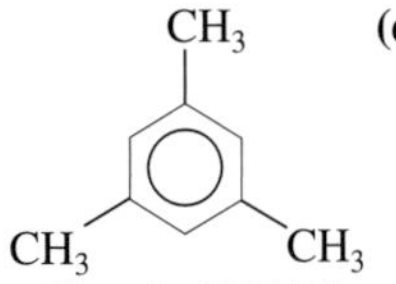

(d) 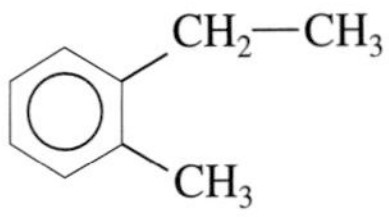

18.46 Give the IUPAC name for each of the following aromatic hydrocarbons using the word benzene in the name and using numbers to locate any alkyl groups present.

(a) 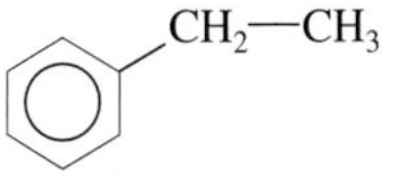

(b) 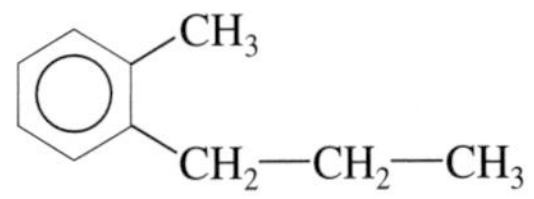

(c) 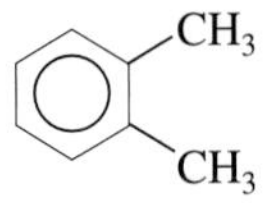

(d) 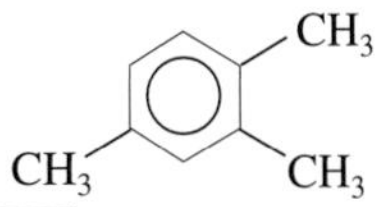

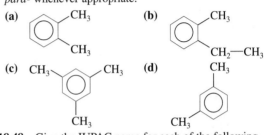

18.47 Give the IUPAC name for each of the following aromatic hydrocarbons, using the prefixes *ortho-*, *meta-*, and *para-* whenever appropriate.

(a) 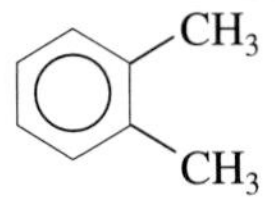

(b) 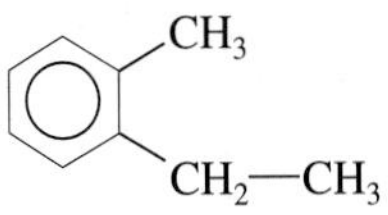

(c) 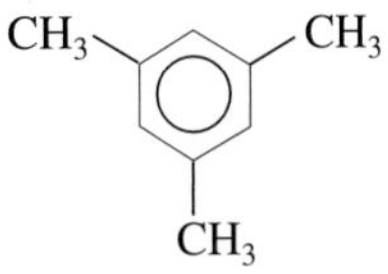

(d) 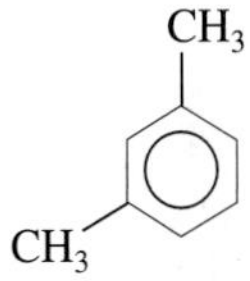

18.48 Give the IUPAC name for each of the following aromatic hydrocarbons, using the prefixes *ortho-*, *meta-*, and *para-* whenever appropriate.

(a) 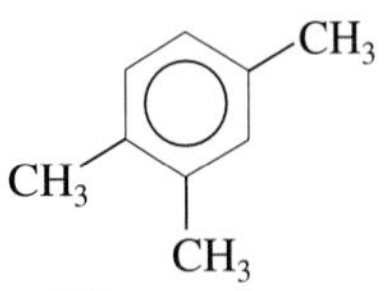

(b) 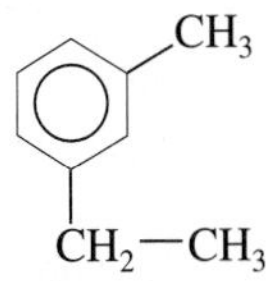

(c) 

(d) 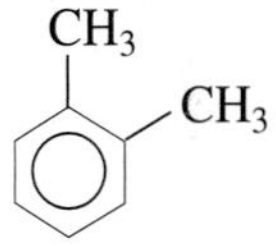

18.49 Draw structural formulas for the following aromatic hydrocarbons.

(a) *m*-ethylmethylbenzene
(b) 1-ethyl-3-propylbenzene
(c) toluene
(d) *p*-xylene

18.50 Draw structural formulas for the following aromatic hydrocarbons.

(a) 1,2-dipropylbenzene
(b) *o*-xylene
(c) *m*-dimethylbenzene
(d) 1-methyl-4-isopropylbenzene

18.51 Indicate whether or not the following pairs of compounds are structural isomers.

(a) ethylbenzene and 1,2-dimethylbenzene

(b) *o*-xylene and *p*-xylene

(c) 1,3-dimethylbenzene and *p*-dimethylbenzene

(d) toluene and propylbenzene

18.52 Indicate whether or not the following pairs of compounds are structural isomers.

(a) propylbenzene and *p*-ethylmethylbenzene

(b) benzene and toluene

(c) *o*-dimethylbenzene and 1,2-dimethylbenzene

(d) *o*-xylene and *m*-dimethylbenzene

18.53 Draw structural formulas for the three isomeric tetramethylbenzenes.

18.54 Draw structural formulas for the six isomeric ethyltrimethylbenzenes.

Halogenated Hydrocarbons (Sec. 18.11)

18.55 Name the following halogenated hydrocarbons using IUPAC rules.

(a) $Cl—CH_2—CH_2—CH_2—Cl$

(b) 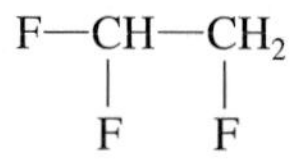

(c) $CH_3—CH_2—Br$

(d) 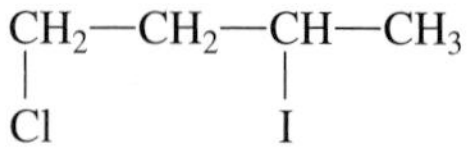

18.56 Name the following halogenated hydrocarbons using IUPAC rules.

(a) 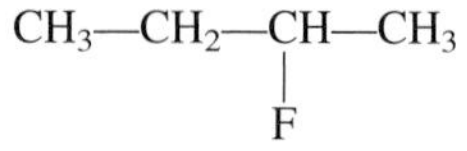 **(b)** 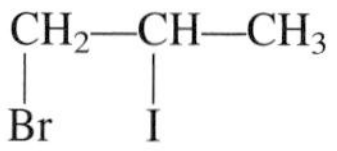

(c) 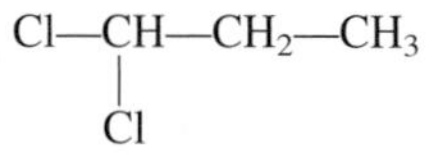 **(d)** 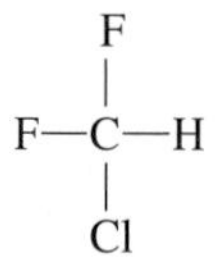

18.57 Name the following halogenated hydrocarbons using IUPAC rules.

(a) 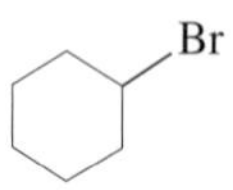 **(b)** 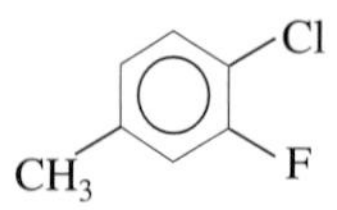

(c) 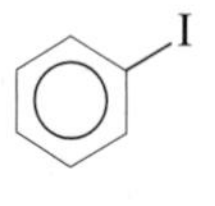 **(d)** 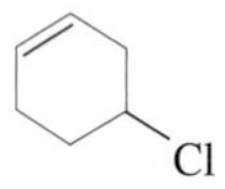

18.58 Name the following halogenated hydrocarbons using IUPAC rules.

(a) 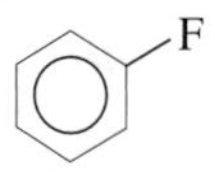 **(b)** 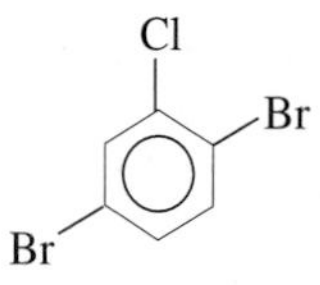

(c) 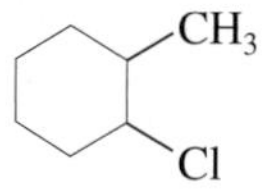 **(d)** 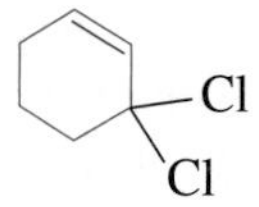

18.59 There are five isomeric chlorinated alkanes with the formula $C_3H_5Cl_3$. Draw structural formulas for and assign IUPAC names to each isomer.

18.60 There are nine isomeric chlorinated alkanes with the formula $C_4H_8Cl_2$. Draw structural formulas for and assign IUPAC names to each isomer.

18.61 There are six isomeric trisubstituted benzenes with the formula $C_6H_3Br_2Cl$. Draw structural formulas for and assign IUPAC names to each isomer.

18.62 There are three isomeric trisubstituted benzenes with the formula $C_6H_3Cl_3$. Draw structural formulas for and assign IUPAC names to each isomer.

Alcohols (Sec. 18.12)

18.63 Name the following alcohols using IUPAC rules.

(a) 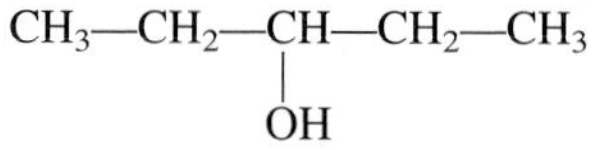

(b) 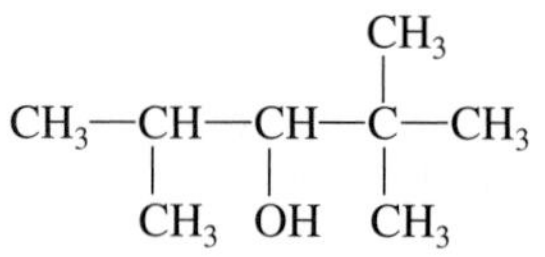

(c) 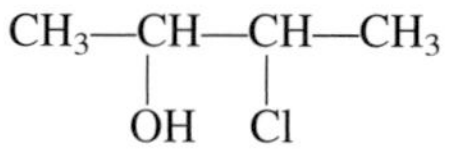

(d) 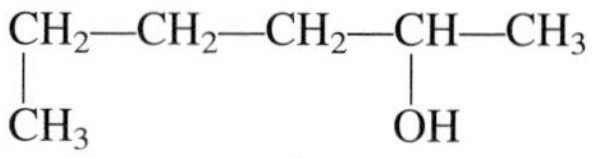

18.64 Name the following alcohols using IUPAC rules.

(a) 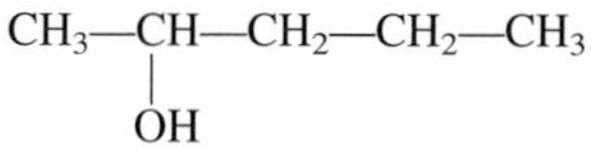

(b) 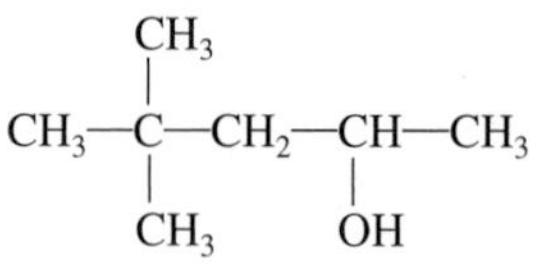

(c) 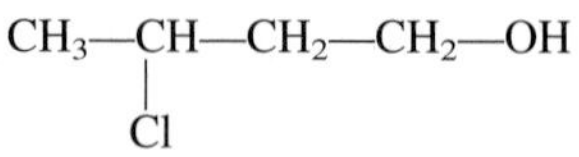

(d) 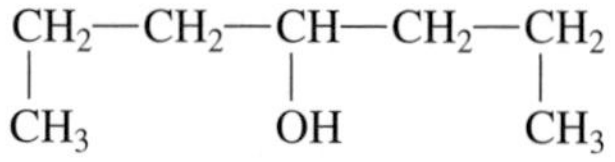

18.65 Draw structural formulas for the following alcohols.

(a) 3-methyl-2-butanol

(b) 2,3-dimethyl-2-hexanol

(c) propyl alcohol

(d) cyclopentanol

18.66 Draw structural formulas for the following alcohols.

(a) 3,3-dimethyl-2-pentanol

(b) 2-methyl-2-butanol

(c) isopropyl alcohol

(d) cyclobutanol

18.67 Assign IUPAC names to the following polyhydric alcohols.

(a) $CH_2—CH_2—CH_2$
OH　　　OH

(b) $CH_2—CH_2—CH—CH_2$
OH　　　OH　OH

(c) $CH_2—CH_2$
OH　OH

(d) $CH_2—CH—CH_2$
OH　CH_3　OH

18.68 Assign IUPAC names to the following polyhydric alcohols.

(a) $CH_3—CH—CH—OH$
OH　CH_3

(b) $CH_3—CH—CH—CH_2—CH_2$
OH　CH_3　　OH

(c) $HO—CH_2—CH—OH$
CH_3

(d) $CH_2—CH—CH_2$
OH　OH　OH

18.69 How many different ways exist (isomers) for the attachment of two hydroxyl groups to an unbranched saturated chain of five carbon atoms?

18.70 How many different ways exist (isomers) for the attachment of two hydroxyl groups and a methyl group to an unbranched saturated chain of four carbon atoms?

Ethers (Sec. 18.13)

18.71 Assign IUPAC names to each of the following ethers.

(a) $CH_3—CH_2—O—CH_2—CH_3$

(b) $CH_3—CH—CH_3$
$O—CH_3$

(c) $CH_3—O—CH_2—CH_2—CH_3$

(d) $CH_3—CH—CH_2—CH_2—O—CH_3$
CH_3

18.72 Assign IUPAC names to each of the following ethers.

(a) $CH_3—CH_2—O—CH_2—CH_2—CH_3$

(b) $CH_3—CH_2—CH—CH_3$
$O—CH_2—CH_3$

(c) $CH_3—CH_2—CH_2—CH_2—O—CH_3$

(d) $CH_3—CH—CH_2—O—CH_3$
CH_3

18.73 Draw structural formulas for the following ethers.

(a) dipropyl ether

(b) isopropyl methyl ether

(c) 2-methoxypropane

(d) 1-methoxypropane

18.74 Draw structural formulas for the following ethers.

(a) diisopropyl ether

(b) ethyl propyl ether

(c) 2-ethoxypentane

(d) 1-ethoxypentane

18.75 Draw the structural formulas (showing only carbon and oxygen atoms) of each of the three isomeric ethers with the formula $C_4H_{10}O$.

18.76 Draw the structural formulas (showing only carbon and oxygen atoms) of each of the six isomeric ethers with the formula $C_5H_{12}O$.

Carboxylic Acids (Sec. 18.14)

18.77 Using IUPAC rules, name the following carboxylic acids.

(a) $CH_3—CH_2—CH_2—C(=O)—OH$

(b) $CH_3—CH_2—CH—CH_2—C(=O)—OH$
CH_3

(c) $CH_3—CH—CH_2—C(=O)—OH$
CH_3

(d) $CH_3—CH—CH—CH_2—CH_2—C(=O)—OH$
CH_3　CH_3

18.78 Using the IUPAC rules, name the following carboxylic acids.

(a) $CH_3—CH_2—CH_2—CH_2—C(=O)—OH$

(b) 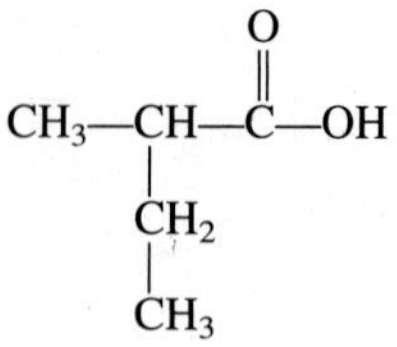

(c) 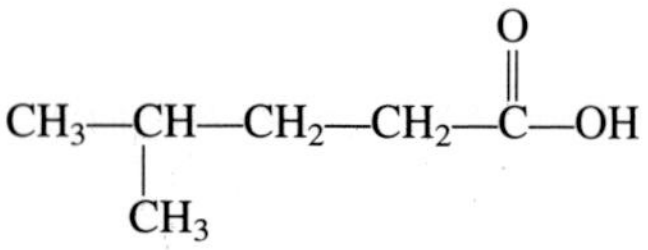

(d) 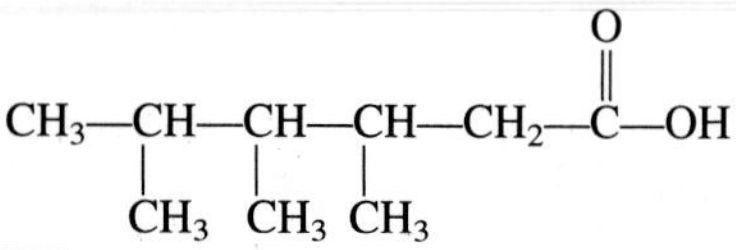

18.79 Draw structural formulas for the following carboxylic acids.

(a) heptanoic acid

(b) lactic acid (common name)

(c) oxalic acid (common name)

(d) propanedioic acid

18.80 Draw structural formulas for the following carboxylic acids.

(a) hexanoic acid

(b) citric acid (common name)

(c) acetic acid (common name)

(d) butanedioic acid

18.81 Draw the structural formula of each of the four isomeric carboxylic acids with the formula $C_5H_{10}O_2$.

18.82 Draw the structural formula of each of the eight isomeric carboxylic acids with the formula $C_6H_{12}O_2$.

Esters (Sec. 18.15)

18.83 Using the IUPAC rules, name the following esters.

(a) 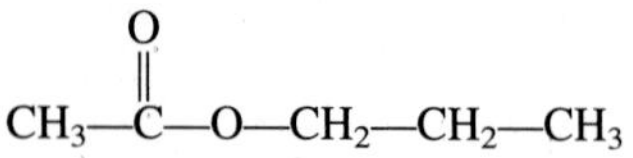

(b) 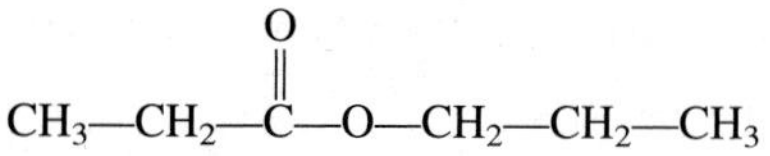

(c) 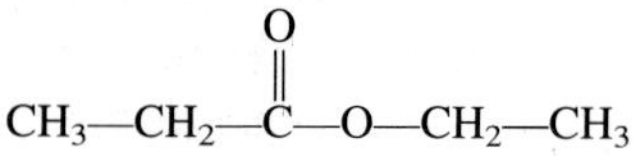

(d) 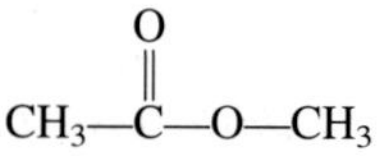

18.84 Using the IUPAC rules, name the following esters.

(a) 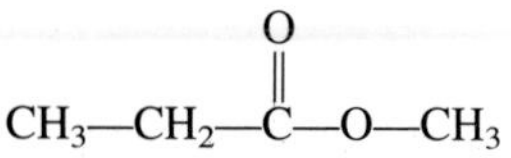

(b) 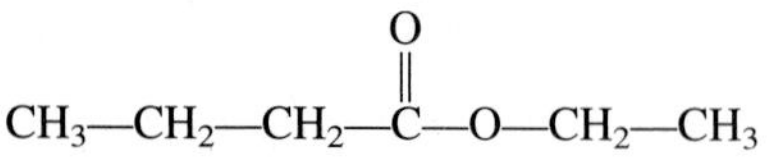

(c) 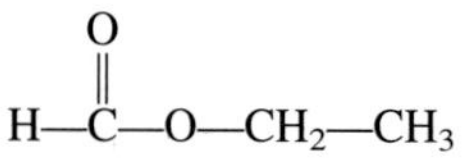

(d) 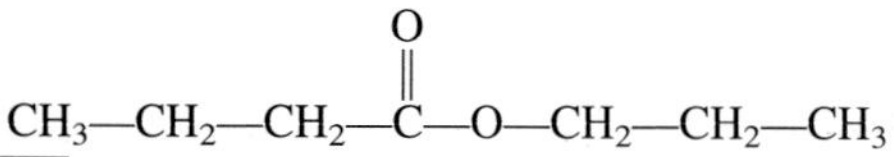

18.85 Draw structural formulas for the following esters.

(a) ethyl acetate

(b) ethyl butanoate

(c) methyl methanoate

(d) propyl methanoate

18.86 Draw structural formulas for the following esters.

(a) propyl acetate

(b) propyl butanoate

(c) ethyl propanoate

(d) methyl propanoate

18.87 Draw the structural formulas of each of the nine isomeric esters with the formula $C_5H_{10}O_2$.

18.88 Draw the structural formulas of each of the four isomeric esters with the formula $C_4H_8O_2$.

ADDITIONAL PROBLEMS

18.89 Using IUPAC rules, name the following hydrocarbon derivatives.

(a) 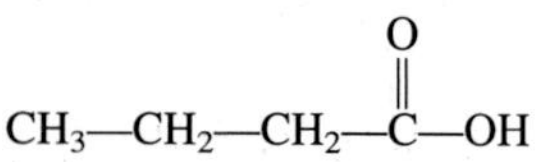

(b) $CH_3—CH_2—OH$

(c) 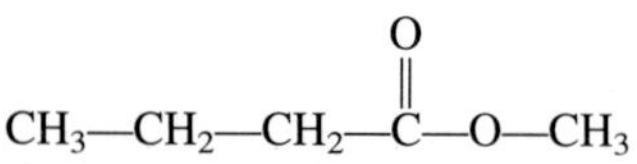

(d) $CH_3—CH_2—CH_2—O—CH_2—CH_3$

18.90 Using IUPAC rules, name the following hydrocarbon derivatives.

(a) $CH_3—CH_2—O—CH_3$

(b) $CH_3—CH_2—Br$

(c) $CH_3—CH_2—\overset{\overset{\displaystyle O}{\|}}{C}—O—CH_2—CH_3$

(d) $CH_3—CH_2—CH_2—\underset{\underset{\displaystyle OH}{|}}{CH}—CH_2—CH_3$

18.91 Draw structural formulas for the following compounds.

(a) the simplest possible ester

(b) the simplest possible ether

(c) the simplest possible cycloalkane

(d) the simplest possible monochlorinated hydrocarbon

18.92 Draw structural formulas for the following compounds.

(a) the simplest possible alcohol

(b) the simplest possible carboxylic acid

(c) the simplest possible alkene

(d) the simplest possible dichlorinated hydrocarbon

18.93 Write the molecular formula for each of the following.

(a) a 20-carbon noncyclic alkane

(b) a 4-carbon noncyclic alkadiene

(c) a 6-carbon cycloalkane

(d) a 7-carbon dichlorinated alkadiene

18.94 Write the molecular formula for each of the following.

(a) a 6-carbon cycloalkene with one multiple bond

(b) a 7-carbon cycloalkane

(c) a 4-carbon noncyclic alkyne with one multiple bond

(d) a 6-carbon dichlorinated cycloalkane

18.95 How many hydrogen atoms are present in a molecule of each of the following compounds? (Assume all carbon chains are saturated.)

(a) a 3-carbon alcohol with two —OH groups

(b) a 3-carbon carboxylic acid

(c) a 4-carbon ether

(d) a 4-carbon ester

18.96 How many hydrogen atoms are present in a molecule of each of the following compounds? (Assume all carbon chains are saturated.)

(a) a 4-carbon alcohol with three —OH groups

(b) a 4-carbon carboxylic acid

(c) a 3-carbon ether

(d) a 3-carbon ester

18.97 Draw structural formulas for all isomers with the following molecular formulas.

(a) C_4H_8 (5 isomers)

(b) C_3H_4 (3 isomers)

(c) C_3H_8O (3 isomers)

(d) $C_4H_8O_2$ (acids and esters only; 6 isomers)

18.98 Draw structural formulas for all isomers with the following molecular formulas.

(a) C_3H_6 (2 isomers)

(b) C_4H_6 (8 isomers)

(c) $C_4H_{10}O$ (7 isomers)

(d) $C_3H_6O_2$ (acids and esters only; 3 isomers)

CUMULATIVE PROBLEMS

18.99 Write a balanced molecular equation for the reaction in which each of the following compounds burns in oxygen to produce carbon dioxide and water vapor.

(a) 3,5-diethyl-2,2-dimethyloctane

(b) 3-ethyl-2-pentanol

18.100 Write a balanced molecular equation for the reaction in which each of the following compounds burns in oxygen to produce carbon dioxide and water vapor.

(a) 3,5-diethyl-4,4-dimethylheptane

(b) 5-methyl-2-octanol

18.101 Draw an electron-dot structure for each of the following hydrocarbons or hydrocarbon derivatives.

(a) ethane (b) ethyne

(c) ethanoic acid (d) methoxyethane

18.102 Draw an electron-dot structure for each of the following hydrocarbons or hydrocarbon derivatives.

(a) ethene (b) ethanol

(c) methyl methanoate (d) cyclohexane

18.103 What are the mass percents of the elements present in THC (the active ingredient in marijuana) whose structure is

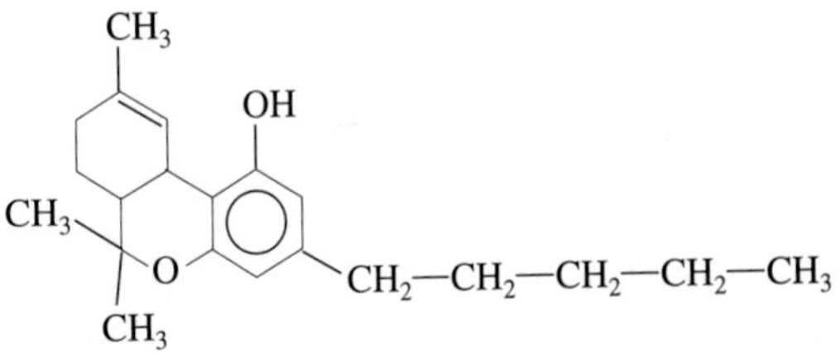

18.104 What are the mass percents of the elements present in cocaine whose structure is

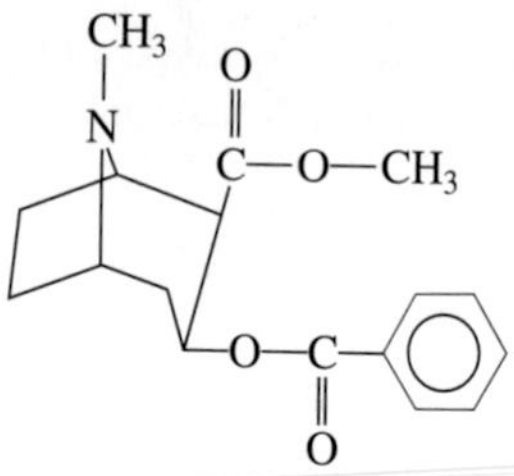

18.105 At 20°C the hydrocarbon hexane is a liquid with a density of 0.66 g/mL. How many liters of O_2 at 20°C and 1.50 atm pressure are needed for the complete combustion of 25 mL of hexane?

18.106 At 20°C the hydrocarbon heptane is a liquid with a density of 0.68 g/mL. How many liters of O_2 at 20°C and 0.75 atm pressure are needed for the complete combustion of 15 mL of heptane?

18.107 Complete combustion of a hydrocarbon sample produces 6.06 g of CO_2 and 3.10 g of H_2O. Another sample of the same hydrocarbon with a mass of 1.14 g occupies a volume of 473 mL at 20°C and 1 atm pressure.

(a) What is the empirical formula of the hydrocarbon?

(b) What is the molecular formula of the hydrocarbon?

18.108 Complete combustion of a hydrocarbon sample produces 67.6 g of CO_2 and 13.8 g of H_2O. Another sample of the same hydrocarbon with a mass of 3.34 g occupies a volume of 875 mL at 225°C and 2 atm pressure.

(a) What is the empirical formula of the hydrocarbon?

(b) What is the molecular formula of the hydrocarbon?

Appendix A

Mathematical Review

A.1 Basic Mathematical Operations

Multiplication *is the addition of a number or quantity to itself a certain number of times.* Multiplying 7 times 3 means 7 *added* 3 *times*, which gives 21. The term *product* is used to denote the answer obtained by multiplying two or more numbers together.

Various ways exist for representing the multiplication process. Each of the following five notations means *3 times b*.

$$3 \times b \qquad 3 \cdot b \qquad 3(b) \qquad (3)(b) \qquad 3b$$

The sign × is the most used symbol for multiplication. However, a dot and parentheses are also in common use. Notice that in the last notation, $3b$, no symbol is present to indicate multiplication.

Multiplication of a series of terms by a given quantity is denoted by parentheses, as in the expression

$$3(a + b + 2c)$$

Each term within the parentheses is to be multiplied by 3, giving as an answer the expression $3a + 3b + 6c$.

Division can be considered to be the reverse process of multiplication. **Division** *is the process of finding out how many times one number or quantity is contained in another.*

Three common notations are in use for indicating division.

$$a \div b \qquad \frac{a}{b} \qquad a/b$$

All mean a divided by b. The result of dividing one number into another is called the *quotient*.

In problems stated in words, the term *per* is an indication of division. For example, density, which is defined as mass *per* unit volume, is written as

$$\text{Density} = \frac{\text{mass}}{\text{volume}}$$

In working problems it is sometimes necessary to *invert* or *take the reciprocal of a* number. Both of these expressions mean the same thing, namely, to divide that number into 1. Thus, the number 3 when inverted becomes $\frac{1}{3}$, and the reciprocal of the number 8 is $\frac{1}{8}$. The reciprocal of $\frac{1}{5}$ is 5.

A.2 FRACTIONS

A **fraction** *is an arithmetic expression showing that one number is to be divided by another number*. Some examples of fractions are $\frac{2}{5}$, $\frac{3}{7}$, $\frac{8}{3}$, and $\frac{1}{10}$.

In a fraction the number above the division line is called the *numerator*, and the number below the line the *denominator*.

$$\text{numerator} \longrightarrow \frac{a}{b} \longleftarrow \text{denominator}$$

Fractions are classified into two categories: proper and improper. A *proper fraction* is one in which the numerator is smaller than the denominator. An *improper fraction* has a larger numerator than its denominator. The fraction $\frac{5}{9}$ is a proper fraction, whereas the fraction $\frac{11}{9}$ is an improper fraction. Improper fractions are often encountered in scientific calculations. There is nothing "wrong" with an improper fraction, and most often it is used "as is" in chemical calculations.

All fractions can be converted to decimal numbers simply by carrying out the implied division. For example,

$$\frac{5}{7} = 5 \div 7 = 0.71428571$$

Example A.1

Classify each of the following fractions as proper or improper, and express it as a decimal number to three significant figures.

(a) $\frac{5}{11}$ **(b)** $\frac{11}{5}$ **(c)** $\frac{17}{12}$ **(d)** $\frac{1}{7}$

Solution

(a) Since the numerator is smaller than the denominator, this is a **proper** fraction. Its decimal value is

$$\frac{5}{11} = 5 \div 11 = 0.45454545 \text{ (calculator answer)}$$
$$= 0.455 \text{ (correct answer)}$$

(b) Since the numerator is larger than the denominator, this is an **improper** fraction. Its decimal value is

$$\frac{11}{5} = 11 \div 5 = 2.2 \text{ (calculator answer)}$$
$$= 2.20 \text{ (correct answer)}$$

(c) This is an **improper** fraction whose decimal value is

$$\frac{17}{12} = 17 \div 12 = 1.4166666 \text{ (calculator answer)}$$
$$= 1.42 \text{ (correct answer)}$$

(d) This is a **proper** fraction whose decimal value is

$$\frac{1}{7} = 1 \div 7 = 0.14285714 \text{ (calculator answer)}$$
$$= 0.143 \text{ (correct answer)}$$

Note that proper fractions always give decimal numbers with values less than 1, and improper fractions always have decimal values greater than 1.

A *ratio* conveys the same information as a fraction. If the ratio of nickels to dimes in a pile of coins is 3 to 2, we are mathematically saying that

$$\frac{\text{Nickels}}{\text{Dimes}} = \frac{3}{2}$$

The product obtained from multiplying two fractions is another fraction whose numerator is the product of the two given numerators and whose denominator is the product of the two given denominators. In general terms we have

$$\frac{a}{b} \times \frac{c}{d} = \frac{a \times c}{b \times d} = \frac{ac}{bd}$$

Division of fractions is most easily handled by converting the division operation into multiplication. This is accomplished by inverting (turning upside down) the fraction to the right of the division sign and then following the rules for multiplication. In general terms, division of fractions proceeds as follows.

$$\frac{a/b}{c/d} = \frac{a}{b} \div \frac{c}{d} = \frac{a}{b} \times \frac{d}{c} = \frac{a \times d}{b \times c} = \frac{ad}{bc}$$

division sign has been changed to multiplication

this fraction has been inverted

Example A.2

Carry out the following multiplications of fractions.

(a) $\frac{3}{4} \times \frac{5}{8}$ **(b)** $\frac{2}{3} \times \frac{6}{7}$ **(c)** $\frac{1}{3} \times \frac{2}{3} \times \frac{4}{5}$

Solution

(a) Fractions are multiplied together by multiplying the numerators together and the denominators together.

$$\frac{3}{4} \times \frac{5}{8} = \frac{3 \times 5}{4 \times 8} = \frac{15}{32}$$

(b) Multiplying the numerators together and then multiplying the denominators together gives

$$\frac{2}{3} \times \frac{6}{7} = \frac{2 \times 6}{3 \times 7} = \frac{12}{21}$$

(c) The product of three fractions is obtained in the same manner as the product of two fractions. All numerators are multiplied together, and all denominators are multiplied together.

$$\frac{1}{3} \times \frac{2}{3} \times \frac{4}{5} = \frac{1 \times 2 \times 4}{3 \times 3 \times 5} = \frac{8}{45}$$

Example A.3

Carry out the following divisions of fractions.

(a) $\frac{3}{7} \div \frac{2}{5}$ **(b)** $\frac{2/3}{3/5}$

Solution

(a) We convert the division operation into a multiplication by inverting the fraction $\frac{2}{5}$ and replacing the division sign with a multiplication sign.

$$\frac{3}{7} \div \frac{2}{5} = \frac{3}{7} \times \frac{5}{2} = \frac{3 \times 5}{7 \times 2} = \frac{15}{14}$$

multiplication sign ↑ ↑ inverted fraction

(b) Inverting the fraction to the right of the division sign, $\frac{3}{5}$ in this case, gives

$$\frac{2/3}{3/5} = \frac{2}{3} \div \frac{3}{5} = \frac{2}{3} \times \frac{5}{3} = 2 \times \frac{5}{3} \times 3 = \frac{10}{9}$$

Adding and subtracting fractions is usually more complicated than multiplying and dividing them. This is because fractions must have the same denominator, called a *common denominator*, before they can be added or subtracted. By analogy, to add values for coins of different denominations, we first must express them as equivalent amounts with a common denomination and then add; for example, 2 dimes and 1 quarter would be expressed as 20 cents and 25 cents, respectively, and then added, obtaining the sum, 45 cents. The actual numerical examples of addition and subtraction of fractions that follow should clarify this "common denominator process."

Example A.4

Carry out the following additions and subtractions of fractions.

(a) $\frac{2}{3} + \frac{3}{4}$ **(b)** $\frac{7}{8} - \frac{1}{4}$ **(c)** $\frac{7}{6} + \frac{2}{3} - \frac{1}{2}$

Solution

(a) Examining the denominators (3 and 4) of the two fractions, we note that the smallest number that each can be divided into evenly is the number 12; that is, each can be converted to 12 by multiplying it by some whole number.

$$3 \times 4 = 12 \qquad 4 \times 3 = 12$$

denominator of first fraction ↑ ↑ denominator of second fraction

Thus, 12 is the common denominator for 3 and 4.

The numerator and denominator of each fraction are both multiplied by the number that will convert the denominator of that particular fraction to the common denominator, which is 12 in this case.

$$\frac{2}{3} \times \frac{4}{4} = \frac{8}{12} \qquad \text{and} \qquad \frac{3}{4} \times \frac{3}{3} = \frac{9}{12}$$

Now that both fractions have the same denominator, we can proceed to add them together by adding numerators and retaining the denominator.

$$\frac{8}{12} + \frac{9}{12} = \frac{17}{12}$$

(b) The common denominator for 8 and 4 is 8.

$$8 \times 1 = 8 \qquad \text{and} \qquad 4 \times 2 = 8$$

Expressing both fractions in eighths, we obtain

$$\frac{7}{8} \times \frac{1}{1} = \frac{7}{8} \qquad \text{and} \qquad \frac{1}{4} \times \frac{2}{2} = \frac{2}{8}$$

Now that both fractions have the same denominator, we can perform the subtraction by subtracting numerators and maintaining the denominator.

$$\frac{7}{8} - \frac{2}{8} = \frac{5}{8}$$

(c) This problem requires that we find a common denominator for three numbers (6, 3, and 2). This common denominator is 6, since both 3 and 2 will divide evenly into 6. Thus, changing all fractions to sixths we have

$$\frac{7}{6} \times \frac{1}{1} = \frac{7}{6}$$
$$\frac{2}{3} \times \frac{2}{2} = \frac{4}{6}$$
$$\frac{1}{2} \times \frac{3}{3} = \frac{3}{6}$$

Carrying out the indicated arithmetic gives

$$\frac{7}{6} + \frac{4}{6} - \frac{3}{6} = \frac{7 + 4 - 3}{6} = \frac{8}{6}$$

A.3 POSITIVE AND NEGATIVE NUMBERS

All numbers have both a magnitude and a sign. The sign can be either positive or negative. If a sign is not shown, it is assumed to be positive.

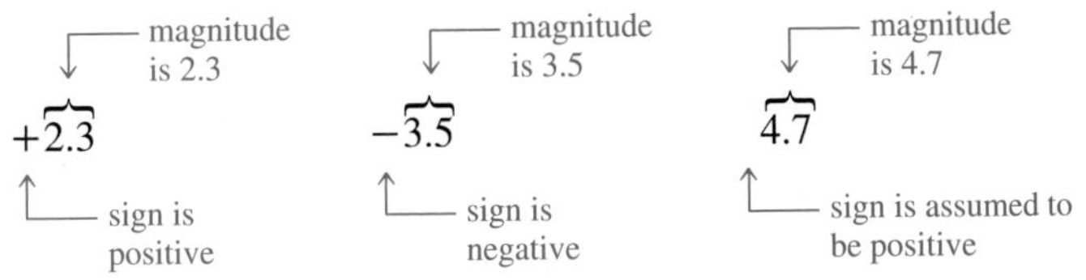

Most students feel comfortable with the use of positive numbers, but are much less certain when negative numbers are encountered. Hence, a review of the use of negative numbers is in order. The effects that negative numbers have on the processes of multiplication, division, addition, and subtraction will be considered.

The sign associated with the product of the multiplication of two signed numbers is positive if both numbers have the same sign and negative if they have opposite signs. Thus,

$$\begin{aligned}\text{Positive} \times \text{positive} &= \text{positive}\\ \text{Positive} \times \text{negative} &= \text{negative}\\ \text{Negative} \times \text{positive} &= \text{negative}\\ \text{Negative} \times \text{negative} &= \text{positive}\end{aligned}$$

In division the same "sign rules" as those for multiplication are used. That is, the quotient of two signed numbers is positive if both numbers have the same sign and negative if they have opposite signs.

Example A.5

Carry out the following multiplications and divisions, each of which involves signed numbers.

(a) $(+3) \times (-4)$ **(b)** $(-2) \times (-3)$ **(c)** $(-4)/(2)$ **(d)** $(+9)/(+3)$

Solution

(a) The sign of the product will be negative because the two numbers have opposite signs.

$$(+3) \times (-4) = -12$$

(b) The product of two negative numbers is a positive number; if both numbers have the same sign the product is positive.

$$(-2) \times (-3) = +6$$

(c) The "sign rules" for division are the same as those for multiplication. A "negative" divided by a "positive" is a "negative."

$$(-4)/(2) = -2$$

(d) Since both numbers have the same sign, the quotient will be positive.

$$(+9)/(+3) = +3$$

When adding numbers, if the signs of the two numbers are the same, add the numbers and keep that same sign. When the signs of the two numbers to be added are not the same, subtract the number of smaller magnitude from the one of larger magnitude and keep the sign of the larger number.

Subtraction of signed numbers is most easily handled by converting the subtraction operation into an addition. This is accomplished by changing the sign of the *number being subtracted* and then following the rules for addition.

sign indicating subtraction
is changed to addition

$$(-3) - (+\,7) = (-3) + (-\,7) = -10$$

sign of number being
subtracted is changed

Example A.6

Carry out the following additions, each of which involves signed numbers.

(a) $(+3) + (+3)$ **(b)** $(-7) + (-2)$

(c) $(-6) + (+9)$ **(d)** $(-4) + (+3)$

Solution

(a) Since both numbers are positive, the sum will be positive.

$$(+3) + (+3) = +6$$

(b) Since both numbers are negative, the sum will be negative.

$$(-7) + (-2) = -9$$

(c) The signs of the two numbers are not the same. Thus, the number of smaller magnitude, 6, is subtracted from the number of larger magnitude, 9, and the sign of the larger number, a plus, is attached to the answer.

$$(-6) + (+9) = +3$$

(d) Again, the signs of the two numbers are not the same. The smaller number, 3, is subtracted from 4, and the sign of the larger number, a minus, is used.

$$(-4) + (+3) = -1$$

Example A.7

Carry out the following subtractions, each of which involves signed numbers.

(a) $(-2) - (-5)$ **(b)** $(+5) - (-4)$ **(c)** $(+2) - (+5)$

Solution

(a) The -5 is changed to a $+5$, and the two numbers are added.

$$(-2) - (-5) = (-2) + (+5) = +3$$

(b) The -4 is changed to a $+4$, and the two numbers are added.

$$(+5) - (-4) = (+5) + (+4) = +9$$

(c) The $+5$ is changed to a -5, and the two numbers are added.

$$(+2) - (+5) = (+2) + (-5) = -3$$

Remember that in adding the two numbers in each of the above examples, the rules for addition shown in Example A.6 were used.

A.4 SOLVING ALGEBRAIC EQUATIONS

A common type of mathematical problem encountered in chemistry involves finding the value of an unknown. Such problems almost always involve manipulating an equation by simple algebraic techniques.

To solve for an unknown in an algebraic equation we must change the equation in such a way that the desired unknown is "isolated" on one side of the equation. This can be accomplished in a number of ways; the method used is determined by the nature of the equation. Equation manipulation procedures include

1. Adding the same quantity to both sides of an equation.
2. Subtracting the same quantity from both sides of an equation.
3. Multiplying both sides of an equation by the same quantity.
4. Dividing both sides of an equation by the same quantity.

All four of these procedures encompass the same fundamental mathematical concept: *The validity of an equation is maintained as long as whatever is done to one side of the equation is also done to the other side of the equation.* Illustrative of this mathematical principle is the procedure of multiplying both sides of the equation.

$$a = b$$

by the number 3 to give the equation

$$3a = 3b$$

This latter equation is "just as valid" as the original equation.

Example A.8 contains five examples of algebraic manipulation of equations by the previously listed procedures either singly or in combination.

Example A.8

Solve each of the following equations for the quantity b.

(a) $ab = 3c$ **(b)** $\frac{a}{c} = \frac{b}{d}$ **(c)** $\frac{3b}{c} = a$

(d) $b + c = 4a$ **(e)** $\frac{3(b - c)}{2} = a$

Solution

(a) Solving the equation

$$ab = 3c$$

for b requires only one step. Divide both sides of the equation by the quantity a.

$$\frac{\cancel{a}b}{\cancel{a}} = \frac{3c}{a}$$

The a's on the left cancel, leaving only b on the left side of the equation.

$$b = \frac{3c}{a}$$

(b) To solve the equation

$$\frac{a}{c} = \frac{b}{d}$$

for b we must eliminate d from the right side of the equation. To do that we multiply both sides of the equation by the quantity d.

$$d\left(\frac{a}{c}\right) = \cancel{d}\left(\frac{b}{\cancel{d}}\right)$$

The d's on the right cancel, giving the equation

$$d\left(\frac{a}{c}\right) = b$$

Note the procedure for removing a quantity from the denominator of one side of an equation. You multiply both sides of the equation by that quantity.

(c) To solve the equation

$$\frac{3b}{c} = a$$

for the quantity b we first multiply both sides of the equation by c.

$$\cancel{c}\left(\frac{3b}{\cancel{c}}\right) = ac$$

The c's on the left cancel, giving

$$3b = ac$$

Next, we divide both sides of the equation by 3:

$$\frac{\cancel{3}b}{\cancel{3}} = \frac{ac}{3}$$

Canceling the 3's on the left gives

$$b = \frac{ac}{3}$$

(d) The equation

$$b + c = 4a$$

is different from the previous three examples in that it contains a "plus" sign. To remove c from the left side of the equation, thus isolating b, requires a new procedure. We subtract the quantity c from both sides of the equation.

$$b + c - c = 4a - c$$

The $+c$ and $-c$ on the left add to give zero, leaving the equation

$$b = 4a - c$$

(e) To isolate the quantity b from the equation

$$\frac{3(b - c)}{2} = a$$

requires three steps. First, we multiply both sides of the equation by 2.

$$\cancel{2}\left[\frac{3(b - c)}{\cancel{2}}\right] = 2a$$

The 2's on the left cancel, giving

$$3(b - c) = 2a$$

Second, we divide both sides of the equation by 3.

$$\frac{\cancel{3}(b - c)}{\cancel{3}} = \frac{2a}{3}$$

Canceling the 3's on the left yields the equation

$$b - c = \frac{2a}{3}$$

Finally, we add the quantity c to both sides of the equation.

$$b - c + c = \frac{2a}{3} + c$$

The c's on the left add to zero, giving the equation

$$b = \frac{2a}{3} + c$$

Answers to Odd-Numbered Problems

CHAPTER 1

1. (a) true (b) false (c) false (d)true **3.** c, b, e, a, d **5.** (a) fact (b) hypothesis (c) fact (d) law **7.** (a) false (b) false (c) true (d) true **9.** While a theory may not be an absolute answer, it is the best answer available. It may be supplanted only if repeated experimental evidence conclusively disproves it and a new theory is developed. **11.** (a) 4 (b) 1 (c) 1 and 4 (d) 1 and 2. **13.** The product of the pressure times the volume is constant, or $P_1V_1 = P_2V_2$. **15.** Scientific laws are discovered by research. Researchers have no control over what the laws turn out to be. Societal laws are arbitrary conventions that can be and are changed by the society when necessary.

CHAPTER 2

1. Student A: low precision, low accuracy; Student B: high precision, high accuracy; Student C: high precision, low accuracy. **3.** (a) 0.1 degree (b) 0.1 fluid ounce (c) 1 milliliter (d) 0.1 millimeter **5.** The uncertainty in the first reading is ±0.1 second, and the uncertainty in the second reading is ±0.01 second. **7.** (a) 42.1°, 61.5° (b) 35.03°, 47.98° **9.** (a) estimated (b) exact (c) estimated (d) exact **11.** (a) 2 (b) 6 (c) 8 (d) 7 **13.** (a) 2 (b) 4 (c) 5 (d) 6 **15.** (a) 6 (b) 5 (c) 6 (d) 6 **17.** (a) yes (b) no (c) yes (d) yes **19.** (a) yes (b) no (c) yes (d) no **21.** (a)±100 (b) ±0.01 (c) ±10 (d) ±1 **23.** (a) 23,000 (b) $23,\bar{0}00$ (c) 23,000.0 (d) 23,000.000 **25.** (a) 3 (b) 4 (c) 2 (d) 5 **27.** 6 **29.** (a) 3.630502 (b) 3.6305 (c) 3.631 (d) 3.63 **31.** (a) 30,427.3 (b) 30,427 (c) 30,430 (d) $3\bar{0},000$ **33.** (a) 0.035 (b) 2.50 (c) 1,500,000 (d) $1\bar{0}0$ **35.** (a) 0.12 (b) 120,000 (c) 12 (d) 0.00012 **37.** (a) 2 (b) 1 (c) 3 (d) 3 **39.** (a) 3 (b) 2 (c) 4 or more (d) 4 or more **41.** (a) 0.0299 (b) 140,000 (c) 1280 (d) 0.988 **43.** (a) 3.9 (b) 4.84 (c) 63 (d) 1 **45.** (a) ±0.1 (b) ±0.1 (c) ±1 (d) ±100 **47.** (a) 162 (b) 9.3 (c) 1261 (d) 20.0 **49.** (a) 957.0 (b) 343 (c) 1200 (d) 132 **51.** 14.6 centimeters **53.** (a) 267.3 (b) 260,000 (c) 201.3 (d) 3.8 **55.** (a) 5^3 (b) 10^4 (c) 3^{-4} (d) 10^{-3} **57.** (a) 10^7 (b) 10^{-1} (c) 10^{-3} (d) 10^{-7} **59.** (a) 4.732×10^2 (b) 1.234×10^{-3} (c) 2.3100×10^2 (d) 2.31×10^8 **61.** (a) 4 (b) 2 (c) 4 (d) 5 **63.** (a) 7×10^4 (b) 6.70×10^4 (c) 6.7000×10^4 (d) 6.700000×10^4 **65.** (a) 0.000170 (b) 573 (c) 0.5550 (d) $11,1\bar{0}0,000,000$ **67.** (a) 10^8 (b) 10^{-8} (c) 10^2 (d) 10^{-2} **69.** (a) 2.992×10^8 (b) 9.1×10^2 (c) 2.7×10^{-9} (d) 2.7×10^{11} **71.** (a) 10^2 (b)10^8 (c) 10^{-8} (d) 10^{-2} **73.** (a) 2.86×10^1 (b) 8.999×10^{17} (c) 3.49×10^{-2} (d) 1.111×10^{-18} **75.** (a) 10^1 (b) 10^{-1} (c) 10^{20} (d) 10^2 **77.** (a) 1.5×10^0 (b) 6.7×10^{-1} (c) 8.51×10^{-19} (d) 8×10^{15} **79.** (a) 4.42×10^3 (b) 9.30×10^{-2} (c) 9.683×10^5 (d) 1.919×10^4 **81.** (a) 7.713×10^7 (b) 8.253×10^7 (c) 8.307×10^7 (d) 8.313×10^7 **83.** (a) 3 (b) 4 (c) 4 (d) 4 **85.** (a) no (b) yes (c) yes (d) yes **87.** (a) yes (b) no (c) no (d) yes **89.** (a) 6.326×10^5 (b) 3.13×10^{-1} (c) 6.300×10^7 (d) 5.000×10^{-1} **91.** (a) 2 (b) 4 or more (c) 3 (d) 3 **93.** (a) 27 (b) 0.9 (c) 1.8×10^2 (d) 0.07 **95.** an exact number has no decimal digits **97.** (a) 3.5×10^2 (b) 3.8×10^2 (c) 0.408 (d) 4.18 **99.** (a) 2.07×10^2, 243, 1.03×10^3 (b) 2.11×10^{-3}, 0.0023, 3.04×10^{-2} (c) 23,000, 9.67×10^4, 2.30×10^5 (d) 0.000014, 0.00013, 1.5×10^{-4}

CHAPTER 3

1. (a) kilo (b) centi (c) micro (d)10^{-9} (e) 10^6 (f) 10^{-3} **3.** (a) μg (b) km (c) cL (d) decimeter (e) milliliter (f) picogram **5.** (a) n (nano) (b) μ (micro) (c) k (kilo) (d) G (giga) **7.** (a) smaller, 100 times (b) smaller, 1000 times (c) larger, 10,000 times (d) smaller, 1,000,000 times **9.** (a) volume (b) length (c) volume (d) area **11.** (a) 5 g (b) 100 kg (c) 2 mm (d) 5 mL **13.** (a) 1 inch (b) 1 meter (c) 1 pound (d) 1 gallon **15.** (a) mm^3 (b) nm^2 (c) km (d) cm **17.** (a) 20.4 cm^2 (b) 32 m^2 (c) 65.88 mm^2 (d) 8.2 mm^2 **19.** (a) 9.5 cm^3 (b) 1.3×10^2 cm^3 (c) 2.8×10^6 mm^3 (d) 3.7×10^2 cm^3

21. (a) 24 hr = 1 day; $\dfrac{1 \text{ day}}{24 \text{ hr}}$; $\dfrac{24 \text{ hr}}{1 \text{ day}}$

(b) 60 sec = 1 min; $\dfrac{1 \text{ min}}{60 \text{ sec}}$; $\dfrac{60 \text{ sec}}{1 \text{ min}}$

(c) 10 decades = 1 century; $\dfrac{1 \text{ century}}{10 \text{ decades}}$; $\dfrac{10 \text{ decades}}{1 \text{ century}}$

(d) 365.25 days = 1 yr; $\dfrac{365.25 \text{ days}}{1 \text{ yr}}$; $\dfrac{1 \text{ yr}}{365.25 \text{ days}}$

23. (a) $\dfrac{1 \text{ kL}}{10^3 \text{ L}}$; $\dfrac{10^3 \text{ L}}{1 \text{ kL}}$ (b) $\dfrac{1 \text{ mg}}{10^{-3} \text{ g}}$; $\dfrac{10^{-3} \text{ g}}{1 \text{ mg}}$

(c) $\frac{1\ \text{cm}}{10^{-2}\ \text{m}}$; $\frac{10^{-2}\ \text{m}}{1\ \text{cm}}$ (d) $\frac{1\ \mu\text{sec}}{10^{-6}\ \text{sec}}$; $\frac{10^{-6}\ \text{sec}}{1\ \mu\text{sec}}$
25. (a) 4 sig fig (b) exact (c) 4 sig fig (d) exact
27. (a) 0.025 g (b) 3.23×10^5 m (c) 2.50×10^2 dL (d) 1.0×10^{10} pg **29.** (a) 2.3×10^2 cL (b) 6.00×10^6 mg (c) 6 nL (d) 2.5×10^{16}nm **31.** (a) 6.0×10^{-4} cm^2 (b) 7.2×10^{-9} m^3 (c) 2.5×10^{-9} dm^3 (d) 2.3×10^{34} nm^3
33. (a) 2.0×10^5 L/hr (b) 5.5×10^{-2} kL/sec (c) 3.3×10^4 dL/min (d) 4.8×10^{-9} mL/day
35. (a) 2.30×10^{-3} mg/L (b) 2.30×10^{-3} μg/mL (c) 2.30×10^{-6} cg/cL (d) 2.30×10^{-6} kg/m^3
37. (a) 1.23 gal (b) 1.234 gal (c) 1.2337 gal (d) 1.23371 gal
39. (a) 91.44 m (b) 9144 cm (c) 0.09144 km (d) 3.600×10^3 in. **41.** (a) 0.079 qt (b) 0.020 gal (c) 2.5 fl oz (d) 75 cm^3 **43.** (a) 6.0×10^{27} g (b) 6.0×10^{24} kg (c) 6.0×10^{36} ng (d) 2.1×10^{26} oz
45. (a) 5.2 cm^2 (b) 0.81 in.^{2} **47.** (a) 51 ft^3
49. (a) 0.789 g/mL (b) 7.18 g/mL (c) 0.916 g/mL (d) 1.49 g/mL **51.** (a) 22.8 g (b) 428 g (c) 28.6 g (d) 22.8 g **53.** (a) 63.2 mL (b) 4.81 mL (c) 4.00×10^4 mL (d) 22.9 mL **55.** (a) 1.039 g/mL **57.** 61 lb
59. 266 g **61.** (a) 12.5% (b) 37.5% (c) 60% (d) 47.5%
63. (a) 95.05% copper (b) 4.95% zinc **65.** 22.3 g water
67. 66.9 g salt **69.** 3 one-ear **71.** (a) 2284°F (b) 73.8°F (c) 28°F (d) −125°F **73.** (a) 1343°C (b) 169°C (c) −12°C (d) −38°C **75.** (a) 504 K (b) 504.9 K (c) 504.89 K (d) 276.1 K **77.** (a) 38°C (b) −361.1°F (c) 1077 K (d) −321°F **79.** −10°C is higher
81. −60°F is lower **83.** (a) light and heat (b) light and heat (c) chemical (d) light and heat **85.** (a) potential (b) kinetic (c) potential (d) potential **87.** (a) 2.29×10^6 J (b) 547 kcal (c) 5.47×10^5 cal (d) 547 Cal **89.** 112 J **91.** 47.7°C
93. 1.58 g **95.** 0.39 J/g·°C **97.** 80.0 g copper is larger
99. 0.382 J/g·°C **101.** 47 g **103.** 198 cm
105. 6.10 cents/shave **107.** 994 g **109.** 0.19 g
111. (a) yes (b) no **113.** 1.7×10^{-3} cm **115.** 6.3 g
117. −160°H **119.** 0.718 J/g·°C **121.** 25 kJ
123. 122 bears/bag **125.** (a) 5.000×10^{-1} g/mL (b) 5.0×10^{-1} g/mL (c) 5.0000×10^{-1} g/mL (d) 5.000×10^{-1} g/mL **127.** (a) 8×10^3 cm^3 (b) 8.0×10^3 cm^3 (c) 8.00×10^3 cm^3 (d) 8.0×10^3 cm^3
129. (a) 3.256×10^3 g (b) 3.34 mg (c) all same precision (d) 3.2500 g

CHAPTER 4

1. (a) indefinite vs definite shape (b) indefinite vs definite volume **3.** b and d **5.** (a) solid (b) solid (c) liquid (d) gas
7. (a) chemical (b) chemical (c) physical (d) physical
9. (a) chemical (b) physical (c) physical (d) physical
11. (a) physical (b) physical (c) chemical (d) chemical
13. (a) physical (b) physical (c) chemical (d) chemical
15. (a) freezing (b) condensation (c) sublimation (d) evaporation **17.** (a) heterogeneous mixture (b) heterogeneous mixture (c) heterogeneous mixture (d) pure substance
19. (a) false (b) true (c) false (d) true **21.** (a) heterogeneous mixture (b) homogeneous mixture (c) homogeneous mixture (d) heterogeneous mixture **23.** (a) homogeneous mixture, 1 phase (b) homogeneous mixture, 1 phase (c) heterogeneous mixture, 2 phases, (d) heterogeneous mixture, 2 phases
25. (a) compound (b) compound (c) no classification possible (d) no classification possible **27.** (a) true (b) false (c) false (d) false **29.** (a) A and B: no classification possible; C: compound (b) D: compound; E,F and G: no classification possible **31.** first box: mixture; second box: compound
33. (a) element and compound (b) element and compound (c) two elements (d) two compounds **35.** (a) 13 (b) 34 (c) 82 **37.** (a) true (b) false (c) false (d) false
39. (a) argon (b) barium (c) calcium (d) fluorine (e) helium (f) potassium (g) carbon (h) copper **41.** (a) Br (b) Cl (c) Fe (d) Hg (e) Li (f) Na (g) H (h) I **43.** (a) fluorine (b) zinc (c) potassium (d) sulfur **45.** (a) iron, Fe (b) tin, Sn (c) sodium, Na (d) gold, Au **47.** (a) Re-Be-C-Ca (b) Ra-Y-Mo-Nd (c) Na-N-C-Y (d) Br-U-Ce or B-Ru-Ce (e) S-H-Ar-O-N (f) Al-I-Ce **49.** (a) colorless gas; odorless gas, colorless, liquid, boils at 43°C (b) toxic to humans
51. (a) compound (b) mixture (c) compound (d) mixture
53. (a) homogeneous mixture (b) element (c) heterogeneous mixture (d) homogeneous mixture **55.** (a) no (b) yes (c) yes (d) no **57.** Bi, Bk, Cf, Co, Cs, Cu, Hf, Ho, Hs, In, Nb, Ni, No, Np, Ns, Os, Pb, Po, Pu, Sb, Sc, Si, Sn, Yb
59. (a) solid (b) not possible (c) gas (d) not possible
61. (a) 73.87% (b) 22.52% (c) 69.37% (d) 11.71%
63. (a) density I. = 1.14 g/mL; density II. = 1.14 g/mL; density III. = 1.14 g/mL. Likely all three same substance. (b) no
65. compound **67.** mixture

CHAPTER 5

1. (a) consistent (b) consistent (c) not consistent **3.** (a) heteroatomic, diatomic compound (b) heteroatomic, triatomic compound (c) homoatomic, diatomic element (d) heteroatomic, triatomic compound **5.** (a) false; molecules must contain two or more atoms (b) true (c) false; compounds may have either molecules or ions as their basic structural unit (d) true (e) true **7.** (a) X_2Q (b) XQZ (c) Z_2Q (d) Z_2
9. (a) compound (b) compound (c) element (d) element
11. (a) $C_{20}H_{30}O$ (b) H_2SO_4 (c) P_4 (d) HCN **13.** (a) same (b) same (c) fewer (d) more **15.** Cs_2 is a diatomic, homoatomic element; CS_2 is a triatomic, heteroatomic compound. **17.** (a) H_3PO_4 (b) $SiCl_4$ (c) NO_2 (d) H_2O_2
19. (a) electron (b) proton (c) proton, neutron (d) neutron
21. (a) false (b) false (c) false (d) true **23.** (a) true (b) false (c) false (d) false **25.** (a) 50 (b) 47 (c) 80 (d) 103
27. (a) 24, 53 (b) 44, 103 (c) 101, 256 (d) 16, 34
29. (a) number of protons or number of electrons (b) number of neutrons **31.** (a) ${}^{11}_{5}B$ (b) ${}^{16}_{8}O$ (c) ${}^{27}_{13}Al$ (d) ${}^{40}_{18}Ar$
33. (a) 16 protons, 16 electrons, 16 neutrons (b) 29 protons, 29 electrons, 34 neutrons (c) 22 protons, 22 electrons, 28 neutrons (d) 92 protons, 92 electrons, 146 neutrons

35. (a) 27 protons, 27 electrons, 32 neutrons (b) 45 protons, 45 electrons, 58 neutrons (c) 69 protons, 69 electrons, 100 neutrons (d) 9 protons, 9 electrons, 10 neutrons **37.** (a) $^{31}_{15}P$ (b) $^{18}_{8}O$ (c) $^{54}_{24}Cr$ (d) $^{197}_{79}Au$ **39.** (a) same total (b) same neutrons (c) same neutrons (d) none of the above **41.** $^{54}_{26}Fe$; $^{56}_{26}Fe$; $^{57}_{26}Fe$; $^{58}_{26}Fe$ **43.** $^{96}_{40}Zr$; $^{94}_{40}Zr$; $^{92}_{40}Zr$; $^{91}_{40}Zr$; $^{90}_{40}Zr$ **45.** (a) isotopes (b) isobars (c) neither (d) isotopes **47.** (a) isotopes (b) isobars (c) neither (d) isotopes **49.** $^{40}_{18}Ar$ $^{40}_{19}K$ $^{40}_{20}Ca$ **51.** (a) Be,F,Na,Al,P (b) O,Ne,Mg,Si,Ar,K (c) He,Li,B,Ar (d) H,He **53.** (a) 32.066 (b) 14.0067 (c) 196.9665 (d) 55.845 **55.** Q = 8.00 bebs; X = 4.00 bebs; Z = 2.00 bebs **57.** (a) Z = 9 amu; X = 27 amu; Q = 108 amu (b) Z = Be; X = Al; Q = Ag **59.** 224.8 lb **61.** 35.46 amu **63.** 47.88 amu **65.** (a) 9.156 (b) 21.8556 **67.** The number 12.0000 amu applies only to the ^{12}C isotope. The number 12.011 applies to naturally occurring carbon, a mixture of ^{12}C, ^{13}C, and ^{14}C isotopes. **69.** (a) 2 (b) 3 (c) 3 (d) 6 **71.** 672 protons **73.** (a) 4 (b) 2 (c) 4 (d) 6 (e) 54 **75.** (a) false (b) false (c) true (d) true **77.** (a) $^{37}_{18}Ar$, $^{39}_{19}K$, $^{42}_{20}Ca$, $^{44}_{21}Sc$, $^{43}_{22}Ti$ (b) $^{44}_{21}Sc$, $^{42}_{20}Ca$, $^{43}_{22}Ti$, $^{39}_{19}K$, $^{37}_{18}Ar$ (c) $^{37}_{18}Ar$, $^{39}_{19}K$, $^{42}_{20}Ca$, $^{44}_{21}Sc$, $^{43}_{22}Ti$ (d) $^{44}_{21}Sc$, $^{43}_{22}Ti$, $^{42}_{20}Ca$, $^{39}_{19}K$, $^{37}_{18}Ar$ **79.** (a) $^{8}_{5}B$ (b) $^{12}_{5}B$ (c) $^{12}_{5}B$ (d) $^{16}_{5}B$ **81.** (a) 29 and 29 (b) 29 and 29 (c) 24 and 26 **83.** (a) 36.756 new amu (b) 67.117 new amu **85.** 122 **87.** 80.9170 amu **89.** two: ^{79}Br and ^{81}Br **91.** ^{39}K **93.** 370 $^{17}_{8}O$ atoms, 2.0×10^{-3} $^{18}_{8}O$ atoms **95.** (a) compound (b) mixture (c) element (d) mixture **97.** 1.90×10^{23} Au atoms **99.** 2.84×10^{9} miles **101.** 0.022% **103.** 2.4 miles

CHAPTER 6

1. (a) Ga (b) Zr (c) Li (d) Cl **3.** b **5.** a,d **7.** (a) group (b) periodic law (c) periodic law (d) period **9.** (a) orbital (b) orbital (c) shell (d) shell **11.** (a) 2 (b) 10 (c) 6 (d) 14 **13.** (a) 2 (b) 2 (c) 2 (d) 2 **15.** (a) 3*d* subshell (b) 2*p* subshell (c) 3*p* subshell (d) third shell **17.** (a) true (b) true (c) true (d) true (e) false **19.** (a) spherical (b) dumbbell (c) cloverleaf (d) spherical **21.** b and c **23.** (a) 3*s* (b) 3*d* (c) 7*s* (d) 4*p* **25.** (a) $1s^22s^22p^63s^23p^1$ (b) $1s^22s^22p^3$ (c) $1s^22s^22p^63s^23p^6$ (d) $1s^22s^22p^63s^2$ **27.** (a) $1s^22s^22p^63s^23p^64s^23d^6$ (b) $1s^22s^22p^63s^23p^64s^23d^{10}4p^65s^1$ (c) $1s^22s^22p^63s^23p^64s^23d^{10}4p^65s^24d^{10}5p^5$ (d) $1s^22s^22p^63s^23p^64s^23d^{10}4p^65s^24d^{10}5p^66s^24f^{14}5d^{10}6p^6$ **29.** (a) $_{10}Ne$ (b) $_{19}K$ (c) $_{22}Ti$ (d) $_{30}Zn$ **31.** (a) 2,8,3 (b) 2,8,8 (c) 2,8,4 (d) 2,2

33.

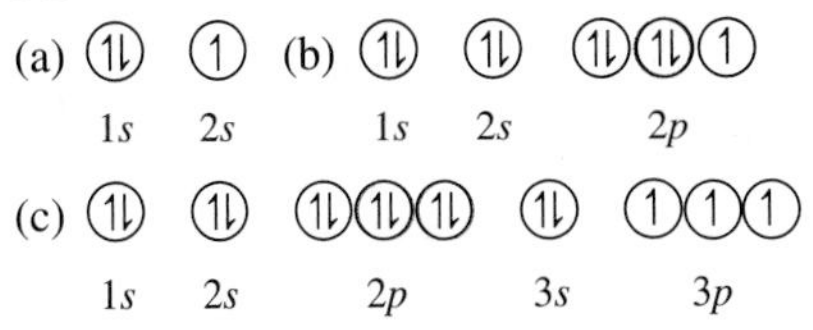

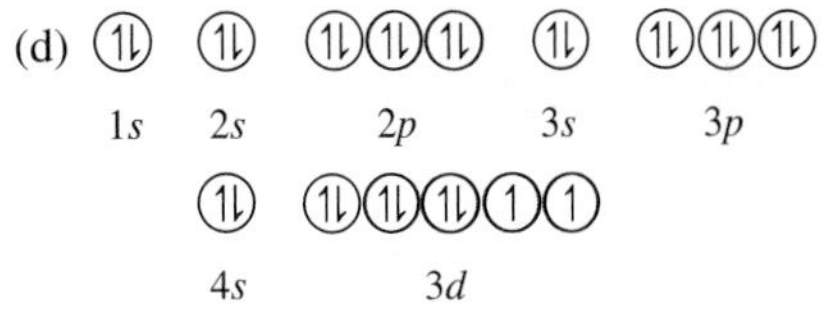

35.

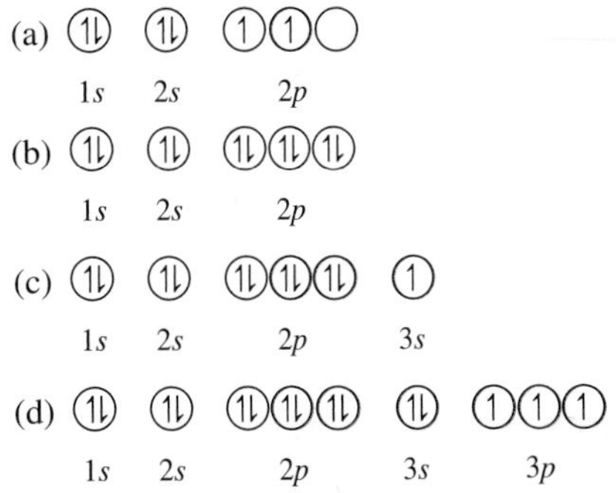

37. (a) one (b) one (c) one (d) none **39.** (a) paramagnetic (b) paramagnetic (c) paramagnetic (d) diamagnetic **41.** (a) no (b) yes (c) no (d) yes **43.** (a) *d* (b) *p* (c) *s* (d) *f* **45.** (a) 4*s* (b) 4*p* (c) 4*d* (d) 5*d* **47.** (a) $_{35}Br$ (b) $_{87}Fr$ (c) $_{40}Zr$ (d) $_{23}V$ **49.** (a) Al (b) Li (c) La (d) Sc **51.** (a) Kr (b) Li (c) K (d) Lu **53.** (a) $1s^22s^22p^63s^23p^64s^23d^{10}4p^3$ (b) $1s^22s^22p^63s^23p^64s^23d^{10}4p^65s^24d^{10}5p^3$ (c) $1s^22s^22p^63s^23p^64s^23d^{10}4p^65s^1$ (d) $1s^22s^22p^63s^23p^64s^23d^{10}4p^65s^24d^{10}$ **55.** (a) 4 (b) 9 (c) 2 (d) 0 **57.** (a) transition (b) representative (c) noble gas (d) inner transition **59.** (a) S (b) P (c) I (d) Cl **61.** (a) Li (b) K (c) Fe (d) Hg **63.** (a) Sc (b) He (c) Li (d) Ce **65.** (a) Ge (b) B (c) Po (d) Te **67.** (a) Mg (b) Au (c) S (d) Rb **69.** (a) F (b) P (c) Zn (d) Cl **71.** (a) N (b) Ga (c) K (d) Rb **73.** (a) The 2*s* subshell has a maximum capacity of 2 electrons. (b) The 2*p* subshell must be filled before the 3*s* subshell. It can hold 6 electrons. (c) The 2*p* subshell is filled after the 2*s*. (d) The 3*p* and 4*s* subshells are filled after the 3*s* and before the 3*d* subshell. **75.** (a) period 3, group IA (b) period 3, group IIIA (c) period 4, group IIIB (d) period 4, group VIIA **77.** (a) x=6, y=2 (b) x=2, y=1 (c) x=2, y=10 (d) x=2, y=6 **79.** (a) $1s^22s^22p^63s^23p^2$(Si) (b) $1s^22s^22p^2$(C) (c) $1s^22s^22p^3$(N) (d) $1s^22s^22p^3$(N) **81.** (a) Be (b) Be (c) Ne (d) Ar **83.** (a) 2 (b) 4 (c) 1 (d) 2 **85.** (a) 7 (b) 2 (c) 4 (d) 3 **87.** (a) groups IVA and VIA (b) group VB and middle column of group VIIIB (c) group IVB and last column of group VIIIB (d) group IA **89.** (a) Po (b) Cr (c) any element 88–111 (d) any element 12–18 **91.** (a) F (b) Ag **93.** (a) $1s^32s^32p^7$ (b) $1s^32s^32p^93s^33p^94s^33d^5$ (c) $1s^32s^32p^93s^1$ (d) $1s^32s^32p^93s^33p^8$ **95.** (a) O,Li,He,B,K (b) O,He,B (c) O,Li,He,B,Sr,K (d) none of them **97.** the same **99.** 54 **101.** 38 (Sr)

CHAPTER 7

1. (a) 3 (b) 7 (c) 1 (d) 3 **3.** (a) 2 (b) 6 (c) 3 (d) 5 **5.** (a) C (b) F (c) Mg (d) P **7.** (a) Be:

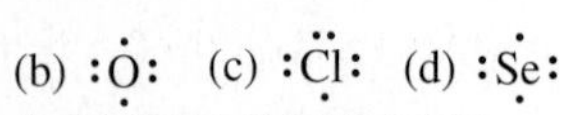

9. (a) B (b) C (c) F (d) Li **11.** (a) Li^+ (b) P^{3-} (c) Br^- (d) Ba^{2+} **13.** (a) $3p,2e$ (b) $7p$, $10e$ (c) $20p$, $18e$ (d) $17p$, $18e$ **15.** (a) Al^{3+} (b) O^{2-} (c) Mg^{2+} (d) Be^{2+} **17.** (a) +2 (b) −2 (c) +1 (d) +2 **19.** (a) 2 lost (b) 1 gained (c) 3 gained (d) 1 lost **21.** (a) loss (b) loss (c) loss (d) gain **23.** (a) $1s^22s^22p^63s^23p^6$ (b) $1s^2$ (c) $1s^22s^22p^6$ (d) $1s^22s^22p^6$ **25.** (a) Cl (b) Na (c) Al (d) S **27.** (a) He (b) Xe (c) Kr (d) Xe **29.** (a) Na^+ (b) F^- (c) O^{2-} (d) Mg^{2+} **31.** (a) yes (b) no (c) yes (d) no

33.

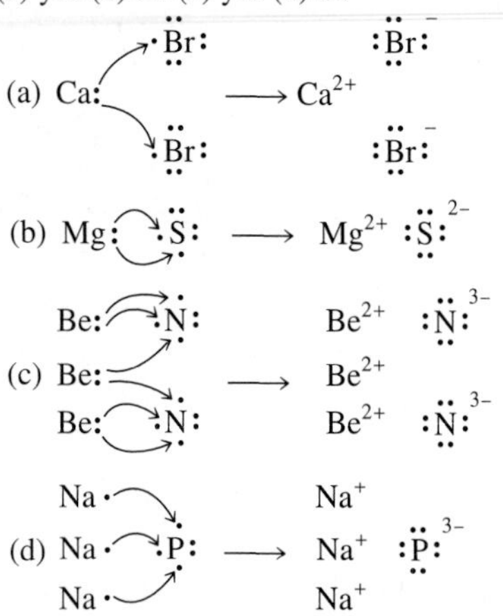

35.

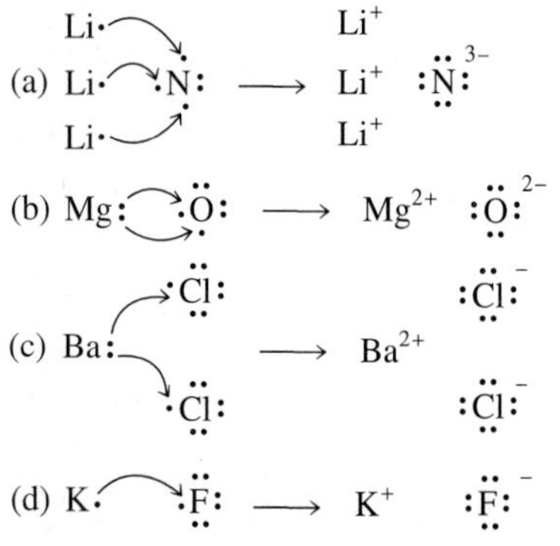

37. (a) $CaCl_2$ (b) BeO (c) AlN (d) K_2S **39.** (a) Mg^{2+}, S^{2-} (b) Al^{3+}, N^{3-} (c) 2 Na^+, O^{2-} (d) 3 Ca^{2+}, 2 N^{3-} **41.** (a) X_2Z (b) XZ_3 (c) X_3Z (d) ZX_2 **43.** (a) $Mg(CN)_2$ (b) $CaSO_4$ (c) $Al(OH)_3$ (d) NH_4NO_3 **45.** (a) $AlPO_4$ (b) $Al_2(CO_3)_3$ (c) $Al(ClO_3)_3$ (d) $Al(C_2H_3O_2)_3$

47.

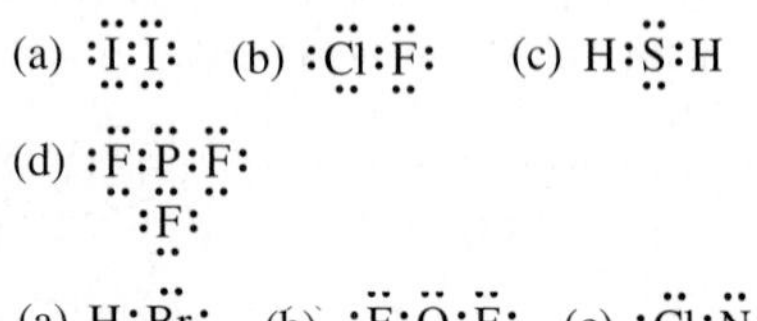

49.

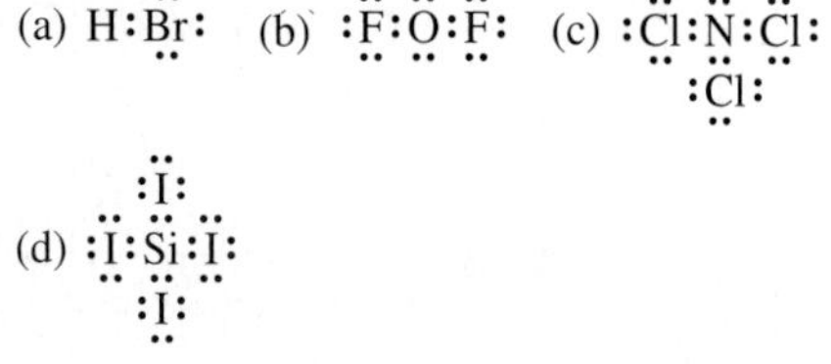

51. (a) 3 bonding, 2 nonbonding (b) 6 bonding, 0 nonbonding (c) 4 bonding, 4 nonbonding (d) 7 bonding, 2 nonbonding **53.** (a) 1 triple (b) 4 single, 1 double (c) 2 double (d) 2 triple, 1 single **55.** The double bond has a shorter bond length and almost twice as great bond energy as a single bond. **57.** a, b, and d **59.** A bond in which both electrons in a shared pair come from one of the two atoms. **61.** (a) the N–O bond (b) none (c) the O–Cl bond (d) the two O–Br bonds **63.** Resonance structures are two or more electron dot structures for a molecule or ion that have the same arrangement of atoms, contain the same number of electrons, and differ only in the location of the electrons.

65.

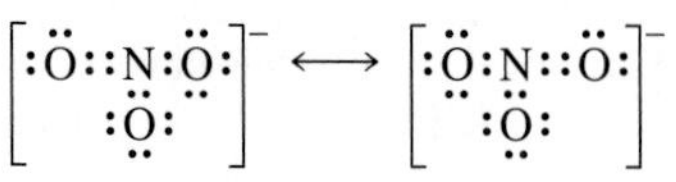

67. (a) 24 (b) 26 (c) 30 (d) 32

69.

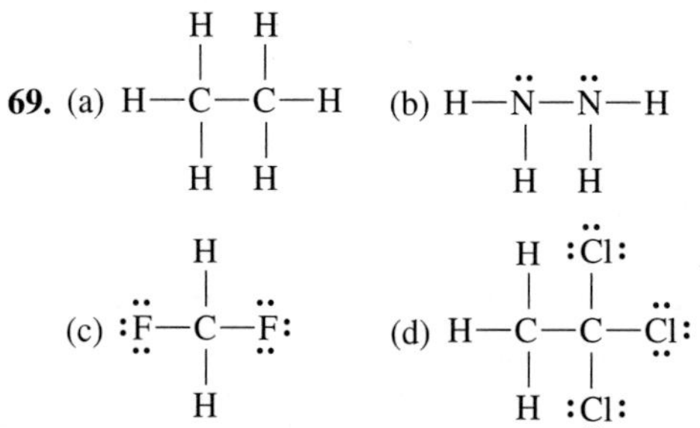

71.

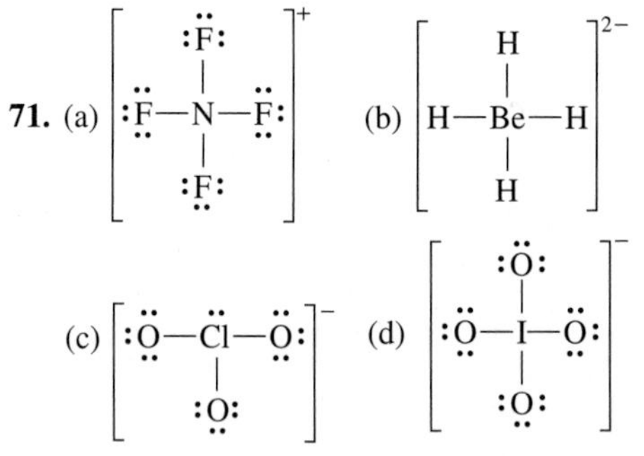

73.

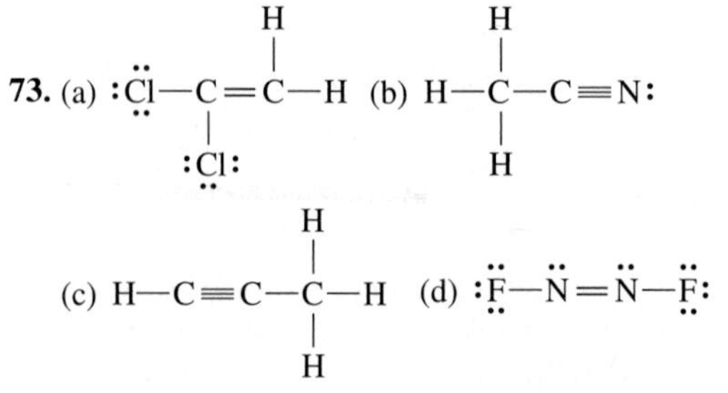

75.

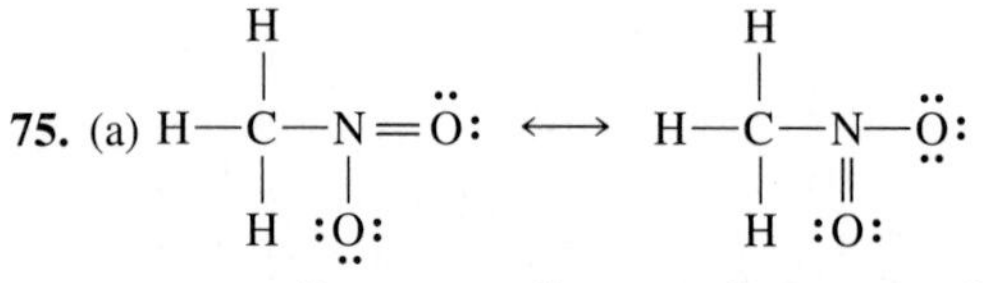

(c) 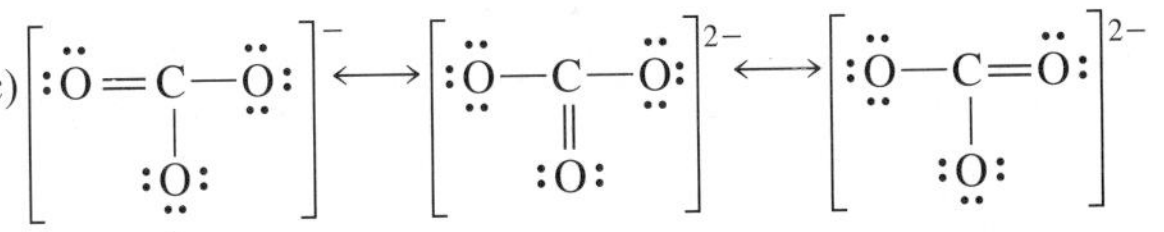

(d) 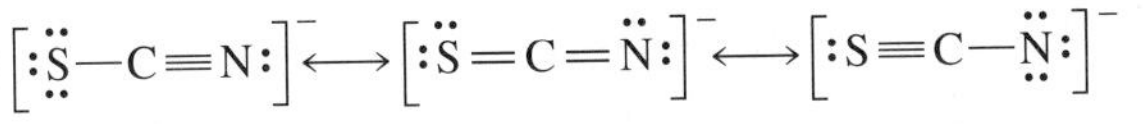

77. (a) 1 bonding, 3 nonbonding (b) 3 bonding, 2 nonbonding (c) 1 bonding, 3 nonbonding (d) 3 bonding, 2 nonbonding **79.** $n = 1$ **81.** (a) angular (b) linear (c) angular (d) linear **83.** (a) angular (b) linear (c) linear (d) angular **85.** (a) trigonal pyramidal (b) linear (c) trigonal planar (d) tetrahedral **87.** (a) tetrahedral (b) linear (c) trigonal pyramidal (d) trigonal pyramidal **89.** (a) Each center has trigonal pyramidal shape. (b) C: tetrahedral; O: angular **91.** (a) O (b) O (c) S (d) Mg **93.** (a) Na,Mg,Al,P (b) I,Br,Cl,F (c) As,P,S,O (d) Ca,Ge,C,O **95.** c **97.** (a) H—O (b) O—Al (c) B—N (d) Al—Cl **99.** (a) polar covalent (b) ionic (c) polar covalent (d) nonpolar covalent **101.** (a) polar (b) polar (c) polar (d) nonpolar **103.** (a) polar bonds, nonpolar molecule (b) polar and nonpolar bonds, polar molecule (c) polar bonds, polar molecule (d) nonpolar bonds, nonpolar molecule **105.** b, c, and d **107.** (a) Mg-$1s^22s^22p^63s^2$, Mg^{2+}-$1s^22s^22p^6$ (b) F-$1s^22s^22p^5$, F^--$1s^22s^22p^6$ (c) N-$1s^22s^22p^3$, N^{3-}-$1s^22s^22p^6$ (d) Ca^{2+}-$1s^22s^22p^63s^23p^6$, S^{2-}-$1s^22s^22p^63s^23p^6$ **109.** (a) too few electron dots (b) too many electron dots **111.** (a) covalent (b) ionic (c) covalent (d) covalent **113.** a and d **115.** (a) O (b) N (c) O (d) F **117.** a, b, and d **119.** BA, CA, DB, and DA **121.** (a) same (b) different (c) different (d) different **123.** (a) Cl (b) Cl

125. (a) 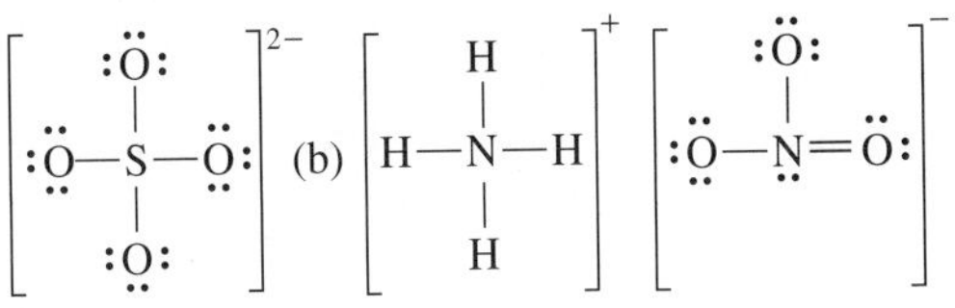

127. (a) tetrahedral electron pairs, tetrahedral molecule (b) tetrahedral electron pairs, trigonal pyramidal molecule (c) linear electron pairs, linear molecule (d) tetrahedral electron pairs, tetrahedral molecule **129.** (a) 109° (b) 120° **131.** (a) HF (b) H_3CF (c) HCN (d) SO_2 **133.** A = Al, D = N **135.** A = Mg, D = S **137.** 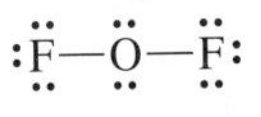 **139.** H_2O

141. A = Be, D = F 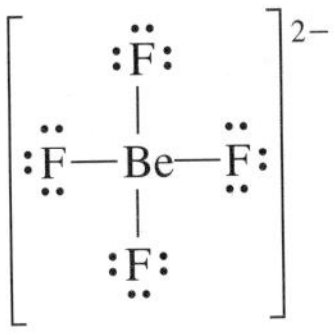

CHAPTER 8

1. a,d **3.** a,d **5.** a,b **7.** (a) I (b) II (c) I (d) II **9.** a,d **11.** (a) II (b) II (c) I (d) I **13.** (a) Cu^{2+} (b) Fe^{3+} (c) Pb^{4+} (d) Sn^{2+} **15.** (a) Na^+ (b) Ag^+ (c) Zn^{2+} (d) Ca^{2+} **17.** (a) Cl^- (b) O^{2-} (c) P^{3-} (d) F^- **19.** (a) magnesium oxide (b) lithium sulfide (c) silver chloride (d) zinc bromide **21.** (a) no (b) no (c) yes (d) yes **23.** (a) +2 (b) +2 (c) +3 (d) +2 **25.** (a) iron(II) bromide, iron(III) bromide (b) copper(I) oxide, copper(II) oxide (c) tin(II) sulfide, tin(IV) sulfide (d) nickel(II) oxide, nickel(III) oxide **27.** (a) aluminum chloride (b) nickel(III) chloride (c) zinc oxide (d) cobalt(II) oxide **29.** (a) plumbic oxide (b) auric chloride (c) iron(III) iodide (d) tin(II) bromide **31.** (a) FeS (b) SnS_2 (c) Li_2S (d) ZnS **33.** b,c **35.** (a) PO_4^{3-} (b) ClO_3^- (c) NO_3^- (d) CN^- **37.** (a) peroxide (b) thiosulfate (c) oxalate (d) permanganate **39.** (a) SO_4^{2-}, SO_3^{2-} (b) PO_4^{3-}, HPO_4^{2-} (c) OH^-, O_2^{2-} (d) CrO_4^{2-}, $Cr_2O_7^{2-}$ **41.** b,c **43.** (a) I (b) I (c) II (d) II **45.** (a) +2 (b) +4 (c) +3 (d) +1 **47.** (a) zinc sulfate (b) barium hydroxide (c) iron(III) nitrate (d) copper(II) carbonate **49.** (a) iron(III) carbonate, iron(II) carbonate (b) gold(I) sulfate, gold(III) sulfate (c) tin(II) hydroxide, tin(IV) hydroxide (d) chromium(III) acetate, chromium(II) acetate **51.** (a) ammonium nitrate (b) ammonium chloride (c) sodium phosphate (d) copper(I) phosphate **53.** (a) Ag_2CO_3 (b) $AuNO_3$ (c) $Cr_2(SO_4)_3$ (d) $NH_4C_2H_3O_2$ **55.** (a) $Fe_2(SO_4)_3$ (b) CuCN (c) $Sn(CO_3)_2$ (d) $Pb(OH)_2$ **57.** (a) 7 (b) 5 (c) 3 (d) 10 **59.** (a) tetraphosphorus decoxide (b) sulfur tetrafluoride (c) carbon tetrabromide (d) chlorine dioxide **61.** (a) ICl (b) NCl_3 (c) SF_6 (d) OF_2 **63.** (a) hydrogen sulfide (b) hydrogen fluoride (c) ammonia (d) methane **65.** (a) PH_3 (b) HBr (c) C_2H_6 (d) H_2Te **67.** (a) no (b) yes (c) yes (d) no **69.** (a) hydrocyanic acid (b) sulfuric acid (c) nitrous acid (d) boric acid **71.** (a) HCN (b) H_2SO_4 (c) HNO_2 (d) H_3BO_3 **73.** (a) nitric acid (b) hydriodic acid (c) hypochlorous acid (d) acetic acid **75.** (a) arsenious acid (b) periodic acid (c) hypophosphorous acid (d) bromous acid **77.** (a) hydrogen bromide (b) hydrocyanic acid (c) hydrogen sulfide (d) hydriodic acid **79.** (a) $HClO_3$ (b) HNO_2 (c) HF (d) $HC_2H_3O_2$ **81.** b,c **83.** c **85.** a,d **87.** (a) 4 (b) 2 (c) 3 (d) 3 **89.** (a) Ca_3N_2 (b) $Ca(NO_3)_2$ (c) $Ca(NO_2)_2$ (d) $Ca(CN)_2$ **91.** (a) K_3P (b) K_3PO_4 (c) K_2HPO_4 (d) KH_2PO_4 **93.** (a) N_2O, CO_2 (b) NO_2, SO_2 (c) SF_2, SCl_2 (d) N_2O_3 **95.** (a) $CaCO_3$, HNO_2 (b) $NaClO_4$, $NaClO_3$ (c) $HClO_2$, HClO (d) Li_2CO_3, Li_3PO_4 **97.** (a) sodium nitrate (b) aluminum sulfide, magnesium nitride, beryllium phosphide (c) iron(III) oxide (d) gold(I) chlorate **99.** (a) $Ni_2(SO_4)_3$ (b) Ni_2O_3 (c) $Ni_2(C_2O_4)_3$ (d) $Ni(NO_3)_3$ **101.** NO_2O_2 **103.** (a) magnesium chloride (b) oxygen difluoride **105.** (a) x = 4; silicon tetrachloride (b) x = 2; magnesium chloride (c) x = 3; potassium nitride (d) x = 3; nitrogen trichloride **107.** beryllium bromate **109.** aluminum nitride **111.** carbon dioxide **113.** beryllium cyanide

CHAPTER 9

1. the same **3.** Both samples have the same %A and %B. **5.** 1 and 3 **7.** 100. g CO **9.** (a) 184.05 amu

(b) 61.84 amu (c) 201.24 amu (d) 375.98 amu **11.** (a) 62.0 amu (b) 60.0 amu (c) 379.3 amu (d) 224.2 amu **13.** 4 **15.** 62.1 amu **17.** (a) 75.75% Sn, 24.25% F (b) 36.76% Fe, 19.81% S, 39.53% O (c) 42.10% C, 6.49% H 51.41% O (d) 47.43% C, 2.56% H, 50.00% Cl **19.** (a) 79.89% Cu, 20.11% O (b) 32.37% Na, 22.57% S, 45.06% O (c) 25.9% N, 74.08% O (d) 33.4% S, 66.6% O **21.** a **23.** both have 92.24% C and 7.76% H **25.** 363.1 g **27.** each has 6.02×10^{23} **29.** (a) 1.50×10^{24} (b) 1.96×10^{24} (c) 1.4×10^{23} (d) 1.875×10^{23} **31.** (a) 9.03×10^{23} (b) 3.01×10^{23} (c) 1.40×10^{24} (d) 6.715×10^{23} **33.** (a) 63.5 g (b) 137 g (c) 28.1 g (d) 238 g **35.** (a) 78.01 g (b) 100.9 g (c) 187.6 g (d) 325.8 g **37.** (a) 79.30 g (b) 105.9 g (c) 115.3 g (d) 222.5 g **39.** (a) 2.00 moles Cu (b) 1.00 mole Br (c) 1.50 moles N_2O (d) 4.87 moles B_2H_6 **41.** 48.0 g/mole **43.** 3.16×10^{-23} g **45.** 30.97 amu **47.** (a) 3.0426 (b) 107.018 (c) 6.74202 (d) 3.482567 **49.** (a) 2.99390 (b) 5.98780 (c) 1.49645 (d) 11.9756 **51.** (a) equal (b) not equal (c) equal (d) equal **53.** (a) more Fe (b) fewer Ni (c) equal (d) more Al **55.** (a) 31.07 g S (b) 8.730 g Be (c) 230.5 g U (d) 27.21 g Si **57.** 3 moles Na = 1 mole Na_3PO_4; 1 mole P = 1 mole Na_3PO_4; 4 moles O = 1 mole Na_3PO_4; 3 moles Na = 1 mole P; 3 moles Na = 4 moles O; 1 mole P = 4 moles O. **59.** a, c, and d **61.** (a) $NaAuBr_4$ (b) $C_2H_2Cl_4$ (c) $Ba(NO_3)_2$ (d) NH_4CN **63.** (a) 1.411×10^{23} (b) 5.012×10^{23} (c) 3.290×10^{22} (d) 2.293×10^{22} **65.** (a) 7.52×10^{23} (b) 4.70×10^{23} (c) 2.35×10^{23} (d)1.53×10^{23} **67.** (a) 107.9 (b) 4.071 (c) 4.8×10^{8} (d) 1.131×10^{-20} **69.** (a) 3.818×10^{-23} (b) 4.035×10^{-23} (c) 9.655×10^{-23} (d) 1.297×10^{-22} **71.** (a) 60.66 (b) 921.93 (c) 9.73 (d) 17.1 **73.** (a) 1.71×10^{23} (b) 3.38×10^{23} (c) 9.84×10^{22} (d) 1.23×10^{23} **75.** (a) 5.3×10^{3} (b) 30.00 (c) 168 (d) 196 **77.** (a) 0.03118 mole (b) 2.347×10^{21} molecules (c) 3.040 g (d) 1.878×10^{22} atoms **79.** (a) 4.213 (b) 1.005×10^{24} (c) 2.544 (d) 4.787×10^{22} **81.** (a) HO (b) C_4H_7 (c) C_3H_8 (d) SN **83.** (a) Na_2S (b) $KMnO_4$ (c) H_2SO_4 (d) $C_2H_3O_5N$ **85.** (a) 3 to 4 (b) 3 to 4 (c) 7 to 8 to 9 (d) 4 to 7 to 6 **87.** (a) P_2O_5 (b) Mg_3N_2 (c) $Na_2S_2O_3$ (d) $Mg_2P_2O_7$ **89.** $NiCl_2$, $NiCl_3$ **91.** C_2H_6S **93.** BeO **95.** (a) CH_4 (b) CH (c) C_2H_5 (d) CH_3 **97.** C_5H_6O **99.** (a) P_4O_{10} (b) S_4N_4 (c) $C_3H_6O_2$ (d) $B_3N_3H_6$ **101.** (a) $C_8H_8O_2$ (b) $C_8H_8O_2$ **103.** (a) $C_3H_6O_3$ (b) $C_3H_6O_3$ **105.** 2.837 g K, 1.163 g S **107.** 37.8 **109.** 4.49 **111.** 7.21×10^{23} **113.** (a) CrO_3 (b) CrO_3 **115.** 216 g **117.** 3.2×10^{18} **119.** (a) Ag (b) B **121.** (a) 8.3% (b) 64.285714% (c) 64.285714% **123.** (a) Na_3AlF_6 (b) SO_2 (c) $BaCO_3$ (d) HClO **125.** 2.43×10^{23} **127.** (a) $C_4H_6O_2$ (b) $C_6H_9O_3$ (c) $C_{12}H_{18}O_6$ (d) $C_4H_6O_2$ **129.** 5.013 **131.** (a) CH_4S (b) 7.221 g **133.** 40.0% NaF, 40.1% $NaNO_3$, 19.9% Na_2SO_4 **135.** 3 **137.** 40.08 **139.** $C_3H_8O_2$ **141.** (a) 142 (b) 141.9 (c) 141.94 (d) 141.943 **143.** 11.3 g/cm^3 **145.** 22.9 cm^3 **147.** (a) 148.3 g (b) 65.02 g (c) 96.69 g (d) 117.5 g **149.** 6.5×10^{2} J **151.** 7×10^{17} **153.** (a) 1.40×10^{24} (b) 2.81×10^{24} (c) 4.21×10^{24} (d) 7.01×10^{24} **155.** 4.51×10^{24} **157.** 6.82×10^{26} **159.** 81.0 mL **161.** 1.39×10^{24} **163.** 6.67×10^{22}

CHAPTER 10

1. 14.33 **3.** 2.2 g **5.** a and d **7.** (a) (s): solid; (aq): water solution (b) (g): gas; (l): liquid; (aq): water solution **9.** (a) balanced (b) balanced (c) not balanced (d) balanced **11.** (a) $2\ Cu + O_2 \rightarrow 2\ CuO$ (b) $2\ Al + N_2 \rightarrow 2\ AlN$ (c) $2\ HgO \rightarrow 2\ Hg + O_2$ (d) $2\ H_2O \rightarrow 2\ H_2 + O_2$ **13.** (a) $BaCl_2 + Na_2S \rightarrow BaS + 2\ NaCl$ (b) $Mg + 2\ HBr \rightarrow MgBr_2 + H_2$ (c) $2\ Co + 3\ HgCl_2 \rightarrow 2\ CoCl_3 + 3\ Hg$ (d) $2\ Na + 2\ H_2O \rightarrow 2\ NaOH + H_2$ **15.** (a) $3\ PbO + 2\ NH_3 \rightarrow 3\ Pb + N_2 + 3\ H_2O$ (b) $2\ NaHCO_3 + H_2SO_4 \rightarrow Na_2SO_4 + 2\ CO_2 + 2\ H_2O$ (c) $TiO_2 + C + 2\ Cl_2 \rightarrow TiCl_4 + CO_2$ (d) $2\ NBr_3 + 3\ NaOH \rightarrow N_2 + 3\ NaBr + 3\ HBrO$ **17.** (a) $Ca(OH)_2 + 2\ HNO_3 \rightarrow Ca(NO_3)_2 + 2\ H_2O$ (b) $BaCl_2 + (NH_4)_2SO_4 \rightarrow BaSO_4 + 2\ NH_4Cl$ (c) $2\ Fe(OH)_3 + 3\ H_2SO_4 \rightarrow Fe_2(SO_4)_3 + 6\ H_2O$ (d) $Na_3PO_4 + 3\ AgNO_3 \rightarrow 3\ NaNO_3 + Ag_3PO_4$ **19.** (a) $AgNO_3 + KCl \rightarrow AgCl + KNO_3$ (b) $CS_2 + 3\ O_2 \rightarrow CO_2 + 2\ SO_2$ (c) $2\ H_2 + O_2 \rightarrow 2\ H_2O$ (d) $2\ Ag_2CO_3 \rightarrow 4\ Ag + 2\ CO_2 + O_2$ **21.** (a) $CH_4 + 2\ O_2 \rightarrow CO_2 + 2\ H_2O$ (b) $2\ C_6H_6 + 15\ O_2 \rightarrow 12\ CO_2 + 6\ H_2O$ (c) $C_6H_{12} + 9\ O_2 \rightarrow 6\ CO_2 + 6\ H_2O$ (d) $C_3H_4 + 4\ O_2 \rightarrow 3\ CO_2 + 2\ H_2O$ **23.** (a) $CH_2O + O_2 \rightarrow CO_2 + H_2O$ (b) $C_3H_6O + 4\ O_2 \rightarrow 3\ CO_2 + 3\ H_2O$ (c) $2\ CH_2O_2 + O_2 \rightarrow 2\ CO_2 + 2\ H_2O$ (d) $C_4H_8O_2 + 5\ O_2 \rightarrow 4\ CO_2 + 4\ H_2O$ **25.** (a) $K_2CO_3 \rightarrow K_2O + CO_2$ (b) $CaCO_3 \rightarrow CaO + CO_2$ (c) $NiCO_3 \rightarrow NiO + CO_2$ (d) $Fe_2(CO_3)_3 \rightarrow Fe_2O_3 + 3\ CO_2$ **27.** (a) $4\ C_2H_7N + 19\ O_2 \rightarrow 8\ CO_2 + 14\ H_2O + 4\ NO_2$ (b) $CH_4S + 3\ O_2 \rightarrow CO_2 + 2\ H_2O + SO_2$ **29.** (a) synthesis (b) synthesis (c) double replacement (d) single replacement **31.** (a) Cu, $Zn(NO_3)_2$; $Zn + Cu(NO_3)_2 \rightarrow Cu + Zn(NO_3)_2$ (b) CaO; $2\ Ca + O_2 \rightarrow 2\ CaO$ (c) $BaSO_4$, KNO_3; $K_2SO_4 + Ba(NO_3)_2 \rightarrow BaSO_4 + 2\ KNO_3$ (d) Ag, O_2; $2\ Ag_2O \rightarrow 4\ Ag + O_2$ **33.** (a) 2 (b) 9 (c) 6 (d) 2

35. $\frac{4 \text{ moles } NH_3}{3 \text{ moles } O_2}$; $\frac{3 \text{ moles } O_2}{4 \text{ moles } NH_3}$; $\frac{4 \text{ moles } NH_3}{2 \text{ moles } N_2}$; $\frac{2 \text{ moles } N_2}{4 \text{ moles } NH_3}$; $\frac{4 \text{ moles } NH_3}{6 \text{ moles } H_2O}$; $\frac{6 \text{ moles } H_2O}{4 \text{ moles } NH_3}$; $\frac{3 \text{ moles } O_2}{2 \text{ moles } N_2}$; $\frac{2 \text{ moles } N_2}{3 \text{ moles } O_2}$; $\frac{3 \text{ moles } O_2}{6 \text{ moles } H_2O}$; $\frac{6 \text{ moles } H_2O}{3 \text{ moles } O_2}$; $\frac{2 \text{ moles } N_2}{6 \text{ moles } H_2O}$; $\frac{6 \text{ moles } H_2O}{2 \text{ moles } N_2}$

37. (a) 2.00 moles NaN_3 (b) 9.00 moles CO (c) 6.00 moles NH_2Cl (d) 2.00 moles $C_3H_5O_9N_3$ **39.** (a) 0.129 mole C_7H_{16} (b) 2.84 moles HCl (c) 0.710 mole Na_2SO_4 (d) 5.68 moles Na_2CO_3 **41.** (a) 6.12 moles (b) 4.38 moles (c) 2.62 moles (d) 4.81 moles **43.** (a) 19.0 moles Cl_2 (b) 0.33 mole HCl (c) 0.575 mole CH_4 (d) 0.308 mole CCl_4 **45.** (a) 1.063 g C (b) 0.4885 g C_3H_8 (c) 31.31 g Cl_2 (d) 0.2312 g H_2O **47.** (a) 30.0 g SiO_2 (b) 76.7 g CO

(c) 66.8 g SiC (d) 20.7 g C **49.** (a) 215 g (b) 238 g (c) 26.6 g (d) 6.48 g **51.** (a) 0.2977 mole Na_2SiO_3 (b) 79.95 g HF (c) 9.864×10^{21} molecules H_2SiF_6 (d) 65.55 g HF **53.** 23.6 g **55.** 65.66 g Cr, 134.3 g Cl_2 **57.** 284 bolts **59.** 213 kits **61.** (a) 3.65 moles H_2 (b) 2.60 moles N_2 (c) 3.00 moles H_2 (d) 55.0 g N_2 **63.** (a) 13.8 g Mg_3N_2 (b) 27.6 g Mg_3N_2 (c) 36.0 g Mg_3N_2 (d) 36.0 g Mg_3N_2 **65.** 175 **67.** 9.6 g O_2 **69.** 2.57 g SF_4, 3.21 g S_2Cl_2, 5.57 g NaCl **71.** 30.8% **73.** (a) 208 g Al_2S_3 (b) 60.1% **75.** 96.44% **77.** 131 g **79.** 31.8 g CO_2 **81.** 54.7 g **83.** 294 g **85.** 22.2 g **87.** 9.00 g **89.** 6.69 g **91.** 8.33 g N_2, 21.4 g H_2O, 45.2 g Cr_2O_3 **93.** 41.53 g H_2S **95.** (a) B (b) A (c) B (d) A **97.** 80.1 g **99.** 44.5% **101.** 92.8% **103.** (a) 75.4% IF_5 **105.** 3.678 tons **107.** 38.3% **109.** (a) $Zn + 2\,AgNO_3 \rightarrow Zn(NO_3)_2 + 2\,Ag$ (b) $HCl + NaOH \rightarrow NaCl + H_2O$ (c) $PCl_3 + Cl_2 \rightarrow PCl_5$ (d) $2\,Cu + O_2 \rightarrow 2\,CuO$ **111.** C_3H_6 **113.** $2\,Cu_2S + 3\,O_2 \rightarrow 2\,Cu_2O + 2\,SO_2$ **115.** $2\,C_6H_6 + 15\,O_2 \rightarrow 12\,CO_2 + 6\,H_2O$ **117.** (a) 45.1 g (b) 0.045061 kg (c) 4.506×10^5 μg (d) 0.09934 lb **119.** (a) 2.039 moles **121.** 7.09 moles **123.** 8.138×10^{24} **125.** 87.3 mL **127.** 0.0789 g **129.** 0.16 ton

CHAPTER 11

1. (a) gas (b) liquid (c) gas (d) gas **3.** a, c, and d **5.** (a) direct (b) potential (c) direct (d) all three **7.** (a) solid (b) liquid (c) solid (d) solid or liquid **9.** (a) The predominant cohesive forces in the solid hold the particles in essentially fixed position. (b) The gas particles are widely separated (disruptive forces). The solid and liquid particles have very little space between them (cohesive forces). The space between the particles can be decreased greatly in gases, but not in solids or liquids. (c) The cohesive forces are dominant enough that changing the temperature has only a small effect on the space between particles. (d) The disruptive forces in a gas are so dominant that each particle can act independently of the others. **11.** b and c **13.** b **15.** a and d **17.** (a) heat of solidification (b) heat of condensation (c) heat of fusion (d) heat of vaporization **19.** (a) 1.96×10^4 J (b) 1.13×10^5 J (c) 1.02×10^4 J (d) 1.13×10^5 J **21.** 6680 J **23.** 5.125 times **25.** 3.83 moles **27.** A is higher by 6 J/g **29.** 7.30×10^2 J **31.**

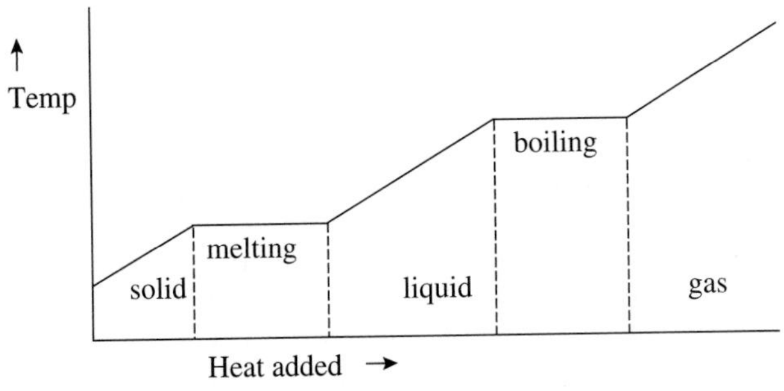

33. (a) 2.4×10^3 J (b) 3.60×10^4 J (c) 2.30×10^5 J (d) 2.33×10^5 J **35.** (a) 7.73×10^3 J (b) 2.36×10^5 J (c) 2.25×10^3 J (d) 8.30×10^4 J **37.** 2.14×10^4 J **39.** (a) boiling point (b) vapor pressure (c) boiling (d) boiling point **41.** (a) Increasing the temperature increases the average kinetic energy of the particles, enabling more molecules to evaporate. (b) The boiling point is lower; reactions occur more slowly at lower temperatures. (c) The boiling point is higher; reactions occur more rapidly at higher temperatures. (d) The particles leaving have higher than average kinetic energy. The particles remaining as liquids have a lower average kinetic energy and lower temperature. **43.** (a) increase (b) no change (c) increase (d) no change **45.** (a) no change (b) decrease (c) no change (d) no change **47.** (a) increase (b) no change (c) no change (d) no change **49.** B must have lower cohesive forces between particles. **51.** CS_2 **53.** Polar molecules must be present. **55.** The stronger the intermolecular forces, the higher the boiling point. **57.** (a) London forces (b) hydrogen bond and dipole-dipole (c) dipole-dipole (d) London forces **59.** b and c **61.** (a) Cl_2 (b) HF (c) NO (d) C_2H_6 **63.** (a) The high heat of fusion, heat of vaporization, and specific heat allow water to absorb (or liberate) a large amount of heat without much change in temperature. (b) If hydrogen bonding were not present in water, its boiling point would be lower than gaseous H_2S (c) Because water expands when it is cooled in the region from 4°C to 0°C, the coldest water is on the surface and ice floats in water. (d) The evaporation of perspiration cools the skin and the circulating blood. **65.** four **67.** (a) true (b) false (c) true (d) false **69.** (a) NaCl (b) SiO_2 (c) Cu (d) MgO **71.** (a) $SnCl_4$ (b) SnI_4 (c) SnI_4 (d) SnI_4 **73.** vaporizing **75.** 9.52 g **77.** 38.3°C **79.** 0.421 J/g · °C **81.** 1.4×10^3 J/g **83.** 0.775 g/mL **85.** 30.1 amu **87.** 9.45 kJ **89.** 2.37×10^3 J **91.** 2.9×10^7 gal/day

CHAPTER 12

1. (a) 4.71×10^3 mm Hg (b) 186 in. Hg (c) 91.0 psi (d) 471 cm Hg **3.** (a) smaller (b) equal (c) larger (d) equal **5.** 999 mm Hg **7.** (a) 1.28 L (b) 5.53 L (c) 4.00 L (d) 81.8 L **9.** (a) 2.6×10^2 mL (b) 16 mL (c) 6.8×10^2 mL (d) 2.9×10^3 mL **11.** 6.6 atm **13.** 1.94×10^3 mm Hg **15.** (a) 6.46 L (b) 4.51 L (c) 4.35 L (d) 20.5 L **17.** (a) 19.3 mL (b) 1.11×10^3 mL (c) 1.52×10^3 mL (d) 1.36×10^4 mL **19.** −125°C **21.** 97°C **23.** (a) 1.34 atm (b) 1.68 atm (c) 2.36 atm (d) 3.37 atm **25.** (a) 3.70 atm (b) 1.23 atm (c) 2.232 atm (d) 0.608 atm **27.** −125°C **29.** 3.1 atm **31.** (a) $T_2 = T_1 \cdot \dfrac{P_2V_2}{P_1V_1}$ (b) $\dfrac{V_2}{P_1} = \dfrac{V_1T_2}{P_2T_1}$ **33.** (a) 1.78 mL (b) 0.900 mL (c) 0.348 mL (d) 0.153 mL **35.** (a) 5.90 L (b) 3.70×10^3 mL (c) 2.11 atm (d) −171°C **37.** (a) 235 mm Hg (b) 727°C **39.** (a) 2.48×10^3°C

(b) −196.7°C (c) 183°C (d) 33°C **41.** (a) 3.15 L (b) 10.5 L (c) 3.64 L (d) 2.63 L **43.** (a) 0.953 L (b) 0.678 L (c) 1.89 L (d) 0.995 L **45.** the second **47.** (a) 0.433 L (b) 0.325 L **49.** 0.781 mole **51.** 1.5 g **53.** 1.33 L **55.** (a) the O_2 (b) the CH_4 (c) the N_2O (d) the CO **57.** NF_3 **59.** (a) 28.0 L (b) 28.0 L (c) 28.0 L (d) 28.0 L **61.** (a) NH_3 (b) O_2 (c) NO_2 (d) NO **63.** (a) 42.2 g (b) 46.6 g (c) 84.7 g (d) 36.0 g **65.** (a) 1.96 g/L (b) 4.11 g/L (c) 2.05 g/L (d) 1.34 g/L **67.** (a) O_3 (b) PH_3 (c) CO_2 (d) F_2 **69.** (a) 44.1 g/mole (b) 28.0 g/mole (c) 16.0 g/mole (d) 71.0 g/mole **71.** 2.51 L **73.** (a) 26.9 L (b) 22.1 L (c) 5.27 L (d) 24.5 L **75.** 0.900 mole **77.** −3°C **79.** (a) 1.2 atm (b) 4.89 atm (c) 0.044 atm (d) 0.175 atm **81.** $0.0824 \frac{\text{atm L}}{\text{mole K}}$ **83.** 36.2 L **85.** (a) 29.5 g (b) 3.19 g (c) 15.7 g (d) 0.363 g **87.** 45 g **89.** CO **91.** (a) 1.24 g/L (b) 1.68 g/L (c) 0.994 g/L (d) 0.829 g/L **93.** (a) 0.938 atm (b) 0.201 atm (c) 0.482 atm (d) 0.597 atm **95.** CO **97.** 18.7 L **99.** 60.9 g **101.** 12.2 L **103.** 260. L **105.** 78.1 L **107.** (a) 167 mm Hg (b) 354 mm Hg (c) 235 mm Hg (d) 86 mm Hg **109.** (a) mole fractions: CO = 0.407, CO_2 = 0.259, H_2S = 0.334 (b) partial pressures (atm): CO = 0.700, CO_2 = 0.445, H_2S = 0.574 **111.** (a) 4.8 atm (b) 4.8 atm (c) 4.8 atm (d) 0.15 atm **113.** (a) 0.60 atm (b) 0.40 atm (c) 0.400 atm (d) 0.400 atm **115.** 0.65 atm **117.** Partial pressures (mm Hg): N_2 = 213, H_2 = 639 **119.** (a) 726 mm Hg (b) 617 mm Hg (c) 722 mm Hg (d) 690. mm Hg **121.** (a) 0.909 (b) 72.7% (c) 18.2% (d) 9.09% **123.** (a) 25% O_2 (b) 37% Ar (c) 37% Ne (d) 0.25 atm **125.** 2.69×10^{22} molecules **127.** 3.3×10^{13} molecules **129.** 3.48 L NH_3 **131.** 0.15 g O_2 **133.** 43.8 mL **135.** 13.5 L **137.** 25% **139.** 50.0% **141.** (a) 1.14:1 (b) 1.14:1 **143.** 1200.0 mm Hg **145.** 1.33 atm **147.** 2.5 atm **149.** 1.24×10^3 mm Hg **151.** steam **153.** 75 amu **155.** 3 **157.** (a) mole fractions: HCl = 0.215, H_2 = 0.260, He = 0.525 (b) partial pressures (atm): HCl = 0.258, H_2 = 0.312, He = 0.630 (c) 10.47 amu (d) 0.467 g/L **159.** C_4H_{10} **161.** $C_2H_4Cl_2$ **163.** 120. L **165.** 1.06 g H_2O **167.** 0.00870 m^3 **169.** 2.0 atm **171.** 0.0147 cent/gram

CHAPTER 13

1. (a) water (b) ethyl alcohol **3.** (a) soluble (b) immiscible **5.** (a) unsaturated (b) saturated (c) saturated (d) unsaturated **7.** (a) soluble (b) very soluble (c) very soluble (d) slightly soluble **9.** (a) concentrated (b) concentrated (c) concentrated (d) dilute **11.** (a) very soluble (b) slightly soluble (c) slightly soluble (d) very soluble **13.** a, c, and d **15.** b **17.** (a) 7.03% (b) 1.81% (c) 19.4% (d) 31% **19.** (a) 0.700 g (b) 4.38 g (c) 135.5 g (d) 3.57 g **21.** (a) 5.18 g (b) 51.8 g (c) 56.9 g (d) 2.3 g **23.** (a) 9.50×10^2 g (b) 9.50×10^2 g (c) 9.50×10^2 g (d) 9.50×10^2 g **25.** (a) 79.1% (b) 38.1% (c) 20.9% (d) 5.14% **27.** (a) 36.06% (b) 66.72% **29.** 11.2 mL **31.** 1.40 gal **33.** (a) 2.67% (b) 1.20% (c) 3.32% (d) 6.6% **35.** (a) 750. mL (b) 33 mL **37.** (a) 11.4 g (b) 3.75×10^3 g **39.** 24% **41.** (a) 1.79 ppm (b) 170. ppm (c) 0.312 ppm (d) 2270 ppm **43.** (a) 1790 ppb (b) 17$\bar{0}$,000 ppb (c) 312 ppb (d) 2,270,000 ppb **45.** yes **47.** 0.0044 mL **49.** (a) 2.6×10^{-3} g (b) 5.4×10^{-6} g (c) 8.7 g **51.** (a) 4.0 M (b) 3.81 M (c) 1.06 M (d) 0.40 M **53.** (a) 13.2 g (b) 0.378 g (c) 2.36×10^4 g (d) 149 g **55.** (a) 33.8 mL (b) 226 mL (c) 1.80×10^3 mL (d) 1.52 mL **57.** (a) 0.877 L (b) 0.725 L (c) 0.386 L (d) 0.323 L **59.** 22.72 M **61.** 16.21% **63.** 3.75 M **65.** (a) 0.0357 *m* (b) 6.92 *m* (c) 0.00801 *m* (d) 0.452 *m* **67.** (a) 10.6 g (b) 4.76 g (c) 8.20 g (d) 8.70 g **69.** (a) 2.7×10^4 g (b) 6.0×10^3 g (c) 1.02×10^3 g (d) 4.9×10^2 g **71.** 0.585 *m* **73.** 0.763 M **75.** 2.58 *m* **77.** (a) 0.200 M (b) 0.120 M (c) 0.0267 M (d) 0.00375 M **79.** (a) 0.601 M (b) 0.638 M (c) 0.694 M (d) 2.27 M **81.** (a) 25 mL (b) 1100 mL (c) 270 mL (d) 0.050 mL **83.** (a) 380. mL (b) 30.0 mL (c) 3070 mL (d) 12.000 L **85.** (a) 3.3 M (b) 2.5 M (c) 8.86 M (d) 0.00909 M **87.** (a) 0.468 M (b) 6.00 M (c) 4.75 M (d) 5.99 M **89.** 1.00 L **91.** 17.3 g **93.** 610 mL **95.** 0.0632 M **97.** 1.7 L **99.** 0.0510 M **101.** (a) $(NH_4)_3PO_4$ (b) $Ca(OH)_2$ (c) $AgNO_3$ (d) CaS, $Ca(NO_3)_2$, $Ca(C_2H_3O_2)_2$ **103.** 80. g **105.** (a) 17.8% (b) 1.05 M **107.** 6.72 g **109.** 1245 g **111.** (a) 563 mL (b) 435 mL (c) 4540 mL (d) 896 mL **113.** (a) 0.245 M Al^{3+}, 0.735 M NO_3^- (b) 0.0735 M $Al(NO_3)_3$, 0.0735 M Al^{3+}, 0.221 M NO_3^- **115.** (a) 0.29 M (b) 0.027 M (c) 0.0866 M **117.** 3.74 mg/Kg **119.** 0.171 M **121.** 8.06 mL **123.** 340 g **125.** 298 g, 266 mL **127.** (a) 11.9% (m/v) (b) 12.2% (m/m) (c)15.1% (v/v) **129.** 1.01 g/mL **131.** (a) AgCl (b) $Ba_3(PO_4)_2$ (c) $PbSO_4$ (d) $BaSO_4$ and CuS **133.** 117 L **135.** 2.32 g **137.** 288 g **139.** 37.1% **141.** Zn **143.** 0.0833 M

CHAPTER 14

1. $HBr \rightarrow H^+ + Br^-$ (b) $HClO_2 \rightarrow H^+ + ClO_2^-$ (c) $LiOH \rightarrow Li^+ + OH^-$ (d) $Ba(OH)_2 \rightarrow Ba^{2+} + 2\,OH^-$ **3.** (a) acid (b) base (c) acid (d) base **5.** (a) $HBr + H_2O \rightarrow H_3O^+ + Br^-$ (b) $H_2O + N_3^- \rightarrow HN_3 + OH^-$ (c) $H_2S + H_2O \rightarrow H_3O^+ + HS^-$ (d) $HClO_4 + NO_2^- \rightarrow HNO_2 + ClO_4^-$ **7.** (a) HSO_3^- (b) HCN (c) HS^- (d) ClO^- **9.** (a) $H_2C_2O_4$ and $HC_2O_4^-$, HClO and ClO^- (b) HSO_4^- and SO_4^{2-}, H_3O^+ and H_2O (c) $H_2PO_4^-$ and HPO_4^{2-}, NH_4^+ and NH_3 (d) H_2CO_3 and HCO_3^-, H_2O and OH^- **11.** a, d **13.** (a) $HS^- + H_3O^+ \rightarrow H_2S + H_2O$; $HS^- + OH^- \rightarrow H_2O + S^{2-}$ (b) $HPO_4^{2-} + H_3O^+ \rightarrow H_2PO_4^- + H_2O$; $HPO_4^{2-} + OH^- \rightarrow H_2O + PO_4^{3-}$ (c) $HCO_3^- + H_3O^+ \rightarrow H_2CO_3 + H_2O$; $HCO_3^- + OH^- \rightarrow H_2O + CO_3^{2-}$ (d) $H_2PO_3^- + H_3O^+ \rightarrow H_3PO_3 + H_2O$; $H_2PO_3^- + OH^- \rightarrow H_2O + HPO_3^{2-}$ **15.** (a) monoprotic (b) triprotic (c) monoprotic (d) diprotic **17.** (a) $H_2C_4H_4O_4 + H_2O \rightarrow H_3O^+ +$

$HC_4H_4O_4^-$; $HC_4H_4O_4^- + H_2O \rightarrow H_3O^+ + C_4H_4O_4^{2-}$ (b) $H_3BO_3 + H_2O \rightarrow H_3O^+ + H_2BO_3^-$; $H_2BO_3^- + H_2O \rightarrow H_3O^+ + HBO_3^{2-}$; $HBO_3^{2-} + H_2O \rightarrow H_3O^+ + BO_3^{3-}$
19. to emphasize the acidic proton **21.** monoprotic; hydrogen atoms bonded to carbon are not acidic. **23.** (a) strong (b) weak (c) weak (d) weak **25.** d **27.** a, b **29.** (a) HNO_3 (b) $HClO_4$ (c) H_3PO_4 (d) HF **31.** (a) salt (b) base (c) salt (d) acid **33.** (a) Na^+ - sodium, PO_4^{3-} - phosphate (b) Li^+ - lithium, NO_3^- - nitrate (c) NH_4^+ - ammonium, Cl^- - chloride (d) K^+ - potassium, CN^- - cyanide **35.** Each salt is soluble. **37.** (a) $NaI \rightarrow Na^+ + I^-$ (b) $BaS \rightarrow Ba^{2+} + S^{2-}$ (c) $Li_2SO_4 \rightarrow 2\ Li^+ + SO_4^{2-}$ (d) $Al(NO_3)_3 \rightarrow Al^{3+} + 3\ NO_3^-$ **39.** (a) molecular (b) net ionic (c) ionic (d) net ionic **41.** (a) $Pb^{2+} + 2\ Br^- \rightarrow PbBr_2$ (b) $Fe^{3+} + 3\ OH^- \rightarrow Fe(OH)_3$ (c) $Zn + 2\ H^+ \rightarrow Zn^{2+} + H_2$ (d) $H_2S + 2\ OH^- \rightarrow S^{2-} + 2\ H_2O$ **43.** (a) $2\ Ag^+ + Pb \rightarrow 2\ Ag + Pb^{2+}$ (b) $Cl_2 + 2\ Br^- \rightarrow 2\ Cl^- + Br_2$ (c) $2\ Al^{3+} + 3\ S^{2-} \rightarrow Al_2S_3$ (d) no net reaction **45.** (a) yes (b) yes (c) no (d) yes **47.** (a) $Ni + 2\ HCl \rightarrow NiCl_2 + H_2$ (b) $Ca + 2\ H_2O \rightarrow Ca(OH)_2 + H_2$ (c) $Mg + 2\ HCl \rightarrow MgCl_2 + H_2$ (d) $Zn + 2\ H_2O \rightarrow Zn(OH)_2 + H_2$ **49.** (a) $H^+ + OH^- \rightarrow H_2O$ (b) $HC_2H_3O_2 + OH^- \rightarrow H_2O + C_2H_3O_2^-$ (c) $2H^+ + Mg(OH)_2 \rightarrow 2\ H_2O + Mg^{2+}$ (d) $H^+ + OH^- \rightarrow H_2O$
51. (a) H_3PO_4, NaOH (b) HCN, KOH (c) HCl, $Be(OH)_2$ (d) $HC_2H_3O_2$, $Ca(OH)_2$ **53.** (a) $Zn + 2\ HCl \rightarrow ZnCl_2 + H_2$ (b) $HCl + NaOH \rightarrow NaCl + H_2O$ (c) $2\ HCl + Na_2CO_3 \rightarrow 2\ NaCl + CO_2 + H_2O$ (d) $HCl + NaHCO_3 \rightarrow NaCl + CO_2 + H_2O$ **55.** (a) no (b) yes (c) no (d) yes **57.** (a) $Fe + Cu^{2+} \rightarrow Fe^{2+} + Cu$ (b) $Sn + 2\ Ag^+ \rightarrow Sn^{2+} + 2\ Ag$ (c) $Zn + Ni^{2+} \rightarrow Zn^{2+} + Ni$ (d) $Cr + Pb^{2+} \rightarrow Cr^{2+} + Pb$
59. (a) insoluble salt formed (b) insoluble salt formed (c) insoluble salt formed (d) gas evolved **61.** (a) $2\ Al^{3+} + 3\ S^{2-} \rightarrow Al_2S_3$ (b) $H^+ + OH^- \rightarrow H_2O$ (c) no reaction (d) no reaction
63. (a) 5.0×10^{-11} M (b) 1.4×10^{-8} M (c) 3.3×10^{-5} M (d) 4.0×10^{-7} M **65.** (a) basic (b) basic (c) acidic (d) acidic **67.** (a) 3.7×10^{-11} M (b) 1.3×10^{-6} M (c) 1.0×10^{-7} M (d) 2.0×10^{-7} M **69.** (a) acidic (b) basic (c) neutral (d) basic **71.** (a) 4.0 (b) 9.0 (c) 5.0 (d) 9.0 **73.** (a) 1.4 (b) 3.2 (c) 9.1 (d) 6.3 **75.** (a) 2.5 (b) 2.52 (c) 2.523 (d) 2.5229 **77.** a **79.** (a) 1×10^{-3} M (b) 1×10^{-5} M (c) 2×10^{-6} M (d) 5×10^{-7} M
81. (a) 3.7×10^{-3} M (b) 3.7×10^{-4} M (c) 3.7×10^{-8} M (d) 3.5×10^{-8} M **83.** (a) solution A (b) solution B
85. (a) 4.199 (b) 3.900 (c) 3.500 (d) 1.500 **87.** (a) 2.20 (b) 0.70 (c) 4.38 (d) 10.37 **89.** (a) PO_4^{3-} (b) CN^- (c) NH_4^+ (d) none **91.** (a) neutral (b) basic (c) acidic (d) neutral **93.** (a) $NH_4^+ + H_2O \rightarrow H_3O^+ + NH_3$ (b) $C_2H_3O_2^- + H_2O \rightarrow HC_2H_3O_2 + OH^-$ (c) $F^- + H_2O \rightarrow HF + OH^-$ (d) $CN^- + H_2O \rightarrow HCN + OH^-$ **95.** (a) no (b) no (c) yes (d) yes **97.** (a) $F^- + H_3O^+ \rightarrow HF + H_2O$ (b) $H_2CO_3 + OH^- \rightarrow HCO_3^- + H_2O$ (c) $NH_3 + H_3O^+ \rightarrow NH_4^+ + H_2O$ (d) $H_3PO_4 + OH^- \rightarrow H_2PO_4^- + H_2O$
99. (a) 2.50 (b) 5.00 (c) 1.50 (d) 0.590 **101.** (a) 0.500 N (b) 0.408 N (c) 5.00 N (d) 0.200 N **103.** (a) 0.390 N (b) 2.30 N (c) 0.090 N (d) 1.00 N **105.** (a) 3.00 M (b) 2.06 M (c) 1.00 M (d) 1.00 M **107.** 0.900 g
109. (a) 35.0 mL (b) 750. mL (c) 25.0 mL (d) 0.22 mL
111. (a) 8.00 L (b) 96.0 mL (c) 270. mL (d) 4.00 mL
113. (a) 0.0705 M (b) 0.176 M (c) 0.0587 M (d) 0.352 M
115. (a) 3 M (b) 6 M (c) 16 M (d) 6 M **117.** (a) strong (b) weak (c) weak (d) strong **119.** 200 times **121.** c, d
123. 1.5 **125.** $Ba(OH)_2$, LiCN, K_2SO_4, NH_4Br, $HClO_4$
127. $HC_2H_3O_2/C_2H_3O_2^-$ and HCO_3^-/CO_3^{2-}
129. (a) 0.348 N (b) 0.00696 equiv. (c) 0.00232 mole (d) 0.116 M **131.** (a) I^-, iodide ion (b) HPO_4^{2-}, hydrogen phosphate ion (c) OH^-, hydroxide ion (d) H_3O^+, hydronium ion **133.** H : C: : : N : and [: C: : : N : $]^-$
135. (a) 0.48 (b) 14.364 (c) 0.22 (d) 0.46
137. (a) 4.0×10^{-4} M (b) 0.0013% **139.** 2.3×10^{16} ions
141. 1.82×10^{23} ions **143.** 65.9%

CHAPTER 15

1. (a) oxidation occurs when an atom loses electrons (b) oxidation occurs when the oxidation number of an atom increases
3. (a) an oxidizing agent gains electrons from another substance (b) an oxidizing agent contains the atom that shows an oxidization number decrease. (c) an oxidizing agent is itself reduced. **5.** (a) oxidized (b) decrease (c) reducing agent (d) loses **7.** (a) N = −3, H = +1 (b) H = +1, S = +4, O = −2 (c) H = +1, N = +3, O = −2 (d) Na = +1, P = +5, O = −2 **9.** (a) P = −3 (b) Mg = +2 (c) N = −3, H = +1 (d) P = +5, O = −2 **11.** (a) +4 (b) +3 (c) −2 (d) $-2\frac{2}{3}$ **13.** (a) +2 (b) +2 (c) +3 (d) +2
15. (a) −2 (b) +2 (c) −1 (d) −2 **17.** (a) −1 (b) +1 (c) +1 (d) −1 **19.** (a) H_2 oxidized, N_2 reduced (b) I^- oxidized, Cl_2 reduced (c) Fe oxidized, Sb in SbO_2 reduced (d) S in H_2SO_3 oxidized, N in HNO_3 reduced **21.** (a) H_2 is reducing agent, N_2 is oxidizing agent (b) I^- is reducing agent, Cl_2 is oxidizing agent (c) Fe is reducing agent, SbO_2 is oxidizing agent (d) H_2SO_3 is reducing agent, HNO_3 is oxidizing agent
23. (a) S in SO_2 (b) HNO_3 (c) HNO_3 (d) SO_2
25. (a) redox, synthesis (b) redox, single replacement (c) non-redox, decomposition (d) non-redox, double replacement
27. (a) redox (b) redox (c) can't classify (d) redox
29. (a) $2\ Cr + 6\ HCl \rightarrow 2\ CrCl_3 + H_2$ (b) $2\ Cr_2O_3 + 3\ C \rightarrow 4\ Cr + 3\ CO_2$ (c) $SO_2 + NO_2 \rightarrow SO_3 + NO$ (d) $BaSO_4 + 4\ C \rightarrow BaS + 4\ CO$ **31.** (a) $Br_2 + 2\ H_2O + SO_2 \rightarrow 2\ HBr + H_2SO_4$ (b) $3\ H_2S + 2\ HNO_3 \rightarrow 3\ S + 2\ NO + 4\ H_2O$ (c) $SnSO_4 + 2\ FeSO_4 \rightarrow Sn + Fe_2(SO_4)_3$ (d) $Na_2TeO_3 + 4\ NaI + 6\ HCl \rightarrow 6\ NaCl + Te + 3\ H_2O + 2\ I_2$ **33.** (a) $I_2 + 5\ Cl_2 + 6\ H_2O \rightarrow 2\ HIO_3 + 10\ Cl^- + 10\ H^+$ (b) $8\ MnO_4^- + 5\ AsH_3 + 24\ H^+ \rightarrow 5\ H_3AsO_4 + 8\ Mn^{2+} + 12\ H_2O$ (c) $2\ Br^- + SO_4^{2-} + 4\ H^+ \rightarrow Br_2 + SO_2 + 2\ H_2O$ (d) $Au + 4\ Cl^- + 3\ NO_3^- + 6\ H^+ \rightarrow AuCl_4^- + 3\ NO_2 + 3\ H_2O$ **35.** (a) $8\ OH^- + S^{2-} + 4\ Cl_2 \rightarrow SO_4^{2-} + 8\ Cl^- + 4\ H_2O$ (b) $5\ H_2O + 3\ SO_3^{2-} + 2\ CrO_4^{2-} \rightarrow 2\ Cr(OH)_4^- + 3\ SO_4^{2-} + 2\ OH^-$ (c) $H_2O + 2\ MnO_4^- + 3\ IO_3^- \rightarrow 2\ MnO_2 + 3\ IO_4^- + 2\ OH^-$ (d) $18\ OH^- + I_2 + 7\ Cl_2 \rightarrow 2\ H_3IO_6^{2-} + 14\ Cl^- + 6\ H_2O$ **37.** (a) MnO_2 +

$4\ H^+ + e^- \rightarrow Mn^{3+} + 2\ H_2O$ (b) $H_3MnO_4 + 5\ H^+ + 5\ e^- \rightarrow Mn + 4\ H_2O$ (c) $MnO_4^- + 8\ H^+ + 5\ e^- \rightarrow Mn^{2+} + 4\ H_2O$ (d) $MnO_4^- + 4\ H^+ + 3\ e^- \rightarrow MnO_2 + 2\ H_2O$
39. (a) $SeO_4^{2-} + 4\ H_2O + 6\ e^- \rightarrow Se + 8\ OH^-$ (b) $Se^{2-} + 6\ OH^- \rightarrow SeO_3^{2-} + 6\ e^- + 3\ H_2O$ (c) $SeO_4^{2-} + H_2O + 2\ e^- \rightarrow SeO_3^{2-} + 2\ OH^-$ (d) $Se + 6\ OH^- \rightarrow SeO_3^{2-} + 4\ e^- + 3\ H_2O$ **41.** (a) $Zn + Cu^{2+} \rightarrow Zn^{2+} + Cu$ (b) $Br_2 + 2\ I^- \rightarrow I_2 + 2\ Br^-$ (c) $S_2O_3^{2-} + 4\ Cl_2 + 5\ H_2O \rightarrow 2\ SO_4^{2-} + 10\ H^+ + 8\ Cl^-$ (d) $6\ Zn + As_2O_3 + 12\ H^+ \rightarrow 6\ Zn^{2+} + 2\ AsH_3 + 3\ H_2O$ **43.** (a) $I_2 + 5\ Cl_2 + 6\ H_2O \rightarrow 2\ HIO_3 + 10\ Cl^- + 10\ H^+$ (b) $8\ MnO_4^- + 5\ AsH_3 + 24\ H^+ \rightarrow 5\ H_3AsO_4 + 8\ Mn^{2+} + 12\ H_2O$ (c) $2\ Br^- + SO_4^{2-} + 4\ H^+ \rightarrow Br_2 + SO_2 + 2\ H_2O$ (d) $Au + 4\ Cl^- + 3\ NO_3^- + 6\ H^+ \rightarrow AuCl_4^- + 3\ NO_2 + 3\ H_2O$ **45.** (a) $2\ NH_3 + ClO^- \rightarrow N_2H_4 + Cl^- + H_2O$ (b) $Cr(OH)_2 + 2\ OH^- + 2\ BrO^- \rightarrow CrO_4^{2-} + 2\ H_2O + 2\ Br^-$ (c) $2\ CrO_2^- + 2\ OH^- + 3\ H_2O_2 \rightarrow 2\ CrO_4^{2-} + 4\ H_2O$ (d) $3\ Sn(OH)_3^- + 3\ OH^- + 2\ Bi(OH)_3 \rightarrow 3\ Sn(OH)_6^{2-} + 2\ Bi$ **47.** (a) $8\ OH^- + S^{2-} + 4\ Cl_2 \rightarrow SO_4^{2-} + 8\ Cl^- + 4\ H_2O$ (b) $5\ H_2O + 3\ SO_3^{2-} + 2\ CrO_4^{2-} \rightarrow 2\ Cr(OH)_4^- + 3\ SO_4^{2-} + 2\ OH^-$ (c) $H_2O + 2\ MnO_4^- + 3\ IO_3^- \rightarrow 2\ MnO_2 + 3\ IO_4^- + 2\ OH^-$ (d) $18\ OH^- + I_2 + 7\ Cl_2 \rightarrow 2\ H_3IO_6^{2-} + 14\ Cl^- + 6\ H_2O$ **49.** (a) $3\ HNO_2 \rightarrow 2\ NO + NO_3^- + H_2O + H^+$ (b) $Cl^- + ClO^- + 2\ H^+ \rightarrow Cl_2 + H_2O$ (c) $3\ S + 6\ OH^- \rightarrow 2\ S^{2-} + SO_3^{2-} + 3\ H_2O$ (d) $3\ Br_2 + 6\ OH^- \rightarrow BrO_3^- + 5\ Br^- + 3\ H_2O$
51. (a) $3\ HNO_2 \rightarrow 2\ NO + NO_3^- + H_2O + H^+$ (b) $Cl^- + ClO^- + 2\ H^+ \rightarrow Cl_2 + H_2O$ (c) $3\ S + 6\ OH^- \rightarrow 2\ S^{2-} + SO_3^{2-} + 3\ H_2O$ (d) $3\ Br_2 + 6\ OH^- \rightarrow BrO_3^- + 5\ Br^- + 3\ H_2O$ **53.** (a) $Pb \rightarrow Pb^{2+} + 2\ e^-$ (b) $Cu^{2+} + 2\ e^- \rightarrow Cu$ (c) anode is Pb; cathode is Cu (d) from anode to cathode
55. anode: $Zn \rightarrow Zn^{2+} + 2\ e^-$; cathode: $2\ MnO_2 + 2\ NH_4^+ + 2\ e^- \rightarrow Mn_2O_3 + 2\ NH_3 + H_2O$ **57.** anode: $Pb + SO_4^{2-} \rightarrow PbSO_4 + 2\ e^-$ cathode: $PbO_2 + 4\ H^+ + SO_4^{2-} + 2\ e^- \rightarrow PbSO_4 + 2\ H_2O$ **59.** The H_2SO_4 in solution is used up and H_2O is produced, lowering the density of the solution. **61.** $2\ Cl^- \rightarrow Cl_2 + 2\ e^-$ **63.** N_2O, NO, N_2O_3, NO_2, N_2O_5 **65.** b, c, d **67.** (a) both reductions (b) reduction and oxidation (c) both oxidations (d) oxidation and reduction **69.** (1) $5\ PH_3 + 4\ MnO_4^- + 12\ H^+ \rightarrow 5\ H_3PO_2 + 4\ Mn^{2+} + 6\ H_2O$ (2) $PH_3 + 2\ SO_4^{2-} + 4\ H^+ \rightarrow H_3PO_2 + 2\ SO_2 + 2\ H_2O$ (3) $5\ As + 3\ MnO_4^- + 3\ H_2O + 9\ H^+ \rightarrow 5\ H_3AsO_3 + 3\ Mn^{2+}$ (4) $2\ As + 3\ SO_4^{2-} + 6\ H^+ \rightarrow 2\ H_3AsO_3 + 3\ SO_2$ **71.** $Zn \rightarrow Zn^{2+} + 2\ e^-$; $NO_3^- + 10\ H^+ + 8\ e^- \rightarrow NH_4^+ + 3\ H_2O$ **73.** (a) redox, HNO_3 is oxidizing agent (b) acid–base, H_2S is acid (c) redox, H_2O_2 is oxidizing agent (d) acid–base, H_2SO_4 is acid **75.** (a) $Sn^{2+} + 2\ Fe^{2+} \rightarrow Sn + 2\ Fe^{3+}$ (b) $PH_3 + 2\ NO_2 \rightarrow H_3PO_4 + N_2$ (c) $S + 3\ H_2O + 2\ Pb^{2+} \rightarrow 2\ Pb + H_2SO_3 + 4\ H^+$ (d) $4\ Zn + 10\ H^+ + NO_3^- \rightarrow 4\ Zn^{2+} + NH_4^+ + 3\ H_2O$
77. (a) $8\ H^+ + 3\ H_2S + Cr_2O_7^{2-} \rightarrow 2\ Cr^{3+} + 3\ S + 7\ H_2O$ (b) $3\ H_2O + 5\ ClO_3^- + 3\ I_2 \rightarrow 6\ IO_3^- + 5\ Cl^- + 6\ H^+$ (c) $8\ OH^- + S^{2-} + 4\ Br_2 \rightarrow SO_4^{2-} + 8\ Br^- + 4\ H_2O$
79. 0.324 M **81.** 30.8 ppm O_3

CHAPTER 16

1. The solute molecules have more motion, allowing more frequent collisions with other reactant molecules throughout the solution. Only those molecules on the surface of a solid can collide with other reactant molecules. **3.** The reaction with the lower activation energy, 45 kJ/mole will have the faster rate. At any given temperature, there will be a greater fraction of collisions with a combined energy equal to or exceeding the lower activation energy. **5.** The activation energy required and the orientation.

7.

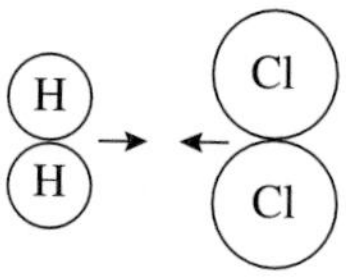

9.

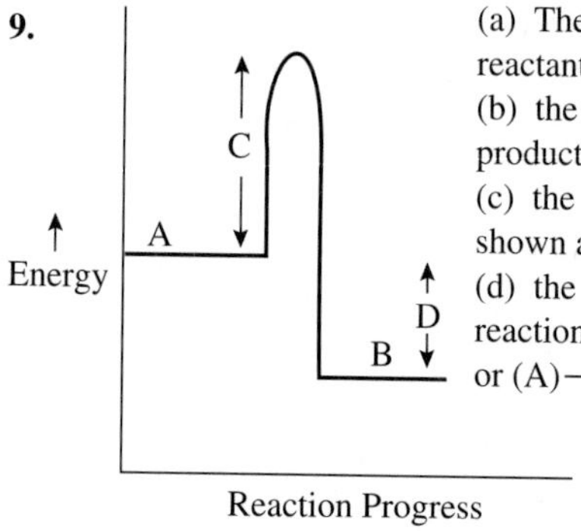

(a) The average energy of the reactants is shown as (A); (b) the average energy of the products is shown as (B); (c) the activation energy is shown as (C); (d) the energy liberated in the reaction is shown as (D), or (A)−(B).

11. Similarities: Both reactions are exothermic to the same extent and the energy difference between the reactants and products is the same. Difference: The activation energy is lower for the reaction occurring at room temperature.

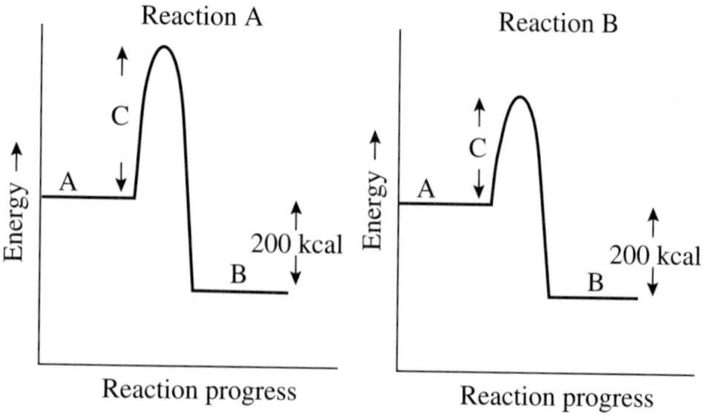

13. At higher temperatures, the faster moving molecules collide more frequently, and a greater fraction of the collisions will have a total energy greater than the activation energy.
15. Increasing the surface area increases the number of collisions on the surface. More collisions result in a higher rate.
17. The coal dust has much more surface exposed than charcoal. Thus the reaction is much faster. **19.** (a) #1, (b) #3,

(c) #4, (d) #3, **21.** The difference is in the magnitude of the activation energy.

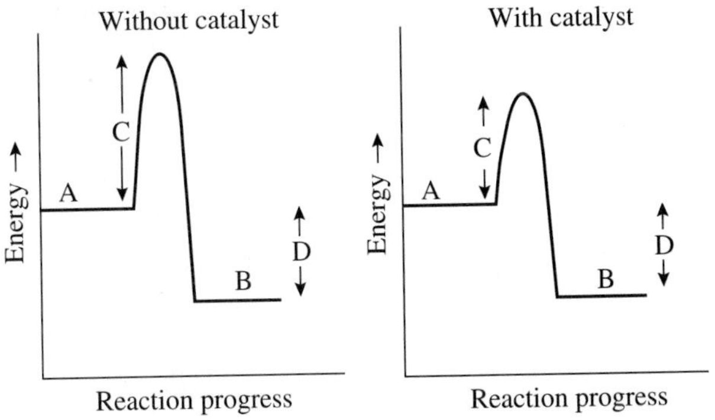

23. The two rates are equal. **25.** A physical equilibrium involves a physical change; a chemical equilibrium involves a chemical change. **27.** SO_3 = 0.194 mole, SO_2 = 0.0058 mole, O_2 = 0.0029 mole **29.** NH_3 = 0.268 mole, N_2 = 0.184 mole, H_2 = 0.137 mole

31. (a) $K = \dfrac{[SO_2Cl_2]}{[SO_2][Cl_2]}$ (b) $K = \dfrac{[N_2][O_2]^2}{[NO_2]^2}$

(c) $K = \dfrac{[CS_2][O_2]^4}{[SO_3]^2[CO_2]}$ (d) $K = \dfrac{[CH_4][H_2S]^2}{[H_2]^4[CS_2]}$

33. (a) $K = [NO_2]^4[O_2]$ (b) $K = [O_2]^3$ (c) $K = 1/[Cl_2]$ (d) $K = [Cl_2]$ **35.** (a) 6.67 (b) 1.04 (c) 3.75 (d) 0.0216 **37.** 0.00730 M **39.** 48.9 **41.** (a) left (b) far right (c) right (d) neither **43.** (a) to right (b) to left (c) to left (d) to right **45.** (a) to right (b) to left (c) to left (d) to right **47.** (a) to right (b) to left (c) no effect (d) to right **49.** c **51.** (a) $N_2 + 3\ H_2 \rightleftarrows 2\ NH_3$ (b) $4\ NH_3 + 3\ O_2 \rightleftarrows 2\ N_2 + 6\ H_2O$ (c) $2\ NO \rightleftarrows N_2 + O_2$ (d) $N_2 + O_2 \rightleftarrows 2\ NO$ **53.** (a) at equilibrium (b) shift to left (c) at equilibrium (d) shift to left **55.** 0.0005 **57.** (a) no effect (b) no effect (c) change value of equilibrium constant (d) no effect **59.** (a) right (b) no effect (c) right (d) right

61. (a) $K = \dfrac{[SO_2]^2[O_2]}{[SO_3]^2}$ (b) $K = \dfrac{[NH_3]^2[H_2O]^4}{[H_2]^7[NO_2]^2}$

(c) $K = \dfrac{[CO_2]}{[CO]}$ (d) $K = [CO_2]$ **63.** 38.8 **65.** 0.0244 **67.** 2.20 atm **69.** 0.119

CHAPTER 17

1. A radioactive nuclide has an unstable nucleus; a nonradioactive nuclide has a stable nucleus **3.** (a) ${}^{9}_{5}B$ or boron-9 (b) ${}^{44}_{19}K$ or potassium-44 (c) ${}^{96}_{45}Rh$ or rhodium-96 (d) ${}^{182}_{73}Ta$ or tantalum-182 **5.** (a) ${}^{4}_{2}\alpha$ (b) ${}^{0}_{-1}\beta$ (c) ${}^{0}_{0}\gamma$ **7.** 2 protons and 2 neutrons **9.** (a) ${}^{192}_{78}Pt \rightarrow {}^{4}_{2}\alpha + {}^{188}_{76}Os$ (b) ${}^{217}_{86}Rn \rightarrow {}^{4}_{2}\alpha + {}^{213}_{84}Po$ (c) ${}^{212}_{85}As \rightarrow {}^{4}_{2}\alpha + {}^{208}_{83}Bi$ (d) ${}^{244}_{96}Cm \rightarrow {}^{4}_{2}\alpha + {}^{240}_{94}Pu$ **11.** (a) ${}^{48}_{21}Sc \rightarrow {}^{0}_{-1}\beta + {}^{48}_{22}Ti$ (b) ${}^{117}_{47}Ag \rightarrow {}^{0}_{-1}\beta + {}^{117}_{48}Cd$ (c) ${}^{92}_{36}Kr \rightarrow {}^{0}_{-1}\beta + {}^{92}_{37}Rb$ (d) ${}^{138}_{55}Cs \rightarrow {}^{0}_{-1}\beta + {}^{138}_{56}Ba$ **13.** The atomic number decreases by 2, and the mass number decreases by 4. **15.** (a) ${}^{0}_{-1}\beta$ (b) ${}^{125}_{52}Te$ (c) ${}^{4}_{2}\alpha$ (d) ${}^{229}_{90}Th$ **17.** (a) ${}^{199}_{79}Au \rightarrow {}^{0}_{-1}\beta + {}^{199}_{80}Hg$ (b) ${}^{120}_{48}Cd \rightarrow {}^{0}_{-1}\beta + {}^{120}_{49}In$ (c) ${}^{152}_{67}Ho \rightarrow {}^{4}_{2}\alpha + {}^{148}_{65}Tb$ (d) ${}^{226}_{88}Ra \rightarrow {}^{4}_{2}\alpha + {}^{222}_{86}Rn$ **19.** (a) 1/32 (b) 1/16 (c) 1/8 (d) 1/256 **21.** (a) 6.0 yr (b) 2.4 yr (c) 1.5 yr (d) 1.2 yr **23.** (a) 2.0 g (b) 0.50 g (c) 0.062 g (d) 0.0078 g **25.** (a) 7.50 g (b) 9.84 g (c) 9.98 g (d) 10.0 g **27.** (a) 16 hr (b) 32 hr (c) 40. hr (d) 56 hr **29.** (a) ${}^{4}_{2}\alpha$ (b) ${}^{2}_{1}H$ (c) ${}^{81}_{34}Se$ (d) ${}^{9}_{4}Be$ **31.** (a) ${}^{9}_{4}Be + {}^{4}_{2}\alpha \rightarrow {}^{1}_{0}n + {}^{12}_{6}C$ (b) ${}^{58}_{28}Ni + {}^{1}_{1}H \rightarrow {}^{4}_{2}\alpha + {}^{55}_{27}Co$ (c) ${}^{113}_{48}Cd + {}^{1}_{0}n \rightarrow {}^{114}_{48}Cd + {}^{0}_{0}\gamma$ (d) ${}^{27}_{13}Al + {}^{4}_{2}\alpha \rightarrow {}^{30}_{15}P + {}^{1}_{0}n$ **33.** (a) ${}^{242}_{96}Cm$ (b) ${}^{238}_{92}U$ (c) ${}^{252}_{98}Cf$ (d) ${}^{209}_{83}Bi$ **35.** nine **37.** 1600 **39.** the atomic number decreases by 1 and the mass number stays the same **41.** beta decay, ${}^{0}_{-1}\beta$ **43.** (a) ${}^{29}_{15}P \rightarrow {}^{0}_{1}\beta + {}^{29}_{14}Si$ (b) ${}^{112}_{51}Sb \rightarrow {}^{0}_{1}\beta + {}^{112}_{50}Sn$ (c) ${}^{46}_{23}V \rightarrow {}^{0}_{1}\beta + {}^{46}_{22}Ti$ (d) ${}^{132}_{58}Ce \rightarrow {}^{0}_{1}\beta + {}^{132}_{57}La$ **45.** (a) ${}^{76}_{36}Kr + {}^{0}_{-1}e \rightarrow {}^{76}_{35}Br$ (b) ${}^{122}_{54}Xe + {}^{0}_{-1}e \rightarrow {}^{122}_{53}I$ (c) ${}^{100}_{46}Pd + {}^{0}_{-1}e \rightarrow {}^{100}_{45}Rh$ (d) ${}^{175}_{73}Ta + {}^{0}_{-1}e \rightarrow {}^{175}_{72}Hf$ **47.** (a) ${}^{0}_{1}\beta$ (b) ${}^{0}_{-1}e$ (c) ${}^{103}_{47}Ag$ (d) ${}^{133}_{55}Cs$ **49.** ${}^{63}_{30}Zn + {}^{0}_{-1}e \rightarrow {}^{63}_{29}Cu$; ${}^{63}_{30}Zn \rightarrow {}^{0}_{1}\beta + {}^{63}_{29}Cu$ **51.** beta emission **53.** (a) nickel-65: 1.32142857; copper-65: 1.24137931 (b) platinum-192: 1.46153846; osmium-188: 1.47368421 (c) thulium-165: 1.39130435; erbium-165: 1.42647059 (d) indium-107: 1.18367347; cadmium-107: 1.22916667 **55.** (a) krypton-74, positron; krypton-87, beta (b) arsenic-68, positron; selenium-84, beta (c) gallium-74, beta; gallium-64, positron (d) niobium-99, beta; palladium-99, positron **57.** stable **59.** ${}^{220}_{86}Rn \rightarrow {}^{4}_{2}\alpha + {}^{216}_{84}Po$; ${}^{216}_{84}Po \rightarrow {}^{4}_{2}\alpha + {}^{212}_{82}Pb$; ${}^{212}_{82}Pb \rightarrow {}^{0}_{-1}\beta + {}^{212}_{83}Bi$; ${}^{212}_{83}Bi \rightarrow {}^{0}_{-1}\beta + {}^{212}_{84}Po$ **61.** An ion pair is an electron and a positive ion produced by the interaction of ionizing radiation with an atom or ion. **63.** A thick sheet of paper stops only the alpha particles. **65.** α: one tenth of the speed of light; β: nine tenths of the speed of light; γ: The speed of light **67.** The α and β particles do not penetrate through the body enough to be detected. **69.** (a) implant cancer therapy (b) diagnostic for intercellular space (c) external beam cancer therapy (d) diagnostic for red blood cell lifetime **71.** (a) fusion (b) both (c) both (d) fusion **73.** (a) fusion (b) neither (c) neither (d) fission **75.** (a) ${}^{228}_{88}Ra$, ${}^{228}_{89}Ac$ (b) ${}^{228}_{89}Ac$, ${}^{228}_{90}Th$, ${}^{220}_{86}Rn$ **77.** six α and four β **79.** ${}^{28}P$ will decay by positron emission or electron capture. ${}^{34}P$ will decay by beta emission. **81.** A = 0; B = 0.250 mole; C = 0; D = 0.750 mole **83.** 6.0 hr **5.** (a) 7.0 days (b) 3.80 g Q **87.** (a) isobars (b) neither (c) isotopes (d) isotopes **89.** 19 days **91.** (a) UF_6 (b) UF_6 **93.** 0.364 L **95.** 3.08×10^6 kg C

CHAPTER 18

1. (a) 8 (b) 18 (c) 24 (d) 38

3. (a) 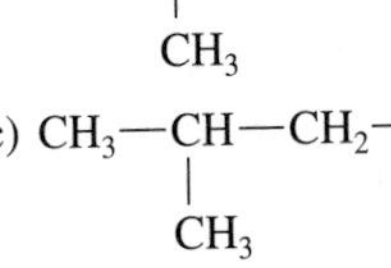

(b) 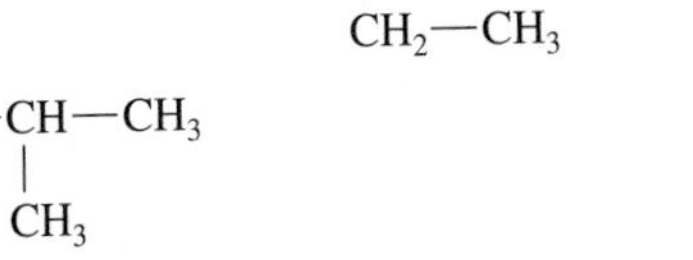

(c)

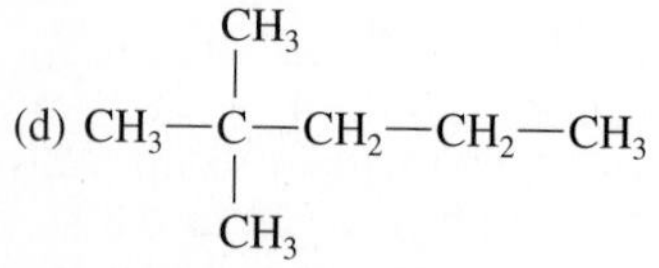

5. (a) 7 (b) 7 (c) 7 (d) 7 **7.** (a) hexane (b) 2,4-dimethylpentane (c) 3,3-dimethylpentane (d) 3-ethylhexane **9.** (a) normal (b) branched (c) branched (d) branched **11.** (a) none (b) 2 (c) 2 (d) 1 **13.** (a) $CH_3—CH_2—CH_2—CH_2—CH_3$

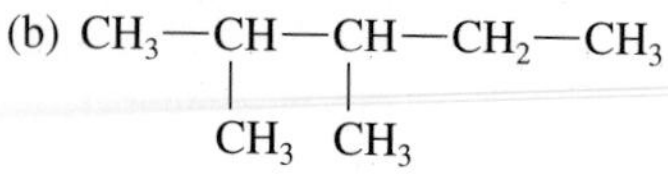

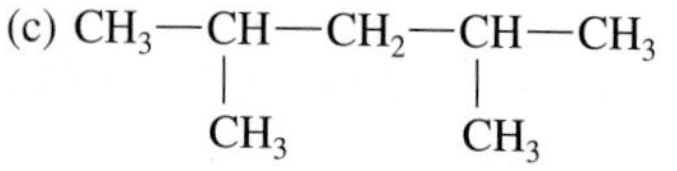

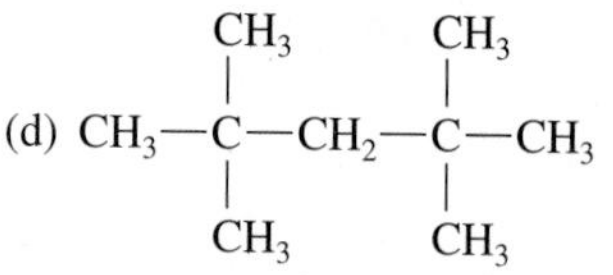

15.

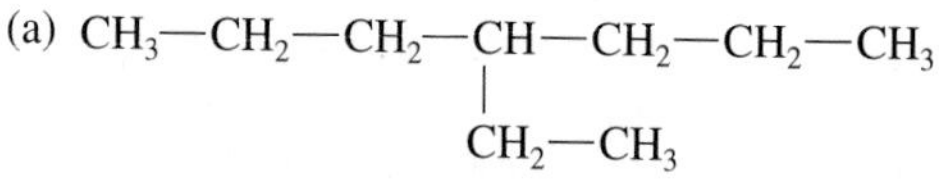

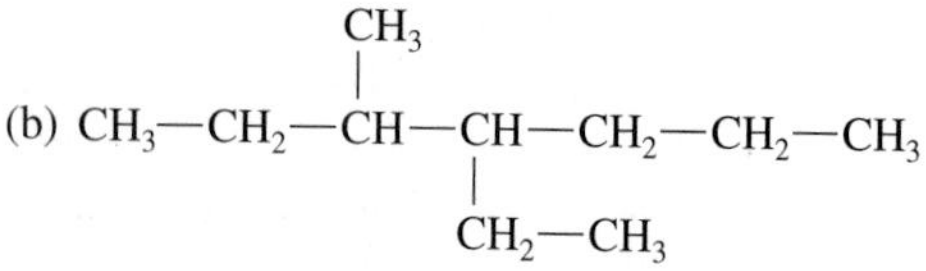

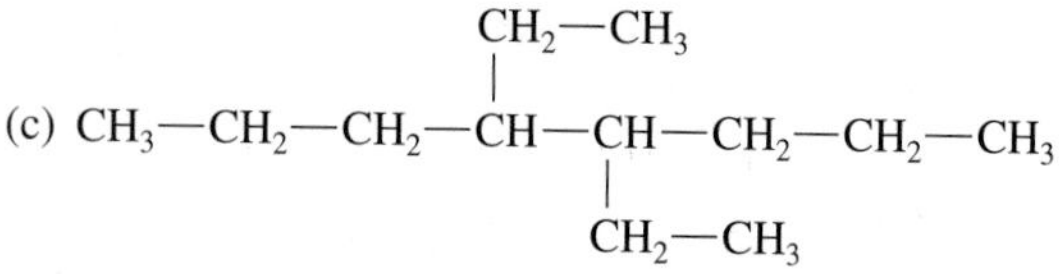

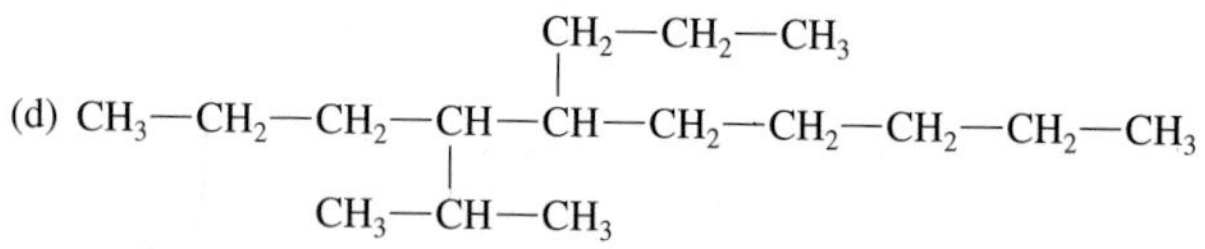

17.

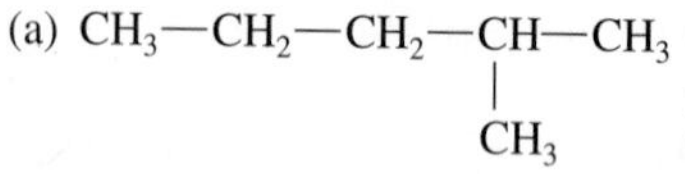

Chain was numbered from the wrong end. The correct name is 2-methylpentane.

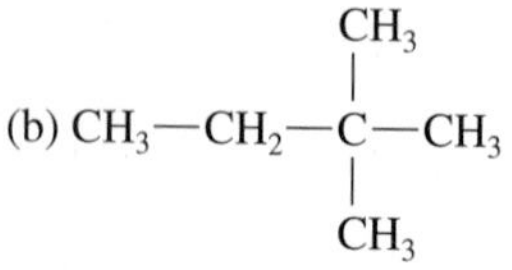

Chain was numbered from the wrong end. The correct name is 2,2-dimethylbutane.

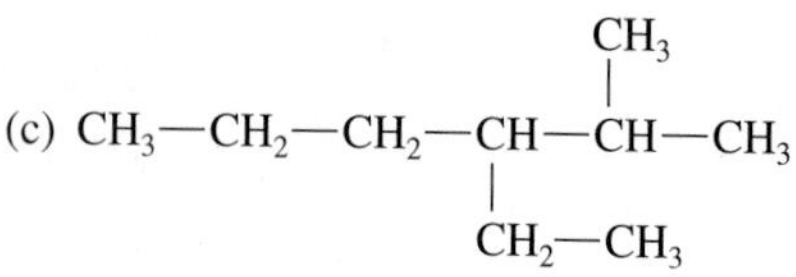

Chain was numbered from the wrong end. The correct name is 3-ethyl-2-methylhexane.

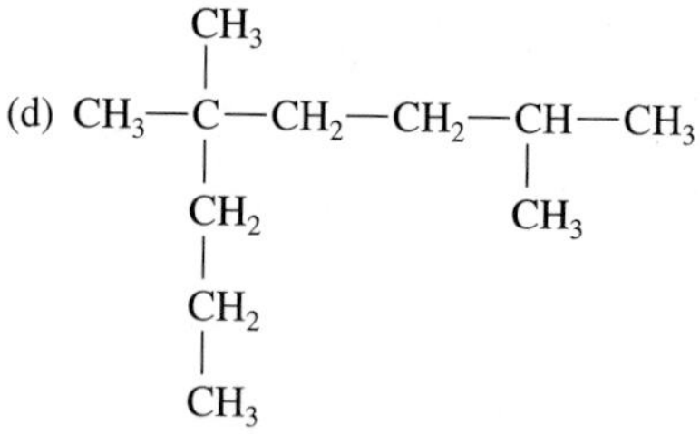

The longest carbon chain was not used as the base for the name. The correct name is 2,5,5-trimethyloctane.

19. (1) C—C—C—C—C—C Hexane

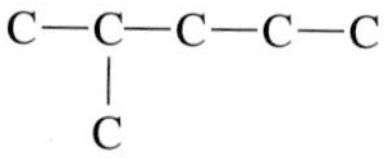

2-methylpentane

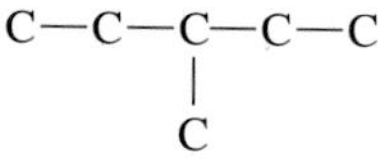

3-methylpentane

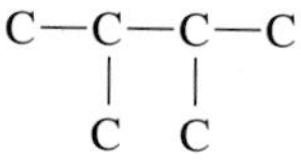

2,3-dimethylbutane

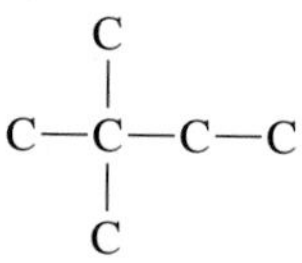

2,2-dimethylbutane

21. (a) yes (b) no (c) no (d) yes **23.** (a) 8 (b) 16 (c) 12 (d) 14 **25.** (a) cyclobutane (b) 1,2-dimethylcyclohexane (c) methylcyclopentane (d) 1-ethyl-2-methylcyclobutane

27.

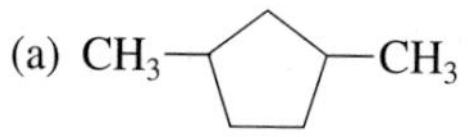

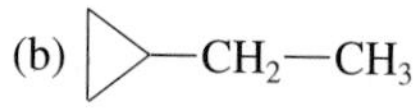

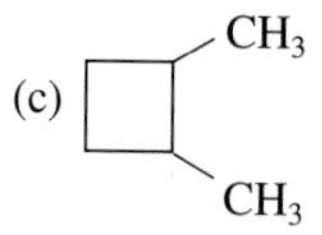

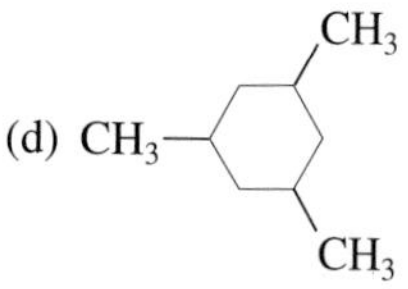

29.

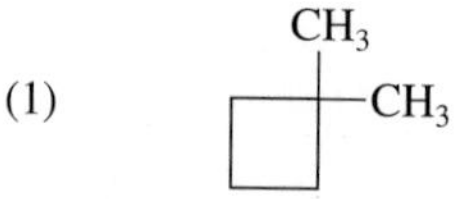

1,1-dimethylcyclobutane

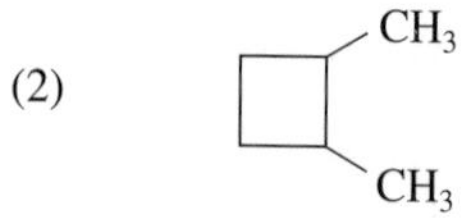

1,2-dimethylcyclobutane

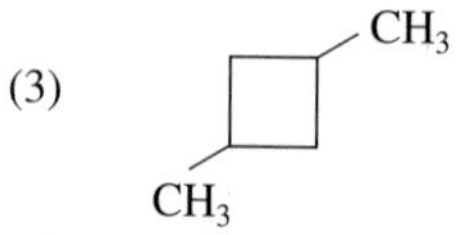

1,3-dimethylcyclobutane

31. (a) unsaturated (b) unsaturated (c) saturated (d) unsaturated **33.** (a) 2-butene (b) 2-butyne (c) 3-methylcyclohexene (d) 3,5-dimethylcyclohexene **35.** (a) 1,3-pentadiene (b) 2-ethyl-1,3-butadiene (c) 1,3-cyclohexadiene (d) 1-methyl-1,3-cyclobutadiene

37.

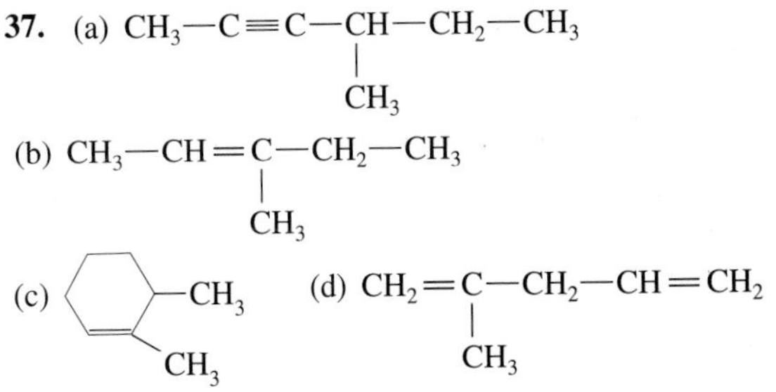

39.

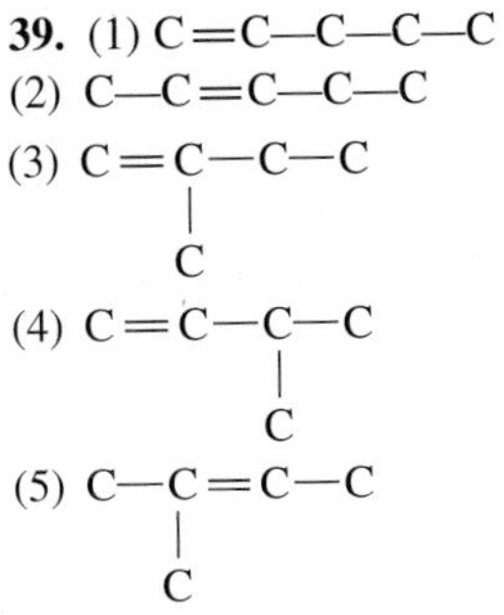

(1) 1-pentene
(2) 2-pentene
(3) 2-methyl-1-butene
(4) 3-methyl-1-butene
(5) 2-methyl-2-butene

41. (a) nonaromatic (b) aromatic (c) nonaromatic (d) aromatic **43.** (a) 10 (b) 6 (c) 22 (d) 8 **45.** (a) benzene (b) 1,3-dimethylbenzene (c) 1,3,5-trimethylbenzene (d) 1-ethyl-2-methylbenzene **47.** (a) *ortho*-dimethylbenzene (b) *ortho*-ethylmethylbenzene (c) 1,3,5-trimethylbenzene (d) *meta*-dimethylbenzene

49.

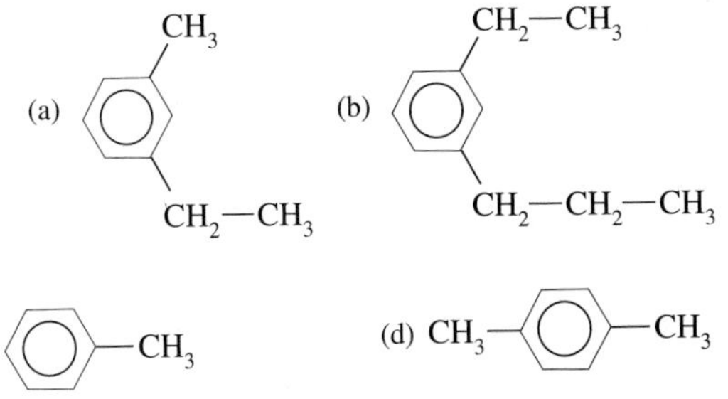

51. (a) yes (b) yes (c) yes (d) no

53.

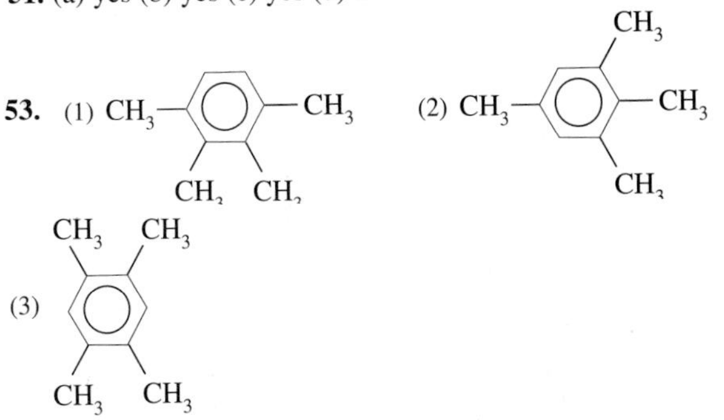

55. (a) 1,3-dichloropropane (b) 1,1,2-trifluoroethane (c) bromoethane (d) 1-chloro-3-iodopropane **57.** (a) bromocyclohexane (b) 4-chloro-3-fluorotoluene (c) iodobenzene (d) 4-chlorocyclohexene

59.

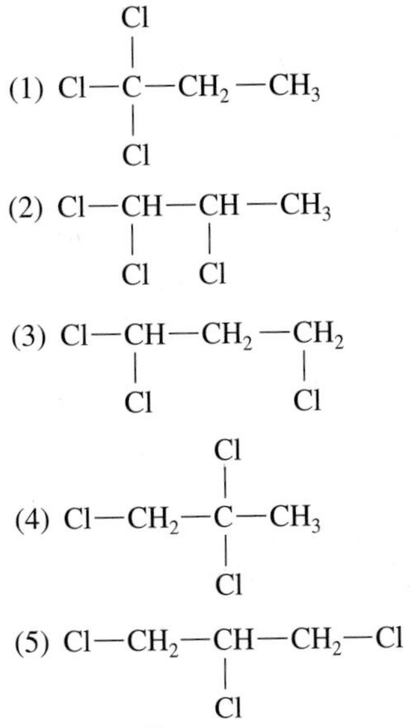

(1) 1,1,1-trichloropropane
(2) 1,1,2-trichloropropane
(3) 1,1,3-trichloropropane
(4) 1,2,2-trichloropropane
(5) 1,2,3-trichloropropane

61.

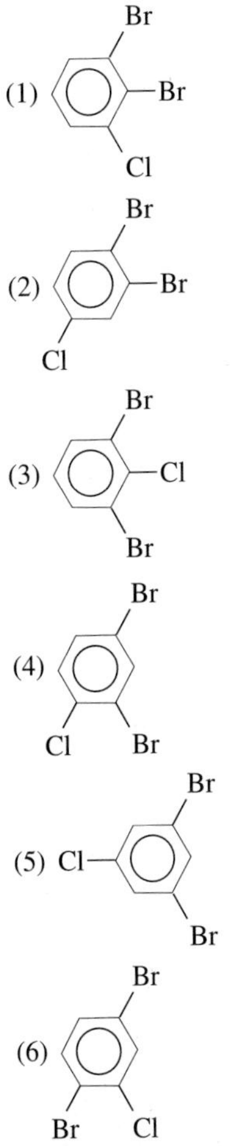

(1) 1,2-dibromo-3-chlorobenzene
(2) 1,2-dibromo-4-chlorobenzene
(3) 1,3-dibromo-2-chlorobenzene
(4) 1,3-dibromo-4-chlorobenzene
(5) 1,3-dibromo-5-chlorobenzene
(6) 1,4-dibromo-2-chlorobenzene

63. (a) 3-pentanol (b) 2,2,4-trimethyl-3-pentanol (c) 3-chloro-2-butanol (d) 2-hexanol

65. 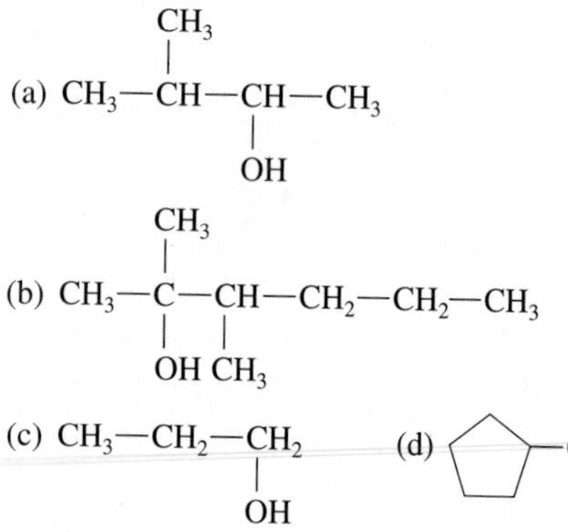

67. (a) 1,3-propanediol (b) 1,2,4-butanetriol (c) 1,2-ethanediol (d) 2-methyl-1,3-propanediol **69.** nine **71.** (a) ethoxyethane (b) 2-methoxypropane (c) 1-methoxypropane (d) 3-methyl-1-methoxybutane

73. (a) CH_3—CH_2—CH_2—O—CH_2—CH_2—CH_3

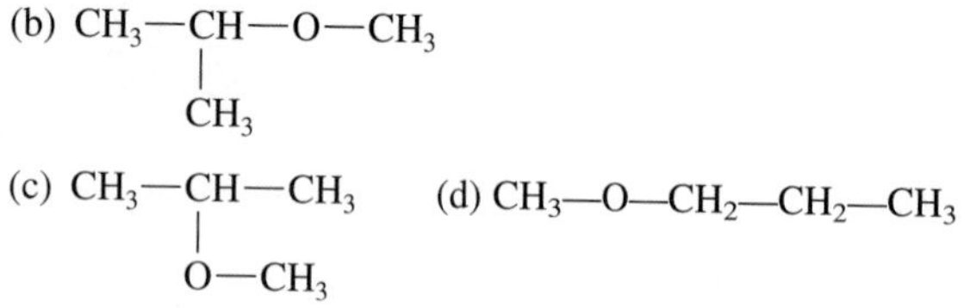

75. (1) C—O—C—C—C 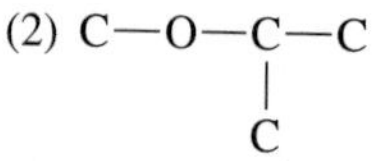

(3) C—C—O—C—C

77. (a) butanoic acid (b) 3-methylpentanoic acid (c) 3-methylbutanoic acid (d) 4,5-dimethylhexanoic acid

79. 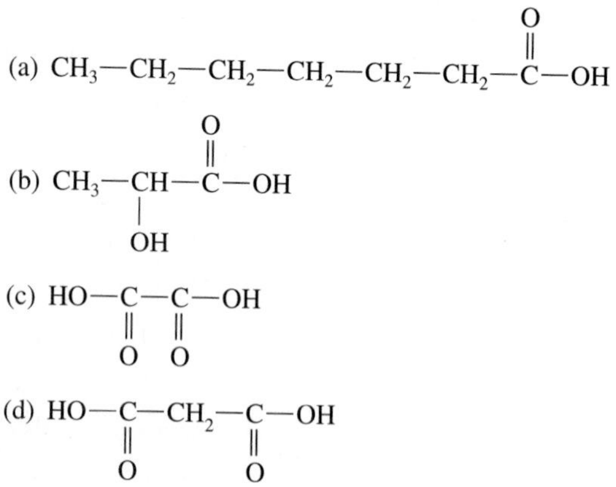

81. 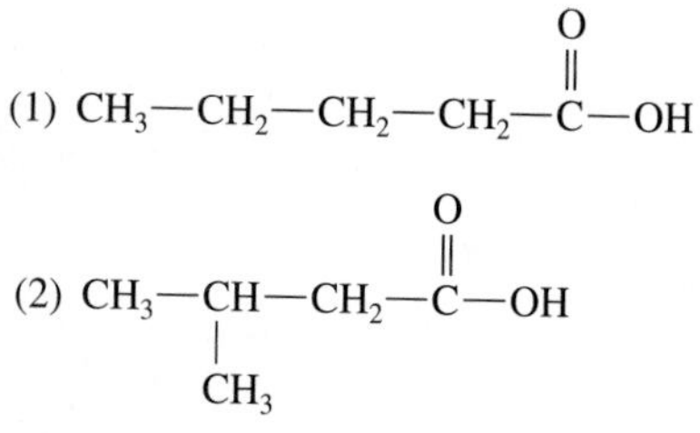

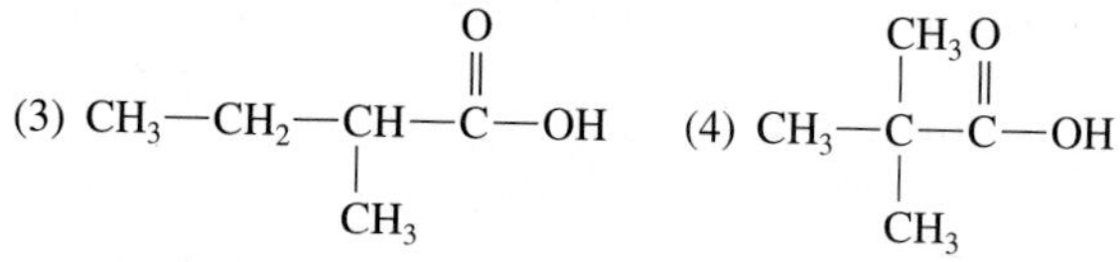

83. (a) propyl ethanoate (b) propyl propanoate (c) ethyl propanoate (d) methyl ethanoate

85. 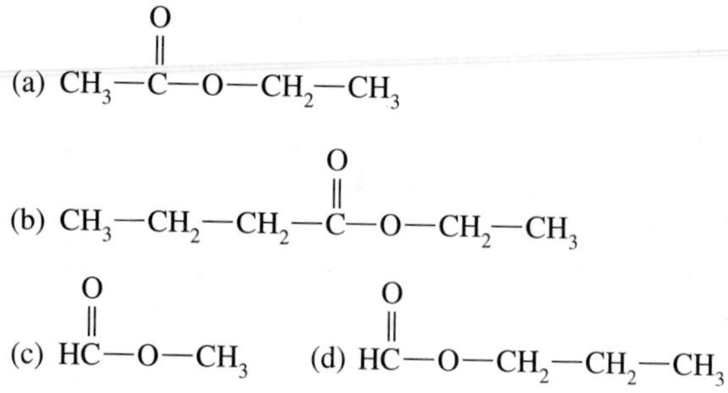

87. 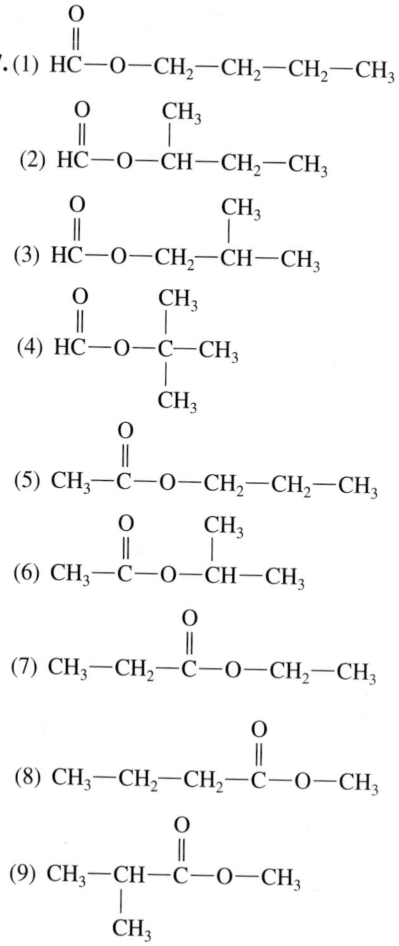

89. (a) butanoic acid (b) ethanol (c) methyl butanoate (d) 1-ethoxypropane

91. (a) $HC(=O)-O-CH_3$ (b) CH_3-O-CH_3

(c) 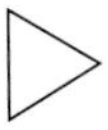 (d) CH_3Cl

93. (a) $C_{20}H_{42}$ (b) C_4H_6 (c) C_6H_{12} (d) $C_7H_{10}Cl_2$

95. (a) 8 (b) 6 (c) 10 (d) 8

97. (a) $CH_2{=}CH-CH_2-CH_3$; $CH_3-CH{=}CH-CH_3$; $CH_2{=}C(CH_3)-CH_3$; 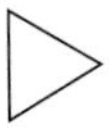 $-CH_3$; 

(b) $CH{\equiv}C-CH_3$; $CH_2{=}C{=}CH_2$; 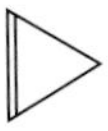

(c) $CH_3-CH_2-CH_2-OH$; $CH_3-CH(CH_3)-OH$; $CH_3-O-CH_2-CH_2$

(d) $CH_3-CH_2-CH_2-C(=O)-OH$; $CH_3-CH_2-C(=O)-O-CH_3$; $CH_3-C(=O)-O-CH_2-CH_3$; $HC(=O)-O-CH_2-CH_2-CH_3$; $CH_3-CH(CH_3)-C(=O)-OH$; $HC(=O)-O-CH(CH_3)-CH_3$

99. (a) $2\ C_{14}H_{30} + 43\ O_2 \longrightarrow 28\ CO_2 + 30\ H_2O$

(b) $2\ C_7H_{16}O + 21\ O_2 \longrightarrow 14\ CO_2 + 16\ H_2O$

101. (a) 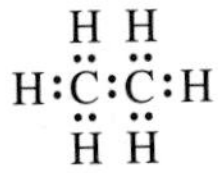 (b) H:C:::C:H (c) 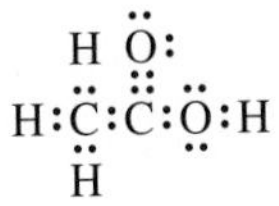

(d) 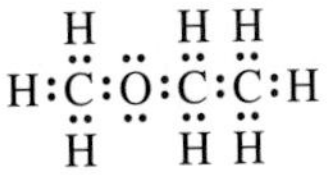

103. 80.22% C; 9.618% H; 10.18% O **105.** 29 L

107. (a) C_2H_5 (b) C_4H_{10}

Index

*A boldfaced term is defined on the indicated page.

Atomic Numbers and Atomic Masses of the Elements

Atomic masses are based on $^{12}_{6}C$. Numbers in parentheses are the mass numbers of the most stable isotopes of radioactive elements.

Element	Symbol	Atomic Number	Atomic Mass	Element	Symbol	Atomic Number	Atomic Mass
Actinium	**Ac**	89	227.0278	Erbium	**Er**	68	167.26
Aluminum	**Al**	13	26.98154	Europium	**Eu**	63	151.96
Americium	**Am**	95	(243)	Fermium	**Fm**	100	(257)
Antimony	**Sb**	51	121.760	Fluorine	**F**	9	18.998403
Argon	**Ar**	18	39.948	Francium	**Fr**	87	(223)
Arsenic	**As**	33	74.9216	Gadolinium	**Gd**	64	157.25
Astatine	**At**	85	(210)	Gallium	**Ga**	31	69.723
Barium	**Ba**	56	137.33	Germanium	**Ge**	32	72.59
Berkelium	**Bk**	97	(247)	Gold	**Au**	79	196.9665
Beryllium	**Be**	4	9.01218	Hafnium	**Hf**	72	178.49
Bismuth	**Bi**	83	208.9804	Hahnium	**Ha**	105	(262)
Boron	**B**	5	10.811	Hassium	**Hs**	108	(265)
Bromine	**Br**	35	79.904	Helium	**He**	2	4.002602
Cadmium	**Cd**	48	112.41	Holmium	**Ho**	67	164.9304
Calcium	**Ca**	20	40.078	Hydrogen	**H**	1	1.00794
Californium	**Cf**	98	(251)	Indium	**In**	49	114.818
Carbon	**C**	6	12.011	Iodine	**I**	53	126.9045
Cerium	**Ce**	58	140.12	Iridium	**Ir**	77	192.217
Cesium	**Cs**	55	132.9054	Iron	**Fe**	26	55.845
Chlorine	**Cl**	17	35.453	Krypton	**Kr**	36	83.80
Chromium	**Cr**	24	51.996	Lanthanum	**La**	57	138.9055
Cobalt	**Co**	27	58.9332	Lawrencium	**Lr**	103	(262)
Copper	**Cu**	29	63.546	Lead	**Pb**	82	207.2
Curium	**Cm**	96	(247)	Lithium	**Li**	3	6.941
Dysprosium	**Dy**	66	162.50	Lutetium	**Lu**	71	174.967
Einsteinium	**Es**	99	(252)	Magnesium	**Mg**	12	24.305
Element 110	—	110	(271)	Manganese	**Mn**	25	54.9380
Element 111	—	111	(272)	Meitnerium	**Mt**	109	(266)

Element	Symbol	Atomic Number	Atomic Mass
Mendelevium	**Md**	101	(260)
Mercury	**Hg**	80	200.59
Molybdenum	**Mo**	42	95.94
Neodymium	**Nd**	60	144.24
Neon	**Ne**	10	20.179
Neptunium	**Np**	93	237.0482
Nickel	**Ni**	28	58.6934
Nielsbohrium	**Ns**	107	(262)
Niobium	**Nb**	41	92.9064
Nitrogen	**N**	7	14.0067
Nobelium	**No**	102	(259)
Osmium	**Os**	76	190.23
Oxygen	**O**	8	15.9994
Palladium	**Pd**	46	106.42
Phosphorus	**P**	15	30.97376
Platinum	**Pt**	78	195.08
Plutonium	**Pu**	94	(244)
Polonium	**Po**	84	(209)
Potassium	**K**	19	39.0983
Praseodymium	**Pr**	59	140.9077
Promethium	**Pm**	61	(145)
Protactinium	**Pa**	91	231.0359
Radium	**Ra**	88	226.0254
Radon	**Rn**	86	(222)
Rhenium	**Re**	75	186.207
Rhodium	**Rh**	45	102.9055
Rubidium	**Rb**	37	85.4678
Ruthenium	**Ru**	44	101.07
Rutherfordium	**Rf**	104	(261)
Samarium	**Sm**	62	150.36
Scandium	**Sc**	21	44.9559
Seaborgium	**Sg**	106	(266)
Selenium	**Se**	34	78.96
Silicon	**Si**	14	28.0855
Silver	**Ag**	47	107.8682
Sodium	**Na**	11	22.98977
Strontium	**Sr**	38	87.62
Sulfur	**S**	16	32.066
Tantalum	**Ta**	73	180.9479
Technetium	**Tc**	43	(98)
Tellurium	**Te**	52	127.60
Terbium	**Tb**	65	158.9254
Thallium	**Tl**	81	204.383
Thorium	**Th**	90	232.0381
Thulium	**Tm**	69	168.9342
Tin	**Sn**	50	118.710
Titanium	**Ti**	22	47.867
Tungsten	**W**	74	183.84
Uranium	**U**	92	238.0289
Vanadium	**V**	23	50.9415
Xenon	**Xe**	54	131.29
Ytterbium	**Yb**	70	173.04
Yttrium	**Y**	39	88.9059
Zinc	**Zn**	30	65.38
Zirconium	**Zr**	40	91.22

Molarity and molality are approximately equal when a solution is dilute and the solution density is about 1 g/mL. Neither of these conditions applies in this problem; hence, molarity (5.87) and molality (8.92) are quite different.

Practice Exercise 13.15

Calculate the molarity of a 2.73 *m* (molal) methyl alcohol (CH_4O) solution whose density is 0.976 g/mL.

Ans. 2.46 M

13.10 DILUTION

A common problem encountered when working with solutions in the laboratory is that of diluting a solution of known concentration (usually called a stock solution) to a lower concentration. **Dilution** *is the process in which more solvent is added to a solution in order to lower its concentration.* Dilution always lowers the concentration of a solution. The same amount of solute is present, but it is now distributed in a larger amount of solvent (the original solvent plus the added solvent).

Since laboratory solutions are almost always liquids, dilution is normally a volumetric procedure. Most often, a solution of a specific molarity must be prepared by adding a predetermined volume of solvent to a specific volume of stock solution.

With molar concentration units, a very simple mathematical relationship exists between the volumes and molarities of the diluted and stock solutions. This relationship is derived from the fact that the same amount of solute is present in both solutions; only solvent is added in a dilution procedure.

$$\text{Moles solute}_{\text{stock solution}} = \text{moles solute}_{\text{diluted solution}}$$

The number of moles of solute in both solutions is given by the expression

$$\text{Moles solute} = \text{molarity } (M) \times \text{liters of solution } (V)$$

(This equation is just a rearrangement of the defining equation for molarity to isolate moles of solute on the left side.) Substitution of this second expression into the first one gives the equation

$$M_s \times V_s = M_d \times V_d$$

In this equation M_s and V_s are the molarity and volume of the stock solution (the solution to be diluted) and M_d and V_d the molarity and volume of the solution resulting from the dilution. Because volume appears on both sides of the equation, any volume unit, not just liters, may be used as long as it is the same on both sides of the equation. Again, the validity of this equation is based on there being no change in the amount of solute present.

Example 13.16

What is the molarity of the solution prepared by diluting 65 mL of 0.95 M sodium sulfate (Na_2SO_4) solution to a final volume of 135 mL?